Essentials of Organic Chemistry

Essentials of Organic Chemistry

For students of pharmacy, medicinal chemistry
and biological chemistry

Paul M Dewick

School of Pharmacy
University of Nottingham, UK

John Wiley & Sons, Ltd

Other Wiley Editorial Offices

John Wiley & Sons Inc., 111 River Street, Hoboken, NJ 07030, USA

Jossey-Bass, 989 Market Street, San Francisco, CA 94103-1741, USA

Wiley-VCH Verlag GmbH, Boschstr. 12, D-69469 Weinheim, Germany

John Wiley & Sons Australia Ltd, 42 McDougall Street, Milton, Queensland 4064, Australia

John Wiley & Sons (Asia) Pte Ltd, 2 Clementi Loop #02-01, Jin Xing Distripark, Singapore 129809

John Wiley & Sons Canada Ltd, 6045 Freemont Blvd, Mississauga, ONT, L5R 4J3, Canada

Wiley also publishes its books in a variety of electronic formats. Some content that appears
in print may not be available in electronic books.

British Library Cataloguing in Publication Data

A catalogue record for this book is available from the British Library

ISBN-13: 978-0-470-01665-7 (HB)   978-0-470-01666-4 (PB)
ISBN-10: 0-470-01665-5 (HB)   0-470-01666-3 (PB)

Typeset in 9.5/11.5pt Times by Laserwords Private Limited, Chennai, India

Contents

Preface

For more years than I care to remember, I have been teaching the new intake of students to the Nottingham pharmacy course, instructing them in those elements of basic organic chemistry necessary for their future studies. During that time, I have also referred them to various organic chemistry textbooks for additional reading. These texts, excellent though they are, contain far too much material that is of no immediate use to pharmacy students, yet they fail to develop sufficiently areas of biological and medicinal interest we would wish to study in more detail. The organic chemistry needs of pharmacy students are not the same as the needs of chemistry students, and the textbooks available have been specially written for the latter group. What I really wanted was an organic chemistry textbook, considerably smaller than the 1000–1500-page tomes that seem the norm, which had been designed for the requirements of pharmacy students. Such a book would also serve the needs of those students on chemistry-based courses, but who are not specializing in chemistry, e.g. students taking medicinal chemistry and biological chemistry. I have wanted to write such a book for a long time now, and this is the result of my endeavours. I hope it proves as useful as I intended it.

Whilst the content is not in any way unique, the selection of topics and their application to biological systems should make the book quite different from others available, and of especial value to the intended readership. It is a combination of carefully chosen material designed to provide a thorough grounding in fundamental chemical principles, but presenting only material most relevant to the target group and omitting that which is outside their requirements. How these principles and concepts are relevant to the study of pharmaceutical and biochemical molecules is then illustrated through a wide range of examples.

I have assumed that readers will have some knowledge of organic chemistry and are familiar with the basic philosophy of bonding and reactivity as covered in pre-university courses. The book then presents material appropriate for the first 2 years of a university pharmacy course, and also provides the fundamental chemical groundwork for courses in medicinal chemistry, biological chemistry, etc. Through selectivity, I have generated a textbook of more modest size, whilst still providing a sufficiently detailed treatment for those topics that are included.

I have adopted a mechanism-based layout for the majority of the book, an approach that best enables the level of detail and selection of topics to be restricted in line with requirements. There is a strong emphasis on understanding and predicting chemical reactivity, rather than developing synthetic methodology. With extensive use of pharmaceutical and biochemical examples, it has been possible to show that the same simple chemistry can be applied to real-life complex molecules. Many of these examples are in self-contained boxes, so that the main theme need not be interrupted. Lots of cross-referencing is included to establish links and similarities; these do not mean you have to look elsewhere to understand the current material, but they are used to stress that we have seen this concept before, or that other uses are coming along in due course.

I have endeavoured to provide a friendly informal approach in the text, with a clear layout and easy-to-find sections. Reaction schemes are annotated to keep material together and reduce the need for textual explanations. Where alternative rationalizations exist,

I have chosen to use only the simpler explanation to keep the reasoning as straightforward as possible. Throughout, I have tried to convince the reader that, by applying principles and deductive reasoning, we can reduce to a minimal level the amount of material that needs be committed to memory. Worked problems showing typical examination questions and how to approach them are used to encourage this way of thinking.

Four chapters towards the end of the book diverge from the other mechanism-oriented chapters. They have a strong biochemical theme and will undoubtedly overlap with what may be taught separately by biochemists. These topics are approached here from a chemical viewpoint, using the same structural and mechanistic principles developed earlier, and should provide an alternative perspective. It is probable that some of the material described will not be required during the first 2 years of study, but it could sow the seeds for more detailed work later in the course.

There is a measure of intended repetition; the same material may appear in more than one place. This is an important ploy to stress that we might want to look at a particular aspect from more than one viewpoint. I have also used similar molecules in different chapters as illustrations of chemical structure or reactivity. Again, this is an intentional strategy to illustrate the multiple facets of real-life complex molecules.

I am particularly grateful to some of my colleagues at Nottingham (Barrie Kellam, Cristina De Matteis, Nick Shaw) for their comments and opinions. I would also like to record the unknowing contribution made by Nottingham pharmacy students over the years. It is from their questions, problems and difficulties that I have shaped this book. I hope future generations of students may benefit from it.

Finally, a word of advice to students, advice that has been offered by organic chemistry teachers many times previously. *Organic chemistry is not learnt by reading*: paper and pencil are essential at all times. It is only through drawing structures and mechanisms that true understanding is attained.

Paul M Dewick
Nottingham, 2005

1

Molecular representations and nomenclature

1.1 Molecular representations

From the beginnings of chemistry, scientists have devised means of representing the materials they are discussing, and have gradually developed a comprehensive range of shorthand notations. These cover the elements themselves, bonding between atoms, the arrangement of atoms in molecules, and, of course, a systematic way of naming compounds that is accepted and understood throughout the scientific world.

The study of carbon compounds provides us with the subdivision 'organic chemistry', and a few simple organic compounds can exemplify this shorthand approach to molecular representations. The primary alcohol propanol (systematically propan-1-ol or 1-propanol, formerly n-propanol, n signifying normal or unbranched) can be represented by a structure showing all atoms, bonds, and lone pair or non-bonding electrons.

Lines are used to show what we call **single bonds**, indicating the sharing of one pair of electrons. In writing structures, we have to remember the number of bonds that can be made to a particular atom, i.e. the **valency** of the atom. In most structures, carbon is tetravalent, nitrogen trivalent, oxygen divalent, and hydrogen and halogens are univalent. These valencies arise from the number of electrons available for bonding. More often, we trim this type of representation to one that shows the layout of the carbon skeleton with attached hydrogens or other atoms. This can be a formula-like structure without

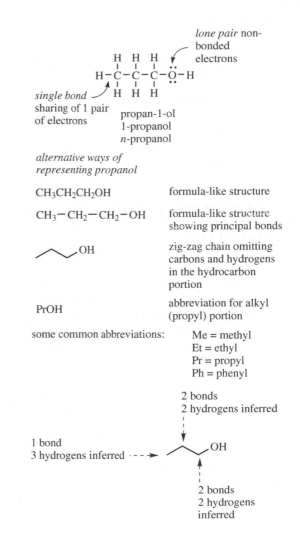

$CH_3CH_2CH_2OH$	formula-like structure
$CH_3-CH_2-CH_2-OH$	formula-like structure showing principal bonds
	zig-zag chain omitting carbons and hydrogens in the hydrocarbon portion
PrOH	abbreviation for alkyl (propyl) portion

some common abbreviations:

Me = methyl
Et = ethyl
Pr = propyl
Ph = phenyl

Essentials of Organic Chemistry   Paul M Dewick
© 2006 John Wiley & Sons, Ltd

bonds, or it can be one showing just the principal bonds, those of the carbon chain.

However, for many complex structures, even these approaches become too tedious, and we usually resort to a shorthand version that omits most, if not all, of the carbon and hydrogen atoms. Propanol is now shown as a **zig-zag chain** with an OH group at one end. The other end of the chain, where it stops, is understood to represent a methyl group; three attached hydrogens have to be inferred. At a point on the chain, two hydrogens are assumed, because two bonds to carbons are already shown. In a structure where three bonds joined, a single additional hydrogen would be assumed (see vinyl chloride, below).

The zig-zag arrangement is convenient so that we see where carbons are located (a long straight line would not tell us how many carbons there are), but it also mimics the low-energy arrangement (conformation) for such a compound (see Section 3.3.1). Note that it is usual to write out the hydroxyl, or some alternative group, in full. This group, the so-called **functional group**, tends to be the reactive part of the molecule that we shall be considering in reactions. When we want an even more concise method of writing the molecule, abbreviations for an alkyl (or aryl) group may be used, in which case propanol becomes PrOH. Some more common abbreviations are given later in Table 1.3.

chloroethene
vinyl chloride

Double bonds, representing the sharing of two pairs of electrons, are inferred by writing a double line. Vinyl chloride (systematically chloroethene) is shown as two different representations according to the conventions we have just seen for propanol. Note that it is customary always to show the reactive double bond, so that CH_2CHCl would not be encountered as an abbreviation for vinyl chloride.

The six-membered cyclic system in **aromatic rings** is usually drawn with alternating double and single bonds, i.e. the **Kekulé form**, and it is usually immaterial which of the two possible versions is used. Aniline (systematically aminobenzene or benzenamine) is shown with and without carbons and hydrogens. It is quite rare to put in any of the ring hydrogens on an aromatic ring, though it is sometimes convenient to put some in on the substituent, e.g. on a methyl, as in toluene (methylbenzene), or an aldehyde group, as in benzaldehyde.

Benzene strictly does not have alternating double and single bonds, but the aromatic sextet of electrons is localized in a π orbital system and bond lengths are somewhere in between double and single bonds

aminobenzene
aniline

the two Kekulé versions of aniline

circle represents
aromatic π
electron sextet

toluene
methylbenzene

benzaldehyde

it is more common
to show hydrogens
in substituents

(see Section 2.9.4). To represent this, a circle may be drawn within the hexagon. Unfortunately, this version of benzene becomes quite useless when we start to draw reaction mechanisms, and most people continue to draw benzene rings in the Kekulé form. In some cases, such as fused rings, it is actually incorrect to show the circles.

two Kekulé versions of naphthalene

each circle must represent six aromatic π electrons

this is strictly incorrect!

Thus, naphthalene has only 10 π electrons, one from each carbon, whereas the incorrect two-circle version suggests it has 12 π electrons.

We find that, in the early stages, students are usually happier to put in all the atoms when drawing structures, following earlier practices. However, you are urged to adopt the shorthand representations as soon as possible. This saves time and cleans up the structures of larger molecules. Even a relatively simple molecule such as 2-methylcyclohexanecarboxylic acid, a cyclohexane ring carrying two substituents, looks a mess when all the atoms are put in. By contrast, the line drawing looks neat and tidy, and takes much less time to draw.

2-methylcyclohexanecarboxylic acid

Do appreciate that there is no strict convention for how you orientate the structure on paper. In fact, we will turn structures around, as appropriate, to suit our needs. For example, the amino acid tyrosine has three functional groups, i.e. a carboxylic acid, a primary amine, and a phenol. How we draw tyrosine will

tyrosine

we might use this version if we were considering reactions of the carboxylic acid group

we might use this version if we were considering reactions of the amine group

we might use this version if we were considering reactions of the phenol group

depend upon what modifications we might be considering, and which functional group is being altered.

You will need to be able to reorientate structures without making mistakes, and also to be able to recognize different versions of the same thing. A simple example is with esters, where students have learnt that ethyl acetate (ethyl ethanoate) can be abbreviated to $CH_3CO_2C_2H_5$. When written backwards, i.e. $C_2H_5OCOCH_3$, the ester functionality often seems less recognizable.

1.2 Partial structures

We have just seen that we can save a lot of time and effort by drawing structures without showing all of the atoms. When we come to draw reaction sequences, we shall find that we are having to repeat large chunks of the structure each time, even though no chemical changes are occurring in that part of the molecule. This is unproductive, so we often end up writing down just that part of the structure that is of interest, i.e. a **partial structure**. This will not cause problems when you do it, but it might when you see one and wish to interpret it.

In the representations overleaf, you can see the line drawing and the version with methyls that stresses the bond ends. Both are satisfactory. When we wish to consider the reactivity of the double bond, and perhaps want to show that reaction occurs irrespective of the alkyl groups attached to the double bond, we put in the abbreviation R (see below), or usually just omit them. When we omit the attached groups, it helps to show what we mean by using wavy lines across the bonds, but in our urge to proceed we tend to omit even these indicators. This may

*a typical line
drawing*

*this is what the line
drawing conveys*

*this version emphasizes
the chain ends*

*using the R abbreviation for
an unspecified alkyl group;
different R groups may be
indicated by R^1, R^2, R^3, etc.,
or R, R', R'', etc.*

*a partial structure; this
shows the double bond that
has four groups attached;
wavy lines indicate bonds
to something else*

*in context, this might mean
the same, but could be
mistaken for a double bond
with four methyls attached*

*this would be better; putting
in the carbons emphasizes
that the other lines represent
bonds, not methyls*

cause confusion in that we now have what looks like a double bond with four methyls attached, not at all what we intended. A convenient ploy is to differentiate this from a line drawing by putting in the alkene carbons.

1.3 Functional groups

The reactivity of a molecule derives from its **functional group** or groups. In most instances the hydrocarbon part of the molecule is likely to be unreactive, and the reactivity of the functional group is largely independent of the nature of the hydrocarbon part. In general terms, then, we can regard a molecule as R–Y or Ar–Y, a combination of a functional group Y with an alkyl group R or aryl group Ar that is not participating in the reaction under consideration. This allows us to discuss reactivity in terms of functional groups, rather than the reactivity of individual compounds. Of course, most of the molecules of interest to us will have more than one functional group; it is this combination of functionalities that provides the reactions of chemical and biochemical importance. Most of the functional groups we shall encounter are included in Table 1.1, which also contains details for their nomenclature (see Section 1.4).

It is particularly important that when we look at the structure of a complex molecule we should visualize it in terms of the functional groups it contains. The properties and reactivity of the molecule can

generally be interpreted in terms of these functional groups. It may sometimes be impossible to consider the reactions of each functional group in complete isolation, but it is valuable to disregard the complexity and perceive the simplicity of the structure. With a little practice, it should be possible to dissect the functional groups in complex structures such as morphine and amoxicillin.

morphine

amoxicillin

Table 1.1 Functional groups and IUPAC nomenclature (arranged in order of decreasing priority)

Functional group	Structure	Suffix	Prefix
Cation			
ammonium	$R_4N^{\oplus}$	-ammonium	ammonio-
phosphonium	$R_4P^{\oplus}$	-phosphonium	phosphonio-
sulfonium	$R_3S^{\oplus}$	-sulfonium	sulfonio-
Carboxylic acid	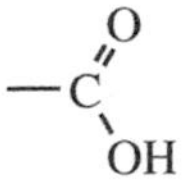 $-CO_2H$	-oic acid	carboxy-
Carboxylic acid anhydride (anhydride)	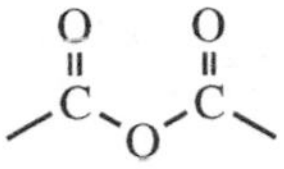	-oic anhydride	
Carboxylic acid ester (ester)	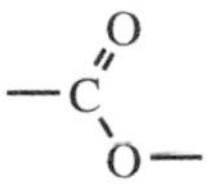 $-CO_2R$	alkyl -oate	alkoxylcarbonyl- (or carbalkoxy-)
Acyl halide	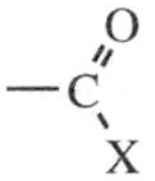 $-COX$	-oyl halide	haloalkanoyl-
Amide			
primary amide	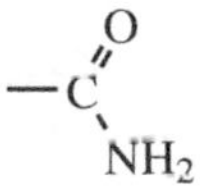 $-CONH_2$	-amide	carbamoyl-
secondary amide	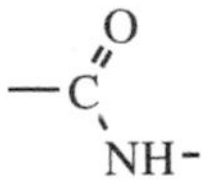 $-CONHR$		
tertiary amide	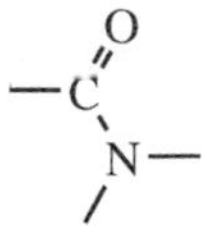 $-CONR_2$		
Nitrile	$-C\equiv N$ $-CN$	-nitrile (or -onitrile)	cyano-
Aldehyde	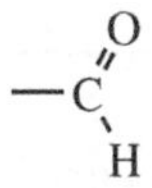 $-CHO$	-al	formyl-
Ketone	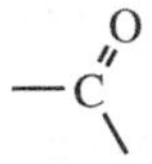 $-COR$	-one	-oxo-

(continued overleaf)

Table 1.1 (*continued*)

Functional group	Structure	Suffix	Prefix
Alcohol			
primary alcohol	$-CH_2OH$	-ol	hydroxy-
secondary alcohol	$\diagdown$CHOH$\diagup$		
tertiary alcohol	$-\overset{\textstyle\vert}{\underset{\textstyle\vert}{C}}-OH$		
phenol	$Ar-OH$		
Thiol (mercaptan)	$-SH$	-thiol	mercapto-
Amine			
primary amine	$-NH_2$	-amine	amino- (or aza-)
secondary amine	$\diagdown$NH$\diagup$ $-NHR$		
tertiary amine	$\diagdown$N$-$$\diagup$ $-NR_2$		
Ether	$-O\diagdown$ $-OR$	(ether)	-oxa- (or alkoxy-)
Sulfide (thioether)	$-S\diagdown$ $-SR$	(sulfide)	alkylthio- (or thia-)
Alkene	$\diagdown$C$=$C$\diagup$	-ene	alkenyl-
Alkyne	$-C\equiv C-$	-yne	alkynyl-
Halides	$-X$	(halide)	halo-
Nitro	$-\overset{\textstyle O}{\underset{\textstyle O\ominus}{N\oplus}}$ $-NO_2$		nitro-
Alkanes	$\diagdown$C$-$C$\diagup$	-ane	alkyl-

1.4   Systematic nomenclature

Organic compounds are named according to the internationally accepted conventions of the International Union of Pure and Applied Chemistry (IUPAC). Since these conventions must cover all eventualities, the documentation required spans a book of similar size to this volume. A very much-abbreviated version suitable for our requirements is given here:

- the functional group provides the suffix name;

- with two or more functional groups, the one with the highest priority provides the suffix name;

- the longest carbon chain containing the functional group provides the stem name;

- the carbon chain is numbered, keeping minimum values for the suffix group;

- side-chain substituents are added as prefixes with appropriate numbering, listing them alphabetically.

The stem names are derived from the names of hydrocarbons. Acyclic and cyclic saturated hydrocarbons (alkanes) in the range C_1-C_{12} are listed in Table 1.2.

Aromatic systems are named in a similar way, but additional stem names need to be used. Parent aromatic compounds of importance are benzene,

Table 1.2 Names of parent hydrocarbons

Acyclic hydrocarbon	Cyclic hydrocarbon
Methane CH_4	
Ethane H_3C-CH_3	
Propane	 Cyclopropane
Butane	 Cyclobutane
Pentane	 Cyclopentane 
Hexane	 Cyclohexane 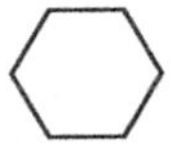
Heptane	 Cycloheptane 
Octane	 Cyclooctane 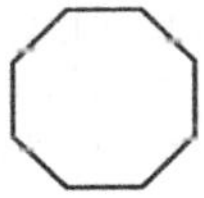
Nonane	 Cyclononane 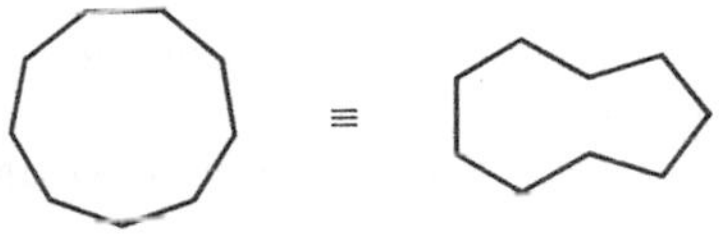
Decane	 Cyclodecane 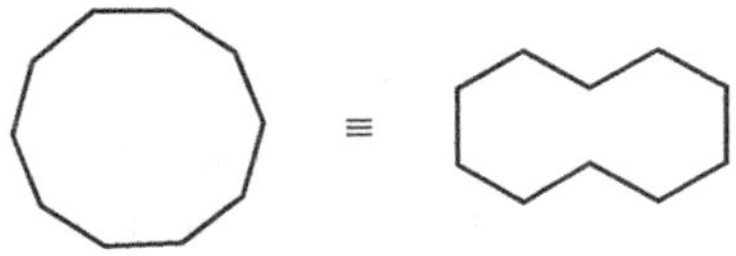
Undecane	 Cycloundecane 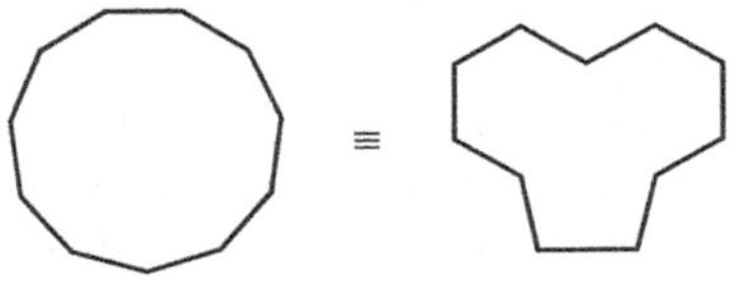
Dodecane	 Cyclododecane 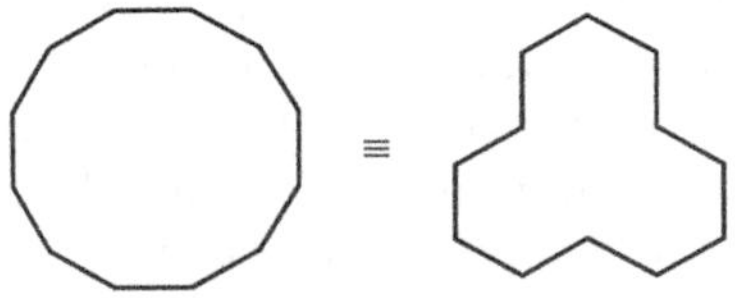

naphthalene, anthracene, and phenanthrene. The last three contain fused rings, and they have a fixed numbering system that includes only those positions at which substitution can take place.

benzene   naphthalene

anthracene   phenanthrene

It is anticipated that readers will already be familiar with many of the general principles of nomenclature and will be able to name a range of simple compounds. It is not the object of this section to provide an exhaustive series of instructions for naming every class of compound. Instead, the examples chosen here (Box 1.1) have been selected to illustrate some of the perhaps less familiar aspects that will be commonly encountered, and to foster a general understanding of the approach to nomenclature.

Alternative names are shown in some cases; this should emphasize that there is often no unique 'correct' name. Sometimes, it can be advantageous to bend the rules a little so as to provide a neat name rather than a fully systematic one. Typically, this might mean adopting a lower priority functional group as the suffix name. It is important to view nomenclature as a means of conveying an acceptable unambiguous structure rather than a rather meaningless scholastic exercise. Other examples will occur in subsequent chapters, and specialized aspects, e.g. **heterocyclic nomenclature**, will be treated in more detail at the appropriate time (see Chapter 11). **Stereochemical descriptors** are omitted here, but will be discussed under stereochemistry (see Sections 3.4.2 and 3.4.3).

Box 1.1

Systematic nomenclature: some examples

6-chloro-5-methylhepta-2,4-diene
6-chloro-5-methyl-2,4-heptadiene

- alkenes have higher priority than halides; suffix is -ene
- longest carbon chain is seven carbons: heptane
- numbering is chosen to give lowest numbers for the double bonds; 2-ene denotes 2,3-double bond, 4-ene denotes 4,5-double bond
- the European system hepta-2,4-diene is less prone to errors than the US system 2,4-heptadiene
- an additional syllable -a- is used but is not obligatory; heptadiene is easier to say than heptdiene

3-methylhex-5-yn-2-ol
3-methyl-5-hexyn-2-ol

- alcohols have higher priority than alkynes; suffix is -ol
- longest carbon chain is six carbons: hexane
- numbering is chosen to give lowest number for alcohol
- the European system hex-5-yn-2-ol keeps numbers and functionalities together

2-amino-4,4-dimethylpentanoic acid

- acids have higher priority than amines; remember 'amino acids'
- suffix is -oic acid
- one of the methyls is part of the five carbon chain, the others are substituents
- note the use of 4,4-, which shows both methyls are attached to the same carbon; 4-dimethyl would not be as precise

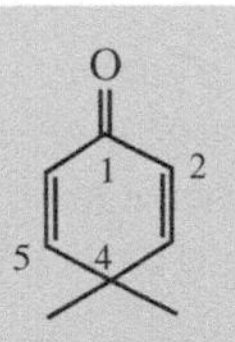

4,4-dimethylcyclohexa-2,5-dienone

- highest priority group is ketone; suffix -one
- longest carbon system is the ring cyclohexane
- numbering is around the ring starting from ketone as position 1
- 2,5-diene conveys 2,3- and 5,6-double bonds
- note 2,5-dienone means two double bonds and one ketone; contrast endione which would be one double bond and two ketones

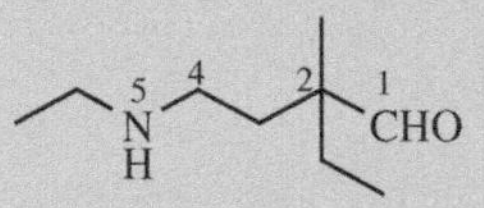

2-ethyl-4-ethylamino-2-methylbutanal
2-ethyl-2-methyl-5-azaheptanal

- highest priority group is aldehyde; suffix -al
- amino group at 4 is also substituted; together they become ethylamino
- the alternative name invokes a seven-carbon chain with one carbon (C-5) replaced by nitrogen; this is indicated by using the extra syllable -aza-, so the chain becomes 5-azaheptane

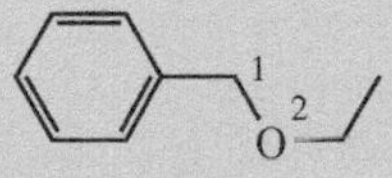

benzyl ethyl ether
benzyloxyethane
1-phenyl-2-oxa-butane

- simple ethers are best named as an alkyl alkyl ether
- the phenylmethyl group is commonly called benzyl
- an acceptable alternative is as an alkoxy alkane: the alternative ethoxytoluene would require an indication of the point of attachment
- the second alternative invokes a three-carbon chain with one carbon replaced by oxygen; this is indicated by using the extra syllable -oxa-, so the chain becomes 2-oxabutane

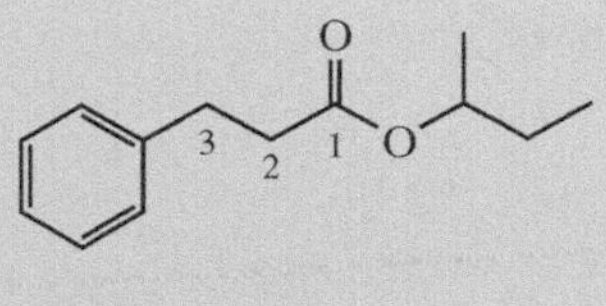

but-2-yl 3-phenylpropanoate

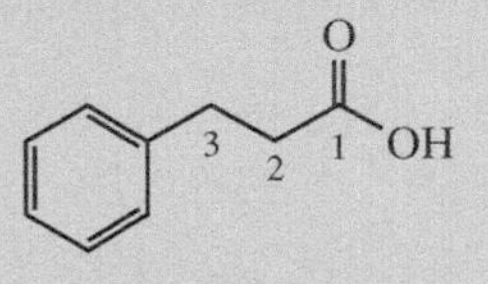

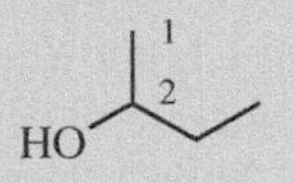

3-phenylpropanoic acid   2-butanol

- esters are named alkyl alkanoate – two separate words with no hyphen or comma
- alkyl signifies the alcohol part from which the ester is constructed, whilst alkanoate refers to the carboxylic acid part
- but-2-yl means the ester is constructed from the alcohol butan-2-ol; 3-phenylpropanoate means the acid part is 3-phenylpropanoic acid
- note the numbers 2 and 3 are in separate words and do not refer to the same part of the molecule

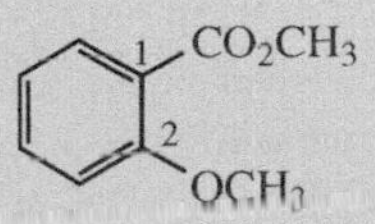

methyl 2-methoxybenzoate

- this is a methyl ester of a substituted benzoic acid; the ring is numbered from the point of attachment of the carboxyl
- the acid portion for the ester is 2-substituted
- the ether group is most easily treated as a methoxy substituent on the benzene ring

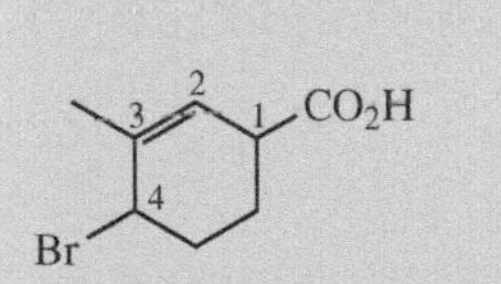

4-bromo-3-methylcyclohex-2-enecarboxylic acid

- the carboxylic acid takes priority; suffix usually -oic acid
- the carboxylic acid is here treated as a substituent on the cyclohexane ring; the combination is called cyclohexanecarboxylic acid

Box 1.1 (continued)

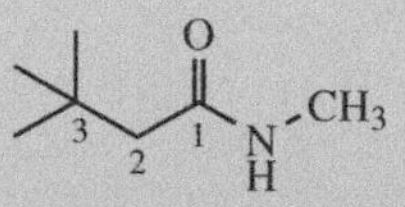

N,3,3,-trimethylbutanamide

- this is a secondary amide of butanoic acid; thus the root name is butanamide
- two methyl substituents are on position 3, and one on the nitrogen, hence N,3,3-trimethyl; the N is given in italics

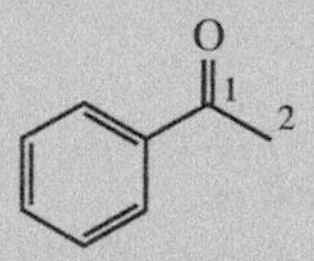

1-phenylethanone
methyl phenyl ketone
acetophenone

- a ketone in which the longest chain is two carbons; thus the root name is ethanone
- the phenyl substituent is on the carbonyl, therefore at position 1
- without the 1-substituent, ethanone is actually an aldehyde, and would be ethanal!
- the alternative methyl phenyl ketone is a neat and easy way of conveying the structure
- this structure has a common name, acetophenone, which derives from an acetyl (CH_3CO) group bonded to a phenyl ring

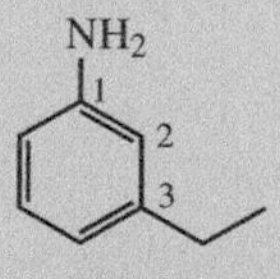

3-ethylaniline
m-ethylaniline
3-ethylphenylamine
3-ethylbenzenamine

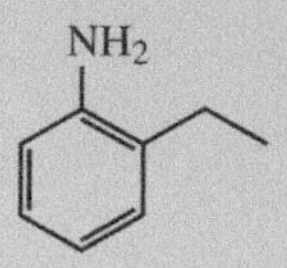

o-ethylaniline

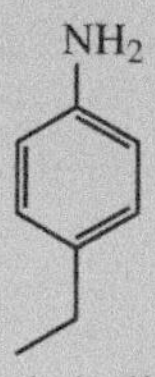

p-ethylaniline

- an amine; suffix usually -amine
- the root name can be phenylamine, as an analogue of methylamine, or the systematic benzenamine; in practice, the IUPAC accepted name is aniline
- the ring is numbered from the point of attachment of the amino group
- the prefixes ortho-, meta-, and para- are widely used to denote 1,2-, 1,3-, or 1,4-arrangements respectively on an aromatic ring; these are abbreviated to o-, m-, and p-, all in italics

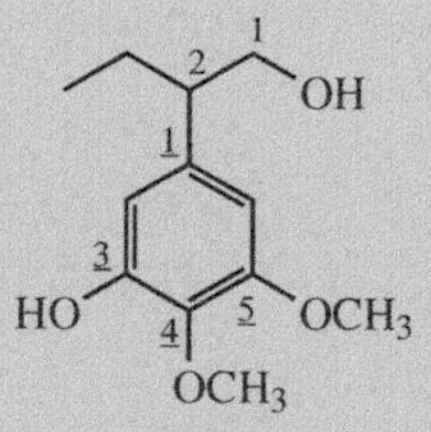

2-(3-hydroxy-4,5-dimethoxyphenyl)butanol

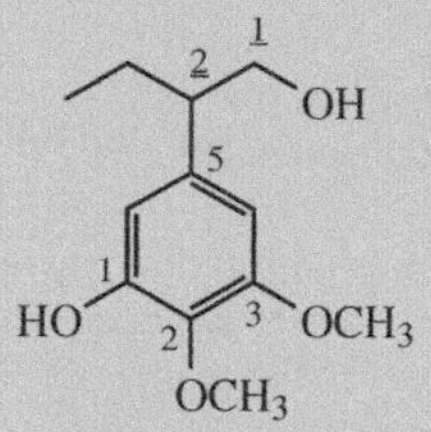

5-(1-hydroxybut-2-yl)-2,3-dimethoxyphenol

- this could be named as an alcohol, or as a phenol
- as an alcohol (butanol), there is a substituted phenyl ring attached at position 2
- note the phenyl and its substituents are bracketed to keep them together, and to separate their numbering (shown underlined) from that of the alcohol chain
- as a phenol, the substituted butane side-chain is attached through its 2-position so has a root name but-2-yl to show the position of attachment; again, this is in brackets to separate its numbering from that of the phenol
- di-, tri-, tetra-, etc. are not part of the alphabetical sequence for substituents; dimethoxy comes under m, whereas trihydroxy would come under h, etc.

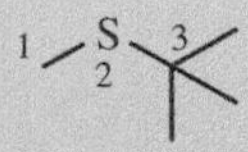

tert-butyl methyl thioether
tert-butyl methyl sulfide
3,3-dimethyl-2-thiabutane

- this is a thioether, which can be named as a thioether or as a sulfide
- an alternative invokes a four-carbon chain with one carbon replaced by sulfur using the extra syllable -thia-; this chain thus becomes 2-thiabutane
- note how the (trimethyl)methyl group is most frequently referred to by its long-established name of tertiary-butyl, abbreviated to tert-butyl, or t-butyl

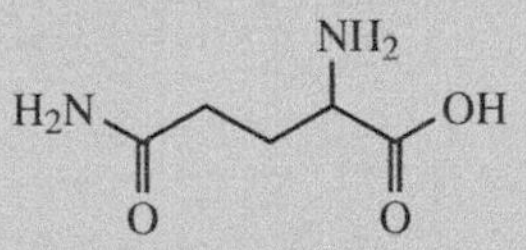

2-amino-4-carbamoylbutanoic acid
2,5-diamino-5-oxo-pentanoic acid
glutamic acid

- this contains an amine, an amide, and a carboxylic acid; the acid takes priority
- the amide group as a substituent is termed carbamoyl; this includes one carbon, so the chain length remaining to name is only four carbons – butane
- it is rather easier to consider the amide as amino and ketone substituents on the five-carbon chain
- the ketone is indicated by oxo-; do not confuse this with -oxa-, which signifies replacement of one carbon by oxygen
- the common name is glutamic acid; it is an amino acid found in proteins

There now follow a number of examples demonstrating how to convert a systematic name into a structure, with appropriate guidance hints (Box 1.2). For added relevance, these are all selected from routinely used drugs. Again, any stereochemical aspects are not included.

Box 1.2

Converting systematic names into structures: selected drug molecules

1-chloro-3-ethylpent-1-en-4-yn-3-ol (ethchlorvynol)

- main chain is pentane (C_5)
  number it

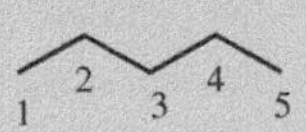

- put in unsaturation
  1-ene (=1,2-ene)
  4-yne (=4,5-yne)

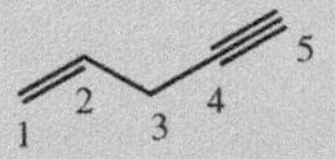

- put in substituents
  1-chloro
  3-ethyl
  3-hydroxy (3-ol)

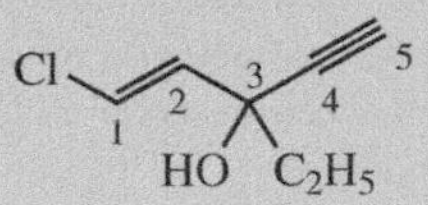

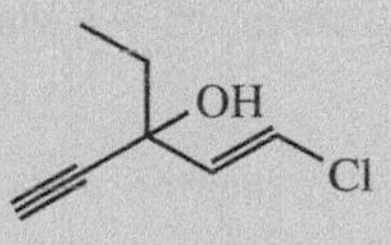

ethchlorvynol

4-aminohex-5-enoic acid (vigabatrin)

- main chain is hexane (C_6)
  number it

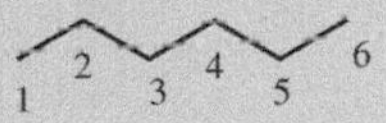

- put in unsaturation
  5-ene (=5,6-ene)

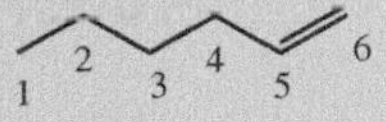

- main functional group is an acid (-oic acid)
  this will be carbon-1

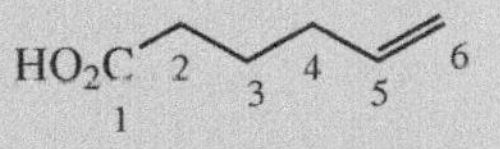

- put in substituent
  4-amino

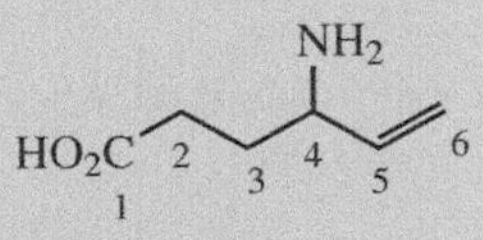

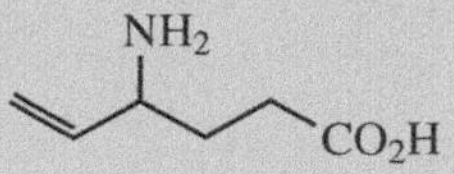

vigabatrin

2-(2-chlorophenyl)-2-methylaminocyclohexanone (ketamine)

- main chain is cyclohexane (C_6) number it

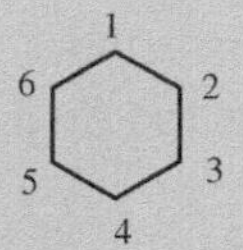

- main functional group is a ketone (-one) this will be carbon-1

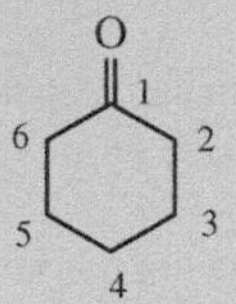

- put in substituents
2-methylamino = 2-amino carrying a methyl (contrast aminomethyl = methyl carrying an amino)
2-(2-chlorophenyl) = 2-chlorophenyl at position 2; the phenyl carries a chloro substituent at its own position 2;
note the use of brackets to separate the two types of numbering

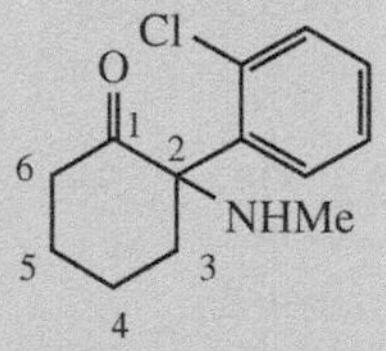

methylamino   aminomethyl   phenyl   2-chlorophenyl

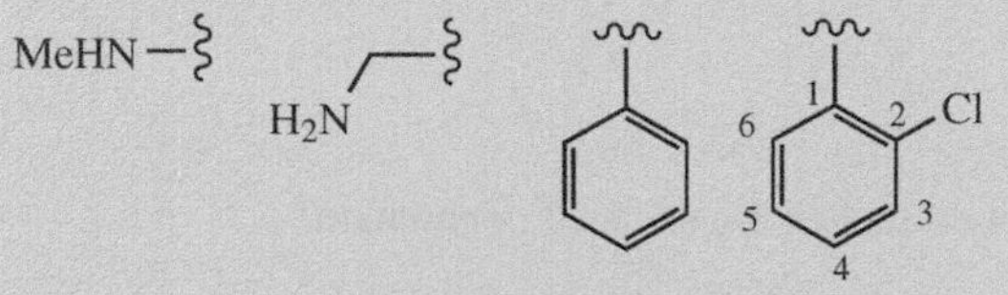

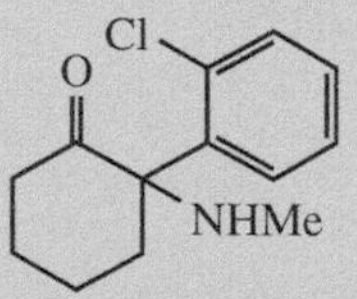

ketamine

5-methyl-2-(2-propyl)-cyclohexanol (menthol)

- main chain is cyclohexane (C_6) number it

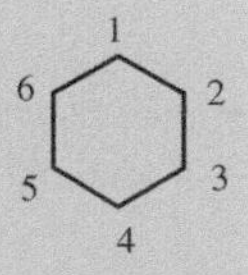

- main functional group is an alcohol (-ol) this will be carbon-1

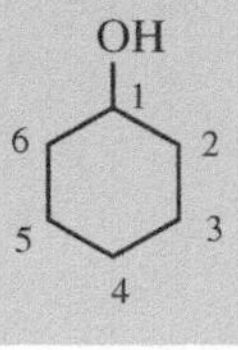

- put in substituents
5-methyl
2-(2-propyl) = 2-propyl at position 2; 2-propyl is a propyl group joined via its 2-position

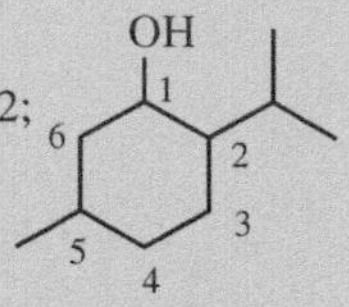

propyl   2-propyl

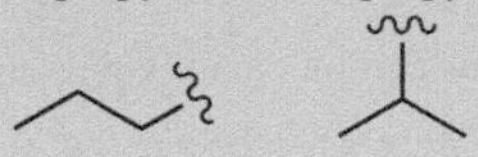

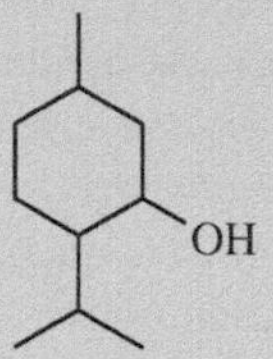

menthol

1-(3,4-dihydroxyphenyl)-2-methylaminoethanol (adrenaline; epinephrine)

- main chain is ethane (C_2) number it

- main functional group is an alcohol (-ol) this will be carbon-1

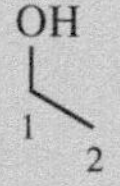

- put in substituents
2-methylamino = 2-amino carrying a methyl
1-(3,4-dihydroxyphenyl) = 3,4-dihydroxyphenyl at position 1;
the phenyl carries hydroxy substituents at its own positions 3 and 4;
note the use of brackets to separate the two types of numbering

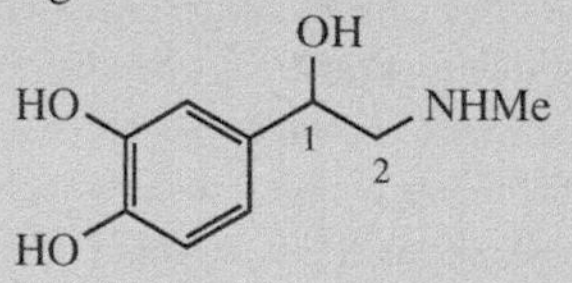

methylamino   phenyl   3,4-dihydroxyphenyl

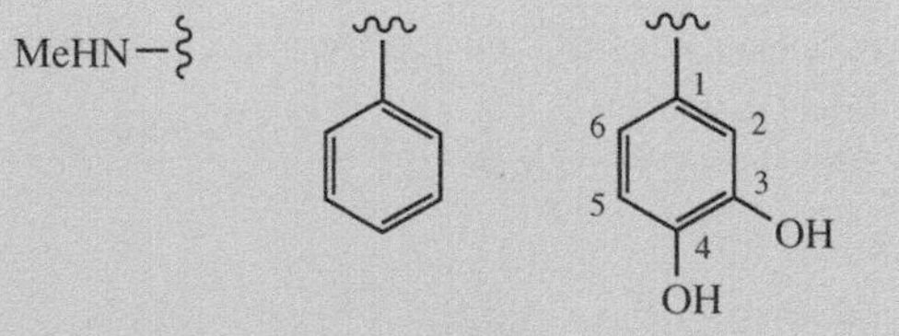

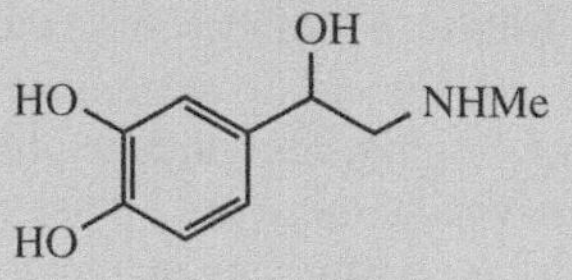

adrenaline

1-benzyl-3-dimethylamino-2-methyl-1-phenylpropyl propionate (dextropropoxyphene)

- this is an ester (two words, -yl -oate)
  the -oate part refers to the acid component, the -yl part to the esterifying alcohol

- main chain of acid is propane (C_3)
  main chain of alcohol is propane (C_3)
  these are numbered separately (the ester has two separate words)

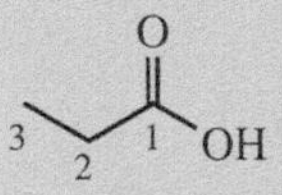

propionic acid   propanol 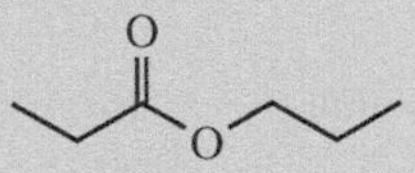   propyl propionate
propanoic acid   propyl propanoate

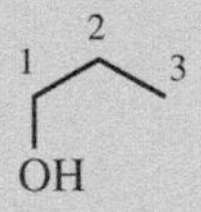

- no substituents on acid component

- put in substituents on alcohol component
  1-phenyl; 1-benzyl; 2-methyl; 3-dimethylamino

phenyl   benzyl 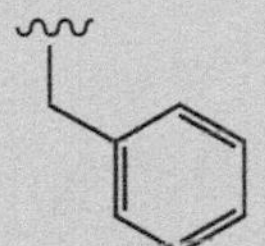   dimethylamino

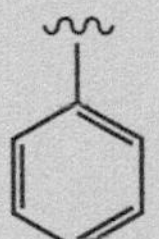

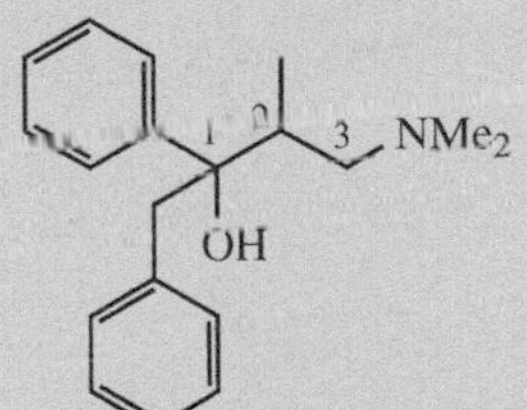

- join with acid component via ester linkage

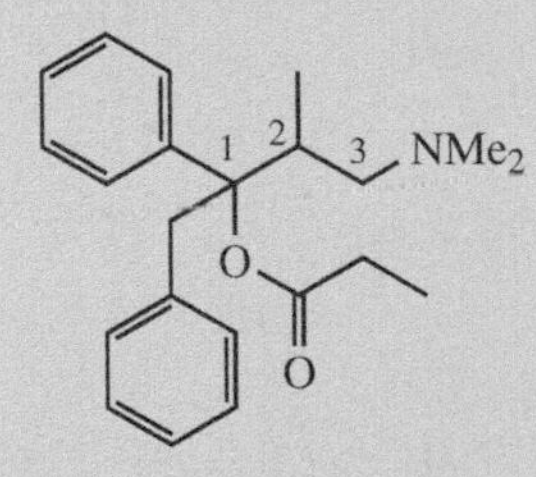

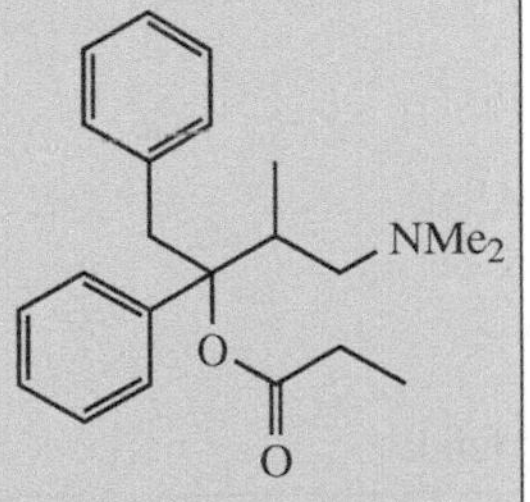

dextropropoxyphene

2-[4-(2-methylpropyl)phenyl]propanoic acid (ibuprofen)

- main chain is propane (C_3)
  number it

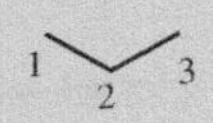

- main functional group is an acid (-oic acid)
  this will be carbon-1

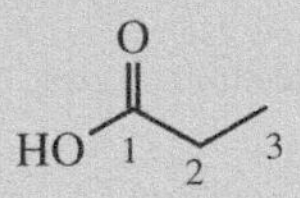

- put in substituents
  consider brackets;
  we have square brackets
  with curved brackets inside
  initially ignore the contents
  of the curved brackets and
  its numbering (4);

2-phenylpropanoic acid

  this reduces to 2-[phenyl]
  propanoic acid, which indicates
  phenyl at position 2 on propanoic acid
  4-(2-methylpropyl)phenyl = 2-methylpropyl
  at position 4 of the phenyl;
  2-methylpropyl = propyl with methyl at position 2
  note the brackets separate different substituents
  and their individual numbering systems

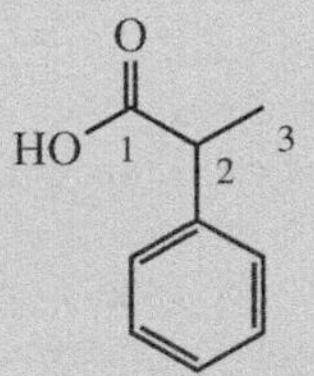

propyl   2-methylpropyl

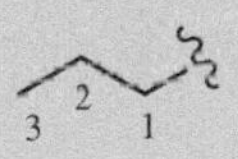

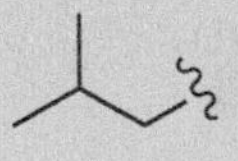

4-(2-methylpropyl)phenyl

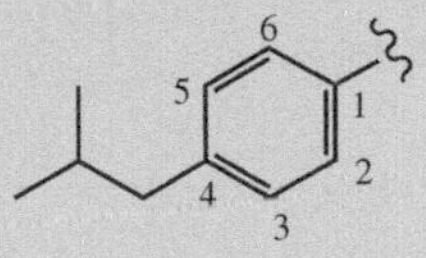

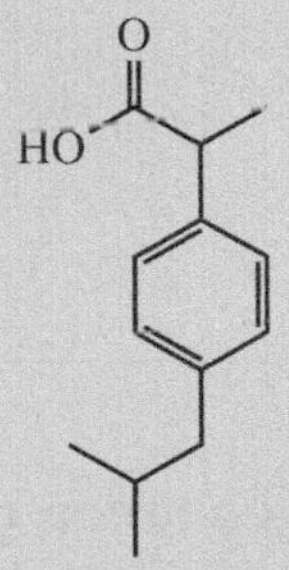

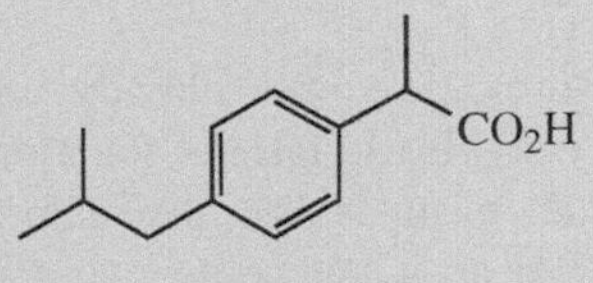

ibuprofen

2-(diethylamino)-*N*-(2,6-dimethylphenyl)acetamide (lidocaine; lignocaine)

- this is an amide; acetamide is the amide of acetic acid (C_2)

- number it; the carbonyl carbon is C-1

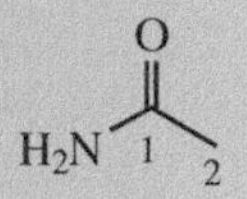

- there are two main substituents, on C-2 and the nitrogen, with brackets to keep the appropriate groups together
  the substituent at C-2 is diethylamino, an amino which is itself substituted with two ethyl groups
  the substituent on the nitrogen is 2,6-dimethylphenyl, a phenyl group substituted at positions 2 and 6 on the phenyl

diethylamino   2,6-dimethylphenyl

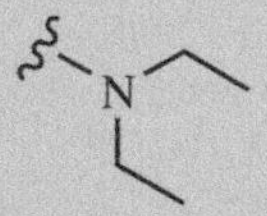

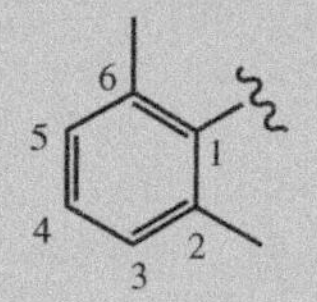

- put in substituents

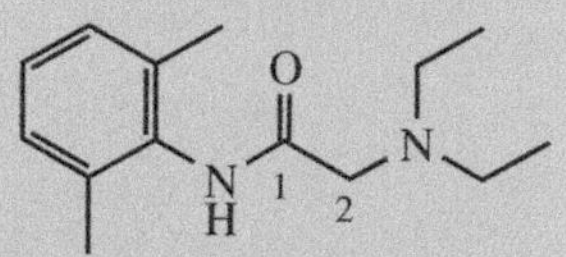

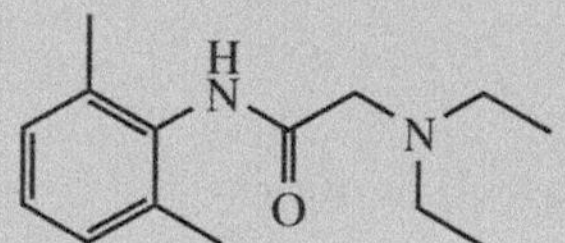

lidocaine

1.5   Common groups and abbreviations

In drawing structures, we are already using a sophisticated series of **abbreviations** for atoms and bonding. Functional groups are also abbreviated further, in that $-CO_2H$ or $-CHO$ convey considerably more information to us than the simple formula does. Other common abbreviations are used to specify particular alkyl or aryl groups in compounds, to speed up our writing of chemistry. It is highly likely that

Table 1.3   Some common structural abbreviations

Group	Abbreviation	Structure
Alkyl	R	
Aryl	Ar	
Methyl	Me	$-CH_3$
Ethyl	Et	$-CH_2CH_3$
Propyl	Pr or *n*-Pr	$-CH_2CH_2CH_3$
Butyl	Bu or *n*-Bu	$-CH_2CH_2CH_2CH_3$
Isopropyl	*i*-Pr or ^{i}Pr	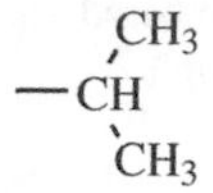
Isobutyl	*i*-Bu or ^{i}Bu	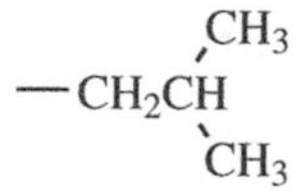
sec-Butyl	*s*-Bu or ^{s}Bu	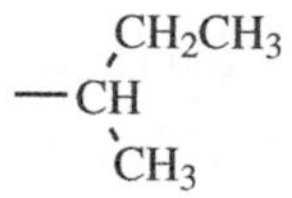
tert-Butyl	*t*-Bu or ^{t}Bu	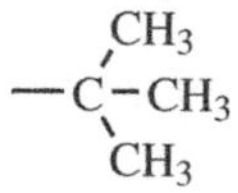
Phenyl	Ph	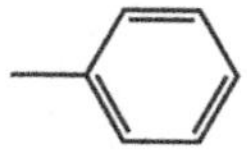
Benzyl	Bn	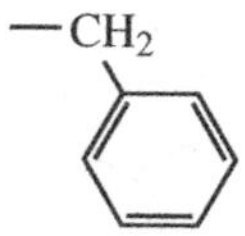
Acetyl	Ac	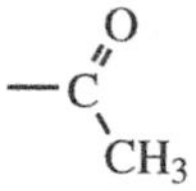
Vinyl		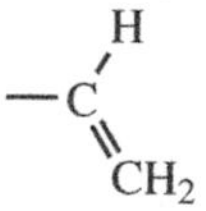
Allyl		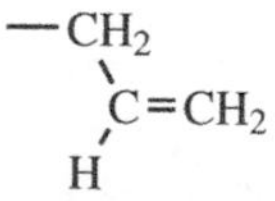
Halide	X	$-F$ $-Cl$ $-Br$ $-I$

some of these are already familiar, such as Me for methyl, and Et for ethyl. Others are included in Table 1.3.

1.6 Common, non-systematic names

Systematic nomenclature was introduced at a relatively late stage in the history of chemistry, and thus **common names** had already been coined for a wide range of chemicals. Because these names were in everyday usage, and familiar to most chemists, a number have been adopted by IUPAC as the approved name, even though they are not systematic. These are thus names that chemists still use, that are used for labelling reagent bottles, and are those under which the chemical is purchased. Some of these are given in Table 1.4, and it may come as a shock to realize that the systematic names school chemistry courses have provided will probably have to be 'relearned'.

The use of the old terminology *n-* (normal) for unbranched hydrocarbon chains, with *i-* (iso), *s-* (secondary), *t-* (tertiary) for branched chains is still quite common with small molecules, and can be acceptable in IUPAC names.

Table 1.4 Common, non-systematic names

Structure	Systematic name	IUPAC approved name
$H_2C{=}CH_2$	Ethene	Ethylene
$HC{\equiv}CH$	Ethyne	Acetylene
HCO_2H	Methanoic acid	Formic acid
$HCO_2CH_2CH_3$	Ethyl methanoate	Ethyl formate
CH_3CO_2H	Ethanoic acid	Acetic acid
$CH_3CO_2CH_2CH_3$	Ethyl ethanoate	Ethyl acetate
CH_3COCH_3	Propan-2-one	Acetone
CH_3CHO	Ethanal	Acetaldehyde
$CH_3C{\equiv}N$	Ethanenitrile	Acetonitrile
⬡—CH₃	Methylbenzene	Toluene
⬡—OH	Hydroxybenzene	Phenol
⬡—NH₂	Benzenamine	Aniline
$H_3C{-}NH_2$	Methanamine	Methylamine
⬡—CO₂H	Benzenecarboxylic acid	Benzoic acid

1.7 Trivial names for complex structures

Biochemical and natural product structures are usually quite complex, some exceedingly so, and fully systematic nomenclature becomes impracticable. Names are thus typically based on so-called **trivial nomenclature**, in which the discoverer of the natural product exerts his or her right to name the compound. The organism in which the compound has been found is frequently chosen to supply the root name, e.g. hyoscyamine from *Hyoscyamus*, atropine from *Atropa*, or penicillin from *Penicillium*. Name suffixes might be -in to indicate 'a constituent of', -oside to show the compound is a sugar derivative, -genin for the aglycone released by hydrolysis of the sugar derivative, -toxin for a poisonous constituent, or they may reflect chemical functionality, such as -one or -ol. Traditionally, -ine is always used for alkaloids (am*ine*s).

Structurally related compounds are then named as derivatives of the original, using standard prefixes, such as hydroxy-, methoxy-, methyl-, dihydro-, homo-, etc. for added substituents, or deoxy-, demethyl-, demethoxy-, dehydro-, nor-, etc. for removed substituents. Homo- is used to indicate one carbon more, whereas nor- means one carbon less. The position of this change is then indicated by systematic numbering of the carbon chains or rings. Some groups of compounds, such as steroids and prostaglandins, are named semi-systematically from an accepted root name for the complex hydrocarbon skeleton. Drug names chosen by pharmaceutical manufacturers are quite random, and have no particular relationship to the chemical structure.

1.8 Acronyms

Some of the common reagent chemicals and solvents are usually referred to by acronyms, a sequence of letters derived from either the systematic name or a trivial name. We shall encounter some of these in due course, and both name and acronym will be introduced when we first meet them. For reference purposes, those we shall meet are also listed in Table 1.5. Far more examples occur with biochemicals. Those indicated cover many, but the list is not comprehensive.

Table 1.5 Some common acronyms

Acronym	Chemical/biochemical name
Reagents and solvents	
DCC	Dicyclohexylcarbodiimide
DMF	Dimethylformamide
DMSO	Dimethylsulfoxide
LAH	Lithium aluminium hydride
LDA	Lithium di-isopropylamide
mCPBA	*meta*-Chloroperoxybenzoic acid
NBS	*N*-Bromosuccinimide
PTSA	*para*-Toluenesulfonic acid
tBOC	*tert*-Butyloxycarbonyl
THF	Tetrahydrofuran
Biochemicals	
ADP	Adenosine diphosphate
AMP	Adenosine monophosphate
ATP	Adenosine triphosphate
CDP	Cytidine diphosphate
CTP	Cytidine triphosphate
DNA	Deoxyribonucleic acid
FAD	Flavin adenine dinucleotide
$FADH_2$	Flavin adenine dinucleotide (reduced)
FMN	Flavin mononucleotide
$FMNH_2$	Flavin mononucleotide (reduced)
GDP	Guanosine diphosphate
GTP	Guanosine triphosphate
NAD^+	Nicotinamide adenine dinucleotide
NADH	Nicotinamide adenine dinucleotide (reduced)
$NADP^+$	Nicotinamide adenine dinucleotide phosphate
NADPH	Nicotinamide adenine dinucleotide phosphate (reduced)
PLP	Pyridoxal 5′-phosphate
RNA	Ribonucleic acid
SAM	*S*-Adenosylmethionine
TPP	Thiamine diphosphate
UDP	Uridine diphosphate
UTP	Uridine triphosphate

1.9 Pronunciation

As you listen to chemists talking about chemicals, you will soon realize that there is no strict protocol for pronunciation. Even simple words like ethyl produce a variety of sounds. Many chemists say 'eethyle', but the Atlantic divide gives us 'ethel' with short 'e's, and continental European chemists often revert to the German pronunciation 'etool'. There is little to guide us in the words themselves, since methane is pronounced 'meethayne' whilst methanol tends to have short 'e', 'a', and 'o', except for occasional cases, mainly European, when it may get a long 'o'. On the other hand, propanol always seems to have the first 'o' long, and the second one short. Vinyl can be 'vinil' or 'vynyl' according to preference, and amino might be 'ameeno' or 'amyno'. Need we go on? Your various teachers will probably pronounce some common words quite differently. Try to use the most commonly accepted pronunciations, and don't worry when a conversation with someone involves differences in pronunciation. As long as there is mutual understanding, it's not really important how we say it. By and large, chemists are a very tolerant group of people.

2

Atomic structure and bonding

2.1 Atomic structure

Atoms are composed of protons, neutrons and electrons. **Protons** are positively charged, **electrons** carry a negative charge, and **neutrons** are uncharged. In a neutral atom, the nucleus of protons and neutrons is surrounded by electrons, the number of which is equal to the number of protons. This number is also the same as the **atomic number** of the atom. If the number of electrons and protons is not equal, the atom or molecule containing the atom will necessarily carry a charge, and is called an **ion**. A negatively charged atom or molecule is termed an **anion**, and a positively charged species is called a **cation**.

The inert or **noble gases**, such as helium, neon, and argon, are particularly unreactive, and this has been related to the characteristic number of electrons they contain, 2 for helium, 10 for neon $(2 + 8)$, and 18 for argon $(2 + 8 + 8)$. They are described as possessing 'filled shells' of electrons, which, except for helium, contain eight electrons, an **octet**. Acquiring a noble gas-like complement of electrons governs the bonding together of atoms to produce molecules. This is achieved by losing electrons, by gaining electrons, or by sharing electrons associated with the unfilled shell, and leads to what we term ionic bonds or covalent bonds. The unfilled shell involved in bonding is termed the **valence shell**, and the electrons in it are termed **valence electrons**.

2.2 Bonding and valency

For many years now, these types of bonding have been represented in chemistry via a shorthand

notation. **Ionic bonds** have been shown as a simple electrostatic interaction of appropriate counter ions, so that sodium chloride and magnesium chloride are conveniently drawn as Na^+Cl^- and Mg^{2+} $2Cl^-$ respectively. It becomes increasingly difficult to remove successive electrons from an atom, and ionic bonding is not usually encountered for some atoms, especially carbon. Organic chemistry, the study of carbon compounds, is dominated by covalent bonding and the sharing of electrons.

A **covalent bond** between atoms involves the sharing of two electrons, one from each atom. The sharing of two electrons is described as a **single bond**, and is indicated in shorthand notation by a single line. Depending upon the number of electrons an atom carries, it is able to form a certain number of bonds, and this number is called the **valency** of the atom. The valency of hydrogen is 1, of oxygen 2, of nitrogen 3, and carbon 4. This means that we can indicate the bonding in simple organic molecules such as methane, methanol, and methylamine via single bonds (see Section 1.1).

single bonds

methane   methanol   methylamine

Carbon is particularly versatile, in that it can sometimes share two of its electrons with a second carbon, with nitrogen, or with oxygen. It can even use three of its four valencies in bonding to another carbon, or to nitrogen. In this way, we generate

Essentials of Organic Chemistry   Paul M Dewick
© 2006 John Wiley & Sons, Ltd

double and **triple bonds**, indicated by two or three adjacent lines in our molecular representations (see Section 1.1).

double bonds

ethylene
(ethene)

formaldehyde
(methanone)

triple bonds

acetylene
(ethyne)

acetonitrile
(ethanenitrile)

These are extensions of **Lewis dot structures**, where bonding electrons associated with each bond are shown as dots. In our simple structures, bonding is associated with eight electrons in the valence shell of the atom, unless it is hydrogen, when two electrons are required for bonding. Whilst we have almost completely abandoned putting in electron dots for bonds, we still routinely show some pairs of electrons not involved in bonding (**lone pairs**) because these help in our mechanistic rationalizations of chemical reactions.

methane

methanol

formaldehyde
(methanone)

This system has its merits and uses – indeed, we shall employ the line notation almost exclusively – but to understand how bonding occurs, and to explain molecular shape and chemical reactivity, we need to use orbital concepts.

2.3 Atomic orbitals

The electrons in an atom surround the nucleus, but are constrained within given spatial limits, defined by atomic orbitals. **Atomic orbitals** describe the probability of finding an electron within a given space. We are unable to pin-point the electron at any particular time, but we have an indication that it will be within certain spatial limits. A farmer knows his cow is in a field, but, at any one time, he does not know precisely where it will be located. Even this is not a good analogy, because electrons do not behave as nice, solid particles. Their behaviour is in some respects like that of waves, and this can best be analysed through mathematics.

Atomic orbitals are actually graphical representations for mathematical solutions to the **Schrödinger wave equation**. The equation provides not one, but a series of solutions termed **wave functions** ψ. The square of the wave function, ψ^2, is proportional to the electron density and thus provides us with the probability of finding an electron within a given space. Calculations have allowed us to appreciate the shape of atomic orbitals for the simplest atom, i.e. hydrogen, and we make the assumption that these shapes also apply for the heavier atoms, like carbon.

Each wave function is defined by a set of **quantum numbers**. The first quantum number, the **principal quantum number** n, generally relates to the distance of the electron from the nucleus, and hence the energy of the electron. It divides the orbitals into groups of similar energies called **shells**. The principal quantum number also defines the row occupied by the atom in the periodic table. It has integral values, $n = 1, 2, 3, 4$, etc. The numerical values are used to describe the shell.

The second quantum number, the **orbital angular momentum quantum number** l, is generally related to the shape of the orbital and depends upon n, taking integral values from 0 to $n - 1$. The different values are always referred to by letters: s for $l = 0$, p for $l = 1$, d for $l = 2$, and f for $l = 3$.

The third quantum number is related to the orientation of the orbital in space. It is called the **magnetic quantum number** m_l, and depends upon l. It can take integral values from $-l$ to $+l$. For p orbitals, suffix letters are used to define the direction of the orbital along the x-, y-, or z-axes. Organic chemists seldom need to consider subdivisions relating to d orbitals.

Finally, there is the **spin quantum number** s, which may have only two values, i.e. $\pm\frac{1}{2}$. This relates to the angular momentum of an electron spinning on its own axis. The magnitude of an electron's spin is constant, but it can take two orientations.

Table 2.1 shows the possible combinations of quantum numbers for $n = 1$ to 3.

For a hydrogen atom, the lowest energy solution of the wave equation describes a spherical region about the nucleus, a **1s atomic orbital**. When the wave equation is solved to provide the next higher energy level, we also get a spherical region of high probability, but this **2s orbital** is further away from the nucleus than the 1s orbital. It also contains a node, or point of zero probability within the sphere

Table 2.1 Quantum number combinations and atomic orbitals

Principle quantum number n	Orbital angular momentum quantum number l	Magnetic quantum number m_l	Spin quantum number s	Atomic orbital designation
1	0	0	$\pm1/2$	1s
2	0	0	$\pm1/2$	2s
2	1	-1	$\pm1/2$	2p
2	1	0	$\pm1/2$	2p
2	1	$+1$	$\pm1/2$	2p
3	0	0	$\pm1/2$	3s
3	1	-1	$\pm1/2$	3p
3	1	0	$\pm1/2$	3p
3	1	$+1$	$\pm1/2$	3p
3	2	-2	$\pm1/2$	3d
3	2	-1	$\pm1/2$	3d
3	2	0	$\pm1/2$	3d
3	2	$+1$	$\pm1/2$	3d
3	2	$+2$	$\pm1/2$	3d

of high probability. Radial probability density plots (Figure 2.1) showing the probability of finding an electron at a particular distance from the nucleus are presented for the 1s and 2s orbitals, to illustrate the node in the 2s orbital.

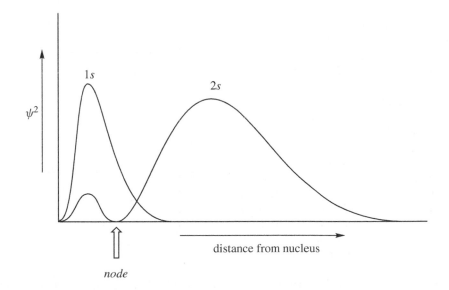

Figure 2.1 Radial probability density plots for 1s and 2s orbitals of hydrogen atom

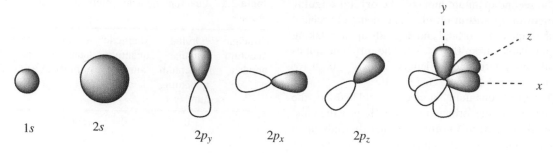

Figure 2.2   Shapes of atomic orbitals

To appreciate the node concept, it is useful to think of wave analogies. Thus, a vibrating string might have no nodes, one node, or several nodes according to the frequency of vibration. We can also realize that the wave has different phases, which we can label as positive or negative, according to whether the lobe is above or below the median line.

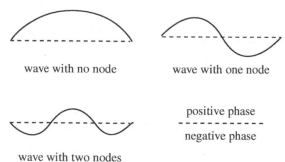

Then follow three additional atomic orbitals, which are roughly dumbbell or propeller-like in appearance. These are aligned along mutually perpendicular axes, and are termed the **$2p_x$, $2p_y$, and $2p_z$ orbitals**. These orbitals possess major probability regions either side of the nucleus, but zero electron probability (a node) at the nucleus. In one lobe of the orbital the **phase sign of the wave function** is positive; in the other it is negative. To avoid confusion with electrical charge, the phase sign of the wave function is usually indicated by shading of the lobes; in everyday usage we may draw them without either sign or shading. These three orbitals are of equal energy, somewhat higher than that of the $2s$ orbital. We use the term **degenerate** to describe orbitals of identical energy. The general appearance of these orbitals is shown in Figure 2.2.

Consideration of $1s$, $2s$, and $2p$ orbitals will allow us to describe the electronic and bonding

characteristics for most of the atoms encountered in organic molecules. Atoms such as sulfur and phosphorus need **$3s$ and $3p$ orbitals** to be utilized, after which five more-complex **$3d$ orbitals** come into play. As the principal quantum number increases, so the average radius of the s orbitals or the length of the lobes of p orbitals increases, and the electrons in the higher orbitals are thus located further from the nucleus. Each subsequent orbital is also at a higher energy level (Figure 2.3). These energy levels can be calculated from the wave function. They may also be measured directly from **atomic spectra**, where lines correspond to electrons moving between different energy levels. As the relative energy levels in Figure 2.3 show, $4s$ orbitals are actually of lower energy than $3d$ orbitals.

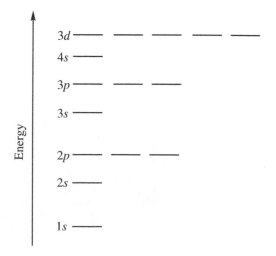

Figure 2.3   Relative energies of atomic orbitals (not to scale)

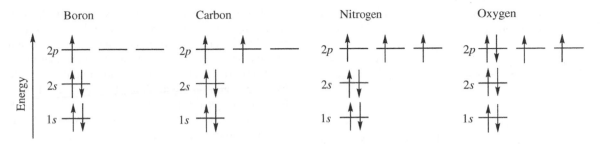

Figure 2.4 Electronic configurations: energy diagrams

2.4 Electronic configurations

Each atomic orbital can accommodate just two electrons, provided these can be paired by virtue of having opposite spin quantum numbers. If the spins are the same, then the electrons must be located in different orbitals. We can now describe the **electronic configuration** for atoms of interest in the first two rows of the periodic table. Electrons are allocated to atomic orbitals, one at a time, so that orbitals of one energy level are filled before proceeding to the next higher level. Where electrons are placed in orbitals of the same energy (degenerate orbitals, e.g. p orbitals) they are located singly in separate orbitals before two electrons are paired. Further, the electronic configuration with the greatest number of parallel spins (same spin quantum number) results in the lowest energy overall.

The electronic configuration can be expressed as a list of those orbitals containing electrons, as shown below. Although it is usual just to indicate the number of electrons in p orbitals, e.g. C: $1s^2 2s^2 2p^2$ as in the first column, it is more informative to use the second column designation, i.e. C: $1s^2 2s^2 2p_x^{\,1} 2p_y^{\,1}$, where the non-pairing of electrons is emphasized. As more $2p$ electrons are allocated, pairing becomes

H	$1s^1$
He	$1s^2$
Li	$1s^2 2s^1$
Be	$1s^2 2s^2$
B	$1s^2 2s^2 2p^1$ or $1s^2 2s^2 2p_x^{\,1}$
C	$1s^2 2s^2 2p^2$ or $1s^2 2s^2 2p_x^{\,1} 2p_y^{\,1}$
N	$1s^2 2s^2 2p^3$ or $1s^2 2s^2 2p_x^{\,1} 2p_y^{\,1} 2p_z^{\,1}$
O	$1s^2 2s^2 2p^4$ or $1s^2 2s^2 2p_x^{\,2} 2p_y^{\,1} 2p_z^{\,1}$
F	$1s^2 2s^2 2p^5$ or $1s^2 2s^2 2p_x^{\,2} 2p_y^{\,2} 2p_z^{\,1}$
Ne	$1s^2 2s^2 2p^6$ or $1s^2 2s^2 2p_x^{\,2} 2p_y^{\,2} 2p_z^{\,2}$

obligatory, e.g. O: $1s^2 2s^2 2p_x^{\,2} 2p_y^{\,1} 2p_z^{\,1}$. There is no hidden meaning in allocating electrons to $2p_x$ first. In any case, we are unable to identify which of these orbitals is filled first.

Alternatively, we can use the even more informative energy diagram (Figure 2.4). Electrons with different spin states are then designated by upward ($\uparrow$) or downward ($\downarrow$) pointing arrows. Note particularly that the **noble gas** neon has enough electrons to fill the orbitals of the '2' shell completely; it has a total of eight electrons in these orbitals, i.e. an **octet**. In the case of helium, the '1' shell orbital is filled with two electrons. The next most stable electronic configurations are those of argon, $1s^2 2s^2 2p^6 3s^2 3p^6$, and then krypton, $1s^2 2s^2 2p^6 3s^2 3p^6 4s^2 3d^{10} 4p^6$. The filled electron shells are especially favourable and responsible for the lack of reactivity of these two elements. Attaining **filled shells** is also the driving force behind bonding.

2.5 Ionic bonding

The simplest type of bonding to comprehend is **ionic bonding**. This involves loss of an electron from one atom, and its transfer to another, with bonding resulting from the strong electrostatic attraction. For this ionic bonding, the electron transfer is from an atom with a low **ionization potential** to an atom with high electron affinity, and the atomic objective is to mimic for each atom the nearest noble gas electronic configuration.

Let us consider sodium and chlorine. Sodium ($1s^2 2s^2 2p^6 3s^1$) has one electron more than neon ($1s^2 2s^2 2p^6$), and chlorine ($1s^2 2s^2 2p^6 3s^2 3p^5$) has one electron less than the noble gas argon ($1s^2 2s^2 2p^6 3s^2 3p^6$). Chlorine has high **electronegativity** (see Section 2.7) and acquires one electron to become a chloride anion Cl^-. Sodium loses one electron to become the cation Na^+.

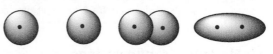

The simplest of the cations we encounter is H^+. This is the result of a hydrogen atom losing an electron, and simple arithmetic tells us that this entity now has no electrons, being composed of just a proton. We thus refer to H^+ as a **proton**, and combination with H^+ as **protonation**.

In favourable circumstances, we may see more than one electron being donated/acquired, e.g. Mg^{2+} O^{2-}, though the more electrons involved the more difficult it is to achieve the necessary ionizations. Molecules such as methane, CH_4, are not obtained through ionic bonding, but through the covalent electron-sharing mechanism.

2.6 Covalent bonding

2.6.1 Molecular orbitals: σ and π bonds

We have used the electronic energy levels for atomic hydrogen to serve as a model for other atoms. In a similar way, we can use the interaction of two hydrogen atoms giving the hydrogen molecule as a model for bonding between other atoms. In its simplest form, we can consider the bond between

two hydrogen atoms originates by bringing the two atoms together so that the atomic orbitals **overlap**, allowing the electrons from each atom to mingle and become associated with both atoms. This sharing of electrons effectively brings each atom up to the noble gas electronic configuration (He, two electrons). Furthermore, it creates a new orbital spanning both atoms in which the two electrons are located; this is called a **molecular orbital**.

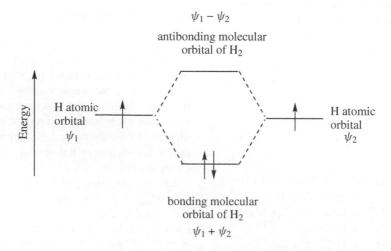

$1s$ atomic orbitals of   overlap of   molecular orbital
two hydrogen atoms   orbitals   of H_2 molecule

Graphically, we can represent this as in Figure 2.5.

There must be some energy advantage by bonding, otherwise it would not occur. The two atomic orbitals, therefore, are used to create a new molecular orbital of rather lower energy, the **bonding molecular orbital**. However, since we are considering mathematical solutions to a wave equation, there is an alternative higher energy solution also possible. Remind yourself that the solution to $x^2 = 1$ is $x = +1$ or -1. The higher energy solution is represented by the **antibonding molecular orbital**. The bonding molecular orbital is where combination of atomic orbitals leads to an increased probability of finding the electrons between the two atoms, i.e. bonding. The antibonding molecular orbital is where combination of atomic orbitals leads to a reduced or negligible

Figure 2.5 Energy diagram: molecular orbitals of hydrogen molecule

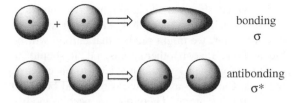

Figure 2.6   Molecular orbitals: the σ bond

probability of finding the electrons between the two atoms, and does not produce bonding (Figure 2.6).

Combining the two atomic orbitals produces two molecular orbitals, here shown as $\psi_1 + \psi_2$, and $\psi_1 - \psi_2$: in the additive mode the electronic probability increases between the atoms, whereas in the subtractive mode the electronic probability between the atoms decreases, i.e. the antibonding situation. Bonding results where we have interaction of orbitals with the same **phase sign of the wave function**, whereas the antibonding orbital originates from interaction of orbitals with different phase signs of the wave function. This approach to molecular orbitals is called a **linear combination of atomic orbitals**: wave functions for the atomic orbitals are combined in a linear fashion, by simple addition or subtraction, to generate new wave functions for molecular orbitals. The number of molecular orbitals formed is the same as the number of atomic orbitals combined. Electrons are allocated to the resultant molecular orbitals as with atomic orbitals. We start with

the lower energy orbital, putting one electron in each degenerate orbital, before we add a second with spin pairing. In the case of hydrogen, therefore, we have two spin-paired electrons in the bonding molecular orbital. The antibonding orbital remains empty in the so-called **ground state** of the molecule, unless we input enough energy to promote one electron to the higher energy state, the **excited state**. This type of transfer gives rise to spectral absorption or emission.

The bonding in the hydrogen molecule formed by overlap of s orbitals is called a **sigma (σ) bond**; the antibonding orbital is designated σ*. It is a term generally applied where orbital overlap gives a bond that is cylindrically symmetrical in cross-section when viewed along the bond axis. All single bonds are sigma bonds. The other important type of bonding in organic molecules is the **pi (π) bond**, the result of side-to-side interaction of p orbitals. Here, we consider the two lobes separately overlapping; the p orbitals have lobes of different phase signs, and for bonding we require overlap of lobes with the same phase sign (Figure 2.7). This produces a bonding π molecular orbital with regions of greatest probability of finding electrons above and below the atomic axis. The π bond thus has a nodal plane passing through the bonded atoms. The antibonding π* orbital can be deduced in a similar manner. Double and triple bonds are characterized by π bonding. π bonds possess an enhanced reactivity not associated with σ bonds.

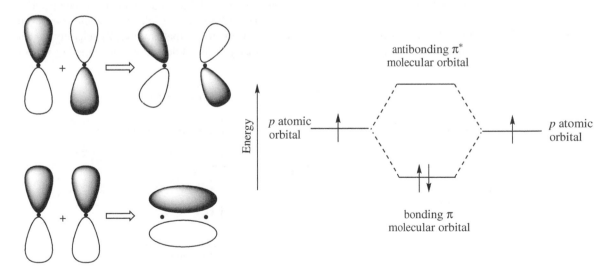

Figure 2.7   Molecular orbitals: the π bond

2.6.2 Hybrid orbitals in carbon

We start this section with a word of caution. Students frequently find **hybridization** a rather difficult concept to understand and appreciate. However, there is no particular reason why this should be so.

Chemistry is an experimental science, and to rationalize our observations we gradually develop and invoke a number of rules and principles. Theories may have to change as scientific data increase, and as old principles cease to explain the facts. All of the foregoing description of atomic and molecular orbitals is a hypothesis for atomic and molecular structure supported by experimental data. So far, the description meets most of our needs and provides a good rationalization of chemical behaviour. However, it falls short in certain ways, and we have to invoke a further modification to explain the facts. Here are three observations based upon sound experimental evidence, which are not accommodated by the above description of bonding:

- The hydrocarbon **methane** (CH_4) is tetrahedral in shape with bond angles of about 109°, and the four C–H bonds are all equivalent and identical in reactivity.

- **Ethylene** (ethene, C_2H_4) is planar, with bond angles of about 120°, and it contains one π bond.

- **Acetylene** (ethyne, C_2H_2) is linear, i.e. bond angles 180°, and it contains two π bonds.

None of these observations follows immediately from the electronic configuration of carbon ($1s^2 2s^2 2p_x^1$

$2p_y^1$), which shows that carbon has two unpaired electrons, each in a $2p$ orbital. From our study of bonding so far, we might predict that carbon will be able to bond to two other atoms, i.e. it should be divalent, though this would not lead to an octet of electrons. Carbon is usually tetravalent and bonds to up to four other atoms. Therefore, we need to modify the model to explain this behaviour. This modification is **hybridization**.

sp^3 hybrid orbitals

Methane is a chemical combination of one carbon atom and four hydrogen atoms. Each hydrogen atom contributes one electron to a bond; so, logically, carbon needs to provide four unpaired electrons to allow formation of four σ bonds. The ability of carbon to bond to four other atoms requires unpairing of the $2s^2$ electrons. We might consider promoting one electron from a $2s$ orbital to the third, as yet unoccupied, $2p$ orbital (Figure 2.8). This would produce an excited-state carbon; since the $2p$ orbital is of higher energy than the $2s$ orbital, the process would require the input of energy. We could assume that the ability to form extra bonds would more than compensate for this proposed change. We now have four unpaired electrons in separate orbitals, and the electronic configuration of carbon has become $1s^2 2s 2p_x^1 2p_y^1 2p_z^1$. Each electron can now form a bond by pairing with the electron of a hydrogen atom.

However, this does not explain why methane is tetrahedral and has four equivalent bonds. The bond that utilizes the $2s$ electron would surely be different from those that involve $2p$ electrons, and the geometry of the molecule should somehow reflect

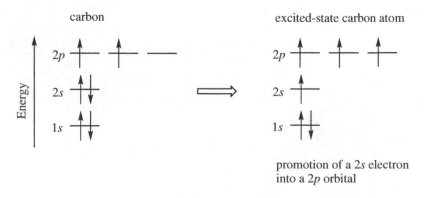

Figure 2.8   Electronic configuration: excited-state carbon atom

that the *p* orbitals are positioned at right angles to each other, whilst the *s* orbital is spherical and might bond in any direction. If all the bonds in methane turn out to be equivalent, they must be some sort of 'hybrid' version of those we have predicted. We can explain many features of organic chemicals, including their reactivity and shape, by a mathematical model in which **hybrid orbitals** for carbon are derived by mixing the one 2*s* orbital and three 2*p* atomic orbitals (Figure 2.9). This generates four equivalent hybrid orbitals, which we designate sp^3, since they are derived from one *s* orbital and three *p* orbitals. The sp^3 **orbitals** will be at an energy level intermediate between those of the 2*s* and 2*p* orbitals, and will have properties intermediate between *s* and *p*, though

with greater *p* character. The mathematical model then provides us with the shape and orientation of these hybrid orbitals (Figure 2.10). For convenience of drawing, we tend to omit the small lobes at the centre of the array.

These new hybrid orbitals are then all equivalent, and spaced to minimize any interaction; this is a tetrahedral array, the best way of arranging four groups around a central point. Each hybrid orbital can now accommodate one electron.

Now we can consider the bonding in **methane**. Using orbital overlap as in the hydrogen molecule as a model, each sp^3 orbital of carbon can now overlap with a 1*s* orbital of a hydrogen atom, generating a bonding molecular orbital, i.e. a σ bond. Four such

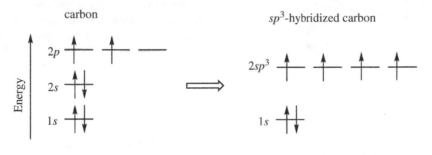

Figure 2.9 Electronic configuration: sp^3-hybridized carbon atom

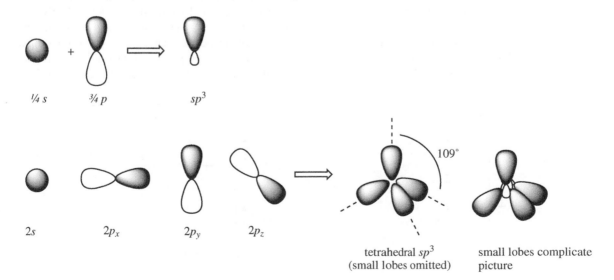

Figure 2.10 sp^3 hybrid orbitals

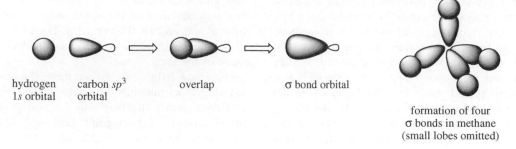

hydrogen
1s orbital

carbon sp^3
orbital

overlap

σ bond orbital

formation of four
σ bonds in methane
(small lobes omitted)

Figure 2.11 Bonding in methane

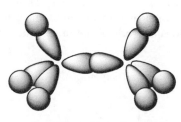

formation of one C−C and
six C−H σ bonds in ethane

Figure 2.12 Bonding in ethane

bonds can be created, and they will be produced in a tetrahedral array (Figure 2.11).

We can also consider C–C σ bonding, as in **ethane** (C_2H_6), by overlap of two carbon sp^3 orbitals. The three remaining sp^3 orbitals of each carbon are used to make C–H σ bonds to hydrogen atoms (Figure 2.12).

It may be argued that we have actually started from the tetrahedral array in methane to propose a tetrahedral array of atomic orbitals in carbon.

This is undoubtedly true, but is part of the process of refining the model as we need to explain new observations. We make models to describe nature; nature merely adopts a minimum energy situation. We gain confidence in the approach by using similar rationale to account for the second of the observations above, that ethylene is planar, with bond angles of about 120°, and contains one π bond.

sp^2 hybrid orbitals

The sp^3 hybrid orbitals of carbon were considered as a mix of the $2s$ orbital with three $2p$ orbitals. To provide a model for ethylene, we now need to consider hybrid orbitals that are a mix of the $2s$ orbital with two $2p$ orbitals, giving three equivalent **sp^2 orbitals**. In this case, we use just three orbitals to create three new hybrid orbitals. Accordingly, we find that the energy level associated with an sp^2 orbital will be below that of the sp^3 orbital: this time, we have mixed just two high-energy p orbitals with the lower energy s orbital (Figure 2.13). The

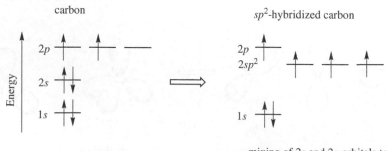

carbon

sp^2-hybridized carbon

Energy

2p

2s

1s

2p
$2sp^2$

1s

mixing of 2s and 2p orbitals to
create sp^2 hybrid orbitals

Figure 2.13 Electronic configuration: sp^2-hybridized carbon atom

four electrons are accommodated one in each hybrid orbital and one in the remaining $2p$ orbital.

The sp^2 hybrid orbitals are distributed in a planar array around the atom; this spacing minimizes any interactions. The $2p$ orbital is then located perpendicular to this plane. Such information is again obtained from the mathematical analysis, but simple logic would lead us to predict that this is the most favourable arrangement to incorporate the components. The sp^2 orbital will be similar in shape to the sp^3 orbital, but somewhat shorter and fatter, in that it has more s character and less p character (Figure 2.14).

The bonding in **ethylene** is based initially on one C–C σ bond together with four C–H σ bonds, much as we have seen in ethane. We are then left with a p orbital for each carbon, each carrying one electron, and these interact by side-to-side overlap to produce a π bond (Figure 2.15). This makes the ethylene molecule planar, with bond angles of 120°, and the π bond has its electron density above and below this plane. The combination of the C–C σ bond and the C–C π bond is what we refer to as a **double bond**; note that we cannot have π bond formation without the accompanying σ bond. You will observe that it becomes progressively more difficult to draw a combination of σ and π molecular orbitals to illustrate the bonding that constitutes a double bond. We often resort to a picture that illustrates the potential overlap of p orbitals by means of a dotted line or similar device. This cleans up the picture, but leaves rather more to the imagination. The properties of an alkene (like ethylene) are special, in that the π bond is more reactive than the σ bond, so that alkenes show a range of properties that alkanes (like ethane) do not (see Chapter 8).

We can only get overlap of the p orbitals if their axes are parallel. If their axes were perpendicular, then there would be no overlap and, consequently, no bonding (Figure 2.16). This situation might arise if we tried to twist the two parts of the ethylene molecule about the C–C link. This is not easily achieved, and would require a lot of energy (see Section 3.4.3). It can be achieved by absorbing sufficient energy to promote an electron to the antibonding π^* orbital. This temporarily destroys the π bond, allows rotation about the remaining σ bond, and the π bond may reform as the electron is restored

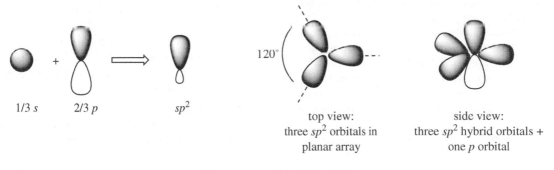

| 1/3 s | 2/3 p | sp^2 |

top view:
three sp^2 orbitals in
planar array

side view:
three sp^2 hybrid orbitals +
one p orbital

Figure 2.14 sp^2 hybrid orbitals

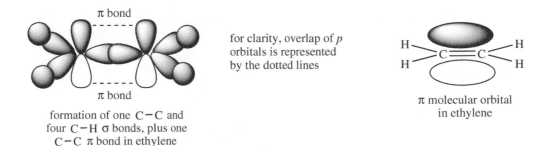

π bond

for clarity, overlap of p orbitals is represented by the dotted lines

π bond

formation of one C–C and
four C–H σ bonds, plus one
C–C π bond in ethylene

π molecular orbital
in ethylene

Figure 2.15   Bonding in ethylene

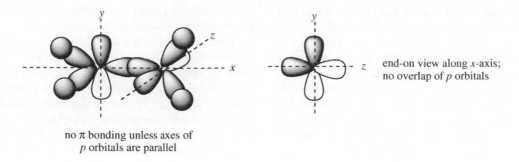

no π bonding unless axes of
p orbitals are parallel

end-on view along *x*-axis;
no overlap of *p* orbitals

Figure 2.16 High-energy state ethylene without π bond

to the bonding orbital. It accounts for a change in configuration (see Section 3.4.3) of the double bond, so-called *cis–trans* isomerism, and we shall see an example shortly (see Box 2.1).

Compounds with π bonds are said to be **unsaturated**, whereas compounds without π bonds containing only σ bonds are referred to as **saturated**.

sp hybrid orbitals

The third observation relates to acetylene (ethyne, C_2H_2), which is linear, i.e. bond angles of 180°, and contains two π bonds. This introduces what we term **triple bonds**, actually a combination of one σ bond and two π bonds. In this molecule, we invoke another type of hybridization for carbon, that of *sp* hybrid orbitals. These are a mix of the 2*s* orbital with one 2*p* orbital, giving two equivalent *sp* orbitals. Each hybrid orbital takes one electron, whilst the remaining two electrons are accommodated in two different 2*p* orbitals (Figure 2.17).

The *sp* hybrid orbitals can be visualized as a straight combination of an *s* and a *p* orbital, so that

it will now be the shortest and fattest of the hybrid orbitals, with most *s* character and least *p* character. Its energy will be above that of the *s* orbital, but below that of sp^2 orbitals, since the *p* contribution is the higher energy component. The atomic orbitals in *sp*-hybridized carbon are going to be two equivalent *sp* orbitals, arranged opposite each other to minimize interaction, plus the two remaining *p* orbitals, which will be at right angles to each other, and also at right angles to the *sp* orbitals (Figure 2.18).

The bonding in **acetylene** has one C–C σ bond together with two C–H σ bonds; the *p* orbitals on each carbon, each carrying one electron, interact by side-to-side overlap to produce two π bonds (Figure 2.19). Note again that the *p* orbitals can only overlap if their axes are parallel.

This makes the acetylene molecule linear, i.e. bond angles of 180°, and there are two π bonds with electron density either side of this axis. The properties of an alkyne, like acetylene, are also special in that the π bonds are again much more reactive than the σ bond.

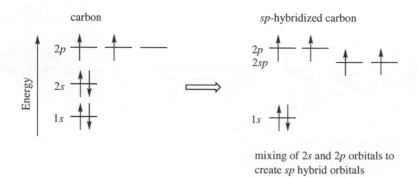

mixing of 2*s* and 2*p* orbitals to
create *sp* hybrid orbitals

Figure 2.17 Electronic configuration: *sp*-hybridized carbon atom

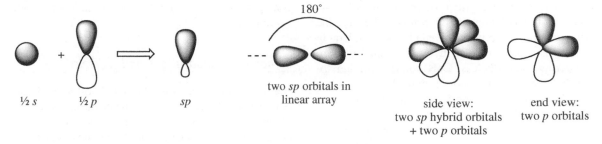

½ s   ½ p   sp   two *sp* orbitals in   side view: end view:
                                  linear array   two *sp* hybrid orbitals   two *p* orbitals
                                                            + two *p* orbitals

Figure 2.18   *sp* hybrid orbitals

π bonds

for clarity, overlap of *p*
orbitals is represented
by the dotted lines

π bonds

formation of one C‒C and
two C‒H σ bonds, plus two
C‒C π bonds in acetylene

π molecular orbitals
in acetylene

Figure 2.19   Bonding in acetylene

Hybridization and bond lengths

We also note that there are significant differences in
bond lengths for single, double, and triple bonds. The
carbon atoms in ethane are further apart (1.54 Å) than
in ethylene (1.34 Å), and those in acetylene are even
closer together (1.20 Å); Å refers to the Ångström
unit, 10^{-10} m. This is primarily a consequence of
the different nature of the σ bonds joining the two

carbons. Because sp^2 hybrid orbitals have less *p*
character than sp^3 hybrid orbitals, they are less
elongated; consequently, a σ bond formed from sp^2
orbitals will be rather shorter than one involving sp^3
orbitals. By similar reasoning, *sp* hybrid orbitals will
be shorter than sp^2 orbitals, because they have even
less *p* character, and will form even shorter C–C
σ bonds.

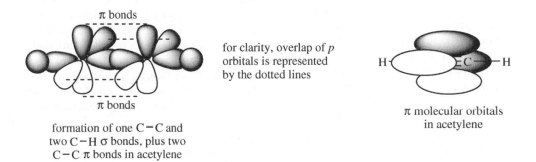

hybridization
in C‒C bond: $sp^3 - sp^3$ $sp^2 - sp^2$

1.10 Å   1.09 Å

1.54 Å   1.34 Å

ethane   ethylene

*a dotted bond indicates it goes   these bonds are in the
behind the plane of the paper*   plane of the paper*

$sp - sp$

1.08 Å

1.20 Å

acetylene

*a wedged bond indicates it comes
in front of the plane of the paper*

depiction of tetrahedral array
using wedged and dotted bonds

There is a similar effect in the length of C–H bonds, but this is less dramatic, primarily because the hydrogen atomic orbital involved ($1s$) is considerably smaller than any of the hybrid orbitals we are considering. Nevertheless, C–H bonds involving sp-hybridized carbon are shorter than those involving sp^2-hybridized carbon, and those with sp^3-hybridized carbon are the longest.

Note how we have resorted to another form of representation of the ethane, ethylene, and acetylene molecules here, representations that are probably familiar to you (see Section 1.1). These **line drawings** are simpler, much easier to draw, and clearly show how the atoms are bonded – we use a line to indicate the bonding molecular orbital. They do not show the difference between σ and π bonds, however. We also introduce here the way in which we can represent the tetrahedral array of bonds around carbon in a two-dimensional drawing. This is to use **wedges** and **dots** for bonds instead of lines. By convention, the wedge means the bond is coming towards you, out of the plane of the paper. The dotted bond means it is going away from you, behind the plane of the paper. We shall discuss stereochemical representations in more detail later (see Section 3.1).

At the beginning of this section we suggested that students often found hybridization a difficult concept to understand. We should emphasize that hybridization is a model that helps us to appreciate molecular structure and predict chemical reactivity. Do not think in terms of atomic orbitals merging to form hybrid orbitals, but consider that such orbitals already exist as the lowest energy arrangement. Hybridization is our modification of the first model, which we saw had its limitations, to an improved model that provides a rationale for experimental observations. As research progresses, we may have to apply even further modifications! At the present, though, the concept of hybrid orbitals provides us with satisfactory explanations for many chemical features. We have already seen that hybridization helps to define features such as bond angles and bond lengths that dictate molecular shape (stereochemistry; see Section 3.1). In later sections we shall see that hybridization gives us good explanations for other aspects of chemistry, such as acidity and basicity (see Sections 4.3.4 and 4.5.3), the relative reactivity of nucleophiles (see Section 6.1.2), and the chemical behaviour of compounds having conjugation (see Section 8.2) or aromatic rings (see Section 8.4).

Carbanions, carbocations and radicals

Before we move on from the hybrid orbitals of carbon, we should take a look at the electronic structure of important reactive species that will figure prominently in our consideration of chemical reactions. First, let us consider **carbanions** and **carbocations**. We shall consider the simplest examples, the methyl anion $CH_3{}^-$ and the methyl cation $CH_3{}^+$, though these are not going to be typical of the carbanions and carbocations we shall be meeting, in that they lack features to enhance their stability and utility.

The **methyl anion** is what would arise if we removed H^+ (a proton) from methane by fission of the C–H bond so that the two electrons are left with carbon. We can immediately deduce that carbon has its full octet of electrons, and that we shall have a tetrahedral array of three bonds and a lone pair of electrons in sp^3 orbitals.

cleavage of C–H bond;
both electrons left with carbon

methyl anion
$^-CH_3$

cleavage of C–H bond;
both electrons removed with hydrogen

methyl carbocation
$^+CH_3$

On the other hand, the **methyl carbocation** is the result of removing a hydride anion (a hydrogen atom and an electron) from methane by fission of the C–H bond so that the two electrons are removed with hydrogen. We can now deduce that carbon has only six electrons in its outer shell. This arrangement is best accommodated by sp^2 hybridization and a vacant p orbital. The alternative of four sp^3 hybrid orbitals with one unfilled does not minimize repulsion between the filled orbitals, and is also a higher energy arrangement. To deduce this, we need to go back to the energy diagrams for sp^3 and sp^2 hybrid orbitals. The lower p character of sp^2 hybrid orbitals means they are of lower energy than sp^3 orbitals; this is because the $2p$ orbitals are of higher energy than the $2s$ orbital. Consequently, we can work out that six electrons in sp^2 orbitals will have a lower energy than six electrons in sp^3 orbitals. Hence, the methyl carbocation is of planar sp^2 nature with an unoccupied p orbital at right angles to this plane. The consequences of this will be developed in Section 6.2.

There is also a third type of reactive species that we shall discuss in detail in Chapter 9, namely radicals. Briefly, **radicals** are uncharged entities that carry an unpaired electron. A methyl radical $CH_3^{\bullet}$ results from the fission of a C–H bond in methane so that each atom retains one of the electrons. In the methyl radical, carbon is sp^2 hybridized and forms three σ C–H bonds, whilst a single unpaired electron is held in a $2p$ orbital oriented at right angles to the plane containing the σ bonds. The unpaired electron is always shown as a dot. The simplest of the radical species is the other fission product, a hydrogen atom.

cleavage of C–H bond; each atom retains one electron

methyl radical

$^{\bullet}CH_3$

hydrogen atom

2.6.3 Hybrid orbitals in oxygen and nitrogen

Hybridization concepts can also be applied to atoms other than carbon. Here, we look at how we can understand the properties of oxygen and nitrogen compounds by considering hybrid orbitals for these atoms.

Let us recap on the electronic configurations of oxygen and nitrogen. Nitrogen has one more electron than carbon, and oxygen has two more. For each atom, we can consider hybrid sp^3 orbitals derived from the $2s$ and $2p$ orbitals as we have seen with carbon (Figure 2.20). We shall then obtain electronic configurations in which nitrogen has two paired electrons in one of these orbitals, whilst the remaining three orbitals each have a single unpaired electron available for bonding. Oxygen has two sets of paired electrons and has two unpaired electrons available for bonding.

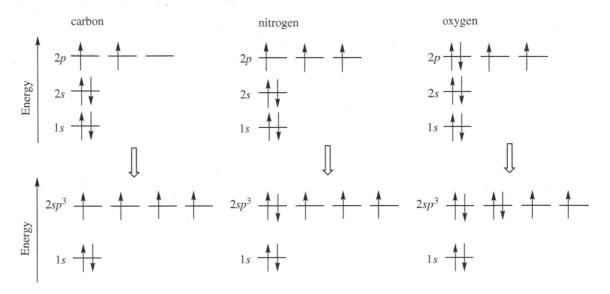

Figure 2.20 Electronic configurations: sp^3-hybridized nitrogen and oxygen

σ bonds

The simplest compounds to consider here are **ammonia** and **water**. It is apparent from the above electronic configurations that nitrogen will be able to bond to three hydrogen atoms, whereas oxygen can only bond to two. Both compounds share part of the tetrahedral shape we saw with sp^3-hybridized carbon. Those orbitals not involved in bonding already have their full complement of electrons, and these occupy the remaining part of the tetrahedral array (Figure 2.21). These electrons are not inert, but play a major role in chemical reactions; we refer to them as **lone pair electrons**.

These orbital pictures tend to get a little confusing, in that we really need to put in the elemental symbol to distinguish it from carbon, and we usually wish to show the lone pair electrons. We accordingly use a compromise representation that employs the cleaner line drawings for part of the structure and shows the all-important orbital with its lone pair of electrons. These are duly shown for ammonia and water.

The tetrahedral geometry resultant from these sp^3-hybridized nitrogen and oxygen atoms is found to exist in both ammonia and water. **Bond angles** in these molecules are not quite the 109° of the perfect tetrahedron, because the electrons in the lone pair atomic orbital are not involved in bonding. They are, therefore, closer to the nucleus than the electrons in the N–H or O–H bond σ molecular orbitals. Lone pairs thus tend to exert a greater electronic repulsive force between themselves, and also towards the bonding electrons, than the σ bonding electrons do to each other. The net result is that bond angles between lone pairs, or between lone pairs and σ bonds, are somewhat greater than between σ bonds, a distortion of the perfect tetrahedral array.

Lone pair electrons may be used in bonding. Since they already have a complement of two electrons, bonds will need to be made to an atom that is electron deficient, e.g. a proton. Thus, the **ammonium cation** and the **hydronium cation** also share tetrahedral geometry, and each possesses a σ bond formed from lone pair electrons.

The hydronium cation still possesses a lone pair of electrons. It does not bond to a second proton for the simple reason that the cation would then be required to take on an unfavourable double positive charge.

π bonds

When we consider double bonds to oxygen, as in **carbonyl groups** (C=O) or to nitrogen, as in **imine** functions (C=N), we find that experimental data are best accommodated by the premise that these atoms are sp^2 hybridized (Figure 2.22). This effectively follows the pattern for carbon–carbon double bonds (see Section 2.6.2). The double bond is again a combination of a σ bond plus a π bond resulting from overlap of p atomic orbitals. The carbonyl

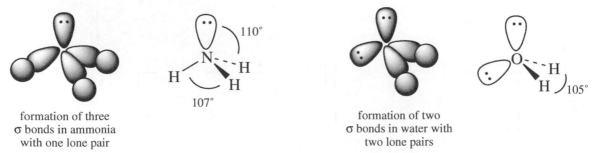

formation of three
σ bonds in ammonia
with one lone pair

formation of two
σ bonds in water with
two lone pairs

Figure 2.21 sp^3 hybrid orbitals: ammonia and water

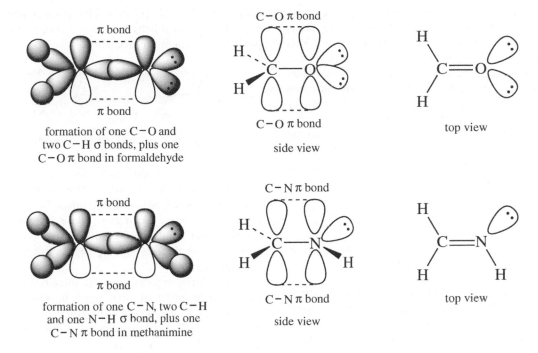

formation of one C−O and
two C−H σ bonds, plus one
C−O π bond in formaldehyde

side view

top view

Figure 2.22 Bonding in formaldehyde and methanimine

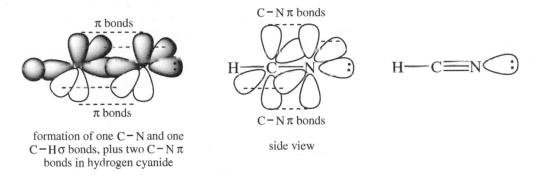

formation of one C−N and one
C−Hσ bonds, plus two C−N π
bonds in hydrogen cyanide

side view

Figure 2.23 Bonding in hydrogen cyanide

oxygen carries two lone pairs in sp^2 orbitals, whereas nitrogen carries one. Thus, the main difference from the alkene structure, apart from the atoms involved, is that lone pairs in atomic orbitals replace one or more of the σ molecular orbitals that constituted the C–H bonds. The atoms around the double bond are in a planar array, just as in an alkene.

Triple bonds are also encountered in **cyanides/nitriles**. We can compare these with alkynes in much the same way (see Section 2.6.2). With sp-hybridized nitrogen, we can form one C–N σ bond and two

C–N π bonds, leaving a lone pair of electrons on nitrogen in an sp atomic orbital (Figure 2.23). The cyanide/nitrile system is linear, just like an alkyne.

2.7 Bond polarity

The nucleus of each atom has a certain ability to attract electrons. This is termed its **electronegativity**. This means that, when it is bonded to another atom, the bonding electrons are not shared equally between the two atoms. Covalent bonds may, therefore, possess a charge imbalance, with one of the atoms taking more than its share of the electrons. This is referred to as **bond polarity**. An atom that is more electronegative than carbon will thus polarize the bond, and we can consider the atoms as being partially charged. This is indicated in a structure by putting partial charges ($\delta+$ and $\delta-$) above the atoms. It can also be represented by putting an arrowhead on the bond, in the direction of electron imbalance. Alternatively, we use a specific dipole arrow at the side of the bond.

bromine is more electronegative than carbon

$\delta+$ $\delta-$		
C—Br	C→Br	C—Br
partial charges	*dipole on bond*	*dipole arrow*

In general, electronegativities increase from left to right across the periodic table, and decrease going down a particular column of the periodic table. The **relative electronegativities** of those atoms most likely to be found in typical organic molecules are included in Table 2.2. The numbers (**Pauling electronegativity values**) are on an arbitrary scale from Li = 1 to F = 4.

From the sequence shown, it is readily seen that hydrogen and carbon are among the least electronegative atoms we are likely to encounter in organic molecules. The relatively small difference in electronegativities between hydrogen and carbon also means there is not going to be much polarity associated with a C–H bond. Most atoms other than hydrogen and carbon when bonded to carbon are going to be electron rich; therefore, bonds may

Table 2.2 Pauling electronegativity values

H						
2.1						
Li	Be	B	C	N	O	F
1.0	1.6	2.0	2.5	3.0	3.5	4.0
Na	Mg	Al	Si	P	S	Cl
0.9	1.2	1.5	1.8	2.1	2.5	3.0
K						Br
0.8						2.8
						I
						2.5

display considerable polarity. This polarity helps us to predict chemical behaviour, and it is crucial to our prediction of chemical mechanisms.

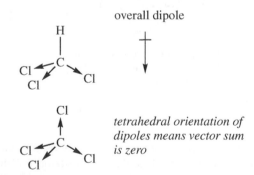

polarity in C—O and C—N bonds

We must also modify our thinking of bonding as being simply ionic (where there is transfer of electrons between atoms) or covalent (where there is equal sharing of electrons). These represent two extremes, but bond polarity now provides a middle ground where there is sharing of electrons, but an unequal sharing.

Bond polarity in a molecule can often be measured by a **dipole moment**, expressed in Debye units (D). However, the physical measurement provides only the overall dipole moment, i.e. the sum of the individual dipoles. A molecule might possess bond polarity without displaying an overall dipole if two or more polar bonds are aligned so that they cancel each other out. The C–Cl bond is polar, but although chloroform ($CHCl_3$) has a dipole moment (1.02 D), carbon tetrachloride (CCl_4) has no overall dipole. Because of the tetrahedral orientation of the dipoles in carbon tetrachloride, the vector sum is zero.

overall dipole

tetrahedral orientation of dipoles means vector sum is zero

Polarization in one bond can also influence the polarity of an adjacent bond. Thus, in ethyl chloride, the polarity of the C–Cl bond makes the carbon more positive ($\delta+$); consequently, electrons in the C–C bond are drawn towards this partial positive charge. The terminal carbon thus also experiences a partial positive charge, somewhat smaller than $\delta+$ and so depicted as $\delta\delta+$.

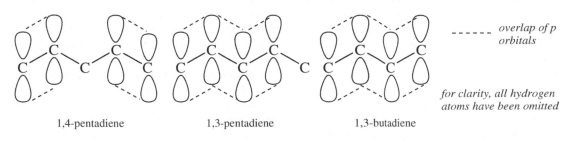

$$\delta- \quad \delta+ \quad \delta\delta+$$
$$Cl \leftarrow C \leftarrow CH_3$$

inductive effect

$$\delta- \quad \delta+ \quad \delta\delta+ \quad \delta\delta\delta+$$
$$X \leftarrow C \leftarrow C \leftarrow CH_3$$

inductive effect decreases as polar group is located further away

$$Cl \leftarrow C \leftarrow CH_3$$

$$Cl \leftarrow C \leftarrow CH_3$$

inductive effects increase as number of polar groups increases

This transmission of polarity through the σ bonds is termed an **inductive effect**. It is relatively short range, decreasing rapidly as the original dipole is located further away. It becomes unimportant after about the third carbon atom. However, the effects will increase with the number of polar groups, so we see increasing polarization effects with 1,1-dichloroethane and 1,1,1-trichloroethane. We shall often need to consider inductive effects when attempting to predict chemical reactivity.

2.8 Conjugation

Double bonds, whether they be C=C, C=O, or C=N, are sites of special reactivity in a molecule. This reactivity may take on different characteristics if we have two or more double bonds in the same molecule, depending upon whether the double bonds are isolated or conjugated. We use the term **conjugated** to describe an arrangement in which double bonds are separated by a single bond. Thus, in 1,3-pentadiene the double bonds are conjugated,

whereas in 1,4-pentadiene they are isolated or non-conjugated. The nomenclature 'diene' indicates two C=C double bonds, the numbers the position in the molecule (see Section 1.4). Conjugated dienes usually display rather different chemical reactivity and spectral properties from non-conjugated dienes (see Section 8.2).

1,3-pentadiene

conjugated double bonds

1,4-pentadiene

isolated double bonds

The differences arise from the nature of the π orbitals in the double bond system. Consider **1,4-pentadiene** first. We may draw this to show overlap of *p* orbitals to create two separate π bonds, and effectively that is all there is that is worthy of note (Figure 2.24). The double bonds are isolated entities that do not interact.

In **1,3-pentadiene**, however, the *p* orbitals are all able to overlap in such a way that a lower energy molecular orbital can be formed. We have more physical data available for **1,3-butadiene**, so let us consider this slightly simpler conjugated system instead.

We have four $2p$ orbitals on four adjacent carbon atoms, and these can overlap to produce four π molecular orbitals. These are as shown, and their relative energies can be visualized from the bonding interactions possible (Figure 2.25). Remember, bonding results from overlap of orbitals that have the same phase sign of the wave function, whereas anti-bonding orbitals originate from interaction of orbitals with different phase signs of the wave function. Thus, ψ_1 has three bonding interactions and no antibonding interactions, ψ_2 has two bonding interactions and one

1,4-pentadiene   1,3-pentadiene   1,3-butadiene

- - - - - *overlap of p orbitals*

for clarity, all hydrogen atoms have been omitted

Figure 2.24 Overlap of *p* orbitals in dienes

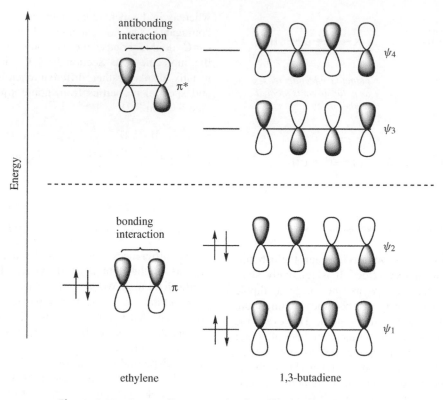

Figure 2.25   Energy diagram: molecular orbitals of 1,3-butadiene

antibonding interaction, ψ_3 has one bonding interaction and two antibonding interactions, and ψ_4 has no bonding interaction and three antibonding interactions. The four electrons will be allocated to ψ_1 and ψ_2.

Conjugation introduces a number of features. We can consider that the π electrons in a conjugated system are no longer associated with specific bonds, but are **delocalized** over those atoms constituting the conjugated system. This has energy implications. The overall energy associated with butadiene is actually less than we might expect. It is lower than that of non-conjugated dienes, e.g. 1,4-pentadiene, and less than what we might estimate from figures for the monounsaturated but-1-ene. Thus, compounds with two conjugated double bonds are thermodynamically more stable (less reactive) than compounds with two isolated double bonds. In due course, we shall see that the double bond reactivity of butadiene is also influenced by conjugation: butadiene behaves differently from compounds with isolated double bonds (see Section 8.2). We also need to appreciate that conjugation, and its influence on reactivity, is

not restricted to alkenes. Any system containing two or more π bonds may be conjugated, so that we can include triple bonds (alkynes), carbonyl groups, imines, and nitriles in this description. In its broadest sense, **conjugation** refers to a system that has a p orbital adjacent to a π bond allowing delocalization of electrons. The adjacent p orbital may be a vacant one, as in a carbocation (see Section 2.6.2), one that contains a single electron, as in a radical (see Section 2.6.2), or may be part of another π bond, as in a conjugated diene. At first glance, a conjugated anion does not fit the broad definition of conjugation, since we would expect the carbanion centre to be sp^3 hybridized (see Section 2.6.2). Nevertheless, there is delocalization of electrons and this system is considered to be conjugated. In this system, delocalization results from accommodating the negative charge in a p orbital rather than an sp^3 orbital, so that we again achieve p orbital overlap.

Conjugated systems also give characteristic **spectral absorptions**, especially in the UV–visible regions. As the extent of conjugation increases, i.e. more than two double bonds separated by single

conjugated diene 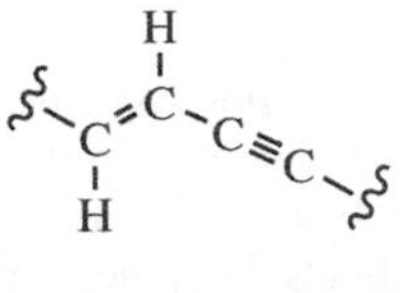   conjugated ene–yne 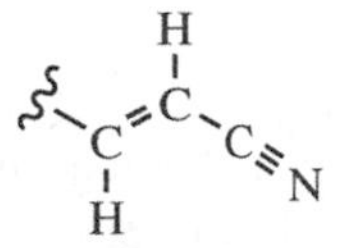   conjugated nitrile 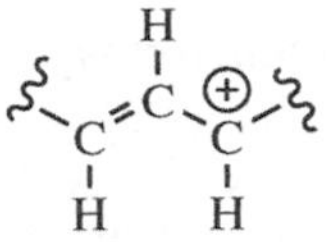   conjugated carbocation

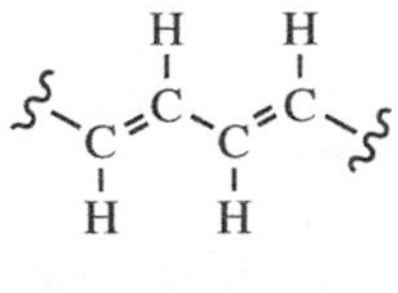

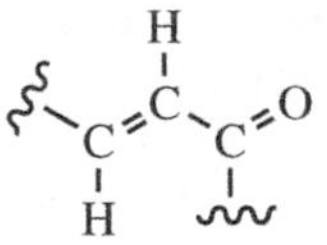

conjugated carbonyl 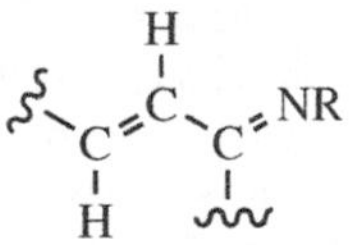   conjugated imine 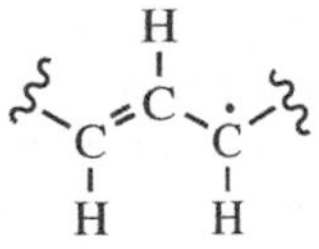   conjugated radical 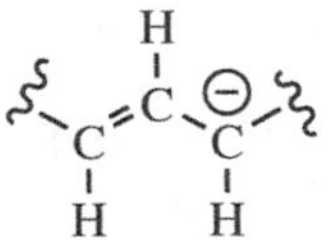   conjugated carbanion

bonds, these compounds have more intense absorptions at longer wavelengths (lower energies). This is because the energy difference between bonding and antibonding molecular orbitals becomes smaller with increasing conjugation. The spectral data arise from the transition of an electron between these energy levels. This means that, with increasing conjugation, the characteristic absorption moves from the UV to the visible region and, typically, the compound becomes coloured. A compound appears coloured to the human eye when it removes by absorption some of the wavelengths from white light.

Box 2.1

Carotenoids, vitamin A, and vision

Carotenoids are a group of natural products found predominantly in plants. They are characterized by an extended chain of conjugated double bonds, giving an extended π electron system. They are highly coloured and contribute to yellow, orange, and red pigmentations in plants. **Lycopene** is the characteristic carotenoid pigment in ripe tomato fruits, and the orange colour of carrots is caused by β-carotene. **Capsanthin** is the brilliant red pigment of capsicum peppers.

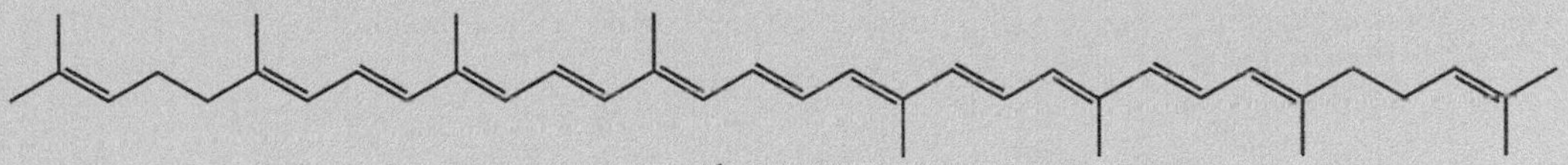

lycopene

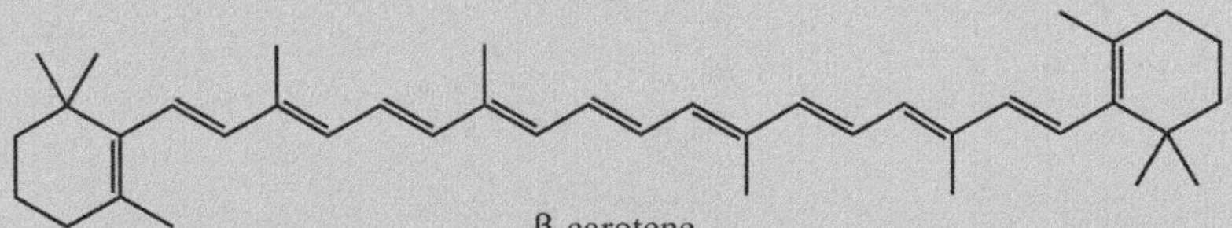

β-carotene

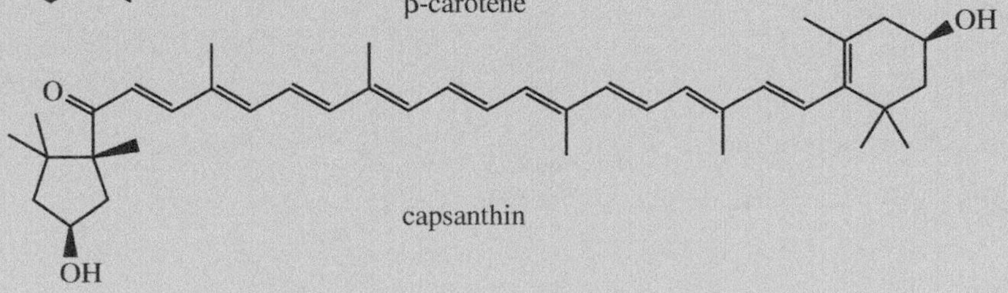

capsanthin

Box 2.1 (continued)

Carotenoids function along with chlorophylls (see Box 11.4) in **photosynthesis** as accessory light-harvesting pigments, effectively extending the range of light that can be absorbed by the photosynthetic apparatus. The absorption maximum of carotenoids is typically between 450 and 500 nm, which indicates that the energy difference between bonding and antibonding molecular orbitals is quite small. This absorption maximum corresponds to blue light, so that with blue light absorbed, the overall impression to the human eye is of a bright yellow−orange coloration. Recent research suggests that carotenoids are important **antioxidant** molecules for humans, helping to remove toxic oxygen-derived radicals, and thus minimizing cell damage (see Box 9.2). The most beneficial dietary carotenoid in this respect is lycopene, with tomatoes featuring as the predominant source.

Vitamin A$_1$ (**retinol**) is derived in mammals by oxidative metabolism of plant-derived dietary carotenoids in the liver, especially β-carotene. Green vegetables and rich plant sources such as carrots help to provide us with adequate levels. Oxidative cleavage of the central double bond of β-carotene provides two molecules of the aldehyde retinal, which is subsequently reduced to the alcohol retinol. Vitamin A$_1$ is also found in a number of foodstuffs of animal origin, especially eggs and dairy products. Some structurally related compounds, including retinal, are also included in the A group of vitamins.

β-carotene

O_2 ↓ cleavage of central double bond generates two molecules of retinal

retinal   NADH →   reduction of aldehyde to alcohol   retinol (vitamin A$_1$)

A deficiency of vitamin A leads to vision defects, including a visual impairment at low light levels, termed night blindness. For the processes of vision, retinol needs to be converted first by oxidation into the aldehyde retinal, and then by enzymic isomerization to *cis*-retinal. *cis*-Retinal is then bound to the protein opsin in the retina via an imine linkage (see Section 7.7.1) to give the red visual pigment rhodopsin.

retinol

↓ NADP$^+$

hydrolytic cleavage of imine

retinal

enzymic trans−cis isomerism of 11,12-double bond

formation of imine with amino group on protein opsin

absorption of light energy restores trans configuration of 11,12-double bond

11-*cis*-retinal H_2N−opsin →   rhodopsin

Rhodopsin is sensitive to light by a process that involves isomerization of the *cis*-retinal portion back to the *trans* form, thus translating the light energy into a molecular change that then triggers a nerve impulse to the brain. The absorption of light energy promotes an electron from a π to a π^* orbital, thus temporarily destroying the double bond character and allowing rotation (see Section 2.6.2). *trans*-Retinal is then subsequently released from the protein by hydrolysis, and the process can continue.

2.9 Aromaticity

Aromatic compounds constitute a special group of conjugated molecules; these are cyclic unsaturated molecules with unusual stability and characteristic properties. The term aromatic originates from the odour displayed by many of the simple examples.

2.9.1 Benzene

The parent compound is benzene. **Benzene**, C_6H_6, contains an array of six sp^2-hybridized carbons, each attached by a σ bond to the adjacent carbons, and by a third σ bond to a hydrogen atom. The six *p* atomic orbitals from carbon are all aligned so that they can overlap to form molecular orbitals, and this is most favourable when the carbons are all in one plane. The lowest energy molecular orbital can be considered as an extended ring-like system with a high electron probability above and below the plane of the ring (Figure 2.26). This is a bonding π molecular orbital in a conjugated system extending over all six atoms. The electrons will be distributed evenly, or **delocalized**, over the whole molecule.

The six *p* atomic orbitals combine to give six molecular orbitals for the π system. The relative energies for these are shown in Figure 2.27. There is one low-energy bonding molecular orbital and two degenerate bonding orbitals at higher energy. There will be an analogous array of antibonding orbitals at higher energy.

The six electrons are assigned to these orbitals as we have seen previously, beginning with the lowest energy level. This leads to the six electrons completely filling the bonding molecular orbitals and providing an extremely favourable arrangement, in that the overall energy is significantly below that of six electrons in the contributing *p* atomic orbitals. The energy stabilization is considerable, and also much more than could be accounted for by simple conjugation. The special stability afforded by this planar cyclic array is what we understand by **aromaticity**. The chemical reactivity associated with aromatic systems will be covered in Chapter 8.

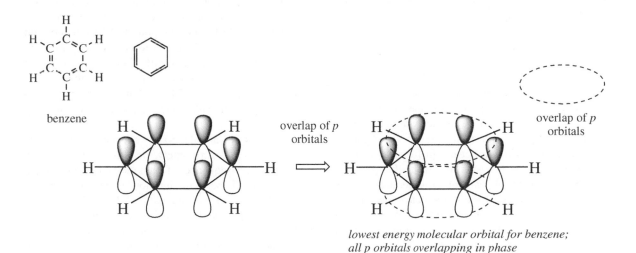

lowest energy molecular orbital for benzene; all p orbitals overlapping in phase

Figure 2.26 Lowest energy molecular orbital for benzene

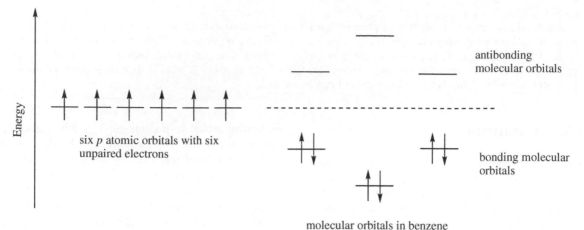

six *p* atomic orbitals with six
unpaired electrons

antibonding
molecular orbitals

bonding molecular
orbitals

molecular orbitals in benzene

Figure 2.27   Energy diagram: molecular orbitals of benzene

2.9.2 Cyclooctatetraene

Let us consider the origins of benzene's aromatic stabilization. Another cyclic hydrocarbon, **cyclooctatetraene** (pronounced cyclo-octa-tetra-ene), certainly
looks conjugated according to our criteria, but chemical evidence shows that it is very much more reactive
than benzene, and does not undergo the same types
of reaction. It does not possess the enhanced aromatic
stability characteristic of benzene.

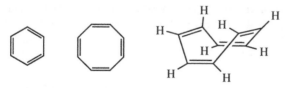

benzene   cyclooctatetraene

Further, cyclooctatetraene has been shown to be
non-planar; it adopts a tub shape. This originates from
bond angles. A regular octagon has internal bond
angles of 135°, quite far from the optimum angle of
120° for sp^2 hybridization. In benzene's hexagon, the
internal angle is 120°, a perfect fit for sp^2 geometry.
Cyclooctatetraene thus distorts from the planar to
relieve this strain. A careful consideration of this
shape may then suggest the immediate consequences.
These are that none of the double bonds are in
the same plane; therefore, there is going to be no
overlap of *p* orbitals between the double bonds. We
cannot get any enhanced stability associated with
conjugation.

2.9.3 Hückel's rule

Cyclooctatetraene has eight π electrons and benzene
has six. The number of π electrons that confer
aromaticity is given by **Hückel's rule**: a planar cyclic
conjugated system will be particularly stable if the
number of π electrons is $4n + 2$, where *n* is an
integer (0, 1, 2, 3, etc). Although the significance
of this will not become apparent until later (see
below), we must stress that $4n + 2$ refers to the
number of π electrons, and not the number of
atoms in the ring. Benzene, therefore, with six π
electrons ($n = 1$, $4n + 2 = 6$), is aromatic; however,
cyclooctatetraene, with eight π electrons, is not
aromatic. Also aromatic would be a system with 10 π
electrons ($n = 2$), or 14 π electrons ($n = 3$). The first
of these would be the compound [10]annulene, and
the second [14]annulene. **Annulene** is a general term
for a carbon ring system with alternating single and
double bonds; the number in brackets is the number
of carbons in the ring. For example, we could call
benzene [6]annulene, though in practice, nobody ever
does.

Whereas [14]annulene shows aromatic properties,
[10]annulene, unfortunately, does not, but we know
this is a consequence of the molecule adopting a
non-planar shape. The interior angle for a planar
10-carbon system would have to be 144°, and this
is too far removed from the sp^2-hybridized angle
of 120° to be feasible. As ring sizes get larger, it
becomes possible to have a cyclic system where all
bond angles can be the ideal 120°. There is a way
of drawing a 10-carbon ring system with angles of

120°, but we must realize that this attempts to place two hydrogens in the same space. This is clearly not feasible; as the hydrogens are pushed away from each other, therefore, this must lead to a non-planar molecule.

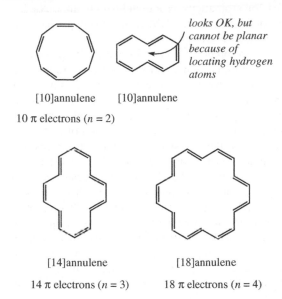

looks OK, but cannot be planar because of locating hydrogen atoms

[10]annulene   [10]annulene

10 π electrons ($n = 2$)

[14]annulene   [18]annulene

14 π electrons ($n = 3$)   18 π electrons ($n = 4$)

Structures that are also aromatic are the cyclopropenyl cation (2 π electrons; $n = 0$) and the cyclopentadienyl anion (6 π electrons; $n = 1$). Although we do not wish to pursue these examples further, they are representative of systems where the number of π electrons is not the same as the number of carbon atoms in the ring.

The stabilization conferred by **aromaticity** results primarily from the much lower energy associated with a set of electrons in molecular orbitals compared

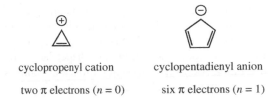

cyclopropenyl cation   cyclopentadienyl anion

two π electrons ($n = 0$)   six π electrons ($n = 1$)

with them being in atomic orbitals. We have seen how this originates in benzene by allocating electrons to the bonding orbitals. We can apply the same procedure to other annulene compounds, and there exists a very neat way of finding the relative energies of molecular orbitals without resource to mathematical calculations. This device, the **Frost circle**, inscribes the appropriate polygon in a circle, with one vertex pointing vertically downwards. The intersections of other vertices with the circle then mark the positions of the molecular orbitals. The position of the horizontal diameter represents the energy of the carbon p orbital; intersections below this are bonding, those above are antibonding, and nonbonding orbitals are on the diameter line. Frost circles for benzene and cyclooctatetraene are drawn in Figure 2.28.

We can immediately see that allocating six electrons into the benzene molecular orbitals fills all three bonding orbitals (a closed shell structure) and there is substantial aromatic stabilization, in that the energy associated with electrons in the molecular orbitals is greatly reduced compared with that of electrons in the six atomic orbitals. For cyclooctatetraene, allocating eight electrons to the molecular orbitals leads to three filled orbitals, but then the remaining two electrons are put singly into each of the degenerate nonbonding orbitals. Cyclooctatetraene does not have a filled shell

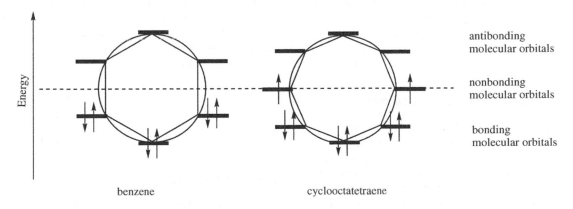

benzene   cyclooctatetraene

Figure 2.28 Relative energies of benzene and cyclooctatetraene molecular orbitals from Frost circles

structure like benzene, but has two nonbonding electrons; it does not have the special stability we see in benzene. As we have seen in Section 2.9.2, cyclooctatetraene also adopts a non-planar shape, lacks the stabilization associated with conjugation, and behaves like four separate normal alkenes.

2.9.4 Kekulé structures

Benzene is usually drawn as a structure with alternating single and double bonds. We can draw it in two ways.

Kekulé representations   benzene; circle represents
      of benzene   delocalized π electrons

These two forms are so-called **Kekulé structures**; but neither is correct, in that benzene does not have single and double bonds. This immediately follows from a measurement of C–C bond lengths. For sp^2-hybridized carbons, we expect the C=C bonds to be about 1.34 Å, whereas the C–C bond length would be about 1.47 Å. Measurements show that all of the carbon–carbon bond lengths are the same, at 1.40 Å. This length is between that of single and double bonds, and suggests that we have C–C bonds that are somewhat between single and double bond in character. From the point of stability, and now also bond lengths, we must view benzene as quite different from cyclohexatriene. To emphasize this, a different representation for the benzene ring has been proposed, i.e. a circle within a hexagon. The circle represents the six π-electron system, and this, therefore, highlights the special nature of the aromatic ring. As we shall see in due course, this representation has considerable limitations, and most chemists, ourselves included, do not use it.

2.9.5 Aromaticity and ring currents

One can demonstrate the particular stability of aromatic compounds by their characteristic chemical reactions. For example, benzene reacts with bromine only with difficulty and gives bromobenzene, a substitution product (see Section 8.4). This leaves the aromatic ring intact. By contrast, a typical alkene reacts readily with bromine by an addition process

to give a dibromo product (see Section 8.1.2). This reaction destroys the π bond. When it comes to compounds such as annulenes, it is not always easy to synthesize sufficient material to demonstrate typical chemical reactivity, and a simple spectroscopic analysis for aromaticity is infinitely preferable. Nuclear magnetic resonance (NMR) spectroscopy has provided such a probe.

The proton NMR signals for hydrogens on a double bond are found in the region δ 5–6 ppm. In contrast, those in benzene are detected at δ 7.27 ppm. This substantial difference is ascribed to the presence of a **ring current** in benzene and other aromatic compounds. A ring current is the result of circulating electrons in the π system of the aromatic compound. Without entering into any discussion on the origins of NMR signals, the ring current creates its own magnetic field that opposes the applied magnetic field, and this affects the chemical shift of protons bonded to the periphery of the ring. Signals are shifted downfield (greater δ) relative to protons in alkenes. Proton NMR spectroscopy can, therefore, be used as a test for aromaticity. In this way, [14]annulene and [18]annulene have been confirmed as aromatic.

2.9.6 Aromatic heterocycles

In due course we shall see that unsaturated cyclic compounds containing atoms other than carbon, e.g. nitrogen, oxygen, or sulfur, can also be aromatic. For example, **pyridine** can be viewed as a benzene ring in which one CH has been replaced by a nitrogen. It is aromatic and, like benzene, displays enhanced stability. **Pyrrole** is a five-membered heterocycle, but also displays aromaticity. Like the cyclopentadienyl anion (see Section 2.9.3), the number of π electrons is not the same as the number of atoms. In pyrrole, nitrogen provides two of the six π electrons. Examples of these molecules are discussed under heterocycles in Chapter 11.

benzene   pyridine   pyrrole

2.9.7 Fused rings

We may also encounter aromatic hydrocarbons that feature fused rings. Thus, **naphthalene** is effectively two benzene rings fused together, and **anthracene** has three fused rings. The heterocycle **quinoline** (see Section 11.8.1) is a fusion of benzene and pyridine. These ring systems are undoubtedly aromatic, and they display the enhanced stability and reactivity associated with simple aromatic compounds like benzene.

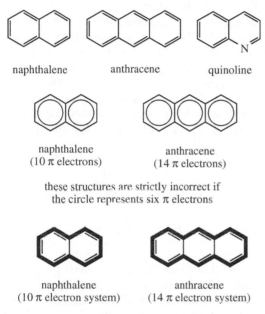

naphthalene   anthracene   quinoline

naphthalene
(10 π electrons)

anthracene
(14 π electrons)

these structures are strictly incorrect if
the circle represents six π electrons

naphthalene
(10 π electron system)

anthracene
(14 π electron system)

the π electron system may involve just
the periphery of the molecule

Molecular orbital calculations suggest that the π electrons in naphthalene are delocalized over the two rings and this results in substantial stabilization. These molecules are planar, and all *p* orbitals are suitably aligned for overlap to form π bonding molecular orbitals. Although we can draw Kekulé structures for these compounds, it is strictly incorrect to use the circle in hexagon notation since the circle represents six π electrons. Naphthalene has 10 carbons, and therefore 10 π electrons, and anthracene has 14 π electrons. The circle notation suggests 12 or

18 π electrons. Note that Hückel's rule applies only to monocyclic compounds, and although 10 π electrons (naphthalene) and 14 π electrons (anthracene) seem to be meet the criteria for aromaticity, there is good evidence to suggest we should consider the aromatic system not as a combination of benzene rings, but as a single ring involving the periphery of the molecule.

2.10 Resonance structures and curly arrows

The molecular orbital picture of benzene proposes that the six π electrons are no longer associated with particular bonds, but are effectively delocalized over the whole molecule, spread out via orbitals that span all six carbons. This picture allows us to appreciate the enhanced stability of an aromatic ring, and also, in due course, to understand the reactivity of aromatic systems. There is an alternative approach based on **Lewis structures** that is also of particular value in helping us to understand chemical behaviour. Because this method is simple and easy to apply, it is an approach we shall use frequently. This approach is based on what we term **resonance structures**.

Let us go back to the two **Kekulé representations** for benzene. The Lewis structure for benzene has alternating single and double bonds, but there are two ways of writing this. In one form, a particular bond is single; in the other form, this bond has become double. Resonance theory suggests that these two structures are both valid representations, and that each contributes to the structure of benzene, but the true structure is something in between, a lower energy hybrid of the two Kekulé forms, a **resonance hybrid**. If this is the case, then each bond is neither single nor double, but, again, something in between. As we have already seen, all C–C bond lengths in benzene are 1.40 Å, which is in between the bond lengths for single (1.47 Å) and double (1.34 Å) bonds.

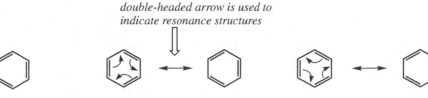

*double-headed arrow is used to
indicate resonance structures*

Kekulé representations
of benzene

*curly arrow represents the
movement of two electrons*

*we could also have written
curly arrows like this*

To indicate resonance forms, we use a **double-headed arrow** between the contributing structures. This arrow is reserved for resonance structures and never used elsewhere. The difference between the two structures is that the electrons in the π bonds have been redistributed, and we can illustrate this by use of another type of arrow, a **curly arrow**. This arrow is used throughout chemistry to represent the movement of two electrons. In the benzene case, a cyclic movement of electrons accounts for the apparent relocation of double bonds, though there are two ways we might show this process; both are equally satisfactory.

Benzene is a nice example to choose to illustrate the concept of resonance. It is not the best example for explaining the rules governing the use of curly arrows, so we must move to some simpler compounds. Being able to draw curly arrows is an essential skill for an organic chemist, and you will see from a cursory glance at the following chapters just how frequently they are employed. We shall use the same curly arrows and precisely the same principles for predicting the outcome of chemical reactions (see Section 5.1). They allow us to follow bond making and bond breaking processes, and provide us with a device we can use to keep track of the electrons.

- The curly arrow represents the movement of two electrons.

- The tail of the arrow indicates where the electrons are coming from, and the arrowhead where they are going to.

- Curly arrows must start from an electron-rich species. This can be a negative charge, a lone pair, or a bond.

- Arrowheads must be directed towards an electron-deficient species. This can be a positive charge, the positive end of a polarized bond, or a suitable atom capable of accepting electrons, i.e. an electronegative atom.

In our brief introduction to Lewis structures (see Section 2.2), we paid particular attention to **valency**, the number of bonds an atom could make to other atoms via the sharing of electrons. We must now broaden this idea to consider atoms in a molecule that are no longer neutral, but which carry a formal positive or negative charge. This means we are considering **cations** and **anions**, as in ionic bonding,

but the atom involved is still part of a molecule, and the molecule consequently also carries a formal charge. We have already met a few such entities in this chapter, e.g. the ammonium and hydronium cations, looking specifically at the molecular orbital descriptions (see Section 2.6.3). As indicated above, the use of curly arrows may involve species with positive or negative charges.

As simple examples, **ammonia** and **water** are neutral molecules. Nitrogen has five valence electrons, and it acquires a stable octet of electrons in making three bonds to hydrogen atoms. Each hydrogen has its stable arrangement of two electrons. The nitrogen in ammonia also carries a lone pair of electrons. Oxygen, with six valence electrons, makes two bonds to hydrogen atoms. Its octet of electrons will carry two lone pairs.

ammonia   water   ammonium   hydronium
                                 cation   cation

We can deduce the charge associated with ammonia and water by simply considering that the component atoms are neutral, that all we have done is share the electrons, so the molecules must also be neutral. The **formal charge** on an individual atom can be assessed more rigorously by subtracting the number of valence electrons assigned to an atom in its bonded state from the number of valence electrons it has as a neutral free atom. Electrons in bonds are considered as shared equally between the atoms, whereas unshared lone pairs are assigned to the atom that possesses them.

$$\begin{matrix} \text{formal} \\ \text{charge} \end{matrix} = \begin{matrix} \text{number of valence} \\ \text{electrons as neutral} \\ \text{free atom} \end{matrix} - \begin{matrix} \text{number of} \\ \text{valence electrons} \\ \text{assigned in} \\ \text{bonded state} \end{matrix}$$

Hence, for nitrogen, the number of valence electrons in a free atom is five. In ammonia, the number of assigned electrons is also five (three in bonds plus a lone pair). Therefore, the formal charge on nitrogen is zero. For hydrogen, the formal charge is also zero, since the number of valence electrons is one, and the number of assigned electrons is one. For oxygen, the number of valence electrons in a free atom is six. In water, the number of assigned electrons is also six

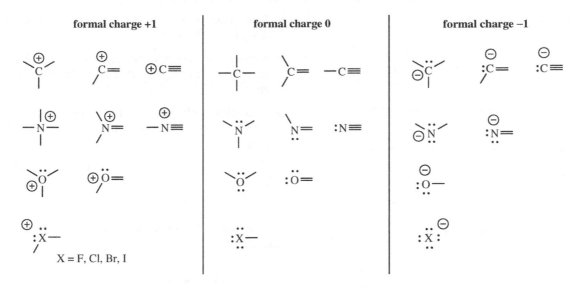

Figure 2.29 Formal charges of common atoms and ions

(two in bonds plus two lone pairs). Therefore, the formal charge on oxygen is zero. The hydrogens are also uncharged, as in ammonia.

Now consider the **ammonium** and **hydronium cations**. In the ammonium system, for nitrogen the formal charge is now +1. This follows from the number of valence electrons, i.e. five, minus the number of assigned electrons, i.e. four (four in bonds). In the hydronium system, the formal charge on oxygen is also +1. This is assessed from the number of valence electrons, i.e. six, minus the number of assigned electrons, i.e. five (three in bonds plus a lone pair). Of course, we already knew that ammonium and hydronium cations were the result of bonding neutral ammonia or water with a proton (charge +1), so an overall charge of +1 comes as no particular surprise (see Section 2.6.3).

Other systems are less familiar, and will therefore have to be assessed carefully. For example, what charge is associated with the structure shown on the left below?

$$H-\overset{\displaystyle ..}{\underset{\displaystyle |}{\overset{\displaystyle |}{C}}}-H \qquad H-\overset{\displaystyle \ominus}{\underset{\displaystyle |}{\overset{\displaystyle |}{C}}}-H$$
$$\quad\;\; H \qquad\qquad\qquad H$$

methyl anion

This is the **methyl anion**, and carries one negative charge. Carbon has four valence electrons, and in this structure the number of assigned electrons is

five (three bonds plus a lone pair). Therefore, the formal charge on carbon is $4 - 5 = -1$. We must always indicate the charge in structures pictured as shown in the right-hand representation; the left-hand representation is incomplete and, therefore, wrong. The most common formal charges we shall meet are summarized in Figure 2.29.

Now let us return to **curly arrows** and **resonance structures**.

- Resonance structures differ only in the position of the electrons; the positions of the atoms do not change.

- Resonance structures can be interconverted by the movement of electrons indicated by curly arrows.

- Three main types of electron movement can be implicated:

  bonding to nonbonding

$$\overset{\backslash}{\underset{/}{C}}=\ddot{O}: \quad\longleftrightarrow\quad \overset{\oplus\backslash}{\underset{/}{C}}-\ddot{O}:^{\ominus}$$

two electrons are moved from the π bond to the electronegative oxygen;
carbon now has formal charge +1, oxygen has formal charge −1;
the molecule still has overall charge of zero

nonbonding to bonding

*the two lone pair electrons are used to make a π bond
to carbon; this effectively assigns one electron to
each atom;
C⁺ was electron deficient but now gains one electron
from the lone pair;
oxygen now loses one electron from the lone pair and
carries formal charge +1;
the new structure still has overall charge +1*

bonding to new bonding

*π electrons from the double bond are used to make a
new double bond;
C⁺ was electron deficient but now gains one electron
from the pair of π electrons;
the carbon at the donor end of the arrow is now
electron deficient and carries formal charge +1;
the new structure still has overall charge +1*

- All structures must be valid Lewis structures. An
  atom may become electron deficient, but, on the
  other hand, it must never be shown with more
  valence electrons than it can accommodate. For
  example, it is not possible to have pentavalent
  carbon.

- The overall charge must remain the same.

- There should be the same number of unpaired
  electrons in each structure.

This redistribution of electrons provides us with
one or more new **resonance structures** (also
called **canonical structures, limiting structures**, or
mesomers). However, some structures are more realistic than others.

- The more covalent bonds a structure has, the more
  stable it is.

- A structure in which all the atoms have the noble
  gas structure is particularly stable.

- Separation of charge in a resonance structure
  decreases stability.

- Structures with charge separation are more stable if
  the negative charge is located on an electronegative
  atom.

- Structures with adjacent like charges are disfavoured, as are those with multiple isolated
  charges.

- The σ bond framework and steric factors must
  permit a planar relationship between contributory
  resonance structures.

This is illustrated for a carbonyl compound and an
alkene.

*most favourable
resonance form*

*charge separation;
less favourable;
negative charge on more
electronegative atom*

*least important;
negative charge on less
electronegative atom*

*most favourable
resonance form*

*charge separation;
less favourable;
positive and negative charges
on same type of atom*

also unfavourable

We then consider the potential relative importance
of the resonance structures we have drawn.

- Equivalent resonance structures contribute equally
  to the hybrid.

- Structures that are not equivalent do not contribute equally; the more stable a structure is, the more it is likely to contribute.

- Highly unstable structures make little contribution and may be ignored.

Acceptable resonance structures can then be imagined as contributing to the overall electronic distribution in the molecule. By considering the properties of contributing structures, we can also predict some of the properties of the molecule. We imagine that the molecule is not fully represented by a single structure, but is better represented as a hybrid of its contributing resonance forms. It is likely that the energy associated with the molecule is actually lower than that of any contributing resonance form; therefore, the **delocalization** of electrons that resonance represents is a stabilizing feature. The larger the number of stable resonance structures we can draw, the greater the extent of delocalization. The difference in energy between the actual molecule and that suggested by the best of the resonance structures is termed the **resonance energy** or **resonance stabilization energy**. This can usually only be an estimated amount.

The resonance terminology and the double-headed arrow may give the impression that the structures are rapidly interconverting. This is not true. We must appreciate right from the start that resonance structures are entirely hypothetical. They are our (sometimes clumsy) attempt to write down on paper what the bonding in the molecule might be like, and they may depict only the extreme possibilities. The molecule is presumably happily going about its business in a form that we cannot easily depict. Nevertheless, resonance structures are extremely useful and do help us to explain chemical behaviour.

Let us look again at the simple examples shown above and the consequences of our hypothetical resonance structures.

We shall see that most of the reactions of simple carbonyl compounds, like **formaldehyde**, are a consequence of the presence of an electron-deficient carbon atom. This is accounted for in resonance theory by a contribution from the resonance structure with charge separation (see Section 7.1). The second example shows the so-called conjugate acid of **acetone**, formed to some extent by treating acetone with acid (see Section 7.1). Protonation in this way typically activates acetone towards reaction, and we

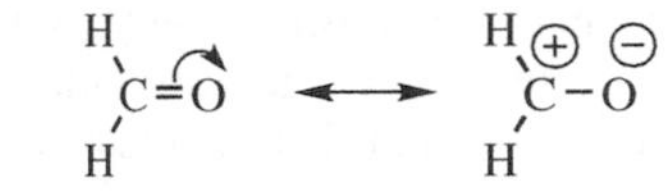

formaldehyde

conjugate acid
of acetone

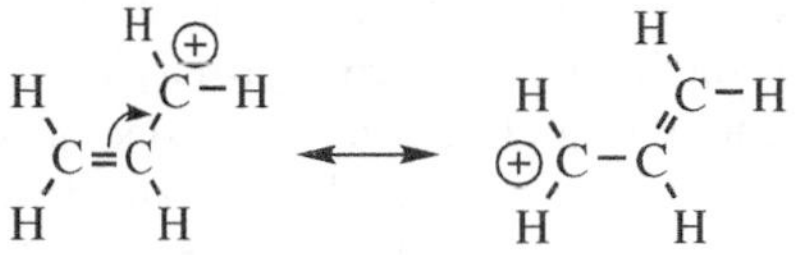

allylic cation

*note that this resonance is only
possible if the atoms are coplanar*

find that electron-rich reagents (**nucleophiles**) attack the carbon atom (see Section 7.1). This is reasonable, since we can show this carbon as positively charged in the right-hand resonance structure. The third example is the **allylic cation**. This is a reasonably stable carbocation, and we attribute this to resonance stabilization; this is particularly favourable in this case, since both contributing resonance forms are identical. We can visualize the allylic cation as an entity in which the positive charge is delocalized over the whole structure (strictly, it is the electrons that are delocalized, but we are one short of a full complement and it is the positive charge that dominates the representation).

2.11 Hydrogen bonding

Hydrogen bonds (H-bonds) describe the weak attraction of a hydrogen atom bonded to an **electronegative atom**, such as oxygen or nitrogen, to the lone pair electrons of another electronegative atom. These bonds are different in nature from the covalent bonds we have described; they are considerably weaker than covalent bonds, but turn out to be surprisingly important in chemistry and biochemistry.

Let us consider a molecule possessing an O–H σ bond. This bond is polar because hydrogen is less electronegative than oxygen (see Section 2.7), and this allows the partially positive hydrogen atom to associate with a centre in another molecule carrying a partial negative charge. This is likely to be the oxygen atom in another molecule.

hydrogen bond

If we consider the interaction between two **water** molecules, then the partial positive charge on hydrogen induced by the electronegativity of oxygen is attracted to the high electron density of the oxygen lone pair in another molecule. This hydrogen is now linked to its original oxygen by a σ bond, and to another oxygen by an electrostatic attraction. The **bond length** of the H–O hydrogen bond is typically about twice the length of the H–O covalent bond, and the hydrogen bond is very much weaker than the covalent bond, though stronger than other interactions between molecules. In water, further hydrogen bonds involving other molecules are formed, leading to a network throughout the entire sample. The extensive hydrogen bonding in water is responsible for some of water's unusual properties, its relatively high boiling point, and its high polarity that makes it a particularly good solvent for ionic compounds. Alcohols also exhibit hydrogen bonding; but, with only a single O–H, the network of bonds cannot be as extensive as in water.

Hydrogen bonds connecting different molecules are termed **intermolecular**, and when they connect groups within the same molecule they are called **intramolecular**. A simple example of an intramolecular hydrogen bond is seen in the enol form of **acetylacetone** (see Section 10.1). **Carboxylic acids**

are known to form hydrogen-bonded 'dimers' in solution through two quite strong intermolecular hydrogen bonds. Hydrogen bonds involving N–H are also common, and we can also meet hydrogen bonding between alcohols and amines.

intramolecular hydrogen bonding
in enol form of acetylacetone

intermolecular hydrogen bonding
in carboxylic acids

Box 2.2

Hydrogen bonds and DNA

The nucleic acids known as deoxyribonucleic acid (**DNA**) are the molecules that store genetic information. This information is carried as a sequence of bases in the polymeric molecule. Remarkably, the interpretation of this sequence depends upon simple **hydrogen bonding** interactions between base pairs. Hydrogen bonding is fundamental to the double helix arrangement of the DNA molecule, and the translation and transcription via ribonucleic acid (**RNA**) of the genetic information present in the DNA molecule.

extensive hydrogen
bonding in water

hydrogen bonding in methanol

In DNA, the base pairs are adenine–thymine and guanine–cytosine. **Adenine** and **guanine** are purine bases, and **thymine** and **cytosine** are pyrimidines (see Section 14.1).

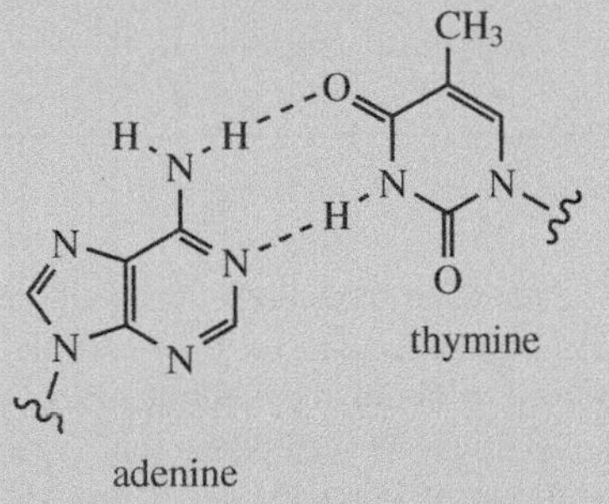

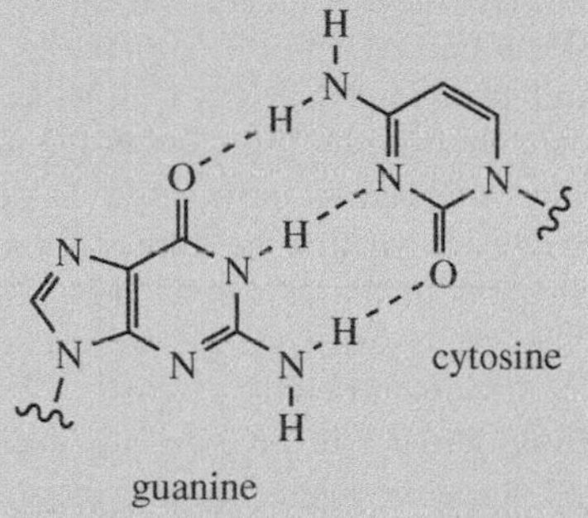

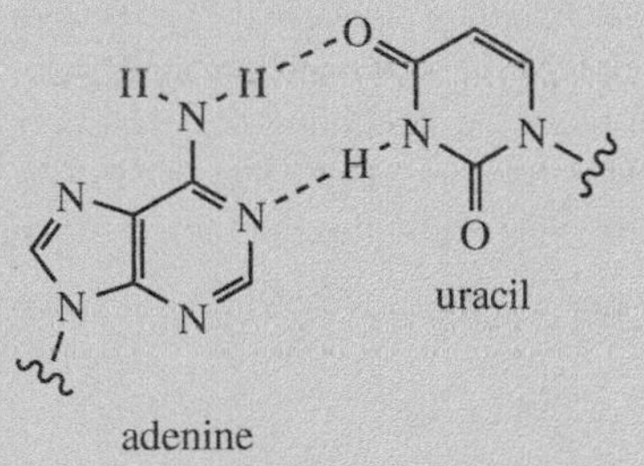

adenine   thymine

guanine   cytosine

adenine   uracil

Thus, each **purine** residue is specifically linked by hydrogen bonding to a **pyrimidine** residue. This may involve either two or three hydrogen bonds, with hydrogen of N–H groups bonding to oxygen or to nitrogen. The result of these interactions is that each base can hydrogen bond only with its complementary partner. The specific base-pairing means that the two strands in the DNA double helix are complementary. Wherever adenine appears in one strand, thymine appears opposite it in the other; wherever cytosine appears in one strand, guanine appears opposite it in the other.

Another pyrimidine base, **uracil**, is found in RNA instead of thymine. Base pairing between adenine and uracil involves two hydrogen bonds and resembles the adenine–thymine interaction. This type of base pairing is of importance in transcription, the synthesis of messenger RNA (see Section 14.2.5).

2.12 Molecular models

We soon come to realize that molecules are not two-dimensional objects as we draw on paper; they are three-dimensional and their overall size and shape can have a profound effect on some of their properties, especially biological properties. We have seen that four single bonds to carbon are distributed in a tetrahedral array, an arrangement that minimizes any steric or electrostatic interactions (see Section 2.6.2). Atoms around double bonds are in a planar array, and angles are 120°. Again, this trigonal arrangement minimizes interactions. A triple bond creates a linear array of atoms. Now, careful measurements of bond angles and also bond lengths in a wide variety of molecules have convinced us that these features are sufficiently constant that we can use them to predict the **shape** and **size** of other molecules.

Bond lengths in molecules usually correlate with one of five kinds, and their typical measurements are shown in Table 2.3. The five types of bond are:

- single bonds between atoms, one of which is hydrogen;
- single bonds between atoms, neither of which is hydrogen;
- double bonds;
- triple bonds;
- bonds in aromatic rings.

Bond angles can be related to **hybridization**, and so can bond lengths. Thus, electrons in sp hybrid orbitals are held closer to the nucleus than electrons in sp^2 orbitals, which are correspondingly closer than electrons in sp^3 orbitals (see Section 2.6.2). The bond lengths below follow this generalization. The shortest bond lengths involve bonds to hydrogen, the smallest atom that utilizes an s orbital in bonding. Note also that aromatic carbon–carbon bonds have a bond length between that of single and double bonds, a feature of aromatic bonding (see Section 2.9.4).

Table 2.3   Typical bond lengths[a]

Bond type	Bond	Length (Å)	Bond	Length (Å)	Bond	Length (Å)
Single H–X	H–C	1.06–1.10	H–N	1.01	H–O	0.96
Single C–X	C–C	1.54	C–N	1.47	C–O	1.43
Double	C=C	1.34	C=N	1.30	C=O	1.23
Triple	C≡C	1.20	C≡N	1.16		
Aromatic	C–C aromatic	1.40				

[a] 1 Å (Angstrom unit) $= 10^{-10}$ m.

With these five typical bond lengths, and the typical bond angles for tetrahedral, trigonal, and linear arrays, it becomes possible to construct molecular models to predict a molecule's size and shape. This may be achieved via a **molecular model kit**, or by **computer graphics**.

Many different molecular model kits have been produced over the years, each varying in their approach to atoms and bonds, and also in their cost. However, there are three main types, which can provide us with three main types of information. These are the framework, ball-and-stick, and space-filling versions (Figure 2.30).

Framework models concentrate on the bonds in the molecule. In the least expensive kits, these are represented by narrow tubes, joined by linker pieces that signify the atom position. The linkers have four, three, or two stubs to fit into the tubes, according to the number of bonds required, and these are also arranged at appropriate angles (tetrahedral 109°, trigonal 120°, linear 180°). The linkers are also coloured differently to show the atom they represent, typically black for carbon, white for hydrogen, red for oxygen, and blue for

nitrogen. The resultant model tells us about the bonding, and the overall size and shape of the molecule, but gives us little indication about the volume taken up by the atoms and electrons. Nevertheless, this type of model is probably the most popular, and it provides a lot of information. It is the three-dimensional equivalent of our two-dimensional line drawings of structures.

In **ball-and-stick models**, balls with holes drilled at appropriate angles are used to represent the atoms, and sticks or springs are used for the bonds to link them. The resultant model is rather like the framework model, but has representations of the atoms. Models tend to look better than the framework type, but in practice tend to be bigger and less user friendly.

Space-filling model kits are even less user friendly. They employ specially shaped atomic pieces that clip together, each representing the volume taken up by the atom and its bonding electrons. This system produces a rather more globular model that indicates the whole bulk of the molecule, including the electron clouds that are involved in bonding. The value of this type of model is that it shows just how big the molecule really is, and

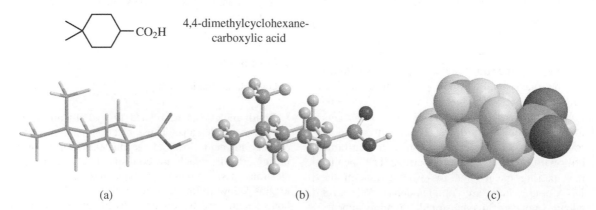

4,4-dimethylcyclohexane-carboxylic acid

(a)   (b)   (c)

Figure 2.30   Molecular models depicting 4,4-dimethylcyclohexanecarboxylic acid: (a) framework; (b) ball-and-stick; (c) space-filling. Note that the size of atoms reflects the electronic charge associated with the atom. Therefore, as seen in models (b) and (c), a hydrogen atom attached to electronegative oxygen appears smaller than a hydrogen atom attached to carbon

not just the atoms and/or bonds. When one is faced with interpreting how a molecule might interact with, say, a receptor protein, then space-filling models are essential. There is a downside, however, because it becomes very difficult indeed to visualize the structure and bonding in the molecule.

These three types of model are illustrated in Figure 2.30, though we have employed computer graphics to generate these pictures. Significantly, **computer modelling** programs all allow generation of images according to the three main types of model; each provides us with subtly different information, so they are not alternatives, but are actually complementary.

Whilst many people still like to handle and view a model, computer programs have allowed us to create and manipulate representations of three-dimensional molecules rapidly with varying degrees of accuracy and sophistication. We can easily view the image from any angle, and can see the effect that any molecular modifications might have. Further, although interactions of groups in a molecule can lead to changes in bond angles, and to a lesser extent in bond lengths, ordinary models may not show this. Computer graphics programs are able to carry out quite sophisticated energy calculations to show the most favourable arrangement of atoms, the interatomic distances, and the bond angles.

3

Stereochemistry

3.1 Hybridization and bond angles

From our discussions of bonding, we have learnt something about the arrangement of bonds around various atoms (see Chapter 2). These concepts are fundamental to our appreciation of the shape of molecules, i.e. **stereochemistry**. Before we delve into these matters, let us recap a little on the disposition of bonds around carbon.

Bonding at four-valent carbon is tetrahedral, with four sp^3-hybridized orbitals mutually inclined at 109.5°. Remember that the tetrahedral array is demonstrated by experimental measurements, and that hybridization is the mathematical model put forward to explain this observation (see Section 2.6.2). We can conveniently represent the tetrahedral arrangement in two dimensions by using a wedge–dot convention. In this convention, single bonds written as normal lines are considered to be in the plane of the paper. Bonds in front of this plane, i.e. coming out from the paper, are then drawn as a wedge, whilst bonds behind the plane, i.e. going into the paper, are drawn as a broken or dotted bond (see Section 2.6.2).

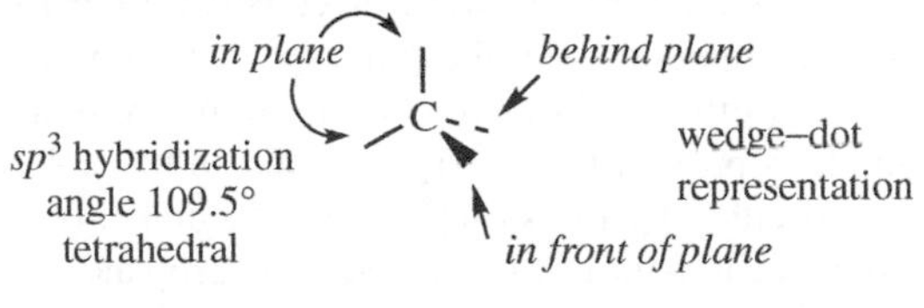

As we get more familiar with this representation, we may begin to abbreviate it by showing either the wedge or the dotted bond, rather than both. Of course, it is important to remember that these abbreviated forms actually represent a tetrahedral array, and not something with three bonds planar plus one other.

drawing stereostructures:

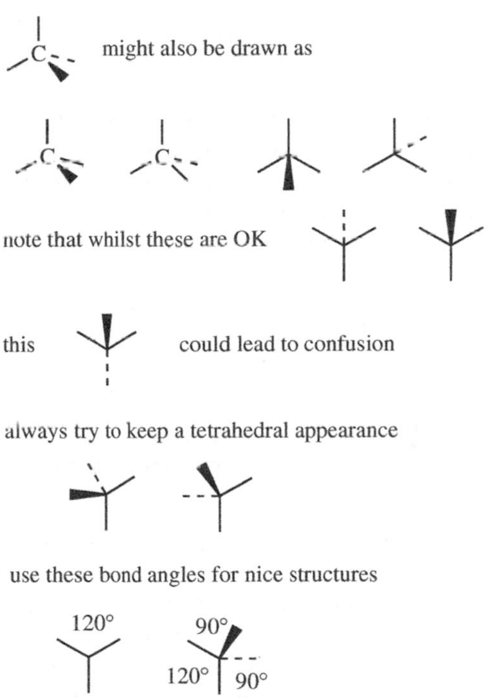

Bonding at three-valent carbon is trigonal planar with bond angles of 120°, an observation that we account for through sp^2 hybridization plus formation of a π bond by overlap of p orbitals (see Section 2.6.2). Thus, an alkene double bond involves electrons in sp^2 hybrid orbitals making σ single bonds, and the remaining electrons in p orbitals overlapping to produce the π-bond component of the double bond. We can draw this as a planar representation, all single bonds in the plane of the

Essentials of Organic Chemistry   Paul M Dewick
© 2006 John Wiley & Sons, Ltd

paper, or show the π bonding in the plane of the paper, so that some bonds now require to be drawn in wedge form and others in dotted form.

sp² hybridization
angle 120°
planar

single bonds in plane of paper, π bond perpendicular to plane

overlap of *p* orbitals generates π bond

π bond in plane of paper

Bonding at two-valent carbon is linear, i.e. bond angles are 180°, and the triple bond comprises two π bonds and a σ single bond formed from *sp* hybrid orbitals (see Section 2.6.2). The two π bonds are at right angles to each other.

sp hybridization
angle 180°
linear

$-C\equiv C-$

overlap of *p* orbitals generates two π bonds

Although most of the atoms in the framework of an organic molecule tend to be carbon, other atoms, such as oxygen and nitrogen, are routinely encountered. We can consider the arrangement of bonds around these atoms as approximately the same as the *sp³*-hybridized tetrahedral array seen with carbon (see Section 2.6.3). One (nitrogen) or two (oxygen) of the *sp³* orbitals will be occupied by lone pair electrons. The consequences of this include the fact that the two single bonds to oxygen are not linear, but are inclined at about 109° (see Section 2.6.3), and the three bonds to nitrogen are similarly not planar.

When oxygen or nitrogen are linked to another atom, e.g. carbon, by double bonds, the arrangement will be equivalent to the trivalent carbon, i.e. trigonal planar with a π bond perpendicular to the plane (see Section 2.6.3). Lone pair electrons (one lone pair for nitrogen, two in the case of oxygen) will occupy nonbonding *sp²* orbitals. A triple bond to nitrogen, as

Bonding at nitrogen and oxygen approximates to that at carbon via lone pairs:

sp³ tetrahedral

sp² trigonal

sp linear

in cyanide, will dictate a linear arrangement, with a nitrogen lone pair occupying a nonbonding *sp* orbital (see Section 2.6.3).

Bond angles depend upon the type of hybridization as just described, but in most molecules they appear to be very similar. There can often be a small degree of variation because of the nature of the precise atoms being bonded, and the presence of lone pair electrons (see Section 2.6.3), but the level of consistency is very high. Similarly, bond lengths are also remarkably consistent, depending mainly on the nature of the atoms bonded and whether bonds are single, double, aromatic, or triple (see Section 2.12). With bond lengths and bond angles being sufficiently consistent between molecules, it is possible to predict the shape and size of a molecule using simple molecular models or computer graphics (see Section 2.12).

3.2   Stereoisomers

For a given molecular formula there is often more than one way of joining the atoms together, whilst still satisfying the rules of valency. Such variants are called **structural isomers** or **constitutional isomers** – compounds with the same molecular formula but with a different arrangement of atoms. A simple example is provided by C_4H_{10}, which can be accommodated either by the straight-chained butane, or by the branched-chain isobutane (2-methylpropane).

structural isomers
constitutional isomers

butane

isobutane
(2-methylpropane)

Stereoisomers, on the other hand, are compounds with the same molecular formula, and the same sequence of covalently bonded atoms, but with a different spatial orientation. Two major classes of stereoisomers are recognized, **conformational isomers** and **configurational isomers**.

Conformational isomers, or **conformers**, interconvert easily by rotation about single bonds. Configurational isomers interconvert only with difficulty and, if they do, usually require bond breaking. We shall study these in turn.

3.3 Conformational isomers

3.3.1 Conformations of acyclic compounds

Let us consider first the simple alkane **ethane**. Since both carbons have a tetrahedral array of bonds, ethane may be drawn in the form of a wedge–dot representation.

Now let us consider rotation of the right-hand methyl group about the C–C bond, and we eventually get to a different wedge–dot representation as shown. This is more easily visualized by looking at the molecule from one end down the C–C bond, and this gives us what is termed a **Newman projection**. The Newman projection shows the hydrogen atoms and their bonds, but the carbons are represented by a circle; since we are looking down the C–C bond, we cannot see the rear carbon. A further feature is that the C–H bonds of the methyl closest to us are shown drawn to the centre of this circle, whilst those of the rear methyl are partially obscured and drawn only to the edge of this circle. We can

draw a similar Newman projection for the second wedge–dot representation, but the C–H bonds of the front and rear methyls will appear to be on top of each other. We therefore draw a slightly modified version showing all bonds, but must remember that this really represents a system where the bonds at the rear are obscured by the bonds at the front.

In the **sawhorse representation**, the molecule is viewed from an oblique angle, and all bonds can be seen.

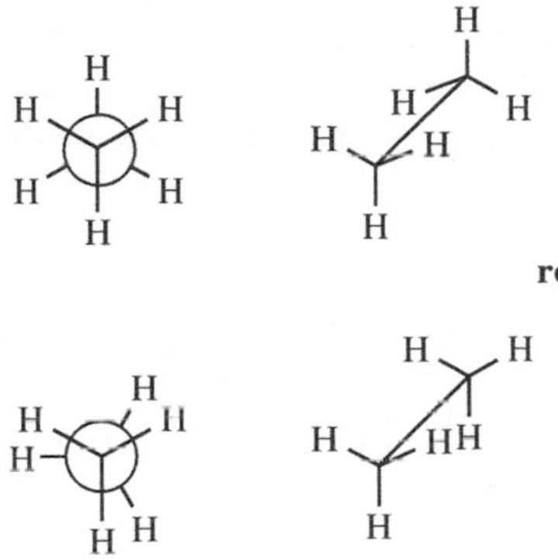

sawhorse representations

The two representations shown here are actually two different conformers of ethane; there will be an infinite number of such conformers, depending upon the amount of rotation about the C–C bond. Although there is fairly free rotation about this bond, there does exist a small energy barrier to rotation of about 12 kJ mol^{-1} due to repulsion of the electrons in the C–H bonds. By inspecting the Newman projections, it can be predicted that this repulsion will be a minimum when the C–H bonds are positioned as far away from each other

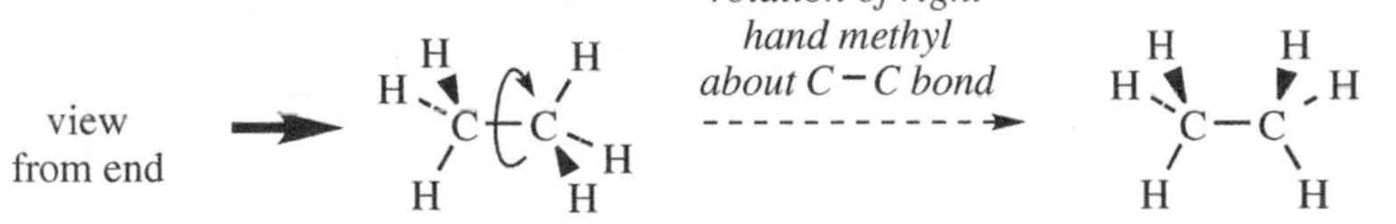

view from end

rotation of right hand methyl about C–C bond

wedge–dot representation

view from end gives **Newman projection**

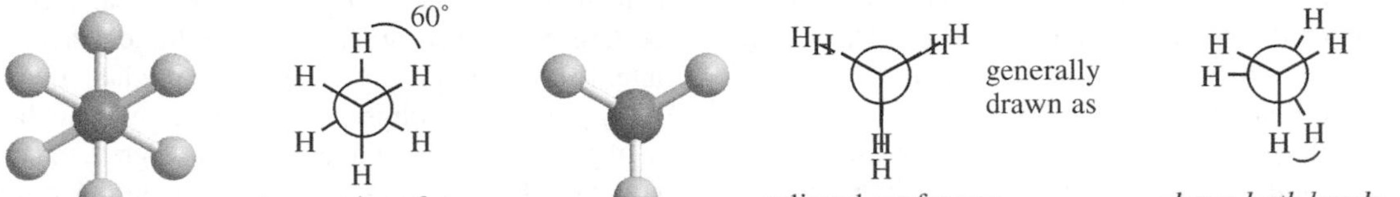

staggered conformer
low energy

eclipsed conformer
high energy

generally drawn as

shows both bonds, but angle is assumed to be 0°

as is possible. This is when the dihedral angle between the C–H bonds of the front and rear methyls is 60°, as exists in the left-hand conformer. This conformation is termed the **staggered conformation**. On the other hand, electronic repulsion will be greatest when the C–H bonds are aligned, as in the right-hand conformer. This conformation is termed the **eclipsed conformation**. In between these two extremes there will be other conformers of varying energies, depending upon the degree of rotation. Energies for these will be greater than that of the staggered conformer, but less than that of the eclipsed conformer. Indeed, if one considers a gradual rotation about the C–C bond, the energy diagram will

take the form of a sine wave, because rotations of either 120° or 240° will produce an indistinguishable conformer of identical energy. This is shown in Figure 3.1.

It follows that the preferred conformation of ethane is a staggered one; but, since the energy barrier to rotation is relatively small, at room temperature there will be free rotation about the C–C bond.

Let us now consider rotation about the central C–C bond in **butane**. Rotation about either of the two other C–C bonds will generate similar results as with ethane above. Wedge–dot, Newman, and sawhorse representations are all shown; use the version that appears most logical to you.

wedge–dot representations

rotation of right-hand group

Newman projections

rotation of rear groups

eclipsed conformer*

staggered conformer
anti
lowest energy

staggered conformer**
gauche

eclipsed conformer
highest energy

staggered conformer**
gauche

eclipsed conformer*

* equal energies

** equal energies

sawhorse representations
rotation of rear groups

C - CH₃ bonds shown in bold

As we rotate the groups, we shall get a series of staggered and eclipsed conformers. The energy barrier to rotation will be larger than the 12 kJ mol⁻¹ seen with ethane. This is because, in addition to the similar electronic repulsion in the bonds, there is now a spatial interaction involving the large methyl groups. It follows that the repulsive energy associated with a methyl–methyl interaction will be larger than

a methyl–hydrogen interaction, which in turn will be larger than that arising from hydrogen–hydrogen interactions. Logically then, we predict that the energy of the eclipsed conformer in which the methyl groups are aligned will be higher than that in which there are methyl–hydrogen alignments, and that there will be two equivalent versions of the latter.

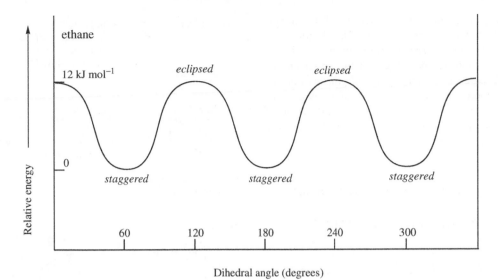

Figure 3.1 Energy diagram: ethane conformations

Similarly, of the low-energy staggered conformers, there will be two equivalent ones where the carbon–methyl bonds are inclined at 60° to each other, and one in which the carbon–methyl bonds are inclined at 180°. We can also predict that the latter conformer, which has the methyl groups as far away from each other as possible, will be of lower energy than the alternative staggered conformers, where there must be at least some spatial interaction between the methyl groups. The staggered conformer with maximum separation of methyl groups is termed the ***anti* conformer** (Greek: *anti* = against), whilst the two other ones are termed ***gauche* conformers** (French: *gauche* = left). The energy diagram observed (Figure 3.2) reflects these predictions, and the energy difference between the

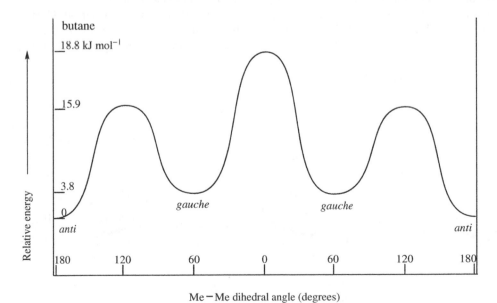

Figure 3.2 Energy diagram: butane conformations

low-energy staggered *anti* conformer and the highest energy eclipsed conformer is about 18.8 kJ mol^{-1}.

There will still be free rotation about C–C bonds in butane at room temperature, but the larger energy barrier compared with that for ethane means that the staggered conformers are preferred, and calculations show that, at room temperature, about 70% of molecules will be in the *anti* conformer and about 15% in each *gauche* conformer.

3.3.2 Conformations of cyclic compounds

Cyclopropane, cyclobutane, cyclopentane, cyclohexane

The practical consequences of conformational isomerism become much more significant when we consider cyclic compounds. The smallest ring system will contain three atoms; in the case of hydrocarbons this will be **cyclopropane**.

Now, simple geometry tells us that the inside angle in cyclopropane must be 60°. This is considerably less than the 109.5° of tetrahedral carbon, and the consequences are that the amount of overlap of the sp^3 orbitals in forming the C–C bonds must be considerably less than in an acyclic system like ethane. With poorer overlap, we get a potentially weaker bond that can be broken more easily. We term this **ring strain**, and although three-membered rings exist and are quite stable, they are frequently subject to ring-opening reactions (see Section 6.3.2).

cyclopropane
- angle 60°
- highest ring strain
- must be planar
- bonds eclipsed

poor orbital overlap in C–C bonds

maximum orbital overlap   *poor orbital overlap*

A further feature of three-membered rings is that they must be planar, and a consequence of this is that, in cyclopropane, all C–H bonds are in the high-energy eclipsed state. There can be no conformational mobility to overcome this.

In **cyclobutane**, the internal angle is 90°. Consequently, there is high ring strain, but this is not so great as in cyclopropane.

If cyclobutane were planar, all C–H bonds would be in the high-energy eclipsed state. It transpires that cyclobutane is not planar, since it can adopt a

cyclobutane
- angle 90°
- high ring strain
- non-planar conformer minimizes eclipsing

equilibrium arrow is used to indicate interconversion of structures

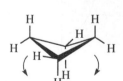 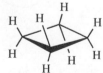 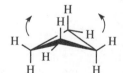

interconversion of equal energy non-planar conformers through planar intermediate

note how we can show perspective by having a full bond at the front, and an incomplete bond at the rear

more favourable conformation in which eclipsing is reduced, and the ring appears puckered. This appears to be achieved by pushing pairs of opposite carbons in different directions; but, in reality, it is only a combination of rotations about C–C bonds as we have seen with the simpler acyclic compounds. It is not possible to achieve the ideal 60° staggered arrangement, but it does produce a lower energy conformer. Of course, there are two alternative ways of doing this, depending on whether pairs of carbons are 'pushed' or 'pulled'. Both conformers will be produced equally, and can interconvert at room temperature because the energy barrier is fairly small at about 5.8 kJ mol^{-1}. The interconversion of the two forms is depicted by the equilibrium arrow, comprised of two half arrows. At equilibrium, both conformers coexist, and in this case in equal amounts since they have the same energy.

The planar form of cyclobutane will be the energy maximum in the interconversion of conformers.

Box 3.1

Compounds with cyclopropane or cyclobutane rings

A cyclopropane ring has the highest level of ring strain in the carbocycles. This means that they are rather susceptible to ring-opening reactions, but it does not mean that they are unstable and cannot exist. Indeed, there are many examples of natural products that contain cyclopropane rings, and these are perfectly stable under normal conditions.

One group of natural cyclopropane derivatives of especial importance is the **pyrethrins**, insecticidal components of pyrethrum flowers, and widely used in agriculture and in the home. These compounds have very high toxicity towards insects without being harmful to animals and man, and are rapidly biodegraded in the environment. The pyrethrins are esters of two acids, **chrysanthemic acid** and **pyrethric acid**, with three alcohols, pyrethrolone, cinerolone, and jasmolone, giving six major ester structures. The acids contain the cyclopropane ring, and this appears essential for the insecticidal activity.

pyrethrins general structure

acids: chrysanthemic acid   pyrethric acid

alcohols: pyrethrolone   cinerolone   jasmolone

semi-synthetic pyrethrins

bioresmethrin   permethrin   phenothrin

Many semi-synthetic esters, e.g. **bioresmethrin, permethrin**, and **phenothrin**, have been produced and these have increased toxicity towards insects and also extended lifetimes. All such esters retain a high proportion of the natural chrysanthemic acid or pyrethric acid structure.

Box 3.1 (continued)

The drugs **naltrexone** and **nalbuphine** are semi-synthetic analogues of the analgesic **morphine**. Morphine is a good painkiller, but has some unpleasant side effects, the most serious of which is the likelihood of becoming addicted.

morphine   nalbuphine   naltrexone

Nalbuphine is a modified structure containing a cyclobutane ring as part of the tertiary amine function. Extending the size of the nitrogen substituent makes the drug larger and allows it to exploit extra binding sites on the receptor that morphine cannot interact with. Nalbuphine is found to be a good analgesic with fewer side effects than morphine. Naltrexone incorporates a cyclopropane ring in the nitrogen substituent. This, together with the other structural modifications, produces a drug that has hardly any analgesic effects, but is a morphine antagonist. Accordingly, it can be used to assist in detoxification of morphine and heroin addicts.

Let us move on to **cyclopentane**, where geometry tells us the internal angle is 108°. This is so close to the tetrahedral angle of 109.5° that cyclopentane can be considered essentially free of ring strain. However, planar cyclopentane would have all its C–H bonds eclipsed, which is obviously not desirable. Accordingly, it adopts a lower energy conformation in which one of the carbon atoms is out of planarity. 'Pushing' this carbon out of the plane is achieved by rotation about C–C bonds, and it reduces eclipsing along all but one of the C–C bonds.

cyclopentane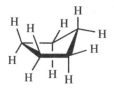
- angle 108°
- little ring strain
- non-planar conformer minimizes eclipsing

envelope conformation

The energy barrier to this conformational change is about 22 kJ mol^{-1}. There is no reason why any one particular carbon should be out of the plane, and at room temperature there is rapid interconversion of all possible variants. Again, a planar form would feature as the energy maximum in the interconversions. The conformation with four carbons in plane and one out of plane is termed an **envelope conformation**. This terminology comes from the similarity to an envelope with the flap open.

For **cyclohexane**, the calculated internal angle is 120° if the molecule were to be planar, but the tetrahedral angle of 109.5° turns out to be perfect if the molecule is non-planar. It is possible to construct a cyclohexane ring from tetrahedral carbons without introducing any strain whatsoever. The ring shape formed in this way is termed a **chair conformation**. There is considerable resemblance to a folding chair having a back rest and leg rest, though the open seat might be regarded as a distinct disadvantage. Not only is the bond angle perfect, but it also turns out that all C–H bonds are in a staggered relationship with adjacent ones. The chair conformation cannot be improved upon.

cyclohexane   • angle 109.5° if non-planar
                • no ring strain
                • no eclipsing in **chair** conformation

Newman projection of chair
conformation looking along
two opposite C−C bonds;
all bonds are staggered

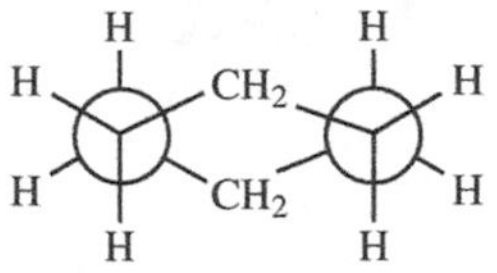

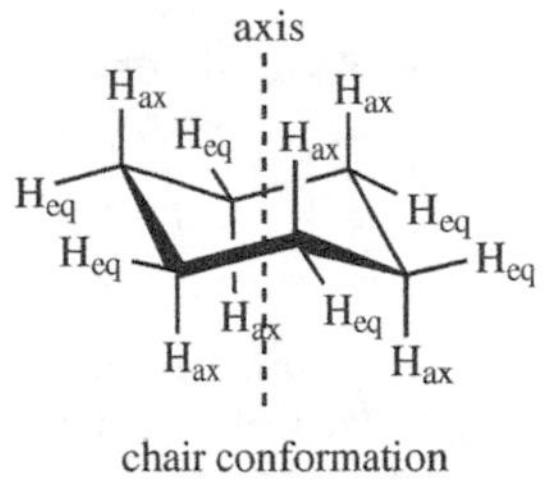

axis

hydrogens are **axial**
or **equatorial**

chair conformation

The total ring strain in various cycloalkanes compared with their strain-free acyclic counterparts has been estimated, as shown in Table 3.1. Thus, small rings like cyclopropane and cyclobutane have considerable ring strain, and cyclohexane is effectively strain free. Larger rings (8–11 atoms) have more ring strain than might be predicted, certainly much more than cyclohexane, but any puckering that reduces ring strain actually creates eclipsing. We shall meet rings containing more than six carbons only infrequently.

Table 3.1   Ring strain[a] in cycloalkanes

Number of atoms in ring	Total ring strain (kJ mol^{-1})	Number of atoms in ring	Total ring strain (kJ mol^{-1})
3	115	8	41
4	110	9	53
5	26	10	51
6	0	11	47
7	26	12	17

[a] Values relative to strain-free acyclic analogue, e.g. cyclobutane and butane.

Box 3.2

How to draw chair conformations of cyclohexane

You can only appreciate stereochemical features if you can draw a representation that correctly pictures the molecule. One of the most challenging is the chair conformation of cyclohexane. Practice makes perfect; so this is how it is done.

Draw two inclined bonds of the same length

Draw two further parallel bonds

Add the two remaining bonds ensuring they are parallel to existing bonds

Put in the axial substituent bonds, up from top points, down from bottom points

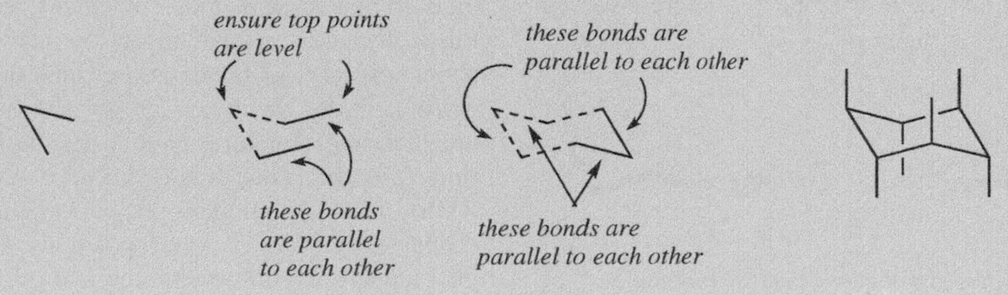

ensure top points are level

these bonds are parallel to each other

these bonds are parallel to each other

these bonds are parallel to each other

Box 3.2 (continued)

Add three pairs of equatorial substituent bonds ensuring they are parallel to existing bonds

these four bonds are all parallel to each other

The end result – perfect!
Put in wedges and bold bond for perspective if required; the lower part of the ring is always at the front

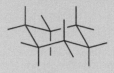

these four bonds are all parallel to each other

these four bonds are all parallel to each other

Note that the wedges and bold bonds help to show how we are looking at the cyclohexane chair. In practice, particularly to speed up the drawing of structures, we tend to omit these. Then, by convention, the lower bonds represent the nearest part of the ring.

for ease of drawing, we usually omit bold bonds and wedges

the lower bonds always represent the nearest part of the ring

When one looks at the hydrogens in the chair conformation of cyclohexane, one can see that they are of two types. Six of them are parallel to the central rotational axis of the molecule, so are termed **axial**. The other six are positioned around the outside of the molecule and are termed **equatorial**. One might imagine, therefore, that these two types of hydrogen would have some different characteristics, and be detectable by an appropriate spectral technique. Such a technique is NMR spectroscopy; but, at room temperature, only one type of proton is detectable. At room temperature, all hydrogens of cyclohexane can be considered equivalent; this is a consequence of conformational mobility, and the interconversion of two chair conformations.

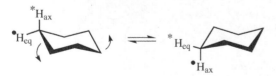

interconversion of conformers via ring flip changes axial / equatorial relationship; the conformers have the same energy

This interconversion may be considered as the simultaneous pushing down/pulling up of carbons on opposite sides of the ring, as indicated in the left-hand structure. As a result, the ring 'flips' into an alternative conformation, also a chair, as in the right-hand structure. This ring flip is actually achieved by rotation about several of the C–C bonds at the same time. The ring flip can be demonstrated with suitable molecular models, and it is possible to feel the resistance in the model to this rotation, which represents the energy barrier to the change. Both conformers have the same energy, but the energy barrier is about 42 kJ mol^{-1}. The energy barrier looks high compared with those in ethane or butane, but this is because the interconversion involves rotations about several C–C bonds at the same time.

Look at the hydrogen atoms shown labelled in the left-hand structure. Note particularly that, after ring flip, the axial hydrogen becomes equatorial, whilst the equatorial hydrogen becomes axial. Similar changes occur at all other positions. With rapidly interconverting conformers, the hydrogens cannot be distinguished by NMR spectroscopy and they all merge to give a single signal. However, as one cools the sample, the energy available to overcome the interconversion energy barrier diminishes, until at a sufficiently low temperature, the interconversion stops, and two types of hydrogen are detectable in the NMR spectrum. This temperature is $-89\,^{\circ}$C. Measurement of this temperature allows the energy barrier to be calculated.

If we look at the two-dimensional hexagon representation for cyclohexane, we could put in the bonds to hydrogens as wedges (up bonds) or dotted lines (down bonds). We now know the cyclohexane ring is not planar, but has a chair conformation. We shall frequently want to use the hexagon representation, and it will be necessary to assign hydrogens or other substituents onto the chair representation with the correct stereochemistry. At this stage it is salutary to look at both the two-dimensional hexagon and the chair representations of cyclohexane. Note particularly that we must not confuse 'up' with axial, and 'down' with equatorial. As the structures show, 'up' hydrogens or substituents will alternate axial and equatorial as we go round the ring positions.

incorrect planar representation

hydrogens shown in bold are 'up' and alternate axial–equatorial around the ring; they are not all axial or all equatorial

The chair is not the only conformation that cyclohexane might adopt. An alternative **boat conformation** is attained if the ring flip-type process is confined to just one carbon. The name boat comes from the similarity to boats formed by paper folding; sea-worthiness is rather questionable. Again, there is no ring strain in this conformation, but it turns out that some of the C–H bonds are eclipsed, as seen in the accompanying Newman projection.

flagpole interaction

chair   boat

- no ring strain
- eclipsing in **boat** conformation
- flagpole interaction in boat
- eclipsing and flagpole interaction reduced in **twist-boat**
- both higher energy than chair

twist-boat

Newman projection of boat conformation looking along the two horizontal bonds; some bonds are eclipsed

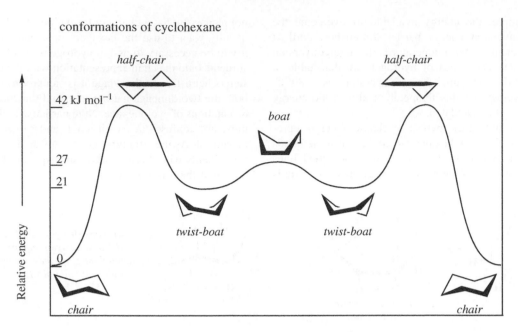

Figure 3.3 Energy diagram: cyclohexane conformations

In addition, the hydrogens at the top of the structure are getting rather close to each other, and there is some interaction, termed a **flagpole interaction**, again from the nautical analogy. Both the eclipsing and the flagpole interactions can be minimized when the boat conformation undergoes further subtle changes by rotation about C–C bonds to form the **twist-boat**. This is a result of twisting the flagpole hydrogens apart. Making a molecular model of the boat conformation immediately shows how easy it is to modify it to the twist-boat variant; the boat conformation is quite floppy compared with the chair, which is very rigid. An energy diagram linking the chair, boat and twist-boat conformations is shown in Figure 3.3. The boat conformation is represented by an energy maximum.

In practice, only the chair conformation is important for cyclohexane, since the energy differences between it and the other conformations make them much less favourable. However, there are plenty of structures where cyclohexane rings are forced into the boat or twist-boat conformation because of other limiting factors. For example, **bornane** is a terpene hydrocarbon where opposite carbons in a cyclohexane ring are bridged by a methylene group. This is stereochemically impossible to achieve with a chair – the carbons are too far apart. However, it is possible with a boat conformation. In such a structure, there are no further possibilities for conformational mobility – the conformation is now fused in and no further changes are possible, even though there may be unfavourable eclipsing interactions.

bornane boat conformation cyclohexene half-chair conformation (planar around double bond, non-planar elsewhere removes eclipsing) tetrahydro-naphthalene

In **cyclohexene**, the double bond and adjacent carbons must all be planar. The remainder of the molecule avoids unfavourable eclipsing interactions by adopting what is termed a **half-chair conformation**. This would also be found in a cyclohexane ring fused onto an aromatic ring (**tetrahydronaphthalene**) or fused to a three-membered ring (see Section 3.5.2). The half-chair conformation in cyclohexane (without the double bond) is thought to be equivalent to the energy maximum in Figure 3.3 that must be overcome in the chair–twist-boat interconversion.

Substituted cyclohexanes

The ring flipping conformational mobility in the un-substituted compound cyclohexane has little practical significance; but, when the ring is substituted, we have to take ring flip into account, because one particular conformation is usually favoured over the other.

Let us look at a simple example, namely **methyl-cyclohexane**.

Ring flip in the case of methylcyclohexane achieves interconversion of one conformer where the methyl group is equatorial into a conformer where this group is axial (compare the hydrogens in cyclohexane). It turns out that the conformer with the equatorial methyl group is favoured over the conformer where the methyl group is axial. The energy difference of these two conformers is estimated to be about 7.1 kJ mol^{-1}; this is the energy difference, not the barrier to interconversion. Because of this energy difference, the equilibrium mixture at room temperature has about 95% of conformers with the equatorial methyl and only 5% where the methyl is axial. We can account for the difference in energy between the two conformers quite easily using the reasoning we applied earlier for the acyclic hydrocarbon butane.

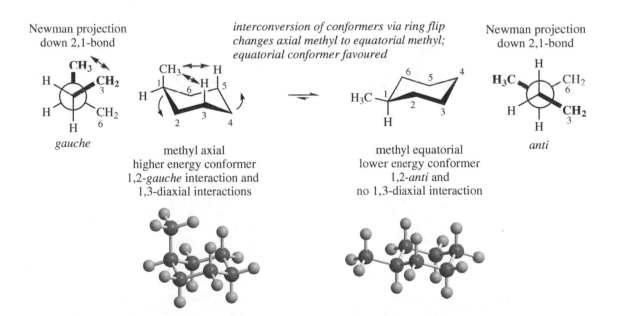

methylcyclohexane

Newman projection
down 2,1-bond

gauche

methyl axial
higher energy conformer
1,2-*gauche* interaction and
1,3-diaxial interactions

*interconversion of conformers via ring flip
changes axial methyl to equatorial methyl;
equatorial conformer favoured*

methyl equatorial
lower energy conformer
1,2-*anti* and
no 1,3-diaxial interaction

Newman projection
down 2,1-bond

anti

We need to consider a Newman projection looking down the 2,1 bond. When the methyl is axial, it can be seen that there will be a *gauche* interaction between this methyl and the ring methylene (C-3); a second, similar interaction will be seen if we looked

down the 6,1 bond. Now, in the conformer where the methyl is equatorial the Newman projection shows the most favourable *anti* arrangement for the methyl and methylene(s); there will be a similar *anti* interaction if we looked down the 6,1 bond. On this basis alone, we can predict that the equatorial conformer is of lower energy and, thus, more favoured. However, there is a further feature that destabilizes the axial conformer, and that is the spatial interaction between the axial methyl and the axial hydrogens at positions 3 and 5, termed a **1,3-diaxial interaction**. Together, they account for the equilibrium mixture consisting mainly of the equatorial conformer. We can indicate this by using arrows of unequal size in the equilibrium equation.

Note that it is not necessary to consider both forms of cyclohexane, where the methyl is either wedged (up) or dotted (down). If the cyclohexane ring were planar, the two structures would be the same, since one merely has to turn the structure over to get the other. Although the cyclohexane ring is not planar, it turns out that the two structures are still identical, because of the ring flip process. This is shown below. One set of conformers is simply the upside-down version of the other.

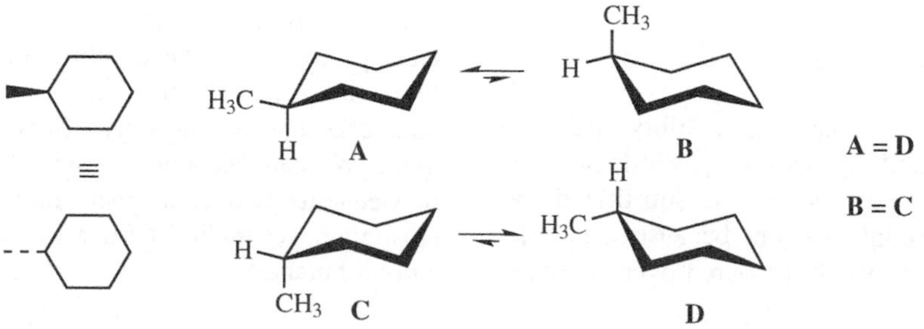

Now, as the substituent gets bigger, the proportion of axial conformer will diminish even further. With a substituent as big as a *tert*-butyl group, the equilibrium is such that essentially all molecules are in the equatorial conformation; in general terms, we can consider that a *tert*-butyl group will never be axial.

tert-butyl group **never** axial

Although analysis of the consequences of ring flip in a monosubstituted cyclohexane is pretty straightforward, the presence of two or more substituents requires careful consideration to decide which conformer, if any, is the more favoured. Let us illustrate the approach using **1,4-dimethylcyclohexane**. Now, two **configurational isomers** of this structure can exist, namely *trans* and

cis. The terms *trans* and *cis* are used to describe the configuration, not conformation, of the isomers; in the *trans* isomer, the two methyl substituents are on opposite sides (faces) of the ring (Latin: *trans* = across), whereas in the *cis* isomer they are on the same side of the ring (Latin: *cis* = on this side). These concepts will become clear when we reach Section 3.4.

1,4-dimethylcyclohexane

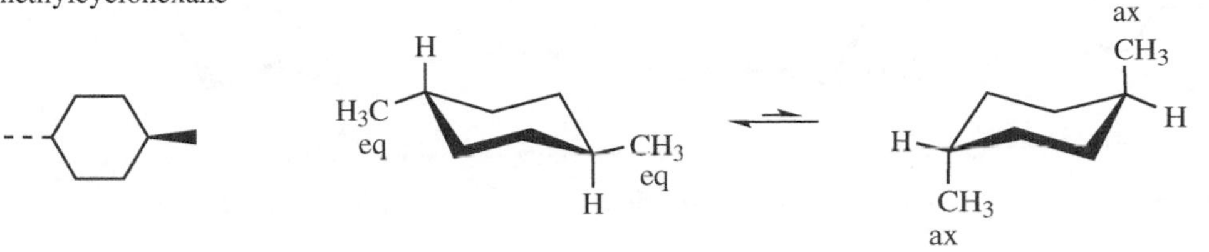

lower energy – both methyl
substituents equatorial

higher energy – both methyl
substituents axial

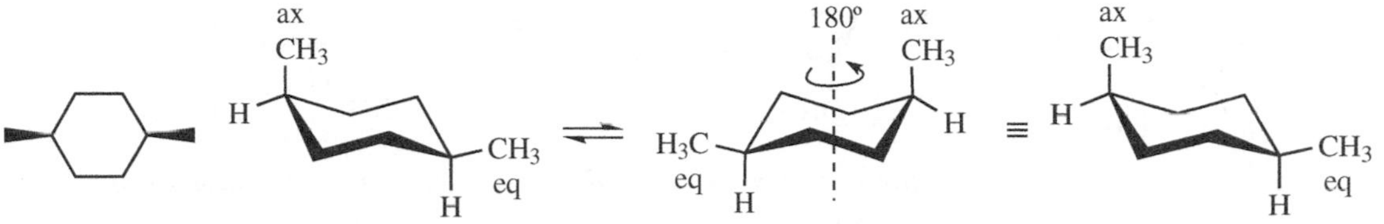

both conformations have same energy –
one axial methyl and one equatorial methyl

In the *trans* isomer, one methyl is written down (dotted bond) whilst the other is written up (wedged bond). If we transform this to a chair conformation, as shown in the left-hand structure, the down methyl will be equatorial and the up methyl will also be equatorial. With ring flip, both of these substituents then become axial as in the right-hand conformer. From what we have learned about monosubstituted cyclohexanes, it is now easily predicted that the diequatorial conformer will be very much favoured over the diaxial conformer.

In the *cis* isomer, both methyls are written with wedges, i.e. up. In the left-hand chair conformation, one methyl is therefore axial and the other is equatorial. With ring flip, the axial methyl becomes equatorial and the equatorial methyl becomes axial. Both conformers have one equatorial methyl and one axial methyl; they must, therefore, be of the same energy, so form a 50 : 50 equilibrium mixture. In fact, it is also easy to see that rotation of either structure about its central axis produces the other structure, a clear illustration that they must be energetically equivalent. Note that the *cis* isomer with both methyls down is actually the same compound viewed from the opposite side.

This type of reasoning may be applied to other **dimethylcyclohexanes**, as indicated in the figure. There is no easy way to predict the result; it must be deduced in each case. One conformer is of much lower energy in the cases of *trans*-1,2-, *cis*-1,3-, and *trans*-1,4-dimethylcyclohexane; both conformers have equal energy in the cases of *cis*-1,2-, *trans*-1,3-, and *cis*-1,4-dimethylcyclohexane.

cis-1,2-dimethylcyclohexane	methyls eq and ax	
trans-1,2-dimethylcyclohexane	methyls both eq or both ax	
cis-1,3-dimethylcyclohexane	methyls both eq or both ax	
trans-1,3-dimethylcyclohexane	methyls eq and ax	
cis-1,4-dimethylcyclohexane	methyls eq and ax	
trans-1,4-dimethylcyclohexane	methyls both eq or both ax	

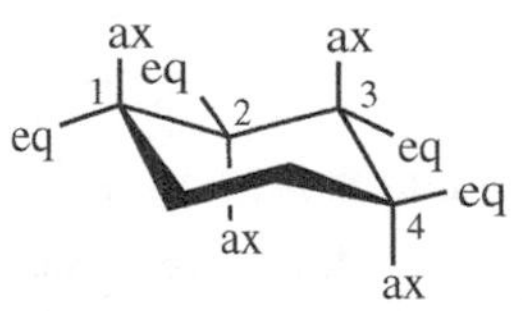

Should the two substituents be different, and especially of different sizes, then the simple reasoning used above with two methyl substituents will need adapting; the larger substituent will prefer to be equatorial. Where we have three or more substituents, the most favoured conformer is going to be the one with the maximum number of equatorial substituents, or perhaps where we have the large substituents equatorial. This is seen in the following examples.

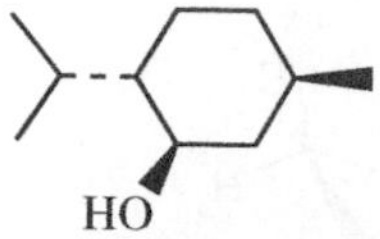

menthol

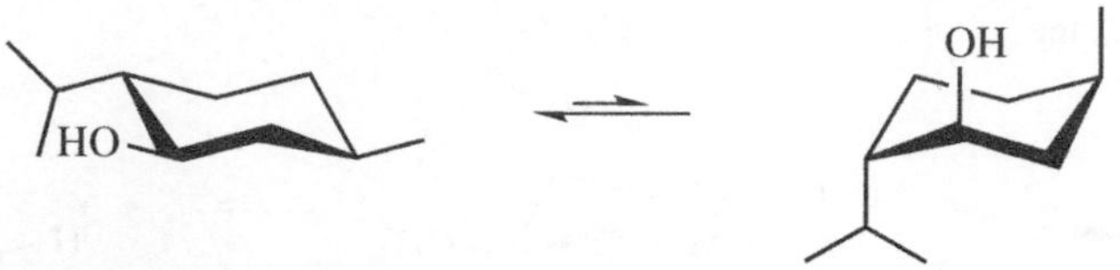

all substituents equatorial;
favoured

all substituents axial

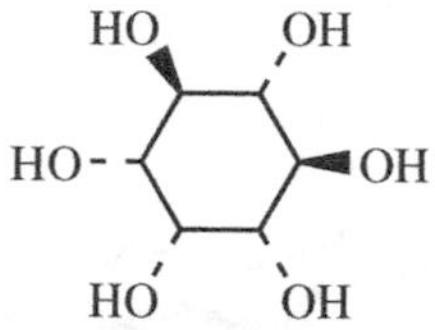

myo-inositol

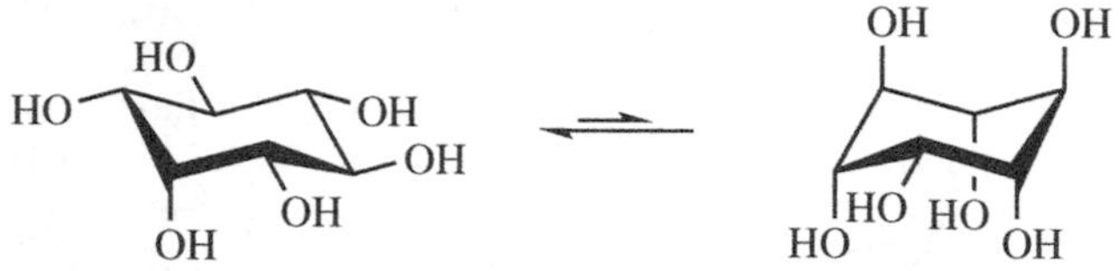

*five substituents equatorial,
one axial;*
favoured

*five substituents axial,
one equatorial*

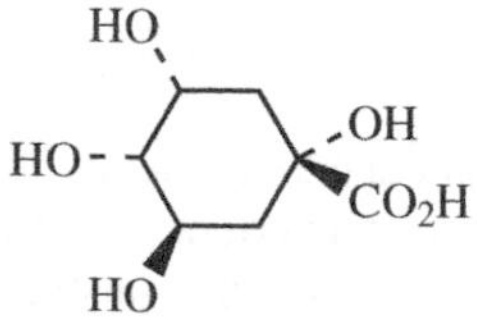

quinic acid

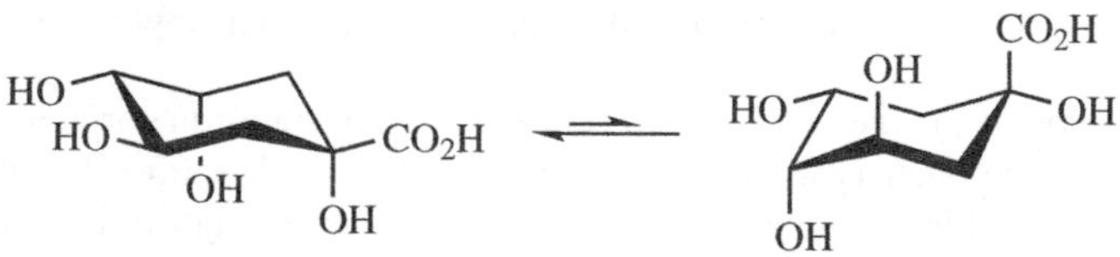

*three substituents equatorial,
two axial;*
favoured

*three substituents axial, two
equatorial*

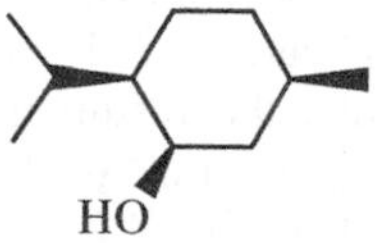

neoisomenthol

*two substituents equatorial
but the large isopropyl group
is axial*

*two substituents axial with large
isopropyl group equatorial;*
favoured

Box 3.3

How to draw conformational isomers and to flip cyclohexane rings

Interpreting a two-dimensional stereochemical structure, converting it into a conformational drawing, and considering the consequences of ring flip can cause difficulties. The process can be quite straightforward if you approach it systematically.

We saw early in Section 3.3.2 that, if we draw cyclohexane in typical two-dimensional form, the bonds to the ring could be described as 'up' or 'down', according to whether they are wedged or dotted. This is how we would see the molecule if we viewed it from the top. When we look at the molecule from the side, we now see the chair conformation; the ring is not planar as the two-dimensional form suggests. Bonds still maintain their 'up' and 'down' relationship, but this means bonds shown as 'up' alternate axial–equatorial around the ring; they are not all axial or all equatorial. Whilst the ring flip process changes equatorial bonds to axial bonds, and vice versa, it does not change the 'up'–'down' relationship.

*hydrogens shown in bold are 'up' and
alternate axial–equatorial around the ring;
they are not all axial or all equatorial*

*ring flip changes equatorial to axial,
and axial to equatorial; it does not
change the 'up'–'down' relationship*

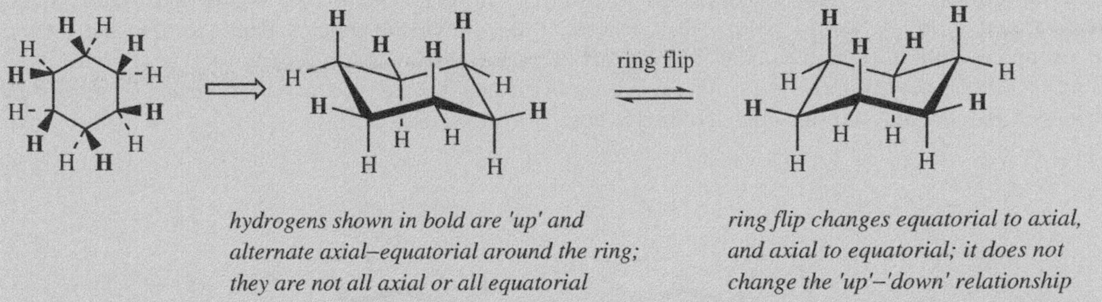

Let us consider the trimethylcyclohexane isomer shown below. All three substituents are 'up'. We need to use one of the carbons as a reference marker; let us choose the top one. I like to make this the left-hand carbon in the chair; to make the process more obvious, we could turn the structure so that our reference carbon is also on the left. It is most important to have this reference carbon, so that as we put the various substituents in we put them on the correct carbons.

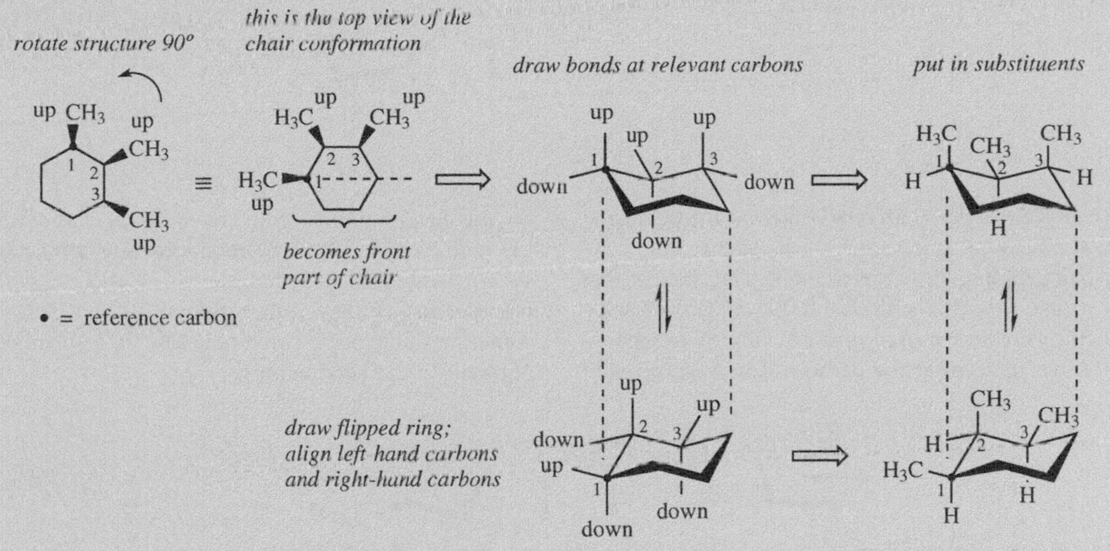

Now draw the two chair conformations of cyclohexane, both having the reference carbon on the left. The carbons opposite our reference point must be furthest right. If we draw the structures one above the other, left-hand carbons and right-hand carbons should be aligned. Draw axial and equatorial bonds at the relevant carbons where we have the substituents and identify them as 'up' or 'down'. Since we are interpreting the structure as though we are looking down on it from the top, the lower part of the ring represents the nearmost part of the conformational drawing. It can also help to number the carbons. Then fill in the substituents as necessary. In this example, our three methyl groups are all 'up', which means that in one conformer the groups will be axial, equatorial, and axial, whereas in the other they will be equatorial, axial, and equatorial. The latter conformer, with the most equatorial substituents, will be the favoured one.

A word of warning is appropriate here. As we shall see in due course (see Box 3.11), merely changing a substituent from, say, equatorial to axial without flipping the ring changes the configuration, and can produce a different molecule. It would also destroy the 'up' or 'down' identifier.

To take this general principle to its extreme, we noted above that *tert*-butyl groups are sufficiently large that they never occupy an axial position. It *is* possible to make di-*tert*-butylcyclohexanes where conformational mobility would predict that one of these groups would have to be axial, namely *cis*-1,2-, *trans*-1,3- or *cis*-1,4-derivatives. As a result, in these cases, we do not see an axial *tert*-butyl, but instead the ring system adopts the less favourable twist-boat conformation. It follows, therefore, that there must be a greater energy difference between chair conformations carrying axial and equatorial *tert*-butyl substituents than there is between chair and twist-boat conformations. These conformational changes are shown for *trans*-1,3-di-*tert*-butyl-cyclohexane.

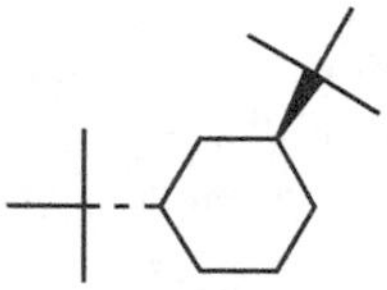

trans-1,3-di-*tert*-butylcyclohexane

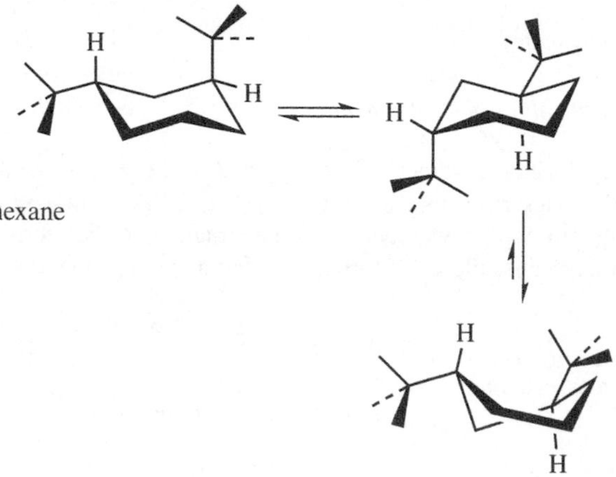

each conformer has one *tert*-butyl group axial

tert-butyl group **never** axial, so chair forced into twist-boat conformation

We noted earlier that bonds around nitrogen and oxygen atoms occupied some of the tetrahedral array, lone pairs taking up other orbitals. This means that we can use essentially the same basic principles for predicting the shape and conformation of **heterocycles** as we have used for carbocycles. A substituent on the heteroatom is considered to be larger than the lone pair electrons. Some common examples are shown below. As we shall see in Section 12.4, the heteroatom may have other influences, and there are sometimes unexpected effects involving a substituent adjacent to the heteroatom.

tetrahydrofuran

piperidine

ethylene oxide (planar)

tetrahydropyran

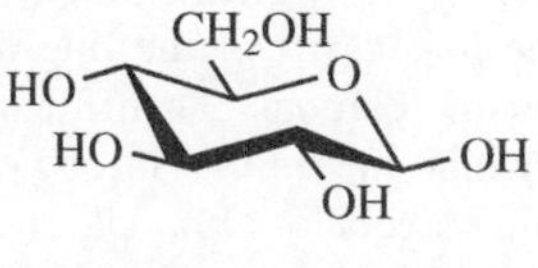

glucose (cyclic hemiacetal form)

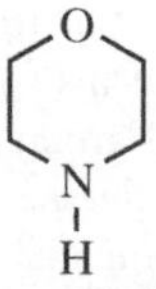

morpholine

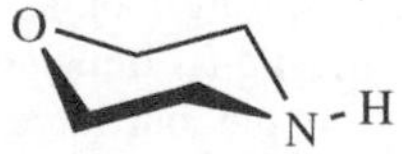

Box 3.4

Conformation of lindane

Chlorination of benzene gives an addition product that is a mixture of stereoisomers known collectively as **hexachlorocyclohexane** (HCH). At one time, this was incorrectly termed benzene hexachloride. The mixture has insecticidal activity, though activity was found to reside in only one isomer, the so-called gamma isomer, γ-HCH. γ-HCH, sometimes under its generic name **lindane**, has been a mainstay insecticide for many years, and is about the only example of the chlorinated hydrocarbons that has not been banned and is still available for general use. Although chlorinated hydrocarbons have proved very effective insecticides, they are not readily degraded in the environment, they accumulate and persist in animal tissues, and have proved toxic to many bird and animal species.

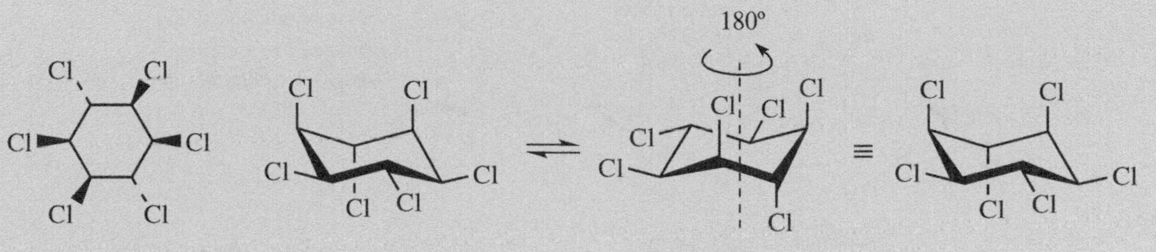

lindane
γ-HCH

The stereochemistry of the γ-isomer is shown in the diagram, and when converted into a conformational stereodrawing it can be seen that there are three axial chlorines and three equatorial ones. Ring flip produces an alternative conformation of equal energy, but it can be seen that this is identical to the first structure; rotation through 180° produces an identical and, therefore, superimposable structure. It can be seen that conformational change will not stop the compound interacting with the insect receptor site.

3.4 Configurational isomers

As we have now seen, conformational isomers interconvert easily by rotation about single bonds. **Configurational isomers**, on the other hand, are isomers that interconvert only with difficulty, and it usually requires bond breaking if they do interconvert.

3.4.1 Optical isomers: chirality and optical activity

If tetrahedral carbon has four different groups attached, it is found that they can be arranged in two different ways. These molecules are not superimposable and they have a mirror image relationship to each other. This is most easily seen with models.

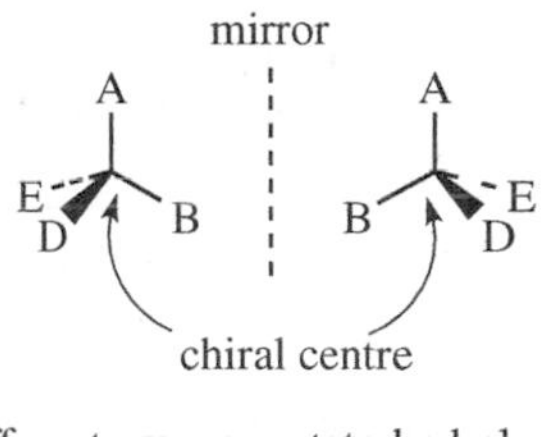

Four different groups on tetrahedral carbon can be arranged in two ways – non-superimposable molecules with a mirror image relationship

Arrangement is described as **chiral**

The two arrangements (non-superimposable mirror images) are called **enantiomers**

Such an arrangement is called **chiral** (Greek: *cheir* = hand), and the carbon atom is termed a **chiral centre** or **stereogenic centre**. Look at your two hands. You will see that they appear identical (allowing for minor blemishes or broken fingernails). However, do what you will, it is not possible to superimpose them, and you should be able to appreciate the mirror image relationship. The two different arrangements – non-superimposable mirror images – are called **enantiomers** (Greek: *enantios* = opposite), and we say that enantiomers have different configurations. The **configuration** is thus the spatial sequence about a chiral centre. It is also apparent that

enantiomers are not going to interconvert readily, and to achieve interconversion we would have to break one of the bonds then remake it so as to get the other configuration.

Note that the enantiomer of a particular compound can be drawn by reversing two of the substituents; this is actually much easier than drawing the mirror image compound, especially in more complicated structures. As an alternative, the wedge–dot relationship could be reversed.

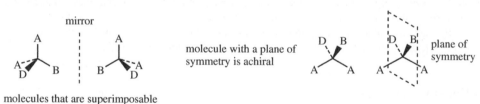

Molecules that *are* superimposable on their mirror images are said to be **achiral**. With tetrahedral carbon, this is typically the case when two or more of the

attached groups are the same. This introduces a plane of symmetry into the molecule; molecules with a plane of symmetry can be superimposed on their mirror images.

Note that chirality is not restricted to tetrahedral carbon; it can also be associated with other

tetrahedral systems, such as quaternary nitrogen compounds.

However, non-quaternary nitrogen, although tetrahedral, is not chiral. There is a rapid inversion that converts one enantiomer into the other; effectively, the lone pair does not maintain its position. The energy barrier to interconversion is about 25 kJ mol^{-1}, which is sufficiently low that inversion

occurs readily at room temperature. This usually makes it impossible to obtain neutral amines in optically active form; quaternization stops this inversion.

We shall later need to introduce a related term, **prochiral**. The concept of **prochirality** is discussed in Section 3.7.

Box 3.5

Manipulating stereostructures

It is not always easy to look at stereostructures – two-dimensional representations of three-dimensional molecules – and decide whether two separate representations are the same or different. To compare structures, it is usually necessary to manipulate one or both so that they can be compared directly. Here are a few demonstrations of how to approach the problem on paper. Of course, constructing models for comparison is the easiest method, but there will always be occasions when we have to figure it out on paper.

Question: A molecule is represented by the Newman projection:

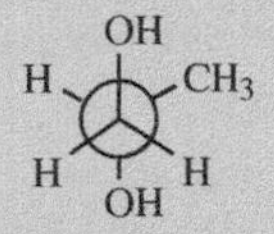

Which of the following are equivalent to the above Newman projection, or to its enantiomer?

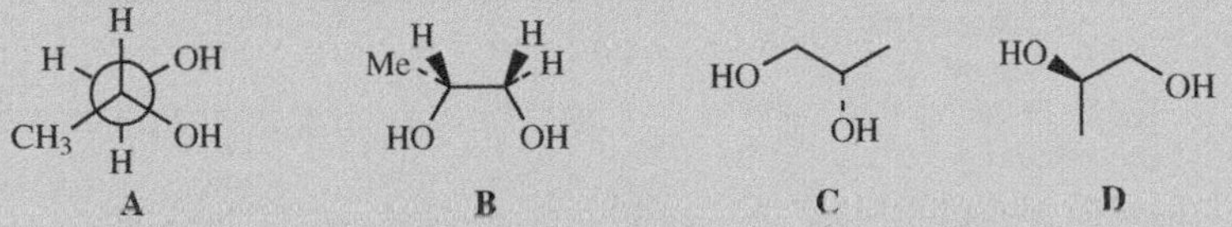

Answer: A and **B** are same as original; **C** and **D** are enantiomers of original

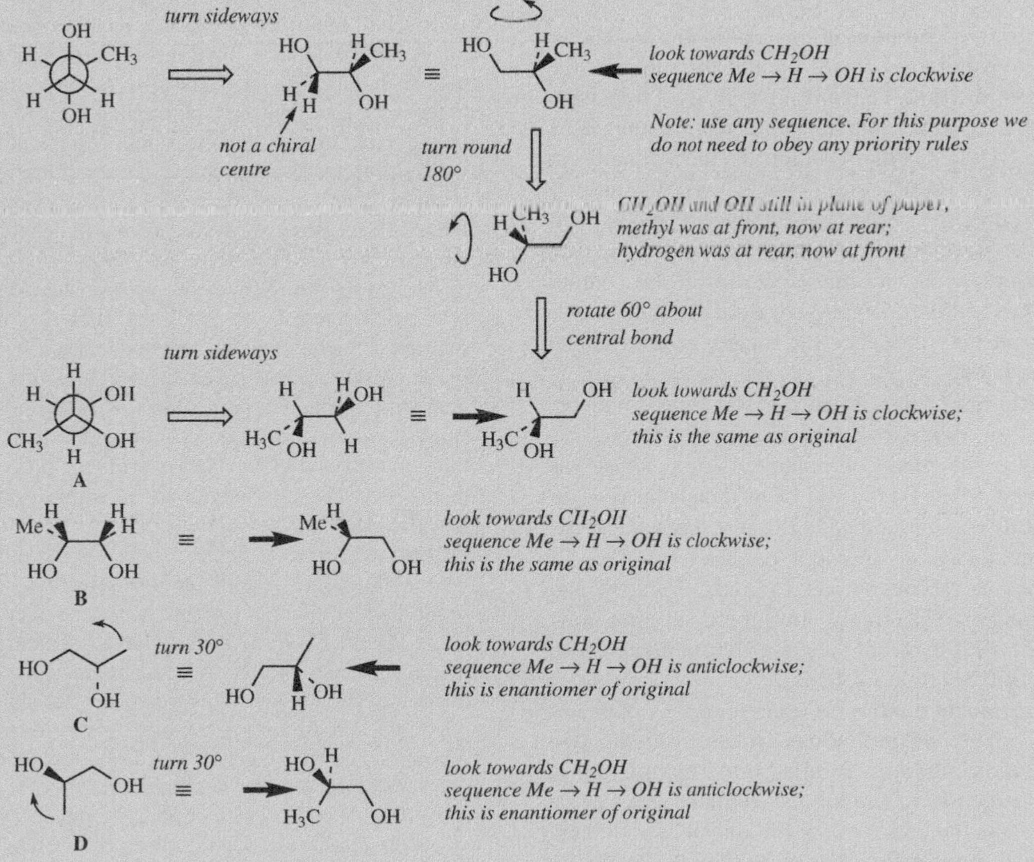

turn sideways

not a chiral centre

turn round 180°

look towards CH_2OH sequence Me → H → OH is clockwise

Note: use any sequence. For this purpose we do not need to obey any priority rules

CH_2OH and OH still in plane of paper, methyl was at front, now at rear; hydrogen was at rear, now at front

rotate 60° about central bond

turn sideways

A

look towards CH_2OH sequence Me → H → OH is clockwise; this is the same as original

B

look towards CH_2OH sequence Me → H → OH is clockwise; this is the same as original

turn 30°

C

look towards CH_2OH sequence Me → H → OH is anticlockwise; this is enantiomer of original

turn 30°

D

look towards CH_2OH sequence Me → H → OH is anticlockwise; this is enantiomer of original

Optical activity is the ability of a compound to rotate the plane of polarized light. This property arises from an interaction of the electromagnetic radiation of polarized light with the unsymmetric electric fields generated by the electrons in a chiral molecule. The rotation observed will clearly depend on the number of molecules exerting their effect, i.e. it depends upon the concentration. Observed rotations are thus converted into specific rotations that are a characteristic of the compound according to the formula below.

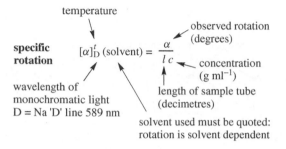

The observed rotation in degrees is divided by the sample concentration (g ml^{-1}) and the sample tube length (decimetres). The unusual units used transform the measured small rotations into more manageable numbers. The specific rotation is then usually in the range 0–1000°; the degree units are strictly incorrect, but are used for convenience. The polarized light must be monochromatic, and for convenience and consistency the D line (589 nm) in the sodium spectrum is routinely employed. Both the temperature and solvent may influence the rotation somewhat, so must be stated.

Enantiomers have equal and opposite rotations. The (+)- or **dextrorotatory** enantiomer is the one that rotates the plane of polarization clockwise (as determined when facing the beam), and the (−)- or **laevorotatory** enantiomer is the one that rotates the plane anticlockwise. In older publications, d and l were used as abbreviations for dextrorotatory and laevorotatory respectively, but these are not now employed, thus avoiding any possible confusion with D and L (see Section 3.4.10).

An equimolar mixture of enantiomers is optically inactive, since the individual effects from the two types of molecule are cancelled out. This mixture is called a **racemic mixture** or **racemate**, and can be referred to as the (±)-form. A mixture of enantiomers in unequal proportions has a rotation numerically less than that of either enantiomer; this measurement

could be used to determine the proportions of each (see Box 3.6). Note that it is not possible to predict the sign or magnitude of the optical activity for a particular enantiomer; it must be measured experimentally. The presence of more than one chiral centre in a molecule results in an optical rotation that reflects a contribution from each centre, though this is unlikely to be a simple summation. It must also be appreciated that a positive contribution from one centre may be reduced, countered, or cancelled out by a negative contribution arising from another centre or centres (see Section 3.4.5).

Box 3.6
Optical purity and enantiomeric excess

A racemic mixture contains equal amounts of the two enantiomeric forms of the compound and has an optical rotation of zero: the optical rotations arising from each of the two types of molecule are cancelled out. It follows that a mixture of enantiomers in unequal proportions will have a rotation that is numerically less than that of an enantiomer. Here, we see how to use the measured optical activity to determine the proportions of each enantiomer in the mixture, and therefore its optical purity. **Optical purity** is a measure of the excess of one enantiomer over the other in a sample of a compound.

There are a number of occasions when optical purity is of interest. We shall see later that many drugs are chiral compounds, and that biological activity often resides in just one enantiomer (see Box 3.7). To minimize potential side effects, it is desirable to supply the drug in a single enantiomeric form. This might be achieved by devising a synthetic procedure that produces a single enantiomer, an **enantiospecific** synthesis. However, syntheses that are enantiospecific can be difficult to achieve, and it is more likely that the procedure is only **enantioselective**, i.e. it produces both enantiomers but with one predominating. Alternatively, it is possible to separate the racemic mixture into the two enantiomers (resolution; see Section 3.4.8). This might not be achieved in a single step. In both cases, it is usually necessary to monitor just how much of the desired enantiomer is present in the product mixture.

To illustrate the calculation of optical purity, we shall consider another type of reaction of interest, **racemization**. This is the conversion of a single enantiomer into a racemic mixture of the two enantiomers. It depends upon the chemical nature of the compound whether this is easily achievable (see Sections 10.1.2 and 10.8). One compound that racemizes readily is **hyoscyamine**, a natural alkaloid found in deadly nightshade, which is used as an anticholinergic drug (see Box 3.7). The natural compound is laevorotatory, $[\alpha]_D^{20} - 21°$ (EtOH), and the enantiomer is almost devoid of biological activity.

Upon heating with dilute base such as 1% NaOH for about an hour, hyoscyamine racemizes, and the solution becomes optically inactive (see Box 10.9). At shorter times, racemization is incomplete and the solution will still be optically active. Consider first a very simple situation in which exactly half of the material has racemized. Half of the material is now optically inactive, consisting of equal amounts of each enantiomer, whilst the other half is still unchanged. Since the concentration of the unchanged part is half of the original concentration, the optical rotation will also have dropped to half its original value. The solution will contain 50% laevorotatory isomer and 50% racemate. However, the racemate is itself a 50 : 50 mixture of the two enantiomers, so the solution actually contains 25% dextrorotatory and 25 + 50% = 75% laevorotatory enantiomers.

Now let us consider when measurements indicate $[\alpha]_D^{20} - 9.2°$. Calculations now tell us that the sample is 56.2% racemic, and contains 71.9% laevorotatory enantiomer and 28.1% dextrorotatory enantiomer. These figures are derived as follows:

$$\text{the optical purity}(\%)$$
$$= \frac{\text{specific rotation of sample}}{\text{specific rotation of pure enantiomer}} \times 100$$
$$= -9.2°/ - 21° \times 100 = 43.8\%$$

The sample thus contains 43.8% of laevorotatory enantiomer and $100 - 43.8\% = 56.2\%$ of racemate, the latter contributing no overall optical activity. The racemate contains equal amounts of laevorotatory and dextrorotatory enantiomers, i.e. it contributes 28.1% of each isomer to the overall mixture. Therefore, we have $43.8 + 28.1 = 71.9\%$ of laevorotatory enantiomer, and 28.1% of dextrorotatory enantiomer in the partially racemized mixture.

Many workers use the equivalent term percentage **enantiomeric excess** rather than optical purity:

$$\% \text{ Enantiomeric excess}$$
$$= \frac{\text{moles of one enantiomer} - \text{moles of other enantiomer}}{\text{total moles of both enantiomers}} \times 100$$

but this is exactly equivalent to optical purity. From the above calculations, one can see that the laevorotatory enantiomer (71.9%) is in excess of the dextrorotatory enantiomer (28.1%) by 43.8%.

The physical properties of enantiomers and racemates, except for optical rotation and melting points, are usually the same. The melting points of (+)- and (−)-enantiomers are the same, though that of the racemate is

---

Box 3.7

Pharmacological properties of enantiomers

Although most physical properties of enantiomers are identical, pharmacological properties may be different. There are examples of compounds where:

• only one enantiomer is active;

• both enantiomers show essentially identical activities;

• both enantiomers have similar activity, but one enantiomer is more active;

• enantiomers show different pharmacological activities.

These observations may reflect the proximity of the chiral centre to the part of the molecule that binds with the receptor site.

Box 3.7 (continued)

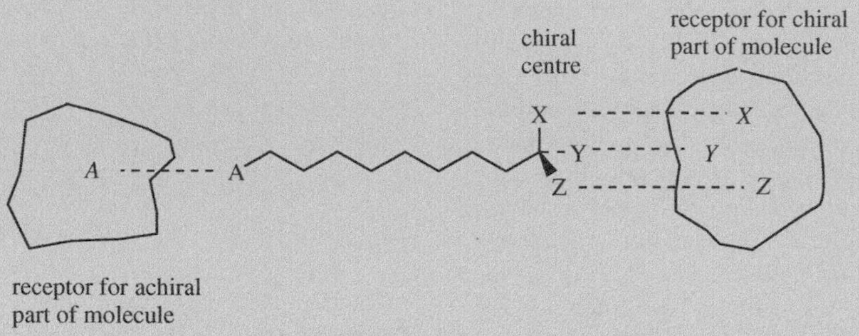

chiral
centre

receptor for chiral
part of molecule

receptor for achiral
part of molecule

If binding to the receptor involves the chiral centre, then we may see activity in only one enantiomer, but if binding does not involves the chiral centre, then there may be similar activities for each enantiomer. Binding close to the chiral centre may cause the same type of activity but of a different magnitude. A different pharmacological activity for each enantiomer almost certainly reflects different receptors.

Further, drug absorption, distribution, and elimination from the body may vary due to differences in protein binding, enzymic modification, etc, since proteins are also chiral entities (see Chapter 13).

Thus, the anticholinergic activity of the alkaloid **hyoscyamine** is almost entirely confined to the (−)-isomer, and the (+)-isomer is almost devoid of activity. The racemic (±)-form, atropine, has approximately half the activity of the laevorotatory enantiomer. An anticholinergic drug blocks the action of the neurotransmitter acetylcholine, and thus occupies the same binding site as acetylcholine. The major interaction with the receptor involves that part of the molecule that mimics acetylcholine, namely the appropriately positioned ester and amine groups. The chiral centre is adjacent to the ester, and also influences binding to the receptor.

the descriptors R, S, and RS
are defined in Section 3.4.2

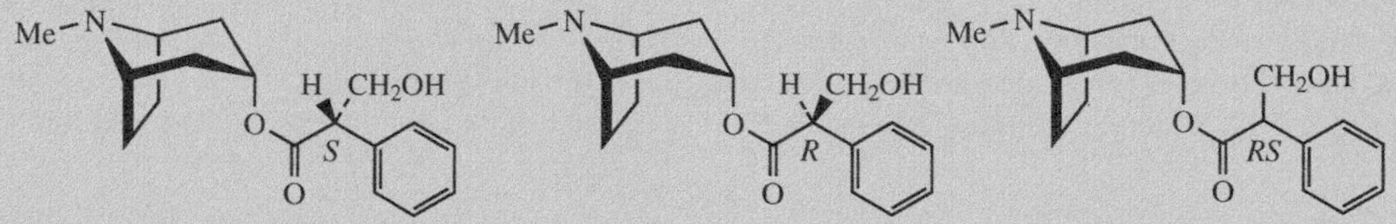

	(−)-hyoscyamine	(+)-hyoscyamine	(±)-hyoscyamine (atropine)
relative anticholinergic activity (%)	100	0	50

The major constituent of caraway oil is (+)-**carvone**, and the typical caraway odour is mainly due to this component. On the other hand, the typical minty smell of spearmint oil is due to its major component, (−)-carvone. These enantiomers are unusual in having quite different smells, i.e. they interact with nasal receptors quite differently. The two enantiomeric forms are shown here in their half-chair conformations.

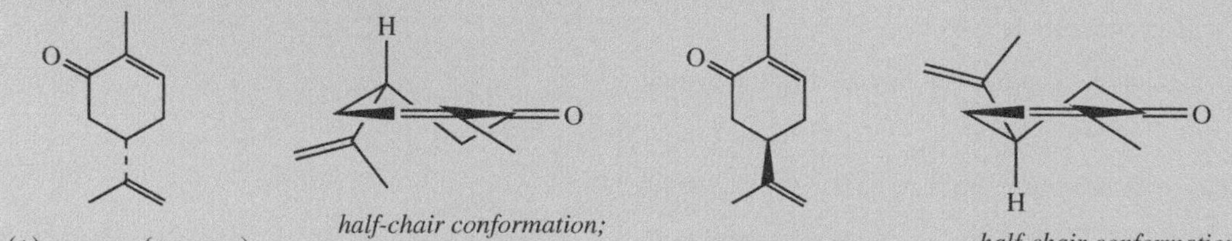

(+)-carvone (caraway)

half-chair conformation; isopropenyl group nearly equatorial

(−)-carvone (spearmint)

half-chair conformation; isopropenyl group nearly equatorial

One of the most notorious and devastating examples of a drug's side effects occurred in the early 1960s, when **thalidomide** was responsible for many thousands of deformities in new-born children. Thalidomide was marketed in racemic form as a sedative and antidepressant, and was prescribed to pregnant women. Although one enantiomer, the (R)-form, has useful antidepressant activity, it was not realized at that time that the (S)-form, thought to be inactive, actually has mutagenic activity and causes defects in the unborn fetus. Furthermore, the (S)-isomer also has antiabortive activity, facilitating retention of the damaged fetus in the womb, so that any natural tendency to abort a damaged fetus was suppressed.

(R)-thalidomide

useful antidepressant activity

(S)-thalidomide

mutagenic activity
antiabortive activity – retained damaged fetus

It is now general policy in the pharmaceutical industry to release new drugs as optically pure isomers, rather than as racemates. It is desirable to minimize the amount of foreign chemical a patient is subjected to, since even the inactive portion of a drug has to be metabolized and removed from the body. Such tragedies as occurred with thalidomide may also be avoided. Where a drug is supplied as a single enantiomer, the optical isomer is often incorporated into the drug name, e.g. dexamfetamine, dexamethasone, levodopa, levomenthol, levothyroxine. Nevertheless, many racemic compounds are currently used as drugs, including atropine, mentioned above, and the analgesic ibuprofen.

ibuprofen

(R)-(−)-isomer inactive

metabolic conversion

(S)-(+)-isomer active

Ibuprofen is an interesting case, in that the (S)-(+)-form is an active analgesic, but the (R)-(−)-enantiomer is inactive. However, in the body there is some metabolic conversion of the inactive (R)-isomer into the active (S)-isomer, so that the potential activity from the racemate is considerably more than 50%. Box 10.11 shows a mechanism to account for this isomerism.

There are two approaches to producing drugs as a single enantiomer. If a synthetic route produces a racemic mixture, then it is possible to separate the two enantiomers by a process known as resolution (see Section 3.4.8). This is often a tedious process and, of course, half of the product is then not required. The alternative approach, and the one now favoured, is to design a synthesis that produces only the required enantiomer, i.e. a chiral synthesis.

Note, the descriptors R and S for enantiomers and RS for racemates are defined in Section 3.4.2.

usually different and can be greater or less than the melting point of the enantiomers. Most spectral properties, e.g. NMR, mass spectrometry, etc., of (+)-, (−)-, and (±)-forms are indistinguishable. However, pharmacological properties are frequently different, because they may depend upon the overall shape of the compound and its interaction with a receptor.

3.4.2   Cahn–Ingold–Prelog system to describe configuration at chiral centres

The arrangement of groups around a chiral atom is called its **configuration**, and enantiomers have different configurations. Therefore, it is necessary for us to have a means of describing configuration so that we are in no doubt about which enantiomer we are talking about. Although enantiomers have equal and opposite optical rotations, the sign of the optical rotation does not tell us anything about the configuration. The system adopted by IUPAC for describing configuration was devised by **Cahn, Ingold, and Prelog**, and is often referred to as the **R,S convention**.

The approach used is as follows:

- Assign an order of priority, 1, 2, 3, and 4, to the substituents on the chiral centre.

- View the molecule through the chiral centre towards the group of lowest priority, i.e. priority 4.

- Now consider the remaining groups in order of decreasing priority. If the sense of decreasing priority 1 → 2 → 3 gives a clockwise sequence, then the configuration is described as R (Latin: *rectus* = right); if the sequence is anticlockwise, then the configuration is described as S (Latin: *sinister* = left).

The remaining part of the procedure is to assign the priorities. The IUPAC **priority rules** form a rather long

document in order to encompass all possibilities. Here is a very short version suitable for our requirements. Note that it applies to both acyclic and cyclic compounds.

- Higher atomic number precedes lower, e.g. Br > Cl > S > O > N > C > H.

- For isotopes, higher atomic mass precedes lower, e.g. T > D > H.

- If atoms have the same priority, then secondary groups attached are considered. If necessary, the process is continued to the next atom in the chain.

e.g. $-CH_2-CH_3$ > $-CH_2-H$

first atom is carbon in both cases; consider the second atom: carbon as second atom has higher priority than hydrogen

first atom is carbon in both cases; consider the second atom: second atom is carbon in both cases; consider the next atom(s): carbon directly bonded to two further carbons has higher priority than carbon directly bonded to just one further carbon

- Double and triple bonds are treated by assuming each atom is duplicated or triplicated.

As simple examples of the approach, let us consider the amino acid (−)-serine and the Krebs cycle intermediate (+)-malic acid.

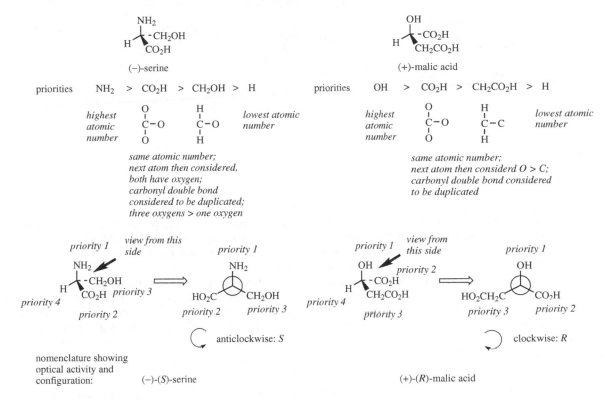

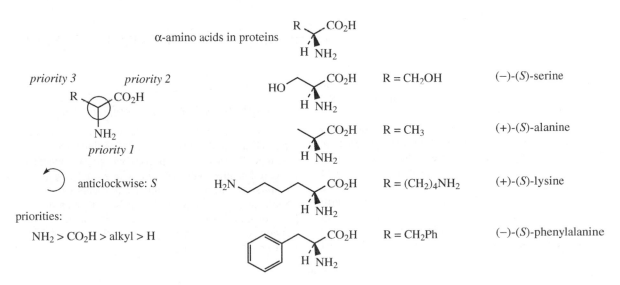

It is now possible to incorporate the configuration of the compound into its nomenclature to give more detail. (−)-Serine becomes (−)-(S)-serine, whilst (+)-malic acid becomes (+)-(R)-malic acid. Because there is no relationship between (+)/(−) and configuration (R)/(S), it is necessary to quote both optical activity and configuration to convey maximum information. The

descriptor (RS) is used to indicate a (±) racemic mixture (see Section 3.4.1).

Note also that the configuration (R) or (S) is defined by the priority rules, and configuration (R) could easily become (S) merely by altering one substituent. For instance, all the amino acids found in proteins can be represented by the formula

priority 2   *priority 3*

HSH$_2$C   CO$_2$H   HS⌒CO$_2$H   R = CH$_2$SH   (+)-(R)-cysteine

NH$_2$ H NH$_2$

priority 1

clockwise: *R*

priorities:

NH$_2$ > CH$_2$SH > CO$_2$H > H

Now all these amino acids that are chiral (glycine, R = H is achiral) have the (S) configuration except for cysteine, which is (R). Just looking at the structures, one might imagine that they would all have the same configuration, and indeed one can consider that they have; they differ only in the nature of the R group, but are all arranged around the chiral centre in the same manner. But since (R) and (S) are only *descriptors* of configuration, the designation depends upon the nature of the R group. In most cases, R is an alkyl or substituted alkyl, so it has a lower priority than the carboxyl. In

the case of cysteine, R = CH$_2$SH, and since S has a higher atomic number than any of the other atoms under consideration, this group will have a higher priority than the carboxyl. The net result is that cysteine is (R)-cysteine.

Configurations in cyclic compounds are considered in the same way as for acyclic compounds. If you cannot get an answer with the first atom, move on to the next, even though this may mean working around the ring system. Consider, for example, the stereoisomer of 3-methylcyclohexanol.

anticlockwise: 1*S*   clockwise: 3*R*

(1*S*,2*R*)-3-methylcyclohexanol

priorities:

OH > CH$_2$CH(CH$_3$)CH$_2$ > CH$_2$CH$_2$CH$_2$ > H

CH$_2$CH(OH)CH$_2$ > CH$_2$CH$_2$CH$_2$ > CH$_3$ > H

This has two chiral centres, C-1 and C-3. It can readily be deduced that this isomer is actually (1*S*,2*R*)-3-methylcyclohexanol.

At both centres, two of the groups under consideration for priority assignment are part of the ring system. These are only differentiable when one comes to the ring substituent, the methyl group when one considers C-1 and the hydroxyl when one considers C-3. In each case, the substituted arm is going to take precedence over the unsubstituted arm. A more interesting example (6-aminopenicillanic acid) containing heterocyclic rings is discussed in Box 3.8.

Box 3.8

Configurations in 6-aminopenicillanic acid

Let us look at the common substructure of the **penicillin** antibiotics, namely 6-aminopenicillanic acid, to illustrate some aspects of working out whether a chiral centre is allocated the R or S configuration.

First of all, there are three chiral centres in this molecule, carbons 3, 5 and 6; note that carbon 2 is not chiral, since two of the groups attached are methyls. Only the three carbons indicated have four different groups attached.

* chiral centre

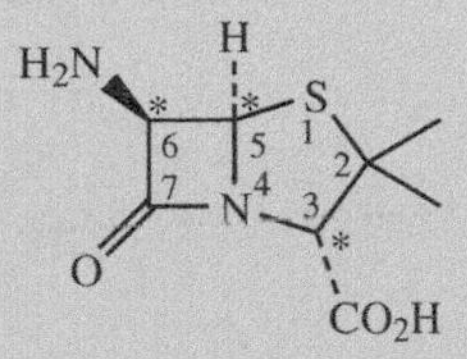

6-aminopenicillanic acid

clockwise: 5R

viewed from the front clockwise

therefore, if viewed from rear, must be anticlockwise: 3S

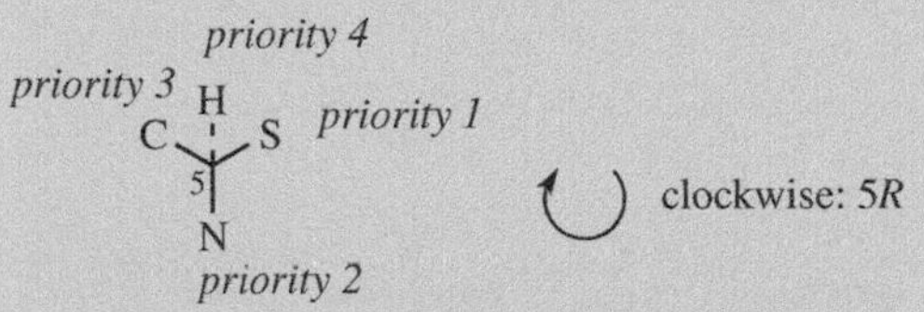

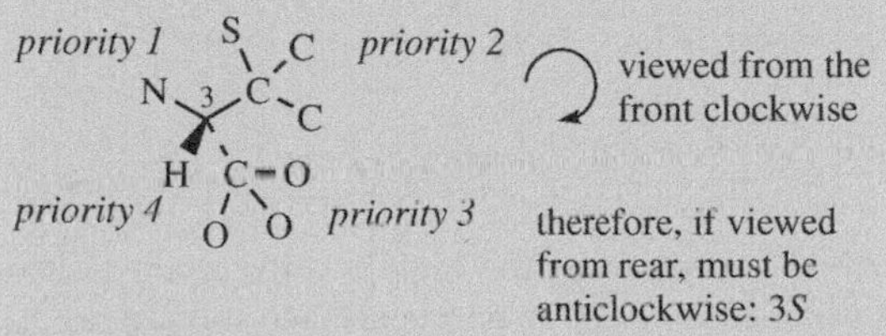

The chirality at C-5 is assigned in the usual way. The groups attached have easily assigned priorities, with S > N > C > H. The configuration is thus 5R. For the chirality at position 3, the priorities are assigned N > C–S > C–O > H. Now a very useful hint. Since the group of lowest priority is wedged/up, it is rather difficult to imagine the sequence when viewed from the rear. Accordingly, view the sequence from the front, which is easy, and reverse it. From the front, the sequence for C-3 looks clockwise, so if viewed from the rear, it must be anticlockwise, and the descriptor is 3S. Note how we consider substituents in the standard way even if they are part of a ring system. If you cannot get an answer with the first atom, move on to the next around the ring system.

3.4.3 Geometric isomers

Restricted rotation about double bonds or due to the presence of ring systems leads to configurational isomers termed **geometric isomers**. Thus, we recognize two isomers of but-2-ene, as shown below, and we term these *cis* and *trans* isomers. We have met these terms earlier (see Section 3.3.2).

Lastly, suppose one is asked to draw a particular configuration at C-6, namely 6R. There is no way one can visualize a particular configuration, so the approach is to draw one and see if it is correct; if it is not correct, then change it by reversing wedged/dotted bonds. And which to try first? Well, always put the group of lowest priority, usually H, away from you, i.e. dotted/down. Then you can see the clockwise/anticlockwise relationship easily from the front. In this case, the version with H down gave the 6R configuration; but, if it were to be wrong, then the alternative configuration at this centre would be the required one, i.e. a wedged bond to the hydrogen.

to draw (6R)-configuration:

first try this:

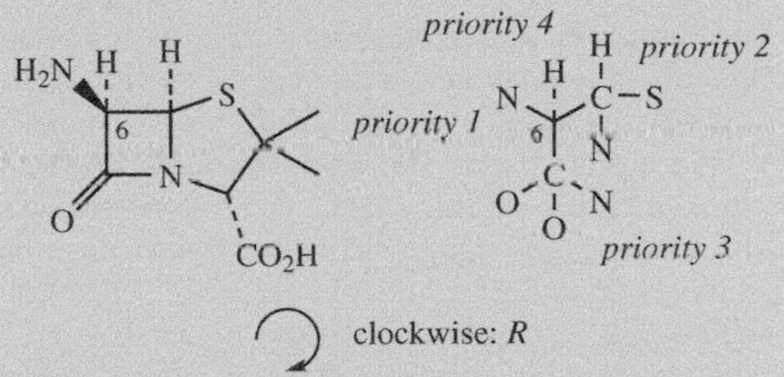

clockwise: R

it is always easier to see clockwise / anticlockwise if the group of lowest priority is at the rear (dotted)

it turns out to be R; if it were incorrect, then the required isomer would be:

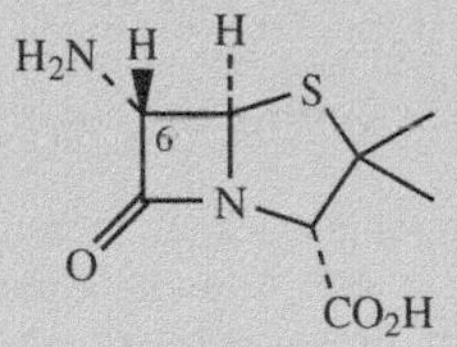

6S configuration   (3S,5R,6R)-6-aminopenicillanic acid

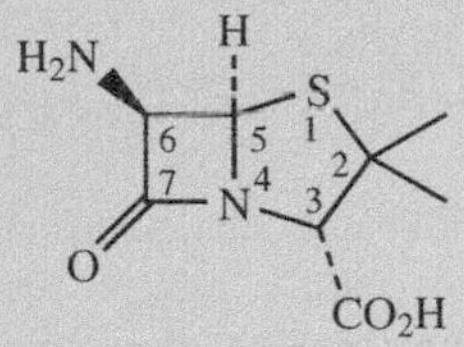

With a double bond, rotation would destroy the π bond that arises from overlap of p orbitals; consequently, there is a very large barrier to rotation. It is of the order of 263 kJ mol^{-1}, which is very much higher than any of the barriers to rotation about single bonds that we have seen for conformational isomerism. Accordingly, *cis* and *trans* isomers do not interconvert under normal conditions. Ring systems can also lead to geometric isomerism, and *cis* and *trans* isomers of cyclopropane-1,2-dicarboxylic acid similarly do not interconvert; interconversion would require the breaking of bonds.

but-2-ene

cis

trans

cyclopropane-1,2-dicarboxylic acid

cis

trans

The terms *cis* and *trans* are used to describe the configuration, which is considered to be the spatial sequence about the double bond or the spatial sequence relative to a ring system. The *cis* isomer has substituents on the same side of the double bond or ring system (Latin: *cis* = on this side), whereas the *trans* isomer has substituents on opposite sides (Latin: *trans* = across).

With simple compounds, like the isomers of but-2-ene, the descriptors *cis* and *trans* are quite satisfactory, but a compound such as 3-methylpent-2-ene causes problems. Do we call the isomer below *cis* because the methyls are on the same side, or *trans* because the main chain goes across the bond?

3-methylpent-2-ene

is this cis or trans?

For double bonds, the configuration is now usually described via the non-ambiguous *E,Z* **nomenclature**, assigned using the **Cahn–Ingold–Prelog priority rules** for substituents on each carbon. First, consider each carbon of the double bond separately, and assign priorities to its two substituents. Then consider the double bond with its four substituents. If the two substituents of higher priority are on the same side of the double bond, the configuration is *Z* (German: *zusammen* = together), whereas if they are on opposite sides, the configuration is *E* (German: *entgegen* = across).

Thus, for the 3-methylpent-2-ene isomer we can see that, for C-2, the substituents are methyl and hydrogen with priorities methyl > hydrogen. For C-3, we have substituents methyl and ethyl, with ethyl having the

Z

E

(*E*)-3-methylpent-2-ene

(*Z*)-3-methylpent-2-ene

higher priority. Thus, the high-priority groups are on opposite sides of the double bond, and this isomer has the *E* configuration. The alternative arrangement with high-priority substituents on the same side of the double bond has the *Z* configuration.

Box 3.9

Configurations of tamoxifen, clomifene and triprolidine

The oestrogen-receptor antagonist **tamoxifen** is used in the treatment of breast cancer, and is highly successful. **Clomifene** is also an oestrogen-receptor antagonist, but is principally used as a fertility drug, interfering with feedback mechanisms and leading to ova release, though this often leads to multiple pregnancies.

configuration (*Z*)

tamoxifen

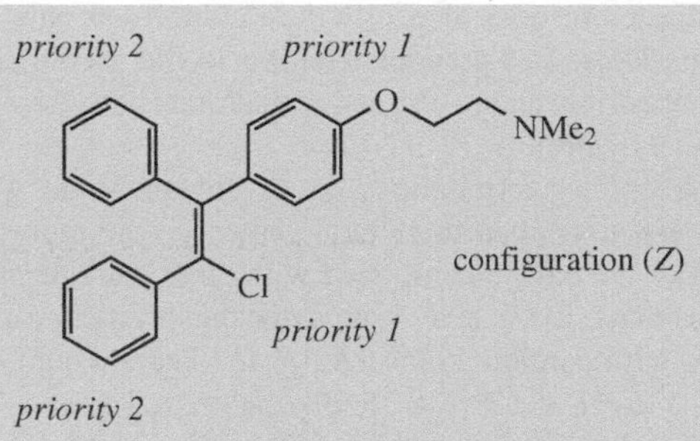

priority 2   priority 1

NMe₂

configuration (Z)

priority 1

priority 2

clomifene

As can be deduced from application of the Cahn–Ingold–Prelog priority rules, high-priority groups are positioned on the same side of the double bond in each case. Note that the substituted aromatic ring has higher priority than the unsubstituted ring. Both tamoxifen and clomifene thus have the Z configuration.

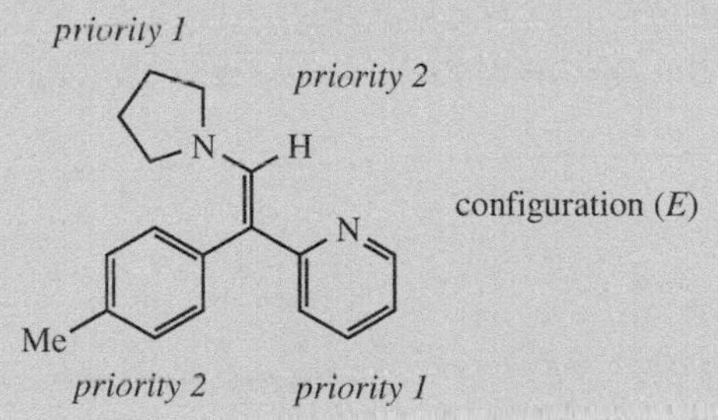

priority 1

priority 2

priority 2   priority 1

triprolidine

The antihistamine drug **triprolidine** has the E configuration; note that the heterocyclic pyridine ring takes priority over the benzene ring, even though the latter has a substituent. Priority is deduced by working along the carbon chain towards the first atom that provides a decision, in this case the nitrogen atom in the pyridine.

3.4.4 Configurational isomers with several chiral centres

Configurational isomerism involving one chiral centre provides two different structures, the two enantiomers. If a structure has more than one chiral centre, then there exist two ways of arranging the groups around each chiral centre. Thus, with n chiral centres in a molecule, there will be a maximum number of 2^n configurational isomers. Sometimes, as we shall see in Section 3.4.5, there are less.

Starting with two chiral centres, there should, therefore, be four stereoisomers, and this is nicely exemplified by the natural alkaloid (−)-ephedrine, which is employed as a bronchodilator drug and decongestant. **Ephedrine** is (1R,2S)-2-methylamino-1-phenylpropan-1-ol, so has the structure and stereochemistry shown.

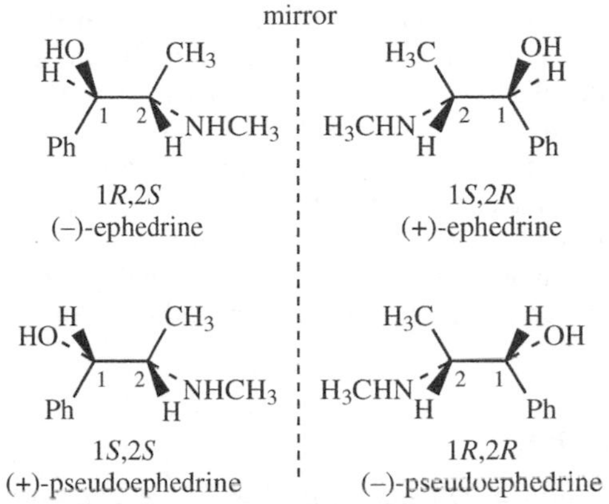

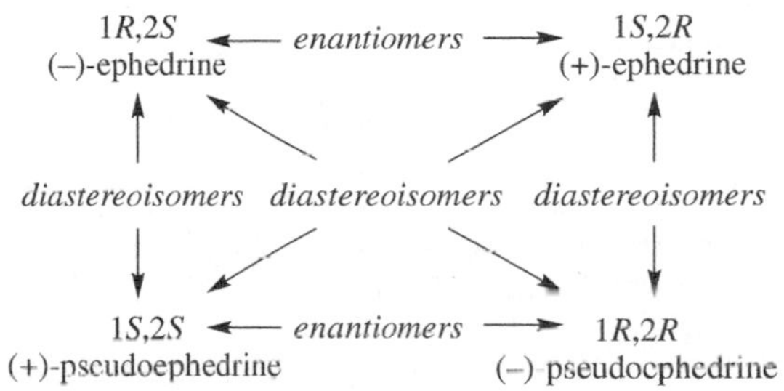

Now the other three of the possible four stereoisomers are the (1S,2S), (1R,2R), and (1S,2R) versions. These are also shown, and mirror image relationships are emphasized. The (1S,2R) isomer is the mirror image of (−)-ephedrine, which has the (1R,2S) configuration. Therefore, it is the enantiomer of (−)-ephedrine, and can be designated (+)-ephedrine. Note that the enantiomeric form has the *opposite configuration at both chiral centres*.

The other two isomers are the (1S,2S) and (1R,2R) isomers, and these two also share a mirror image relationship, have the opposite configuration at both chiral centres, and are, therefore, a pair of enantiomers. From a structure with two chiral centres, we thus have four stereoisomers that consist of two pairs of enantiomers. Stereoisomers that are not enantiomers we term **diastereoisomers**, or sometimes **diastereomers**. Thus, the (1S,2S) and (1R,2R) isomers are diastereoisomers of the (1R,2S) isomer. Other enantiomeric or diastereomeric relationships between the various isomers are indicated in the figure.

We have seen earlier that enantiomers are chemically identical except in optical properties, although biological properties may be different (see Box 3.7). On the other hand, diastereoisomers have different physical and chemical properties, and probably different biological properties as well. As a result, they are considered a completely different chemical entity, and are often given a different chemical name. The $(1S,2S)$ and $(1R,2R)$ isomers are thus known as $(+)$-pseudoephedrine and $(-)$-pseudoephedrine respectively. Interestingly, $(+)$-pseudoephedrine has similar biological properties to $(-)$-ephedrine, and it is used as a bronchodilator and decongestant drug in the same way as ephedrine.

One more useful piece of terminology can be introduced here. This is the term **epimer**. An epimer is a diastereoisomer that differs in chirality at only one centre. Thus, $(-)$-pseudoephedrine is the 2-epimer of $(-)$-ephedrine, and $(+)$-pseudoephedrine is the 1-epimer of $(-)$-ephedrine.

The epimer terminology is of greater value when there are more than two chiral centres in the molecule. Suppose we have a compound with three chiral centres, at positions 2, 3, and 4 in some unspecified carbon chain, with configurations $2R,3R,4S$. There would thus exist a total of $2^3 = 8$ configurational isomers. The enantiomer would have the configuration $2S,3S,4R$, i.e. changing the configuration at all centres. The $2S,3R,4S$ diastereoisomer we could then refer to as 'the 2-epimer', and the $2R,3S,4S$ diastereoisomer as 'the 3-epimer', since we have changed the stereochemistry at just one centre, keeping other configurations the same.

Box 3.10

Drawing enantiomers and epimers: 6-aminopenicillanic acid

The structure of the natural isomer of 6-aminopenicillanic acid is shown. You are asked to draw the structure of its enantiomer and its 6-epimer.

$(3S,5R,6R)$-
6-aminopenicillanic acid

enantiomer

$(3R,5S,6S)$-
6-aminopenicillanic acid

it is easier to change wedges/dots than to draw the mirror image; reverse the configuration at all centres

change the configuration at one centre only, by reversing wedge/dots

6-epimer

$(3S,5R,6S)$-
6-aminopenicillanic acid

The enantiomer will have the configuration changed at all chiral centres, whereas the 6-epimer retains all configurations except for that at position 6. Note that it is not necessary to draw the mirror image compound for the enantiomer, just reverse the wedge–dot relationship for the bonds at each chiral centre. This is much easier and less prone to errors whilst transcribing the structure.

Now for a rather important point. In a compound such as (−)-ephedrine there are going to be many different conformations as a result of rotation about the central C–C bond; three of them are shown here, the energetically most favourable staggered conformer with all large groups *anti*, a less favourable staggered conformer, and a high-energy eclipsed version.

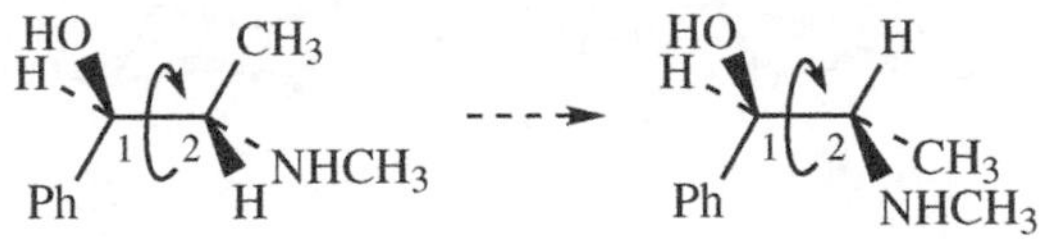

1R,2S
(−)-ephedrine

*favourable
staggered conformer:
large groups all anti*

1R,2S
(−)-ephedrine

*less favourable
staggered conformer*

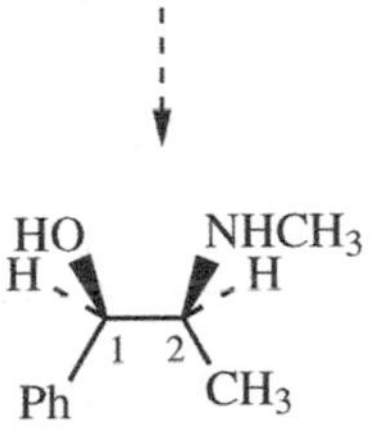

a change in **conformation** does not affect **configuration**

1R,2S
(−)-ephedrine

*unfavourable
eclipsed conformer*

However, note carefully that changing the conformation does *not* affect the spatial sequence about the chiral centres, i.e. it does not change the configuration at either chiral centre. This seems a trivial and rather obvious statement, and indeed it probably is in the case of acyclic compounds. It is when we move on to cyclic compounds that we need to remember this fundamental concept, because a common mistake is to confuse conformation and configuration (see Box 3.11).

The same stereochemical principles are going to apply to both acyclic and cyclic compounds. With simple cyclic compounds that have little or no conformational mobility, it is easier to follow what is going on. Consider a disubstituted cyclopropane system. As in the acyclic examples, there are four different configurational stereoisomers possible, comprising two pairs of enantiomers. No conformational mobility is possible here.

2-methylcyclopropanecarboxylic acid

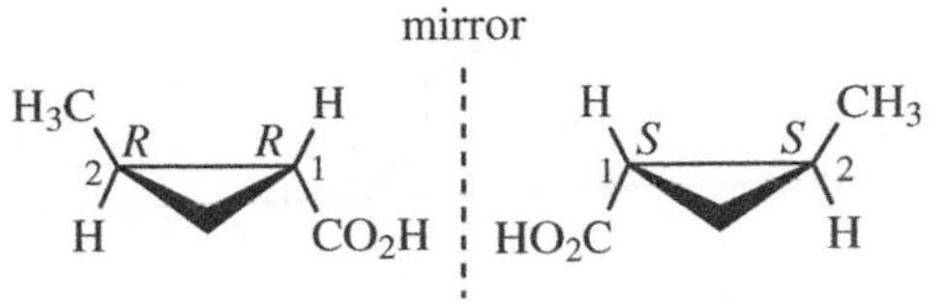

(+)- and (−)-*trans* enantiomers

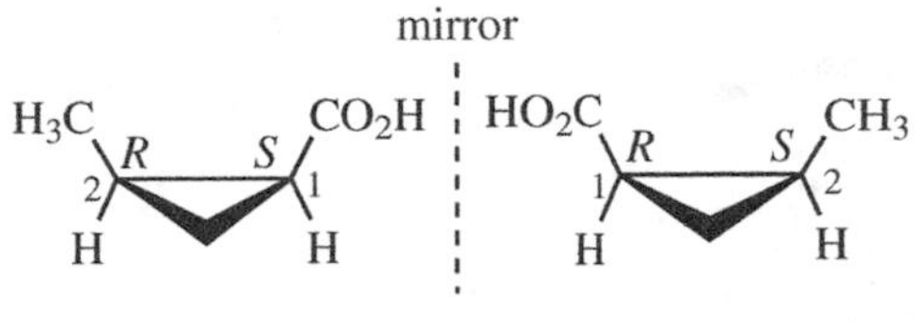

(+)- and (−)-*cis* enantiomers

However, in a cyclohexane system we also need to consider the conformational mobility that generates two different chair forms of the ring (see Section 3.3.2). Let us consider 3-methylcyclohexanecarboxylic acid. This has two chiral centres, and thus there are four configurational stereoisomers. These are the enantiomeric forms of the *trans* and *cis* isomers.

3-methylcyclohexanecarboxylic acid

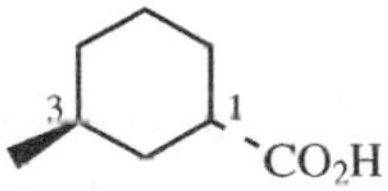

trans

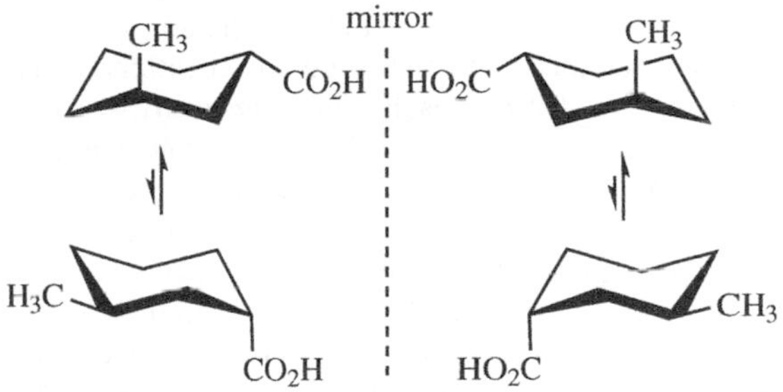

(+)- and (−)-*trans* enantiomers; two chair conformations are shown for each, the favoured one is likely to have the larger carboxylic acid group equatorial − note that the mirror image relationship is readily apparent in both conformers

Care: this shows two interconvertible conformers for each of the two non-interconvertible enantiomers

3-methylcyclohexanecarboxylic acid

cis

CO₂H label: CO_2H

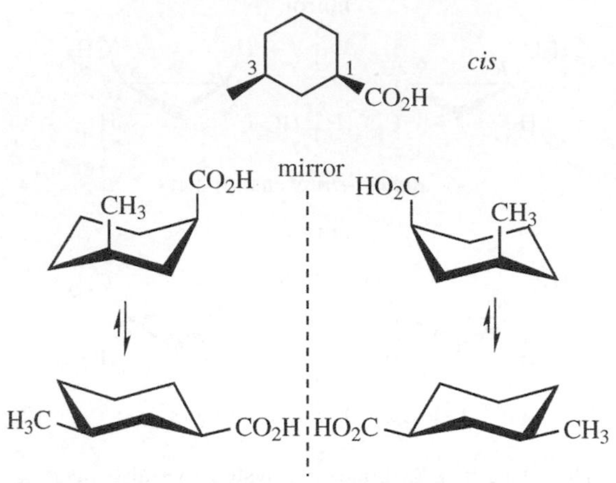

(+)- and (−)-*cis* enantiomers; two chair conformations are shown for each, the favoured diequatorial and the unfavoured diaxial — note that the mirror image relationship is readily apparent in both conformers

Care: this shows two interconvertible conformers for each of the two non-interconvertible enantiomers

4-methylcyclohexanecarboxylic acid

trans

- - -◁- - -CO_2H- - - plane of symmetry

cis

- - -◁- - -CO_2H- - - plane of symmetry

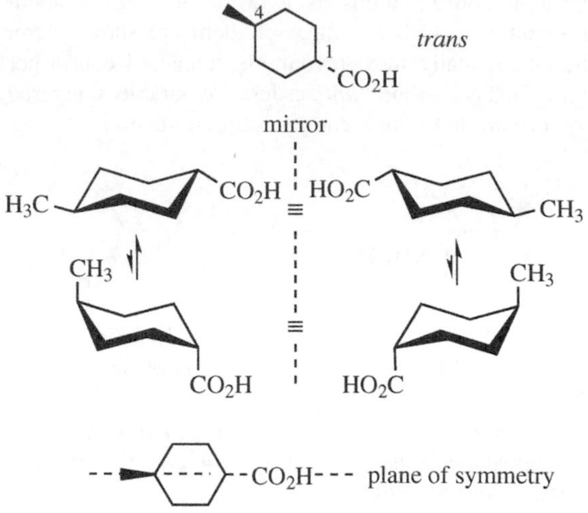

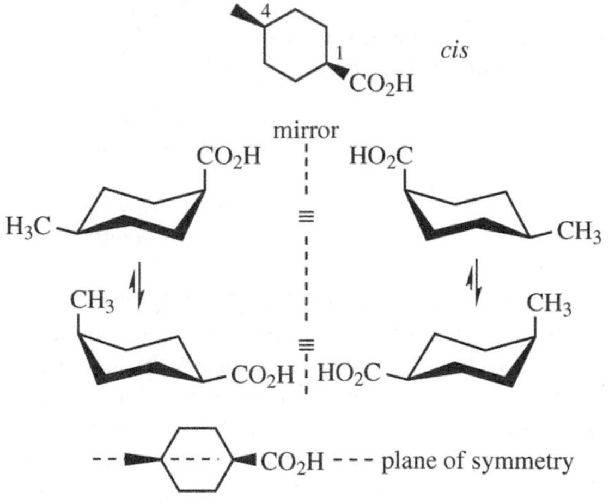

Each isomer can also adopt a different chair conformation as a consequence of ring flip (see Section 3.3.2). We thus can write down eight possible stereoisomers, comprised of two interconvertible conformers for each of the four non-interconvertible configurational isomers. Put another way, there are four configurational isomers ($2^2 = 4$), but each can exist as two possible conformational isomers. Note that you can also see the mirror image relationship in the conformational isomers. Of course, in practice, some conformers are not going to be energetically favourable. The *cis* compound has favoured diequatorial and unfavoured diaxial conformers. The *trans* compound has one equatorial and one axial substituent; we can assume that the larger carboxylic acid group will prefer to be equatorial.

Do appreciate that cyclohexane rings with 1,2- or 1,3-substitution fit into the above discussions; however, if we have 1,4-substitution there are no chiral centres in the molecule, since two of the groups are the same at each possible site! However, *cis* and *trans* forms still

exist; these are geometric isomers (see Section 3.4.3) and can still be regarded as diastereoisomers.

We can spot this type of situation by looking for symmetry in the molecule. Both *cis*- and *trans*-4-methylcyclohexanecarboxylic acid isomers have a plane of symmetry, and, as we saw for simple tetrahedral carbons (see Section 3.4.1), this symmetry means the molecule is achiral.

Box 3.11

Configurations and conformations: avoiding confusion

At this stage, a word of caution: *do not confuse conformation with configuration*. Different conformations interconvert easily; different configurations do not interconvert without some bond-breaking process. We commented above that changing the conformation did *not* affect the spatial sequence about chiral centres, and used ephedrine as a rather trivial and obvious example. Rotation about single bonds did not change the configuration at either chiral centre.

To emphasize this point, look at the following relationships for *trans*-3-methylcyclohexyl bromide.

don't confuse conformation with configuration

ring flip

these do not interconvert

conformer: has same configuration at each centre

ring flip

enantiomer: has different configuration at each centre

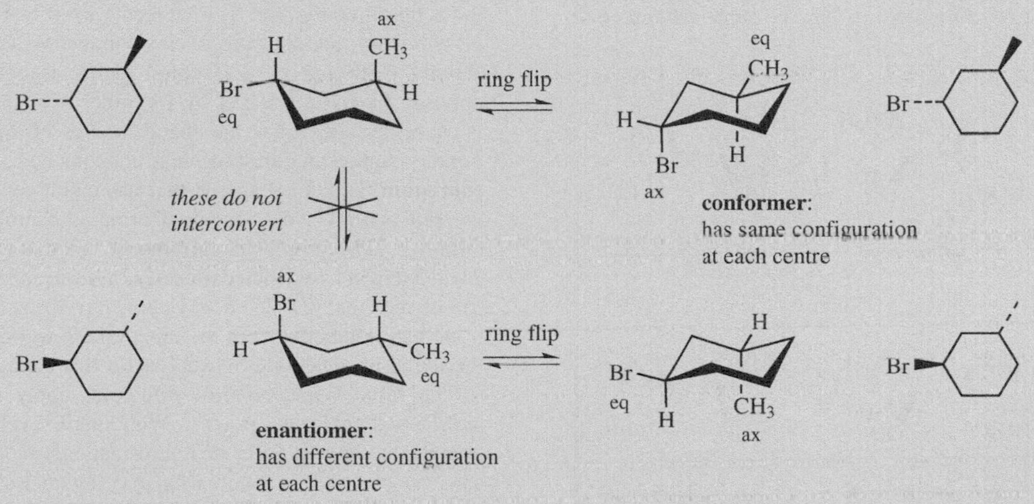

Ring flip of the upper left structure produces an alternative conformer. Ring flip does not change the configuration. The axial–equatorial relationship (conformation) is modified, but the up–down relationship (configuration) is still there. The enantiomer of this structure has the alternative configuration at both chiral centres, but it cannot be produced from the first structure by any simple isomerization process. However, it is still conformationally mobile. The figure thus shows the conformational isomerism for two different configurational isomers, the enantiomeric pair.

A common mistake that can be made when one is trying to draw the different conformers that arise from ring flip in a cyclohexane compound (see Box 3.3) is to remember vaguely that axial groups become equatorial, and vice versa, and to apply this change without flipping the ring. Of course, as can be seen from looking at the compounds below, transposing the equatorial bromine to axial and the axial methyl to equatorial changes the configuration at both centres, so we have produced the enantiomer. This is a configurational isomer and not a conformer.

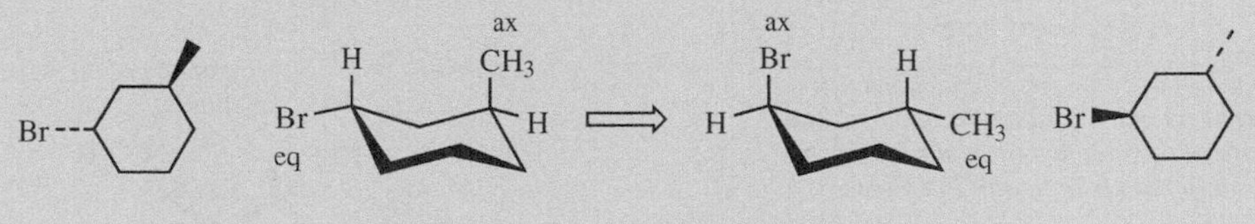

changing axial to equatorial and vice versa without ring flip creates the enantiomer, not a conformer

3.4.5 Meso compounds

Now for a rather unexpected twist. We have seen that if there are n chiral centres there should be 2^n configurational isomers, and we have considered each of these for $n = 2$ (e.g. ephedrine, pseudoephedrine). It transpires that if the groups around chiral centres are the same, then the number of stereoisomers is less than 2^n. Thus, when $n = 2$, there are only three stereoisomers, not four. As one of the simplest examples, let us consider in detail **tartaric acid**, a component of grape juice and many other fruits. This fits the requirement, since each of the two chiral centres has the same substituents.

2S,3S
(−)-tartaric acid

2R,3R
(+)-tartaric acid

2R,3S
meso-tartaric acid

2S,3R
meso-tartaric acid

these two structures are superimposable; this is more easily seen by considering the eclipsed conformer

meso-tartaric acid
eclipsed conformer

because of the symmetry, optical activity conferred by one chiral centre is equal and opposite to that conferred by the other; this **meso** compound is optically inactive

We can easily draw the four predicted isomers, as we did for the ephedrine–pseudoephedrine group, and two of these represent the enantiomeric pair of (−)-tartaric acid and (+)-tartaric acid. Now let us consider the other pair of isomers, and we shall see the consequences of the substituent groups being the same, because these two structures are actually superimposable and, therefore, only represent a single compound. This is not so easily seen with the staggered conformers drawn, so it is best to rotate these about the 2,3-bond to give an eclipsed conformer. They can both be rotated to give the same structure, so they represent only a single compound. This is called *meso*-tartaric acid (Greek: *mesos* = middle). Furthermore, since we have superimposable mirror images, there can be no optical activity.

We can see why a compound with chiral centres should end up optically inactive by looking again at the eclipsed conformer. The molecule itself has a plane of symmetry, and because of this symmetry the optical activity conferred by one chiral centre is equal and opposite to that conferred by the other and, therefore, is cancelled out. It has the characteristics of a racemic mixture, but as an intramolecular phenomenon. A ***meso* compound** is defined as one that has chiral centres but is itself achiral. Note that numbering is a problem in tartaric acid because of the symmetry, and that positions 2 and 3 depend on which carboxyl is numbered as C-1. It can be seen that $(2R,3S)$ could easily have been $(3R,2S)$ if we had numbered from the other end, a warning sign that there is something unusual about this isomer.

The same stereochemical principles apply to both acyclic and cyclic compounds. With simple cyclic compounds that have little or no conformational mobility, it can even be easier to follow what is going on. Let us first look at cyclopropane-1,2-dicarboxylic acid. These compounds were considered in Section 3.4.3 as examples of geometric isomers, and *cis* and *trans* isomers were recognized.

cyclopropane-1,2-dicarboxylic acid

(+)- and (−)-*trans* enantiomers

cis isomer is an optically inactive *meso* compound

$n = 2$, but only three isomers

This is essentially the same as the tartaric acid example, without the conformational complication. Thus, there are two chiral centres, and the groups around each centre are the same. Again, we get only three stereoisomers rather than four, since the *cis* compound is an optically inactive *meso* compound. There is a plane of symmetry in this molecule, and it is easy to see that one chiral centre is mirrored by the other, so that we lose optical activity.

Conformational mobility, such as we get in cyclohexane rings, makes the analysis more difficult, and manipulating molecular models provides the clearest vision of the relationships. Let us look at 1,2-dimethylcyclohexane as an example. Again, we have met the *cis* and *trans* isomers when we looked at conformational aspects (see Section 3.3.2). Here, we need to consider both configuration and conformation.

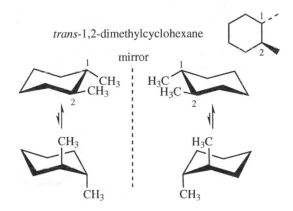

trans-1,2-dimethylcyclohexane

(+)- and (−)- *trans* enantiomers; two chair conformations are shown for each, the favoured diequatorial and the unfavoured diaxial – note that the mirror image relationship is readily apparent in both conformers

Care: this shows two interconvertible conformers for each of the two non-interconvertible enantiomers

In the *trans* compound, two mirror image enantiomeric forms can be visualized. These will be the (+)- and (−)-*trans* isomers. Note particularly that conformational changes may also be considered, but these *do not change configuration*, so we are only seeing different conformers of the same compound. The above scheme thus shows two interconvertible conformers (upper and lower structures) for each of the two non-interconvertible enantiomers (left and right structures).

The *cis* compound provides the real challenge, however. If we draw version A, together with its mirror image C, they do not look capable of being

cis-1,2-dimethylcyclohexane

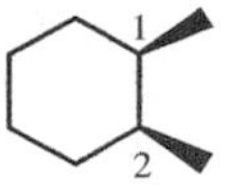

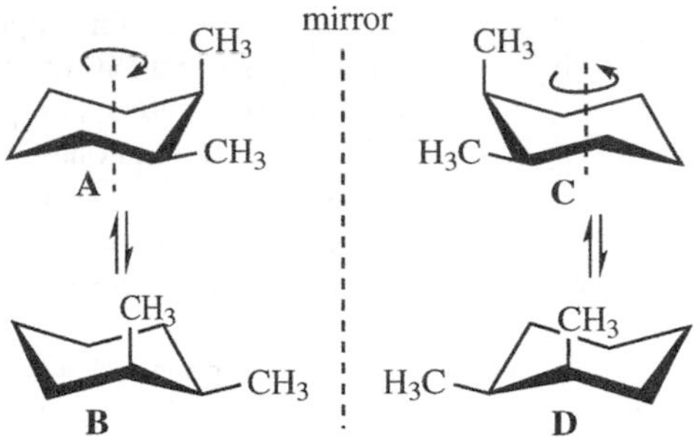

this is the difficult one!

the *cis* isomer is an optically inactive *meso* compound

the picture shows mirror images of the equal-energy interconvertible conformers

however, consider a 120° rotation of A about the central axis which produces D; 120° rotation of C produces B; therefore, they are all the same compound, but different conformers

superimposed. However, conformer A may be ring-flipped to an equal-energy conformer B, and this will have a corresponding mirror image version D. Now consider a 120° rotation of version A about the central axis; this will give D. A similar 120° rotation of version C about the central axis will give B. It follows, therefore, that if simple rotation of one structure about its axis gives the mirror image of a conformational isomer, then we cannot have enantiomeric forms but must have the same compound. These are thus two different conformers of an optically inactive *meso* compound. It may require manipulation of models to really convince you about this!

Now, although the cyclohexane ring is not planar, the overall consequences for *trans*- and *cis*-dimethylcyclohexane can be predicted by looking at the two-dimensional representations.

the *meso* nature of *cis*-1,2-dimethylcyclohexane can be deduced from the plane of symmetry in the 2D representation:

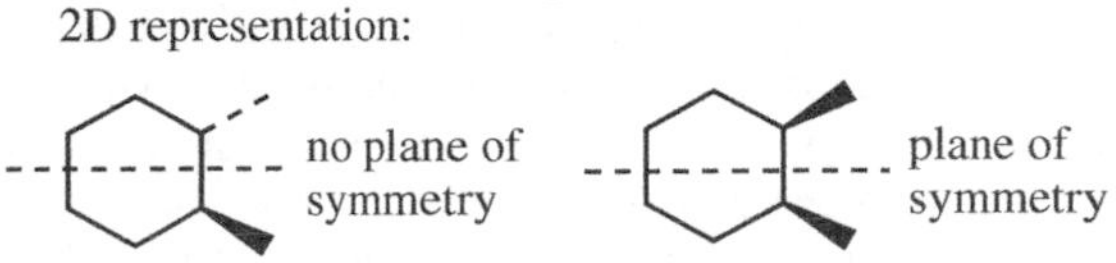

no plane of symmetry

plane of symmetry

trans

cis

It is clear that this representation of *cis*-dimethyl-cyclohexane shows a plane of symmetry, and we can deduce it to be a *meso* compound. No such plane of symmetry is present in the representation of *trans*-dimethylcyclohexane. Why does this approach work? Simply because the transformation of planar cyclohexane (with eclipsed bonds) into a non-planar form (with staggered bonds) is a conformational change achieved by rotation about single bonds. The fact that cyclohexane is non-planar means we may have to invoke the conformational mobility to get the three-dimensional picture.

Our consideration of *meso* compounds leads us to generalize:

- a molecule with one chiral centre is chiral;

- a molecule with more than one chiral centre may be chiral or achiral.

Now let us extend this generalization with a further statement:

- a molecule may be chiral without having a chiral centre.

This is the subject of the next section.

3.4.6 *Chirality without chiral centres*

We shall restrict discussions here to three types of compound. In the first we get what is termed **torsional asymmetry**, where chirality arises because of restricted rotation about single bonds. The commonest examples involve two aromatic rings bonded through a single bond (**biphenyls**). If large groups are present in the *ortho* positions, these prevent rotation about the inter-ring single bond, and the most favourable arrangement to minimize interactions is when the aromatic rings are held at right angles to each other. As a result, two enantiomeric forms of the molecule can exist. Because of the size of the *ortho* groups, it is not possible to interconvert these stereoisomers merely by rotation. Even when we only have two different types of substituent, as shown, we get two enantiomeric forms.

large *ortho* groups prevent rotation
two enantiomeric forms exist

chirality via restricted rotation – **torsional asymmetry**

The second type of compound is called an **allene**; these compounds contain two double bonds involving the same carbon. These compounds exist, but are often difficult to prepare and are very reactive. It is the concept of chirality which is more important here than the chemistry of the compounds. If a carbon atom is involved in two double bonds, it follows that the π bonds created must be at right angles to each other. The consequence of this is that the substituents on the other carbons of the allene are also held at right angles to each other. Again, two enantiomeric forms of the molecule can exist.

chirality in allenes

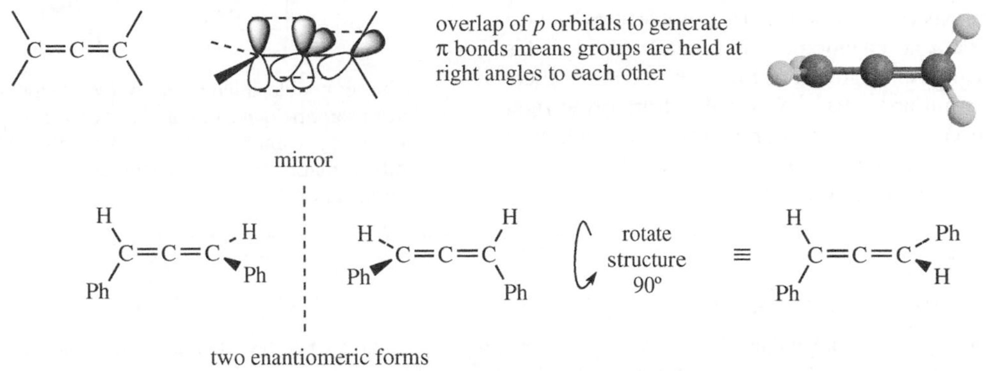

overlap of *p* orbitals to generate π bonds means groups are held at right angles to each other

mirror

rotate structure 90°

≡

two enantiomeric forms

The third example of chirality without a chiral centre is provided by *spiro* **compounds**, which we shall meet later when we consider the stereochemistry of polycyclic systems (see Section 3.5.1), but at this stage it is worth noting that they provide a third example of chirality without a chiral centre. *Spiro* compounds contain two ring systems that have one carbon in common, and it is easy to see this carbon could be chiral if four different groupings are present. A nice natural example, the antibiotic griseofulvin, is shown here.

spiro compounds

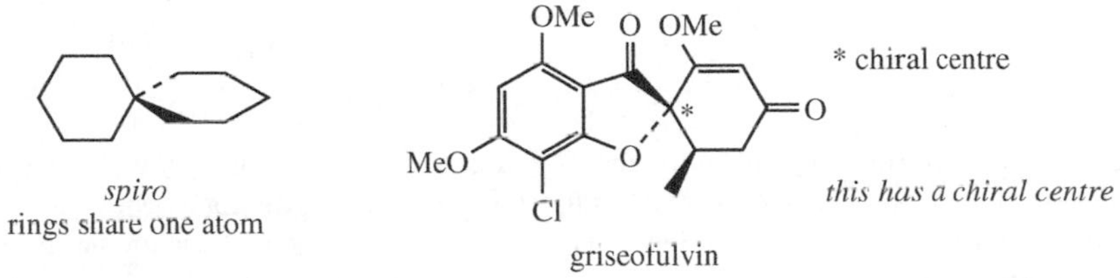

spiro
rings share one atom

* chiral centre

this has a chiral centre

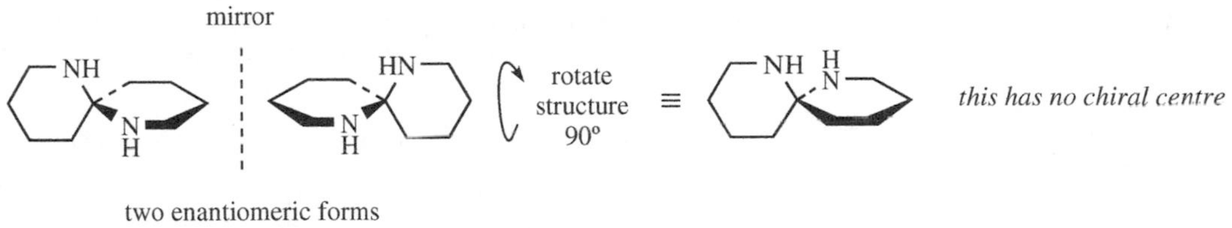

griseofulvin

mirror

rotate structure 90°

≡

this has no chiral centre

two enantiomeric forms

However, it is also possible to visualize *spiro* compounds with groupings that are not all different, where enantiomeric forms exist because mirror image compounds are not superimposable. The diamine shown is chiral, in that the mirror image forms are not superimposable, even though only two types of substituent are attached to the *spiro* centre. Both rings in this compound will have the chair conformation, but it is not easy to draw these because one ring will always be viewed face on. The solution is to ensure the *spiro* centre is not on the left or right tip of either ring.

it is difficult to show the chair conformation for both rings

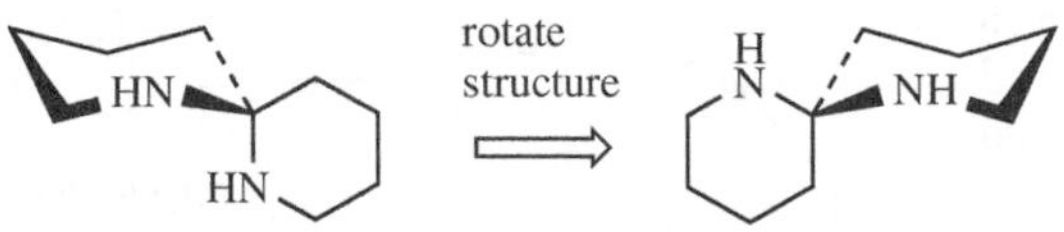

rotate structure

the solution is to ensure the spiro centre is not on the tip of either ring

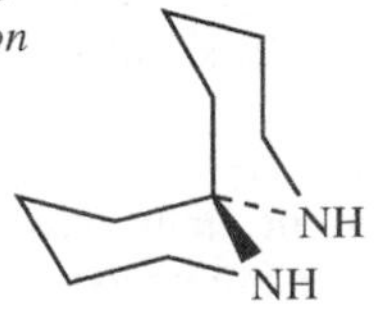

With biphenyls, allenes, and *spiro* compounds, groups are held at right angles by a rigid system, and this feature allows the existence of non-superimposable mirror image stereoisomers, i.e. enantiomers. It is useful to think of this arrangement as analogous to a simple chiral centre, where the tetrahedral array also holds pairs of groups at right angles. In contrast to tetrahedral carbon, it is not even necessary for all the groups to be different to achieve chirality, as can be seen in the examples above.

with biphenyls, allenes, and *spiro* compounds, groups are held at right angles by a rigid system; the arrangement produces non-superimposable mirror images and is thus analogous to a chiral centre

---

Box 3.12

Torsional asymmetry: gossypol

The concept of torsional asymmetry is not just an interesting abstract idea. Some years ago, fertility in some Chinese rural communities was found to be below normal levels, and this was traced back to the presence of **gossypol** in dietary cottonseed oil. Gossypol acts as a male contraceptive, altering sperm maturation, spermatozoid motility, and inactivation of sperm enzymes necessary for fertilization. Extensive trials in China have shown the antifertility effect is reversible after stopping the treatment, and it has potential, therefore, as a contraceptive for men.

(+)-gossypol   (−)-gossypol

Gossypol is chiral due to restricted rotation, and only the (−)-isomer is pharmacologically active as an infertility agent. The (+)-isomer has been found to be responsible for some toxic symptoms. Most species of cotton (*Gossypium*) produce both enantiomers of gossypol in unequal amounts, with the (+)-enantiomer normally predominating over the (−)-isomer. It has proved possible to separate racemic (±)-gossypol from this type of mixture – the racemate complexes with acetic acid, whereas the separate enantiomers do not. The racemic form can then be resolved (see Section 3.4.8) to give the useful biologically active (−)-isomer.

3.4.7 Prochirality

Enantiotopic groups

We have defined chirality in terms of 'handedness', such that mirror image stereoisomers are not superimposable. In the case of tetrahedral carbon, chirality is a consequence of having four different groups attached to it. If two or more groups were the same, then the compound would be termed achiral (see Section 3.4.1). Now we introduce another term, **prochiral**. Achiral molecules that can become chiral by one simple change are called prochiral. The simplest example we could include under this definition would be an achiral molecule in which two groups are the same. The two like groups are termed **enantiotopic**, in that separate replacement of each would generate enantiomers.

Molecules that are superimposable on their mirror images are **achiral**

achiral molecules, that can become chiral by one simple change are called **prochiral**; the A groups are termed **enantiotopic**

enantiomers

This seems an unnecessary complication. Why do we want to call an achiral centre prochiral? What benefits are there? Well, remember that the Cahn–Ingold–Prelog system allowed us to describe a particular chiral arrangement of groups at a chiral centre; prochirality now allows us to distinguish between the two like groups at an achiral centre. When might we want to do that? The following example from biochemistry shows the type of occasion when we might need to identify one or other of the like groups.

The enzyme alcohol dehydrogenase oxidizes ethanol to acetaldehyde, passing the hydrogen to the coenzyme nicotinamide adenine dinucleotide NAD^+ (see Section 15.1.1). This is the enzyme that restores normal service after excessive consumption of alcoholic drinks. By specifically labelling each hydrogen in turn, then observing whether the substrate loses or retains label in the enzymic reaction, it has been determined which hydrogen is lost from the methylene group of ethanol.

ethanol is prochiral

How then, in unambiguous fashion, can we describe which hydrogen is lost? We define the two hydrogens as *pro-R* and *pro-S*, by considering the effect of increasing their effective priorities according to the Cahn–Ingold–Prelog system; this is simply achieved if we consider having deuterium instead of protium (normal hydrogen). Then, if replacing a particular hydrogen with deuterium produces a chiral centre with the R configuration, that hydrogen is termed the *pro-R* hydrogen. Similarly, increasing the priority of the other hydrogen should generate the S configuration, so that that hydrogen is termed the *pro-S* hydrogen. We can also label hydrogens in a structure as H_R and H_S according to this procedure.

We can thus deduce that alcohol dehydrogenase stereospecifically removes the *pro-R* hydrogen from the prochiral methylene.

use *pro-R* and *pro-S* descriptors to distinguish enantiotopic hydrogens/groups

pro-S   *pro-R*

pro-R

increasing the priority of the pro-R hydrogen creates R configuration

pro-S

increasing the priority of the pro-S hydrogen creates S configuration

the enzyme is stereospecific; it removes the pro-R hydrogen

This example is from biochemistry. It is a feature of biochemical reactions that enzymes almost always catalyse reactions in a completely stereospecific manner. They are able to distinguish between enantiotopic hydrogens because of the three-dimensional nature of the binding site (see Section 13.3.2). There are also occasions where chemical reactions are stereospecific; refer to the stereochemistry of E2 eliminations for typical examples (see Section 6.4.1).

Box 3.13

Citric acid has three prochiral centres

The **Krebs cycle** is a process involved in the metabolic degradation of carbohydrate (see Section 15.3). It is also called the citric acid cycle, because **citric acid** was one of the first intermediates identified. Once formed, citric acid is modified by the enzyme aconitase through the intermediate cis-aconitic acid to give the isomeric isocitric acid. This is not really an isomerization, but the result of a dehydration followed by a rehydration. Both steps feature stereospecific *anti* processes, i.e. groups are removed or added from opposite sides of the molecule (see Sections 6.4.1 and 8.1.2).

citric acid has three prochiral centres;
it is also prochiral at the central carbon

First, let us look closely at the structure of citric acid. It has three prochiral centres. Two of these are the methylenes, but note that the central carbon is also prochiral. It has two groups the same, namely the $-CH_2CO_2H$ groups. The loss of water from citric acid is an *anti* elimination, so that the hydroxyl is lost together with one of the methylene hydrogens. The hydrogen lost has been found to be the *pro-R* hydrogen from the *pro-R*–CH_2CO_2H group.

This is followed by an *anti* addition reaction in which water is added to the new double bond, but in the reverse sense. The hydrogen retained throughout the process is shown with an asterisk. Note that we

can only label this hydrogen as *pro-S* in citric acid; in *cis*-aconitic acid and isocitric acid, it is no longer attached to a prochiral centre, and we must resort to some other labelling system, namely the asterisk.

This is a nice example of **enzymic stereospecificity**. It involves specific removal of one hydrogen atom from a substrate that appears to have four equivalent hydrogens. Because of the three-dimensional characteristics of both the enzyme and the substrate, the apparently equivalent side-chains on the central carbon are going to be positioned quite differently and the enzyme is able to distinguish between them. Further, it also distinguishes between the two hydrogens of a methylene group. An interesting consequence of this stereospecificity is that, because only one of the citric acid side-chains is modified in the aconitase reaction, it takes further turns of the cycle before material entering the cycle (acetyl-CoA) is actually degraded (see Section 15.3).

A reaction that gives a mixture of isomeric products with one isomer predominating would be termed **stereoselective**.

Enantiotopic faces

We have thus seen that there could be a need to distinguish between two similar groups attached to tetrahedral carbon, and have exploited the Cahn–Ingold–Prelog priorities to label the separate groups. We also need to consider another way in which a chiral centre might be generated, and that is by addition of a group to a planar system. For example, if we reduce a simple ketone that has two different R groups with lithium aluminium hydride we shall produce a racemic alcohol product (see Section 7.5). This is because hydride can be delivered to either face of the planar carbonyl group with equal probability.

addition from either face of
planar carbonyl group

In marked contrast, nature's reducing agent, reduced nicotinamide adenine dinucleotide (NADH), delivers hydride in a stereospecific manner because it is a cofactor in an enzyme-catalysed reaction. For example, reduction of pyruvic acid to lactic acid in vertebrate muscle occurs via attack of hydride to produce just one enantiomer, namely (*S*)-lactic acid.

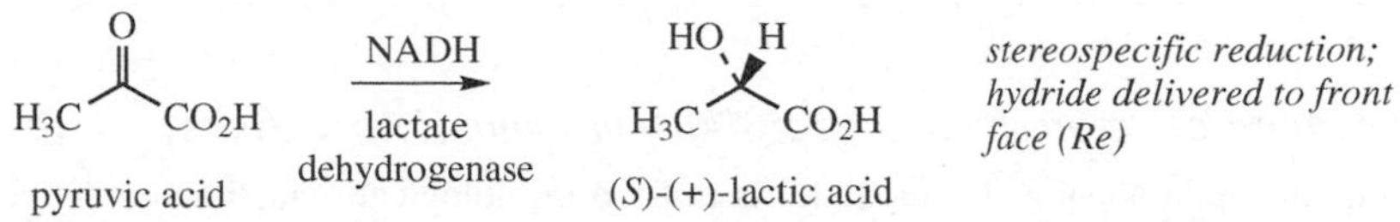

pyruvic acid → (S)-(+)-lactic acid

stereospecific reduction; hydride delivered to front face (Re)

We can see from the diagram that hydride must be delivered from the front face as shown, but it makes sense to have a more precise descriptor for faces than front or back. Once again, the Cahn–Ingold–Prelog system can help us out. We assign priorities to the three groups attached to the planar carbon.

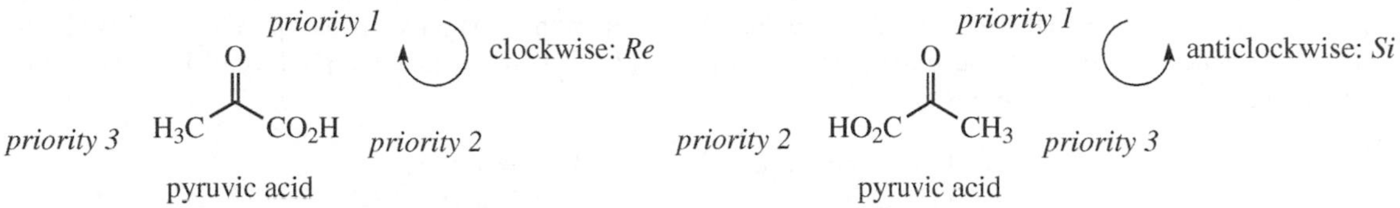

We then consider the descending sequence and decide whether this is clockwise or anticlockwise; the face that provides a clockwise sequence is then labelled *Re* and the face that provides an anticlockwise sequence is labelled *Si*. These are simply variants on *R* and *S*, in fact the first two letters of *rectus* and *sinister*.

Note that there is no correlation between *Re* or *Si* and the chirality *R* or *S* of the tetrahedral product formed.

It can now be seen that, in the enzymic reduction of pyruvic acid to lactic acid, hydride is delivered to the *Re* face of the pyruvic acid.

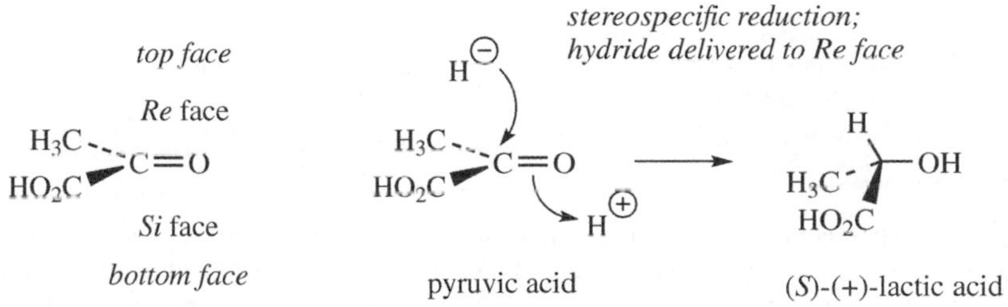

stereospecific reduction; hydride delivered to Re face

A molecule such as pyruvic acid is said to have two **enantiotopic** faces. Attack of a reagent onto the *Re* face yields one enantiomer, whereas attack onto the *Si* face will produce the other enantiomer.

The *Re* and *Si* descriptors are similarly applied to the carbon atoms making up C=C bonds. This gets a little more complex, in that a C=C bond generates four faces to be considered, two at each carbon. It is necessary to systematically deduce the descriptor for each, as shown below.

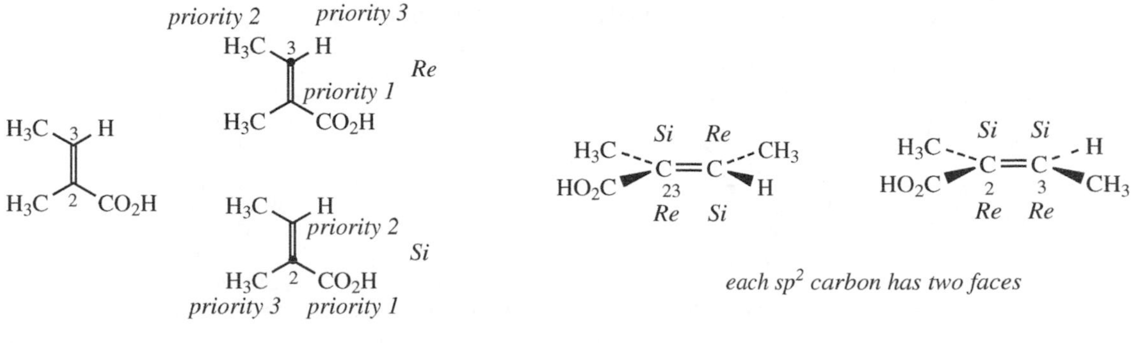

each sp² carbon has two faces

Box 3.14

NADH delivers hydride from a prochiral centre; NAD$^+$ has enantiotopic faces

NADH (reduced nicotinamide adenine dinucleotide) is utilized in biological reductions to deliver hydride to an aldehyde or ketone carbonyl group (see Box 7.6). A proton from water is used to complete the process, and the product is thus an alcohol. The reaction is catalysed by an enzyme called a **dehydrogenase**. The reverse reaction may also be catalysed by the enzyme, namely the oxidation of an alcohol to an aldehyde or ketone. It is this reverse reaction that provides the dehydrogenase nomenclature.

During the reduction sequence, NADH transfers a hydride from a prochiral centre on the dihydropyridine ring, and is itself oxidized to **NAD$^+$** (nicotinamide adenine dinucleotide) that contains a planar pyridinium ring. In the oxidation sequence, NAD$^+$ is reduced to NADH by acquiring hydride to an enantiotopic face of the planar ring. The reactions are completely stereospecific.

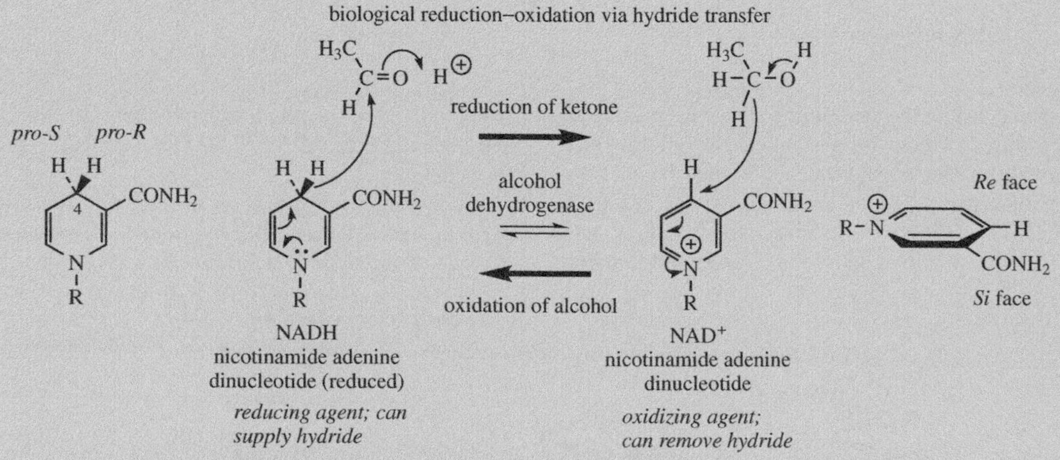

biological reduction–oxidation via hydride transfer

The stereospecificity depends upon the enzyme in question. Let us consider the enzyme alcohol dehydrogenase, which is involved in the ethanol to acetaldehyde interconversion. It has been deduced that the hydrogen transferred from ethanol is directed to the *Re* face of NAD$^+$, giving NADH with the 4R configuration. In the reverse reaction, it is the 4-*pro-R* hydrogen of NADH that is transferred to acetaldehyde.

Note also that transfer of hydride to the carbonyl compound is also stereospecific, as is removal of hydrogen from the prochiral centre of ethanol in the reverse reaction (see Section 3.4.7).

We should note that prochiral molecules have the potential to become chiral if we make certain changes, and we have used the term **enantiotopic** to identify the groups at sp^3-hybridized carbon or the faces of sp^2-hybridized carbon where alternative changes lead to the production of enantiomers. However, if there is also a chiral centre in the molecule, then the same changes would lead to the formation of diastereoisomers, not enantiomers. Such groups or faces are now correctly termed **diastereotopic**.

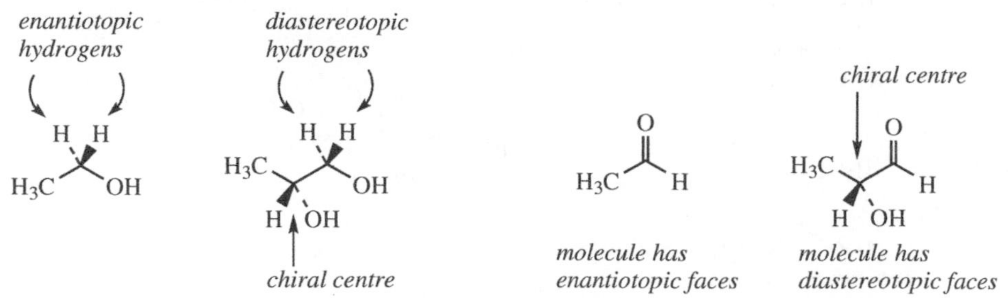

3.4.8 Separation of enantiomers: resolution

We saw in Section 3.4.1 that **enantiomers** have the same physical and chemical properties, except for optical activity, and thus they behave in exactly the same manner. We also saw, however, that this generalization did not extend into biological properties, and that there were compelling reasons for administering drugs as a single enantiomer rather than a racemate (see Box 3.7). At some stage, therefore, it might be necessary to have the means of separating individual enantiomers from a racemic mixture. This is termed **resolution**. The traditional method has been to convert enantiomers into **diastereoisomers**, because diastereoisomers have different physical and chemical properties and can, therefore, be separated by various methods (see Section 3.4.4). Provided one can convert the separated diastereoisomers back to the original compound, this offers a means of separating or resolving enantiomers.

The simplest method has been to exploit salt formation by reaction of a racemic acid (or base) with a chiral base (or acid). For example, treating a racemic acid with a chiral base will give a mixture of two salts that are diastereoisomeric. Although there is no covalent bonding between the acid and base, the ionic bonding is sufficient that the diastereoisomeric salts can be separated by some means, typically fractional crystallization. Although fractional crystallization may have to be repeated several times, and, therefore, is tedious, it has generally been an effective means of separating the diastereoisomeric salts. Finally, the salts can separately be converted back to the acid, completing the resolution.

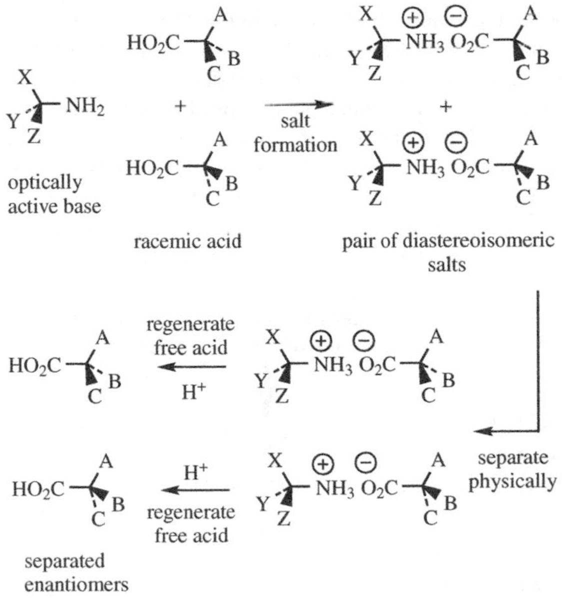

The bases generally employed in such resolutions have been natural alkaloids, such as strychnine, brucine, and ephedrine. These alkaloids are more complex than the general case shown in the figure, in that they contain several chiral centres (ephedrine is shown in Section 3.4.4). Tartaric acid (see Section 3.4.5) has been used as an optically active acid to separate racemic bases. Of course, not all materials contain acidic or basic groups that would lend themselves to this type of resolution. There are ways of introducing such groups, however, and a rather neat one is shown here.

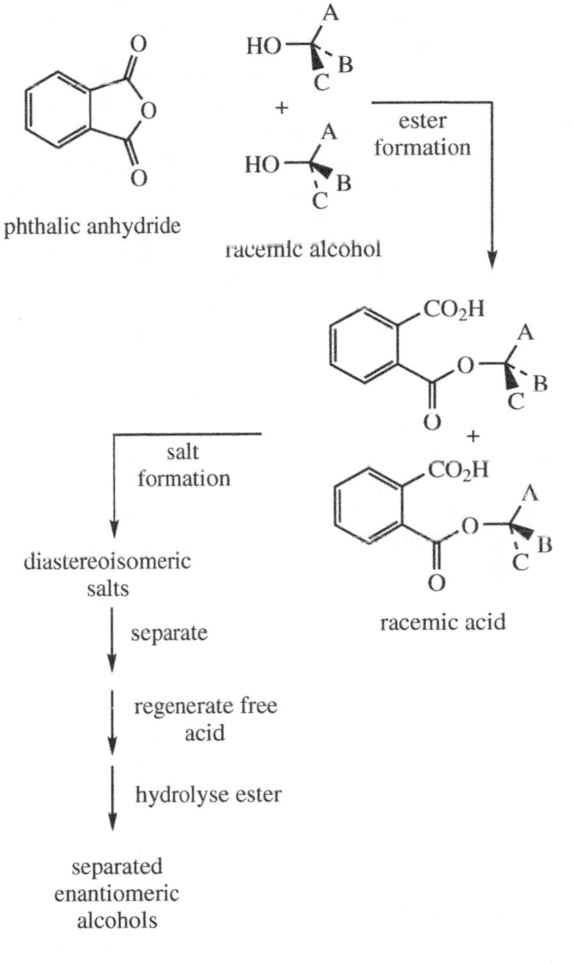

A racemic alcohol may be converted into a racemic acid by reaction with one molar equivalent of phthalic anhydride; the product is a half ester of a dicarboxylic acid (see Section 7.9.1). This can now be subjected to the resolution process for acids and, in due course, the alcohols can be regenerated by hydrolysis of the ester.

A significant improvement on the fractional crystallization process came with the introduction of chiral phases for column chromatography. This allows simple chromatographic separation of enantiomers. In practice it is effectively the same principle, that of forming diastereoisomeric complexes with the chiral material comprising the column. One enantiomer binds more tightly than the other and, therefore, passes through the column at a different rate. The two enantiomers thus emerge from the column as separate fractions.

It has also proved possible to exploit the enantiospecific properties of enzymes to achieve resolution of a racemic mixture during a chemical synthesis. Enzymes (see Section 13.4) are proteins that catalyse biochemical reactions with outstanding efficiency and selectivity. This is a consequence of the size and shape of the enzyme's binding site, a feature that is determined by the sequence of amino acid residues in the protein (see Section 13.3.2). The selectivity of enzymes means that they carry out reactions on one functional group in the presence of others that might be affected by a chemical reagent. It also means that they can be stereoselective, either performing reactions in a stereospecific manner or only reacting with substrates with a particular chirality. As a simple example, racemic ester structures may be resolved by the use of ester hydrolysing enzymes called lipases.

With the appropriate choice of enzyme, it has been found that only one enantiomer of the racemic mixture is hydrolysed, whilst the other remains unreacted. It is then a simple matter to separate the unreacted ester from the alcohol. The unreacted ester may then be hydrolysed chemically, thus achieving resolution of the enantiomeric alcohols.

3.4.9 Fischer projections

Fischer projections provide a further approach to the two-dimensional representations of three-dimensional formulae. They become particularly useful for molecules that contain several chiral centres, and are most frequently encountered in discussions of sugars (see

racemic ester   only one enantiomer is hydrolysed

Section 12.2). To start, though, let us consider just one chiral centre, and choose the amino acid we met earlier (see Section 3.4.2), $(-)$-(S)-serine.

The **Fischer projection** is drawn with groups on horizontal and vertical lines, but without showing the chiral carbon atom. Should you put in this carbon atom, it can no longer be considered that you are representing stereochemistry. The Fischer projection then implies that horizontal bonds are wedged, whilst vertical bonds are dotted, and it thus speeds up the drawing of stereochemical features. For $(-)$-(S)-serine, the wedge–dot version is what one would see if one looked down on the right-hand stereostructure

Fischer projection

carbon with highest oxidation state at top

carbon not shown – intersection of lines

longest carbon chain vertical

horizontal lines *above* plane
vertical lines *below* plane

Fischer projection is equivalent to viewing molecule from the top

$(-)$-(S)-serine

as indicated. Accordingly, we can now transform stereostructures into Fischer projections, and vice versa.

The only significant restrictions are

- we should draw the longest carbon chain vertical;
- we should place the carbon of highest oxidation state at the top.

However, when we come to manipulate Fischer projections, we may need to disregard these restrictions in the interests of following the changes.

Manipulations we can do to a Fischer projection may at first glance appear confusing, but by reference to a model of a tetrahedral array, or even a sketch of the representation, they should soon become quite understandable, perhaps even obvious. The molecular manipulations shown are given to convince you of the reality of the following statements.

- Rotation of the formula by 180° gives the same molecule.

180°

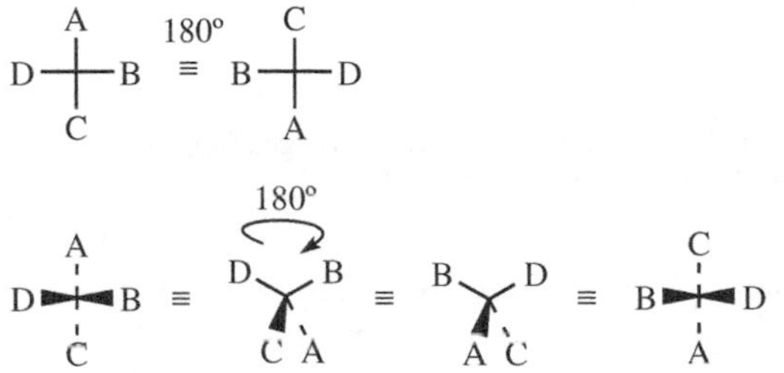

- Rotation of any three groups clockwise or anticlockwise gives the same molecule.

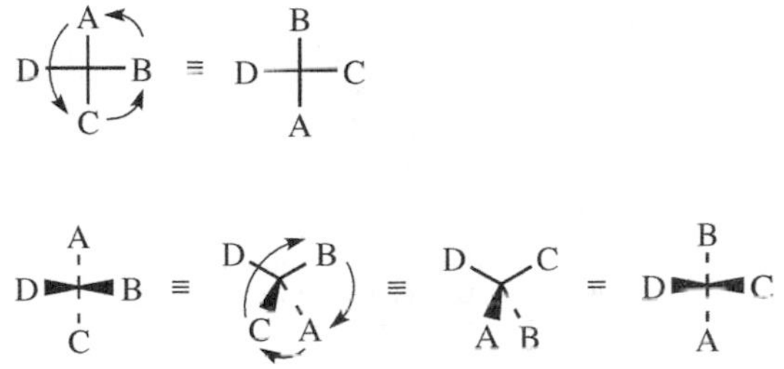

- Exchange of any two groups gives the enantiomer.

original isomer
mirror
enantiomer
enantiomer
exchange of two groups gives enantiomer
mirror image of enantiomer is original isomer

- Rotation of the formula by 90° gives the enantiomer.

original isomer
mirror
90°
enantiomer
enantiomer
one exchange gives enantiomer, second exchange restores original isomer

It is also surprisingly easy to assign *R* or *S* configurations to chiral carbons in the Fischer projections; but, because horizontal lines imply wedged bonds (towards you) and vertical lines imply dotted bonds (away from you), there are important guidelines to remember:

- if the group of lowest priority is on the vertical line, a clockwise sequence gives the *R* configuration;
- if the group of lowest priority is on the horizontal line, a clockwise sequence gives the *S* configuration.

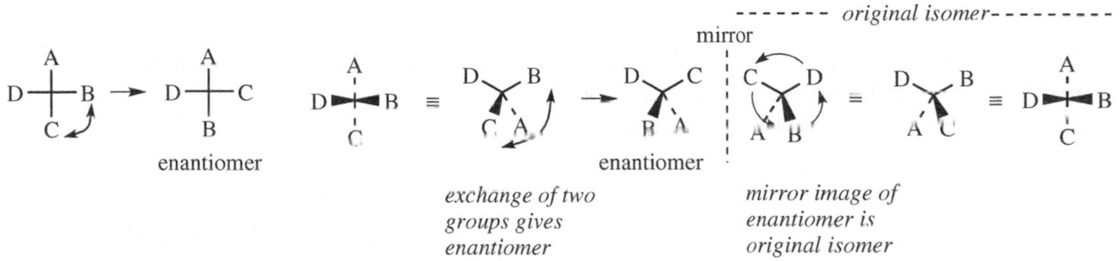

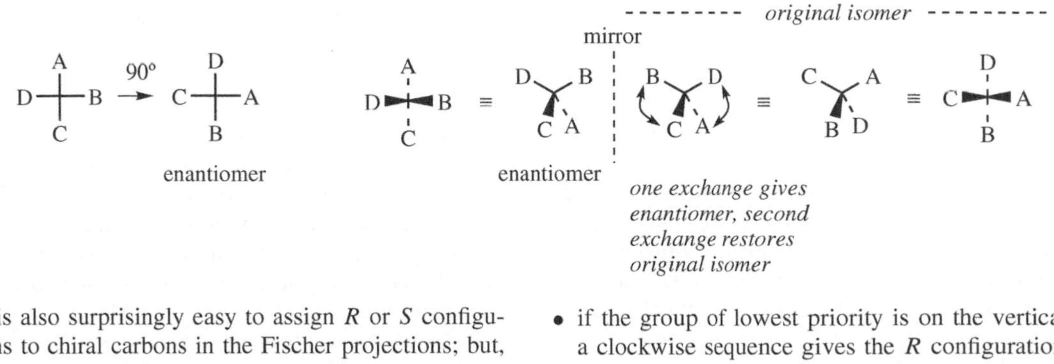

These do not represent a different set of rules from the clockwise = R, anticlockwise = S conventions we already use (see Section 3.4.2). It is merely a consequence of the lowest priority group being down (dotted bond) on the vertical line, but up (wedged) on the horizontal line. We have noted (see Box 3.8) that, if the lowest priority group is wedged, it is easier to look at the sequence from the front, then reverse it to give us the sequence as viewed from the rear, i.e. towards the group of lowest priority.

rotation by 180°
gives same molecule

interchange of two groups gives enantiomer

numbers refer to assigned priorities

if group of lowest priority is on the vertical line, a clockwise sequence gives the R configuration

if group of lowest priority is on the horizontal line, a clockwise sequence gives the S configuration

relate this to horizontal bonds implying wedged (up) and vertical bonds implying dotted (down)

hydrogen down, clockwise = R

hydrogen up, must view from rear; alternatively, front view clockwise needs reversing = S

Let us apply these principles to tartaric acid. This compound has two chiral centres; but, as we saw previously, only three stereoisomers exist, since there is an optically inactive *meso* compound involved (see Section 3.4.5).

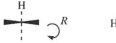

H on horizontal anticlockwise = R
(2R,3R)-(+)-tartaric acid

H on horizontal clockwise = S
(2S,3S)-(−)-tartaric acid

(2R,3S)-*meso*-tartaric acid

We can draw these three stereoisomers as Fischer projections, reversing the configurations at both centres to get the enantiomeric stereoisomers, whilst the Fischer projection for the third isomer, the *meso* compound, is characterized immediately by a plane of symmetry. For (+)-tartaric acid, the configuration is (2R,3R), and for (−)-tartaric acid it is (2S,3S). For both chiral centres, the group of lowest priority is hydrogen, which is on a horizontal line. In fact, this is the case in almost all Fischer projections, since, by convention, the vertical line is the longest carbon chain. Thus, we have to reverse our normal configurational thinking: a clockwise sequence of priorities gives S and an anticlockwise sequence gives R. The configuration of the *meso* isomer can be deduced by abstracting the appropriate portions from the other two structures and assigning equivalent configurations.

It should be appreciated that a Fischer projection involving more than one chiral centre actually depicts an eclipsed conformer, which is naturally a high-energy state, and is normally an unlikely arrangement of atoms (see Section 3.3.1). We need to bear this in mind when we transpose Fischer projections into wedge–dot stereochemical drawings, with further manipulations necessary to give lower energy staggered conformers. This is illustrated here with the five-carbon sugar (−)-ribose.

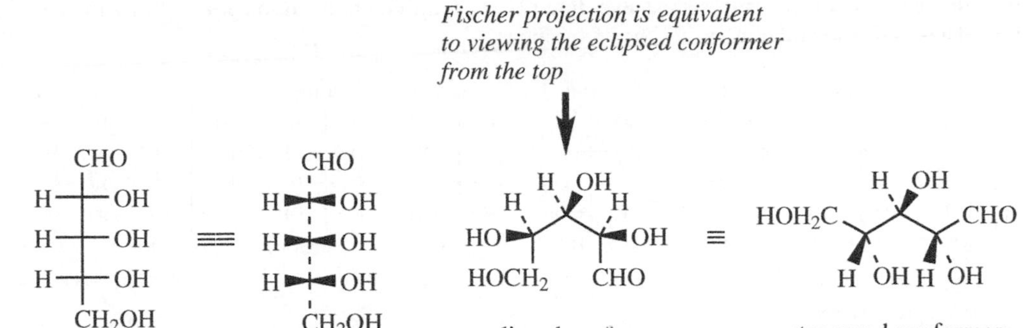

Fischer projection is equivalent to viewing the eclipsed conformer from the top

CHO / H—OH / H—OH / H—OH / CH₂OH

(−)-ribose
Fischer projection

≡

implied stereochemical relationship

eclipsed conformer   ≡   staggered conformer

However, as we shall see shortly, Fischer-projection-derived eclipsed conformers are particularly useful in deducing the stereochemistry in cyclic forms of sugars (see Box 3.16).

3.4.10 D and L configurations

The concept of D and L as configurational descriptors is well established, particularly in amino acids and sugars; frankly, however, we could live without them and save ourselves a lot of confusion. Since they are so widely used, we need to find out what they mean, but in most cases the information conveyed is less valuable than sticking with R and S.

D and L sugars

The simplest of the sugars is glyceraldehyde, which has one chiral centre. Long before R and S were adopted as descriptors, the two enantiomers of glyceraldehyde were designated as D and L. D-(+)-Glyceraldehyde is equivalent to (R)-(+)-glyceraldehyde, the latter configuration being fully systematic. Configurations in other compounds were then related to the configurations of D- and L-glyceraldehyde by direct comparison of Fischer projections. For example, (+)-glucose (= dextrose) is represented by a Fischer projection that defines the configuration at all four chiral centres.

(R)-(+)-glyceraldehyde
=
D-(+)-glyceraldehyde

(S)-(−)-glyceraldehyde
=
L-(−)-glyceraldehyde

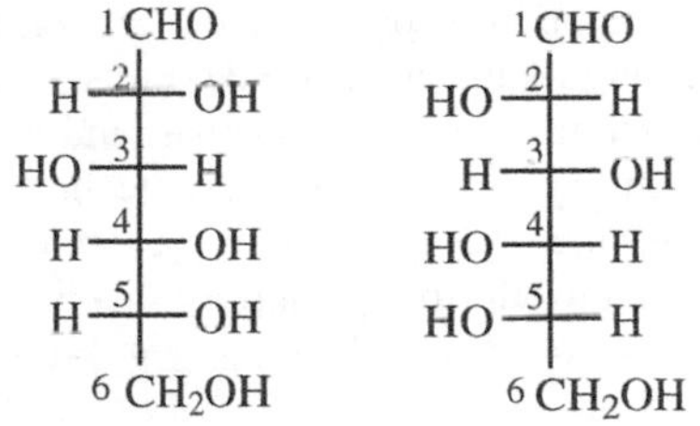

D-(+)-glucose   L-(−)-glucose

Since the configuration at position 5 in (+)-glucose can be directly related to that in D-(+)-glyceraldehyde, (+)-glucose is said to have the D configuration, and is thus termed D-(+)-glucose. By similar reasoning, the enantiomer of glucose has the L configuration, and is termed L-(−)-glucose. Now the limitations of this system become obvious when one realizes that D and L refer to the configuration at just one centre, by convention the highest numbered chiral centre, and the remaining configurations are not specified, except by the name of the sugar (see Box 3.15).

Box 3.15

Fischer projections of glucose and stereoisomers

The sugar **glucose** has four chiral centres; therefore, $2^4 = 16$ different stereoisomers of this structure may be considered. These are shown below as Fischer projections.

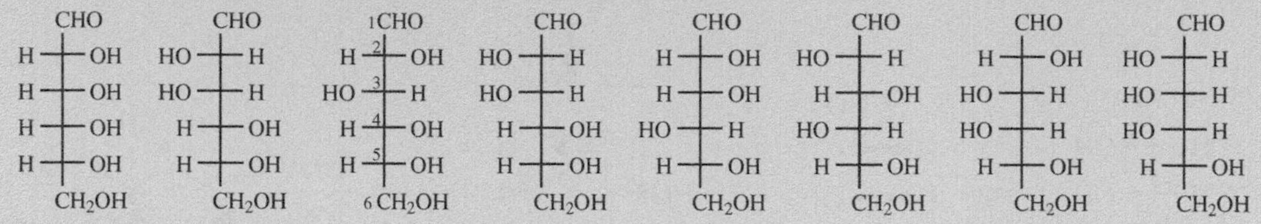

| D-(+)-allose | D-(+)-altrose | D-(+)-glucose | D-(+)-mannose | D-(-)-gulose | D-(+)-idose | D-(+)-galactose | D-(+)-talose |

(C-3 epimer of D-glucose) (C-2 epimer of D-glucose)   (C-4 epimer of D-glucose)

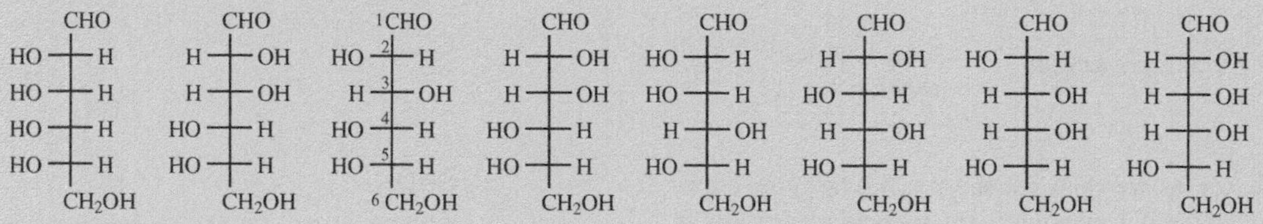

| L-(-)-allose | L-(-)-altrose | L-(-)-glucose | L-(-)-mannose | L-(+)-gulose | L-(-)-idose | L-(-)-galactose | L-(-)-talose |

(C-5 epimer of D-glucose)

The 16 stereoisomers are divided into D and L groups, which reflect only the configuration at the highest numbered chiral centre, namely C-5. The chirality at other centres is defined solely by the name given to the sugar, so we have eight different names for particular configurational combinations. Note that although D and L strictly refer to the configuration at only one centre, L-glucose is the enantiomer of D-glucose and, therefore, must have the opposite configuration at all chiral centres. A change in configuration at only one centre produces a diastereoisomer that has different chemical properties, and is accordingly given a different name.

Whilst this system of nomenclature has some obvious shortcomings, it is analogous to the ephedrine and pseudoephedrine example where we were considering just two chiral centres (see Section 3.4.4). A more systematic approach (though not one that is used) might give all the above sugars the same name, e.g. hexose, but specify the chirality at each centre, e.g. D-(+)-glucose would be $(+)$-$(2R,3S,4R,5R)$-hexose and L-(-)-galactose would become $(-)$-$(2S,3R,4R,5S)$-hexose. Instead, we have the eight different names in two configurational classes, D and L.

We can also use the term epimer to describe the relationship between isomers, where the difference is in the configuration at just one centre (see Section 3.4.4). This is shown for the four epimers of D-(+)-glucose. An interesting observation with the 16 stereoisomers is that optical activity of a particular isomer does not appear to relate to the configuration at any particular chiral centre.

Box 3.16

Stereochemistry in hemiacetal forms of sugars from Fischer projections

In solution, aldehyde sugars normally exist as cyclic hemiacetals through reaction of one of the hydroxyls with the aldehyde group, giving a strain-free six- or five-membered ring (see Section 3.3.2). The Fischer projection for the sugar is surprisingly useful in predicting the configuration and conformation of the cyclic form.

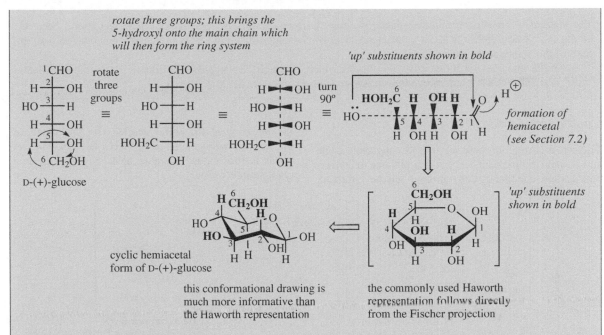

this conformational drawing is
much more informative than
the Haworth representation

the commonly used Haworth
representation follows directly
from the Fischer projection

The approach is straightforward. Since cyclic hemiacetal formation requires a hydroxyl group as the nucleophile to attack the protonated carbonyl (see Section 7.2), we put this hydroxyl group on the vertical, thus getting all the ring atoms onto the vertical. This requires rotation of three groups attached to the appropriate atom, C-5 in the case of D-(+)-glucose. Such rotation does not affect the configuration at C-5. Then put in the stereochemistry implied by the Fischer projection, using wedges and dots. This structure should then be turned on its side, and the ring formation considered by joining up the C-5 hydroxyl and the carbonyl at the rear of the structure. Note that, as drawn, this eclipsed conformer from the Fischer projection actually has these atoms quite close together, so that ring formation is easily achieved and, most importantly, easily visualized (see Section 3.4.9).

The net result is a cyclic system looking like the Haworth representation that is commonly used, especially in biochemistry books. The Haworth representation nicely reflects the up–down relationships of the various substituent groups, but is uninformative about whether these are equatorial or axial. The last step, therefore, is to transcribe this representation into a chair conformation, as shown, so that we see the conformational consequences.

The alternative chair conformation, should we draw it instead, would be less favoured than that shown because of the increased number of axial substituents. The conformation of D-glucose is the easily remembered one, in that all the substituents are equatorial.

A similar procedure is shown for D-(−)-ribose, which, although it is capable of forming a six-membered cyclic form, is found to exist predominantly as a five-membered ring (see Section 12.2.2).

D and L amino acids

There is a correlation between D- and L-glyceraldehyde and D- and L-amino acids, in that it is possible to convert one system chemically into another without affecting the integrity of the chiral centre. The fine detail of the transformations need not concern us here. The net result is that D- and L-amino acids have the general configurations shown.

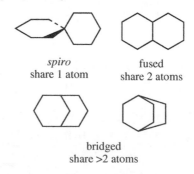

the common way of
presenting L-amino acids

Note that all the amino acids found in proteins are of the L configuration (excepting the achiral glycine); D-amino acids are found in some polypeptide antibiotics (see Section 13.1). As we pointed out in Section 3.4.2, this brings up an apparent anomaly in nomenclature. In all protein L-amino acids, except for cysteine, this represents an S configuration; cysteine, because of its high-priority sulfur atom has the R configuration. One can consider they all have the same configuration based on the L descriptor, but the priority rules lead to a different label.

One further point; as mentioned in Section 3.4.1, the now obsolete descriptors d and l are abbreviations for dextrorotatory (+) and laevorotatory (−) respectively. They do not in any way relate to D and L.

3.5 Polycyclic systems

Many molecules of biological or pharmaceutical importance contain polycyclic ring systems, and we have already met some examples in other contexts, e.g. penicillins (see Box 3.8). There are three main ways in which rings can be joined together, according to whether they share one atom, two atoms, or more than two atoms. These are termed *spiro*, **fused**, or **bridged** systems respectively. Examples are shown where six-membered rings are joined in the various ways, but the concepts apply equally to rings of other sizes.

spiro
share 1 atom

fused
share 2 atoms

bridged
share >2 atoms

3.5.1 Spiro systems

Spiro systems have two rings sharing a single carbon atom, and since this has essentially a tetrahedral array of bonds, the bonds starting the two rings must be arranged perpendicular to each other. If there is appropriate substitution on the rings, then this can lead to the *spiro* centre becoming chiral (see Section 3.4.6).

Box 3.17

Natural spiro compounds

Spiro compounds are exemplified by several natural product structures. One of these is the antifungal agent **griseofulvin** produced by cultures of the mould *Penicillium griseofulvum*. Griseofulvin is the drug of choice for many fungal infections, but it is ineffective when applied topically, so is administered orally. Griseofulvin has two chiral centres, one of which is the *spiro* centre, so there are potentially four configurational isomers for the structure. Natural griseofulvin has the configurations shown.

griseofulvin

solasodine

tomatidine

diosgenin

Solasodine and **tomatidine** are steroidal alkaloids produced by potatoes (*Solanum tuberosum*) and tomatoes (*Lycopersicon esculente*) respectively. These compounds, as glycosides (see Section 12.4), are responsible for the toxic properties of the foliage and green fruits of these plants. They are not present in potato tubers, unless green, or in ripe tomato fruits. Both compounds contain a *spiro* system, a nitrogen analogue of a ketal (see Section 7.2). A **spiroketal** is present in **diosgenin** from *Dioscorea* species, a raw material used for the semi-synthesis of steroidal drugs. Note that solasodine and tomatidine demonstrate the different configurations at the *spiro* centre; all natural spiroketals have the same stereochemistry at the *spiro* centre as in diosgenin.

3.5.2 Fused ring systems

Fused ring systems are particularly common. It is logical to suppose that fusing on one or more additional ring systems is going to have stereochemical consequences, in particular that the conformational changes seen with single ring systems are likely to be significantly modified. Initially, let us consider two cyclohexane rings fused together, giving a bicyclic system called **decalin**.

trans ring fusion

trans-decalin

cis ring fusion

cis-decalin

Two configurational isomers exist, *trans*- and *cis*-decalin, according to the stereochemistry of ring fusion. The *trans* or *cis* relationship is most easily seen with the hydrogens at the ring fusion carbons, but it also follows that the bonds forming part of the second ring can be considered to share a *trans* or *cis* relationship to each other. It is usual practice to show the stereochemistry in the former way, via the ring fusion substituents.

The situation is in many ways analogous to *trans*- and *cis*-1,2-dimethylcyclohexane (see Section 3.3.2), and these afford useful comparisons as we consider conformational changes.

H
eq
eq
H

trans-decalin

ax
H
H
ax

cannot achieve bonding within a six-membered ring

relate to
eq
eq

ax
ax

Now, ***trans*-decalin** forms a rather rigid system, and it transpires that the only conformational mobility possible is ring flip of chairs to very much less favourable boats. Since both bonds of the second ring are equatorial with respect to the first ring, any other type of conformational change would require these to become axial. It is impossible to join the two axial bonds into a ring system as small as six carbons; hence, there is no conformational mobility.

ax
H
eq
H

cis-decalin

ax
H
eq
H

rotate 60°
≡
eq
ax
H
H

relate to
ax
eq

ax
eq

rotate 60°
≡
eq
ax

On the other hand, ***cis*-decalin** is conformationally mobile, and a simultaneous flipping in both rings produces a new conformer of equal energy. This is not easy to visualize. In the scheme, the middle conformer has one ring viewed face on, so that we have resorted to rotation of the structure to get an appreciation of the new conformer with its rings in chair form. It is best to have models to appreciate this conformational flexibility. It is quite clear, though, that an axial bond becomes equatorial and an equatorial one becomes axial, just as with substituents in the *cis*-1,2-dimethylcyclohexane analogue (see Section 3.3.2). However, it is probably reassuring to appreciate that this conformational flexibility in two *cis*-fused cyclohexane rings is lost when a third ring is fused on, and in many of the fused ring systems of interest to us it becomes of no further consequence.

Since the second ring in *trans*-decalin effectively introduces two equatorial substituents to the first ring, whilst in *cis*-decalin it provides one equatorial and one axial substituent, it is logical to predict that *trans*-decalin should have a lower energy than *cis*-decalin. This is indeed the case, the energy difference being about 12 kJ mol^{-1}.

When we considered *trans*- and *cis*-1,2-dimethyl-cyclohexane, we found that only three configurational isomers exist, enantiomeric forms of the *trans* isomer, together with the *cis* isomer, which is an optically inactive *meso* compound (see Section 3.4.5). The *meso* relationship could be deduced from the plane of symmetry in the hexagon representation.

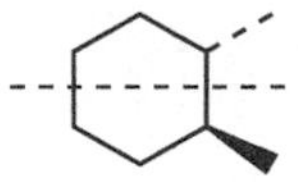

no plane of
symmetry

trans
two enantiomers

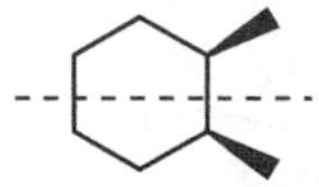

plane of symmetry

cis
meso

three configurational isomers

When we look at the structures of *trans*- and *cis*-decalin, it is apparent that a further plane of symmetry, through the ring fusion, is present in both structures. This means that each isomer is superimposable on its mirror image; consequently, there are only two configurational isomers of decalin, one *trans* and one *cis*.

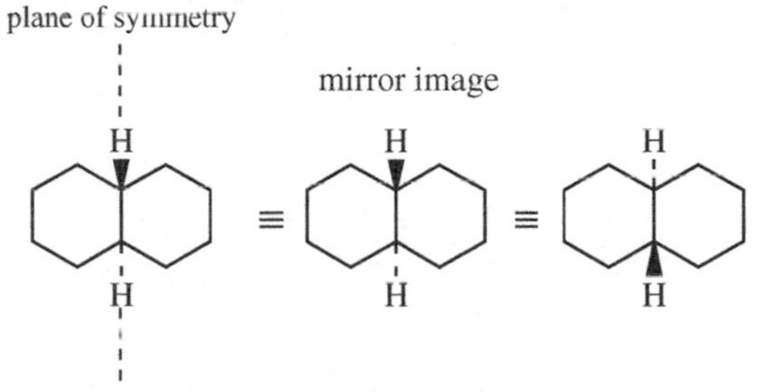

plane of symmetry

mirror image

trans-decalin

planes of symmetry

mirror image

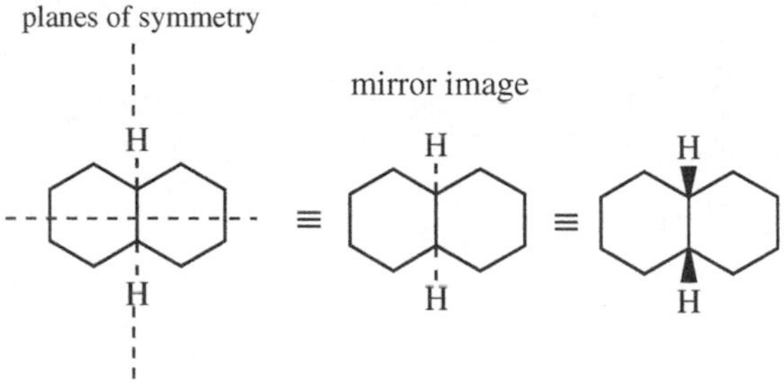

cis-decalin

only two configurational isomers

The situation in *trans*- and *cis*-decalin is complicated by the symmetry elements. If this symmetry is destroyed, e.g. by introducing dimethyl substituents, we get back to reassuringly familiar territory in which two chiral centres lead to four configurational isomers. The same is true in the *trans*- and *cis*-1,2-dimethylcyclohexane series.

destroy symmetry

trans
two enantiomers

cis
two enantiomers

four configurational isomers

dimethyl substitution removes symmetry without adding a new chiral centre

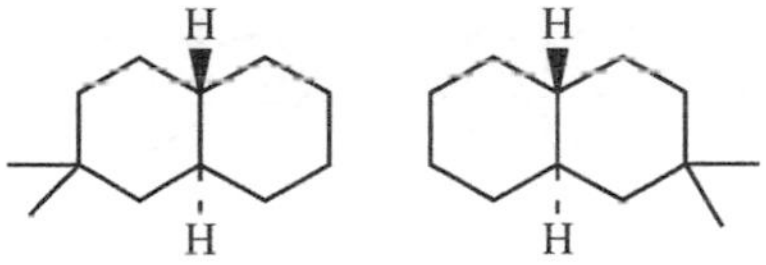

trans
two enantiomers

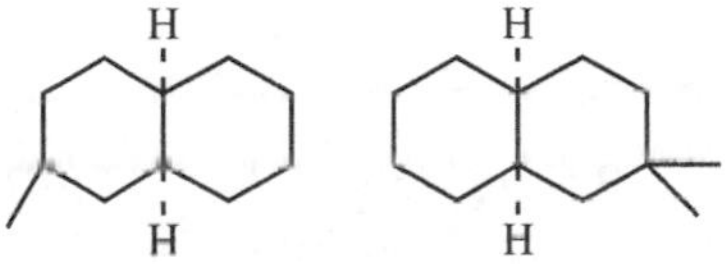

cis
two enantiomers

four configurational isomers

Fusing rings of different sizes can produce significant restraints, especially when rings of less than six carbons are involved. However, the characteristics of these fused systems can be deduced logically by applying our knowledge of single ring systems.

Fusion of a five-membered ring to a six-membered ring gives a **hydrindane** system, and, as with decalins, *cis* and *trans* forms are possible. Because the cyclopentane ring is more planar than a cyclohexane ring (see Section 3.3.2), this causes deformation and increases strain at the ring fusion. This deformation is more easily accommodated with the *cis*-fusion than the *trans*-fusion, and, in contrast to the decalins, the *cis* isomer has a lower energy than the *trans* isomer (by about $1 \, kJ \, mol^{-1}$). As in the decalins though, the *cis* form is conformationally mobile, whereas the *trans* form is fixed.

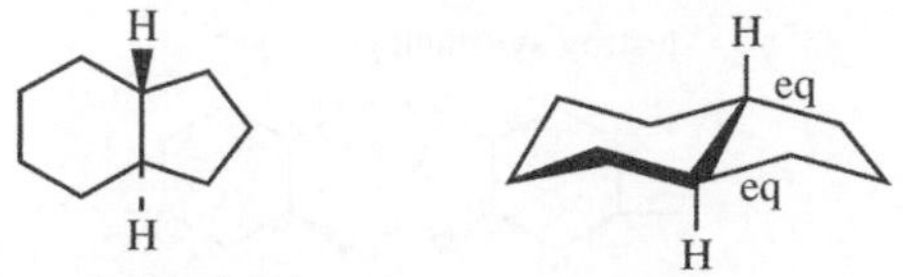

trans-hydrindane

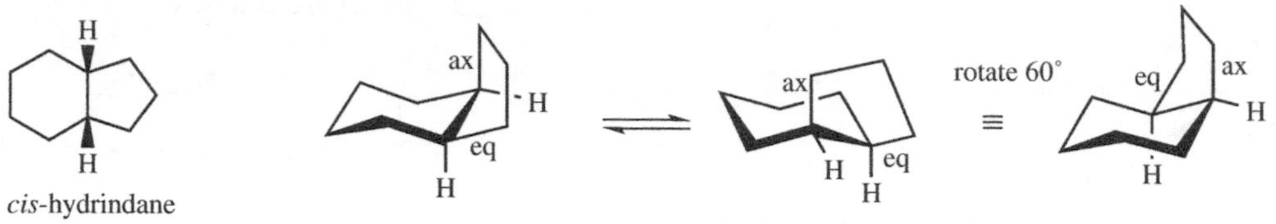

cis-hydrindane

The fusion of rings of different sizes reduces symmetry in the structures; instead of the rather unusual situation with the decalins, where there are only two configurational isomers, the hydrindanes exist in the anticipated three isomeric forms, two enantiomeric *trans* isomers and a *meso cis* isomer (compare 1,2-dimethylcyclohexane).

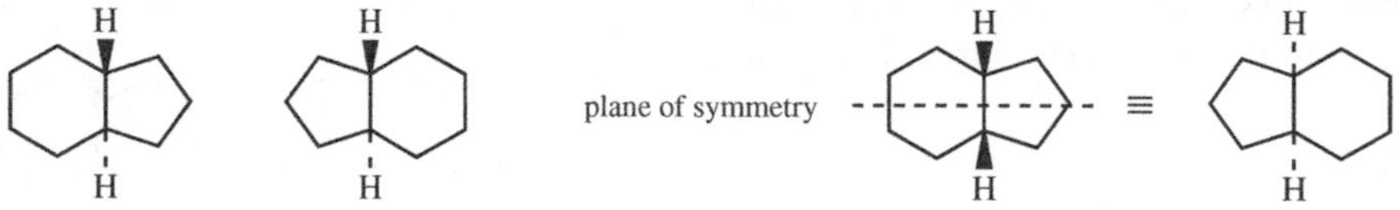

trans
two enantiomers

plane of symmetry

cis
meso

Box 3.18

Isomerizations influenced by ring fusions

Epimerization of *cis*-decalone   If *cis*-decalone is treated with mild base, it is predominantly isomerized to *trans*-decalone. This can be rationalized by considering stereodrawings of the two isomers.

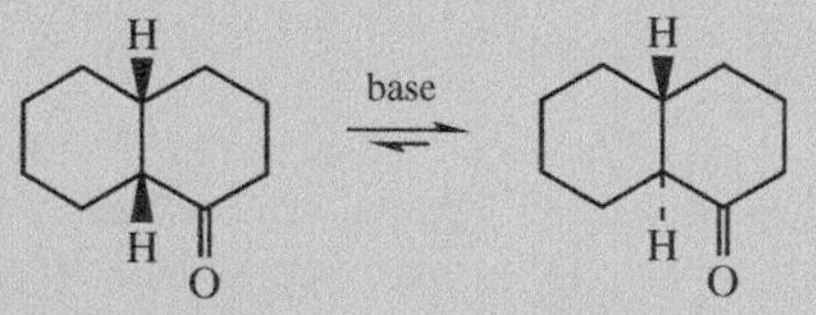

cis-decalone   *trans*-decalone

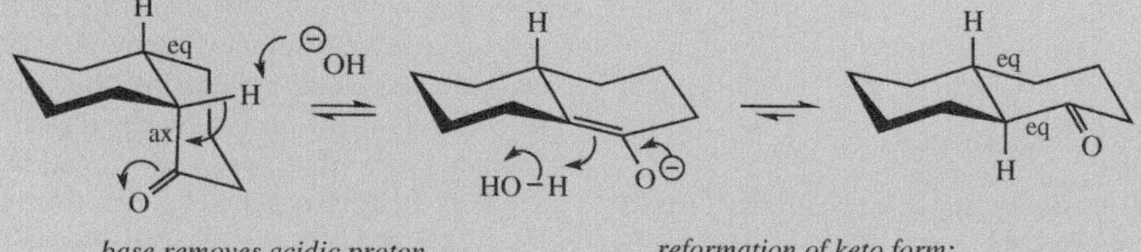

base removes acidic proton
α to carbonyl and generates
enolate anion

reformation of keto form;
proton is acquired on lower face
to produce more stable isomer

The ring fusion in *cis*-decalone means that bonds forming the second ring have a relationship to the first ring in which one bond is equatorial and one axial. In contrast, both such bonds in *trans*-decalone are equatorial to the first ring. We can predict, therefore, that *trans*-decalone has a lower energy than *cis*-decalone. The isomerization is brought about because the carbonyl group is adjacent (α) to the hydrogen at the ring fusion. This hydrogen is relatively acidic and may be removed by base, generating the enolate anion (see Sections 4.3.5 and 10.1 for detail of this reaction). The enolate anion must now be planar around the site of ring fusion and, by a reversal of the process, may pick up a proton from either side of the double bond. However, instead of getting a 1 : 1 mixture of the two possible isomers, this reaction very much favours the *trans* isomer because of its lower thermodynamic energy. The equilibrium mixture contains principally *trans*-decalone.

Epimerization of etoposide

The anticancer agent **etoposide** contains a five-membered lactone function that is significantly strained because it is *trans*-fused. This material is readily converted into a relatively strain-free *cis*-fused system by treating with very mild alkali, e.g. traces of detergent, and produces an epimer (see Section 3.4.4) called picroetoposide. This isomer has no significant biological activity.

etoposide   picroetoposide

The epimerization can be formulated as involving an enolate anion, as above (see Section 10.8). However, in contrast to the decalin example above, *cis*-hydrindane is of lower energy than *trans*-hydrindane. In this particular case, on reverting back to a carbonyl compound, the planar enolate anion is presented with the alternatives of receiving a proton from one face to form a strained *trans*-fused system, or from the other face to form a strain-free *cis*-fused system. The latter is very much preferred, so much so that the conversion of etoposide into its epimer is almost quantitative. Although we can rationalize this behaviour simply by considering the hydrindane-type rings, the fusion of this system to an aromatic ring causes additional distortion (see below), and the effect becomes even more pronounced in favour of the *cis*-fused system.

This behaviour contrasts with the racemization of hyoscyamine to atropine, which also involves an enolate anion derived from an ester system (see Section 10.8). As the term racemization implies, atropine is a 50 : 50 mixture of the two enantiomers. It shows how the proportion of each epimer formed can be influenced by other stereochemical factors.

The fusion of a three-membered ring onto a six-membered ring has much more serious limitations. A three-membered ring must be planar, so it will distort the ring it is being fused to, and this restricts stereochemical possibilities. For example, **epoxycyclohexane** can, therefore, only be *cis*-fused, and the six-membered ring is forced to adopt the half-chair conformation we saw with cyclohexene (see Section 3.3.2). There will be conformational mobility in this ring provided that there are no other ring fusions to prevent this.

epoxycyclohexane

6-membered ring adopts half-chair conformation

Note that, in situations where a ring fusion produces chiral centres, we can find the number of configurational isomers possible is less than that predicted from the 2^n guidelines. This may be the consequence of symmetry, in that an isomer is the same as its mirror image, as we have seen above. However, it can also be the result of restrictions caused by the ring fusion, so that one centre effectively defines the chirality of another, thus reducing the number of combinations. In epoxycyclohexanes, no *trans*-fused variants can exist.

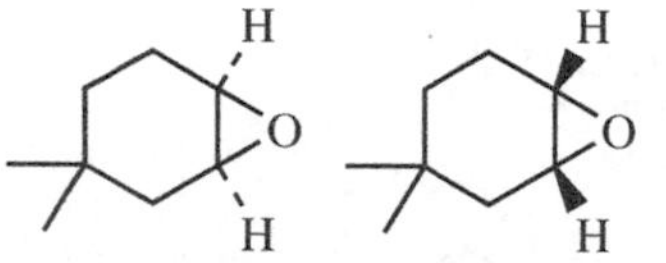

Note that a cyclohexane system will be forced into a similar half-chair conformation by fusing a planar aromatic ring onto a cyclohexane ring (a tetrahydronaphthalene system).

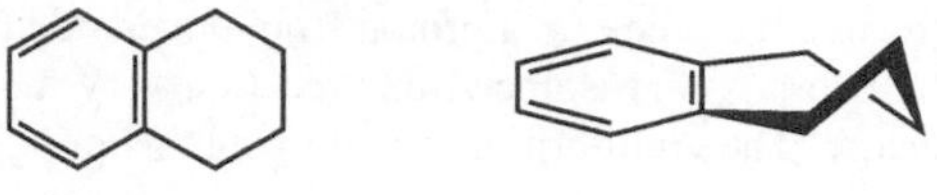

tetrahydronaphthalene

dimethyl substitution removes symmetry without adding a new chiral centre

two chiral centres, but only two configurational isomers; no trans isomers can exist

cyclohexene ring adopts half-chair conformation

Box 3.19

Shapes of steroids

Steroids all contain a tetracyclic ring system comprised of three six-membered rings and one five-membered ring fused together. **Cholesterol** is the best known of the steroids. It is an essential structural component of animal cells, though the presence of excess cholesterol in the blood is definitely associated with the incidence of heart disease and heart attacks.

Whilst cholesterol typifies the fundamental structure, further modifications to the side-chain and the ring system help to create a wide range of biologically important natural products, e.g. sterols, steroidal saponins, cardioactive glycosides, bile acids, corticosteroids, and mammalian sex hormones. Because of the profound biological activities encountered, many natural steroids, together with a considerable number of synthetic and semi-synthetic steroidal compounds, are routinely employed in medicine. The markedly different biological activities observed emanating from compounds containing a common structural skeleton is, in part, ascribed to the functional groups attached to the steroid nucleus and, in part, to the overall shape conferred on this nucleus by the stereochemistry of ring fusions.

Let us start with **cholestane**, which is the basic hydrocarbon skeleton of cholesterol. This structure has all ring fusions *trans*, and by logical extension of *trans*-decalin and *trans*-hydrindane can be deduced to have approximately the shape illustrated. Because of the *trans* fusions, there is no conformational mobility except for the unlikely flipping of ring A into a boat form, which we can ignore. The overall shape of cholestane is a rather rigid and flattish structure. The rings are designated A–D as indicated.

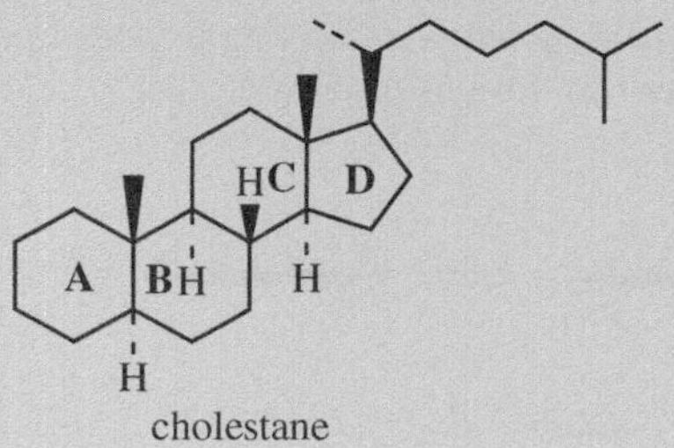

cholestane

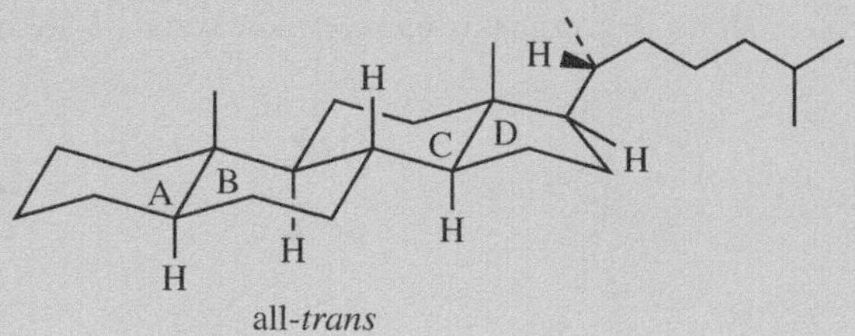

all-*trans*

Cholesterol has a double bond in ring B at the A–B ring fusion, so this distorts the rings by demanding that the arrangement around the double bond is planar. It is not possible to depict this perfectly in a typical two-dimensional representation.

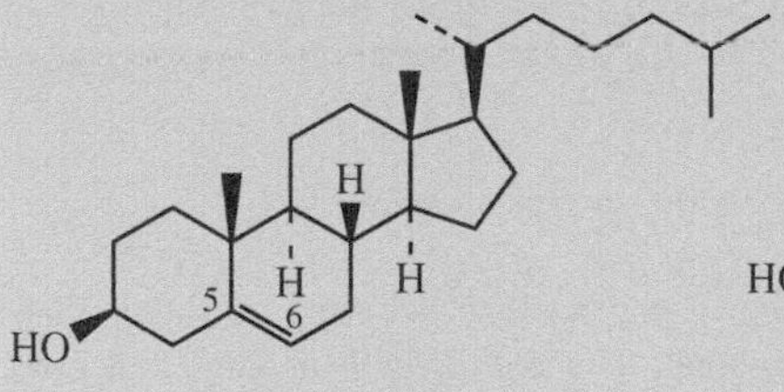

cholesterol

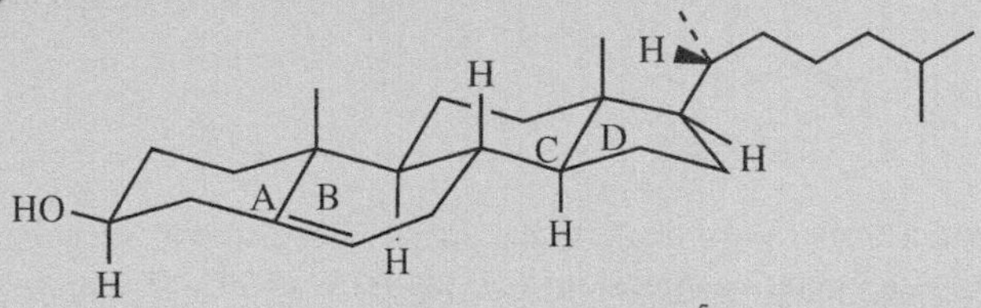

Δ^5-unsaturation

Note: Δ^5 is a neat way of indicating that there is a double bond at position 5

The natural progestogen hormone **progesterone** also has a double bond at the A–B ring fusion, but this time in ring A, so a similar distortion in ring A is required.

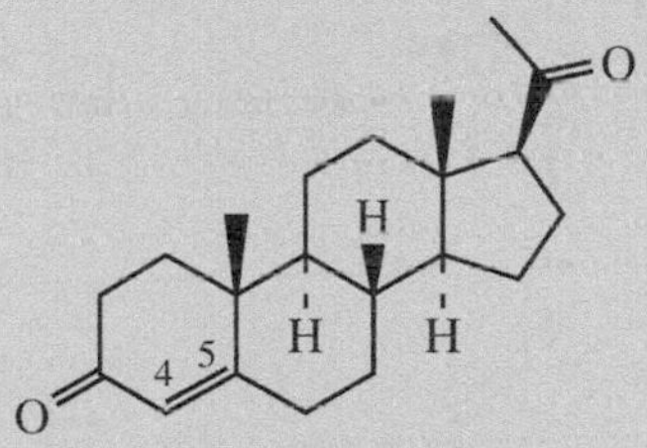

progesterone

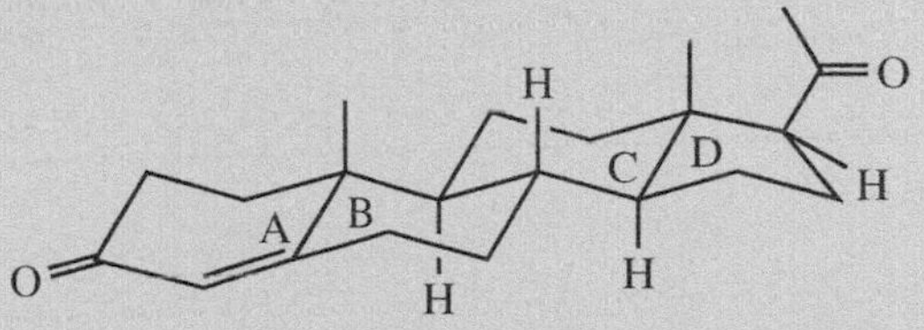

Δ^4-unsaturation

The fungal sterol **ergosterol** has double bonds at positions 5 and 7, both in the B ring, which consequently should become essentially planar. The picture shown is a rough approximation. The antifungal effect of polyene antibiotics, such as amphotericin and nystatin (see Box 7.14), depends upon their ability to bind strongly to ergosterol in fungal membranes. They do not bind significantly to cholesterol in mammalian cells, so this provides selective toxicity. The binding to ergosterol is very much influenced by the changes in shape conferred by the extra double bond in ring B.

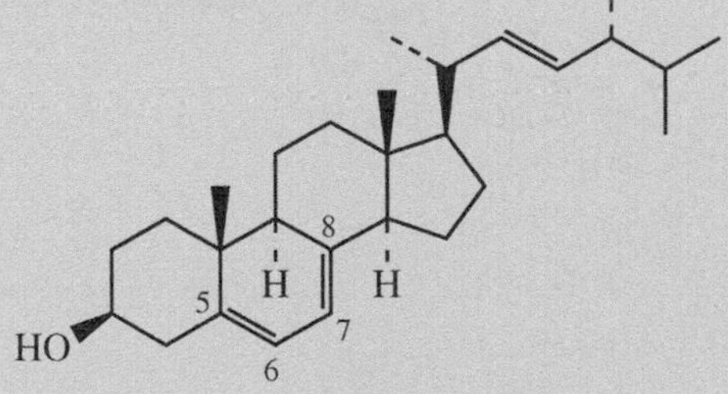

ergosterol

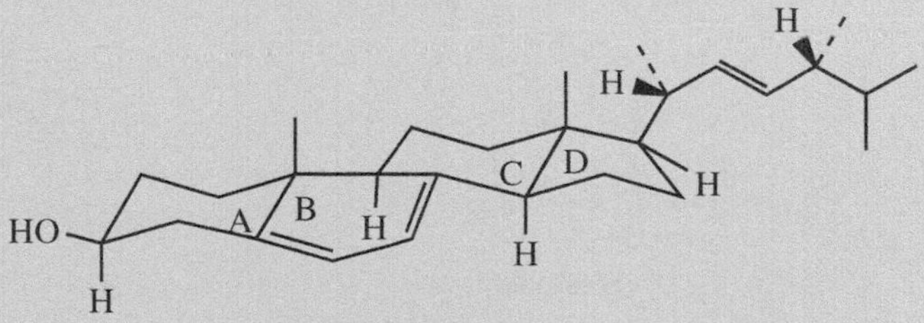

$\Delta^{5,7}$-unsaturation

In oestrogens, such as **estradiol**, the A ring is aromatic. Consequently, this ring is planar and distorts ring B accordingly; again, it is difficult to draw this perfectly. The stereochemical outcome makes oestrogens seem rather more flattened than the original all-*trans* arrangement in cholestane.

Box 3.19 (continued)

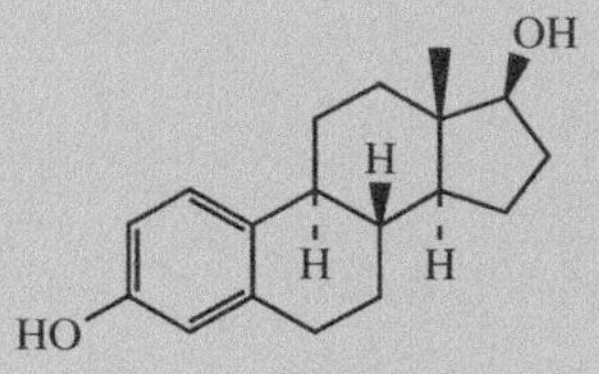

estradiol

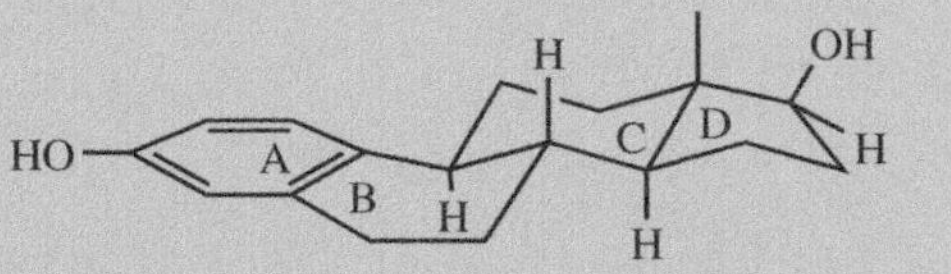

A ring aromatic

More dramatic changes are made to the shape of the steroid skeleton if ring fusions become *cis* rather than *trans*. The most important examples involve the A–B and C–D ring fusions. It is not difficult to work out how the modified skeleton looks after these changes. The approach is to start from the all-*trans* system and to delete the appropriate ring, though retaining the bonds to the unchanged part as a guide to putting in the new ring. This provides us with three of the bonds in the new ring, and it is just necessary to fill in the rest, using earlier decalin or hydrindane templates.

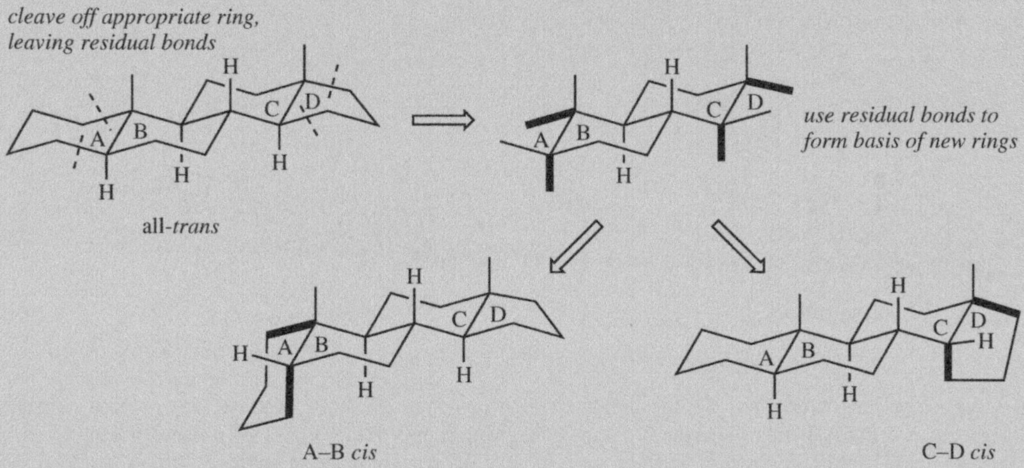

cleave off appropriate ring, leaving residual bonds

all-*trans*

use residual bonds to form basis of new rings

A–B *cis*

C–D *cis*

The approach is used to show the shape of **cholic acid**, one of the bile acids secreted into the gut to emulsify fats and encourage digestion. Cholic acid is characterized by a *cis* fusion of rings A and B.

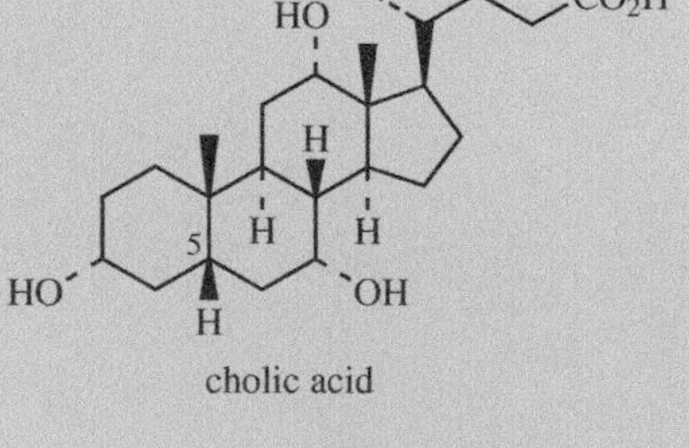

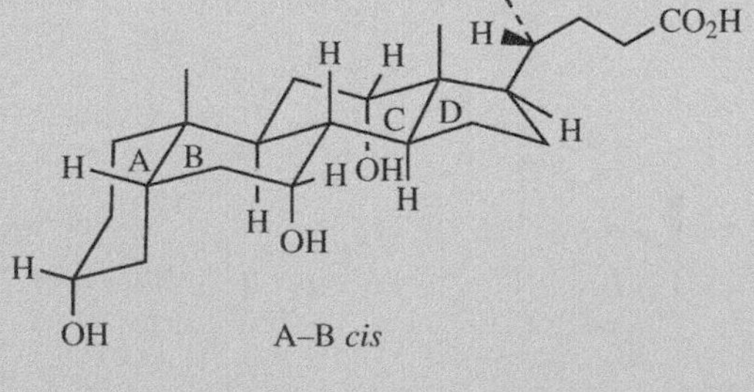

cholic acid

A–B *cis*

Digitoxigenin has *cis* fusions for both A–B and C–D rings. Glycosides of digitoxigenin are the powerful heart drugs found in the foxglove, *Digitalis purpurea*. Note how a *cis* ring fusion changes the more-or-less flat molecule of cholestane into a molecule with a significant 'bend' in its shape; digitoxigenin has two such 'bends'. These features are important in the binding of steroids to their receptors, and partially explain why we observe quite different biological activities from compounds containing a common structural skeleton.

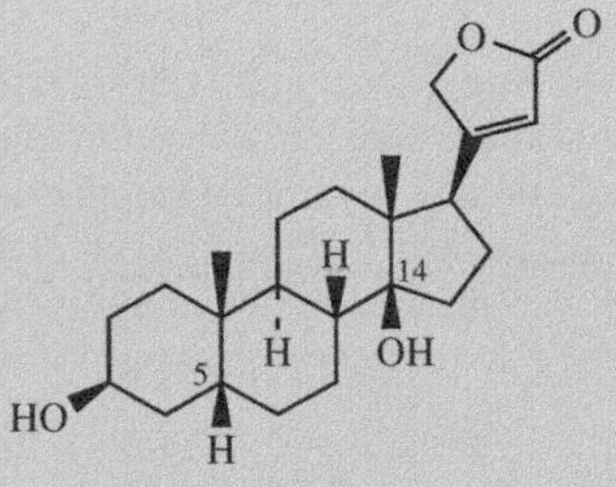

digitoxigenin

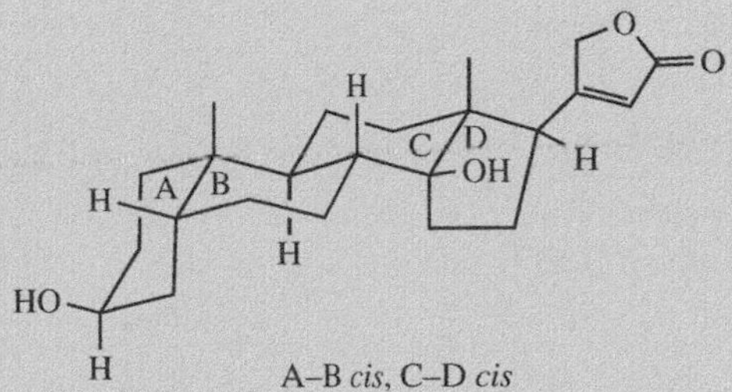

A–B *cis*, C–D *cis*

Most natural steroids have the stereochemical features seen in cholesterol, though, as we have seen, there may be some variations, particularly with respect to ring fusions affecting the A and D rings. Note that *trans* fusion at the hydrindane C–D ring junction is energetically less favourable than a *cis* fusion (see Section 3.5.2), but most natural steroid systems actually have this *trans* fusion.

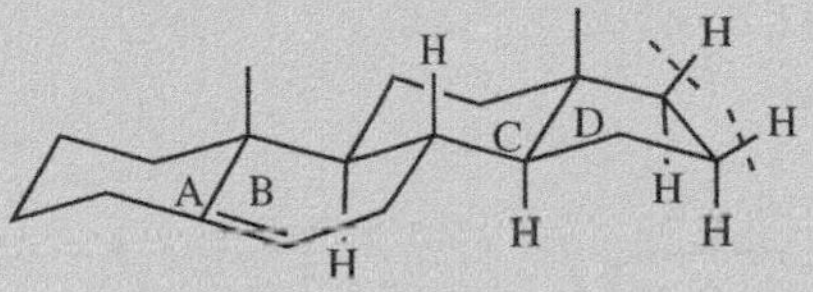

Δ⁵-unsaturation

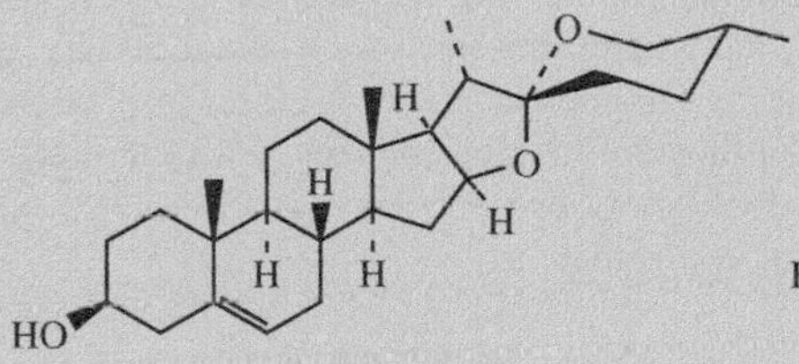

diosgenin

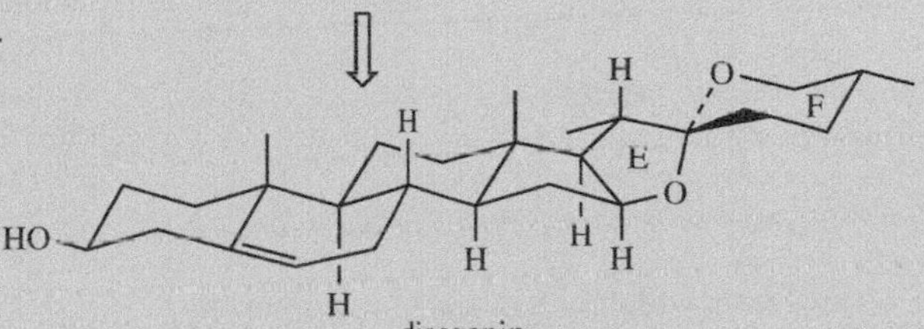

diosgenin

We have met **diosgenin** as an example of a natural *spiro* compound (see Box 3.17), and further examination of the structure shows the Δ⁵ double bond as in cholesterol, a second five-membered ring *cis*-fused onto the five-membered ring D, as well as the *spiro* fusion of a six-membered ring. Before this structure dismays you, take it slowly and logically. It should not be too difficult to end up with the stereodrawing shown here.

Box 3.20

The shape of penicillins

Penicillins are the most widely used of the clinical antibiotics. They contain in their structures an unusual fused ring system in which a four-membered β-lactam ring is fused onto a five-membered thiazolidine. Both rings are heterocyclic, and one of the ring fusion atoms is nitrogen. These heteroatoms do not alter our understanding of molecular shape, since we can consider that they also have an essentially tetrahedral array of bonds or lone pair electrons (see Section 2.6.3).

We have seen that, in cyclobutane and cyclopentane, a lower energy conformation is attained if the rings are not planar (see Section 3.3.2). If one fuses a five-membered ring onto a four-membered ring, models demonstrate that it is only possible to have a *cis* fusion in such a structure, and that conformational freedom in the four-membered ring disappears if we are to achieve this bonding; the four-membered ring reverts to a more planar shape. It is still possible to have the five-membered ring non-planar, thereby reducing eclipsed interactions.

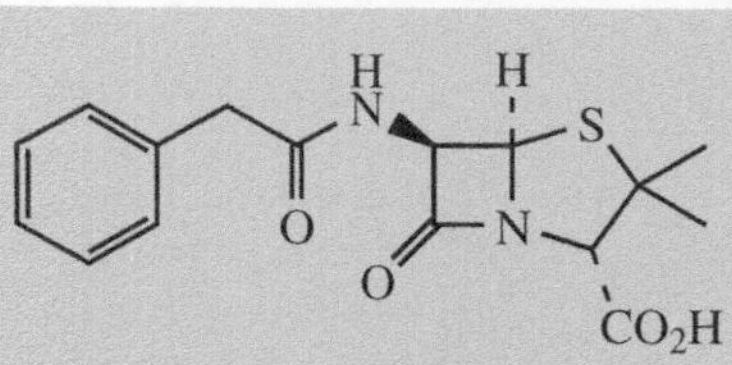

benzylpenicillin
(penicillin G)

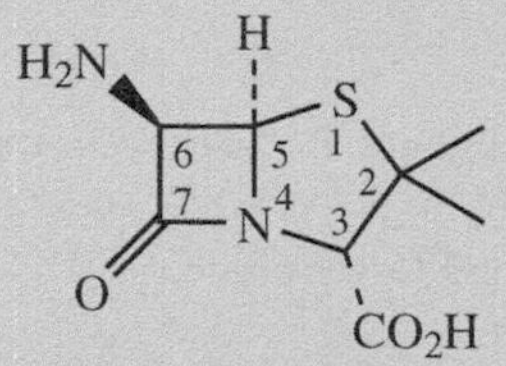

6-aminopenicillanic acid

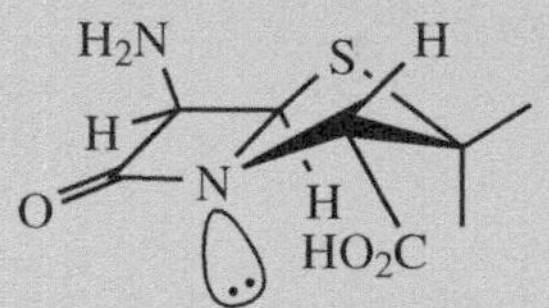

can only be cis-fused
four-membered ring is planar
five-membered ring is non-planar
ring fusion forces N into one configuration

The *cis* fusion in which one of the fusion atoms is nitrogen merely indicates that the nitrogen lone pair electrons occupy the remaining part of the tetrahedral array. It does, however, mean that inversion at the nitrogen atom (see Section 3.4.1) is not possible, since that would hypothetically result in formation of the impossible *trans*-fused system. The ring fusion has thus frozen the nitrogen atom into one configuration.

Fusion of a four-membered ring onto a six-membered ring is also only possible with a *cis* fusion; **cephalosporins** provide excellent examples of such compounds, and the comments made above for penicillins are equally valid for these compounds.

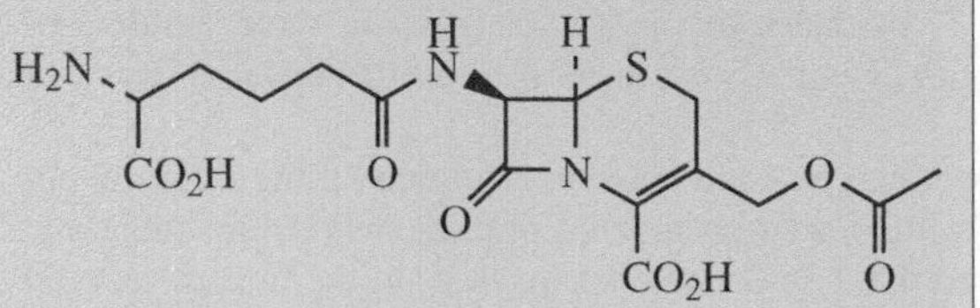

cephalosporin C

3.5.3   *Bridged ring systems*

In bridged ring compounds, rings share more than two atoms, and the bridge can consist of one or more atoms. We have already met an example in **bornane** (see Section 3.3.2), which we used as an illustration of how a cyclohexane ring can be forced into a boat conformation to achieve the necessary bonding.

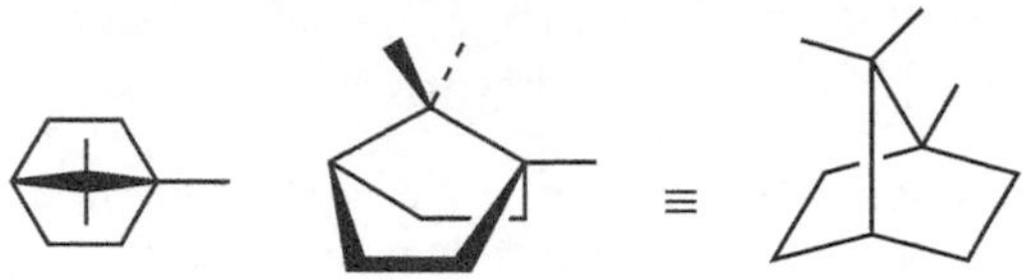

bornane   *cyclohexane ring*
                 forced into boat
                 conformation

If we inspect the ring system of bornane, omitting the methyl groups, we can see that there are actually several bridges of different lengths spanning the bridgehead atoms, depending upon which atoms are considered. This is used in nomenclature, as illustrated below, including in square brackets all the bridges, listed in decreasing lengths. Numbering, when necessary, always starts from a bridgehead atom. A closer inspection of the shape of bicyclo[2,2,2]octane (best with a model), which has two-carbon bridges, shows that each ring system has the boat conformation.

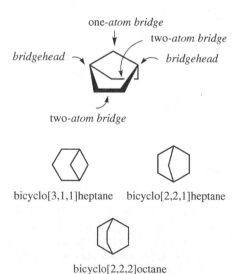

one-*atom bridge*

two-*atom bridge*

bridgehead   *bridgehead*

two-*atom bridge*

bicyclo[3,1,1]heptane   bicyclo[2,2,1]heptane

bicyclo[2,2,2]octane

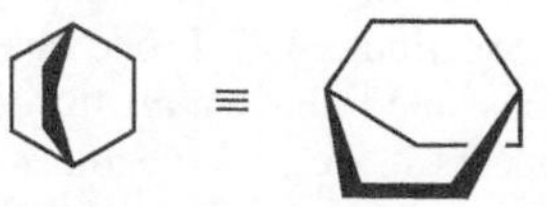

bicyclo[2,2,2]octane   ≡   *all rings boat*

Note that the ring systems with small bridges illustrated here can have no conformational mobility, and are quite fixed. Bornane also has no configurational isomers. If we are going to bridge a cyclohexane ring with a one-carbon bridge, there is only one way to achieve this; in other words, the configuration at the second bridgehead is fixed by that chosen at the first. A similar situation confronted us with fused rings, in that, in order to achieve the fusion of a small ring, only a *cis* fusion was feasible (see Section 3.5.2). Furthermore, bornane has a plane of symmetry and can be superimposed on its mirror image, so only one configurational isomer can exist.

mirror image

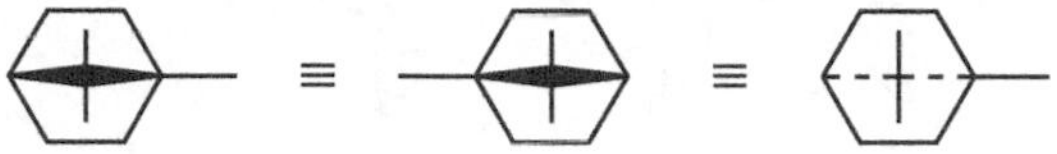

bornane

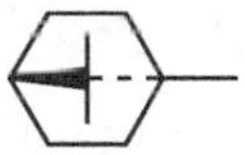

this type of bridging
is stereochemically
impossible

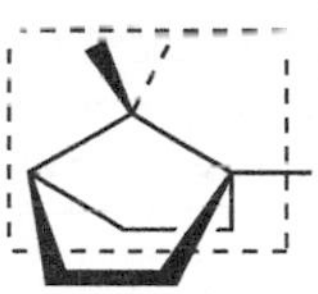

We should compare this system with a 1,4-disubstituted cyclohexane such as 4-methylcyclohexanecarboxylic acid (see Section 3.4.4). There is a plane of symmetry in this molecule, so there are no chiral centres; but geometric isomers exist, allowing *cis* and *trans* stereoisomers. The restrictions imposed by bridging have now destroyed any possibility of geometric isomerism.

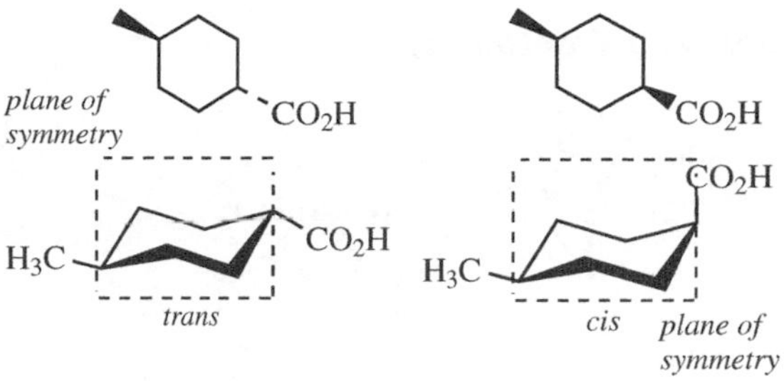

When we move on to **camphor**, a ketone derivative of bornane, we find this can exist in two enantiomeric forms because the plane of symmetry has been destroyed. Nevertheless, there are only two configurational isomers despite the presence of two chiral centres; bridging does not allow the other two variants to exist.

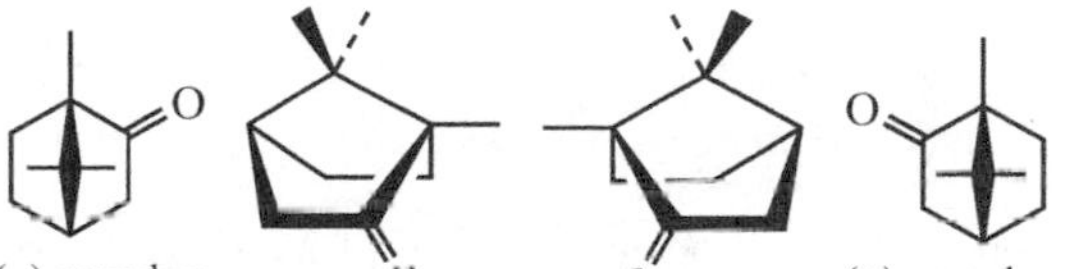

(−)-camphor   (+)-camphor

*two chiral centres
only two stereoisomers*

β-Pinene is representative of a bicyclo[3,1,1]heptane system. This natural product has two chiral centres, but can exist only in the (+)- and (−)-enantiomeric forms shown.

(−)-β-pinene   (+)-β-pinene

*two chiral centres
only two stereoisomers*

Box 3.21

Stereochemistry of tropane alkaloids

The tropane alkaloids (−)-**hyoscyamine** and (−)-**hyoscine** are found in the toxic plants deadly nightshade (*Atropa belladonna*) and thornapple (*Datura stramonium*) and are widely used in medicine. Hyoscyamine, usually in the form of its racemate atropine, is used to dilate the pupil of the eye, and hyoscine is employed to control motion sickness. Both alkaloids are esters of (−)-tropic acid.

The alcohol portion in hyoscyamine is **tropine**; in hyoscine it is the epoxide **scopine**. Tropine is an example of an azabicyclo[3,2,1]octane system with a nitrogen bridge, whereas scopine is a tricyclic system with a three-membered epoxide ring fused onto tropine. Note that systematic nomenclature considers an all-carbon ring system with one carbon replaced by nitrogen; hence, tropane is an azabicyclooctane (see Section 1.4).

There are several interesting stereochemical features accommodated within these structures. First, both tropine and scopine are optically inactive *meso* compounds; despite the chiral centres, two for tropine and four for

Box 3.21 (continued)

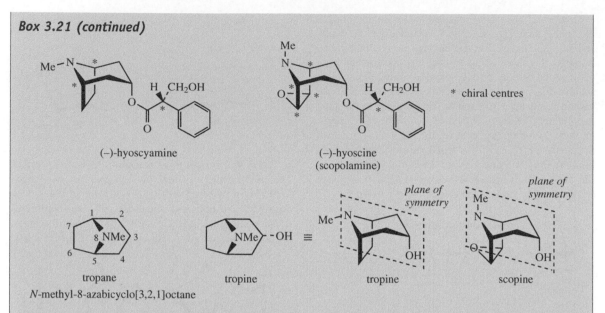

scopine, both compounds have a plane of symmetry, so that optical activity conferred by one centre is cancelled out by its mirror image centre. The optical activities of hyoscyamine and hyoscine are derived entirely from the chiral centre in the tropic acid portion. Atropine, the racemic form of hyoscyamine, is the ester of tropine with (±)-tropic acid (see Box 10.9).

nitrogen inversion can occur: the methyl group is preferentially equatorial in tropine but axial in scopine (minimizes interaction with epoxide)

Note also that, although we normally see rapid inversion at a nitrogen atom, the *N*-methyl group in hyoscyamine is preferentially in the lower energy equatorial position of the chair-like piperidine ring, as would be predicted. However, in hyoscine, the *N*-methyl group has been found to be axial, not the expected equatorial. This seems to arise to minimize interaction with the extra epoxide ring in scopine.

When we look at another tropane alkaloid, **cocaine**, we get a different scenario. Cocaine is obtained from the coca plant *Erythroxylum coca*, and is a powerful local anaesthetic, but now known primarily as a drug of abuse. There is no chiral centre in the acid portion, which is benzoic acid, but the optical activity of cocaine comes from the alcohol **methylecgonine**. Because of the ester function in methylecgonine, the tropane system is no longer symmetrical, and the four chiral centres all contribute towards optical activity.

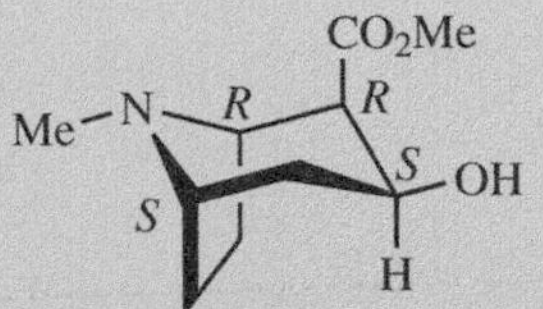

(−)-methylecgonine 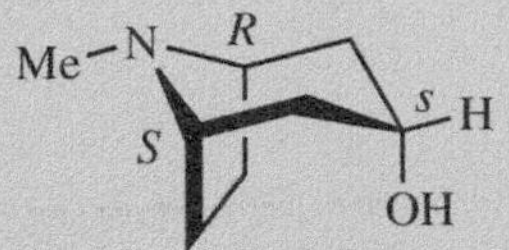   tropine 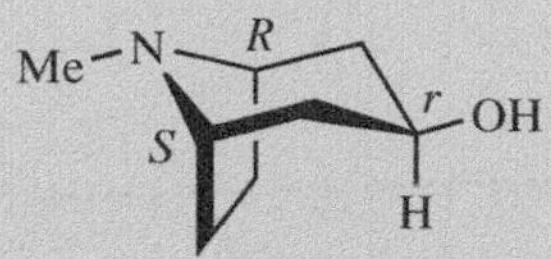   pseudotropine

Now, you may have noticed that the hydroxyl group in methylecgonine is oriented differently from that in tropine. In methylecgonine it is easy to define the position of the hydroxyl, since this is a chiral centre and we can use the *R/S* nomenclature. An alternative stereoisomer of tropine exists, and this is called **pseudotropine**. How can we define the configuration for the hydroxyl when the plane of symmetry of the molecule goes through this centre and means this centre is not chiral but can exist in two different arrangements?

This is a situation allowed for in the IUPAC nomenclature rules, because if we are faced with two groups which are the same but have opposite chiralities, then the group with *R* chirality has a higher priority than the group with *S* chirality. Applying this rule, tropine would have the *S* configuration and pseudotropine the *R* configuration at this centre. Because of the plane of symmetry, these atoms are not strictly chiral, and this is taken into account by using lower-case letters; tropine is *s* and pseudotropine is *r*.

4

Acids and bases

4.1 Acid–base equilibria

A particularly important concept in chemistry is that associated with proton loss and gain, i.e. **acidity** and **basicity. Acids** produce positively charged hydrogen ions H^+ (protons) in aqueous solution; the more acidic a compound is, the greater the concentration of protons it produces. In water, protons do not have an independent existence, but become strongly attached to a water molecule to give the stable **hydronium ion** H_3O^+. In the **Brønsted–Lowry definition**:

- an acid is a substance that will donate a proton;

- a base is a substance that will accept a proton.

Thus, in water, the acid HCl ionizes to produce H_3O^+ and Cl^- ions.

H_2O : $H{-}Cl$ ⇌ $H_3O^{\oplus}$: $Cl^{\ominus}$

base	acid	conjugate acid	conjugate
(proton acceptor)	(proton donor)	of H_2O	base of HCl

H_3O^+ is termed the **conjugate acid** (of the base H_2O) and Cl^- is termed the **conjugate base** (of the acid HCl). In general terms, cleavage of the H–A bond in an acid HA is brought about by a base, generating the conjugate acid of the base, together with the conjugate base of the acid. You may wish to read that sentence again!

$B:$ $H{-}A$ ⇌ $\overset{\oplus}{B}{-}H$: $A^{\ominus}$

base	acid	conjugate acid	conjugate base

The **Lewis definition of acids and bases** is rather more general than the Brønsted–Lowry version (which refers to systems involving proton transfer) in that:

- an acid is an electron-pair acceptor;

- a base is an electron-pair donor.

Thus, Lewis acids include such species as boron trifluoride, which is able to react with trimethylamine to form a salt.

$Me_3N:$ BF_3 ⇌ $\overset{\oplus}{Me_3N}{-}\overset{\ominus}{BF_3}$

Lewis base	Lewis acid	

There is no fundamental difference between trimethylamine acting as a Brønsted base or as a Lewis base, except that in the Brønsted concept it donates its electrons to a proton electrophile, whereas as a Lewis base it donates its electrons to a Lewis acid electrophile.

$R_3N:$ $H^{\oplus}$ ⇌ $\overset{\oplus}{R_3N}{-}H$

Brønsted base	conjugate acid

$R_3N:$ $E^{\oplus}$ ⇌ $\overset{\oplus}{R_3N}{-}E$

Lewis base	

Essentials of Organic Chemistry   Paul M Dewick
© 2006 John Wiley & Sons, Ltd

4.2 Acidity and pK_a values

For the ionization of the acid HA in water

$$H_2O \ + \ HA \ \underset{}{\overset{K}{\rightleftharpoons}} \ H_3O^{\oplus} \ + \ A^{\ominus}$$

the equilibrium constant K is given by the formula

$$K = \frac{[A^-][H_3O^+]}{[HA][H_2O]}$$

where [HA] signifies the concentration of HA, etc.

However, because the concentration of water is essentially constant in aqueous solution, a new equilibrium constant K_a is defined as

$$K_a = \frac{[A^-][H_3O^+]}{[HA]}$$

K_a is termed the **acidity constant**, and its magnitude allows us to classify acids as **strong acids** (a large value for K_a and, consequently, a high H_3O^+ concentration) or **weak acids** (a small value for K_a and, thus, a low H_3O^+ concentration). For example, the strong acid HCl has $K_a = 10^7$. However, for weak acids, the amount of ionization is much less and, consequently, the value of K_a is rather small. Thus, acetic acid CH_3CO_2H has $K_a = 1.76 \times 10^{-5}$. To avoid using such small numbers as these, K_a is usually expressed in the logarithmic form **pK_a** where

$$pK_a = -\log_{10} K_a$$

Accordingly, the pK_a for acetic acid is 4.75:

$$pK_a = -\log(1.76 \times 10^{-5}) = -(-4.75) = 4.75$$

The pK_a for hydrochloric acid can similarly be calculated to be −7:

$$pK_a = -\log(10^7) = -7$$

This means there is an inverse relationship between the strength of an acid and pK_a:

- a strong acid has a large K_a and, thus, a small pK_a, i.e. A^- is favoured over HA;
- a weak acid has a small K_a and, thus, a large pK_a, i.e. HA is favoured over A^-.

Or, put another way:

- **the smaller the value of pK_a, the stronger is the acid;**
- **the larger the value of pK_a, the weaker is the acid.**

We find that pK_a values range from about −12 to 52, but it must be appreciated right from the start that a difference of one pK_a unit actually represents a 10-fold difference in K_a and, thus, a 10-fold difference in H_3O^+ concentration. A twofold difference in acidity would be indicated by a pK_a difference of just 0.3 units (log 2 = 0.3). Accordingly, a difference of n pK_a units indicates a 10^n-fold difference in acidity, so that the range −12 to 52 actually represents a huge factor of 10^{64}. A compound with p$K_a < 5$ is regarded as a reasonably strong acid, and those with p$K_a < 0$ are very strong acids. At first glance, negative pK_a values seem rather strange, but this only means that the equilibrium lies heavily towards ionization; K_a is large and, therefore, p$K_a = -\log K_a$ becomes negative.

$$H_2O \ + \ HA \ \underset{}{\overset{}{\rightleftharpoons}} \ H_3O^{\oplus} \ + \ A^{\ominus}$$

$$K_a = \frac{[A^-][H_3O^+]}{[HA]} \qquad pK_a = -\log_{10} K_a$$

$K_a = 0.01$	$K_a = 0.1$	$K_a = 1$	$K_a = 10$	$K_a = 100$

$$\Longrightarrow$$

increasing acid strength

$pK_a = 2$	$pK_a = 1$	$pK_a = 0$	$pK_a = -1$	$pK_a = -2$

As we use pK_a values, we shall find that, in most cases, relative, rather than specific, values are all we need to consider to help us predict chemical behaviour and reactivity. Thus, from pK_a values, we can see that acetic acid (pK_a 4.75) is a weaker acid than hydrochloric acid (pK_a − 7).

pK_a values for a wide variety of different compounds are given in Tables 4.1–4.6. Compounds are listed in order of increasing acidity. Although pK_a values included extend from about 52 to −10, values in the middle of the range are known most accurately. This is because they can be measured readily in aqueous solution. Outside of the range from about 2 to 12, pK_a values have to be determined in other solvents,

Table 4.1 pK_a values of H–X acids

Acid	Conjugate base	pK_a
CH_4	$CH_3^{\ominus}$	48
NH_3	$NH_2^{\ominus}$	38
H_2	$H^{\ominus}$	35
H_2O	$HO^{\ominus}$	15.7
H–C≡N	$^{\ominus}$C≡N	9.1
H_2S	$HS^{\ominus}$	7
HF	$F^{\ominus}$	3.2
H_3PO_4	$H_2PO_4^{\ominus}$	2.1
HNO_3	$NO_3^{\ominus}$	−1.4
H_2SO_4	$HSO_4^{\ominus}$	−3.0
HCl	$Cl^{\ominus}$	−7
HBr	$Br^{\ominus}$	−9
HI	$I^{\ominus}$	−10

Table 4.2 pK_a values of C–H acids

Acid	Conjugate base	pK_a
(cyclohexane CH₂)	(cyclohexyl anion)	52
$H_3C–CH_3$	$H_3C–CH_2^{\ominus}$	50
CH_4	$CH_3^{\ominus}$	48
$H_2{=}CH_2$	$H_2{=}CH^{\ominus}$	44
(benzene–H)	(phenyl anion)	44
$H_2C{=}CH–CH_3$	$H_2C{=}CH–CH_2^{\ominus}$	43
(toluene CH₃)	(benzyl anion $CH_2^{\ominus}$)	41
$Ph_3C–H$	$Ph_3C^{\ominus}$	32
$H_3C–C{\equiv}N$	$^{\ominus}H_2C–C{\equiv}N$	25
HC≡C–H	HC≡$C^{\ominus}$	25

Table 4.3 pK_a values of N–H, O–H, and S–H acids

Acid	Conjugate base	pK_a
NH_3	$NH_2^{\ominus}$	38
$(CH_3)_2CH$ N–H $(CH_3)_2CH$	$(CH_3)_2CH$ $N^{\ominus}$ $(CH_3)_2CH$	36
$CH_3CH_2NH_2$	$CH_3CH_2NH^{\ominus}$	35
$Ph–NH_2$	$Ph–NH^{\ominus}$	28
CH_3CONH_2	$CH_3CONH^{\ominus}$	15
$(CH_3)_3C–OH$	$(CH_3)_3C–O^{\ominus}$	19
CH_3OH	$CH_3O^{\ominus}$	15.5
CH_3CH_2OH	$CH_3CH_2O^{\ominus}$	16
H_2O	$HO^{\ominus}$	15.7
$Ph–OH$	$Ph–O^{\ominus}$	10
CH_3SH	$CH_3S^{\ominus}$	10.5
H_2S	$HS^{\ominus}$	7
$Ph–SH$	$Ph–S^{\ominus}$	6.5

Table 4.4 pK_a values of CO_2H and SO_3H acids

Acid	Conjugate base	pK_a
CH_3CO_2H	$CH_3CO_2^{\ominus}$	4.8
$Ph–CO_2H$	$Ph–CO_2^{\ominus}$	4.2
HCO_2H	$HCO_2^{\ominus}$	3.7
$ClCH_2CO_2H$	$ClCH_2CO_2^{\ominus}$	2.9
Cl_2CHCO_2H	$Cl_2CHCO_2^{\ominus}$	1.3
Cl_3CCO_2H	$Cl_3CCO_2^{\ominus}$	0.7
F_3CCO_2H	$F_3CCO_2^{\ominus}$	−0.3
$Me–SO_3H$	$Me–SO_3^{\ominus}$	−1.2
H_3C—(benzene)—SO_3H	H_3C—(benzene)—$SO_3^{\ominus}$	−1.3

or even by indirect methods; results are then extrapolated to give the value in water. The figures presented in Tables 4.1–4.6 have been intentionally rounded to stress that a high level of accuracy is usually inappropriate.

The range of pK_a values that can be measured in water is determined by the ionization of water itself, i.e. −1.74 (the pK_a of H_3O^+) to 15.74 (the pK_a of H_2O); see Box 4.1. Acids that are stronger than H_3O^+ simply protonate water, whereas bases that are stronger than HO^- remove protons from water.

However, the fact that they do have to be measured means that as you look in the literature for the pK_a of a particular compound you may find slightly different values can be presented. Do not let this confuse you. As mentioned above, relative, rather than specific, values are our main concern.

We have chosen to present the pK_a values as a series of tables, rather than in a single one. This should help you to locate a particular compound

Table 4.5 pK_a values of CH–CO, CH–CN, and CH–NO$_2$ acids

Acid	Conjugate base	pK_a
$CH_3CO_2CH_2$	$^{\ominus}CH_3CO_2CH_3$	24
CH_3COCH_3	$CH_3COCH_2^{\ominus}$	19
$CH_3CH{=}O$	$^{\ominus}CH_2CH{=}O$	17
$CH_3O_2C{-}CH_2{-}CO_2CH_3$	$CH_3O_2C{-}\overset{\ominus}{C}H{-}CO_2CH_3$	13
$CH_3COCH_2CO_2CH_3$	$CH_3CO\overset{\ominus}{C}HCO_2CH_3$	11
$CH_3COCH_2COCH_3$	$CH_3CO\overset{\ominus}{C}HCOCH_3$	9
$H_3C{-}C{\equiv}N$	$^{\ominus}H_2C{-}C{\equiv}N$	25
CH_3NO_2	$^{\ominus}CH_2NO_2$	10

Table 4.6 pK_a values of N$^+$, O$^+$, and S$^+$ acids

Acid	Conjugate base	pK_a
$(H_2N)_2\overset{\oplus}{C}{=}NH_2$	$(H_2N)_2C{=}NH$	13.6
$CH_3\overset{\oplus}{N}H_3$	CH_3NH_2	10.6
$(CH_3)_2\overset{\oplus}{N}H_2$	$(CH_3)_2NH$	10.7
$(CH_3)_3\overset{\oplus}{N}H$	$(CH_3)_3N$	9.8
$\overset{\oplus}{N}H_4$	NH_3	9.2
pyridinium ($\overset{\oplus}{N}$H)	pyridine (N)	5.2
$Ph{-}\overset{\oplus}{N}H(CH_3)_2$	$Ph{-}N(CH_3)_2$	5.1
$Ph{-}\overset{\oplus}{N}H_3$	$Ph{-}NH_2$	4.6
$Ph_2\overset{\oplus}{N}H_2$	Ph_2NH	0.8
$H_3C{-}C{\equiv}\overset{\oplus}{N}H$	$H_3C{-}C{\equiv}N$	−10
$H_3C{-}\underset{NH_2}{\overset{\overset{\oplus}{O}H}{C}}$	CH_3CONH_2	−1.4

(continues)

Table 4.6   *(continued)*

Acid	Conjugate base	pK_a
$H_3\overset{\oplus}{O}$	H_2O	−1.7
$Ph{-}\underset{NH_2}{\overset{\overset{\oplus}{O}H}{C}}$	$PhCONH_2$	−2.2
$CH_3\overset{\oplus}{O}H_2$	CH_3OH	−2.2
$CH_3CH_2\overset{\oplus}{O}H_2$	CH_3CH_2OH	−2.4
$(CH_3)_3C{-}\overset{\oplus}{O}H_2$	$(CH_3)_3C{-}OH$	−3.8
$(H_3C)_2\overset{\oplus}{O}{-}H$	$(H_3C)_2O$	−3.8
$H_3C{-}\underset{OH}{\overset{\overset{\oplus}{O}H}{C}}$	$H_3C{-}\underset{OH}{\overset{O}{C}}$	−6.1
$H_3C{-}\underset{OCH_3}{\overset{\overset{\oplus}{O}H}{C}}$	$H_3C{-}\underset{OCH_3}{\overset{O}{C}}$	−6.5
$Ph{-}\overset{\oplus}{O}H_2$	$Ph{-}OH$	−6.7
$PhCH{=}\overset{\oplus}{O}H$	$PhCH{=}O$	−7
$(CH_3)_2C{=}\overset{\oplus}{O}H$	$(CH_3)_2C{=}O$	−7.2
$CH_3CH{=}\overset{\oplus}{O}H$	$CH_3CH{=}O$	−8
$(H_3C)_2\overset{\oplus}{S}{-}H$	$(H_3C)_2S$	−5.4
$CH_3\overset{\oplus}{S}H_2$	CH_3SH	−6.8

according to its functional group, and we hope that this will also emphasize similarities and differences in related structures. It also means that you may find some examples turning up in more than one table.

As we consider different aspects of chemical reactivity in subsequent chapters, we shall see how pK_a

values can be used to predict whether a reagent is a good or a poor nucleophile, whether it can function as a good leaving group, and how easy it is to generate anionic nucleophiles. We shall also find that pK_a values can tell us how much of a compound or a drug is ionized under particular conditions and, therefore, whether or not it can be produced in a soluble form.

It is now appropriate to consider some of the electronic and structural features that influence pK_a so that we can rationalize and predict relative acidities.

4.3 Electronic and structural features that influence acidity

4.3.1 Electronegativity

The more electronegative an element is, the more it helps to stabilize the negative charge of the conjugate base. For example, the acidities of compounds of second-row elements in the periodic table increase as the atom to which hydrogen is attached becomes more electronegative:

- pK_a values for CH_4, NH_3, H_2O and HF are about 48, 38, 16 and 3, respectively, i.e. we have increasing acidity left to right as the electronegativity of the atom attached to hydrogen increases.

4.3.2 Bond energies

Within a single column of the periodic table, acidities increase as one descends the column: pK_a values for HF, HCl, HBr, and HI are about 3, −7, −9, and −10 respectively, i.e. we have increasing acidity on descending the group.

This is the reverse of what might be expected simply based on electronegativities, but relates to the increasing size of the atom and the corresponding improved ability to disperse the negative charge over the atom. We are seeing a weakening in bond strengths on descending the group.

Similarly, although sulfur is less electronegative than oxygen, thiols (RSH) are more acidic than alcohols (ROH). For example, pK_a values for methanethiol and methanol are 10.5 and 16 respectively.

4.3.3 Inductive effects

Electron-donating and electron-withdrawing groups influence acidity by respectively destabilizing or stabilizing the conjugate base. This **inductive effect**, a charge polarization transmitted through σ bonds (see Section 2.7), causes a shift in electron density, and its influence may easily be predicted.

$$X-A-H \rightleftharpoons X-A^{\ominus} \quad H^{\oplus}$$

$$\overset{\longleftarrow}{X-A^{\ominus}} \qquad \overset{\longrightarrow}{X-A^{\ominus}}$$

electron-withdrawing
inductive effect
stabilizing

electron-donating
inductive effect
destabilizing

Thus, **electron-withdrawing groups** increase acidity:

- pK_a values for the simple **carboxylic acid** acetic acid and its halogenated derivatives chloroacetic acid, dichloroacetic acid, and trichloroacetic acid are about 4.8, 2.9, 1.3, and 0.7 respectively, the inductive effects of the chlorine atoms spreading the charge of the conjugate base and thus stabilizing it.

acetic acid
pK_a 4.8

chloroacetic acid
pK_a 2.9

dichloroacetic acid
pK_a 1.3

trichloroacetic acid
pK_a 0.7

acidity increases as the number of electron-withdrawing substituents increases

- Increasing the number of halogen atoms increases this effect, with a consequent increase in acidity. Note that introduction of one chlorine atom increases acidity by a factor of almost 100, and trichloroacetic acid is a strong acid.

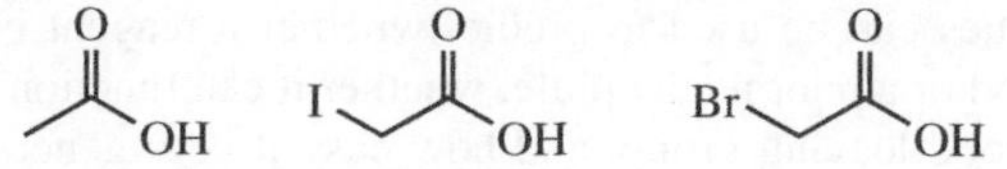

- Because of the different electronegativities of the various halogens, we can also predict that fluorine will have a greater effect than chlorine, which in turn will increase acidity more than bromine or iodine. This is reflected in the observed acidities of monohalogenated acetic acids, though the increased acidity of chloroacetic acid (pK_a 2.87) over bromoacetic acid (pK_a 2.90) is not apparent because of the rounding-up process.

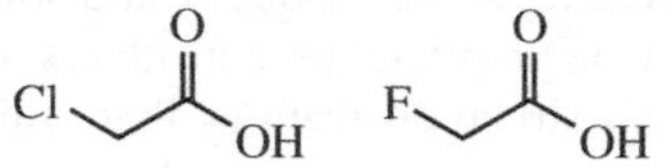

- The inductive effect is a rather short-range effect, and its influence decreases rapidly as the

acetic acid   iodoacetic acid   bromoacetic acid

pK_a 4.8 pK_a 3.2 pK_a 2.9

chloroacetic acid   fluoroacetic acid

pK_a 2.9 pK_a 2.6

*acidity increases as
substituent becomes
more electronegative*

butanoic acid   2-chlorobutanoic acid   3-chlorobutanoic acid   4-chlorobutanoic acid

pK_a 4.9 pK_a 2.9 pK_a 4.1 pK_a 4.5

*effect of electronegative
substituent decreases as
it is located further away
from acidic group*

Table 4.7 Inductive effects from functional groups

Electron-withdrawing groups			Electron-donating groups
—F	—CO$_2$H	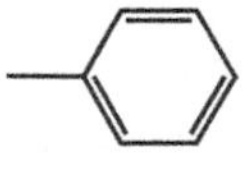	—O$^{\ominus}$
—Cl	—CO$_2$R		—CH$_3$
—Br	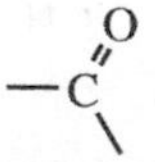	 $\overset{\oplus}{-N-}$	—CO$_2{}^{\ominus}$
—I	$-C\overset{O}{\underset{\backslash}{\parallel}}$		
—OR		$\overset{\oplus}{-S-}$	
—OH	—C≡N		
$-N\overset{/}{\underset{\backslash}{}}$	—NO$_2$		
—SR	—SO$_2$—		
—SH			

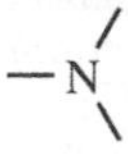

substituent in question is located further away from the site of negative charge, because it has to be transmitted through more bonds. Thus, the effect on the acidity in butanoic acid derivatives can be seen to diminish with distance. 2-Chlorobutanoic acid (pK_a 2.9) shows a significant enhancement in acidity over butanoic acid (pK_a 4.9), whereas 3-chlorobutanoic acid (pK_a 4.1) and 4-chlorobutanoic acid (pK_a 4.5) show rather more modest changes.

- Other electron-withdrawing groups that increase the acidity of acids include, listed in decreasing order of their effect: $-NO_2$, $-N^+R_3$, $-CN$, $-CO_2R$, $-CO-$, $-OR$ and $-OH$. A more extensive list is given in Table 4.7.

Electron-donating groups will have the opposite effect, destabilizing the conjugate base by increasing electron density, and thus produce weaker acids. The most common electron-donating groups encountered are going to be alkyl groups, though the effect from alkyl groups is actually rather small. Indeed, it is not immediately apparent why there should be any inductive effect at all, since substitution of hydrogen by alkyl should not lead to any bond polarization. At this point, we should merely note that alkyl groups have a weak electron-donating effect – it may not be strictly an inductive effect (see Section 6.2.1).

- The pK_a value for formic acid (pK_a 3.7) makes it more acidic than acetic acid (pK_a 4.8). The electron-donating effect of the methyl group is most marked on going from formic acid to acetic acid, since the acidity of propionic acid (pK_a 4.9) and butanoic acid (pK_a 4.8) vary little from that of acetic acid. The electron-donating effect from alkyl substituents is relatively small, being considerably smaller than inductive effects from most electron-withdrawing groups, and also rapidly diminishes along a carbon chain.

- **Alcohols** are much less acidic than carboxylic acids; but, as one progresses through the sequence methanol, ethanol, isopropanol, and *tert*-butanol, pK_a values gradually increase from 15.5 to 19, a substantial decrease in acidity. Although this was originally thought to be caused by the inductive effects of methyl groups, it is now known to be

formic acid
pK_a 3.7

acetic acid
pK_a 4.8

propionic acid
pK_a 4.9

butanoic acid
pK_a 4.8

electron-donating effect of alkyl groups is most marked on going from formic acid to acetic acid

primarily related to solvation effects. In solution, the conjugate base anion is surrounded with polar solvent molecules. This solvation helps to stabilize the conjugate base, and thus increases the acidity of the alcohol. As we get more alkyl groups, solvation of the anion is diminished because of the increased steric hindrance they cause, and observed acidity also decreases.

methanol
pK_a 15.5

ethanol
pK_a 16.0

isopropanol
pK_a 17

tert-butanol
pK_a 19

alkyl groups hinder approach of solvation molecules

4.3.4 Hybridization effects

The acidity of a C–H bond is influenced by the
hybridization state of the carbon atom attached to the
acidic hydrogen. Dissociation of the acid generates
an anion whose lone pair of electrons is held in a
hybridized orbital. We can consider sp orbitals to
have more s character than sp^2 orbitals, and similarly
sp^2 orbitals to have more s character than sp^3 orbitals
(see Section 2.6.2). Since s orbitals are closer to
the nucleus than p orbitals, it follows that electrons
in an sp-hybridized orbital are held closer to the
nucleus than those in an sp^2 orbital; those in an
sp^2 orbital are similarly closer to the nucleus than
those in an sp^3 orbital. It is more favourable for the
electrons to be held close to the positively charged
nucleus, and thus an sp-hybridized anion is more
stable than an sp^2-hydridized anion, which is more
stable than an sp^3-hybridized anion. Thus, the acidity
of a C–H bond decreases as the s character of the
bond decreases.

• The pK_a of the **hydrocarbon** ethane is about 50,
that of ethylene about 44, and that of acetylene
is about 25. The hybridization of the C–H bond
in ethane is sp^3 (25% s character), in ethylene it

is sp^2 (33% s character), and in acetylene it is sp
(50% s character). This makes alkynes (acetylenes)
relatively acidic for hydrocarbons. It is also a
contributing factor in the acidity of HCN (pK_a
9.1), where the conjugate base cyanide is an sp-
hybridized anion, though additional stabilization
comes from the electronegative nitrogen atom.

So far we have considered the hybridization state
of the orbital associated with the anionic charge.
However, the hybridization state elsewhere in the
molecule may also affect acidity. The more s character an orbital has, the closer the electrons are held
to the nucleus, and this effectively makes the atom
more electronegative. This may be explained in terms
of hybridization modifying inductive effects, such
that sp-hybridized carbons are effectively more electronegative than sp^2-hybridized carbons, and similarly, sp^2-hybridized carbons are more electronegative than sp^3-hybridized carbons.

• The pK_a values for the following acids illustrate
that, as the carbon atom adjacent to the **carboxylic
acid** group changes from sp^3 to sp^2 to sp
hybridization, the acidity increases, in accord with
the electronegativity explanation above. Note that
benzoic acid (sp^2 hybridization) has a similar pK_a
to acrylic acid (propenoic acid), which also has sp^2
hybridization.

ethane
pK_a 50

sp^3 orbital

ethylene
pK_a 44

sp^2 orbital

acetylene
pK_a 25

sp orbital

hydrocyanic acid
pK_a 9.1

sp orbital

propionic acid
pK_a 4.9

acrylic acid
(propenoic acid)
pK_a 4.2

benzoic acid
pK_a 4.2

propiolic acid
(propynoic acid)
pK_a 1.8

*inductive effect resulting
from hybridization*

4.3.5 Resonance/delocalization effects

Delocalization of charge in the conjugate base anion through **resonance** is a stabilizing factor and will be reflected by an increase in acidity. Drawing resonance structures allows us to rationalize that the negative charge is not permanently localized on a particular atom, but may be dispersed to other areas of the structure. We should appreciate that a better interpretation is that the electrons are contained in a molecular orbital that spans several atoms.

However, drawing resonance structures provides a simple and convenient way of predicting stability through delocalization (see Section 2.10).

The pK_a of ethanol is 16, and that of acetic acid is 4.8. The increased acidity of acetic acid relative to ethanol can be rationalized in terms of delocalization of charge in the acetate anion, whereas in ethoxide the charge is localized on oxygen. Even more delocalization is possible in the methanesulfonate anion, and this is reflected in the increased acidity of methanesulfonic acid ($pK_a - 1.2$).

ethanol
pK_a 16

ethoxide anion
localized charge

acetic acid
pK_a 4.8

acetate anion
delocalized charge

*delocalization of charge
is sometimes depicted
via partial bonds*

methanesulfonic acid
$pK_a - 1.2$

methanesulfonate anion
delocalized charge

*delocalization depicted
via partial bonds*

We have also shown some representations of acetate and methanesulfonate anions that have been devised to emphasize resonance delocalization; these include partial bonds rather than double/single bonds. Although these representations are valuable, they can lead to some confusion in interpretation. It is important to remember that there is a double bond in these systems. Therefore, we prefer to draw out the contributing resonance structures.

The alkane propane has pK_a 50, yet the presence of the double bond in propene means the methyl protons in this alkene have pK_a 43; this value is similar to that of ethylene (pK_a 44), where increased acidity was rationalized through sp^2 hybridization effects. 1,3-Pentadiene is yet more acidic, having pK_a 33 for the methyl protons. In each case, increased acidity in the unsaturated compounds may be ascribed to delocalization of charge in the conjugate base. Note that we use the term allyl for the propenyl group.

propane
pK_a 50

propene
pK_a 43

resonance stabilized
allyl anion

*delocalization depicted
via partial bonds*

1,3-pentadiene
pK_a 33

resonance stabilized
pentadienyl anion

*delocalization depicted
via partial bonds*

Resonance stabilization is also responsible for the increased acidity of a C–H group situated adjacent to a carbonyl group. The anion is stabilized through delocalization of charge, similar to that seen with the allyl anion derived from propene; but this system is even more favourable, in that delocalization allows the charge to be transferred to the electronegative oxygen atom. As a result, acetone (pK_a 19) is significantly more acidic than propene (pK_a 43). Anions of this type, termed **enolate anions**, are some of the most important reactive species used in organic chemistry (see Chapter 10).

acetone
pK_a 19

resonance-stabilized
enolate anion

*favoured – charge on
electronegative oxygen*

*delocalization depicted
via partial bonds*

The acidity of a C–H is further enhanced if it is adjacent to two carbonyl groups, as in the 1,3-diketone acetylacetone. The enolate anion is stabilized by delocalization, and both carbonyl oxygens can participate in the process. This is reflected in the pK_a 9 for the protons between the two carbonyls, whereas the terminal protons adjacent to just a single carbonyl have pK_a 19, similar to acetone above. It is clear that increased delocalization has a profound effect on the acidity. These two values should be compared with that of the hydrocarbon propane (pK_a 50).

pK_a 9 pK_a 19

acetylacetone

resonance-stabilized
enolate anion

*delocalization depicted
via partial bonds*

Aromatic rings are themselves excellent examples of resonance and delocalization of electrons (see Section 2.10). They also influence the acidity of appropriate substituent groups, as seen in **benzoic acids**. Benzoic acid (pK_a 4.2) is a stronger acid than acetic acid (pK_a 4.8), and it is also stronger than its saturated analogue cyclohexanecarboxylic acid (pK_a 4.9). The phenyl group exerts an electron-withdrawing effect because the hybridization of the ring carbons is sp^2; consequently, electrons are held closer to the carbon atom than in an sp^3-hybridized orbital. This polarizes the bond between the aromatic ring and the carboxyl. The pK_a of phenylacetic acid (pK_a 4.3), compared with acetic acid (pK_a 4.8), demonstrates the inductive effect of a benzene ring. However, we might then expect benzoic acid to be a rather stronger acid than it actually is, since the phenyl group is closer to the carboxyl group than in phenylacetic acid. We attribute the lower acid strength to an additional resonance effect in the carboxylic acid that is not favourable in the anion, where it would lead to a carboxylate carrying a double negative charge; therefore, the resonance effect weakens the acid strength.

CO$_2$H
|
CH$_3$

acetic acid
pK_a 4.8

cyclohexane-
carboxylic acid
pK_a 4.9

benzoic acid
pK_a 4.2

*inductive effect
resulting from
hybridization*

phenylacetic acid
pK_a 4.3

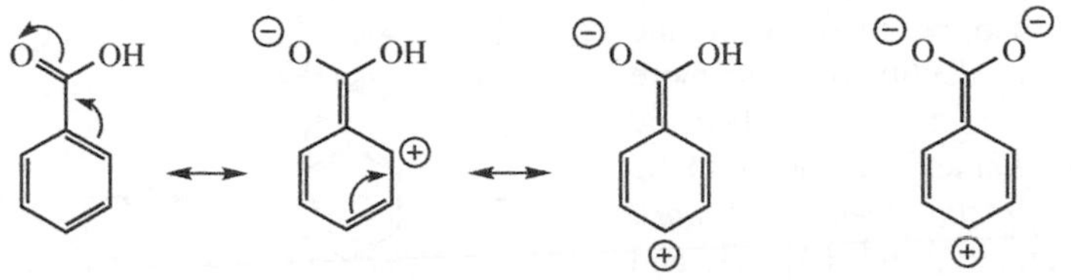

resonance favours non-ionized benzoic acid   *resonance unfavourable in anion*

Further inductive effects from other substituents enhance or counter these effects with predictable results. Thus, a halogen such as chlorine, with a strong inductive effect, produces stronger acids, especially in the case of the *ortho* derivative. Here, the extra inductive effect is correspondingly closer to the carboxyl group, and it will help to stabilize the conjugate base. The acid-weakening resonance effects are also diminished by the inductive effects of halogens; it is not favourable to have an electron-withdrawing substituent close to a positive charge.

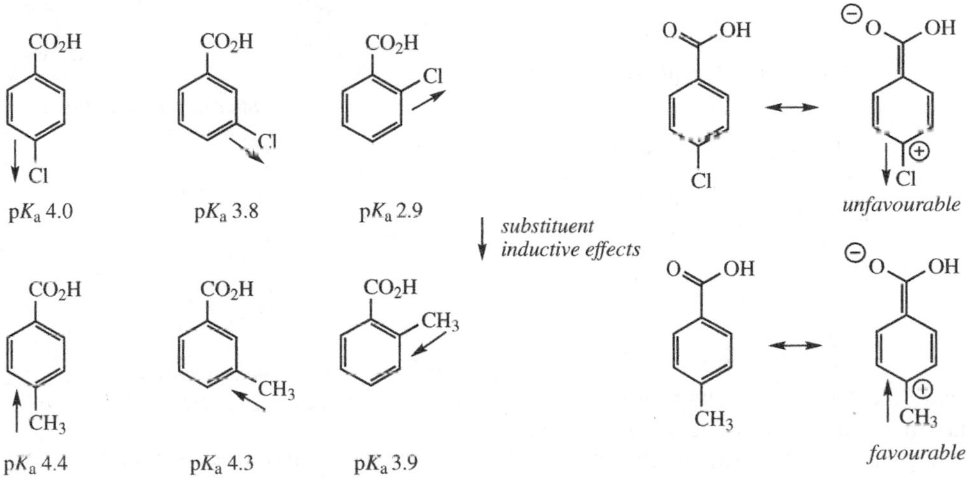

pK$_a$ 4.0   pK$_a$ 3.8   pK$_a$ 2.9

substituent inductive effects

unfavourable

pK$_a$ 4.4   pK$_a$ 4.3   pK$_a$ 3.9

favourable

On the other hand, methyl substituents have a weak electron-donating effect opposing that of the aromatic ring. This also favours resonance in the non-ionized acid. There is only a modest effect on acidity, except when the methyl is in the *ortho* position, where the effect is closer to the carboxyl group. However, *ortho* substituents add a further dimension that is predominantly steric. Large groups in the *ortho* position can have an influence on the carboxyl group, forcing it out of the plane of the ring. The result is that resonance is now inhibited because the orbitals of the carbonyl group are no longer coplanar with the benzene ring. In almost all cases, the *ortho*-substituted benzoic acid tends to be the strongest acid of the three isomers.

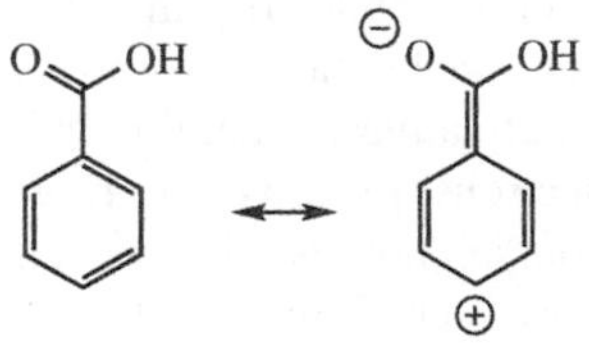

resonance requires coplanarity of carbonyl with benzene ring

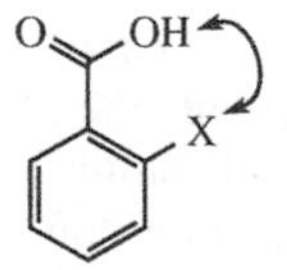

steric hindrance distorts carboxyl from coplanarity with benzene ring and inhibits resonance

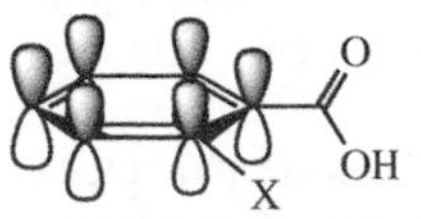

e.g. rotation of carboxyl group; carbonyl now at right angles to benzene ring and orbitals cannot overlap

When substituents can also be involved in the resonance effects, changes in acidity become more marked. Consider hydroxy- and methoxy-benzoic acid derivatives. The pK_a values are found to be 3.0, 4.1, and 4.6 for the *ortho, meta*, and *para* hydroxy derivatives respectively, and 4.1, 4.1, and 4.5 respectively for the corresponding methoxy derivatives.

(pK_a values for CO_2H group)

pK_a (CO_2H) 4.6   pK_a (CO_2H) 4.1   pK_a (CO_2H) 3.0

pK_a 4.5   pK_a 4.1   pK_a 4.1

Let us ignore the figure for *ortho*-hydroxybenzoic acid for the moment, since there is yet another feature affecting acidity. We then see that the *para* derivatives are rather less acidic than we might predict merely from the inductive effect of the OH or OMe groups. In fact, pK_a values show that these compounds are less acidic than benzoic acid, whereas the inductive effect would suggest they should be more acidic. This is because of a large resonance effect emanating from the substituent in which electronic charge is transmitted through the conjugated system of the aromatic ring into the carboxyl group.

The electron-donating effect originates from the lone pair electrons on oxygen, with overlap into the π electron system. This electron donation will stabilize the non-ionized acid via electron delocalization, but would destabilize the conjugate base by creating a double charge in the carboxylate system. The net result is lower acidity.

This electron-donating effect from lone pair electrons is simply a resonance effect, but is often termed a **mesomeric effect**. A mesomer is another term for a

*resonance stabilizes
the non-ionized acid*

*resonance destabilizes the
conjugate base*

*resonance delocalizes
electrons only to ring carbons*

resonance structure (see Section 2.10). We shall use **'resonance effect'** rather than 'mesomeric effect' to avoid having the alternative terminologies.

We can write a similar delocalization picture for the *ortho*-substituted compounds, but this is countered by the opposing inductive effect close to the carboxyl. However, the steric effect, as described above, means large groups in the *ortho* position can force the carboxyl group out of the plane of the ring. This weakens the resonance effect, since delocalization is dependent upon coplanarity in the conjugate system.

Resonance stabilization is not as important for the *meta* derivatives, where it is only possible to donate electrons towards the ring carbons, which are, of course, not as electronegative as oxygen. In fact, *meta* substitution is the least complicated, in that groups placed there exert their influence almost entirely through inductive effects. It should be noted that, where we have opposing resonance and inductive effects, the resonance effect is normally of much greater magnitude than the inductive effect, and its contribution predominates (but see below for chlorine).

The relatively high acidity of *ortho*-hydroxybenzoic acid (salicylic acid), compared with the other derivatives just considered, is ascribed to **intramolecular hydrogen bonding**, which is not possible in the other compounds, even with *ortho*-methoxybenzoic acid.

| favourable H-bonding stabilizes anion | H-bonding in non-ionized acid |

Hydrogen bonding involves a favourable six-membered ring and helps to stabilize the conjugate base. Although some hydrogen bonding occurs in the non-ionized acid, the effect is much stronger in the carboxylate anion.

It should be noted that the electron-donating resonance effects just considered are the result of lone pair electrons feeding in to the π electron system. Potentially, any substituent with a lone pair might do the same, yet we did not invoke such a mechanism with chlorine substituents above. As the size of the atom increases, lone pair electrons will be located in orbitals of higher level, e.g. $3p$ rather than $2p$ as in carbon. Consequently, the ability to overlap the lone pair orbital with the π electron system of the aromatic ring will diminish, a simple consequence of how far from the atom the electrons are mostly located. Chlorine thus produces a low resonance effect but a high inductive effect, and the latter predominates.

strong inductive effect > *weak resonance effect*

strong resonance effect > *weak inductive effect*

Resonance can also influence the acidity of hydroxyl groups, as seen in **phenols**. Cyclohexanol has pK_a 16, comparable to that of ethanol. On the other hand, phenol has pK_a 10, making it considerably more acidic than a simple alcohol, though less so than a carboxylic acid. This increased acidity is explained in terms of delocalization of the negative charge into the aromatic ring system, with resonance structures allowing ring carbons *ortho* and *para* to the original phenol group to become electron rich. Although the aromatic ring acts as an acceptor of electrons, and may be termed an electron sink, charge is dispersed towards carbon atoms, which is going to be less favourable than if it can be dispersed towards more electronegative atoms such as oxygen.

| cyclohexanol pK_a 16 | phenol pK_a 10 | phenoxide conjugate base | charge delocalized towards ortho and para carbons |

A good illustration of this concept is seen in a series of nitrophenols. The **nitro group** itself has to be drawn with charge separation to accommodate the electrons and our rules of bonding. However, resonance structures suggest that there is electron delocalization within the nitro group.

nitro group

With substituted phenols, there can be similar delocalization of charge into the aromatic ring as with phenol, but substituents will introduce their own effects, be it inductive or resonance related. It can be seen that the nitro group allows further delocalization of the negative charge of the phenoxide conjugate base if it is situated in the *ortho* or *para* positions. This increases acidity relative to phenol, and both compounds have essentially the same pK_a of 7.2.

phenol	*o*-nitrophenol	*m*-nitrophenol	*p*-nitrophenol
pK_a 10	pK_a 7.2	pK_a 8.4	pK_a 7.2

2,4-dinitrophenol
pK_a 4.1

2,4,6-trinitrophenol
(picric acid)
pK_a 0.4

*resonance effects
stabilize anions*

*inductive effect helps
to stabilize anion*

The effect is magnified considerably if there are nitro groups both *ortho* and *para*, so that the pK_a for 2,4-dinitrophenol is 4.1. A third nitro group, as in 2,4,6-trinitrophenol, confers even more acidity, and this compound has pK_a 0.4, making it a strong acid. This is reflected in its common name, picric acid.

Note that *m*-nitrophenol has pK_a 8.4, and is a lot less acidic than *o*-nitrophenol or *p*-nitrophenol. We can draw no additional resonance structures here, and the nitro group cannot participate in further electron delocalization. The increased acidity compared with phenol can be ascribed to stabilization of resonance structures with the charge on a ring carbon through the nitro group's inductive effect.

From the above, it should not be difficult to rationalize the effects of other types of substituent on the acidity of phenols. Thus electron-donating groups, e.g. alkyl, reduce acidity, and electron-withdrawing groups, e.g. halogens, increase acidity. With strongly electron-withdrawing groups, such as cyano and nitro, the acid-strengthening properties can be quite pronounced. A summary list of **resonance effects** emanating from various groups is shown in Table 4.8. We should also point out that these very same principles will be used to rationalize aromatic

Table 4.8 Resonance effects from functional groups

Electron-donating groups		Electron-withdrawing groups
—F	$-N\diagup\diagdown$	—C≡N
—Cl		
—Br	—SR	$-C\diagdown^{O}$
—I	—SH	—SO₂—
—O$^{\ominus}$	—CH₃	—NO₂
—OR		
—OH		
—OCOR		

substitution reactions in Chapter 8, and this is why we have purposely discussed the acidity of aromatic derivatives in some detail.

4.4 Basicity

We have already defined a base as a substance that will accept a proton by donating a pair of electrons. Just as we have used pK_a to measure the strength of an acid, we need a system to measure the strength of a base. Accordingly, a basicity scale based on pK_b was developed in a similar way to pK_a.

For the ionization of the base B in water

$$B \ + \ H_2O \ \underset{}{\overset{K}{\rightleftharpoons}} \ BH^{\oplus} \ + \ HO^{\ominus}$$

the equilibrium constant K is given by the formula

$$K = \frac{[HO^-][BH^+]}{[B][H_2O]}$$

and since the concentration of water will be essentially constant, the equilibrium constant K_b and the logarithmic pK_b may be defined as

$$K_b = \frac{[HO^-][BH^+]}{[B]}$$

with

$$pK_b = -\log_{10} K_b$$

This system has been almost completely dropped in favour of using pK_a throughout the **acidity–basicity scale**. To measure the strength of a base, we use the pK_a of its conjugate acid, i.e. we consider the equilibrium

$$\underset{\substack{\text{conjugate} \\ \text{acid}}}{BH^{\oplus}} \ + \ H_2O \ \underset{}{\overset{K}{\rightleftharpoons}} \ H_3O^{\oplus} \ + \ B$$

for which

$$K_a = \frac{[B][H_3O^+]}{[BH^+]}$$

It follows that

- a strong base has a small K_a and thus a large pK_a, i.e. BH^+ is favoured over B;

- a weak base has a large K_a and thus a small pK_a, i.e. B is favoured over BH^+.

Or, put another way:

- **the larger the value of pK_a, the stronger is the base;**

- **the smaller the value of pK_a, the weaker is the base.**

The relationship between pK_a and pK_b can be deduced as follows:

$$K_b = \frac{[HO^-][BH^+]}{[B]}$$

$$K_a = \frac{[B][H_3O^+]}{[BH^+]}$$

$$K_a \times K_b = \frac{[B][H_3O^+]}{[BH^+]} \times \frac{[HO^-][BH^+]}{[B]}$$

$$= [H_3O^+][HO^-]$$

Thus, $K_a \times K_b$ reduces to the ionization constant for water K_w.

$$\underset{\substack{\text{base} \\ \textit{accepts} \\ \textit{proton}}}{H_2O} \ + \ \underset{\substack{\text{acid} \\ \textit{donates} \\ \textit{proton}}}{H_2O} \ \underset{}{\overset{K}{\rightleftharpoons}} \ H_3O^{\oplus} \ + \ HO^{\ominus}$$

In this reaction, one molecule of water is acting as a base and accepts a proton from a second water molecule. This second water molecule, therefore, is acting as an acid and donates a proton. The equilibrium constant K for this reaction is given by the formula

$$K = \frac{[H_3O^+][HO^-]}{[H_2O][H_2O]}$$

and because the concentration of water is essentially constant in aqueous solution, the new equilibrium constant K_w is defined as

$$K_w = [HO^-][H_3O^+]$$

For every hydronium ion produced, a hydroxide anion must also be formed, so that the concentrations of hydronium and hydroxide ions must be equal. In pure water at 25 °C, this value is found to be 10^{-7} M.

$$K_w = [HO^-][H_3O^+] = 10^{-7} \times 10^{-7} = 10^{-14}$$

Box 4.1

pKa values for water

Acting as an acid: pK_a of H_2O Water is a very weak acid and can undergo self-ionization as follows:

$$H_2O \;+\; H_2O \;\rightleftharpoons\; H_3O^{\oplus} \;+\; HO^{\ominus}$$
$$\text{base} \qquad\qquad \text{acid}$$

Thus, one molecule of water is acting as an acid and donating a proton to a second water molecule, whilst the other acts as a base accepting a proton from the other water molecule. In pure water at 25 °C, the concentrations of hydronium ions and hydroxide ions are equal and found to be 10^{-7} M. The concentration of pure water is $1000/18 = 55.5$ M.

Therefore

$$K_a = \frac{[HO^-][H_3O^+]}{[H_2O]} = \frac{10^{-7} \times 10^{-7}}{55.5} = 1.8 \times 10^{-16}$$

Hence, $pK_a = 15.7$.

Acting as a base: pK_a of H_3O^+ Here, we need to consider the pK_a for ionization of the conjugate acid:

$$H_3O^{\oplus} \;+\; H_2O \;\rightleftharpoons\; H_2O \;+\; H_3O^{\oplus}$$
$$\text{conjugate} \qquad \text{base}$$
$$\text{acid}$$

Obviously, the two sides of this equation are identical, and K must therefore be 1. However, one of the water concentrations is already assimilated into K_a.

This makes

$$K_a = \frac{[H_2O][H_3O^+]}{[H_3O^+]} = [H_2O] = 55.5$$

and $pK_a = -1.74$.

These are the two figures seen for water in the tables of pK_a values. Water acting as an acid, i.e. losing a proton, has pK_a 15.7. Water acting as a base, i.e. accepting a proton, has $pK_a - 1.74$.

We now have the relationship that $K_a \times K_b = 10^{-14}$, or

$$pK_a + pK_b = 14$$

4.5 Electronic and structural features that influence basicity

Basicity relates to the ability of a compound to use its nonbonding electrons to combine with a proton. We have already seen that features such as inductive or delocalization effects can make an acid stronger. They increase the stability of the conjugate base, and consequently favour loss of a proton from an acid. It follows that features that stabilize a conjugate base are going to discourage its protonation, i.e. they are going to make it a weaker base. Thus, a compound in which the electrons are delocalized will be less basic than one in which the electrons are localized. For example, carboxylate anions (delocalized charge) are going to be weaker bases than alkoxide ions (localized charge).

Anionic (charged) bases are naturally going to be more ready to donate electrons to a positively charged proton than a neutral base (uncharged) that uses lone pair electrons. Most of our organic bases are not anionic, so we need to look at features that affect basicity, just as we have done for acids. Nitrogen compounds are good examples of organic bases and the ones we shall meet most frequently, though oxygen systems will feature prominently in our mechanistic rationalizations.

4.5.1 Electronegativity

The acidity of an acid HX increases as X becomes more electronegative. Conversely, basicity will decrease as an atom becomes more electronegative. Ammonia (pK_a 9.2) is a stronger base than water ($pK_a - 1.74$). These figures relate to release of a proton from the conjugate acid, namely ammonium ion and hydronium ion respectively. This is sometimes confusing; we talk about the pK_a of a base when we really mean the pK_a of its conjugate acid. We cannot avoid this, because it becomes too complicated to use the name of the conjugate acid, but we shall endeavour to show the conjugate acid in structures.

$$NH_4^{\oplus} \rightleftharpoons H^{\oplus} + NH_3$$

ammonium ion

$$H_3O^{\oplus} \rightleftharpoons H^{\oplus} + H_2O$$

hydronium ion

Oxygen is more electronegative than nitrogen, so its electrons are less likely to be donated to a proton. Neutral oxygen bases are generally very much weaker than nitrogen bases, but as we shall see later, protonation of an oxygen atom is important and the first step in many acid-catalysed reactions, especially carbonyl compounds.

4.5.2 Inductive effects

Electron-donating groups on nitrogen are going to increase the likelihood of protonation, and help to stabilize the conjugate acid. They thus increase the basic strength.

The pK_a values for the **amines** ammonia, methylamine, dimethylamine, and trimethylamine are 9.2, 10.6, 10.7, and 9.8 respectively. The electron-donating effect of the methyl substituents increases the basic strength of methylamine over ammonia by about 1.4 pK_a units, i.e. by a factor of over 25 ($10^{1.4} = 25.1$). However, the introduction of a second methyl substituent has a relatively small effect, and the introduction of a third methyl group, as in trimethylamine, actually reduces the basic strength to nearer that of methylamine.

ammonium
pK_a 9.2

methylammonium
pK_a 10.6

dimethylammonium
pK_a 10.7

trimethylammonium
pK_a 9.8

electron-donating effects of alkyl groups stabilize positive charge

This apparent anomaly is a consequence of measuring pK_a values in aqueous solution, where there is more than ample opportunity for hydrogen bonding with water molecules. Hydrogen bonding helps to stabilize a positive charge on nitrogen, and this effect will decrease as the number of alkyl groups increases. Therefore, the observed pK_a values are a combination of increased basicity with increasing alkyl groups (as predicted via electron-donating effects) countered by a stabilization of the cation through hydrogen bonding, which decreases with increasing alkyl groups. Note that we saw solvent molecules influencing the acidity of alcohols by stabilizing the conjugate base (see Section 4.3.3).

H-bonding stabilizes cations

When pK_a values are measured in the gas phase, where there are no hydrogen bonding effects, they are found to follow the predictions based solely on electron-donating effects. In water, mono-, di-, and tri-alkylated amines all tend to have rather similar pK_a values, typically in the range 10–11.

Electron-withdrawing groups will have the opposite effect. They will decrease electron density on the nitrogen, destabilize the conjugate acid, and thus make it less likely to pick up a proton, so producing a weaker base. Refer back to Table 4.7 for a summary list of inductive effects from various groups.

For example, groups with a strong electron-withdrawing inductive effect, such as trichloromethyl, decrease basicity significantly.

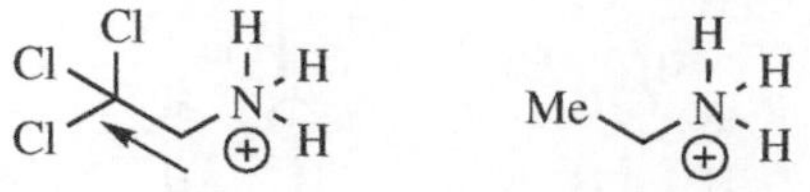

pK_a 5.5 pK_a 10.7

We have already seen that water is a much weaker base than ammonia, because oxygen is more electronegative than nitrogen and its electrons are thus less likely to be donated to a proton. Neutral oxygen bases are also generally very much weaker than nitrogen bases. Nevertheless, protonation of an oxygen atom is a critical first step in many acid-catalysed reactions.

Oxygen is so electronegative that inductive effects from substituents have rather less influence on basicity than they would in similar nitrogen compounds. **Alcohols** are somewhat less basic than water, with **ethers** weaker still.

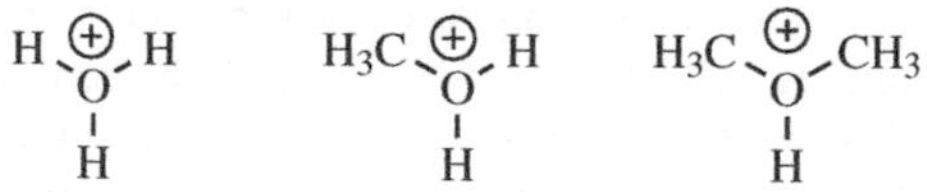

pK_a −1.7 pK_a −2.2 pK_a −3.8

This is precisely opposite to what would be expected from the inductive effects of alkyl groups, and the observations are likely to be the result primarily of solvation (hydrogen bonding) effects. Note, the cations shown all have negative pK_a values. In other words, they are very strong acids and will lose a proton readily. Conversely, the non-protonated compounds are weak bases.

4.5.3 Hybridization effects

We have seen above that acidity is influenced by the hybridization of the atom to which the acidic hydrogen is attached. The acidity of a C–H bond was found to increase as the *s* character of the bond increased. The more *s* character in the orbital, the closer the electrons are held to the nucleus. Similar reasoning may be applied to basicity. If the lone pair is in an sp^2 or sp orbital, it is held closer to the nucleus and is more difficult to protonate than if it is in an sp^3 orbital.

Accordingly, we find that a nitrile nitrogen (lone pair in an sp orbital) is not at all basic (pK_a about −10), though ethylamine (lone pair in an sp^3 orbital)

has pK_a 10.7. Imines (lone pair in an sp^2 orbital) are less basic than amines. Cyclohexanimine (the imine of cyclohexanone) has pK_a 9.2, and is less basic than cyclohexylamine (pK_a 10.6).

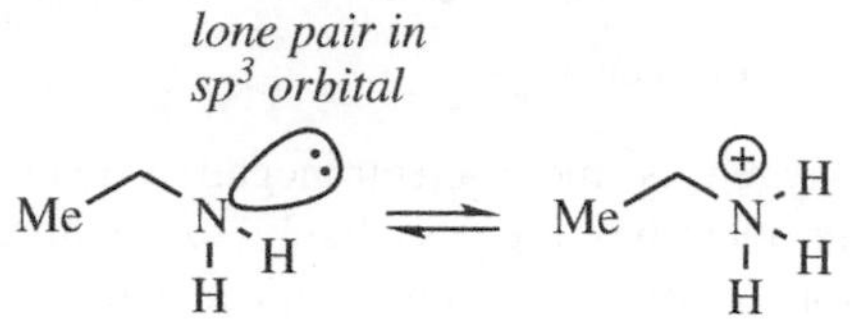

lone pair in sp^3 orbital

ethylamine pK_a 10.7

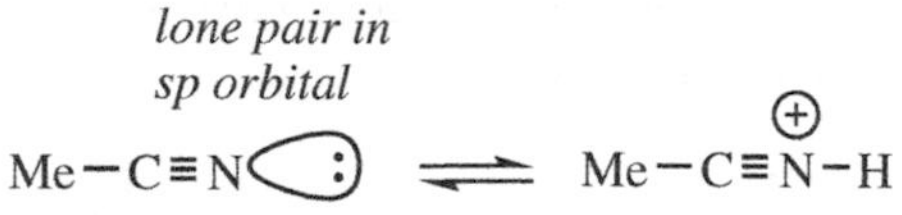

lone pair in sp orbital

acetonitrile pK_a −10

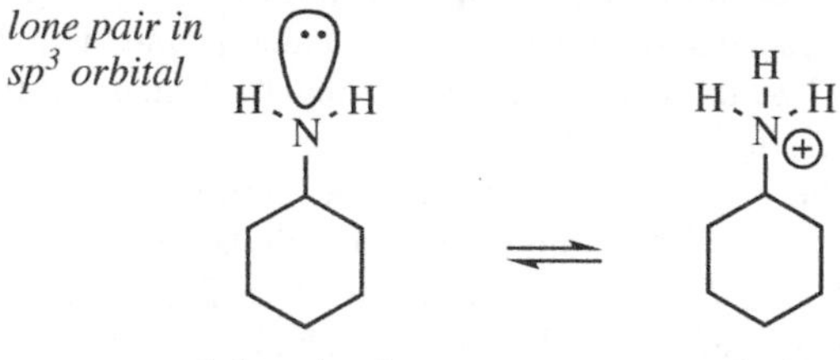

lone pair in sp^3 orbital

cyclohexylamine pK_a 10.6

lone pair in sp^2 orbital

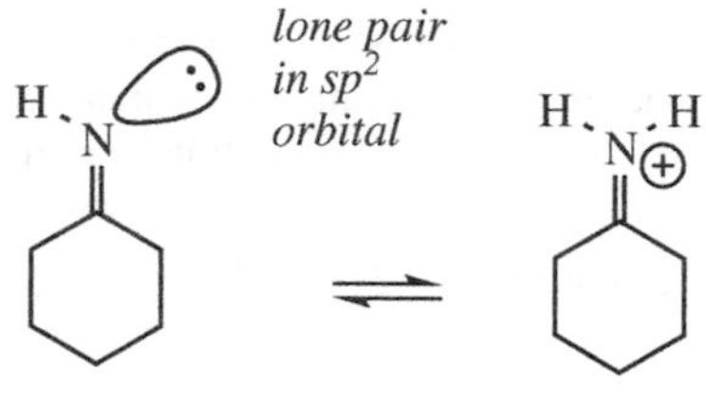

cyclohexanimine pK_a 9.2

Similarly, alcohols (sp^3 hybridization), although they are themselves rather weak bases, are going to be more basic than aldehydes and ketones (sp^2 hybridization)

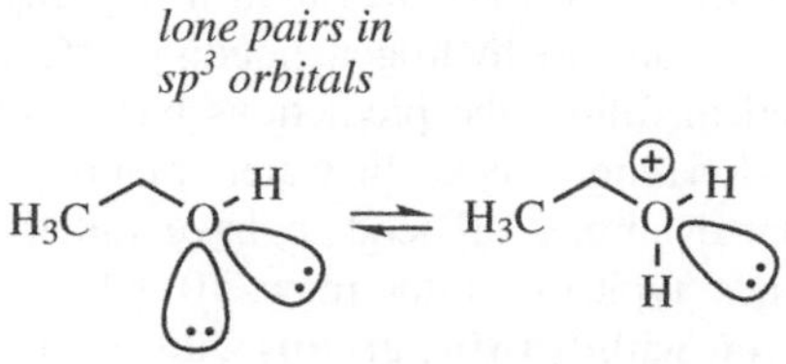

lone pairs in sp^3 orbitals

ethanol pK_a −2.4

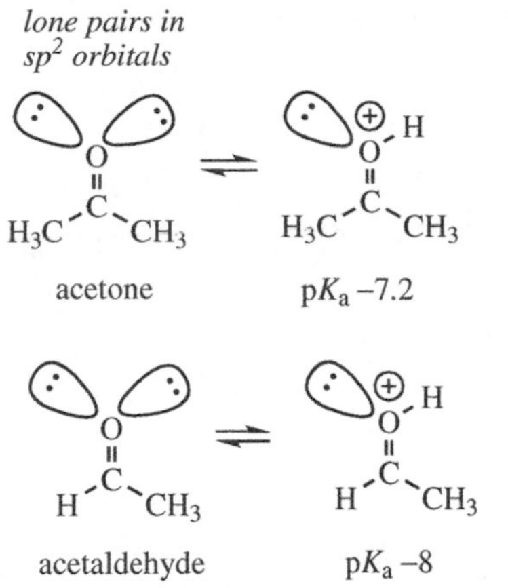

lone pairs in
sp² orbitals

acetone $pK_a -7.2$

acetaldehyde $pK_a -8$

4.5.4 Resonance/delocalization effects

Delocalization of charge in the conjugate base anion contributes to stabilization of the anion, and thus ionization of the acid is enhanced. Delocalization effects in bases are more likely to stabilize the base

rather than the conjugate acid, and thus tend to reduce the basicity. Refer again to Table 4.8 for a summary of various groups that may contribute resonance effects.

Pre-eminent amongst examples is the case of **amides**, which do not show the typical basicity of amines. Acetamide, for example, has $pK_a - 1.4$, compared with a pK_a 10.7 in the case of ethylamine. This reluctance to protonate on nitrogen is caused by delocalization in the neutral amide, in which the nitrogen lone pair is able to overlap into the π system. This type of resonance stabilization would not be possible with nitrogen protonated, since the lone pair is already involved in the protonation process. Indeed, if amides do act as bases, then protonation occurs on oxygen, not on nitrogen. Resonance stabilization is still possible in the *O*-protonated amide, whereas it is not possible in the *N*-protonated amide. Note that resonance stabilization makes the *O*-protonated amide somewhat less acidic than the hydronium ion ($pK_a - 1.7$); the amide oxygen is more basic than water.

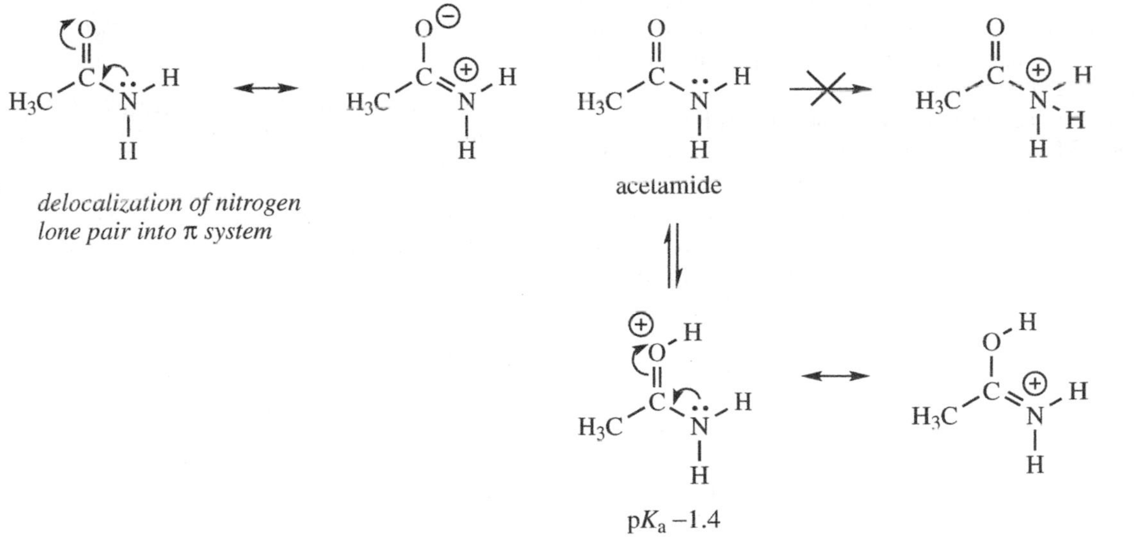

delocalization of nitrogen
lone pair into π system

acetamide

$pK_a -1.4$

The carbonyl oxygen of **aldehydes** and **ketones** is less basic than that of an alcohol by several powers of 10. We have just seen above that this arises because the lone pair electrons of the carbonyl oxygen are in orbitals that are approximately sp^2 in character, and are more tightly held than the alcohol lone pairs in sp^3 orbitals. The neutral carbonyl group is thus

favoured, and the conjugate acid is correspondingly more acidic.

On the other hand, protonated **carboxylic acids** and **esters** are shown with the proton on the carbonyl oxygen, despite this oxygen having sp^2 hybridization, whereas the alternative oxygen has sp^3 hybridization.

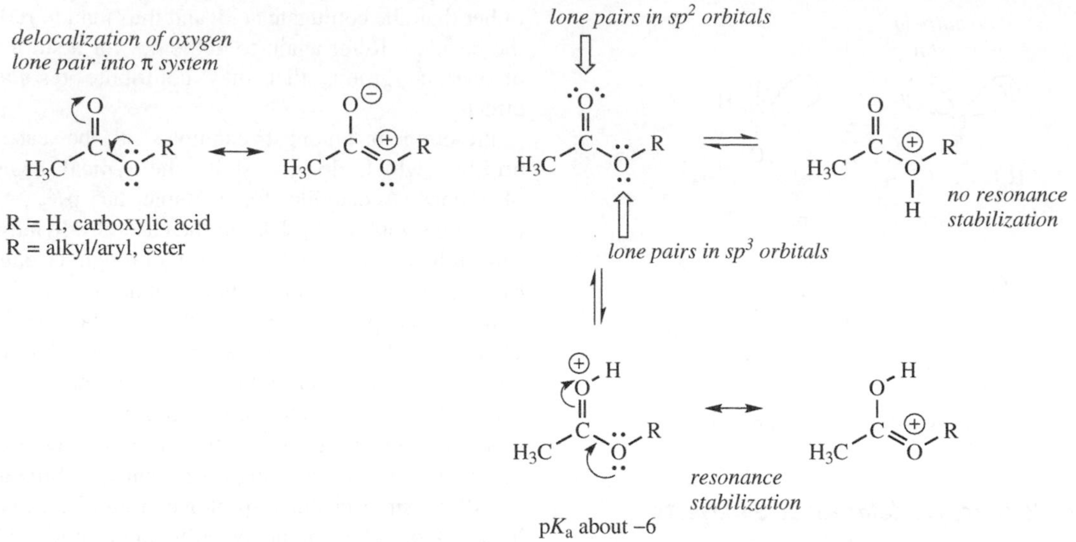

*delocalization of oxygen
lone pair into π system*

R = H, carboxylic acid
R = alkyl/aryl, ester

lone pairs in sp² orbitals

*no resonance
stabilization*

lone pairs in sp³ orbitals

*resonance
stabilization*

pK_a about −6

This is a consequence of delocalization, with resonance stabilization being possible when the carbonyl oxygen is protonated, but not possible should the OR oxygen become protonated. This additional resonance stabilization is not pertinent to aldehydes and ketones, which are thus less basic than the carboxylic acid derivatives. However, these oxygen derivatives are still very weak bases, and are only protonated in the presence of strong acids.

In the case of the sulfur analogues **thioesters** and **thioacids**, this delocalization is much less favourable. In the oxygen series, delocalization involves overlap between the oxygen sp^3 orbital and the π system of the carbonyl, which is composed of $2p$ orbitals. Delocalization in the sulfur series would require

overlap between a sulfur $3p$ orbital and a carbon $2p$ orbital, which is much less likely because of the size difference between these orbitals.

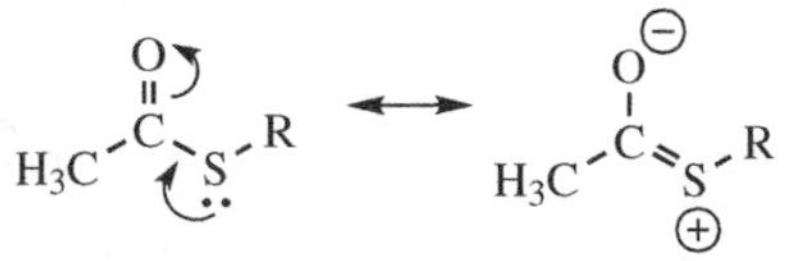

resonance of this type is less favourable in the sulfur esters and acids due to the larger S atom, and less orbital overlap

Amidines are stronger bases than amines. The pK_a for acetamidine is 12.4.

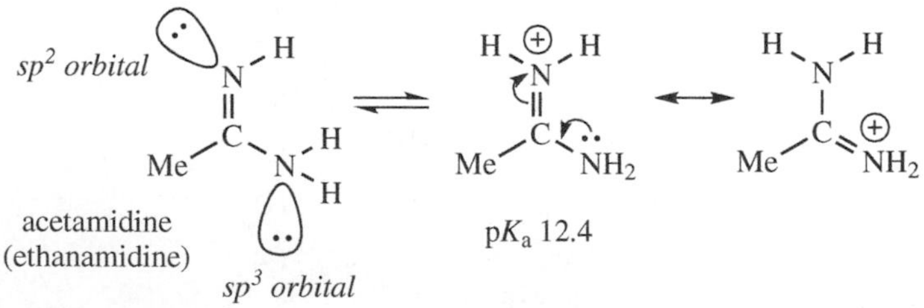

sp^2 orbital

acetamidine
(ethanamidine)

sp^3 orbital

pK_a 12.4

Amidines are essentially amides where the carbonyl oxygen has been replaced with nitrogen, i.e. they are nitrogen analogues of amides. It is the nitrogen replacing the oxygen that becomes protonated. This is easily rationalized, even though the hydridization here is sp^2, which in theory should be less basic than the sp^3-hybridized nitrogen. Protonation of the imine nitrogen allows resonance stabilization in the cation,

which could not happen if the amide nitrogen were protonated. In addition, the two resonance structures both have charge on nitrogen, and in fact are identical. We have a similar situation in the carboxylate anion. Amidines, therefore, are quite strong bases, with the potential for electron delocalization being a greater consideration than the hybridization state of the orbital housing the lone pair.

Now let us look at **guanidines**, which are even stronger bases. Guanidine itself has pK_a 13.6. It can be seen that there is delocalization of charge in the conjugate acid, such that in each resonance structure the charge is favourably associated with one of the three nitrogen atoms. No such favourable delocalization is possible in the neutral molecule, so guanidines are readily protonated and, therefore, are strong bases.

guanidine pK_a 13.6

Should you need further convincing that resonance stabilization is an important criterion in acidity and basicity, it is instructive to consider the bond lengths in the carboxylate anion and in the amidinium and guanidinium cations. Now we would expect double bonds to be shorter than single bonds (see Section 2.6.2), and this is true in the corresponding non ionized systems. However, bond length measurements for the carboxylate anion tells us that the bonds in question (C–O/C=O) are actually the same length, being somewhere between the expected single and double bond lengths. The same is true of the C–N/C=N bonds in amidinium and guanidinium cations. This fits in nicely with the concept that the actual ion is not a mixture of the various resonance forms that we can draw, but something in between. Compare this with the fact that the C–C bond lengths in benzene are somewhere between single and double bonds, and thus do not correspond to either of the Kekulé resonance structures (see Section 2.9.4).

If we look at the pK_a values for the conjugate acids of cyclohexylamine and the **aromatic amine** aniline, we see that aniline is the weaker base. Cyclohexylamine has pK_a 10.6, whereas aniline has pK_a 4.6. There is an inductive effect in aniline because the phenyl ring is electron withdrawing. The carbon atoms of the aromatic ring are sp^2 hybridized, and more electronegative than sp^3-hybridized carbons of alkyl groups. We might, therefore, expect some reduction in basicity. However, a more prominent effect arises from resonance, which can occur in the uncharged amine, but not in the protonated conjugate acid. This makes the unprotonated amine favourable, and aniline is consequently a very weak base.

cyclohexylamine pK_a 10.6   aniline pK_a 4.6

Hint: if you draw the alternative Kekulé form, you can push electrons around the ring

resonance stabilization possible in the base, but not the conjugate acid

This effect is increased if there is a suitable electron-withdrawing group in the *ortho* or *para* position on the aromatic ring. Thus, *p*-nitroaniline and *o*-nitroaniline have pK_a 1.0 and −0.3 respectively. These aromatic amines are thus even weaker bases than aniline, a result of improved delocalization in the free base. The increased basicity of the *ortho* isomer is a result of the very close inductive effect of the nitro group; the *meta* isomer has only the inductive effect, and its pK_a is about 2.5.

p-nitroaniline

Of course, those groups that can act as electron-donating groups through resonance will produce the opposite effect, and increase the basicity. Through resonance, groups such as hydroxyl and methoxyl can distribute negative charge towards the amino substituent, facilitating its protonation. The pK_a values for *o*-methoxyaniline and *p*-methoxyaniline are about 4.5 and 5.4 respectively, and that for *m*-methoxyaniline is about 4.2. The electron-donating resonance effect is countered by the electron-withdrawing inductive effects of these electronegative substituents, so that predictions about basicity become a little more complex.

stabilizing resonance effect
destabilizing inductive effect

stabilizing resonance effect
destabilizing inductive effect

As we pointed out after our considerations of acidity in aromatic derivatives, we wish to emphasize that the very same principles will be used when we consider aromatic substitution reactions in Chapter 8. The methods used to understand the basicity of aromatic derivatives will be applied again in a different format.

A word of warning is now needed! Some compounds may have pK_a values according to whether they are acting as acids or as bases. For example, CH_3OH has pK_a 15.5 and −2.2; the first figure refers to methanol acting as an acid via loss of a proton and giving CH_3O^-, and the second value refers to methanol acting as a base, i.e. the conjugate acid losing a proton. Similarly, CH_3NH_2 has pK_a values of 35 and 10.6, again referring to acid and base behaviour.

It is important to avoid confusion in such cases, and this requires an appreciation of typical pK_a values for simple acids and bases. There is no way we would encourage memorizing of pK_a values, but two easily remembered figures can be valuable for comparisons. These are pK_a around 5 for a typical aliphatic carboxylic acid, and pK_a around 10 for a typical aliphatic amine. These then allow us to consider whether the compound in question is more acidic, more basic, etc.

It then becomes fairly easy to decide that methanol is not a strong acid, like nitric acid say, so that the pK_a − 2.2 is unlikely to refer to its acid properties. Methylamine ought to be basic rather like ammonia, so the pK_a value of 35 would appear well out of the normal range for bases and must refer to its acidic properties. In such cases, there appear to be very good reasons for continuing to use pK_b values for bases; unfortunately, however, this is not now the convention.

4.6 Basicity of nitrogen heterocycles

Our discussions of the basicity of organic nitrogen compounds have concentrated predominantly on simple amines in which the nitrogen atom under consideration is part of an acyclic molecule. Many biologically important compounds, and especially drug molecules, are based upon systems in which nitrogen is part of a heterocycle. We shall consider the properties of heterocyclic compounds in more detail in Chapter 11; here, we mainly want to show how our rationalizations of basicity can be extended to a few commonly encountered **nitrogen heterocycles**.

The basicities of the simple heterocycles **piperidine** and **pyrrolidine** vary little from that of a secondary amine such as dimethylamine. pK_a values for the conjugate bases of these three compounds are 11.1, 11.3, and 10.7 respectively.

piperidine   piperidinium cation   pyrrolidine   pyrrolidinium cation pK_a 10.7

pK_a 11.1 pK_a 11.3

However, **pyridine** and **pyrrole** are significantly less basic than either of their saturated analogues. The pyridinium cation has pK_a 5.2, making pyridine a much weaker base than piperidine, whereas the pyrrolium cation ($pK_a - 3.8$) can be considered a very strong acid, and thus pyrrole is not at all basic.

Although the nitrogen atom in these systems carries a lone pair of electrons, these electrons are not able to accept a proton in the same way as a simple amine. The dramatic differences in basicity are a consequence of the π electron systems, to which the nitrogen contributes (see Section 2.9.6).

protonation on α-carbon; aromaticity destroyed, but resonance stabilization of cation

pyridine   pyridinium cation   pyrrole   pyrrolium cation

pK_a 5.2   *N lone pair is part of aromatic π electron system* pK_a -3.8

protonation on N not favoured; destroys aromaticity

Pyridine, like benzene, is an aromatic system with six π electrons (see Section 11.3). The ring is planar, and the lone pair is held in an sp^2 orbital. The increased s character of this orbital, compared with the sp^3 orbital in piperidine, means that the lone pair electrons are held closer to the nitrogen and, consequently, are less available for protonation. This hybridization effect explains the lower basicity of pyridine compared with piperidine. **Pyrrole** is also aromatic, but there is a significant difference, in that both of the lone pair electrons are contributing to the six-π-electron system. As part of the delocalized π electron system, the lone pairs are consequently not available for bonding to a proton. Protonation of the nitrogen in pyrrole is very unfavourable: it would destroy the aromaticity. It is possible to protonate pyrrole using a strong acid; but, interestingly, protonation occurs on the α-carbon and not on the nitrogen. Although this still destroys aromaticity, there is some favourable resonance stabilization in the conjugate acid.

Let us consider just one more nitrogen heterocycle here, and that is **imidazole**, a component of the amino acid histidine (see Box 11.6). The imidazolium cation has pK_a 7.0, making imidazole less basic than a simple amine, but more basic than pyridine. Imidazole has two nitrogen atoms in its aromatic ring system. One of these nitrogens contributes its lone pair to make up the aromatic sextet, but the other has a free lone pair that is available for protonation. As with pyridine, this lone pair is in an sp^2 orbital, but the increased basicity of imidazole compared with pyridine is a result of additional resonance in the conjugate acid.

imidazole   imidazolium
                          cation
                        pK_a 7.0

The basicity of some other heterocyclic systems will be considered in Chapter 11.

4.7 Polyfunctional acids and bases

We have so far considered acids and bases with a single ionizable group, and have rationalized the measured pK_a values in relation to structural features in the molecule. This additional structural feature could well have its own acidic or basic properties, and we thus expect that such a compound will be characterized by more than one pK_a value.

Before we consider polyfunctional organic compounds, we should consider the inorganic acids sulfuric acid and phosphoric acid. **Sulfuric acid** is termed a dibasic acid, in that it has two ionizable groups, and **phosphoric acid** is a tribasic acid with three ionizable hydrogens. Thus, sulfuric acid has two pK_a values and phosphoric acid has three.

sulfuric acid   hydrogensulfate   sulfate   resonance
                         (bisulfate)   stabilization

phosphoric acid   dihydrogenphosphate   hydrogenphosphate   phosphate   resonance
                                                                                  stabilization

Both acids give rise to resonance-stabilized conjugate bases (compare carboxylate) and are strong acids. Sulfuric acid is the stronger, owing to improved resonance possibilities provided by the two S=O functions, as against just one P=O in phosphoric acid. Note particularly, though, that the pK_a values for the second and third ionizations are higher than the first. This indicates that loss of a further proton from an ion is much less favourable than loss of the first proton from the non-ionized acid. Nevertheless, the sulfate dianion is sufficiently well

resonance stabilized via the two S=O functions that hydrogensulfate (bisulfate) is still a fairly strong acid.

We can generalize that it is going to be more difficult to lose a proton from an anion than from an uncharged molecule. This is also true of polyfunctional acids, such as **dicarboxylic acids**. However, it is found that this effect diminishes as the negative centres become more separated. Thus, pK_a values for some simple aliphatic dicarboxylic acids are as shown, loss of the first proton being represented by pK_{a1} and loss of the second by pK_{a2}.

oxalic acid   malonic acid   succinic acid   glutaric acid   acetic acid

pK_{a1} 1.3 pK_{a1} 2.9 pK_{a1} 4.2 pK_{a1} 4.3 pK_a 4.8
pK_{a2} 3.8 pK_{a2} 5.7 pK_{a2} 5.6 pK_{a2} 5.4

It can be seen that the difference between the first and second pK_a values diminishes as the number of methylene groups separating the carboxyls increases, i.e. it becomes easier to lose the second proton as

the other functional group is located further away. It can also be seen that since malonic acid is a stronger acid than acetic acid, then the extra carboxyl is an electron-withdrawing substituent that is stabilizing the conjugate base. Again, this effect diminishes rapidly as the chain length increases, as anticipated for an inductive effect. Of course, ionization of a carboxylic acid group to a carboxylate anion reverses the inductive effect, in that the carboxylate will be electron donating, and will destabilize the dianion. This is reflected in the pK_{a2} values for malonic, succinic and glutaric acids all being larger than the pK_a for acetic acid. Oxalic acid appears anomalous in this respect, and this appears to be a result of the high charge density associated with the dianion and subsequent solvation effects.

In the aromatic benzenedicarboxylic acid derivatives, the pattern is not dissimilar, especially since we have no oxalic acid-like anomaly. Carboxylic acid groups are electron withdrawing, and all three diacids are stronger acids than benzoic acid. On the other hand, the carboxylate group is electron donating, and this weakens the second ionization. This makes the second acid a weaker acid than benzoic acid. The effects are greatest in the *ortho* derivative, where there are also going to be steric factors (see Section 4.3.5).

phthalic acid	isophthalic acid	terephthalic acid	benzoic acid
pK_{a1} 2.9	pK_{a1} 3.7	pK_{a1} 3.5	pK_a 4.2
pK_{a2} 5.4	pK_{a2} 4.6	pK_{a2} 4.3	

Compounds with two basic groups, e.g. **diamines**, can be rationalized in a similar manner. Here, we must appreciate that both amino groups and

ammonium cations are electron withdrawing, the positively charged entity having the greater effect. pK_a values for a series of aliphatic diamines are shown.

1,2-diaminoethane	1,3-diaminopropane	1,4-diaminobutane	ethylamine
pK_{a1} 9.9	pK_{a1} 10.6	pK_{a1} 10.8	pK_a 10.7
pK_{a2} 6.9	pK_{a2} 8.9	pK_{a2} 9.6	

As the distance between the amino groups increases, the effect of the NH_2 on the first protonation diminishes, so that pK_{a1} values for the 1,3- and 1,4-diamino compounds are very similar to that of ethylamine. Only in 1,2-diaminoethane do we see the electron-withdrawing effects of the second amino group decreasing basicity. However, for the second protonation, it is clear that an ionized

amino group has a much larger effect than a non-ionized one. The effects fall off as the separation increases, but persist further. Thus, pK_{a2} values for the 1,3- and 1,4-diamino compounds are now rather different.

The aromatic diamines present a much more complex picture, and we do not intend to justify the observed pK_a values in detail.

1,2-diaminobenzene	1,3-diaminobenzene	1,4-diaminobenzene	aniline
pK_{a1} 4.6	pK_{a1} 5.1	pK_{a1} 6.3	pK_a 4.6
pK_{a2} 0.8	pK_{a2} 2.5	pK_{a2} 3.0	

There are going to be a number of effects here, with some that provide opposing influences. An amino group has an electron-withdrawing inductive effect, but has an electron-donating resonance effect that tends to be greater in magnitude than the inductive effect. A protonated amino group also has an electron-withdrawing inductive effect that is greater than that of an uncharged amino group. On the other hand, it no longer supplies the electron-donating resonance effect. As with other disubstituted benzenes, the *ortho* compound also experiences steric effects that may reduce the benefits of resonance. Both the *meta* and *para* diamines are stronger bases than aniline, and protonation of the first amine in all three compounds considerably inhibits the second protonation.

4.8 pH

The acidity of an aqueous solution is normally measured in terms of **pH**. pH is defined as

$$pH = -\log_{10}[H_3O^+]$$

The lower the pH, the more acidic the solution; the higher the pH, the more basic the solution. The pH scale only applies to aqueous solutions, and is only a measure of the acidity of the solution. It does not indicate how strong the acid is (that is a function of pK_a) and the pH of an acid will change as we alter its concentration. For instance, dilution will decrease the H_3O^+ concentration, and thus the pH will increase.

In water, the hydronium ion concentration arises by the self-dissociation equilibrium (see Section 4.4):

$$H_2O \ + \ H_2O \ \rightleftharpoons \ H_3O^{\oplus} \ + \ HO^{\ominus}$$

In this reaction, one molecule of water is acting as a base, accepting a proton from a second water molecule. The second molecule is acting as an acid and donating a proton. For every hydronium ion produced, a hydroxide anion must also be formed, so that the concentrations of hydronium and hydroxide ions must be equal. In pure water at 25 °C, this value is found to be 10^{-7} M.

The equilibrium constant K is given by the formula

$$K = \frac{[H_3O^+][HO^-]}{[H_2O][H_2O]}$$

and because the concentration of water is essentially constant in aqueous solution, the new equilibrium constant K_w (the ionization constant for water) is defined as

$$K_w = [HO^-][H_3O^+] = 10^{-7} \times 10^{-7} = 10^{-14}$$

This means that the pH of pure water at 25 °C is therefore

$$pH = -\log 10^{-7} = 7$$

pH 7 is regarded as neither acidic, nor basic, but neutral. It follows that acids have pH less than 7 and bases have pH greater than 7.

Box 4.2

K_w and pH of neutrality at different temperatures

We rapidly become accustomed to the idea that the pH of water is 7.0, and that this represents the pH of neutrality. Unfortunately, this is only true at 25 °C; at other temperatures, the amount of ionization varies, so that K_w will consequently be different. We find that the amount of ionization increases with temperature and the pH of neutrality decreases accordingly. A few examples are shown in Table 4.9.

Table 4.9 K_w and pH of neutrality at different temperatures

Temperature (°C)	K_w	pH of neutrality
0	0.12×10^{-14}	7.97
25	1.00×10^{-14}	7.00
37	2.51×10^{-14}	6.80
40	2.95×10^{-14}	6.77
75	16.9×10^{-14}	6.39
100	48.0×10^{-14}	6.16

Box 4.3

Calculation of pH: strong acids and bases

A **strong acid** is considered to be completely ionized in water, so that the hydronium ion concentration is the same as its molarity.

Thus, a 0.1 M solution of HCl in water has $[H_3O^+] = 0.1$, and pH $= -\log 0.1 = 1$.
Similarly, a 0.01 M solution has $[H_3O^+] = 0.01$ and pH $= -\log 0.01 = 2$, and a 0.001 M solution has $[H_3O^+] = 0.001$ and pH $= -\log 0.001 = 3$.

It follows from this that, because we are using a logarithmic scale, a pH difference of 1 corresponds to a factor of 10 in hydronium ion concentration.

If the pH is known, then we can calculate the hydronium ion concentration. Since

$$pH = -\log[H_3O^+]$$

the hydronium ion concentration is given by

$$[H_3O^+] = 10^{-pH}$$

For example, if the pH $= 4$, $[H_3O^+] = 10^{-4} = 0.0001$ M.

When we have a **strong base**, our calculations need to invoke the ionization constant for water

$$K_w = [H_3O^+][HO^-] = 10^{-14}$$

Thus, the pH of a 0.1 M solution of NaOH in water is calculated from $[HO^-] = 0.1$, and since $[H_3O^+][HO^-] = 10^{-14}$, $[H_3O^+]$ must be 10^{-13}.

Hence, the pH of a 0.1 M solution of NaOH in water will be $-\log 10^{-13} = 13$. A 0.01 M solution of NaOH will have $[HO^-] = 10^{-2}$ and pH $= -\log 10^{-12} = 12$, and a 0.001 M solution has $[HO^-] = 10^{-3}$ and pH $= -\log 10^{-11} = 11$.

Weak acids are not completely ionized in aqueous solutions, and the amount of ionization, and thus hydronium ion concentration, is governed by the equilibrium

$$HA + H_2O \rightleftharpoons H_3O^+ + A^-$$

and the equilibrium constant K_a we defined above:

$$K_a = \frac{[A^-][H_3O^+]}{[HA]}$$

However, since $[H_3O^+]$ must be the same as $[A^-]$, we can write

$$K_a = \frac{[H_3O^+]^2}{[HA]}$$

and therefore

$$[H_3O^+] = \sqrt{K_a[HA]}$$

If we take negative logarithms of both sides, we get

$$-\log[H_3O^+] = -\tfrac{1}{2}\log K_a - \tfrac{1}{2}\log[HA]$$

which becomes

$$pH = \tfrac{1}{2}pK_a - \tfrac{1}{2}\log[HA]$$

Note: this is simply a variant of the Henderson–Hasselbalch equation below, when $[A^-] = [H_3O^+]$.

The calculation of the pH of a **weak base** may be approached in the same way. The equilibrium we need to consider is

$$B + H_2O \rightleftharpoons BH^+ + HO^-$$

and the equilibrium constant K_b will be defined as

$$K_b = \frac{[HO^-][BH^+]}{[B]}$$

However, since $[HO^-]$ must be the same as $[BH^+]$, we can write

$$K_b = \frac{[HO^-]^2}{[B]}$$

and therefore

$$[HO^-] = \sqrt{K_b[B]}$$

Now we need to remember that

$$K_w = [HO^-][H_3O^+]$$

so that we can replace $[HO^-]$ with $K_w/[H_3O^+]$; this leads to

$$\frac{K_w}{[H_3O^+]} = \sqrt{K_b[B]}$$

and hence

$$[H_3O^+] = \frac{K_w}{\sqrt{K_b[B]}}$$

If we now take negative logarithms of both sides, we get

$$-\log[H_3O^+] = -\log K_w + \tfrac{1}{2}\log K_b + \tfrac{1}{2}\log[B]$$

which becomes

$$pH = pK_w - \tfrac{1}{2}pK_b + \tfrac{1}{2}\log[B]$$

Box 4.4

Calculation of pH: weak acids and bases

Consider a 0.1 M solution of the **weak acid** acetic acid ($K_a = 1.76 \times 10^{-5}$; $pK_a = 4.75$). Since the degree of ionization is small, the concentration of undissociated acid may be considered to be approximately the same as the original concentration, i.e. 0.1. The pH can be calculated using the equation

$$pH = \tfrac{1}{2}pK_a - \tfrac{1}{2}\log[HA]$$

Thus

$$pH = 2.38 - 0.5 \times \log 0.1$$

$$= 2.38 - (-0.5)$$

$$= 2.88$$

The calculation of the pH of a **weak base** can be achieved in a similar way; but again, since we have a base, our calculations need to invoke the ionization constant for water

$$K_w = [H_3O^+][HO^-] = 10^{-14}$$

and $pK_a + pK_b = 14$. Thus, for a 0.1 M solution of ammonia (conjugate acid $pK_a = 9.24$)

$$pH = pK_w - \tfrac{1}{2}pK_b + \tfrac{1}{2}\log[B]$$

and pK_b is thus $14 - 9.24 = 4.76$. This leads to

$$pH = 14 - 2.38 + 0.5 \times \log 0.1$$

$$= 14 - 2.38 + 0.5(-1)$$

$$= 11.12$$

Alternatively, we could use

$$pH = \tfrac{1}{2}pK_w + \tfrac{1}{2}pK_a + \tfrac{1}{2}\log[B]$$

to get the same result:

$$pH = 7 + 4.62 + 0.5 \times \log 0.1$$

$$= 11.12$$

These calculations are for the pH of **weak acids** and **weak bases**. It is well worth comparing the figures we calculated above for strong acids and bases. Thus, a 0.1 M solution of the strong acid HCl had pH 1, and a 0.1 M solution of the strong base NaOH had pH 13.

Although this produces a similar type of expression to that for the pH of a weak acid above, it does employ pK_b rather than pK_a. To keep to a 'pK_a only' concept, we need to incorporate the $pK_a + pK_b = pK_w$ expression. Then we get the alternative formula

$$pH = pK_w - \tfrac{1}{2}(pK_w - pK_a) + \tfrac{1}{2}\log[B]$$

or

$$pH = \tfrac{1}{2}pK_w + \tfrac{1}{2}pK_a + \tfrac{1}{2}\log[B]$$

Box 4.5

The pH of salt solutions

It should be self-evident that solutions comprised of equimolar amounts of a **strong acid**, e.g. HCl, and a **strong base**, e.g. NaOH, will be neutral, i.e. pH 7.0 at 25 °C. We can thus deduce that a solution of the salt NaCl in water will also have pH 7.0.

However, salts of a **weak acid and strong base** or of a **strong acid and weak base** dissolved in water will be alkaline or acidic respectively. Thus, aqueous sodium acetate is basic, whereas aqueous ammonium chloride is acidic. pH values may be calculated from pK_a as follows.

Consider the ionization of sodium acetate in water; this leads to an equilibrium in which AcO^- acts as base

$$^\ominus OAc + H_2O \;\rightleftharpoons\; HOAc + HO^\ominus$$

We can treat this equilibrium in exactly the same way as the ionization of a weak base, where we deduced the pH to be

$$pH = \tfrac{1}{2}pK_w + \tfrac{1}{2}pK_a + \tfrac{1}{2}\log[B]$$

Thus, for a 0.1 M solution of sodium acetate in water, where pK_a for the conjugate acid HOAc is 4.75

$$pH = \tfrac{1}{2}pK_w + \tfrac{1}{2}pK_a + \tfrac{1}{2}\log[B]$$

$$= 7 + 2.38 + 0.5 \times \log 0.1$$

$$= 7 + 2.38 + 0.5 \times (-1)$$

$$= 8.88$$

If we now consider a 0.1 M solution of ammonium chloride in water, where pK_a for the conjugate acid NH_4^+ is 9.24, we have the equilibrium

$$NH_4^{\oplus} + H_2O \rightleftharpoons NH_3 + H_3O^{\oplus}$$

in which the ammonium ion is acting as an acid.

For the ionization of a weak acid, we calculated above that the pH is given by the equation

$$pH = \tfrac{1}{2}pK_a - \tfrac{1}{2}\log[HA]$$

Thus

$$pH = 4.62 - 0.5 \times \log 0.1$$
$$= 4.62 - 0.5 \, (-1)$$
$$= 5.12$$

In both cases, we are making the assumption that the concentration of the ion (either AcO^- or NH_4^+) is not significantly altered by the equilibrium and can, therefore, be considered to be equivalent to the molar concentration

4.9 The Henderson–Hasselbalch equation

K_a for the ionization of an **acid** HA has been defined as

$$K_a = \frac{[A^-][H_3O^+]}{[HA]}$$

and this can be rearranged to give

$$[H_3O^+] = K_a \times \frac{[HA]}{[A^-]}$$

Taking negative logarithms of each side, this becomes

$$-\log[H_3O^+] = -\log K_a + \log\frac{[A^-]}{[HA]}$$

or

$$pH = pK_a + \log\frac{[A^-]}{[HA]}$$

This is referred to as the **Henderson–Hasselbalch equation**, and it is sometimes written as

$$pH = pK_a + \log\frac{[base]}{[acid]}$$

Using this relationship, it is possible to determine the degree of ionization of an acid at a given pH.

An immediate outcome from this expression is that the pK_a of an acid is the pH at which it is exactly half dissociated. This follows from

$$pH = pK_a + \log\frac{[A^-]}{[HA]}$$

But when the concentrations of acid HA and conjugate base A^- are equal, then

$$\log\frac{[A^-]}{[HA]} = \log 1 = 0$$

so that

$$pH = pK_a$$

This means we can determine the pK_a of an acid by measuring the pH at the point where the acid is half neutralized. As we increase the pH, the acid becomes more ionized; as we lower the pH, the acid becomes less ionized.

For a **base**, K_a is defined as

$$K_a = \frac{[B][H_3O^+]}{[BH^+]}$$

which can be rearranged to give

$$[H_3O^+] = K_a \times \frac{[BH^+]}{[B]}$$

so that the Henderson–Hasselbalch equation is written

$$pH = pK_a + \log\frac{[B]}{[BH^+]}$$

or, as previously

$$pH = pK_a + \log\frac{[base]}{[acid]}$$

Again, we can see that the pK_a of a base is the pH at which it is half ionized. As we increase the pH, the base becomes less ionized; as we lower the pH, the base becomes more ionized.

A further useful generalization can be deduced from the Henderson–Hasselbalch equation. This relates to the ratio of ionized to non-ionized forms as the pH varies. A shift in pH by one unit to either side of the pK_a value must change the ratio of ionized to non-ionized forms by a factor of 10. Every further shift of pH by one unit changes the ratio by a further factor of 10.

Thus, for example, if the pK_a of a base is 10, at pH 7 the ratio of free base to protonated base is $1:10^3$. An acid with pK_a 2 at pH 7 would produce a ratio of acid to anion of $10^5:1$.

Box 4.6

Calculation of percentage ionization

Using the Henderson–Hasselbalch equation, we can easily calculate the amount of ionized form of an acid or base present at a given pH, provided we know the pK_a.

For example, consider aqueous solutions of acetic acid (pK_a 4.75) first at pH 4.0 and then at pH 6.0. Since

$$pH = pK_a + \log \frac{[A^-]}{[HA]}$$

at pH 4.0

$$\log \frac{[A^-]}{[HA]} = pH - pK_a = 4.0 - 4.75 = -0.75$$

Thus

$$\frac{[A^-]}{[HA]} = 10^{-0.75} = 0.18$$

If we consider $[A^-] = I$, the fraction ionized, then [HA] is the fraction non-ionized, i.e. $1 - I$, and $I/(1 - I) = 0.18$, from which I may be calculated to be about 0.15 or 15%.

At pH 6.0, $pH - pK_a = 1.25$, and the calculation yields $I/(1 - I) = 10^{1.25} = 17.8$, so that $I = 0.95$, i.e. 95% ionized.

With a base such as ammonia (pK_a 9.24), the percentages ionized at pH 8.0 and 10.0 are calculated as follows:

$$pH = pK_a + \log \frac{[B]}{[BH^+]}$$

At pH 8.0

$$\log \frac{[B]}{[BH^+]} = pH - pK_a = 8.0 - 9.24 = -1.24$$

Thus

$$\frac{[B]}{[BH^+]} = 10^{-1.24} = 0.057$$

Now with bases, [B] is the non-ionized fraction $1 - I$ and $[BH^+]$ is the ionized fraction I, so $(1 - I)/I = 0.057$, and therefore $I = 0.95$, i.e. 95% ionized.

At pH 10.0, the calculation yields $(1 - I)/I = 10^{0.76} = 5.75$, so that I = 0.15, i.e. 15% ionized.

Box 4.7

The ionization of amino acids at pH 7

Peptides and proteins are composed of **α-amino acids** linked by amide bonds (see Section 13.1). Their properties, for example the ability of enzymes to catalyse biochemical reactions, are dependent upon the degree of ionization of various acidic and basic side-chains at the relevant pH. This aspect will be discussed in more detail in Section 13.4, but, here, let us consider a simple amino acid dissolved in water at pH 7.0. An α-amino acid has an acidic carboxylic acid group and a basic amine group. Both of these entities need to be treated separately.

$$R \diagdown\!\!\!\diagup CO_2H$$
$$|$$
$$NH_2$$

α-amino acid

The carboxylic acid groups of amino acids have pK_a values in a range from about 1.8 to 2.6 (see Section 13.1). Let us consider a typical **carboxylic acid group** with pK_a 2.0. Using the Henderson–Hasselbalch equation

$$pH = pK_a + \log \frac{[RCO_2^-]}{[RCO_2H]}$$

we can deduce that

$$\log \frac{[RCO_2^-]}{[RCO_2H]} = pH - pK_a = 7.0 - 2.0 = 5.0$$

Thus

$$\frac{[RCO_2^-]}{[RCO_2H]} = 10^5 = 10\,000 : 1$$

Therefore, the carboxylic acid group of an amino acid can be considered to be completely ionized in solution at pH 7.0.

Now let us consider the **amino group** in α-amino acids. The pK_a values of the conjugate acids are found to range from about 8.8 to 10.8. We shall consider a typical group with pK_a 10.0. From

$$pH = pK_a + \log \frac{[RNH_2]}{[RNH_3^+]}$$

$$\log \frac{[RNH_2]}{[RNH_3^+]} = pH - pK_a = 7.0 - 10.0 = -3.0$$

Thus

$$\frac{[RNH_2]}{[RNH_3^+]} = 10^{-3} = 1 : 1000$$

Therefore, as with the carboxylic acid group, we find that the amino group of an amino acid is effectively ionized completely, i.e. fully protonated, in solution at pH 7.0.

Therefore, we can deduce that α-amino acids in solution at pH 7.0 exist as dipolar ions; these are called **zwitterions** (German; *zwitter* = hybrid) (see Section 13.1).

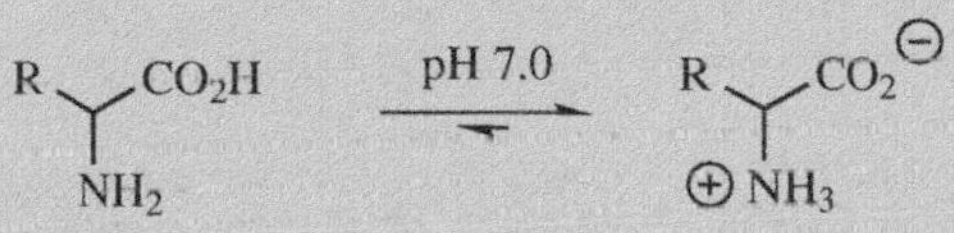

α-amino acid   zwitterion

Some amino acids have additional ionizable groups in their side-chains. These may be acidic or potentially acidic (aspartic acid, glutamic acid, tyrosine, cysteine), or basic (lysine, arginine, histidine). We use the term 'potentially acidic' to describe the phenol and thiol groups of tyrosine and cysteine respectively; under physiological conditions, these groups are unlikely to be ionized. It is relatively easy to calculate the amount of ionization at a particular pH, and to justify that latter statement.

Similar calculations as above for the basic side-chain groups of arginine (pK_a 12.48) and lysine (pK_a 10.52), and the acidic side-chains of aspartic acid (pK_a 3.65) and glutamic acid (pK_a 4.25) show essentially complete ionization at pH 7.0. However, for cysteine (pK_a of the thiol group 10.29) and for tyrosine (pK_a of the phenol group 10.06) there will be negligible ionization at pH 7.0.

For **cysteine** at pH 7.0, the Henderson–Hasselbalch equation leads to

$$\log \frac{[A^-]}{[HA]} = pH - pK_a = 7.0 - 10.29 = -3.29$$

and

$$\frac{[A^-]}{[HA]} = 10^{-3.29} = 5.1 \times 10^{-4}$$

i.e. no significant ionization.

Box 4.7 (continued)

Interestingly, the heterocyclic side-chain of **histidine** is partially ionized at pH 7.0. This follows from

$$\log \frac{[B]}{[BH^+]} = pH - pK_a = 7.0 - 6.00 = -1.0$$

and

$$\frac{[B]}{[BH^+]} = 10^{-1.0} = 10$$

which translates to approximately 9.1% ionization.

We shall see that this modest level of ionization is particularly relevant in some enzymic reactions where histidine residues play an important role (see Section 13.4.1). Note, however, that when histidine is bound in a protein structure, pK_a values for the imidazole ring vary somewhat in the range 6–7 depending upon the protein, thus affecting the level of ionization.

The ionic states at pH 7.0 of these amino acids with ionizable side-chains are shown below.

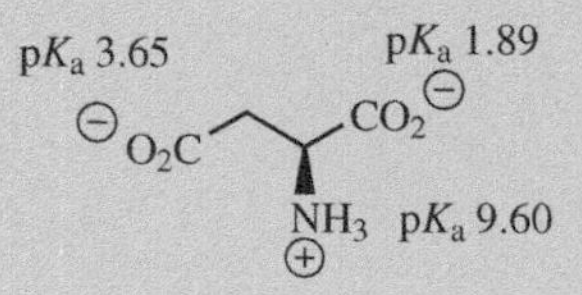

pK_a 3.65 pK_a 1.89
$\ominus O_2C$ $CO_2 \ominus$
NH_3 pK_a 9.60
$\oplus$

aspartic acid

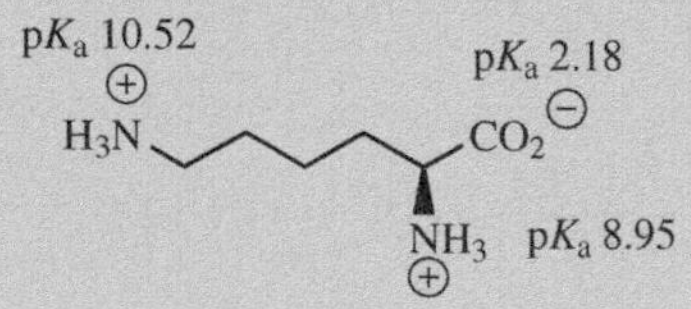

pK_a 10.52
$\oplus$
H_3N $CO_2 \ominus$ pK_a 2.18
NH_3 pK_a 8.95
$\oplus$

lysine

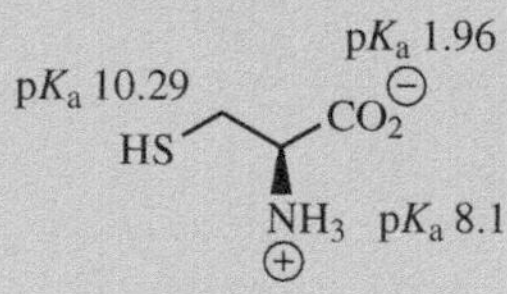

pK_a 10.29 pK_a 1.96
HS $CO_2 \ominus$
NH_3 pK_a 8.18
$\oplus$

cysteine

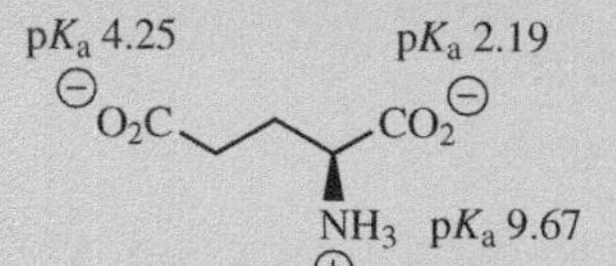

pK_a 4.25 pK_a 2.19
$\ominus O_2C$ $CO_2 \ominus$
NH_3 pK_a 9.67
$\oplus$

glutamic acid

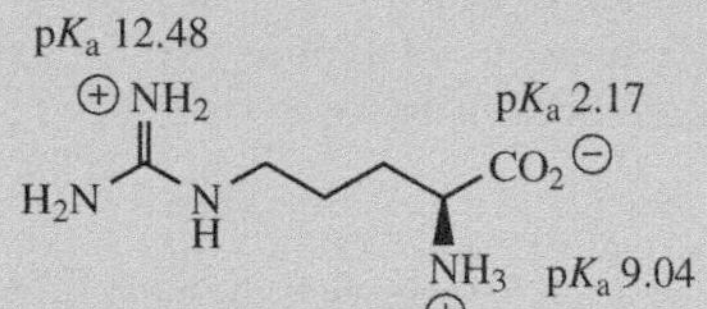

pK_a 12.48
$\oplus NH_2$
H_2N N $CO_2 \ominus$ pK_a 2.17
H
NH_3 pK_a 9.04
$\oplus$

arginine

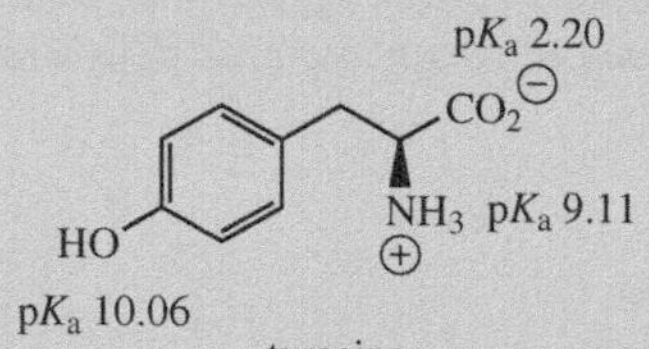

pK_a 2.20
$CO_2 \ominus$
HO NH_3 pK_a 9.11
pK_a 10.06 $\oplus$

tyrosine

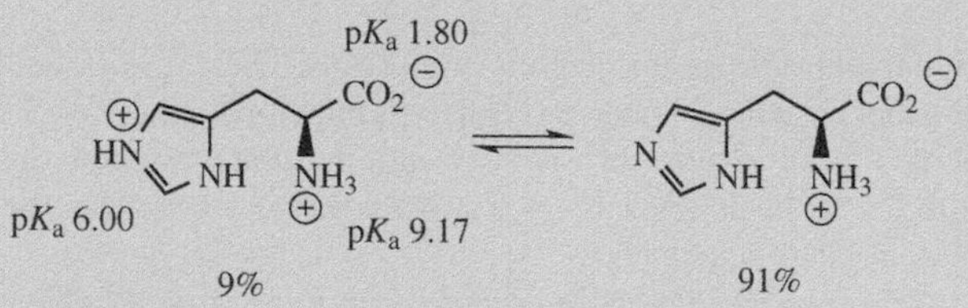

pK_a 1.80
$\oplus$ $CO_2 \ominus$
HN   NH NH_3 N   NH $CO_2 \ominus$
pK_a 6.00 $\oplus$ NH_3
pK_a 9.17 $\oplus$

9%   91%

histidine

pK_a values adjacent to carboxylate functions refer to conjugate acid

4.10 Buffers

A **buffer** is a solution that helps to maintain a reasonably constant pH environment by countering the effects of added acids or bases. They are used extensively for the handling of biochemicals, especially enzymes, as well as in chromatography and drug extractions.

The simplest type of buffer is composed of a weak acid–strong base combination or a weak base–strong acid combination. This may be prepared by combining the weak acid (or base) together with its salt. For example, the sodium acetate–acetic acid combination is one of the most common buffer systems. Although tabulated data are available for the preparation of buffer solutions, a sodium acetate–acetic acid buffer could be prepared simply by adding sodium hydroxide to an acetic acid solution until the required pH is obtained. For maximum efficiency, this pH needs to be within about 1 pH unit either side of the pK_a of the weak acid or base used.

Since acetic acid is only weakly dissociated, the concentration of acetic acid will be almost the same as the amount put in the mixture. On the other hand, the sodium acetate component will be almost completely dissociated, so the acetate ion concentration can be considered the same as that of the sodium acetate used for the solution.

Addition of an acid such as HCl to the buffer solution provides H^+, which combines with the acetate ion to give acetic acid. This has a twofold effect: it reduces the amount of acetate ion present and, by so doing, also increases the amount of undissociated acetic acid. Provided the amount of acid added is small relative to the original concentration of base in the buffer, the alteration in base: acid ratio in the Henderson–Hasselbalch equation is relatively small and has little effect on the pH value.

ionization of weak acid

$$HOAc + H_2O \rightleftharpoons H_3O^{\oplus} + {}^{\ominus}OAc$$

addition of acid

$${}^{\ominus}OAc + HCl \longrightarrow HOAc + Cl^{\ominus}$$

base reduced   *non-ionized acid increased*

addition of base

$$HOAc + HO^{\ominus} \longrightarrow H_2O + {}^{\ominus}OAc$$

non-ionized acid decreased   *base increased*

Similar considerations apply if a base such as NaOH is added to the buffer solution. This will decrease the amount of undissociated acid, and increase the amount of acetate ion present.

The Henderson–Hasselbalch equation may be employed in calculations relating to the properties and effects of buffer solutions (see Box 4.8).

Box 4.8

Preparation of a buffer

One litre of 0.1 M sodium acetate buffer with a pH 4.9 is required. The pK_a of acetic acid is 4.75. From the Henderson–Hasselbalch equation

$$pH = pK_a + \log \frac{[A^-]}{[HA]}$$

Therefore

$$4.9 = 4.75 + \log \frac{[A^-]}{[HA]}$$

so that

$$\log \frac{[A^-]}{[HA]} = 0.15$$

and

$$\frac{[A^-]}{[HA]} = 10^{0.15} = \frac{1.41}{1}$$

This means that the buffer solution requires 1.41 parts sodium acetate to 1 part acetic acid. Therefore, this can be prepared by mixing $1.41/2.41 = 0.585$ l of 0.1 M sodium acetate with $1/2.41 = 0.415$ l of 0.1 M acetic acid.

The amount of sodium acetate in 1 l of solution will thus be 0.0585 M, and the amount of acetic acid will be 0.0415 M.

Buffering effect

If 1 ml of 1 M HCl is added to this sodium acetate buffer solution, the pH change may be calculated as follows. Again, we require the Henderson–Hasselbalch equation:

$$pH = pK_a + \log \frac{[A^-]}{[HA]}$$

We are adding an additional $[H_3O^+]$ of 0.001 M, and this reacts

$${}^{\ominus}OAc + HCl \longrightarrow HOAc + Cl^{\ominus}$$

so effectively reducing the amount of acetate base by 0.001 M and also increasing the amount of acetic acid by 0.001 M. We can ignore the small change in volume arising from addition of the acid.

The Henderson–Hasselbalch equation becomes

$$pH = 4.75 + \log \frac{0.0585 - 0.001}{0.0415 + 0.001}$$

so

$$pH = 4.75 + \log \frac{0.0575}{0.0425} = 4.75 + \log 1.35$$
$$= 4.75 + 0.13 = 4.88$$

It can be seen, therefore, that the effect of addition of the acid is to change the pH value from 4.90 to 4.88, i.e. by just 0.02 of a pH unit.

Box 4.8 (continued)

This contrasts with the effect of adding 0.001 M of HCl to 1 litre of water (pH 7). The new $[H_3O^+]$ of 0.001 M gives $pH = -\log[H_3O^+] = -\log 0.001 = 3$, i.e. a change of four pH units.

If 1 ml of 1 M NaOH was added to this buffer solution, the pH change may be calculated similarly.

We are adding an additional $[HO^-]$ of 0.001 M, and this reacts

$$HOAc \;+\; HO^{\ominus} \longrightarrow H_2O \;+\; {}^{\ominus}OAc$$

effectively increasing the amount of acetate base by 0.001 M and also decreasing the amount of acetic acid by 0.001 M.

The Henderson–Hasselbalch equation becomes

$$pH = 4.75 + \log \frac{0.0575 + 0.001}{0.0415 - 0.001}$$

so

$$pH = 4.75 + \log \frac{0.0585}{0.0405} = 4.75 + \log 1.44$$

$$= 4.75 + 0.16 = 4.91$$

Again, the pH change is minimal, which is the whole point of a buffer solution.

Note also that the pH of a buffer solution is essentially independent of dilution. An unbuffered solution of an acid or base would suffer a pH change on dilution because pH relates to hydronium ion concentration (see Section 4.8). Dilution of a buffered solution does not affect pH because any such changes are accommodated in the $\log([A^-]/[HA])$ component and, therefore, cancel out. We have made certain approximations in deriving the equations, and at very high dilutions the pH does begin to deviate.

The sodium acetate–acetic acid combination is one of the most widely used buffers, and is usually referred to simply as **acetate** buffer. Other buffer combinations commonly employed in chemistry and biochemistry include **carbonate–bicarbonate** (sodium carbonate–sodium hydrogen carbonate), **citrate** (citric acid–trisodium citrate), **phosphate** (sodium dihydrogen phosphate–disodium hydrogen phosphate), and **tris** [tris(hydroxymethyl)amino-methane–HCl].

Box 4.9

The buffering effect of blood plasma

In humans, the **pH of blood** is held at a remarkably constant value of 7.4 ± 0.05. In severe diabetes, the pH can drop to pH 7.0 or below, leading to death from acidotic coma. Death may also occur at pH 7.7 or above, because the blood is unable to release CO_2 into the lungs. The pH of blood is normally controlled by a buffer system, within rather narrow limits to maintain life and within even narrower limits to maintain health.

The buffering system for blood is based on carbonic acid (H_2CO_3) and its conjugate base bicarbonate (HCO_3^-):

$$H_2CO_3 \rightleftharpoons H^{\oplus} \;+\; HCO_3^{\ominus}$$

From the Henderson–Hasselbalch equation

$$pH = pK_a + \log \frac{[HCO_3^-]}{[H_2CO_3]}$$

we can see that maintaining the pH depends upon the ratio of bicarbonate to carbonic acid concentrations. Large quantities of acid formed during normal metabolic processes react with bicarbonate to form carbonic acid. This, however, dissociates and rapidly loses water to form CO_2 that is removed via the lungs.

$$H_2CO_3 \rightleftharpoons H_2O \;+\; CO_2$$

The pH is maintained, therefore, in that a reduction in $[HCO_3^-]$ is countered by a corresponding decrease in $[H_2CO_3]$. The increase in metabolic acid is compensated by a corresponding increase in CO_2.

If the pH of blood rises, $[HCO_3^-]$ temporarily increases. The pH is rapidly restored when atmospheric CO_2 is absorbed and converted into H_2CO_3. It is a reservoir of CO_2 that enables the blood pH to be maintained so rigidly. This reservoir of CO_2 is large and can be altered quickly via the breathing rate.

4.11 Using pK_a values

4.11.1 Predicting acid–base interactions

With a knowledge of pK_a values, or a rough idea of relative values, one can predict the outcome of acid–base interactions. This may form an essential preliminary to many reactions, or provide us with an understanding of whether a compound is ionized under particular conditions, and whether or not it is in a soluble form.

As a generalization, acid–base interactions result in the formation of the weaker acid and the weaker base, a consequence of the most stable species being favoured at equilibrium. Thus, a carboxylic acid such as acetic acid will react with aqueous sodium hydroxide to form sodium acetate and water.

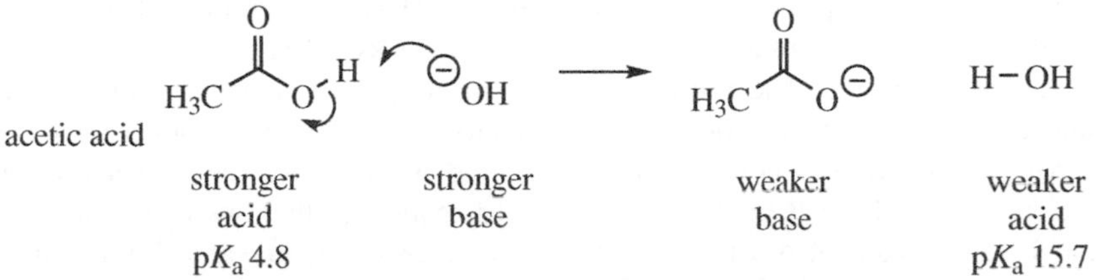

acetic acid

stronger acid pK_a 4.8   stronger base   weaker base   weaker acid pK_a 15.7

Consider the pK_a values. Acetic acid (pK_a 4.8) is a stronger acid than water (pK_a 15.7), and hydroxide is a stronger base than acetate. Accordingly, hydroxide will remove a proton from acetic acid to produce acetate and water, the weaker base–weaker acid combination. Because of the large difference in pK_a values, the position of equilibrium will greatly favour the products, and we can indicate this by using a single arrow and considering the reaction to be effectively irreversible. Even acids that are not particularly soluble in water, e.g. benzoic acid, will participate in this reaction, because the conjugate base benzoate is a water-soluble anion.

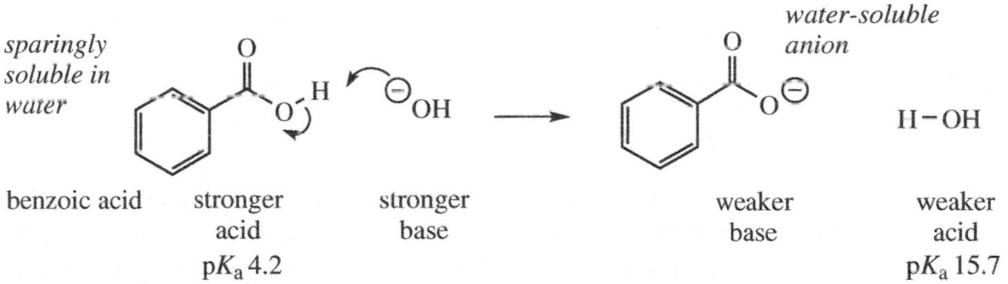

sparingly soluble in water

water-soluble anion

benzoic acid

stronger acid pK_a 4.2   stronger base   weaker base   weaker acid pK_a 15.7

Bases can be considered in the same way. Thus, methylamine will react with aqueous HCl to produce methylammonium chloride and water.

Hydronium is a stronger acid than methylammonium, and methylamine is a stronger base than water, so methylamine will become protonated in aqueous acid. Again, there is a large difference in pK_a values, so the position of equilibrium is well over to the right-hand side. Bases that are not particularly soluble in water, e.g. aniline, can easily be made soluble by conversion to the salt form.

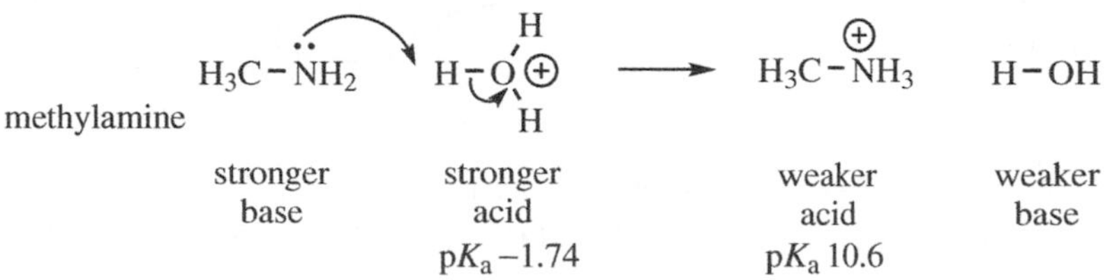

methylamine

stronger base   stronger acid pK_a −1.74   weaker acid pK_a 10.6   weaker base

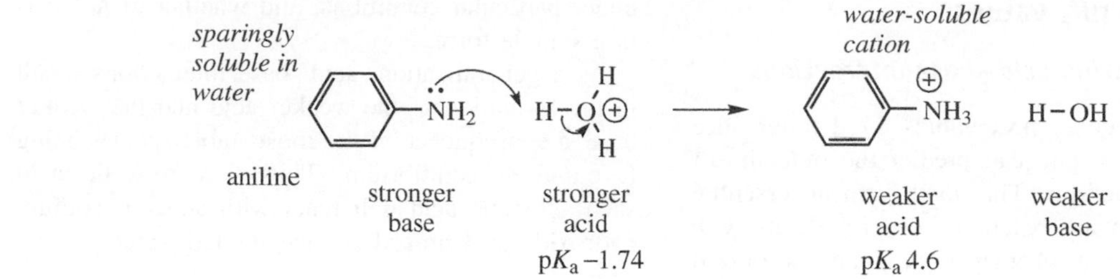

aniline — *sparingly soluble in water*

stronger base

stronger acid
pK_a −1.74

water-soluble cation

weaker acid
pK_a 4.6

weaker base

However, attempts to make an aqueous solution of the base sodium amide would result in the formation of sodium hydroxide and ammonia. The amide ion is a strong base and abstracts a proton from water, a weak acid. The reverse reaction is not favoured, in that hydroxide is a weaker base than the amide ion, and ammonia is a weaker acid than water. Take care with the terminology 'amide': the amide

anion H_2N^- is quite different from the amide molecule $RCONH_2$.

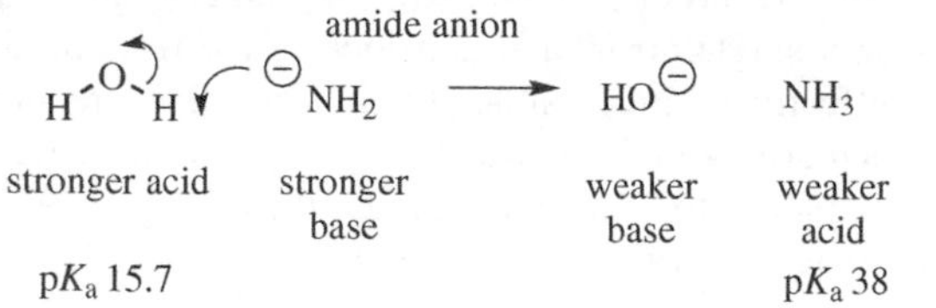

In general, in aqueous solutions, water can donate a proton to any base stronger than the hydroxide ion. If we wish to use bases that are stronger than the hydroxide ion, then we must employ a solvent that is a weaker acid than water. For example, hydrocarbons (pK_a about 50), ethers (pK_a about 50), or even liquid ammonia (pK_a 38). Since these are all extremely weak acids, they will not donate a proton even to a strong base such as amide ion.

Thus, in liquid ammonia, the amide ion may be used to convert an acetylene to its acetylide ion conjugate base (see Section 6.3.4).

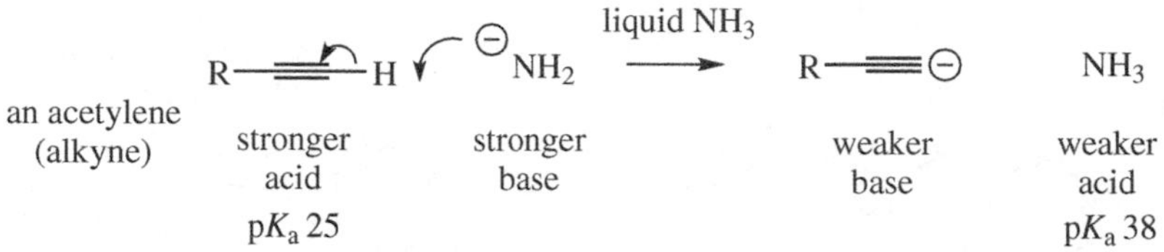

amide anion

stronger acid
pK_a 15.7

stronger base

weaker base

weaker acid
pK_a 38

an acetylene (alkyne)

stronger acid
pK_a 25

stronger base

liquid NH_3

weaker base

weaker acid
pK_a 38

Similarly, the amide ion could be used to abstract a proton from a ketone to produce an enolate anion (see Section 10.2) in an essentially irreversible reaction, since the difference in acidities of the ketone and ammonia is so marked. However, if the base chosen were ethoxide, then enolate anion formation would

be dependent on an equilibration reaction, since the two acids are more comparable in acidity. As we shall see, this latter system may be used, provided the equilibrium can be disturbed in favour of the products (see Section 10.3).

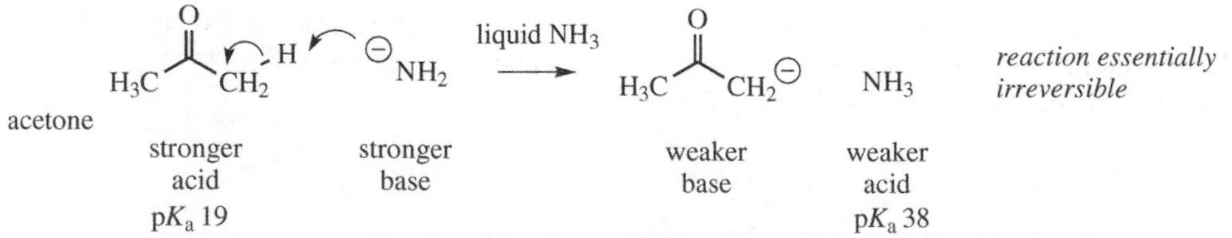

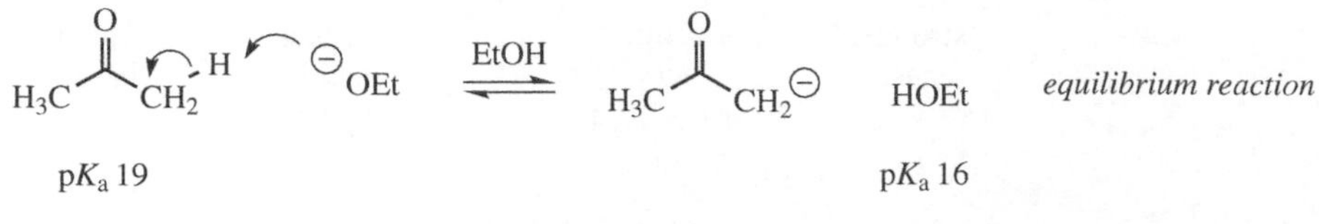

acetone

stronger acid
pK_a 19

stronger base

liquid NH_3

weaker base

weaker acid
pK_a 38

reaction essentially irreversible

pK_a 19

EtOH

pK_a 16

equilibrium reaction

Sodium ethoxide may be produced by treating ethanol with sodium hydride. Again, hydride is the strong base, the conjugate base of the weak acid hydrogen, so the reaction proceeds readily. Sodium

tert-butoxide (from sodium hydride treatment of *tert*-butanol) is a stronger base than sodium ethoxide, since *tert*-butanol (pK_a 19) is less acidic than ethanol (pK_a 16).

ethanol

$$CH_3CH_2O-H \quad \ominus H \longrightarrow CH_3CH_2O^{\ominus} \quad H_2$$

stronger acid	stronger base		weaker base	weaker acid
pK_a 16				pK_a 35

ethoxide

tert-butanol

pK_a 19

tert-butoxide

Some bases and acids that are commonly used as reagents to initiate reactions are listed in Table 4.10, in decreasing order of basicity or acidity.

4.11.2 Isotopic labelling using basic reagents

By definition, an acid will donate a proton to a base, and it is converted into its conjugate base. Conversely, a base will accept a proton from a

suitable donor, generating its conjugate acid. We can utilize these properties to label certain compounds with isotopes of hydrogen, namely deuterium (^{2}H or D) and tritium (^{3}H or T). To differentiate normal hydrogen (^{1}H) from deuterium and tritium, it is sometimes referred to as protium. Because these isotopes are easily detectable and measurable by spectroscopic methods, labelling allows us to follow the fate of particular atoms during chemical reactions, or during metabolic studies.

Table 4.10 Common basic and acidic reagents

Base	Name		Acid	Name	
NaH	Sodium hydride		$FSO_3H \cdot SbF_5$	Antimony pentafluoride–	
t-BuLi	*tert*-Butyl lithium			fluorosulfonic acid	
n-BuLi	*n*-Butyl lithium		CF_3SO_3H	Trifluoromethanesulfonic acid	
(*i*-Pr)$_2$NLi	Lithium	Decreasing basicity	FSO_3H	Fluorosulfonic acid	Decreasing acidity
	diisopropylamide		HCl	Hydrochloric acid	
NaNH$_2$	Sodium amide		H_2SO_4	Sulfuric acid	
	(sodamide)		BF$_3$	Boron trifluoride	
Ph$_3$CNa	Triphenylmethyl sodium		p-CH$_3$C$_6$H$_4$SO$_3$H	p-Toluenesulfonic acid	
(CH$_3$)$_3$COK	Potassium *tert*-butoxide		CF_3CO_2H	Trifluoroacetic acid	
C$_2$H$_5$ONa	Sodium ethoxide		CH_3CO_2H	Acetic acid	
NaOH	Sodium hydroxide				
Et$_3$N	Triethylamine				
Pyridine	Pyridine				
CH$_3$CO$_2$Na	Sodium acetate				

In general, the labelling process is one of **exchange labelling**, removing protium from an acid, and allowing the conjugate base to accept isotopic hydrogen from a suitable donor, most conveniently and cheaply supplied as labelled water. If the labelled compound is going to be of use, say in metabolic studies, labelling must be achieved at a position that does not easily exchange again in an aqueous environment. This rules out hydrogens attached to oxygen or nitrogen that can exchange through simple acid–base equilibria.

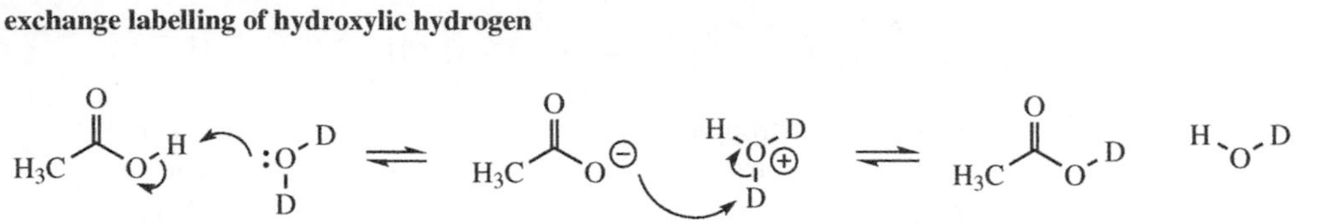

exchange labelling of hydroxylic hydrogen

Thus, dissolving acetic acid in deuteriated water will rapidly give deuteriated acetic acid by acid–base equilibria. However, if the deuteriated acetic acid were then dissolved in normal water, the reverse process would wash out the label equally rapidly.

Useful labelled compounds containing deuterium or tritium normally require the isotopic hydrogen to be attached to carbon, so the acid–base equilibrium will require cleavage of a C–H bond, where acid strength is usually very weak. This will necessitate the use of a very strong base to achieve formation of the conjugate base.

A simple example follows from the reactions considered in Section 4.11.1. We saw that we needed to use a strong base such as the amide ion to form the conjugate base of an acetylene. This reaction was favoured, in that the products were the weaker base acetylide and the weaker acid ammonia.

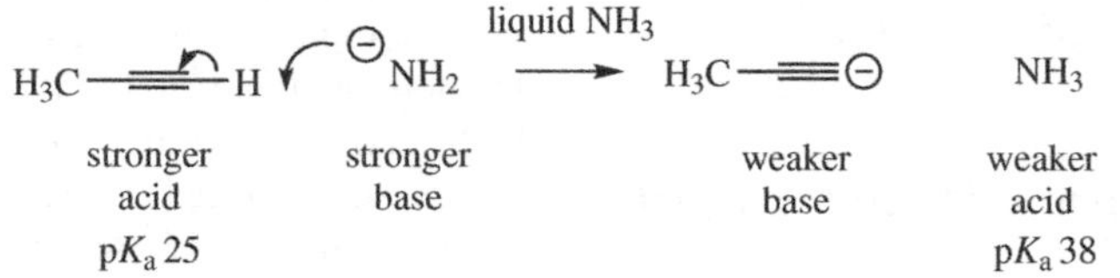

Upon completion of this ionization, we can then add labelled water D_2O. Under these conditions, labelling occurs through abstraction of a deuteron $^2H^+$ from D_2O. This is feasible because acetylide is a stronger base than hydroxide and water is a stronger acid than the acetylene.

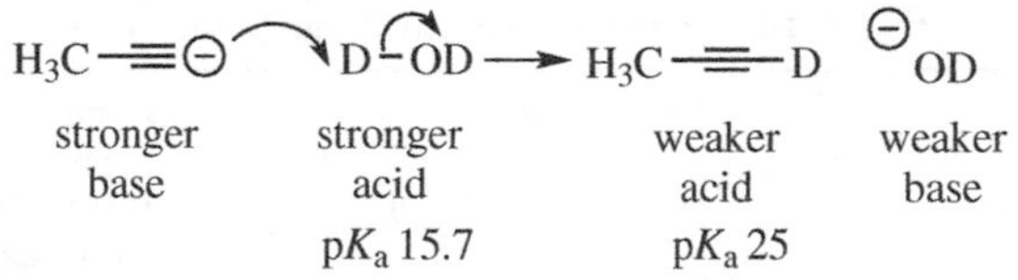

It is also possible to produce deuterium-labelled acetaldehyde by an exchange reaction with D_2O and NaOD. This results in exchange of all three α-hydrogens for deuterium and depends upon generation of the conjugate base of acetaldehyde under basic conditions (see Section 10.1.1).

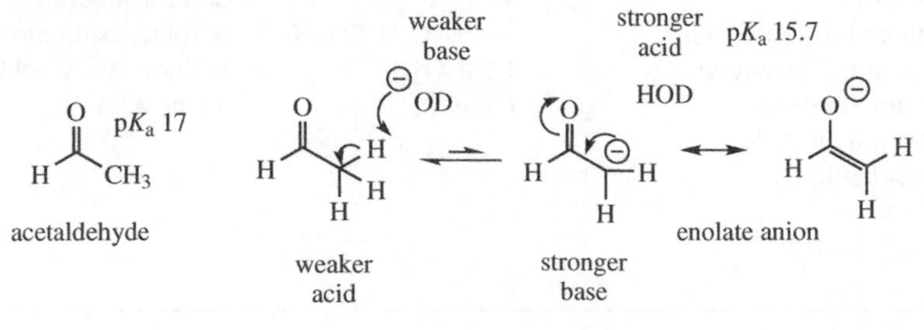

Unlike the example of the acetylene above, the feature of this process is that it is an equilibrium. This is because acetaldehyde is a weak acid (pK_a 17), a weaker acid in fact than water (pK_a 15.7). In other words, the conjugate base of acetaldehyde (the enolate anion) is a stronger base than hydroxide. Treatment of acetaldehyde with hydroxide thus generates an equilibrium mixture, with only a small amount of enolate anion present. Nevertheless, since we get a small amount of conjugate base, this is able to abstract a deuteron from the solvent D_2O in the reverse reaction. Provided excess D_2O is available, equilibration allows exchange of all three hydrogens in the methyl group of acetaldehyde. The label introduced is unfortunately not stable enough, in that similar treatment with strong base and H_2O will reverse the process and incorporate ¹H. Note that the aldehydic hydrogen is not acidic and, therefore, not removed by base. This follows from consideration of the conjugate base, which has no stabilizing features.

The amount of enolate anion present at equilibrium may be calculated from the pK_a values:

$$K = \frac{[\text{enolate}][H_2O]}{[\text{acetaldehyde}][HO^-]}$$

$$= \frac{[\text{enolate}][H^+]}{[\text{acetaldehyde}]} \times \frac{[H_2O]}{[HO^-][H^+]} = 10^{-17}\frac{1}{10^{-15.7}}$$

$$= 10^{-1.3} = 0.05$$

i.e. the proportion of enolate is about 5%.

Despite the unfavourable equilibrium, this type of reaction works surprisingly well under relatively mild conditions. By using a much stronger base, e.g. sodium hydride or lithium diisopropylamide (see Section 10.2), generation of the conjugate base would be essentially complete. Treating the enolate anion with D_2O would give a deuterium-labelled acetaldehyde, but only one atom of deuterium would be introduced. It is the equilibration process that allows exchange of all three hydrogens.

4.11.3 Amphoteric compounds: amino acids

Amphoteric compounds are compounds that may function as either acid or base, depending upon conditions. We have already met this concept in Section 4.5.4, where simple alcohols and amines have two pK_a values according to whether the compound loses or gains a proton. Of course, with alcohols and amines, acidity and basicity involve the same functional group. Other amphoteric compounds may contain separate acidic and basic groups. Particularly important examples of this type are the **amino acids** that make up proteins.

The carboxylic acid groups of protein amino acids have pK_a values about 2, ranging from about 1.8 to 2.6, making them significantly more acidic than simple alkanoic acids (pK_a about 5). For the amino groups, the pK_a values of the conjugate acids are found to range from about 8.8 to 10.8, with most of them in the region 9–10. These values are thus much closer to those of simple amines (pK_a about 10).

Protein amino acids are α-amino acids, the amino and carboxylic acid groups being attached to the same carbon. Thus, the groups are close and will exert maximum inductive effects. The increased acidity of the carboxylic group, therefore, reflects the electron-withdrawing inductive effect of the amino group, or, more correctly, the ammonium ion. This is because we should not consider the amino acid as the non-ionized structure, but as the doubly charged form termed a zwitterion (German: *zwitter* = hybrid) (see Box 4.7).

Consider the two pK_a values. The carboxylic acid group (pK_a 2) is a stronger acid than the protonated NH_2 group (pK_a 9). Thus, the carboxylic acid will

protonate the amino group, and, in pure water (pH 7), amino acids having neutral side-chains will exist predominantly as the doubly charged zwitterion. This may be considered an internal salt, and could be compared to ammonium acetate, the salt formed when ammonia (pK_a 9.2) reacts with acetic acid (pK_a 4.8).

At low pH (acidic solution), an amino acid will exist as the protonated ammonium cation, and at high pH (basic solution) as the aminocarboxylate anion. The intermediate zwitterion form will predominate at pHs between these extremes. The uncharged amino acid has no real existence at any pH. It is ironic that we are so familiar with the terminology amino acid, yet such a structure has no real existence! Amino acids are ionic compounds, solids with a high melting point.

We can appreciate that ionization of the carboxylic acid is affected by the electron-withdrawing inductive effect of the ammonium residue; hence the increased acidity when compared with an alkanoic acid. Similarly, loss of a proton from the ammonium cation of the zwitterion is influenced by the electron-donating inductive effect from the carboxylate anion, which should make the amino group more basic than a typical amine. That this is not the case is thought to be a solvation effect (compare simple amines).

The pH at which the concentration of the zwitterion is a maximum is equal to the **isoelectric point** pI, strictly that pH at which the concentrations of cationic and anionic forms of the amino acid are equal. With a simple amino acid, this is the mean of the two pK_a values:

$$pH = \frac{pK_{a1} + pK_{a2}}{2}$$

This is deduced from

$$K_{a1} = \frac{[H^+][\text{zwitterion}]}{[\text{cation}]} \text{ and } K_{a2} = \frac{[H^+][\text{anion}]}{[\text{zwitterion}]}$$

It follows, therefore, that $pH = pK_{a1}$ when [cation] = [zwitterion], and that $pH = pK_{a2}$ when [zwitterion] = [anion].

At the isoelectric point, [cation] = [anion]; thus

$$[\text{cation}] = \frac{[H^+][\text{zwitterion}]}{K_{a1}}$$

$$= [\text{anion}] = \frac{K_{a2}[\text{zwitterion}]}{[H^+]}$$

Therefore

$$K_{a1} \times K_{a2} = [H^+]^2$$

and by taking negative logarithms, it follows that

$$pH = \frac{pK_{a1} + pK_{a2}}{2}$$

For alanine, $pK_{a1} = 2.34$ and $pK_{a2} = 9.69$, so pI = 6.02. Most amino acids have a pI around this figure.

zwitterion

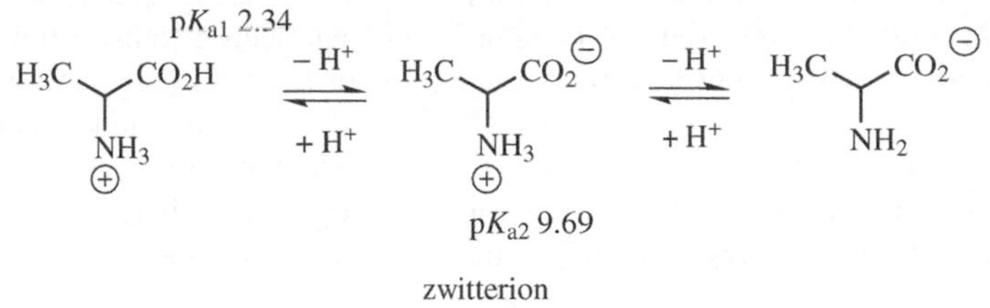

Note that a number of the protein amino acids also have ionizable functions in the side-chain R group. These may be acidic or potentially acidic (aspartic acid, glutamic acid, tyrosine, cysteine), or basic (lysine, arginine, histidine). These amino acids are thus characterized by three pK_a values. We have used the term 'potentially acidic' to describe the phenol and thiol groups of tyrosine and cysteine respectively; under physiological conditions, these groups are unlikely to be ionized (see Box 4.7).

Let us consider lysine, which has a second amino group in its side-chain. The pK_a values for lysine

are 2.18 (CO_2H), 8.95 (α-NH_2), and 10.52 (ε-NH_2). Note that the ε-amino group has a typical amine pK_a value, and is a stronger base than the α-amino group. In strongly acidic solution, lysine will be present as a di-cation, with both amino groups protonated. As the pH is raised, the most acidic proton will be lost first, and this is the carboxyl proton (pK_a 2.18). As the pH increases further, protons will be lost first from the more acidic α-ammonium cation (pK_a 8.95), and lastly from the least acidic ε-ammonium cation (pK_a 10.52). Again, there is no intermediate with uncharged functional groups.

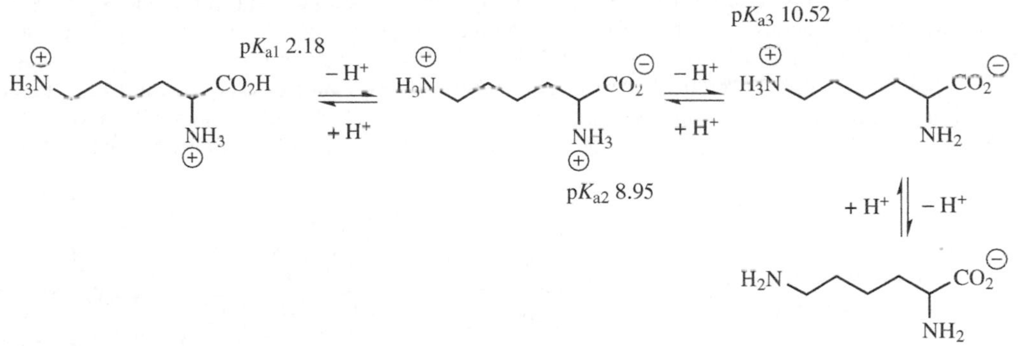

The isoelectric point for lysine is that pH at which the compound is in an electrically neutral form, and this will be the average of pK_{a2} (the cation) and pK_{a3} (the dipolar ion). For lysine, $pK_{a2} = 8.95$ and $pK_{a3} = 10.52$, so pI = 9.74.

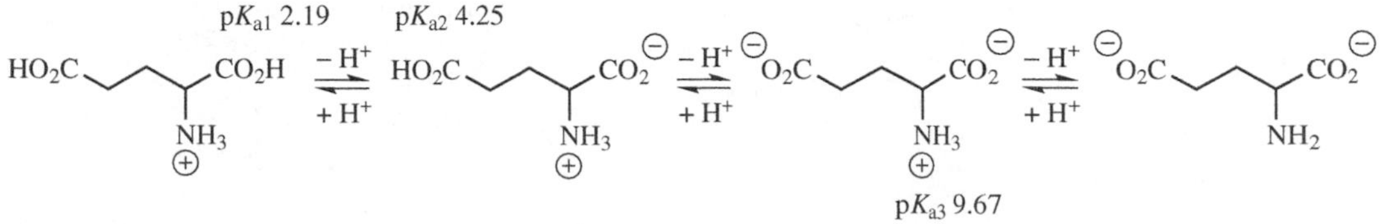

Glutamic acid is an example of an amino acid with an acidic side-chain. The pK_a values are 2.19 (CO_2H), 4.25 (γ-CO_2H), and 9.67 (NH_2). Here, the γ-carboxyl is more typical of a simple carboxylic acid. In strongly acidic solution, glutamic acid will be present as a cation, with the amino group protonated. As the pH is raised, a proton will be lost from the most acidic group, namely the 1-carboxyl (pK_a 2.19), followed by the γ-carboxyl (pK_a 4.25). Lastly, the proton from the α-ammonium cation (pK_a 9.67) will be removed. Yet again, there is no intermediate with uncharged functional groups.

The isoelectric point for glutamic acid is that pH at which the compound is in an electrically neutral form, and this will be the average of pK_{a1} (the

cation) and pK_{a2} (the dipolar ion). For glutamic acid, pK_{a1} = 2.19 and pK_{a2} = 4.25, so pI = 3.22.

Isoelectric points are useful concepts for the separation and purification of amino acids and proteins using electrophoresis. Under the influence of an electric field, compounds migrate according to their overall charge. As we have just seen for amino acids, this very much depends upon the pH of the solution. At the isoelectric point, there will be no net charge, and, therefore, no migration towards either anode or cathode.

Box 4.10

Ionization of morphine: extraction from opium

Morphine is the major alkaloid in **opium**, the dried latex obtained from the opium poppy, *Papaver somniferum*. About 25% of the mass of opium is composed of alkaloids, with morphine constituting about 12–15%. Morphine is a powerful analgesic, and remains one of the most valuable for relief of severe pain. However, most of the morphine extracted from opium is processed further to give a range of semi-synthetic drugs, with enhanced or improved properties. A means of extracting morphine from the other alkaloids in opium is thus desirable.

Alkaloids are found mainly in plants, and are nitrogenous **bases**, typically primary, secondary, or tertiary amines. The basic properties facilitate their isolation and purification. Water-soluble salts are formed in the presence of mineral acids (see Section 4.11.1), and this allows separation of the alkaloids from any other compounds that are neutral or acidic. It is a simple matter to take a plant extract in a water-immiscible organic solvent, and to extract this solution with aqueous acid. Salts of the alkaloids are formed, and, being water soluble, these transfer to the aqueous acid phase. On basifying the acid phase, the alkaloids revert back to an uncharged form, and may be extracted into fresh organic solvent.

Opium contains over 40 different alkaloids, all of which will be extracted from opium by the procedure just described. It then remains to separate morphine from this mixture. Of the main opium alkaloids, only morphine displays some acidic properties as well as basic properties. Although a tertiary amine, morphine also contains a

morphine   codeine   thebaine   papaverine

noscapine
(narcotine)

Typical alkaloid composition of opium

morphine	4–21%
codeine	0.8–2.5%
thebaine	0.5–2.0%
papaverine	0.5–2.5%
noscapine	4–8%

phenolic group. The acidity of this group can be exploited for the preferential extraction of morphine from an organic solvent by partitioning with aqueous base.

Thus, if a solution of opium alkaloids in an organic solvent, e.g. dichloromethane, is shaken with aqueous NaOH, only morphine will ionize at this pH, and it will form the water-soluble phenolate anion. The other alkaloids will remain non-ionized and stay in the organic layer, allowing their separation from the aqueous morphine phenolate fraction. By adding acid to the aqueous fraction, the phenolate will become protonated to give the non-ionized phenol, which may be extracted by shaking with organic solvent. Care is needed during the acidification, since addition of too much acid would ionize the amine and create another water-soluble ion, the protonated amine. This would stay in the aqueous phase and not be extracted by an organic solvent.

The optimum pH will be the isoelectric point as described under amino acids (see Section 4.11.3). This is the pH at which the concentrations of cationic and anionic forms of morphine are equal, and is the mean of the two pK_a values. Morphine has pK_a (phenol) 9.9, and pK_a (amine) 8.2, so that pI = 9.05. Note that the protonated amine is a stronger acid than the phenol, so that the intermediate between the two ionized forms will be the non-ionized alkaloid.

phenolate anion   *morphine*   *ammonium cation*

We can then use the Henderson–Hasselbalch equation

$$pH = pK_a + \log \frac{[base]}{[acid]}$$

to calculate the relative amounts of the ionic forms at this pH.

(a) For the phenol, pK_a 9.9

$$\log \frac{[base]}{[acid]} = pH - pK_a = 9.05 - 9.9 = -0.85$$

thus [base]/[acid] = $10^{-0.85}$ = 0.14 or about 1:7.

(b) For the amine, pK_a 8.2

$$\log \frac{[base]}{[acid]} = pH - pK_a = 9.05 - 8.2 = 0.85$$

thus [base]/[acid] = $10^{0.85}$ = 7.08 or about 7:1.

At pH 9.05, the phenol–phenolate equilibrium favours the phenol by a factor of 7 : 1, and the amine–ammonium ion equilibrium favours the amine by a factor of 7 : 1. In other words, the non-ionized morphine predominates, and this can thus be extracted into the organic phase. What about the amounts in ionized form; are these not extractable? By solvent extraction of the non-ionized morphine, we shall set up a new equilibrium in the aqueous phase, so that more non-ionized morphine is produced at the expense of the two ionized forms. A second solvent extraction will remove this, and we shall effectively recover almost all the morphine content. A third extraction would make certain that only traces of morphine were left as ionized forms.

4.11.4 pK$_a$ and drug absorption

Cell membranes are structures containing lipids and proteins as their main components. Many drug molecules are weak acids or bases and can, therefore, exist as ionized species, depending upon their pK$_a$ values and the pH of the environment. One of the more important concepts relating to drug absorption is that ionized species have very low lipid solubility, and are unable to permeate through membranes. Only the non-ionized drug is usually able to cross membranes. A range of pK$_a$ values covered by some common drugs is shown in Table 4.11.

If we invoke the Henderson–Hasselbalch equation

$$pH = pK_a + \log \frac{[base]}{[acid]}$$

where pH = pK$_a$ for the drug, then, for a **weak acid**

$$\log \frac{[base]}{[acid]} = \log \frac{[ionized]}{[non\text{-}ionized]} = 0$$

In other words, [ionized] = [non-ionized] = 50%.

As we saw in Section 4.9, a shift in pH by one unit to either side of the pK$_a$ value must change the ratio of ionized to non-ionized forms by a factor of 10.

For pH = pK$_a$ + 1

$$\log \frac{[ionized]}{[non\text{-}ionized]} = 1 \text{ and therefore } \frac{[ionized]}{[non\text{-}ionized]} = 10$$

For pH = pK$_a$ − 1

$$\log \frac{[ionized]}{[non\text{-}ionized]} = -1 \text{ and therefore } \frac{[ionized]}{[non\text{-}ionized]} = 0.1$$

For a **weak base**, we have a similar relationship, though

$$\log \frac{[base]}{[acid]} = \log \frac{[non\text{-}ionized]}{[ionized]}$$

The human body has a number of different pH environments. For example, blood plasma has a rigorously controlled pH of 7.4 (see Box 4.9), the gastric juice is usually strongly acidic (pH from about 1 to 7), and urine can vary from about 4.8 to 7.5. It is possible to predict the qualitative effect of pH changes on the distribution of weakly acidic and basic drugs, especially in relation to gastric absorption and renal excretion:

Table 4.11 pK$_a$ values of some common drugs

Weak acids	pK$_a$		Weak bases	pK$_a$	
Levodopa	2.3	strong	Dapsone	1.3	weak
Cromoglycic acid	2.5		Diazepam	3.3	
Penicillins	2.5–2.8		Quinidine	4.1	
Probenecid	3.4		Chlordiazepoxide	4.5	
Aspirin	3.5		Ergometrine	6.8	
Ascorbic acid	4.0		Trimethoprim	7.2	
Warfarin	5.0		Lidocaine	7.8	
Sulfamethoxazole	5.6		Morphine	8.2	
Methotrexate	5.7		Noradrenaline (norepinephrine)	8.6	
Sulfadiazine	6.5				
Thiopental	7.6		Adrenaline (epinephrine)	8.7	
Phenobarbital	7.4		Dopamine	8.8	
Pentobarbital	8.0		Chlorpromazine	9.3	
Phenytoin	8.3		Propranolol	9.5	
Theophylline	8.7		Amphetamine	9.9	
Paracetamol (acetaminophen)	9.5	weak	Atropine	10.2	
			Chloroquine	10.8	
			Guanethidine	11.4	strong

- Very weak acids with pK_a values greater than 7.5 will be essentially non-ionized at all pH values in the range 1–8, so that absorption will be largely independent of pH.

- Acids with pK_a values in the range 2.5–7.5 will be characterized by significant changes in the proportion of non-ionized drug according to the pH. As the pH rises, the percentage of non-ionized drug decreases, and absorption therefore also decreases.

- Absorption of stronger acids (pK_a less than about 2.5) should also depend upon pH, but the fraction that is non-ionized is going to be very low except under the most acidic conditions in the stomach. Absorption is typically low, even under acidic conditions.

- Basic drugs will not be absorbed from the stomach, where the pH is strongly acidic.

- Excretion of drugs will be affected by the pH of the urine. If the urine is acidic, weak bases are ionized and there will be poor re-absorption. With basic urine, weak bases are non-ionized and there is more re-absorption. The pH of the urine can be artificially changed in the range 5–8.5: oral administration of sodium bicarbonate ($NaHCO_3$) increases pH values, whereas ammonium chloride (NH_4Cl) lowers them. Thus, urinary acidification will accelerate the excretion of weak bases and retard the excretion of weak acids. Making the urine alkaline will facilitate the excretion of weak acids and retard that of weak bases.

5

Reaction mechanisms

A **reaction mechanism** is a detailed step-by-step description of a chemical process in which reactants are converted into products. It consists of a sequence of bond-making and bond-breaking steps involving the movement of electrons, and provides a rationalization for chemical reactions. Above all, by following a few basic principles, it allows one to predict the likely outcome of a reaction. On the other hand, it must be appreciated that there will be times when it can be rather difficult to actually 'prove' the mechanism proposed, and in such instances we are suggesting a reasonable mechanism that is consistent with experimental data.

The basic layout of this book classifies chemical reactions according to the type of reaction mechanism involved, not by the reactions undergone within

any specific group of compounds. As we proceed, we shall meet several types of general reaction mechanism. Initially, however, reactions can be classified as **ionic** or **radical**, according to whether bond-making and bond-breaking processes involve two electrons or one electron respectively.

5.1 Ionic reactions

As the name implies, ionic reactions involve the participation of charged entities, i.e. ions. Bond-making and bond-breaking processes in ionic reactions are indicated by **curly arrows** that represent the movement of **two electrons**. The tail of the arrow indicates where the electrons are coming from, the arrowhead where they are going to.

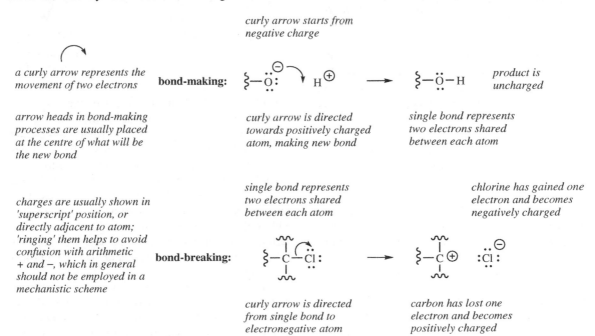

a curly arrow represents the movement of two electrons

bond-making:

curly arrow starts from negative charge

arrow heads in bond-making processes are usually placed at the centre of what will be the new bond

curly arrow is directed towards positively charged atom, making new bond

single bond represents two electrons shared between each atom

product is uncharged

single bond represents two electrons shared between each atom

charges are usually shown in 'superscript' position, or directly adjacent to atom; 'ringing' them helps to avoid confusion with arithmetic + and −, which in general should not be employed in a mechanistic scheme

bond-breaking:

curly arrow is directed from single bond to electronegative atom

chlorine has gained one electron and becomes negatively charged

carbon has lost one electron and becomes positively charged

Essentials of Organic Chemistry Paul M Dewick
© 2006 John Wiley & Sons, Ltd

Lone pairs, originally nonbonding electrons, can also be used in bond-making processes.

curly arrow starts from lone pair electrons

oxygen has donated electron; product carries positive charge on oxygen

$$\{-\overset{\cdot\cdot}{\underset{|}{O}}: \quad H^{\oplus} \longrightarrow \{-\overset{\oplus}{\underset{|}{O}}-H$$

curly arrow is directed towards positively charged atom, making new bond

single bond represents two electrons shared between each atom

curly arrow starts from lone pair electrons

oxygen has donated electron; product carries positive charge on oxygen

$$\{-\overset{\cdot\cdot}{\underset{\cdot\cdot}{O}}-\overset{\oplus}{\underset{|}{C}}-\{ \quad \longrightarrow \quad \{-\overset{\oplus}{O}=\underset{|}{C}-\{$$

curly arrow is directed towards positively charged atom, making new bond

generation of double bond from single bond; new bond represents another pair of electrons shared between each atom

These simple examples illustrate the basic rules for mechanism and the use of **curly arrows**. The concepts are no different from those we have elaborated for drawing resonance structures (see Section 2.10):

- Curly arrows must start from an electron-rich species. This can be a negative charge, a lone pair, or a bond.

- Arrowheads must be directed towards an electron-deficient species. This can be a positive charge, the positive end of a polarized bond, or a suitable atom capable of accepting electrons, e.g. an electronegative atom or Lewis acid.

If we are to draw sensible mechanisms, putting in the correct number of bonds and assigning the correct charges, then it is vital that we know the number of electrons around any particular atom. We have already considered how to assess the **formal charge** on an atom in Section 2.10; the following résumé covers those occasions that we are most likely to meet.

Carbon has four bonding electrons and can attain a stable octet of electrons by bonding to four other atoms, i.e. it has a valency of four.

$$\{-\underset{|}{\overset{|}{C}}-\{$$

tetravalent carbon

$$\{-\underset{|}{\overset{X}{C}}-\{ \rightarrow \{-\overset{\oplus}{\underset{|}{C}}-\{ \quad X^{\ominus}$$

carbocation

sp^2 + unfilled p

carbocation is planar

$$\{-\underset{|}{\overset{H}{C}}-\{ \longrightarrow \{-\overset{\ominus}{\underset{|}{C}}-\{ \quad H^{\oplus}$$

carbanion

sp^3

carbanion is tetrahedral

Carbon can also bond to just three other atoms by donating a pair of electrons from the octet to one of the atoms originally bonded, in so doing breaking the bond. It will then carry a positive charge; it has effectively donated its single electron contribution from the shared pair comprising the single bond. This positively charged carbon is called a **carbocation** (in older nomenclature a carbonium ion). Note that, with only six electrons involved in bonding, the carbocation is a planar entity, having two electrons in each of three sp^2 orbitals and with an unfilled p orbital.

Alternatively, carbon can carry a negative charge if it accepts both electrons from one of the original bonds, leaving the other group electron deficient and positively charged. It has effectively gained a single electron, and is termed a **carbanion**. In this case, carbon carries a full octet of electrons and is tetrahedral, as if it had four single bonds. The lone pair electrons occupy the fourth sp^3 orbital.

Remember, *carbon cannot form more than four bonds*!

Nitrogen has three bonding electrons and a lone pair; it can bond to three atoms, i.e. it has a valency of three. However, it can also bond to four atoms by donating its lone pair, in which case it will then carry a positive charge.

trivalent nitrogen   an ammonium
                                           cation

an amide anion

Nitrogen can also bond to just two atoms. Here, it carries a negative charge, since the octet is made up by acquiring one electron.

Oxygen has two bonding electrons and two lone pairs. It can bond to two other atoms, and is usually divalent. It can also bond to one atom in a negatively charged form, or to three atoms in a positively-charged form. The oxonium cation produced still carries a lone pair, but these electrons will not participate

in further bonding, since this would necessitate an unfavourable double-charged oxygen.

divalent oxygen   an oxide anion

an oxonium cation

Hydrogen has one bonding electron and can bond to one other atom; it is monovalent. The electrons in this bond can be donated to hydrogen, giving the **hydride anion**, or can be donated to the other atom, generating a **proton**.

monovalent hydrogen   hydride

proton

The proton thus contains no electrons. This seems a rather unnecessary statement, but it means a proton can only be an acceptor of electrons, and can never donate any. *Curly arrows may be directed towards protons, but can never start from them!* This would be a serious mechanistic error. Nevertheless, most students seem to make this error at some time or other.

this is impossible!
a proton has no electrons

**bond making and
bond breaking**

carbon already has an octet of electrons;
making a new bond means another bond
must be broken

curly arrow represents
movement of two electrons
to make new bond to carbon

iodine has gained one
electron and becomes
negatively charged

curly arrows represent making and
breaking of bonds until electrons reside
on electronegative oxygen; this involves
breaking and making of double bonds

Counting the number of electrons on a particular atom becomes even more important when mechanisms become a little more complex and involve the making and breaking of bonds at the same atom. This is going to be routine at carbon atoms, and the statement above, that 'carbon cannot form more than four bonds', becomes an important guiding principle. Any mechanism that adds electrons to a carbon atom that is already carrying its full octet of electrons will also require the breaking of a bond and the removal of the excess electrons.

Initially, it is good idea to show nonbonding electrons in a mechanism, so that the number of electrons can be assessed and the correct charges defined. In due course, it is quicker to draw mechanisms without all the lone pairs, and it is normal practice to use representations showing just charges and only the lone pairs involved in subsequent bonding. The following mechanisms omit the lone pairs not involved in bonding, but are perfectly acceptable.

5.1.1 Bond polarity

The concept of bond polarity has been discussed in some detail in Chapter 2 (see Section 2.7). Because different atomic nuclei have a particular ability to attract electrons, bonds between unlike atoms may not be shared equally. This leads to a charge imbalance, with one of the atoms taking more than its share of the electrons. We refer to this as **bond polarity**. An atom that is more electronegative than carbon will thus polarize the bond, and we can consider the atoms as being partially charged. This is indicated in a structure by putting partial charges ($\delta+$ and $\delta-$) above the atoms. It can also be represented by putting an arrowhead on the bond in the direction of electron excess.

$$\overset{\delta+ \quad \delta-}{C—Br} \qquad\qquad C \rightarrow Br$$

bromine is more   *polarity shown*
electronegative   *as arrow*
than carbon

The relatively small difference in electronegativities between hydrogen and carbon means there is not going to be much polarity associated with a C–H bond. Most atoms other than hydrogen and carbon when bonded to carbon are going to be electron rich and bonds may therefore display considerable polarity. This is illustrated for carbon–oxygen and carbon–nitrogen single bonds. Double bonds show even greater polarity (see Section 7.1). This polarity helps us to predict chemical behaviour, and is crucial to our prediction of chemical mechanisms.

typical mechanisms where lone pairs,
apart from those involved in
subsequent bonding, are omitted

polarity in C–O and C–N bonds

polarity in C=O and C=N bonds

5.1.2 Nucleophiles, electrophiles, and leaving groups

Reagents are classified as nucleophiles or electrophiles.

Nucleophiles are electron-rich, nucleus-seeking reagents, and typically have a negative charge (anions) or a lone pair.

charged nucleophiles (anions)

$$RO\overset{..}{\underset{..}{:}}^{\ominus} \qquad :\overset{..}{\underset{..}{Br}}:^{\ominus} \qquad \overset{\xi}{\underset{|}{-}}\overset{..}{\underset{..}{C}}\overset{\ominus}{-\xi} \qquad ^{\ominus}:C{\equiv}N:$$

uncharged nucleophiles (lone pair)

$$H_2\overset{..}{O}: \qquad R\overset{..}{N}H_2 \qquad R\overset{..}{\underset{..}{S}}H$$

Compounds with multiple bonds, e.g. alkenes, alkynes, aromatics, can also act as nucleophiles in so-called electrophilic reactions (see Chapter 8).

Electrophiles are electron-deficient, electron-seeking reagents, and typically have a positive charge (cations) or are polarizable molecules that can develop an electron-deficient centre.

charged electrophiles (cations)

$$H^{\oplus} \qquad \overset{\xi}{\underset{|}{-}}\overset{\oplus}{\underset{|}{C}}-\xi \qquad :\overset{..}{O}{=}\overset{\oplus}{N}{=}\overset{..}{O}:$$

uncharged electrophiles (polarization)

$$\overset{\delta+}{:\overset{..}{Cl}}{\rightarrow}\overset{\delta-}{\overset{..}{O}H} \qquad \overset{\xi}{\underset{|}{-}}\overset{\delta+}{\underset{|}{C}}{\rightarrow}\overset{\delta-}{:\overset{..}{Br}:} \qquad \overset{\delta+}{C}{=}\overset{\delta-}{\overset{..}{O}:}$$

polarization is a consequence of atoms with different electronegatives

$$\overset{\delta+}{:\overset{..}{Br}}{\rightarrow}\overset{\delta-}{\overset{..}{Br}:}$$

polarization is brought about by the proximity of a nucleophile

Many reactions will involve both nucleophiles and electrophiles. These may then be classified as nucleophilic if the main change to the substrate involves attack of a nucleophile, or electrophilic if the principal change involves attack of the substrate onto an electrophile. This distinction will become clearer in due course (see Section 7.1). The electron-rich species is always regarded as the attacking agent.

$$Nu^{\ominus} \overset{\frown}{} E^{\oplus} \longrightarrow Nu{-}E$$

nucleophile   electrophile

Leaving group is the terminology used for ions or neutral molecules that are displaced from a reactant as part of a mechanistic sequence. Frequently, this displacement is the consequence of a nucleophile attacking an electrophile, and where the electrophile carries a suitable leaving group.

$$\underset{\text{nucleophile}}{Nu^{\ominus}} \quad \overset{\xi}{\underset{|}{-}}\overset{|}{\underset{|}{C}}{\frown}L \longrightarrow Nu{-}\overset{|}{\underset{|}{C}}{-}\xi \quad L^{\ominus}$$

electrophile   leaving group

Good leaving groups are those that form stable ions or neutral molecules after they leave the substrate.

We shall frequently need to write mechanisms involving general nucleophiles, electrophiles or leaving groups. Standard abbreviations are Nu^- or Nu: for a nucleophile (charged or uncharged), E^+ for an electrophile, and L^- or L: for a leaving group. In many instances, an electrophile containing a leaving group would simply be represented by C–L.

5.2   Radical reactions

Radicals (sometimes termed free radicals) are uncharged high-energy species with an unpaired electron, and may contain one or more atoms:

$$H^{\bullet} \qquad :\overset{..}{Cl}^{\bullet} \equiv Cl^{\bullet} \qquad H_3C^{\bullet}$$

H atom   Cl atom   methyl radical

the unpaired electron is always shown

For clarity, nonbonding electrons are usually omitted, though in order to propose meaningful mechanisms it is important to remember how many electrons are associated with each atom. The unpaired electron must always be shown.

In the formation of radicals, a bond is broken and each atom takes one electron from the pair constituting the bond. Bond-making and bond-breaking processes are indicated by single-headed (fishhook) curly arrows representing the movement of **one electron**.

radicals

*fish-hook curly arrow
representing the movement
of one electron*

A radical mechanism sequence requires three distinct types of process: initiation, propagation, and termination. Initiation is the formation of two radical species by bond fission, whereas propagation involves reaction of a radical with a neutral molecule, a process that leads to generation of a new radical. Because radicals are so reactive, the propagation process may continue as long as reagent molecules are available. Finally, the reaction is brought to a

*fission of single bond (two electrons)
creates two radicals each containing
one unpaired electron)*

initiation

*new bond formed by combination
of one electron from radical, and
one electron from single bond*

*this creates a
new radical*

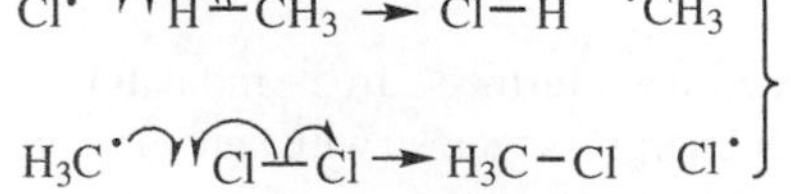

propagation

*the new radical reacts further,
generating another radical*

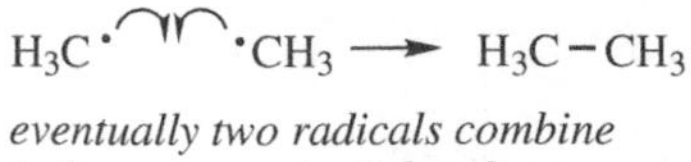

termination
(radical
pairing)

*eventually two radicals combine
to form a new single bond*

conclusion by the combination of two radical species, so that the unpaired electrons, one from each species, are combined into a new single bond. The radical-pairing termination step is analogous to a reversal of the initiation step. It occurs readily because of the reactivity of radicals; it follows, therefore, that the initiation step will require the input of a considerable amount of energy in order to dissociate the single bond.

In the propagation steps shown above, the radical propagates a further radical by causing fission of a single bond in the substrate. Many important radical reactions actually involve compounds with double bonds as substrates, and the π bond is cleaved during the radical addition reaction.

radical addition

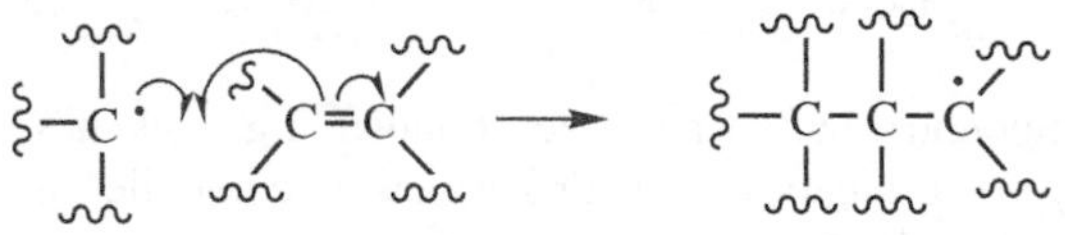

cleavage of π bond

*creation of a new
radical*

It makes good sense to draw free-radical mechanisms in the manner shown by these examples. However, shorter versions may be encountered in which not all of the arrows are actually drawn. These versions bear considerable similarity to two-electron curly arrow mechanisms, in that a fishhook arrow is shown attacking an atom, and a second fishhook arrow is then shown leaving this atom. The other electron movement is not shown, but is implicit. This type of representation is quite clear if the complement of electrons around a particular atom is counted each time; but, if in any doubt, use all the necessary fishhook arrows.

alternative representation of radical
addition omitting some curly arrows

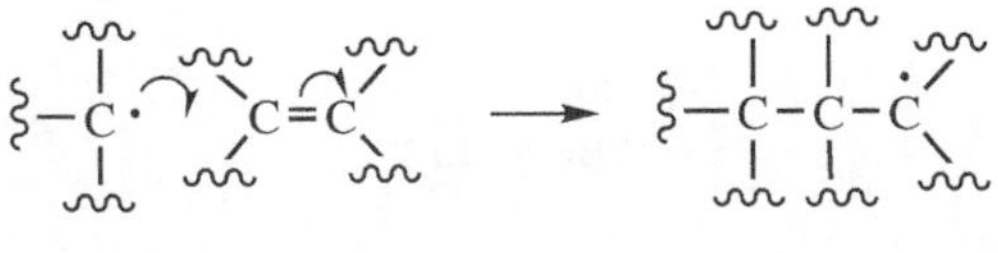

*electron movement gives
carbon nine electrons;
therefore, one must be lost
by transfer to next atom*

5.3 Reaction kinetics and mechanism

One of the ways in which we can obtain information about a mechanistic sequence is to study the rate of reaction. The dependence of the reaction rate on the concentration of reagents and other variables indicates the number and nature of the molecules involved in the **rate-determining step** of the reaction. The rate-determining step is defined as the slowest transformation in the sequence, all other transformations proceeding much faster than this. Consider a turnstile at a football match. This limits the rate at which spectators enter the ground. How rapidly people walk towards the turnstile or away from it once they are in the ground cannot influence the rate at which they get through the turnstile.

The **rate of reaction** is given by the equation

concentrations of A, B, ...

$$\text{Rate} = k \, [\text{A}][\text{B}]...$$

rate constant

in which k is the rate constant, and A, B, etc. are the variables on which the rate depends. Square brackets are used to indicate concentrations. It is rare for more than two variables to be involved, and often it is only one. The most common types of rate expression are given in Table 5.1.

In first-order reactions, the rate expression depends upon the concentration of only one species, whereas second-order reactions show dependence upon two species, which may be the same or different. The **molecularity**, or number of reactant molecules involved in the rate-determining step, is usually equivalent to the kinetic reaction order, though there can be exceptions. For instance, a bimolecular reaction can appear to be first order if there is no apparent dependence on the concentration of one of the

Table 5.1 Rate expressions, reaction order, and molecularity

Rate expression	Reaction order	Probable reaction	Molecularity
$k[\text{A}]$	first	$\text{A} \rightarrow$	unimolecular
$k[\text{A}][\text{B}]$	second	$\text{A} + \text{B} \rightarrow$	bimolecular
$k[\text{A}]^2$	second	$\text{A} + \text{A} \rightarrow$	bimolecular

reagents. Such a situation might occur when the solvent was also one of the reagents.

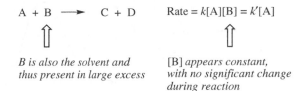

B is also the solvent and thus present in large excess

[B] *appears constant, with no significant change during reaction*

Despite occasional apparent anomalies such as this, the rate expression gives us valuable information about the likely reaction mechanism. If the reaction is unimolecular, the rate-determining step involves just one species, whereas the rate-determining step involves two species if it is bimolecular. As indicated in Table 5.1, we can then deduce the probable reaction, and our proposed mechanism must reflect this information. The kinetic rate expressions will be considered further as we meet specific types of reaction.

5.4 Intermediates and transition states

Any realistic mechanism will include a number of postulated structures, perhaps charged structures or radicals, which lie on the pathway leading from reactants to products. Some of these intervening structures are termed intermediates, and others transition states. These are differentiated by their stability, and whether they can be detected by appropriate analytical methods. A diagram that follows the energy change during the reaction can illustrate their involvement. The x-coordinate is usually termed 'reaction coordinate', and in many cases equates to time, though the possibility that the reaction is reversible prevents us from showing this as a simple time coordinate.

Consider the energy profile in Figure 5.1, in which reactants are converted into products. The difference between the energy of the reactants and products is called the **standard free energy change** $\Delta G°$ for the reaction. As shown, the change in energy is negative, so that the reaction liberates energy and is potentially favourable. It does not occur spontaneously, however, since the reactants need to acquire sufficient energy to collide and react. This energy is termed the **activation energy** – even gunpowder needs a match to set off the explosion! The high-energy peak in the curve is termed the **transition state** or sometimes,

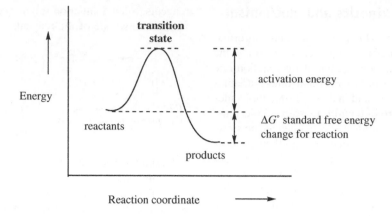

Figure 5.1 Energy profile diagram: transition state

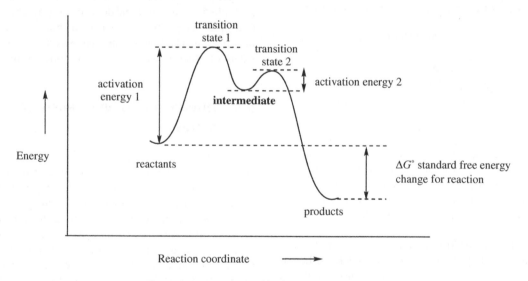

Figure 5.2 Energy profile diagram: intermediate

activated complex. This material cannot be isolated, or even detected.

In an alternative scenario, again with a negative energy change, the energy profile may appear different, as shown in Figure 5.2. In this case, there is again an activation energy required to set the reaction off, but this energy maximum is then followed by an energy minimum. The energy minimum represents an **intermediate** in the reaction pathway. It is converted into the products by overcoming a second activation energy, though this is likely to be considerably less than the first activation energy. Because the intermediate is at an energy minimum, this material may be stable and can be isolated, or it may be reactive

and short-lived, but nevertheless detectable. The two energy maxima represent different transition states.

The energy diagrams shown here are merely generalized examples. We shall meet some specific examples as we consider various reaction mechanisms.

5.5 Types of reaction

At first glance, there appear to be an infinite number of different chemical reactions, all of which will have to be remembered. A cursory look through any textbook of organic chemistry does little to dispel this fear. However, the beauty and strength of mechanism is that it allows us to predict chemical behaviour without having to remember lots of chemical reactions. A further reassuring fact is that virtually all of the chemical reactions can be classified according to a reaction type, and the number of distinct reaction types is actually rather few. We only need to consider reaction types according to what is achieved in the conversion, namely **substitution, elimination, addition**, or **rearrangement**. In general terms these may be represented as follows:

substitution

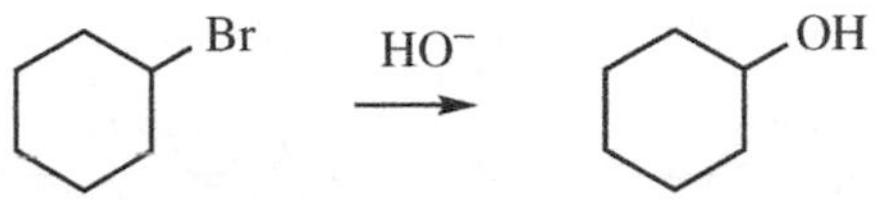

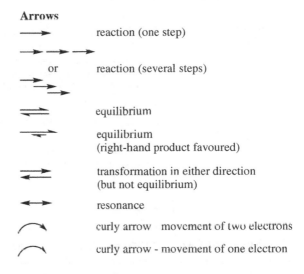

Br is substituted by OH

elimination

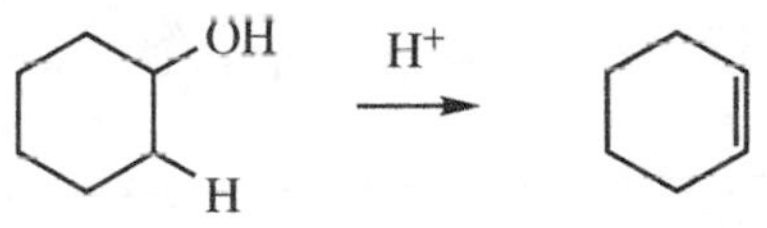

H_2O is eliminated to form double bond

addition

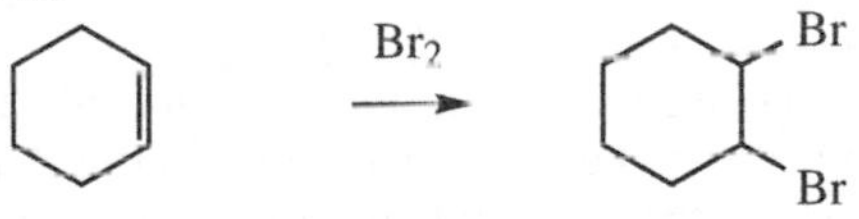

Br_2 is added across double bond

rearrangement

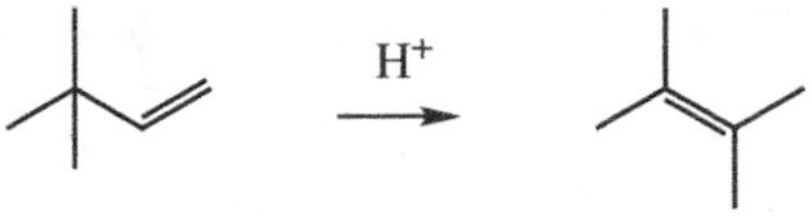

the molecular skeleton has been rearranged

We will then subdivide these reaction types according to the type of reagent that brings about the change, in order to rationalize typical reactions further. For example, addition reactions can be subdivided into **nucleophilic** addition, **electrophilic** addition, or **radical** addition. Whilst this does increase the number of permutations, we shall see that it is necessary to do this, and it is also perfectly logical for our understanding of how reactions occur.

5.6 Arrows

We have now encountered a number of different types of arrow routinely used in chemistry to convey particular meanings. We have met curly arrows used in mechanisms, double-headed resonance arrows, equilibrium arrows, and the simple single arrows used for reactions. This is a convenient point to bring together the different types and provide a checklist for future reference. We are also showing how additional information about a reaction may be presented with the arrow.

Arrows

→	reaction (one step)
→→→ or →→→	reaction (several steps)
⇌	equilibrium
⇌	equilibrium (right-hand product favoured)
⇌	transformation in either direction (but not equilibrium)
↔	resonance
⤴	curly arrow movement of two electrons
↷	curly arrow - movement of one electron

Information on arrows

A $\xrightarrow{\ a\ }$ B   reaction with **a** converts A into B

A $\xrightarrow{\ a/b\ }$ B   reaction with **a** in the presence of **b** converts A into B

A $\xrightarrow{\ a/\text{solvent}\ }$ B   reaction with **a** in suitable solvent converts A into B

A $\xrightarrow[t°,\,h\ \text{hr}]{\ a\ }$ B   reaction with **a** at t °C, for h hours converts A into B

A $\xrightarrow{\ a\ }$ $\xrightarrow{\ b\ }$ B

    or

A $\xrightarrow[\text{(ii)}\,b]{\text{(i)}\,a}$ B

   reaction with **a** first, then with **b** converts A into B

A $\underset{b}{\overset{a}{\rightleftarrows}}$ B   reagent **a** achieves conversion A → B, reagent **b** achieves conversion B → A

Box 5.1

Some common mistakes in drawing mechanisms

Experience tells us that whilst many students find mechanisms easy and logical, others despair and are completely bewildered. We cannot guarantee success for all, but we hope that by showing a few of the common mistakes we may help some of the latter group join the former. In order to make the examples chosen as real as possible, these have all been selected from students' examination answers. The mechanisms relate to reactions we have yet to meet, but this is not important. At this stage, it is the manipulation of curly arrows that is under consideration. You may wish to return to this section later.

Mistakes with valencies   As electrons are moved around via curly arrows, it is imperative to remember how many electrons are associated with a particular atom, and not to exceed the number of bonds permitted. The usual clanger is five-valent carbon, typically the result of making a new bond to a fully substituted carbon (four bonds, eight electrons) without breaking one of the old bonds. This is the case in the example shown.

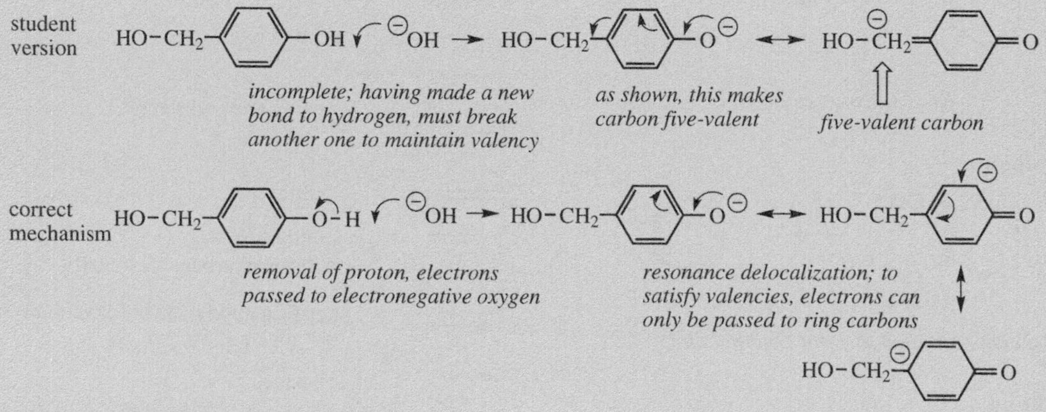

Mistakes with formal charges   It is also important when counting electrons to assign any formal charge as necessary. It is all too common to see hydroxide presented with a lone pair, but without any charge. Unfortunately, subsequent ionic reactions then just do not 'balance'. If one considers that hydroxide is derived by ionization of NaOH, or by loss of a proton from H_2O, this problem should not arise.

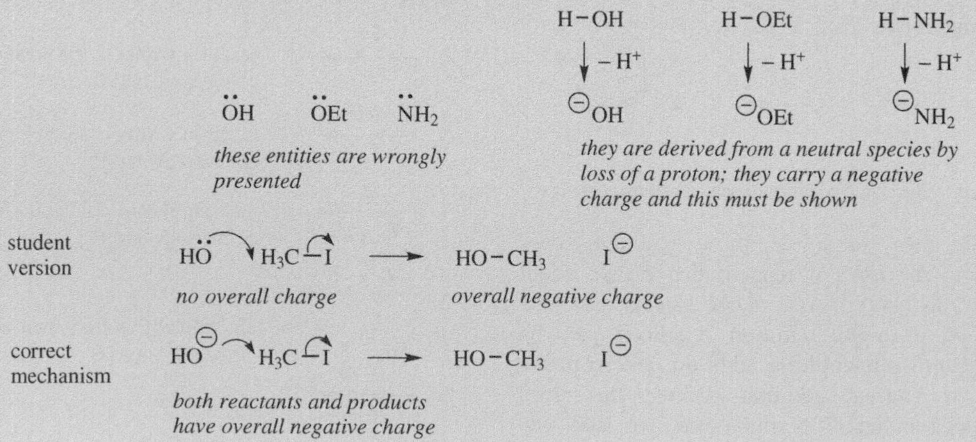

Arrows from protons   Ask yourself how many electrons are there in a proton? We trust the answer is none, and you will thus realize that arrows representing movement of electrons can never ever start from a proton. It seems that this mistake is usually made because, if one thinks of protonation as addition of a proton, it is tempting to show the proton being put on via an arrow. With curly arrows, we must always think in terms of electrons.

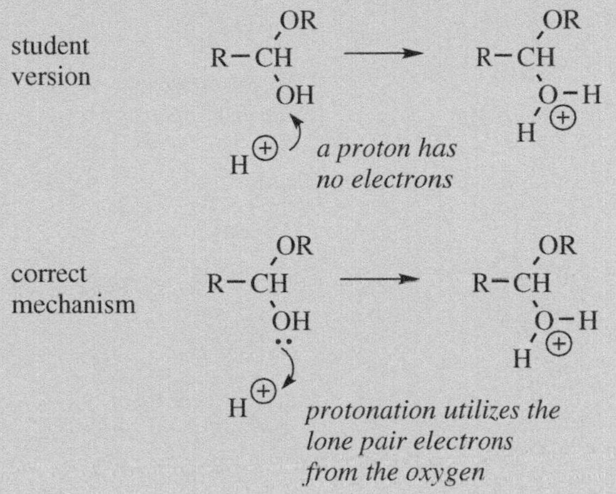

We were even less keen on the second example, where, in the resonance delocalization step, an arrow is shown taking electrons away from a positive charge and creating a new positive centre.

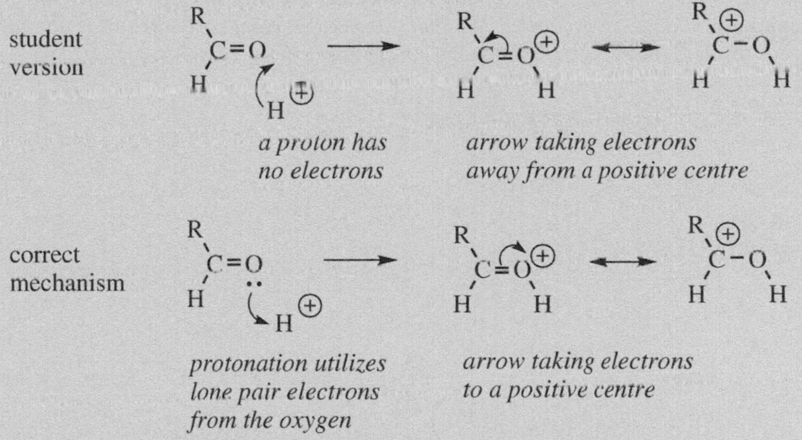

Vague arrows   Some mechanisms have arrows going in all sorts of directions. Arrows must 'flow' from start to finish; they should not veer off in different directions. Many of the arrows do not represent electron movements, and it would appear that, as a last resort, students have tried to memorize the mechanism rather than rationalizing it. This is both dangerous and really rather unnecessary. The logical approach gets the right answer, requires relatively little effort, and cuts out the need to learn the mechanism. Mechanisms should not be learnt; they should be deduced.

Box 5.1 (continued)

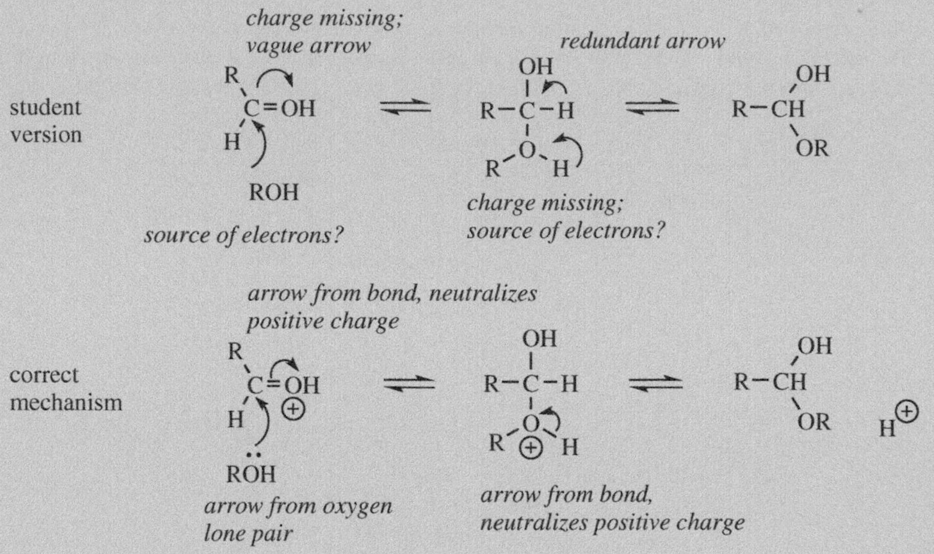

student version

*charge missing;
vague arrow*

redundant arrow

source of electrons?

*charge missing;
source of electrons?*

correct mechanism

*arrow from bond, neutralizes
positive charge*

*arrow from oxygen
lone pair*

*arrow from bond,
neutralizes positive charge*

In this example, the student remembered that a series of curly arrows was required, and they are generally in the right places, but not coming from electron-rich species, and not flowing in the right direction. This is typical of trying to remember a mechanism, which then fails to obey the general rules.

Too many steps at once   It is tempting to draw a mechanism with a series of curly arrows leading to the product via the minimum number of structures. We can often use several curly arrows in the same structure, but only provided we do not destroy the rationale for the mechanism.

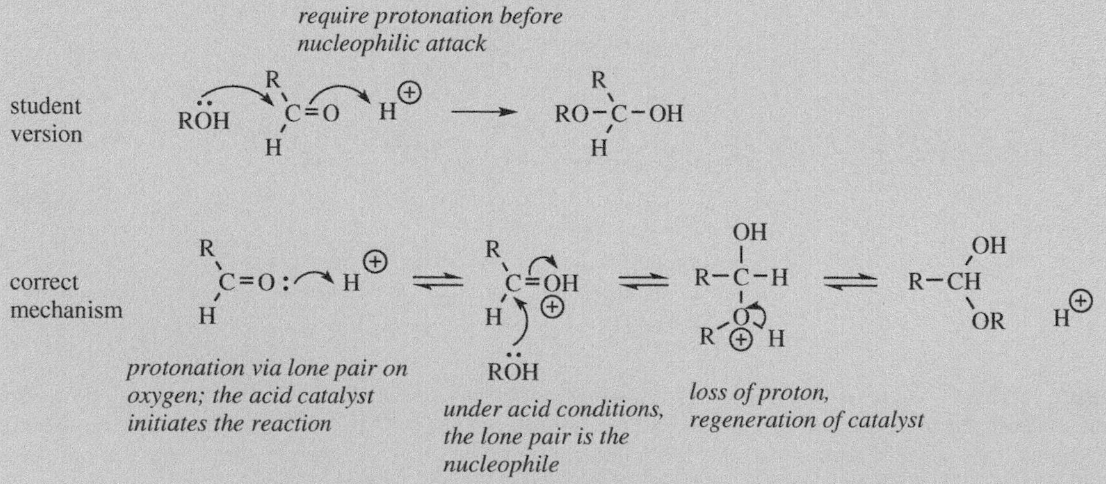

student version

*require protonation before
nucleophilic attack*

correct mechanism

*protonation via lone pair on
oxygen; the acid catalyst
initiates the reaction*

*under acid conditions,
the lone pair is the
nucleophile*

*loss of proton,
regeneration of catalyst*

In the example shown, the two curly arrows suggest a concerted interaction of three entities. This is improbable, and does not tell us why the reaction should actually take place. Using the longer sequence, we see that the acid catalyst activates the carbonyl group towards nucleophilic attack, and is later regenerated.

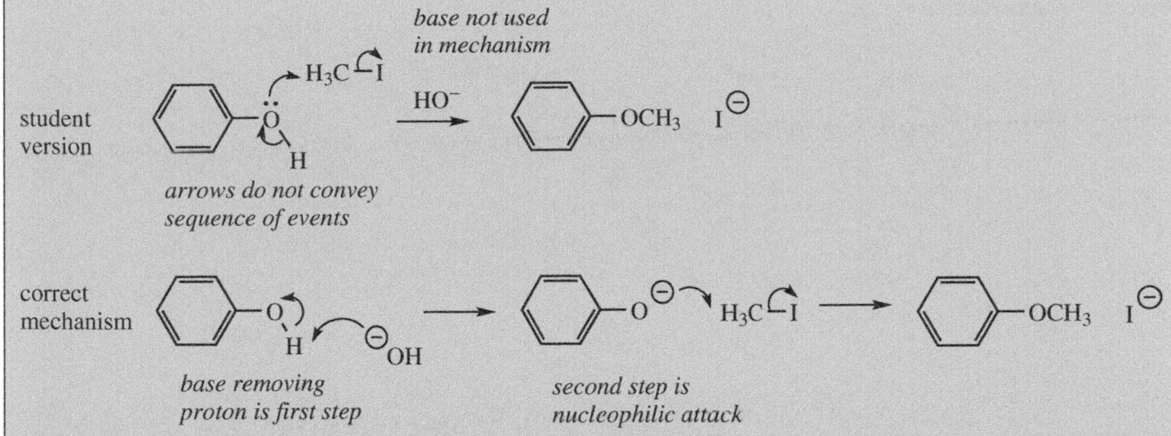

student version

base not used in mechanism

H_3C—I

HO^- ⟶ ⬡—OCH_3 $I^{\ominus}$

arrows do not convey sequence of events

correct mechanism

base removing proton is first step

second step is nucleophilic attack

The second example also emphasizes that base is needed to generate the nucleophile, the charged phenoxide being a better nucleophile than the phenol.

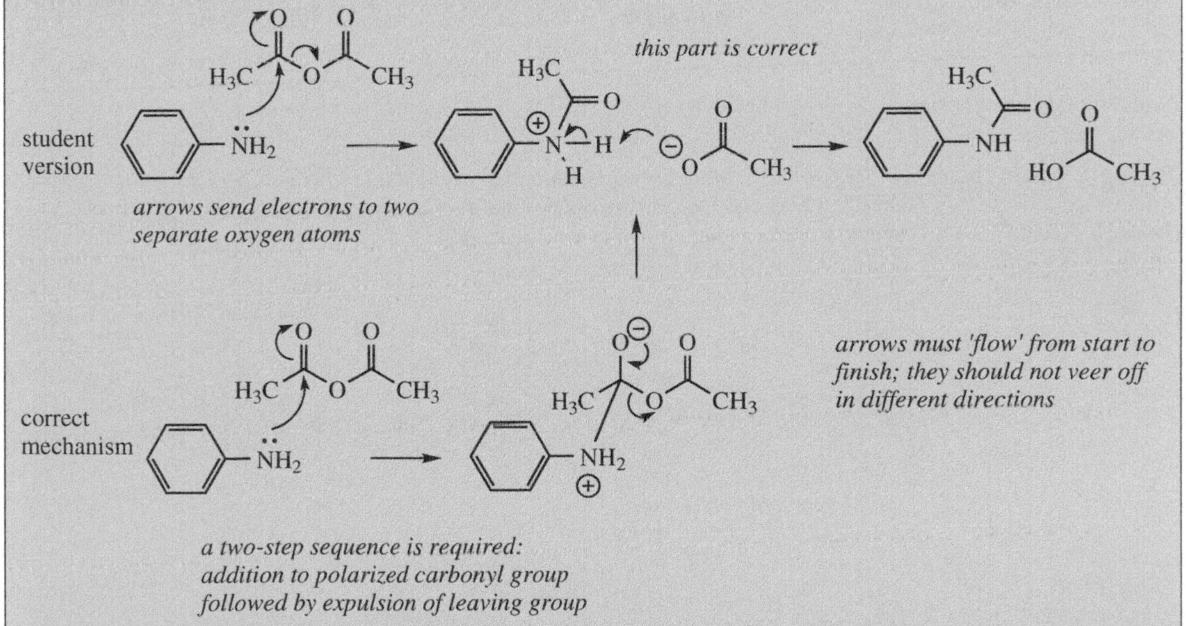

student version

arrows send electrons to two separate oxygen atoms

this part is correct

correct mechanism

arrows must 'flow' from start to finish; they should not veer off in different directions

a two-step sequence is required: addition to polarized carbonyl group followed by expulsion of leaving group

In the third example, arrows veer off in different directions, rather than flowing smoothly from start to finish. This mechanism is wrong in that an intermediate has been omitted.

Unrealistic ionizations   It is often necessary to ionize one of the reagents to initiate a reaction, and this requires careful consideration if the mechanism is to be realistic. For example, we should not attempt to protonate substrates under basic conditions, and we are unlikely to generate anionic species under acidic conditions. These are fairly obvious limitations, but are frequent mistakes.

Box 5.1 (continued)

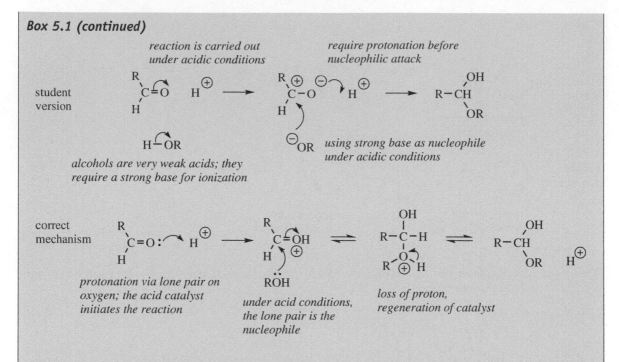

Some thought about relative acidity and basicity is also sensible; ionization of alcohols does not occur without a strong base, as suggested in the example.

Primary carbocations   Should you wish to use carbocations in a reaction mechanism, you must consider the relative stability of these entities. Tertiary carbocations are OK, and in many cases so are secondary carbocations. Primary carbocations are just not stable enough, unless there is the added effect of resonance, as in benzylic or allylic systems.

Arrows curled the wrong way   Can arrows curl the wrong way? Yes, they can, as the example shows. You should always understand that the arrowhead is depositing electrons between the start of the arrow and the atom it is approaching, so that the new bond is formed at the inside of the curl.

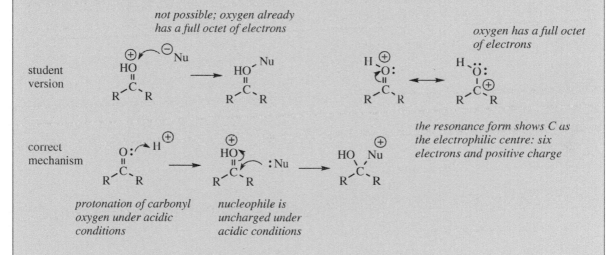

The electrophilic addition to alkenes is one of the occasions when the direction of the curl matters and can convey formation of different products. Although the product shown is correct, the curly arrow is wrong.

Making bonds to O^+ or N^+

It is tempting to consider O^+ and N^+ as electron-deficient species and, therefore, open to attack by nucleophiles. Here, we must count electrons to appreciate the true nature of these charged systems.

Both O^+ bonded to three atoms and N^+ bonded to four atoms are isoelectronic with tetravalent carbon, in other words, they have a full octet of electrons. Despite the positive charge, these atoms are not electron-deficient and are unable to make a new bond with the electron-rich nucleophiles.

6

Nucleophilic reactions: nucleophilic substitution

As the term suggests, a substitution reaction is one in which one group is substituted for another. For nucleophilic substitution, the reagent is a suitable nucleophile and it displaces a leaving group. As we study the reactions further, we shall see that mechanistically related competing reactions, eliminations and rearrangements, also need to be considered.

6.1 The S_N2 reaction: bimolecular nucleophilic substitution

The abbreviation S_N2 conveys the information 'substitution–nucleophilic–bimolecular'. The reaction is essentially the displacement of one group, a leaving group, by another group, a nucleophile. It is a bimolecular reaction, since kinetic data indicate that two species are involved in the rate-determining step:

$$\text{Rate} = k[\text{RL}][\text{Nu}]$$

where Nu is the nucleophile, RL the substrate containing the leaving group L, and k is the rate constant.

In general terms, the reaction can be represented as below

Differences in electronegativities (see Section 2.7) between carbon and the leaving group atom lead to bond polarity. This confers a partial positive charge on the carbon and facilitates attack of the nucleophile. As the nucleophile electrons are used to make a new bond to the carbon, electrons must be transferred away to a suitable acceptor in order to maintain carbon's octet. The suitable acceptor is the electronegative leaving group.

The nucleophile attacks from the side opposite the leaving group – electrostatic repulsion prevents attack in the region of the leaving group. This results in an **inversion process** for the other groups on the carbon centre under attack, rather like an umbrella turning inside out in a violent gust of wind. The process is **concerted**, i.e. the bond to the incoming nucleophile is made at the same time as the bond to the leaving group is being broken. As a consequence, the mechanism involves a high-energy **transition state** in which both nucleophile and leaving group are partially bonded, the Nu–C–L bonding is linear, and the three groups X, Y, and Z around carbon are in a planar array. This is the natural arrangement to

S_N2 reaction

nucleophile

positive end of polarized C–Br bond

leaving group

partially bonded transition state

$sp^2 + p$

note inversion of configuration due to rearside attack

this is a concerted reaction

Essentials of Organic Chemistry Paul M Dewick
© 2006 John Wiley & Sons, Ltd

minimize steric interactions if we wish to position five groups around an atom, and will involve three sp^2 orbitals and a p orbital as shown. The p orbital is used for the partial bonding; note that we cannot have five full bonds to a carbon atom. The energy profile for the reaction (Figure 6.1) proceeds from reactants to products via a single high-energy transition state (see Section 5.4).

Figure 6.1   Energy profile: S_N2 reaction

The rate of an S_N2 reaction depends upon several variables. These are:

- the nature of the substituents bonded to the atom attacked by the nucleophile;

- the nature of the nucleophile;

- the nature of the leaving group;

- solvent effects.

We can consider these in turn.

6.1.1   The effect of substituents

The S_N2 mechanism requires attack of a nucleophile at the rear of the leaving group, and consequently the size of the groups X, Y, and Z will influence the ease of approach of the nucleophile. Experimental evidence shows the relative rates for S_N2 reactions of halides are as shown in Table 6.1. This is primarily a result of steric hindrance increasing as

Box 6.1

S_N2 reactions: the racemization of 2-iodobutane

The inversion in an S_N2 reaction can be demonstrated in a rather simple experiment. If $(+)$-(R)-2-iodobutane is heated in acetone solution, it is recovered unchanged. However, when sodium iodide is added to the mixture, there is no apparent chemical change, but the optical activity gradually diminishes until it becomes zero, i.e. racemic $(\pm)$-(RS)-2-iodobutane has been formed (see Section 3.4.1).

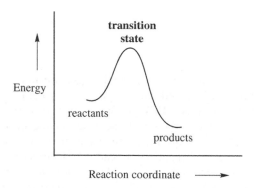

In this reaction, an equilibrium is set up. The nucleophile, iodide, is the same as the leaving group. Therefore, inversion of configuration merely converts the $(+)$-isomer into the $(-)$-isomer. As a result, the optical activity gradually disappears and ultimately becomes zero as the mixture becomes the racemic $(\pm)$-form. We are never going to get complete conversion of the $(+)$- into $(-)$-enantiomer because the reverse reaction will also occur. This is mechanistically identical to the forward reaction, so either $(+)$- or $(-)$-2-iodobutane as starting material would give racemic product, i.e. it is a racemization reaction.

This is an unusual reaction, in that the energy of the products will be identical to the energy of the reactants, though the interconversion of isomers involves an activation energy that must be overcome by the application of heat.

Table 6.1 Effect of structure on rates of S$_N$2 reactions

Halide	Relative rate of reaction	Class of halide
CH$_3$–	30	primary
CH$_3$CH$_2$–	1	primary
(CH$_3$)$_2$CH–	0.03	secondary
(CH$_3$)$_3$C–	0	tertiary

one goes from primary to secondary to tertiary compounds. With the *tert*-butyl group, approach of the nucleophile is hindered by three methyl groups, so much so that the S$_N$2 reaction is not normally possible.

Nu$^{\ominus}$ ----→ C—X

nucleophile

H$_3$C, H$_3$C, H$_3$C

leaving group

*methyl groups
hinder approach
of nucleophile*

In general terms then, the S$_N$2 reaction is only important for primary and secondary substrates, and the rate of reaction for primary substrates is considerably greater than that for secondary substrates. Should a reaction be attempted with tertiary substrates, one does not usually get substitution, but alternative side-reactions occur (see Section 6.4).

If the potential leaving group is attached to unsaturated carbon, as in vinyl chloride or phenyl chloride, attack by nucleophiles is also extremely difficult, and these compounds are very unreactive in S$_N$2 reactions compared with simple alkyl halides. In these cases, the reason is not so much steric but electrostatic, in that the nucleophile is repelled by the electrons of the unsaturated system. In addition, since the halide is attached to carbon through an sp^2-hybridized bond, the electrons in the bond are considerably closer to carbon than in an sp^3-hybridized bond of an alkyl halide (see Section 2.6.2). Lastly, resonance stabilization in the halide gives some double bond character to the C–Hal bond. This effectively strengthens the bond and makes it harder to break. This lack of reactivity is also true for S$_N$1 reactions (see Section 6.2).

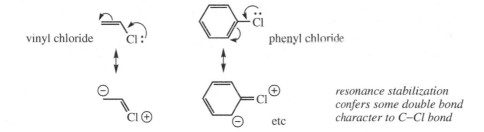

vinyl chloride   Cl:

phenyl chloride

Cl$^{\ominus}$

Cl$^{\oplus}$

Cl$^{\oplus}$

$^{\ominus}$ etc

*resonance stabilization
confers some double bond
character to C–Cl bond*

6.1.2 Nucleophiles: nucleophilicity and basicity

The S$_N$2-type reaction can be considered simply as being initiated by attack of a nucleophile onto the electron-deficient end of a polarized bond X–Y.

Nu :$^{\ominus}$ ⤳ $\overset{\delta+}{X}$–$\overset{\delta-}{Y}$ B :$^{\ominus}$ ⤳ $\overset{\delta+}{H}$–$\overset{\delta-}{Y}$

nucleophilic attack   acidity

If X = H, then this equates to removal of a proton and we would consider the nucleophile to be a base (see Section 4.1). It follows that there is going to be

a close relationship between a group's capacity to act as a nucleophile, i.e. **nucleophilicity**, and its ability to act as a base, i.e. **basicity**. Thus, the hydroxide ion can act as a nucleophile or as a base.

HO$^{\ominus}$ ⤳ $\overset{\delta+}{C}$–$\overset{\delta-}{Br}$ HO$^{\ominus}$ ⤳ H–O–C

*hydroxide acting
as nucleophile*   *hydroxide
acting as base*

In many cases, nucleophilicity can be correlated with basicity, and this forms a helpful way of predicting how good a potential nucleophile may be. The sequences of relative basicity given in Table 6.2

Table 6.2 Nucleophilicity and basicity with N and O nucleophiles

Base (pK_a conjugate acid)		Base (pK_a conjugate acid)	
H_2N^- (38)	Decreasing basicity and decreasing nucleophilicity ↓	$C_2H_5O^-$ (16)	Decreasing basicity and decreasing nucleophilicity ↓
$C_2H_5NH_2$ (10.6)		HO^- (15.7)	
H_3N (9.2)		PhO^- (10.0)	
$PhNH_2$ (4.6)		$CH_3CO_2^-$ (4.8)	
		H_2O (-1.7)	

are also reflected in relative nucleophilicities. The approximation works best for comparisons where the identity of the attacking atom is the same, e.g. N, or O, as illustrated in Table 6.2.

The correlation is useful but not exact. This is because basicity is a measure of the position of equilibrium between a substrate and its conjugate acid (see Section 4.4), whereas nucleophilicity relates to a rate of reaction. The above relationship breaks down when one looks at atoms in the same column of the periodic table. As atomic number increases, basicity decreases, whilst nucleophilicity actually increases (Table 6.3). This originates from the size of the atom, so that electrons associated with larger atoms become less localized, consequently forming weaker bonds with protons (see acidity of HX, Section 4.3.2). On

the other hand, electrons in the larger atoms are more easily polarizable, and it becomes easier for them to be donated to an electrophile; this leads to greater nucleophilicity.

Despite these inconsistencies, there are two important features worth remembering

- an anion is a better nucleophile than the uncharged conjugate acid;

- strong bases are good nucleophiles.

Table 6.4 summarizes relative nucleophilicity for some common reagents.

Table 6.3 Nucleophilicity and basicity for atoms in same column of periodic table

		Base (pK_a conjugate acid)	
Decreasing basicity ↑	Decreasing nucleophilicity ↓	RS^- (10.5)	I^- (-10)
		RO^- (16)	Br^- (-9)
			Cl^- (-7)
			F^- (3.2)

Table 6.4 Nucleophilicity of common reagents

	Nucleophile	Name	
Very good nucleophiles	NC^-	Cyanide	Decreasing nucleophilicity ↓
	HS^-	Thiolate	
	I^-	Iodide	
Good nucleophiles	HO^-	Hydroxide	
	Br^-	Bromide	
	NH_3	Ammonia	
Reasonable nucleophiles	Cl^-	Chloride	
	$CH_3CO_2^-$	Acetate	
	F^-	Fluoride	
	CH_3OH	Methanol	
	H_2O	Water	

Box 6.2

Selective alkylation of morphine to codeine and pholcodine

Opium is a crude exudate obtained from the opium poppy *Papaver somniferum*, and it provides several medicinally useful alkaloids. One of these is **codeine**, which is widely used as a moderate analgesic. Opium contains only relatively small amounts of codeine (1–2%), however, and most of the codeine for drug use is obtained by semi-synthesis from morphine, which is the major component (12–20%) in opium. Conversion of morphine into codeine requires selective methylation of the phenolic hydroxyl. This can be achieved by an S$_N$2 reaction under basic conditions.

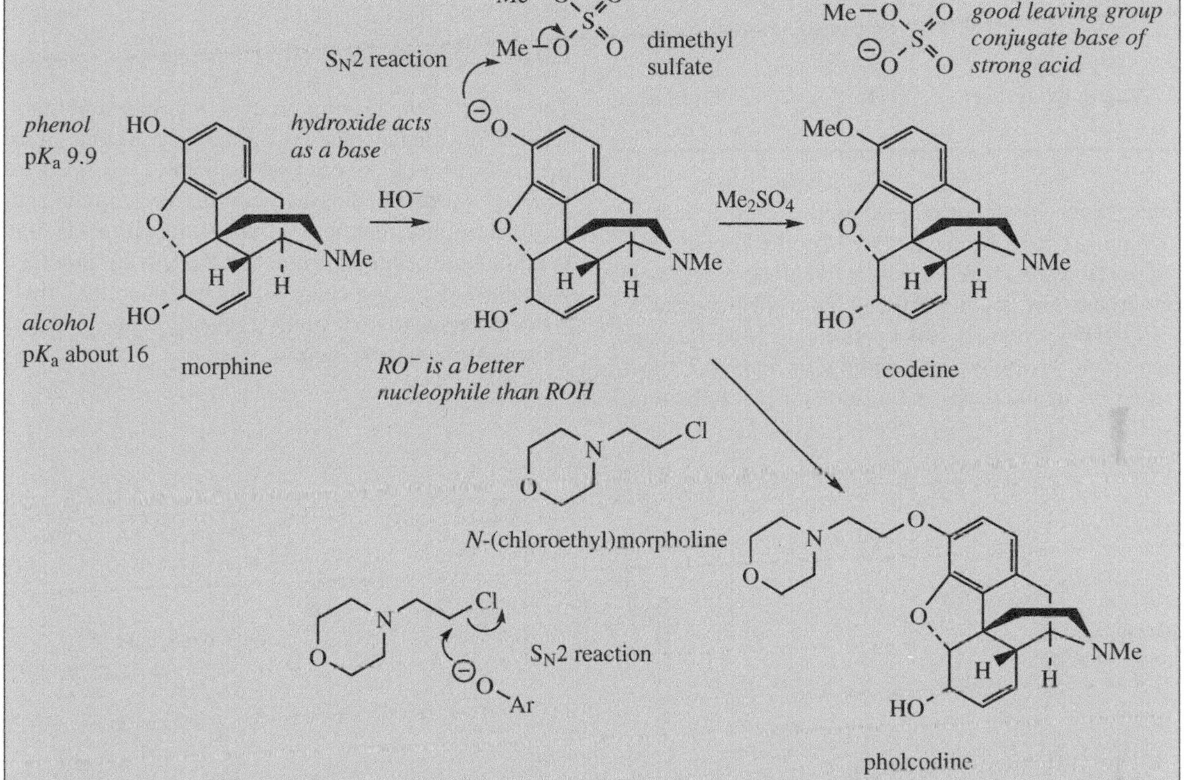

dimethyl sulfate

good leaving group conjugate base of strong acid

phenol pK$_a$ 9.9

S$_N$2 reaction

hydroxide acts as a base

alcohol pK$_a$ about 16   morphine

RO$^-$ is a better nucleophile than ROH

codeine

N-(chloroethyl)morpholine

S$_N$2 reaction

pholcodine

Morphine has two hydroxyls, but one is a phenol and the other is an alcohol. Because phenols (pK$_a$ about 10) are considerably more acidic than alcohols (pK$_a$ about 16), only the phenol will become ionized under mild basic conditions (see Box 4.10). Since the phenolate anion (charged) will then be a much better nucleophile than the alcohol hydroxyl (uncharged), the S$_N$2 reaction will selectively involve the phenolate group. The alcohol group does not react under these conditions. The methylating agent (electrophile) used in this reaction is dimethyl sulfate (see Section 7.13.1); the leaving group is the anion of a sulfuric acid ester, and is the conjugate base of a strong acid.

The same type of reasoning allows production of pholcodine, an effective cough suppressant, from morphine. In this semi-synthesis, the electrophile is *N*-(chloroethyl)morpholine, and the leaving group is chloride.

6.1.3 Solvent effects

Nucleophilicities are affected by solvent, and any correlations with basicity can break down in **protic solvents** like methanol or ethanol. This is because anions are stabilized by hydrogen bonding, and become solvated. These solvation molecules must be lost before the anion can attack as a nucleophile. Accordingly,

better solvents for nucleophilic substitution reactions are the so-called **aprotic polar solvents**, which contain no protons that allow hydrogen bonding to occur (Table 6.5). Anions, consequently, become more nucleophilic in aprotic polar solvents than they are in protic solvents.

As an example, the S$_N$2 reaction of chloride with methyl iodide leading to methyl chloride is some 10^6

Table 6.5 Aprotic polar solvents

Name	Formula	Abbreviation
Acetone	Me_2CO	
Acetonitrile	MeCN	
Dimethylformamide	$HCONMe_2$	DMF
Dimethylsulfoxide	Me_2SO	DMSO
Hexamethylphosphoric triamide	$(Me_2N)_3PO$	HMPT

times faster in dimethylformamide (DMF) than in methanol. This is because there is no hydrogen bonding possible in DMF. In sharp contrast, reaction in the structurally similar solvent N-methylformamide (HCONHMe), which still contains an N–H that can participate in hydrogen bonding, is only 45 times as fast as in methanol. Chloride ions actually form

stronger hydrogen bonds with methanol than with N-methylformamide, so there is an increase in reactivity, but hardly as dramatic as with the aprotic solvent DMF.

6.1.4 Leaving groups

The nature of the leaving group is a further important feature of nucleophilic substitution reactions. For the S_N2 reaction to proceed smoothly, we need to generate strong bonding between the nucleophile and the electrophilic carbon, at the same time as the bonding between this carbon and the leaving group is weakened. The high-energy transition state may thus be considered to require the general characteristics shown in the scheme below.

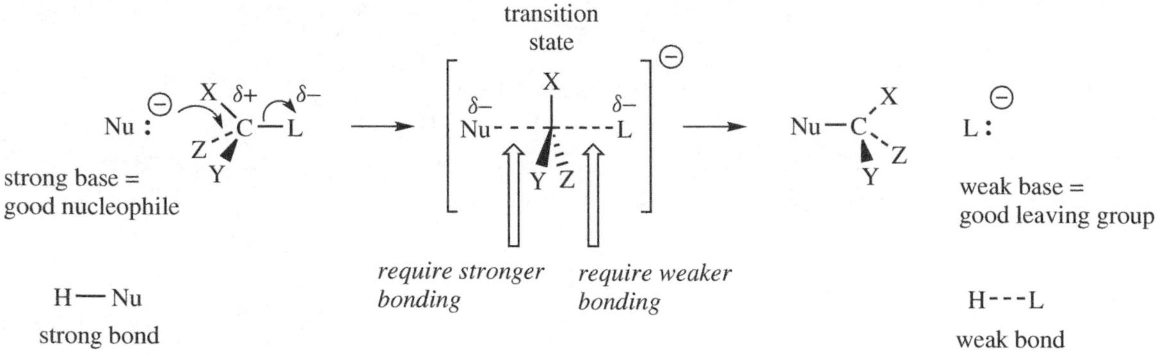

Good leaving groups are those that form stable ions or neutral molecules after they leave the substrate. Consequently, the capacity of a substituent to act as a leaving group can also be related to basicity. Strong bases (the conjugate bases of weak acids) are poor leaving groups; but, as we have seen above, they are good nucleophiles. On the other hand, weak bases (the conjugate bases of strong acids) are good leaving groups, but they make poor nucleophiles (Table 6.6).

We can now understand and predict why some nucleophilic substitution reactions are favoured and others are not. Thus, it is easy to convert methyl bromide into methanol by the use of hydroxide as nucleophile. On the other hand, it is not feasible to convert methanol into methyl bromide merely by using bromide as the nucleophile.

$$CH_3Br + HO^{\ominus} \longrightarrow CH_3OH + Br^{\ominus}$$

strong base   *weak base*
good nucleophile   *good leaving group*

$$CH_3OH + Br^{\ominus} \xrightarrow{\quad\times\quad} CH_3Br + HO^{\ominus}$$

weak base   *strong base*
poor nucleophile   *poor leaving group*

The difference here is primarily due to the nature of the leaving groups. Bromide is a weak base and a good leaving group, whereas hydroxide is a strong base and, therefore, a poor leaving group. Nevertheless, the latter transformation can be achieved by improving the ability of the leaving group to depart by carrying out the reaction under acidic conditions.

Table 6.6 Leaving groups and acidity of conjugate acid

	Leaving group	pK_a of conjugate acid
Good leaving groups	I$^-$	-10
	Br$^-$	-9
	Cl$^-$	-7
	Me$_2$S	-5.4
	CH$_3$SO$_3^-$ (MsO$^-$)	-2.6
	p-CH$_3$(C$_6$H$_4$)SO$_3^-$ (TsO$^-$)	-1.3
	H$_2$O	-1.7
	RCO$_2^-$	4.8
Moderate leaving groups	HS$^-$	7
	CN$^-$	9.1
	NH$_3$	9.2
	RNH$_2$	10.6
	RS$^-$	10.5
Poor leaving groups	F$^-$	3.2[a]
	HO$^-$	15.7
	RO$^-$	16
Very poor leaving groups	NH$_2^-$	38
	R$_2$N$^-$	36
	H$^-$	35
	R$^-$	50

[a]The C–F bond is one of the strongest, and fluoride is a poor leaving group. On pK_a values alone, this appears out of sequence.

methanol acts as base; protonation of oxygen

conjugate acid

nucleophilic substitution

weak base and neutral molecule, good leaving group

Thus, protonation of the substrate via the oxygen lone pair produces the conjugate acid. This now has greater polarization favouring nucleophilic attack and, most importantly, changes the leaving group from hydroxide (a strong base) to water (a weak base). The reaction is now facilitated and proceeds readily.

Chemical modification of poor leaving groups into good leaving groups may also be considered as a way of enhancing the ease of substitution reactions. Two important reagents that may be used with alcohols are p-toluenesulfonyl chloride (tosyl chloride) and methanesulfonyl chloride (mesyl chloride) (see Section 7.13.1). Both anions p-toluenesulfonate (**tosylate**) and methanesulfonate (**mesylate**) are excellent leaving groups, being the conjugate bases of strong acids (pK_a < 0).

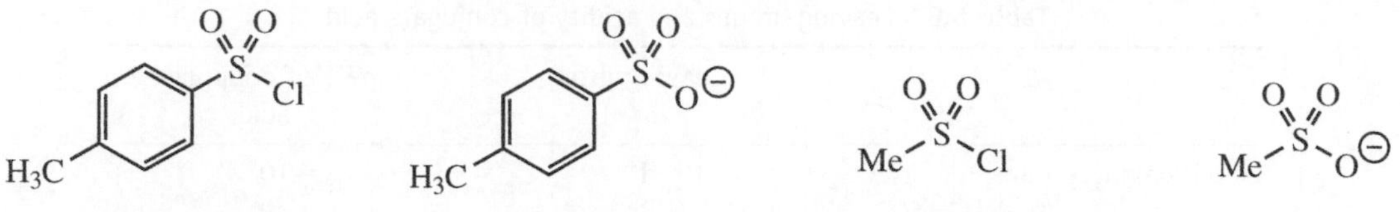

tosyl chloride
(*p*-toluenesulfonyl chloride)
TsCl

tosylate
TsO⁻

mesyl chloride
(methanesulfonyl chloride)
MsCl

mesylate
MsO⁻

Typically, these sulfonyl chlorides would be used to convert an alcohol into a sulfonate ester (see Section 7.13.1), and this would then be the substrate used for the nucleophilic substitution reaction.

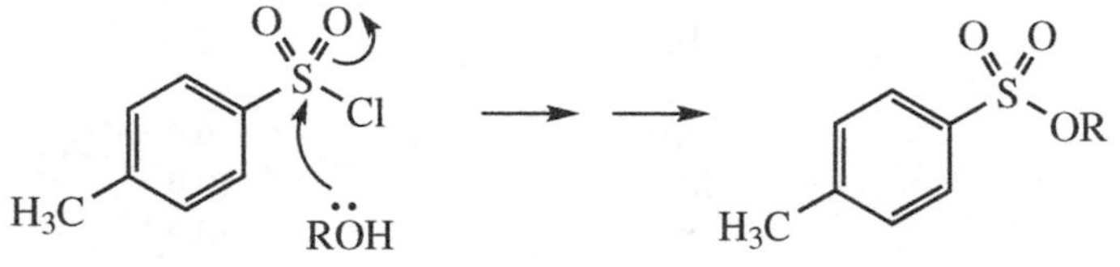

tosyl chloride
TsCl

tosyl ester
R—OTs

formation of sulfonate ester

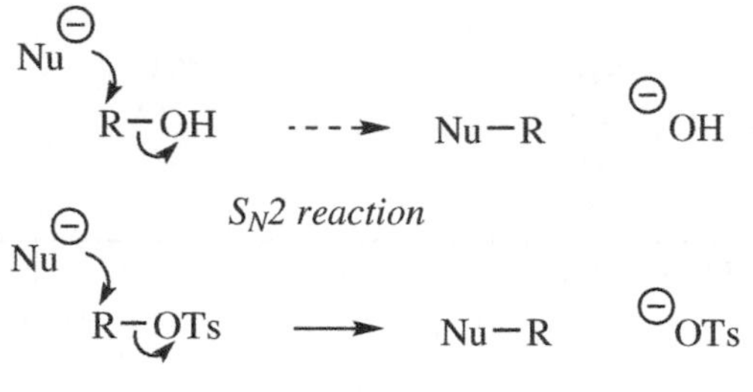

not favoured; hydroxide poor leaving group

S_N2 reaction

favoured; tosylate excellent leaving group

6.1.5 S_N2 reactions in cyclic systems

The inversion process accompanying S_N2 reactions may have particular significance in cyclic compounds. Thus, if we consider the disubstituted cyclopentane derivative shown undergoing an S_N2

reaction, we observe that the substituents were arranged in a *cis* relationship in the original compound and the consequence of inversion is formation of a *trans* product.

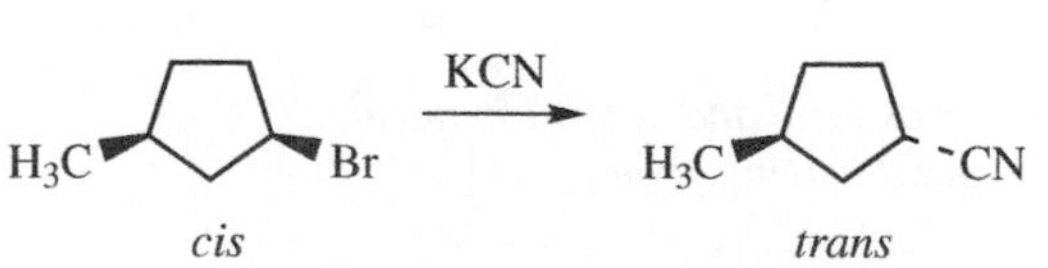

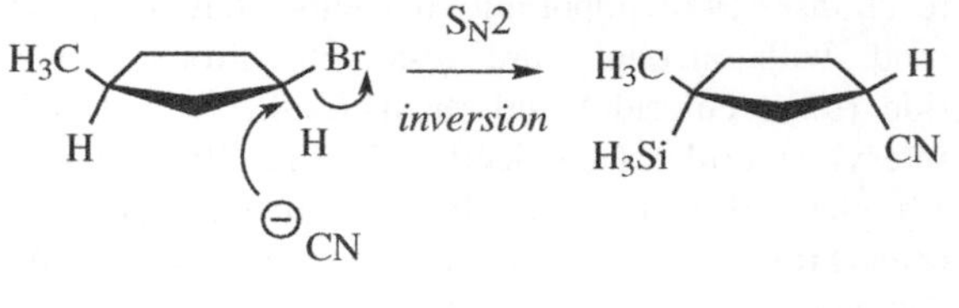

However, it is found that cyclic substrates tend to react much more slowly than do similar acyclic compounds. In small rings this is a consequence of ring strain; the S_N2 transition state requires the three groups other than the nucleophile and leaving group to be spaced 120° apart (see Section 6.1). This would be a severe problem for three- and four-membered rings (angles 60° and 90° respectively). It is not a problem for five-membered rings, where this is the normal bond angle in the ring, and such compounds react just as readily as acyclic compounds. Cyclohexyl derivatives react some 100-fold less readily than acyclic compounds, however, and ring strain cannot be an important factor: the 109° tetrahedral angles are the same as in an acyclic compound. In cyclohexyl compounds, the rate of reaction is apparently slowed by steric interactions with axial hydrogens.

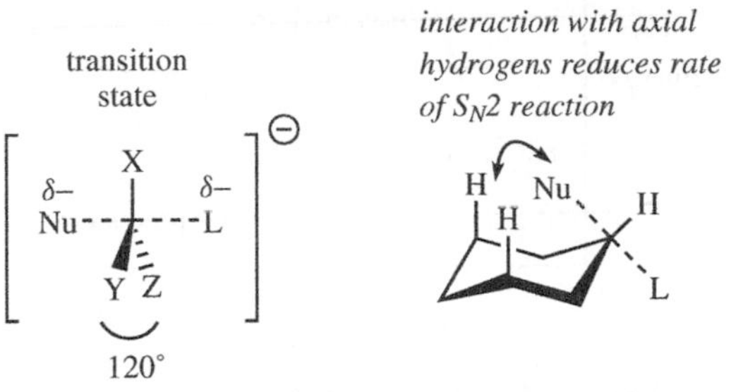

transition state

interaction with axial hydrogens reduces rate of S_N2 reaction

A consequence of the low rate of reaction in S_N2 reactions is that side-reactions in cyclohexane derivatives, especially elimination reactions (see Section 6.4.1), may often dominate over substitution.

6.2 The S_N1 reaction: unimolecular nucleophilic substitution

The abbreviation S_N1 conveys the information 'substitution–nucleophilic–unimolecular'. The reaction achieves much the same result as the S_N2 reaction, i.e. the replacement of a leaving group by a nucleophile, but is mechanistically different. It is unimolecular, since kinetic data indicate that only one species is involved in the rate-determining step:

$$\text{Rate} = k[\text{RL}]$$

where RL the substrate containing the leaving group L and k is the rate constant. Note that the nucleophile Nu does not figure in the rate equation.

In general terms, the reaction can be represented as below.

S_N1 reaction

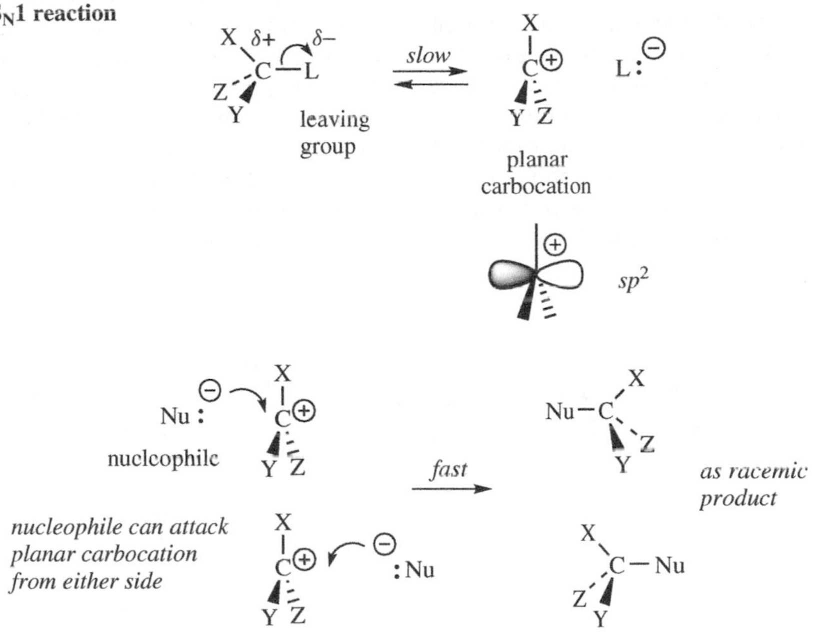

The first step of the reaction is loss of the leaving group, transforming the initial polarization ($\delta + /\delta -$) in the molecule into complete charge separation. To achieve this, we need a good leaving group as with S_N2 reactions, but also a structure in which the positively charged carbon, a **carbocation**, is suitably stabilized. This ionization step constitutes the slow part of the sequence, the **rate-determining step**, and,

since only one molecular species is involved, it is responsible for the observed kinetic data. Once the reactive carbocation is formed, it is rapidly attacked by a suitable nucleophilic species, thus generating the final product.

In S_N1 reactions, the nucleophilicity of the nucleophile is relatively unimportant. Because of the high reactivity of the carbocation, any nucleophile, charged or uncharged, will rapidly react. Therefore, as the rate equation shows, the nucleophile plays no part in controlling the overall reaction rate. We have shown carbocation formation as reversible; it would be if the leaving group recombined with the carbocation. If there is an excess of an alternative nucleophile, however, we shall get the required product.

The carbon atom of the carbocation has only six bonding electrons, and is a planar entity. The bonding electrons are in sp^2 orbitals, and there is also an unoccupied p orbital. The attacking nucleophile is able to attack from either face of this planar species; so, when X, Y, and Z are different, the product will turn out to be a mixture of two possible stereoisomers. As there is usually an equal probability of attack at each face, the product will be a **racemic mixture**. This is in marked contrast to the product from an S_N2 reaction, where there would be inversion of configuration and formation of a single enantiomer. The carbocation is an intermediate in the reaction sequence (Figure 6.2), and corresponds to a minimum in the energy profile (see Section 5.4). Its formation depends upon overcoming an activation energy, corresponding to that required for fission of the bond to the leaving group. Since the carbocation is very reactive, there will be a very much smaller activation energy for reaction with the nucleophile.

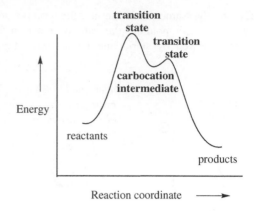

Figure 6.2   Energy profile: S_N1 reaction

Thus, *tert*-butanol reacts readily with HBr to give the corresponding bromide. This reaction could not proceed via the S_N2 mechanism because steric crowding prevents access of the nucleophile (see Section 6.1.1). Instead, an S_N1 mechanism can be formulated. The initial step would be protonation of the alcohol group to improve the nature of the leaving group, i.e. water rather than hydroxide, and allowing formation of the carbocation. Loss of the leaving group would be the slow, rate-determining step, but the following step, attack of the nucleophile onto the carbocation, would then be rapid.

tert-butanol
protonation of alcohol provides better leaving group

slow
rate-determining step

loss of leaving group

attack of nucleophile onto carbocation

tert-butyl bromide

Box 6.3

Why some S_N1 reactions do not lead to racemic products

Notwithstanding the remarks above concerning the equal probability of a nucleophile attacking either face of the planar carbocation and, therefore, producing a racemic product, many S_N1 reactions result in varying degrees of inversion and racemization. This can be rationalized in terms of preferential attack of the nucleophile from the face opposite the leaving group simply because, as the leaving group departs, it actually hinders attack from that side.

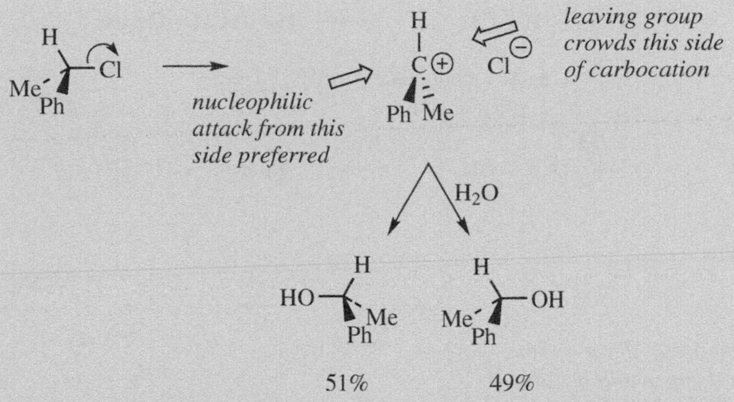

In the example shown, there is slightly more of the 'inverted' product in the reaction mixture, though the effect is not especially large. In other recorded examples, up to about 80% of the product might be the inverted form. It follows that the S_N2 process is accompanied by complete inversion, whereas an S_N1 process will involve racemization or partial inversion.

6.2.1 The effect of substituents

The S_N1 mechanism requires initial loss of the leaving group to generate a reactive carbocation.

Table 6.7 Effect of structure on rates of S_N1 reactions

Halide	Relative rate of reaction	Class of halide
CH_3-	1	primary
CH_3CH_2-	1	primary
$(CH_3)_2CH-$	12	secondary
$(CH_3)_3C-$	1.2×10^6	tertiary

Experimental evidence concerning the relative rates for S_N1 reactions of halides is listed in Table 6.7. The differences in reactivity reflect structural features that stabilize the intermediate carbocation. **Carbocations are stabilized** by the **electron-donating effect** of alkyl groups, which help to disperse the positive charge. We have noted that alkyl groups have a modest electron-donating effect (see Section 4.3.3). In carbocations, this is not a simple inductive effect, but results from overlap of the σ C–H (or C–C) bond into the vacant p orbital of the carbocation. This leads to a favourable delocalization of the positive charge.

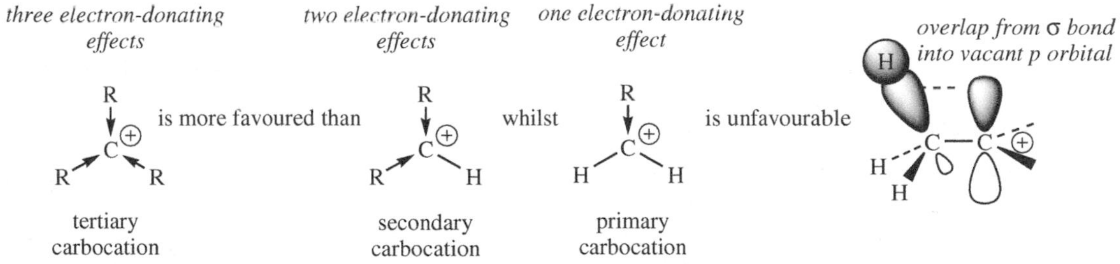

Accordingly, tertiary carbocations benefit from three such effects and are favoured over secondary carbocations with two effects, whilst the single effect in primary carbocations is insufficient to provide significant stabilization. Thus, S_N1 reactions are highly favoured at tertiary carbon, and very much

disfavoured at primary carbon. However, in addition, carbocations may be stabilized by resonance. Simple examples of this are met with the **allyl** and **benzyl cations**, so that allyl chloride and benzyl chloride react via S_N1 reactions, although superficially these appear to involve primary carbocations.

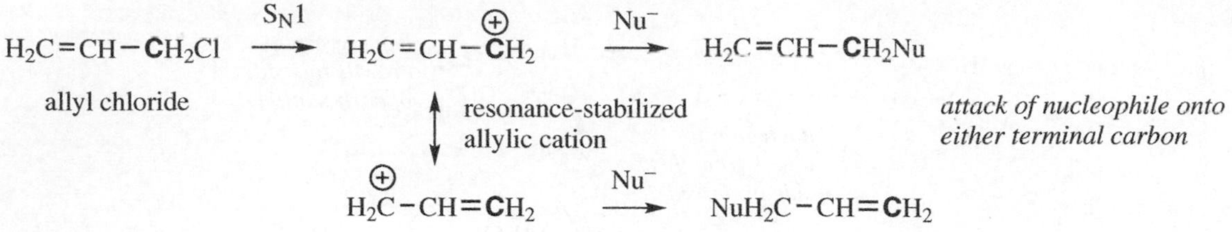

$$H_2C=CH-CH_2Cl \xrightarrow{S_N1} H_2C=CH-\overset{\oplus}{C}H_2 \xrightarrow{Nu^-} H_2C=CH-CH_2Nu$$

allyl chloride

resonance-stabilized allylic cation

attack of nucleophile onto either terminal carbon

$$\overset{\oplus}{H_2C}-CH=CH_2 \xrightarrow{Nu^-} NuH_2C-CH=CH_2$$

with a non-symmetrical allylic system, two different products would be formed

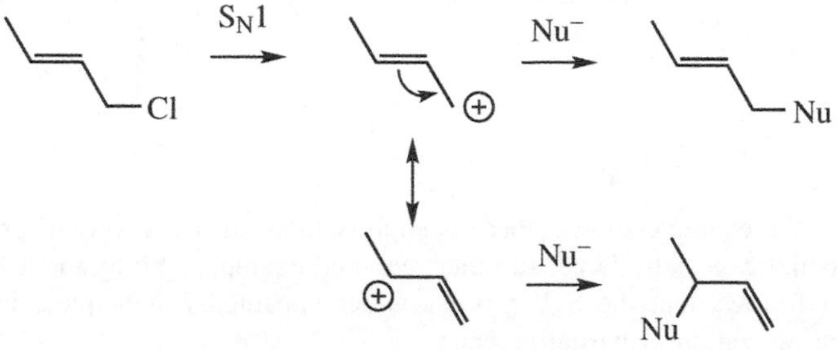

Note that with allyl derivatives there is potential for the nucleophile to react with the different resonance forms, perhaps leading to a mixture of products. This is not the case with the benzylic substrates, since only the benzylic product is formed; addition to the ring would destroy the stability conferred by aromaticity.

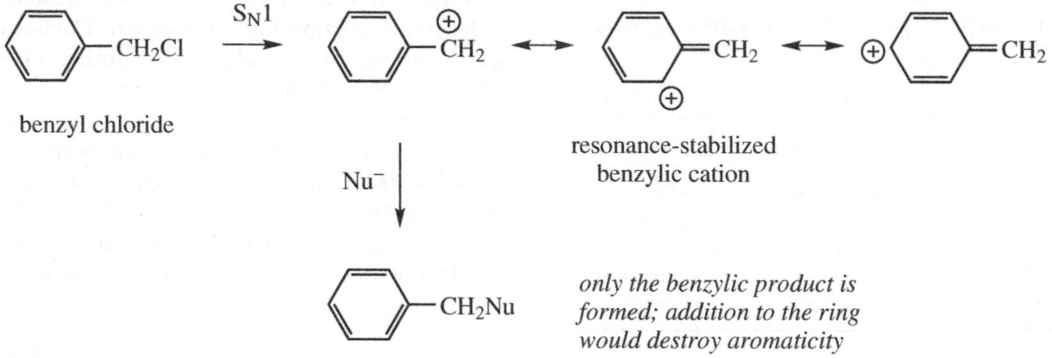

benzyl chloride

resonance-stabilized benzylic cation

only the benzylic product is formed; addition to the ring would destroy aromaticity

One of the most stable carbocation structures is the triphenylmethyl cation (**trityl cation**). In this structure, the positive charge is stabilized by resonance employing all three rings. Trityl chloride ionizes readily, and can capture an available nucleophile.

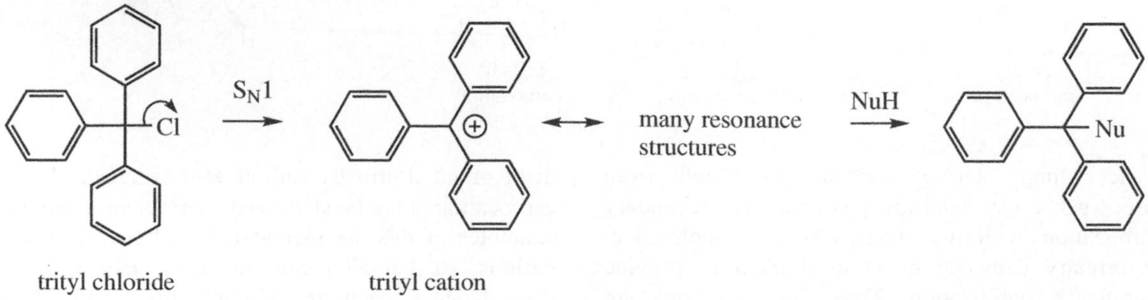

trityl chloride   trityl cation

many resonance structures

6.2.2 S$_N$1 reactions in cyclic systems

We noted above that the inversion of configuration that accompanied S$_N$2 reactions was particularly apparent in cyclic systems, and that *cis* derivatives would be converted into *trans* products in disubstituted rings, and vice versa (see Section 6.1.5). Should

an S$_N$1 reaction occur in a similar sort of cyclic system, then there may be stereochemical consequences, though these are easily predicted. Thus, should the dimethylcyclohexanol shown below participate in an S$_N$1 reaction, then we can deduce that the carbocation will be attacked from either face by the nucleophile, but not necessarily to the same extent.

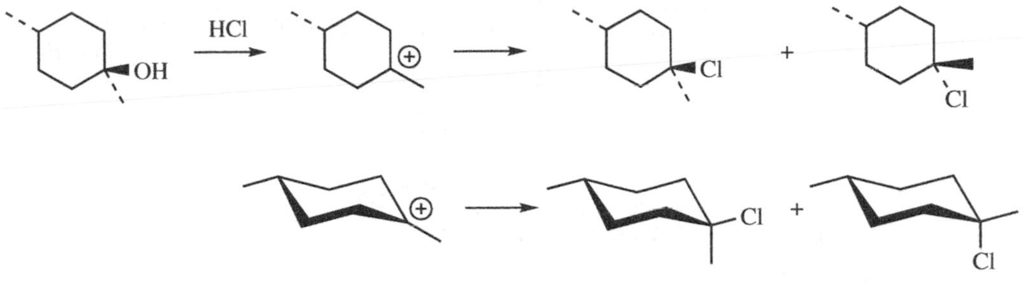

diastereoisomers

The net result is that the product mixture consists of two diastereoisomers.

6.2.3 S$_N$1 or S$_N$2?

As we have just seen, S$_N$1 reactions are highly favoured at tertiary carbon, and very much disfavoured at primary carbon. This is in marked contrast to S$_N$2 reactions, which are highly favoured at primary carbon and not at tertiary carbon. With S$_N$2 reactions, consideration of steric hindrance rationalized the results observed. This leads to the generalizations for nucleophilic substitutions shown in Table 6.8, with secondary substrates being able to participate in either type of process.

The most distinguishing feature of the S$_N$1 mechanism is the intermediate carbocation. Formation of the carbocation is the rate-determining step, and this is more favourable in **polar solvents** that are able to assist in facilitating the charge separation/ionization. A useful, though not always exact,

Table 6.8 Occurrence of S$_N$1 or S$_N$2 reactions according to substrate

Class of substrate	S$_N$1	S$_N$2
Tertiary	Common	Never
Secondary	Sometimes	Sometimes
Primary	Never	Common

guide is that S$_N$1 reactions are going to be favoured by an acidic/positive environment, and are less likely to occur under basic/negative conditions. Since good nucleophiles are often also strong bases, this does tend to limit the applicability of S$_N$1 reactions. Indeed, under strongly basic conditions, side-reactions such as elimination (see Section 6.4.1) are more likely to occur than nucleophilic substitution reactions. However, all is not lost, because the carbocation is a particularly good electrophile and can be used with relatively poor nucleophiles. This is illustrated in the following examples.

S$_N$2 reactions

$$RCH_2Cl \xrightarrow[\text{very slow}]{H_2O} RCH_2OH$$

hydroxide is a much better nucleophile than water

$$RCH_2Cl \xrightarrow[\text{fast}]{HO^-} RCH_2OH$$

S_N1 reaction

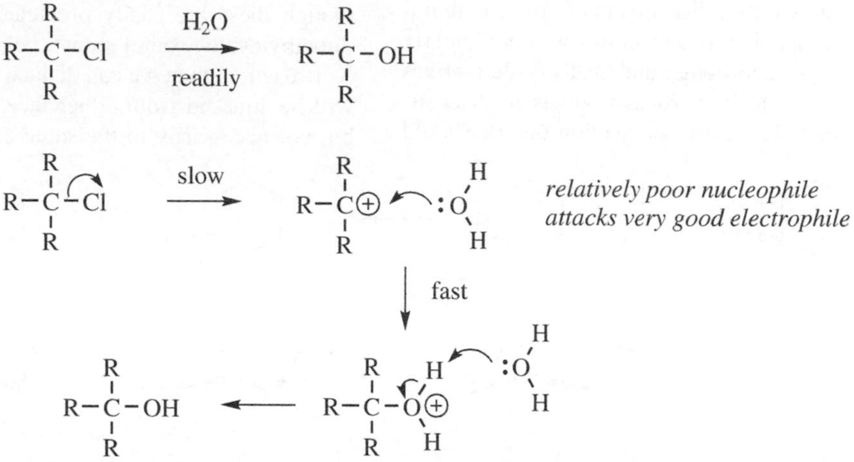

Finally, do appreciate that, depending upon conditions, it is quite possible that both S_N1 and S_N2 mechanisms might be operating at the same time, with each contributing its own stereochemical characteristics upon the product.

Box 6.4
Biological S_N1 reactions involving allylic cations

The leaving groups most commonly employed in nature are **phosphates** and **diphosphates**. These good leaving groups are anions of the strong acids phosphoric (pK_a 2.1) and diphosphoric (pK_a 1.5) acids respectively. The pK_a values given refer to the first ionization of these polyfunctional acids (see Section 4.7).

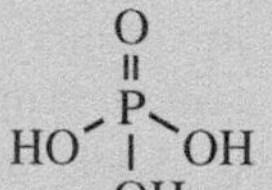

phosphoric acid

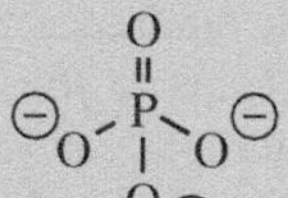

phosphate

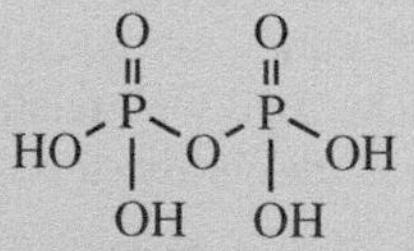

diphosphoric acid
(pyrophosphoric acid)

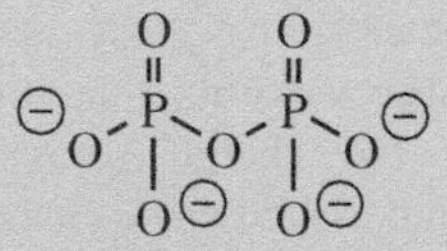

diphosphate
(pyrophosphate)

The compound **dimethylallyl diphosphate** provides an excellent example of a natural product with a diphosphate leaving group that can be displaced in a nucleophilic substitution reaction. Suitable nucleophiles are hydroxyl groups, e.g. a phenol, though frequently an electron-rich nucleophilic carbon is employed. Dimethylallyl diphosphate is a precursor of many natural products that contain in their structures branched-chain C_5 subunits termed isoprene units.

diphosphate is a good leaving group

PPO — dimethylallyl diphosphate (DMAPP)

S$_N$1 reaction

resonance-stabilized allylic carbocation

OPP (diphosphate)

Although both S$_N$2 and S$_N$1 mechanisms might be formulated for such reactions, all the available evidence favours an S$_N$1 process. This is rationalized in terms of formation of a favourable resonance-stabilized **allylic cation** by loss of the leaving group. In the majority of natural product structures, the nucleophile has attacked the allylic system on the same carbon that loses the diphosphate, but there are certainly examples of nucleophilic attack on the alternative tertiary carbon.

geranyl diphosphate (GPP)

farnesyl diphosphate (FPP)

Geranyl diphosphate and **farnesyl diphosphate** are analogues of dimethylallyl diphosphate that contain two and three C$_5$ subunits respectively; they can undergo exactly the same S$_N$1 reactions as does dimethylallyl diphosphate. In all cases, a carbocation mechanism is favoured by the resonance stabilization of the allylic carbocation. Dimethylallyl diphosphate, geranyl diphosphate, and farnesyl diphosphate are precursors for natural terpenoids and steroids.

The possibility of nucleophilic attack on different carbons in the resonance-stabilized carbocation facilitates another modification exploited by nature during terpenoid metabolism. This is a change in double-bond stereochemistry in the allylic system. The interconversions of **geranyl diphosphate, linalyl diphosphate**, and **neryl diphosphate** provide neat but satisfying examples of the chemistry of simple allylic carbocations.

Thus, geranyl diphosphate ionizes to the resonance-stabilized geranyl carbocation; in nature, this can recombine with the diphosphate anion in two ways, reverting to geranyl diphosphate or forming linalyl diphosphate. In linalyl diphosphate, the original double bond from geranyl diphosphate has now become a single bond, and free rotation is possible. Ionization of linalyl diphosphate occurs, giving a resonance-stabilized neryl carbocation, one form of which now has a Z double bond. Recombination of this with diphosphate leads to neryl diphosphate, a geometric configurational isomer of geranyl diphosphate. It is normally very difficult to change the configuration of a double bond. Nature achieves it easily in this allylic system via carbocation chemistry.

Box 6.4 (continued)

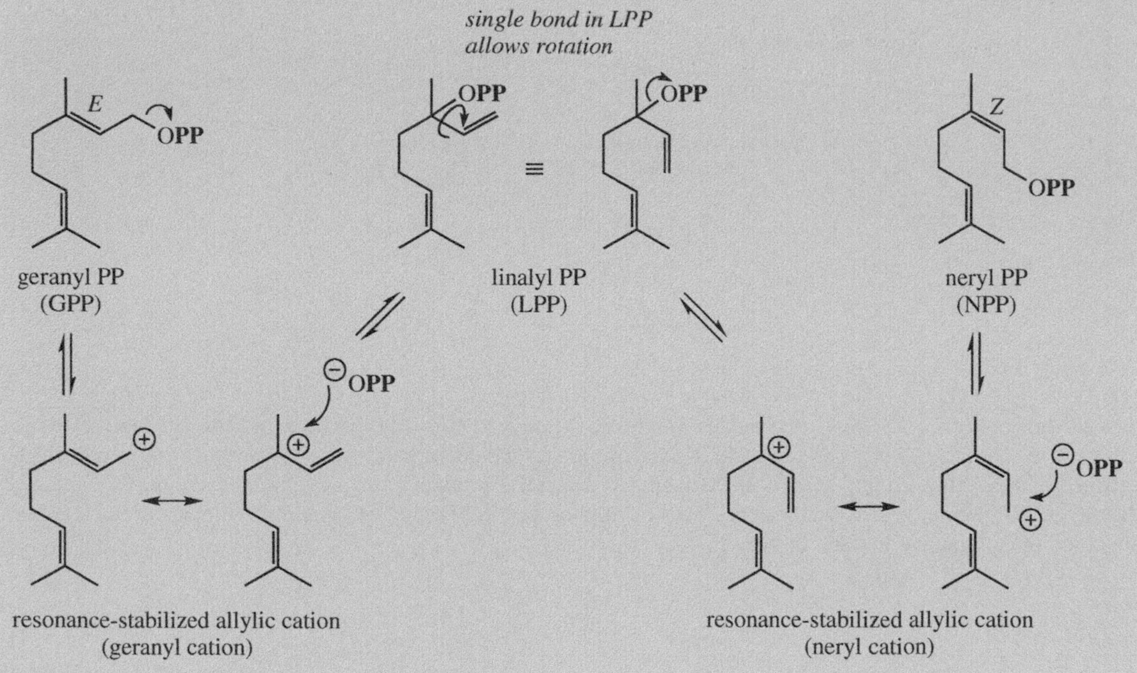

single bond in LPP
allows rotation

geranyl PP
(GPP)

linalyl PP
(LPP)

neryl PP
(NPP)

resonance-stabilized allylic cation
(geranyl cation)

resonance-stabilized allylic cation
(neryl cation)

6.3 Nucleophilic substitution reactions

6.3.1 *Halide as a nucleophile: alkyl halides*

Halide can be employed as a nucleophile in either S_N2 or S_N1 reactions to generate an alkyl halide. However, note that, in the general example shown, protonation by the acidic reagent HBr is required to improve the leaving group (see Section 6.1.4).

$$ROH \xrightarrow{HBr} RBr \xrightarrow{Nu} RNu$$

The utility of this simple transformation is often to increase the reactivity of the substrate, in that halide is a good leaving group and so can participate in other nucleophilic substitution reactions.

6.3.2 *Oxygen and sulfur as nucleophiles: ethers, esters, thioethers, epoxides*

Alkyl halides can react with water or alcohols by either S_N2 or S_N1 mechanisms to give alcohols or ethers respectively.

$$RBr \xrightarrow{H_2O} ROH$$

$$RBr \xrightarrow{R'OH} ROR'$$
ether

$$RBr \xrightarrow{R'O^-} ROR'$$

It is often preferable to use basic conditions with hydroxide or alkoxide as a better nucleophile, though this may lead to elimination and alkene formation as a competing reaction (see Section 6.4).

$$RCO_2^{\ominus} \xrightarrow{R'Br} RCO_2R'$$
ester

Although a carboxylate anion is only a relatively modest nucleophile (see Section 6.1.2), it is possible to exploit an S_N2 reaction to prepare esters from carboxylic acids as an alternative to the usual esterification methods (see Section 7.9). Such methods might be useful, depending upon the nature and availability of starting materials.

$$RBr \xrightarrow{R'SH} RSR'$$

thioether

$$RBr \xrightarrow{R'S^-} RSR'$$

Sulfur nucleophiles behave similarly to oxygen compounds. Again, the anion will be a better nucleophile than the thiol; and since thiols are more acidic than alcohols (see Section 4.3.2), the conjugate bases are more easily generated.

Note also that ring-opening nucleophilic substitution reactions may be possible, and that these will give a product with two functional groups, since the leaving group is still attached to the original molecule through another bond.

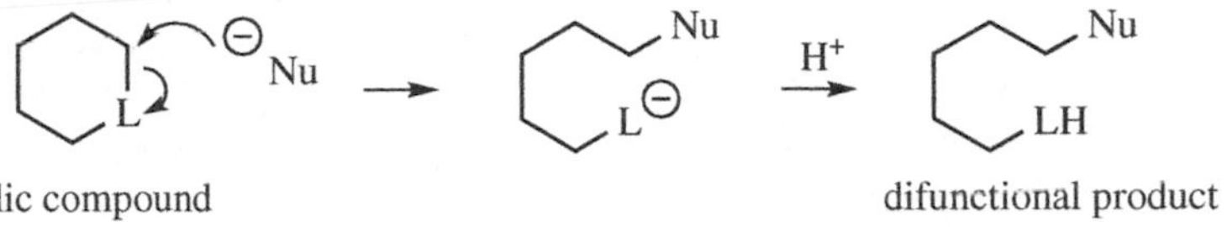

cyclic compound   difunctional product

A simple example of a ring-opening substitution reaction is the acid-catalysed hydrolysis of **epoxides**. In the example shown, protonation of the epoxide oxygen improves the leaving group, and an S_N2 reaction may then proceed using water as the nucleophile. Three-membered rings must of necessity be *cis*-fused (see Section 3.5.2), and the inversion process, therefore, generates a *trans*-1,2-diol. This is true even if the other end of the epoxide ring system is attacked, though it will produce the enantiomeric product. Since both reactions can occur with equal probability, the product here is racemic.

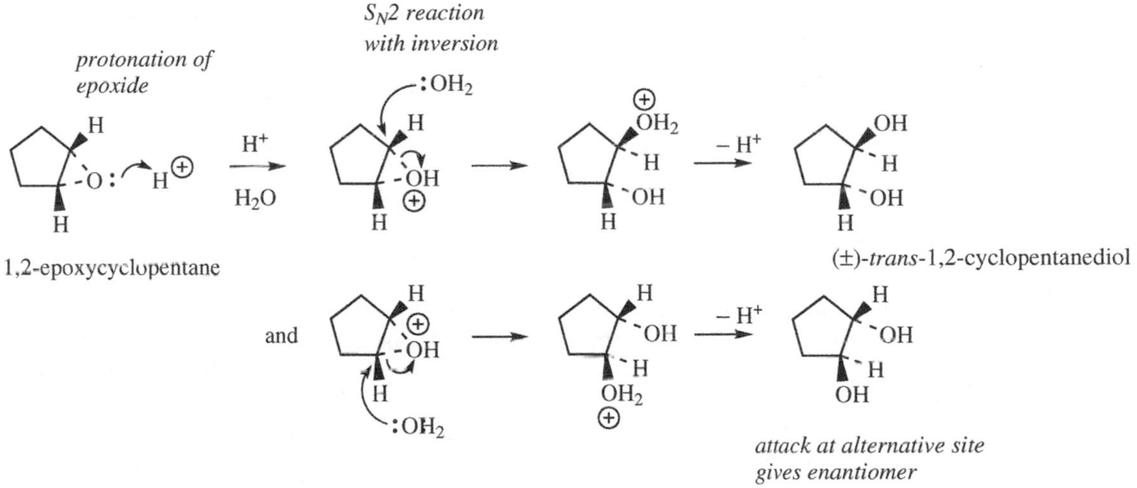

1,2-epoxycyclopentane

($\pm$)-*trans*-1,2-cyclopentanediol

attack at alternative site gives enantiomer

Box 6.5

S-Adenosylmethionine in biological methylation reactions

In biological methylation, the *S*-methyl group of the amino acid L-methionine is used to methylate suitable O, N, S, and C nucleophiles. First, methionine is converted into the methylating agent **S-adenosylmethionine (SAM)**. SAM is nucleoside derivative (see Section 14.3). Both the formation of SAM and the subsequent methylation reactions are nice examples of biological S_N2 reactions.

Box 6.5 (continued)

formation of SAM

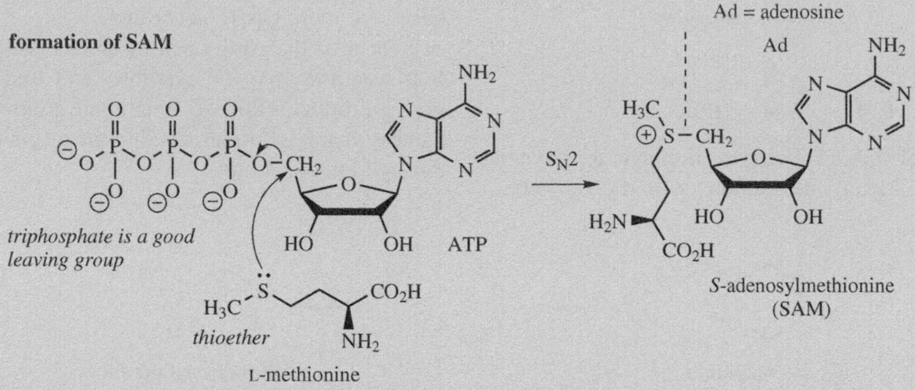

Ad = adenosine

triphosphate is a good leaving group

thioether

ATP

L-methionine

S_N2

S-adenosylmethionine (SAM)

L-Methionine is a thioether that acts as a sulfur nucleophile to react with **adenosine triphosphate** (**ATP**); see Box 7.25. Sulfur is a good nucleophile, and ATP contains a good leaving group, the triphosphate moiety. The leaving group is at a primary position, favouring an S_N2 reaction, the product of which is SAM. This can be regarded as similar to a protonated alcohol in nucleophilic substitution, in that it now contains a good leaving group that is a neutral molecule, in this case the thioether *S*-adenosylhomocysteine. Subsequent S_N2 reactions with appropriate nucleophiles (alcohols, phenols, amines, etc.) produce the methylated compounds.

O- and *N*-alkylation using SAM

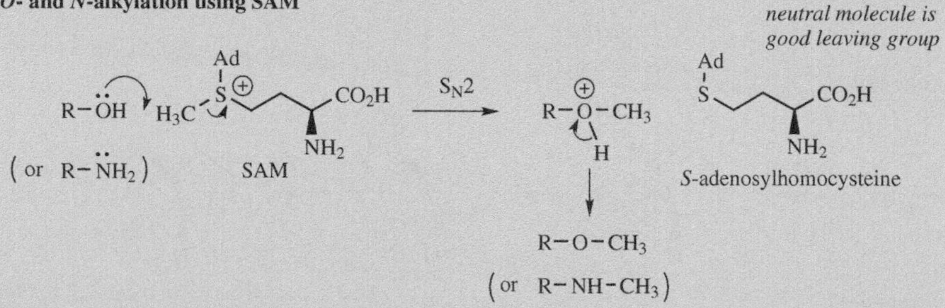

neutral molecule is good leaving group

$R-\overset{..}{O}H$

(or $R-\overset{..}{N}H_2$)

SAM

S_N2

$R-\overset{+}{O}-CH_3$

S-adenosylhomocysteine

$R-O-CH_3$

(or $R-NH-CH_3$)

Note that in nature, these are all enzyme-catalysed reactions. This makes the reactions totally specific. It means possible competing S_N2 reactions involving attack at either of the two methylene carbons in SAM are not encountered. It also means that where the substrate contains two or more potential nucleophiles, reaction occurs at only one site, dictated by the enzyme. The enzymes are usually termed **methyltransferases**. Thus, in animals an *N*-methyltransferase is responsible for SAM-dependent *N*-methylation of noradrenaline (norepinephrine) to adrenaline (epinephrine), whereas an *O*-methyltransferase in plants catalyses esterification of salicylic acid to methyl salicylate.

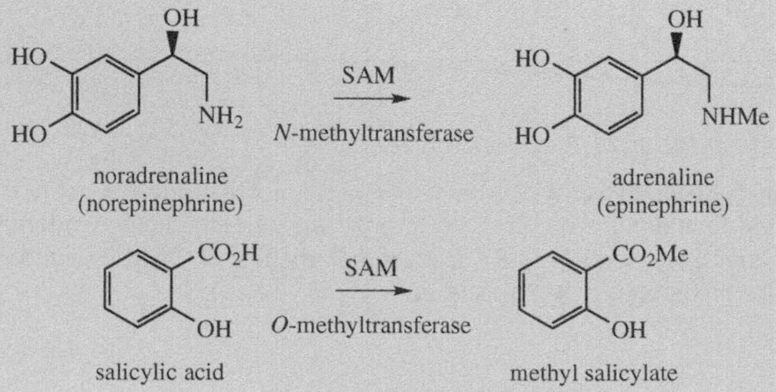

noradrenaline (norepinephrine)

SAM
N-methyltransferase

adrenaline (epinephrine)

salicylic acid

SAM
O-methyltransferase

methyl salicylate

Box 6.6

Glutathione as a sulfur nucleophile in the metabolism of foreign compounds

Glutathione is a tripeptide containing a thiol grouping, which is part of the amino acid cysteine. This SH group plays an important role as a nucleophile in the metabolism of potentially dangerous foreign compounds taken in by the body. The potential of SH as a nucleophile is exploited in metabolic reactions catalysed by enzymes termed **glutathione S-transferases**, which conjugate the foreign compounds, i.e. bind them to glutathione. Conjugation markedly reduces the biological activity of the compound, and most conjugates are actually inactive. In addition, conjugation usually increases the polarity of the substrate, thus increasing its water solubility and its potential to be excreted. There may be further modification to the glutathione part of the conjugate before the foreign compound is finally excreted. Care: this is not the structural "conjugation" of Section 2.8.

Glutathione is able to react with many potentially toxic electrophiles, including halides and epoxides that react via simple S_N2 reactions.

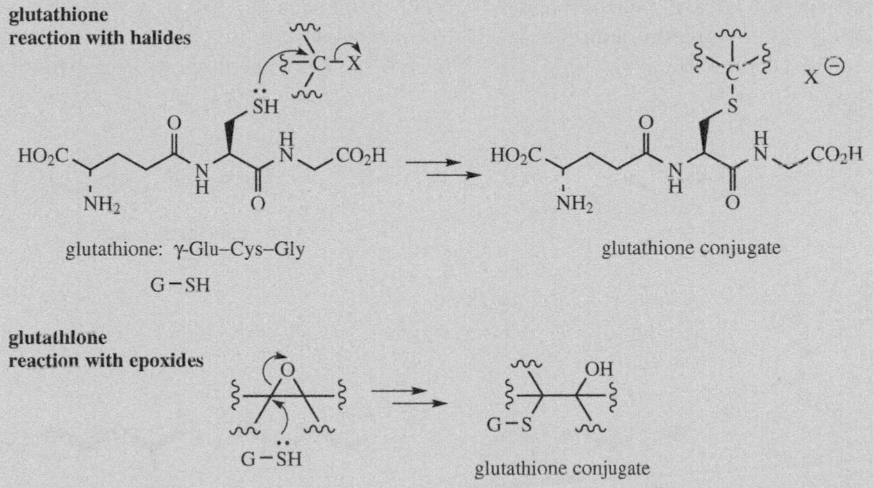

A specific example involving aflatoxins is shown in Box 6.8. We shall see other examples of glutathione reacting as a nucleophile in detoxification reactions, where conjugation is not the result of nucleophilic substitution. For example, it might be nucleophilic addition to an electrophile such as an unsaturated carbonyl compound (see Box 10.20).

6.3.3 Nitrogen as a nucleophile: ammonium salts, amines

Amines react with alkyl halides to give initially **ammonium salts**, from which an amine product is liberated in the presence of base, typically an excess of the amine. However, this is not always a useful reaction, in that the product amine is usually just as nucleophilic as the starting amine, allowing further S_N2 reactions to occur. Depending upon conditions, mixtures of amines together with the quaternary salt may be produced.

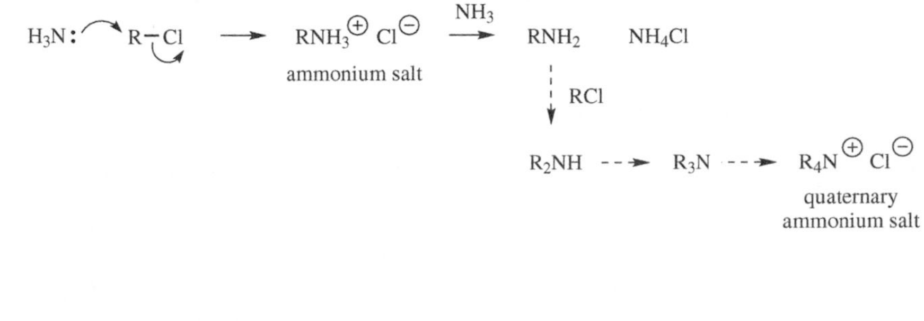

Nevertheless, it offers a convenient route to amino acids, both natural and unnatural, since the amino group in amino acids is less basic (pK_a about 9.8) than a simple amine (pK_a about 10.6) and is consequently rather less nucleophilic (see

Section 4.11.3). Steric hindrance also reduces the chance of multiple alkylations.

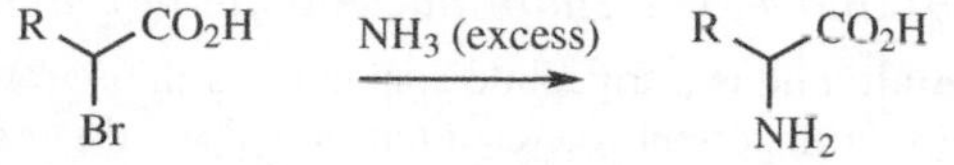

Box 6.7

Curare-like muscle relaxants: quaternary ammonium salts

The production of a quaternary ammonium salt from a tertiary amine and an alkyl halide forms the synthetic route to **decamethonium**, the first of a range of synthetic muscle relaxants having an action like the natural materials found in the arrow-poison **curare**. Decamethonium is actually a di-quaternary salt, as are more modern analogues, such as suxamethonium. **Suxamethonium** superseded decamethonium as a drug because it has a shorter and more desirable duration of action in the body. This arise because it can be metabolized by ester-hydrolysing enzymes (esterases) (see also Box 6.9).

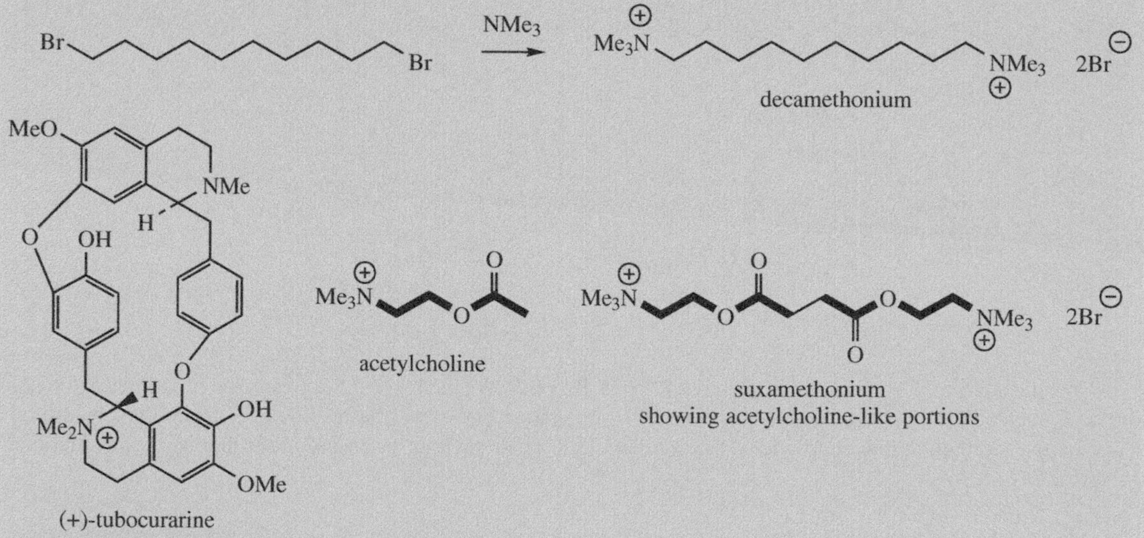

decamethonium

(+)-tubocurarine

acetylcholine

suxamethonium
showing acetylcholine-like portions

Curare-like muscle relaxants act by blocking **acetylcholine** receptor sites, thus eliminating transmission of nerve impulses at the neuromuscular junction. There are two acetylcholine-like groupings in the molecules, and the drugs, therefore, probably span and block several receptor sites. The neurotransmitter acetylcholine is also a quaternary ammonium compound. The natural material present in curare is **tubocurarine**, a complex alkaloid that is a mono-quaternary salt. Under physiological conditions, the tertiary amine will be almost completely protonated (see Section 4.9), and the compound will similarly possess two positively charged centres.

Box 6.8

Aflatoxins and DNA damage

The **aflatoxins** are rather unpleasant fungal toxins. At high levels they can cause severe liver damage in animals and humans, and at lower levels they are implicated in liver cancer. These toxins are produced by the fungus *Aspergillus flavus*, a common contaminant on nuts and grains. Aflatoxin B_1 is the most commonly encountered example, and it is also one of the most toxic. We now know that the toxicity is initiated by oxidative metabolism of the toxin in the body, converting aflatoxin B_1 into an electrophilic epoxide (see Section 6.3.2). This **epoxide** is attacked in an S_N2 reaction by a nitrogen atom in a guanine residue of DNA. This leads to irreversible binding of the toxin to DNA, inhibition of DNA replication and RNA synthesis (see Section 14.2), and initiation of mutagenic activity.

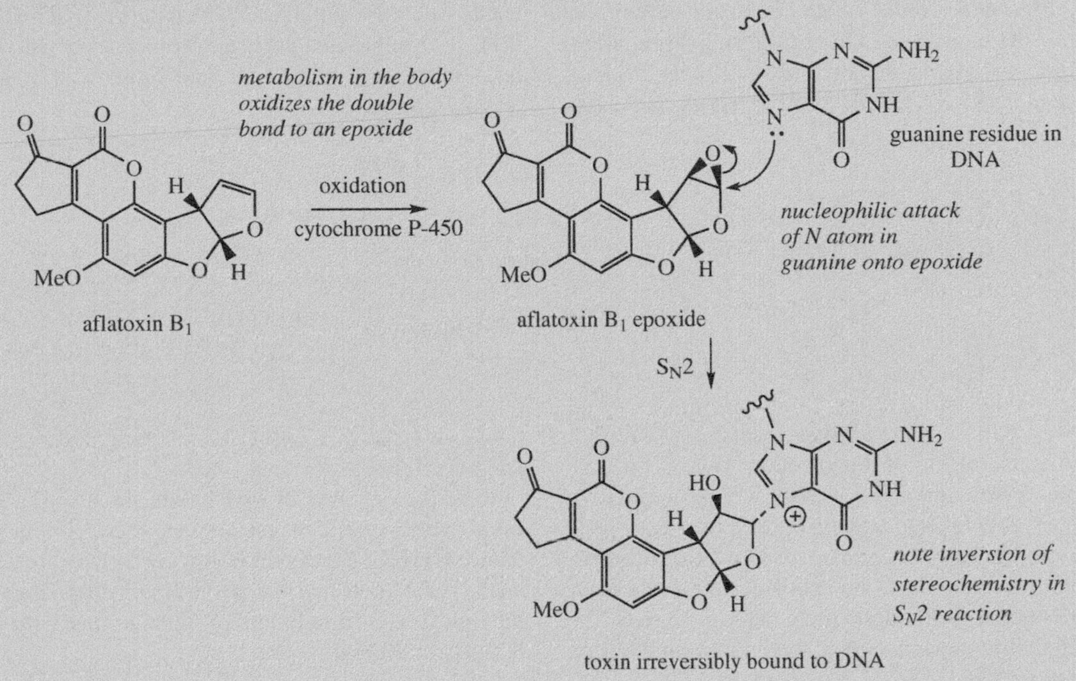

aflatoxin B_1

metabolism in the body oxidizes the double bond to an epoxide

oxidation
cytochrome P-450

aflatoxin B_1 epoxide

guanine residue in DNA

nucleophilic attack of N atom in guanine onto epoxide

S_N2

note inversion of stereochemistry in S_N2 reaction

toxin irreversibly bound to DNA

Fortunately, nature provides an alternative nucleophile whose role is to mop up dangerous electrophiles such as aflatoxin B_1 epoxide before they can do damage, and to remove them from the body. This compound is **glutathione** (see Box 6.6), a tripeptide composed of glutamic acid, cysteine, and glycine.

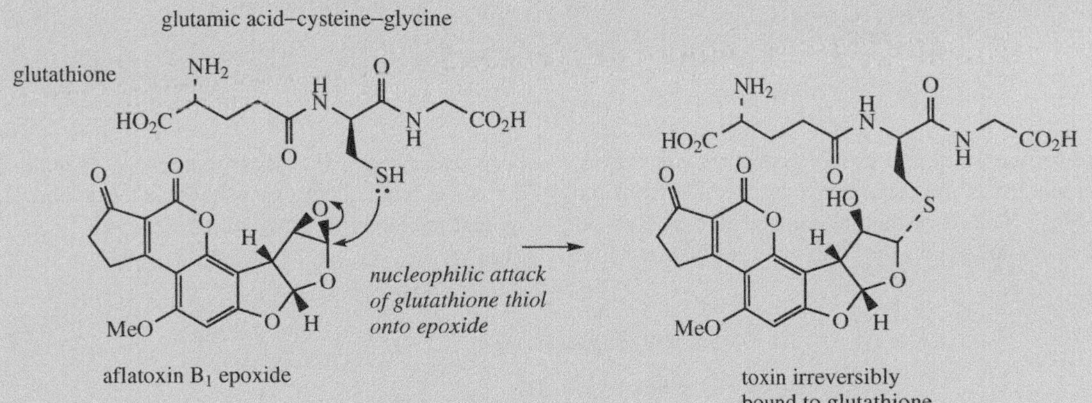

glutamic acid–cysteine–glycine

glutathione

aflatoxin B_1 epoxide

nucleophilic attack of glutathione thiol onto epoxide

toxin irreversibly bound to glutathione

It is the thiol grouping that acts as a nucleophile, attacking the epoxide function of the toxin (see Box 6.6). In this way, the toxin becomes irreversibly bound to glutathione, and the additional polar functionalities in the adduct mean that the product becomes water soluble. The glutathione–toxin adduct can thus be excreted from the body.

6.3.4   Carbon as a nucleophile: nitriles, Grignard reagents, acetylides

Nucleophilic substitution reactions employing carbon as a nucleophile are important in synthetic chemistry in that they create a new C–C bond. A carbon nucleophile, of course, must be in the form of anionic carbon, or its equivalent. One of the simplest sources of anionic carbon is the **cyanide** anion. HCN is a weak acid (pK_a 9.1) and forms a series of stable

salts. Sodium and potassium cyanides are convenient sources of cyanide, which in many reactions behaves similarly to a halide nucleophile. Thus, reaction of an alkyl halide with cyanide creates a nitrile, and extends the carbon chain in the substrate by one carbon. It is easy to rationalize why cyanide is able to displace a halide such as bromide: HCN is a weak acid (pK_a 9.1), so cyanide is a good nucleophile, whereas HBr is a strong acid (pK_a − 9), and bromide is a good leaving group.

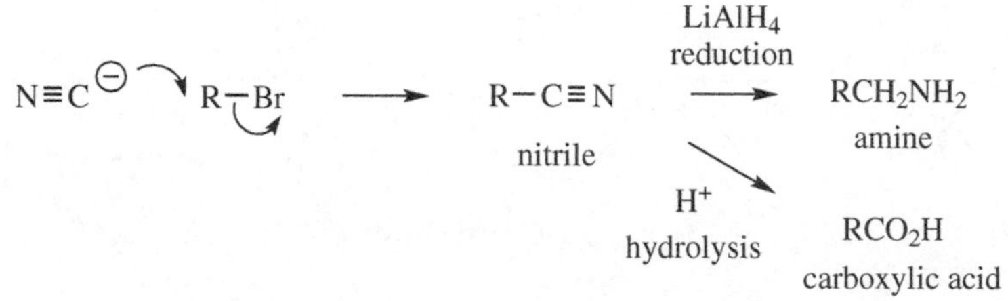

As we shall see later, other reactions of nitriles extend the usefulness of this reaction. Thus, reduction of nitriles gives amines (see Section 7.6.1), whereas hydrolysis generates a carboxylic acid (see Box 7.9).

Organometallic reagents also provide carbon nucleophiles that can be considered to behave as carbanions. Although there are a variety of organometallic reagents available, we include here

only two types of reagent, namely Grignard reagents and acetylides.

Reacting an alkyl or aryl halide, usually the bromide, with metallic magnesium in ether solution produces **Grignard reagents**. An exothermic reaction takes place in which the magnesium dissolves, and the product is a solution of the Grignard reagent RMgBr or ArMgBr.

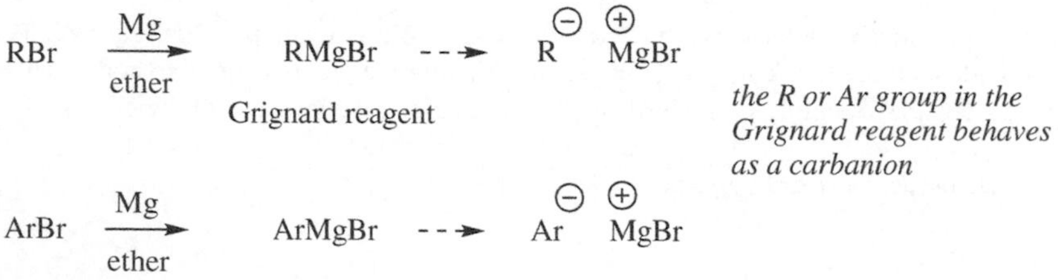

the R or Ar group in the Grignard reagent behaves as a carbanion

The formation of this product need not concern us, but its nature is important. We can deduce from the ions Mg^{2+} and Br^- that it contains the equivalent of R^- or Ar^-, i.e. the alkyl or aryl group has

been transformed into its carbanion equivalent. This carbanion equivalent can behave as a nucleophile in typical nucleophilic substitution reactions.

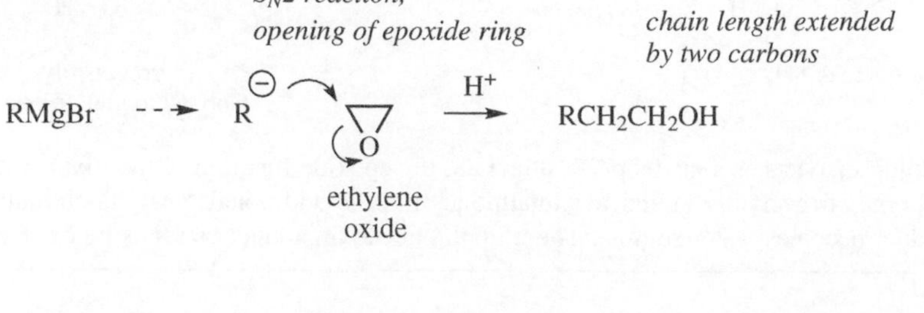

In the example shown, reaction of a Grignard reagent with the epoxide electrophile ethylene oxide proceeds as expected, and after acidification results in formation of an alcohol that is two carbons longer than the original nucleophile.

The carbanion equivalent from a Grignard reagent is also a strong base. pK_a values for alkanes are typically about 50, and for aromatics about 44. Not surprisingly, a Grignard reagent reacts readily with

Acetylides are formed by treating terminal acetylenes with a strong base, sodium amide in liquid ammonia being the one most commonly employed. Acetylenes with a hydrogen atom attached to the triple bond are weakly acidic (pK_a about 25) due to the stability of the acetylide anion (see Section 4.3.4),

$$R-C{\equiv}C-H \quad \xrightarrow[\text{liquid NH}_3]{\text{NaNH}_2} \quad R-C{\equiv}C^{\ominus} \; \overset{\oplus}{\text{Na}}$$
acetylide

$$R-C{\equiv}C^{\ominus} \quad R'{-}Br \longrightarrow R-C{\equiv}C-R'$$
S_N2 reaction

water to form the hydrocarbon, so these reactions must be conducted under anhydrous conditions.

$$\overset{\ominus}{R} \quad \overset{\oplus}{\text{MgBr}} \quad \xrightarrow{\text{H}_2\text{O}} \quad R-H$$

$$\overset{\ominus}{\text{Ar}} \quad \overset{\oplus}{\text{MgBr}} \quad \xrightarrow{\text{H}_2\text{O}} \quad \text{Ar}-H$$

and this anion can thus act as a nucleophile. It reacts with appropriate electrophiles, e.g. alkyl halides, in the manner expected. This reaction extends a carbon chain by two or more atoms, depending on the acetylide used.

$$R-CH_2-CH_2-\underline{H} \qquad pK_a \; 50$$
$$R-C{\equiv}C-\underline{H} \qquad pK_a \; 25$$

alkynes are considerably
more acidic than alkanes

Probably the most significant examples of carbon nucleophiles are enolate anions. These can participate in a wide variety of important reactions, and simple nucleophilic substitution reactions are included amongst these. However, we shall consider these reactions at a later stage, when the nature and formation of enolate anions is discussed (see Chapter 10).

6.3.5 Hydride as nucleophile: lithium aluminium hydride and sodium borohydride reductions

A number of **complex metal hydrides** such as **lithium aluminium hydride** (LiAlH$_4$, abbreviated to LAH) and **sodium borohydride** (NaBH$_4$) are able to deliver hydride in such a manner that it appears to act as a nucleophile. We shall look at the nature of these reagents later under the reactions of carbonyl compounds (see Section 7.5), where we shall see that the complex metal hydride never actually produces hydride as a nucleophile, but the aluminium hydride anion has the ability to effect transfer of hydride. Hydride itself, e.g. from sodium hydride, never acts as a nucleophile; owing to its small size and high charge density it always acts as a base. Nevertheless, for the purposes of understanding the transformations,

we shall consider hydride as a nucleophile that participates in a typical S_N2 process. This achieves replacement of a leaving group by hydrogen and, therefore, is a reduction of the substrate.

$$H^{(-)} \quad R{-}Br \quad \xrightarrow{\text{LiAlH}_4} \quad H-R$$
'hydride' acting
as a nucleophile

In the example shown overleaf where hydride attacks the epoxide function, the product is an alcohol, the reaction being completed by supplying a proton source, usually water.

Lithium aluminium hydride reacts violently with water, liberating hydrogen, and the heat of reaction usually ignites the hydrogen. LAH must, therefore, be used in rigorously anhydrous conditions, usually in ether solution. In fact, any solvent containing OH or NH groups would destroy the reagent by acting as a proton donor for hydride. The addition of water as a proton source has to be carried out with considerable caution, since any unreacted LAH will react violently with this water. In the laboratory, safe removal of excess LAH may be achieved by adding small amounts of an ester such

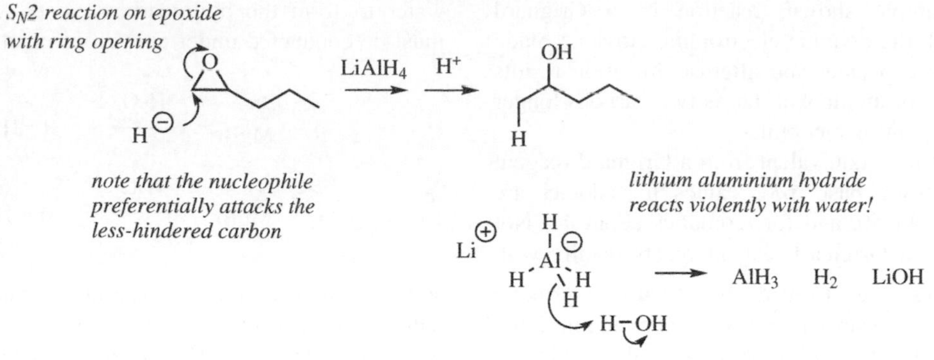

S$_N$2 reaction on epoxide with ring opening

note that the nucleophile preferentially attacks the less-hindered carbon

lithium aluminium hydride reacts violently with water!

$$Li^{\oplus} \quad \overset{\overset{H}{|}}{\underset{\underset{H}{|}}{\overset{H}{Al}}}{}^{\ominus}{}_{H} \quad \longrightarrow \quad AlH_3 \quad H_2 \quad LiOH$$

H–OH

as ethyl acetate (see Section 7.5). Note that LAH is a powerful reducing agent and will attack many other functional groups, especially carbonyl groups (see Section 7.11).

An analogous series of reactions is involved when sodium borohydride is used as the reducing agent. Sodium borohydride is considerably less reactive than LAH, and reactions proceed much more slowly. This reagent may be used in alcoholic or even aqueous solution, so there are no particular hazards associated with its use.

base removes proton from alcohol (not particularly favourable)

intramolecular S$_N$2 reaction

an epoxide

ethylene oxide

alternative intramolecular S$_N$2 reaction

intramolecular S$_N$2 reaction

pyrrolidine

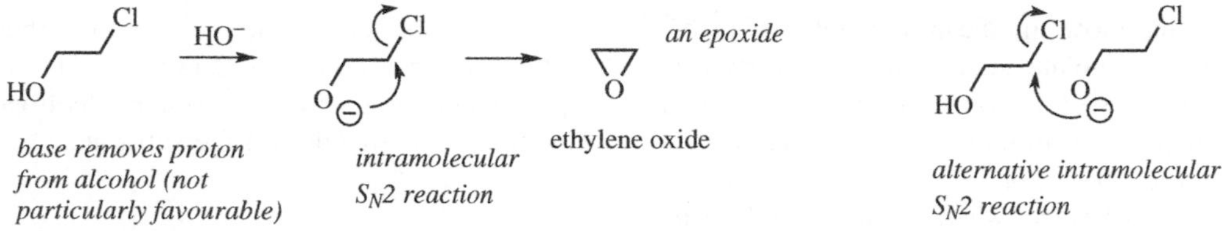

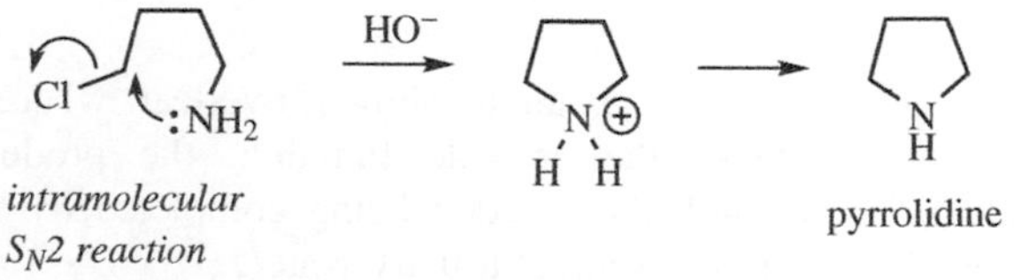

Simple examples shown above are the base-catalysed formation of oxygen- and nitrogen-containing ring systems. We have shown base-initiated ionization of the alcohol to an alkoxide anion in epoxide formation; the anion is a better nucleophile than the alcohol. For pyrrolidine synthesis, the amino group is sufficiently nucleophilic for reaction to occur, but base is required to remove a proton from the first-formed intermediate.

6.3.6   Formation of cyclic compounds

In substrates where there is a good leaving group in the same molecule as the nucleophile, one may get an **intramolecular process** and create a ring system. It is usually necessary to find conditions that favour an intramolecular process over the alternative intermolecular reaction. This is typically achieved by carrying out the reaction at relatively higher dilutions, thereby minimizing the intermolecular processes.

6.4   Competing reactions: eliminations and rearrangements

When nucleophilic substitution reactions are attempted, the expected product may often be accompanied by one or more additional products that arise from competing reactions. Since these competing reactions share features of the nucleophilic substitution mechanism, they are readily rationalized, and it is possible to devise conditions to minimize or maximize the formation of such products. The most common alternative reactions are **eliminations** and **rearrangements**, which we shall consider in turn.

6.4.1 Elimination reactions

The E2 reaction: bimolecular elimination

The abbreviation E2 conveys the information 'elimination–bimolecular'. The reaction is a concerted process in which a nucleophile removes an electrophile at the same time as a leaving group departs. It is bimolecular, since kinetic data indicate that two species are involved in the rate-determining step:

$$\text{Rate} = k[\text{RL}][\text{Nu}]$$

where Nu is the nucleophile, RL is the substrate containing the leaving group L, and k is the rate constant.

The electrophile removed is usually hydrogen, so we can consider that the nucleophile is acting as a base. We have seen above the close relationship between **basicity** and **nucleophilicity** (see Section 6.1.2), so the E2 mechanism provides an example of how the alternative property of nucleophiles may come into play and lead to different products. To achieve an S_N2 reaction, the nucleophile must approach to the rear of the leaving group and then displace it (see Section 6.1). If a rear-side approach is hindered by adjacent groups, or perhaps because the nucleophile is rather large, it becomes energetically easier for the nucleophile to act as a base and remove a proton from the substrate.

E2 mechanism

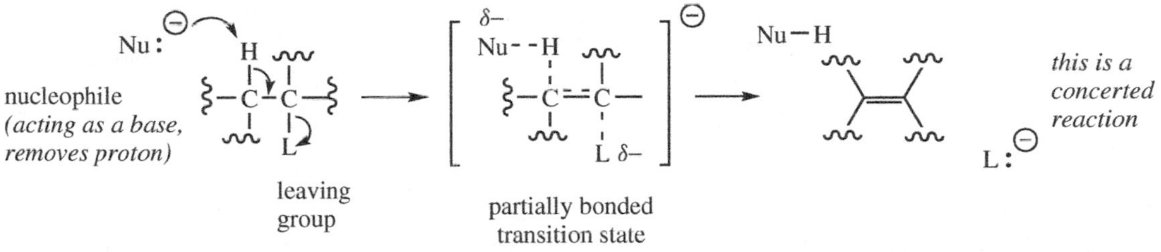

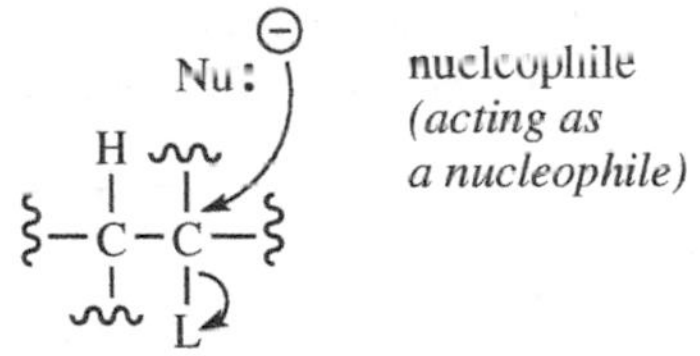

As the proton is removed, electrons that were involved in bonding the proton to the substrate are then used to form the double bond; however, to maintain the octet of electrons on the neighbouring carbon, the electrons will have to be transferred to a suitable acceptor, in this case the leaving group. As with the S_N2 mechanism, the reaction is **concerted** and proceeds through a high-energy **transition state**, in which partial bonds have been established. The energy profile will look the same as that of an S_N2 reaction (see Section 6.1). The elimination reaction generates a new π bond in a planar **alkene**. Since the π bond is perpendicular to the plane of the alkene, we can predict that the most favourable way to achieve the new π bonding is to start with the H–C–C–L atoms in a planar array. This will line up the orbitals and allow easy development of the π bond.

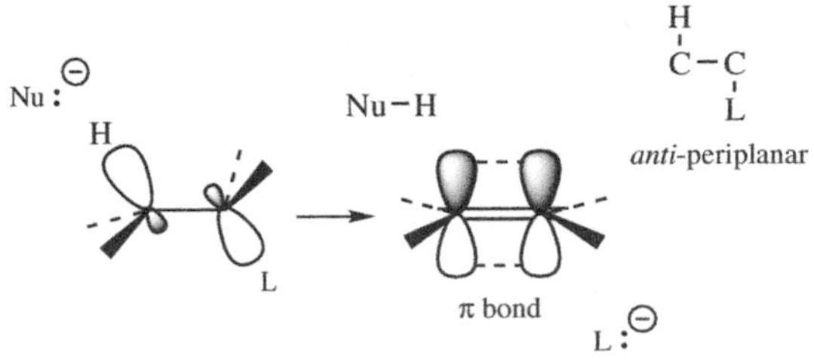

The nucleophile approaches from the side opposite the electronegative leaving group – electrostatic repulsion discourages attack in the region of the leaving group. With the substrate in the favoured staggered conformation, we describe this arrangement of atoms as ***anti*-periplanar**.

The requirement for the proton electrophile and the leaving group involved in the elimination to be *anti* to each other is demonstrated by the nature of the product obtained from a suitable substrate, e.g.

(1*R*,2*R*)-1-bromo-1,2-diphenylpropane, when treated with base. The only product formed is (*Z*)-1,2-diphenylprop-1-ene. This is the product from an *anti* elimination of H and Br when the substrate is in a staggered conformer. If H and Br were positioned on the same side of the conformer, then it would need to be in an unfavourable eclipsed conformer to line up the orbitals. Elimination of H and Br in this fashion is termed a *syn* elimination, and would lead to the *E*-product. However, this is not the product formed, and, in general, *syn* eliminations are very rare.

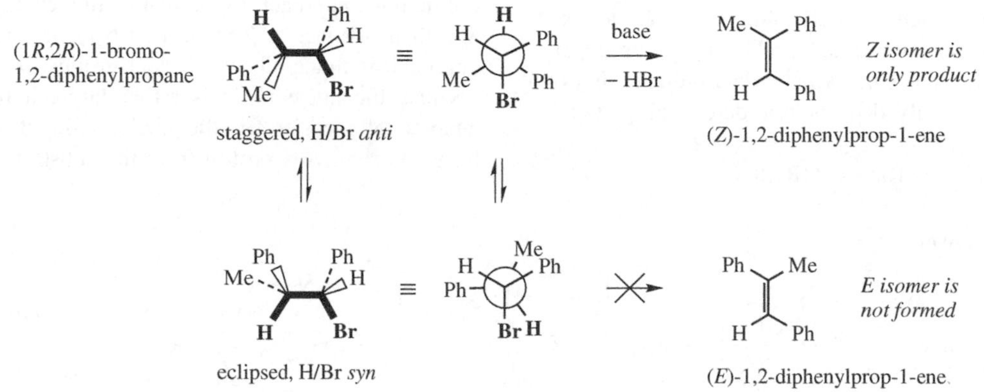

(1*R*,2*R*)-1-bromo-1,2-diphenylpropane

staggered, H/Br *anti*

base
– HBr

Z isomer is only product

(*Z*)-1,2-diphenylprop-1-ene

eclipsed, H/Br *syn*

E isomer is not formed

(*E*)-1,2-diphenylprop-1-ene

It is particularly evident that the *anti* stereochemical relationship is obligatory by observing elimination reactions in suitable cyclohexane derivatives. The only way to achieve a planar arrangement of

the H–C–C–L atoms is when H and L are both axial and, consequently, *trans* to each other (*trans*-diaxial). Thus, consider menthyl chloride and neomenthyl chloride, which are stereoisomers differing in configuration at just one centre.

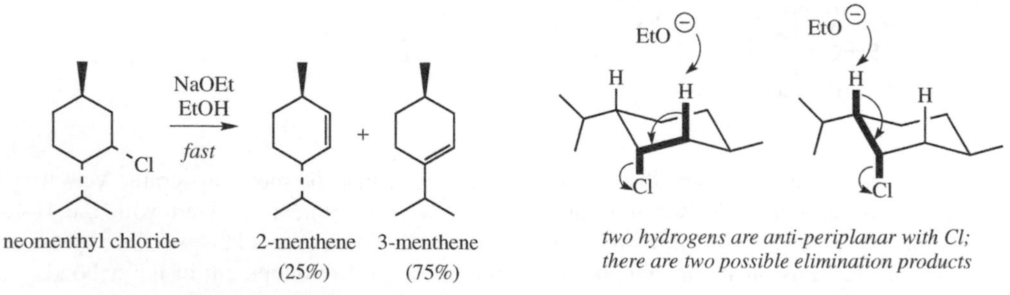

neomenthyl chloride

NaOEt
EtOH
fast

2-menthene (25%)   +   3-menthene (75%)

two hydrogens are anti-periplanar with Cl; there are two possible elimination products

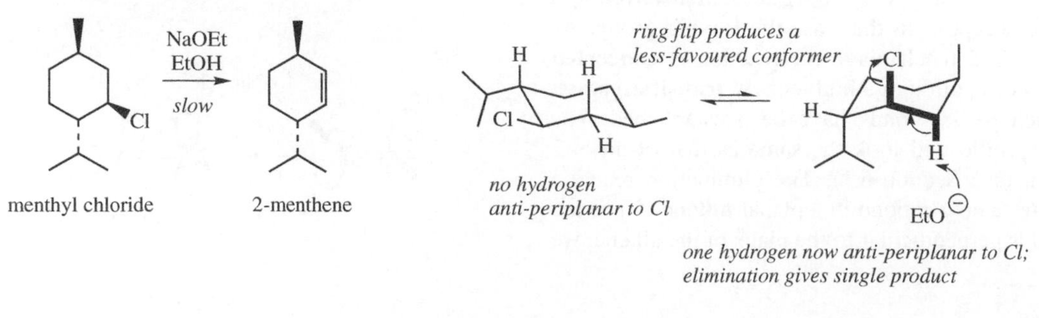

menthyl chloride

NaOEt
EtOH
slow

2-menthene

no hydrogen anti-periplanar to Cl

ring flip produces a less-favoured conformer

one hydrogen now anti-periplanar to Cl; elimination gives single product

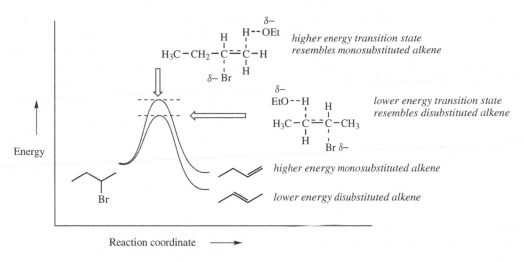

Figure 6.3   Energy profile: E2 reaction to more- or less-substituted alkenes

Treatment of neomenthyl chloride with base rapidly produces two different alkenes, i.e. 2-menthene and 3-menthene. If one considers the three-dimensional shape of neomenthyl chloride, it can be seen that, in the preferred conformer with the two alkyl groups equatorial (see Section 3.3.2), the chlorine is an axial substituent. This means there are two different hydrogen atoms adjacent that are also axial and *anti*-periplanar to the chlorine. As a consequence, two different E2 eliminations can occur; hence the two observed products. That the two products are not formed in equal amounts will be considered in the next section.

On the other hand, menthyl chloride is only slowly converted by treatment with base, and into a single product, i.e. 2-menthene. In the preferred conformation of menthyl chloride, all three substituents are equatorial, and no adjacent hydrogen is in a planar relationship to the chlorine leaving group. The fact that slow elimination occurs at all is a result of conformational isomerism into the less-favoured conformer that has all three substituents axial. In this conformer, there is a single hydrogen *anti*-periplanar with the chlorine, so elimination occurs giving just one product. The conformational equilibrium is slowly disturbed because the elimination removes the small concentration of unfavoured conformer.

Direction of elimination

The E2 elimination of HCl from neomenthyl chloride described above produced two products, namely 2-menthene and 3-menthene in a ratio of about 1 : 3.

It is a general observation that, where different alkene products can arise through E2 elimination, the more-substituted alkene predominates. 2-Menthene contains a double bond with two alkyl substituents, whereas the double bond in 3-menthene has three substituents. The more-substituted alkene is termed the **Saytzeff product**; the less-substituted alkene is termed the **Hofmann product**. We recommend you disregard the proper names, and think of the products in terms of 'more-substituted alkene' and 'less-substituted alkene'.

A further example of the more-substituted alkene predominating is found in the elimination of HBr from 2-bromobutane. The major product is the more-substituted alkene but-2-ene, which predominates over the less-substituted alkene but-1-ene by a factor of 4 : 1. The reasoning for this direction of elimination is twofold. The more-substituted alkene is actually of lower energy than the less-substituted alkene because of the stabilizing electron-donating effect of alkyl groups (see Section 4.3.3), and a similar effect will occur in the transition state where the double bond is developing. This is seen in the energy profile for the reaction (Figure 6.3).

The stabilizing effect of alkyl groups appears to involve overlap of σ C–H (or C–C) orbitals with the π system of the alkene, rather as we have seen with carbocations (see Section 6.2.1). The more alkyl groups attached, the more stabilization the alkene derives.

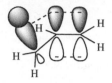

overlap from σ bond with π system

This effect is relatively small and both products are formed, usually with one predominating. The more-substituted Saytzeff product typically predominates when the leaving group is small, e.g. halide. On the other hand, when there is a large leaving group present, e.g. quaternary ammonium, then steric effects

become more important than the stabilizing effects of alkyl groups. This is exemplified by the heat-initiated decomposition of the quaternary ammonium salt below. The elimination is now governed by which is the more favourable conformer of the substrate where a hydrogen atom is positioned *anti* to the quaternary ammonium substituent. Two such possibilities can be considered. It is apparent that the conformer set up for 1,2-elimination is more favourable than the conformer for 2,3-elimination, since the latter conformer would necessitate a less favourable *gauche* interaction (see Section 3.3.1). An alternative conformer for 2,3-elimination has two unfavourable *gauche* interactions. Thus, it is the large leaving group that now dictates the direction of elimination, and the less-substituted alkene (Hofmann product) predominates. Again, it should be noted that both products are actually obtained – the effect is not sufficiently great to produce one product exclusively.

N,N,N-trimethyl-
2-butylammonium
hydroxide

HO⁻ (⊖)

counter-ion acts as base

heat ⟶

(5%)   *more-substituted alkene*

(95%)   *less-substituted alkene*

looking along
C-1,2 bond

we are showing only one configuration at C-2; the conformational consequences are the same in the enantiomer

looking along
C-3,2 bond

and

less favourable conformer; gauche interaction

this conformer with two gauche interactions is even less favourable

Note that with some cyclic substrates, the leaving group may remain as part of the product alkene. Elimination reactions played an important role in early structural analysis of alkaloids (typically cyclic amines). Combination of *N*-methylation followed by elimination may be used to open up nitrogen heterocycles, as shown with piperidine.

successive nucleophilic attacks of amine onto methyl iodide generate quaternary ammonium salt

base-initiated elimination

repeat methylation and elimination sequence

piperidine

MeI ⟶ − H⁺ ⟶   MeI ⟶ − H⁺ ⟶   HO⁻   E2 heat ⟶   MeI HO⁻ ⟶

One further consideration relating to the nature of the products in eliminations is the stereochemistry about the double bond. For instance, base-catalysed elimination of HBr from 2-bromopentane gives three products.

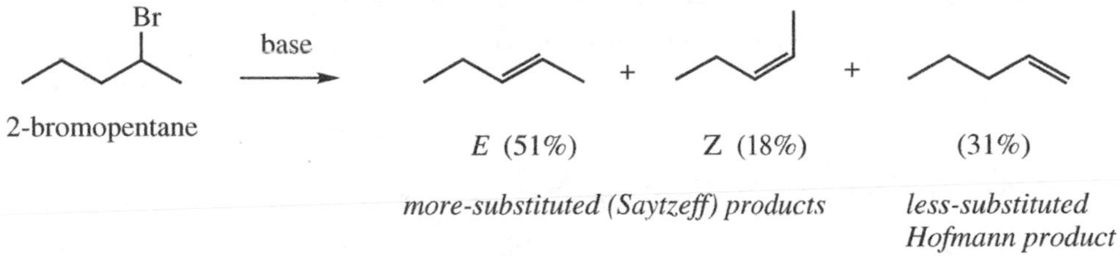

2-bromopentane

E (51%) + Z (18%) + (31%)

more-substituted (Saytzeff) products   *less-substituted Hofmann product*

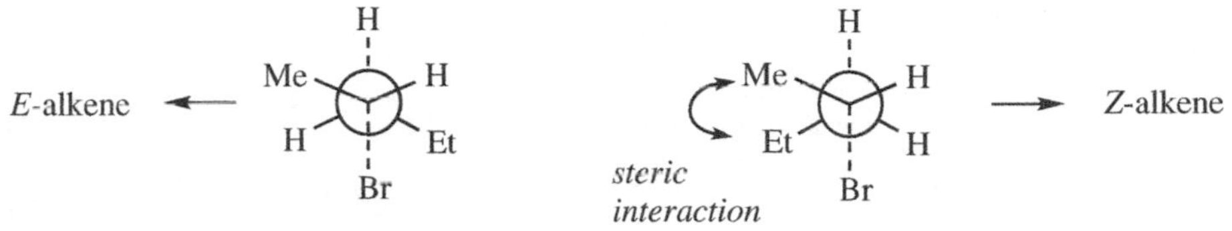

E-alkene   *steric interaction*   Z-alkene

transition states

This elimination involves a small leaving group, so the more-substituted alkene predominates. However, E and Z isomers of this Saytzeff product are produced, and in unequal amounts. That the major product is the E-alkene can be rationalized in terms of minimizing steric repulsion during the transition state.

Note the terminology that can be used to describe product distribution in this type of reaction. Reactions are termed **regiospecific** where one product is formed exclusively, or **regioselective** where one product predominates.

Box 6.9

Atracurium, a curare-like muscle relaxant that is metabolized via an elimination reaction

We have seen above that the muscle relaxant properties of curare and synthetic analogues result from competing with **acetylcholine** at receptors, thus blocking nerve impulses at the neuromuscular junction (see Box 6.7). As diquaternary ammonium salts, there are two well-separated acetylcholine like groupings in the molecules, and the drugs probably span and block several receptor sites. These agents work rapidly, and are of considerable value in surgery. However, artificial respiration is required until the agent is metabolized, and thus broken down by the patient.

Recent developments have led to agents with a built-in functional group that allows more rapid metabolism. Initially, the presence of ester groupings, as in **suxamethonium**, allowed fairly rapid metabolism in the body via esterase enzymes that hydrolyse these linkages. The enzyme involved appears to be a non-specific serum acetylcholinesterase (see Box 13.4). Even better is the inclusion of functionalities that allow additional degradation via an elimination reaction. Such an agent is **atracurium**.

In addition to enzymic ester hydrolysis, atracurium is also degraded in the body by a non-enzymic elimination reaction that is independent of liver or kidney function. Normally, this elimination would require strongly alkaline conditions and a high temperature, but the presence of the carbonyl group increases the acidity of the proton (see Section 4.3.5) and thus facilitates its removal. Elimination can proceed readily under physiological conditions, giving atracurium a half-life of about 20 minutes. This is particularly valuable where patients have low or atypical esterase enzymes. Atracurium contains four chiral centres (including the quaternary nitrogens) and is supplied as a mixture of stereoisomers; a single isomer cisatracurium has now been introduced. This isomer is more potent than the mixture, has a slightly longer duration of action, and produces less cardiovascular side-effects.

Box 6.9 (continued)

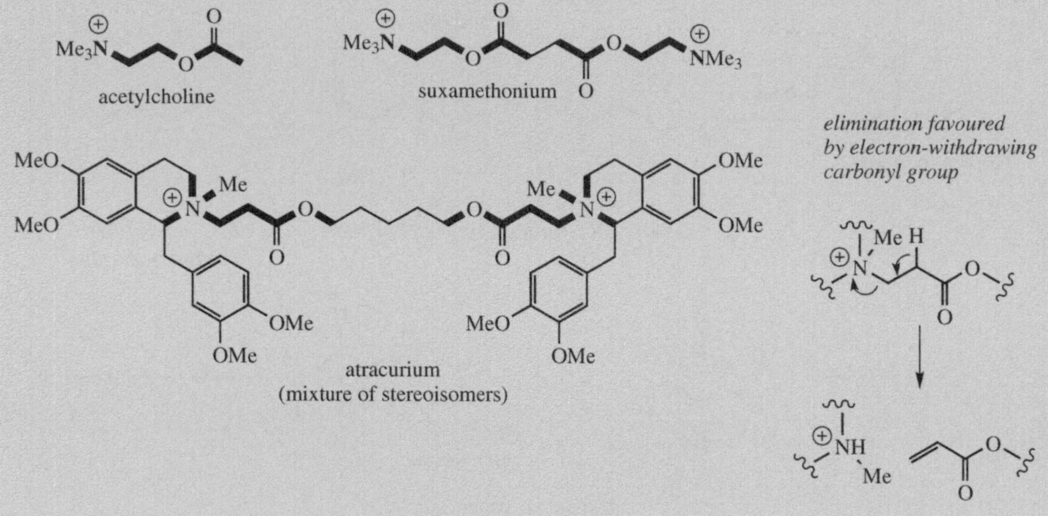

acetylcholine

suxamethonium

elimination favoured by electron-withdrawing carbonyl group

atracurium
(mixture of stereoisomers)

The E1 reaction: unimolecular elimination

The abbreviation E1 conveys the information 'elimination–unimolecular'. The reaction achieves the same result as the E2 reaction, but is mechanistically different in that it involves a **carbocation** intermediate. It is unimolecular, since kinetic data indicate that only one species is involved in the rate-determining step:

$$\text{Rate} = k[\text{RL}]$$

where RL is the substrate containing the leaving group L and k is the rate constant. The nucleophile Nu does not figure in the rate equation.

Just as the E2 mechanism shares features of the S_N2 mechanism, the E1 mechanism shares features of the S_N1 reaction. The initial step is formation of a carbocation intermediate through loss of the leaving group. This slow step becomes the **rate-determining step** for the whole reaction, i.e. the E1 mechanism is unimolecular. In general terms, the reaction can be represented as follows.

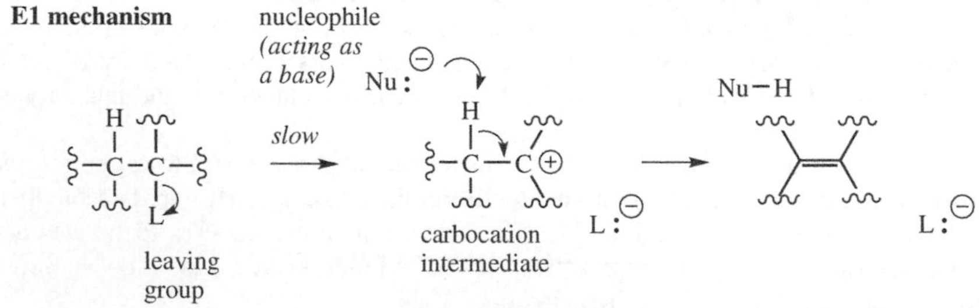

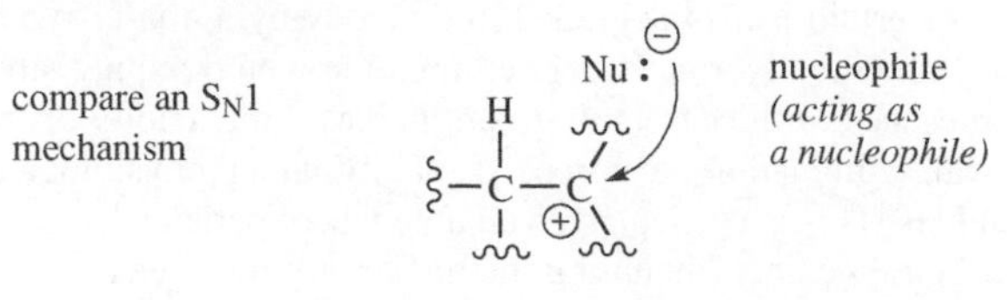

E1 mechanism

nucleophile
(acting as a base)

leaving group

carbocation intermediate

Nu—H

compare an S_N1 mechanism

nucleophile
(acting as a nucleophile)

Once formed, the carbocation could be attacked by a nucleophile – the S_N1 reaction. However, if the nucleophile acts as a base, then it removes a proton from a position adjacent to the positive centre and

p orbital should be parallel to C–H bond

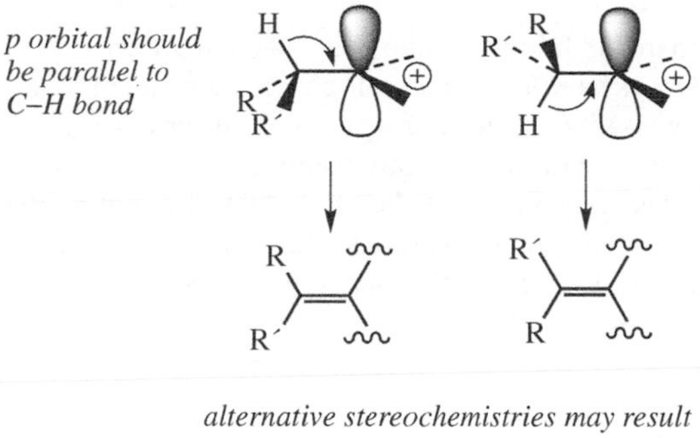

alternative stereochemistries may result

the original bonding electrons are used to discharge the positive charge and make a new double bond. A stereochemical consequence of this is that the proton lost should be perpendicular to the plane of the carbocation to achieve maximum overlap with the unfilled *p*-orbital during formation of the π bond.

We do not have the same strict stereochemical requirements as in the E2 mechanism, and isomeric alkenes may well be produced. If several hydrogens are available for elimination, then the preferred product formed is the more-substituted Saytzeff alkene.

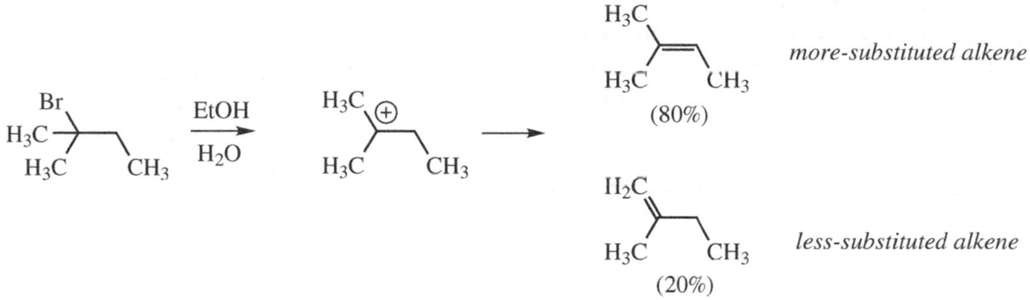

more-substituted alkene

(80%)

less-substituted alkene

(20%)

Box 6.10

E1 elimination in the synthesis of tamoxifen

We have already employed the tamoxifen structure as an example of defining the configuration about double bonds (see Box 3.9). Tamoxifen is a highly successful oestrogen-receptor antagonist used in the treatment of breast cancer. It may be synthesized by the following sequence:

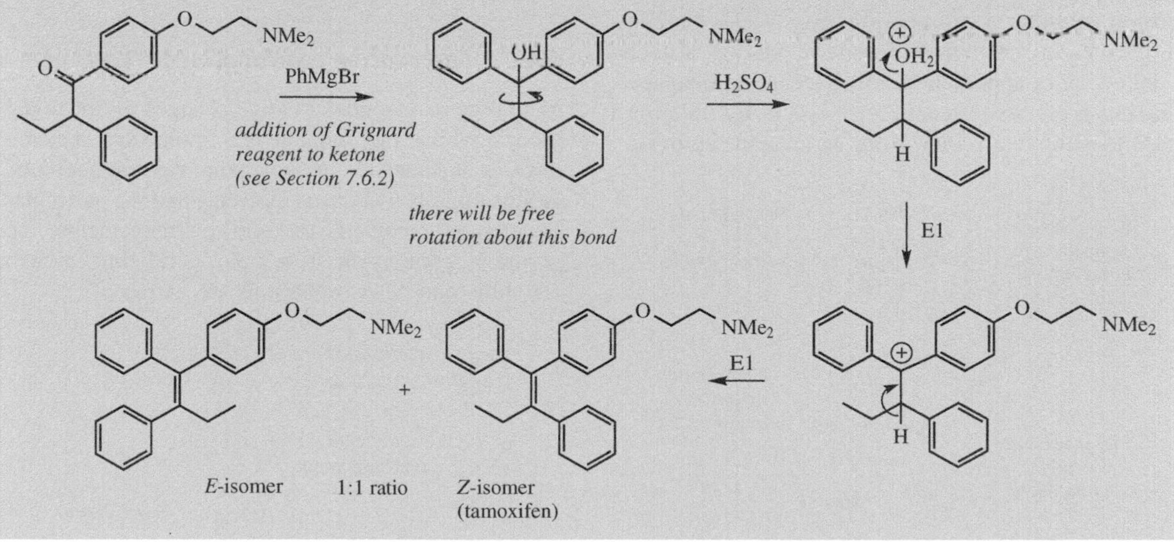

PhMgBr

addition of Grignard reagent to ketone (see Section 7.6.2)

there will be free rotation about this bond

H_2SO_4

E1

E1

E-isomer   1:1 ratio   *Z*-isomer
                             (tamoxifen)

+

The main skeleton of the drug is constructed by a Grignard addition reaction (see Section 7.6.2) on the appropriate ketone using phenyl magnesium bromide. This produces a tertiary alcohol. It now remains to eliminate water from this structure. This is achieved under acid conditions. An E1 mechanism is involved: protonation of the tertiary alcohol allows loss of water as the leaving group and generation of a carbocation, which is favoured since it is both tertiary and benzylic (see Section 6.2.1). However, completion of the elimination by proton loss gives a 1 : 1 mixture of the *E*- and *Z*-alkenes, since there is no stereocontrol at this stage – free rotation about the C–C bond in the alcohol and subsequent structures until the double bond is actually formed means both stereochemistries will be produced. The drug material tamoxifen is the *Z*-isomer.

E1 or E2?

We have seen above that the structure of the substrate is the most important feature that dictates the mechanism of substitution reactions. Thus, the S_N2 mechanism is favoured when the reaction takes place at a primary centre, whereas an S_N1 mechanism is preferred at tertiary centres, or where stable intermediate carbocations can be produced (see Section 6.2.3).

We can use similar reasoning to predict that an E2 mechanism might be preferred when the leaving group departs from a primary centre, and that an E1 mechanism is likely when structural features facilitate carbocation formation. By structural features we mean tertiary, allylic, or benzylic centres (see Section 6.2.1). When a secondary centre is involved, then either E1 or E2 might occur, depending upon reaction conditions. In general, these predictions are found to be sound. However, there is an apparent anomaly, in that E2 reactions also frequently occur with tertiary substrates. If we think a little deeper, we shall discover that it is not unreasonable for this to be so. The E2 reaction is initiated by base removing a proton, and this is still possible even where there is a tertiary centre. Although, for steric reasons, a nucleophile cannot approach a tertiary centre to displace a leaving group (S_N2 reaction), it is still feasible for a base to remove a proton from an adjacent carbon.

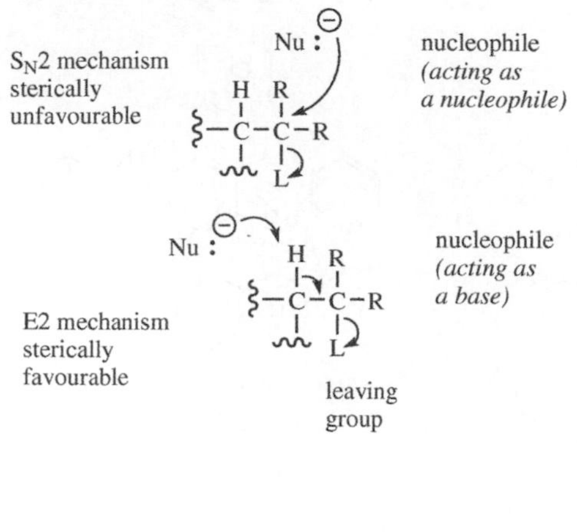

Accordingly, the E2 mechanism becomes relatively favourable, even with tertiary substrates, when we use a strong base or more concentrated base. We are thus more likely to get an E1 mechanism when we have a tertiary centre, and weak bases or bases in low concentration. Obviously, polar solvents are also going to be conducive to carbocation mechanisms (see Section 6.2.3). Just as acidic conditions help to favour S_N1 reactions, they also going to favour E1 reactions.

Elimination or substitution?

Elimination can be a troublesome side-reaction during substitution reactions. In general terms:

- strong bases favour elimination;
- large bases favour elimination;
- steric crowding in the substrate favours elimination;
- high temperatures and low solvent polarity favour elimination.

6.4.2 Carbocation rearrangement reactions

Most organic reactions involve changes to functional groups whilst the fundamental molecular skeleton remains unchanged. In molecular **rearrangements**, groups migrate within the molecule and the molecular skeleton is modified. In most rearrangements, the groups migrate to the next atom, a 1,2-shift, though 1,3-shifts and other migrations are known.

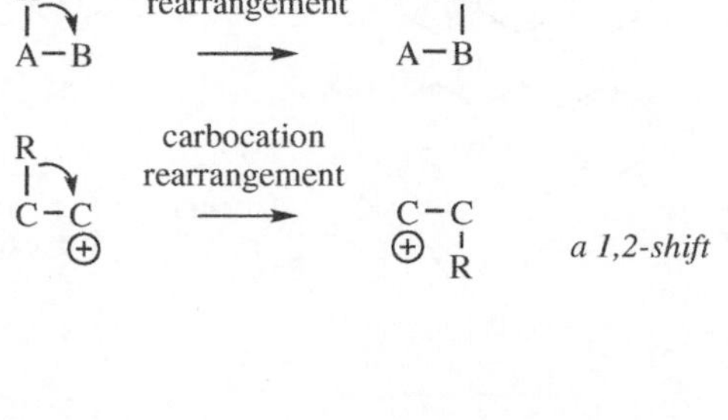

The most common examples of rearrangements involve an electron-deficient atom, and pre-eminent amongst these are carbocations. Since carbocations are a feature of the S_N1 and E1 mechanisms, it follows that rearrangements can be side-reactions of these types of transformation. The driving force in **carbocation rearrangements** is to form a more stable carbocation.

Consider a proposed nucleophilic substitution reaction on the secondary alcohol shown using aqueous HBr. As a secondary alcohol, either S_N2 or S_N1 mechanisms are possible (see Section 6.2.3), but S_N1 is favoured because of the acidic environment and the large *tert*-butyl group hindering approach of the nucleophile. The expected S_N1 bromide product is formed, together with a smaller amount of the E1-derived alkene in a competing reaction.

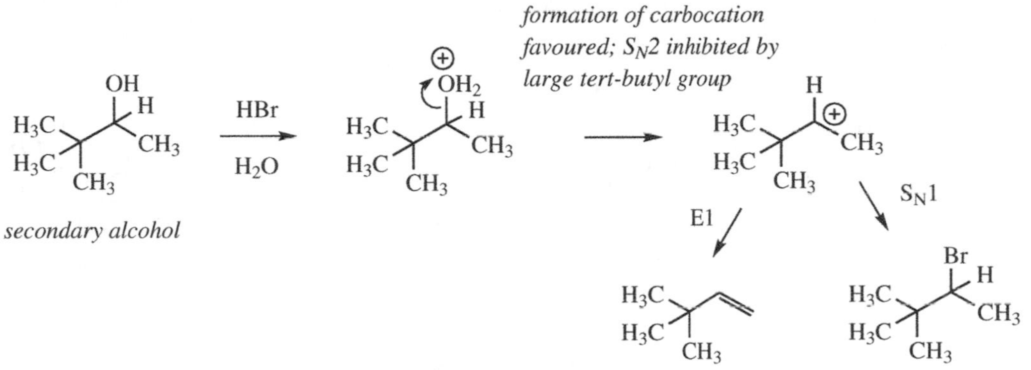

formation of carbocation favoured; S_N2 inhibited by large tert-butyl group

secondary alcohol

However, other products are also produced. These are isomers of the above products and have a rearranged carbon skeleton. Their formation is rationalized as follows:

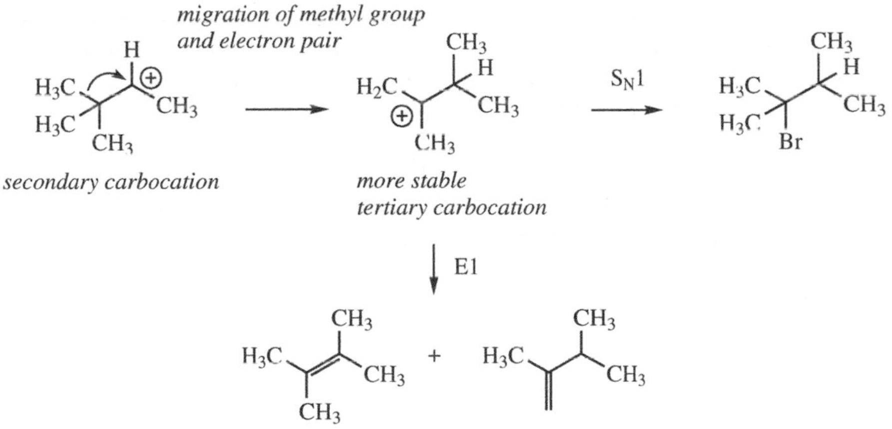

migration of methyl group and electron pair

secondary carbocation

more stable tertiary carbocation

The first-formed carbocation is secondary. It is possible for this carbocation to become a more stable tertiary carbocation via rearrangement, in which a methyl group with its pair of electrons migrates from one carbon to the adjacent positive centre. Now the rearranged tertiary carbocation can yield S_N1- and E1-type products in much the same manner as the original secondary carbocation. A rearranged bromide is formed, together with two alkenes from an E1

process, with both more-substituted Saytzeff and less-substituted Hofmann alkenes being produced. The formation of such rearranged products proves that this unexpected transformation must occur.

These carbocation rearrangements are termed **Wagner–Meerwein rearrangements**. They are most commonly encountered with secondary carbocations where rearrangement produces a more stable tertiary carbocation. They are less common with tertiary carbocations, which are already stabilized by the maximum number of alkyl groups, and where any rearrangement would tend to produce only a less stable secondary carbocation. Wagner–Meerwein

rearrangements are not restricted to methyl migrations, and we may also see transfer of hydrogen with an electron pair, i.e. a **hydride migration**.

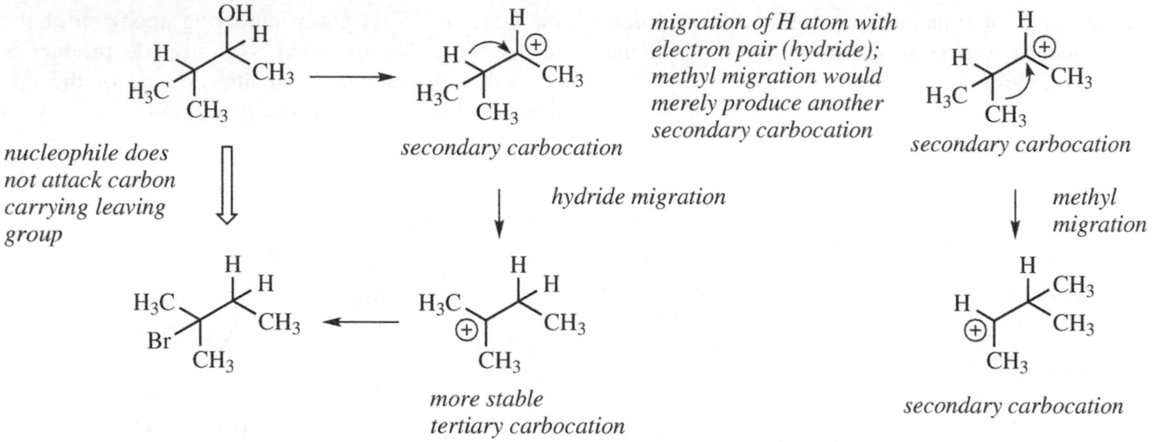

nucleophile does not attack carbon carrying leaving group

secondary carbocation

migration of H atom with electron pair (hydride); methyl migration would merely produce another secondary carbocation

secondary carbocation

hydride migration

methyl migration

more stable tertiary carbocation

secondary carbocation

This is observed in the case of the secondary alcohol illustrated, where a secondary carbocation would be generated. A methyl migration would merely lead to another secondary carbocation, and this serves no stabilizing effect. However, a hydride migration produces a tertiary carbocation, so this process will stabilize the system. This is what actually happens, and the major product is a bromide where the halogen appears to have attacked the wrong position,

i.e. different from that which originally carried the leaving group. This is the pointer to something unusual occurring. Again, the driving force is the conversion of a secondary carbocation into a more stable tertiary carbocation.

Hydride migration also accounts for one of the observed products from treatment of the cyclohexenol tosylate with acetic acid.

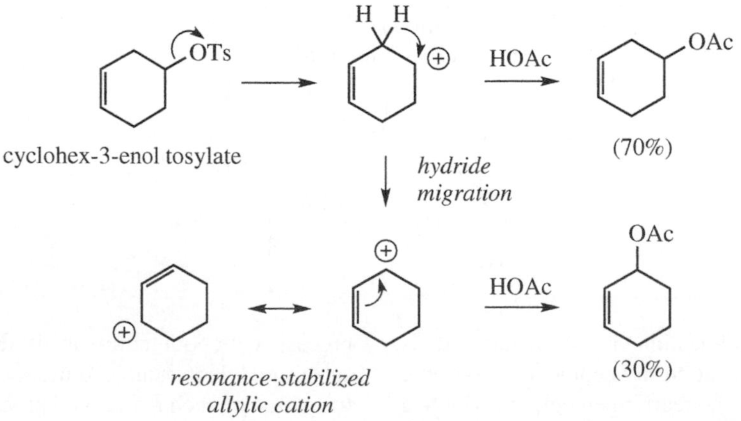

cyclohex-3-enol tosylate

(70%)

hydride migration

resonance-stabilized allylic cation

(30%)

Although the predominant product is the corresponding acetate (one could formulate either S_N1 or S_N2 mechanisms for formation of this product), about 30% of the alternative acetate is formed. This can be

rationalized as arising from a carbocation that rearranges by hydride migration. This is favoured because the resultant carbocation is an **allylic cation**, and stabilized by resonance (see Section 2.10).

In most cases, the driving force for a rearrangement is the conversion of a secondary carbocation into a more stable tertiary carbocation. Surprisingly, there are examples of where a tertiary carbocation is transformed into a secondary carbocation, but there needs to be some more powerful driving force to achieve this. Relief of ring strain is a particular case.

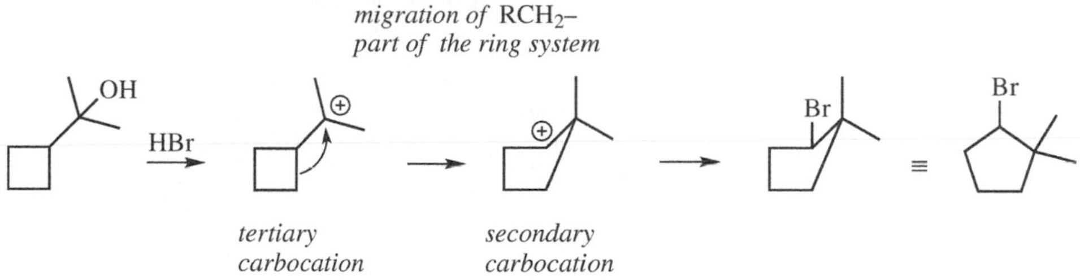

*migration of RCH₂–
part of the ring system*

*tertiary
carbocation*

*secondary
carbocation*

The cyclobutane-ring-containing alcohol can yield a tertiary carbocation, but the product from an S_N1 reaction with HBr contains a cyclopentane ring. Its formation is rationalized via a Wagner–Meerwein rearrangement in which ring expansion occurs. This is represented as equivalent to a methyl migration, but the methylene group is part of the carbon chain. There is significant relief of ring strain in going from a four-membered ring to a five-membered ring (see Section 3.3.2), which is obviously more than enough to make up for the energy change in going from a tertiary carbocation to the less stable secondary carbocation.

Carbocations also feature as intermediates in **electrophilic addition reactions** (see Section 8.1) and in **Friedel–Crafts alkylations** (see Section 8.4.1).

Rearrangements may also be observed in these carbocations if they have the appropriate structural features. It does not matter how the carbocation is produced, subsequent transformations will be the same as we have seen where rearrangements are competing reactions in nucleophilic substitution. Thus, electrophilic addition of HCl to 3,3-dimethylbut-1-ene proceeds via protonation of the alkene, and leads to the preferred secondary rather than primary carbocation (see Section 8.1.1). However, this carbocation may then undergo a methyl migration to produce the even more favourable tertiary carbocation. Finally, the two carbocations are quenched by reaction with chloride ions. The product mixture is found to contain predominantly the chloride from the rearranged carbocation.

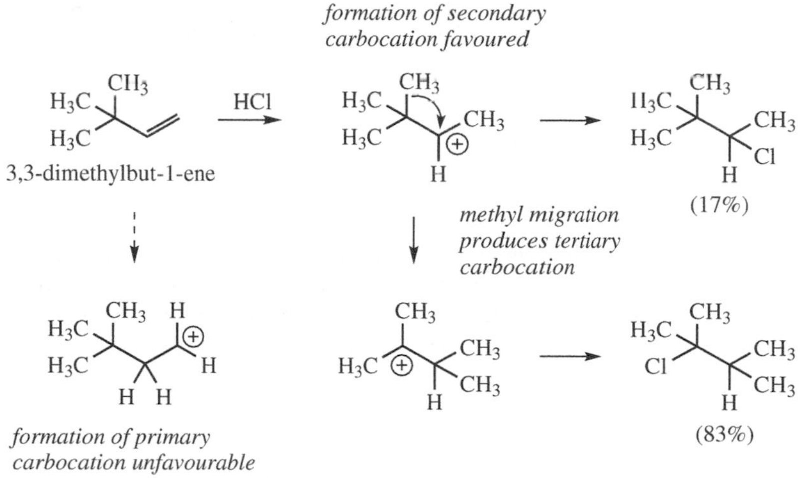

Similar reaction of 3-methylbut-1-ene with HCl gives roughly equal amounts of two isomeric chlorides. One of these is the result of a carbocation rearrangement where hydride is the migrating group.

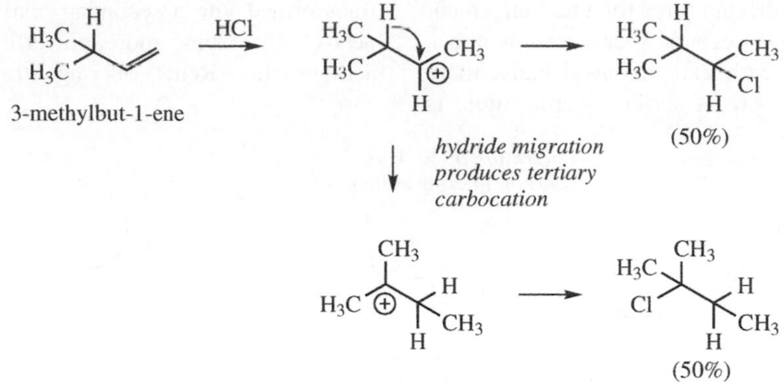

3-methylbut-1-ene

hydride migration produces tertiary carbocation

(50%)

(50%)

The enhanced stability of benzylic carbocations is nicely illustrated by the addition of HBr to the two alkenes shown below. In the case of 2-phenylbut-1-ene, protonation of the alkene leads to a carbocation that is both tertiary and benzylic, and is significantly favoured over an alternative primary carbocation.

Quenching with the bromide nucleophile gives the tertiary bromide. On the other hand, 3-phenylprop-1-ene is protonated to a secondary carbocation. In this case, rearrangement by hydride migration leads to a more favourable benzylic carbocation, and a benzylic bromide is the observed product.

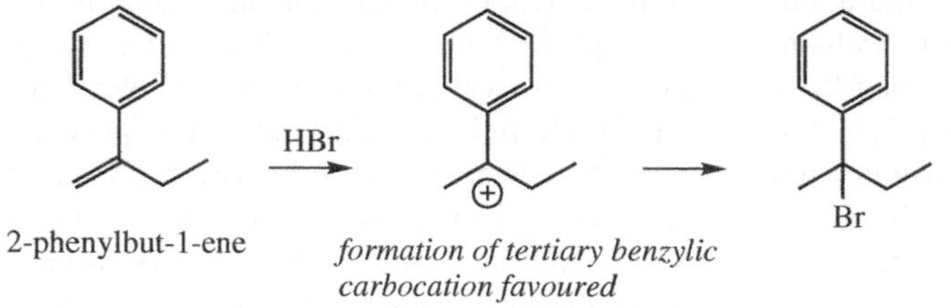

2-phenylbut-1-ene

formation of tertiary benzylic carbocation favoured

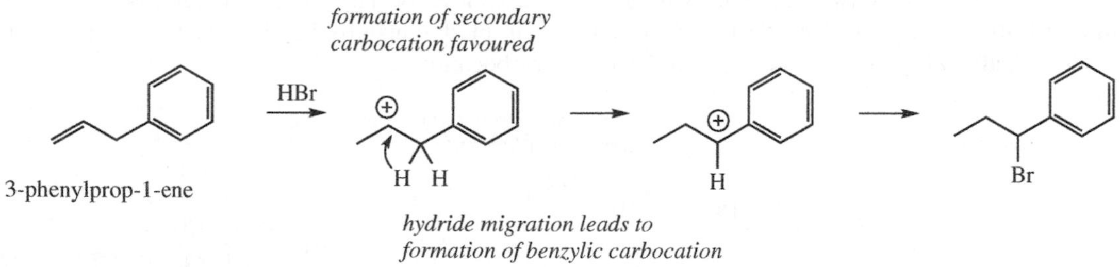

formation of secondary carbocation favoured

3-phenylprop-1-ene

hydride migration leads to formation of benzylic carbocation

Rearrangements seem to provide us with an unexpected complication to ruin our carefully thought-out plans for interconverting chemicals. It is sometimes difficult to predict when they might occur, but we should recognize occasions when they might become

a nuisance, e.g. look at the structure of any proposed carbocation intermediate. In most cases, we shall be more concerned with rationalizing such transformations, rather than trying to predict their possible occurrence.

Box 6.11

Carbocation rearrangements: synthesis of camphor from α-pinene

Although the monoterpene camphor occurs naturally, substantial amounts are produced semi-synthetically from α-pinene, a component in turpentine. Treatment of α-pinene with aqueous HCl protonates the double bond by an electrophilic addition (see Section 8.1.1) and generates the more favoured tertiary carbocation. Rather than simply being attacked by a nucleophile, this carbocation rearranges.

protonation of alkene gives carbocation; more stable tertiary cation formed

alkyl shift; relief of ring strain

H_2O:

α-pinene → HCl → *tertiary carbocation* → *secondary carbocation* → H_2O → isoborneol → CrO_3 → camphor

α-pinene   camphor

The tertiary carbocation contains a strained four-membered ring, and an alkyl shift allows relief of ring strain, generating five-membered rings and a secondary carbocation. It would appear that the relief of ring strain more than compensates for the loss of tertiary character in the carbocation. Thus, it is the secondary carbocation that interacts with a nucleophile. In this case, the nucleophile is water, the major component of the aqueous HCl. The product is thus isoborneol.

Camphor is then obtained from isoborneol by oxidation of the secondary alcohol to a ketone.

Box 6.12

Carbocation rearrangements in nature: biosynthesis of lanosterol

Many examples of carbocation rearrangements can be found in nature, particularly in the biosynthesis of terpenoids and steroids. Nature generates carbocations in three main ways. The first of these is loss of a leaving group, with diphosphate being the most common leaving group (see Box 6.4). Protonation of an alkene also produces a carbocation, and, as we would predict, this tends to form the more-substituted and thus more stable carbocation (see Section 8.1.1). Also encountered is ring opening of an epoxide group (see Section 6.3.2), which may be considered to be acid initiated.

Generation of carbocations in nature

loss of leaving group; L is usually diphosphate   *protonation of alkene*   *protonation and ring opening of epoxide*

Perhaps the most spectacular of the natural carbocation rearrangements is the concerted sequence of 1,2-methyl and 1,2-hydride Wagner–Meerwein shifts that occurs during the formation of **lanosterol** from squalene. Lanosterol is then the precursor of the steroid **cholesterol** in animals.

Carbocation formation is initiated by epoxide ring opening in squalene oxide, giving a tertiary carbocation, and this is transformed into the four-ring system of the protosteryl cation by a series of electrophilic addition reactions (see Box 8.3).

The resultant protosteryl cation has a tertiary carbocation in the side-chain, and a hydride shift generates another tertiary cation. A second hydride shift follows, then two methyl shifts, each time generating a new tertiary cation.

Box 6.12 (continued)

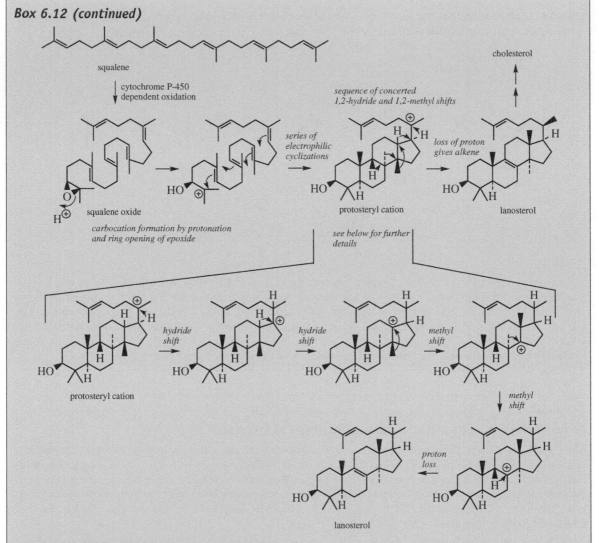

Lastly, the positive charge is neutralized via loss of a proton, giving the alkene lanosterol. There is no obvious energy advantage in such tertiary-to-tertiary cation changes, but it must be appreciated that this is an enzyme-catalysed reaction, and the enzyme plays a crucial role in the reactions that occur. These hydride and methyl migrations definitely do occur, as demonstrated by isotopic labelling studies.

Further, it is noted that most of them involve inversion of stereochemistry at the particular centre, a feature of the concerted nature of these rearrangements, so that as one group leaves another approaches from the rear. Thus, we have the features of S_N2 reactions in a carbocation mechanism.

This is a complicated series of reactions, but includes impressive examples of carbocation rearrangements. The electrophilic cyclization sequence is also quite striking, and this is discussed in more detail in Box 8.3.

7

Nucleophilic reactions of carbonyl groups

7.1 Nucleophilic addition to carbonyl groups: aldehydes and ketones

The carbon–oxygen double bond C=O is termed a carbonyl group, and represents one of the most important reactive functional groups in chemistry and biochemistry. Since oxygen is more electronegative than carbon, the electrons in the double bond are not shared equally and the carbon–oxygen bond is polarized, with the oxygen atom attracting more of the electron density (see Section 2.7). This polarization may be represented via the resonance structures A and B, where A is uncharged and B has full charge separation (see Section 2.10).

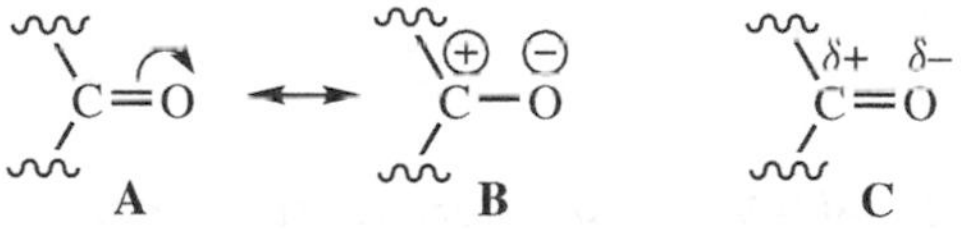

However, since the contribution from B is smaller than that from A, the charge distribution is often presented as in C. The partial charges $\delta+$ and $\delta-$ indicate the imbalance in electron density.

Ketones have two alkyl or aryl groups attached to the carbonyl group; in **aldehydes** one or both of these groups is hydrogen (the simplest aldehyde is formaldehyde $H_2C{=}O$). The carbon atom is sp^2 hybridized, so that the carbonyl group and the directly bonded atoms are in one plane (see Section 2.6.2). The two pairs of nonbonding electrons, i.e. lone pairs, on the carbonyl oxygen tend to be omitted in most representations; but, as we shall see, these frequently play an important part in mechanisms involving carbonyl groups. For convenience, or out of sheer laziness, we may show only one of these lone pairs even when they do play a role.

Because of the polarization, it is possible for the carbonyl group to be involved in both **nucleophilic addition** reactions and **electrophilic addition** reactions.

nucleophilic addition:

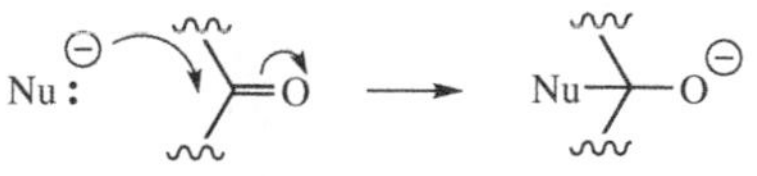

nucleophile attacks
$\delta+$ centre

electrophilic addition:

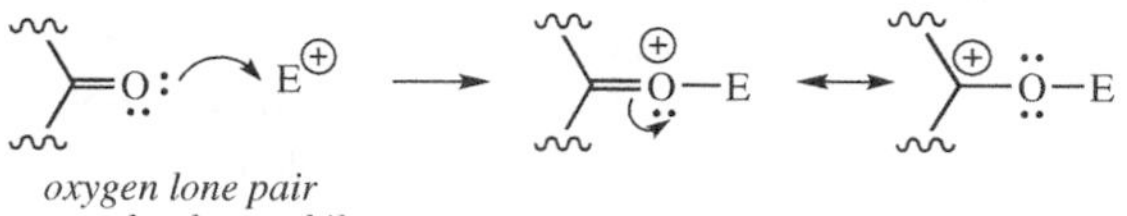

oxygen lone pair
attacks electrophile

Essentials of Organic Chemistry   Paul M Dewick
© 2006 John Wiley & Sons, Ltd

Most addition reactions actually involve both steps, but the order in which these occur depends on the nature of the reagent and the reaction conditions. Under basic conditions, the nucleophile attacks the carbonyl first, and the reaction is completed by abstraction of an electrophile, a proton, from the solvent.

proton abstracted from solvent

Under acidic conditions, electrophilic addition occurs first, namely protonation of the carbonyl and formation of the conjugate acid. The conjugate acid, with a full positive charge, is now a more reactive electrophile than the original uncharged carbonyl group, which only has partial charge separation due to polarization. As a result, addition can now occur with less-reactive nucleophiles, and typically these are uncharged and attack via their lone pair electrons.

protonation leads to conjugate acid

nucleophilic attack on to conjugate acid

loss of proton from nucleophile

In most of the reactions that we shall encounter there will be attack of a charged nucleophile on to an uncharged carbonyl or, alternatively, attack of lone pair electrons in an uncharged nucleophile onto a charged conjugate acid. An uncharged nucleophile tends to be insufficiently reactive for addition reactions to occur with an uncharged carbonyl. At the other extreme, the combination of charged nucleophile and charged carbonyl is not usually favourable. Since negatively charged nucleophiles are also bases, an acidic environment will not permit their involvement.

The most significant change in these reactions is the formation of the carbon–nucleophile bond; so, in both types of mechanism, the reaction is termed a **nucleophilic addition**. It should be noted that the polarization in the carbonyl group leads to nucleophilic addition, whereas the lack of polarization in the C=C double bond of an alkene leads to electrophilic addition reactions (see Chapter 8). Carbonyl groups in carboxylic acid derivatives undergo a similar type of reactivity to nucleophiles, but the presence of a leaving group in these compounds leads to substitution reactions rather than addition (see Section 7.8).

7.1.1 Aldehydes are more reactive than ketones

The reactions undergone by **aldehydes** and **ketones** are essentially the same, but aldehydes are more reactive than ketones. There are two rational reasons for this. Alkyl groups have an electron-donating inductive effect (see Section 4.3.3) and the presence of two such groups in ketones against just one in aldehydes means the magnitude of $\delta+$ is reduced in ketones. Put another way, the carbonyl group in aldehydes is more electrophilic than that in ketones. It should also be noted that aromatic aldehydes, such as benzaldehyde, are less reactive than alkyl aldehydes. This is because the aromatic ring allows electron delocalization via a resonance effect that also reduces the positive charge on the carbonyl carbon.

aldehydes are more reactive than ketones

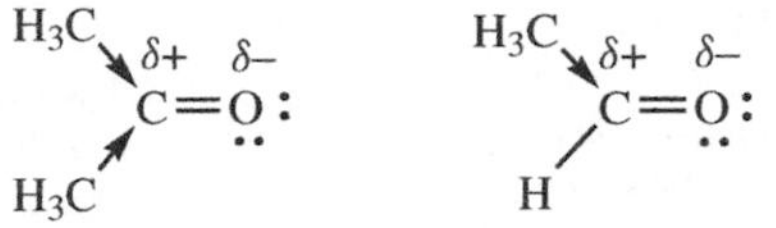

the aldehyde carbon is more electrophilic than the ketone carbon

aromatic aldehydes are less reactive than aliphatic aldehydes

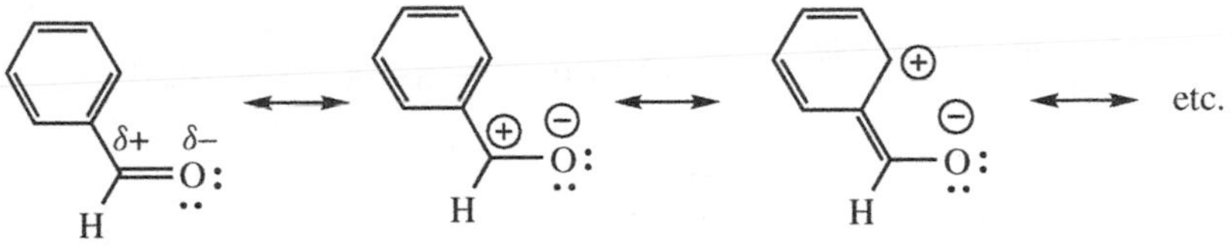

etc.

aromatic compounds are less reactive; the aromatic ring delocalizes positive charge away from carbonyl carbon

The second feature is a steric consideration. During nucleophilic addition, the planar sp^2 system of the carbonyl compound (bond angle 120°) is converted into a tetrahedral sp^3 system in the product (bond angle 109°) creating more steric crowding, i.e. the groups are brought closer together.

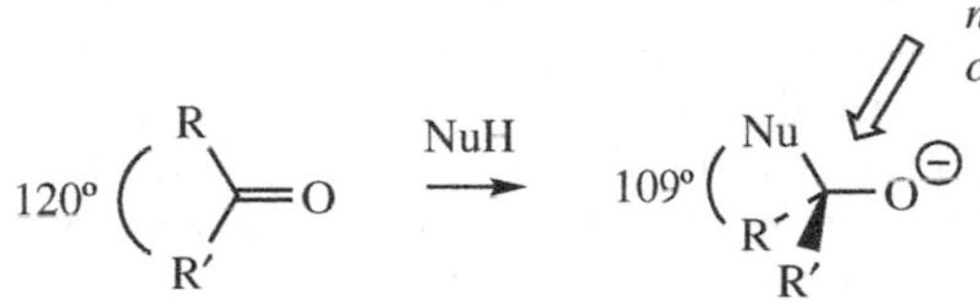

note formation of chiral centre

the addition reaction increases steric crowding

formation of new bond creates more steric crowding

This crowding is more severe with two alkyl substituents (from ketones) than with one alkyl and the much smaller hydrogen (from aldehydes). A consequence of this change is that the planar aldehyde or ketone can be attacked from either side of the plane with essentially equal probability. If the substituents are all different, then this will result in the creation of a chiral centre; but, since both enantiomers will be formed in equal amounts, the product will be an optically inactive racemate (assuming no other chiral centres are present in the R groups); see Section 3.4.1.

7.1.2 Nucleophiles and leaving groups: reversible addition reactions

In principle, all carbonyl addition reactions could be reversible; but, in practice, many are essentially irreversible. Let us consider mechanisms for the reverse of the nucleophilic addition reactions given above. For the base-catalysed reaction, we would invoke the following mechanism:

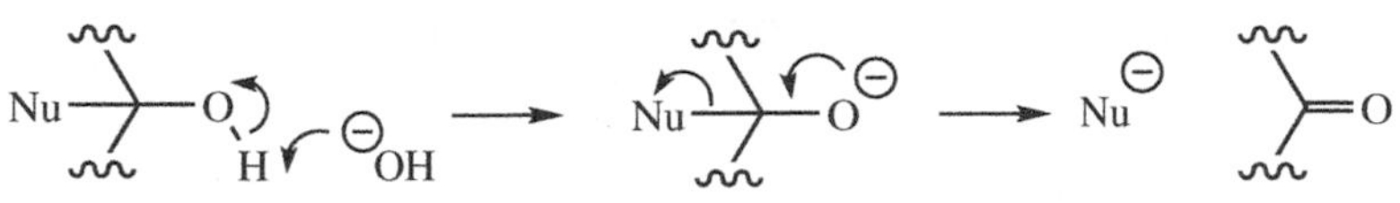

base removes proton

carbonyl formation with loss of leaving group

nucleophile as leaving group

For the acid-catalysed reaction, we would write

nucleophile as leaving group

loss of proton from carbonyl conjugate acid

protonation of original nucleophilic species   *loss of leaving group*

resonance stabilization

we normally combine the above steps:

loss of leaving group and formation of protonated carbonyl

It becomes clear that, in the reverse reactions, we need the original nucleophile to behave as a good leaving group, either as Nu⁻ in base-catalysed reactions or as Nu–H in acid-catalysed situations. Conversely, if the nucleophile cannot act as a leaving group, then the reverse reaction is going to be unfavourable and the addition will be essentially irreversible. By appreciating this fundamental concept, we shall be able to rationalize the various carbonyl addition reactions of importance described below. We shall also be able to link in easily the behaviour of carboxylic acid derivatives, where the presence of an alternative leaving group needs to be considered (see Section 7.8).

Reversible reactions include addition of water, alcohols, thiols, HCN, and amines. **Irreversible reactions** include addition of hydride and organometallics. In the latter cases, hydride H⁻ and carbanions such as Me⁻ are going to be very poor leaving groups, predictable from the pK_a values of H_2 (35) and MeH (48). We have seen in Section 6.1.4 that good leaving groups are the conjugate bases of strong acids.

We can also rationalize why some addition reactions simply do not occur, e.g. halide ions do not add to carbonyl groups. Although we know that a halide such as bromide can act as an effective nucleophile in S_N1 and S_N2 reactions (see Section 6.1.2), it is also a very good leaving group (pK_a value for HBr −9). This means that the reverse reaction becomes very much more favourable than the forward reaction. In cases where both forward and reverse reactions are feasible, we can often usefully disturb the equilibrium by using an excess of one reagent (see below).

7.2 Oxygen as a nucleophile: hemiacetals, hemiketals, acetals and ketals

The addition of 1 mol of an alcohol to an aldehyde gives a **hemiacetal**, and to a ketone a **hemiketal**. However, most chemists do not now differentiate between hemiacetals and hemiketals; these are both termed hemiacetals. This reaction is usually catalysed

acid-catalysed formation of hemiacetals

formation of conjugate acid

nucleophilic attack of alcohol onto conjugate acid

ketone   *hemiketal*

hemiacetal

acid catalyst regenerated

by acids, but may also be achieved in the presence of base. The reactions follow the general mechanisms given above. The main difference between the acid-catalysed and base-catalysed mechanisms is that acid increases the electrophilicity of the carbonyl group by protonation, whereas base increases the nucleophilicity of the alcohol via ionization to the conjugate base.

base-catalysed formation of hemiacetals

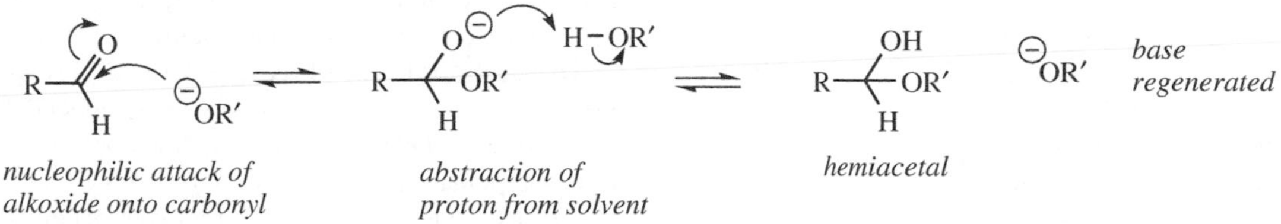

nucleophilic attack of alkoxide onto carbonyl   *abstraction of proton from solvent*   *hemiacetal*

Each reaction is reversible, and by extrapolating from the forward reaction it is relatively easy to propose a mechanism for the reverse reaction. Mechanisms are shown here for both acid- and base-catalysed decomposition of hemiacetals.

acid-catalysed decomposition of hemiacetals

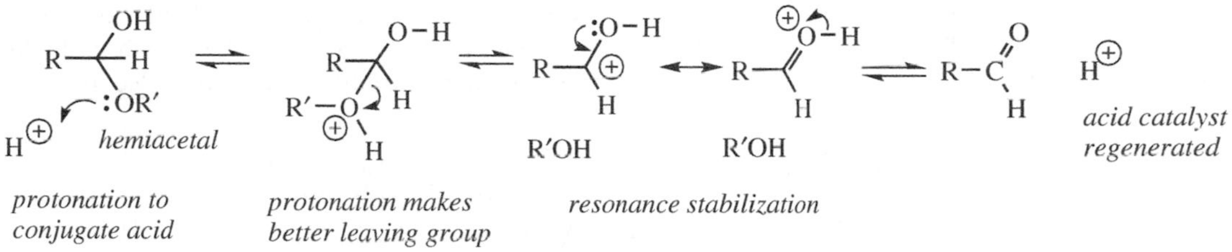

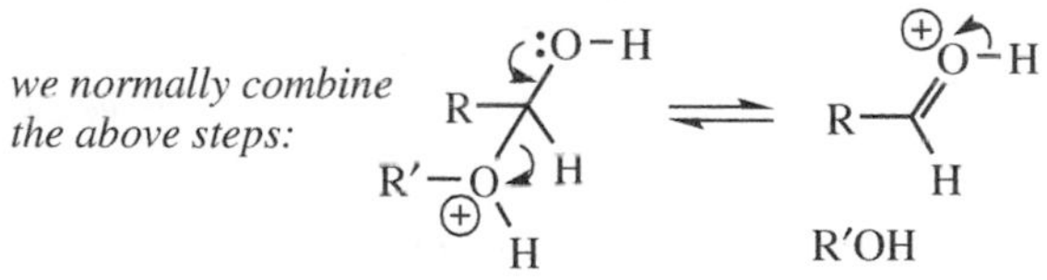

protonation to conjugate acid   *protonation makes better leaving group*   *resonance stabilization*   *acid catalyst regenerated*

we normally combine the above steps:

base-catalysed decomposition of hemiacetals

hemiacetal
base removes proton generating conjugate base   *formation of carbonyl with loss of alkoxide leaving group*

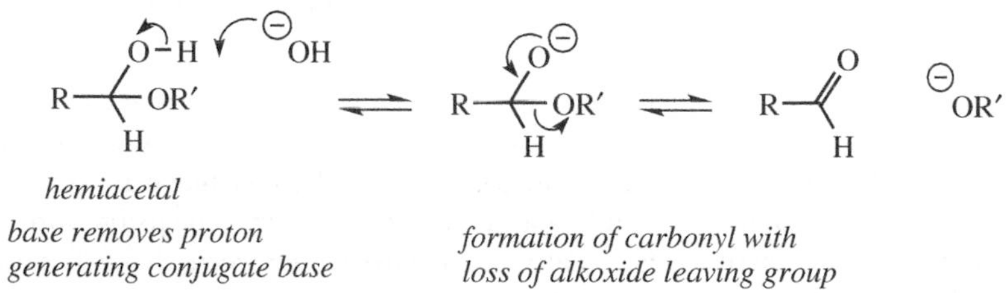

The position of equilibrium, i.e. whether the carbonyl compound or the addition product is favoured, depends on the nature of the reagents. The equilibrium constant is often less than 1, so that the product is not favoured, and many simple hemiacetals and hemiketals are not sufficiently stable to be isolated. However, stable cyclic hemiacetals and hemiketals can be formed. A cyclic product arises if the alcohol function is in the same molecule as the carbonyl, allowing an **intramolecular reaction** rather than an intermolecular one. When these functional groups are separated by three or four carbons, this results in the generation of stereochemically favourable five- or six-membered rings respectively (see Section 3.3.2).

Table 7.1 Hydroxyaldehyde–cyclic hemiacetal equilibria

Hydroxyaldehyde	Hemiacetal	Ring size	% Hemiacetal at equilibrium
		4	0
		5	89
		6	94
		7	15

However, as seen in Table 7.1, rings other than five- and six-membered are either not formed or are not particularly favoured.

A vast range of natural **sugars** exemplify these cyclic addition products. A typical sugar exists predominantly in the form of a hemiacetal or hemiketal in solution, although this is an equilibrium reaction, and the open chain carbonyl form is always present to a small extent (<1%). The formation of a six-membered cyclic hemiacetal from glucose is achieved by attack of the C-5 hydroxyl onto the protonated carbonyl (conjugate acid).

The cyclic form of **glucose** is termed **glucopyranose**, since the new ring system is a reduced form of the oxygen heterocycle pyran. Nucleophilic attack onto the planar carbonyl may occur from either of its two faces, generating two different stereochemistries at this new chiral centre, designated as α or β. This new chiral centre is termed the **anomeric centre**. Since there are other chiral centres in the molecule, the mixture of α- and β-anomeric forms is not a racemate, but a mixture of diastereoisomers (see Section 3.4.4). The mixture does not contain 50% of each anomer (see below). Although both forms are produced, the β form with the equatorial hydroxyl is thermodynamically favoured (see Section 3.3.2).

D-(+)-glucose *Fischer projection*

conjugate acid

wavy bond indicates either stereochemistry

pyranose form

hemiacetal

Note that these are equilibria; sugars display carbonyl reactions

β-D-glucopyranose   α-D-glucopyranose   pyran

Further, the two forms can also equilibrate via the open-chain carbonyl form of the sugar, so that the single isomers in solution are rapidly transformed into the equilibrium mixture (see Box 7.1). Since there are two anomeric forms, and these are often in equilibrium via the acyclic carbonyl compound, we use a new type of bond to indicate that the configuration is not specified, and could be of either stereochemistry. This is the wavy or wiggly bond; and to emphasize our indecision further, it is usually sited halfway between the two possible positions.

Box 7.1

The mutarotation of glucose

It is possible to separate the two **anomeric forms** of **glucose** by careful recrystallization from water. The two forms have different specific optical rotations (see Section 3.4.1), $[\alpha]_D + 112°$ for α-D-glucopyranose, and $[\alpha]_D + 18.7°$

for β-D-glucopyranose. If either of these forms is dissolved in water, the optical rotation slowly changes to yield the same final rotation, $[\alpha]_D + 52.7°$. Because this process produces a change in rotation from that of either pure substance, it is termed **mutarotation**. The final product is an equilibrium mixture of the α and β anomeric forms, and their interconversion involves the open-chain aldehyde form as shown.

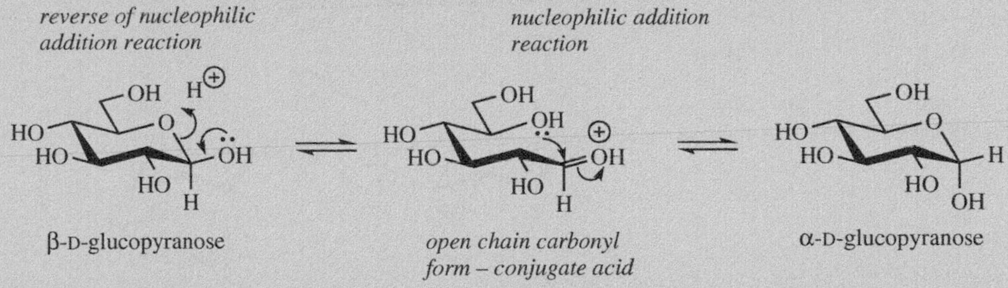

*reverse of nucleophilic
addition reaction*

*nucleophilic addition
reaction*

β-D-glucopyranose

*open chain carbonyl
form – conjugate acid*

α-D-glucopyranose

A similar transformation is common to all aldehyde-containing hexoses (aldohexoses) and pentoses (aldopentoses); see Section 12.2.3.

From these data, it is also easy to calculate the proportions of the two forms in the equilibrium mixture. Since we are dealing with isomeric compounds, we can relate specific rotations to the amount of each isomer. If the fraction of the α form is x and that of the β form is $1 - x$, then

$$112x + 18.7(1 - x) = 52.7$$

From this one can calculate that $x = 0.36$; there is thus 36% α anomer and 64% β anomer in the equilibrium mixture.

The approximately 2 : 1 preference for the β anomer is consistent with our knowledge of conformations in six-membered rings (see Section 3.3.2); this anomer has the variable hydroxyl in a more favourable equatorial position. Note that the difference in thermodynamic stability is not sufficient to force the equilibrium completely in favour of the β anomer, and it is perhaps unexpected that there is quite so much of the α anomer present at equilibrium. We shall return to this topic, the **anomeric effect**, in Section 12.4.

Glucose is also capable of forming a five-membered hemiacetal ring by attack of the 4-hydroxyl onto the carbonyl, though this is much less favourable than formation of the six-membered ring just discussed (see Section 12.2.2). Five-membered rings are termed **furanose** rings by comparison with the oxygen heterocycle furan, with the most prominent example being that from the five-carbon sugar **ribose**.

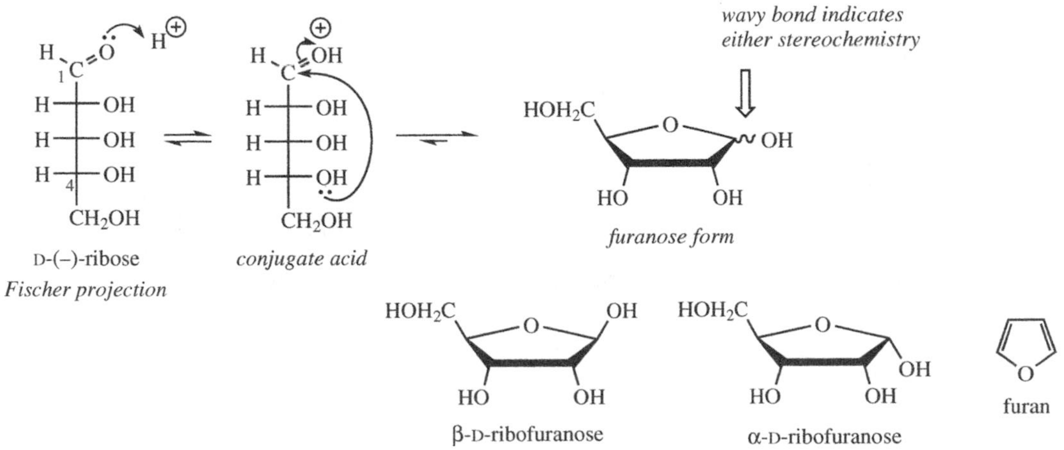

D-(–)-ribose

Fischer projection

conjugate acid

*wavy bond indicates
either stereochemistry*

furanose form

β-D-ribofuranose

α-D-ribofuranose

furan

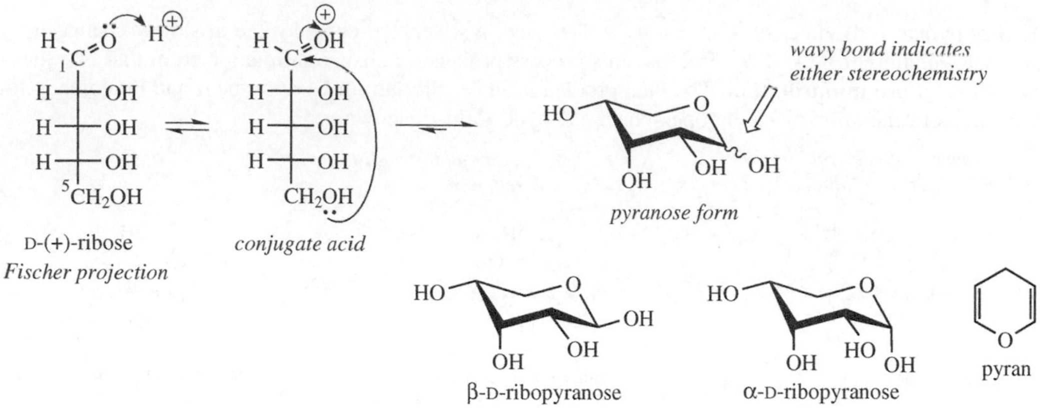

D-(+)-ribose
Fischer projection

conjugate acid

wavy bond indicates either stereochemistry

pyranose form

β-D-ribopyranose

α-D-ribopyranose

pyran

The two anomeric forms are called α- and β-ribofuranose. Again, in solution there exists an equilibrium between the open chain carbonyl form and the two anomeric hemiacetal forms. Ribose also forms six-membered pyranose anomers, and an aqueous solution actually contains about 76% pyranose forms to 24% furanose forms. In the vast majority of cases, ribose is found in nature combined in the β-furanose form (see Box 7.2).

Box 7.2
Some biologically important ribose derivatives

Nature has exploited **ribose** derivatives for a number of crucially significant biochemicals. Many of these contain a heterocyclic base attached to the β-anomeric position of D-ribofuranose, and are termed **nucleosides**. Adenosine, guanosine, cytidine, and uridine are fundamental components of **ribonucleic acids** (**RNA**; see Section 14.1),

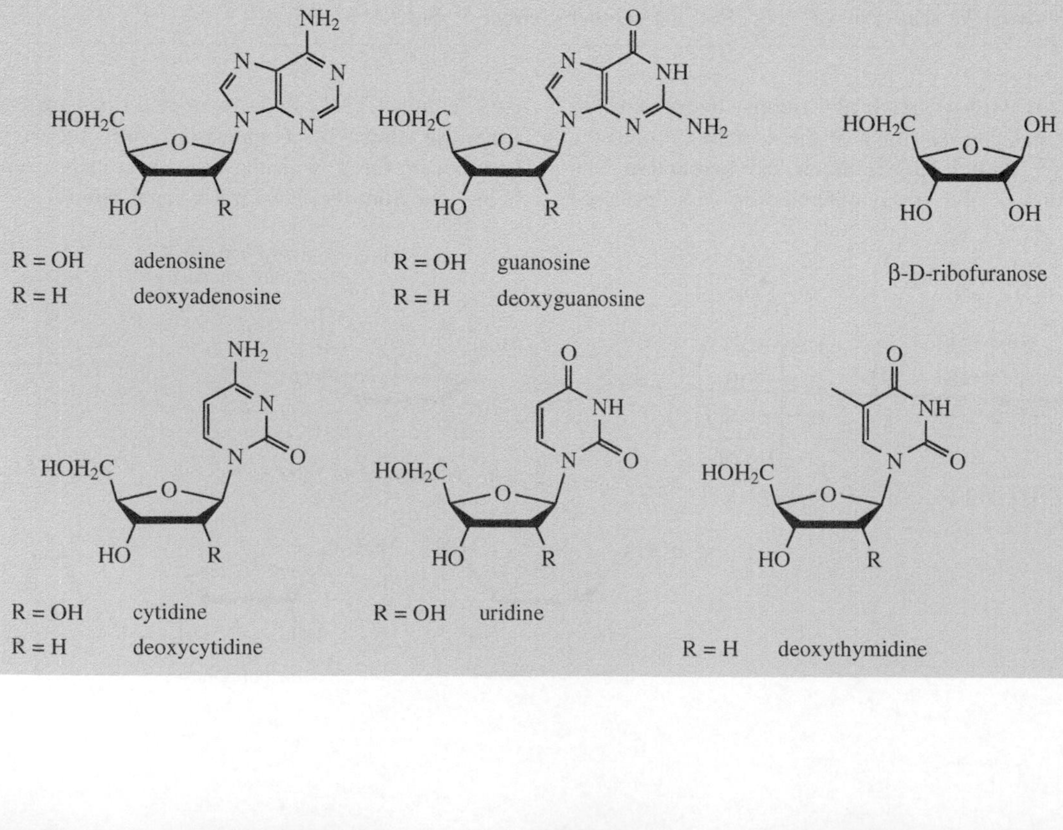

R = OH   adenosine
R = H   deoxyadenosine

R = OH   guanosine
R = H   deoxyguanosine

β-D-ribofuranose

R = OH   cytidine
R = H   deoxycytidine

R = OH   uridine

R = H   deoxythymidine

and similar derivatives of 2-deoxy-β-D-ribofuranose, the deoxynucleosides deoxyadenosine, deoxyguanosine, deoxycytidine, and deoxythymidine, are the building blocks of **deoxyribonucleic acids** (**DNA**; see Section 14.1).

Nucleosides are also encountered in the structures of **adenosine triphosphate** (**ATP**) and coenzyme A (**HSCoA**). ATP provides nature with its currency unit for energy. Hydrolysis of ATP to adenosine diphosphate (ADP) liberates energy, which can be coupled to energy-requiring processes in biochemistry, and synthesis of ATP from ADP can be coupled to energy-releasing processes (see Box 7.25).

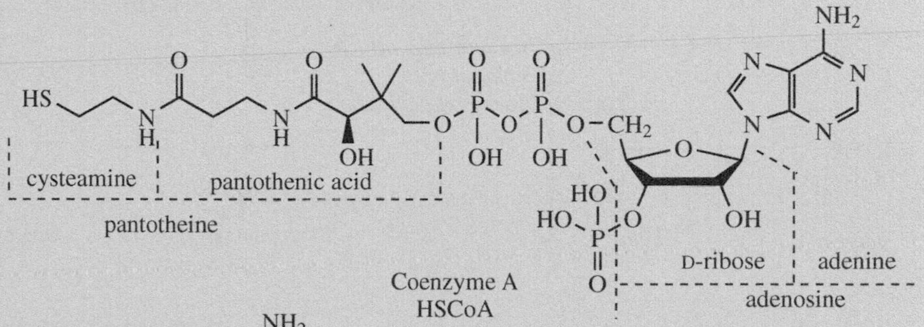

Coenzyme A
HSCoA

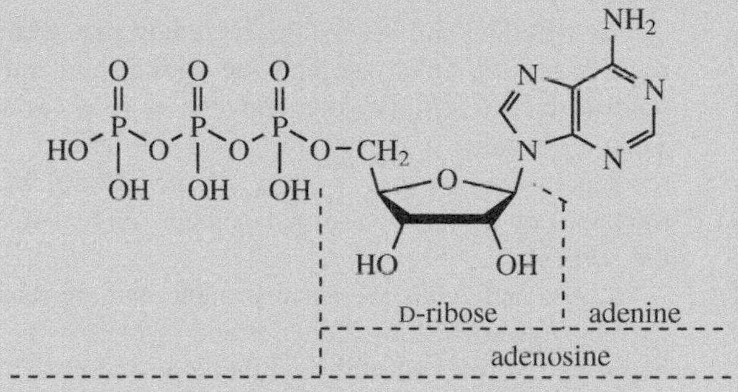

adenosine triphosphate; ATP

Coenzyme A is used as the alcohol part of thioesters, which are more reactive than oxygen esters (see Section 7.9.3) and are thus exploited in biochemistry in a wide range of reactions, e.g. fatty acid biosynthesis and metabolism (see Section 15.5).

The addition of 2 mol of an alcohol to an aldehyde gives an **acetal**, and to a ketone a **ketal**. Most chemists do not now differentiate between acetals and ketals; these are both termed acetals. These products are formed by further reaction of the hemiacetal or hemiketal with a second molecule of alcohol. In contrast to hemiacetal and hemiketal formation, this reaction is catalysed only by acids, not by base.

acid-catalysed formation of acetals

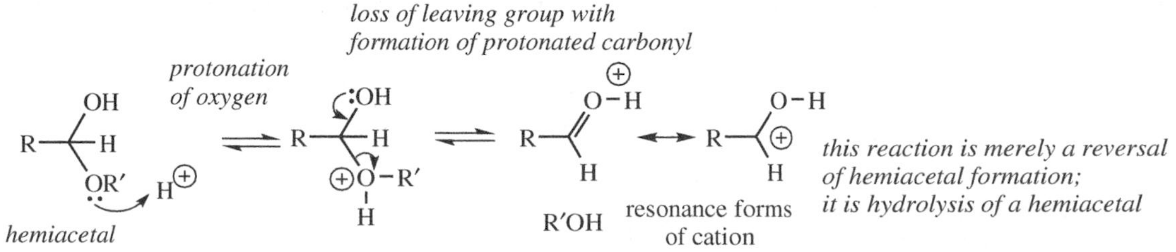

this reaction is merely a reversal of hemiacetal formation; it is hydrolysis of a hemiacetal protonation of oxygen   *loss of leaving group with formation of protonated carbonyl*   *resonance forms of cation*   *this is merely a further example of*

nucleophilic attack of alcohol onto carbonyl equivalent

acid catalyst regenerated

ketone   *ketal*   *acetal*

Initially, the reaction involves protonation of one of the oxygen atoms, followed by loss of this group as a neutral molecule and formation of a resonance-stabilized carbocation. If the oxygen protonated were that of the alkoxy group, then the product would merely be the protonated aldehyde, and the reaction becomes a reversal of hemiacetal formation. Only when the oxygen of the hydroxyl is protonated can the reaction lead to an acetal, and this requires nucleophilic attack of the second alcohol molecule on to the alternative resonance-stabilized carbocation.

This is a further example of a carbonyl–electrophile complex, and equivalent to the conjugate acid, so that the subsequent nucleophilic addition reaction parallels that in hemiacetal formation. Loss of the leaving group occurs first in an S_N1-like process with the cation stabilized by the neighbouring oxygen; an S_N2-like process would be inhibited sterically. It is also possible to rationalize why base catalysis does not work. Base would simply remove a proton from the hydroxyl to initiate hemiacetal decomposition back to the aldehyde – what is needed is to transform the hydroxyl into a leaving group (see Section 6.1.4), hence the requirement for protonation.

The reactions are reversible, including that which regenerates the aldehyde, and the equilibrium must be disturbed to achieve good conversion to an acetal. This is usually achieved by using an excess of the alcohol; if we are using a simple alcohol like methanol or ethanol, we might perhaps employ it as the solvent.

Acetals and ketals are usually stable, but are readily hydrolysed back to aldehydes and ketones by acid hydrolysis, a reversal of the synthetic procedure. This makes acetal or ketal formation a valuable means of protecting an aldehyde or ketone carbonyl from reaction with other reagents being used during a synthetic procedure. For instance, protection of a ketone group may be achieved by forming a cyclic ketal with an excess of ethylene glycol; when protection is no longer required, this **protecting group** may be removed by acid-catalysed hydrolysis using an excess of water. By having the two alcohol functions in the same molecule, formation of a ketal from the intermediate hemiketal becomes favourable, since it requires an intramolecular reaction rather than an intermolecular one. In the example shown, it is feasible under mild acidic conditions to carry out these reactions on a ketoester without affecting the ester function.

ethyl 4-oxocyclohexane carboxylate   ethylene glycol (ethan-1,2-diol)   *cyclic ketal*

Of course, we can use the same strategy to protect a diol. We would convert this into an acetal or ketal using a suitable aldehyde or ketone.

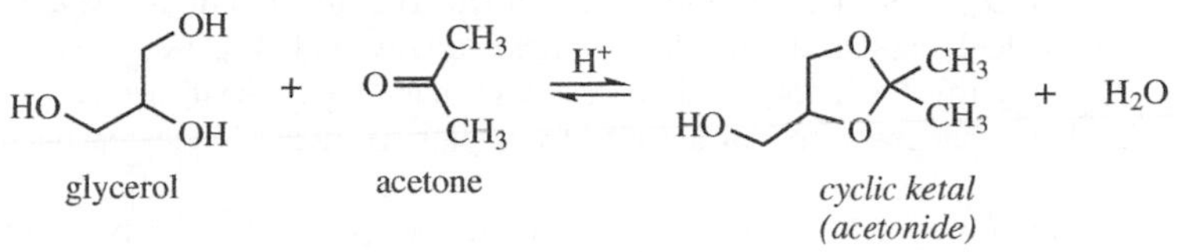

Acetal and ketal linkages are widely found in natural sugars and polysaccharides. The structure of sucrose is a splendid example. **Sucrose** is a disaccharide, composed of two linked monosaccharide units, glucose in pyranose ring form and fructose in furanose ring form. As we have seen above, glucopyranose is a hemiacetal derived from the aldehyde-containing sugar glucose. In sucrose, it is present as its α anomer. On the other hand, fructose is a ketone-containing sugar, and it forms a hemiketal furanose ring by reaction of the C-5 hydroxyl with the ketone group.

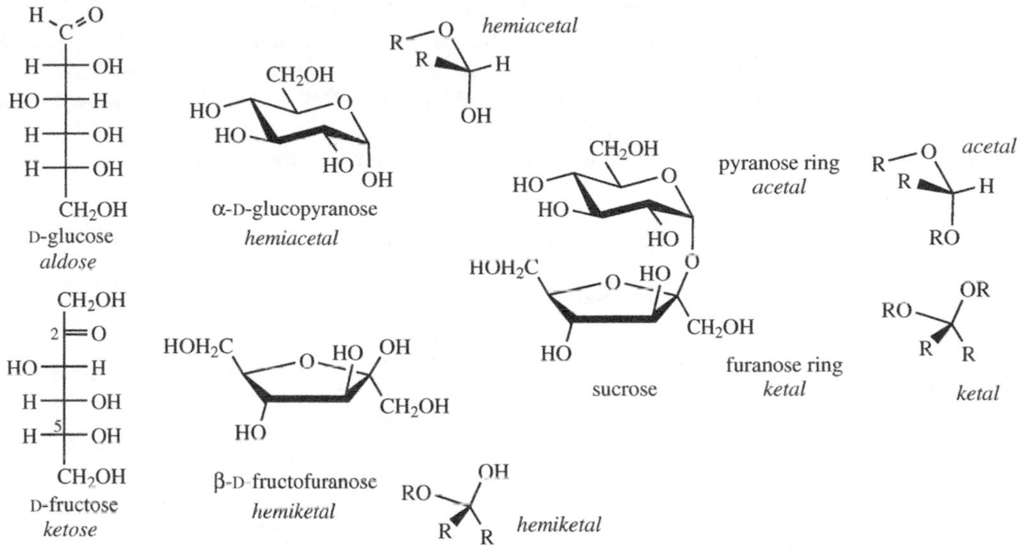

In sucrose, fructose is present as the β anomer. Now, one of these sugars has acted as an alcohol to make a bond to the other sugar. We can look at this in two ways. Either fructose acts as an alcohol to react with the hemiacetal glucose to form an acetal, or alternatively, glucose is the alcohol that reacts with the hemiketal fructose to form a ketal. In sucrose, the pyranose ring is an acetal, whilst the furanose ring is a ketal. This all seems rather complicated at first – look carefully at the structures whilst considering the text.

In aqueous solution, both glucose (hemiacetal) and fructose (hemiketal) exist as equilibrium mixtures of cyclic and open-chain carbonyl forms. Sucrose, however, is a single stable substance (acetal and ketal), and conversion back to glucose and fructose requires more rigorous hydrolytic conditions, such as heating with aqueous acid.

Box 7.3

Invert sugar

Invert sugar is the name given to an equimolar mixture of glucose and fructose, obtained from sucrose by hydrolysis with acid or alternatively using the enzyme **invertase**. During the process, the optical activity changes from that of sucrose, $[\alpha]_D + 66.5°$, to that resulting from an equal mixture of glucose and fructose.

The equilibrium mixture of α and β isomers of glucose has $[\alpha]_D + 52.7°$ (see Box 7.1), and that of fructose is strongly laevorotatory with $[\alpha]_D - 92.4°$. Since glucose and fructose are structural isomers and have the same molecular weight, it can be calculated that the resultant optical activity will be $[\alpha]_D - 19.85°$. This derives from $(-92.4) \times 0.5 + (+52.7) \times 0.5$, the 0.5 factor being necessary because 1 g of the disaccharide sucrose gives 0.5 g of each monosaccharide. This change in optical activity from plus to minus is the reason for the terminology 'invert'.

The high sweetness of fructose combined with that of glucose means invert sugar is sweeter than sucrose, so it provides a cheaper, less calorific sweetener than sucrose. The relative sweetness figures for sucrose, glucose and fructose are 1.0, 0.7 and 1.7 respectively. Honey is also composed mainly of invert sugar.

Box 7.4

Polysaccharides: starch, glycogen and cellulose are polyacetals of glucose

Polysaccharides fulfil two main functions in living organisms, as food reserves and as structural elements. Plants accumulate **starch** as their main food reserve, a material that is composed entirely of glucopyranose units but in two types of molecule. **Amylose** is a linear polymer containing some 1000–2000 glucopyranose units linked through α1 → 4 acetal groups. This terminology means that the 1-position of the first ring is linked to the 4-position of the second ring, and the configuration at the anomeric centre in the first ring is α.

amylopectin
(up to 10^6 residues; branching about every 20 residues)

glycogen
(>10^6 residues; branching about every 10 residues)

acetal linkages α1 → 4 and α1 → 6

amylose
(1000–2000 residues)
acetal linkages α1 → 4

Amylopectin is a much larger molecule than amylose (the number of glucose residues varies widely, but may be as high as 10^6), and it is a branched-chain molecule. In addition to α1 → 4 acetal linkages, amylopectin has branches at about every 20 units through α1 → 6 acetal linkages, i.e. similar acetal bonding but to the 6-hydroxyl of another glucose residue. These branches continue with α1 → 4 linkages, but then may have subsidiary branching giving a tree-like structure. The mammalian carbohydrate storage molecule is **glycogen**, which is analogous to amylopectin in structure, but is larger and contains more frequent branching, about every 10 residues.

cellulose
(~8000 residues)

acetal linkages β1 → 4

Cellulose is reputedly the most abundant organic material on Earth, being the main constituent in plant cell walls. It is composed of glucopyranose units linked β1 → 4 in a linear chain, i.e. this time the configuration at the anomeric centre is β. Alternate residues are found to be 'rotated' in the structure, allowing hydrogen bonding between adjacent molecules, and construction of the strong fibres characteristic of cellulose, as for example in cotton.

There is further discussion of polysaccharides in Chapter 12.

Box 7.5

Acetal linkages in etoposide

Etoposide is an effective anticancer drug used in the treatment of small-cell lung cancer, testicular cancer and lymphomas. It is a semi-synthetic modification of the natural lignan podophyllotoxin, and contains three acetal linkages. Can you identify them?

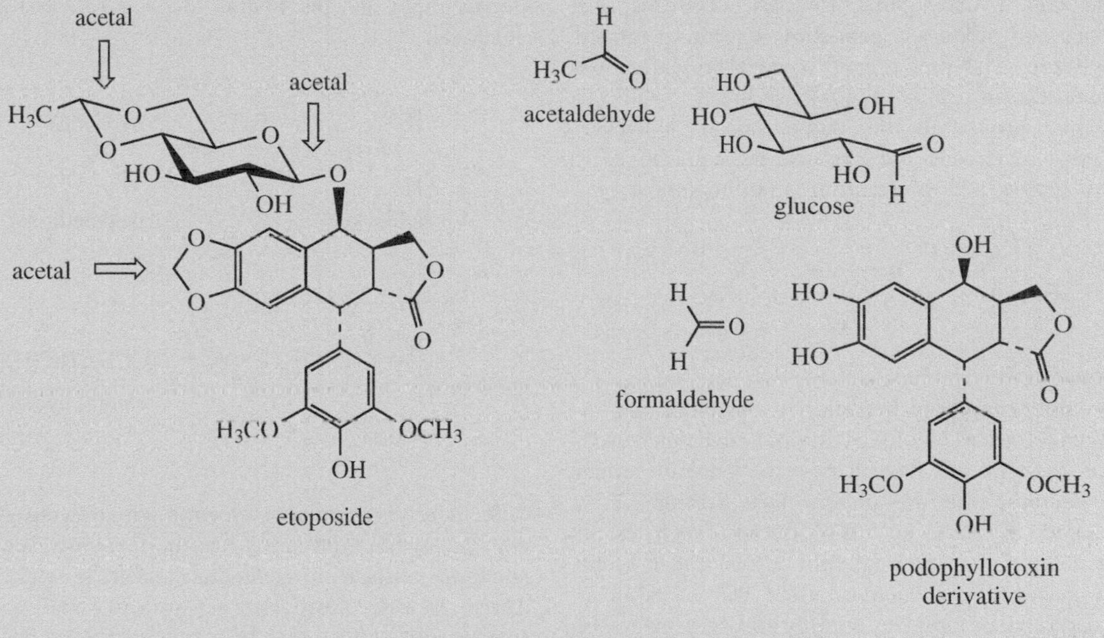

etoposide

glucose

formaldehyde

acetaldehyde

podophyllotoxin derivative

Initially, it is possible to see the cyclic form of glucose as a component of the structure. This is normally a hemiacetal, but here is further bound to an alcohol derived from the podophyllotoxin derivative through an acetal linkage. Secondly, it should be noted that two of the hydroxyl groups of glucose are also bound as a cyclic acetal to acetaldehyde; this linkage can be formed because the two hydroxyls of glucose are suitably positioned and allow a favourable six-membered ring to be constructed (see Section 12.5).

The third acetal linkage is not so obvious, and it is in the five-membered ring fused onto the aromatic ring of the podophyllotoxin derivative. This is called a methylenedioxy group, and it is a common bidentate substituent on many natural aromatic structures. However, it can be formally regarded as an acetal of formaldehyde.

Nature does not actually make a methylenedioxy group using formaldehyde. Instead, it modifies an existing ortho-hydroxy-methoxy arrangement. Enzymic hydroxylation of the methoxy methyl converts this substituent into what is identical to a hemiacetal of formaldehyde, and then acetal formation follows in a process analogous to a chemical synthesis. The hydroxylating enzyme involved is a cytochrome P-450 mono-oxygenase (see Box 11.4).

Box 7.5 (continued)

ortho-hydroxy-   hemiacetal of   methylenedioxy
methoxy derivative   formaldehyde   derivative

7.3 Water as a nucleophile: hydrates

Water, as the simplest alcohol, should also be able to act as a nucleophile towards aldehydes and ketones and produce a **gem-diol**, sometimes termed a **hydrate**. The prefix *gem* is an abbreviation for geminal (Latin *gemini*: twins); we use it to indicate two like groups on the same carbon. However, for most aldehydes and ketones, the equilibrium is unfavourable, and the reaction is not important.

gem-diol
(hydrate)

The equilibrium only becomes favourable if the $\delta+$ charge on the carbon of the carbonyl can be increased. Since alkyl groups have a positive inductive effect and decrease the $\delta+$ charge (see Section 7.1.1), we need to have no alkyl groups, such as in formaldehyde, or alternatively a functional group with a negative inductive effect that destabilizes the carbonyl group. The equilibrium percentages of hydrate for formaldehyde, acetaldehyde, and acetone in water are found to be about 100, 58, and 0 respectively.

	H	H₃C	H₃C	Cl₃C
% hydrate				
at equilibrium	100	58	0	100

Formaldehyde is normally a gas at room temperature, but dissolves in water. In aqueous solution, formaldehyde exists almost entirely as the *gem*-diol; a 37% solution is called formalin and is used for preserving biological tissues. However, the hydrate cannot be isolated, since the reverse reaction is rapid and the hydrate decomposes to formaldehyde.

formaldehyde   methandiol

chloral   chloral hydrate

If there is a suitable electron-withdrawing substituent, hydrate formation may be favoured. Such a situation exists with trichloroacetaldehyde (chloral). Three chlorine substituents set up a powerful negative inductive effect, thereby increasing the $\delta+$ charge on the carbonyl carbon and favouring nucleophilic attack. Hydrate formation is favoured, to the extent that chloral hydrate is a stable solid, with a history of use as a sedative.

These observations emphasize the fact that *gem*-diols are usually unstable and decompose to carbonyl compounds. However, it can be demonstrated that hydrate formation does occur by exchange labelling of simple aldehyde or ketone substrates with ¹⁸O-labelled water. Thus, after equilibrating acetone with labelled water, isotopic oxygen can be detected in the ketone's carbonyl group.

exchange labelling demonstrates intermediacy of hydrate

7.4 Sulfur as a nucleophile: hemithioacetals, hemithioketals, thioacetals and thioketals

The reaction of thiols with aldehydes and ketones parallels that of alcohols. However, the reactions are more favourable because sulfur is a better nucleophile than oxygen (see Section 6.1.2). Electrons in larger atoms are more easily polarizable and it becomes easier for them to be donated to an electrophile.

As a consequence, thiols are preferred to alcohols for the protection of aldehyde and ketone groups in synthetic procedures. Thioacetals and thioketals are

hemithioacetal   *thioacetal*

excellent protecting groups. They are more readily formed, and are more stable to hydrolytic conditions than acetals and ketals.

cyclohexanone   propan-1,3-dithiol   *cyclic thioketal*

7.5 Hydride as a nucleophile: reduction of aldehydes and ketones, lithium aluminium hydride and sodium borohydride

The carbonyl group of aldehydes and ketones may be reduced to an alcohol group by a nucleophilic addition reaction that appears to involve hydride as the nucleophile. The reduction of the carbonyl group may be interpreted as nucleophilic attack of hydride onto the carbonyl carbon, followed by abstraction of a proton from solvent, usually water.

This is not strictly correct, in that hydride, from say sodium hydride, never acts as a nucleophile, but because of its small size and high charge density it always acts as a base. Nevertheless, there are a number of **complex metal hydrides** such as **lithium aluminium hydride** (LiAlH$_4$; LAH) and **sodium borohydride** (NaBH$_4$) that deliver hydride in such a manner that it appears to act as a nucleophile. We have already met these reagents under nucleophilic substitution reactions (see Section 6.3.5). Hydride is also a very poor leaving group, so hydride reduction reactions are also irreversible (see Section 7.1.2).

nucleophilic addition of hydride on to carbonyl

abstraction of proton from water

this is an oversimplification of the process

alcohol

Whilst the complex metal hydride is conveniently regarded as a source of hydride, it never actually produces hydride as a nucleophile, and it is the aluminium hydride anion that is responsible for

transfer of the hydride. Then, the resultant negatively charged intermediate complexes with the residual Lewis acid AlH$_3$.

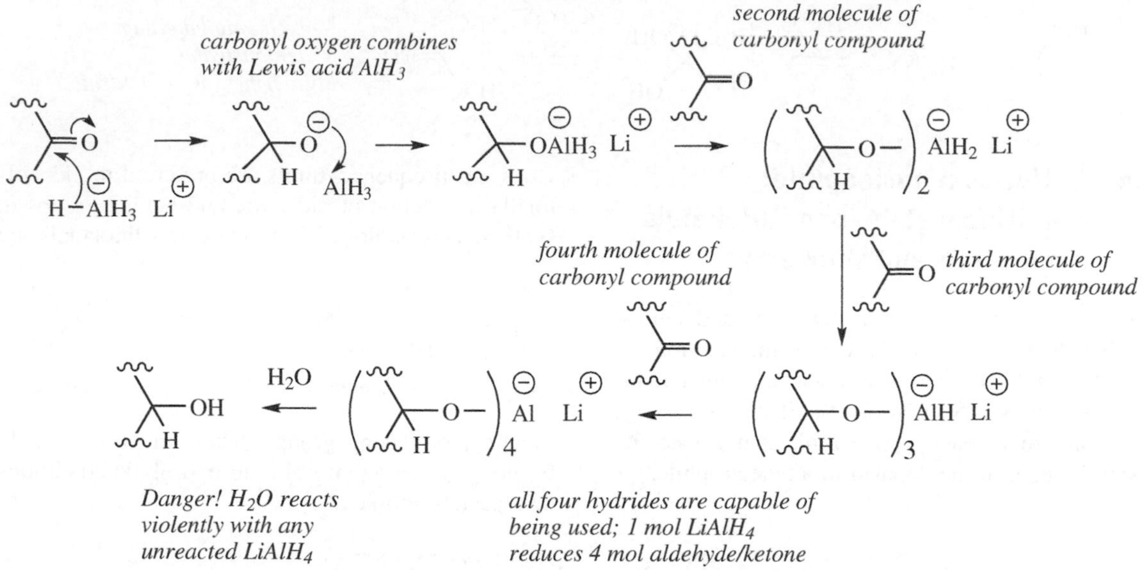

*carbonyl oxygen combines
with Lewis acid AlH₃*

*second molecule of
carbonyl compound*

*fourth molecule of
carbonyl compound*

*third molecule of
carbonyl compound*

*Danger! H₂O reacts
violently with any
unreacted LiAlH₄*

*all four hydrides are capable of
being used; 1 mol LiAlH₄
reduces 4 mol aldehyde/ketone*

This complex can also transfer hydride to another molecule of the carbonyl compound in a similar manner, and the process continues until all four hydrides have been delivered. Since all four hydrogens in the complex metal hydride are capable of being used in the reduction process, 1 mol of reducing agent reduces 4 mol of aldehyde or ketone. Finally, the last complex is decomposed by the addition of water as a proton source.

Lithium aluminium hydride reacts violently with water, liberating hydrogen, and must therefore be used in rigorously anhydrous conditions, usually in ether solution. In fact, any solvent containing OH or NH groups would destroy the reagent by acting as a proton donor for hydride. In the case of LAH reductions, the addition of water as the proton source has to be carried out with considerable caution, since any unreacted LAH will react violently with this

water. In the laboratory, safe removal of excess LAH may be achieved by initially adding small amounts of an ester, such as ethyl acetate (see Section 7.11).

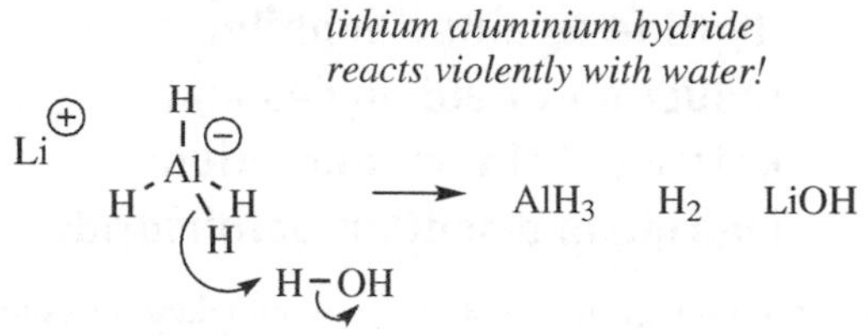

*lithium aluminium hydride
reacts violently with water!*

An analogous series of reactions is involved when sodium borohydride is used as the reducing agent. Sodium borohydride is considerably less reactive than LAH, and may be used in alcoholic or even aqueous solution, so there are no particular problems associated with its use.

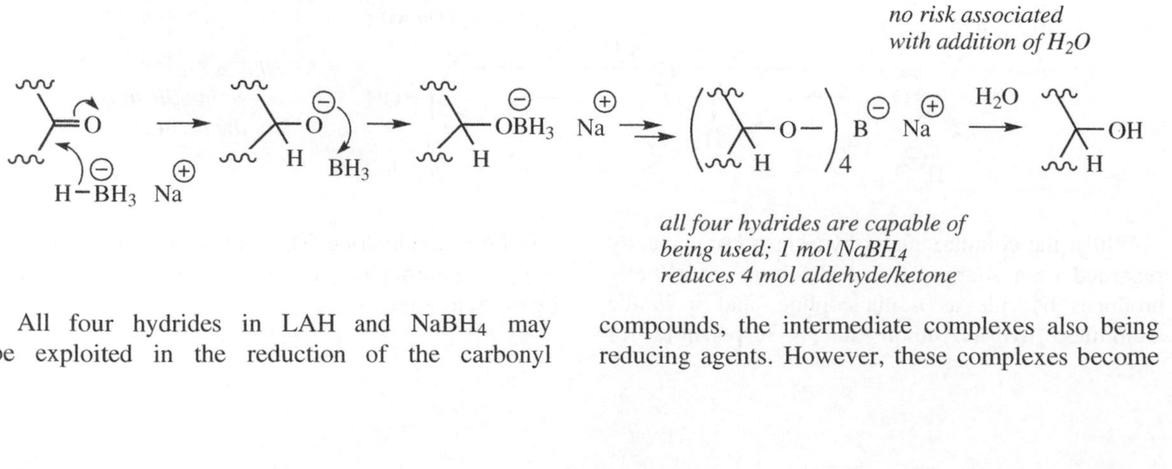

*no risk associated
with addition of H₂O*

*all four hydrides are capable of
being used; 1 mol NaBH₄
reduces 4 mol aldehyde/ketone*

All four hydrides in LAH and NaBH₄ may be exploited in the reduction of the carbonyl

compounds, the intermediate complexes also being reducing agents. However, these complexes become sequentially less reactive than the original reagent, and this has led to the development of other **complex metal hydride reducing agents** that are less reactive and, consequently, more selective than LAH. They are produced by treating LAH with various amounts of an alcohol ROH, giving compounds with the general formulae $(RO)MH_3^-$, $(RO)_2MH_2^-$, and $(RO)_3MH^-$ as their anionic component. These provide a range of reducing agents with different activities. LAH itself is a powerful reducing agent and will react with a number of other functional groups (see Sections 7.7.1 and 7.11).

Note that LAH does not reduce carbon–carbon double bonds; these double bonds lack the charge separation that distinguishes the carbonyl group, and there is no electrophilic character to allow nucleophilic attack. An effective way of reducing C=C is catalytic hydrogenation (see Section 9.4.3).

Box 7.6

Nicotinamide adenine dinucleotide as reducing agent

Biological reduction of aldehydes and ketones is catalysed by an appropriate enzyme, a dehydrogenase or reductase, and most of these use a pyridine nucleotide, such as the reduced form of **nicotinamide adenine dinucleotide** (**NADH**), as the cofactor. This cofactor may be considered as the reducing agent, capable of supplying hydride in a similar manner to lithium aluminium hydride or sodium borohydride (see Section 7.5). NADH is a complex molecule (see Box 11.2), and only the dihydropyridine ring part of the structure is considered here. Some reactions employ the alternative phosphorylated cofactor **NADPH**; the phosphate does not function in the reduction step, but is merely a recognition feature helping to bind the compound to the enzyme.

biological reduction via hydride transfer

NADH
nicotinamide adenine
dinucleotide (reduced)

reducing agent;
can supply hydride

NAD⁺
nicotinamide adenine
dinucleotide

oxidizing agent;
can remove hydride

Hydride may be transferred from NADH to the carbonyl compound because of the electron releasing properties of the ring nitrogen; this also results in formation of a favourable aromatic ring, a pyridinium system since the nitrogen already carries a substituent. The cofactor becomes oxidized to NAD⁺. The reaction is then completed by abstraction of a proton from water.

There is a rather important difference between chemical reductions using complex metal hydrides and enzymic reductions involving NADH, and this relates to stereospecificity. Thus, chemical reductions of a simple aldehyde or ketone will involve hydride addition from either face of the planar carbonyl group, and if reduction creates a new chiral centre, this will normally lead to a racemic alcohol product. Naturally, the aldehyde → primary alcohol conversion does not create a chiral centre.

Box 7.6 (continued)

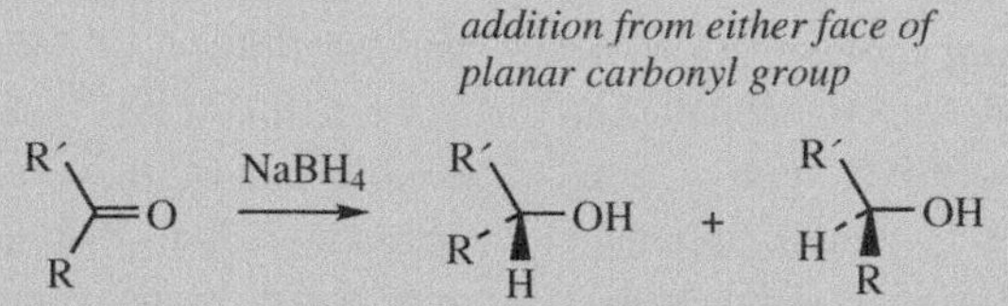

addition from either face of planar carbonyl group

In contrast, an enzymic reduction utilizing NADH will be executed stereospecifically, with hydride attaching to one particular face of the planar carbonyl. Which face is attacked depends upon the individual enzyme involved. For example, reduction of pyruvic acid to lactic acid in vertebrate muscle occurs via attack of hydride from the *Re* face (see Section 3.4.7), and produces the single enantiomer (*S*)-lactic acid. Hydride addition onto the alternative *Si* face is a feature of some microbial dehydrogenase enzymes.

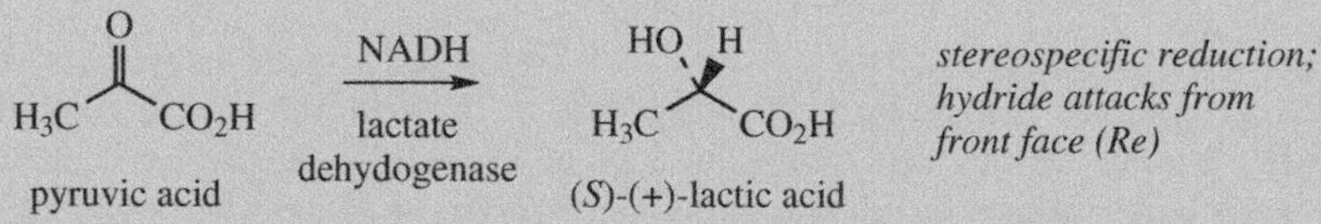

stereospecific reduction; hydride attacks from front face (Re)

These enzymes often also catalyse the reverse reaction, oxidation of an alcohol to an aldehyde or ketone (see Box 11.2). In such reactions, the cofactor NAD^+ abstracts hydride from the alcohol, and may thus be regarded as an oxidizing agent; hence the dehydrogenase terminology for some enzymes, even when they are carrying out a reduction.

7.6 Carbon as a nucleophile

7.6.1 Cyanide: cyanohydrins

Aldehydes and ketones react with HCN to give 2-hydroxynitriles, compounds that are generally termed **cyanohydrins**. HCN is only a weak acid (pK_a 9.1), and proton availability is insufficient to initiate a typical acid-catalysed reaction via the conjugate acid of the carbonyl compound. Instead, the partial ionization of HCN provides a source of cyanide anions, which then react as a nucleophile towards the carbonyl compound. The reaction is terminated by the strongly basic alkoxide ion abstracting a proton from solvent or a further molecule of HCN. To avoid the use of HCN, which is a highly toxic gas, aqueous sodium or potassium cyanides in buffered acid solution are usually employed in the reaction.

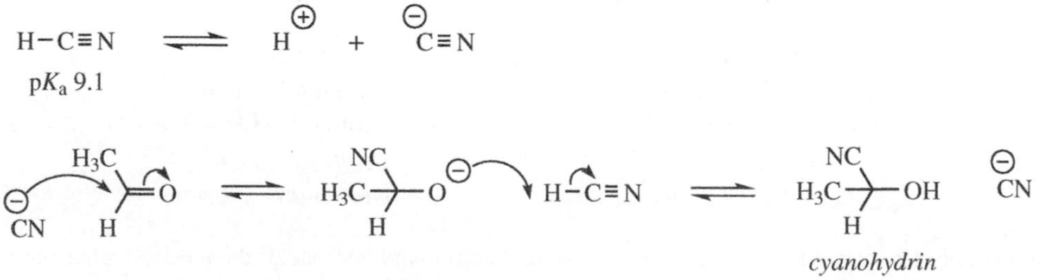

pK_a 9.1

cyanohydrin

The reaction is reversible, and cyanohydrin formation is more favourable with aldehydes than with ketones, as with other addition reactions. The reverse reaction is easily effected by treating a cyanohydrin with aqueous base, since cyanide is a reasonable leaving group (see Section 6.1.4).

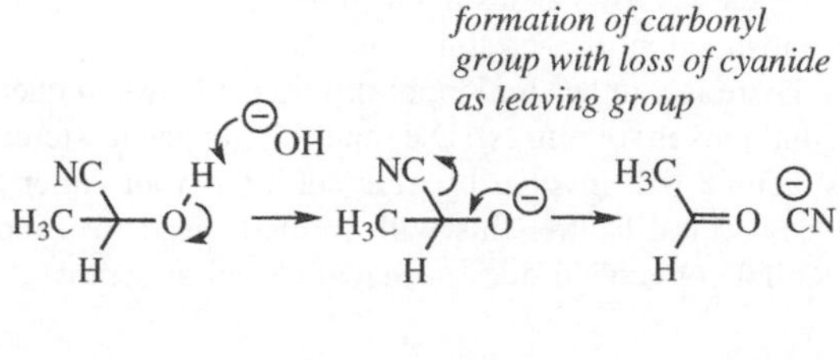

formation of carbonyl group with loss of cyanide as leaving group

Cyanohydrin formation is a useful synthetic reaction, in that it utilizes a simple reagent, cyanide, to create a new C–C bond. The cyano (nitrile) group may easily be modified to other functions, e.g. carboxylic acids via hydrolysis (see Box 7.9) or amines by reduction.

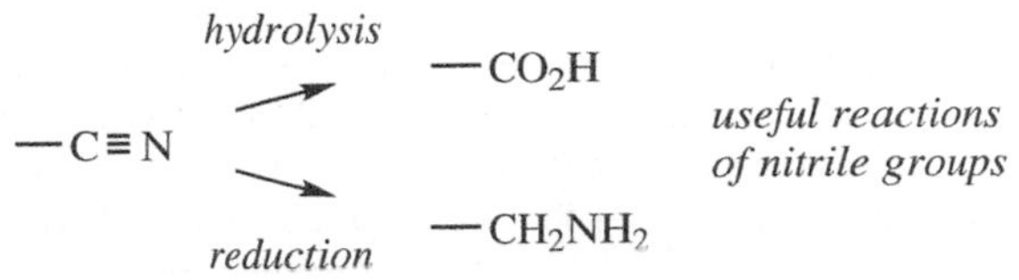

$$-C \equiv N \quad \overset{hydrolysis}{\nearrow} \quad -CO_2H$$
$$\underset{reduction}{\searrow} \quad -CH_2NH_2$$

useful reactions of nitrile groups

Box 7.7

Natural cyanohydrins and cyanogenic glycosides

Natural cyanohydrins feature as toxic constituents in a number of plants, e.g. laurel, bitter almonds, and cassava. In the plant, the **cyanohydrin** is bound through an acetal linkage to a sugar, usually glucose, to produce what is termed a **cyanogenic glycoside**. Cyanogenic means cyanide-producing, because, upon hydrolysis, the glycoside breaks down to the sugar, the carbonyl compound, and HCN. When a plant tissue containing a cyanogenic glycoside is crushed, hydrolytic enzymes also in the plant, but usually located in different cells, are brought into contact with the glycoside and begin to hydrolyse it. Alternatively, hydrolysis may be brought about by ingesting the plant material. Either way, it leads to the production of HCN, which is extremely toxic to humans. The glycoside itself is not especially toxic, and toxicity depends on the hydrolysis reaction.

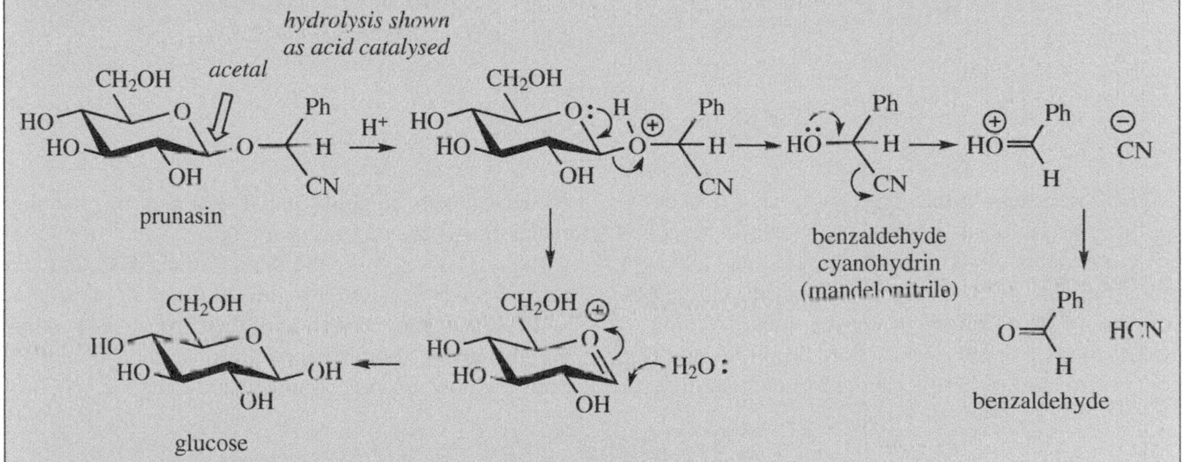

The main cyanogenic glycoside in laurel is **prunasin**, the β-D-glucoside of benzaldehyde cyanohydrin. The enzymic hydrolysis of prunasin may be visualized as an acid-catalysed process, first of all hydrolysing the acetal linkage to produce glucose and the cyanohydrin. Further hydrolysis results in reversal of cyanohydrin formation, giving HCN and benzaldehyde.

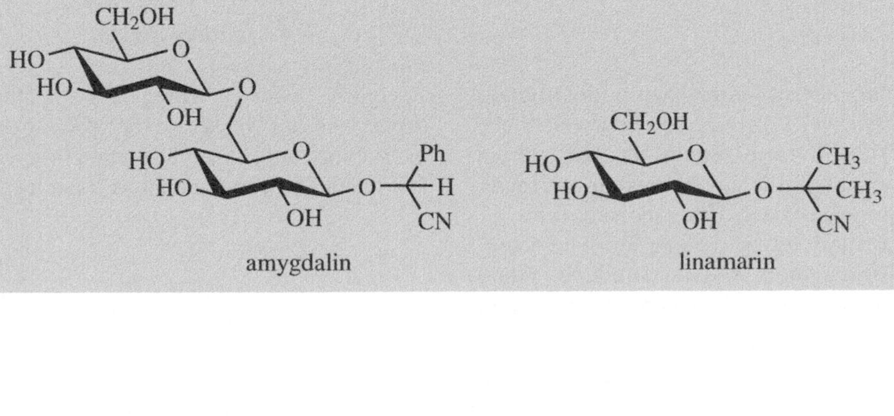

Bitter almonds contain **amygdalin**, which is the β-D-glucoside of prunasin, so it hydrolyses sequentially to the same products. Cassava, which is used in many parts of the world as a food plant, contains **linamarin**, which is the β-D-glucoside of acetone cyanohydrin. Preparation of the starchy tuberous roots of cassava for food involves prolonged hydrolysis and boiling to release and drive off the HCN before they are suitable for consumption.

7.6.2 Organometallics: Grignard reagents and acetylides

The use of **organometallic reagents** as nucleophiles towards carbonyl compounds is also synthetically important, since it results in the formation of new C–C bonds, building up the size and complexity of the molecule. For carbon to act as a nucleophile, we require a negative charge on carbon, i.e. a carbanion or equivalent. Although there are a variety of organometallic reagents available, we include here only two types of reagent, Grignard reagents and acetylides. We have met these organometallic reagents earlier (see Section 6.3.4)

Reacting an alkyl or aryl halide, usually bromide, with metallic magnesium in ether solution, produces **Grignard reagents** (see Section 6.3.4). An exothermic reaction takes place in which the magnesium dissolves, and the product is a solution of the Grignard reagent RMgBr or ArMgBr.

$$RBr \xrightarrow[\text{ether}]{Mg} \underset{\text{Grignard reagent}}{RMgBr} \dashrightarrow \overset{\ominus}{R} \; \overset{\oplus}{MgBr}$$

the R or Ar group in the Grignard reagent behaves as a carbanion

$$ArBr \xrightarrow[\text{ether}]{Mg} ArMgBr \dashrightarrow \overset{\ominus}{Ar} \; \overset{\oplus}{MgBr}$$

The formation of this product need not concern us, but its nature is important, in that it contains the equivalent of R^- or Ar^-, i.e. the alkyl or aryl group has been transformed into its carbanion. Addition of an aldehyde or ketone to the solution of the Grignard reagent allows a nucleophilic addition reaction to occur. The reaction resembles that of reduction with complex metal hydrides, in that the metal forms a complex with the oxygen from the carbonyl; and to complete the addition, this complex must be decomposed by the addition of a proton source through acidification of the mixture. Reactions are also going to be irreversible, since the carbanions are very poor leaving groups (see Section 6.1.4).

$$HCHO \longrightarrow \text{primary alcohol}$$
$$RCHO \longrightarrow \text{secondary alcohol}$$
$$R_2CO \longrightarrow \text{tertiary alcohol}$$

It should be noted that, on reaction with Grignard reagents, aldehydes will produce secondary alcohols, whereas ketones will form tertiary alcohols. Often forgotten is the possibility of synthesizing primary alcohols by using formaldehyde as the substrate.

Acetylides are formed by treating terminal acetylenes with a strong base, sodium amide in liquid ammonia being that most commonly employed.

Acetylenes with a hydrogen atom attached to the triple bond are weakly acidic (pK_a about 25) due to the stability of the acetylide anion (see Section 4.3.4), and this anion can then act as a nucleophile.

$$R-C\equiv C-H \xrightarrow[\text{liquid NH}_3]{NaNH_2} R-C\equiv C^{\ominus} \; \overset{\oplus}{Na}$$

acetylide

It reacts with aldehydes and ketones in the manner expected, and after acidification yields an alcohol. This reaction also extends the carbon chain by two or more atoms, depending on the acetylide used, inserting a triple bond for further modification.

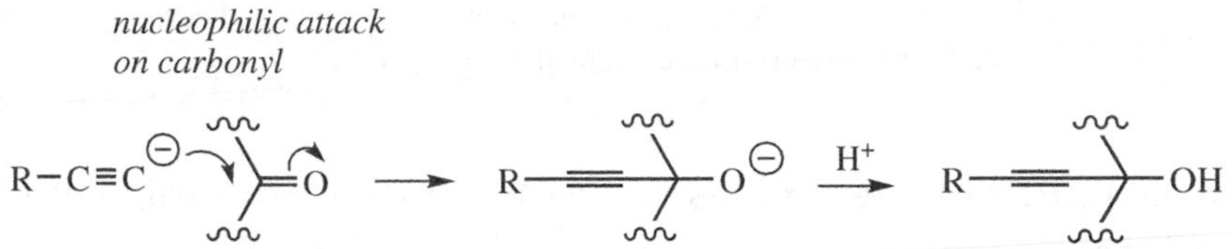

nucleophilic attack on carbonyl

Box 7.8

Synthesis of the oral contraceptive ethinylestradiol

A number of steroidal drugs are produced by procedures that include nucleophilic attack of sodium acetylide onto a ketone, particularly that at position 17 on the five-membered ring of the steroid (see Box 3.19).

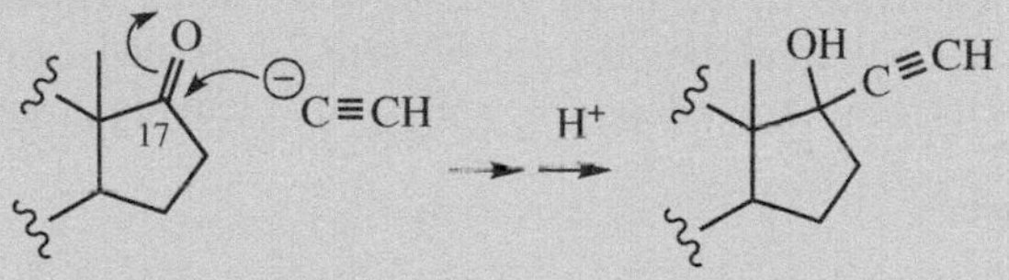

Thus, the natural oestrogen estrone can be converted into the drug **ethinylestradiol** by nucleophilic addition. Ethinylestradiol is some 12 times more effective than estradiol when taken orally, and is widely used in oral contraceptive preparations. In drug nomenclature, the systematic name ethynyl– for the HC≡C– group is usually presented as ethinyl–.

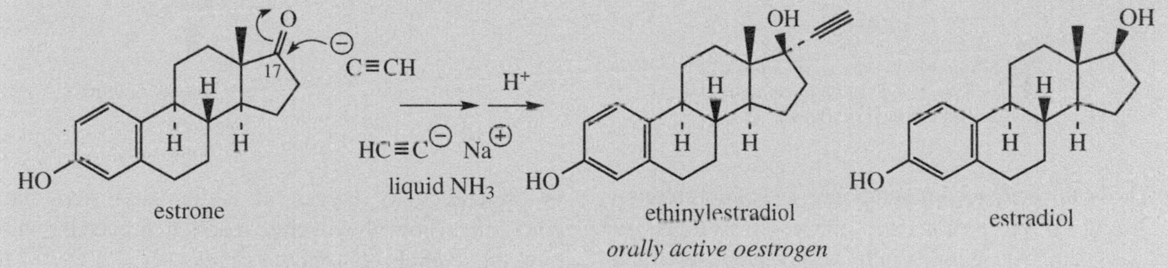

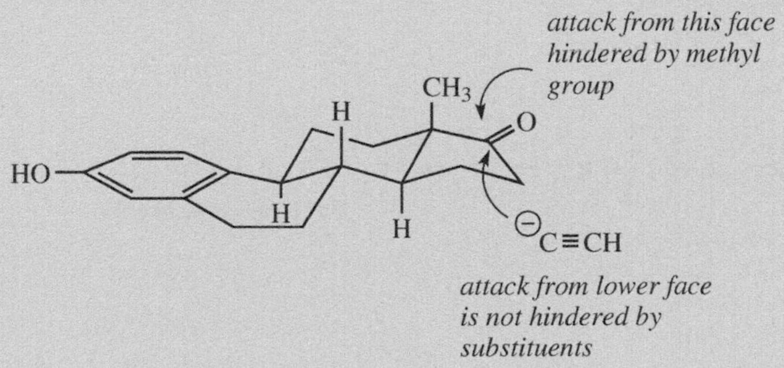

estrone   ethinylestradiol   estradiol
                   orally active oestrogen

attack from this face hindered by methyl group

attack from lower face is not hindered by substituents

With a simple aldehyde or ketone substrate, there is an equal probability that the nucleophile will attack the carbonyl carbon from each face of the planar system, thus producing a racemic product, assuming that there are no other chiral carbons in the starting material. Estrone contains the complex steroidal ring system with four fused rings (see Box 3.19), and the product ethinylestradiol is formed as just one of the two possible epimers at C-17.

By considering the three-dimensional shape of estrone (see Box 3.19), we can appreciate that nucleophilic attack from the upper face is hindered by the methyl group adjacent to the ketone. Therefore, the nucleophile can only approach from the lower face, and the product is formed stereospecifically.

7.7 Nitrogen as a nucleophile: imines and enamines

7.7.1 Imines

The addition of **primary amines** to the carbonyl group of aldehydes and ketones is generally followed by elimination of water (dehydration), and the product is called an **imine** or **Schiff base**.

The elimination reaction is catalysed by acid, but the initial nucleophilic attack depends upon the presence of a lone pair on the nitrogen. Accordingly,

imine formation occurs only within a very limited pH range, typically pH 4–6. At lower pH values, the amine is extensively protonated and is therefore non-nucleophilic. **Addition** followed by **elimination** is not a feature encountered with the other nucleophiles considered in this chapter, but we shall see quite similar processes in aldol reactions where the nucleophile is an enolate anion (see Section 10.3). How are we going to explain this rather different behaviour? It all depends upon leaving groups (see Section 7.1.2), and particularly the presence of two alternative leaving groups.

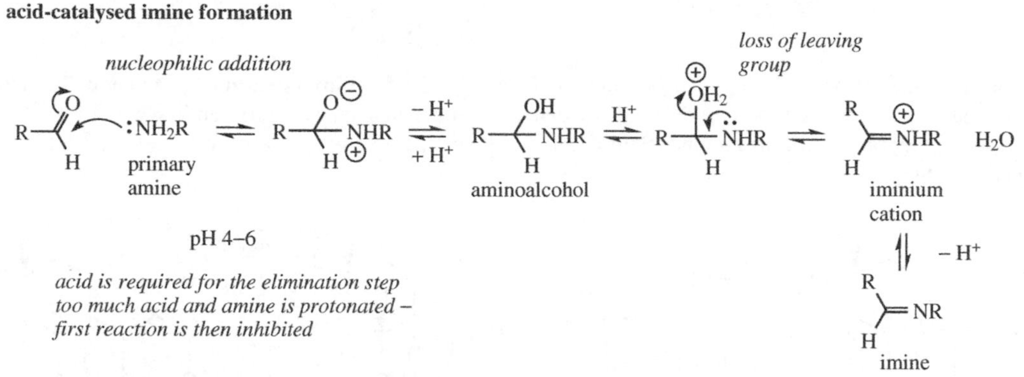

acid-catalysed imine formation

The intermediate aminoalcohol in acidic solution is going to be protonated, either on nitrogen or on oxygen. An equilibrium will be set up, and although we would expect nitrogen to be protonated

in preference to oxygen, it is the next step that determines how the overall reaction proceeds, and how this equilibrium plays its part.

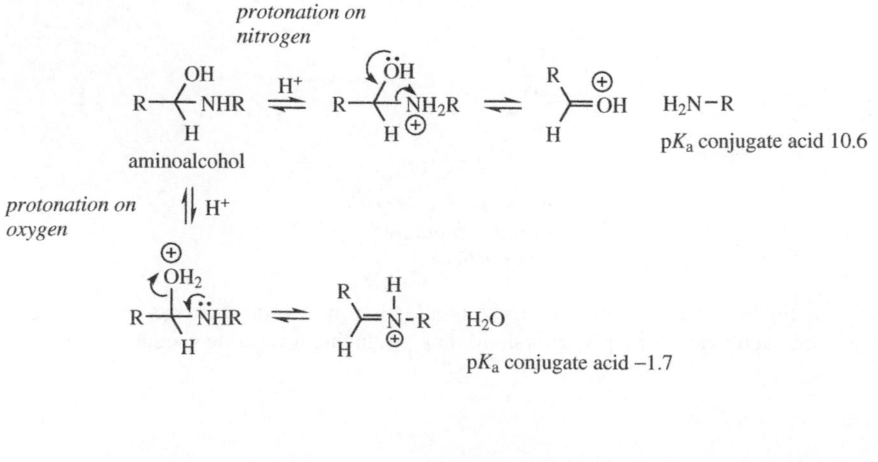

We now have two possibilities for loss of a leaving group. With nitrogen protonated, an amine is lost and the protonated carbonyl reforms, so that we end up with the reverse reaction. With oxygen protonated, water is lost, and an iminium cation forms by an analogous type of electron movement from the nitrogen lone pair. Water (pK_a of conjugate acid -1.7) is a better leaving group than the amine (pK_a of conjugate acid about 10.6), so loss of water is favoured,

and the aminoalcohol protonation equilibrium is disturbed accordingly. The product then is the imine. The elimination reaction is mechanistically an alternative version of the reverse reaction, but involves the second, more favourable, leaving group.

Further, and just to stress that this is not new or novel chemistry, let us go back and compare imine formation with acetal formation from hemiacetals (see Section 7.2).

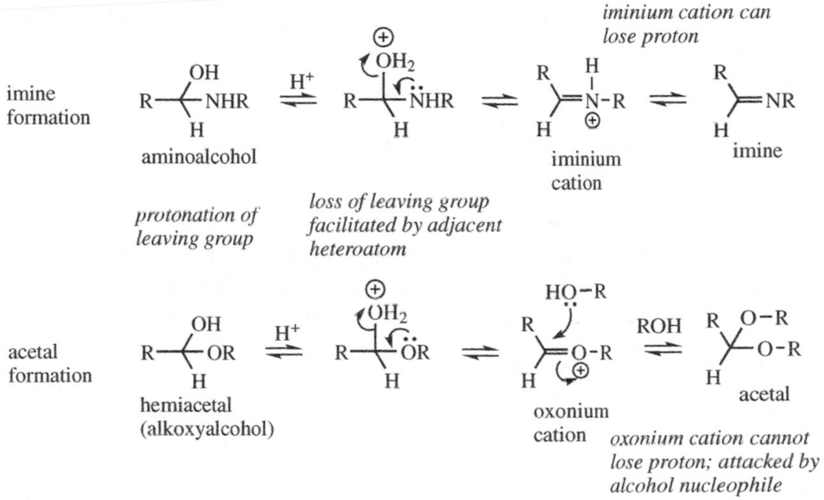

imine formation

aminoalcohol

protonation of leaving group

loss of leaving group facilitated by adjacent heteroatom

iminium cation can lose proton

iminium cation

imine

acetal formation

hemiacetal (alkoxyalcohol)

oxonium cation

oxonium cation cannot lose proton; attacked by alcohol nucleophile

acetal

The amino alcohol intermediate is analogous to the hemiacetal, and both undergo protonation and loss of water, facilitated by the heteroatom. The iminium cation can then lose a proton, but the oxonium cation has no proton to lose; instead, it is attacked by a nucleophile, namely a second molecule of alcohol.

Imines are most conveniently visualized as **nitrogen analogues of carbonyl groups**, since many of the reactions they undergo are paralleled in aldehyde

and ketone chemistry.

behaves as

As a simple example, we need only consider the reverse of imine formation – imines are readily hydrolysed back to carbonyl compounds. In fact, because of this, many imines are somewhat unstable.

hydrolysis of imines ⟶ aldehyde / ketone

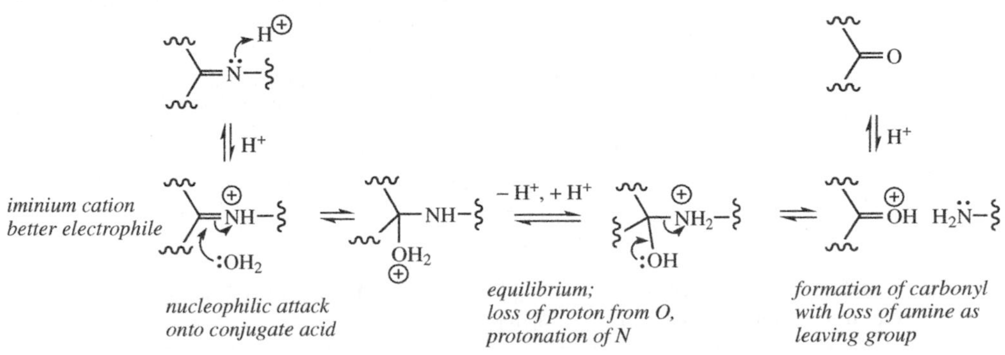

iminium cation better electrophile

nucleophilic attack onto conjugate acid

equilibrium; loss of proton from O, protonation of N

formation of carbonyl with loss of amine as leaving group

Protonation to the conjugate acid (iminium cation) increases the potential of the imine to act as an electrophile (compare carbonyl; see Section 7.1), and this is followed by nucleophilic attack of water. The protonated product is in equilibrium with the other mono-protonated species in which the nitrogen carries the charge. We shall meet this mechanistic feature from time to time, and it is usually represented in a mechanism simply by putting '$-H^+$, $+H^+$' over the equilibrium arrows. Do not interpret this as an internal transfer of a proton; such transfer would not be possible, and it is necessary to have solvent to supply and remove protons.

Protonation of nitrogen allows loss of the amine leaving group and formation of the conjugate acid of the carbonyl compound. Despite the comments made above regarding alternative leaving groups, imine formation and hydrolysis are reversible, though it will usually be necessary to disturb the equilibrium, as required, by using an excess of the appropriate reagent.

In Section 10.6 we shall meet the Mannich reaction, where an imine or iminium ion acts as an electrophile for nucleophiles of the enolate anion type.

Box 7.9

Hydrolysis of nitriles to carboxylic acids

Just as imines may be viewed as nitrogen analogues of carbonyl compounds, the C≡N group may also be viewed as carbonyl-like for interpretation of some of its reactions. For instance, **nitriles** are readily hydrolysed in acid to give **carboxylic acids** (see Section 7.6.1). This process begins in a similar manner to hydrolysis of imines.

In acid-catalysed hydrolysis, we usually invoke protonation of the nitrogen to the conjugate acid to increase the potential of the nitrile to act as an electrophile, though the nitrile nitrogen is actually a very weak base (see Section 4.5.3). This is followed by nucleophilic attack of water. Loss of a proton from the product cation generates a hydroxy-imine, which is a **tautomer** of a carboxylic acid amide (compare keto–enol tautomerism, Section 10.1). The keto-like tautomer (amide) is the more favoured, and is subsequently hydrolysed under the acidic conditions to a carboxylic acid (see Section 7.9.2).

Hydrolysis under basic conditions is mechanistically similar, also proceeding through a hydroxy-imine. The tautomeric amide then undergoes basic hydrolysis (see Section 7.9.2).

The main theme to be appreciated here is that nucleophilic attack onto the nitrile triple bond can be interpreted mechanistically by extrapolation from carbonyl chemistry.

Box 7.10

Nucleophilic addition of carbon to imines: the Strecker synthesis of amino acids

A nice example of the chemical similarity between imines and carbonyl compounds is the **Strecker synthesis of amino acids**. This involves reaction of an aldehyde with ammonia and HCN (usually in the form of ammonium chloride plus KCN) to give an intermediate α-aminonitrile. Hydrolysis of the α-aminonitrile then produces the α-amino acid.

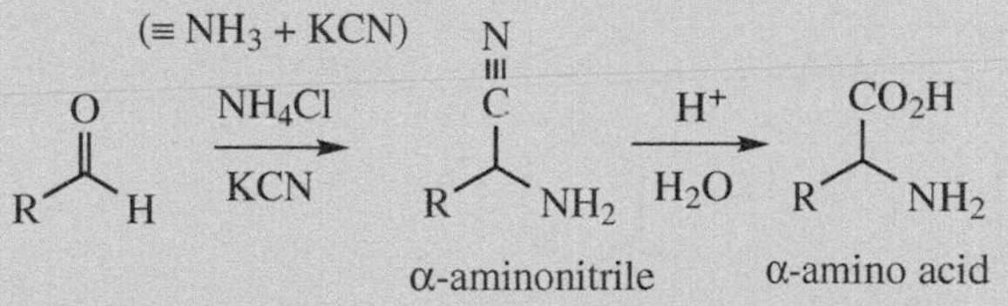

The sequence can be rationalized mechanistically as involving nucleophilic attack of ammonia onto the aldehyde to produce an **imine**, which then acts as the electrophile for further nucleophilic attack, this time by the cyanide ion (see Section 7.7.1). The racemic amino acid is then formed by acid-catalysed hydrolysis of the nitrile function, as above (Box 7.9).

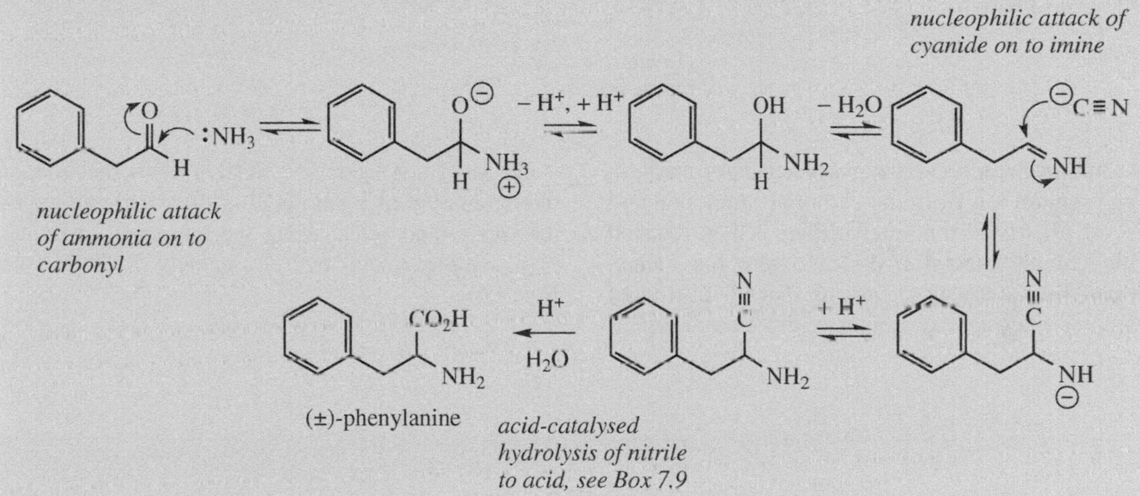

This synthesis is fairly general, and can be used for many amino acids, provided the R side-chain contains no other functional group that is sensitive to the reagents (see Section 13.1). R groups containing –NH$_2$, for example, would require appropriate protection measures. There is also considerable scope for making labelled amino acids via the use of ^{14}C-labelled cyanide.

Imine formation is an important reaction. It generates a C–N bond, and it is probably the most common way of forming heterocyclic rings containing nitrogen (see Section 11.10). Thus, cyclization of 5-aminopentanal to Δ^1-piperideine is merely **intramolecular imine** formation. A further property of imines that is shared with carbonyl groups is their susceptibility to reduction via complex metal hydrides (see Section 7.5). This allows imines to be reduced to amines, such as piperidine.

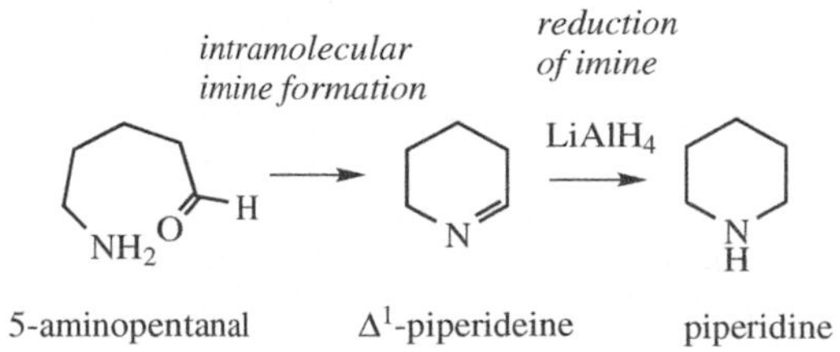

A combined reductive amination sequence has been developed as a useful way of synthesizing amines, with **sodium cyanoborohydride** as the reducing agent of choice. This **complex metal hydride** is a less reactive version of sodium

borohydride (see Section 7.5), since the electron-withdrawing cyano group lowers the ability to transfer hydride. Consequently, sodium cyanoborohydride is rather selective, in that it will reduce iminium systems but does not reduce carbonyl compounds.

The combined reaction thus involves initial formation of the iminium ion from the carbonyl compound and amine at pH 6, and this intermediate is then reduced by the complex metal hydride to give the amine. This can also be a way of making methyl-substituted amines via intermediate imines with formaldehyde.

We shall see later (see Section 15.6) that reductive amination of a keto acid is the way nature synthesizes amino acids, using the biological analogue of a complex metal hydride, namely NADPH (see Box 7.6).

Box 7.11

Pyridoxal and pyridoxamine: vitamins that participate via imine formation

The terminology **vitamin B$_6$** covers a number of structurally related compounds, including **pyridoxal** and **pyridoxamine** and their 5′-phosphates. **Pyridoxal 5′-phosphate** (PLP), in particular, acts as a coenzyme for a large number of important enzymic reactions, especially those involved in amino acid metabolism. We shall meet some of these in more detail later, e.g. transamination (see Section 15.6) and amino acid decarboxylation (see Section 15.7), but it is worth noting at this point that the biological role of PLP is absolutely dependent upon imine formation and hydrolysis. Vitamin B$_6$ deficiency may lead to anaemia, weakness, eye, mouth, and nose lesions, and neurological changes.

Pyridoxal 5′-phosphate is an aldehyde, and this grouping can react with the amino group of α-amino acids to form an imine; since an aldehyde is involved, biochemists often refer to this product as an aldimine. This imine undergoes changes in which the heterocyclic ring plays an important role (see Section 15.6), changes that lead to the double bond of the imine ending up on the other side of the nitrogen atom. This is in many ways similar to rearrangement in an allylic system (see Section 8.2), and at its simplest can be represented as shown. Conjugation with the heterocyclic ring system facilitates loss of a proton to start off the electron redistribution.

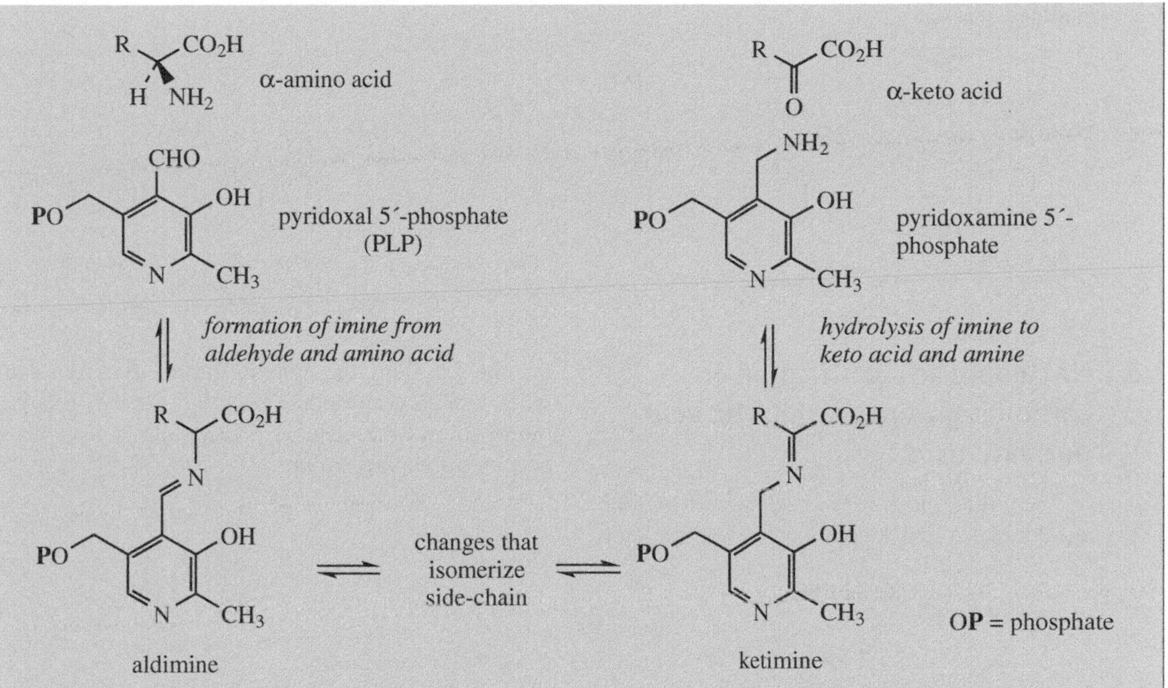

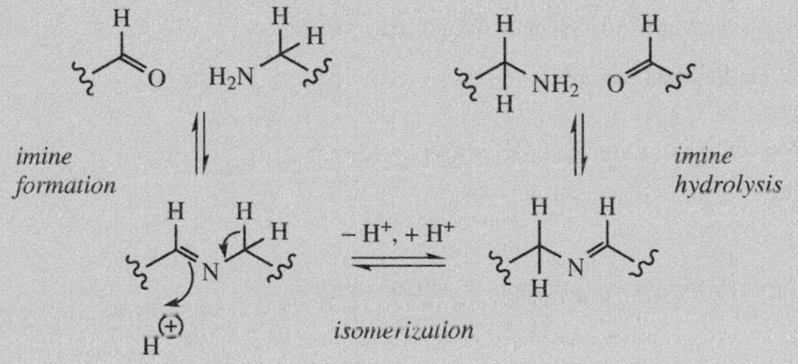

Hydrolysis of the new imine then allows formation of a ketone as part of an α-keto acid, and an amine which is the previously mentioned pyridoxamine 5'-phosphate. Since this imine is the product from an amine and a ketone, it is termed a ketimine. These reactions are reversible in nature, allowing amino acids to be converted into keto acids, and keto acids to be converted into amino acids (see Section 15.6).

7.7.2 Enamines

Secondary amines react with aldehydes and ketones via addition reactions, but instead of forming imines, produce compounds known as **enamines**. Initially, there is the same type of nucleophilic attack of the amine onto the carbonyl system, followed by acid-catalysed dehydration; but, since the amine is secondary, the product of dehydration is an iminium ion rather than an imine. This needs to lose a proton to become neutral, and since there is no available proton on nitrogen, one is lost from the nearest carbon atom, which is β to the nitrogen atom. This produces an enamine (ene-amine). We shall see later (see Section 10.5) that enamines are valuable synthetic intermediates, and are essentially nitrogen analogues of enols.

enamine formation

secondary amine

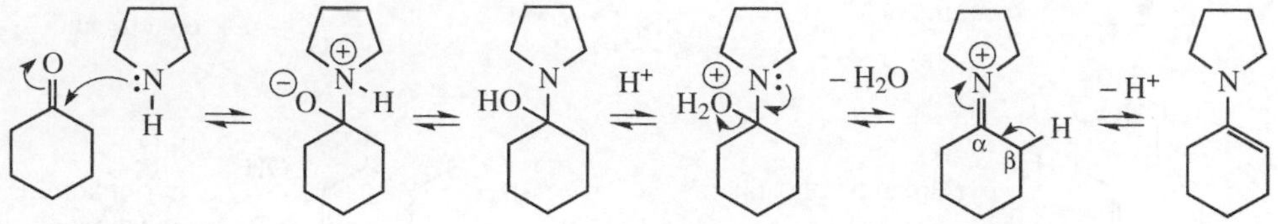

nucleophilic attack   *loss of leaving*   *no proton on N;*   enamine
on to carbonyl   *group*   *therefore, lose proton*
                                                     from β-position

7.8   Nucleophilic substitution on carbonyl groups: carboxylic acid derivatives

We have seen that most of the addition reactions involving aldehydes and ketones are reversible (see

Section 7.1.2). If the anionic intermediate does not react with an electrophile, then the carbonyl group is reformed and the original nucleophile is lost, i.e. it becomes a leaving group.

nucleophilic addition

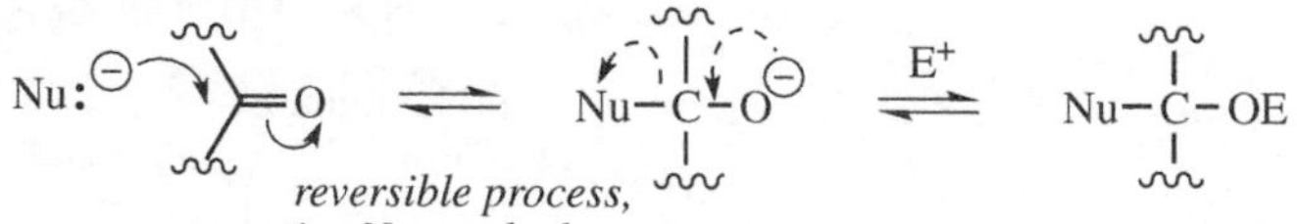

reversible process,
i.e. Nu can be lost

if there is another leaving group in the molecule, can get

nucleophilic substitution

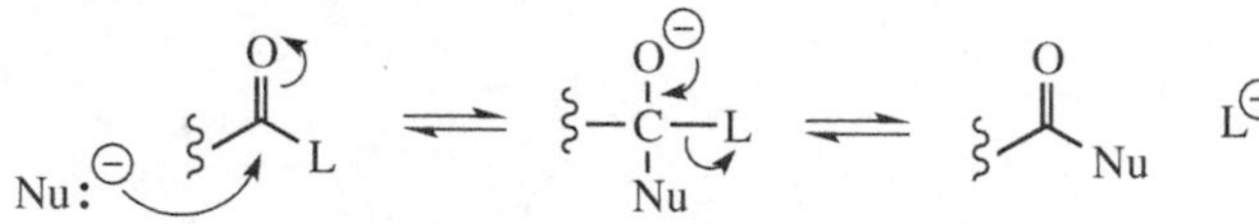

tetrahedral anionic
intermediate

this is really an **addition–elimination** sequence;

it may also be considered as **acylation** of the nucleophile

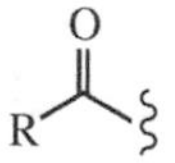

= acyl

However, if there is another leaving group in this molecule, then this may be lost instead, so that overall the reaction becomes a **nucleophilic substitution**, though really it should be regarded as an **addition–elimination** sequence, since the

carbonyl group is essential for this reactivity. This is readily appreciated if one compares the reactivity towards water of acetyl chloride and ethyl chloride.

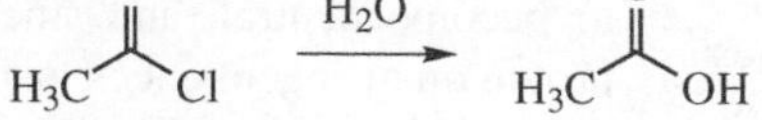

occurs readily at
room temperature

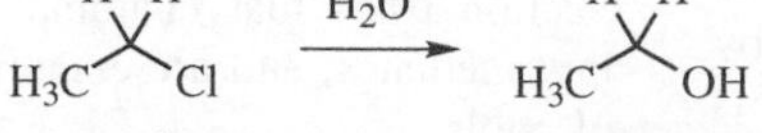

acetyl chloride

no reaction at room
temperature; requires base
and high temperature

ethyl chloride

Acetyl chloride must always be stored under anhydrous conditions, because it readily reacts with moisture and becomes hydrolysed to acetic acid. On the other hand, if one wanted to convert ethyl chloride into ethanol, this nucleophilic substitution reaction would require hydroxide, with its negative charge a better nucleophile than water, and an elevated temperature (see Section 6.3.2). It is clear, therefore, that the carbonyl group is responsible for the increased reactivity, and we must implicate

this in our mechanisms. Although it is easier to draw a mechanism as an S$_N$2 style substitution reaction, this is lazy and wrong, and the two-stage addition–elimination should always be shown. There is ample experimental evidence to show that a tetrahedral addition intermediate does participate in these reactions. The S$_N$1 style mechanism is also incorrect, in that few reactions, and certainly none that we shall consider, actually follow this pathway.

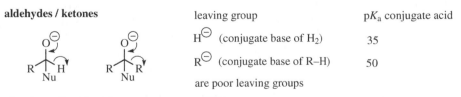

A shorthand addition–elimination mechanism sometimes encountered is also shown. This employs a double-headed curly arrow to indicate the flow of electrons to and from the carbonyl oxygen; we prefer and shall use the longer two-step mechanism to emphasize the addition intermediate.

The reaction may also be considered as acylation of the nucleophile, since an acyl group RCO– is effectively added to the nucleophile; this description, however, conceals the fact that the electron-rich nucleophile is actually the attacking species in the reaction.

Since this reaction, an overall substitution, depends upon the presence of a suitable **leaving group** in the substrate, it is not surprising to find that the level of reactivity depends very much upon the nature of the leaving group. We have already

seen that weak bases, the conjugate bases of strong acids, make good leaving groups (see Section 6.1.4). Conversely, strong bases, the conjugate bases of weak acids, are poor leaving groups. We can now see why aldehydes and ketones react with nucleophiles to give addition products. This is because the tetrahedral anionic intermediate has no satisfactory leaving group apart from the original nucleophile. The alternative possibilities, hydride in the case of aldehydes or an alkyl carbanion in the case of ketones, are both very poor leaving groups; both are the conjugate bases of very weak acids, namely molecular hydrogen (pK_a 35) or an alkane (pK_a 50) respectively. Therefore, in the forward reaction, the alkoxide intermediate instead reacts with an electrophile, usually by abstraction of a proton from solvent, and the overall reaction is addition.

aldehydes / ketones

	leaving group	pK_a conjugate acid
	H$^{\ominus}$ (conjugate base of H$_2$)	35
	R$^{\ominus}$ (conjugate base of R–H)	50
	are poor leaving groups	

*therefore, alkoxide picks up a proton, get **addition***

Much better leaving groups are encountered in carboxylic acid derivatives. Acyl halides possess a good leaving group in chloride, the conjugate base of HCl (pK_a − 7), so react very readily with nucleophiles in overall substitution reactions.

acyl halides

RCOCl $\longrightarrow$

leaving group

pK_a conjugate acid

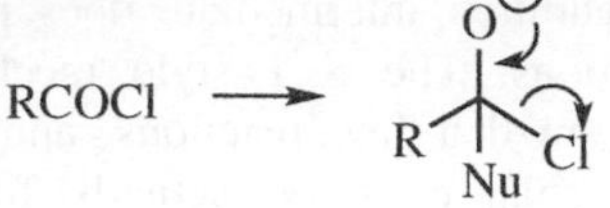

Cl$^{\ominus}$ (conjugate base of HCl)   −7

is good leaving group

carboxylic acids

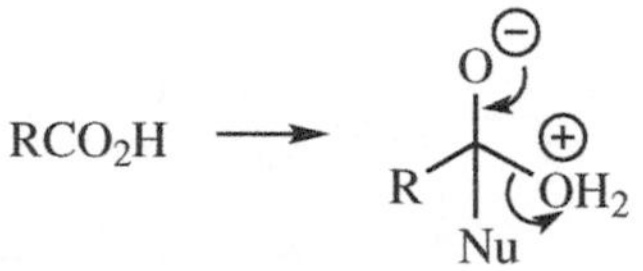

RCO$_2$H $\longrightarrow$

H$_2$O (via protonation)   −1.7

is good leaving group

Where the leaving group is less satisfactory, the reactivity can be improved by carrying out the reaction under acidic conditions. Thus, reaction of carboxylic acids with nucleophiles would require loss of hydroxide as leaving group, and this is the conjugate base of the weak acid water (pK_a 15.7). This is not particularly favourable, but reactivity can be increased by protonation, leading to the expulsion of the neutral molecule water (pK_a conjugate acid −1.7) as a good leaving group (see Section 6.1.4). Purists will dislike the intermediate shown above that has both negatively charged and positively charged oxygens as an unlikely species under acidic conditions. As we shall see, the intermediate actually formed under acidic conditions carries only a positive charge on the potential leaving group.

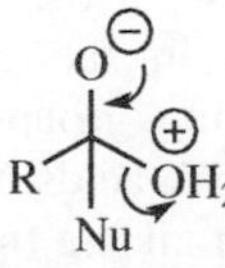

this is an unlikely species under acidic conditions

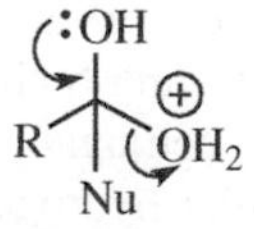

we are likely to see this type of elimination under acidic conditions

Accordingly, the reactivity of compounds in this type of reaction can now be predicted by our appreciation of leaving-group tendencies (Table 7.2).

Reactive substrates are those with a good leaving group, such as halide (in **acyl halides**), hydrosulfide (in **thioacids**), alkyl thiolate or alkyl mercaptide (in **thioesters**), and carboxylate (in **anhydrides**).

Acids, esters and amides are only moderately reactive, in that their leaving groups cannot be classified as good until they become protonated to the conjugate acid. Under acidic conditions, the leaving group then becomes a stable neutral molecule. As we have already seen, aldehydes and ketones have no satisfactory leaving group and undergo addition reactions rather than substitution reactions.

Table 7.2 Leaving groups and reactivity in carboxylic acid derivatives

Compound	Leaving group	pK_a conjugate acid
Reactive: good leaving group		
Acyl halides RCOX	Cl$^-$, Br$^-$	−7,−9
Thioacids RCOSH	HS$^-$	7
Thioesters RCOSR	RS$^-$	11
Anhydrides (RCO)$_2$O	RCO$_2^-$	5
Moderately reactive: good leaving group via protonation		
Acids RCO$_2$H	H$_2$O	−1.7
	[HO$^-$]	[15.7]
Esters RCO$_2$R	ROH	−2.5
	[RO$^-$]	[16]
Amides RCONH$_2$	NH$_3$	9
	[NH$_2^-$]	[38]
RCONHR	RNH$_2$	11
RCONR$_2$	R$_2$NH	11
Unreactive: poor leaving group		
Aldehydes RCHO	H$^-$	35
Ketones R$_2$CO	R$^-$	50

Box 7.12

Synthesis of anhydrides and acyl halides

As one of the most reactive groups of carboxylic acid derivatives, **acyl halides** are very useful substrates for the preparation of the other classes of derivatives. For example, **anhydrides** may be synthesized by the reaction of carboxylic acid salts with an acyl halide. In this reaction, the carboxylate anion acts as the nucleophile, eventually displacing the halide leaving group.

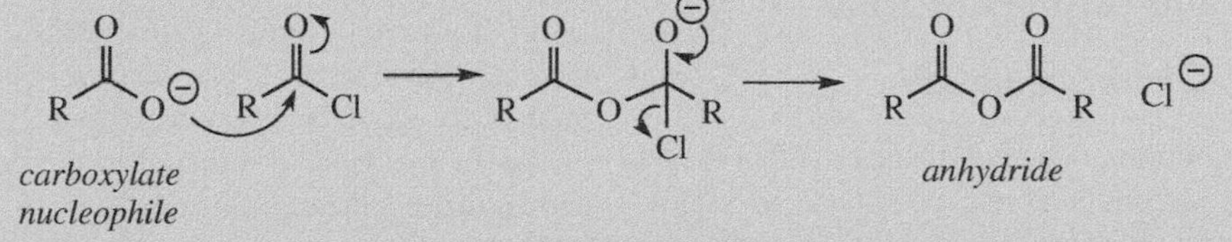

carboxylate nucleophile   *anhydride*

Pyridine is often used as a solvent in such reactions, since it is also functions as a weak base. As an aromatic base, pyridine will promote ionization of the carboxylic acid to carboxylate, and also react with the other product of the reaction, namely HCl. Without removal of HCl, the anhydride formed might be hydrolysed under the acid conditions generated.

Pyridine has another useful attribute, in that it behaves as a nucleophilic catalyst, forming an intermediate **acylpyridinium ion**, which then reacts with the nucleophile. Pyridine is more nucleophilic than the carboxylate anion, and the acylpyridinium ion has an excellent leaving group (pK_a pyridinium 5.2). The reaction thus becomes a double nucleophilic substitution.

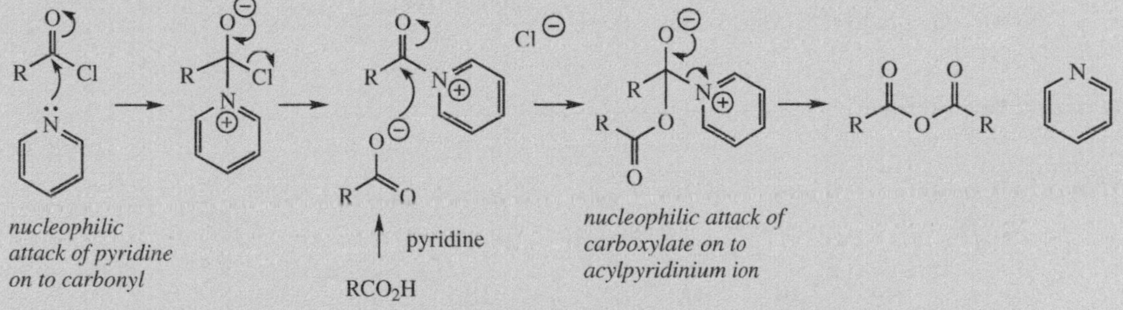

nucleophilic attack of pyridine on to carbonyl   *pyridine*   *nucleophilic attack of carboxylate on to acylpyridinium ion*

RCO_2H

Acyl chlorides themselves may be synthesized by a similar type of reaction, in which we invoke nucleophilic attack of an acid onto thionyl chloride as shown. As we shall see later (see Section 7.13.1), the S=O group behaves as an electrophile in the same way as a C=O group.

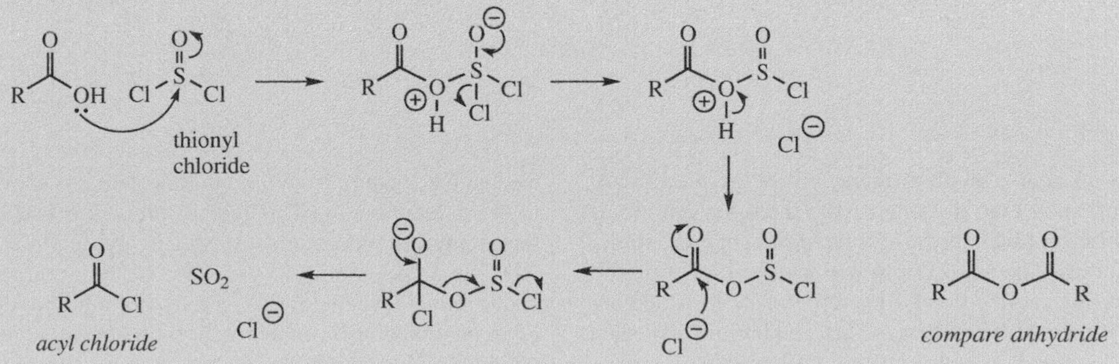

thionyl chloride

acyl chloride SO_2 *compare anhydride*

The leaving group is chloride, and after proton loss, we generate what may be considered a mixed anhydride having both C=O and S=O functionalities. The C=O group in this mixed anhydride is then attacked by chloride, and the good leaving group ($-SO_2Cl$) this time dissociates into sulfur dioxide and chloride, as shown.

7.9 Oxygen and sulfur as nucleophiles: esters and carboxylic acids

7.9.1 Alcohols: ester formation

A well-known reaction of carboxylic acids is that they react with alcohols under acidic conditions to yield **esters**.

$$CH_3CO_2H + EtOH \xrightleftharpoons{H^+} CH_3CO_2Et + H_2O$$
ester

the equilibrium constant is not particularly favourable – may have to remove water or use excess of one reagent, e.g. use the alcohol as solvent

This reaction is, in fact, an equilibrium that often does not favour the product. Thus, to make it a useful procedure for the synthesis of esters, one has to disturb the equilibrium by either removing the water as it is formed, or by using an excess of one reagent, typically the alcohol. It is an easy way to make simple methyl or ethyl esters, where one can employ an excess of methanol or ethanol to act both as solvent and to disturb the equilibrium.

The reaction may be rationalized mechanistically as below, beginning with protonation of the carbonyl oxygen using the acid catalyst. We are using an uncharged nucleophile, i.e. the lone pair of the alcohol oxygen atom acts as a nucleophile, so it is advantageous to increase the electrophilicity of the carbonyl. This is achieved by protonation, which introduces a positive charge. The product from the nucleophilic attack is a protonated tetrahedral addition species.

acid-catalysed esterification

protonation of carbonyl oxygen via lone pair electrons

carboxylic acid

nucleophilic attack on to protonated carbonyl

formation of carbonyl via resonance effect with loss of leaving group

ester *regeneration of acid catalyst*

In an acidic medium there will be an equilibrium set up such that any one of the three oxygen atoms may be protonated; they all have the same or similar basicities. The equilibrium will involve loss of proton to the solvent, followed by reprotonation of another oxygen from the solvent. This equilibrium will then be disturbed as one of the protonated species is removed by further reaction. We shall meet this mechanistic feature from time to time, and it is shown in more detail below. This type of process is usually represented in a mechanism simply by putting '$-H^+$, $+H^+$' over the equilibrium arrows; we also met this under imines (see Section 7.7.1). Do not interpret this as an internal transfer of a proton; such transfer would not be possible, and it is necessary to have solvent to supply and remove protons.

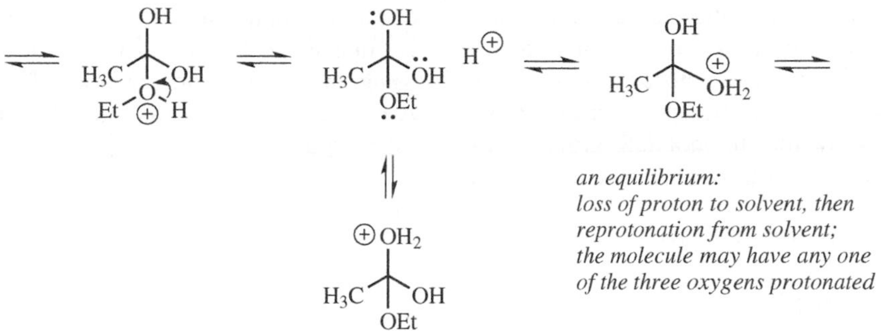

an equilibrium:
loss of proton to solvent, then
reprotonation from solvent;
the molecule may have any one
of the three oxygens protonated

For ester formation to occur, one of the two hydroxyls needs to be protonated, so that it can be lost as a water leaving group; protonation of the ethoxy would lead to loss of ethanol, and reversal of the reaction. Both water and ethanol are good leaving groups, and the reaction is freely reversible. Loss of the leaving group is facilitated by a resonance effect from the other hydroxyl, and leads to regeneration of the carbonyl group in protonated form. Formation of the uncharged carbonyl regenerates the acid catalyst. Note that a base-catalysed process for ester

formation from acid and alcohol is not feasible, since base would immediately ionize the carboxylic acid substrate and a nucleophile would not be able to attack the negatively charged carboxylate anion.

The equilibrium limits the practical applicability of this reaction, and other methods would normally be employed if one were working with uncommon or expensive reagents that could not be used in excess. Esters are actually more conveniently prepared using the more reactive acyl halides or anhydrides, i.e. derivatives with better leaving groups.

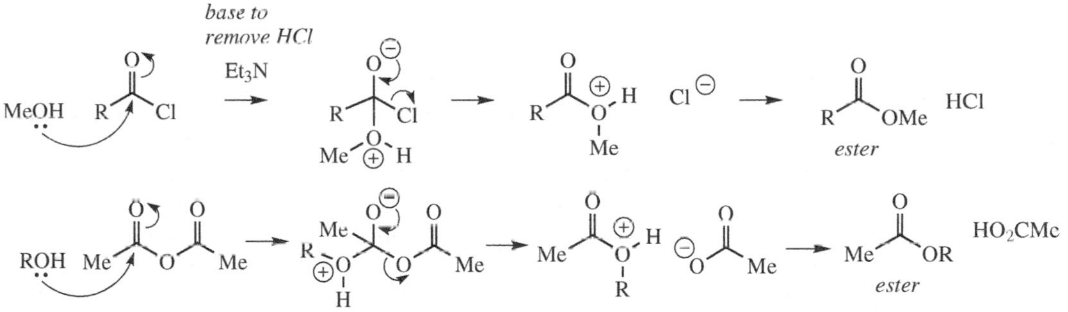

Note the use of a weak base to scavenge the HCl formed as a by-product in the acyl chloride reaction. The aromatic base pyridine is often used for this purpose, though it has other useful attributes. It functions as a good solvent for the reaction, but it

also behaves as a nucleophilic catalyst, forming an intermediate acylpyridinium ion, which then reacts with the alcohol (see Box 7.12). Pyridine helps to catalyse reactions with anhydrides in a similar manner.

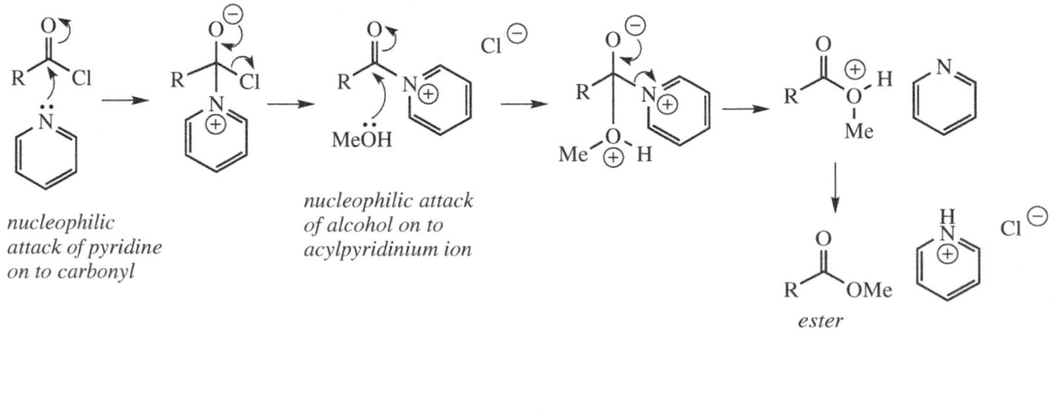

nucleophilic attack of pyridine on to carbonyl

nucleophilic attack of alcohol on to acylpyridinium ion

ester

An interesting extension of the acid-catalysed equilibrium reaction is the process termed **transesterification**. If an ester is treated with an excess of an alcohol and an acid catalyst, then the ester OR group becomes replaced by the alcohol OR group. This reaction proceeds through a tetrahedral intermediate containing both types of OR group, and the product thus depends upon disturbing the equilibrium by using an excess of one or other of the two alcohols. A base-catalysed process may also be used in transesterification.

acid-catalysed transesterification

methyl ester

alcohol in excess

ethyl ester

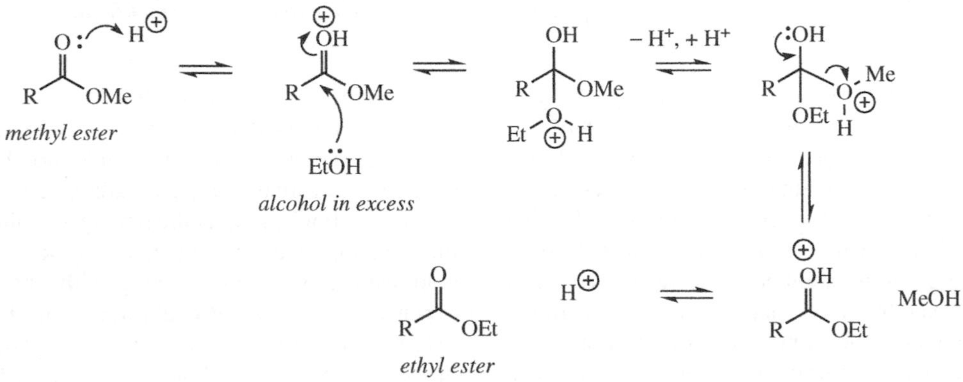

Box 7.13

Transesterification: aspirin as an acetylating agent

The mode of action of the analgesic aspirin is now known to involve a transesterification process. **Aspirin** (acetylsalicylic acid) exerts its action by acetylating the enzyme **cyclooxygenase (COX)** that is involved in the biosynthesis of **prostaglandins** (see Box 9.3). Prostaglandins are modified C_{20} fatty acids found in small quantities in animal tissues and they affect a wide variety of physiological processes, such as blood pressure, gastric secretion, smooth muscle contraction and platelet aggregation. Inflammation is a condition that occurs as a direct result of increased prostaglandin synthesis, and many of the non-steroidal anti-inflammatory drugs, such as aspirin and ibuprofen, exert their beneficial effects by reducing prostaglandin formation. Aspirin is able to do this by specifically acetylating the hydroxyl of a serine residue in COX, thus inactivating the enzyme and stopping the biosynthetic pathway to prostaglandins:

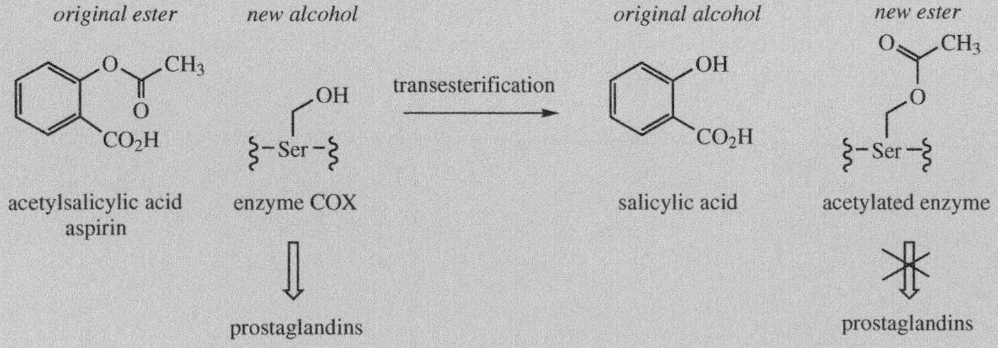

We shall meet another important example of transesterification in the action of the enzyme acetylcholinesterase (see Box 13.4).

Cyclic esters (**lactones**) are formed when the carboxyl and hydroxyl groups are in the same molecule, and are most favoured when this results in the generation of strain-free five- or six-membered rings. Thus, 4-hydroxybutyric acid may form a five-membered lactone, which is termed a γ-lactone, its name coming from the alternative nomenclature γ-hydroxybutyric acid for the acyclic compound. Similarly, six-membered lactones are termed δ-lactones. It is generally easier to use the fully systematic -oxa- nomenclature (see Section 1.4) for the oxygen heterocycle in more complex lactones. This approach considers nomenclature as if we had a carbocyclic ring, and uses an -oxa- syllable to indicate replacement of a carbon with the oxygen heteroatom.

Lactonization, like esterification, is an equilibrium process. γ-Lactones and δ-lactones are so readily formed that the carboxylic acid itself can provide the required acidic catalyst, and substantial amounts of the lactone are typically present in solutions of 4- or 5-hydroxy acids respectively (Table 7.3). Interestingly, the proportion of lactone is usually higher for five-membered rings than for six-membered rings,

$$\underset{\substack{4\ \ 3\ \ 2\ \ 1}}{\overset{\gamma\quad\beta\quad\alpha}{HOCH_2CH_2CH_2CO_2H}} \xrightarrow{\ H^+\ }$$

4-hydroxybutyric acid
γ-hydroxybutyric acid

2-oxacyclopentanone
γ-lactone

$$\underset{\substack{5\ \ 4\ \ 3\ \ 2\ \ 1}}{\overset{\delta\quad\gamma\quad\beta\quad\alpha}{HOCH_2CH_2CH_2CH_2CO_2H}} \xrightarrow{\ H^+\ }$$

5-hydroxypentanoic acid

2-oxacyclohexanone
δ-lactone

though other substituents, if they are present, also affect the equilibrium proportions.

Larger lactones do not exist to any appreciable extent in equilibrium with the free hydroxy acids, but they may be prepared under appropriate conditions. These may include removal of water to disturb the equilibrium, and carrying out the reaction at quite high dilutions in order to minimize intermolecular esterification.

Table 7.3 Hydroxyacid–lactone equilibria

Hydroxyacid	Lactone	Ring size	Equilibrium composition (%)	
			Hydroxyacid	Lactone
HO–CO₂H		4	100	0
HO–CO₂H		5	27	73
HO–CO₂H		6	91	9
HO–CO₂H		7	100	0

Box 7.14

Large ring lactones: erythromycin and amphotericin

Very large ring lactones are called **macrolides**, and are found in the natural **macrolide antibiotics**. Typically, these may have 12-, 14-, or 16-membered lactone rings, though other sizes are encountered. **Erythromycin** is a good example. This antibiotic is prescribed for patients who are allergic towards penicillins, and is the antibiotic of choice for infections of *Legionella pneumophila*, the cause of legionnaire's disease.

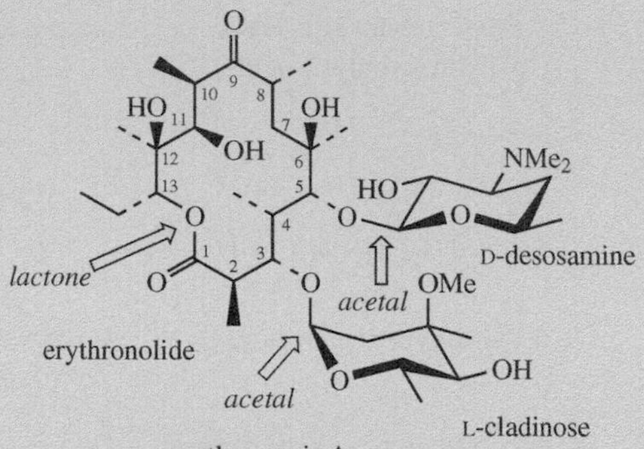

erythromycin A

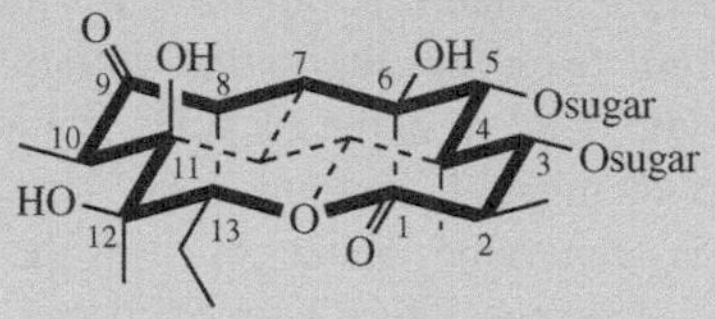

shape of macrolide
ring in erythromycin

Erythromycin is a mixture of at least three structurally similar compounds, the major component of which is erythromycin A. This has a 14-membered lactone ring, with a range of additional substituents. The ring system in erythromycin adopts a conformation that approximates to the periphery of four fused chair-like rings. Note that erythromycin contains two uncommon sugar rings, cladinose and desosamine, the latter being an aminosugar. Both of these sugars are bound to the lactone-containing ring through acetal linkages (see Section 7.2).

Polyene macrolides have even larger lactone rings, typically from 26–38 atoms, which also accommodates a conjugated polyene of up to seven *E* double bonds. **Amphotericin** from cultures of *Streptomyces nodosus* provides a typical example, and is used clinically as an antifungal agent. It is administered intravenously for treating potentially life-threatening fungal infections. Amphotericin is a mixture of compounds, the main and most active component being amphotericin B. The ring size in amphotericin B is 36 atoms, but is contracted from a potential 38 by cross-linking through a hemiketal function (see Section 7.2). An unusual amino sugar, D-mycosamine, is bound to the system through an acetal linkage.

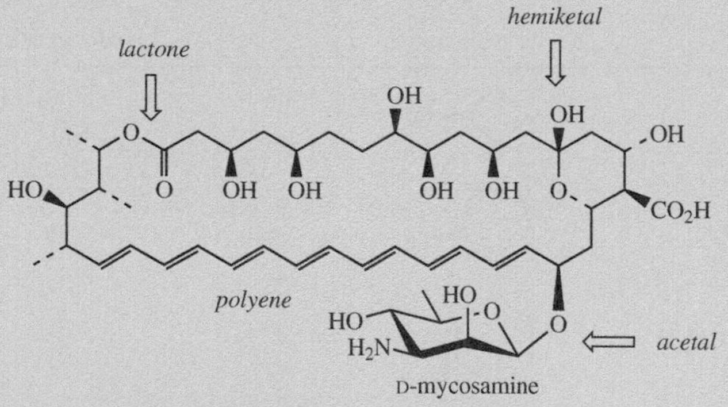

amphotericin B

7.9.2 Water: hydrolysis of carboxylic acid derivatives

All carboxylic acid derivatives are hydrolysed to **carboxylic acids** by the action of water as nucleophile. **Acyl halides** and **anhydrides** of low molecular weight are hydrolysed quite vigorously. **Esters** and **amides** react much more slowly, and hydrolysis normally requires acid or base catalysis. This is nicely

exemplified by the need to store and use compounds such as acetyl chloride and acetic anhydride under anhydrous conditions. On the other hand, the ester ethyl acetate is routinely used for solvent extractions of organic products from aqueous solutions.

Hydrolysis of an **ester** can be achieved by either base- or acid-catalysed reactions, and the nucleophilic substitution mechanisms follow processes that should now be becoming familiar to us. However, there are significant differences between the two types of process, as we shall see. In **acid-catalysed hydrolysis of esters**, the process is analogous to the acid-catalysed formation of esters: it is merely the reverse reaction.

acid-catalysed hydrolysis of esters

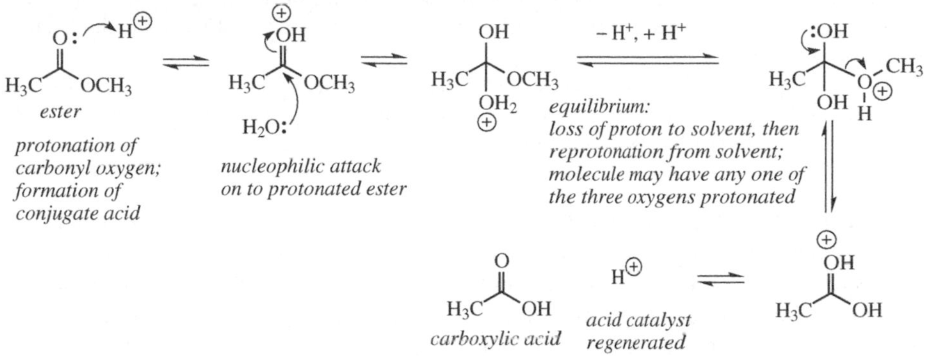

acid is **catalyst** and is regenerated;
equilibrium reaction

The reaction begins with protonation of the carbonyl oxygen to give the conjugate acid, which increases the electrophilicity of the carbonyl. This is necessary because we are using a neutral nucleophile. Nucleophilic attack follows, giving a protonated tetrahedral intermediate. In an acidic medium, an equilibrium will again be set up such that any one of the three oxygen atoms may be protonated. This equilibrium also involves loss of protons to the solvent, followed by reprotonation of another oxygen using the solvent. As in esterification (see Section 7.9.1), the process is not internal transfer of a proton, but requires solvent molecules. For hydrolysis to occur, the methoxyl needs to be protonated, so that it can be lost as a methanol leaving group. Loss of the leaving group is again facilitated by the resonance effect from a hydroxyl, leading to regeneration of the carbonyl group in protonated form. Formation of the uncharged carboxylic acid regenerates the acid catalyst. As with acid-catalysed ester formation, the reaction is an equilibrium, and this equilibrium needs to be disturbed for complete hydrolysis, typically by using an excess of water, i.e. aqueous acid.

Box 7.15

Autolysis of aspirin

The analgesic **aspirin**, acetylsalicylic acid, is an ester. In this compound, the alcohol part is actually a phenol, salicylic acid. Aspirin is synthesized from salicylic acid by treatment with acetic anhydride.

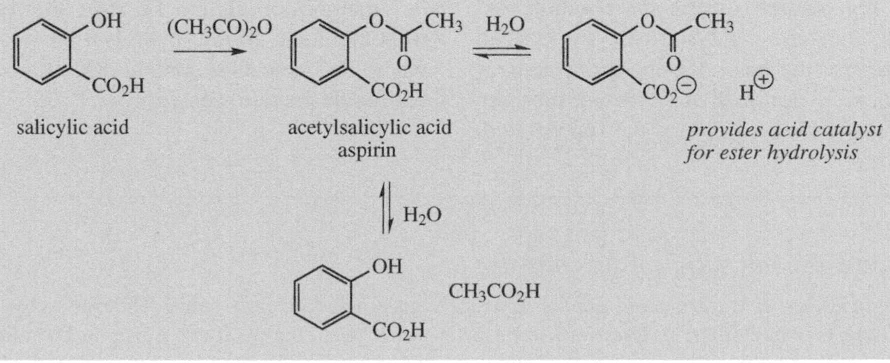

Aspirin is an ester, but it still contains a carboxylic acid function (pK_a 3.5). In aqueous solution, there will thus be significant ionization. However, this ionization now provides an acid catalyst for ester hydrolysis and initiates autolysis (autohydrolysis). The hydrolysis product salicylic acid (pK_a 3.0) is also acidic; both aspirin and salicylic acid are aromatic acids and are rather stronger acids than aliphatic compounds such as acetic acid (pK_a 4.8) (see Section 4.3.5). An aqueous solution of aspirin has a half-life of about 40 days at room temperature. In other words, after about 40 days, half of the material has been hydrolysed, and the biological activity will have deteriorated similarly.

Even aspirin tablets that have been stored under less than ideal conditions and, therefore, have absorbed some water from the atmosphere, are likely to have suffered partial hydrolysis. The characteristic odour of acetic acid (vinegar) from a bottle of aspirin tablets will be an indicator that some hydrolysis has occurred.

In the **base-catalysed hydrolysis of esters**, the nucleophile is hydroxide, a charged species that is able to attack the uncharged carbonyl. The carbonyl group is restored by loss of alkoxide as leaving group. However, alkoxide is a strong base, a poor leaving group, and the reaction seems unlikely to be favourable (see Section 7.8). It does occur, however, and this is because the strong base leaving group is able to abstract a proton from the carboxylic acid product, generating an alcohol and the carboxylate anion. Although the early steps of the reaction are reversible, this last step, ionization of the carboxylic acid, is essentially irreversible and so disturbs the equilibrium reaction. The ionization is not reversible: the carboxylate anion is far too weak a base to ionize an alcohol.

base hydrolysis of esters

nucleophilic attack of hydroxide on to carbonyl

loss of leaving group, and reformation of carbonyl

leaving group (strong base) abstracts proton from acid

base is **reactant** and is consumed; reaction becomes essentially **irreversible** because of formation of the carboxylate anion

carboxylate anion

We can now distinguish differences between acid-catalysed and base-catalysed hydrolysis of esters. The acid-catalysed reaction is an equilibrium, and the equilibrium needs to be disturbed by use of excess reagent (water). The acid used is a true catalyst: it is regenerated during the reaction. On the other hand, the base-catalysed reaction goes to completion, because the basic leaving group ionizes the product and, in so doing, disturbs the equilibrium. This means that the base catalyst is not regenerated, but is actually consumed during the reaction. The description 'base-catalysed hydrolysis' is generally used, but it is strictly incorrect, since the base is a reagent rather than a catalyst; a better terminology is 'base hydrolysis'. Basic hydrolysis of esters is usually the method of choice because the reaction goes to completion. Acidic hydrolysis would be selected where the molecules contain other functional groups that might be base sensitive.

Box 7.16

Ester hydrolysis: saponification of fats and oils

Fats and oils are esters of the trihydric alcohol glycerol with long-chain fatty acids. The descriptor fat or oil is applied according to whether the material is a solid or liquid at room temperature; it has no chemical meaning. All three fatty acids in the ester may be the same, or they may be different. Common saturated fatty acids encountered are stearic acid (C_{18}) and palmitic acid (C_{16}), especially in animal fats, and the unsaturated acids oleic acid (C_{18}) and linoleic acid (C_{18}) in plant oils.

RCO_2H = fatty acid

OCOR
|
—OCOR $\xrightarrow{\text{NaOH}}$ $RCO_2^{\ominus}\ Na^{\oplus}$
| *soap*
OCOR

fat

OH
|
—OH
|
OH

glycerol

base hydrolysis of esters is often termed **saponification**

e.g. $CH_3(CH_2)_{16}CO_2H$ $CH_3(CH_2)_{14}CO_2H$
      stearic acid   palmitic acid

$CH_3(CH_2)_7CH=CH(CH_2)_7CO_2H$
      oleic acid

$CH_3(CH_2)_4CH=CHCH_2CH=CH(CH_2)_7CO_2H$
      linoleic acid

Base hydrolysis of fats with sodium or potassium hydroxide liberates glycerol and the salt of the carboxylic acid(s). This reaction was the basis of soap making; the salt, or mixture of salts, is a soap with characteristic detergent properties. The relationship of ester hydrolysis to soap-making remains, in that base hydrolysis of esters is still commonly referred to as saponification.

Amides may be hydrolysed to carboxylic acids by either acids or bases, though hydrolysis is considerably slower than with esters. Although amines are bases and become protonated on nitrogen via the lone pair electrons, we know that amides are not basic (see Section 4.5.4). This is because the lone pair on the nitrogen in amides is able to overlap into the carbonyl π system, thus creating resonance stabilization in the neutral amide. This effect also diminishes the reactivity of the carbonyl towards nucleophilic attack, since the resonance contribution actually means less carbonyl character and more carbon–nitrogen double bond character.

electron overlap from nitrogen lone pair allows resonance stabilization of amide

protonation on nitrogen does not occur; destroys resonance stabilization

oxygen is more electronegative than nitrogen; electron donation from oxygen is less than from nitrogen

protonation on oxygen allows resonance stabilization of cation

Note that we can write a similar resonance picture for esters, and we shall actually need to invoke this when we discuss enolate anions (see Section 10.7). However, electron donation from oxygen is not as effective as from the less electronegative nitrogen. We shall also see that this resonance effect in amides has other consequences, such as increased acidity of the amide hydrogens (see Section 10.7) and stereochemical aspects of peptides and proteins (see Section 13.3). In addition, the amide derivatives have poorer leaving groups than the corresponding esters, and this also contributes to the lower reactivity of amides.

Although protonation does not occur on nitrogen in an amide, protonation can occur on the carbonyl oxygen, because this still allows the same type of resonance stabilization. Accordingly, **acid hydrolysis of amides** proceeds through nucleophilic attack of water onto the protonated carbonyl, giving a tetrahedral protonated intermediate.

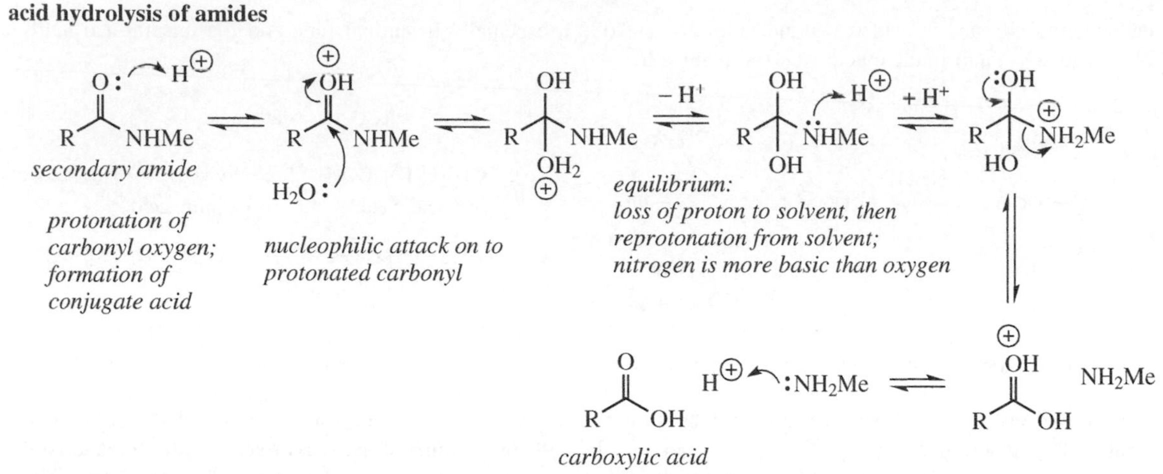

acid hydrolysis of amides

secondary amide

*protonation of
carbonyl oxygen;
formation of
conjugate acid*

*nucleophilic attack on to
protonated carbonyl*

*equilibrium:
loss of proton to solvent, then
reprotonation from solvent;
nitrogen is more basic than oxygen*

carboxylic acid

Loss of a proton from this allows reprotonation on nitrogen; the nitrogen atom is no longer attached to a carbonyl, so it is basic, more basic than the oxygen atoms. The amine molecule is now a satisfactory leaving group, and this allows regeneration of the carbonyl. Of course, under acid conditions, the amine will be rapidly protonated and become non-nucleophilic, so this will help to disturb the equilibrium and discourage the reverse reaction. It also means that acid is used up in the hydrolysis, and we do not have true acid catalysis. Primary, secondary and tertiary amides all undergo similar hydrolytic reactions, though

hydrolysis does require heating with quite concentrated acid.

Base hydrolysis of amides also requires quite vigorous conditions, but mechanistically it is exactly equivalent to base hydrolysis of esters. After nucleophilic attack of hydroxide on to the carbonyl, the tetrahedral anionic intermediate is able to lose either an amide anion (care with nomenclature here, the amide anion is quite different from the amide molecule) or hydroxide. Although loss of hydroxide is preferred, since the amide anion is a stronger base than hydroxide, this would merely reverse the reaction.

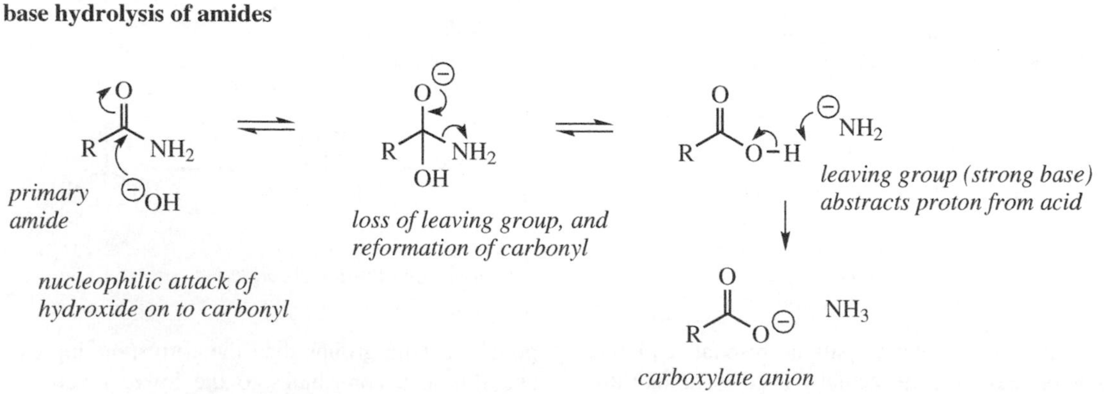

base hydrolysis of amides

*primary
amide*

*nucleophilic attack of
hydroxide on to carbonyl*

*loss of leaving group, and
reformation of carbonyl*

*leaving group (strong base)
abstracts proton from acid*

carboxylate anion

The reaction progresses because the amide anion, once a small amount is released, abstracts a proton from the carboxylic acid product. Again, we have an analogy with the last step in the base hydrolysis of esters, and the ionization becomes an essentially

irreversible step. Furthermore, hydroxide is again consumed as a reagent.

Base hydrolysis of secondary and tertiary amides is less readily achieved than with primary amides, and may require stronger basic conditions.

Box 7.17
Amide hydrolysis: peptides and proteins

Proteins are fundamentally polymers of α-amino acids linked by amide linkages (see Section 13.1). It is a pity that biochemists refer to these amide linkages as peptide bonds; remember, a peptide is a small protein (less than about 40 amino acid residues), whereas a peptide bond is an amide. Therefore, peptides and proteins may be hydrolysed to their constituent amino acids by either acid or base hydrolysis. The amide bond is quite resistant to hydrolytic conditions (see above), an important feature for natural proteins.

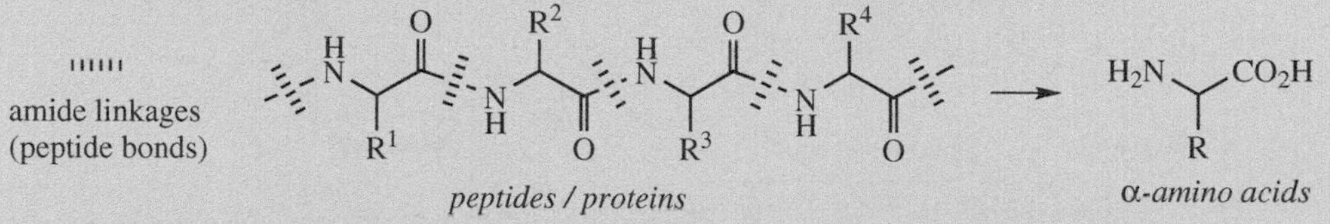

amide linkages (peptide bonds)

peptides / proteins

α-amino acids

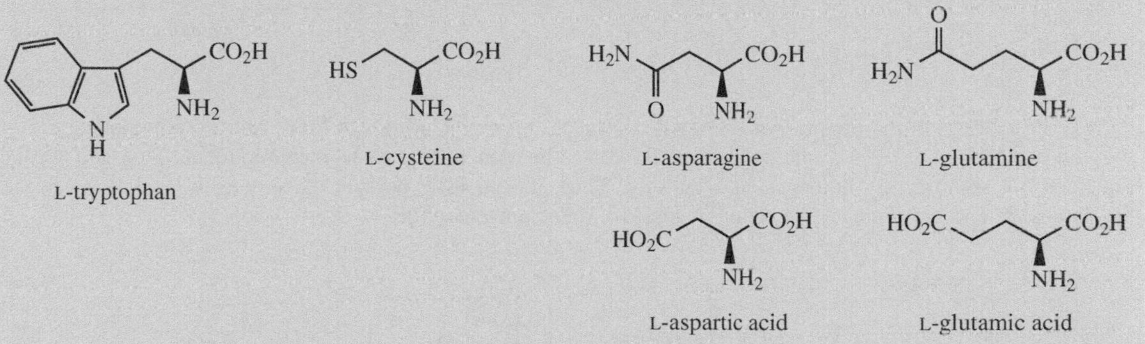

L-tryptophan

L-cysteine

L-asparagine

L-glutamine

L-aspartic acid

L-glutamic acid

Neither acid nor base hydrolysis is ideal, since some of the constituent amino acids are found to be sensitive to the reagents because of the nature of their R side-chains. Acid hydrolysis is the preferred method because it causes less degradation. Nevertheless, the indole system of tryptophan is known to be largely degraded in acid, and the sulfur-containing amino acid cysteine is also unstable. Those amino acids containing amide side-chains, e.g. asparagine and glutamine, will be hydrolysed further, giving the corresponding structures with acidic side-chains, namely aspartic acid and glutamic acid.

7.9.3 Thiols: thioacids and thioesters

Thiols undergo the same types of nucleophilic reaction with carboxylic acid derivatives as do alcohols. However, reactivity tends to be increased for two reasons. First, sulfur, because of its larger size, is a better nucleophile than oxygen (see Section 6.1.2); second, RS^- is a better leaving group than RO^- (see Section 6.1.4), again because of size and the less localized electrons. Simple nucleophilic reactions with H_2S parallel those with H_2O, and those with RSH parallel those with ROH. This gives rise to carboxylic acid derivatives containing sulfur, such as **thioacids** and **thioesters**.

$(RCO)_2O$ + H_2S ⟶ *anhydride* → thioacid

RCOCl + R′SH ⟶ *acyl chloride* → thioester

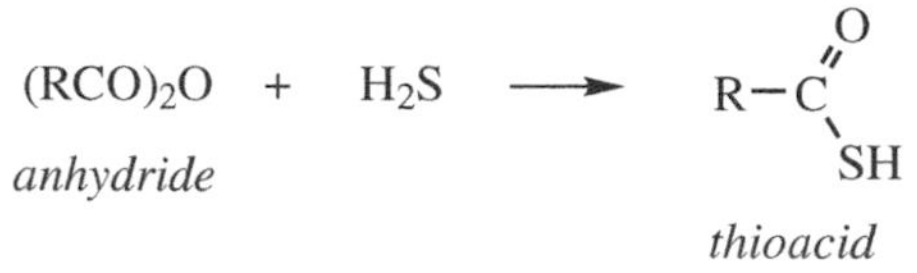

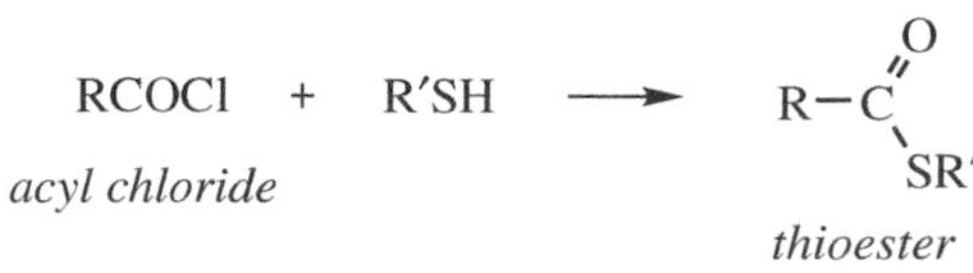

Box 7.18

Thioesters: coenzyme A

Thioesters are more reactive towards nucleophilic substitution than oxygen esters, and are widely employed in natural biochemical processes because of this property. **Coenzyme A** is a structurally complex thiol, and functions in the transfer of acetyl groups via its thioester acetyl coenzyme A (acetyl-CoA; $CH_3CO–SCoA$).

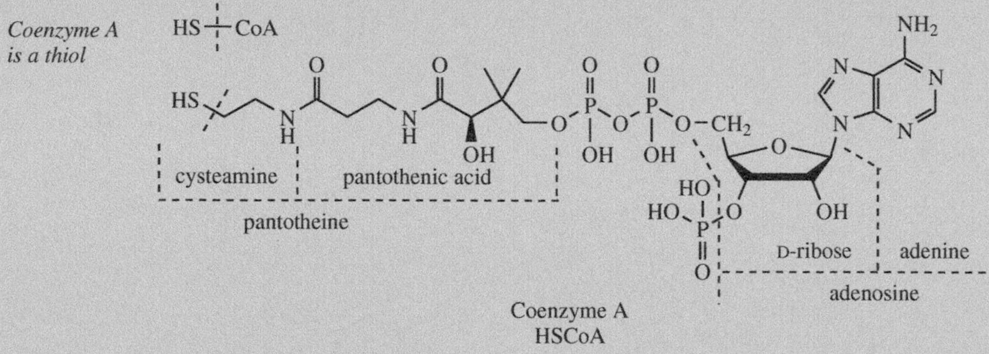

Coenzyme A is a thiol   HS—CoA

Coenzyme A
HSCoA

We can understand the function of coenzyme A merely by appreciating that it is a thiol; the remaining part of the complex structure (it is a nucleotide derivative; see Section 14.3) aids its enzymic recognition and binding but does not significantly influence its reactivity. Thus, the thioester acetyl-CoA is a good acylating agent in biochemistry, and can transfer the acetyl group to a suitable nucleophile.

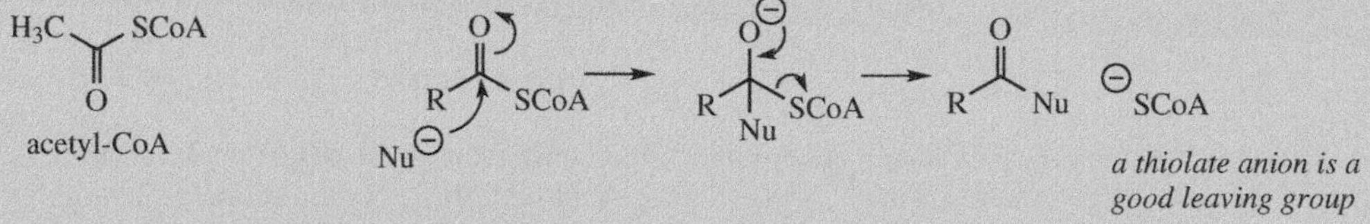

acetyl-CoA

a thiolate anion is a good leaving group

This familiar reaction is effective under physiological conditions because it is enzyme mediated and employs a good leaving group in the form of a thiolate anion $CoAS^-$. It is rather interesting to note that nature uses this reaction to make acetate esters from ROH nucleophiles. Nature thus uses the more reactive combination of thioester plus alcohol, rather than the acid plus alcohol combination we might initially consider to make an ester.

We shall see some other important biological reactions of thioesters in Box 10.8.

7.10   Nitrogen as a nucleophile: amides

Ammonia, primary amines, and **secondary amines** all react with carboxylic acids to give **amides**. However, all of these reagents are bases, and salt formation with the carboxylic acid occurs first, basicity prevailing over nucleophilicity. The negatively charged carboxylate is correspondingly unreactive towards nucleophiles.

Nucleophilic attack only occurs upon heating the ammonium salt, resulting in overall dehydration of the salt. Consequently, it is usual to prepare amides by using a more favourable substrate than the carboxylic acid, one that is more reactive towards nucleophiles by virtue of possessing a better leaving group, and where salt formation does not hinder the reaction.

RCO_2H + H_2NR' ⟶ RCO_2^- $^+NH_2R'$ $\xrightarrow[-H_2O]{heat}$ $RCONHR'$

primary amine   *ammonium salt*   *secondary amide*

Acyl halides and **anhydrides** are the most reactive class of carboxylic acid derivatives, and readily react with amines to give amides. It should be noted that in both cases the leaving group is a conjugate base that, upon protonation during the reaction, will become an acid. Consequently, this acid forms a salt with the amine reagent, and the reaction will tend to stop. For success, the reaction thus requires the use of 2 mol of amine, or some alternative base must be added.

$$RCOCl \ + \ NHR'_2 \ \longrightarrow \ RCONR'_2 \ + \ HCl$$

secondary amine   *tertiary amide*

need extra mole of amine or base to take up HCl

acyl halide

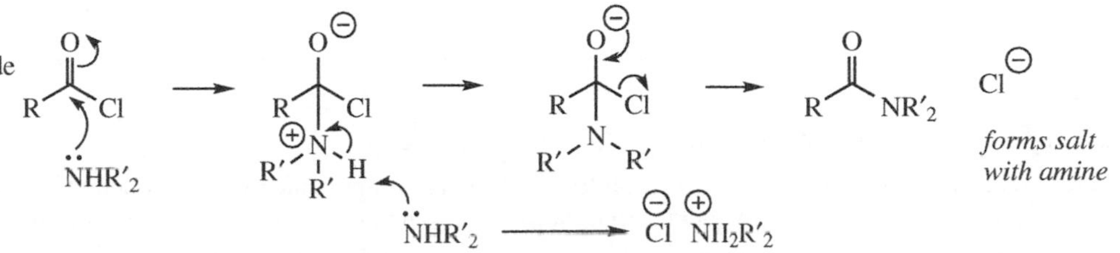

forms salt with amine

anhydride

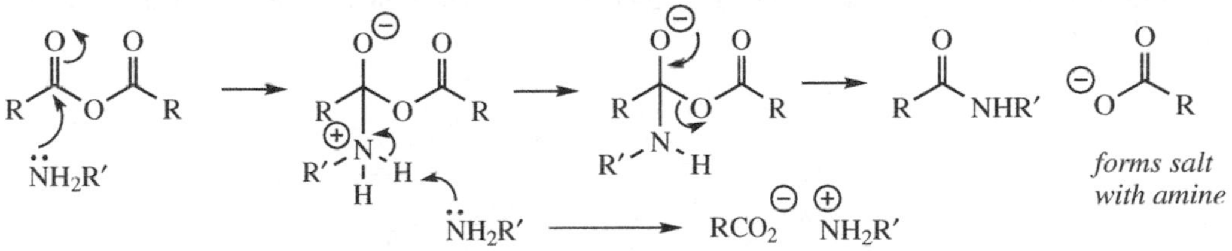

forms salt with amine

Esters also react smoothly with amines, which is a useful reaction if the corresponding acyl halides or anhydrides are not easily available. The reaction proceeds through the anticipated tetrahedral anionic intermediate. There are two possible leaving groups in this tetrahedral intermediate: alkoxide anion or amide anion. Since ammonia is a considerably weaker acid than an alcohol, the preferred leaving group is the weaker base alkoxide.

$$RCO_2Et \ + \ NH_3 \ \longrightarrow \ RCONH_2 \ + \ EtOH$$

primary amide

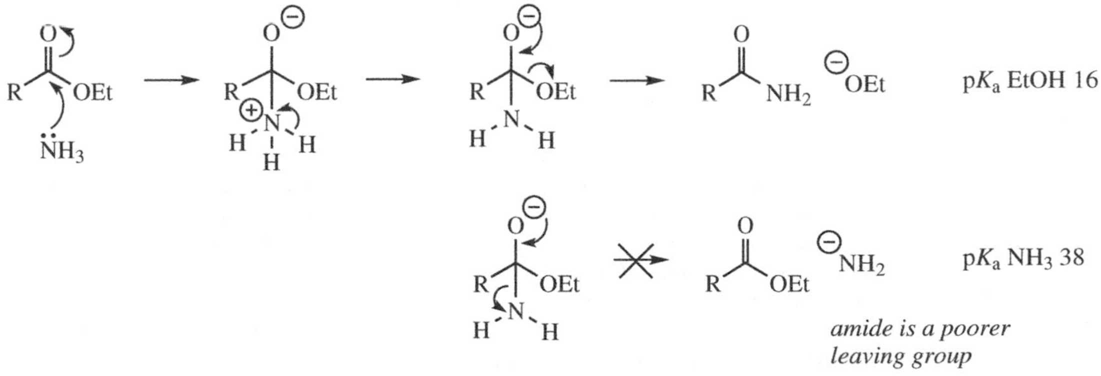

pK_a EtOH 16

pK_a NH$_3$ 38

amide is a poorer leaving group

The consequence of this is that an ester can react with ammonia to give an amide, but the reverse reaction does not occur; simply treating amides with alcohols does not produce esters. Production of esters from amides requires acid or base catalysis. It should also be noted that amines are better nucleophiles than alcohols (see Section 6.1.2), and that addition of nitrogen to the carbonyl compound does not usually require acid or base catalysis. Indeed, acid conditions would protonate the amine and destroy its nucleophilicity (see Section 7.7.1).

Note that, in all of these reaction mechanisms, a proton needs to be removed from the nitrogen nucleophile. Hence ammonia, primary amines, and secondary amines, but not **tertiary amines**, can function as nucleophiles. Where we appeared to exploit the nucleophilicity of a tertiary amine (pyridine) towards carboxylic acid derivatives forming an acylpyridinium ion, the amine was subsequently lost as a leaving group (see Section 7.9.1). Pyridine behaved as a nucleophilic catalyst, and a permanent N–C bond was not produced.

Box 7.19

Synthesis of paracetamol: an example of selective reactivities

The different reactivities associated with nucleophiles and leaving groups is nicely exemplified in the synthesis of the analgesic drug **paracetamol** (USA: **acetaminophen**) from 4-aminophenol. If 4-aminophenol is treated with an excess of acetic anhydride, acetylation of both amino and phenol groups is observed, and the product is the diacetate. Paracetamol is the *N*-acetate of 4-aminophenol, so how might mono-acetylation be achieved? There are two approaches.

One method is to treat 4-aminophenol with just one molar equivalent of acetic anhydride. The main product is paracetamol, which is produced almost selectively since –NH$_2$ is a better nucleophile than –OH. We can predict this from their pK_a values as bases, about 5 for the conjugate acid of a typical aromatic amine, and about −7 for a phenol, i.e. the amine is the stronger base. Although the heteroatoms are not the same (see Section 6.1.2), the pK_a values are significantly different and allow us to predict that the amine is also going to be the better nucleophile. The higher the pK_a of the conjugate acid, the better the nucleophile.

However, the second method of synthesizing paracetamol is to hydrolyse the diacetate of 4-aminophenol carefully using aqueous NaOH. In this case, hydrolysis of the amide is slower than that of the ester, since the ArNH$^-$ ion is a poorer leaving group than ArO$^-$. Again, this can be predicted from pK_a values. ArOH (pK_a 10) is a stronger acid than ArNH$_2$ (pK_a 28). The lower the pK_a of the conjugate acid, the better the leaving group.

We observed that cyclic esters (lactones) may be formed when the carboxyl electrophile and hydroxyl nucleophile are in the same molecule (see Section 7.9.1). Similarly, cyclic amides are produced when carboxyl and amine groups are in the same molecule, and are again most favoured when this results in the generation of strain-free five- or six-membered rings. Cyclic esters are termed lactones, whereas cyclic amides are in turn called **lactams**. The nomenclature of lactams is similar to that used for lactones. Thus, 4-aminobutyric acid may form a five-membered lactam, which is termed a γ-lactam, its name coming from the alternative nomenclature γ-aminobutyric acid for the acyclic compound. Similarly, six-membered lactams are termed δ-lactams. The fully systematic -aza-nomenclature for the nitrogen heterocycle in lactams is much easier to use. In practice, we would probably name lactams as ketone derivatives of heterocycles.

$$H_2NCH_2CH_2CH_2CO_2H$$

4-aminobutyric acid

2-azacyclopentanone
γ-lactam
(pyrrolidin-2-one)

2-azacyclohexanone
δ-lactam
(piperidin-2-one)

Box 7.20

Amides and β-lactams: semi-synthesis and hydrolysis of penicillins and cephalosporins

Penicillin and **cephalosporin** antibiotics possess an unusual and highly strained four-membered lactam ring. These and related antibiotics are commonly called **β-lactam antibiotics**. In both penicillins and cephalosporins, the β-lactam ring is fused through the nitrogen and adjacent carbon to a sulfur-containing ring. This is a five-membered thiazolidine ring in penicillins, and a six-membered dihydrothiazine ring in cephalosporins.

β-lactam

penicillin
β-lactam–thiazolidine

cephalosporin
β-lactam–dihydrothiazine

The penicillins are the oldest of the clinical antibiotics and are still the most widely used. Early examples, such as **benzylpenicillin**, were decomposed by gastric acid and, consequently, could not be administered orally. Modern penicillins have been developed to overcome this sensitivity towards acid by changing the nature of the carboxylic acid that features in the acyclic amide linkage. It was found that introducing electron-withdrawing heteroatoms into the side-chain significantly inhibited sensitivity to acid hydrolysis. Benzylpenicillin, produced in fermentors by cultures of the fungus *Penicillium chrysogenum*, is converted into **6-aminopenicillanic acid** by a suitable bacterial enzyme system. This enzyme selectively hydrolyses the acyclic amide, without affecting the cyclic amide.

Box 7.20 (continued)

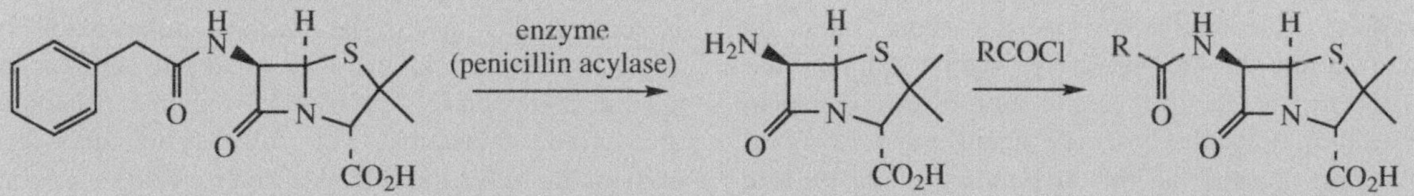

benzylpenicillin → (enzyme (penicillin acylase)) → 6-aminopenicillanic acid → (RCOCl) → semi-synthetic penicillin

This selectivity is not achievable by simple chemical hydrolysis, since the strained β-lactam ring is much more susceptible to nucleophilic attack than the unstrained side-chain amide function. Normally, the electron-donating effect from the lone pair of the adjacent nitrogen stabilizes the carbonyl against nucleophilic attack (see Section 7.9.2); this is not possible with the β-lactam ring because of the geometric restrictions (see Box 3.20).

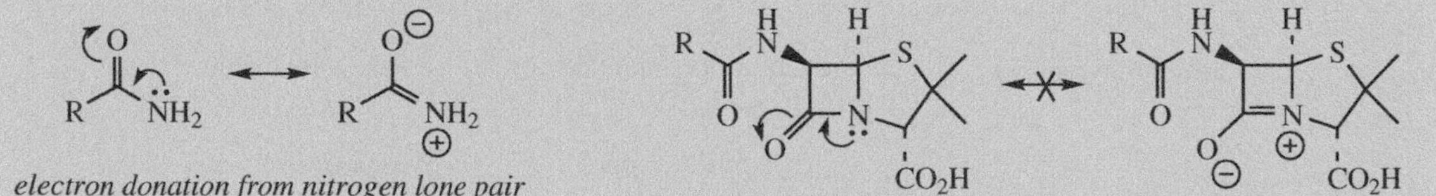

electron donation from nitrogen lone pair allows resonance stabilization of amide; bond angles are approximately 120°

resonance involving the amide lone pair normally decreases carbonyl character and thus stabilizes the carbonyl against attack by nucleophilic reagents

this resonance form is sterically impossible; the β-lactam carbonyl function is thus reactive towards nucleophilic reagents

It is feasible to convert benzylpenicillin into 6-aminopenicillanic acid chemically, but by a procedure involving several steps.

After removal of the carboxylic acid portion of the original amide, a new amide linkage is generated, e.g. by reaction with a suitable acyl chloride. One of the first commercial semi-synthetic penicillins, **methicillin**, was produced as shown. Other agents, e.g. **ampicillin**, may be produced by similar means, though sensitive functional groups in the new side-chain will need suitable protection.

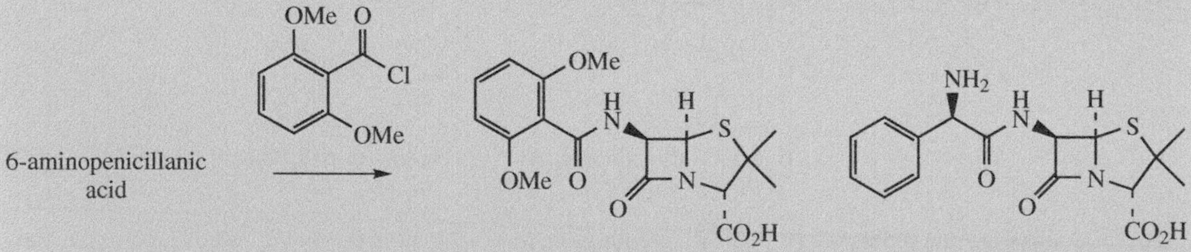

6-aminopenicillanic acid → methicillin   ampicillin

An additional disadvantage with many penicillin and cephalosporin antibiotics is that bacteria have developed resistance to the drugs by producing enzymes capable of hydrolysing the β-lactam ring; these enzymes are called **β-lactamases**. This type of resistance still poses serious problems. Indeed, methicillin is no longer used, and antibiotic-resistant strains of the most common infective bacterium *Staphylococcus aureus* are commonly referred to as **MRSA** (methicillin-resistant *Staphylococcus aureus*). The action of β-lactamase enzymes resembles simple base hydrolysis of an amide.

the common substructure
in β-lactam antibiotics

*antibiotic inactivated
by ring opening and
binding to enzyme*

*hydrolysis
of β-lactam
amide bond*

CO_2H

§– Ser –§ Enzyme

§– Ser –§ Enzyme

$H_2O:$

§– Ser –§ Enzyme

*hydrolysis of
ester function,
release from
enzyme*

HO CO_2H

NH_2

Serine

penicilloic acid
derivative – inactive

HO_2C HN

CO_2H

OH

§– Ser –§ Enzyme

§– Ser –§ Enzyme

It is known that the nucleophilic species in a β-lactamase enzyme is the hydroxyl group of a serine residue in the protein, and that this attacks the β-lactam carbonyl, followed by loss of the leaving group and consequent opening of the four-membered ring. The ring-opened penicillin (or cephalosporin) becomes bound to the enzyme through an ester linkage and is no longer active. The ester linkage is subsequently hydrolysed to release an inactive penicilloic acid derivative and regenerate the functional β-lactamase enzyme.

Note again that the strained β-lactam ring is more susceptible to nucleophilic attack than the unstrained side-chain amide function. However, by increasing the steric bulk of the side-chain, the approach of a β-lactamase enzyme to the β-lactam ring is hindered in the semi-synthetic antibiotic, giving it more resistance to enzymic hydrolysis.

7.11 Hydride as a nucleophile: reduction of carboxylic acid derivatives

We have already noted the ability of **complex metal hydrides** like **lithium aluminium hydride** and **sodium borohydride** to reduce the carbonyl group of aldehydes and ketones, giving alcohols (see Section 7.5). These reagents deliver hydride in such a manner that it appears to act as a nucleophile. However, as we have seen, the aluminium hydride anion is responsible for transfer of the hydride and

the hydride itself never acts as a nucleophile because of its small size and high charge density.

Acyl halides, anhydrides, esters and **acids** all react with LAH to give a **primary alcohol**. Amides (see later) behave differently.

The initial reaction is effectively the same as with an aldehyde or ketone, in that hydride is transferred from the reducing agent, and that the tetrahedral anionic intermediate then complexes with the Lewis acid aluminium hydride. However, the typical reactivity of the carboxylic acid derivatives arises because of the presence of a leaving group.

carbonyl oxygen combines with Lewis acid AlH₃

aldehyde reacts further with LAH; more reactive than carboxylic acid derivative

nucleophilic transfer of hydride on to carbonyl

loss of leaving group, regeneration of carbonyl

aldehyde

primary alcohol

The first-formed product is an **aldehyde**, resulting from loss of the leaving group and regeneration of the carbonyl. It is not normally possible to isolate this aldehyde product, because it reacts rapidly with the reducing agent, more rapidly in fact than the original carboxylic acid derivative. As a result, the aldehyde is further reduced, and after treatment with a proton source is converted into a primary alcohol. The Lewis acid aluminium hydride released during regeneration of the carbonyl will complex with the leaving group and continue as a source of hydride (see Section 7.5).

Although LAH will reduce **carboxylic acids**, it is not usually employed for this purpose, since salt formation can interfere with the reduction process. LAH is a strong base, and the lithium salt of the carboxylic acid typically precipitates out from solution. The usual approach to reducing carboxylic acids is to employ a two-stage process, first making an ester and then reducing this derivative. A feature of ester reduction is that it generates two molecules of alcohol, one from the acyl group and one from the leaving group.

$$RCO_2R' \xrightarrow{\text{LAH}} RCH_2OH + HOR'$$ *reduction of ester gives two alcohols*

Box 7.21

Selective reduction of carbonyl groups

Sodium borohydride is a weaker hydride donor than **lithium aluminium hydride** (see Section 7.5) and it is only really effective for reducing acyl halides, the most reactive of the carboxylic acid derivatives. However, this difference in reactivity of the reducing agents can be very useful, allowing selectivity and reduction of one group in the presence of other susceptible groups. For example, $NaBH_4$ will reduce the more reactive aldehyde and ketone groups but not reduce the less reactive ester group.

LAH reduces both carbonyls

NaBH₄ reduces only the ketone carbonyl

reduction of ketone

reduction of both groups

methyl 4-oxo-cyclohexanecarboxylate

ketal formation protects the ketone carbonyl

regeneration of ketone from ketal

only one carbonyl available for LAH reduction

reduction of ester

Therefore, it is possible to reduce both carbonyl groups in the ketoester methyl 4-oxocyclohexanecarboxylate using LAH. Sodium borohydride will reduce only the ketone, giving the hydroxyester. However, by using a ketal as protecting group (see Section 7.2) it is possible to reduce just the ester with LAH, and the original ketone can be regenerated by hydrolysis of the ketal.

Amides behave differently towards LAH than the other carboxylic acid derivatives, and the overall reaction observed is reduction of the carbonyl to a methylene group, with retention of the amino group.

$$RCONH_2 \xrightarrow{\text{LAH}} RCH_2NH_2$$

primary amide   *primary amine*

This unusual behaviour may be explained simply as a consequence of alternative leaving groups being present in the addition intermediate. After transfer of hydride to the carbonyl with formation of a tetrahedral anionic complex, there are two potential leaving groups, i.e. R_2N^- and the aluminate anion $(OAlH_3)^{2-}$. The aluminate anion is a better leaving group than the amide, and this leads to formation of an iminium ion. This behaviour can thus be seen to be analogous to the dehydration of hydroxyamines to imines during reaction of aldehydes or ketones with amines (see Section 7.7.1). There, we also had an alternative leaving group present.

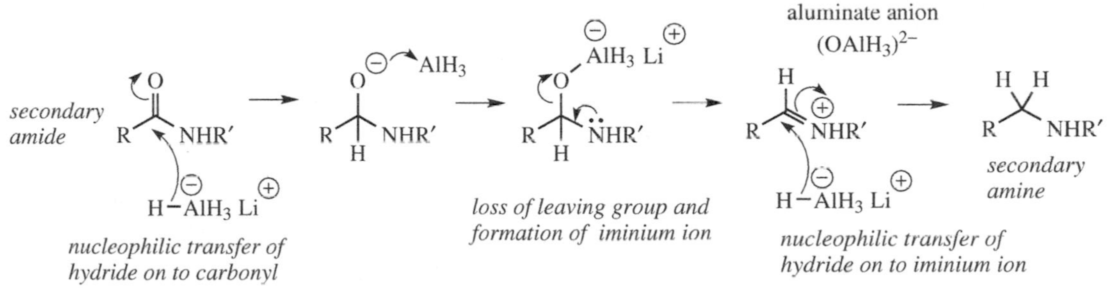

secondary amide — *nucleophilic transfer of hydride on to carbonyl* — *loss of leaving group and formation of iminium ion* — aluminate anion $(OAlH_3)^{2-}$ — *nucleophilic transfer of hydride on to iminium ion* — *secondary amine*

In the LAH reduction sequence, the C=N double bond in the iminium ion now behaves just as the C=O bond of a carbonyl (see Section 7.7.1) and is also reduced by transfer of hydride from a further equivalent of LAH. The final product is thus an amine.

These reactions also provide us with a convenient way of making secondary and tertiary amines. Thus, a primary amine may be converted into an amide by reaction with an acyl chloride, then LAH reduction

leads to a secondary amine. We are effectively introducing an RCH$_2$– group via the corresponding RCO– acyl group.

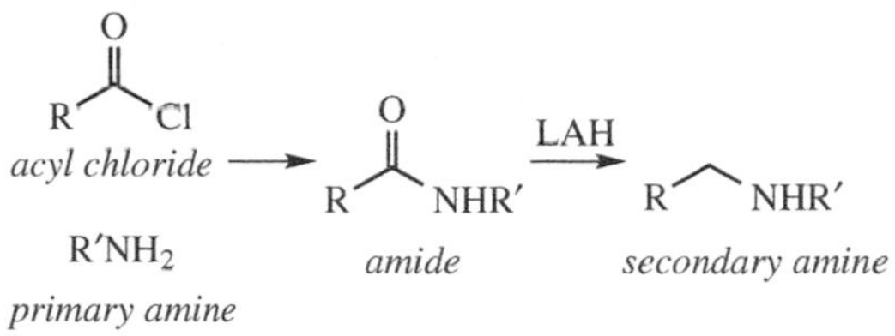

acyl chloride + R'NH$_2$ *primary amine* → *amide* → $\xrightarrow{\text{LAH}}$ *secondary amine*

Box 7.22

The importance of leaving groups: linking up the chemistry of amides with imine formation

Although we have discussed all of the following reactions, they have been covered in separate sections, and this box is included to draw them together and demonstrate that they can all be rationalized via a common theme, namely the nature of **leaving groups**. A fundamental consideration is that addition to the carbonyl group is reversible if the nucleophile can subsequently be lost from the addition product as a leaving group. This explains why halides do not react as nucleophiles towards carbonyl compounds; halides are such good leaving groups that the reverse reaction always predominates. The forward reaction is completed by protonation of the oxyanion in the case of **aldehydes** and **ketones**, or loss of a leaving group for **carboxylic acid derivatives**.

Box 7.22 (continued)

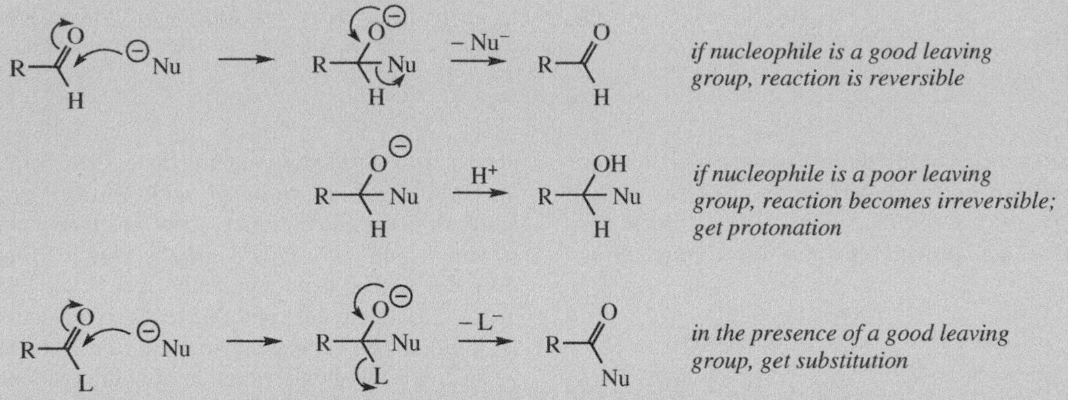

if nucleophile is a good leaving group, reaction is reversible

if nucleophile is a poor leaving group, reaction becomes irreversible; get protonation

in the presence of a good leaving group, get substitution

We saw that reaction of amines with aldehydes or ketones led to **imine** formation, rather than the simple aminoalcohol addition product (see Section 7.7.1). This was because, in acidic solution, the protonated aminoalcohol had two possible leaving groups, and water rather than the amine was the better leaving group. Dehydration occurs, leading to the imine.

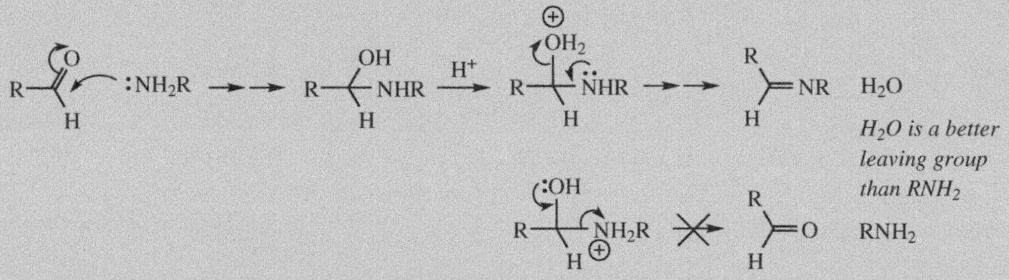

H_2O is a better leaving group than RNH_2

Amide formation involved the same considerations. Thus, esters are readily converted into amides by treatment with ammonia (see Section 7.10). The intermediate anion has two potential leaving groups, alkoxide RO^- and amide NH_2^-, and alkoxide is the better leaving group. The converse of this is that treatment of an amide with an alcohol does not lead to an amide; we generate the same intermediate anion, so the reverse reaction, loss of alkoxide, predominates.

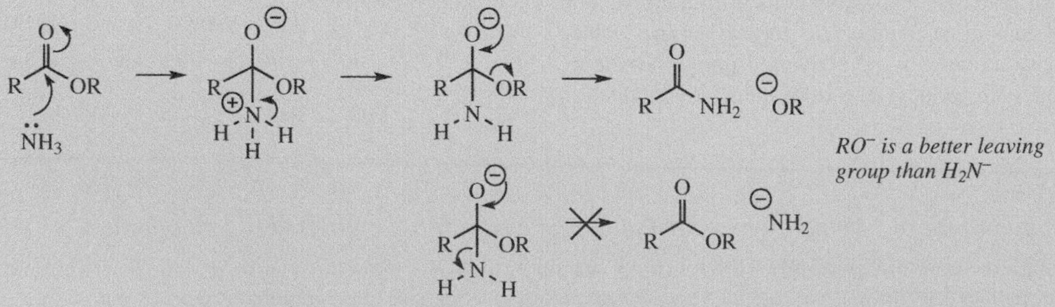

RO^- is a better leaving group than H_2N^-

Reduction of **aldehydes** and **ketones** with a **complex metal hydride** gives an alcohol (see Section 7.5). Such reactions are not reversible because hydride is a very poor leaving group, so we eventually get protonation of the alkoxide system. **Acyl derivatives** generally have a good leaving group and this is lost, restoring the carbonyl group, and producing an **aldehyde**. Of course, this reacts further with reducing agent, and the final product is a **primary alcohol** (see Section 7.11).

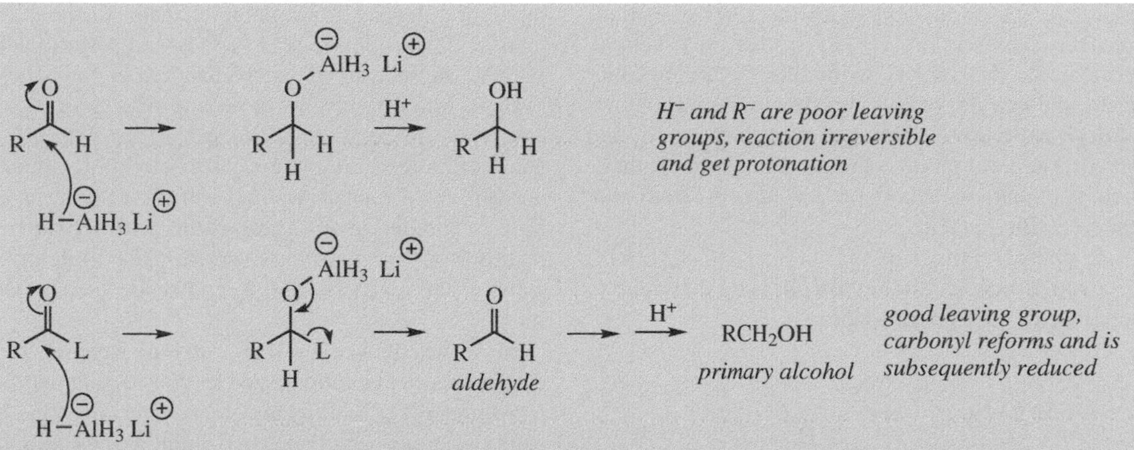

H⁻ and R⁻ are poor leaving groups, reaction irreversible and get protonation

good leaving group, carbonyl reforms and is subsequently reduced

aldehyde

RCH₂OH

primary alcohol

Amides seem to behave differently, with complex metal hydride reduction giving an **amine**, effectively converting the carbonyl group to a methylene (see Section 7.11).

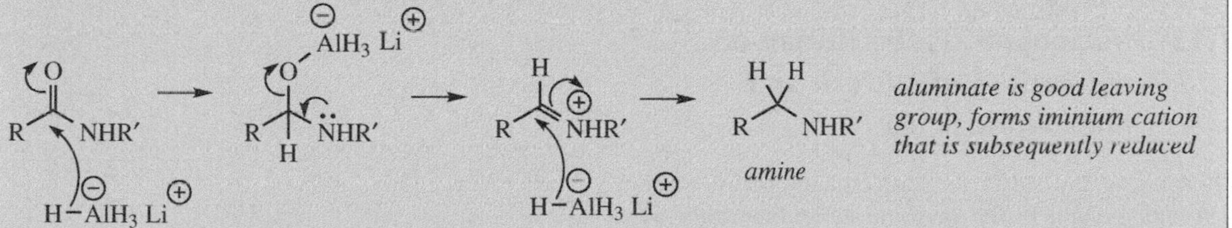

aluminate is good leaving group, forms iminium cation that is subsequently reduced

amine

This behaviour results from initial formation of an intermediate with two potential leaving groups, an amide anion R_2N^- and the aluminate anion $(OAlH_3)^{2-}$. Aluminate is the better leaving group, and its loss produces an iminium cation that is also subject to further reduction. This gives us the **amine** product.

Although at first glance the behaviour of some of these carbonyl compounds towards nucleophiles might seem anomalous, closer consideration shows there is a logical explanation for the reactions observed. Furthermore, if we understand the underlying mechanisms, these reactions become predictable.

7.12 Carbon as a nucleophile: Grignard reagents

The reaction of carbon nucleophiles derived from organometallics with carboxylic acid derivatives follows closely the reactions we have already encountered in Sections 6.3.2 and 7.6.2. Organometallics such as **Grignard reagents** are conveniently regarded as sources of carbanion equivalents, and these add to the carbonyl, followed by loss of the leaving group. As with other examples, a tetrahedral anionic complex with the metal is likely to be produced. Regeneration of the carbonyl with loss of the leaving group produces an intermediate **ketone**.

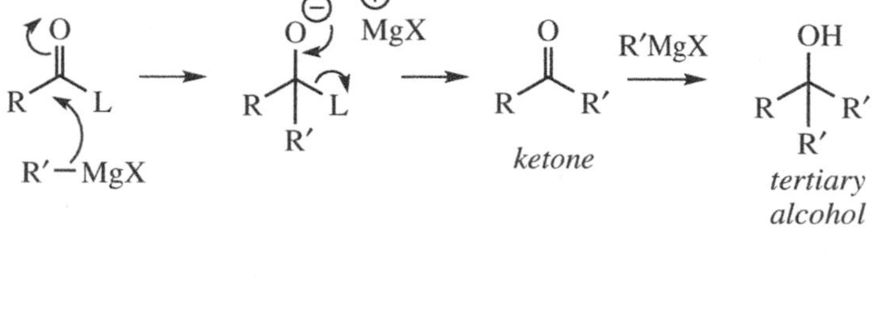

ketone

tertiary alcohol

Now we see an analogy with the LAH reduction sequence (see Section 7.11), in that this ketone intermediate also reacts with the organometallic reagent, rather more readily than the initial carboxylic acid derivative, so that this ketone cannot usually be isolated. The final product is thus a **tertiary alcohol**, which contains two alkyl or aryl groups from the organometallic reagent.

Note that derivatives of formic acid, HCO_2H, will be converted into **secondary alcohols** by this double reaction with a Grignard reagent.

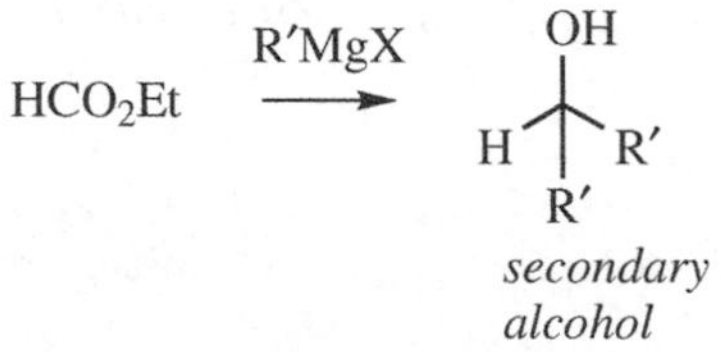

$$HCO_2Et \xrightarrow{R'MgX} \begin{array}{c} OH \\ | \\ H - C - R' \\ | \\ R' \end{array}$$

secondary alcohol

7.13   Nucleophilic substitution on derivatives of sulfuric and phosphoric acids

Phosphorus and sulfur are immediately below nitrogen and oxygen in the periodic table; therefore, we might expect them to have properties akin to nitrogen and oxygen. This is true in principle, so that PH_3 and H_2S are going to be analogues of NH_3 and H_2O. We have already met a number of sulfur derivatives, and have seen how thiols can be considered to behave in much the same way as alcohols (see Sections 6.3.2 and 7.4). However, a major difference that is encountered with phosphorus and sulfur arises from the fact that both have the ability to accommodate more than eight electrons in the outer electron shell. They are able to make use of d orbitals in bonding, and this leads to a greater versatility in bonding and a range of valencies is available to them. In the common acids phosphoric acid H_3PO_4 and sulfuric acid H_2SO_4, phosphorus is pentavalent and sulfur is hexavalent. Interestingly, we now find that these atoms have more in common with carbon than with nitrogen and oxygen. The chemical reactivity of organic derivatives of phosphoric and sulfuric acids in most aspects parallels that of carboxylic acid derivatives, so this is a particularly convenient place to describe some of the reactions, and to emphasize the similarities.

Most reactions of sulfuric and phosphoric acid derivatives can be rationalized by considering that the S=O and P=O functionalities are equivalent to the carbonyl group, and that polarization in these groups allows similar nucleophilic reactions to occur. Initial nucleophilic addition will then be followed by loss of an appropriate leaving group and regeneration of the S=O or P=O.

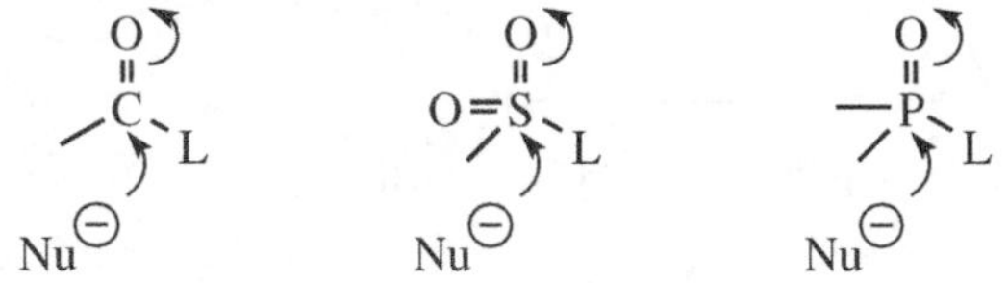

7.13.1   Sulfuric acid derivatives

Sulfuric acid can form ester derivatives with alcohols, though since it is a dibasic acid ($pK_a - 3, 2$) it can form both mono- and di-esters. Thus, acid-catalysed reaction of methanol with sulfuric acid gives initially methyl hydrogen sulfate, and with a second mole of alcohol the diester dimethyl sulfate. Though not shown here, the mechanism will be analogous to the acid-catalysed formation of carboxylic acid esters (see Section 7.9).

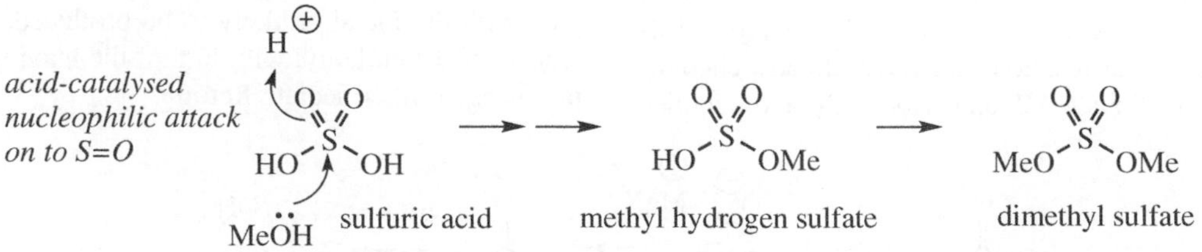

acid-catalysed nucleophilic attack on to S=O

sulfuric acid   methyl hydrogen sulfate   dimethyl sulfate

The simple diesters **dimethyl sulfate** and **diethyl sulfate** are convenient and useful reagents for alkylation reactions. As derivatives of sulfuric acid, the alkyl sulfate anions are also the conjugate bases of strong acids, and are consequently good leaving groups (see Section 6.1.4).

S$_N$2 nucleophilic attack

dimethyl sulfate

alkylation of nucleophile

good leaving group through resonance stabilization

Table 7.4   Comparisons of carboxylic and sulfonic acid derivatives

	Carboxylic acid derivative		Sulfonic acid derivative
RCO.OH	Carboxylic acid	RSO$_2$.OH	Sulfonic acid
RCO.Cl	Acyl chloride	RSO$_2$.Cl	Sulfonyl chloride
RCO.OR'	Carboxylate ester	RSO$_2$.OR'	Sulfonate ester
RCO.NH$_2$	Amide	RSO$_2$.NH$_2$	Sulfonamide

Compounds that are even better analogues of carboxylic acids are produced when an alkyl or aryl group replaces one of the hydroxyls in sulfuric acid. This provides compounds called **sulfonic acids**, which in turn give rise to a range of derivatives exactly comparable to those we have met as carboxylic acid derivatives (Table 7.4).

As with the carboxylic acid group, the reactivity of these sulfonic acid derivatives may be predicted from the properties of the leaving group, and sulfonyl chlorides are the most reactive (see

Section 7.8). Other classes of derivatives are thus most conveniently prepared from the sulfonyl chloride. Reaction with an alcohol leads to formation of a sulfonate ester. Two common sulfonyl chloride reagents employed to make sulfonate esters from alcohols are ***p*-toluenesulfonyl chloride**, known as **tosyl chloride**, and **methanesulfonyl chloride**, known as **mesyl chloride** (see Section 6.1.4). Note the nomenclature tosyl and mesyl for these groups, which may be abbreviated to Ts and Ms respectively.

tosyl chloride
(*p*-toluenesulfonyl chloride)
TsCl

tosyl (Ts)

mesyl chloride
(methanesulfonyl chloride)
MsCl

mesyl (Ms)

The reaction of tosyl chloride with an alcohol is easily represented by the standard nucleophilic

substitution sequence, and gives a sulfonate ester called a **tosyl ester**.

tosyl chloride
(*p*-toluenesulfonyl chloride)
TsCl

tosyl ester
R–OTs

Tosyl esters are good alkylating agents, rather like dimethyl sulphate above, and for the same reason, i.e. the presence of a good resonance-stabilized leaving group, the tosylate anion. This is the conjugate base of *p*-toluenesulfonic acid, a strong acid (pK_a − 1.3).

Thus, tosyl chloride may be used to facilitate **nucleophilic substitutions**.

Hydroxide is a poor leaving group, and nucleophilic reactions on alcohols are not particularly favourable unless acidic conditions are used to protonate the hydroxyl and produce a better leaving group, namely water (see Section 6.1.4). An alternative is to convert the alcohol into its tosylate ester using tosyl chloride and then carry out the

S_N2 reaction

not favoured;
hydroxide poor leaving group

favoured;
tosylate excellent leaving group

E2 reaction

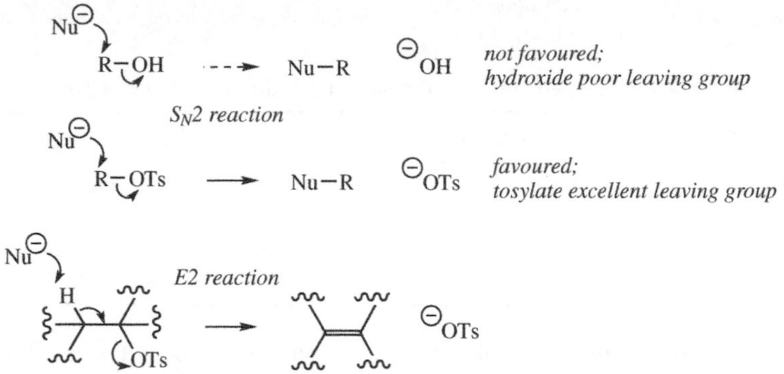

Mesyl chloride may be employed in exactly the same manner as tosyl chloride. Methanesulfonic acid is also a strong acid ($pK_a - 1.2$).

Using amines as nucleophiles, sulfonyl chlorides are readily converted into sulfonamides, exemplified here by the formation of *p*-aminobenzenesulfonamide (sulfanilamide).

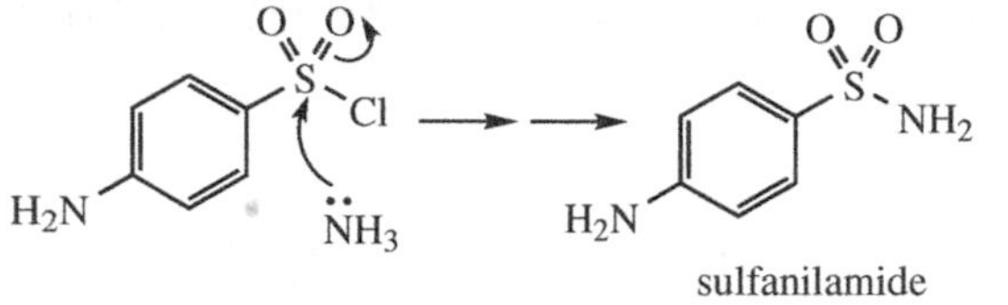

sulfanilamide

It is relatively easy to predict many properties of sulfonamides just by thinking about the

nucleophilic substitution on this ester, where there is now an excellent leaving group in the tosylate anion. This strategy may also be used to facilitate **elimination reactions** by providing a better leaving group.

corresponding amides. For example, just as amide nitrogens are not basic, so the sulfonamide nitrogen is not basic; indeed, it is more likely to be acidic (see Section 4.5.4). This is because of resonance stabilization involving the nitrogen lone pair feeding back towards the oxygen. Resonance is also responsible for stabilizing the anion resulting from loss of a proton from this nitrogen. Thus, pK_a values for sulfanilamide are 2.0 and 10.5. The 2.0 value relates to the aromatic amino group, which is somewhat less basic than aniline (pK_a 4.6; see Section 4.5.4) due to the contribution from the electron-withdrawing *para* sulfonyl group. The sulfonamide amine is rather more acidic than a carboxylic amide (pK_a about 18; see Section 10.7), a feature of the enhanced resonance stabilization conferred by two S=O systems.

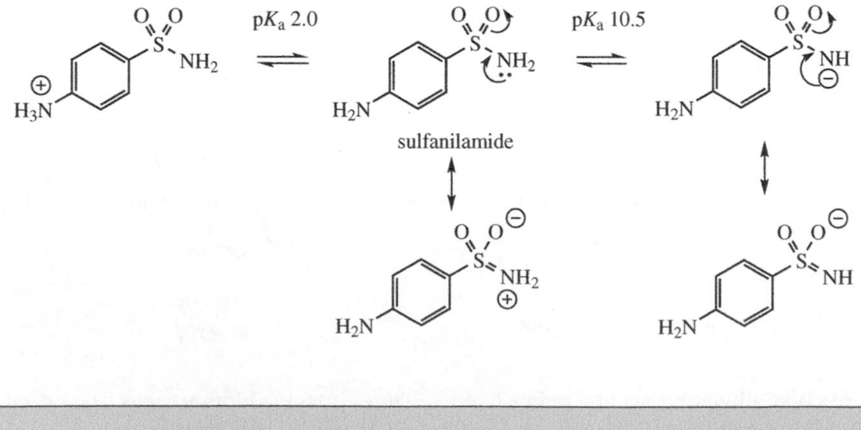

pK_a 2.0 pK_a 10.5

sulfanilamide

Box 7.23
Sulfonamide antibiotics and diuretics

Sulfanilamide was the first of a range of synthetic antibacterial drugs known collectively as **sulfa drugs**. These agents are antibacterial because they mimic in size, shape, and polarity the carboxylic acid *p*-aminobenzoic acid.

p-Aminobenzoic acid is used by bacteria for the synthesis of **folic acid**, and sulfanilamide acts as a competitive inhibitor of an enzyme involved in folic acid biosynthesis (see Box 11.13).

sulfanilamide is an inhibitor of the enzyme that incorporates p-aminobenzoic acid into the folic acid structure

sulfanilamide

a pteridine

p-amino-
benzoic acid
(PABA)

L-glutamic acid

folic acid

Folic acid is vital for both humans and bacteria. Bacteria synthesize this compound, but humans are unable to synthesize it and, consequently, obtain the necessary amounts from the diet, principally from green vegetables and yeast. This allows selectivity of action. Therefore, sulfa drugs are toxic to bacteria because folic acid biosynthesis is inhibited, whereas they produce little or no ill effects in humans. The structural relationships between carboxylic acids and sulfonic acids that we have observed in rationalizing chemical reactivity are now seen to extend to some biological properties.

The use of sulfa drugs as antibacterial drugs has diminished over the years as even better agents have been discovered, but in their time they were crucial to medical health. An interesting additional property of many sulfa drugs has been developed further, however. Many sulfonamides display diuretic activity, and sulfonamide diuretics are still a major drug group. Two widely used examples are **furosemide** and **bendroflumethazide**. Furosemide (frusemide) is an aromatic sulfonamide that can be seen to be a structural variant on sulfanilamide. Bendroflumethazide (bendrofluazide) also contains an aromatic sulfonamide grouping, but in addition contains a second sulfonamide group as part of a ring system (compare lactams, section 7.10). This drug is a member of the thiazide diuretics, so named because of the ring system containing this cyclic sulfonamide.

acyclic sulfonamide *cyclic sulfonamide*

furosemide / frusemide
(sulfonamide diuretic)

bendroflumethiazide / bendrofluazide
(thiazide diueretic)

7.13.2 Phosphoric acid derivatives

Derivatives of phosphoric acid are of particular significance in biochemical reactions, in that many metabolic intermediates are phosphates. The phosphate group introduces polarity, makes the compound water soluble, and provides a group that facilitates binding to proteins, especially enzymes. Phosphoric acid (pK_a 2.1) is a weaker acid than sulfuric acid ($pK_a - 3$), but stronger than a typical carboxylic acid (pK_a 5). It is also a tribasic acid; ionizations and pK_a values are as shown. It follows that, at pH 7, there will be considerable ionization, and by application of the Henderson–Hasselbalch equation (see Section 4.9) the major species at pH 7 can also be determined.

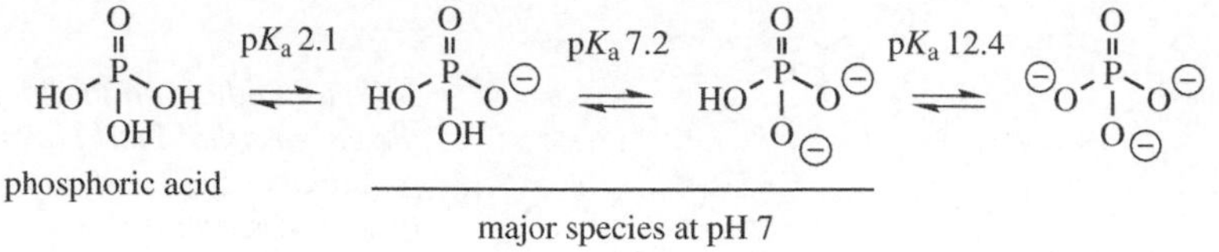

phosphoric acid

major species at pH 7

As a tribasic acid it has three replaceable hydroxyls, so that mono-, di-, and tri-substituted derivatives are possible.

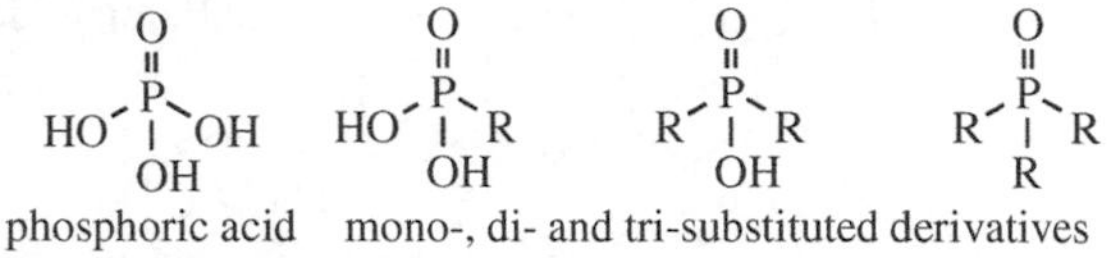

phosphoric acid   mono-, di- and tri-substituted derivatives

Box 7.24

The nucleic acids DNA and RNA feature diesters of phosphoric acid

Whilst many biochemicals are mono-esters of phosphoric acid, the nucleic acids DNA and RNA (see Section 14.2) provide us with good examples of diesters. A short portion of one strand of a DNA molecule is shown here; the most significant difference in RNA is the use of ribose rather than deoxyribose as the sugar unit.

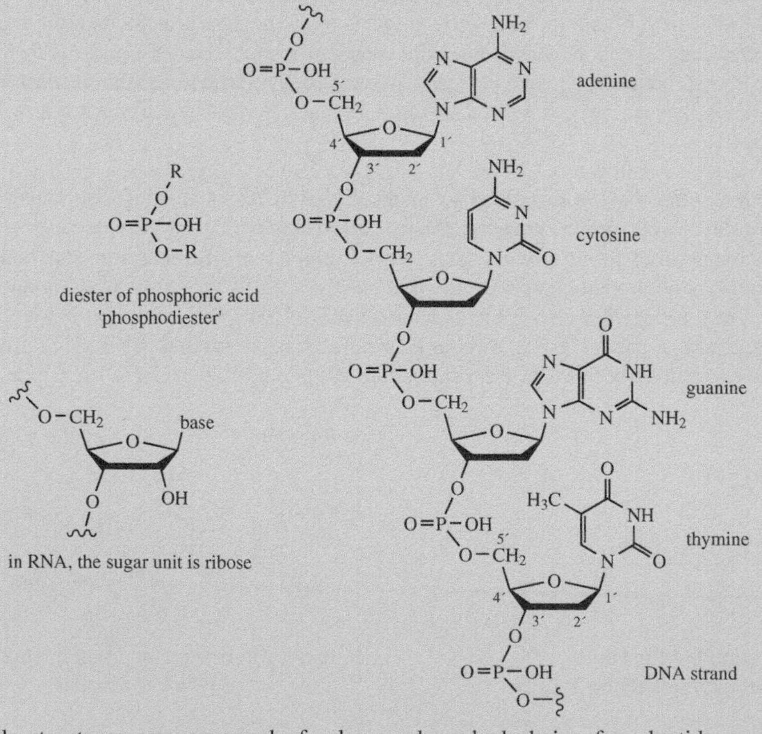

diester of phosphoric acid
'phosphodiester'

in RNA, the sugar unit is ribose

adenine

cytosine

guanine

thymine

DNA strand

These remarkable structures are composed of a long unbranched chain of nucleotide monomeric units (see Section 14.1). A nucleotide is a combination of three parts, a heterocyclic base, a sugar, and phosphate. The nucleotides are linked together via the phosphate group, which joins the sugar units through ester linkages, usually referred to as phosphodiester bonds. The phosphodiester bond links the 5′ position of one sugar with the 3′ position of the next.

A feature of phosphoric acid is that it forms a series of polymeric anhydrides that resemble carboxylic acid anhydrides in structure and reactivity. **Diphosphoric acid** (formerly called pyrophosphoric

acid) and **triphosphoric acid** are the simplest examples, and derivatives of these are the ones we meet in biochemistry.

phosphoric acid

diphosphoric acid
(pyrophosphoric acid)

carboxylic acid anhydride

triphosphoric acid

Since phosphoric acid, diphosphoric acid, and triphosphoric acid are reasonably strong acids, their anions are good leaving groups, and biochemical reactions frequently exploit this leaving group capacity. Phosphate derivatives retaining one or more unsubstituted hydroxyls will usually be significantly ionized at physiological pHs, so that these compounds will be water soluble, which is an important property for substrates in metabolic processes. When we draw the structures of phosphate derivatives in metabolic transformations, we should strictly show these compounds as anions, but, in general, the additional negative charges complicate the structures and interfere with our understanding of mechanistic electron movements. As a result, non-ionized acids may be shown in order to simplify structures and mechanisms and avoid the need for counter-ions; this is the convention we shall use. It is also very common to see abbreviations for phosphate-based structures, such as **OP** for phosphate, and **OPP** for diphosphate, which are convenient to use when mechanisms do not involve the P=O system. When writing such phosphates, drawing a ring round the P is a speedy and accepted way of abbreviating the structure.

a phosphate

a diphosphate (pyrophosphate)

ROP

ROPP

RO(P)

RO(P)(P)

Nucleophilic reactions on phosphate derivatives follow the general mechanisms seen with carboxylic acid derivatives, namely initial attack on to the P=O double bond followed by loss of the leaving group. In the following example, we employ a diphosphate system as the electrophile. Note that there are two types of linkage in this diphosphate, i.e. an anhydride and an ester. Nucleophilic attack results in cleavage of the anhydride bond (phosphate is a good leaving group) and not the ester bond (RO⁻ is a poor leaving group). Nucleophilic attack followed by cleavage of the anhydride bond could also result if the alternative P=O was the electrophile. This is an equally valid mechanism, but it is not as common in enzyme-controlled reactions as attack on the terminal phosphate.

nucleophilic attack
on to P=O

phosphorylation of
nucleophile

resonance stabilization
of leaving group

Box 7.25

Adenosine triphosphate

One of the most important molecules in biochemical metabolism is **adenosine triphosphate (ATP)**. Hydrolysis of ATP to adenosine diphosphate (ADP) liberates energy, which can be coupled to energy-requiring processes. Alternatively, synthesis of ATP from ADP can be coupled to energy-releasing processes. ATP thus provides nature with a molecule for energy storage; we also consider it to be the currency unit for energy (see Section 15.1.1).

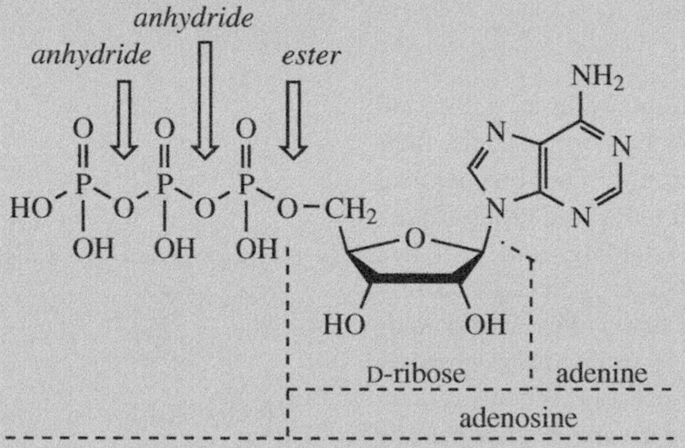

adenosine triphosphate; ATP

Hydrolysis of ATP to ADP is rationalized as nucleophilic attack of water on to the terminal P=O double bond, followed by cleavage of the anhydride bond and expulsion of ADP as the leaving group.

Ad = adenosyl

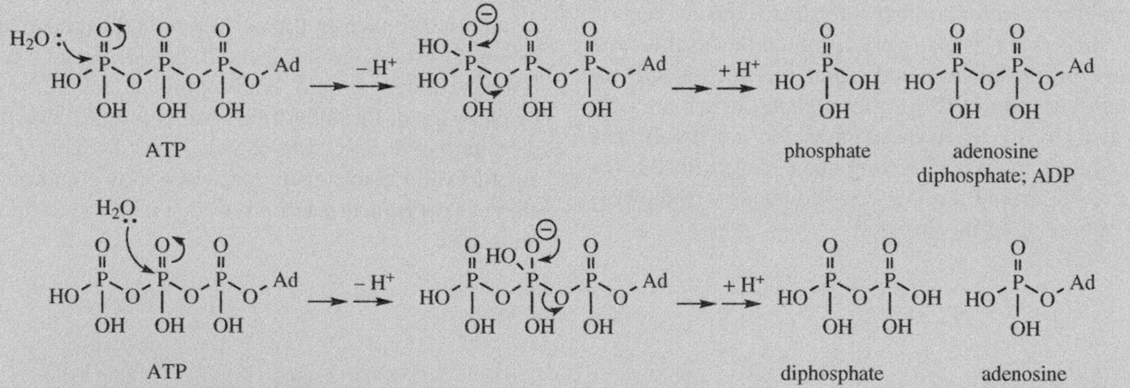

Note that there are two anhydride linkages in ATP, and one ester linkage. We know that hydrolysis of anhydride bonds is more favourable than hydrolysis of ester bonds because of the nature of the leaving group (see Section 7.8). In the enzyme-controlled reaction, nucleophilic attack usually occurs on the terminal P=O (hydrolysis of ATP to ADP), but very occasionally we encounter attack on the central P=O (hydrolysis of ATP to adenosine monophosphate, AMP). Both reactions yield the same amount of energy, $\Delta G = -34$ kJ mol^{-1}. This is not surprising, since the same type of bond is being hydrolysed in each case. The further hydrolysis of AMP to adenosine breaks an ester linkage and would liberate only a fraction of the energy, $\Delta G = -9$ kJ mol^{-1}, so this reaction is not biochemically important.

Box 7.26
Inhibitors of acetylcholinesterase

The neurotransmitter **acetylcholine** is both a quaternary ammonium compound (see Box 6.7) and an ester. After interaction with its receptor, acetylcholine is normally degraded by hydrolysis in a reaction catalysed by the enzyme **acetylcholinesterase**. This enzyme contains a serine residue that acts as the nucleophile, hydrolysing the ester linkage in acetylcholine (see Box 13.4). This effectively acetylates the serine hydroxyl, and is an example of transesterification (see Section 7.9.1). For continuation of acetylcholine degradation, the original form of the enzyme must be regenerated by a further ester hydrolysis reaction.

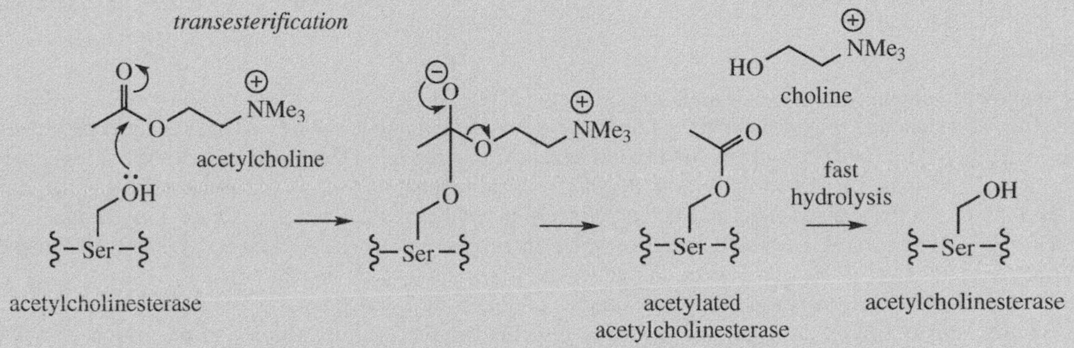

Acetylcholinesterase is a remarkably efficient enzyme; turnover has been estimated as over 10,000 molecules per second at a single active site. This also makes it a key target for drug action, and acetylcholinesterase inhibitors are of considerable importance. Some natural and synthetic toxins also function by inhibiting this enzyme. The natural alkaloid **physostigmine** (eserine) and its synthetic analogue **neostigmine** inhibit acetylcholinesterase by forming a covalent intermediate that is hydrolysed very much more slowly than is the normal substrate. These drugs are carbamoyl esters rather than acetyl esters.

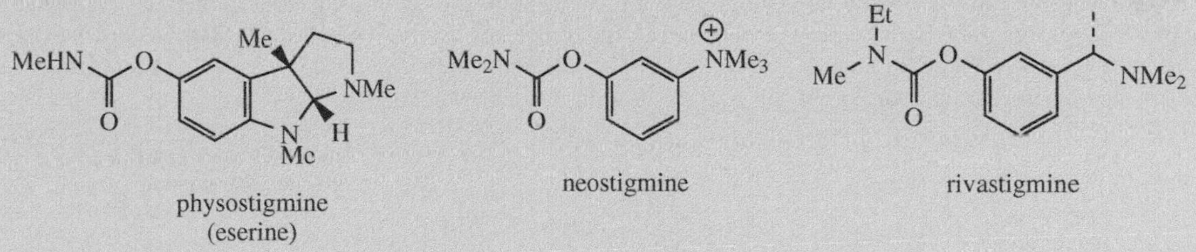

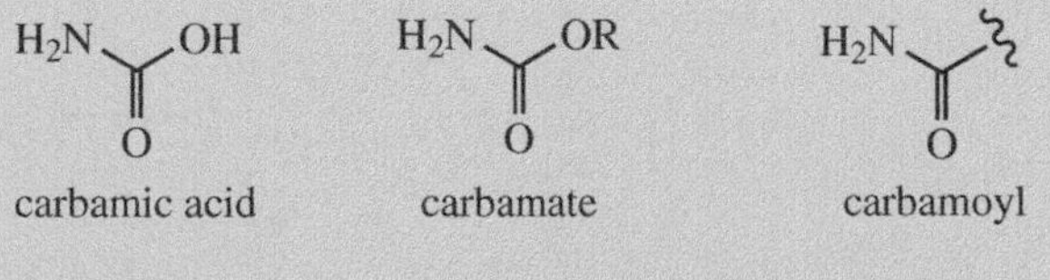

The carbamoyl group is transferred to the serine hydroxyl in the enzyme, but the resultant carbamoyl–enzyme intermediate then hydrolyses only very slowly (minutes rather than microseconds), effectively blocking the active site for most of the time. The slower rate of hydrolysis of the serine carbamate ester is a consequence of decreased carbonyl character resulting from resonance stabilization, as shown.

Box 7.26 (continued)

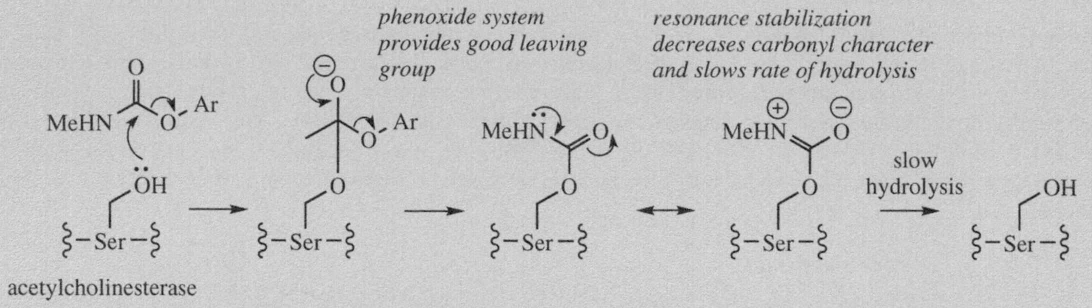

phenoxide system provides good leaving group

resonance stabilization decreases carbonyl character and slows rate of hydrolysis

acetylcholinesterase

By markedly slowing down the degradation of acetylcholine, these drugs are used to prolong the effects of endogenous acetylcholine. Physostigmine and neostigmine have ophthalmic use as a miotic, contracting the pupil of the eye, often to combat the effects of mydriatics such as atropine. They could be used as an antidote to anticholinergic poisons, such as atropine (see Box 10.9), or to reverse the effects of muscle relaxants that block acetylcholine receptors, such as tubocurarine and atracurium (see Boxes 6.7 and 6.9). Other acetylcholinesterase inhibitors, e.g. rivastigmine, are found to be of value in treating Alzheimer's disease, by increasing memory function.

Many phosphorus derivatives function as irreversible inhibitors of acetylcholinesterase, and are thus potentially toxic. These include a range of **organophosphorus insecticides**, such as **malathion** and **parathion**, and nerve gases such as **sarin**.

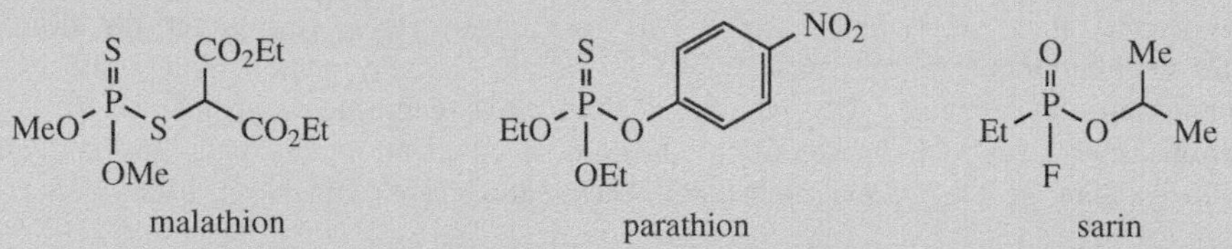

malathion   parathion   sarin

In contrast to the inhibitors such as neostigmine and related compounds described above, where the intermediate complexes hydrolyse slowly, these toxic compounds form complexes that do not hydrolyse. The enzyme becomes irreversibly bound to the toxin and, as a result, ceases to function. These agents all have leaving groups that can be displaced by the serine hydroxyl of the enzyme, leading to stable addition products.

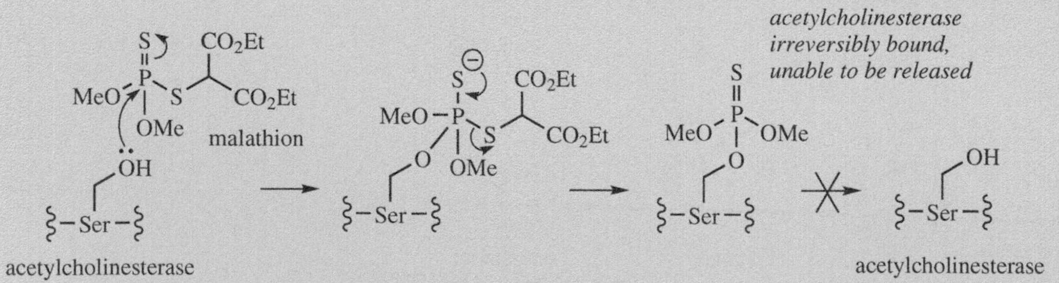

acetylcholinesterase irreversibly bound, unable to be released

acetylcholinesterase   acetylcholinesterase

Malathion and parathion contain a P=S grouping, exemplifying a further carbonyl analogue, in which phosphorus replaces carbon, and sulfur replaces oxygen. Nevertheless, the same type of chemistry occurs, in which the serine hydroxyl of the insect's acetylcholinesterase attacks this P=S electrophile, followed by expulsion of the leaving group, here a thiolate. The esterified enzyme, however, is not hydrolysed back to the original form of the enzyme, and its action is thus totally inhibited. The insect becomes subjected to a build up of acetylcholine that eventually proves fatal. Malathion is much less toxic to mammals because of the two carboxylic ester functions that allow metabolism through hydrolysis to inactive products.

Even more reactive towards acetylcholinesterase are the organophosphorus derivatives developed as chemical warfare nerve agents, e.g. **sarin**. Such compounds react readily with the enzyme and form very stable addition intermediates. It is unusual to see fluoride as a leaving group, as in sarin, but its presence provides a huge inductive effect, thus accelerating the initial nucleophilic addition step (see also Section 13.7).

Box 7.27

Acylphosphates: mixed anhydrides of phosphoric and carboxylic acids

We are now familiar with anhydrides of carboxylic acids, e.g. acetic anhydride, and of phosphoric acid, e.g. ATP. In each case we can rationalize their reactivity by considering nucleophilic attack onto the C=O or P=O, followed by loss of a leaving group, carboxylate or a phosphate derivative.

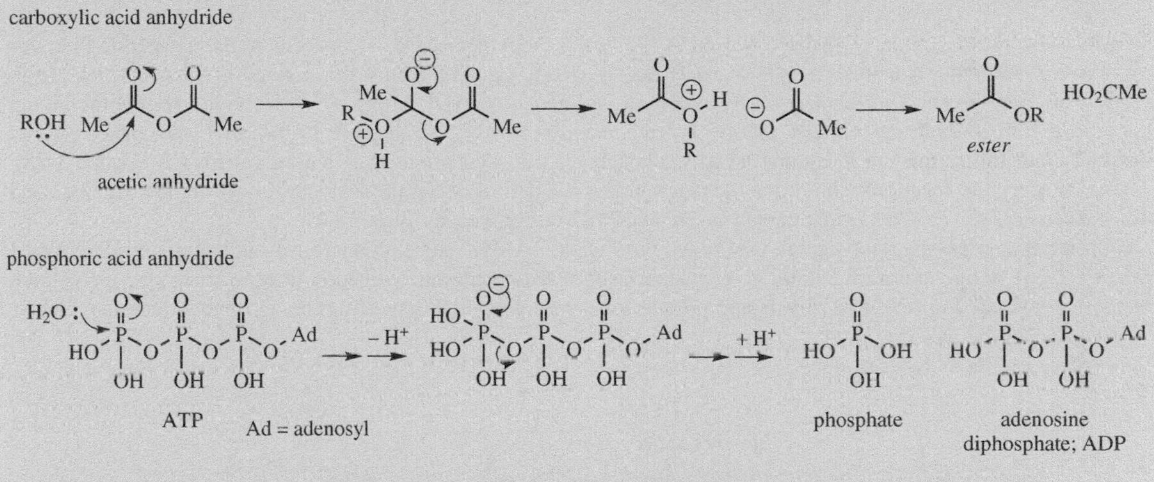

carboxylic acid anhydride

acetic anhydride

phosphoric acid anhydride

ATP   Ad = adenosyl

phosphate   adenosine diphosphate; ADP

As we consider biochemical processes in Chapter 15, we shall encounter some metabolic intermediates that are **acylphosphates**, i.e. hybrid mixed anhydrides of phosphoric and carboxylic acids. The simplest example of an acylphosphate is acetylphosphate, employed by some bacteria as an energy-rich metabolite.

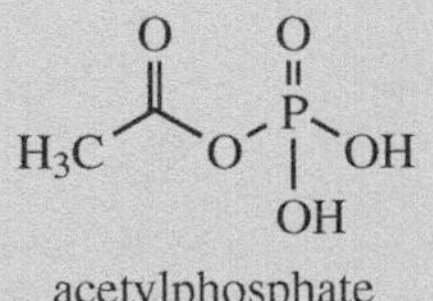

acetylphosphate

Examples of considerably more consequence are 1,3-diphosphoglyceric acid in the glycolytic pathway, and succinyl phosphate in the Krebs cycle. These compounds should not trouble us, since their reactivity is easily explained in terms of the above processes.

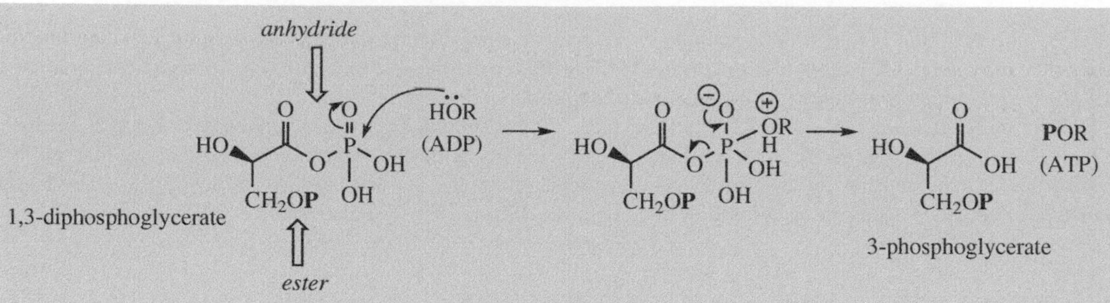

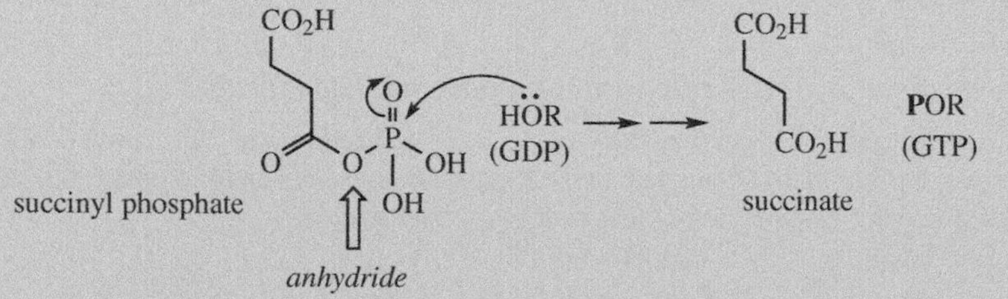

1,3-diphosphoglycerate

3-phosphoglycerate

succinyl phosphate   succinate

In both cases, the mixed anhydride is used to synthesize ATP from ADP. Hydrolysis of the anhydride liberates more energy than the hydrolysis of ATP to ADP and, therefore, can be linked to the enzymic synthesis of ATP from ADP. This may be shown mechanistically as a hydroxyl group on ADP acting as nucleophile towards the mixed anhydride, and in each case a new phosphoric anhydride is formed. In the case of succinyl phosphate, it turns out that GDP rather than ADP attacks the acyl phosphate, and ATP production is a later step (see Section 15.3). These are enzymic reactions; therefore, the reaction and the nature of the product are closely controlled. We need not concern ourselves why attack should be on the P=O rather than on the C=O.

Further examples of acylphosphates are found in fatty acyl-AMPs (see Section 15.4.1) and aminacyl-AMPs (see Section 13.5), activated intermediates in the metabolism of fatty acids and formation of peptides respectively. Each of these is attacked on the C=O by an appropriate S or O nucleophile, displacing the phosphate derivative AMP.

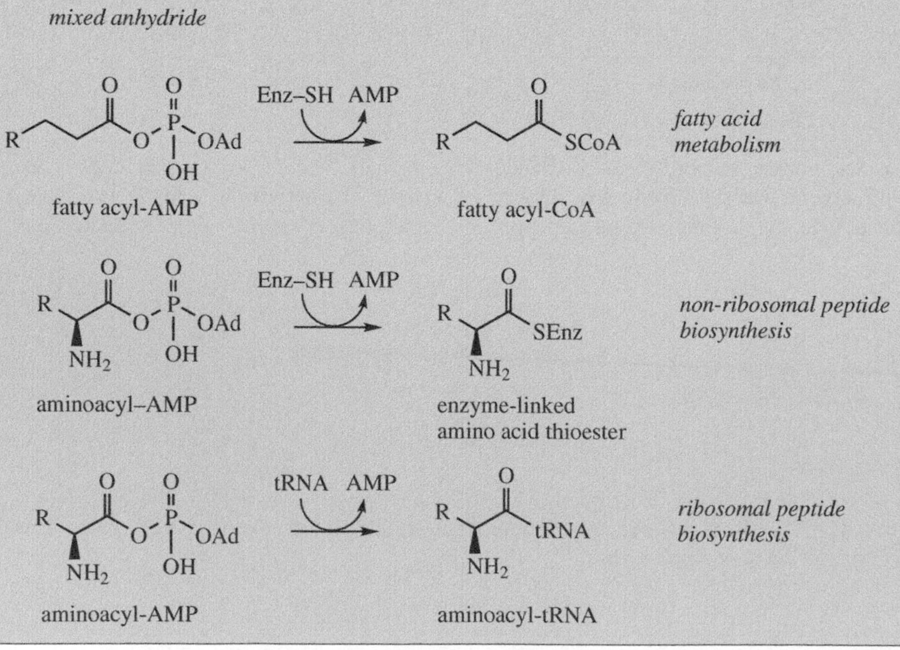

fatty acyl-AMP   fatty acyl-CoA

aminoacyl–AMP   enzyme-linked
                                                 amino acid thioester

aminoacyl-AMP   aminoacyl-tRNA

8

Electrophilic reactions

In the preceding chapters we have seen how new bonds may be formed between nucleophilic reagents and various substrates that have electrophilic centres, the latter typically arising as a result of uneven electron distribution in the molecule. The nucleophile was considered to be the reactive species. In this chapter we shall consider reactions in which electrophilic reagents become bonded to substrates that are electron rich, especially those that contain multiple bonds, i.e. alkenes, alkynes, and aromatics. The π electrons in these systems provide regions of high electron density, and electrophilic reactions feature as the principal reactivity in these classes of compounds. We term the reactions electrophilic rather than nucleophilic, since it is the electrophile that provides the reactive species.

8.1 Electrophilic addition to unsaturated carbon

The characteristic reaction of **alkenes** is **electrophilic addition**, in which the carbon–carbon π bond is replaced by two σ bonds.

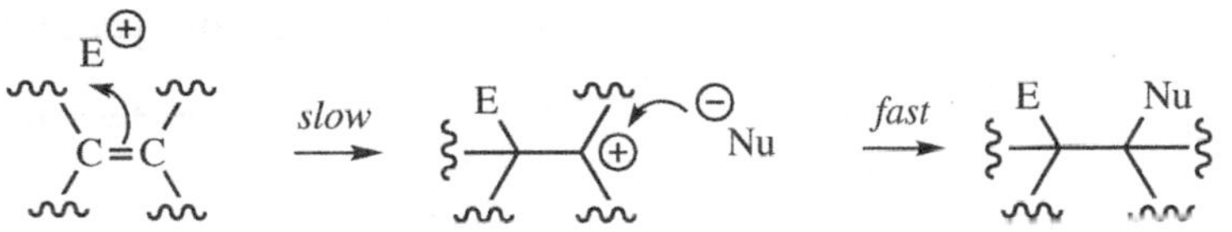

π electrons flow towards electrophile forming σ bond

resultant carbocation quenched by addition of nucleophile

addition of ENu across double bond

The π bond of an alkene results from overlapping of *p* orbitals and provides regions of increased electron density above and below the plane of the molecule. These electrons are less tightly bound than those in the σ bonds, so are more polarizable and can interact with a positively charged electrophilic reagent. This forms the first part of an electrophilic addition, in which the electrons are used to form a σ bond with the electrophile and leave the other carbon of the double bond electron deficient, i.e. it becomes a **carbocation**. This carbocation is then rapidly captured by a nucleophile, which donates its electrons to form the second new σ bond. This latter step is very much faster than the first step, and thus carbocation formation becomes the **rate-determining step** in this **bimolecular** reaction.

$$\text{Rate} = k[\text{alkene}][\text{ENu}]$$

where ENu is the electrophilic reagent and k is the rate constant.

The carbocation is an intermediate in the reaction sequence and corresponds to a minimum in the energy profile (see Section 5.4). Its formation

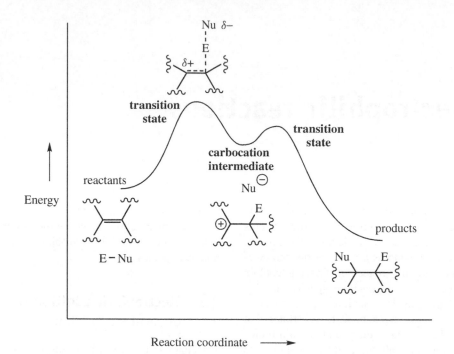

Figure 8.1   Energy profile: electrophilic addition reaction

depends upon producing a high-energy transition state in which there is substantial development of positive charge on carbon (Figure 8.1). The reactive carbocation requires a much smaller activation energy for reaction with the nucleophile.

8.1.1   Addition of hydrogen halides to alkenes

The addition of a **hydrogen halide**, e.g. HCl, to an **alkene** is a simple example of an electrophilic addition.

This type of synthetic reaction requires the use of gaseous hydrogen halide. If **aqueous acid** were employed, then, although water is not such a good nucleophile as halide, it is going to be the most abundant available nucleophile, and the predominant product will be the corresponding alcohol.

Since HCl will be completely dissociated in water, the electrophile in this case will be the hydronium ion, although the same carbocation will be produced. The reaction is completed by nucleophilic attack of water, followed by loss of a proton, thus regenerating the acid catalyst. The overall conversion thus becomes **hydration** of the alkene. This is an important industrial process, typically employing sulfuric acid, but it is seldom used in the laboratory because the yields are very dependent upon the conditions used, and better routes to alcohols are available.

Attack of the nucleophile onto a planar carbocation may take place from either face with equal probability, so that it is easy to see that, when a new chiral centre results, a **racemic product** will be formed, a similarity with S_N1 processes (see Section 6.2).

attack from upper face

(E)-but-2-ene will give the same product as (Z)-but-2-ene

HCl gas

(Z)-but-2-ene

attack from lower face

(E)-but-2-ene

Also, since the same carbocation intermediate will be formed, it can be deduced that the nature of the product will not be dependent upon the configuration of the double bond.

In discussing simple electrophilic additions, the example chosen was but-2-ene. This choice was deliberate, in that it is a symmetrical substrate and it makes no difference which carbon becomes bonded to the electrophile or nucleophile. When the substrate is not symmetrical, we must then consider the relative stabilities of the two carbocations that might be involved in the addition reaction.

2-methylbut-2-ene

favourable tertiary carbocation

major product

less favourable secondary carbocation

minor product

If we consider protonation of 2-methylbut-2-ene, then two different carbocations might be formed. One of these is tertiary, and thus favourable, because three electron-donating alkyl groups help to stabilize the cation by dispersing the charge (see Section 6.2.1). The alternative carbocation intermediate is less favourable, in that it is secondary, with just two alkyl groups helping to stabilize the carbocation. It follows that the tertiary carbocation is more likely to be formed, and that the predominant product will be the result of nucleophilic attack on this intermediate. It is likely that some of the alternative product will be produced, since secondary carbocations are reasonably stabilized and frequently produced in reactions.

Naturally, if the protonation step could lead to either a tertiary or a very unfavourable primary carbocation, then we would expect the product to be almost entirely the result of tertiary carbocation involvement.

favourable tertiary carbocation   *essentially sole product*

2-methylpropene

highly unfavourable primary carbocation

Long before any reaction mechanism had been deduced, **Markovnikov's rule** had been utilized to predict the **regiochemistry** for addition of HX to an unsymmetrical alkene. Markovnikov's rule states that addition of HX across a carbon–carbon multiple bond proceeds in such a way that the proton adds to the less-substituted carbon atom, i.e. that already bearing the greater number of hydrogen atoms. Since we now know that carbocation stability controls the regiochemistry of electrophilic addition, it is recommended that the more favoured product be predicted simply from an inspection of the possible carbocation intermediates. Alternatively, Markovnikov's rule should be restated in mechanistic terms, in that the electrophile adds to the double bond to form the more stable carbocation. In some circumstances, this generalization has appeared incorrect, and so-called anti-Markovnikov addition has been observed. Careful analysis of the reagents has shown that abnormal anti-Markovnikov addition of HX is the result of a radical reaction brought about by the presence of peroxides as radical initiators. This will be discussed further in Section 9.4.

The relative ease with which hydrogen halides react with alkenes is in the order HI > HBr > HCl > HF. This is the same as their relative acidities (see Section 4.3.2) and indicates that protonation of the alkene is the rate-limiting step for the addition reaction.

8.1.2 Addition of halogens to alkenes

Halogens such as **chlorine** (Cl_2) and **bromine** (Br_2) react readily with alkenes to produce 1,2-dihalogen derivatives. Although the halogen–halogen bond of Cl_2 and Br_2 is non-polar, it becomes polarized as it approaches the π electrons of the double bond. The electrons in the halogen–halogen σ bond become unequally shared, and are disturbed towards the atom furthest away from the polarizing double bond. As a result, the dihalogen functions as an electrophile, in much the same way as does HX.

π bond

π electrons cause polarization of dihalogen | *electrophilic addition with loss of bromide* | *lone pair of bromine atom interacts with resultant carbocation giving cyclic bromonium ion* | *rearside attack of bromide nucleophile* | *anti addition of two bromine atoms*

or

As the reactants get closer, there is a flow of electrons from the π bond to the nearer halogen, followed by departure of the further halogen as halide. This results in formation of a carbocation. In the next step, we see a significant difference in mechanism when compared with the addition of HX. Instead of the carbocation being quenched by attack of nucleophile, there is formation of a **cyclic halonium ion**. This is achieved by bonding of a lone pair of electrons from the large halogen atom to the carbocation, and it helps stabilize the cation by transferring the charge to the halogen.

However, the bridging halogen atom now blocks any further attack on the halogen-bonded face of the original double bond, so that when a nucleophile attacks it has to be from the opposite face. This means that there now has to be rearside attack to open the cyclic halonium ion, in a process resembling an S_N2 mechanism (see Section 6.1). Of course, either carbon might be attacked by the nucleophile, but the consequences are the same. The net result is formation of a 1,2-dihalo system, and, stereochemically, the halogen atoms have been inserted onto opposite faces of the double bond. This is described as **anti addition** (Greek: *anti* = against).

A mechanism in which groups become attached to the same face of the double bond would be termed

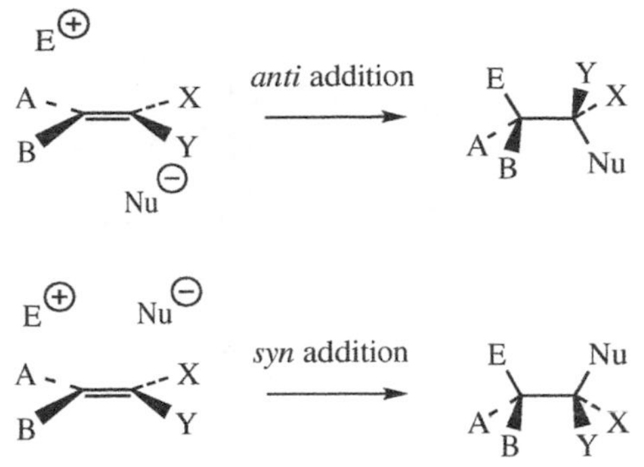

syn addition (Greek: *syn* = with). The observed addition of halogen with *anti* stereochemistry is thus different from the simpler addition of HX, where the initially formed carbocation may be attacked from either face by the nucleophile.

Bromine and chlorine both react via cyclic halonium cations, which we term **bromonium** and **chloronium cations** respectively. Fluorine and iodine are hardly ever used for halogenations; iodine is a rather unreactive halogenating agent, whereas at the other extreme, fluorine is too vigorous to give controllable reactions.

The **stereochemical consequences** of the electrophilic addition of, say, bromine to certain alkenes can be predicted as follows:

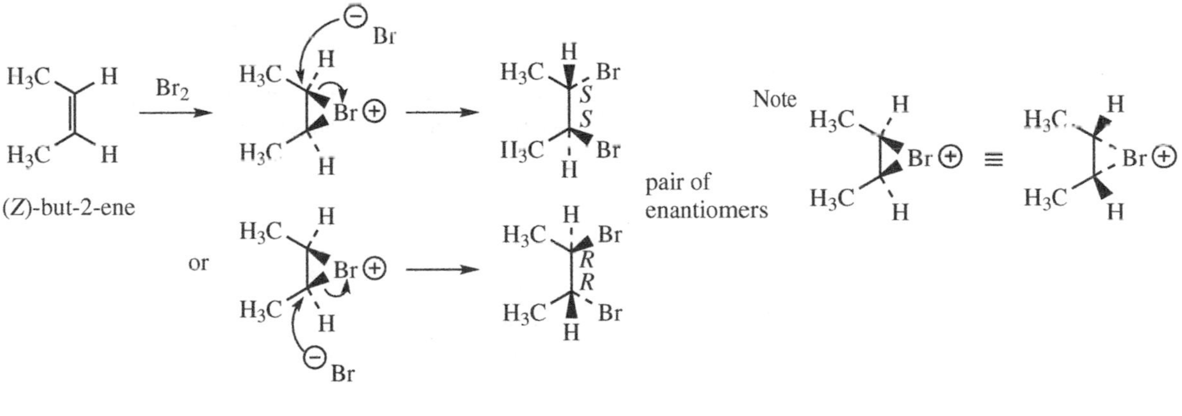

Thus, (Z)-but-2-ene will react to give 2,3-dibromobutane as a pair of enantiomers, R,R and S,S, a result of the *anti* addition. A racemic product will thus be formed, because there is equal probability of

nucleophilic attack at the two possible centres. Because of the symmetry in the molecules, it is only necessary to consider one bromonium ion, since the mirror image version is actually identical.

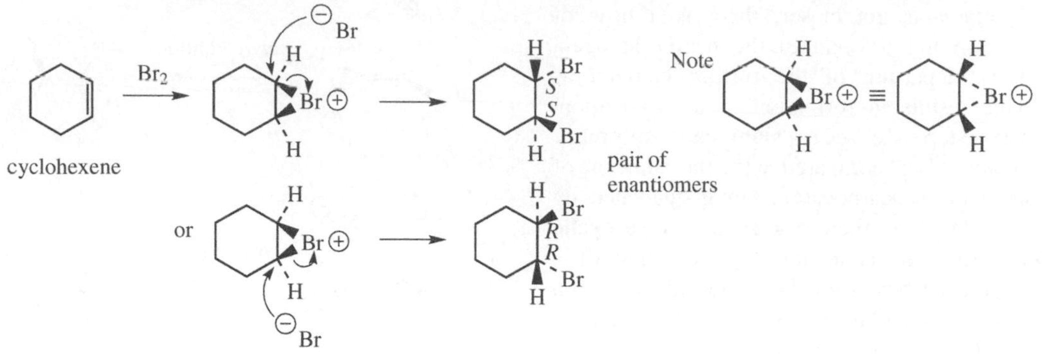

cyclohexene

pair of enantiomers

Note

Similarly, cyclohexene will form 1,2-dibromo-cyclohexane as a racemic product, again *R,R* and *S,S*. Note that the three-membered ring of the bromonium ion must be planar and can only be *cis*-fused to the cyclohexane ring (see Section 3.5.2).

When one considers bromination of (*E*)-but-2-ene, the product turns out to be the *meso R,S* isomer, i.e. a single product.

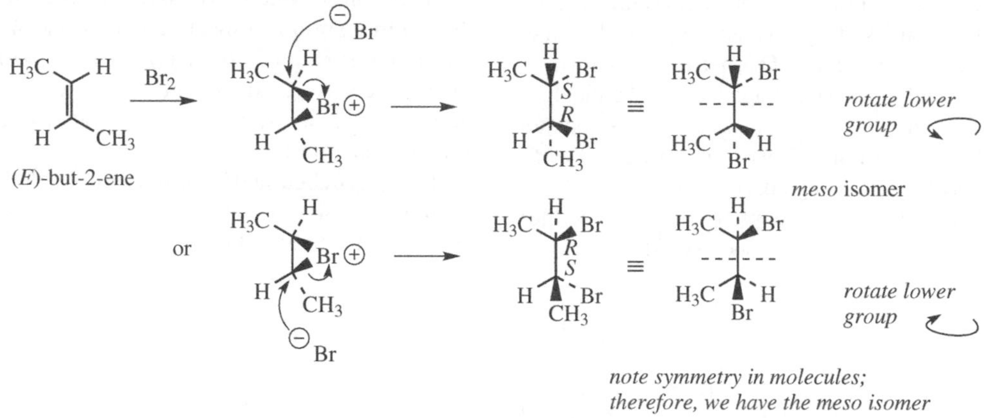

(*E*)-but-2-ene

rotate lower group

meso isomer

rotate lower group

note symmetry in molecules;
therefore, we have the meso isomer

Intriguingly, although there are going to be two different and enantiomeric bromonium ions for an unsymmetrical substrate such as (*Z*)-pent-2-ene, a

pair of enantiomeric products results, due to the two types of nucleophilic attack – try it!

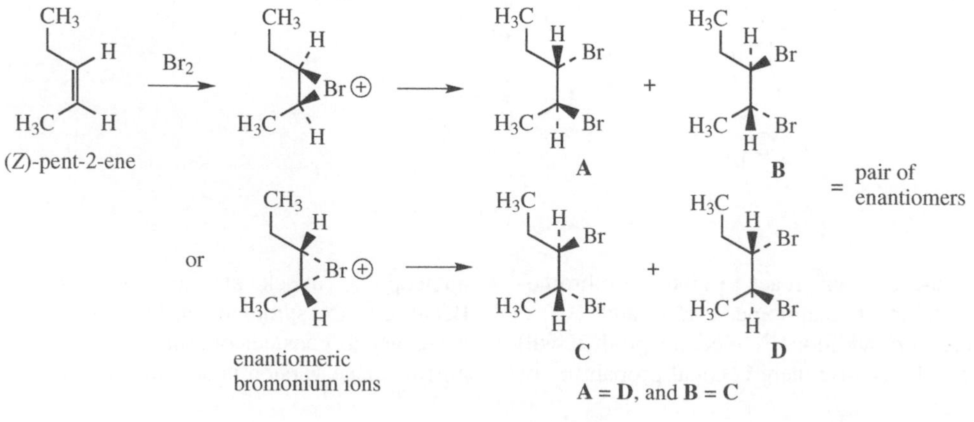

(*Z*)-pent-2-ene

enantiomeric bromonium ions

A   **B**   = pair of enantiomers

C   **D**

A = D, and **B = C**

For the HX additions above, we noted that, in aqueous solution, water would be the most abundant nucleophile, and the predominant product would thus be an alcohol derivative. A similar situation holds if we use aqueous bromine or chlorine, for example. The product is going to be a halo alcohol (**halohydrin**), and overall we are seeing the electrophilic *anti* addition of X^+ and HO^-. This is sometimes considered as the addition of a hypohalous acid HOX, but is much more easily rationalized in terms halogenation in the presence of water as the predominant nucleophile.

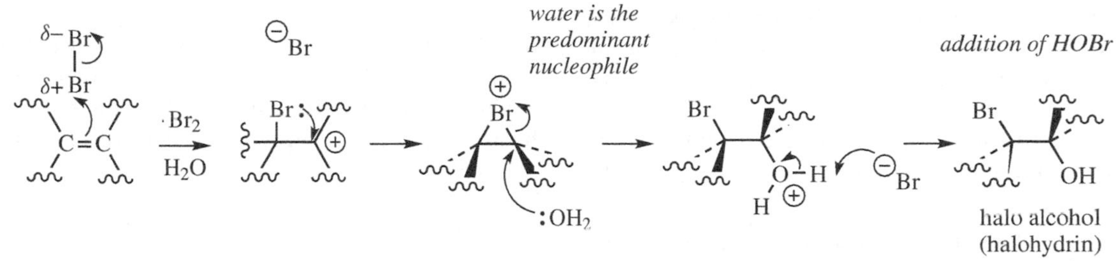

One of the properties of halo alcohols formed in this way is that they can be used to make **epoxides**, three-membered oxygen heterocycles. This is achieved by treatment with base, and an intramolecular S_N2 mechanism is involved (see Section 6.3.2).

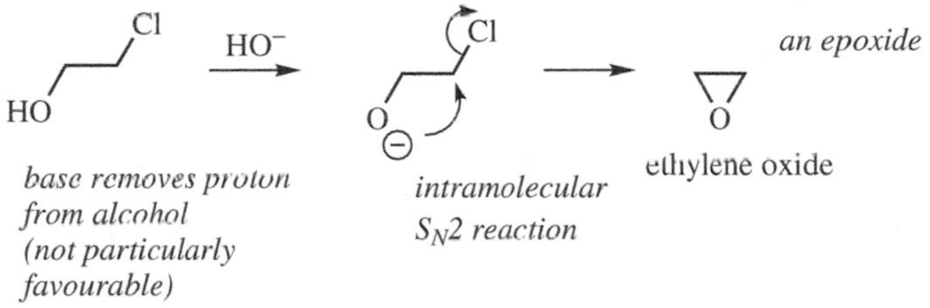

Note that the bromination–hydroxylation sequence is going to produce *anti* addition, and the groups are ideally set up for the S_N2 reaction with the necessary rearside attack (see section 6.1).

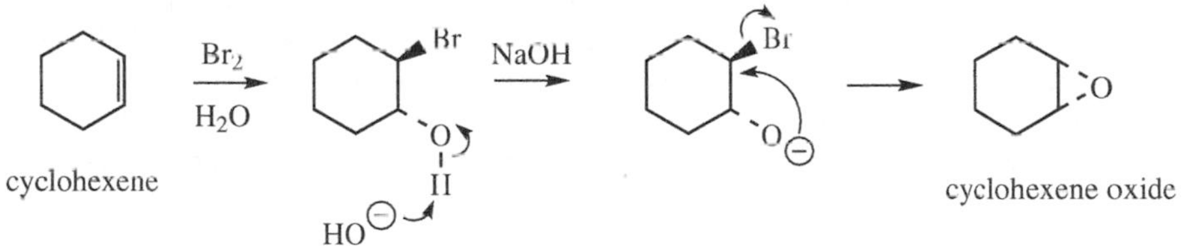

cyclohexene

cyclohexene oxide

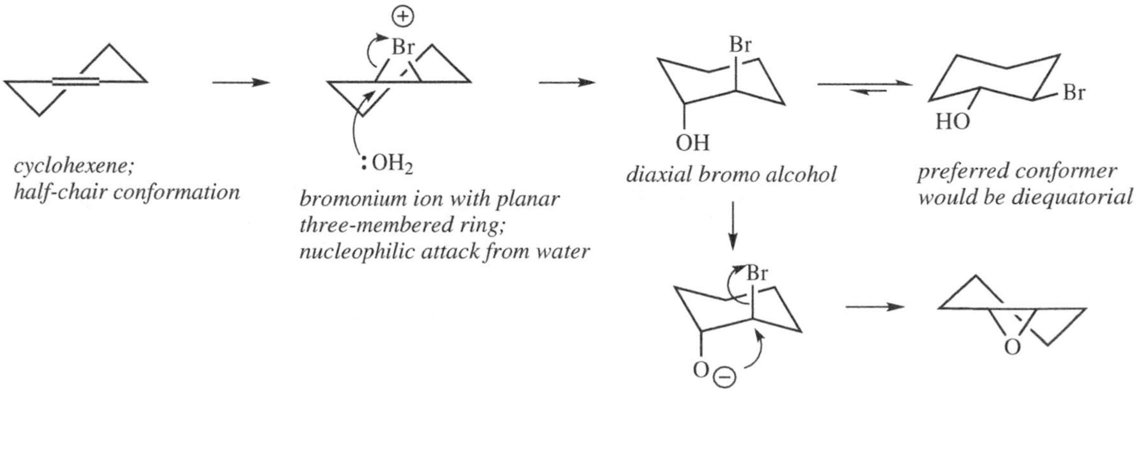

cyclohexene;
half-chair conformation

bromonium ion with planar
three-membered ring;
nucleophilic attack from water

diaxial bromo alcohol

preferred conformer
would be diequatorial

In the synthesis of cyclohexene oxide from cyclohexene shown, this does implicate the less favourable
diaxial conformer in the epoxide-forming step. Cyclohexene oxide contains a *cis*-fused ring system, the
only arrangement possible, since the three-membered
ring is necessarily planar (see Section 3.5.2).

Another method of making **epoxides** is the electrophilic reaction of alkenes with a peroxy acid such
as **peroxyacetic acid** (sometimes simply **peracetic
acid**). Thus, cyclohexene may be converted into the
epoxide in a single reaction.

Mechanistically, this is an electrophilic attack
involving the π electron system of the alkene and the

polarized O–O bond in the peroxy acid. One could
realistically suggest a potential carbocation intermediate, followed by nucleophilic attack of the
hydroxyl oxygen, using as precedent the examples
seen above.

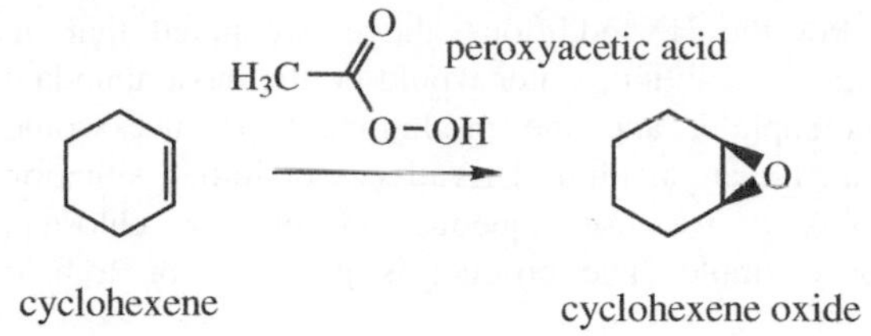

a logical, but apparently incorrect mechanism:

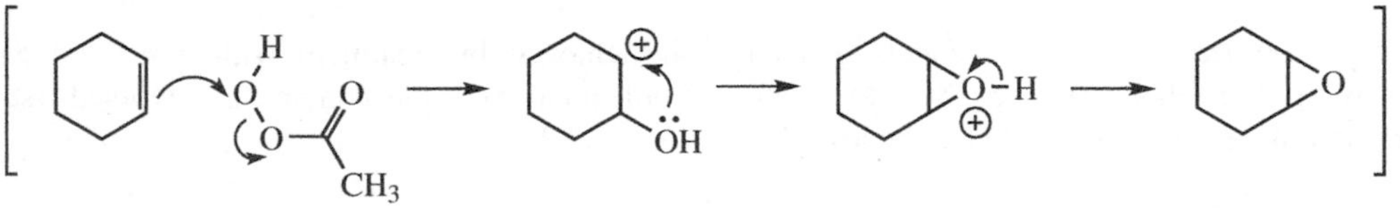

proposed cyclic mechanism

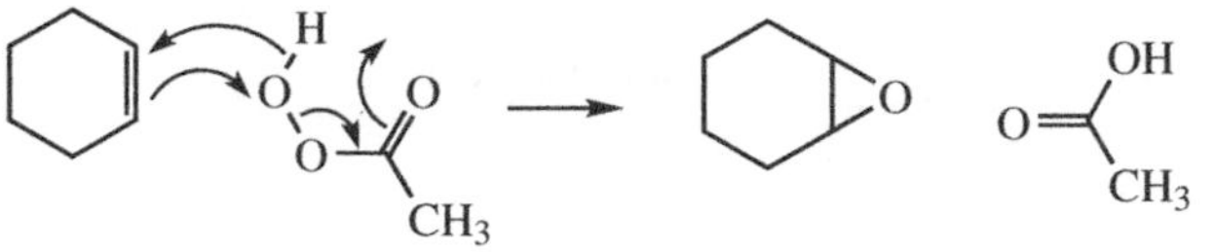

*the hydroxyl proton from
peroxyacetic acid actually ends
up in the acetic acid by-product*

However, since it is found that the hydroxyl proton
from peroxyacetic acid actually ends up in the acetic
acid by-product, a messy-looking cyclic mechanism
has been proposed. This starts with the nucleophilic
π bond attacking the peroxy acid oxygen, breaking
of the O–O bond to form a new carbonyl, with the
original carbonyl picking up the hydroxyl's hydrogen.
The remaining electrons from the hydroxyl are then
used to bond to the electrophilic carbon from the
original double bond.

Epoxides, like cyclic halonium ions, undergo
ring opening through rearside attack of nucleophiles
(see Section 6.3.2). Two mechanisms are shown,
for both basic and acidic conditions. Under acidic
conditions, protonation of the epoxide oxygen occurs
first. The epoxidation–nucleophilic attack sequence
also adds substituents to the double bond in an
anti sense.

basic conditions

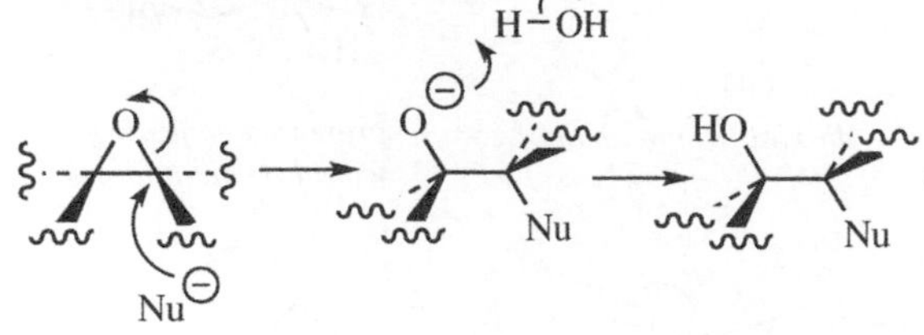

acidic conditions

*ring opening via rearside
nucleophilic attack*

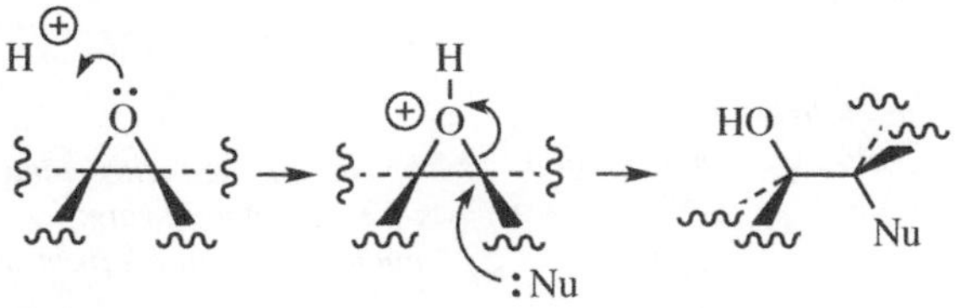

*protonation of epoxide oxygen
precedes rearside nucleophilic attack*

Box 8.1

Halohydrins: biological activity of semi-synthetic corticosteroids

Corticosteroids are produced by the adrenal glands, and display two main types of biological activity. **Glucocorticoids** are concerned with the synthesis of carbohydrate from protein and the deposition of glycogen in the liver. They also play an important role in inflammatory processes. **Mineralocorticoids** are concerned with the control of electrolyte balance, promoting the retention of Na^+ and Cl^-, and the excretion of K^+. Synthetic and semi-synthetic corticosteroid drugs are widely used in medicine. Glucocorticoids are primarily used for their antirheumatic and anti-inflammatory activities, and mineralocorticoids are used to maintain electrolyte balance where there is adrenal insufficiency.

The two groups of steroids share considerable structural similarity, and it is difficult to separate entirely the two types of activity in one molecule. Extensive synthetic effort has been applied to optimize anti-inflammatory activity, whilst minimizing the mineralocorticoid activity, which tends to produce undesirable side-effects. One modification that has proved particularly successful has been to create 9α-halo-11β-hydroxy compounds. The 11β-hydroxyl is present in all glucocorticoids and is known to be essential for activity; the introduction of the halogen atom at position 9 was a major development in this group of drugs. Halogenation was achieved as shown below.

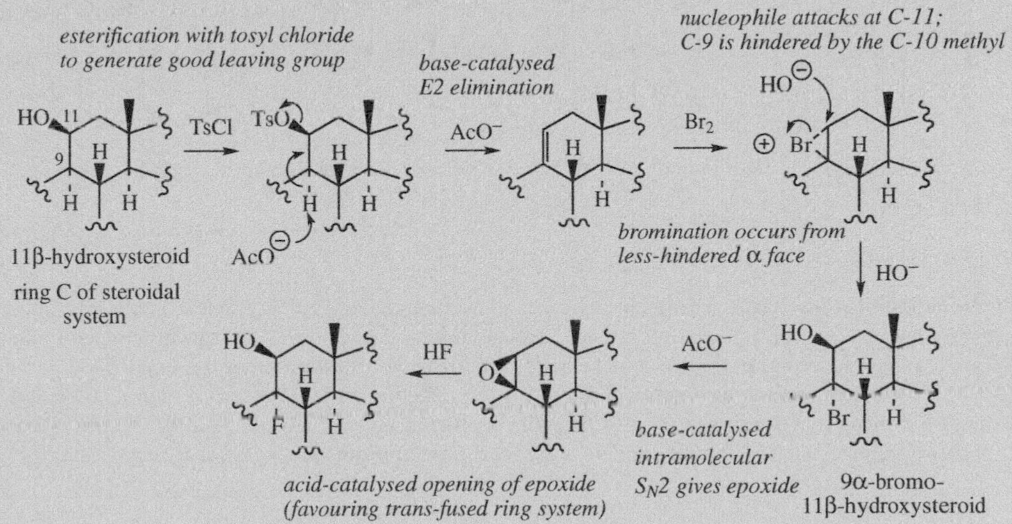
esterification with tosyl chloride to generate good leaving group
base-catalysed E2 elimination
nucleophile attacks at C-11; C-9 is hindered by the C-10 methyl
HO
11
9
H
TsCl
TsO
AcO⁻
Br₂
Br
11β-hydroxysteroid
ring C of steroidal system
AcO
bromination occurs from less-hindered α face
HO⁻
HF
AcO⁻
F
Br
acid-catalysed opening of epoxide (favouring trans-fused ring system)
base-catalysed intramolecular S$_N$2 gives epoxide
9α-bromo-11β-hydroxysteroid

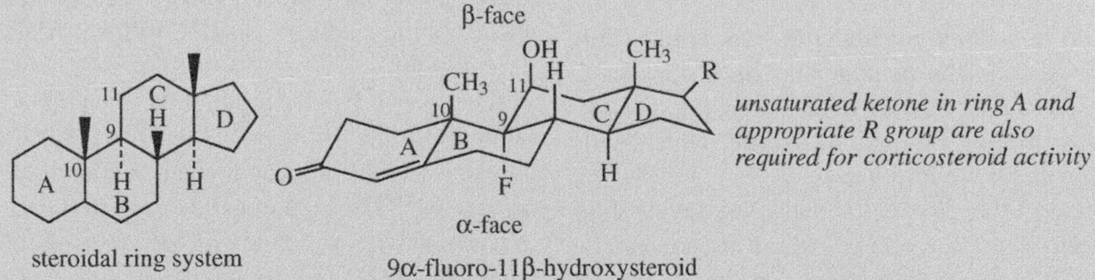
β-face
OH
CH₃
R
CH₃
11
10
9
A
B
C
D
H
F
O
α-face
unsaturated ketone in ring A and appropriate R group are also required for corticosteroid activity

steroidal ring system

9α-fluoro-11β-hydroxysteroid

Treatment of the 11β-hydroxysteroid with tosyl chloride produces a tosylate ester, providing a good leaving group for a base-catalysed E2 elimination (see Section 6.4.1). The favoured product is the more-substituted 9, 11-alkene (see Section 6.4.1). A consideration of the steroid shape (see Box 3.19) shows that the 9α-proton and the 11β-tosylate are both axial and, therefore, *anti* to each other; they are thus ideally positioned for an elimination reaction (see Section 6.4.1). Bromination of the alkene under aqueous alkaline conditions then leads to formation of the bromohydrin. Interestingly, this reaction is quite stereospecific. Only one bromonium cation is formed, because the upper face of the steroid is sterically hindered by the methyl groups and approach of the large bromine molecule occurs from the lower less-hindered α-face. Ring opening by nucleophilic attack of hydroxide occurs from the upper β-face, and at C-11, since the methyl groups, particularly that at C-10, again hinder attack at C-9.

The natural glucocorticoid is **hydrocortisone (cortisol)**. Semi-synthetic 9α-bromohydrocortisone 21-acetate was found to be less active as an anti-inflammatory agent than hydrocortisone 21-acetate by a factor of three, and 9α-iodohydrocortisone 21-acetate was also less active by a factor of 10. However, 9α-fluorohydrocortisone 21-acetate (**fludrocortisone acetate**) was discovered to be about 11 times more active than hydrocortisone acetate. Although the bromination sequence shown is equally applicable to chlorine and iodine compounds, fluorine must be introduced indirectly by the β-epoxide formed by base treatment of the 9α-bromo-11β-hydroxy analogue.

| hydrocortisone (cortisol) | 9α-bromohydrocortisone acetate | fludrocortisone acetate (9α-fluorohydrocortisone acetate) | betamethasone |

The introduction of a 9α-fluoro substituent increases anti-inflammatory activity, but it increases mineralocorticoid activity even more (300×). Fludrocortisone acetate is of little value as an anti-inflammatory, but it is employed as a mineralocorticoid. On the other hand, additional modifications may be employed. Introduction of a 1,2-double bond increases glucocorticoid activity over mineralocorticoid activity, and a 16-methyl group reduces mineralocorticoid activity without affecting glucocorticoid activity. A combination of these three structural modifications gives valuable anti-inflammatory drugs, e.g. **betamethasone**, with hardly any mineralocorticoid activity.

8.1.3 Electrophilic additions to alkynes

Electrophilic reactions of **alkynes** can readily be predicted, based on the mechanisms outlined above for alkenes. Of course, the main extension is that addition will initially produce an alkene, which will then undergo further addition.

Protonation of the alkyne is actually less favourable than protonation of an alkene, because the resulting vinyl cation is *sp* hybridized, having σ bonds to just two substituents, a π bond, and a vacant *p* orbital. A vinyl cation is thus less stable than a comparable trigonal sp^2-hybridized carbocation, since *sp*-hybridization brings bonding electrons closer to carbon; it thus becomes less tolerant of positive charge. Protonation, when it occurs, will be on the less-substituted carbon, a secondary vinyl cation being preferred over a primary vinyl cation. Thus, electrophilic addition of HX follows Markovnikov orientation.

The vinyl halide product is then able to react with a further mole of HX, and the halide atom already present influences the orientation of addition in this step. The second halide adds to the carbon that already carries a halide. In the case of the second addition of HX to RC≡CH, we can see that we are now considering the relative stabilities of tertiary and primary carbocations. The halide's inductive effect actually destabilizes the tertiary carbocation. Nevertheless, this is outweighed by a favourable stabilization from the halide by overlap of lone pair electrons, helping to disperse the positive charge.

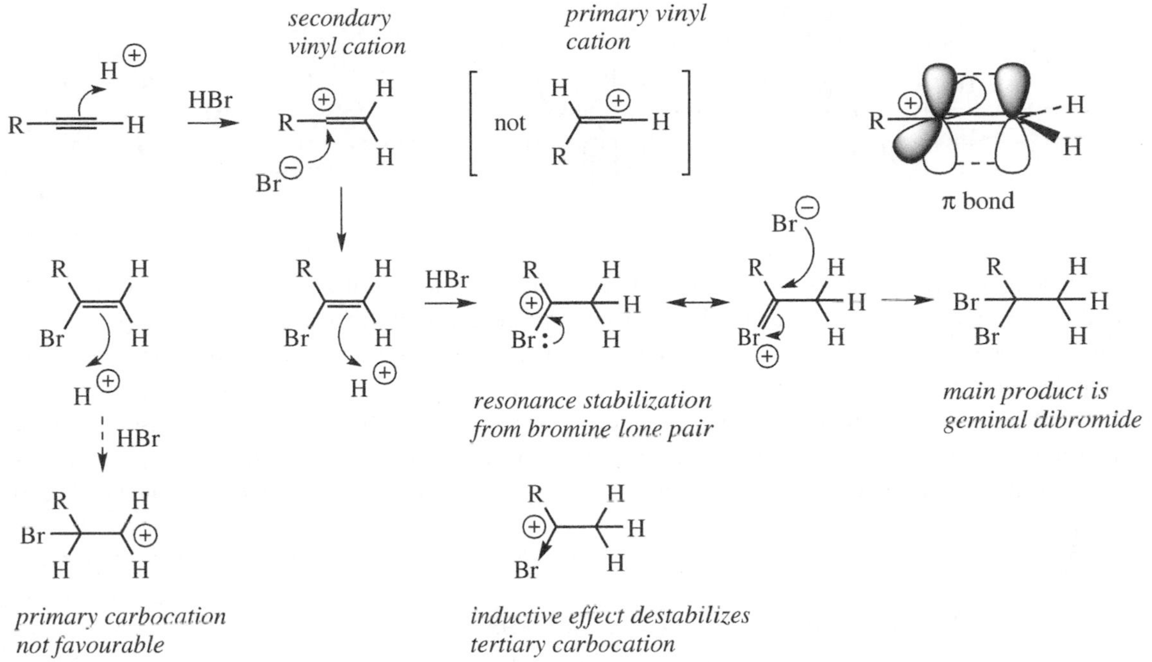

secondary vinyl cation

primary vinyl cation

not

π bond

R—≡—H   HBr

resonance stabilization from bromine lone pair

main product is geminal dibromide

HBr

primary carbocation not favourable

inductive effect destabilizes tertiary carbocation

In the case of electrophilic addition of HX to RC≡CR, it is possible to see even more clearly the role of the first halide atom. After addition of the first mole of HX, further protonation would give either a secondary carbocation or the tertiary carbocation with a destabilizing inductive effect. The resonance contribution from the first halide atom still defines the position for the second protonation, and thus the nature of the major product, the *gem*-dihalide.

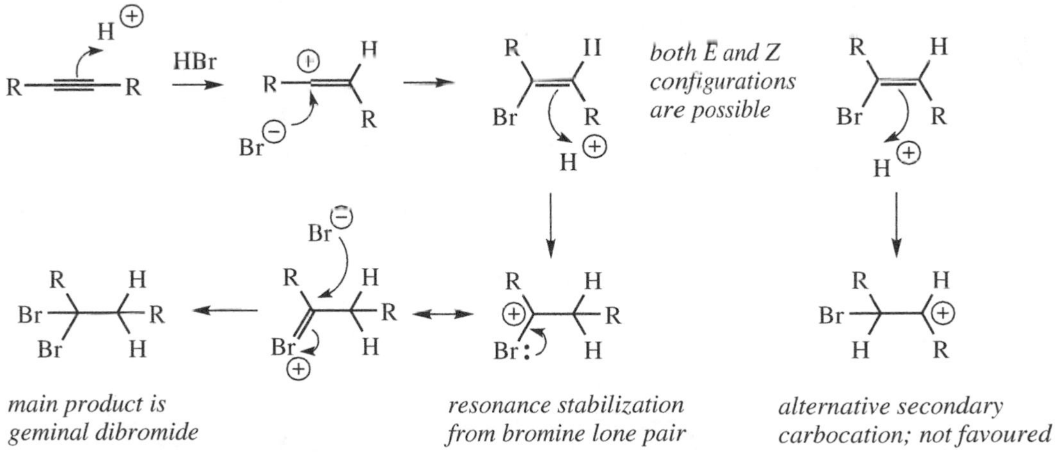

R—≡—R   HBr

both E and Z configurations are possible

main product is geminal dibromide

resonance stabilization from bromine lone pair

alternative secondary carbocation; not favoured

Note also that, if 2 mol of HX add to an alkyne, it is of no consequence whether the first addition produces an alkene with *E* or *Z* stereochemistry, since the orientation of addition means the final product has no potential chiral centres.

Predicting the outcome of electrophilic additions to alkynes from an extension of alkene reactivity usually works well, and can be applied to halogenations and hydrations. **Hydration** of an alkyne has a subtle twist, however; the product is a ketone! This can still be rationalized quite readily, though.

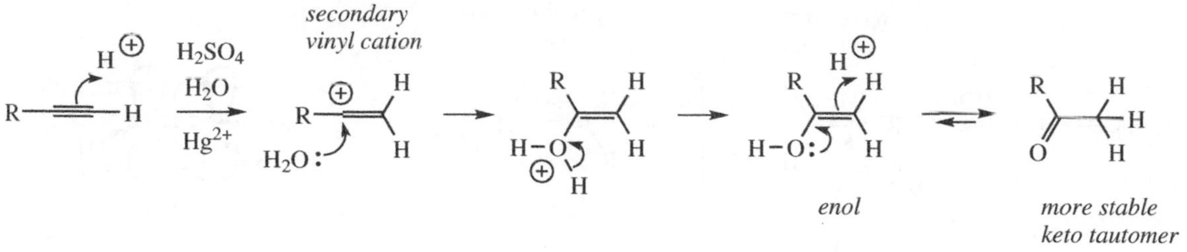

Protonation of the alkyne produces the more favourable secondary vinyl cation, which is then attacked by water, since water is the predominant nucleophile available. Loss of a proton from this produces an enol, which is transformed into a more stable isomeric form, the ketone. This transformation is termed **tautomerism**, and we shall meet it later (see Section 10.1) as an important consideration in the reactivity of many organic compounds. This reaction involves only one electrophilic addition, and, although it does not occur readily using simply aqueous acid, it can be achieved by the use of mercuric salts as catalyst. The mercuric ion may function as a Lewis acid to facilitate formation of the vinyl cation.

Box 8.2

Electrophilic alkylation in steroid side-chain biosynthesis

Polyene macrolide drugs such as **amphotericin** and **nystatin** have useful antifungal activity but no antibacterial action (see Box 7.14). Their activity is a result of binding to **sterols** in the eukaryotic cell membrane; they display no antibacterial activity because bacterial cells do not contain sterol components. Binding to sterols modifies cell wall permeability and leads to pores in the membrane and loss of essential cell components. Fungal cells are also attacked in preference to mammalian cells, since the antibiotics bind much more strongly to **ergosterol**, the major fungal sterol, than to **cholesterol**, the main animal sterol component. This selective action allows these compounds to be used as drugs, though a limited amount of binding to cholesterol is responsible for side-effects of the drugs.

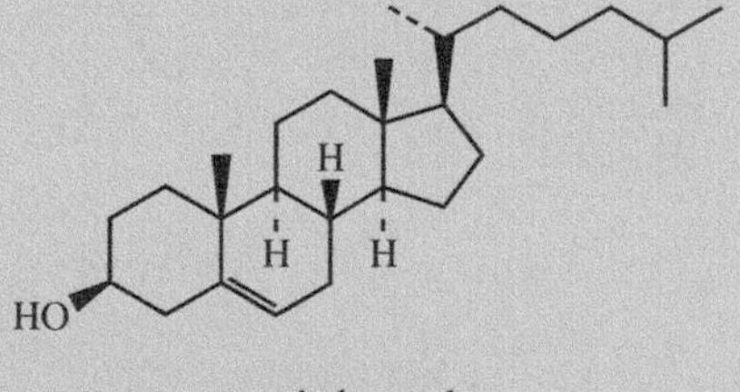

cholesterol
(animals)

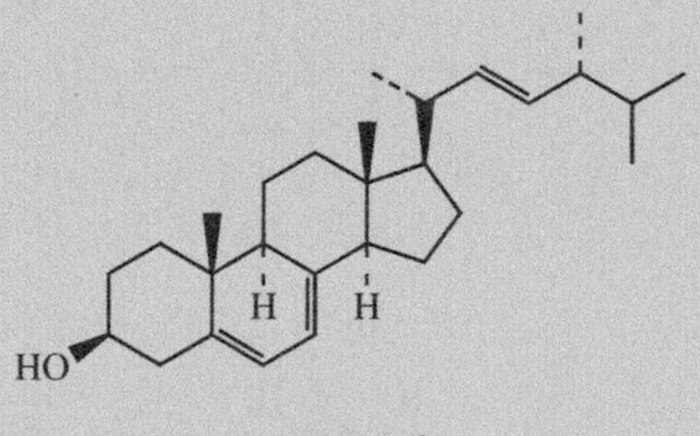

ergosterol
(fungi)

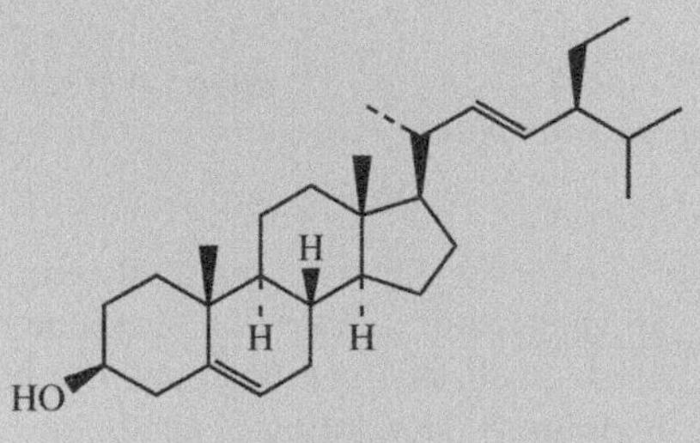

stigmasterol
(plants)

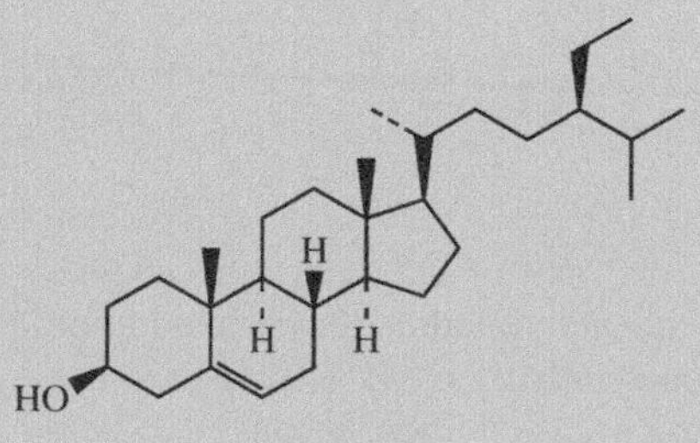

sitosterol
(plants)

A principal structural difference between ergosterol and cholesterol that affects binding to polyene drugs is the extra methyl group on the side-chain in ergosterol. This extra methyl is known to be introduced in nature by an electrophilic alkylation of a double-bond system, and it employs S-adenosylmethionine (SAM) as the electrophilic agent. We have already met SAM as a biochemical alkylating agent through S_N2 reactions (see Box 6.5). The role of SAM in these electrophilic reactions is similar. It possesses a good leaving group in the neutral molecule S-adenosylhomocysteine, and the methyl group can be donated to the alkene nucleophile in an electrophilic addition.

electrophilic addition; C-methylation using SAM

Ad = adenosine

S-adenosyl-homocysteine

second electrophilic addition involving SAM

1,2-hydride shift followed by loss of proton

e.g. lanosterol

NADPH

e.g. cholesterol

NADPH

dehydrogenation

dehydrogenation

e.g. sitosterol

e.g. stigmasterol

e.g. ergosterol

Cholesterol and ergosterol share a common biosynthetic pathway from squalene oxide as far as lanosterol (see Box 6.12), but then subsequent modifications vary. Part of the route to cholesterol involves reduction of the side-chain double bond, an enzymic process utilizing the hydride donor NADPH as reducing agent (see Box 7.6). During ergosterol biosynthesis, the side-chain double bond is involved in an electrophilic reaction with SAM, addition yielding the anticipated tertiary carbocation. This carbocation then undergoes a Wagner–Meerwein 1,2-hydride shift (see Section 6.4.2), an unexpected change, and subsequently loses a proton from the SAM-derived methyl group to generate a new alkene. NADPH reduction of this double bond leads to the C-methylated side-chain, as found in ergosterol, though further unsaturation needs to be introduced via an enzymic dehydrogenation reaction.

Plant sterols such as **stigmasterol** typically contain an extra ethyl group when compared with cholesterol. Now this is not introduced by an electrophilic ethylation process; instead, two successive electrophilic methylation processes occur, both involving SAM as methyl donor. Indeed, it is a methylene derivative like that just seen in ergosterol formation that can act as the alkene for further electrophilic alkylation. After proton loss, the product has a side-chain with an ethylidene substituent; the side-chains of the common plant sterols stigmasterol and **sitosterol** are then related by repeats of the reduction and dehydrogenation processes already seen in ergosterol formation.

8.1.4 Carbocation rearrangements

Carbocations are highly reactive intermediates, and
are notorious for their ability to rearrange into more
stable variants. We have already met this concept
when we considered **carbocation rearrangements**
as competing reactions during S_N1 nucleophilic substitutions (see Section 6.4.2). Since carbocations are
also involved in electrophilic reactions, we must
expect that analogous rearrangements might occur
in these. This is indeed the case. In Section 6.4.2,
we included examples of rearrangements in carbocations formed during electrophilic additions, because
identical processes are involved, and it was more
appropriate to consider these topics together, rather
than separately.

As a simple example, note that the major products
obtained as a result of addition of HBr to the alkenes
shown below are not always those initially expected.
For the first alkene, protonation produces a particularly favourable carbocation that is both tertiary and
benzylic (see Section 6.2.1); this then accepts the bromide nucleophile. In the second alkene, protonation
produces a secondary alkene, but hydride migration
then leads to a more favourable benzylic carbocation.
As a result, the nucleophile becomes attached to a
carbon that was not part of the original double bond.
Further examples of carbocation rearrangements will
be met under electrophilic aromatic substitution (see
Section 8.4.1).

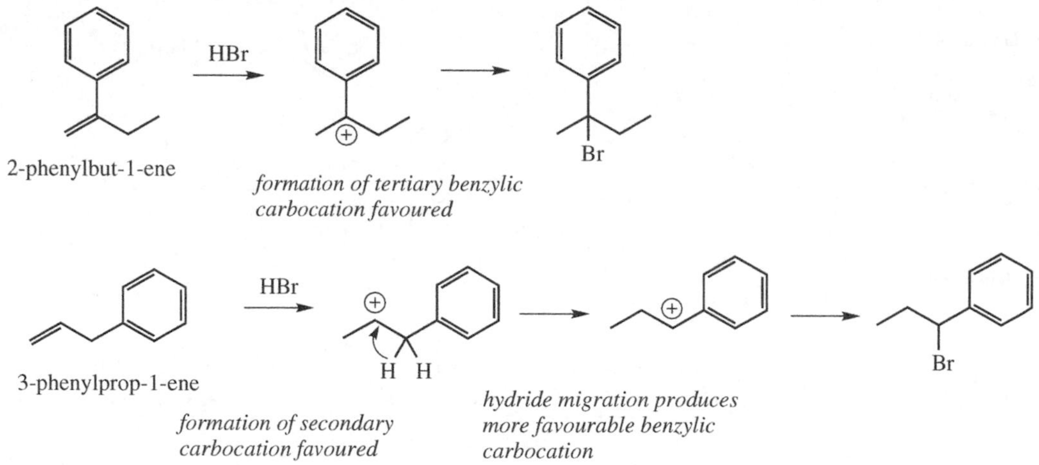

2-phenylbut-1-ene

formation of tertiary benzylic carbocation favoured

3-phenylprop-1-ene

formation of secondary carbocation favoured

hydride migration produces more favourable benzylic carbocation

Rearrangements are an unexpected complication,
and it is sometimes difficult to predict when they
might occur. We need to look carefully at the
structure of any proposed carbocation intermediate
and consider whether any such rearrangements are
probable. In most cases we shall only need to
rationalize such transformations, and will not be
trying to predict their possible occurrence.

8.2 Electrophilic addition to conjugated systems

The first step in the reaction of HX with an alkene
is protonation to yield the more stable cation. If
we extend this principle to a conjugated diene, e.g.
buta-1,3-diene, then we can see that the preferred
carbocation will be produced if protonation occurs

on a terminal carbon atom; protonation of either
C-2 or C-3 would produce an unfavourable primary
carbocation.

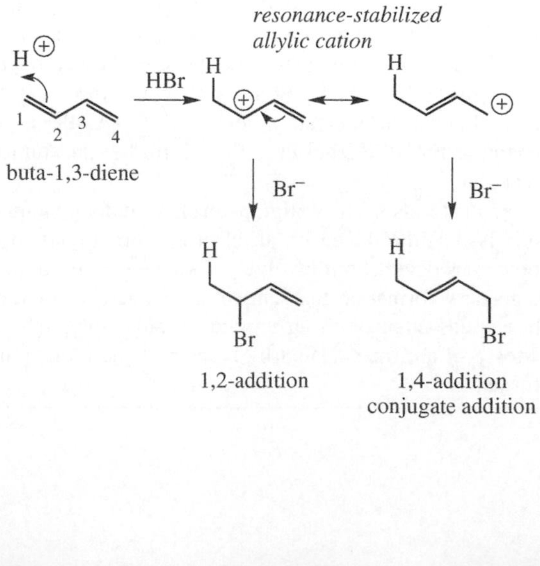

resonance-stabilized allylic cation

buta-1,3-diene

1,2-addition

1,4-addition
conjugate addition

At first glance, this appears to be a secondary carbocation, but on further examination one can see that it is also an **allylic cation**. Allylic carbocations are stabilized by resonance, resulting in dispersal of the positive charge (see Section 6.2.1). From these two resonance forms, we can predict that both carbons 2 and 4 will be electron deficient. Now this has particular consequences when we consider subsequent attack of the nucleophile X^- on to the carbocation. There are two possible centres that may be attacked, resulting in two different products. The products are the result of either 1,2-addition or 1,4-addition. The addition across the four-carbon

system as in the 1,4-adduct is termed **conjugate addition**.

Now the allylic cation has two limiting structures, one of which is a secondary carbocation and the other a primary carbocation. We would expect the secondary carbocation to contribute more than the primary carbocation, and this is usually reflected in the proportions of the two products actually obtained from the reaction carried out at low temperatures. Why the rider about low temperatures? Well, the product ratio is different if the same reaction is carried out at higher temperatures, and typically the 1,2-adduct now predominates.

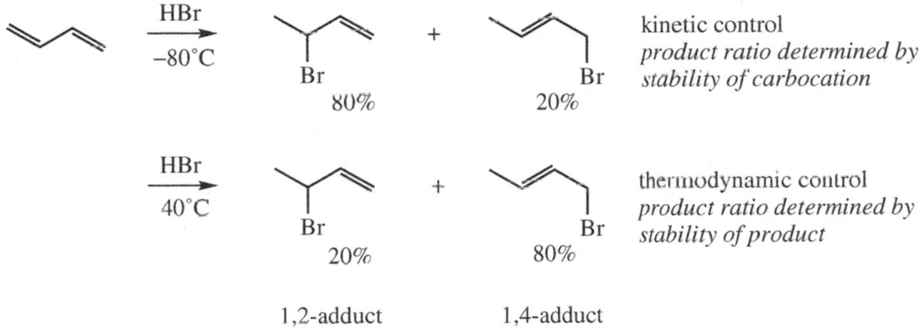

At the higher temperature, the thermodynamic stability of the product is the important consideration, with the 1,4-adduct, a disubstituted alkene, being more stable than the 1,2-adduct, which is a monosubstituted alkene. An essential part of the reasoning is

that the reaction is reversible, so that the product can lose halide to regenerate the allylic cation. Thus, the product mixture from the lower temperature reaction is converted upon heating into the product mixture corresponding to the higher temperature.

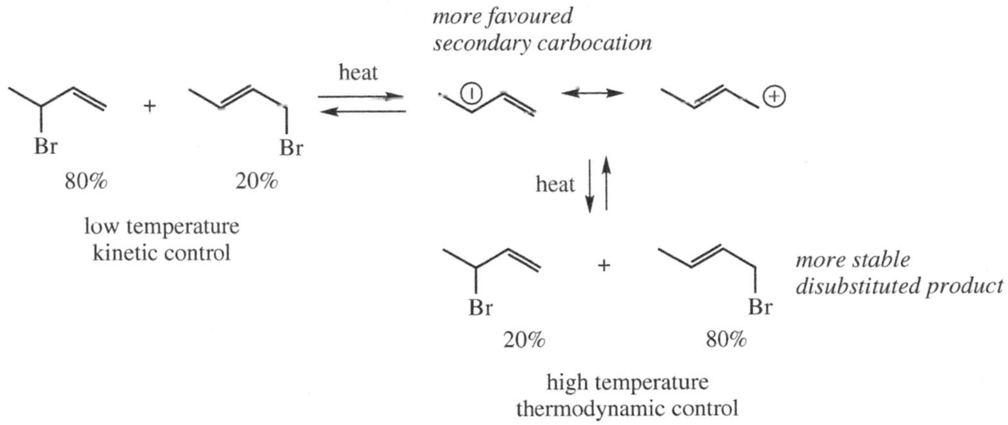

These concepts are termed **kinetic control** and **thermodynamic control**. At the lower temperature, the product ratio is determined by the relative

importance of the carbocations, with the predominant one reacting faster. At the higher temperature, the product ratio is determined by the stability of the product, with the more stable one predominating. These products are, of course, the result of addition of just 1 mol of HX to the conjugated system. The second double bond could also react if further HX were available, with regiochemistry now following the principles already established in Section 8.1.

The energy diagram for kinetic versus thermodynamic control is shown in Figure 8.2. This may be interpreted as follows. The 1,4-addition product is of lower energy, i.e. more stable, than the 1,2-addition product. The critical step, however, is the interaction of the bromide nucleophile with the allylic cation. The activation energy leading to the 1,2-addition product is lower than that leading to the 1,4-addition product. Therefore, at lower temperatures the 1,2-adduct is formed faster, and becomes the dominant product. At the lower temperature, though, there is insufficient energy available to overcome the much larger energy barrier for the reverse reactions, so neither reaction is reversible. Both products are formed, but do not revert back to the allylic cation. Therefore, we have kinetic control: the product ratio depends upon which product is formed faster. At

higher temperatures there is now sufficient energy available to overcome both activation energies with ease, and, more importantly, the reverse reactions become feasible. We can also see that the less stable 1,2-addition product will revert back to the allylic cation faster than the 1,4-addition product simply because the energy barrier is that much less. The dominant equilibrium product will thus become the more stable material, i.e. the 1,4-addition product; we now have thermodynamic control.

Similar observations emerge from addition of halogens to butadiene. Thus, low-temperature bromination gives predominantly the 1,2-adduct. At higher temperatures, the 1,4-adduct is the main product, and the mixture from the lower temperature reaction equilibrates to the same product ratio. The 1,4-product is the thermodynamically more stable; it has the more-substituted double bond, and the two large bromine atoms are further apart in this isomer. Mechanisms for formation and equilibration of the products can be written as shown, using bromonium cation intermediates. It is perhaps less easy to see why the 1,2-adduct should be the kinetically controlled product, until

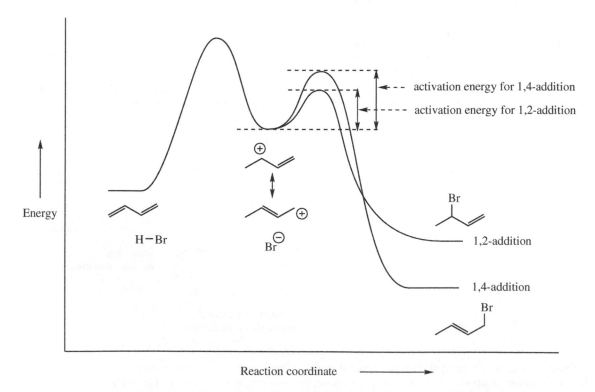

Figure 8.2 Energy profile: 1,2 and 1,4 electrophilic addition to conjugated diene

we consider that the bromonium cation is actually a stabilized carbocation (see Section 8.1.2). We can then use the same type of rationalization as with the carbocation intermediates in HBr addition.

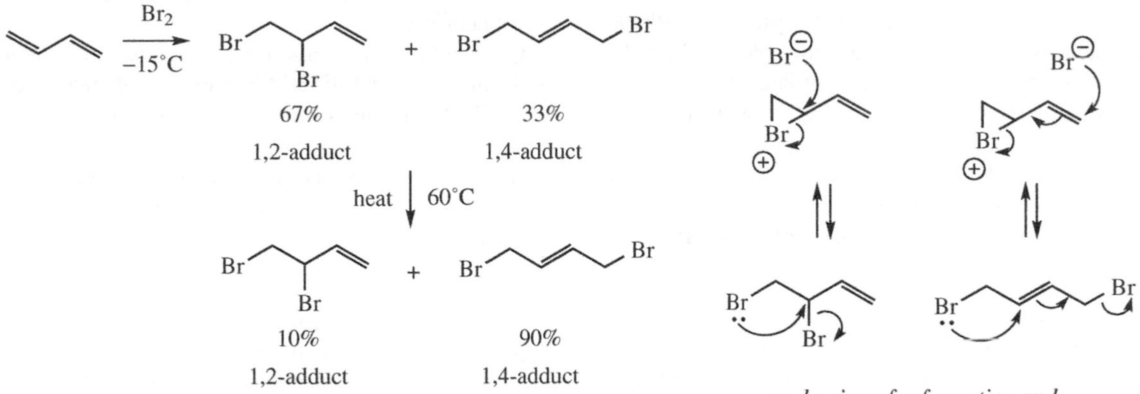

Br₂ / −15°C reaction scheme:

Br with Br (1,2-adduct) 67%   +   Br...Br (1,4-adduct) 33%

heat | 60°C

Br with Br (1,2-adduct) 10%   +   Br...Br (1,4-adduct) 90%

mechanisms for formation and equilibration of 1,2- and 1,4-adducts

8.3 Carbocations as electrophiles

As we have seen in Section 8.1, reaction of an alkene with an electrophile produces a carbocation that is subsequently attacked by a nucleophile. However, the carbocation is itself an electrophilic species, and is vulnerable to attack by another alkene molecule, provided that the alkene concentration is sufficiently high. For example, protonation of styrene leads to a secondary carbocation that is favoured by being benzylic (see Section 6.2.1).

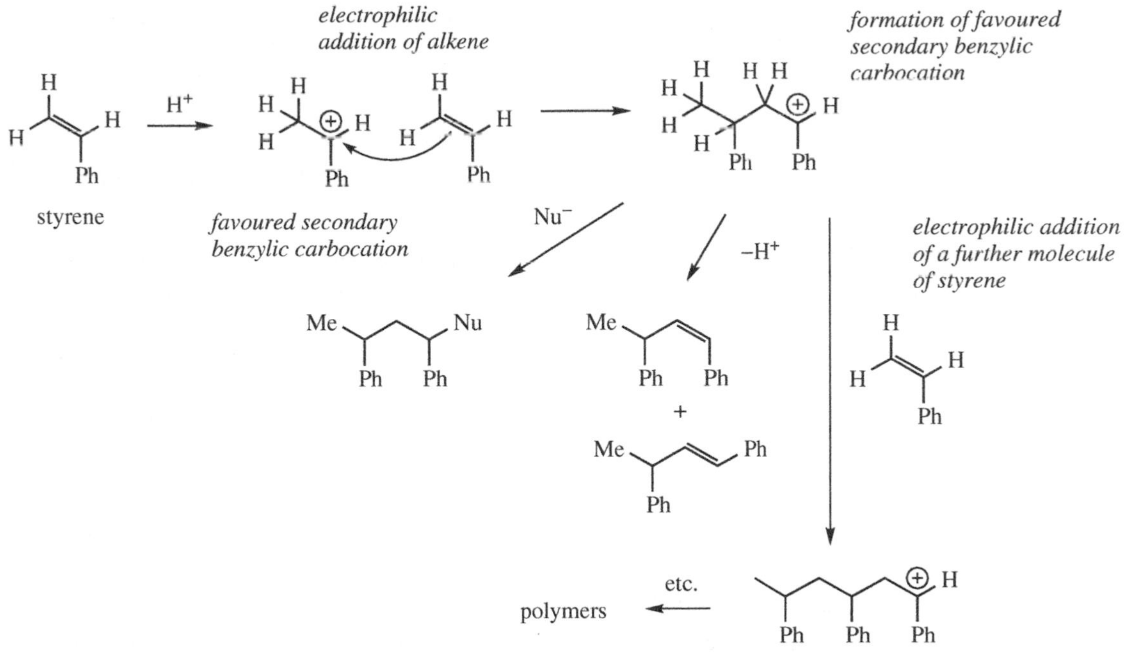

electrophilic addition of alkene

formation of favoured secondary benzylic carbocation

styrene   H⁺

favoured secondary benzylic carbocation

Nu⁻

−H⁺

electrophilic addition of a further molecule of styrene

Me—...—Nu (Ph, Ph)

Me—...— (Ph, Ph)
+
Me—...—Ph (Ph)

etc.

polymers ←

(Ph, Ph, Ph)

This carbocation now becomes the electrophile, and may be attacked by the π electrons of a second styrene molecule, the regiochemistry of attack being the same as with the original protonation, i.e. giving the secondary benzylic carbocation. Now this carbocation may suffer several fates. It may be attacked by a nucleophilic species or, more likely, it may lose a proton to yield an alkene. Alternatively, it may act as the electrophile for reaction with a further styrene molecule, generating yet another carbocation. It can be seen that this type of process may then continue, giving polymeric products: polystyrene. The final carbocation will be discharged most probably by loss of a proton. The process is termed **cationic polymerization**. In practice, the process is more useful for generating dimers and trimers than polymers, and industrial polymers are usually produced by radical processes (see Section 9.4.2).

Cationic polymerization is, of course, an intermolecular electrophilic addition process. **Intramolecular electrophilic addition** involving two double bonds in the same molecule may be used to generate a cyclic system. Thus, the trienone shown is converted into a mixture of cyclic products when treated with sulfuric acid.

protonation to favourable tertiary carbocation

electrophilic addition to give favourable tertiary carbocation

proton loss generates more-substituted double bond and favourable conjugated system

6,10-dimethyl-undeca-3,5,9-trien-2-one

β-ionone

alternative products not favoured
less-substituted double bond; not conjugated

This is easily rationalized by protonation of the terminal alkene, yielding the preferred tertiary carbocation. The carbocation is then attacked by π electrons from the neighbouring double bond, creating a new σ bond and a ring system. Note that this results in a favourable tertiary carbocation and a favourable strain-free six-membered ring (see Section 3.3.2).

The products are then formed by loss of a proton from this carbocation, with a choice of protons that may be lost, so that a mixture of products in varying proportions results. β-Ionone is the predominant product. This is the most substituted alkene, and has the added stability conferred by extending conjugation with the unsaturated ketone (see Section 2.8).

Box 8.3

Electrophilic additions to carbocations in terpenoid and steroid biosynthesis

Terpenoids and **steroids** account for a huge group of natural products, and provide us with many useful materials, including flavouring and perfumery agents, aromatherapy oils, some vitamins, steroidal hormones and a range of drugs. Although spanning a vast range of chemical structures, these compounds all derive from two simple precursors, **dimethylallyl diphosphate** and **isopentenyl diphosphate**.

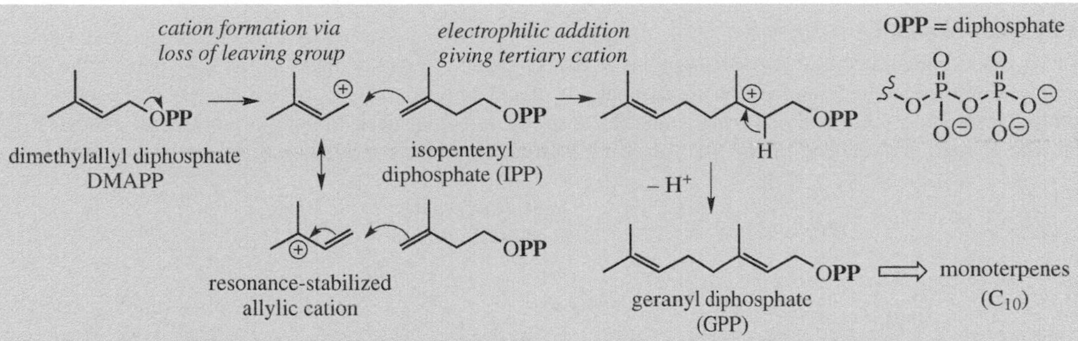

Dimethylallyl diphosphate is responsible for generating the carbocation. Loss of diphosphate as the leaving group produces a resonance-stabilized allylic cation. An intermolecular electrophilic addition follows, with isopentenyl diphosphate as the source of π electrons. Addition to the cationic species takes place at the terminal carbon that is sterically less congested – at first glance, we appear to be invoking the less favourable resonance form, but an alternative addition through the double bond onto the tertiary cation could be drawn. At this stage, it is important to appreciate that these reactions are enzyme controlled, so that we can have two different species reacting in a highly specific manner. The carbocation product is the more stable tertiary carbocation, as we might predict, and this loses a proton to form the more-substituted alkene product, **geranyl diphosphate**.

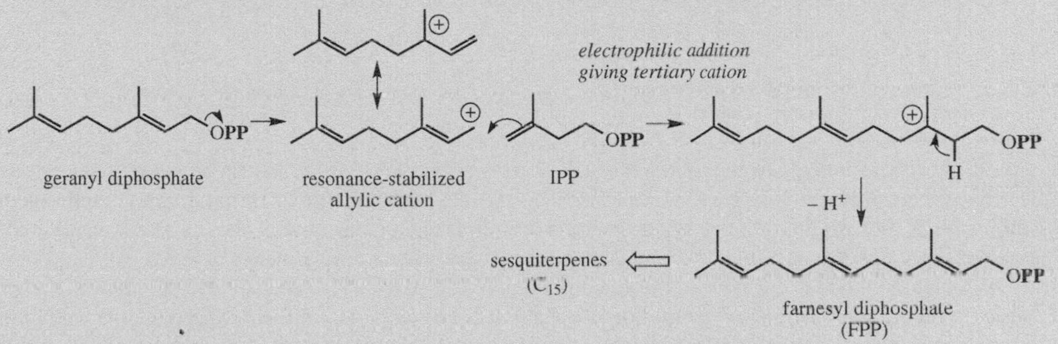

An exactly analogous process can then occur, in which geranyl diphosphate provides the allylic cation, and a further molecule of isopentenyl diphosphate adds on, giving **farnesyl diphosphate**; this can subsequently yield **geranylgeranyl diphosphate**.

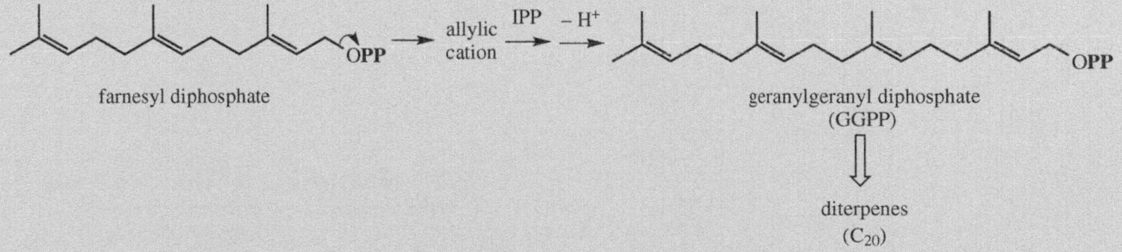

The compounds geranyl diphosphate, farnesyl diphosphate, and geranylgeranyl diphosphate are biochemical precursors of monoterpenes, sesquiterpenes, and diterpenes respectively, and virtually all subsequent modifications of these precursors involve initial formation of an allylic cation through loss of diphosphate as the leaving group.

The formation of cyclic terpenoids involves intramolecular electrophilic addition, and this can be exemplified by the following monoterpene structures, again with all reactions being enzyme controlled.

Box 8.3 (continued)

Before cyclization can occur, however, there has to be a change in stereochemistry at the 2,3-double bond, from *E* in geranyl diphosphate to *Z*, as in neryl diphosphate. It should be reasonably clear that geranyl diphosphate cannot possibly cyclize to a six-membered ring, since the carbon atoms that need to bond are not close enough to each other. The change in stereochemistry is achieved through allylic cations and linalyl diphosphate (see Box 6.4).

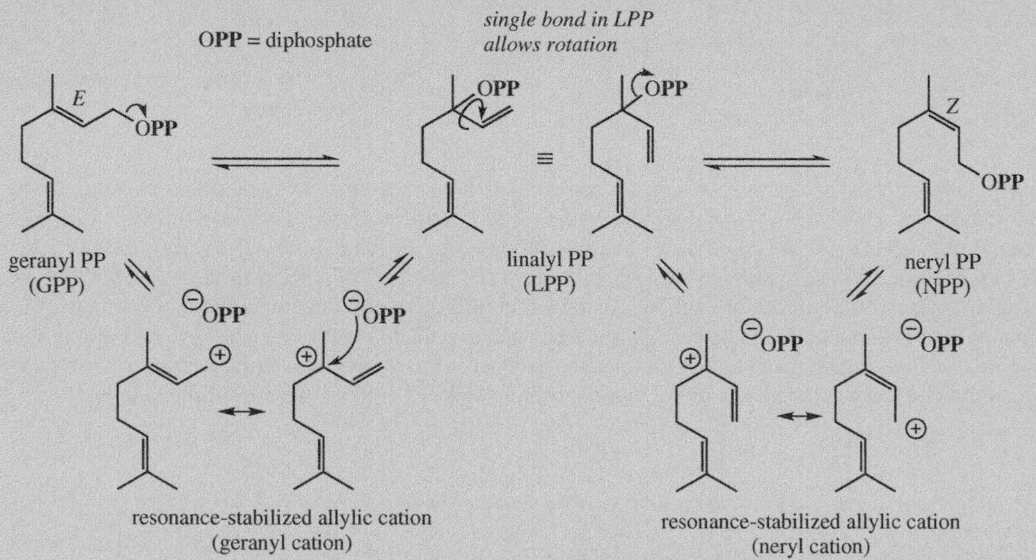

OPP = diphosphate

single bond in LPP allows rotation

geranyl PP (GPP)   linalyl PP (LPP)   neryl PP (NPP)

resonance-stabilized allylic cation (geranyl cation)

resonance-stabilized allylic cation (neryl cation)

Geranyl diphosphate ionizes to the resonance-stabilized geranyl carbocation, which, in nature, can recombine with the diphosphate anion in two ways, either reverting to geranyl diphosphate or forming linalyl diphosphate. In linalyl diphosphate, the original double bond from geranyl diphosphate has now become a single bond, and free rotation is possible. Ionization of linalyl diphosphate then occurs, giving a resonance-stabilized neryl carbocation, one form of which now has a *Z* double bond. Recombination of this with diphosphate leads to neryl diphosphate, a geometric configurational isomer of geranyl diphosphate. It is normally very difficult to change the configuration of a double bond. Nature achieves it easily in this allylic system via carbocation chemistry, and, in metabolic processes, geranyl diphosphate can be isomerized through linalyl diphosphate to neryl diphosphate.

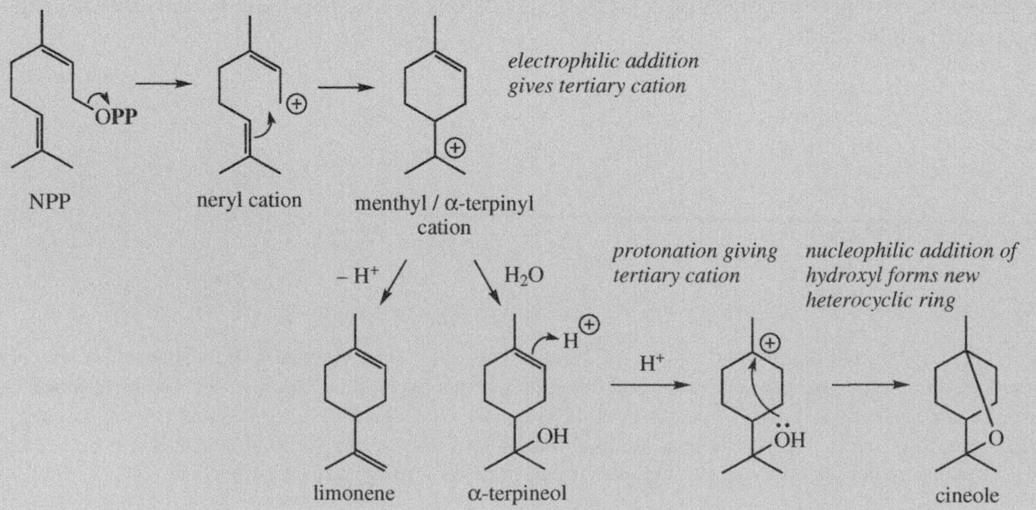

NPP   neryl cation   menthyl / α-terpinyl cation

electrophilic addition gives tertiary cation

– H⁺   H₂O

protonation giving tertiary cation   *nucleophilic addition of hydroxyl forms new heterocyclic ring*

H⁺

limonene   α-terpineol   cineole

Cyclization involves the neryl cation with electrophilic attack from the double bond giving the favoured tertiary carbocation and a favourable six-membered ring. Loss of a proton from this cation results in formation of **limonene**, actually the less-substituted alkene, so this where enzyme control takes over. Alternatively, discharge of the cation by addition of water as a nucleophile leads to α-terpineol. By a similar sequence, α-terpineol may be transformed into cineole. This requires generation of a carbocation by protonation of the double bond; the proton is added so that the favoured tertiary cation is formed. **Cineole** formation then involves nucleophilic attack from the alcohol group with generation of a further ring system, this time a heterocyclic ring. Limonene is a major constituent of lemon oil, α-terpineol is found in pine oil, and cineole is the principal component of eucalyptus oil.

By far the most impressive example of electrophilic addition in natural product formation is in the biosynthesis of **steroids**. The substrate squalene oxide is cyclized to **lanosterol** in a process catalysed by a single enzyme. Lanosterol is then converted into the primary animal-steroid cholesterol. Squalene oxide comes from squalene, which is itself formed through a combination of two molecules of farnesyl diphosphate.

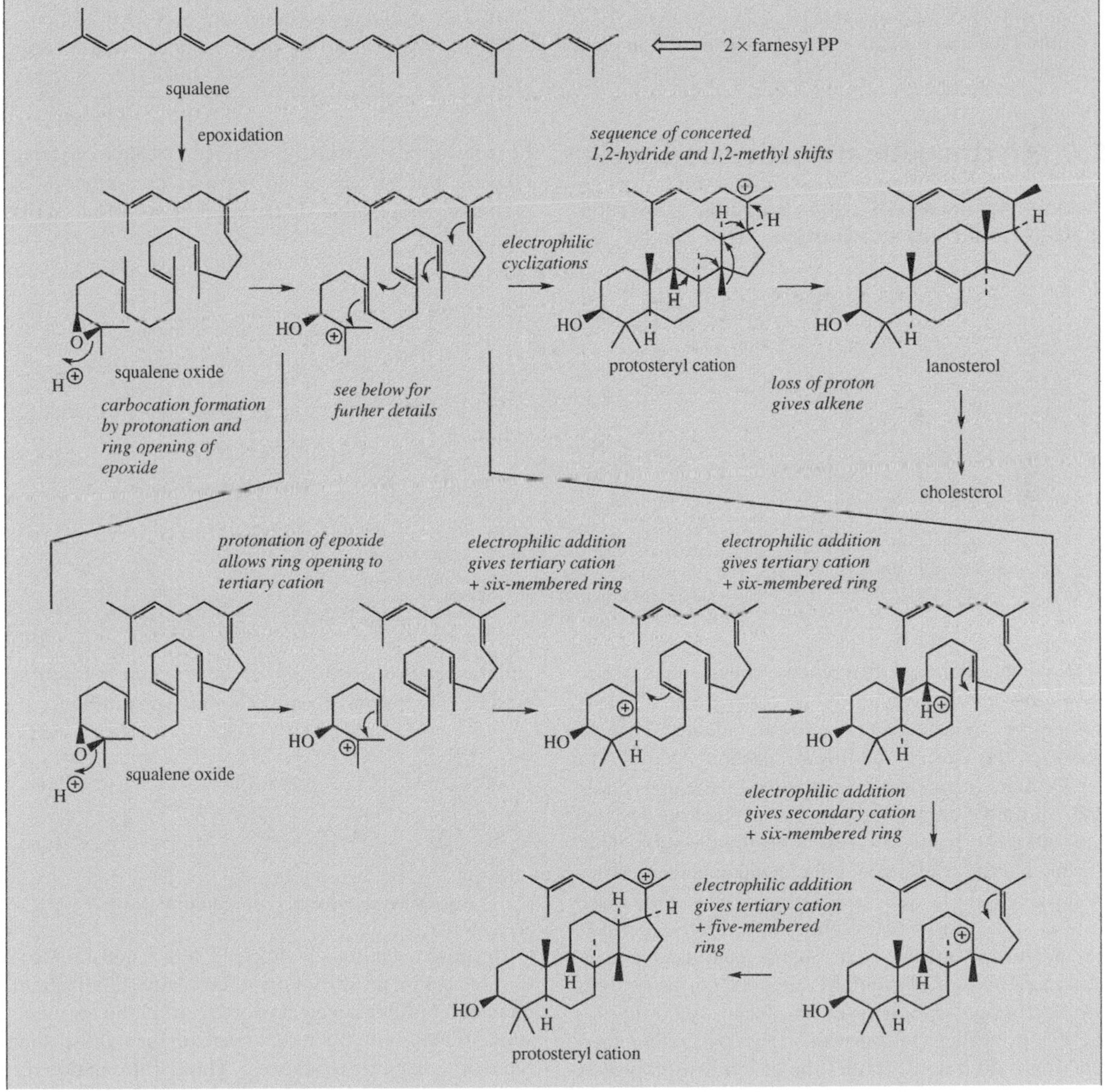

Box 8.3 *(continued)*

With squalene oxide suitably positioned and folded onto the enzyme surface, a series of electrophilic cyclizations can be used to rationalize formation of the polycyclic structure. The cyclizations are carbocation-mediated and proceed in a stepwise sequence. Thus, protonation of the epoxide group will allow opening of this ring and generation of the preferred tertiary carbocation (see Box 6.12). This is suitably placed to allow electrophilic addition to a double bond, formation of a six-membered ring and production of a new tertiary carbocation. This process continues twice more, generating a new carbocation, until the protosteryl cation is formed. This is then followed by a sequence of concerted Wagner–Meerwein migrations of methyls and hydrides leading to lanosterol. It is not appropriate to discuss these migrations in this chapter, but this aspect is studied further in Box 6.12.

Note that the preferred tertiary carbocation (Markovnikov addition) is produced in all of the cyclizations, except in one case, the third ring, which appears to be formed in an anti-Markovnikov sense. The latest studies now show that the reaction as illustrated above is not quite correct. The third ring is first produced as a five-membered one, reassuringly by Markovnikov addition via the predicted tertiary carbocation, and it is subsequently expanded to a six-membered ring through a Wagner–Meerwein 1,2-alkyl shift. This should not be thought of as a complication; simply note the formation of a biochemically important polycyclic ring system through a series of electrophilic additions.

8.4 Electrophilic aromatic substitution

Electrophilic reactions with **aromatic substrates** tend to result in **substitution**. This should not be viewed as markedly different behaviour from alkenes, but merely as an obvious consequence of aromatic stabilization dictating the fate of the initial carbocation.

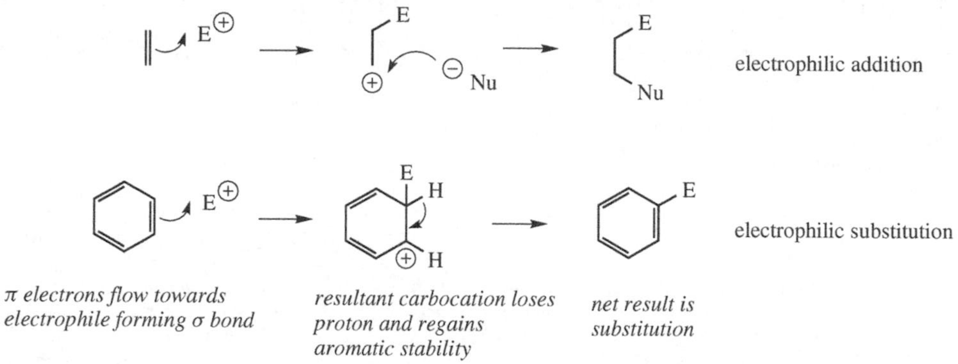

electrophilic addition

electrophilic substitution

π electrons flow towards electrophile forming σ bond

resultant carbocation loses proton and regains aromatic stability

net result is substitution

However, there are differences, in that electrophilic attack on to an aromatic ring is energetically less favourable than attack on to an alkene. This is because the initial addition reaction leading to carbocation formation uses up one of the p orbitals that normally contributes to the π electron system, and thereby creates an sp^3-hybridized centre. This means that the π electron delocalization characteristic of an aromatic system is destroyed. However, there is also some good news: the carbocation generated (an **arenium cation**) is resonance stabilized and is considerably more favourable than the corresponding simple trigonal cation from an alkene. Accordingly, the electrophilic addition can occur; but, rather than reacting with a nucleophile, the cation loses a proton and this leads to restoration of the aromatic π electron system. The overall reaction is thus substitution.

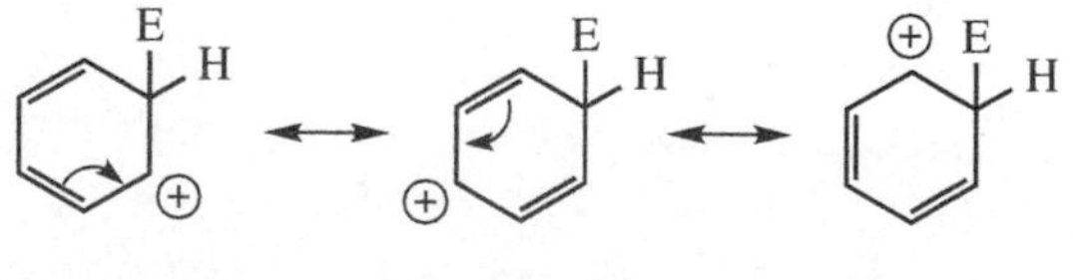

resonance stabilization of arenium cation

Because the initial electrophilic attack and carbocation formation results in loss of aromatic stabilization, the **electrophiles** necessary for electrophilic aromatic substitution must be more reactive than those that typically react with alkenes. Thus, chlorination or bromination generally occurs only in the presence of a Lewis acid, which allows a greater fraction of the positive charge to develop on the electrophilic atom. The role of the Lewis acid $AlCl_3$ in the chlorination of benzene is illustrated below; we can consider the electrophilic species as Cl^+.

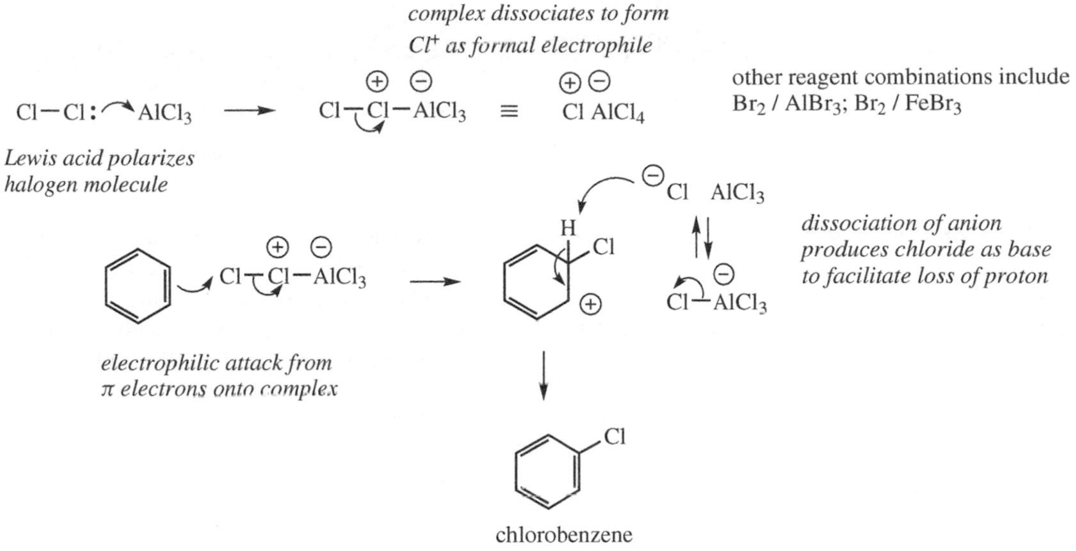
complex dissociates to form
Cl+ as formal electrophile
Cl−Cl: AlCl3
Cl−Cl−AlCl3
Cl AlCl4
other reagent combinations include
Br2 / AlBr3; Br2 / FeBr3
Lewis acid polarizes
halogen molecule
Cl AlCl3
H
Cl
dissociation of anion
produces chloride as base
to facilitate loss of proton
Cl−Cl−AlCl3
Cl−AlCl3
electrophilic attack from
π electrons onto complex
Cl
chlorobenzene

Nitration of an aromatic ring using nitric acid also requires the presence of sulfuric acid. Nitric acid is protonated by the stronger sulfuric acid, leading to loss of water and production of the **nitronium ion** as electrophile.

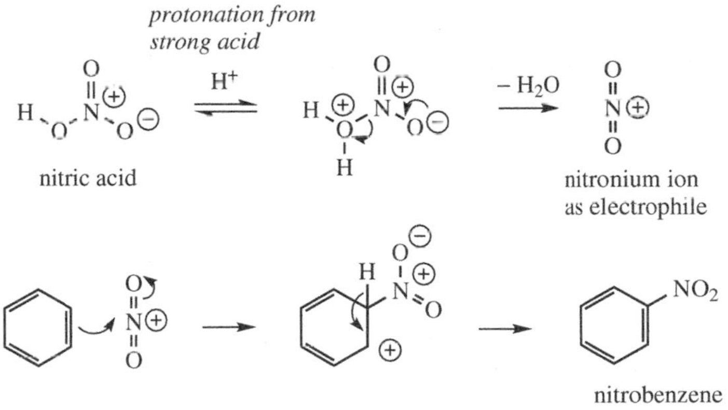
protonation from
strong acid
H+
− H2O
nitric acid
nitronium ion
as electrophile
NO2
nitrobenzene

Aromatic **sulfonation** occurs with fuming sulfuric acid, where the electrophile is **sulfur trioxide**. This is present in concentrated sulfuric acid as a result of the equilibrium shown.

$$2H_2SO_4 \rightleftharpoons SO_3 + H_3O^{\oplus} + HSO_4^{\ominus}$$

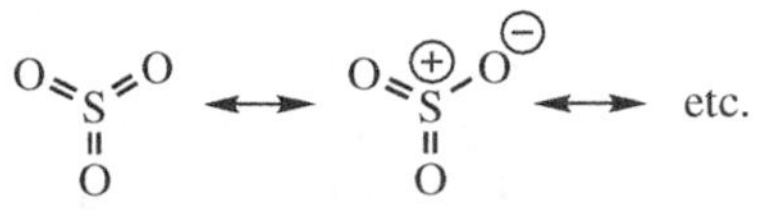
etc.

resonance structures predict the sulfur atom in SO_3 is electron deficient

The product of electrophilic aromatic substitution is a **sulfonic acid** (see Section 7.13.1). Unusually, sulfonation is found to be reversible; it is possible to replace an SO_3H group attached to an aromatic ring with hydrogen by heating the sulfonic acid with steam.

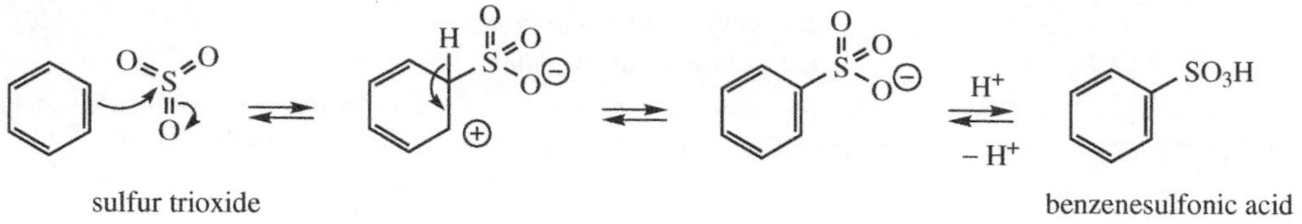

sulfur trioxide   benzenesulfonic acid

8.4.1 Electrophilic alkylations: Friedel–Crafts reactions

Particularly useful reactions result from Friedel–Crafts alkylations and acylations, in which the electrophile is developed from either an alkyl halide or an acyl halide in the presence of a Lewis acid. The alkylation reaction is mechanistically similar to the halogenation process above, with the Lewis acid increasing polarization in the alkyl halide.

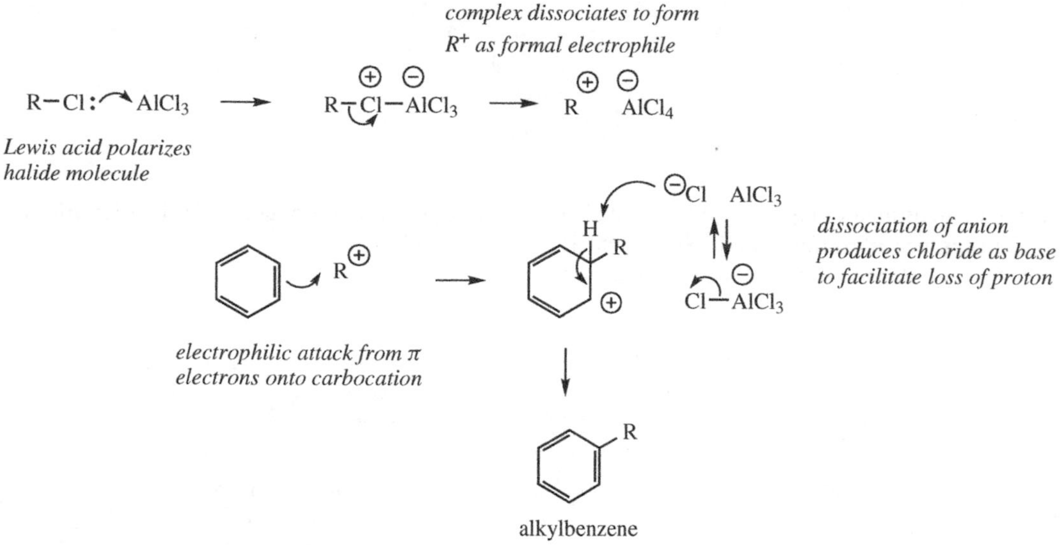

complex dissociates to form
R^+ as formal electrophile

Lewis acid polarizes
halide molecule

dissociation of anion
produces chloride as base
to facilitate loss of proton

electrophilic attack from π
electrons onto carbocation

alkylbenzene

However, although we invoked a Lewis acid complex to provide the halonium electrophile, there is considerable evidence that, where appropriate, the electrophile in **Friedel–Crafts alkylations** is actually the dissociated carbocation itself. Of course, a simple methyl or ethyl cation is unlikely to be formed, so there we should assume a Lewis acid complex as the electrophilic species. On the other hand, if we can get a secondary or tertiary carbocation, then this is probably what happens. There are good stereochemical reasons why a secondary or tertiary complex cannot be attacked. Just as we saw with S_N2 reactions (see Section 6.1), if there is too much steric hindrance, then the reaction becomes S_N1 type.

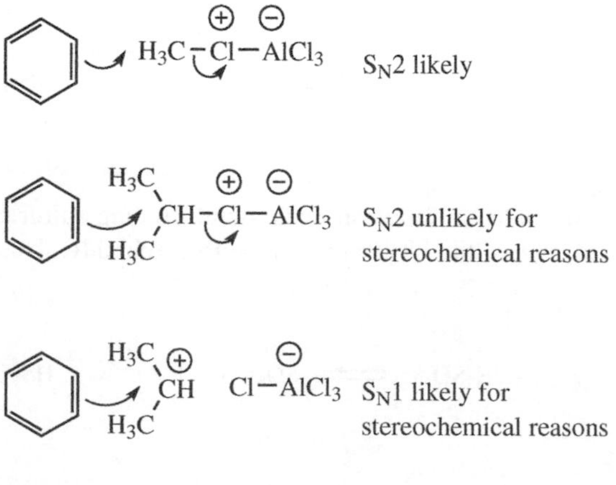

S_N2 likely

S_N2 unlikely for stereochemical reasons

S_N1 likely for stereochemical reasons

Indeed, we can also achieve alkylation of an aromatic ring by using any system that generates a carbocation. In effect, we are paralleling the concept of carbocations as electrophiles as in Section 8.3,

but using an aromatic substrate. Thus, an alkene in strongly acidic conditions, or an appropriate alcohol in acid, may be used to generate a carbocation and achieve electrophilic substitution.

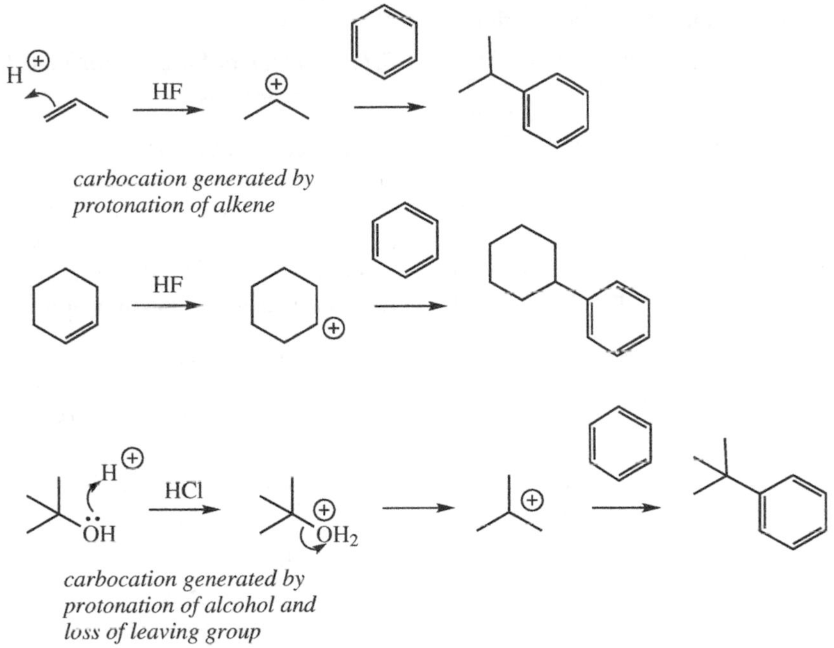

carbocation generated by protonation of alkene

carbocation generated by protonation of alcohol and loss of leaving group

This involvement of carbocations actually limits the utility of Friedel–Crafts alkylations, because, as we have already noted with carbocations, **rearrangement reactions** complicate the anticipated outcome (see Section 6.4.2). For instance, when a Lewis acid

is used to generate what would be a primary carbocation, rearrangement to a secondary carbocation is likely to occur. Both types of cationic species can then bond to the aromatic ring, and a mixture of isomeric products is formed.

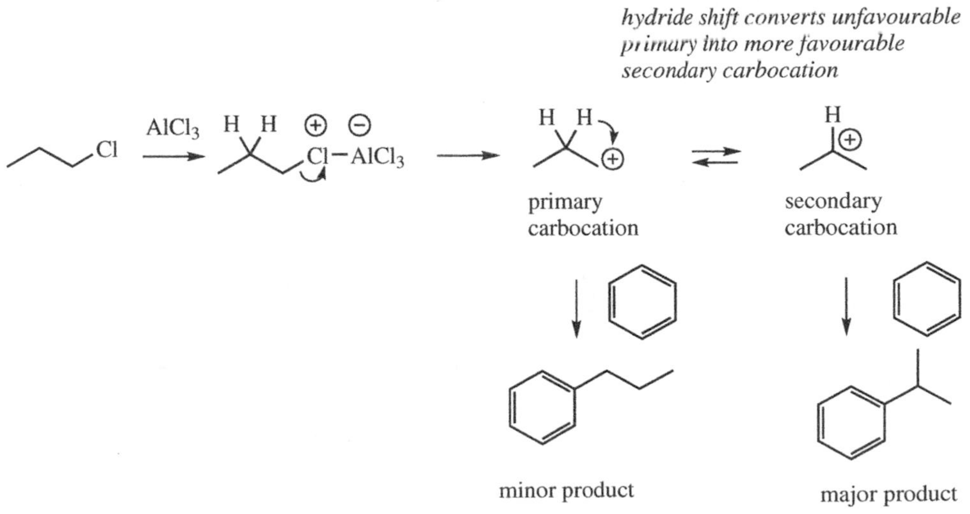

hydride shift converts unfavourable primary into more favourable secondary carbocation

minor product   major product

To be really satisfactory, a Friedel–Crafts alkylation requires one relatively stable secondary or tertiary carbocation to be formed from the alkyl halide by interaction with the Lewis acid, i.e. cases where there is not going to be any chance of rearrangement. Note also that we are unable to generate carbocations from an aryl halide – aryl cations (also vinyl cations, see Section 8.1.3) are unfavourable – so that we cannot use the Friedel–Crafts reaction to join aromatic groups. There is also one further difficulty, as we shall see below. This is the fact that introduction of an alkyl substituent on to an aromatic ring activates the ring towards further electrophilic substitution. The result is that the initial product from Friedel–Crafts alkylations is more reactive than the starting material, with the consequence that di-, tri-, and poly-alkylated products also tend to be formed. It may be possible to minimize this if the starting material is readily available, e.g. benzene, and can thus be used in large excess.

8.4.2 Electrophilic acylations: Friedel–Crafts reactions

In **Friedel–Crafts acylations**, an acyl halide, almost always the chloride, in the presence of a Lewis acid is employed to acylate an aromatic ring. The process is initiated by polarization of the carbon–chlorine bond of the acyl chloride, resulting in formation of a resonance-stabilized **acylium ion**.

lone pair can back-bond to transfer charge to oxygen, thus delocalizing positive charge

resonance-stabilized acylium ion

The acylium ion is now our electrophile, and aromatic substitution proceeds in the predicted manner.

The intermediate cation is subsequently deprotonated to yield the acylated product. However, this acyl derivative is actually a ketone, which can also complex with the Lewis acid. Accordingly, the final complex must be decomposed by treatment with water, and the significant consequence is that the Lewis acid has to be supplied in stoichiometric amounts, in sharp contrast to Friedel–Crafts alkylations, where only catalytic amounts need to be used.

electrophilic attack from π electrons onto acylium ion

dissociation of anion produces chloride as base to facilitate loss of proton

acylbenzene (ketone)

Lewis acid complex

ketone

A similar problem of complex formation may be encountered if either amino or phenol groups are present in the substrate, and the reaction may fail. Under such circumstances, these groups need to be blocked (protected) by making a suitable derivative. Nevertheless, Friedel–Crafts acylations tend to work very well and with good yields, uncomplicated by multiple acylations, since the acyl group introduced deactivates the ring towards further electrophilic substitution. This contrasts with Friedel–Crafts alkylations, where the alkyl substituents introduced activate the ring towards further substitution (see Section 8.4.3).

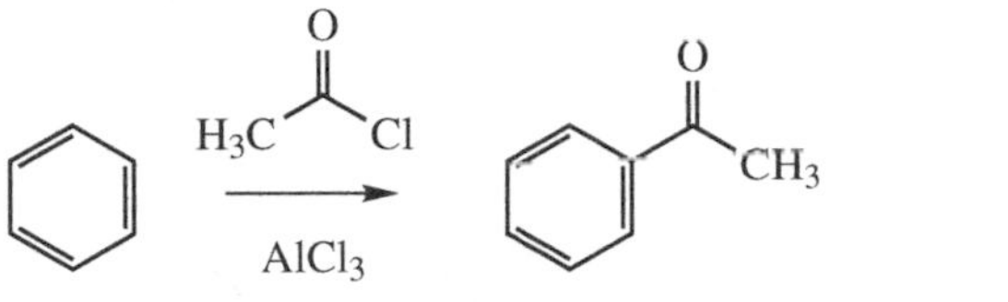

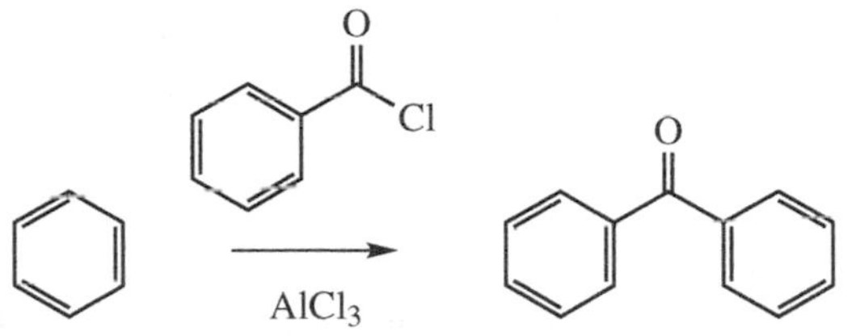

A useful extension of Friedel–Crafts acylation is an **intramolecular reaction** leading to cyclic products. Thus, five- and six-membered rings are readily and efficiently created by use of an appropriate aryl acyl chloride, as shown below.

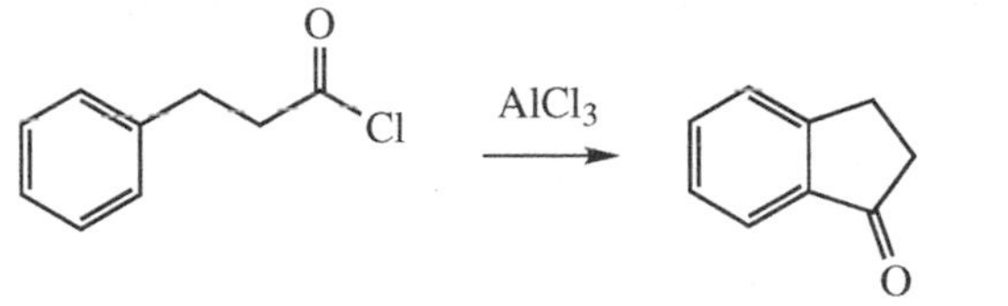

1-hydrindanone
(2,3-dihydroinden-1-one)

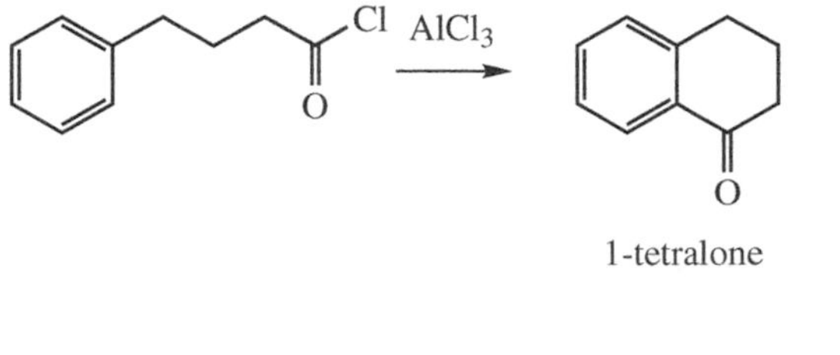

1-tetralone

8.4.3 Effect of substituents on electrophilic aromatic substitution

Substituents already bonded to an aromatic ring influence both the rate of electrophilic substitution and the position of any further substitution. The effect of a particular substituent can be predicted by a consideration of the relative stability of the first-formed **arenium cation**, formation of which constitutes the rate-limiting step. In general, substituents that are electron releasing activate the ring to further substitution – they help to stabilize the arenium ion. Substituents that are electron withdrawing destabilize the arenium ion, therefore, are deactivating and hinder further substitution.

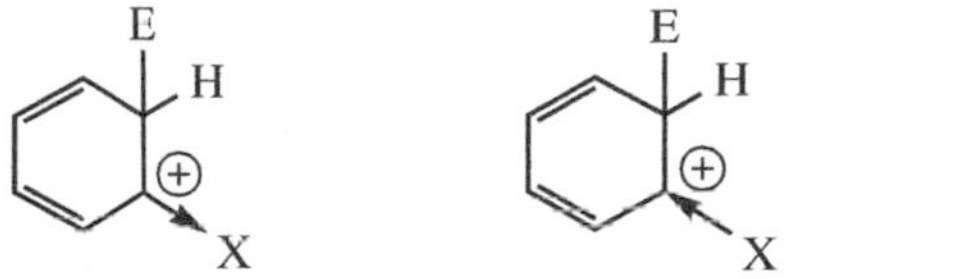

electron-withdrawing effect destabilizes carbocation

electron-donating effect stabilizes carbocation

For the position of further substitution, we also need to consider resonance forms of the arenium ion.

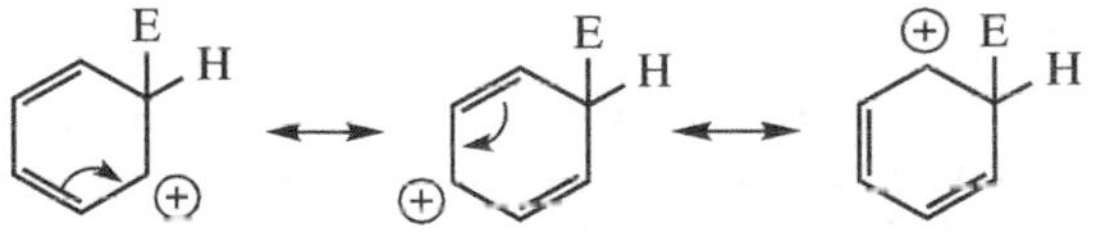

resonance stabilization of arenium cation:
ortho and *para* positions are electron deficient

From these resonance forms we can deduce that positions *ortho* and *para* to the position of attack are electron deficient. This means that any pre-existing substituent will produce maximum effect if it is located in any of these positions. We normally think in terms of the existing substituent directing the attack of the electrophile to a position that optimizes the stability of the arenium ion. Electron-releasing substituents are thus *ortho* and *para* directing because they help to stabilize the arenium ion; electron-withdrawing substituents destabilize the arenium ion more if they are *ortho* or *para*, and, consequently, they are found to be *meta* directing. The observed electron-releasing and electron-withdrawing properties of various groups are summarized in Table 8.1, though to understand these

Table 8.1   Directing effects of substituents in electrophilic aromatic substitution

Electron-releasing groups: *ortho* and *para* directors

Strong: $-\ddot{N}H_2$ $-\ddot{O}H$ $-\ddot{O}\colon^{\ominus}$ $-\ddot{O}R$ $-\ddot{N}R_2$

Moderate: (structures showing $-N(H)-C(=\ddot{O})R$ amide, $-N(H)-SO_2R$, and $-\ddot{O}-C(=\ddot{O})R$ ester)

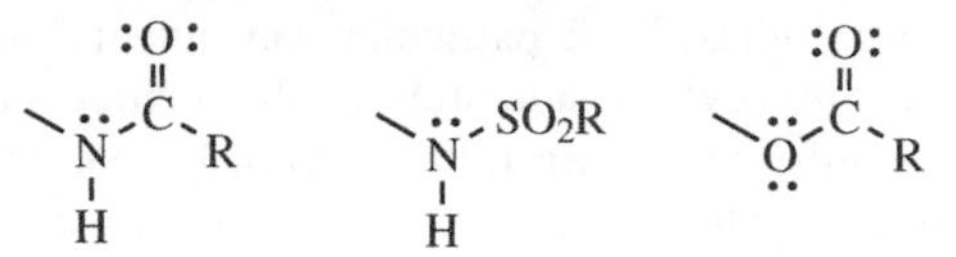

Weak: $-R$ $-Ar$

Electron-withdrawing groups: *meta* directors

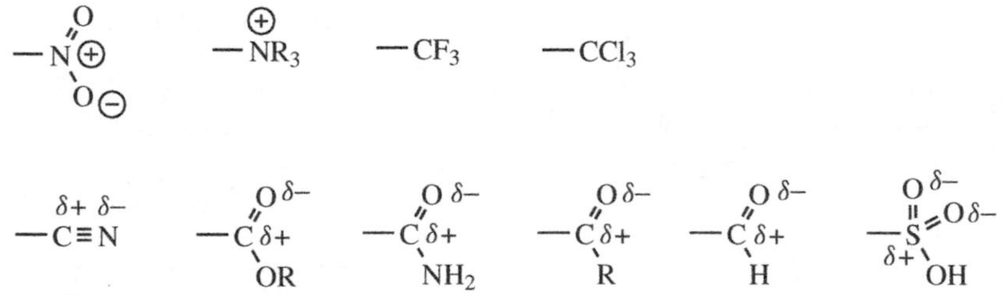

Strong: $-\overset{O}{\underset{O_{\ominus}}{\overset{\|}{N}}}{}^{\oplus}$ $-\overset{\oplus}{N}R_3$ $-CF_3$ $-CCl_3$

Moderate: $\overset{\delta+ \;\; \delta-}{-C\equiv N}$ $-\underset{OR}{\overset{O\,\delta-}{\overset{\|}{C}}}{}^{\delta+}$ $-\underset{NH_2}{\overset{O\,\delta-}{\overset{\|}{C}}}{}^{\delta+}$ $-\underset{R}{\overset{O\,\delta-}{\overset{\|}{C}}}{}^{\delta+}$ $-\underset{H}{\overset{O\,\delta-}{\overset{\|}{C}}}{}^{\delta+}$ $-\underset{\delta+\;\;OH}{\overset{O\,\delta-}{\overset{\|}{S}}}{=}O\,\delta-$

Halogens (σ withdrawers, π donors): *ortho* and *para* directors

    $-F$ $-Cl$ $-Br$ $-I$

fully we need to consider the properties in terms of both **inductive effects** and delocalization or **resonance effects**. We must also appreciate that the term 'directing' indicates the major product(s) formed, and not the exclusive product. Mixtures of products are the norm.

Stabilization of the arenium ion through electron-donating effects is typical of alkyl substituents. This is not strictly an inductive effect (see Section 4.3.3), but is derived from overlap of σ orbitals with the aromatic π orbital system. Thus, with both *ortho* and *para* addition there is one resonance form of the arenium ion that is particularly favourable, in that the positive charge is adjacent to the electron-donating alkyl group, which thus helps to disperse the charge, as with a carbocation (see Section 6.2.1). There is no particularly favourable resonance form resulting from *meta* addition.

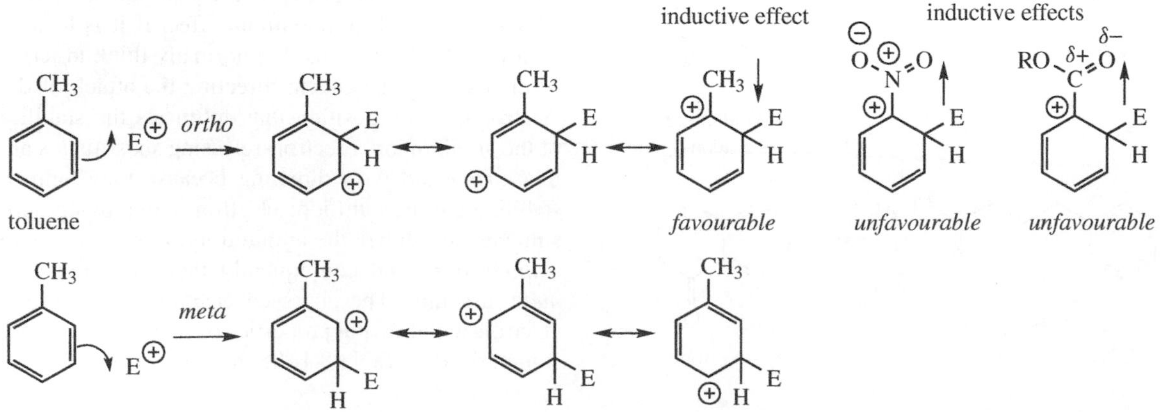

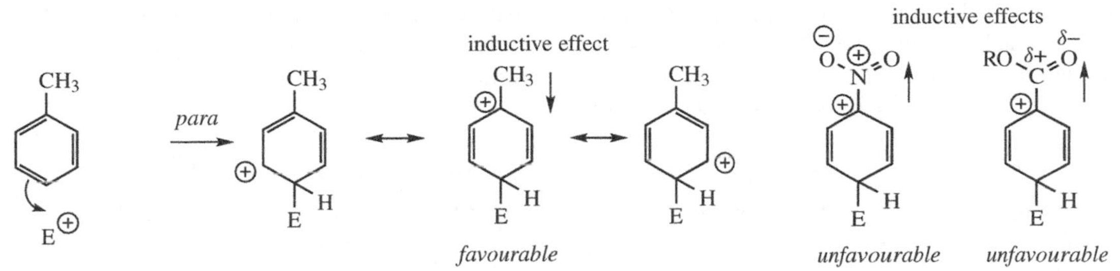

inductive effects

inductive effect

para

favourable unfavourable unfavourable

Nitration of toluene gives approximately 59% *o*-nitrotoluene, 37% *p*-nitrotoluene, and only 4% *m*-nitrotoluene. On the other hand, nitration of nitrobenzene gives about 93% *m*-dinitrobenzene, 6% *o*-dinitrobenzene, and 1% *p*-dinitrobenzene. Since nitro is an electron-withdrawing group, those resonance forms stabilized by the presence of an alkyl substituent are going to be seriously destabilized when alkyl is replaced by an electron-withdrawing group, and so *meta* substitution predominates. This can be appreciated even more readily when one looks at the representation of the nitro group or, say, an ester group: the unfavourable resonance forms have the positive charge positioned adjacent to a full or partial positive charge. However, we must also realize that toluene undergoes electrophilic substitution some 25 times as readily as benzene because of the beneficial inductive effect, whereas nitrobenzene undergoes nitration only about 10^{-4} times as readily as benzene because the inductive effect withdraws electrons from the ring.

Groups that are particularly strong electron releasers do not achieve this by an inductive effect, but they have heteroatoms with lone pair electrons that are able to stabilize resonance structures by transferring the charge to the heteroatom, i.e. an electron-releasing **resonance effect**. An amino group is typical of this type of substituent.

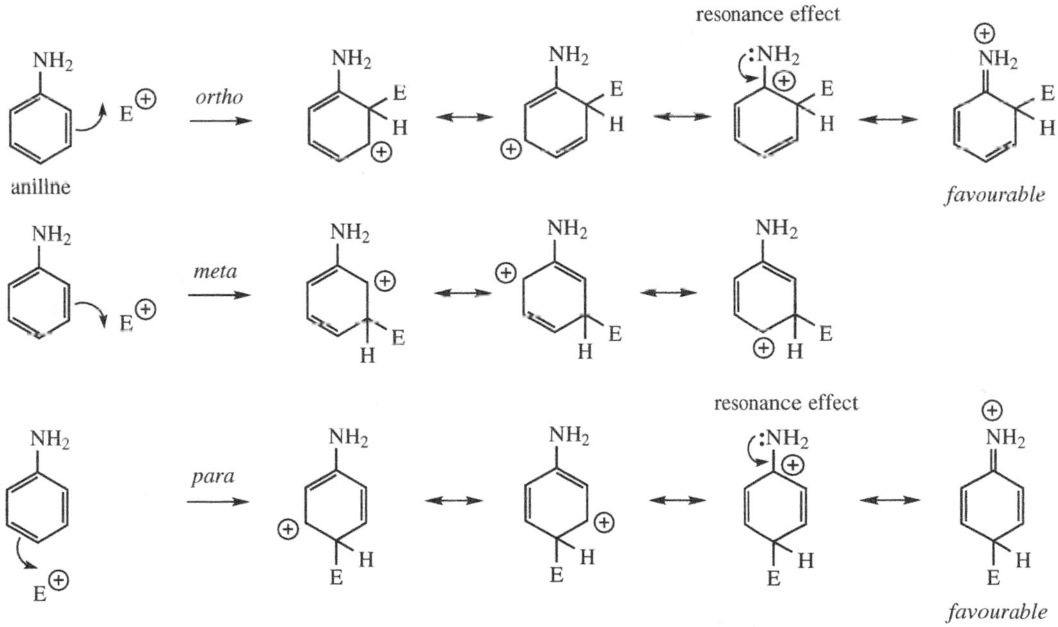

resonance effect

aniline

ortho

favourable

meta

resonance effect

para

favourable

It can be seen that lone pair donation creates another favourable resonance form only when electrophilic attack is *ortho* or *para* to the amino group. There is a small electron-withdrawing inductive effect for an amino group due to the heteroatom, but this effect is vastly overpowered by the resonance effect, and an amino group is strongly activating and gives rise to *ortho* and *para* substitution. In fact, aniline reacts readily with bromine in water, without any need for a catalyst, giving 2,4,6-tribromoaniline in nearly quantitative yield. The same is true for phenol, which rapidly gives 2,4,6-tribromophenol, because of

the powerful activation provided by the phenol group. The activation is so great that all three positions are brominated.

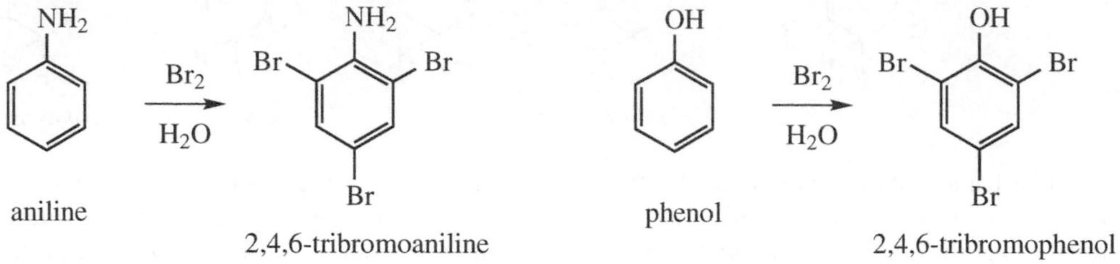

aniline

2,4,6-tribromoaniline

phenol

2,4,6-tribromophenol

Groups such as amides and esters, where the heteroatom is bonded to the aromatic ring, might be expected to behave similarly; this is true, but the level of activation is markedly less than for amines and phenols. We can understand this because of

two conflicting types of resonance behaviour in such molecules. These groups do activate the ring towards electrophilic attack, and are *ortho* and *para* directing, but activation is considerably less than with simple amino and phenolic groups.

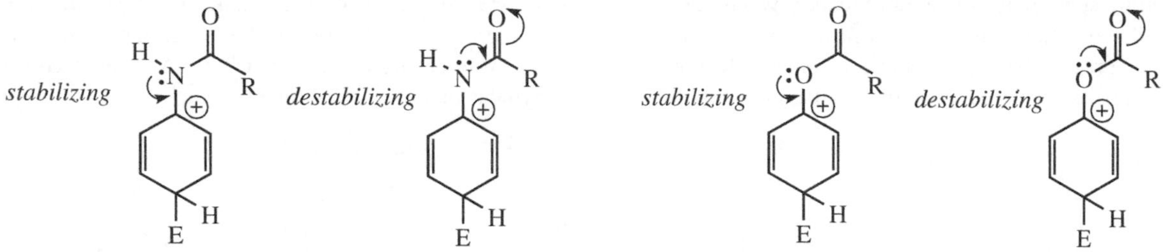

For most substituents, electron-donating ones activate the ring towards electrophilic attack and also direct *ortho* and *para*. Conversely, electron-withdrawing substituents are deactivating and direct substitution *meta*. This appears so straightforward a concept that we must have an exception; this is found with halogen substituents. Thus, chlorobenzene is nitrated about 50 times more slowly than benzene, but yields *o*- and *p*-nitro products. However, the explanation is simple, and does not alter our reasoning. It turns out that, because of the high electronegativity of halogen atoms, we have

a very strong electron-withdrawing inductive effect and, consequently, significant deactivation towards electrophilic substitution. However, because there is an electron-donating resonance effect we get *ortho* and *para* substitution. This lone pair donation is not nearly as effective as with oxygen and nitrogen, however, because in the larger atom the orbitals are less able to overlap effectively. So we have the conflicting trends, deactivation from a strong inductive effect through σ bonds, but *ortho* and *para* directing because of a weak resonance effect through the π bond system.

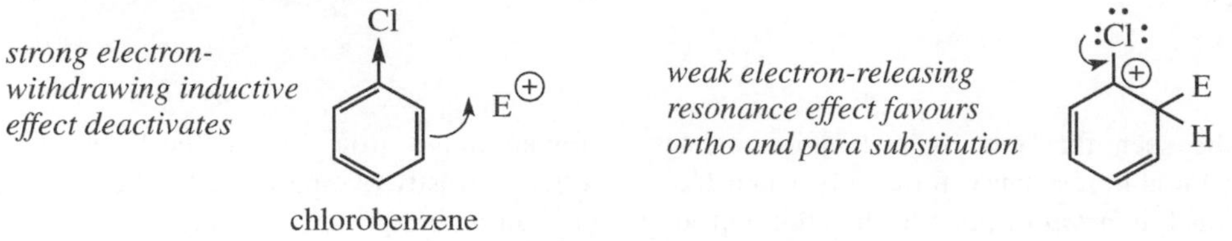

strong electron-withdrawing inductive effect deactivates

chlorobenzene

weak electron-releasing resonance effect favours ortho and para substitution

An understanding of electron-donating and electron-withdrawing substituent effects is crucial to designing the synthesis of aromatic derivatives. For example, from the electrophilic substitution reactions we have studied, there are two potential approaches for the synthesis of *m*-nitroacetophenone:

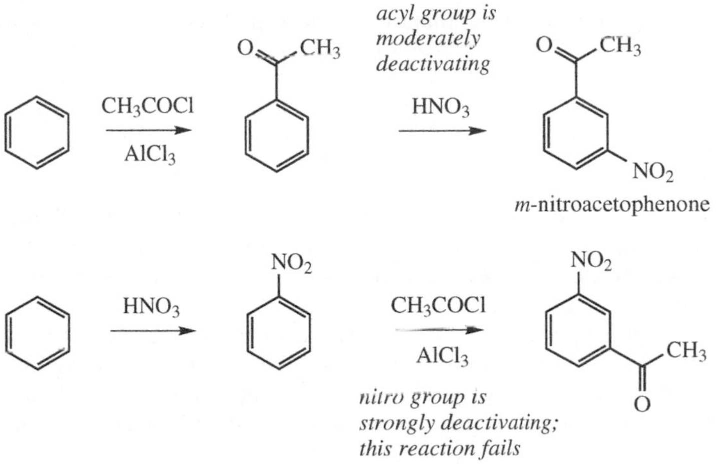

m-nitroacetophenone

acyl group is moderately deactivating

nitro group is strongly deactivating; this reaction fails

Only the first of these is effective, because strongly deactivating groups such as nitro almost completely inhibit Friedel–Crafts acylation (or alkylation), and the alternative sequence shown will fail at the second step. Accordingly, the workable route inserts the less effective deactivating group, the acyl group, first, so that the second electrophilic substitution can proceed, even though it tends to be fairly slow.

Although electron-donating substituents activate the ring towards electrophilic attack, they are both *ortho* and *para* directing, and an electrophilic substitution reaction can be expected to yield a mixture of products that must be separated. In practice, this problem can be minimal because of steric considerations. When the original substituent is large, or the incoming substituent is large, the steric interaction will be considerably less with *para* substitution than with *ortho*. Thus, both nitration of acetanilide and acylation of toluene give predominantly the *para* product. Note that small amounts of the *meta* product are inevitably formed as well as the *ortho* and *para* products; these reactions are only **regioselective**.

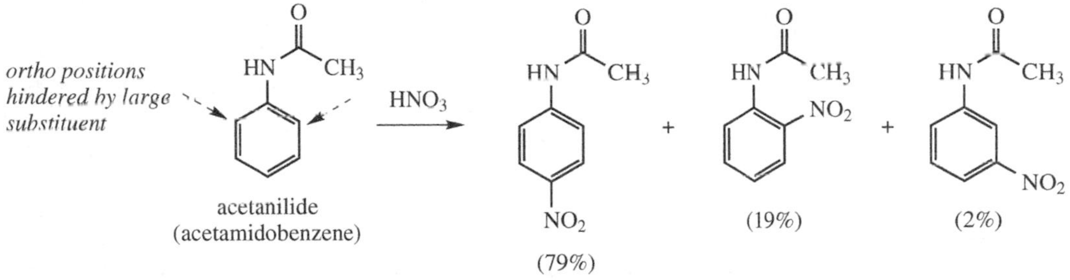

ortho positions hindered by large substituent

acetanilide
(acetamidobenzene)

(79%)   (19%)   (2%)

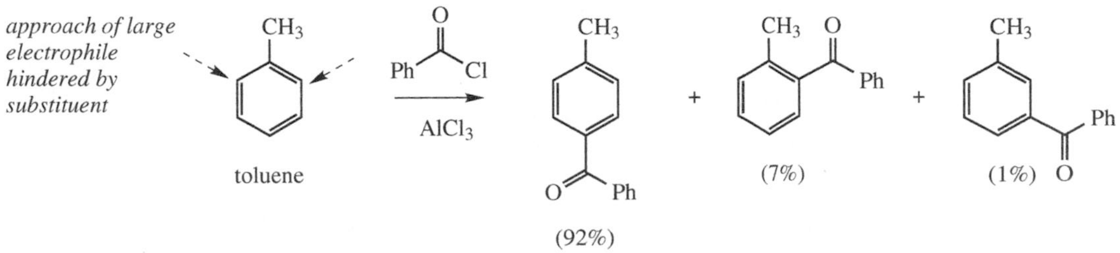

approach of large electrophile hindered by substituent

toluene

(92%)   (7%)   (1%)

This is a good time to have another brief look at Sections 4.3.5 and 4.5.4, and compare how we used similar reasoning to consider the likely stability, or otherwise, of anions and cations in order to predict the acid–base properties of aromatic amines and phenols. The rationalizations are essentially identical.

The effect of heteroatoms on electrophilic aromatic substitution, e.g. the reactions of pyridine, will be considered separately in Chapter 11.

Box 8.4

Synthesis of ibuprofen

There are several approaches to the synthesis of the **analgesic** anti-inflammatory drug **ibuprofen**. Here is one that employs a relatively simple sequence of reactions, beginning with a **Friedel–Crafts acylation** of isobutylbenzene. The alkyl substituent is weakly electron releasing, and thus activates the ring towards electrophilic substitution. It also directs further substitution to the *ortho* and *para* positions. As in most Friedel–Crafts acylations, the *para* product predominates strongly over the *ortho*, a consequence of the relatively large size of the electrophilic reagent (see Section 8.4.3). In this case, we also have a quite large alkyl substituent, again disfavouring the *ortho* product. The subsequent steps are relatively straightforward. Sodium borohydride reduction of the ketone gives an alcohol (see Section 7.5), then the alcohol is converted into a nitrile by successive nucleophilic substitution reactions.

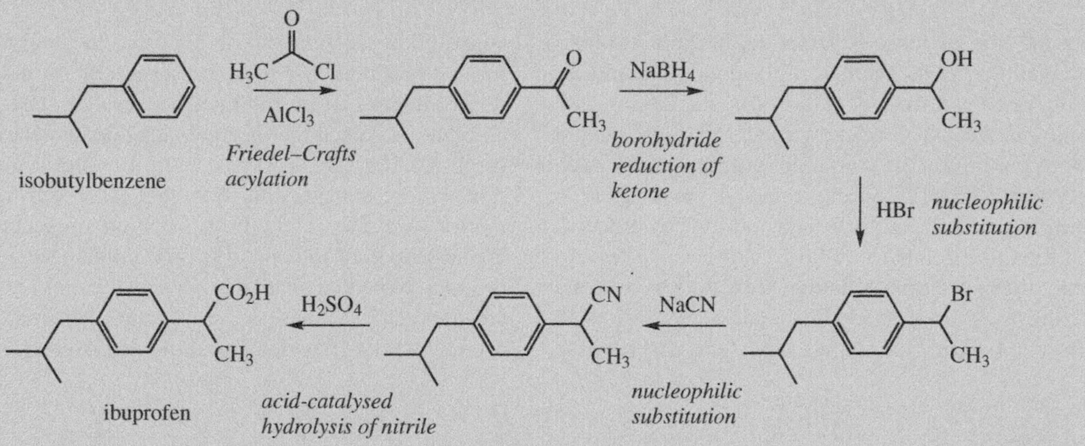

Note that a two-stage process is involved. Since hydroxide is a poor leaving group, nucleophilic substitution requires acidic conditions to protonate the hydroxyl to provide a better leaving group (see Section 6.1.4). We can formulate an S_N1 conversion, since this would involve a favourable benzylic carbocation. HCN is a weak acid (pK_a 9.1), so it is not very effective in protonating the hydroxyl group. Thus, the two-stage process is used, with displacement of hydroxyl via bromide, then subsequent displacement of bromide by cyanide, the latter step usually being an S_N2 process. Lastly, the nitrile group is hydrolysed to a carboxylic acid (see Box 7.9).

The starting material, isobutylbenzene, is readily available, but could be synthesized by exploiting another Friedel–Crafts reaction.

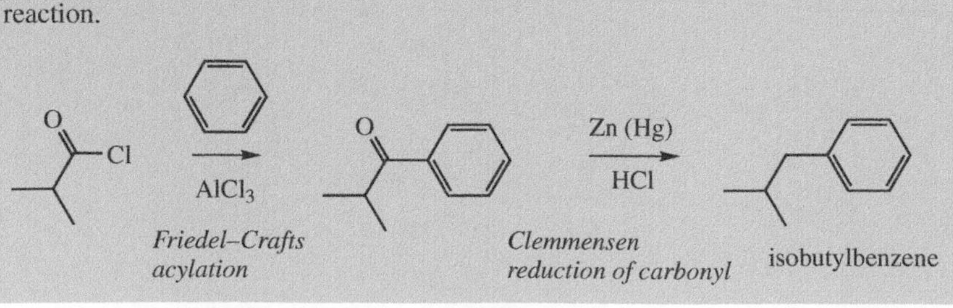

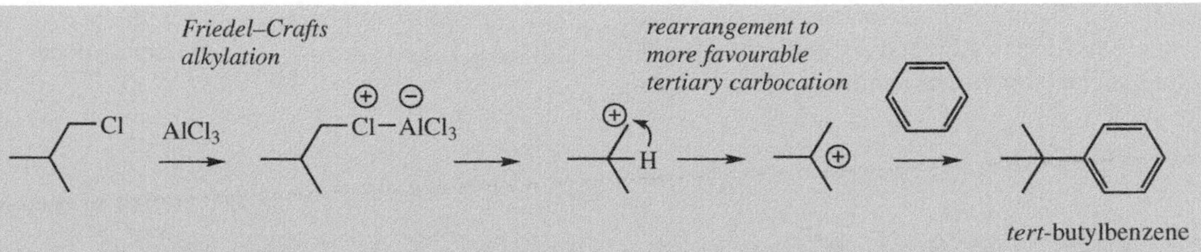

tert-butylbenzene

Note that a Friedel–Crafts alkylation is not a good idea. There is too much chance of rearrangement occurring, since we are trying to generate the equivalent of a primary carbocation. We might expect that rearrangement of the primary carbocation to a tertiary carbocation by hydride migration would occur, so that the product would turn out to be *tert*-butylbenzene rather than isobutylbenzene. The approach then is to use Friedel–Crafts acylation, then reduce the carbonyl group by an appropriate method, here a Clemmensen reduction (see below).

The substituents that can be introduced by electrophilic substitution appear somewhat limited, but there exist standard chemical processes for converting these into other functional groups, thereby extending significantly the scope for this type of process. A few of these are shown below, though they will not be elaborated upon here.

Some useful functional group transformations

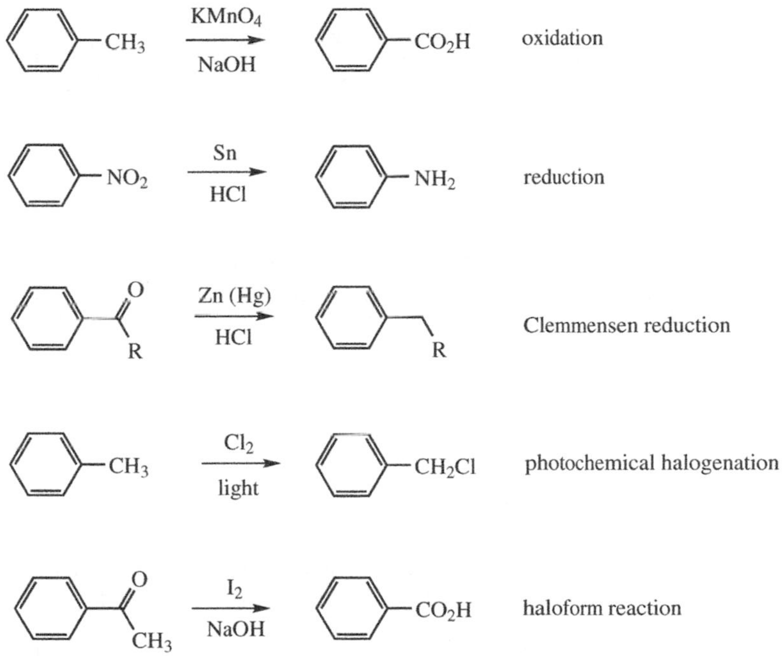

8.4.4 *Electrophilic substitution on polycyclic aromatic compounds*

Fused-ring cyclic hydrocarbons such as naphthalene and anthracene display the enhanced stability and reactivity associated with simple aromatic compounds like benzene. We have briefly looked at the π electron systems and their aromatic status in Chapter 2. In this short section, we wish to demonstrate how the principles developed above for rationalizing the behaviour of benzene compounds can be extended to the more complex ring systems.

We observe that nitration of naphthalene using nitric acid–sulfuric acid gives predominantly 1-nitronaphthalene (sometimes α-nitronaphthalene), and that Friedel–Crafts acylation with acetyl chloride–AlCl₃ gives mainly 1-acetylnaphthalene.

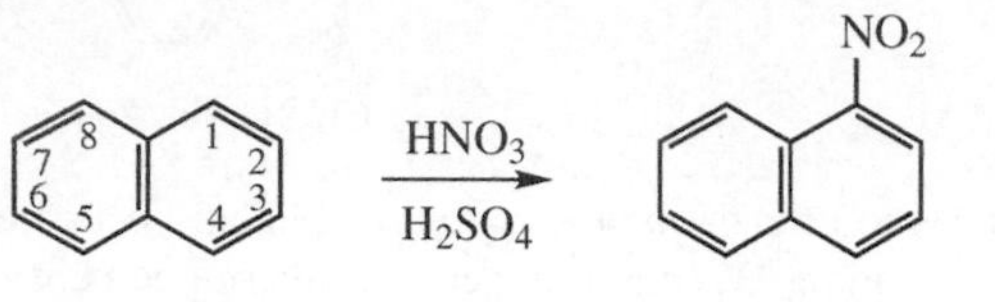

naphthalene   1-nitronaphthalene 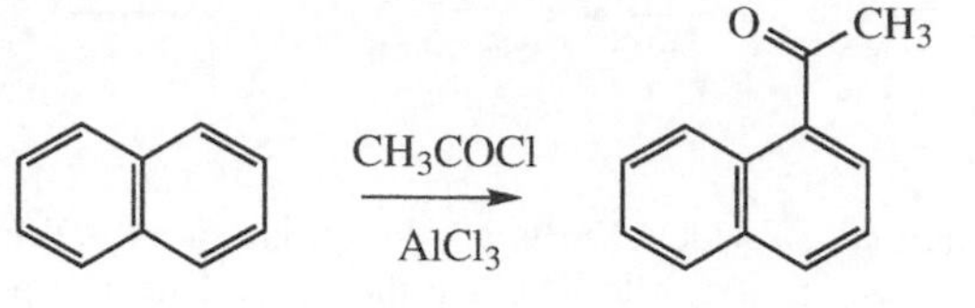   1-acetylnaphthalene

This behaviour can readily be explained. Let us simply consider the resonance structures for the intermediate cation following attack of electrophile at position 1 (α position) or at position 2 (β position). On drawing these out, we find that two of the five structures retain a benzene ring if attack occurs at position 1. For attack at position 2, only one resonance structure has a benzene ring. We know that a benzene ring has special stability (see Section 2.9.1), so we can predict that the intermediate cation with more benzenoid resonance structures should be the more stable. This fits with the observation that electrophilic substitution occurs predominantly at position 1.

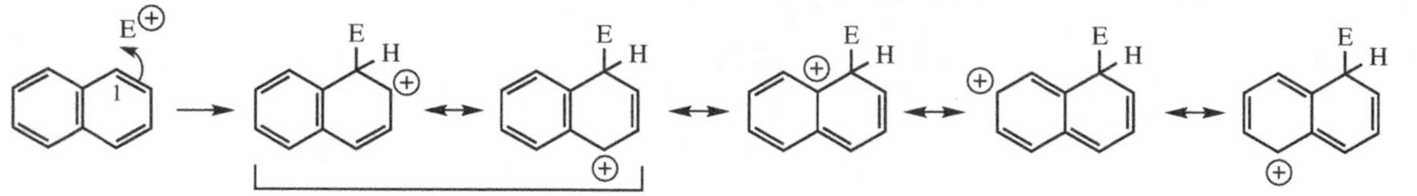

*two resonance structures
retain aromatic benzene ring*

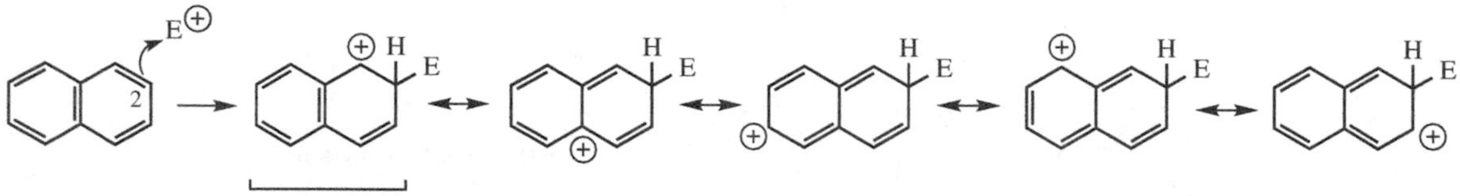

*only resonance
structure that retains
aromatic benzene ring*

Using the same reasoning, it is not difficult to see why anthracene becomes substituted on the central ring. The intermediate cation then benefits from the stability of two benzene rings, which is actually substantially more than for a single naphthalene ring. Anthracene undergoes aromatic substitution more readily than naphthalene, and can frequently lead to disubstitution, with both substituents on the central ring.

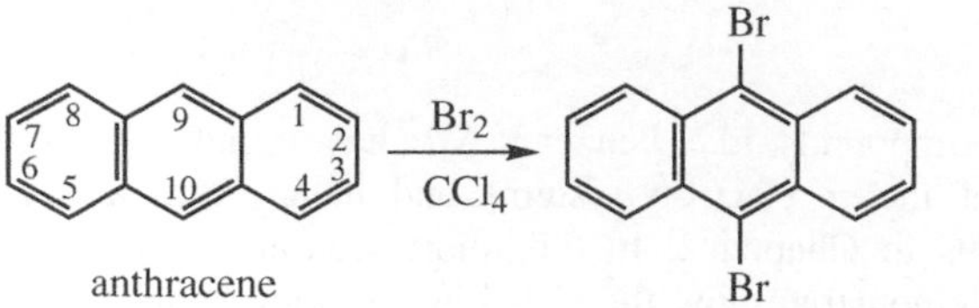

We can also rationalize how a substituent on naphthalene will direct further substitution. If we have an activating group at position 1, electrophilic attack will occur on the same ring and at positions 2 or 4. Consideration of resonance structures shows that the benzene ring can be retained whilst providing favourable structures in which an electron-releasing group minimizes the charge. Further, those groups

anthracene   9,10-dibromoanthracene

that are electron-releasing through lone pair donation, e.g. NH$_2$ or OH, are ideally placed to delocalize charge. This is shown in the case of *para* and *ortho* attack.

X = electron releasing substituent

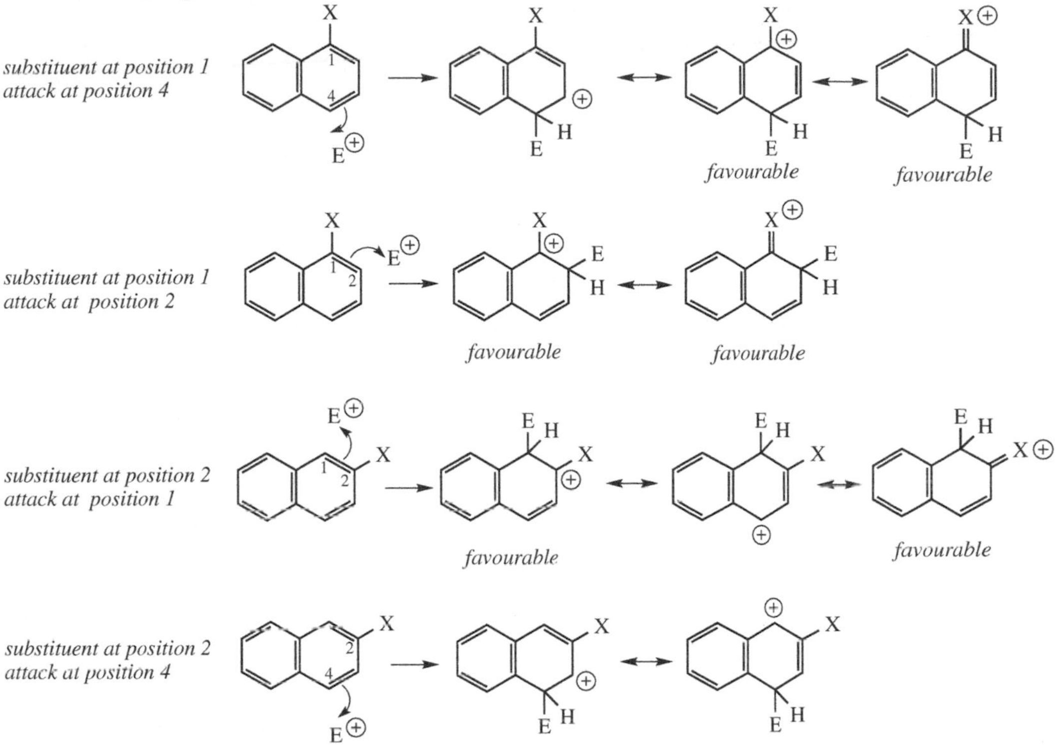

substituent at position 1
attack at position 4

favourable   *favourable*

substituent at position 1
attack at position 2

favourable   *favourable*

substituent at position 2
attack at position 1

favourable   *favourable*

substituent at position 2
attack at position 4

Should the activating substituent be at position 2, further substitution will be almost exclusively at position 1; this follows from consideration of resonance structures, where the 2-substituent has minimal effect if attack occurs at position 4. Of course, this would equate to *meta* attack, which we know is unfavourable for an *ortho* and *para* director (see Section 8.4.3).

Deactivating (electron-withdrawing) substituents do just that: they deactivate the ring to which they are attached and hinder any further attack. Hence, further electrophilic substitution occurs on the other ring whether the substituent is at C-1 or C-2. Further substitution occurs at positions 5 or 8, the positions most susceptible to attack.

These trends are summarized below, though we recommend deducing the reactivity rather than committing it to memory.

position of further electrophilic substitution

substituent activating

substituent deactivating

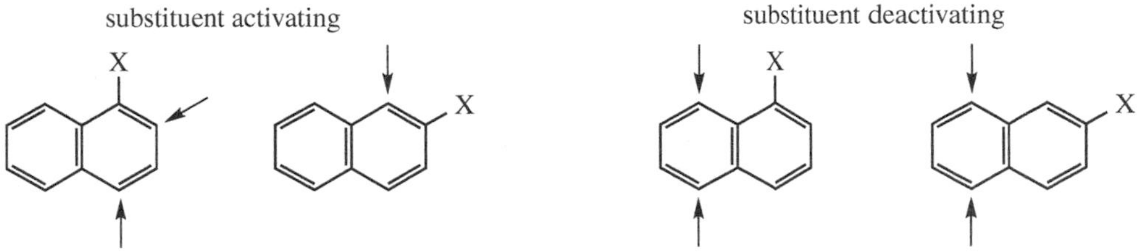

9

Radical reactions

9.1 Formation of radicals

The ionization of HBr distributes the two electrons of the single H–Br bond so that the electronegative bromine accepts electrons whilst hydrogen loses electrons, and the resultant ions are thus H^+ and Br^-. This process is termed **heterolytic cleavage**, in that the two atoms of the bond suffer different fates and that the two electrons are distributed unevenly. In marked contrast, it is possible for the two electrons of the single bond to be distributed evenly, so that one electron becomes associated with each atom. This is termed **homolytic cleavage**, and it generates radicals (often termed free radicals). A **radical** may be defined as a high-energy species carrying an unpaired electron. Note that, to indicate movement of just one electron, we use a **fish-hook curly arrow** in mechanisms (see Section 5.2) rather than the normal curly arrow, which denotes movement of two electrons.

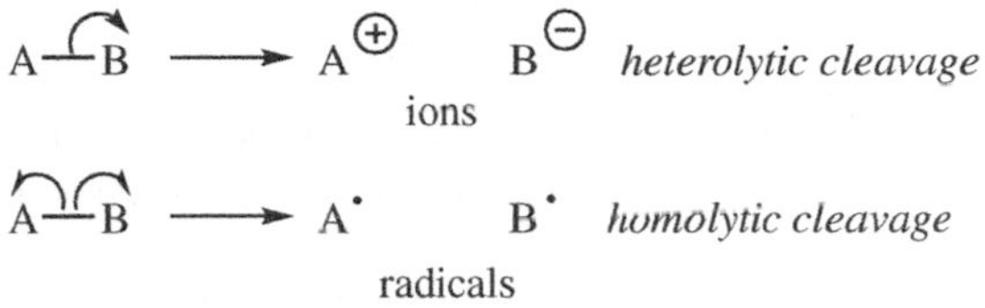

A—B ⟶ A⊕ B⊖ *heterolytic cleavage*
ions

A—B ⟶ A• B• *homolytic cleavage*
radicals

Radicals may be generated in two general ways:

• by homolysis of weak bonds;

• by reaction of molecules with other radicals.

Homolytic cleavage of most σ bonds may be achieved if the compound is subjected to a sufficiently high temperature, typically about 200 °C. However, some weak bonds will undergo homolysis at temperatures little above room temperature. Bonds of peroxy and azo compounds fall in this category, and such compounds may be used to initiate a radical process. Di-*tert*-butyl peroxide, dibenzoyl peroxide

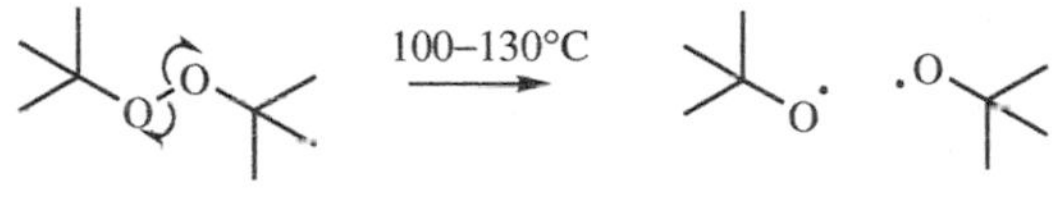

di-*tert*-butyl peroxide   100–130°C   *tert*-butoxyl radicals

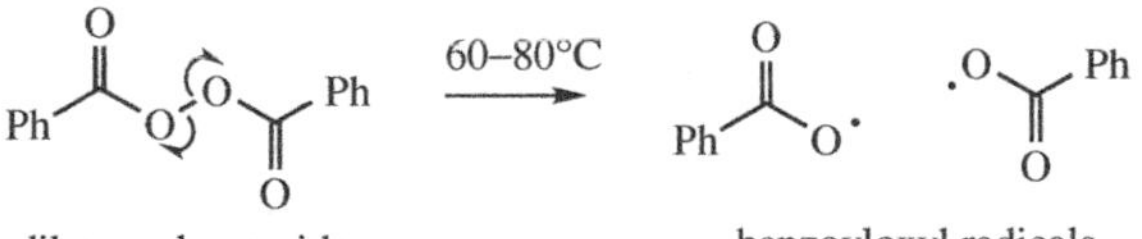

dibenzoyl peroxide   60–80°C   benzoyloxyl radicals

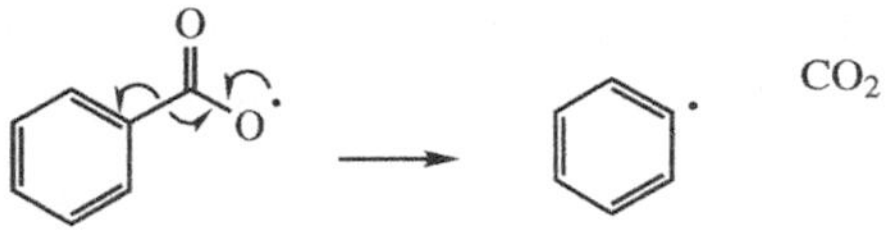

benzoyloxyl radical   phenyl radical CO_2

Essentials of Organic Chemistry   Paul M Dewick
© 2006 John Wiley & Sons, Ltd

and azoisobutyronitrile (AIBN) are good sources of radicals under typical reaction conditions.

At increased temperatures, the peroxide bond is cleaved homolytically, giving radicals. Dibenzoyl peroxide is a diacyl peroxide and cleaves rather more readily than the dialkyl peroxide, but further decomposition then occurs in which carbon dioxide is lost, and the phenyl radical is produced. This displacement is favoured by the inherent stability of carbon dioxide.

Homolytic cleavage of diazo compounds such as AIBN is also driven by the stability of a neutral molecule, this time molecular nitrogen, and two alkyl radicals are produced.

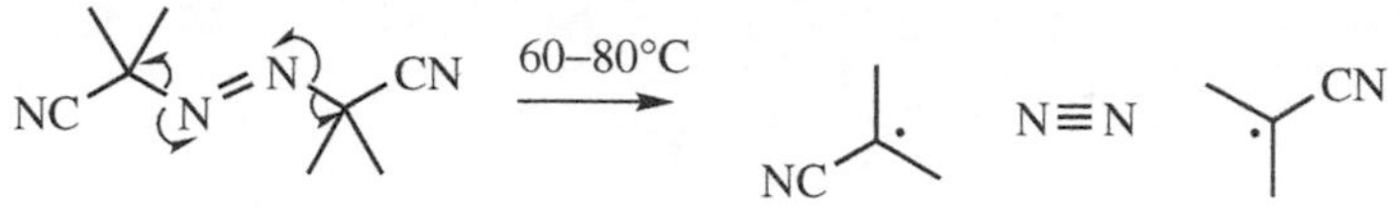

azoisobutyronitrile (AIBN)   2-cyano-2-propyl radical

An alternative approach to homolytic cleavage is **photolysis**, the absorption of light energy, especially UV radiation. Thus, halogen molecules are easily photolysed to generate halogen radicals.

homolytic cleavage of halogen molecules

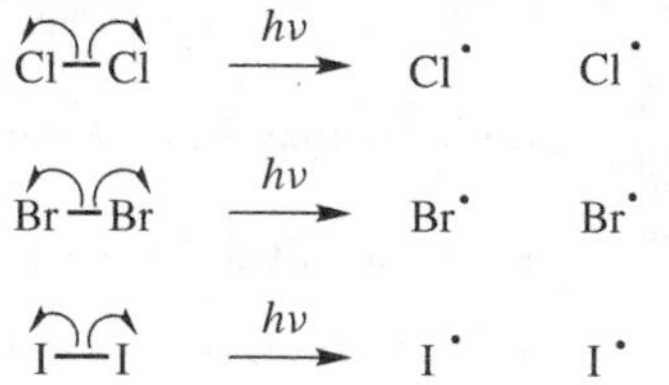

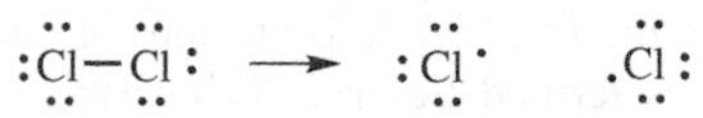

seven electrons in outer shell

$h\nu$ is the accepted abbreviation for electromagnetic radiation

The halogen molecule is comprised of two halogen atoms each with seven electrons in their outer shell. Sharing of the unpaired electrons creates a stable molecule in which each atom has now acquired an octet of electrons in its outer shell. By absorbing energy, we have removed this stabilization and effectively generated halogen atoms, which are our radicals.

Radicals formed in one of these initiation reactions may themselves be the means of producing other radicals, by reacting with another molecular species. Abstraction of a hydrogen atom is a particularly common reaction leading to a new radical.

radical abstracts hydrogen atom, generating a new radical

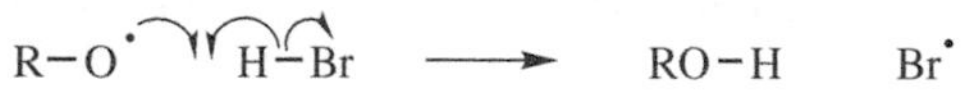

instead of showing all the electron movements, we could write the mechanism like this:

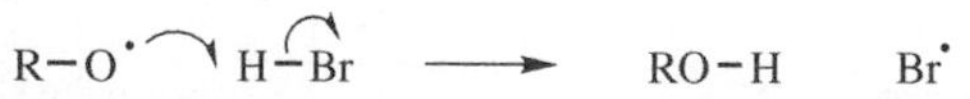

Thus, **abstraction of a hydrogen atom** from HBr generates a bromine radical. Note that, for convenience, we tend not to put in all of the electron movement arrows. This simplifies the representation, but is more prone to errors if we do not count electrons. Our attacking radical has an unpaired electron, and it abstracts the proton plus one of the electrons comprising the H–Br σ bond, i.e. a hydrogen atom, and the remaining electron from the bond now resides with the bromine in the form of a bromine radical. This is shown as a one-electron mechanism, and should be compared with the analogous two-electron mechanisms that account for acidity and S_N2 reactions. The only difference is in the number of electrons involved, which we indicate by the fish-hook or normal curly arrow.

Compare:

one-electron mechanism   R–O• ⌢ H–Br ⟶ RO–H   Br•   *hydrogen atom abstraction*

two-electron mechanism   R–O⊖ ⌢ H–Br ⟶ RO–H   Br⊖   *proton removal – acidity*

two-electron mechanism   R–O⊖ ⌢ H₃C–Br ⟶ RO–CH₃   Br⊖   *S_N2 reaction*

Another possibility is that we can get **radical addition** to an unsaturated molecule, e.g. an alkene. Writing in all the electron movement arrows, we have one of the double bond π electrons being used to make the new σ bond with the original radical species, whilst the second π electron becomes located on the other end of the double bond, and is now the unpaired electron of the new radical. The original radical could potentially have attacked at either end of the double bond; the regiochemistry of addition is governed by the stability of the radical generated (see below).

radical addition to alkene

more favoured tertiary radical

we could have written the mechanism in either of these ways:

the first version is perhaps more commonly used, in that it considers the radical as the attacking species;

however, compare the second one with the electrophilic addition mechanism

electrophilic addition to alkene

Note that if we choose not to put in all the curly arrows, we could write the mechanism in two ways: either considering the radical as the attacking species or the double bond as the electron-rich species. The first version is perhaps more commonly used, but it is much more instructive to compare the second one with an electrophilic addition mechanism (see Section 8.1). The rationalization for the regiochemistry of addition parallels that of carbocation stability (see Section 8.2).

9.2 Structure and stability of radicals

Most radicals have a planar or nearly planar structure. Carbon is sp^2 hybridized in the methyl radical, giving three σ C–H bonds, and the single electron is held in a $2p$ orbital that is oriented at right angles to the plane of the radical.

•CH₃

methyl radical

planar structure with unpaired electron in p orbital

Although a radical is neutral, it is an electron-deficient species that will be very reactive as it attempts to pair off the odd electron. Because radicals are electron deficient, electron-releasing groups such as alkyl groups tend to provide a stabilizing effect. The more electron-releasing groups there are, the more stable the radical. Thus, tertiary radicals are more stable than secondary radicals, which in turn are more stable than primary radicals.

relative stabilities:

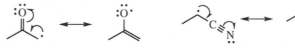

tertiary radical   secondary radical   primary radical   methyl radical

overlap from σ bond into
singly occupied p orbital

The order of stability is thus the same as with car-bocations, another electron-deficient species, and for the same reason. There is favourable delocalization of the unpaired electron through overlap of the σ C–H (or C–C) bond into the singly occupied *p* orbital of

the radical (see Section 6.2.1). The similarity continues, in that **resonance delocalization** also helps to stabilize a radical, so that the **allyl radical** and the **benzyl radical** are more stable than an alkyl radical (compare Section 6.2.1).

allyl radical stabilized by
resonance delocalization

benzyl radical stabilized by
resonance delocalization

Electron-donating functional groups, e.g. ethers, also stabilize radicals via their lone pair orbitals. However, electron-withdrawing groups can also sta-bilize radicals, so that radicals next to carbonyl or nitrile are more stable than even tertiary alkyl radi-cals. This is because these groups possess a π electron system and the unpaired electron can take advantage

of this (compare carbanions, Section 10.4). It tran-spires that features that stabilize an anion, e.g. an electron-withdrawing group, features that stabilize a carbocation, e.g. electron-donating groups, or features such as conjugation that may stabilize either, will all stabilize a radical.

electron-donating group   *electron-withdrawing groups*

radical adjacent to ether   radical adjacent to carbonyl   radical adjacent to nitrile

There is a significant difference between carboca-tions and radicals when we are thinking about sta-bility, however. One of the more confusing aspects relating to carbocations was their ability to rearrange, either by migration of an alkyl group or of hydride, when a more stable system might be attained by this means (see Section 6.4.2). We related this trend to the enhanced stability of, say, a tertiary or allylic carbo-cation over secondary or primary carbocations. Now, although we also find tertiary or allylic radicals are more favourable than secondary or primary radicals, we do not encounter rearrangements with radicals, even if the product radical is more stable. This comes from an increased energy barrier to rearrangement in radicals compared with carbocations, which in turn

relates to the extra unpaired electron in the radical, which has to occupy a higher energy orbital in the transition state.

9.3 Radical substitution reactions: halogenation

Halogenation reactions of alkanes provide good examples of radical processes, and may also be used to illustrate the steps constituting a **radical chain reaction**. Alkanes react with chlorine in the presence of light to give alkyl chlorides, e.g. for cyclohexane the product is cyclohexyl chloride.

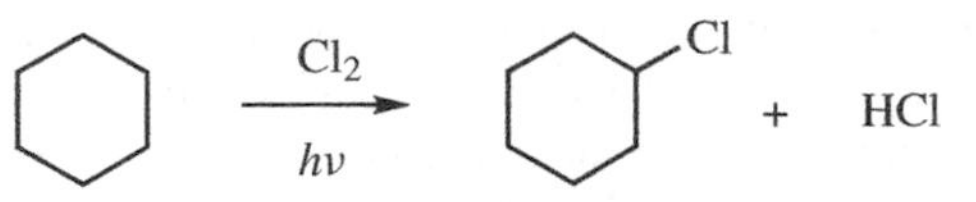

cyclohexane   cyclohexyl chloride

The **initiation step** is the light-induced formation of chlorine atoms as the radicals. Only a few chlorine molecules will suffer this fate, but these highly reactive radicals then rapidly interact with the predominant molecules in the system, namely cyclohexane.

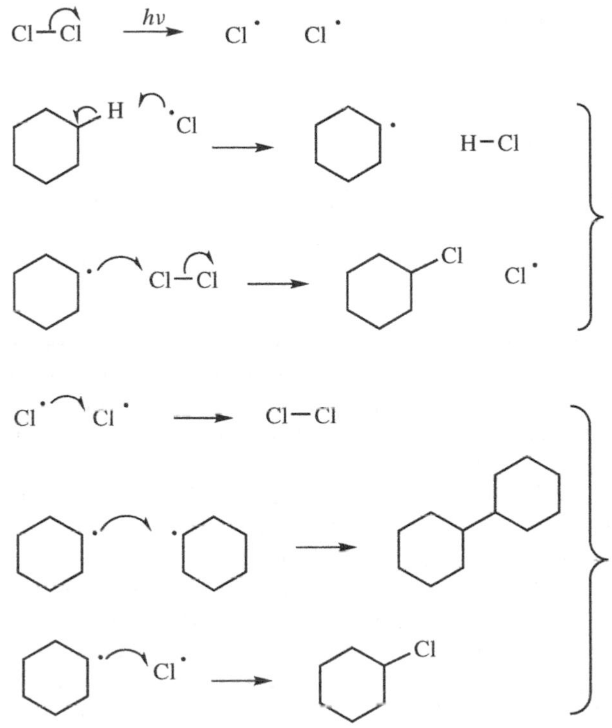

initiation step

propagation steps

termination steps

The chlorine radicals abstract hydrogen atoms from the cyclohexane substrate, producing new radicals, i.e. cyclohexyl radicals. These, in turn, cause further dissociation of chlorine molecules and the production of more chlorine radicals. The cyclohexyl radical reacts with a chlorine molecule rather than, say, a further molecule of cyclohexane simply because bond energies dictate it is easier to achieve fission of the Cl–Cl bond than the C–H bond. This results in production of cyclohexyl chloride and a further chlorine radical. The chlorine radical can now abstract hydrogen from another cyclohexane substrate, and we get a repeat of the same reaction sequence, the so-called **propagation steps** of this chain reaction. During the propagation steps, one radical is used to generate another, so that only one initiation reaction is required to generate a large number of product molecules.

Finally, when we are running out of cyclohexane, the process terminates by the interaction of two radical species, e.g. two chlorine atoms, two cyclohexyl radicals, or one of each species. The combination of two chlorine atoms is probably the least likely of the **termination steps**, since the Cl–Cl bond would be the weakest of those possible, and it was light-induced fission of this bond that started off the radical reaction. Of course, once we have formed cyclohexyl chloride, there is no reason why this should not itself get drawn into the radical propagation steps, resulting in various dichlorocyclohexane products, or indeed polychlorinated compounds. Chlorination of an alkane will give many different products, even when the amount of chlorine used is limited to molar ratios, and in the laboratory it is not going to be a particularly useful process.

However, it is instructive to consider radical chlorination of alkanes just a little further, to appreciate the mechanistic concepts. If we carry out light-induced chlorination of propane, then we obtain two different monochlorinated products, but not in equal amounts. There will also be other products containing more than one chlorine atom. A similar situation pertains if we chlorinate 2-methylpropane.

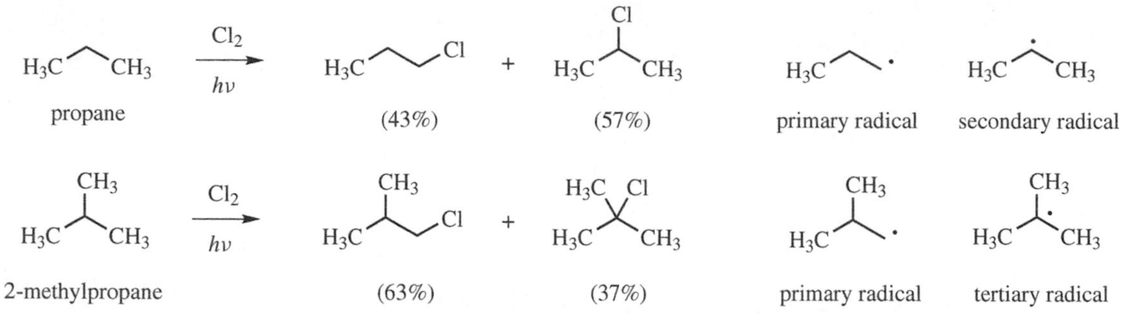

propane	
(43%)	(57%)
primary radical	secondary radical

2-methylpropane	
(63%)	(37%)
primary radical	tertiary radical

The proportion of each product formed can be rationalized by considering a number of factors. First, the products from propane are the result of generating either primary or secondary radicals. We know that tertiary radicals are more favourable than secondary radicals, which in turn are more favourable than primary radicals. It is also true that tertiary C–H bonds are slightly weaker than secondary C–H bonds, which in turn are slightly weaker than primary C–H bonds. It is thus rather easier to break tertiary C–H bonds by the hydrogen abstraction reaction, followed by secondary C–H bonds, and then primary C–H bonds. On the other hand, there is a statistical factor, in that there are six primary hydrogens in propane and only two secondary ones, so it is more likely that a primary C–H bond is attacked by the very reactive radical. The net result of these two opposite trends is the slight excess of the secondary halide product. With 2-methylpropane, the statistical factor is even more pronounced (only one tertiary hydrogen to nine primary hydrogens), and hence we get rather more primary product in the reaction mixture, even though tertiary radicals are the more stable and a tertiary C–H bond is the weakest. In fact, because the chlorine radical is so reactive, the variation in bond strengths is not an especially important factor. It can readily be

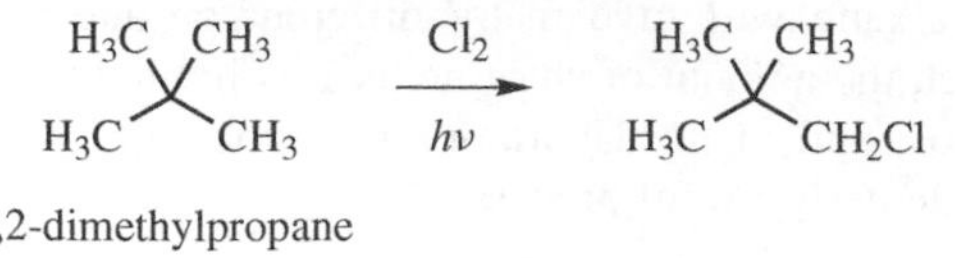

appreciated that, even under conditions in which we can maximize monochlorination, it is highly desirable if there is no chance of forming isomers that have to be separated. Substrates that meet these criteria include cyclohexane and 2,2-dimethylpropane.

Bromine will also halogenate alkanes, but in this case we find that bromine is considerably less reactive than chlorine. As a result, the reaction becomes much more selective, and the product ratios are more distinctive. In fact, bromination of alkanes is so selective that it is a feasible laboratory process to make alkyl bromides from alkanes.

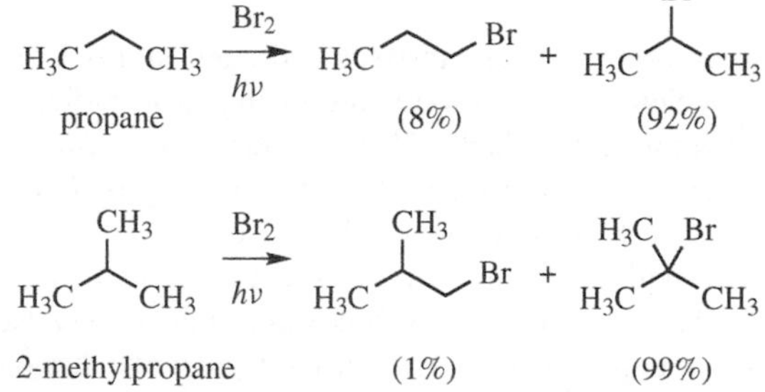

propane	
(8%)	(92%)

2-methylpropane	
(1%)	(99%)

The product ratios for bromination of propane and 2-methylpropane are quite different from those seen above in the chlorination reaction, in that the more-favoured products by far are the secondary and tertiary halides respectively. Abstraction of a hydrogen atom by a bromine atom is now much more difficult than with a chlorine atom. The favoured product may be rationalized in terms of the relative strength of the C–H bond being broken, and the

2,2-dimethylpropane

relative stability of the radical produced, though this is an oversimplification and we ought to consider relative energies of transition states.

9.3.1 Stereochemistry of radical reactions

The planarity of a radical (see Section 9.2) means that, when it reacts with a reagent, there is an equal probability that it can form a new bond to either side of the radical. In many cases this is of no consequence; but, should the formation of the product generate a chiral centre, we are going to get an equimolar mixture of both possible configurations, i.e. formation of a racemic mixture. This outcome has already been noted when a carbocation, another planar system, reacts to produce a chiral centre (see Section 6.2).

Thus, if we consider radical chlorination of butane, we expect to get a mixture of products, including the monochlorinated compounds 1-chlorobutane and 2-chlorobutane.

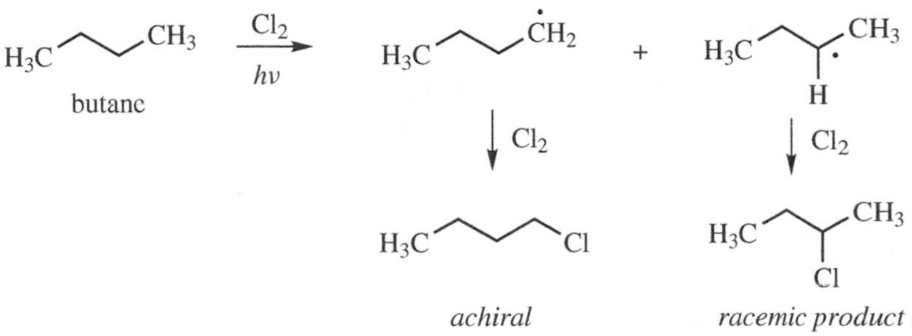

butane

achiral

racemic product

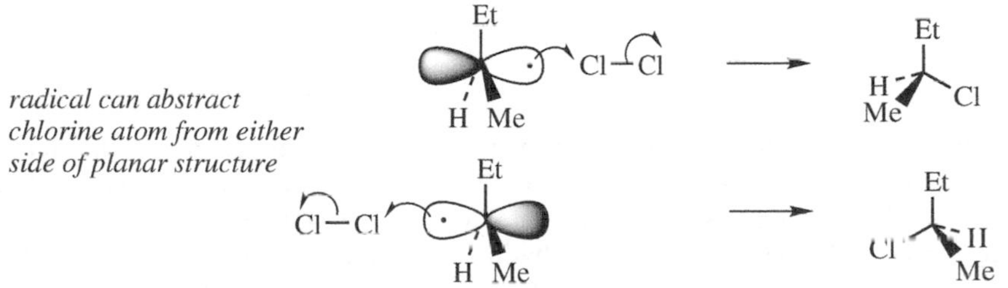

radical can abstract chlorine atom from either side of planar structure

In the formation of 1-chlorobutane, an intermediate primary radical is involved, and there are no stereochemical consequences. However, the secondary radical involved in 2-chlorobutane formation is planar, and when it abstracts a chlorine atom from a chlorine molecule it can do so from either side with equal probability. The result is formation of a racemic product, (±)-2-chlorobutane.

9.3.2 Allylic and benzylic substitution: halogenation reactions

The selectivity of radical bromination reactions depends, in part, on the increased stability of secondary or tertiary radical intermediates compared with primary radicals. In Section 9.2 we noted that **allyl** and **benzyl radicals** were especially

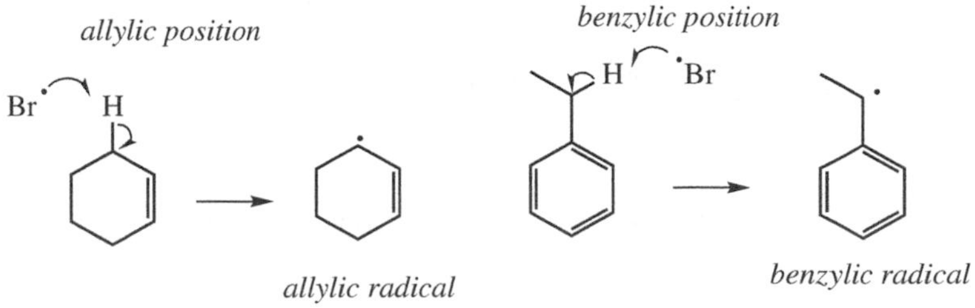

allylic position

allylic radical

benzylic position

benzylic radical

stabilized by resonance delocalization; indeed, they are even more stable than tertiary radicals. In the presence of a suitable initiator, bromine dissociates to bromine atoms that will selectively abstract an allylic or a benzylic hydrogen from a suitable substrate, generating the corresponding allyl and benzyl radicals.

In the case of cyclohexene, this leads to a resonance-stabilized **allylic radical** that then reacts with bromine to give the allylic bromide, plus a further bromine atom to continue the chain propagation steps. The symmetry in cyclohexene means that the two resonance structures are identical. It does not matter which allylic radical picks up bromine, we get the same product. It is not difficult to appreciate that a mixture of brominated products must result if we start with a non-symmetrical substrate.

resonance-stabilized allylic radical

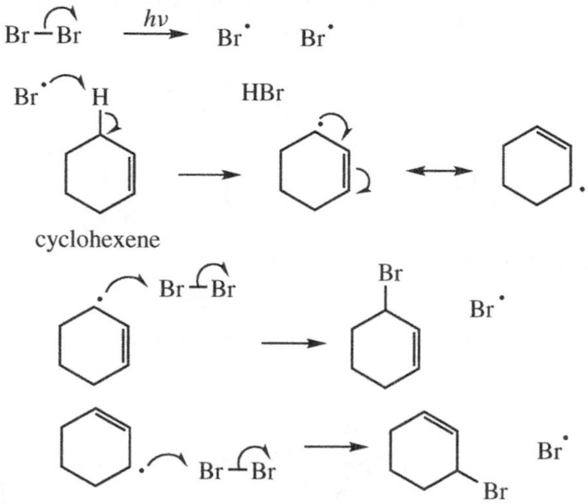

For example, radical allylic bromination of pent-2-ene must produce a mixture of three products. There are two allylic positions in the substrate, and either can suffer hydrogen abstraction. If hydrogen is abstracted from the methylene, then the two contributing resonance structures for the allylic radical are equivalent, and one product results when this captures a bromine atom. Abstraction of hydrogen from the terminal methyl gives an allylic radical for which the resonance structures are not equivalent, and hence two different brominated products may be formed. The net result will be a mixture of all three products. If we want to exploit allylic bromination, this means we must choose the substrate carefully if we prefer to get a single product.

equivalent resonance structures

same product

non-equivalent resonance structures

different products

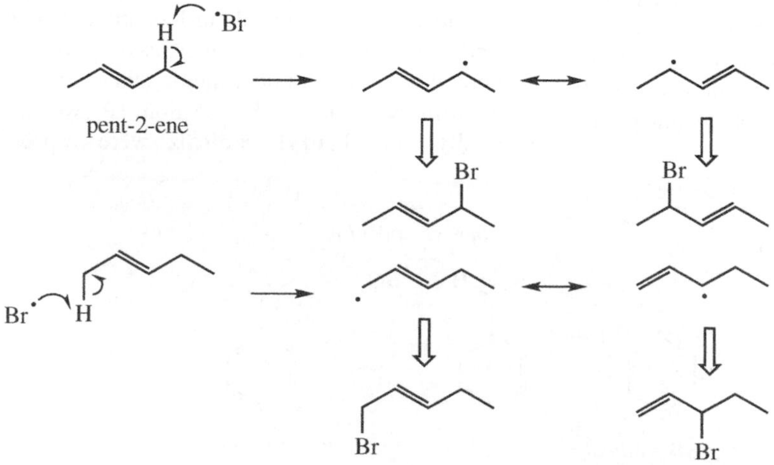

Of course, we have also seen that bromine can react with a double bond via electrophilic addition (see Section 8.1.2); further, it can add to a double bond via a radical mechanism.

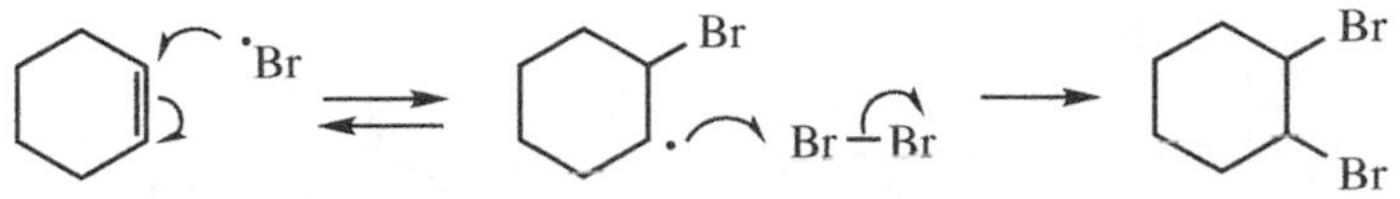

This could complicate an allylic bromination reaction, and it is necessary to choose conditions that minimize any addition to the double bond. This is achieved by carrying out the reaction in a solvent of low polarity, e.g. CCl_4, which suppresses the possibility of the polar electrophilic addition, whilst keeping the concentration of bromine very low to suppress radical addition.

There is, however, a much better reagent than bromine to brominate at an allylic position selectively. This reagent is **N-bromosuccinimide** (NBS), and it also reacts via a radical mechanism. The weak N–Br bond in NBS is susceptible to homolytic dissociation initiated either by light or a chemical initiator, such as a peroxide. This produces a small amount of bromine radicals, which can then abstract hydrogen from an allylic position on the substrate. The chain reaction continues via a small concentration of molecular bromine, which is generated by an ionic mechanism from NBS and the HBr released as a consequence of the hydrogen abstraction. Accordingly, the broad overall reaction is just the same as if we were employing molecular bromine as the reagent. The difference is in the use of NBS to maintain a very low concentration of bromine. Under the conditions used, i.e. in a non-polar solvent in which NBS is not very soluble, and with the very low concentration of bromine produced, there is almost exclusive allylic bromination and very little addition to the double bond.

N-bromosuccinimide
(NBS)

HBr now reacts with NBS

allylic radical

generation of Br₂ from HBr and NBS (ionic mechanism)

Br₂ allows radical chain reaction to continue

protonation of carbonyl oxygen

enol-like tautomer

carbonyl tautomer

succinimide

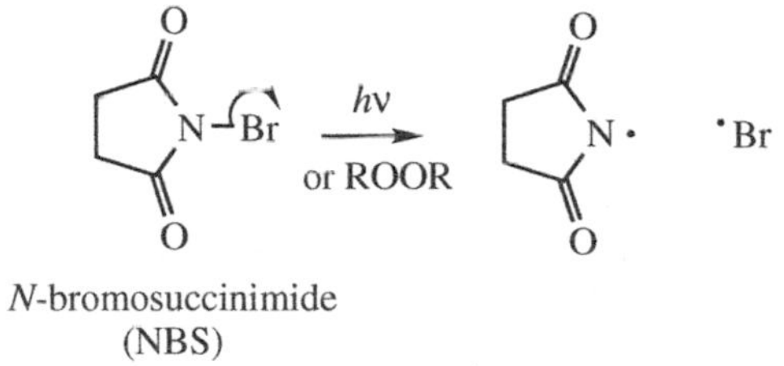

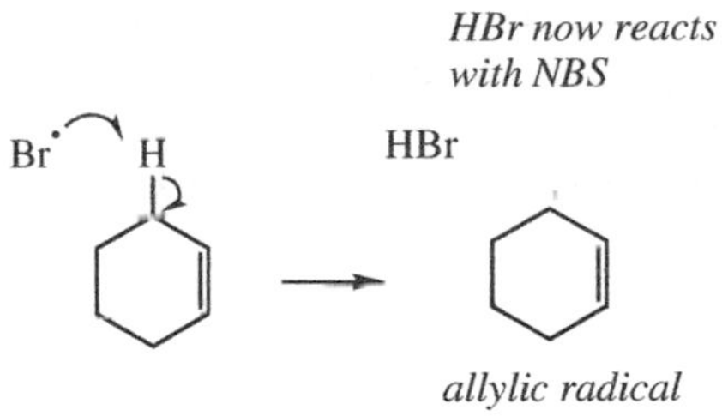

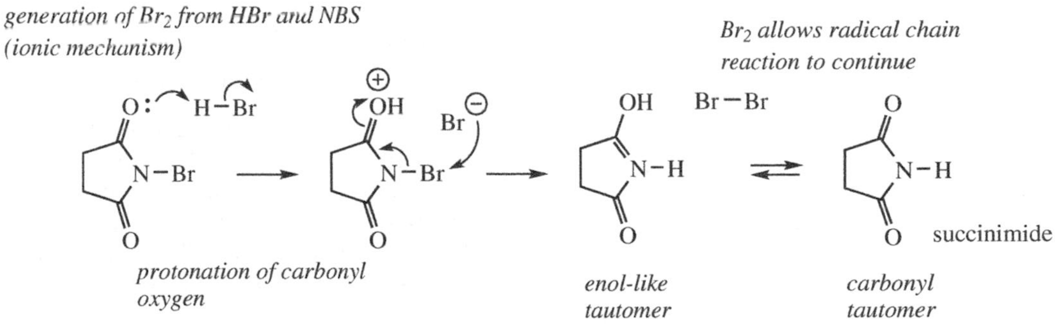

A **benzylic radical** is generated if a compound like toluene reacts with bromine or chlorine atoms. Hydrogen abstraction occurs from the side-chain methyl, producing a resonance-stabilized radical. The dissociation energy for the C–H bonds of the aromatic ring system is considerably more than that for the side-chain methyl, and relates to the stability of the radical produced.

toluene

benzyl radical stabilized by resonance delocalization

benzyl bromide

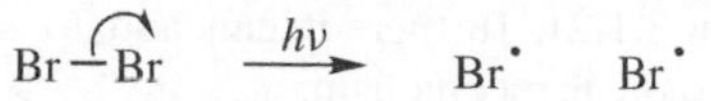

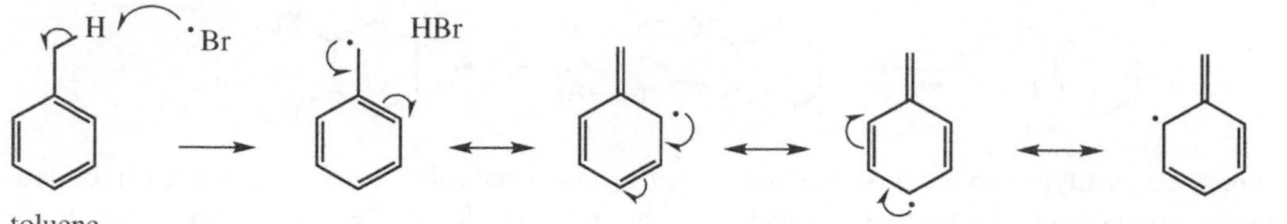

The typical propagation steps now follow, although all halogenation proceeds in the side-chain; addition to the ring would destroy the aromaticity and produce a higher energy product.

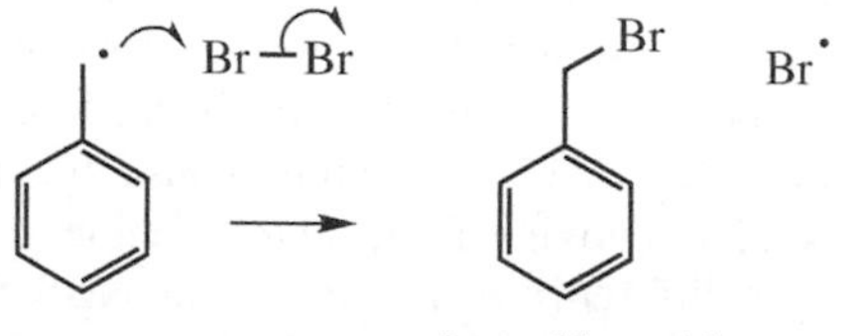

Benzyl chloride undergoes further chlorination to give di- and tri-chloro derivatives, though it is possible to control the extent of chlorination by restricting the amount of chlorine used. As indicated above, it is easier to mono-brominate than it is to mono-chlorinate. The particular stabilization conferred on the benzylic radical by resonance is underlined by the reaction of ethylbenzene with halogens.

(56%) (44%)

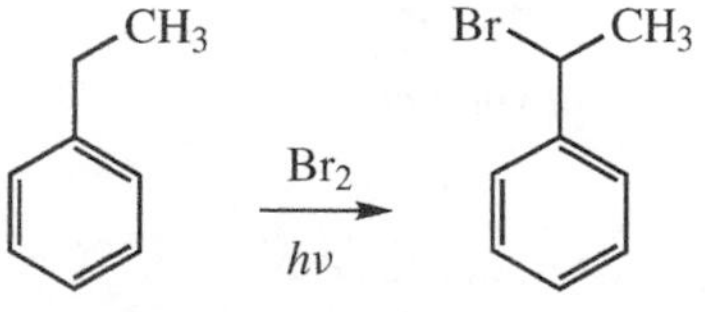

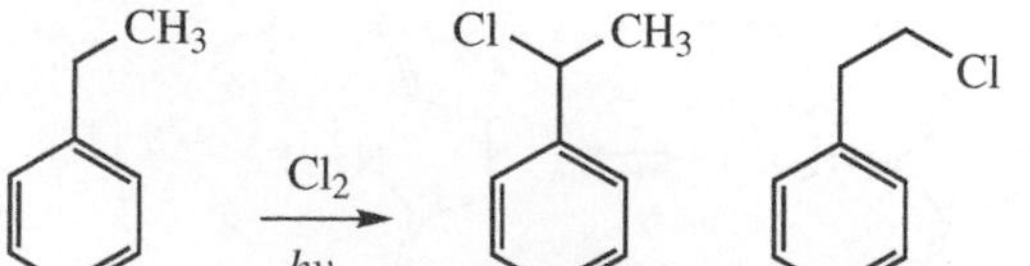

Bromination occurs exclusively at the benzylic position, i.e. adjacent to the benzene ring. The radical formed at this position is resonance stabilized, whereas no such stabilization is available to the primary radical formed by abstraction of one of the methyl hydrogens.

resonance structures

no resonance structures

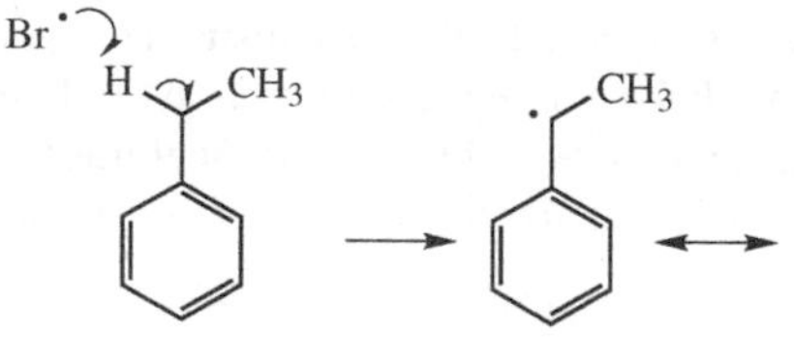

However, with the more reactive chlorine, chlorination can occur at either position, though the major product is the benzylic halide. Benzylic bromination is also efficiently achieved by the use of *N*-bromosuccinimide as the halogenating species.

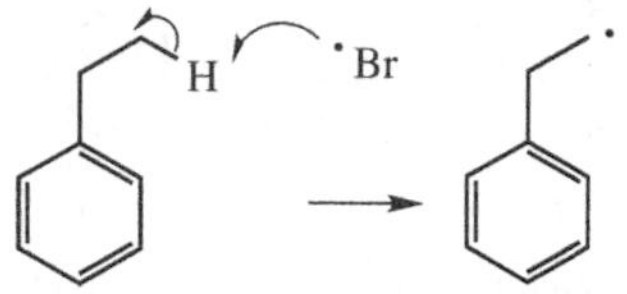

9.4 Radical addition reactions: addition of HBr to alkenes

The radical addition of halogen to an alkene has been referred to briefly in Section 9.3.2. We saw an example of bromination of the double bond in cyclohexene as an unwanted side-reaction in some allylic substitution reactions. The mechanism is quite straightforward, and follows a sequence we should now be able to predict.

More relevant to our consideration now is the radical addition of hydrogen bromide to an alkene. Radical formation is initiated usually by homolysis of a peroxide, and the resultant alkoxyl radical may then abstract a hydrogen atom from HBr.

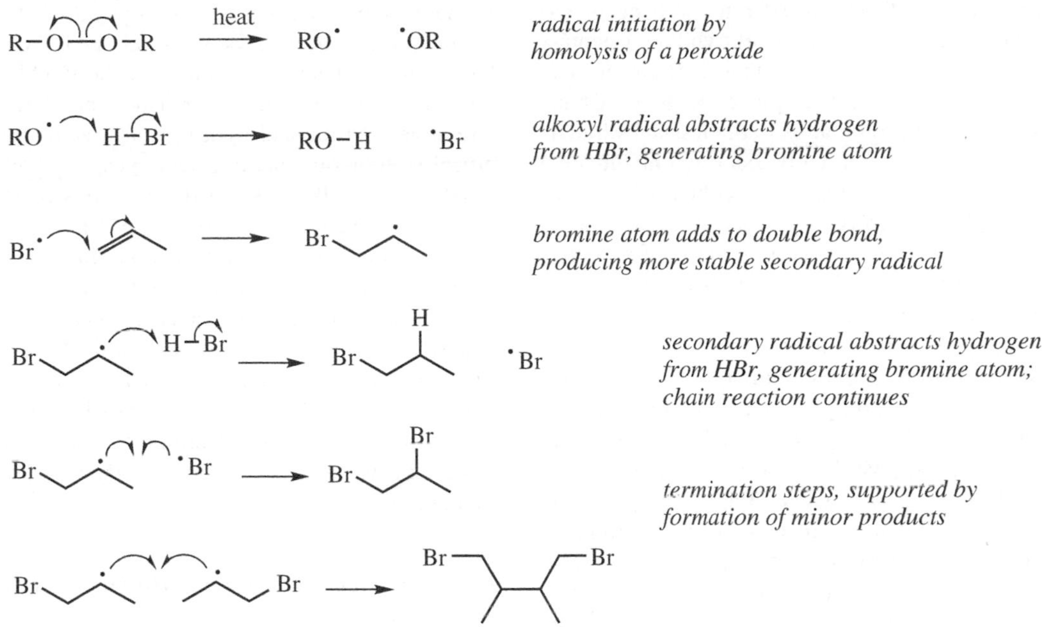

R–O–O–R $\xrightarrow{\text{heat}}$ RO•   •OR   *radical initiation by*
homolysis of a peroxide

RO• ⌢ H–Br $\longrightarrow$ RO–H   •Br   *alkoxyl radical abstracts hydrogen*
from HBr, generating bromine atom

Br• ⌢ [CH₂=CH–CH₃] $\longrightarrow$ Br⌢⌢•   *bromine atom adds to double bond,*
producing more stable secondary radical

Br⌢⌢• H–Br $\longrightarrow$ Br⌢⌢ (with H) •Br   *secondary radical abstracts hydrogen*
from HBr, generating bromine atom;
chain reaction continues

Br⌢⌢• •Br $\longrightarrow$ Br⌢⌢ (with Br)   *termination steps, supported by*
formation of minor products

Br⌢⌢• •⌢⌢Br $\longrightarrow$ Br–CH₂–CH(CH₃)–CH(CH₃)–CH₂–Br

The bromine atom then adds to the alkene, generating a new carbon radical. In the case of propene, as shown, the bromine atom bonds to the terminal carbon atom. In this way, the more stable secondary radical is generated. This is preferred to the primary radical generated if the central carbon were attacked. The new secondary radical then abstracts hydrogen from a further molecule of HBr, giving another bromine atom that can continue the chain reaction.

The main product is thus the result of addition of HBr to the alkene. Minor products detected are consistent with the proposed chain-termination steps.

This looks quite logical and consistent with what we know about radical reactions. However, remind yourself of the addition of HBr to an alkene, as we discussed under electrophilic reactions in Section 8.1.1. There is a significant difference in the nature of the product.

electrophilic addition of HBr

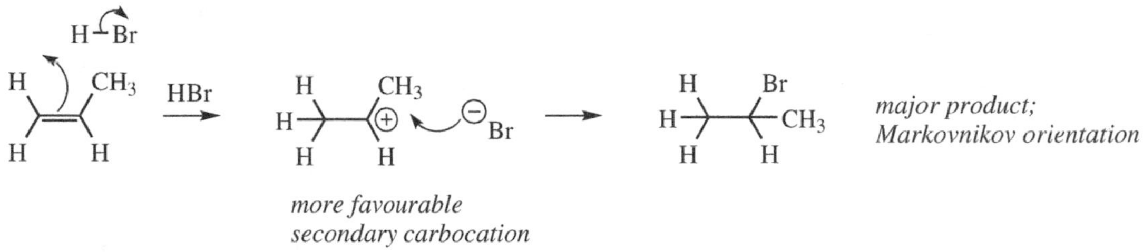

more favourable
secondary carbocation

major product;
Markovnikov orientation

radical addition of HBr

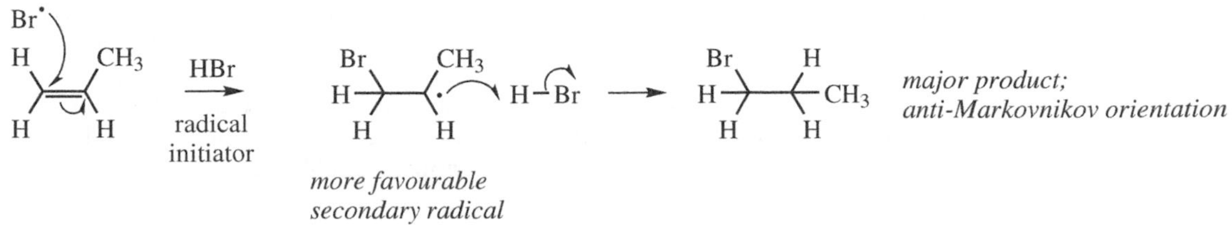

more favourable
secondary radical

major product;
anti-Markovnikov orientation

Electrophilic addition of HBr to propene gives predominantly the so-called Markovnikov orientation; Markovnikov's rule states that addition of HX across a carbon–carbon multiple bond proceeds in such a way that the proton adds to the less-substituted carbon atom, i.e. that already bearing the greater number of hydrogen atoms (see Section 8.1.1). We rationalized this in terms of formation of the more favourable carbocation, which in the case of propene is the secondary carbocation rather than the alternative primary carbocation.

Now, just the same sort of rationalization can be applied to the **radical addition**, in that the more favourable secondary radical is predominantly produced. This, in turn, leads to addition of HBr in what is the anti-Markovnikov orientation. The apparent difference is because the electrophile in the ionic mechanism is a proton, and bromide then quenches the resultant cation. In the radical reaction, the attacking species is a bromine atom, and a hydrogen atom is then used to quench the radical. This is effectively a reverse sequence for the addition process; but, nevertheless, the stability of the intermediate carbocation or radical is the defining feature. The terminologies Markovnikov or anti-Markovnikov orientation may be confusing and difficult to remember; consider the mechanism and it all makes sense.

This radical anti-Markovnikov addition of HX to alkenes is restricted to HBr; both HI and HCl add in a Markovnikov fashion by an ionic

mechanism, because the radical propagation steps are not favoured. The C–I bond is relatively weak, so that addition of an iodine atom to the double bond is not favoured. On the other hand, the H–Cl bond is relatively strong and hydrogen abstraction using a radical is unfavourable. For many years, the addition of HBr to an alkene seemed quite mysterious and erratic, with Markovnikov or anti-Markovnikov orientation occurring apparently at random. Eventually, the problem was solved and traced to the purity of the compounds used. Impure reagents containing traces of peroxides led to addition with anti-Markovnikov orientation, and we can now see that this is the consequence of a radical reaction. Reagents free from peroxides react via the ionic electrophilic addition mechanism, and we thus get predominantly Markovnikov orientation.

9.4.1 Radical addition of HBr to conjugated dienes

Radical addition of HBr to an alkene depends upon the bromine atom adding in the first step so that the more stable radical is formed. If we extend this principle to a **conjugated diene**, e.g. buta-1,3-diene, we can see that the preferred secondary radical will be produced if halogenation occurs on the terminal carbon atom. However, this new radical is also an **allylic radical**, and an alternative resonance form may be written.

A hydrogen atom is abstracted from HBr in the following step of the chain reaction to produce the addition product. Depending upon which resonance

structure is involved, we shall get different products, the results of 1,2- and 1,4-addition. The 1,4-addition is termed **conjugate addition**.

This is comparable to the electrophilic addition of HBr to butadiene (see Section 8.2), though the addition is in the reverse sense overall, in that Br adds before H in the radical reaction, whereas H adds before Br in the ionic mechanism. As with the electrophilic addition, we shall usually obtain a mixture of the two products.

9.4.2 Radical polymerization of alkenes

The addition of a radical on to an alkene generates a new radical, which potentially could add on to a further molecule of alkene, and so on, eventually giving a **polymer**. This becomes an obvious extension of the radical mechanisms we have already studied, and is the basis for the production of many commercial polymers.

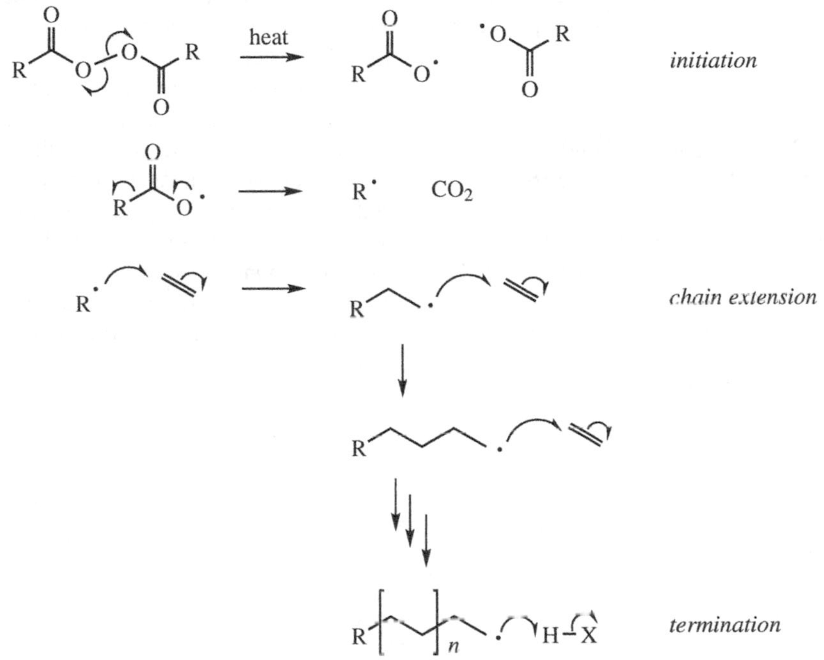

initiation

chain extension

termination

The radical initiator is usually a diacyl peroxide (see Section 9.1) that dissociates to radicals that in turn add on to the alkene. This starts the chain reaction, which is terminated by hydrogen abstraction from some suitable substrate, e.g. another polymeric radical that consequently becomes an alkene. In this general fashion, polymers such as polyethylene (polythene), polyvinyl chloride (PVC), polystyrene, and polytetrafluoroethylene (PTFE) may be manufactured.

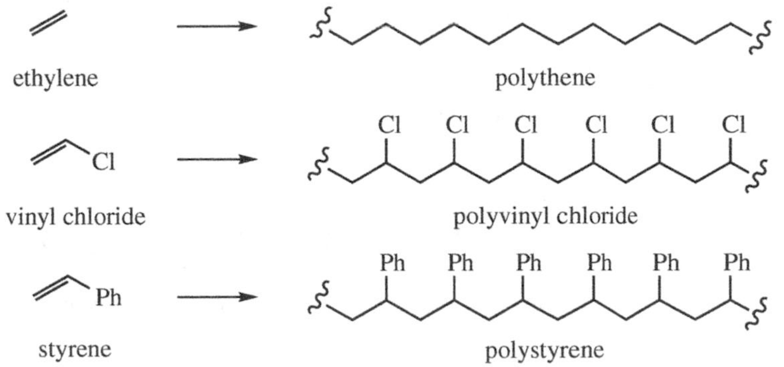

ethylene → polythene

vinyl chloride → polyvinyl chloride

styrene → polystyrene

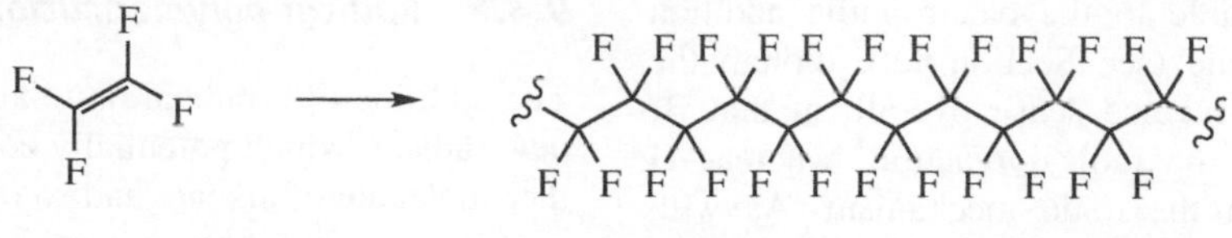

tetrafluoroethylene   polytetrafluoroethylene (PTFE; Teflon®)

We met a rather similar process, cationic polymerization, under electrophilic reactions in Section 8.3. In practice, **radical polymerization** is more effective than cationic polymerization, and industrial polymers are usually produced by radical processes.

9.4.3 Addition of hydrogen to alkenes and alkynes: catalytic hydrogenation

The addition of hydrogen to carbon–carbon multiple bonds (reduction) may be achieved using gaseous hydrogen in the presence of a finely divided noble metal catalyst. This is termed **catalytic hydrogenation**. It is not a radical reaction as we have seen

above, and does not feature initiation, propagation and termination steps. However, since it appears to involve atomic hydrogen, it has much more in common with radical reactions than ionic ones, and we consider it here for convenience.

The catalyst used is typically platinum, palladium, rhodium, or ruthenium, or sometimes an appropriate derivative. Precise details of the reaction remain vague, but we believe the catalyst surface binds to both the substrate, e.g. an **alkene**, and hydrogen, weakening or breaking the π bond of the alkene and the σ bond of hydrogen. Sequential addition of hydrogen atoms to the alkene carbons then occurs and generates the alkane, which is then released from the surface.

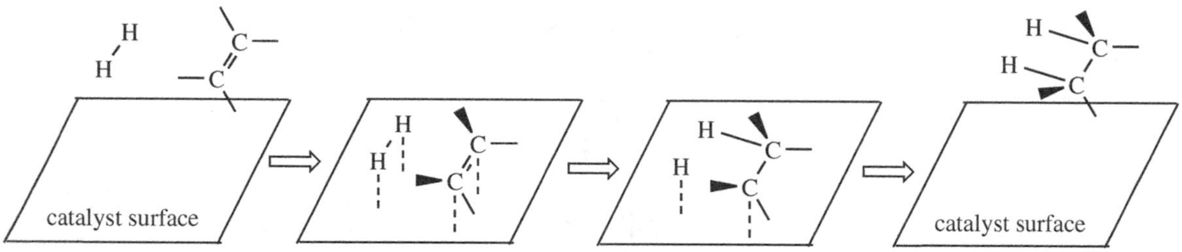

catalyst surface

hydrogen and alkene bonded to catalyst

first hydrogen atom bonds to alkene

second hydrogen atom bonds; alkane released

catalyst surface

Catalytic hydrogenation delivers hydrogen to one face of the alkene; the consequence is *syn* **addition** of hydrogen. This is a departure from our usual observations with ionic mechanisms, where the groups typically add to a double bond with *anti* stereochemistry (see Section 8.1.2).

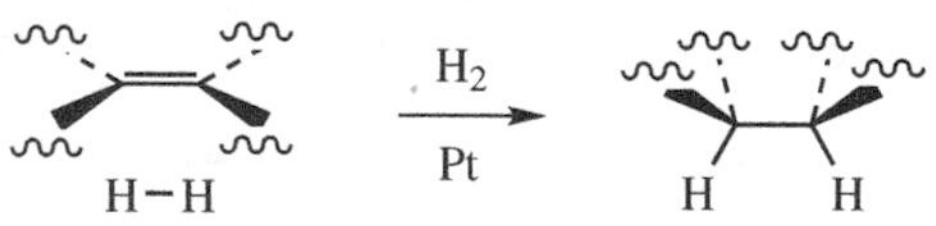

The stereochemical consequences of this are illustrated in the following examples.

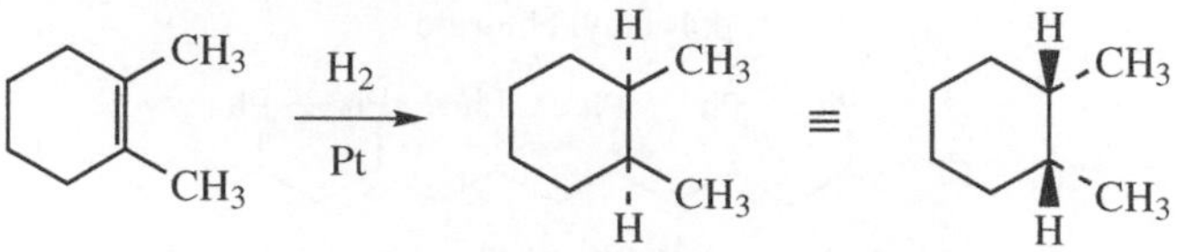

syn addition of hydrogen

1,2-dimethylcyclohexene   *cis*-1,2-dimethylcyclohexane

syn addition from either side of the double bond creates the same product

trans-2,3-diphenylbut-2-ene   (±)-2,3-diphenylbutane

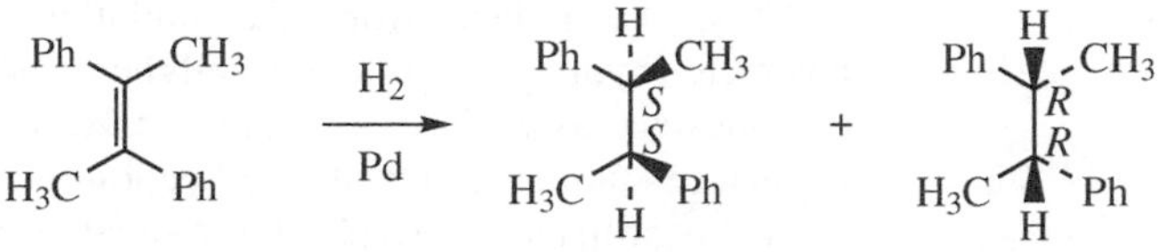

syn addition from either side of the double bond creates a pair of enantiomers

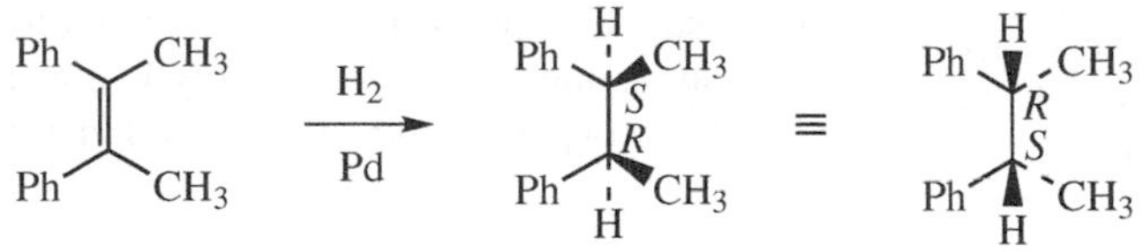

cis-2,3-diphenylbut-2-ene   meso-2,3-diphenylbutane

syn addition from either side of the double bond creates the meso isomer

Alkynes may also be hydrogenated, initially to alkenes, and then further to alkanes. By suitable modification of the catalyst, it has proved possible to stop the reaction at the intermediate alkene. Typically, platinum or palladium catalysts partially deactivated (poisoned) with lead salts are found to be suitable for reduction of alkynes to alkenes. Again, *syn* addition is observed.

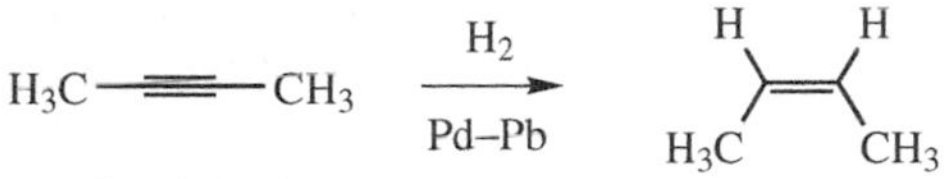

but-2-yne   cis-but-2-ene

Isolated double and triple bonds are reduced readily, whereas conjugated alkenes and aromatic systems are difficult to hydrogenate. Carbonyl double bonds react only very slowly, if at all, so it is possible to achieve selective reduction of C=C double bonds in the presence of aromatic and carbonyl functions.

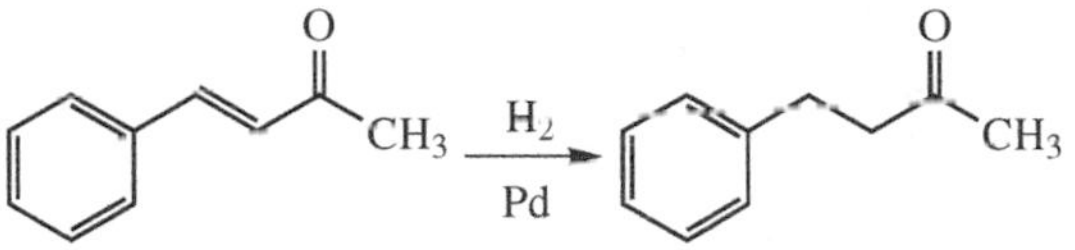

(E)-4-phenylbut-3-en-2-one   4-phenylbutan-2-one

9.5 Radical addition of oxygen: autoxidation reactions

The slow spontaneous oxidation of compounds in the presence of oxygen is termed **autoxidation** (auto-oxidation). This radical process is responsible for a variety of transformations, such as the drying of paints and varnishes, the development of rancidity in foodstuff fats and oils, the perishing of rubber, air oxidation of aldehydes to acids, and the formation of peroxides in ethers.

Unsaturated hydrocarbons undergo autoxidation because allylic hydrogens are readily abstracted by radicals (see Section 9.2). Molecular oxygen in its low-energy arrangement is a diradical, with only one bond between the atoms, and consequently an unpaired electron on each atom. Thus, oxygen can abstract hydrogen atoms like other radicals, though it is not a particularly good hydrogen abstractor. Instead, sequences are initiated by light or by other promoters that generate radicals, and oxygen is involved in the propagation steps.

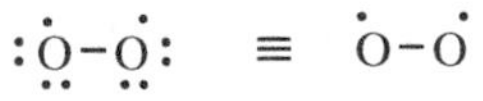

oxygen as diradical

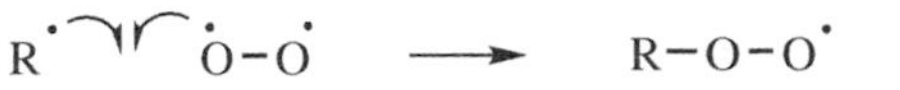

$R \cdot \quad O-O \longrightarrow R-O-O\cdot$

peroxyl radical

formation of peroxyl radical

$R-O-O\cdot \quad H-R \longrightarrow R-O-O-R \quad R\cdot$

propagation step

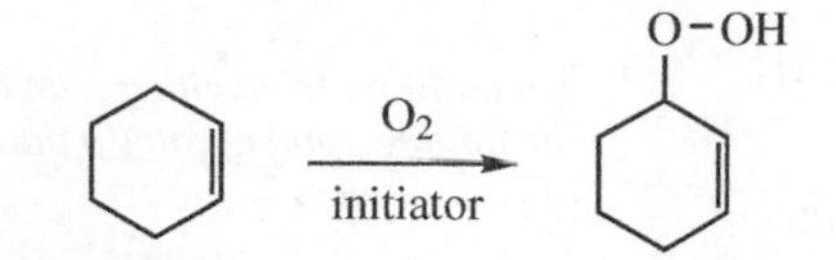

cyclohexene   3-cyclohexenyl
                     hydroperoxide

The processes that occur when cyclohexene reacts with oxygen in the presence of an initiator to give the allylic hydroperoxide exemplify this nicely.

Thus, the radical from the initiation reaction abstracts hydrogen from the allylic position of cyclohexene, as we have seen previously, to give the resonance-stabilized radical (see Section 9.2).

In the propagation steps, this radical then reacts with oxygen, producing a peroxyl radical, which then abstracts hydrogen from a further molecule of the substrate. The product is thus the hydroperoxide, reaction having occurred at the allylic position of the alkene. Two possible chain-termination steps might

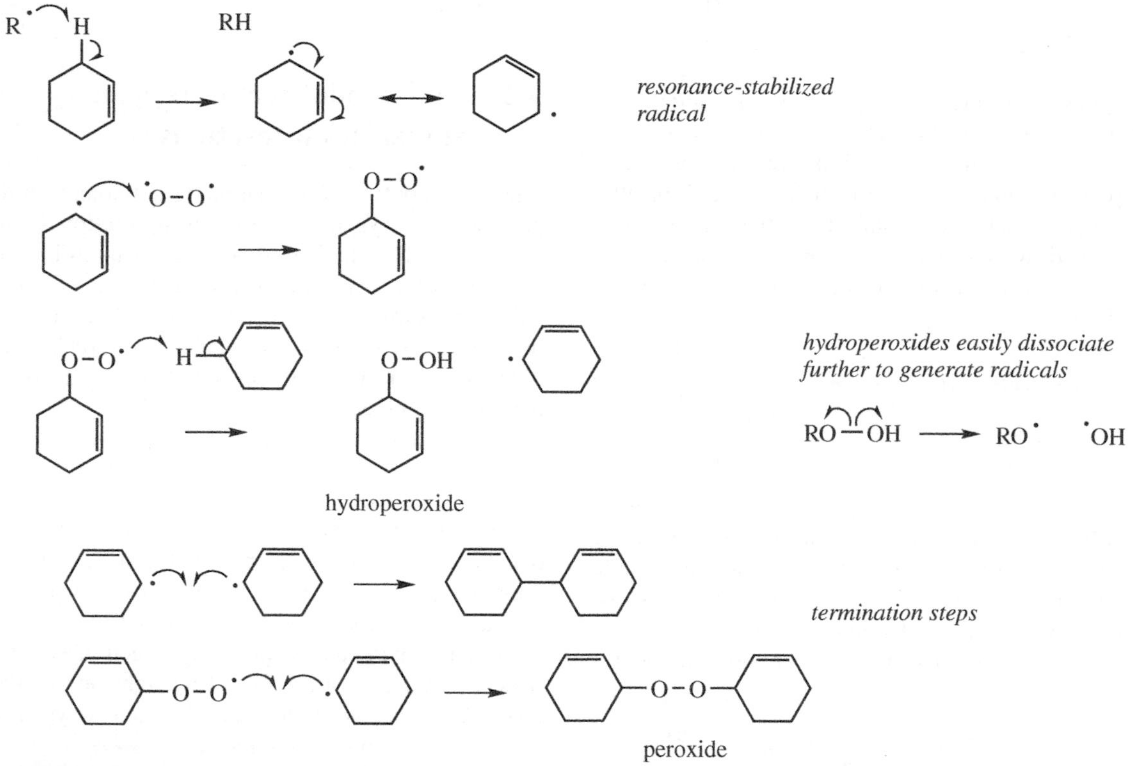

resonance-stabilized radical

hydroperoxides easily dissociate further to generate radicals

hydroperoxide

termination steps

peroxide

be the combination of two cyclohexenyl radicals or the formation of a peroxide, as shown. The hydroperoxide itself can easily dissociate to produce radicals that may then initiate other chain reactions. Peroxyl radicals are not particularly reactive, and thus

tend to be highly selective. They tend to abstract hydrogen atoms most readily from tertiary, allylic and benzylic C–H bonds. These are systems with the weakest bonds and that have maximum stabilization in the radical produced.

Box 9.1

Autoxidation in fats and oils: the origins of rancidity

Oxygen-mediated **autoxidation** can occur with unsaturated acid components of fats and oils, which are esters of fatty acids with glycerol (see Box 7.16). This leads initially to hydroperoxides that decompose further to produce low molecular weight carboxylic acids. These are the cause of **rancidity**, the unpleasant odour and taste associated with badly stored fats. **Linoleic acid** is a typical unsaturated fatty acid component, and hydrogen abstraction will occur from the methylene between the two non-conjugated double bonds. The radical thus produced benefits from extensive delocalization, as shown by the resonance forms that can be drawn.

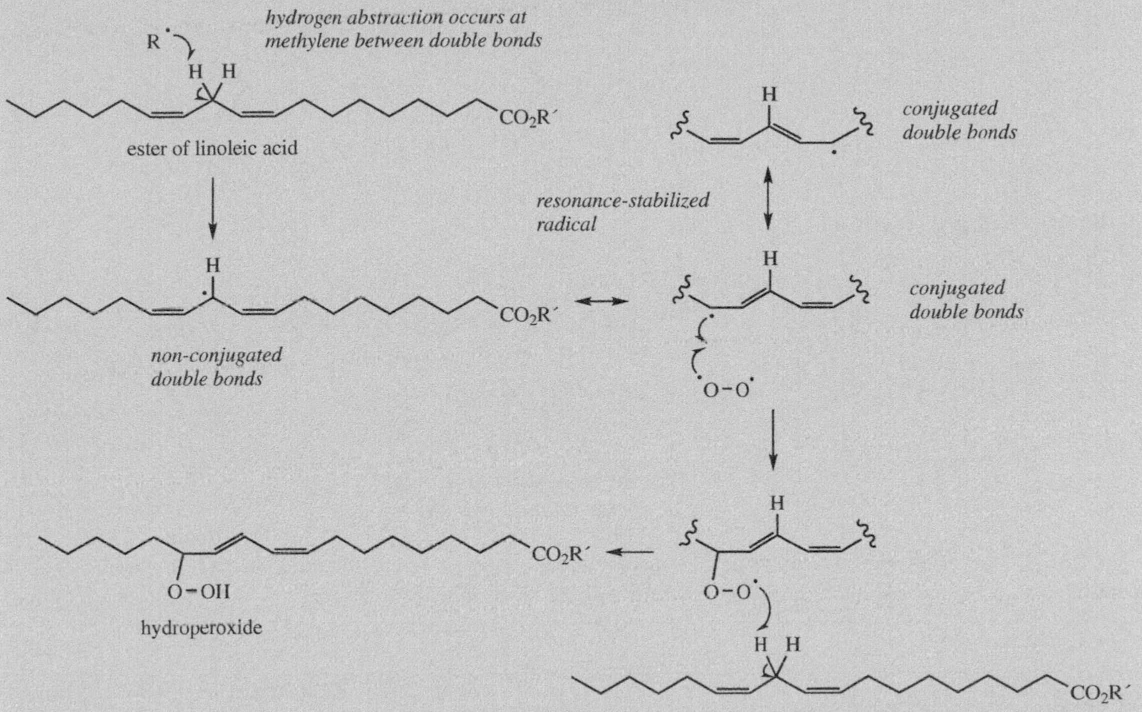

However, the resonance forms in which the double bonds are conjugated are inherently more stable than that with the unconjugated double bonds (see Section 9.2). Accordingly, the hydroperoxide subsequently formed upon reaction with oxygen will have conjugated double bonds. Abstraction of a hydrogen atom to form the hydroperoxide is part of the chain propagation process.

Fragmentation of the hydroperoxide can then lead to chain shortening, as illustrated.

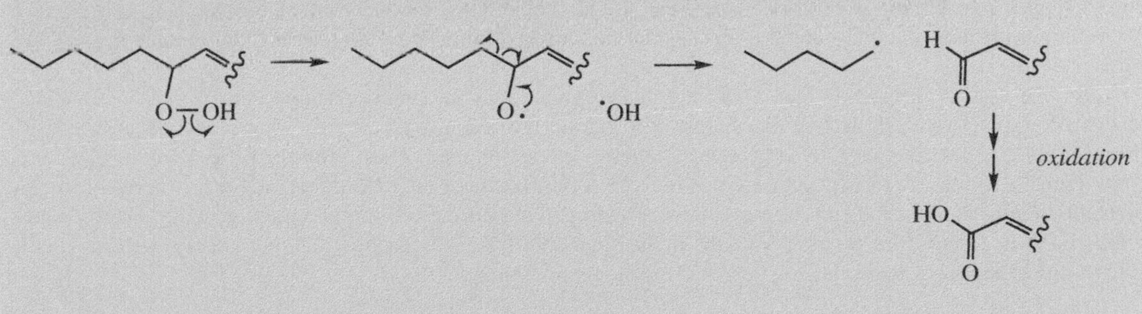

Acidic products result from further oxidation of aldehydes (or ketones), again by a radical process. Oxidation of an aldehyde to a carboxylic acid in the presence of air involves a peroxy acid (compare peroxyacetic acid, Section 8.1.2). Finally, a reaction between the peroxy acid and a molecule of aldehyde yields two carboxylic acid molecules; this is not a radical reaction, but is an example of a **Baeyer–Villiger oxidation**. Baeyer–Villiger

Box 9.1 (continued)

reactions are valuable for converting a ketone into an ester, in which case we see a rearrangement involving migration of an alkyl group.

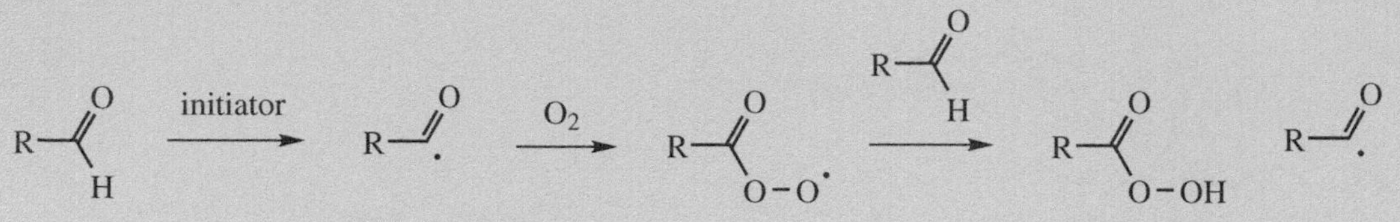

peroxycarboxylic acid

Baeyer–Villiger reaction

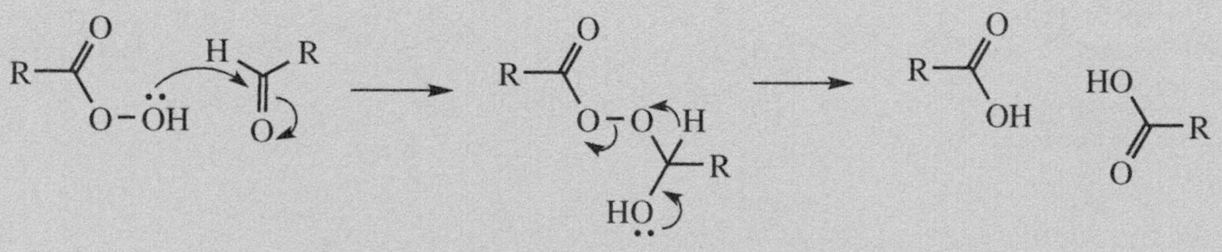

*rearrangement involving
migration of hydride*

In Box 9.2 we shall see how vitamin E is used commercially to retard rancidity in fatty materials in food manufacturing; it reduces autoxidation by reacting with peroxyl radicals.

Box 9.2

Antioxidants and health

The human body is continually exposed to radicals, either from external sources such as pollutants, or from endogenous sources because reactive oxygen species are involved in the natural processes used to detoxify chemicals and invading organisms. Although enzyme systems are present to provide protection from radical production and damage, such systems cannot be completely efficient. There is growing evidence that several disease states can be linked to radical damage. Lipid membranes, proteins, and DNA are all susceptible to interaction with radicals, and natural molecules termed **antioxidants** provide an important defence against such damage.

   Antioxidants are compounds that inhibit autoxidation reactions by rapidly reacting with radical intermediates to form less-reactive radicals that are unable to continue the chain reaction. The chain reaction is effectively stopped, since the damaging radical becomes bound to the antioxidant. Thus, **vitamin E (α-tocopherol)** is used commercially to retard rancidity in fatty materials in food manufacturing. Its antioxidant effect is likely to arise by reaction with peroxyl radicals. These remove a hydrogen atom from the phenol group, generating a resonance-stabilized radical that does not propagate the radical reaction. Instead, it mops up further peroxyl radicals. In due course, the tocopheryl peroxide is hydrolysed to α-tocopherylquinone.

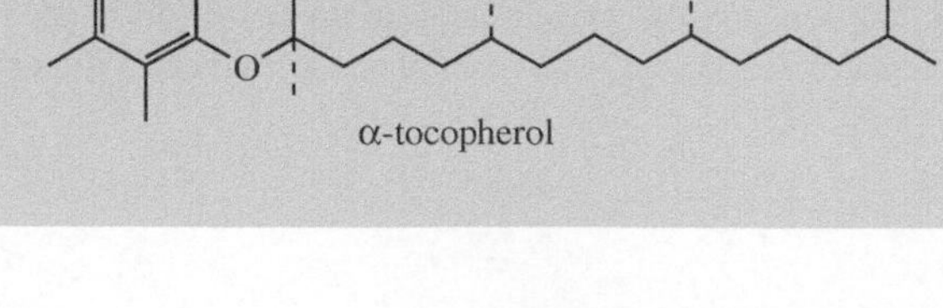

α-tocopherol

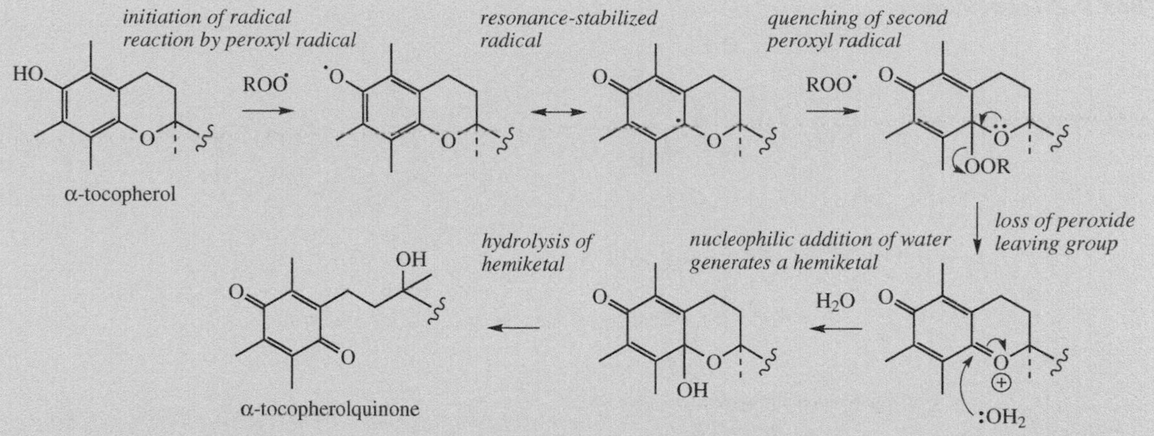

initiation of radical
reaction by peroxyl radical

resonance-stabilized
radical

quenching of second
peroxyl radical

α-tocopherol

hydrolysis of
hemiketal

nucleophilic addition of water
generates a hemiketal

loss of peroxide
leaving group

α-tocopherolquinone

Vitamin E in the diet is known to provide valuable antioxidant properties for humans, preventing the destruction of cellular materials, e.g. unsaturated fatty acids in biological membranes, and also helping to prevent heart disease. Other materials are similarly known to have beneficial antioxidant properties, and we are encouraged to incorporate sufficient levels of antioxidant-rich foods into our diets to minimize the risks of cardiovascular disease, cell degradation, and cancer.

Carotenoids are plant chemicals that function along with chlorophylls in photosynthesis as accessory light-harvesting pigments, effectively extending the range of light absorbed by the photosynthetic apparatus (see Box 11.4). They also serve as important protectants for plants and algae against photo-oxidative damage, by quenching toxic oxygen species. Recent research also suggests that carotenoids are important antioxidant molecules in humans, quenching peroxyl radicals, minimizing cell damage and affording protection against some forms of cancer. The most significant dietary carotenoid in this respect is **lycopene**; tomatoes and processed tomato products feature as the predominant source. The extended conjugated system allows radical addition reactions and hydrogen abstraction from positions allylic to the double bond system.

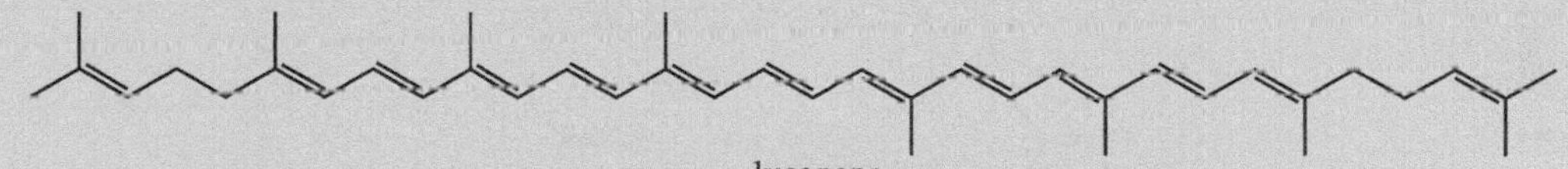

lycopene
(carotenoid)

Considerable quantities of natural polyphenolic compounds are consumed daily in our vegetable diet, and there is growing belief that some **flavonoids** are particularly beneficial, acting as antioxidants and giving protection against cardiovascular disease, certain forms of cancer, and, it is claimed, age-related degeneration of cell components. Their polyphenolic nature enables them to scavenge injurious radicals by hydrogen abstraction from phenol groups, as in α-tocopherol above (see also phenolic oxidative coupling, Section 9.6).

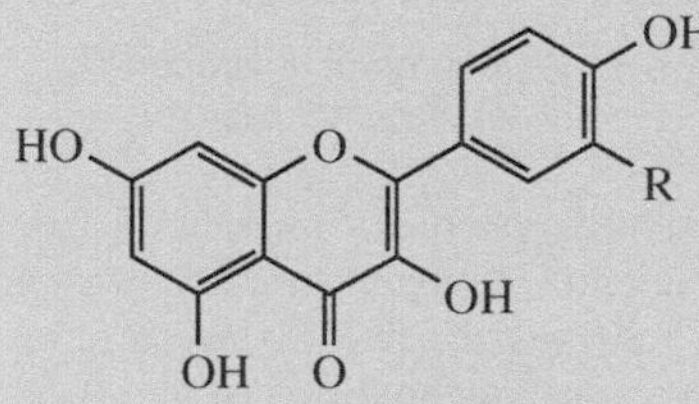

R = H, kaempferol
R = OH, quercetin
(flavonols)

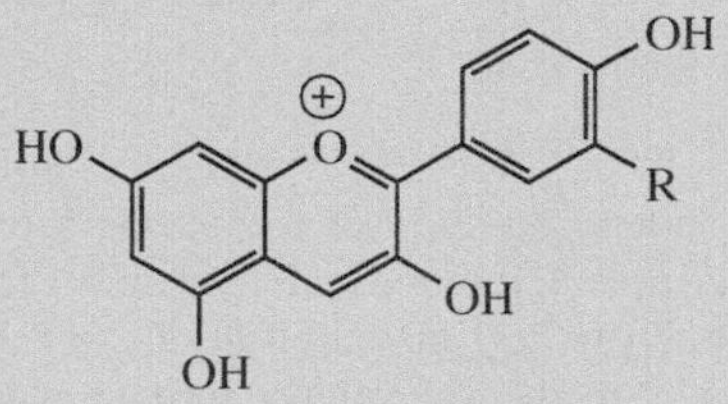

R = H, pelargonidin
R = OH, cyanidin
(anthocyanidins)

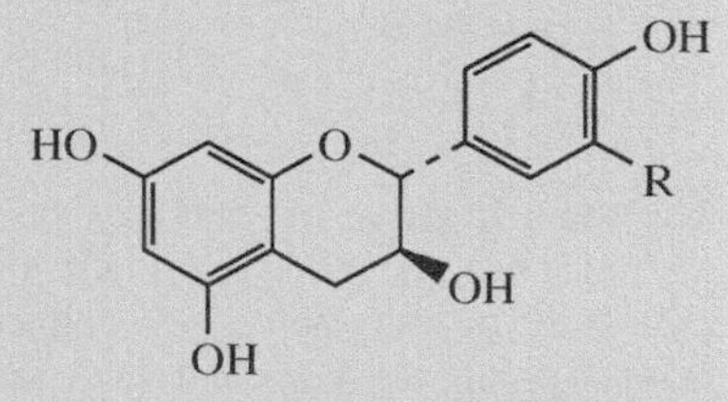

R = H, afzelechin
R = OH, (+)-catechin
(catechins)

Box 9.2 (continued)

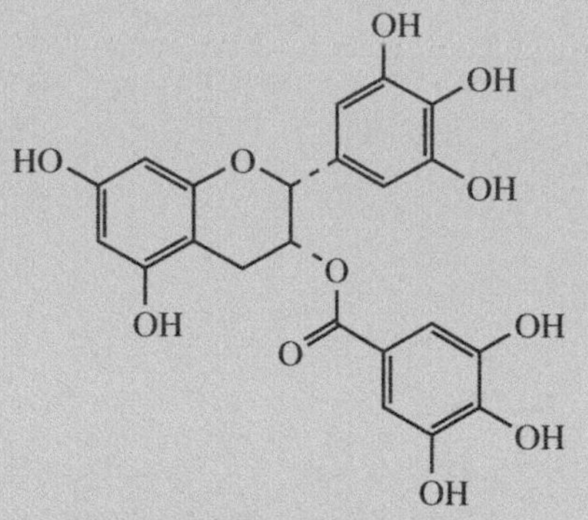

epigallocatechin gallate

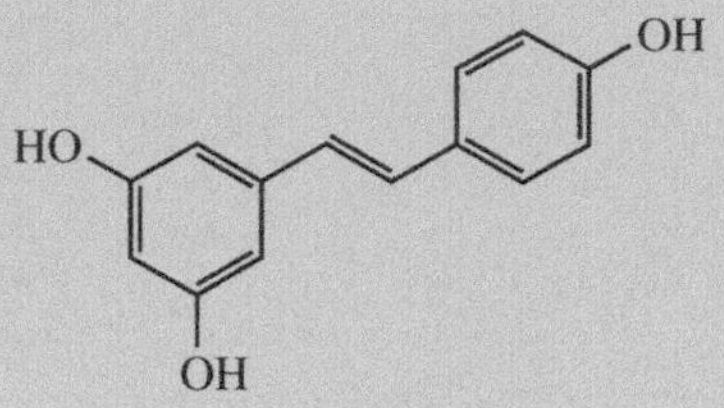

resveratrol
(stilbene)

Quercetin, in particular, is almost always present in substantial amounts in plant tissues, and is a powerful antioxidant, chelating metals, scavenging radicals and preventing oxidation of low-density lipoprotein. Flavonoids in red wine (quercetin, kaempferol, and anthocyanidins) and in tea (catechins and catechin gallate esters) are also demonstrated to be effective antioxidants. A particularly efficient agent in green tea is **epigallocatechin gallate**.

Resveratrol is another type of polyphenol, a stilbene derivative, that has assumed greater relevance in recent years as a constituent of grapes and wine, as well as other food products, with antioxidant, anti-inflammatory, anti-platelet, and cancer preventative properties. Coupled with the cardiovascular benefits of moderate amounts of alcohol, and the beneficial antioxidant effects of flavonoids, red wine has now emerged as an unlikely but most acceptable medicinal agent.

Vitamin C (ascorbic acid) is also a well-known antioxidant. It can readily lose a hydrogen atom from one of its enolic hydroxyls, leading to a resonance-stabilized radical. Vitamin C is acidic (hence ascorbic acid) because loss of a proton from the same hydroxyl leads to a resonance-stabilized anion (see Box 12.8). However, it appears that vitamin C does not act as an antioxidant in quite the same way as the other compounds mentioned above.

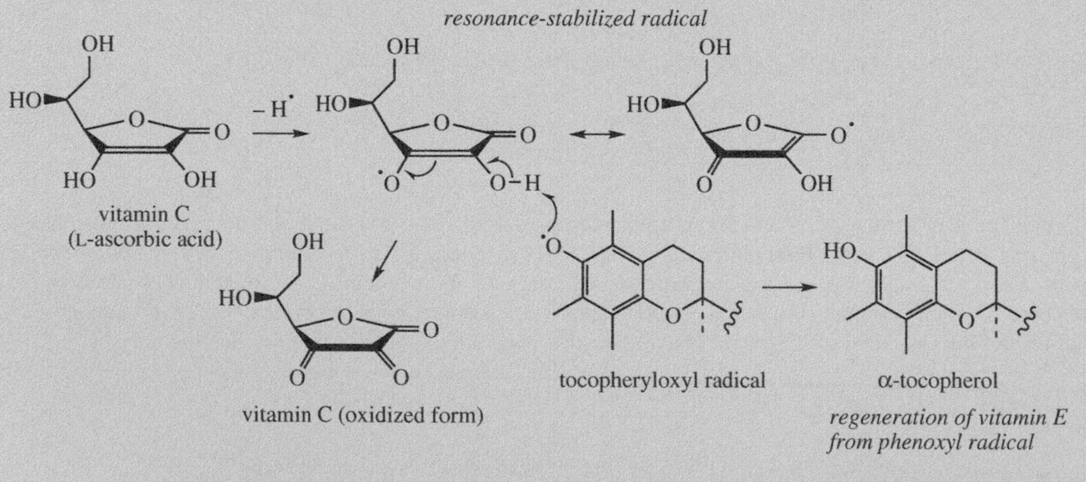

resonance-stabilized radical

vitamin C
(L-ascorbic acid)

vitamin C (oxidized form)

tocopheryloxyl radical

α-tocopherol

*regeneration of vitamin E
from phenoxyl radical*

The main function of vitamin C as a radical producer is to provide a regenerating system for tocopherol (see above). Thus, tocopheryloxyl radicals are able to remove hydrogen atoms from vitamin C to regenerate functioning molecules of tocopherol. A tocopheryloxyl radical may well be the agent that removes the first hydrogen atom from vitamin C. A second such radical can then abstract a further hydrogen atom and produce the oxidized tricarbonyl form of vitamin C.

Box 9.3
Radical oxidations in prostaglandin biosynthesis: cyclooxygenase

In Box 7.13 we saw that the widely used analgesic aspirin exerted its action by acetylating the enzyme **cyclooxygenase (COX)** which is involved in the production of prostaglandins. **Prostaglandins** are modified C_{20} fatty acids synthesized in animal tissues and they affect a wide variety of physiological processes, such as

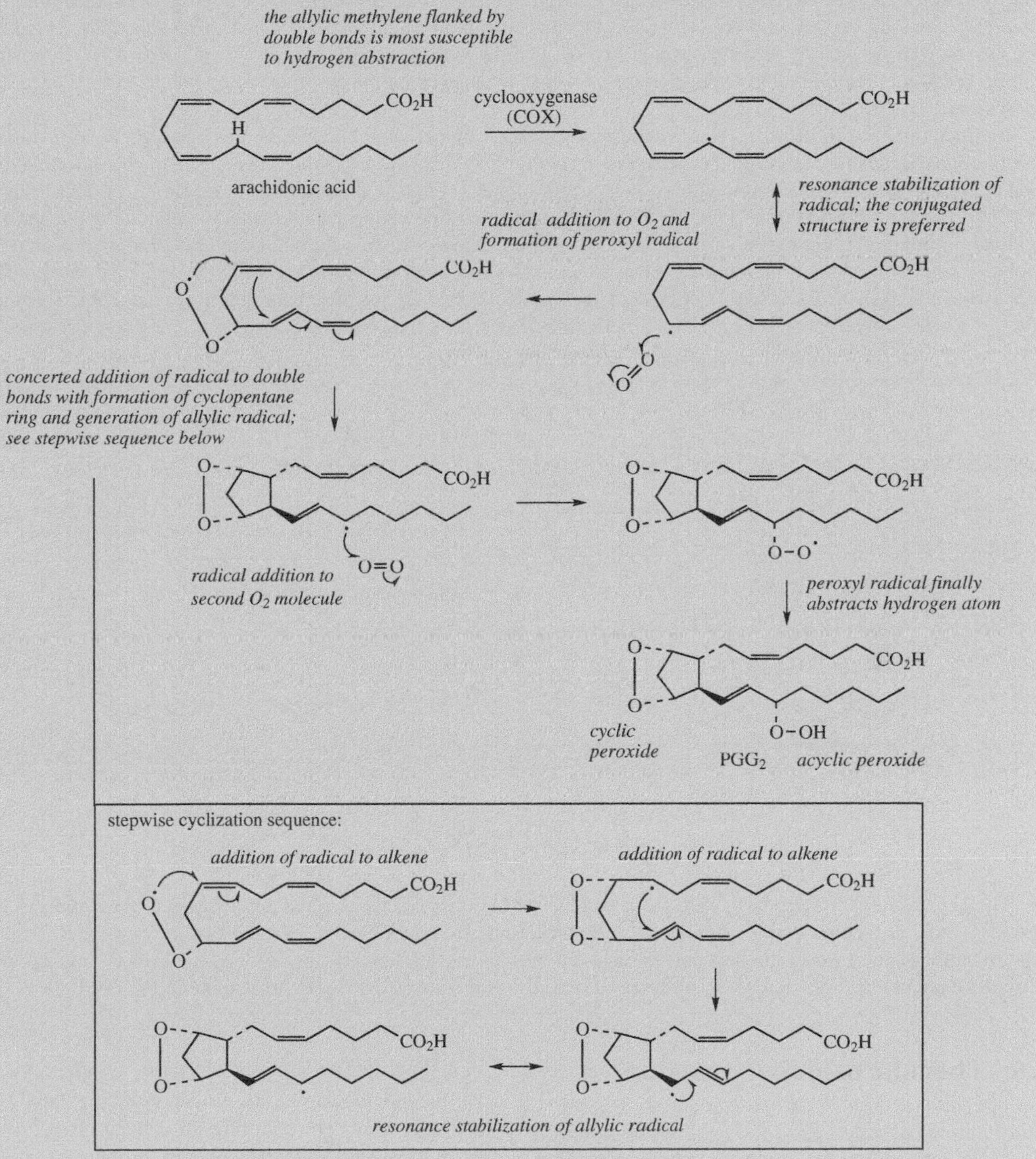

the allylic methylene flanked by double bonds is most susceptible to hydrogen abstraction

arachidonic acid

cyclooxygenase (COX)

resonance stabilization of radical; the conjugated structure is preferred

radical addition to O_2 and formation of peroxyl radical

concerted addition of radical to double bonds with formation of cyclopentane ring and generation of allylic radical; see stepwise sequence below

radical addition to second O_2 molecule

peroxyl radical finally abstracts hydrogen atom

cyclic peroxide

PGG_2 *acyclic peroxide*

stepwise cyclization sequence:

addition of radical to alkene

addition of radical to alkene

resonance stabilization of allylic radical

Box 9.3 (continued)

blood pressure, gastric secretion, smooth muscle contraction and platelet aggregation. Inflammation is a condition that occurs as a direct result of increased prostaglandin synthesis, and many of the non-steroidal anti-inflammatory drugs (NSAIDs), such as aspirin and ibuprofen, exert their beneficial effects by reducing prostaglandin formation.

Prostaglandin biosynthesis from the unsaturated fatty acid arachidonic acid looks very complicated. Breaking the process down into separate steps should reassure us that we have actually met these reactions already. In the reaction catalysed by COX, arachidonic acid is converted into prostaglandin G_2 (PGG_2) by incorporating two molecules of oxygen, and producing a compound with both cyclic and acyclic peroxide functions. This may be rationalized by radical reactions essentially identical to those we have seen above (see Box 9.1). The major difference is that the initiation reaction giving a radical is achieved by the enzyme, rather than by typical chemical processes.

In arachidonic acid, the allylic methylene group flanked by two double bonds is most susceptible to hydrogen abstraction, because of the resonance stabilization conferred. There are two such positions in arachidonic acid, but the enzyme is selective. Reaction with oxygen occurs so that a conjugated diene results (see Box 9.1). This leads to a peroxyl radical. Formation of PGG_2 is then depicted as a concerted cyclization reaction, initiated by the peroxyl radical, through addition to the various double bonds, the enzyme holding the substrate in the required manner to achieve ring formation. It is definitely easier to consider this cyclization via the stepwise sequence shown. The resultant radical then reacts with a second oxygen molecule, which abstracts hydrogen from a suitable substrate and generates a hydroperoxide, giving the structure PGG_2. It is likely that the hydrogen atom donor is another molecule of arachidonic acid, thus continuing the chain reaction.

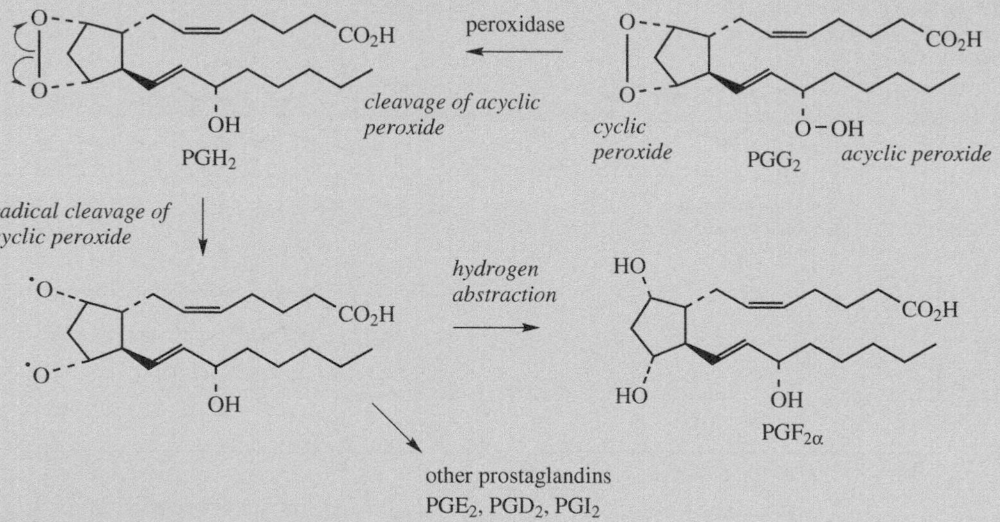

peroxidase

cleavage of acyclic peroxide

PGH_2

cyclic peroxide PGG_2 acyclic peroxide

radical cleavage of cyclic peroxide

hydrogen abstraction

$PGF_{2\alpha}$

other prostaglandins
PGE_2, PGD_2, PGI_2

The acyclic peroxide group in PGG_2 is then cleaved by a peroxidase enzyme and hydrogen abstraction yields prostaglandin H_2 (PGH_2), which occupies a central role and can be modified in several different ways. These further modifications can be rationally accommodated by initial cleavage of the cyclic peroxide to a diradical. For example, simple quenching of the radicals by abstraction of hydrogen atoms gives rise to prostaglandin $F_{2\alpha}$ ($PGF_{2\alpha}$).

9.6   Phenolic oxidative coupling

Many natural products are produced by the coupling of two or more phenolic systems, in a process readily rationalized by means of radical reactions. The reactions can be brought about by oxidase enzymes, including peroxidase and laccase systems, known to be radical generators. Other enzymes catalysing phenolic oxidative coupling have been characterized as cytochrome P-450-dependent proteins, requiring NADPH and O_2 cofactors, though no oxygen is incorporated into the substrate (see Box 11.4). Hydrogen abstraction from a phenol (a one-electron oxidation) gives the radical, and the unpaired electron can then be delocalized via resonance forms in which the free electron is dispersed to positions *ortho* or *para* to the

original oxygen function. We have already seen this property in the antioxidant effect of α-tocopherol and other phenolics (see Box 9.2).

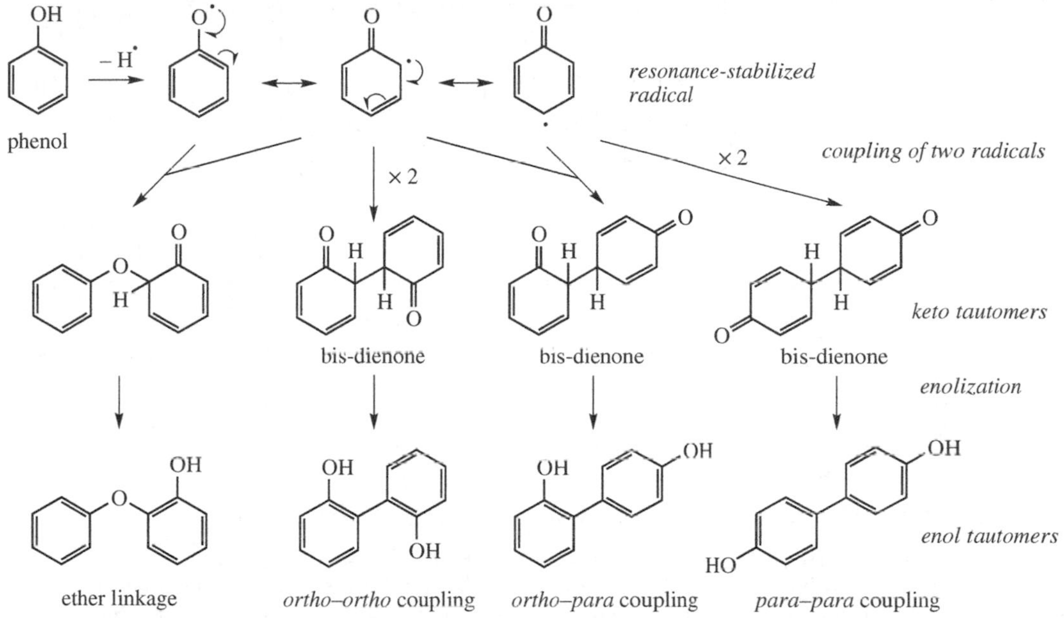

In phenolic oxidative coupling reactions, these phenol-derived radicals do not propagate a radical chain reaction; instead, they are quenched by coupling with other radicals. Thus, coupling of two of these resonance structures in various combinations gives a range of dimeric systems, as shown. The

final products indicated are then derived by enolization, which restores aromaticity to the rings. We shall discuss the concept of enolization in some detail in Section 10.1; for the moment, a simple acid-catalysed mechanism is shown below.

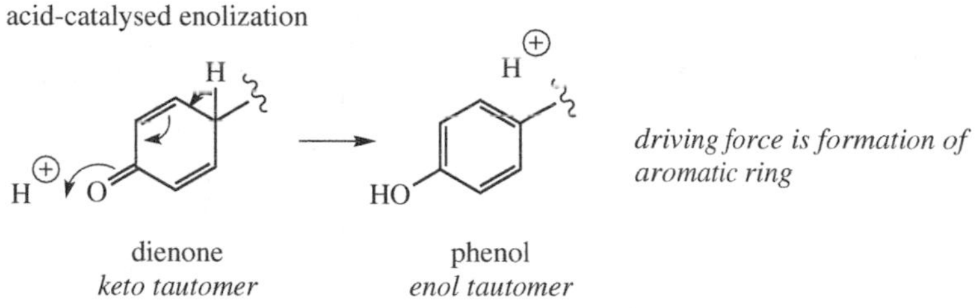

Accordingly, carbon–carbon bonds involving positions *ortho* or *para* to the original phenols, or ether linkages may be formed. The reactive dienone systems formed as intermediates may, in some cases, be attacked by other nucleophilic groupings (see Section 10.10), extending the range of structures ultimately derived from this basic reaction sequence.

The phenolic oxidative coupling process can also be demonstrated in laboratory experiments. Thus, treatment of 1-naphthol with alkaline potassium

ferricyanide yields a mixture of products, including those shown overleaf. As an oxidizing agent, potassium ferricyanide, $K_3Fe(CN)_6$, undergoes a change in oxidation state from Fe^{3+} to Fe^{2+}, i.e. a one-electron change. This makes it capable of initiating radical reactions by removal of one electron from the phenolate anion (hence the requirement for alkaline conditions). Thus, the formation of 1-naphthol dimers having *ortho–ortho*, *ortho–para*, and *para–para* coupling modes is easily accommodated.

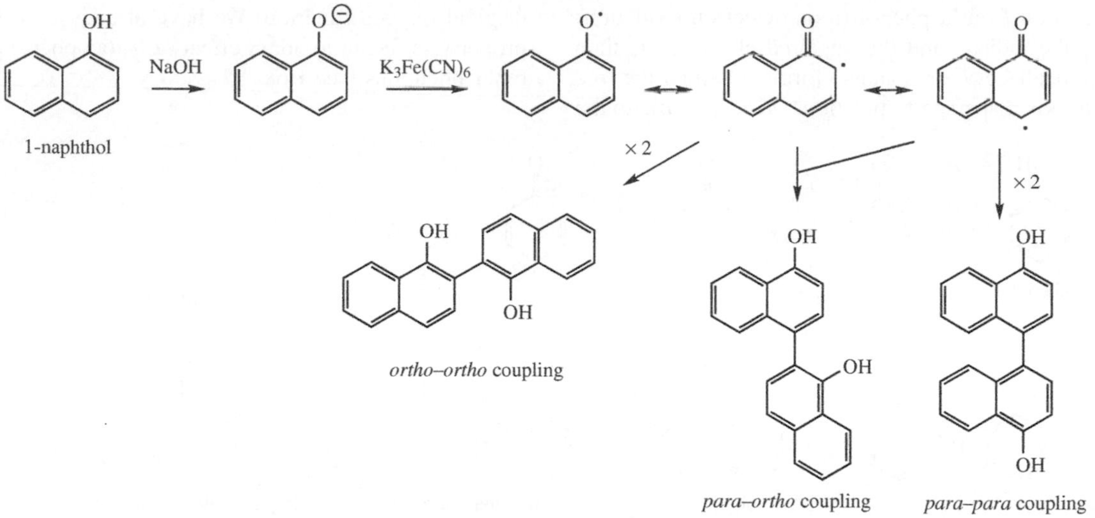

NaOH
$K_3Fe(CN)_6$
1-naphthol
× 2
× 2
ortho–ortho coupling
para–ortho coupling
para–para coupling

Box 9.4

Phenolic oxidative coupling: the biosynthesis of tubocurarine and morphine

Tubocurarine is the principal component of some varieties of **curare**, the arrow poison of the South American Indians. Curare is prepared by extracting the bark of several different plants, then concentrating the extract to a brown glutinous mass. Curare kills by paralysing muscles, particularly those associated with breathing. It achieves this by competing with acetylcholine at nicotinic receptor sites, thus blocking nerve impulses at the neuromuscular junction. Curare, and then tubocurarine, have found considerable use as muscle relaxants in surgery, but synthetic analogues have improved characteristics and are now preferred over the natural product (see Boxes 6.7 and 6.9).

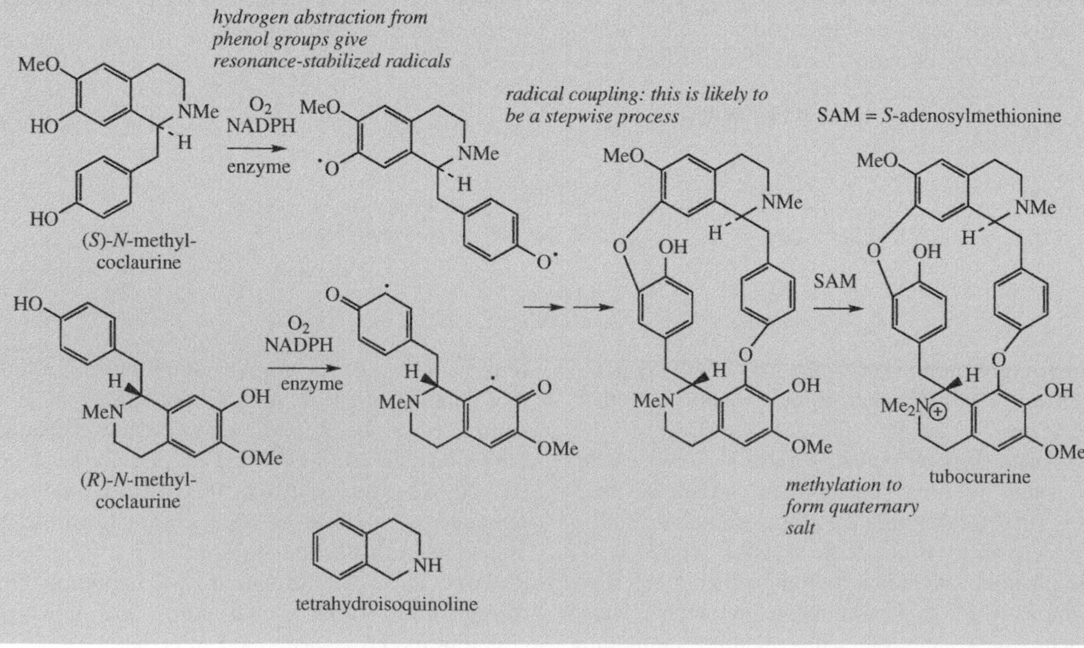

hydrogen abstraction from phenol groups give resonance-stabilized radicals
O_2 NADPH enzyme
(*S*)-*N*-methyl-coclaurine
(*R*)-*N*-methyl-coclaurine
radical coupling: this is likely to be a stepwise process
SAM = *S*-adenosylmethionine
SAM
methylation to form quaternary salt
tubocurarine
tetrahydroisoquinoline

The structure of tubocurarine has two benzyltetrahydroisoquinoline alkaloid units linked together, and this linking is achieved through phenolic oxidative coupling. We shall meet tetrahydroisoquinoline alkaloids as a product of biochemical Mannich-like reactions (see Box 10.7). Tubocurarine is formed in nature from two molecules of *N*-methylcoclaurine, one of each configuration. The coupling enzyme is a cytochrome P-450-dependent mono-oxygenase. Radical coupling is readily rationalized. The two diradicals, formed by hydrogen abstraction from the phenol group in each ring, couple to give ether bridges. This would be a consequence of the free electrons being localized on carbon in one system and on oxygen in the other. It is not proven, but more likely, that the radical coupling is a stepwise process involving simple monoradicals rather than the diradicals shown in the scheme. Tubocurarine is finally elaborated by enzymic methylation of one nitrogen atom to form the quaternary ammonium salt. This involves the participation of SAM as the methyl donor (see Box 6.5).

In natural alkaloids, the coupling of two benzyltetrahydroisoquinoline molecules by ether bridges, as in tubocurarine above, is rather less frequent than that involving carbon–carbon bonding between aromatic rings. The principal **opium alkaloids** morphine, codeine, and thebaine are derived by this type of process, though the subsequent reduction of one aromatic ring to some extent disguises their benzyltetrahydroisoquinoline origins. (*R*)-Reticuline is firmly established as the precursor of the morphine-like alkaloids.

Both morphine and codeine are valuable analgesics. **Morphine** is extracted from opium, the dried latex of the opium poppy, and **codeine** is usually obtained from morphine by semi-synthesis (see Box 6.2), since the amounts in opium are rather small. **Thebaine** is a valuable raw material for semi-synthesis of a wide range of morphine-like drugs.

(*R*)-Reticuline, turned over and rewritten as in the scheme, is the substrate for hydrogen abstractions via the phenol group in each ring, giving the diradical.

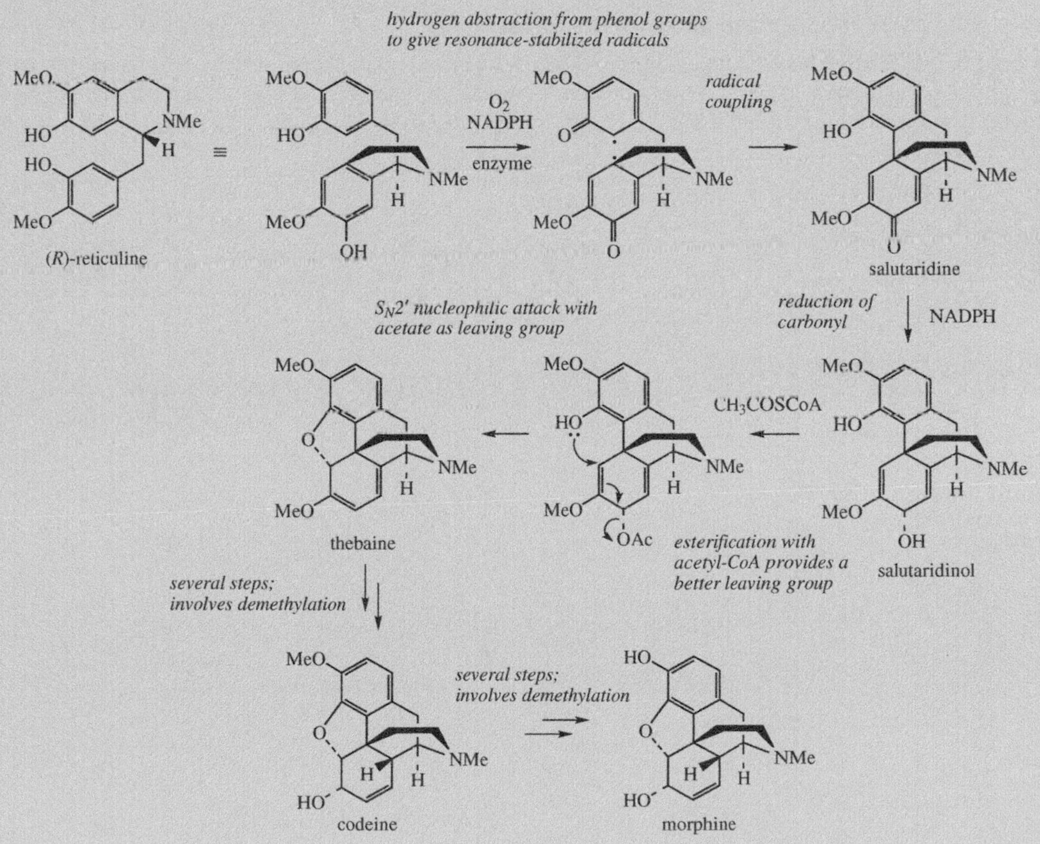

Box 9.4 *(continued)*

Coupling *ortho* to the phenol group in the tetrahydroisoquinoline and *para* to the phenol in the benzyl substituent then yields the dienone salutaridine, found as a minor alkaloid constituent in the opium poppy. Only the original benzyl aromatic ring can be restored to aromaticity, since the tetrahydroisoquinoline fragment becomes coupled *para* to the phenol function, a position that is already substituted.

The alkaloid thebaine is obtained by way of salutaridinol, formed from salutaridine by stereospecific reduction of the carbonyl group involving NADPH as reducing agent (see Box 7.6). Ring closure to form the ether linkage in thebaine would be the result of nucleophilic attack of the phenol group on to the dienol system and subsequent displacement of the hydroxyl (termed an S_N2' reaction). This cyclization step can be demonstrated chemically by treatment of salutaridinol with acid. *In vivo*, however, an additional reaction is used to improve the nature of the leaving group, and this is achieved by acetylation with acetyl-CoA. The cyclization then occurs readily, and without any enzyme participation.

Subsequent reactions involve conversion of thebaine into morphine by way of codeine, a process that most significantly removes two *O*-methyl groups. The involvement of these *O*-demethylation reactions is rather unusual; metabolic pathways tend to increase the complexity of the product by adding methyls rather than removing them. In this pathway, it is convenient to view the methyl groups in reticuline as protecting groups, which reduce the possible coupling modes available during the oxidative coupling process; these groups are then removed towards the end of the synthetic sequence.

Box 9.5

Phenolic oxidative coupling: the biosynthesis of thyroxine

The thyroid hormone **thyroxine** is necessary for the development and function of cells throughout the body. It increases protein synthesis and oxygen consumption in almost all types of body tissue. Excess thyroxine causes hyperthyroidism, with increased heart rate, blood pressure, overactivity, muscular weakness, and loss of weight.

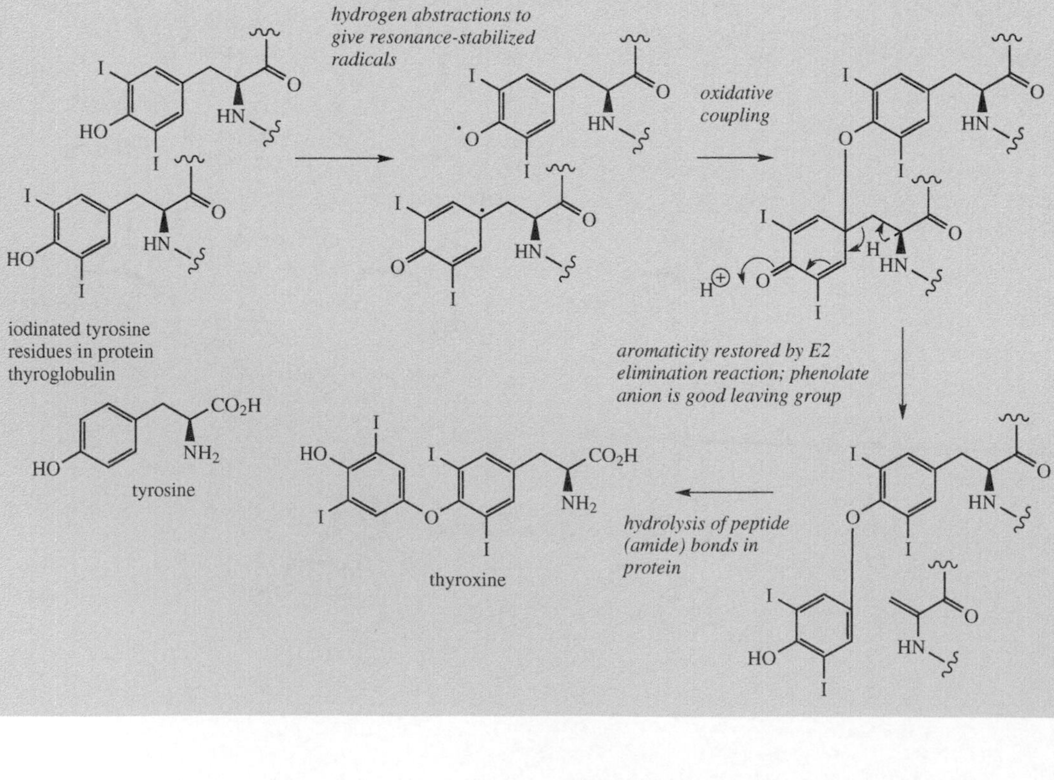

Too little thyroxine may lead to cretinism in children, with poor growth and mental deficiency, or myxoedema in adults, resulting in a slowing down of all body processes.

Thyroxine is actually a simple derivative of the aromatic amino acid tyrosine (see Section 13.1), but is believed to be derived by degradation of a larger protein molecule containing tyrosine residues. One hypothesis for their formation invokes suitably placed tyrosine residues in the protein thyroglobulin being iodinated to di-iodotyrosine. These residues then react together by phenolic oxidative coupling.

Coupling allows formation of an ether linkage, but since the position *para* to the original phenol is already substituted, it does not allow rearomatization through simple keto–enol tautomerization. Instead, rearomatization is achieved by an E2 elimination reaction in the side-chain of one residue, resulting in cleavage of the ring from the side-chain. This is feasible, since the phenolate anion is a good leaving group. Thyroxine is then released from the protein by hydrolytic cleavage of peptide (amide) bonds (see Box 13.5).

10

Nucleophilic reactions involving enolate anions

10.1   Enols and enolization

Aldehydes and ketones, and other carbonyl compounds having hydrogen atoms on the α-carbon, exist in solution as equilibrium mixtures of two or more isomeric forms. These isomers are termed the keto form, which is how we normally represent a carbonyl compound, and the enol form, which takes its name from the combination of double bond and alcohol.

keto form   *enol form*

The interconversion of keto and enol forms is termed **enolization**, or **keto–enol tautomerism**. The two isomeric structures are not resonance forms, but are termed **tautomers**. Resonance forms have the same arrangement of atoms, but the electrons are distributed differently (see Section 2.10). Tautomers have the atoms arranged differently, and tautomerism is an equilibrium reaction between the isomeric forms. Thus, in the general case shown, the α-hydrogen in the keto tautomer disappears and the oxygen atom gains hydrogen to produce the hydroxyl of the enol system.

To indicate the importance of enolization, equilibrium constants for a number of substrates are shown in Table 10.1. These equilibrium constants are only approximate, and they do depend very much on the solvents employed. Nevertheless, we can see that the equilibrium constant $K = $ [enol]/[keto] is very small for substrates like acetaldehyde, acetone, and cyclohexanone, with only a few molecules in every million existing in the enol form. However, in ethyl acetoacetate, enol concentrations are measured in percentages, and in acetylacetone the equilibrium constant indicates the enol form can be distinctly favoured over the normal keto form. In hexane solution, only 8% of acetylacetone molecules remain in the keto form.

Normally then, the keto form we have traditionally written for carbonyl compounds is very much favoured over the enol tautomer. The high contribution of enol forms in equilibrium mixtures of the 1,3-dicarbonyl compounds such as ethyl acetoacetate and acetylacetone is ascribed principally to additional stability conferred by formation of a conjugated enone system, with further stabilization coming from the establishment of hydrogen bonding in a favourable six-membered ring. At the other extreme, as in the case of cyclohexadienone, the enol tautomer is really the only contributing tautomer, since the enol form (phenol) benefits from the stabilization conferred by the aromatic ring system.

Essentials of Organic Chemistry   Paul M Dewick
© 2006 John Wiley & Sons, Ltd

Table 10.1   Keto–enol equilibria

	Keto tautomer	Enol tautomer	$K = \dfrac{[\text{enol}]}{[\text{keto}]}$	% Enol
Acetaldehyde	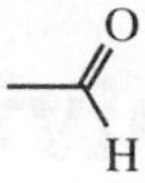	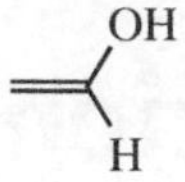	2×10^{-5} (water)	2×10^{-3}
Acetone	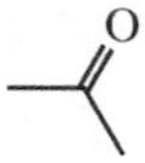	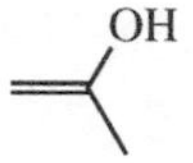	2.5×10^{-6} (water)	2.5×10^{-4}
Cyclohexanone	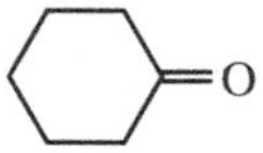	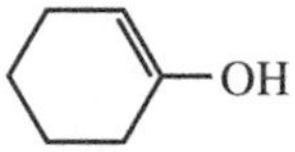	2×10^{-4} (water)	0.02
Ethyl acetoacetate	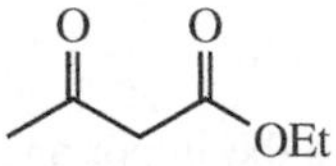	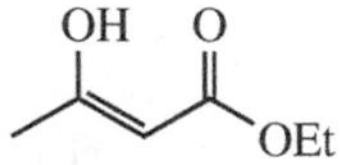	4×10^{-3} (water) 8.7×10^{-2} (liquid) 0.85 (hexane)	0.4 46 8
Acetylacetone	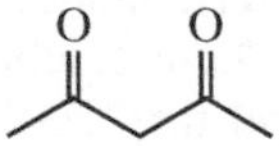	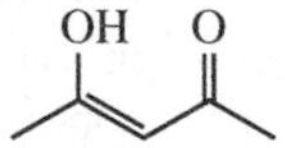	0.25 (water) 11.5 (hexane) 3.2 (liquid)	20 92 76
Phenol	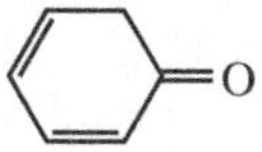	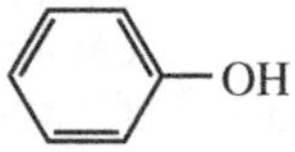	$>10^{13}$ (water)	100

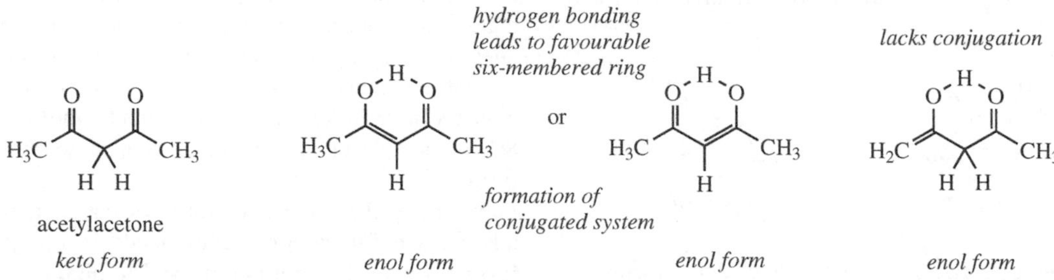

acetylacetone

keto form   *enol form*   *enol form*   *enol form*

It is important to note that, in 1,3-dicarbonyl compounds such as acetylacetone, enolization involves loss of the α-hydrogen between the two carbonyl groups, and not the terminal α-hydrogens. Enolization involving the latter α-hydrogens would not generate conjugation stabilization; and despite the possibility of hydrogen bonding, this enol form is not favoured relative to the alternatives. Conjugation can only be achieved if the central α-hydrogens, those sandwiched between the two carbonyls, are involved.

The **interconversion of keto and enol forms** may be catalysed by both acid and by base. In acid, this may be rationalized by a mechanism in which protonation of the carbonyl to give the conjugate acid is followed by loss of the α-proton.

acid-catalysed tautomerism

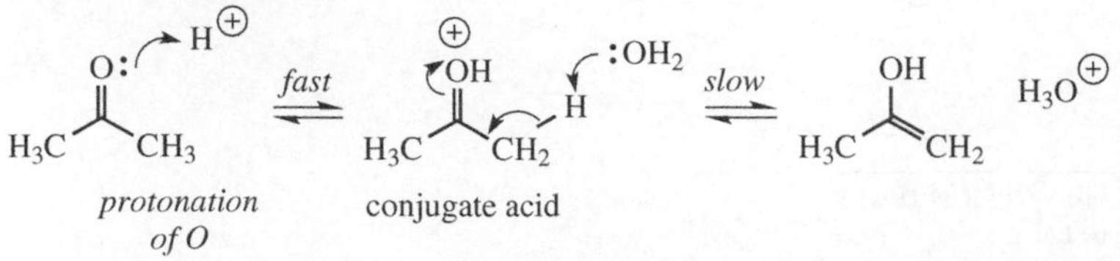

protonation   conjugate acid
of O

It is important to appreciate the role of the solvent in this transformation, removing and supplying protons, and to understand that tautomerism is not merely transfer of a proton from the α-carbon to the carbonyl oxygen. The rate-determining step in tautomerism will be removal of the α-hydrogen; protonation of the carbonyl (formation of the conjugate acid) can be considered rapid.

In base, slow abstraction of the α-hydrogen by the base will be the first step, followed by rapid protonation of the conjugate base, again making use of the solvent for the removal and supply of protons.

base-catalysed tautomerism

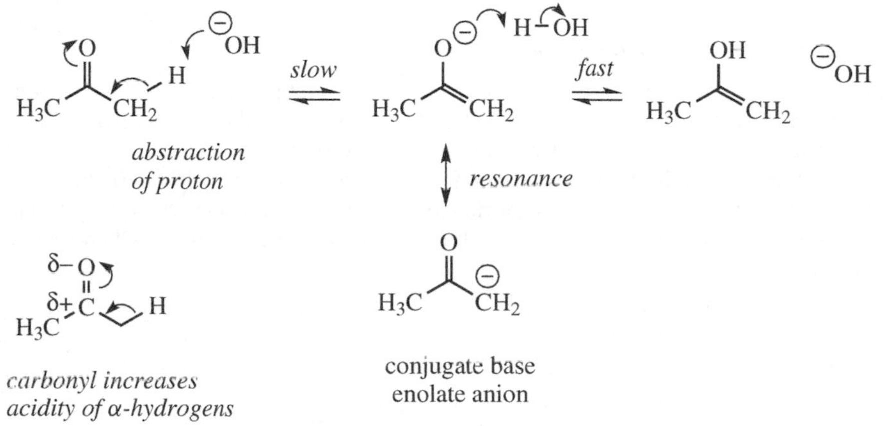

abstraction of proton

resonance

carbonyl increases acidity of α-hydrogens

conjugate base enolate anion

This process is thus exploiting the acidity associated with the α-hydrogens (pK_a 19), which is considerably greater than that of the corresponding alkane (pK_a 50). The effect of the adjacent carbonyl is to increase the acidity of the α-hydrogens (see Section 4.3.5). This is a direct consequence of the polarization of the carbonyl arising from the electronegativity of the oxygen atom. The conjugate base in this process is called an **enolate anion**, and is stabilized by resonance.

Of the two resonance forms of the enolate anion, that with the charge on the electronegative oxygen will be preferred over that with charge on the carbon. Note the distinct difference between resonance as shown here, a redistribution of electrons, and tautomerism, as described above. Tautomers are isomers in equilibrium and have the atoms arranged differently.

resonance forms: electrons distributed differently

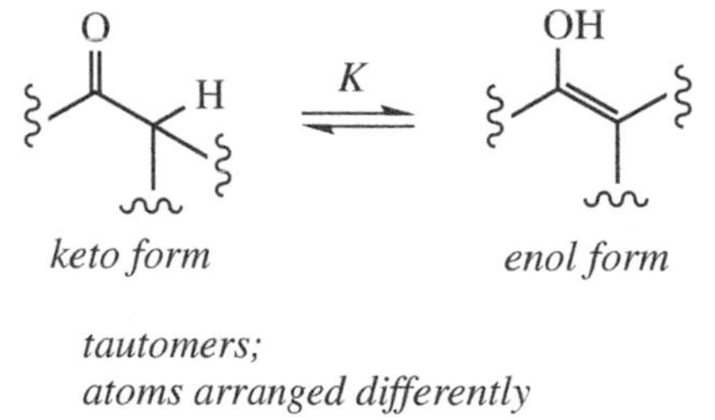

preferred – charge on O

keto form

enol form

tautomers; atoms arranged differently

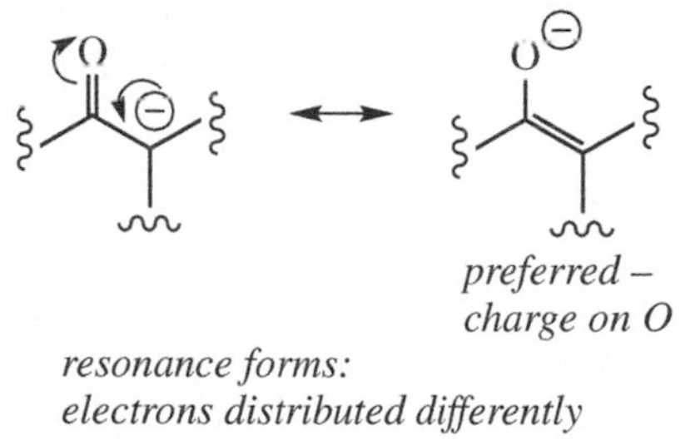

In 1,3-dicarbonyl compounds such as acetylacetone, the protons between the two carbonyls will be even more acidic (pK_a 9), since there are now two carbonyl groups exerting their combined influence. It can also be seen that resonance in the enolate anion is even more favourable with two carbonyl groups. This increased stability is not achieved by removal of the terminal α-hydrogens, and in acetylacetone these have pK_a 20, comparable to that in acetone. Put another way, treatment of acetylacetone with base preferentially removes a proton from the central methylene.

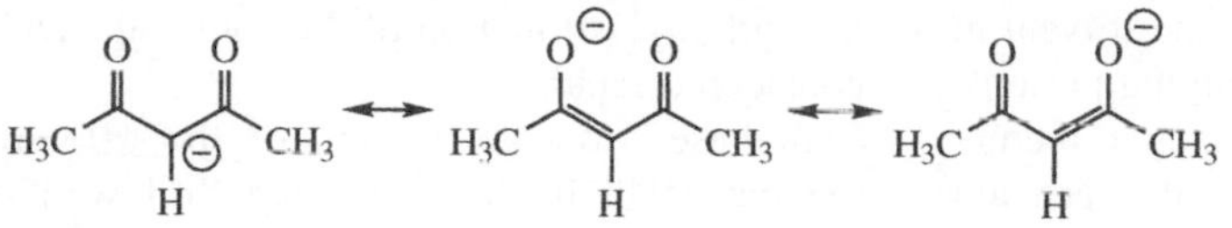

increased stabilization in enolate anion
from 1,3-dicarbonyl compounds

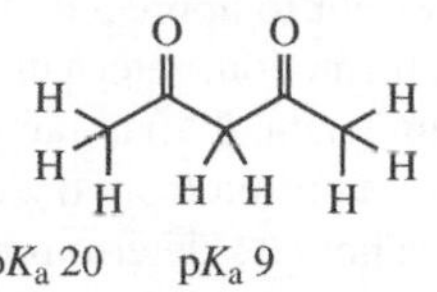

pK_a 20 pK_a 9

Box 10.1

Enols and enolization in the glycolytic pathway

Enols and enolization feature prominently in some of the basic biochemical pathways (see Chapter 15). Biochemists will be familiar with the terminology **enol** as part of the name **phosphoenolpyruvate**, a metabolite of the **glycolytic pathway**. We shall here consider it in non-ionized form, i.e. phosphoenolpyruvic acid. As we have already noted (see Section 10.1), in the enolization between pyruvic acid and enolpyruvic acid, the equilibrium is likely to favour the keto form pyruvic acid very much. However, in phosphoenolpyruvic acid the enol hydroxyl is esterified with phosphoric acid (see Section 7.13.2), effectively freezing the enol form and preventing tautomerism back to the keto form.

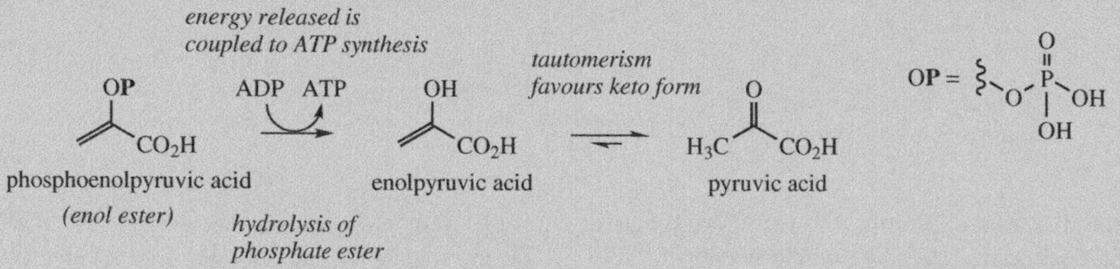

energy released is coupled to ATP synthesis

ADP ATP

tautomerism favours keto form

phosphoenolpyruvic acid
(enol ester)

enolpyruvic acid

hydrolysis of phosphate ester

pyruvic acid

Once the phosphate ester is hydrolysed, there is an immediate rapid tautomerism to the keto form, which becomes the driving force for the metabolic transformation of phosphoenolpyruvic acid into pyruvic acid, and explains the large negative free energy change in the transformation. This energy release is coupled to ATP formation (see Box 7.25).

Tautomerism occurs elsewhere in the glycolytic pathway (see Section 15.2). The transformation of glyceraldehyde 3-phosphate into dihydroxyacetone phosphate involves two such keto–enol tautomerisms, and proceeds through an enediol.

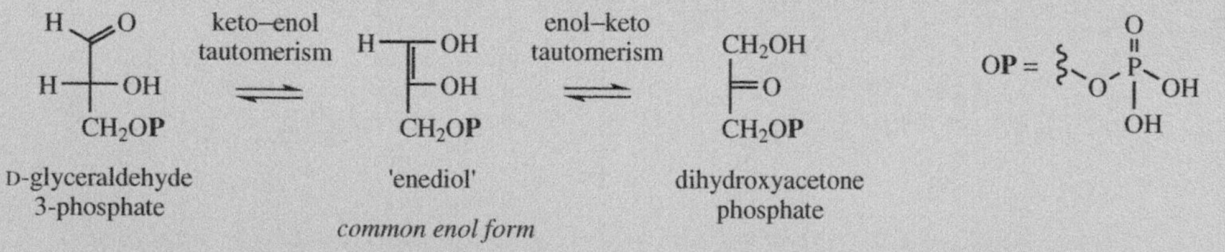

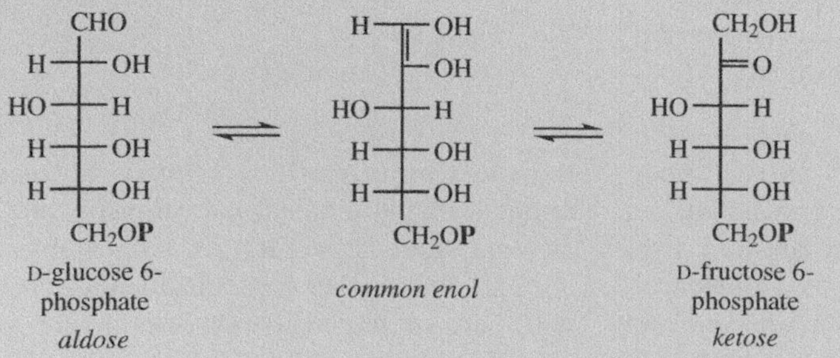

keto–enol tautomerism

enol–keto tautomerism

D-glyceraldehyde 3-phosphate

'enediol'
common enol form

dihydroxyacetone phosphate

these sugar derivatives are shown as Fischer projections to represent stereochemistry

D-glucose 6-phosphate
aldose

common enol

D-fructose 6-phosphate
ketose

This enediol can be regarded as a common enol tautomer for two different keto structures. In other words, there are two ways in which this enediol can tautomerize back to a keto form, and the reaction thus appears to shift the position of the carbonyl group. The reaction is enzyme catalysed, which allows the normal equilibrium processes to be disturbed.

It is nice to see this series of reactions being repeated in the glycolytic pathway, this time accounting for the transformation of glucose 6-phosphate into fructose 6-phosphate. Although the substrates are different, the reacting portion of the molecules is exactly the same as that in the glyceraldehyde 3-phosphate to dihydroxyacetone phosphate transformation. Again, this is an enzyme-catalysed reaction.

10.1.1 Hydrogen exchange

The intermediacy of enols or enolate anions may be demonstrated by **hydrogen exchange reactions** (see Section 4.11.2). Both acid-catalysed and base-catalysed tautomerism mechanisms involve removal of a proton from the α-carbon and supply of a proton from solvent to the carbonyl oxygen. Accordingly, this removal/supply of protons can be observed using isotopes of hydrogen, either radioactive tritium or the stable deuterium, which can be detected easily via NMR techniques.

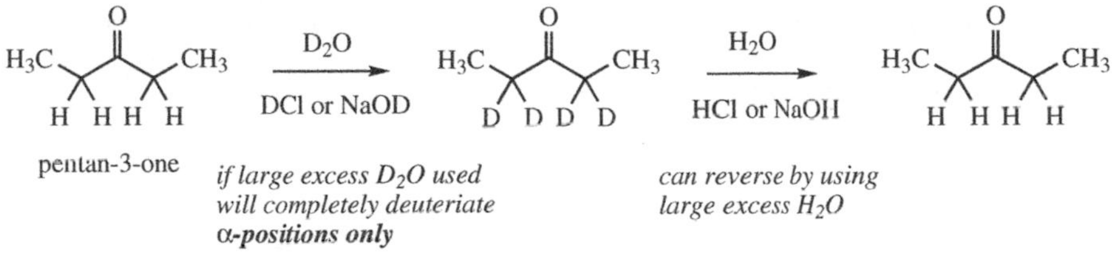

pentan-3-one
if large excess D₂O used will completely deuteriate **α-positions only**

can reverse by using large excess H₂O

Thus, pentan-3-one can be deuteriated using a large excess of D_2O, with either acid (DCl) or base (NaOD) catalyst; the acid or base catalyst should also be deuteriated to minimize dilution of label. After suitable equilibration, usually requiring prolonged heating, the α-positions will become completely labelled with deuterium.

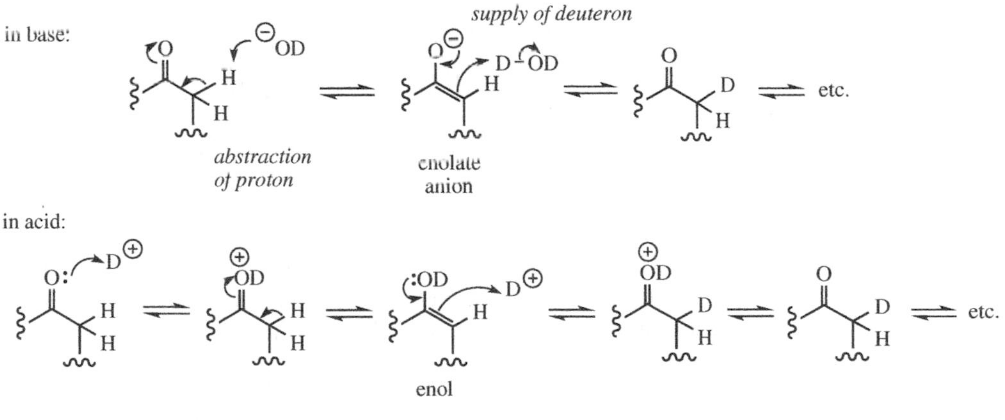

Two mechanisms are shown above. The base-catalysed mechanism proceeds through the enolate anion. The acid-catalysed process would be formulated as involving an enol intermediate. Note that the terminal hydrogens in pentan-3-one are not exchanged, since they do not participate in the enolization process. Of course, it is also possible to re-exchange the labelled hydrogens by a similar process using an excess of ordinary water, a process that might be exploited to determine or confirm the position of labelling in a deuterium-labelled substrate.

Although this section has been termed hydrogen exchange, it is important to realize that we could also visualize this simply as an **enolate anion acting as a base**. This is also true of the next section, and in some of the following sections we shall encounter enolate anions acting as nucleophiles.

10.1.2 Racemization

The process of hydrogen exchange shown above has implications if the α-carbon is chiral and has a hydrogen attached. Removal of the proton will generate a planar enol or enolate anion, and regeneration of the keto form may then involve supply of protons from either face of the double bond, so changing a particular enantiomer into its racemic form. Reacquiring a proton in the same stereochemical manner that it was lost will generate the original substrate, but if it is acquired from the other face of the double bond it will give the enantiomer, i.e. together making a racemate. Note that removal and replacement of protons at the other α-carbon, i.e. the methyl, will occur, but has no stereochemical consequences.

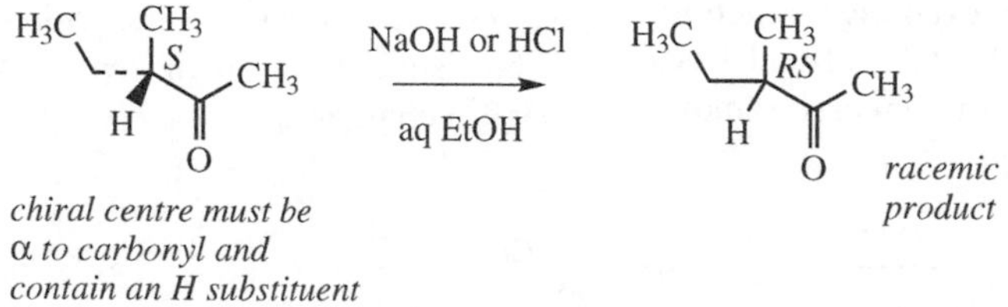

chiral centre must be
α to carbonyl and
contain an H substituent

racemic
product

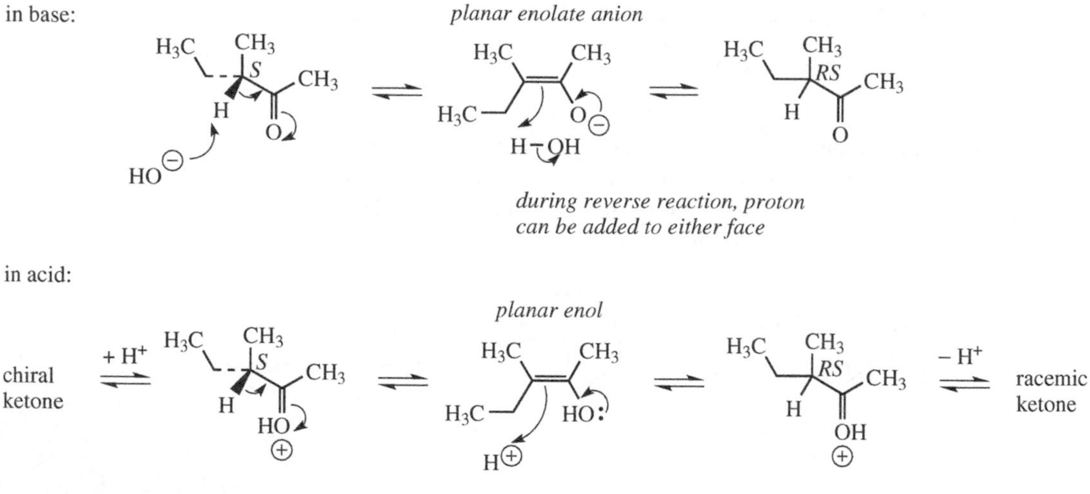

in base:

planar enolate anion

during reverse reaction, proton
can be added to either face

in acid:

planar enol

chiral
ketone

racemic
ketone

The chiral centre must be α to the carbonyl and must contain a hydrogen substituent. If there is more than one chiral centre in the molecule with only one centre α to the carbonyl, then the other centres will not be affected by enolization, so the product will be a mixture of **diastereoisomers** of the original compound rather than the racemate.

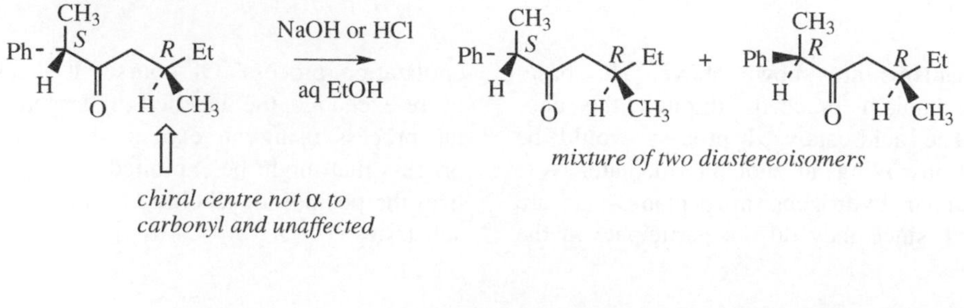

chiral centre not α to
carbonyl and unaffected

mixture of two diastereoisomers

Sometimes, other features in the molecule may facilitate formation of the enol or enolate. Thus, in the ketone shown below, conjugation of the enol double bond with the aromatic ring system helps to stabilize the enol tautomer; therefore, enolization and racemization occur more readily.

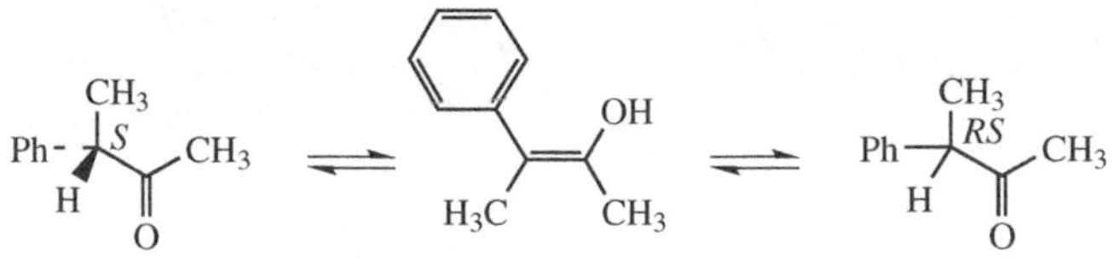

enol stabilized by conjugation with aromatic ring

It should be noted that the rate of racemization (or the rate of hydrogen exchange in Section 10.1.1) is exactly the same as the rate of enolization, since the reprotonation reaction is fast. Hence, the rate is typical of a bimolecular process and depends upon two variables, the concentration of carbonyl compound and the concentration of acid (or base).

$$\text{Rate} = k[\text{C=O}][\text{acid}]$$

or

$$\text{Rate} = k[\text{C=O}][\text{base}]$$

where C=O is the carbonyl substrate and k is the rate constant.

Box 10.2

Interconversion of monoterpene stereoisomers through enolization

On heating with either acid or base, the monoterpene ketone **isodihydrocarvone** is largely converted into one product only, its stereoisomer **dihydrocarvone**.

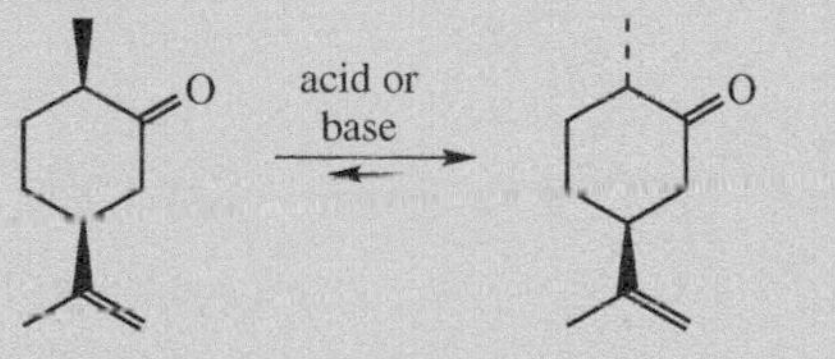

(−)-isodihydrocarvone   (−)-dihydrocarvone

There are two chiral centres in isodihydrocarvone, but only one of these is adjacent to the carbonyl group and can participate in enolization. Under normal circumstances, we might expect to generate an equimolar mixture of two diastereoisomers. This is because two possible configurations could result from the chiral centre α to the carbonyl, whereas the other centre is going to stay unchanged (see Section 3.4.4). We might thus anticipate formation of a 50:50 mixture of isodihydrocarvone and dihydrocarvone. That the product mixture is not composed of equal amounts of isodihydrocarvone and dihydrocarvone can be rationalized by considering stereochemical factors, particularly the conformations adopted by the two compounds, which turn out to favour the product over the starting material.

The favoured conformation of isodihydrocarvone has the large isopropenyl substituent equatorial. On forming the enol (or enolate anion), it will adopt the conformation in which both substituents are equatorial (or equatorial-like). To revert back to a keto tautomer might then involve acquiring a proton from either side of the planar enol/enolate. However, there is going to be a distinct preference for forming the more favoured product that has two equatorial substituents. This is dihydrocarvone. The equilibrium mixture set up thus contains predominantly dihydrocarvone, rather than an equal mixture of two diastereoisomers. The second chiral centre contains a large group, and its stereochemical preference effectively dictates the chirality at the second centre, and thus the nature of the product.

Box 10.2 (continued)

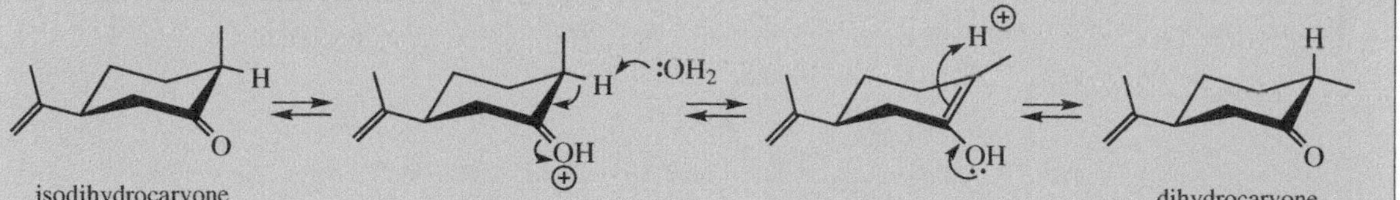

isodihydrocarvone

favoured conformer; large substituent equatorial

acid-catalysed enol formation

enol function planar

dihydrocarvone

favoured conformer; both substituents equatorial

10.1.3 Conjugation

When an enol tautomer reverts back to a keto tautomer, it must acquire a proton, and we have already seen that it may be acquired from different faces of the double bond, giving two types of stereochemistry. In the example described in Box 10.2, the stereochemistry of the product was effectively dictated by the existing chirality at a second centre. Now we can see a further variant, in that the stability of the product dictates that an alternative carbon in the enol tautomer actually receives the proton. This relates to **conjugation** in the product.

A β,γ-unsaturated carbonyl compound exposed to acid or base is usually converted rapidly into an α,β-unsaturated carbonyl derivative. This isomerization is easily interpreted by considering enolization.

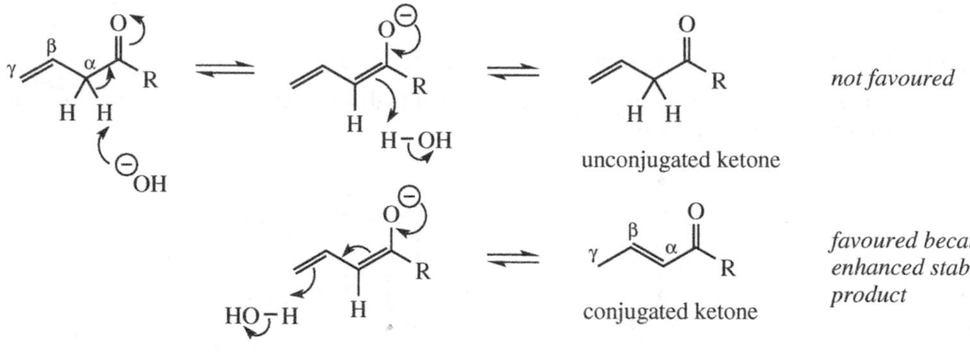

not favoured

unconjugated ketone

favoured because of enhanced stability of product

conjugated ketone

Removal of an α-proton from a β,γ-unsaturated ketone generates an enolate anion, and this might be transformed back to the β,γ-unsaturated compound by reprotonation at the α-position. However, this does not occur because the enolate anion now has conjugated double bonds, and we can propose an alternative mechanism for reprotonation, invoking the conjugation and protonating at the γ-position. This protonation is preferred, in that the product is now a conjugated ketone and, therefore, energetically favoured over the non-conjugated ketone. Since all the reactions are equilibria, eventually the more stable product will result.

Box 10.3

Conversion of pregnenolone into progesterone

An important transformation in **steroid biochemistry** is the conversion of **pregnenolone** into **progesterone**. Progesterone is a female sex hormone, a progestogen, but this reaction is also involved in the production of corticosteroids such as hydrocortisone and aldosterone. The reaction also occurs in plants, and features in the formation of cardioactive glycosides, such as digitoxin in foxglove.

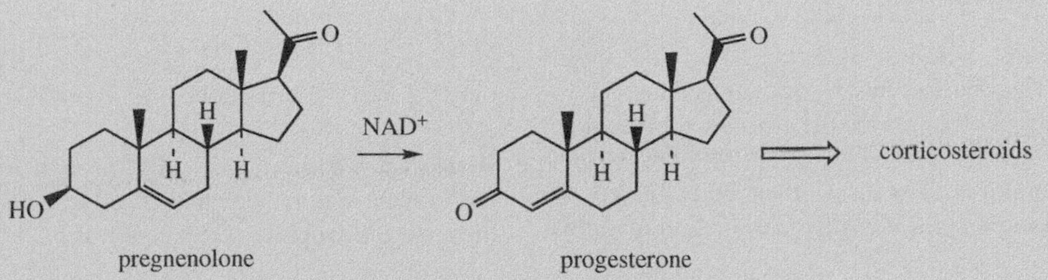

pregnenolone   progesterone

This enzymic conversion involves two enzymes, a dehydrogenase and an isomerase. The dehydrogenase component oxidizes the hydroxyl group on pregnenolone to a ketone, and requires the oxidizing agent cofactor NAD⁺ (see Box 11.2). The isomerase then carries out two tautomerism reactions, enolization to a dienol followed by production of the more stable conjugated ketone.

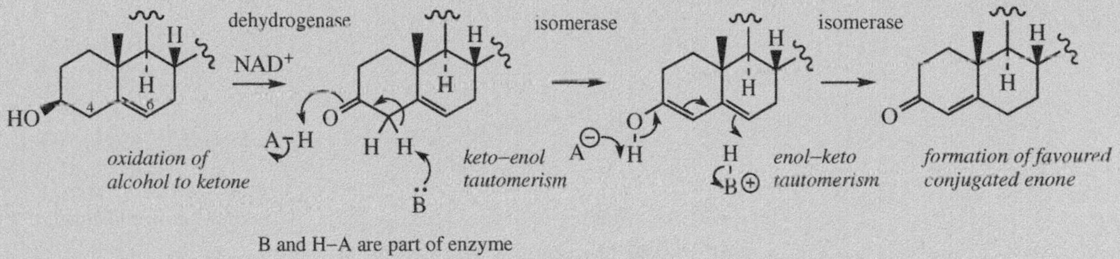

oxidation of   *keto–enol*   *enol–keto*   *formation of favoured*
alcohol to ketone   *tautomerism*   *tautomerism*   *conjugated enone*

B and H–A are part of enzyme

The enzyme provides a base (B:) and an acid (A–H) via appropriate amino acid side-chains on the enzyme (see Section 13.4) to facilitate proton removal and supply. A fascinating aspect is that the proton removed from the methylene (steroid position 4) by the base is then donated back to position 6. The base is suitably positioned to serve both sites in the steroid.

An exactly analogous enzymic transformation is encountered during the formation of oestrogen and androgen sex hormones, e.g. estradiol and testosterone respectively, where dehydroepiandrosterone is oxidized to androstenedione.

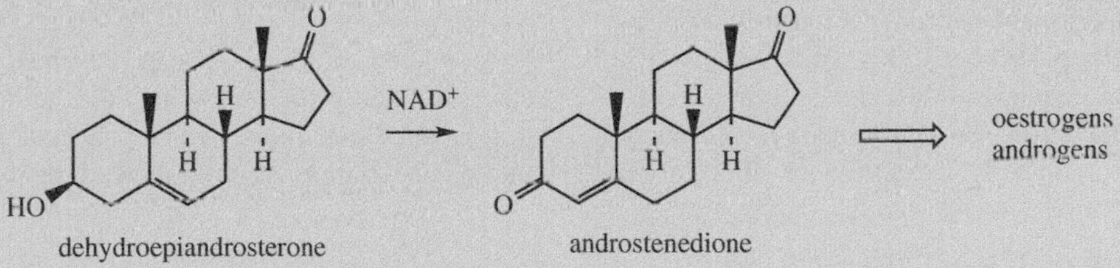

dehydroepiandrosterone   androstenedione

The isomerization reaction is also encountered in chemical manipulations of steroids. Thus, many natural steroids contain a 5-en-3-ol combination of functionalities, e.g. cholesterol. Treatment of cholesterol with an oxidizing agent (aluminium isopropoxide is particularly suitable) leads to cholest-4-en-3-one, the tautomerism occurring spontaneously under the reaction conditions.

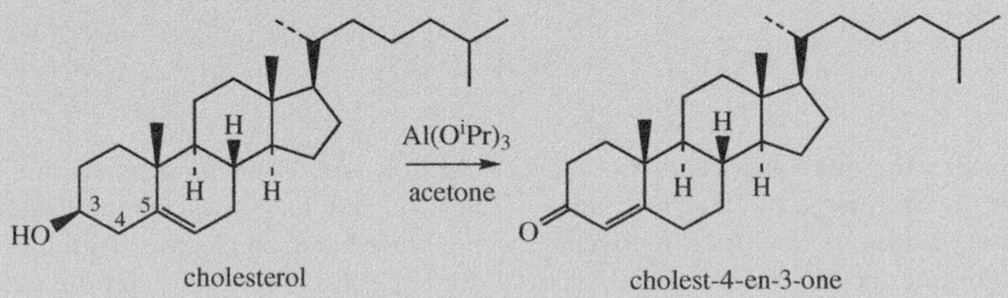

cholesterol   cholest-4-en-3-one

10.1.4   Halogenation

Aldehydes and ketones undergo acid- and base-catalysed **halogenation** in the α position. This is also dependent on enolization or the formation of enolate anions.

Thus, bromination of acetone may be achieved by using bromine in sodium hydroxide solution, and this

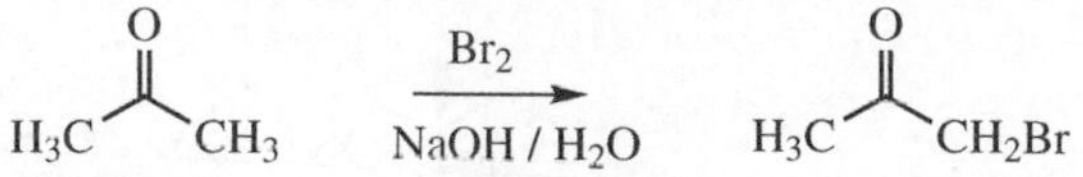

is rationalized mechanistically through formation of the enolate anion, which then attacks the polarized bromine electrophile (see Section 8.1.2).

in base:

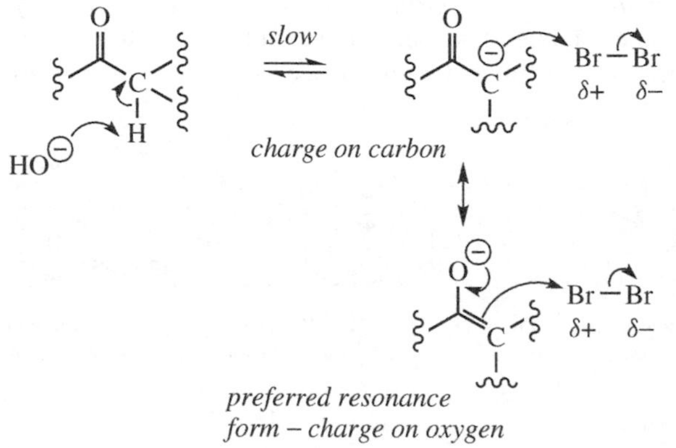

charge on carbon

preferred resonance
form – charge on oxygen

$$\text{Rate} = k \,|\, C{=}O \,|\, |\, HO^- |$$

rate-controlling step is
enolate anion formation

There are two ways of representing this, according to which resonance form of the enolate anion is used. Although the preferred resonance form (charge located on the oxygen atom) should be used as the nucleophile, because carbon is acting as the nucleophile and a new C–Br bond is formed, the less-favoured resonance form is frequently employed in mechanistic pathways. This makes mechanism drawing rather easier, but is technically incorrect.

Kinetic data show us that the rate of reaction is dependent upon two variables, i.e. the carbonyl

substrate concentration and the concentration of base. These are the two components necessary for formation of the enolate anion, which is the slow step in the sequence. After formation of the enolate anion, nucleophilic attack on bromine is rapid; therefore, the bromine concentration does not figure in the rate equation.

A related mechanism can be drawn for acid-catalysed halogenation. Again, the halogen concentration does not figure in the rate equation, and the rate of enolization controls the rate of reaction.

in acid:

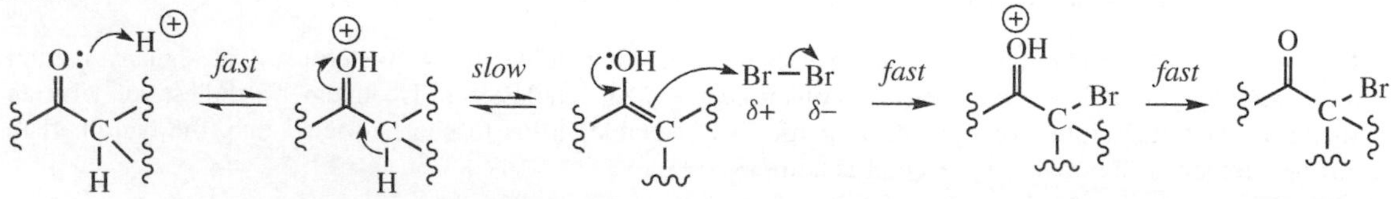

$$\text{Rate} = k \,|\, C{=}O \,|\, |\, H^+ |$$

rate-controlling step
is enolization

If we wish to synthesize a monohalogenated product, then we have to use an acid-catalysed reaction; base catalysis leads to multiple halogenation. This relates to the acidity of intermediates. Thus, each successive halogenation introduces an

electron-withdrawing substituent, which increases acidity and facilitates enolate anion formation. On the other hand, an electron-withdrawing halogen substituent destabilizes the protonated carbonyl compound, and consequently disfavours enolization.

base-catalysed reaction: consider acidity and enolate anion formation

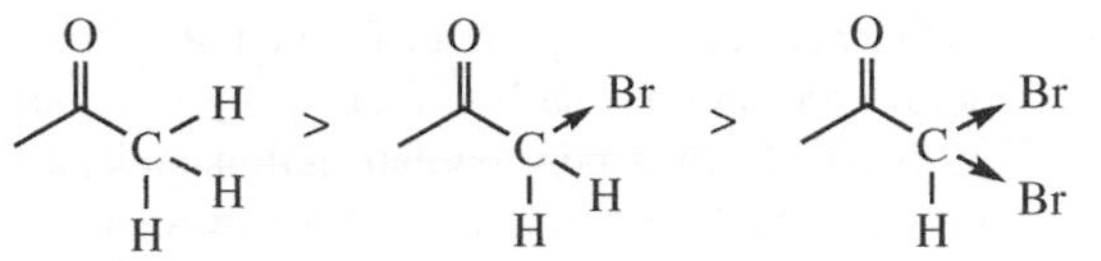

acidity of α-protons increases

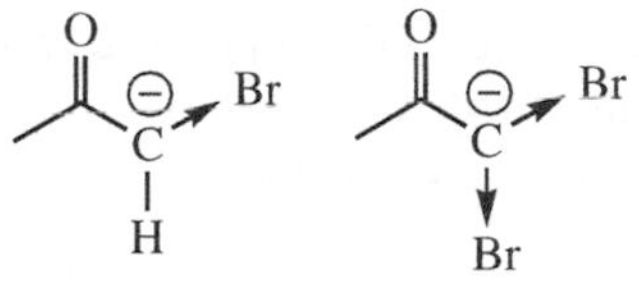

acid-catalysed reaction: consider basicity and conjugate acid formation

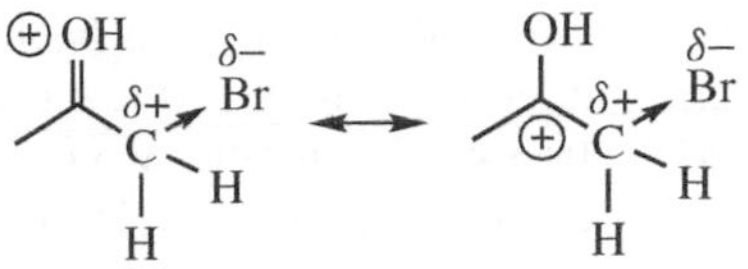

halogenation destabilizes protonated carbonyl

relate to increased stability of enolate anions

10.2   Alkylation of enolate anions

Though this topic is treated here under a separate heading, alkylation of enolate anions is nothing other than **enolate anions acting as carbanion nucleophiles** in S_N2 reactions. We deferred this topic

from Section 6.3.4, since at that stage we had not encountered the concept of enols and enolate anions.

By treating the 1,3-dicarbonyl compound acetylacetone with methyl iodide in the presence of potassium carbonate, one observes alkylation at the central carbon.

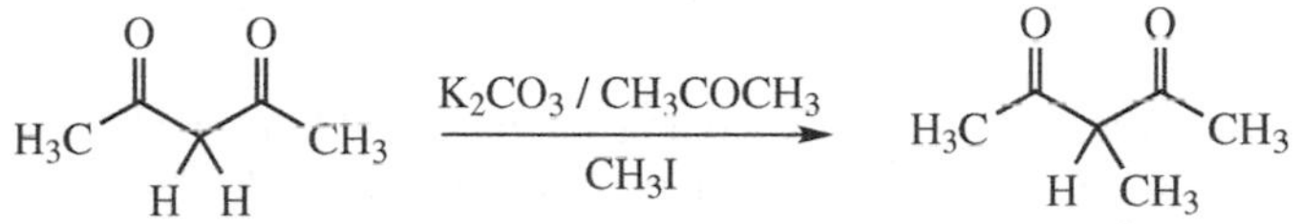

acetylacetone

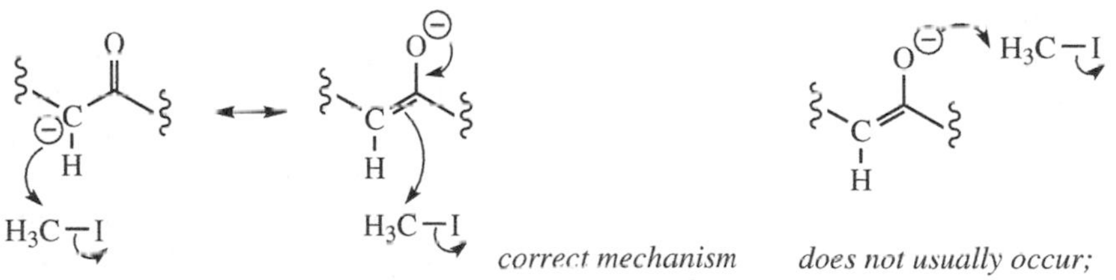

simple mechanism generally used, though strictly incorrect

S_N2 reaction; enolate as nucleophile

correct mechanism using preferred resonance form

does not usually occur; almost always get alkylation on carbon with retention of carbonyl

This is easily rationalized via initial formation of an enolate anion under the basic conditions, followed by an S_N2 reaction on the methyl iodide. The enolate anion is the nucleophile and iodide is displaced as the leaving group. The enolate anion could be drawn with charge on carbon or oxygen; the latter is preferred, as discussed above (see Section 10.1), in that the charge is preferentially located on the electronegative oxygen atom. It is feasible, therefore, that either carbon or oxygen could be the nucleophilic atom, and we might expect more chance of oxygen

participating. Despite this, it is observed that, in almost all cases, alkylation occurs on carbon, not on oxygen, so it does not present a problem. Two mechanisms could be drawn for the reaction, depending on whether the enolate anion has charge on the carbon or oxygen. Since carbon is eventually the nucleophilic centre, it is permissible to use the carbanion version of the enolate (as, in general, we shall do), though this is strictly not correct, and purists would use the alternative version starting with charge on the oxygen.

Now for some interesting features of the reaction, though they become fairly obvious with a little thought. First, the central methylene contains the more acidic protons (pK_a 9) since it is flanked by two carbonyls, so the enolate anion formed involves this carbon (see Section 4.3.5). In other words, alkylation occurs on the central carbon of acetylacetone, not on the terminal carbons. Second, it is possible to use carbonyl compounds such as acetone as a solvent without these reacting under the reaction conditions. Acetone will have similar acidity (pK_a 19) to the acetyl groups of acetylacetone, so likewise will not

form an enolate anion under conditions that only ionize the central methylene of a 1,3-dicarbonyl compound.

Furthermore, the product formed still contains an acidic proton on a carbon flanked by two carbonyls, so it can form a new enolate anion and participate in a second S_N2 reaction. The nature of the product will thus depend on electrophile availability. With 1 mol of methyl iodide, a monomethylated compound will be the predominant product, whereas with 2 mol of methyl iodide the result will be mainly the dimethylated compound.

1 mol MeI gives
monoalkylated product

more acidic   *acidic*
hydrogens   *hydrogen*

2 mol MeI gives
dialkylated product

A further twist is that it is possible to use this reaction to insert two different alkyl groups. This requires treating first with 1 mol of an alkylating agent, allowing the reaction to proceed, then supplying 1 mol of a second, but different, alkylating agent.

use successively 1 mol of
two different alkyl halides

Of course, minor products might be produced, including monoalkylated products and dialkylated products (in which the two alkyl groups are the

same), depending on the conditions and how near to completion the reaction proceeds. Note that we cannot use aryl halides in these reactions; rearside attack is impossible and we do not get S_N2 reactions at sp^2-hybridized carbon (see Section 6.1.1).

1,3-Dicarbonyl compounds, like acetylacetone, are reasonably acidic (pK_a 9) and formation of enolate anions is achieved readily. Potassium carbonate is basic enough to ionize acetylacetone in the above example. However, if we are presented with a substrate having only a single carbonyl group, e.g. acetone (pK_a 19), then it follows that we must use a stronger base to remove the correspondingly less acidic protons. Strong bases that might be used include sodium hydride and sodium amide.

These compounds ionize and act as sources of hydride and amide ions respectively, which are able to remove α-protons from carbonyl compounds. These ions are actually the conjugate bases of hydrogen and ammonia respectively, compounds that are very weak acids indeed. What becomes important here is that enolate anion formation becomes essentially irreversible; the enolate anion formed is insufficiently basic to be able to remove a proton from either hydrogen or ammonia. This is in marked contrast to the earlier examples of enolate anion formation that were reversible. We now have a means of preparing the enolate anion, rather than relying upon an equilibrium reaction. Accordingly, reactions are usually done in two stages, preparation of the enolate anion followed by addition of the alkylating agent electrophile.

no α-hydrogens

In the example shown, alkylation of the ketone is readily accomplished using such a two-stage process with 1 mol of alkyl halide. Note that the specificity of this reaction relies on one of the α-carbons having no acidic hydrogens, so that only one enolate anion can be formed.

Another strong base routinely employed in synthetic procedures to prepare enolate anions is **lithium diisopropylamide** (**LDA**). The diisopropylamide anion is formed by removing a proton from diisopropylamine using the organometallic derivative *n*-butyllithium. Because of the highly reactive nature of *n*-butyllithium (it reacts explosively with air) this reaction has to be conducted in an oxygen-free atmosphere and at very low temperature. The ionization works because although the acidity of diisopropylamine is not great (pK_a 36), the other product formed, i.e. butane, is significantly less acidic (pK_a 50). The reaction is essentially irreversible.

diisopropylamine   *n*-butyllithium   LDA   butane

pK_a 36   *strong base*   pK_a 50
                                                          poor nucleophile

When the carbonyl compound is added to this base, abstraction of a proton and formation of the enolate anion follow, as seen with sodium hydride or sodium amide above. Again, this reaction is essentially irreversible because the other product is the weak base diisopropylamine (pK_a 36). So far, there does not seem any particular advantage in using LDA rather than sodium hydride or sodium amide, and the manipulations required are very much more difficult and dangerous. The real benefit is that LDA is a very strong base, and because of its quite large size it is also a relatively poor nucleophile. This reduces the number of competing reactions that might occur where nucleophilicity competes with basicity (see Section 6.4.1).

cyclohexanone

In symmetrical structures such as cyclohexanone, ionization at α-positions occurs readily and allows the preparation of alkylated products. In unsymmetrical structures, the sheer size of LDA as a base may allow selectivity by preferential removal of certain α-protons. Thus, the ketone pentan-2-one will undergo preferential removal of a proton from the terminal methyl in the generation of an enolate anion. This allows selective alkylation to be achieved.

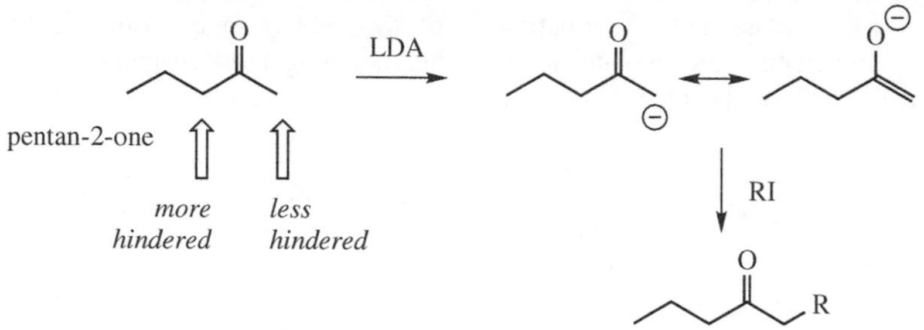

10.3   Addition–dehydration: the aldol reaction

We now have examples of the generation of enolate anions from carbonyl compounds, and their potential as nucleophiles in simple S_N2 reactions. However, we must not lose sight of the potential of a carbonyl compound to act as an electrophile. This section, the **aldol reaction**, is concerned with enolate anion nucleophiles attacking carbonyl electrophiles to give addition compounds (see Section 7.1), though it is usual for such addition compounds to then lose water, i.e. **addition–dehydration**.

The namesake aldol reaction is the formation of an addition compound, aldol, from two molecules of acetaldehyde, when this aldehyde is treated with aqueous sodium hydroxide. The terminology aldol comes from the functional groups in the product, aldehyde and alcohol.

aldol reaction

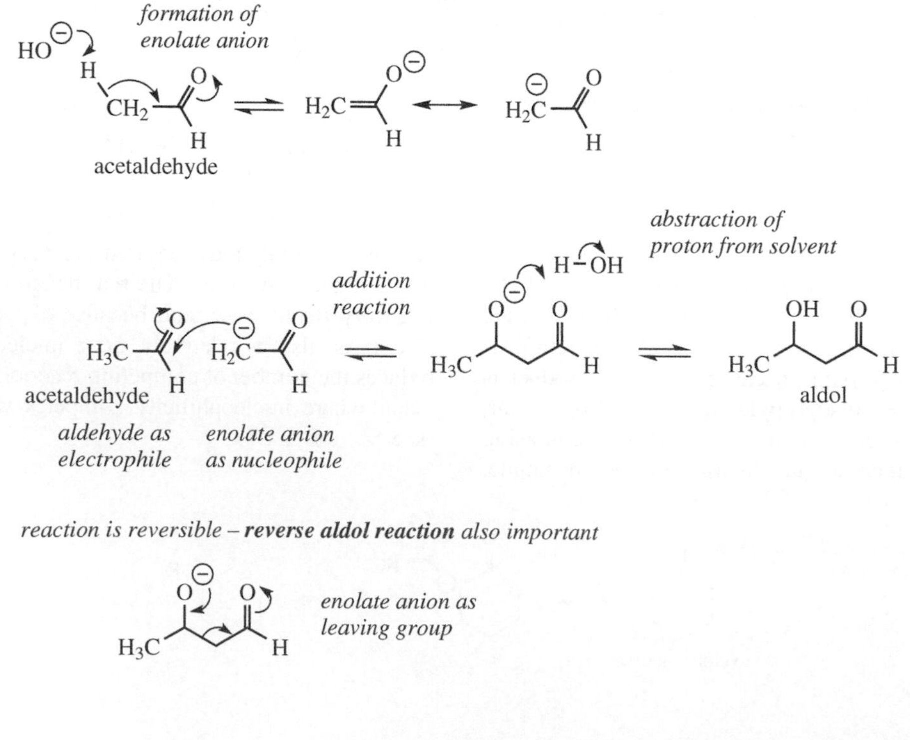

This is easily formulated as production of an enolate anion followed by nucleophilic attack of this anion on to the carbonyl group of a second molecule of acetaldehyde. Aldol is then produced when the addition anion abstracts a proton from solvent. The reaction is reversible, and it is usually necessary to disturb the equilibrium by some means. Removal of product is possible, but, as seen below, the dehydration part of the sequence may be responsible for pushing the reaction to completion.

In the reverse reaction, the addition anion reforms the carbonyl group by expelling the enolate anion as leaving group. This **reverse aldol reaction** is sufficiently important in its own right, and we shall meet examples. Note that, as we saw with simple aldehyde and ketone addition reactions, aldehydes are better electrophiles than ketones (see Section 7.1.1). This arises from the extra alkyl group in ketones, which provides a further inductive effect and extra steric hindrance. Accordingly, the aldol reaction is more favourable with aldehydes than with ketones. With ketones, it is absolutely essential to disturb the equilibrium in some way.

The aldol reaction as formulated above involves two molecules of the starting substrate. However, by a consideration of the mechanism, one can see that different carbonyl compounds might be used as nucleophile or electrophile. This would be termed a **mixed aldol reaction** or crossed aldol reaction. However, if one merely reacted, say, two aldehydes together under basic conditions, one would get a

rather messy mixture of products containing at least four different components. This is because both starting materials might feature as nucleophile or as electrophile.

mixed aldol reaction

$$RCH_2CHO \ + \ R'CH_2CHO \ \longrightarrow \ 4 \ products$$

nucleophile		electrophile
RCH_2CHO	+	RCH_2CHO
RCH_2CHO	+	$R'CH_2CHO$
$R'CH_2CHO$	+	RCH_2CHO
$R'CH_2CHO$	+	$R'CH_2CHO$

For the mixed aldol reaction to be of value in synthetic work, it is necessary to restrict the number of combinations. This can be accomplished as follows. First, if one of the materials has no α-hydrogens, then it cannot produce an enolate anion, and so cannot function as the nucleophile. Second, in aldehyde plus ketone combinations, the aldehyde is going to be a better electrophile, so reacts preferentially in this role. A simple example of this approach is the reaction of benzaldehyde with acetone under basic conditions. Such reactions are synthetically important as a means of increasing chemical complexity by forming new carbon–carbon bonds.

mixed aldol reaction can be of value if one reagent has no α-hydrogens and thus cannot form an enolate anion

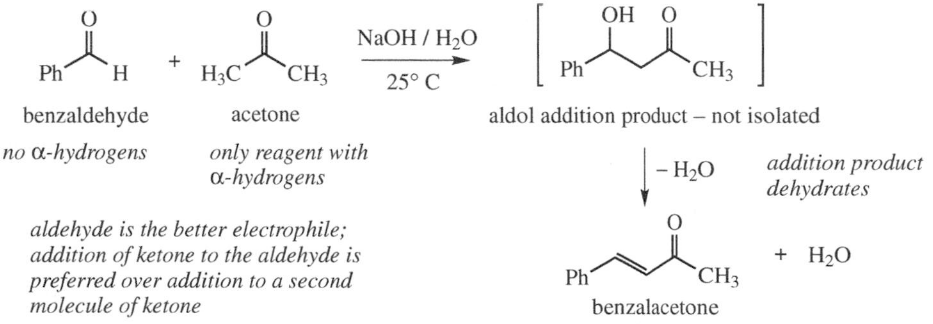

benzaldehyde   acetone

no α-hydrogens   *only reagent with α-hydrogens*

aldehyde is the better electrophile; addition of ketone to the aldehyde is preferred over addition to a second molecule of ketone

aldol addition product – not isolated

addition product dehydrates

benzalacetone

Benzaldehyde has no α-hydrogens, so it cannot be converted into an enolate anion to become a nucleophile. Acetone has α-hydrogens, so it can form an enolate anion and become the nucleophile.

We now have two possible electrophiles, i.e. one an aldehyde and the other a less reactive ketone. The preferred reaction is thus acetone as enolate anion nucleophile, with benzaldehyde as preferred electrophile, giving the addition product shown. This is not actually isolated, since it readily dehydrates to give the unsaturated ketone benzalacetone (see below).

The addition product from aldol reactions frequently dehydrates by heating in acid or in base to give the corresponding α,β-unsaturated carbonyl compound. Under basic conditions, this occurs readily, even though hydroxide is poor leaving group, because of the acidity of the α-proton and the conjugation stabilization in the product.

base-catalysed E2 elimination

aldol

acid-catalysed E2 elimination

− H$_2$O

favoured by formation of conjugated system

benzalacetone

conjugation extends into aromatic ring

E1cb mechanism

formation of enolate anion

loss of leaving group

There is evidence that this is not an E2 mechanism under basic conditions, but a so-called **E1cb mechanism**. This stands for elimination–unimolecular–conjugate base, and proceeds via initial removal of the acidic proton to give the conjugate base (enolate anion). The reaction is unimolecular because it is the loss of a leaving group from the conjugate base that is the rate-determining step. Removal of the acidic proton is actually faster than loss of the hydroxide ion. Since E1cb reactions are rare (this is the only one we shall consider), we deliberately chose not to include it under general elimination reactions in Chapter 6.

The conditions of the reaction are often sufficient to cause dehydration of the addition product as it is formed, and it is normally extremely difficult to isolate the addition product. It turns out that the addition reaction (equilibrium) is slow, whereas the elimination reaction (non-reversible) is faster. This usually disturbs the equilibrium in an aldol reaction, especially if the product is stabilized by even further conjugation, as in the case of benzalacetone above, where the benzene ring also forms part of the conjugated system.

An alternative approach to mixed aldol reactions, and the one usually preferred, is to carry out a two-stage process, forming the enolate anion first using a strong base like LDA (see Section 10.2). The first step is essentially irreversible, and the electrophile is then added in the second step. An aldol reaction between butan-2-one and acetaldehyde exemplifies this approach. Note also that the large base LDA selectively removes a proton from the least-hindered position, again restricting possible combinations (see Section 10.2).

butan-2-one

more hindered *less hindered*

LDA   H$_2$O

Box 10.4

Aldol and reverse aldol reactions in biochemistry: aldolase, citrate synthase

Both the **aldol** and **reverse aldol reactions** are encountered in carbohydrate metabolic pathways in biochemistry (see Chapter 15). In fact, one reversible transformation can be utilized in either carbohydrate biosynthesis or carbohydrate degradation, according to a cell's particular requirement. D-Fructose 1,6-diphosphate is produced during carbohydrate biosynthesis by an aldol reaction between dihydroxyacetone phosphate, which acts as the enolate anion nucleophile, and D-glyceraldehyde 3-phosphate, which acts as the carbonyl electrophile; these two starting materials are also interconvertible through keto–enol tautomerism, as seen earlier (see Section 10.1). The biosynthetic reaction may be simplified mechanistically as a standard mixed aldol reaction, where the nature of the substrates and their mode of coupling are dictated by the enzyme. The enzyme is actually called **aldolase**.

During carbohydrate metabolism in the **glycolytic pathway** (see Section 15.2), fructose 1,6-diphosphate is cleaved to give dihydroxyacetone phosphate and glyceraldehyde 3-phosphate. This is a reverse aldol reaction, in which a carbonyl group is formed at the expense of carbon–carbon bond cleavage with expulsion of an enolate anion leaving group.

The additional functional groups present in the substrates would seriously limit any base-catalysed chemical aldol reaction between these substrates, but this reaction is enzyme mediated, allowing reaction at room temperature and near-neutral conditions. The aldol and reverse aldol reactions just described accommodate the chemical changes observed, though we now know that nature uses a slightly different approach via enamines (see Box 10.5). This does not significantly alter our understanding of the reactions, but it does remove the requirement for a strong base, and also accounts for the bonding of the substrate to the enzyme.

A similar aldol reaction is encountered in the **Krebs cycle** in the reaction of acetyl-CoA and oxaloacetic acid (see Section 15.3). This yields citric acid, and is catalysed by the enzyme **citrate synthase**. This intermediate provides the alternative terminology for the Krebs cycle, namely the **citric acid cycle**. The aldol reaction is easily rationalized, with acetyl-CoA providing an enolate anion nucleophile that adds to the carbonyl of oxaloacetic acid. We shall see later that esters and thioesters can also be converted into enolate anions (see Section 10.7).

We should also consider occasions when there are two carbonyl groups in the same molecule. We then have the possibility of an **intramolecular aldol reaction**, and this offers a convenient way of synthesizing ring systems. Rings with five or six carbons are particularly favoured (see Section 3.3.2). Thus, treatment of octan-2,7-dione with base gives good yields of the cyclopentene derivative shown.

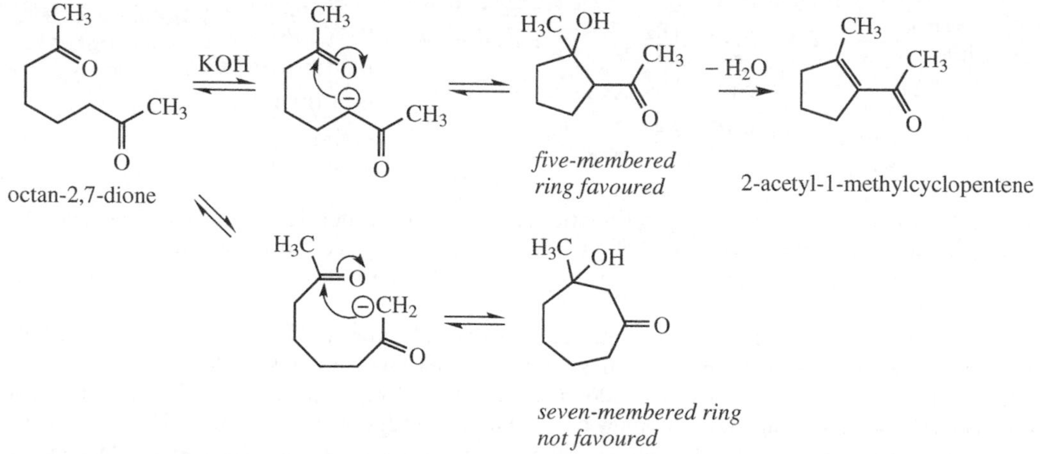

octan-2,7-dione

five-membered ring favoured

2-acetyl-1-methylcyclopentene

seven-membered ring not favoured

The reaction is readily formulated. Note that there are two potential products from the aldol addition, one of which is five-membered and the other seven-membered. The five-membered product is more favourable than the seven-membered one simply based on ring strain. However, if both products form, they will be in equilibrium as shown. It is the next step, the dehydration, that drives the reaction giving the more stable product, the cyclopentene. Any seven-membered addition product can then equilibrate to give more of the five-membered compound. A similar reaction with heptan-2,6-dione would lead to the methylcyclohexenone product, and not the sterically unfavourable four-membered ring alternative.

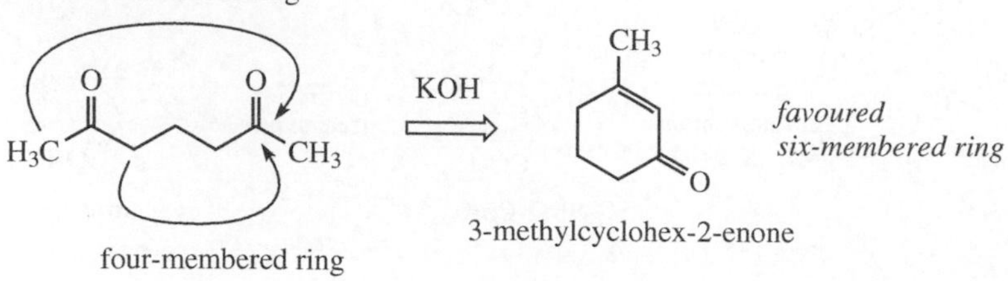

six-membered ring

four-membered ring

favoured six-membered ring

3-methylcyclohex-2-enone

Note also that if the substrate has both aldehyde and ketone functions the aldehyde will act as the electrophile. The ketoaldehyde shown forms the one product in good yield, there now being restrictions on preferred ring size and the regiochemistry of the mixed aldol reaction.

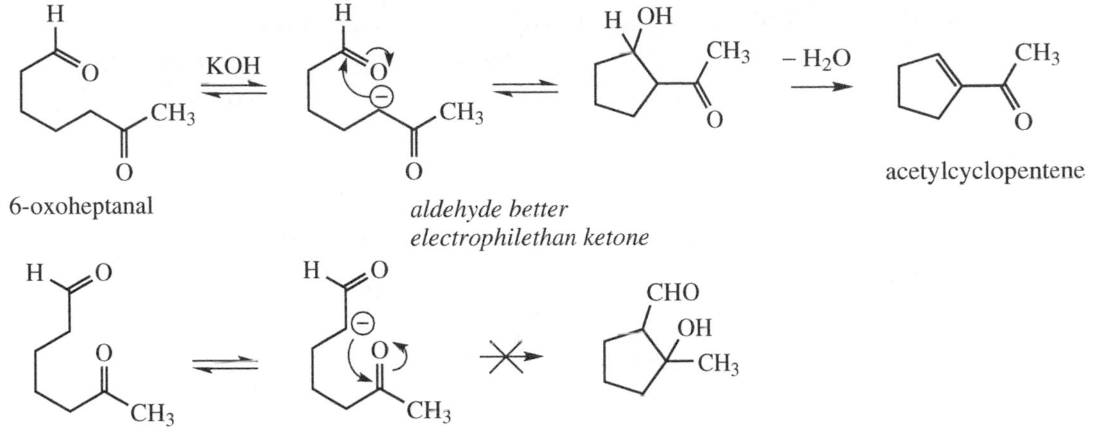

6-oxoheptanal

aldehyde better electrophilethan ketone

acetylcyclopentene

If a five- or six-membered ring can form, then intramolecular aldol reactions usually occur more rapidly than the corresponding intermolecular reactions between two molecules of substrate. This provides a very useful route to cyclic compounds (see Box 10.19).

10.4　Other stabilized anions as nucleophiles: nitriles and nitromethane

An enolate anion behaves as a carbanion nucleophile, the carbonyl group stabilizing the anion by delocalization of charge. Both cyano (nitrile) and nitro groups can fulfil the same role as a carbonyl by stabilizing a carbanion, so we see similar enhanced acidity of α-protons in simple nitrile and nitro compounds. pK_a values for nitriles are about 25, whereas aliphatic nitro compounds have pK_a about 10. Nitro compounds are thus considerably more acidic than aldehydes and ketones (pK_a about 20).

Accordingly, it is possible to generate analogues of enolate anions containing cyano and nitro groups, and to use these as nucleophiles towards carbonyl electrophiles in **aldol-like processes**. Simple examples are shown.

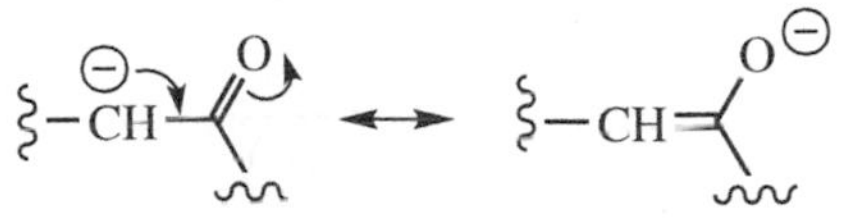

enolate anion; carbonyl stabilizes carbanion by delocalization

cyano (nitrile) and nitro are also able to stabilize carbanions

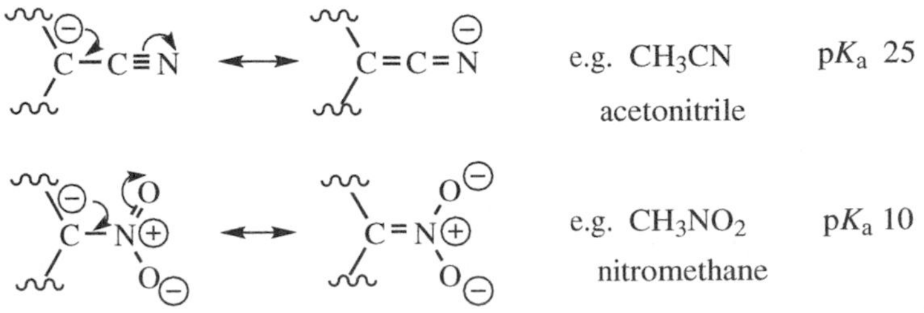

e.g. CH_3CN　　pK_a 25
acetonitrile

e.g. CH_3NO_2　　pK_a 10
nitromethane

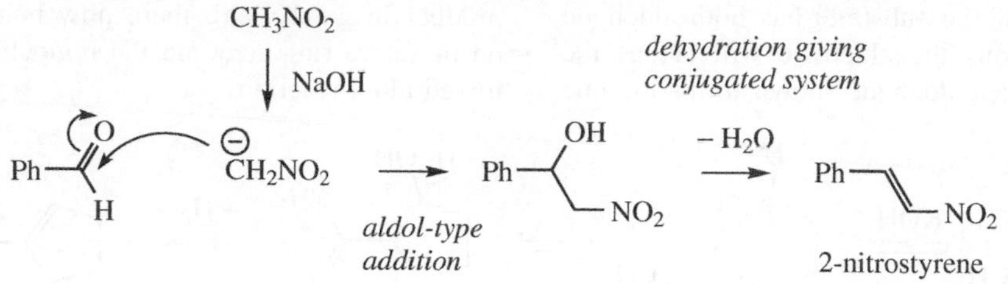

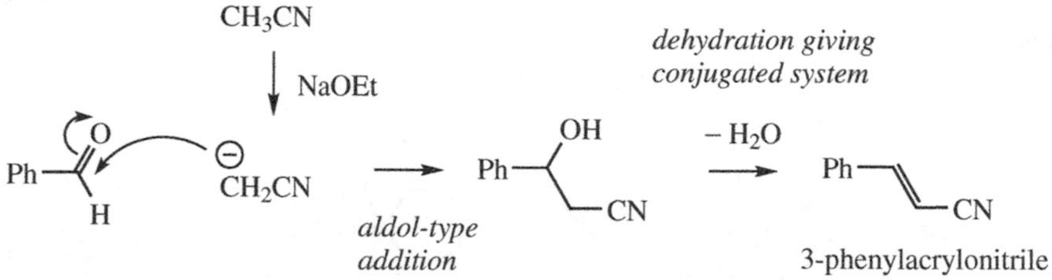

CH₃NO₂

↓ NaOH

dehydration giving conjugated system

Ph—CHO ... ⊖CH₂NO₂ ⟶ Ph—CH(OH)—CH₂—NO₂ $\xrightarrow{-H_2O}$ Ph—CH=CH—NO₂

aldol-type addition

2-nitrostyrene

CH₃CN

↓ NaOEt

dehydration giving conjugated system

Ph—CHO ... ⊖CH₂CN ⟶ Ph—CH(OH)—CH₂—CN $\xrightarrow{-H_2O}$ Ph—CH=CH—CN

aldol-type addition

3-phenylacrylonitrile

As with many aldol reactions, addition is usually followed by elimination of water, generating a conjugated system with the cyano or nitro group. The presence of extended conjugation through aromatic substituents enhances this process.

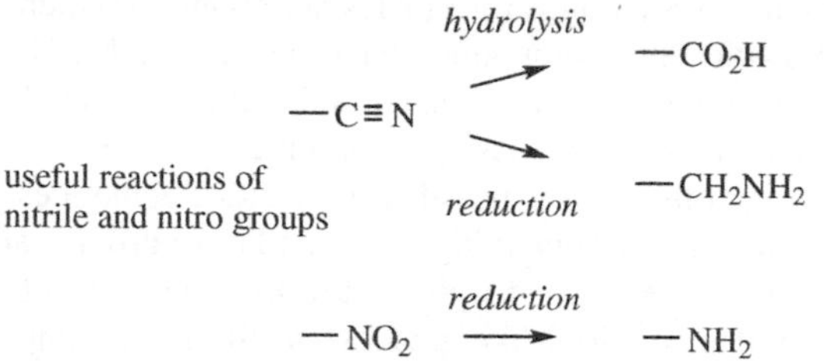

These reactants introduce either nitrile or nitro groups into the product. These groups may be converted into carboxylic acids or amines, as shown.

10.5 Enamines as nucleophiles

In Section 7.7.2 we met enamines as products from addition–elimination reactions of secondary amines with aldehydes or ketones. **Enamines** are formed instead of imines because no protons are available on nitrogen for the final deprotonation step, and the nearest proton that can be lost from the iminium ion is that at the β-position.

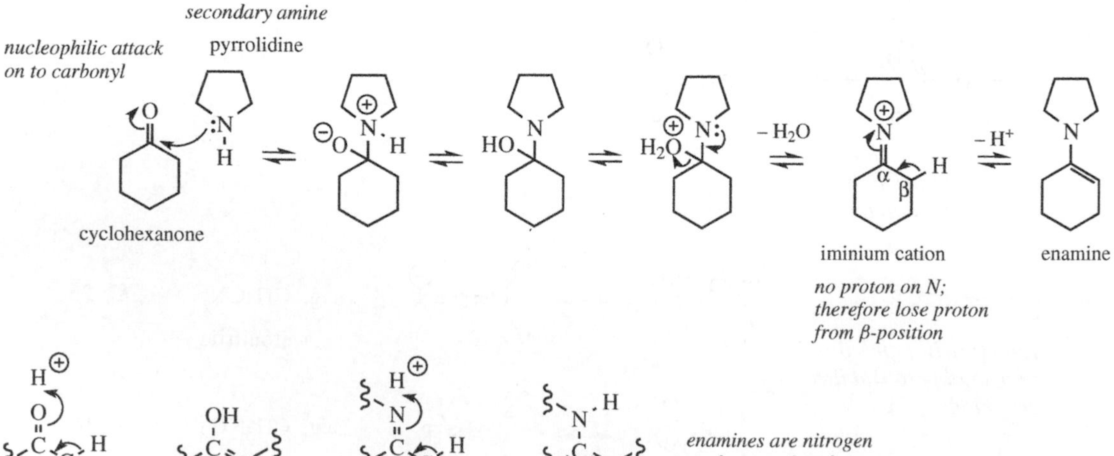

useful reactions of nitrile and nitro groups

$$—C≡N \xrightarrow{hydrolysis} —CO_2H$$
$$\xrightarrow{reduction} —CH_2NH_2$$

$$—NO_2 \xrightarrow{reduction} —NH_2$$

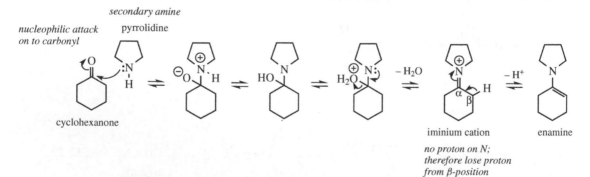

nucleophilic attack on to carbonyl

secondary amine pyrrolidine

cyclohexanone

iminium cation

enamine

no proton on N; therefore lose proton from β-position

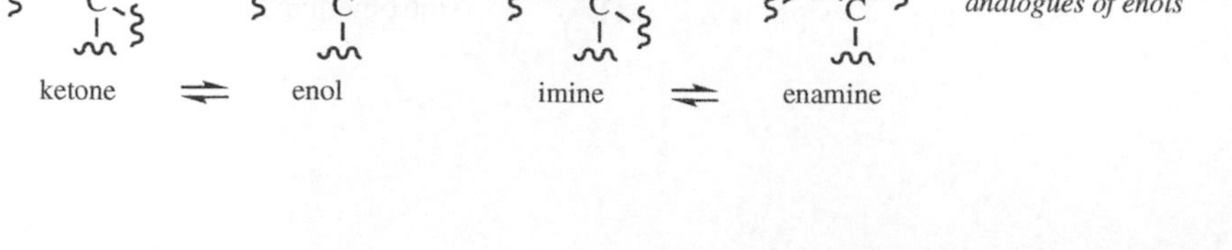

ketone ⇌ enol   imine ⇌ enamine

enamines are nitrogen analogues of enols

There is a distinct relationship between keto–enol tautomerism and the iminium–enamine interconversion; it can be seen from the above scheme that enamines are actually **nitrogen analogues of enols**. Their chemical properties reflect this relationship. It also leads us to another reason why enamine formation is a property of secondary amines, whereas primary amines give imines with aldehydes and ketones (see Section 7.7.1). Enamines from primary amines would undergo rapid conversion into the more stable imine tautomers (compare enol and keto tautomers); this isomerization cannot occur with enamines from secondary amines, and such enamines are, therefore, stable.

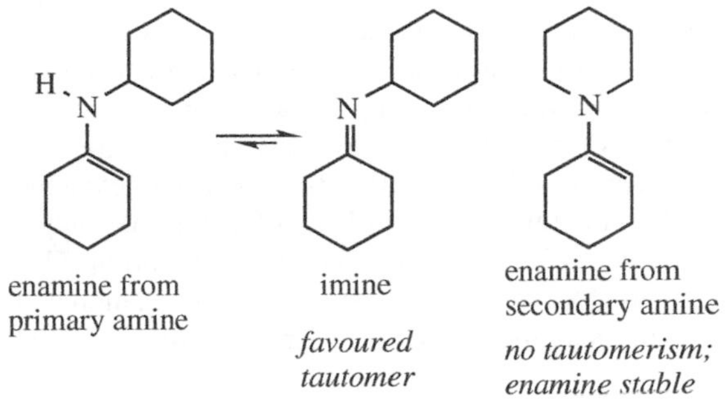

The most prominent property of enamines is that the β-carbon can behave as a **carbon nucleophile**.

enamine from primary amine	imine	enamine from secondary amine
	favoured tautomer	*no tautomerism; enamine stable*

This is a consequence of resonance; overlap of lone pair electrons from the nitrogen provides an iminium system, with the negative counter-charge on the β-carbon.

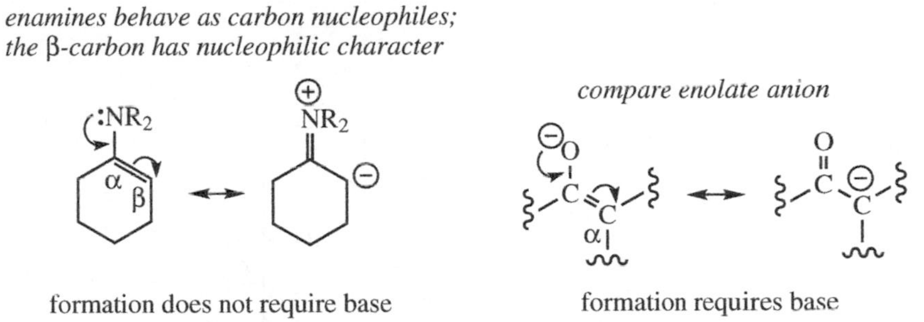

enamines behave as carbon nucleophiles; the β-carbon has nucleophilic character

formation does not require base

compare enolate anion

formation requires base

This resonance form can then act as a nucleophile, in much the same way as an enolate anion can. However, there is a marked difference, and this is what makes enamines such useful synthetic intermediates. Generation of an enolate anion requires the treatment of a carbonyl compound with a base, sometimes a very strong base (see Section 10.2).

The formation of the enamine resonance form is a property of the enamine, and requires no base.

A simple S_N2 alkylation reaction serves as example. As we have already seen, treating cyclohexanone with LDA gives the enolate anion, which can then be allowed to react with methyl iodide to give 2-methylcyclohexanone.

cyclohexanone   *enamine formation*   *nucleophilic substitution*   2-methylcyclohexanone

enamine route produces same product, but under mild conditions without the use of strong base

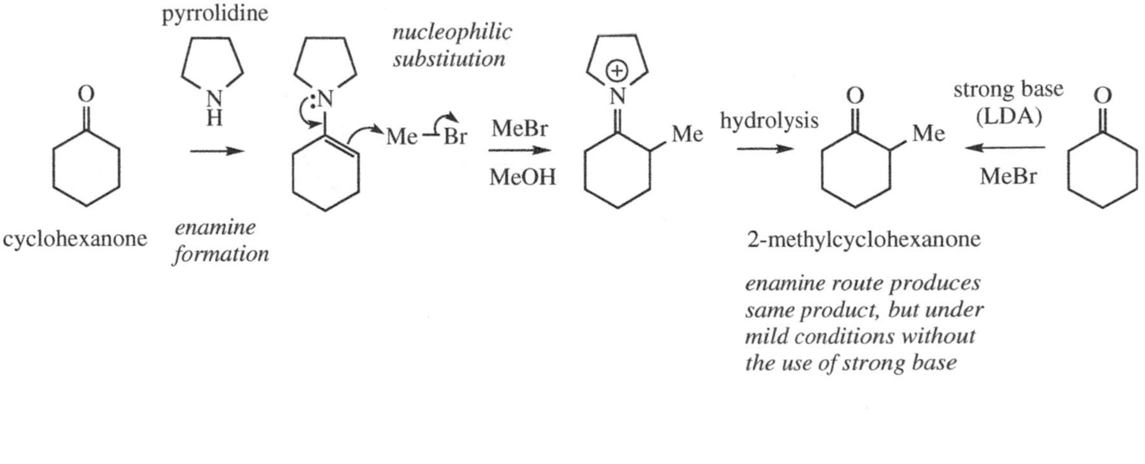

Alternatively, cyclohexanone may initially be transformed into an enamine with a secondary amine, here pyrrolidine. This intermediate enamine can act as a nucleophile and can be alkylated at the β-position using methyl iodide. Finally, 2-methylcyclohexanone may be generated by hydrolysis of the iminium system, effectively a reversal of enamine formation. This gives us two routes to 2-methylcyclohexanone, a short process using the very strong base LDA and

a longer route that involves no strong base and relatively mild conditions. The latter synthesis may well be preferred, depending upon the nature of any other functional groups in the starting substrate.

The essential feature of enamines is that they are **nitrogen analogues of enols** and behave as enolate anions. They effectively mask a carbonyl function while activating the compound towards nucleophilic substitution.

Box 10.5

Enamine reactions in biochemistry: aldolase

In Box 10.4 we saw that an aldol-like reaction could be used to rationalize the biochemical conversion of dihydroxyacetone phosphate (nucleophile) and glyceraldehyde 3-phosphate (electrophile) into fructose 1,6-diphosphate by the enzyme **aldolase** during carbohydrate biosynthesis. The reverse reaction, used in the glycolytic pathway for carbohydrate metabolism, was formulated as a reverse aldol reaction.

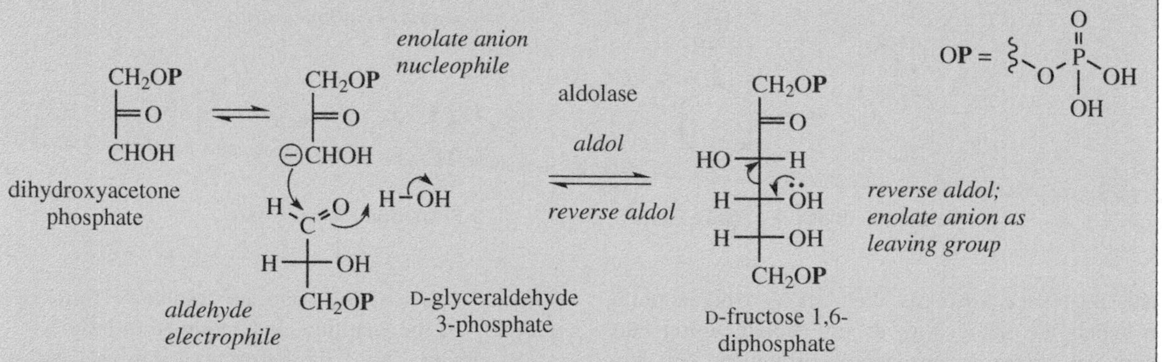

In a postscript, we noted that nature avoided the use of strong base to catalyse the reaction by involving an enzyme. Here, we see how this is achieved through an **enamine**.

Enzymes are very sophisticated systems that apply sound chemical principles. The side-chains of various amino acids are used to supply the necessary bases and acids to help catalyse the reaction (see Section 13.4). Thus, the enzyme **aldolase** binds the dihydroxyacetone phosphate substrate by reacting the ketone group with an amine, part of a lysine amino acid residue. This forms an imine that becomes protonated under normal physiological conditions.

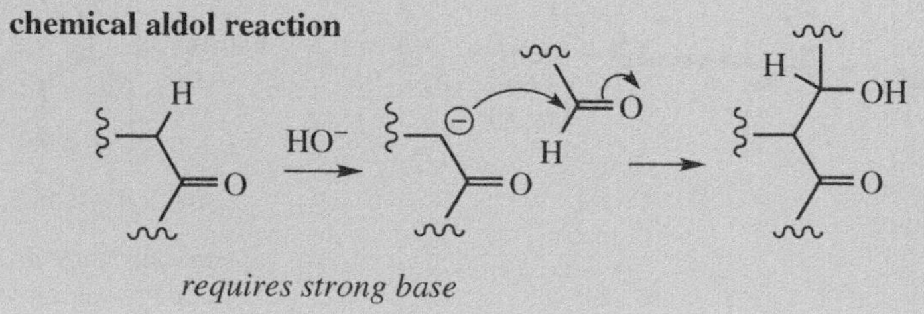

chemical aldol reaction

requires strong base

enzymic aldol reaction

A basic group removes a proton from the β-carbon of the iminium and forms the enamine. This enamine then reacts as a nucleophile towards the aldehyde group of glyceraldehyde 3-phosphate in a simple addition reaction, and the proton necessary for neutralizing the charge is obtained from an appropriately placed amino acid residue. Finally, the iminium ion loses a proton and hydrolysis releases the product from the enzyme.

The reaction is exactly analogous to the chemical aldol reaction (also shown), but it utilizes an enamine as the nucleophile, and it can thus be achieved under typical enzymic conditions, i.e. around neutrality and at room temperature. There is one subtle difference though, in that the enzyme produces an enamine from a primary amine. We have indicated that enamine formation is a property of secondary amines, whereas primary amines react with aldehydes and ketones to form imines (see Section 7.7.1). Thus, a further property of the enzyme is to help stabilize the enamine tautomer relative to the imine.

10.6 The Mannich reaction

We saw in Section 7.7.1 that imines and iminium ions could act as carbonyl analogues and participate in nucleophilic addition reactions.

iminium ion acting as carbonyl analogue for nucleophilic addition reaction

One simple example was the hydrolysis of imines back to carbonyl compounds via nucleophilic attack of water. The **Mannich reaction** is only a special case of **nucleophilic addition to iminium ions**, where the nucleophile is an enol system, the equivalent of an enolate anion. We have to say 'the equivalent of an enolate anion' because conditions that favour iminium cations are not going to allow the participation of negatively charged nucleophiles.

The Mannich reaction is best discussed via an example. A mixture of dimethylamine, formaldehyde and acetone under mild acidic conditions gives N,N-dimethyl-4-aminobutan-2-one. This is a two-stage process, beginning with the formation of an iminium cation from the amine and the more reactive of the two carbonyl compounds, in this case the aldehyde. This iminium cation then acts as the electrophile for addition of the nucleophile acetone. Now it would be nice if we could use the enolate anion as the nucleophile, as in the other reactions we have looked at, but under the mild acidic conditions we cannot have an anion, and the nucleophile must be portrayed as the enol tautomer of acetone. The addition is then unspectacular, and, after loss of a proton from the carbonyl, we are left with the product.

*the **Mannich reaction** proceeds via an intermediate iminium cation*
and utilizes an enolate anion equivalent as the nucleophile

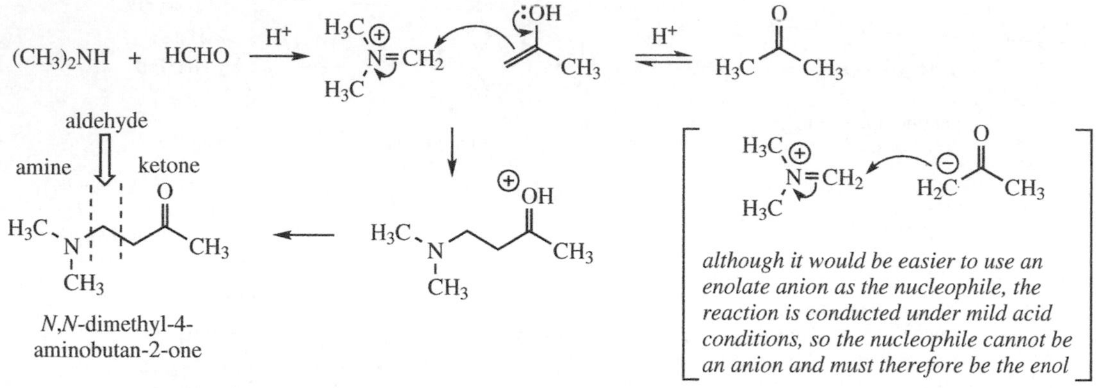

N,N-dimethyl-4-
aminobutan-2-one

although it would be easier to use an
enolate anion as the nucleophile, the
reaction is conducted under mild acid
conditions, so the nucleophile cannot be
an anion and must therefore be the enol

general Mannich reaction:

amine
aldehyde (usually HCHO) ⟶ β-aminoketone
enolizable ketone

This is a fairly general reaction, and requires an amine plus an aldehyde (usually, but not necessarily, formaldehyde) together with an enolizable ketone, which together generate a β-aminoketone via an iminium system. The Mannich reaction is surprisingly important in biochemical processes, especially in the biosynthetic formation of alkaloids (see Box 10.7). We shall also see several examples in heterocyclic chemistry (see Chapter 11).

Box 10.6

Mannich reaction: the synthesis of tropine

The **Mannich reaction** was used for the first synthesis of **tropine**, the parent alcohol of the **tropane alkaloids**. One of the natural tropane alkaloids used medicinally is hyoscyamine, sometimes in its racemic form atropine. **Hyoscyamine** is an anticholinergic, competing with acetylcholine for the muscarinic site of the parasympathetic nervous system, and thus preventing the passage of nerve impulses.

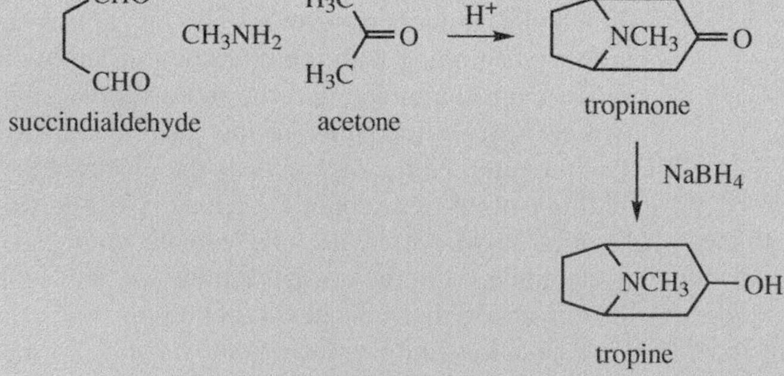

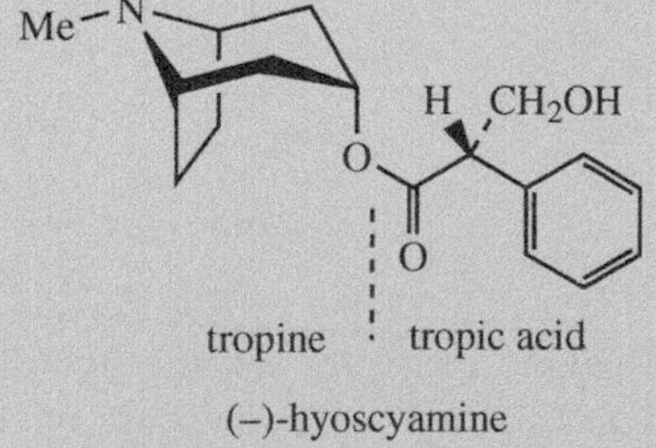

succindialdehyde   acetone   tropinone

tropine

tropine : tropic acid

(−)-hyoscyamine

The synthesis involved reaction of methylamine, succindialdehyde and acetone under mild acid conditions, and although yields were poor, tropinone was formed. This could then be reduced with sodium borohydride to give tropine.

It is instructive to formulate a mechanism for this reaction; note that two Mannich reactions are involved. The scheme below shows the sequence of events, though not all the steps are shown.

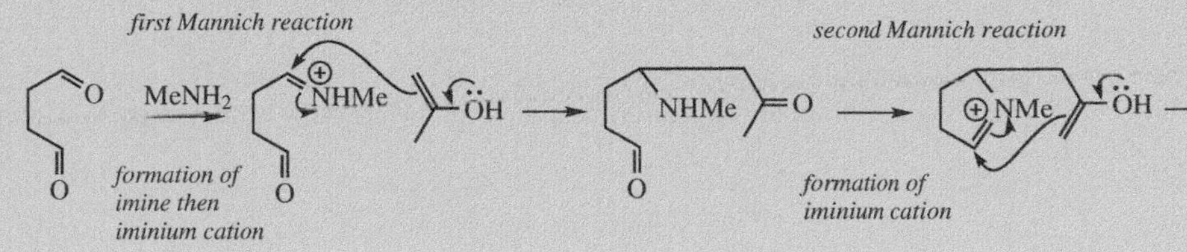

Box 10.7

Biosynthesis of tetrahydroisoquinolines

Mannich and Mannich-like reactions are widely used for the chemical synthesis of heterocycles, and in **alkaloid biosynthesis** in plants. One such reaction important in nature is a biological equivalent of the **Pictet–Spengler tetrahydroisoquinoline synthesis** (see Section 11.10.4), and offers a slight twist, in that the enol nucleophile is actually a phenol.

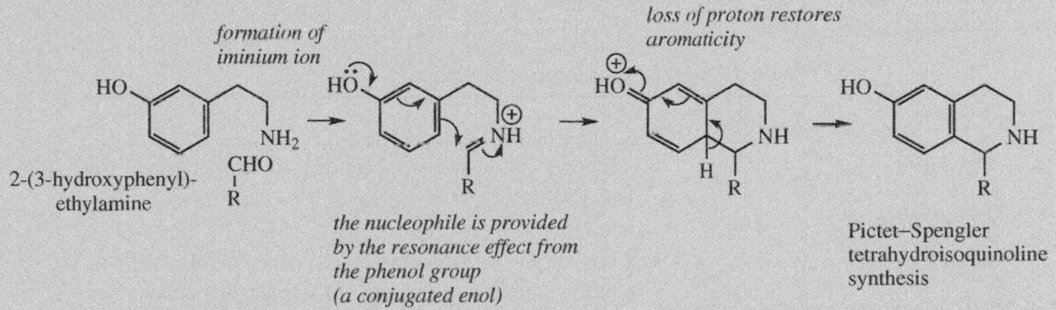

Thus, reaction of 2-(3-hydroxyphenyl)ethylamine with an aldehyde generates initially an imine that will become protonated to an iminium ion. The resonance effect from the phenol group will increase electron density at the *ortho* and *para* positions in the aromatic ring (see Section 4.3.5). With the *para* resonance form, this is equivalent to having a nucleophile located adjacent to the iminium ion, and allows formation of a favourable six-membered ring via the Mannich-like reaction, the nucleophile attacking the C=N. Alternatively, we may consider the phenol to be simply a conjugated enol that is participating in a Mannich reaction. The final step is loss of a proton, and this comes from the position *para* to the oxygen substituent, because this allows regeneration of the aromatic ring and phenol group. In a chemical reaction, a racemic product will be formed, but enzyme-controlled biochemical reactions normally produce just one enantiomer.

For a simple specific example, the tetrahydroisoquinoline alkaloid salsolinol is found in some plants, and it can also be detected in the urine of humans as a product from dopamine and acetaldehyde.

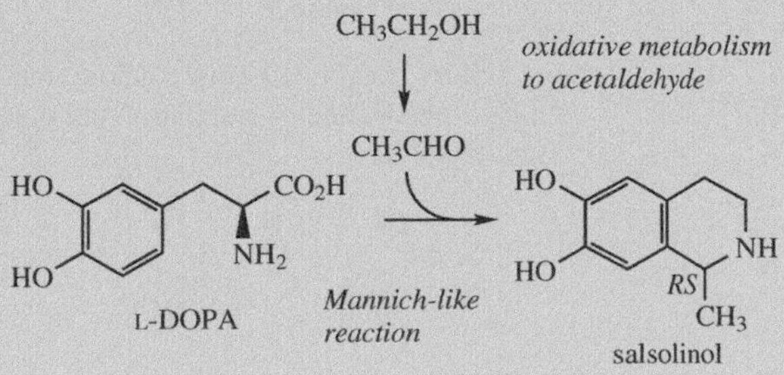

Box 10.7 (continued)

Acetaldehyde is typically formed after ingestion of alcohol (see Section 3.4.7). Since the urine product is racemic, it would appear that a chemical Pictet–Spengler synthesis is being observed here rather than an enzymic one.

In Box 9.4, we saw that tetrahydroisoquinoline alkaloids with appropriate phenol substituents could be involved in radical coupling processes. The complex alkaloids tubocurarine and morphine are derived in nature from simpler tetrahydroisoquinoline alkaloids.

10.7 Enolate anions from carboxylic acid derivatives

The α-hydrogens of **carboxylic acid derivatives** show enhanced acidity, as do those of aldehydes and ketones, and for the same reasons, that the carbonyl group stabilizes the conjugate base. Thus, we can generate **enolate anions** from carboxylic acid derivatives and use these as nucleophiles in much the same way as we have already seen with enolate anions from aldehydes and ketones.

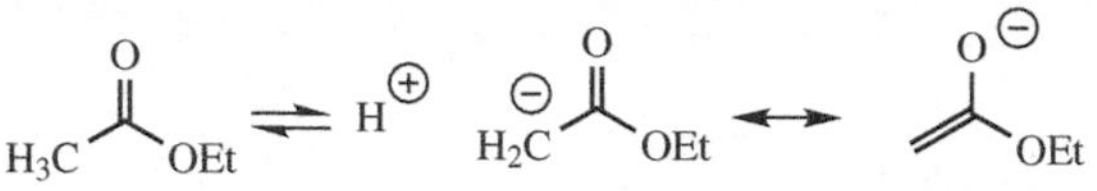

resonance-stabilized enolate anion

Unfortunately, there are some limitations in the carboxylic acid group of compounds, and the derivatives most often used to form enolate anions are **esters**. However, esters are less acidic than the corresponding aldehydes or ketones (Table 10.2).

Table 10.2 pK_a values for carboxylic acid derivatives[a]

	pK_a	
CH$_3$CHO	17	• resonance stabilization of enolate anion (conjugate base) same in all
CH$_3$COCH$_3$	19	
CH$_3$CO$_2$CH$_3$	24	• lower acidity of ester due to resonance stabilization in neutral ester – less carbonyl character, less tendency to lose proton and give enolate
CH$_3$CO$_2$H	4.8	

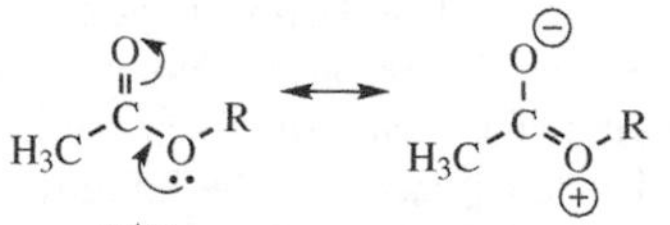

ester

CH$_3$COSCH$_3$	20	• resonance of this type is less favourable in the sulfur ester due to the larger S atom, and less orbital overlap

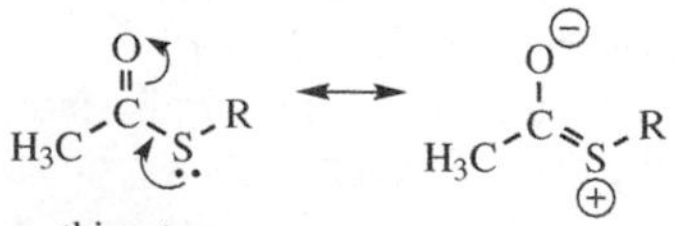

thioester

CH$_3$CONH$_2$	15	• in amides, the N–H is more acidic than the α-hydrogens, due to resonance stabilization of the conjugate base.
CH$_3$CONHMe	18	
CH$_3$CONMe$_2$	30	

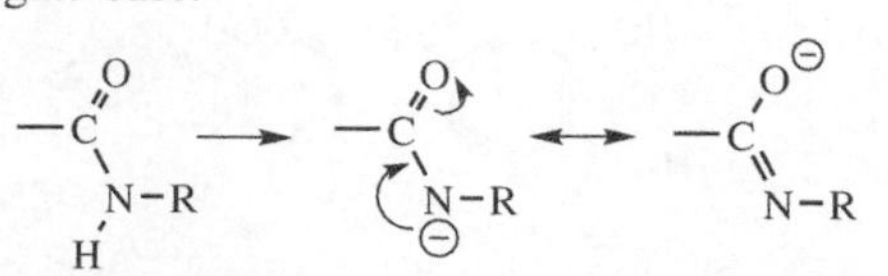

---

[a]pK_a refers to loss of proton underlined

Whereas the pK_a for the α-protons of aldehydes and ketones is in the region 17–19, for esters such as ethyl acetate it is about 25. This difference must relate to the presence of the second oxygen in the ester, since resonance stabilization in the enolate anion should be the same. To explain this difference, overlap of the non-carbonyl oxygen lone pair is invoked. Because this introduces charge separation, it is a form of resonance stabilization that can occur only in the neutral ester, not in the enolate anion. It thus stabilizes the neutral ester, reduces carbonyl character, and there is less tendency to lose a proton from the α-carbon to produce the enolate. Note that this is not a new concept; we used the same reasoning to explain why amides were not basic like amines (see Section 4.5.4).

The α-hydrogens in thioesters are more acidic than in oxygen esters, comparable in fact to those in the equivalent ketone. This can be rationalized from the larger size of sulfur. The sulfur lone pair is located in a $3p$ orbital, whereas oxygen lone pairs are in $2p$ orbitals; there is consequently less overlap of orbitals. There can be relatively little contribution from this type of resonance stabilization in thioesters. Accordingly, normal enolate anion stabilization is not affected.

Note that acids, and primary and secondary amides cannot be employed to generate enolate anions. With acids, the carboxylic acid group has pK_a of about 3–5, so the carboxylic proton will be lost much more easily than the α-hydrogens. In primary and secondary amides, the N–H (pK_a about 18) will be removed more readily than the α-hydrogens. Their acidity may be explained because of resonance stabilization of the anion. Tertiary amides might be used, however, since there are no other protons that are more acidic.

Box 10.8

Coenzyme A and acetyl-CoA

The increased acidity associated with **thioesters** is one of the reasons that biochemical reactions tend to involve thioesters rather than oxygen esters. The most important thiol encountered in such thioesters is **coenzyme A** (see Box 7.18).

This is a complex molecule, made up of an adenine nucleotide (ADP-3′-phosphate), pantothenic acid (vitamin B_5), and cysteamine (2-mercaptoethylamine), but for mechanism purposes can be thought of as a simple **thiol**, HSCoA. Pre-eminent amongst the biochemical thioesters is the thioester of acetic acid, **acetyl-coenzyme A** (acetyl-CoA). This compound plays a key role in the biosynthesis and metabolism of fatty acids (see Sections 15.4 and 15.5), as well as being a building block for the biosynthesis of a wide range of natural products, such as phenols and macrolide antibiotics (see Box 10.4).

Acetyl-CoA is a good biochemical reagent for two main reasons. First, the α-protons are more acidic than those in ethyl acetate, comparable in fact to a ketone, and this increases the likelihood of generating an enolate anion. As explained above, this derives from sulfur being larger than oxygen, so that electron donation from the lone pair that would stabilize the neutral ester is considerably reduced. This means it is easier for acetyl-CoA to lose a proton and become a nucleophile. Second, acetyl-CoA is actually a better electrophile than ethyl acetate,

Box 10.8 (continued)

in that it has a better leaving group; thiols (pK_a 10–11) are stronger acids than alcohols (pK_a 16). Acetyl-CoA is thus rather well suited to participate in aldol and Claisen reactions.

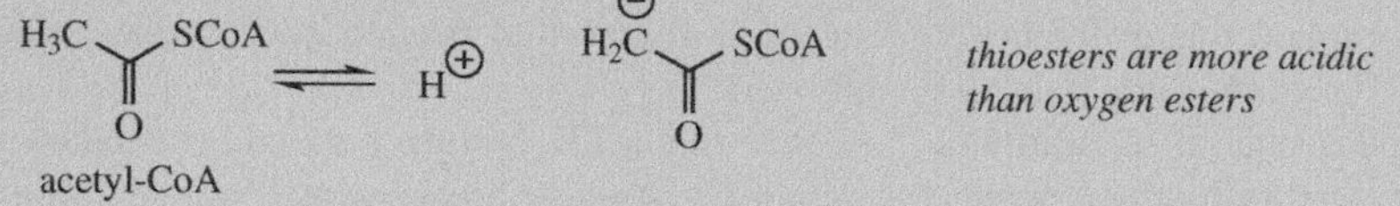

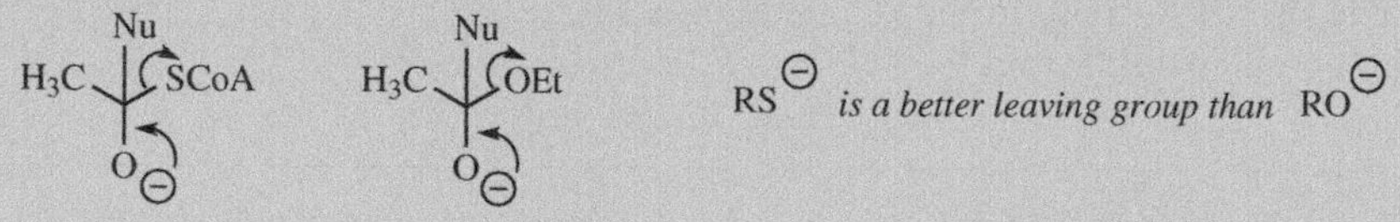

thioesters are more acidic than oxygen esters

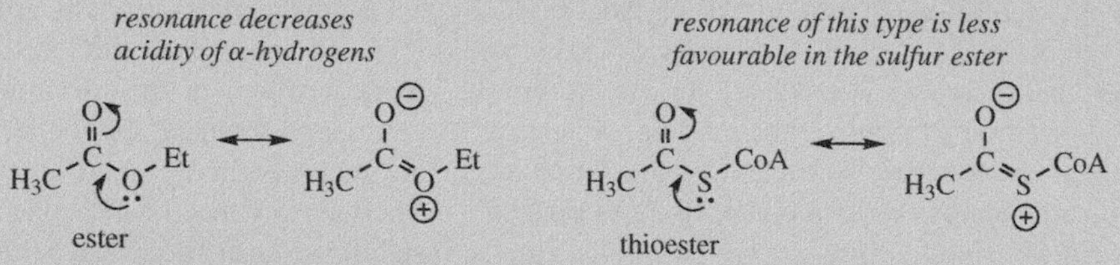

resonance decreases acidity of α-hydrogens

resonance of this type is less favourable in the sulfur ester

ester   thioester

RS$^\ominus$ *is a better leaving group than* RO$^\ominus$

We shall see later (see Box 10.17) that nature can employ yet another stratagem to increase the acidity of the α-protons in thioesters, by converting acetyl-CoA into malonyl-CoA (see Section 15.9).

An enolate anion generated from a carboxylic acid derivative may be used in the same sorts of nucleophilic reactions that we have seen with aldehyde and ketone systems. It should be noted, however, that the base used to generate the enolate anion must be chosen carefully. If sodium hydroxide were used, then hydrolysis of the carboxylic derivative to the acid (see Section 7.9.2) would compete with enolate anion formation. However, the problem is avoided by using the same base, e.g. ethoxide, as is present in the ester function, so that the ester is not hydrolysed. Larger bases, e.g. *tert*-butoxide, may also be valuable, in that they can remove α-protons but tend to be too large to add to the carbonyl group and form a tetrahedral intermediate.

Using ethoxide as base, we can get **hydrogen exchange** by equilibration in a labelled solvent (see Section 10.1.1); but, because of the lower acidity of the α-protons compared with aldehydes and ketones, this process is less favourable.

hydrogen exchange in α-position

Should the α-position be a chiral centre containing hydrogen, it is possible to racemize at that centre (compare Section 10.1.1). Again, **racemization** is less likely to occur with esters than with aldehydes and ketones, and ready racemization may require the contribution of other favourable factors in the enolate anion (see Box 10.9).

racemization

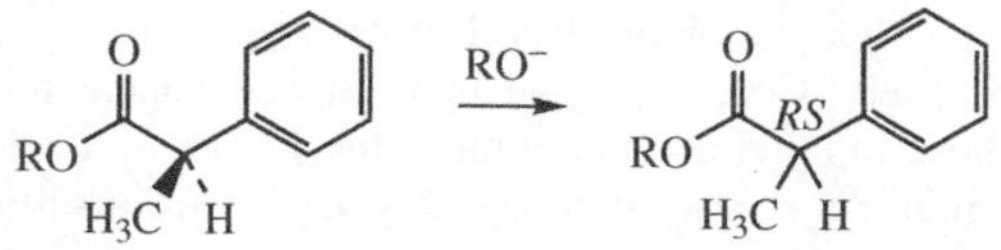

Box 10.9

Racemization of hyoscyamine to atropine

The base-catalysed racemization of the alkaloid (−)-**hyoscyamine** to (±)-hyoscyamine (**atropine**) is an example of enolate anion participation. Alkaloids are normally extracted from plants by using base, thus liberating the free alkaloid bases from salt combinations. (−)-Hyoscyamine is found in belladonna (*Atropa belladonna*) and stramonium (*Datura stramonium*) and is used medicinally as an anticholinergic. It competes with acetylcholine for the muscarinic site of the parasympathetic nervous system, thus preventing the passage of nerve impulses. However, with careless extraction using too much base the product isolated is atropine, which has only half the biological activity of (−)-hyoscyamine, since the enantiomer (+)-hyoscyamine is essentially inactive.

The racemization process involves removal of the α-hydrogen to form the enolate anion, which is favoured by both the enolate anion resonance plus additional conjugation with the aromatic ring. Since the α-protons in esters are not especially acidic, the additional conjugation is an important contributor to enolate anion formation. The proton may then be restored from either side of the planar system, giving a racemic product.

base-catalysed enolate anion formation

(−)-hyoscyamine

double bond of enolate and aromatic ring in conjugation

(+)-hyoscyamine

base-catalysed or heat-initiated keto–enol tautomerism

(−)-hyoscyamine

double bond of enol and aromatic ring in conjugation

(+)-hyoscyamine

atropine

Note that the alcohol portion of hyoscyamine, namely tropine, also contains two chiral centres, but it is a symmetrical molecule and is optically inactive; it can be considered as a *meso* structure (see Box 3.21). Thus, the optical activity of hyoscyamine stems entirely from the chiral centre in the acid portion, tropic acid.

Racemization of hyoscyamine may also be brought about by heating, and it is probable that, under these conditions, there is involvement of the enol form, rather than the enolate anion. The enol is also stabilized by the additional conjugation that the aromatic ring provides. The importance of this additional conjugation is emphasized by the observation that **littorine**, an alkaloid from *Anthocercis littorea*, is not readily racemized by either heat or base. The esterifying acid in littorine is phenyl-lactic acid, and the aromatic ring would not be in conjugation with the double bond of the enol or enolate anion. Racemization depends entirely on the acidity associated with the isolated ester function.

Box 10.9 (continued)

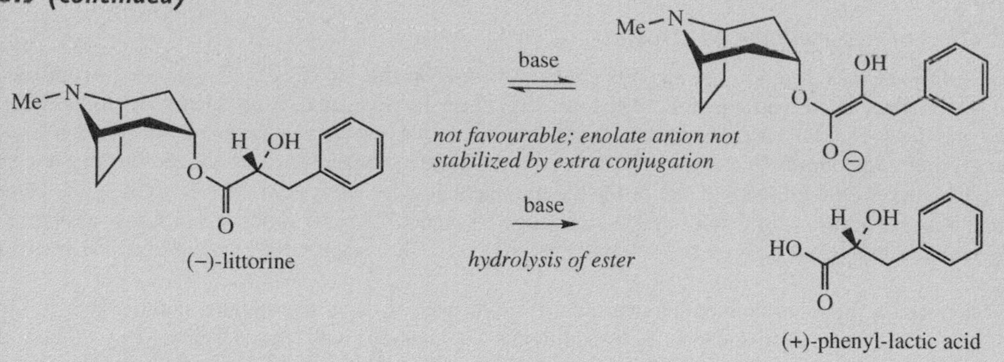

not favourable; enolate anion not stabilized by extra conjugation

base → *hydrolysis of ester*

(−)-littorine

(+)-phenyl-lactic acid

Additionally, note that base hydrolysis of hyoscyamine gives (±)-tropic acid and tropine, with racemization preceding hydrolysis. Base hydrolysis of littorine gives optically pure phenyl-lactic acid, so we deduce that hydrolysis is a more favourable process than racemization.

Box 10.10

Epimerization of L-amino acids to D-amino acids during peptide biosynthesis

Many natural peptide structures, especially the **peptide antibiotics** such as dactinomycin and ciclosporin (see Box 13.10), contain one or more D-amino acids along with L-amino acids in their structures. This contrasts with most proteins, where all the amino acid constituents are of the L-configuration (see Section 13.1). It is now known that the biosynthetic precursors of the D-amino acids are actually the corresponding L-analogues, and that an enzymic epimerization process through an enol-type intermediate is involved. However, this does not appear to involve epimerization of the free L-amino acid followed by incorporation of the D-amino acid into the growing peptide chain. There are good reasons for this. Enolization in base does not occur, since ionization of the carboxylic acid group predominates (see Section 10.7). Enolization in acid is also prevented, because the basic amino group would be protonated rather than the carbonyl (see Section 4.11.3). In fact, epimerization appears to take place after the L-amino acid has been incorporated into the peptide, and is thus occurring on an amide substrate.

A simple example is the tripeptide precursor of the **penicillin antibiotics**, called ACV, an abbreviation for δ-(L-α-aminoadipyl)-L-cysteinyl-D-valine. The amino acid precursors for ACV are L-α-aminoadipic acid (an unusual amino acid derived by modification of L-lysine), L-cysteine, and L-valine (not D-valine).

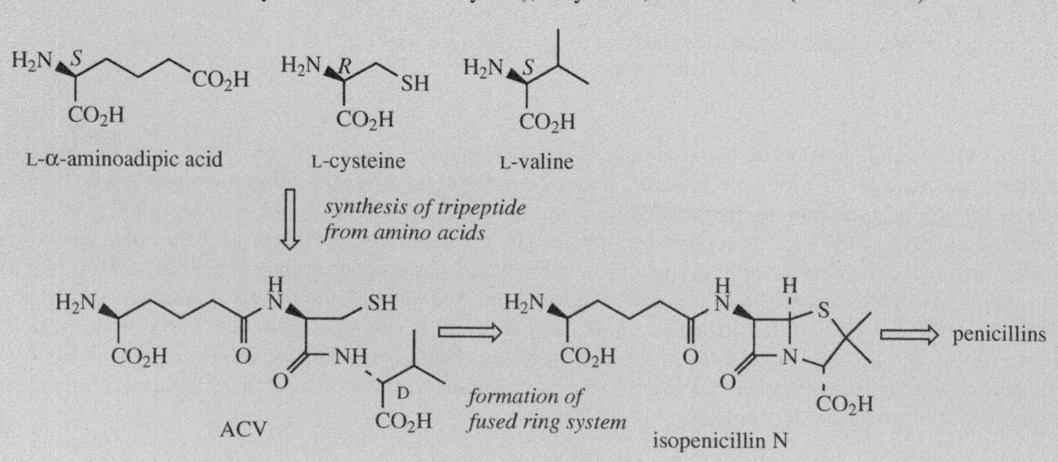

L-α-aminoadipic acid   L-cysteine   L-valine

synthesis of tripeptide from amino acids

ACV   *formation of fused ring system*   isopenicillin N   ⇒ penicillins *epimerization through an intermediate enol-like tautomer*

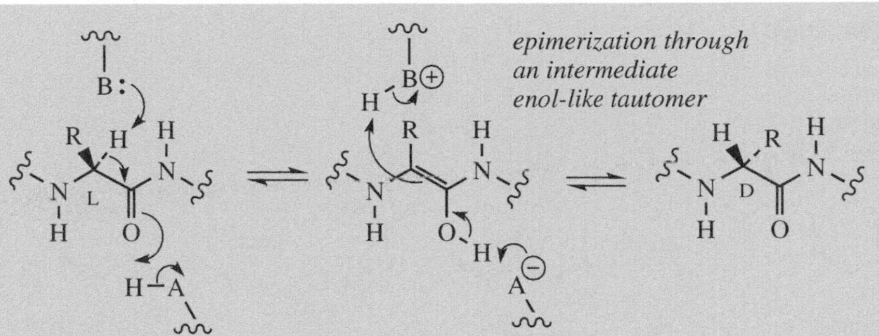

During ACV formation, the stereochemistry of the valine component is changed. ACV is the linear tripeptide that leads to isopenicillin N, the first intermediate with the fused ring system found in the penicillins. Note, we are using the D and L convention for amino acid stereochemistry rather than the fully systematic R and S (see Section 3.4.10). This is one occasion where use of D and L is advantageous, in that the sulfur atom in L-cysteine means this compound has the R configuration, whereas the other L-amino acids have the S configuration.

Evidence points to the most likely explanation for the epimerization of L- to D-amino acids being the involvement of an enol-like intermediate. The carbonyl form is an amide in this example; but, from the comments made earlier (see Section 10.7), such a transformation could not be achieved chemically in solution, since the N–H proton would be more acidic and would, therefore, be preferentially removed using a base. However, this is an enzymic reaction, thus allowing selectivity determined by the functional groups at the enzyme's binding site. A basic residue is responsible for removing the α-hydrogen to generate the enol-like structure, and then a reverse process allows it to be delivered back, though from the opposite side of the planar structure. Since this is an enzymic reaction, the product is also produced in just one configuration, rather than as an equimolar mixture of the two configurations typical of a chemical process.

Box 10.11

Metabolic racemization of ibuprofen

The analgesic **ibuprofen** is supplied for drug use in its racemic form. However, only the (S)-(+)-enantiomer is the biologically active species; the (R)-(−)-form is inactive.

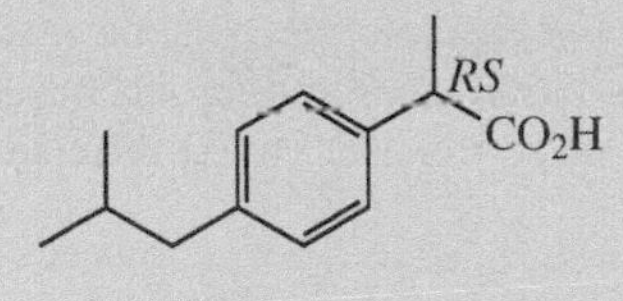

ibuprofen

(S)-(+)-isomer active
(R)-(−)-isomer inactive

some metabolic conversion of R → S via racemization

Nevertheless, the racemate provides considerably more analgesic activity than that expected, since in the body there is some metabolic conversion of the inactive (R)-isomer into the active (S)-isomer. This can be rationalized readily through an enolization mechanism. As we have indicated under D-amino acid formation above, a simple base-catalysed chemical conversion is ruled out by preferential ionization of the carboxylic acid group, though this may have little bearing on a metabolic process. An enzyme-mediated process may possibly involve both basic and acidic amino acid side-chains (see D-amino acid formation above), and we could consider the biological transformation as either base catalysed or acid catalysed, as shown below. Either would generate a planar enediol intermediate, and the reverse process would account for racemization. The enediol also benefits from favourable conjugation with the aromatic ring.

Box 10.11 (continued)

base-catalysed conversion:

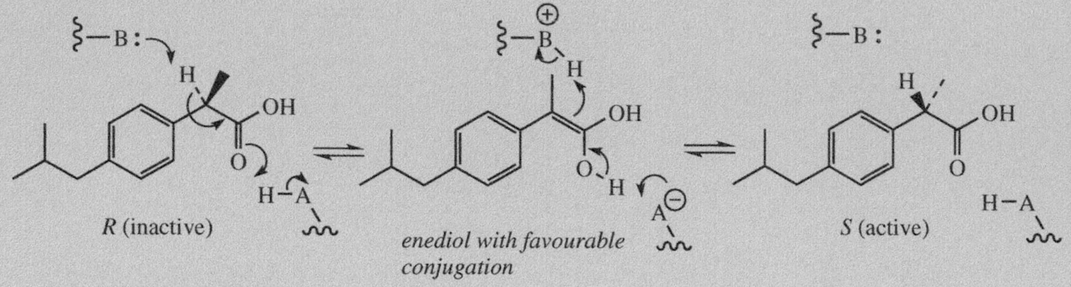

R (inactive)   *enediol with favourable conjugation*   S (active)

acid-catalysed conversion:

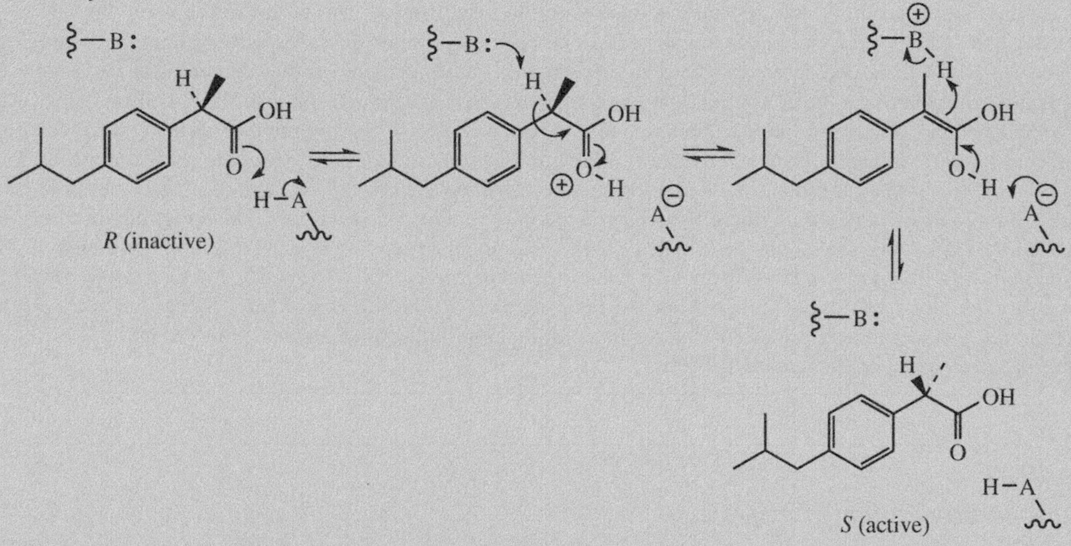

R (inactive)

S (active)

Thus, when (±)-ibuprofen is supplied to the body, the active (+)-isomer can be utilized, with the remaining (−)-isomer then being racemized to provide more of the active isomer. Theoretically, almost all of the (−)-isomer could be converted as the (+)-isomer is gradually removed by the body. For example, since the racemate contains 50% inactive isomer, racemization of this provides another 25% active isomer, then further racemization of the remaining 25% inactive would leave 12.5%, and so on. In practice, transport and excretion differences do not allow total usage of all the material.

Alkylation of the α-position of suitable carboxylic acid derivatives may be achieved using the enolate anion as nucleophile in a typical S_N2 reaction (compare Section 10.2). In the example shown, the base used is LDA. This is a strong base that easily removes the weakly acidic α-proton, but because of its size it is a poor nucleophile and so does not affect the ester function (see Section 10.2).

alkylation

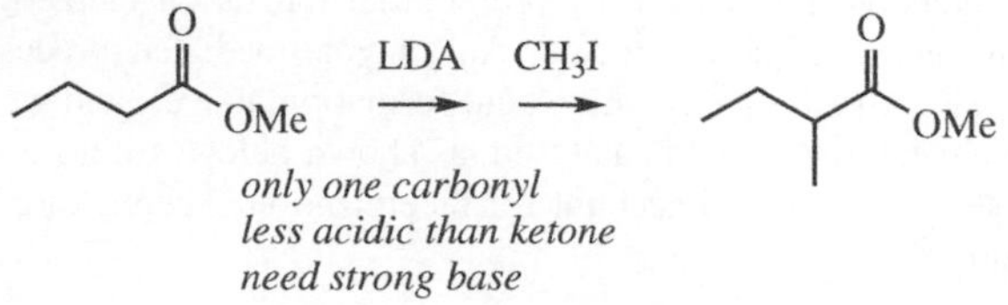

*only one carbonyl
less acidic than ketone
need strong base*

Nucleophilic addition of an enolate anion from a carboxylic acid derivative onto an aldehyde or ketone is simply an **aldol-type reaction** (see Section 10.3).

A simple example is shown; again, LDA is used to generate the enolate anion, and addition to the ketone is carried out as a second step (see Section 10.2).

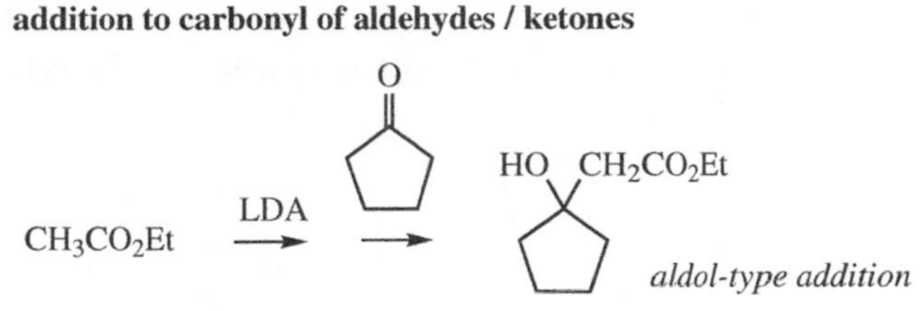

addition to carbonyl of aldehydes / ketones

$$CH_3CO_2Et \xrightarrow{LDA} \longrightarrow$$

aldol-type addition

10.8 Acylation of enolate anions: the Claisen reaction

In the aldol reaction, we saw an enolate anion acting as a nucleophile leading to an addition reaction with aldehydes and ketones.

However, if there is a leaving group present, then instead of the intermediate alkoxide anion abstracting a proton from solvent giving the aldol product, the leaving group may be expelled with regeneration of the carbonyl group.

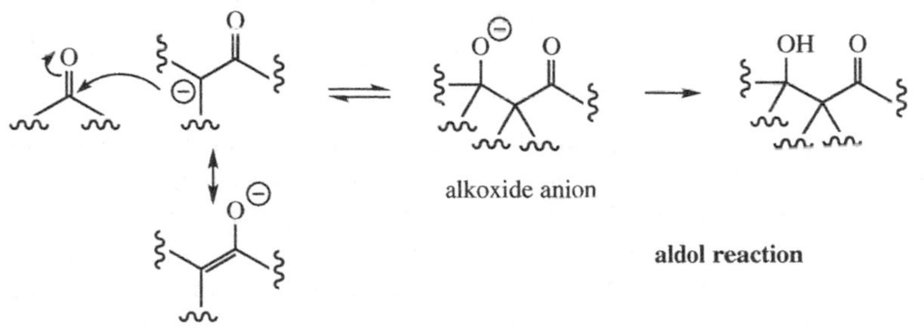

alkoxide anion

aldol reaction

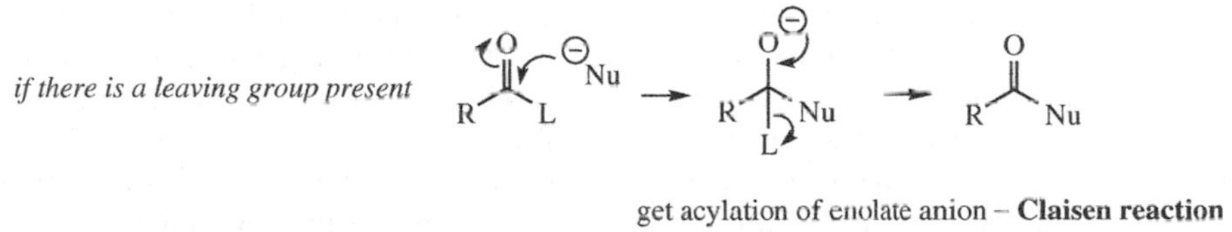

if there is a leaving group present

get acylation of enolate anion – **Claisen reaction**

Now this is exactly the same situation we encountered when we compared the reactivity of aldehydes and ketones with that of carboxylic acid derivatives (see Section 7.8). The net result here is acylation of the nucleophile, and in the case of **acylation of enolate anions**, the reaction is termed a **Claisen reaction**. It is important not to consider aldol and Claisen reactions separately, but to appreciate that the initial addition is the same, and differences in products merely result from the absence or presence

of a leaving group. This is just how we rationalized the different reactions of aldehydes and ketones compared with carboxylic acid derivatives (see Section 7.8).

The Claisen reaction (sometimes Claisen condensation) is formally the base-catalysed reaction between two molecules of ester to give a **β-ketoester**. Thus, from two molecules of ethyl acetate the product is ethyl acetoacetate.

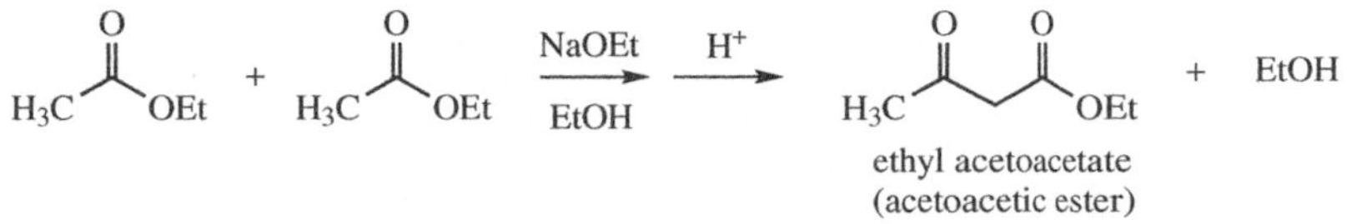

ethyl acetoacetate
(acetoacetic ester)

To participate in this sort of reaction, the carboxylic acid derivative acting as nucleophile must have α-hydrogens in order to generate an enolate anion. In practice, esters are most commonly employed in Claisen-type reactions.

The Claisen reaction may be visualized as initial formation of an enolate anion from one molecule of ester, followed by nucleophilic attack of this species on to the carbonyl group of a second molecule. The addition anion then loses ethoxide as leaving group, with reformation of the carbonyl group.

Claisen reaction

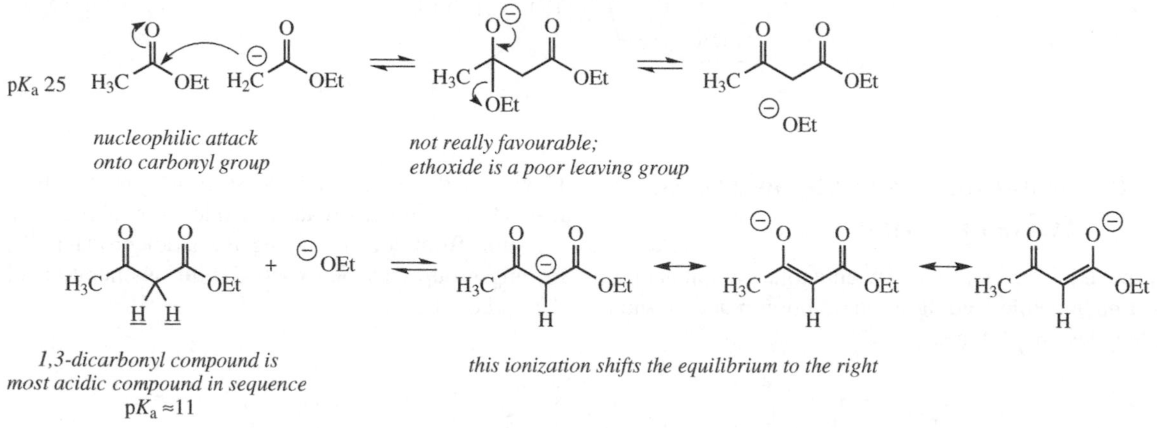

*nucleophilic attack
onto carbonyl group*

*not really favourable;
ethoxide is a poor leaving group*

*1,3-dicarbonyl compound is
most acidic compound in sequence
$pK_a \approx 11$*

this ionization shifts the equilibrium to the right

However, the reaction is not quite that simple, and to understand and utilize the Claisen reaction we have to consider pK_a values again. Loss of ethoxide from the addition anion is not really favourable, since ethoxide is not a particularly good leaving group. This is because ethoxide is a strong base, the conjugate base of a weak acid (see Section 6.1.4). So far then, the reaction will be reversible. What makes it actually proceed further is the fact that ethoxide *is* a strong base, and able to ionize acids. The ethyl acetoacetate product is a 1,3-dicarbonyl compound and has relatively acidic protons on the methylene between the two carbonyls (see Section 10.1). With

a pK_a of about 11, this makes ethyl acetoacetate the most acidic compound in the sequence. Ionization of ethyl acetoacetate, generating a resonance-stabilized enolate anion, removes product from the reaction mixture and shifts the equilibrium to the right. This also explains why, in the simple equation above, two reagents are shown on the arrows, first base and then acid. The acid is required in the workup to liberate the β-ketoester from the enolate anion.

The importance of ionization of the β-ketoester product can be illustrated by the attempted Claisen reaction between two molecules of ethyl 2-methyl-propionate.

ethyl 2-methylpropionate

*no acidic hydrogen
between carbonyls*

$pK_a \approx 20$

*can drive the equilibrium to right
only by removing the γ-hydrogen;
this requires a much stronger base,
e.g. NaH, NaNH$_2$, LDA*

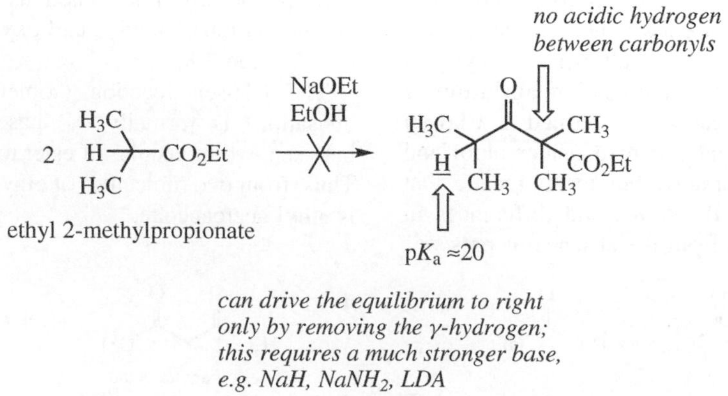

Using sodium ethoxide as base, the reaction does not proceed. This can be ascribed to the nature of the β-ketoester product, which contains no protons sandwiched between two carbonyls and, therefore, no protons that are sufficiently acidic for the final equilibrium-disturbing step. The reaction can be made to proceed, however, and the solution is simple: use a stronger base. In this way, the base used is sufficiently powerful to remove a less acidic proton from the product, removing it from the reaction mixture and disturbing the equilibrium. Any of the strong bases sodium hydride, sodium amide, or LDA might be employed. Although such bases will produce the enolate anion irreversibly (see Section 10.2), it is still necessary to ionize the product to overcome the effect of the poor leaving group. In the β-ketoester product, the pK_a of the only acidic proton is about 20, so this requires a strong base to achieve an equilibrium-disturbing ionization.

Box 10.12

Claisen and aldol reactions in nature: HMG-CoA and mevalonic acid

In nature, the biologically active form of acetic acid is **acetyl-coenzyme A** (acetyl-CoA) (see Box 7.18). Two molecules of acetyl-CoA may combine in a Claisen-type reaction to produce acetoacetyl-CoA, the biochemical equivalent of ethyl acetoacetate. This reaction features as the start of the sequence to **mevalonic acid** (MVA), the precursor in animals of the sterol **cholesterol**. Later, we shall see another variant of this reaction that employs malonyl-CoA as the nucleophile (see Box 10.17).

Three molecules of acetyl-CoA are used to form MVA, a third molecule being incorporated via a stereospecific aldol addition to give the branched-chain ester β-hydroxy-β-methylglutaryl-CoA (**HMG-CoA**). This third acetyl-CoA molecule appears to be bound to the enzyme via a thiol group (see Section 13.4.3), and this linkage is subsequently hydrolysed to form the free acid group of HMG-CoA.

It should be noted that, on purely chemical grounds, acetoacetyl-CoA is the more acidic substrate in this reaction, and might be expected to act as the nucleophile rather than the third acetyl-CoA molecule. The enzyme thus achieves what is a less favourable reaction. There is a rather similar reaction in the Krebs cycle, where acetyl-CoA adds on to oxaloacetate via an aldol reaction, again with the enzymic reaction employing the less acidic substrate as the nucleophile (see Box 10.4).

The subsequent conversion of HMG-CoA into MVA involves a two-step reduction of the thioester group to a primary alcohol (see Section 7.11), and provides an essentially irreversible and rate-limiting transformation. Drug-mediated inhibition of this enzyme, **HMG-CoA reductase** (HMGR), can be used to regulate the biosynthesis of the steroid cholesterol. High levels of blood cholesterol are known to contribute to the incidence of coronary heart disease and heart attacks.

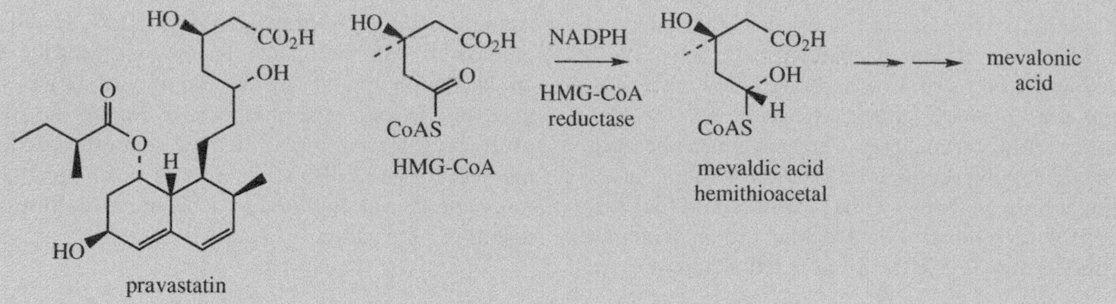

pravastatin

HMG-CoA

mevaldic acid
hemithioacetal

mevalonic
acid

The **statins**, e.g. pravastatin, are a group of HMGR inhibitors that possess functionalities that mimic the half-reduced substrate mevaldate hemithioacetal. The affinity of these agents towards HMG-CoA reductase is some 10^4-fold more than the natural substrate, making them extremely effective inhibitors of the enzyme, and powerful drugs in coronary care.

Should there be two ester functions in the same molecule, then it is possible to achieve an **intramolecular Claisen reaction**, particularly if this results in a favourable five- or six-membered ring. This reaction is usually given a separate name, a **Dieckmann reaction**, but should be thought of as merely an intramolecular extension of the Claisen reaction. As we have seen previously (see Section 7.9.1), intramolecular reactions are favoured over intermolecular reactions when the reaction is carried out at high dilution, conditions that minimize the interaction of two separate molecules. A simple example involving the transformation of diethyl adipate into a cyclic β-ketoester is shown.

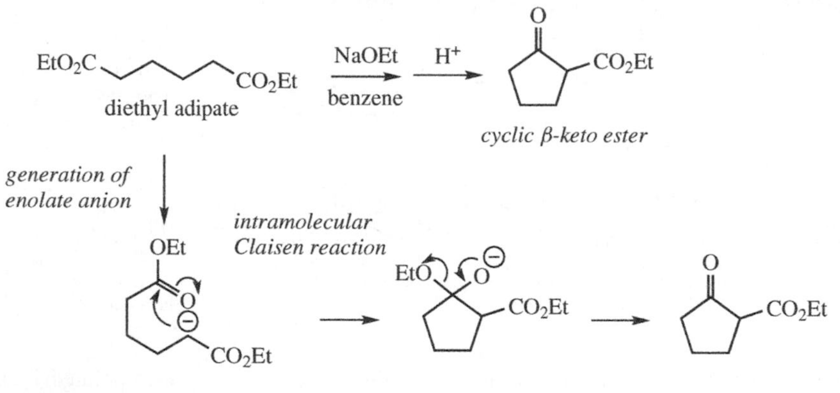

diethyl adipate

cyclic β-keto ester

generation of
enolate anion

intramolecular
Claisen reaction

We saw the possibilities for a mixed aldol reaction above, in which the reaction could become useful if we restricted the number of couplings possible (see Section 10.3). The same considerations can be applied to the Claisen reaction. Thus, it is possible to have four products from two esters, depending on which ester became the nucleophile and which was acting as the electrophile.

mixed Claisen reaction

RCH_2CO_2Et + $R'CH_2CO_2Et$ ⟶ 4 products

nucleophile		electrophile
RCH_2CO_2Et	+	RCH_2CO_2Et
RCH_2CO_2Et	+	$R'CH_2CO_2Et$
$R'CH_2CO_2Et$	+	RCH_2CO_2Et
$R'CH_2CO_2Et$	+	$R'CH_2CO_2Et$

To be synthetically useful, a **mixed Claisen reaction** (crossed Claisen reaction) needs one ester with no α-hydrogens, so that it cannot become the nucleophile. Such reactants include oxalate, formate and benzoate esters. An example is shown below.

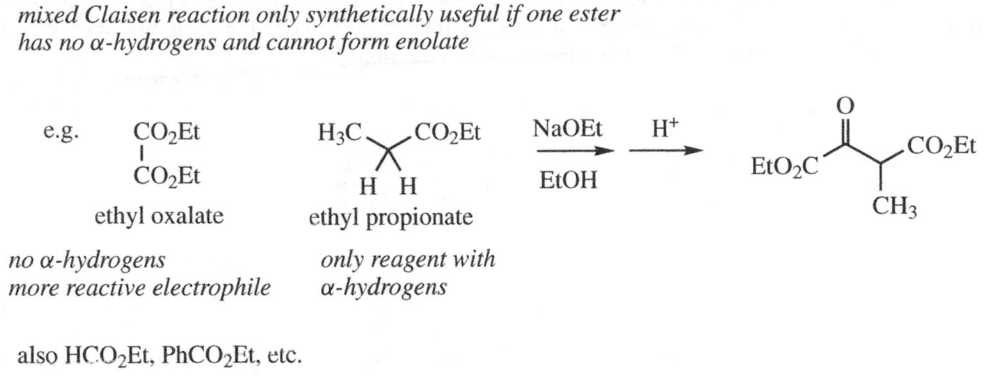

mixed Claisen reaction only synthetically useful if one ester has no α-hydrogens and cannot form enolate

e.g.

ethyl oxalate

no α-hydrogens
more reactive electrophile

ethyl propionate

only reagent with α-hydrogens

also HCO_2Et, $PhCO_2Et$, etc.

However, one might expect that the product from two molecules of ethyl propionate could also be formed. In practice, ethyl oxalate, because of its second electron-withdrawing carboxylate group, is a more reactive electrophile, so the major product is as shown. Formates are also more susceptible to nucleophilic attack; they lack the electron-donating inductive effect of an alkyl group and provide no steric hindrance (see Section 7.1.1). Benzoates are not as reactive as formates and oxalates, but the phenyl ring is electron withdrawing and they also lack α-hydrogens. To minimize self-condensation of the nucleophilic reagent, it helps to add this gradually to the electrophilic species, so that the latter is always present in excess.

Alternatively, and much more satisfactory from a synthetic point of view, it is possible to carry out a two-stage process, forming the enolate anion first. We also saw this approach with a mixed aldol reaction (see Section 10.3). Thus, ethyl acetate could be converted into its enolate anion by reaction with the strong base LDA in a reaction that is essentially irreversible (see Section 10.2).

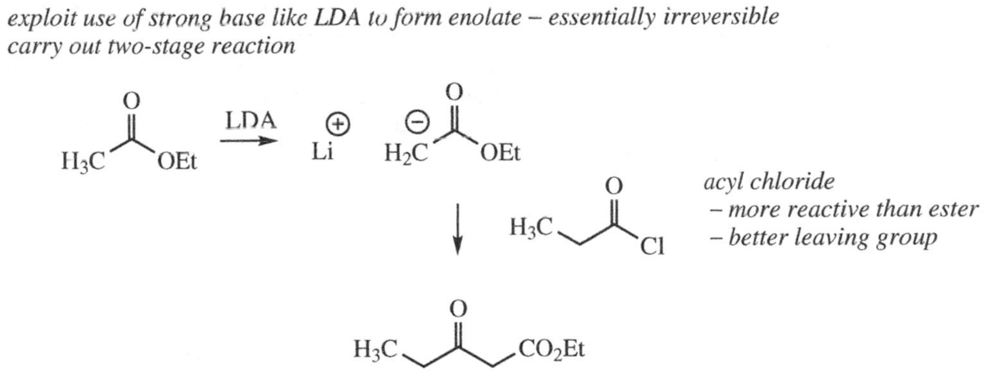

exploit use of strong base like LDA to form enolate – essentially irreversible carry out two-stage reaction

acyl chloride
– more reactive than ester
– better leaving group

This nucleophile can then be treated with the electrophile. This could be a second ester, but there is an even better idea. If one is going to use a two-stage process, one can now employ an electrophile with a better leaving group than ethoxide, and also get over the final ionization problem. It would not be possible to use an acyl halide in a one-pot reaction, because it would be quickly attacked by base. An acyl halide could be used in a two-stage reaction, as shown here.

Box 10.13

Ester–ketone condensations: predicting the product

Let us use a systematic approach to consider what product is most likely to result when a mixture of an **ester** and a **ketone**, both capable of forming **enolate anions**, is treated with base. For example, consider an ethyl acetate–acetone mixture treated with sodium hydride in ether solution.

consider a mixed reaction between ester and ketone

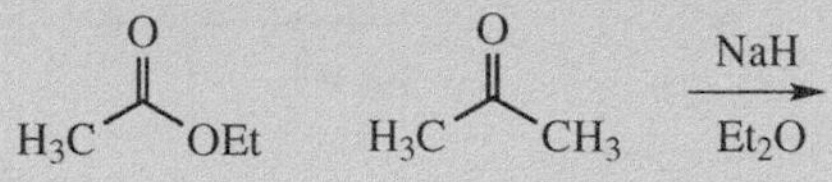

with four possible products

ketone as nucleophile + ketone as electrophile	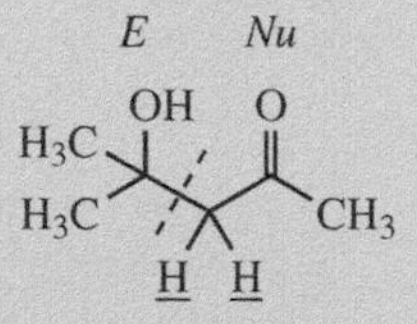	*aldol reaction*
ester as nucleophile + ester as electrophile	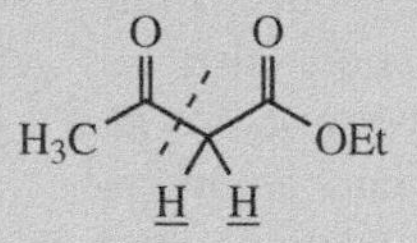	*Claisen reaction*
ester as nucleophile + ketone as electrophile	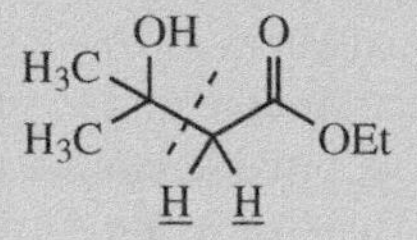	*aldol reaction*
ketone as nucleophile + ester as electrophile	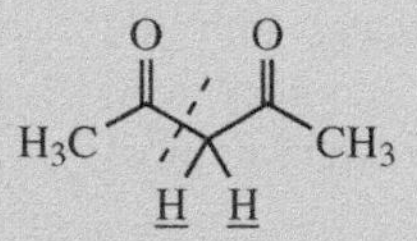	*Claisen reaction*

Four reactions and products can be considered, involving either ketone or ester as the nucleophile, with ketone as the electrophile (**aldol reactions**) or ester as the electrophile (**Claisen reactions**).

Both aldol and Claisen reactions are equilibria, and product formation is a result of disturbing these equilibria. This would be **dehydration** in aldol reactions and **ionization** in Claisen reactions. Ionization would be the more immediate determinant. On that basis, it is obvious that the 1,3-dicarbonyl products from Claisen reactions are going to be more acidic than the aldol products, which possess just one carbonyl group.

Now let us look at the ease of forming the enolate anion nucleophiles. Ketones are more acidic than esters (see Section 10.7). Taken together, these factors mean the more favoured product is going to be the β-diketone (acetylacetone), formed from a ketone nucleophile by a Claisen reaction with an ester. This is the reaction observed.

product is β-diketone acetylacetone

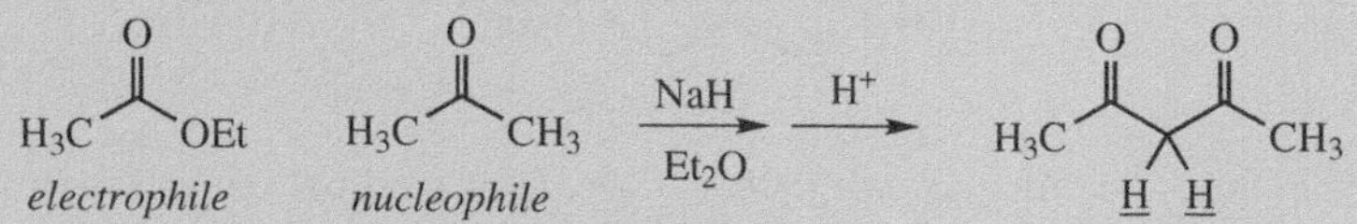

- ketone is more acidic than ester – ketone enolate favoured
- β-diketone is most acidic of four possible products

Box 10.14

Aldol and Claisen reactions in the biosynthesis of phenols

Many natural aromatic compounds are produced from the cyclization of poly-β-keto chains by **enzymic aldol** and **Claisen reactions**. Examples include simple structures like **orsellinic acid** and **phloracetophenone**, and more complex highly modified structures of medicinal interest, such as **mycophenolic acid**, used as an immunosuppressant drug, the antifungal agent **griseofulvin**, and antibiotics of the tetracycline group, e.g. **tetracycline** itself.

orsellinic acid   phloracetophenone   mycophenolic acid

griseofulvin   tetracycline

The more complex structures are inappropriate for consideration here, but the two compounds **orsellinic acid** and **phloracetophenone** exemplify nicely the enolate anion mechanisms we have been considering, as well as the concept of keto–enol tautomerism.

A multifunctional enzyme complex is responsible for producing a poly-β-keto chain via a sequence of several Claisen reactions, together with subsequent reactions that achieve cyclization and aromatization. The C_8 poly-β-keto chain shown is bonded to the enzyme through a thioester linkage (see Section 13.4.3). Because of the number of functional groups in this molecule, it is very reactive, and the enzyme plays a significant role in stabilizing it and preventing any unwanted chemical reactions. In addition, the enzyme binds the substrate in a folded conformation, allowing the atoms to be held in positions approximating to those occupied in the desired product. There are various possibilities for undergoing **intramolecular aldol** or **Claisen reactions**, dictated by the nature of the enzyme and how the substrate is folded on the enzyme surface.

Methylenes flanked by two carbonyl groups are the more acidic, allowing the formation of enolate anions. These may then participate in intramolecular reactions with ketone or ester carbonyl groups, with a natural tendency to form strain-free six-membered rings. To produce the compounds orsellinic acid and phloracetophenone, we can envisage the same substrate being folded in two different ways. Which folding occurs will be dependent on the organism and the enzyme it contains.

With folding A, ionization of the α-methylene allows aldol addition onto the carbonyl six carbons distant along the chain, giving the tertiary alcohol. Dehydration occurs as in most chemical aldol reactions, giving the conjugated system, and enolization follows to attain the stability conferred by the aromatic ring. The thioester bond is then hydrolysed to produce orsellinic acid, at the same time releasing the product from the enzyme. Alternatively, folding B allows a Claisen reaction to occur, which, although mechanistically analogous to the aldol reaction, is terminated by expulsion of the leaving group and direct release from the enzyme. Enolization of the cyclohexatrione produces phloracetophenone.

Box 10.14 (continued)

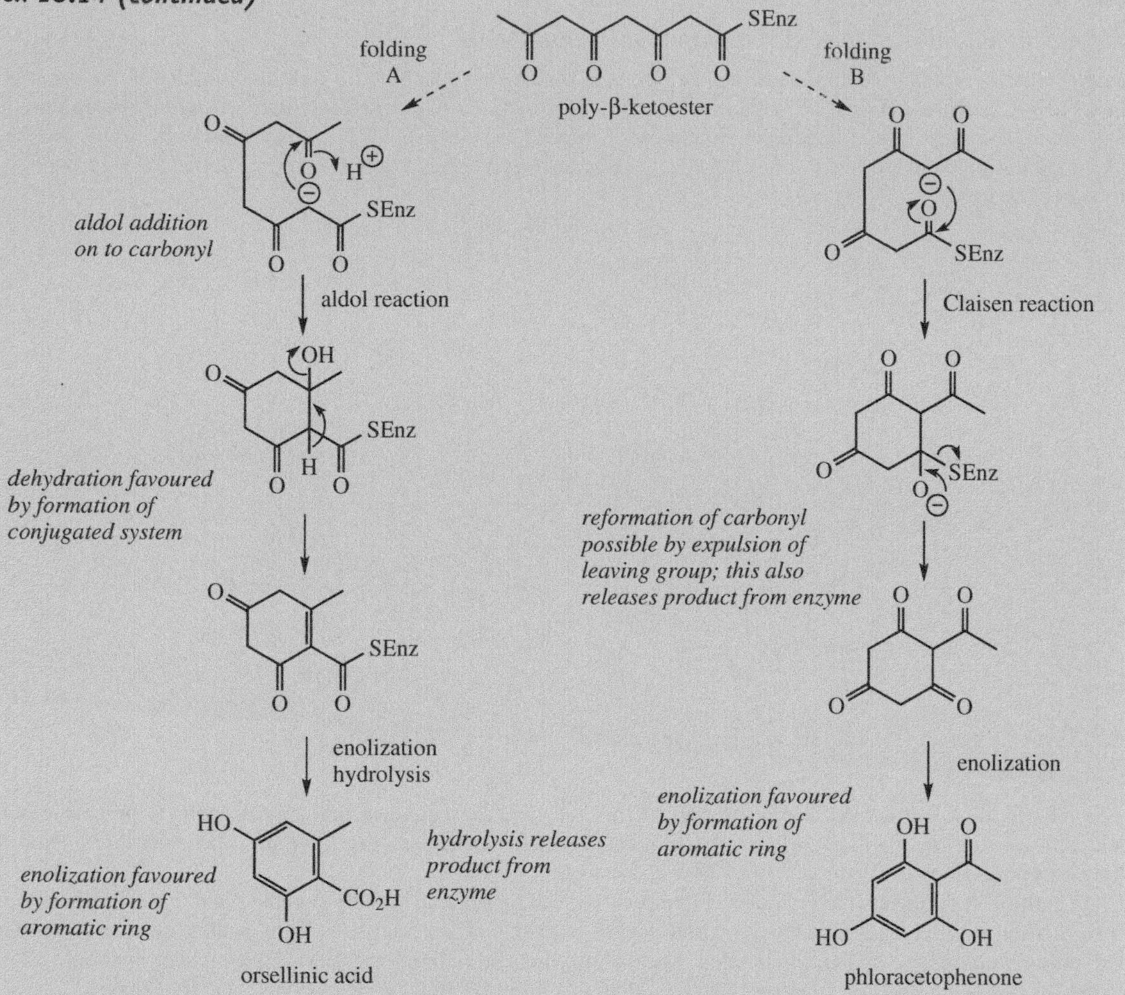

folding A

poly-β-ketoester

folding B

aldol addition on to carbonyl

aldol reaction

Claisen reaction

dehydration favoured by formation of conjugated system

reformation of carbonyl possible by expulsion of leaving group; this also releases product from enzyme

enolization hydrolysis

enolization

enolization favoured by formation of aromatic ring

hydrolysis releases product from enzyme

enolization favoured by formation of aromatic ring

orsellinic acid

phloracetophenone

Essentially the same sort of enolate anion aldol and Claisen reactions occur in the production of the more complex structures mycophenolic acid, griseofulvin, and tetracycline. However, the final structure is only obtained after a series of further modifications.

10.8.1 Reverse Claisen reactions

The driving force for the Claisen reaction is formation of the enolate anion of the β-ketoester product. If this cannot form, the reverse reaction controls the equilibrium.

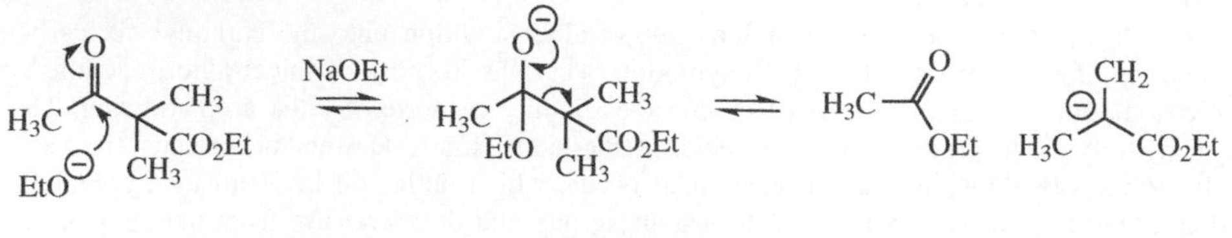

i.e. β-ketoester / base ⟶ reverse Claisen reaction

This means that a **reverse Claisen reaction** can occur if a β-ketoester is treated with base. This is most likely to occur if we attempt to hydrolyse the β-ketoester to give a β-ketoacid using aqueous base. Note that the alcoholic base used for the Claisen reaction does not affect the ester group, since the nucleophile is the same as the leaving group (see Section 10.7). Aqueous base treatment of a β-ketoester will, however, result in both ester hydrolysis and a reverse Claisen reaction, and poses a problem if one only wants to hydrolyse the ester.

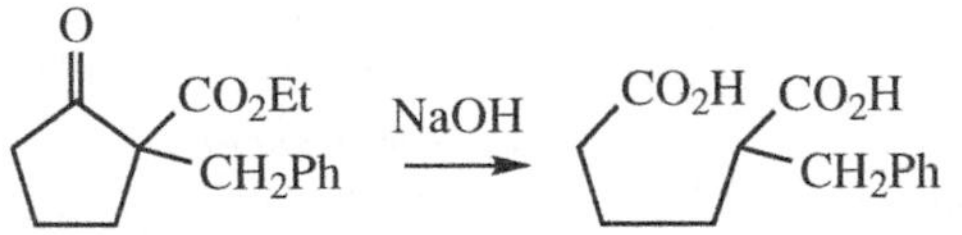

base causes reverse Claisen reaction as well as ester hydrolysis

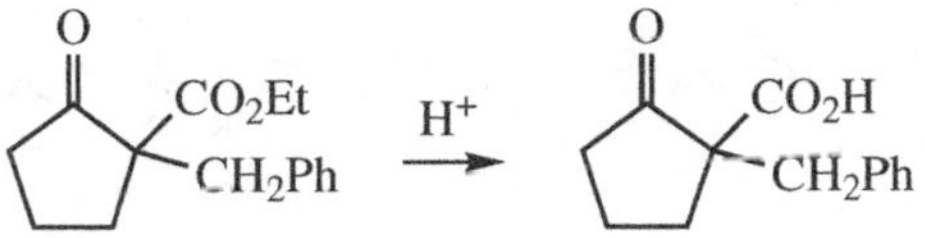

β-ketoester   *β-ketoacid*

acid only hydrolyses ester

The reverse Claisen reaction is common, especially with cyclic β-ketoesters, such as one gets from the Dieckmann reaction (see Section 10.8). If one only wants to hydrolyse the ester, it thus becomes necessary to use the rather less effective acid-catalysed hydrolysis method (see Section 7.9.2).

Cleavage of β-diketones, the products of a mixed Claisen reaction between an ester electrophile and a ketone nucleophile (see Box 10.13), behave similarly towards base, and a reverse Claisen reaction ensues. Again, this is prevalent with cyclic systems.

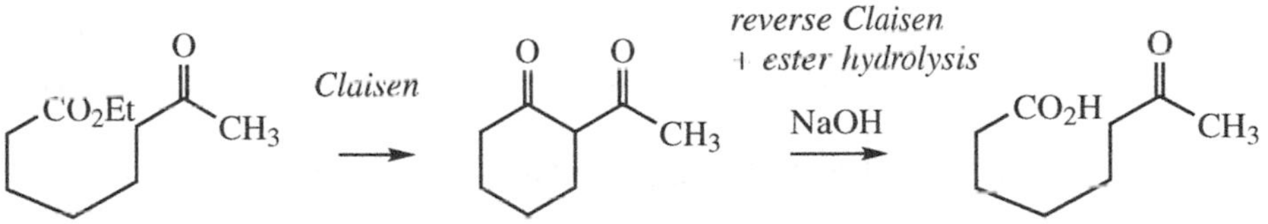

β-diketone

Nevertheless, as we shall see in Section 10.9, it is also possible to exploit the reverse Claisen reaction to achieve useful transformations.

10.9 Decarboxylation reactions

Hydrolysis of the ester function of the β-ketoester Claisen product under acidic conditions yields a β-ketoacid, but these compounds are especially susceptible to loss of carbon dioxide, i.e. **decarboxylation**. Although β-ketoacids may be quite stable, decarboxylation occurs readily on mild heating, and is ascribed to the formation of a six-membered hydrogen-bonded transition state. Decarboxylation is represented as a cyclic flow of electrons, leading to an enol product that rapidly reverts to the more favourable keto tautomer.

Box 10.15

Reverse Claisen reaction in biochemistry: β-oxidation of fatty acids

Perhaps the most important example of the **reverse Claisen reaction** in biochemistry is that involved in the **β-oxidation of fatty acids**, used to optimize energy release from storage fats, or fats ingested as food (see Section 15.4). In common with most biochemical sequences, thioesters rather than oxygen esters are utilized (see Box 10.8).

β-oxidation of fatty acids

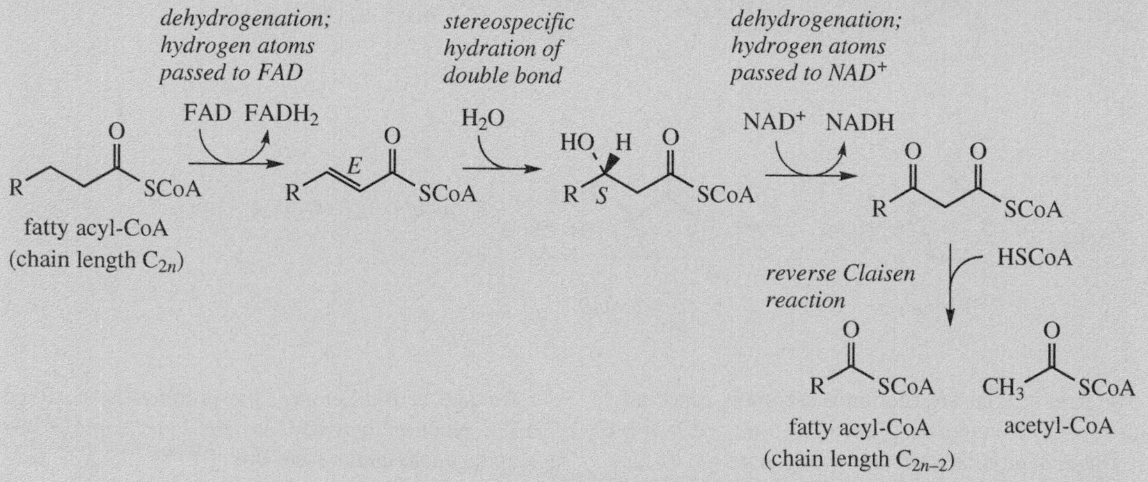

The β-oxidation sequence involves three reactions, dehydrogenation, hydration, then oxidation of a secondary alcohol to a ketone, thus generating a β-ketothioester from a thioester. We shall study these reactions in more detail later (see Section 15.4.1). The β-ketothioester then suffers a reverse Claisen reaction, initiated by nucleophilic attack of the thiol coenzyme A (see Box 10.8).

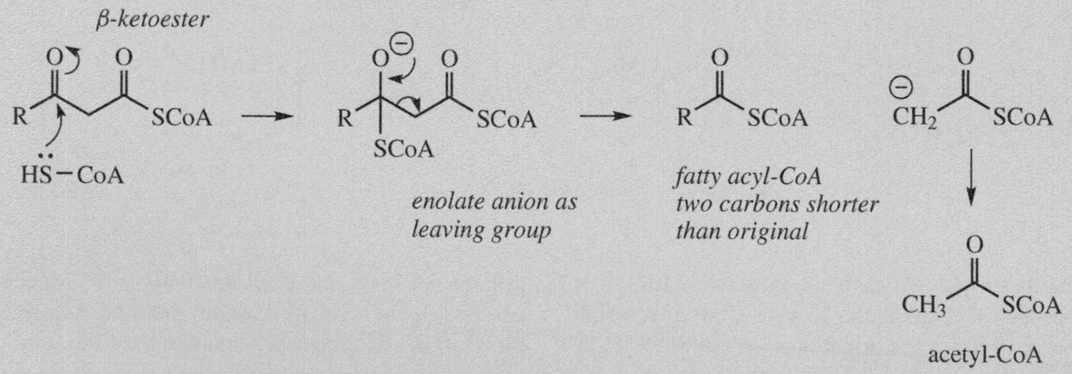

The leaving group is the enolate anion of acetyl-CoA, and the reaction thus cleaves off a two-carbon fragment from the original fatty acyl-CoA. Since the nucleophile is coenzyme A, the other product is also a coenzyme A ester. In fact, the reaction generates a new fatty acyl-CoA, shorter by two carbons, which can re-enter the β-oxidation cycle. Most natural fatty acids have an even number of carbons, so the process continues until the original fatty acid chain is cleaved completely to acetyl-CoA fragments.

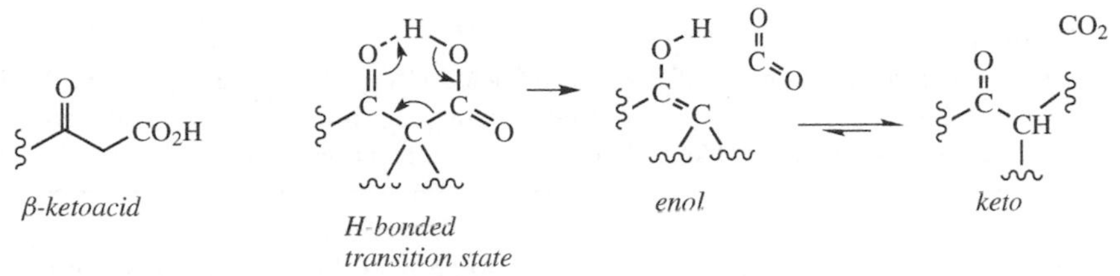

β-ketoacid   H-bonded transition state   enol   keto

We can now see how a number of the reactions recently studied fit together.

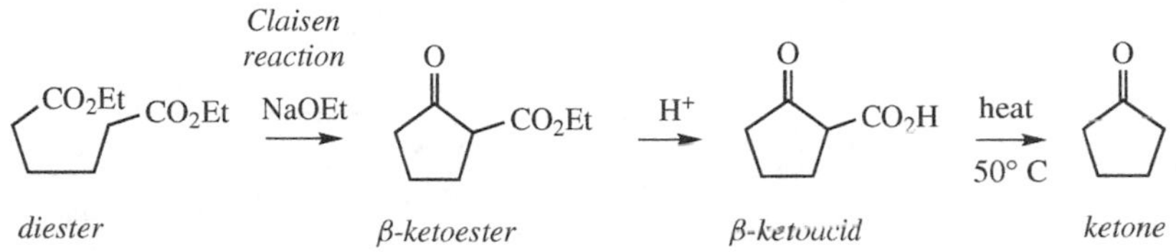

diester   β-ketoester   β-ketoacid   ketone

Box 10.16

Decarboxylation of β-ketoacids in biochemistry: isocitrate dehydrogenase

The enzyme **isocitrate dehydrogenase** is one of the enzymes of the **Krebs** or **citric acid cycle**, a major feature in carbohydrate metabolism (see Section 15.3). This enzyme has two functions, the major one being the dehydrogenation (oxidation) of the secondary alcohol group in isocitric acid to a ketone, forming oxalosuccinic acid. This requires the cofactor NAD^+ (see Section 11.2). For convenience, we are showing non-ionized acids here, e.g. isocitric acid, rather than anions, e.g. isocitrate.

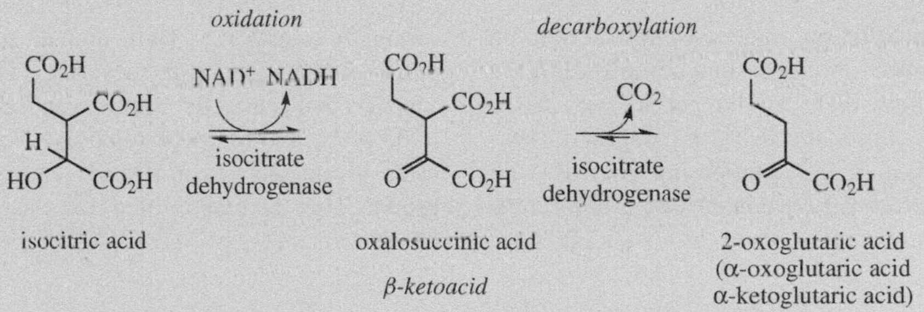

isocitric acid   oxalosuccinic acid β-ketoacid   2-oxoglutaric acid (α-oxoglutaric acid α-ketoglutaric acid)

The second function, and the one pertinent to this section, is the decarboxylation of oxalosuccinic acid to 2-oxoglutaric acid. This is simply a biochemical example of the ready decarboxylation of a β-ketoacid, involving an intramolecular hydrogen-bonded system. This reaction could occur chemically without an enzyme, but it is known that isocitric acid, the product of the dehydrogenation, is still bound to the enzyme isocitrate dehydrogenase when decarboxylation occurs.

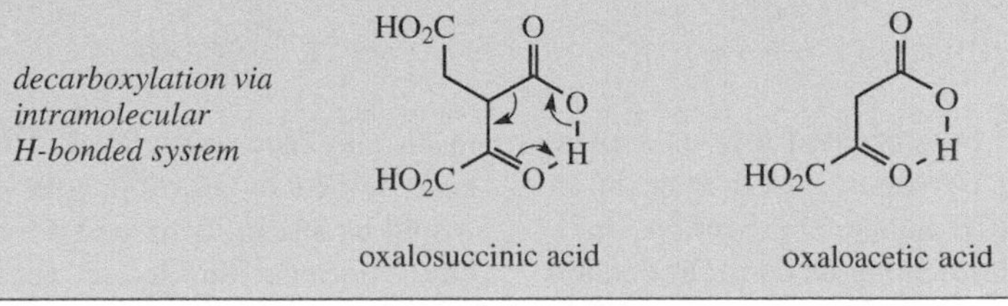

decarboxylation via intramolecular H-bonded system

oxalosuccinic acid   oxaloacetic acid

Box 10.16 (continued)

It is appropriate here to look at the structure of oxaloacetic acid, a critical intermediate in the Krebs cycle, and to discover that it too is a β-ketoacid. In contrast to oxalosuccinic acid, it does not suffer decarboxylation in this enzyme-mediated cycle, but is used as the electrophile for an aldol reaction with acetyl-CoA (see Box 10.4).

Decarboxylation of 1,1-diacids (**gem-diacids**) is a similar reaction involving a hydrogen-bonded transition state. 1,1-Diacids may be stable entities, e.g. malonic acid, but they are susceptible to decarboxylation upon heating; malonic acid decarboxylates at 150 °C.

gem-diacid   *enol*   *keto*

gem-Diacids are typical products that might be obtained from synthetic sequences using esters of malonic acid, e.g. **diethyl malonate**, a 1,3-dicarbonyl compound. Since the methylene group in diethyl malonate is sandwiched between two carbonyls, the protons are considerably more acidic than those in ethyl acetate. The pK_a is of the order of 13, compared with about 24 for ethyl acetate, so it becomes much easier to form the enolate anion.

malonic acid   diethyl malonate

more acidic than CH_3CO_2Et (pK_a 24) enolate anion stabilized by two carbonyls; therefore, better nucleophile

These decarboxylation reactions must not be viewed as unwanted processes that complicate reactions, but reactions that can be put to very good use. There were hints in the last paragraph. Two carbonyl groups in a 1,3-relationship increase the acidity of the α-protons between the two groups compared with protons adjacent to just one carbonyl group. It is easier to form enolate anions and then carry out nucleophilic reactions. Therefore, since we may subsequently remove an ester function by hydrolysis and decarboxylation, we can view an ester group as a useful and temporary activating group. This is exemplified by the two sequences below.

same product as from use of CH_3CO_2Et, but enolate anion formation occurs more readily

Diethyl malonate can be converted into its enolate anion, which may then be used to participate in an S_N2 reaction with an alkyl halide (see Section 10.7). Ester hydrolysis and mild heating leads to production of an alkylated acetic acid. The same product might be obtained by starting with ethyl acetate, but this would be less efficient and possibly require a stronger base, because the lower acidity of the α-protons makes generation of the enolate anion less effective. One of the ester groups in diethyl malonate can thus be regarded as a temporary activating group to increase acidity of the α-protons.

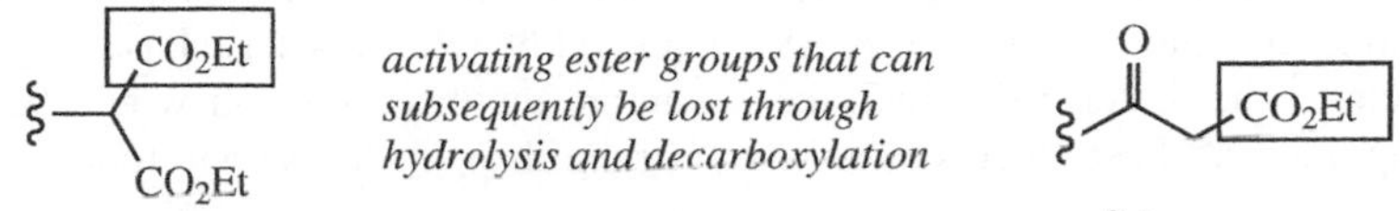

activating ester groups that can subsequently be lost through hydrolysis and decarboxylation

gem-diester

β-ketoester

The same viewpoint can taken for the ester function in a β-ketoester such as ethyl acetoacetate. Again, acidity of the α-protons is increased because there are two carbonyl groups, and generation of an enolate anion is facilitated. Although mono- or di-alkylation of a ketone might be achieved through enolate anions (see Section 10.2), it would be easier to use the more acidic β-ketoester and follow this by hydrolysis and decarboxylation.

mono- or di-alkylation; R groups may be the same or different

same product as from use of ketone, but enolate anion formation occurs more readily

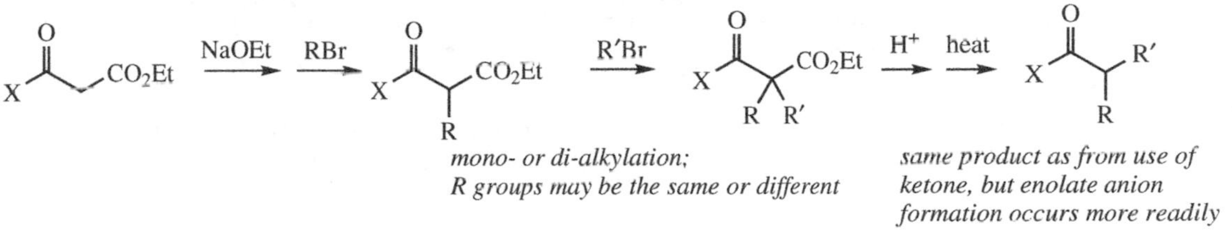

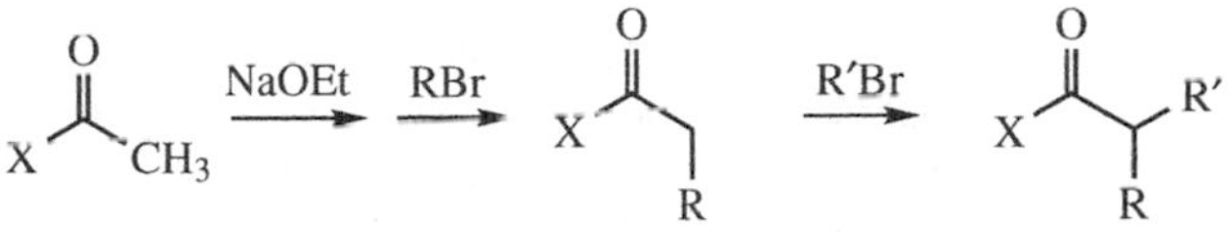

In general terms, a β-ketoester like ethyl acetoacetate can be considered as a pathway to substituted ketones, and diethyl malonate is a source of substituted acids.

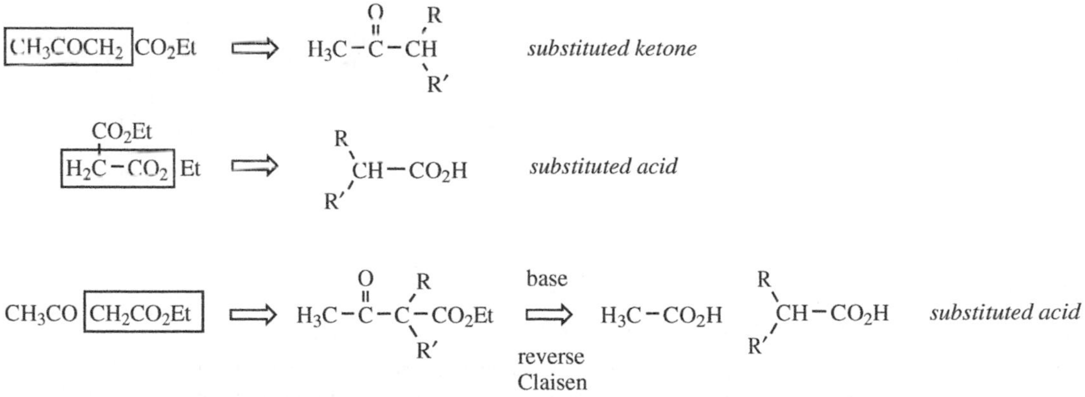

substituted ketone

substituted acid

substituted acid

Note also that we can even make good use of the **reverse Claisen reaction**. Thus, alkylation of ethyl acetoacetate followed by suitable base treatment to effect a reverse Claisen reaction would also generate a substituted acid. Alcoholic base would be used for the enolate anion chemistry, whereas aqueous base would initiate the reverse Claisen reaction and ester hydrolysis. In this sequence, we are using the acyl group as a temporary activating group.

Back in Section 10.5 we saw two methods of synthesizing 2-methylcyclohexanone, i.e. by direct alkylation of the enolate anion derived from cyclohexanone and by using an enamine derivative as the nucleophilic species. The latter route had the advantage of not using a strong base to generate the nucleophile. We can now add a further approach for synthesis of the same compound, via a β-ketoester. This also has the advantage of proceeding smoothly and, although it does use base to generate the enolate anion, the base required would be considerably less strong than for the ketone route.

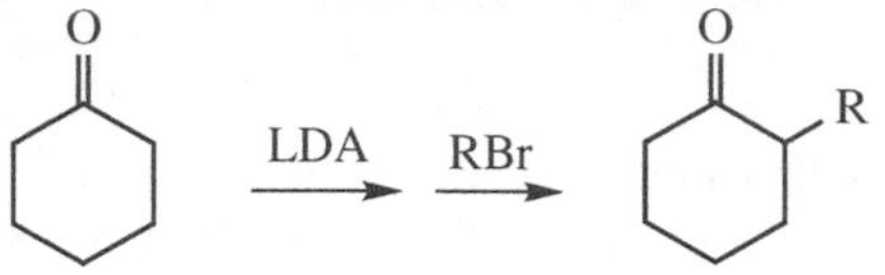

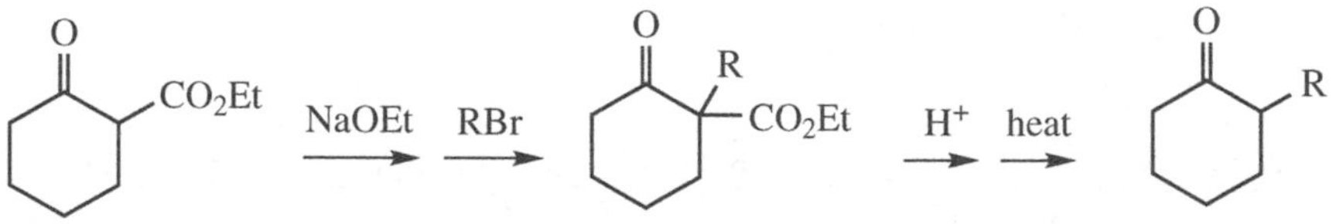

this route will proceed more readily, and uses a less strong base

On a number of occasions (see Sections 10.2, 10.7 and 10.8) we have noted that reactions involving enolate anions could be improved significantly by utilizing strongly basic reagents, such as sodium hydride, sodium amide, or LDA, and carrying out the reaction in two stages. This stratagem removed the constrictions imposed by unfavourable equilibria, by preparing the enolate anion in an essentially irreversible reaction, then adding the electrophile that could have a more reactive leaving group. This is further exemplified by the synthesis of a β-diketone from a β-ketoester, as shown below, again exploiting a decarboxylation reaction.

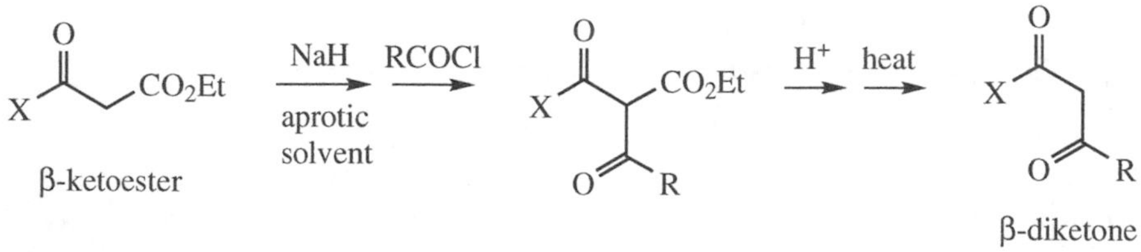

β-ketoester

β-diketone

Box 10.17

Claisen reactions in nature involving malonyl-CoA

In Box 10.12 we saw that nature employs a Claisen reaction between two molecules of acetyl-CoA to form acetoacetyl-CoA as the first step in the biosynthesis of mevalonic acid and subsequently cholesterol. This was a direct analogy for the Claisen reaction between two molecules of ethyl acetate. In fact, in nature, the formation of acetoacetyl-CoA by this particular reaction using the enolate anion from acetyl-CoA is pretty rare.

We have just seen that diethyl malonate can be used instead of ethyl acetate as a nucleophile. The second ester group is effectively used to activate the system for producing a nucleophile, and then is removed when the required reaction has been achieved. Would it surprise you to know that, with respect to this strategy, nature got there first?

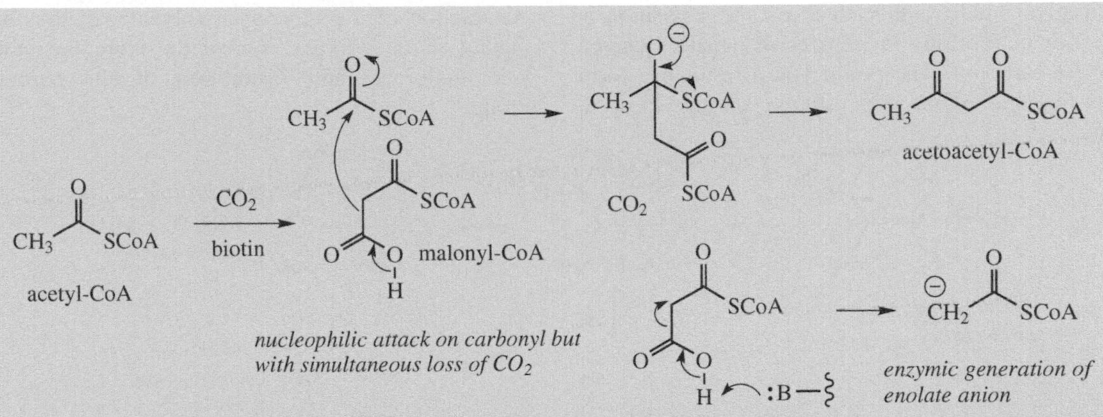

nucleophilic attack on carbonyl but
with simultaneous loss of CO_2

enzymic generation of
enolate anion

The nucleophile in biological Claisen reactions that effectively adds on acetyl-CoA is almost always malonyl-CoA. This is synthesized from acetyl-CoA by a reaction that utilizes a biotin–enzyme complex to incorporate carbon dioxide into the molecule (see Section 15.9). This has now flanked the α-protons with two carbonyl groups, and increases their acidity. The enzymic Claisen reaction now proceeds, but, during the reaction, the added carboxyl is lost as carbon dioxide. Having done its job, it is immediately removed. In contrast to the chemical analogy, a carboxylated intermediate is not formed. Mechanistically, one could perhaps write a concerted decarboxylation–nucleophilic attack, as shown. An alternative rationalization is that decarboxylation of the malonyl ester is used by the enzyme to effectively generate the acetyl enolate anion without the requirement for a strong base.

Malonyl-CoA is used as the nucleophilic species in the biosynthesis of fatty acids (see Section 15.5) and a whole host of other natural products, including the aromatic compounds seen in Box 10.14.

10.10 Nucleophilic addition to conjugate systems: conjugate addition and Michael reactions

We are familiar with the concept that the reactivity of a carbonyl group can be ascribed to the difference in electronegativity between carbon and oxygen, and the resultant unequal sharing of electrons. The polarization δ+/δ− can be considered as a contribution from the resonance form having full charge separation.

Now let us go a step further, and conjugate the carbonyl group with a double bond. If we polarize the carbonyl as before, then conjugation allows another resonance form to be written, in which the β-carbon now carries a positive charge. Thus, as well as the carbonyl carbon being electrophilic, the β-carbon is also an electrophilic centre.

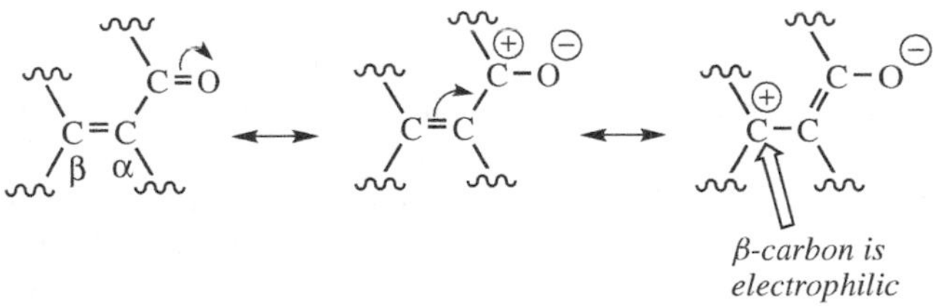

β-carbon is
electrophilic

Conjugation of the carbonyl with a double bond transfers the electronic characteristics δ+/δ− of the carbonyl group along the carbon chain. The alkene would normally be nucleophilic and react with electrophiles (see section 8.1). When conjugated with a carbonyl, it now becomes electrophilic and reacts with nucleophiles.

A typical nucleophilic attack on the β-position is now shown, resulting in transfer of negative charge onto the carbonyl. The product is a resonance form of an enolate anion with charge on the oxygen.

Abstraction of a proton from solvent will thus ultimately result in production of the more favourable keto tautomer, and restoration of the carbonyl group.

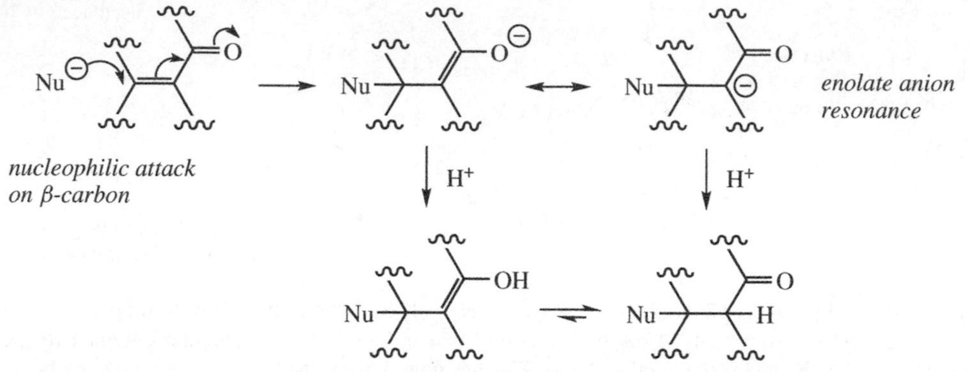

nucleophilic attack on β-carbon

enolate anion resonance

It is possible to get either the typical addition reaction on to the carbonyl group, termed a **1,2-addition**, or this form of **conjugate addition**, termed **1,4-addition**, terminology that is understandable if the enol tautomer is considered as the product formed first.

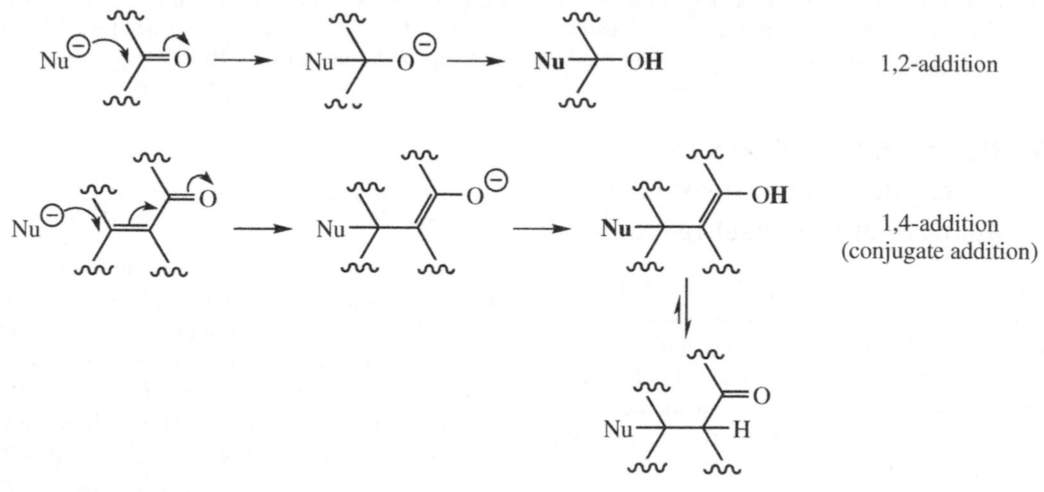

1,2-addition

1,4-addition
(conjugate addition)

Addition to the carbonyl, i.e. 1,2-addition, may be favoured with some nucleophilic reagents; but, more frequently, conjugate addition is the preferred mode of attack. Sometimes the product mixture is a result of both types of reaction.

Simple 1,2-addition is often favoured with good nucleophiles, and conjugate addition with weaker nucleophiles. This can partly be related to reversibility of addition reactions (see Section 7.1.2). Direct attack on the carbonyl will be faster, because this carbon carries rather greater positive charge, so 1,2-addition is favoured kinetically. On the other hand, the 1,4-addition product with the carbonyl group is thermodynamically more stable.

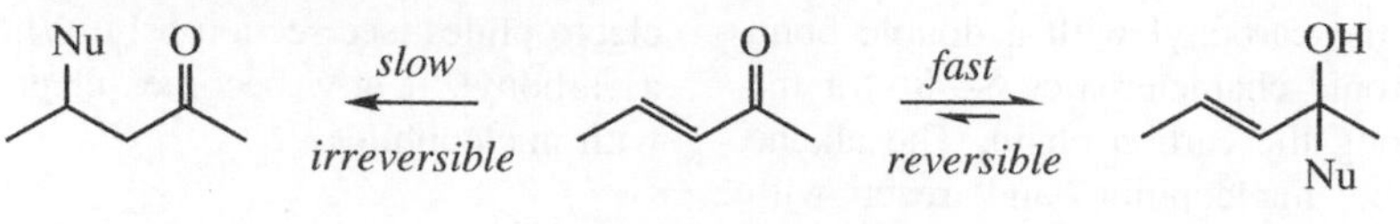

slow

irreversible

fast

reversible

If the 1,2-addition is reversible (the nucleophile is a good leaving group), then we get **thermodynamic control** and the conjugate addition product predominates. When the 1,2-addition is not reversible (the nucleophile is a poor leaving group), we get **kinetic control** and simple addition. Stereochemical considerations are also partly responsible, since it will be easier for larger nucleophiles, especially enolate anions, to attack the C=C double bond that is less hindered, particularly if this is a H_2C=group.

The following examples illustrate typical additions to conjugated systems. Although conjugate addition is more common, **Grignard reagents** (see Section 7.6.2) and **lithium aluminium hydride** (see Section 7.5) are more likely to add directly to the carbonyl.

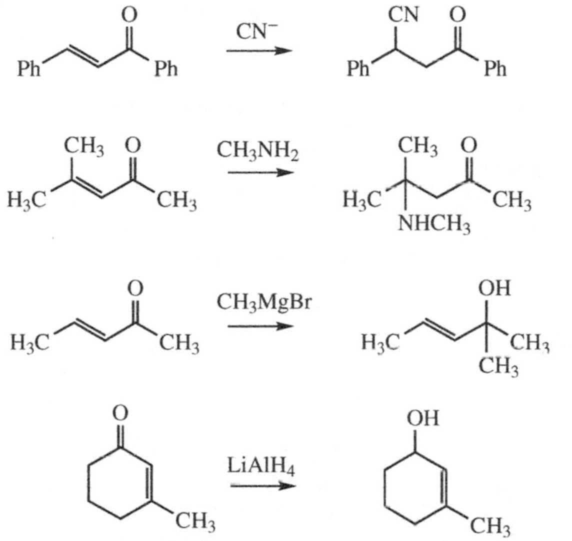

The less-reactive **sodium borohydride** may reduce unsaturated aldehydes by 1,2-addition, whereas unsaturated ketones tend to undergo conjugate addition. This allows selective reduction processes to be exploited. For example, in the unsaturated ketone shown, we may achieve reduction of the carbonyl using LAH, reduction of the double bond via catalytic hydrogenation (see Section 9.4.3), or conjugate reduction using sodium borohydride.

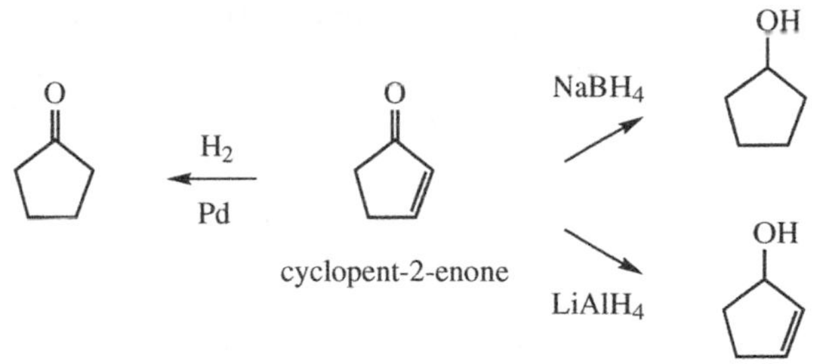

cyclopent-2-enone

The conjugate addition of a thiol, methanethiol, to the α,β-unsaturated aldehyde acrolein may be used in the synthesis of the amino acid methionine. Under basic conditions, the nucleophile will be the thiolate anion, and 1,4-addition leads to the thia-aldehyde. Methionine may then be obtained via the Strecker synthesis (see Box 7.10), a sequence that involves imine formation, then nucleophilic attack of cyanide on this imine/carbonyl analogue. The reaction is completed by acidic hydrolysis of the nitrile function to a carboxylic acid (see Box 7.9).

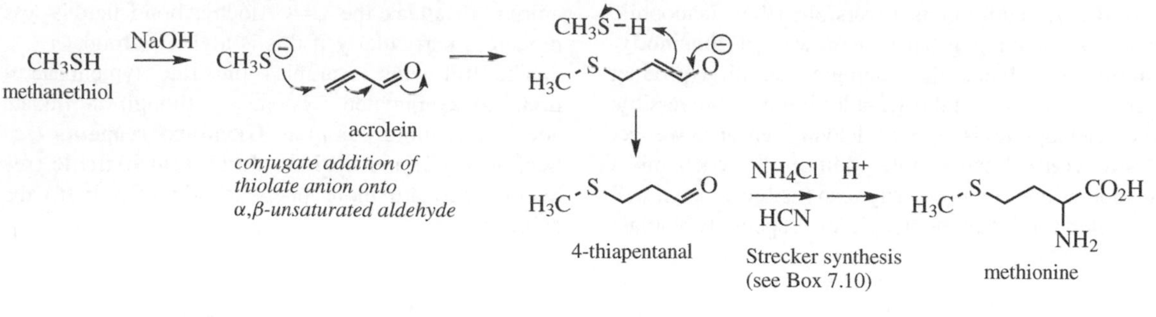

conjugate addition of
thiolate anion onto
α,β-unsaturated aldehyde

acrolein

4-thiapentanal

Strecker synthesis
(see Box 7.10)

methionine

It should also be noted, as we have seen earlier, that other electron-withdrawing groups, e.g. esters and nitriles, can achieve the same end as aldehydes or ketones (see Section 10.4). Conjugate addition can be observed when groups such as these are conjugated with a double bond.

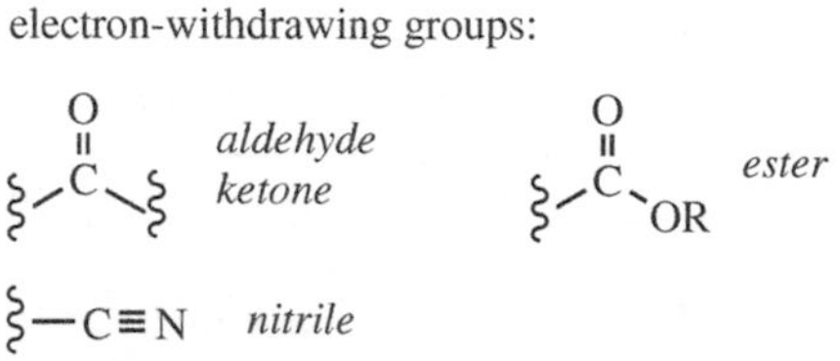

electron-withdrawing groups:

aldehyde
ketone

ester

nitrile

Box 10.18

Flavonoids: conjugate addition and heterocyclic ring formation

Flavonoids are natural plant phenols containing a six-membered oxygen heterocyclic ring. Considerable quantities of flavonoids are consumed daily in our vegetable diet, and there is growing belief that they have beneficial properties, acting as **antioxidants** (see Box 9.2) and giving protection against cardiovascular disease, and perhaps cancer. Their polyphenolic nature enables them to scavenge injurious free radicals, such as superoxide and hydroxyl radicals, which can cause serious cell damage. In particular, flavonoids in red wine and in tea have been demonstrated to be effective antioxidants.

One of the simplest natural flavonoids is the flavanone **liquiritigenin**, a material that contributes to the bright yellow colour of liquorice root. Liquiritigenin may be synthesized readily, as shown, by a two-stage process starting from the phenolic ketone and aldehyde.

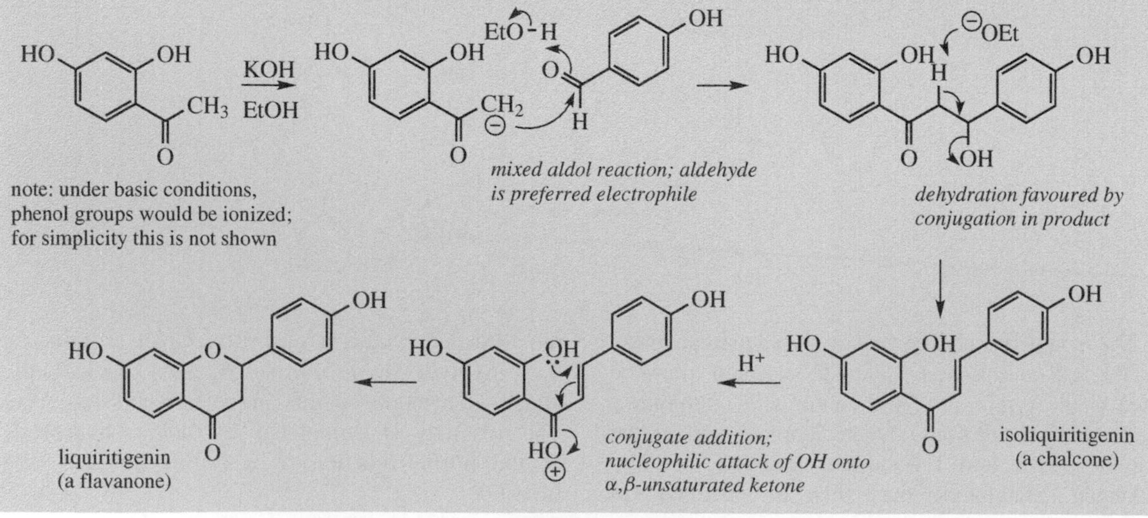

note: under basic conditions,
phenol groups would be ionized;
for simplicity this is not shown

mixed aldol reaction; aldehyde
is preferred electrophile

dehydration favoured by
conjugation in product

isoliquiritigenin
(a chalcone)

conjugate addition;
nucleophilic attack of OH onto
α,β-unsaturated ketone

liquiritigenin
(a flavanone)

The base-catalysed aldol reaction involves the enolate anion from the ketone adding preferentially to the aldehyde (see Section 10.3). Under the reaction conditions, the addition product dehydrates to give the unsaturated ketone (see Section 10.3), favoured because of the extended conjugation afforded in the product. The product is a member of the chalcone class of flavonoids and is called isoliquiritigenin. This material, when heated with acid, is converted into the corresponding flavanone liquiritigenin. This is the result of a conjugate addition reaction, in which the phenol group acts as nucleophile towards the unsaturated ketone, facilitated by protonation of the carbonyl. Formation of a six-membered ring is sterically favourable.

reverse reaction:

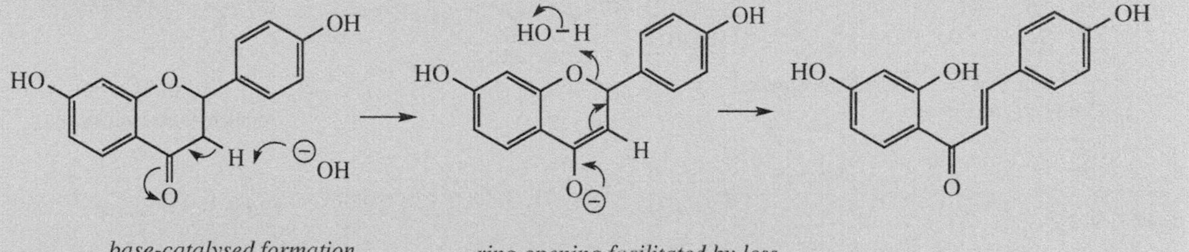

base-catalysed formation   ring opening facilitated by loss
of enolate anion   of phenolate as leaving group

The heterocyclic ring can be opened up again if the flavanone product is heated with alkali. Under these conditions, an enolate anion would be produced, and the addition is reversed, favoured by the phenolate anion as a leaving group. Whereas both isoliquiritigenin and liquiritigenin are stable in neutral solution, isomerizations to the other compound can be initiated by acid or base, as appropriate.

The **conjugate addition of enolate anions** onto α,β-unsaturated systems is an important synthetic reaction, and is termed the **Michael reaction**, though this terminology may often be used in the broader context for the other conjugate additions considered above. A typical example of the Michael reaction is the base-catalysed reaction of ethyl acetoacetate with the α,β-unsaturated ester ethyl acrylate.

Michael reaction

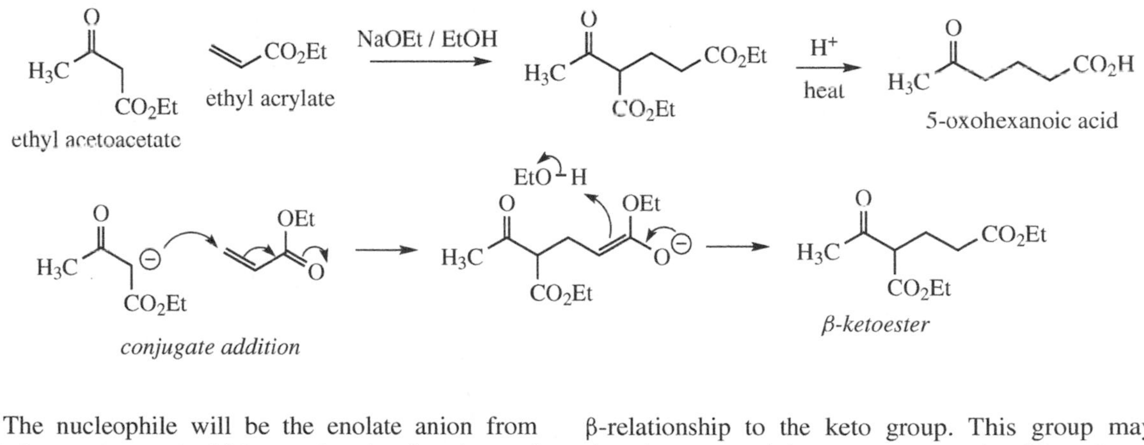

conjugate addition

The nucleophile will be the enolate anion from ethyl acetoacetate, which attacks the β-carbon of the electrophile, generating an addition complex that then acquires a proton at the α-position with restoration of the carbonyl group. The product is a δ-ketoester with an ester side-chain that has a β-relationship to the keto group. This group may thus be removed by a sequence of acid-catalysed hydrolysis, followed by thermal decarboxylation (see Section 10.9). The final product in this sequence is therefore a δ-ketoacid, i.e. a 1,5-dicarbonyl compound.

Other examples of the Michael reaction are shown below. Note the relatively mild bases that are employed in these reactions where the nucleophiles are 1,3-dicarbonyl compounds.

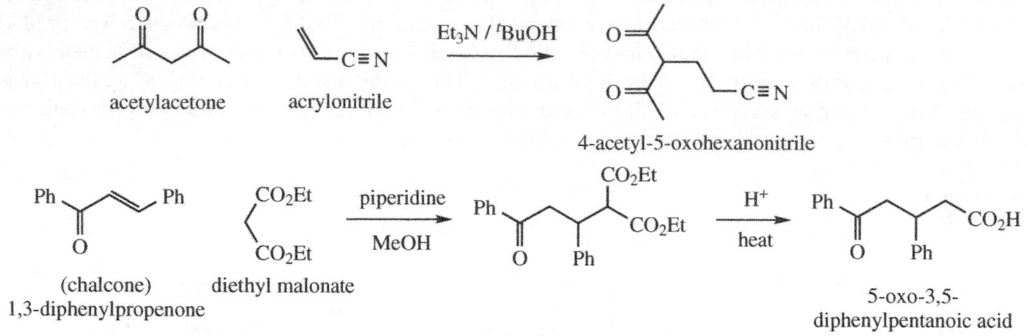

acetylacetone   acrylonitrile

4-acetyl-5-oxohexanonitrile

(chalcone)
1,3-diphenylpropenone   diethyl malonate

5-oxo-3,5-diphenylpentanoic acid

Box 10.19

Michael reaction: the Robinson annulation

A rather nice example of enolate anion chemistry involving the **Michael reaction** and the **aldol reaction** is provided by the **Robinson annulation**, a ring-forming sequence used in the synthesis of steroidal systems (Latin: *annulus*, ring).

In the partial synthesis shown, there are two reagents, the α,β-unsaturated ketone methyl vinyl ketone and the 1,3-diketone 2-methylcyclohexa-1,3-dione.

Robinson annulation

methyl vinyl ketone

acidic protons

2-methylcyclohexa-1,3-dione

NaOEt / EtOH
Michael reaction

EtO—H

acidic protons

diketone is more acidic substrate

NaOEt / EtOH

aldol reaction

testosterone

– H₂O

dehydration produces conjugated system

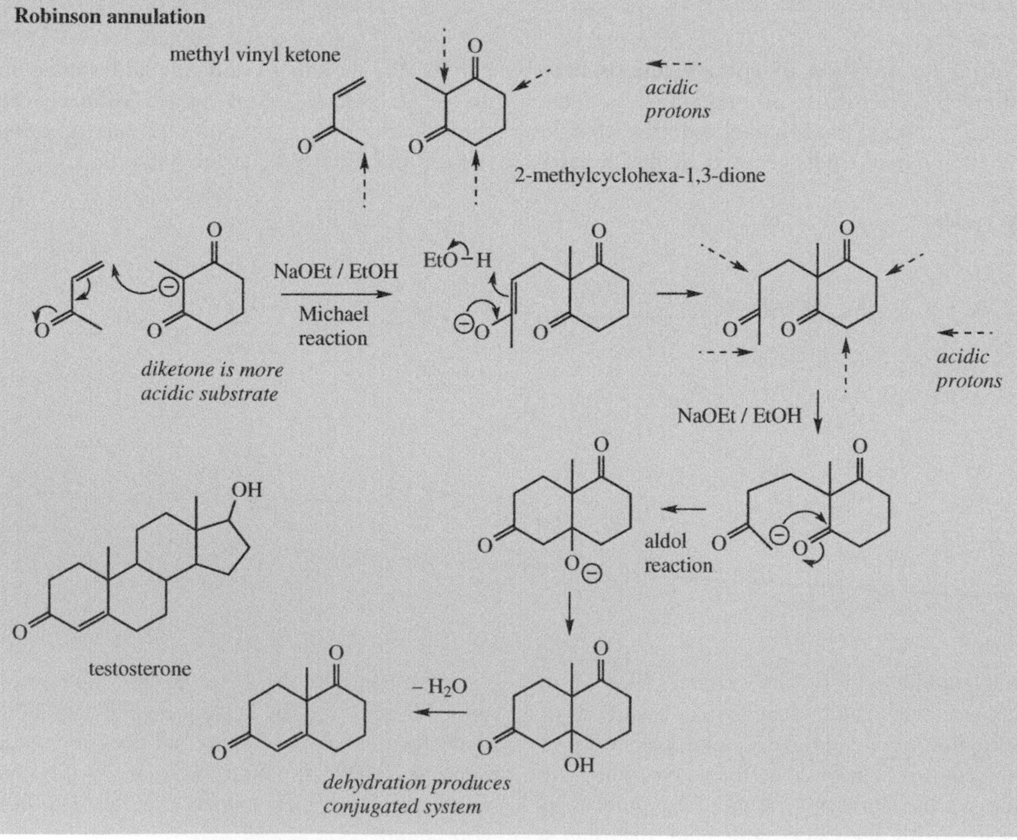

These are reacted together in basic solution. It can be deduced that the 1,3-diketone is more acidic than the monoketone substrate, so will be ionized by removal of a proton from the carbon between the two carbonyls to give the enolate anion as a nucleophile. This attacks the α,β-unsaturated ketone in a Michael reaction. It is understandable that this large nucleophile prefers to attack the unhindered β-position rather than the more congested ketone carbonyl.

The product from the Michael reaction will be a triketone. Now this substrate has four potential sites for proton removal, all flanked by a single ketone group, and thus all hydrogens are of similar acidity. The reaction that occurs is the intramolecular reaction that generates a strain-free six-membered ring system. This involves generating an enolate anion through loss of a proton from the terminal methyl of the side-chain, followed by an aldol reaction involving the appropriate ring carbonyl as electrophile. Dehydration follows to generate the conjugated system, and it is this dehydration that disturbs the equilibrium (see Section 10.3).

This annulation process was of considerable value in early approaches to **steroid synthesis**. The structural relationship of the bicyclic product obtained here to the male sex hormone **testosterone** is immediately apparent. Further, the non-conjugated carbonyl is now activating the adjacent carbon that subsequently features in building up the third ring system.

Box 10.20

Michael acceptors can be carcinogens

The **Michael reaction** involves conjugate addition of a nucleophile onto an α,β-unsaturated carbonyl compound, or similar system. Such reactions take place in nature as well, and some can be potentially dangerous to us. For example, the α,β-unsaturated ester ethyl acrylate is a cancer suspect agent. This electrophile can react with biological nucleophiles and, in so doing, bind irreversibly to the nucleophile, rendering it unable to carry out its normal functions. A particularly important enzyme that can act as a nucleophile is **DNA polymerase**, which is responsible for the synthesis of strands of DNA, especially as part of a DNA repair mechanism (see Section 14.2.2). The nucleophilic centre is a thiol grouping, and this may react with ethyl acrylate as shown.

e.g. DNA polymerase

ethyl acrylate
cancer suspect agent

inactivated enzyme

All is not doom and gloom, however, in that nature has provided in our bodies an alternative nucleophile to react with stray electrophiles like Michael acceptors. This rather important compound is the tripeptide **glutathione**, a combination of glutamic acid, cysteine, and glycine (see Box 6.6).

glutathione

glutamic acid–cysteine–glycine

glutathione

carcinogen

inactivated carcinogen

Box 10.20 (continued)

It is the thiol group in glutathione that reacts with a carcinogenic α,β-unsaturated carbonyl compound in exactly the same way as did the thiol group of DNA polymerase. As a result, the carcinogen becomes irreversibly bound to glutathione, and can no longer interact with other biochemicals. Furthermore, as a result of the amino acid functionalities, the inactivated carcinogen now has increased polarity compared with the original compound. This compound is likely to be water soluble, and can thus be excreted from the body. We have also seen glutathione inactivating other electrophiles, e.g. toxic epoxides (see Box 6.8).

Glutathione is also implicated in the removal of toxic metabolites from the analgesic **paracetamol** (USA: **acetaminophen**). Oxidative metabolism of paracetamol produces an *N*-hydroxy derivative, and this readily loses water to generate a reactive and toxic quinone imine, which interacts with proteins to cause cell damage.

Glutathione normally deactivates this reactive electrophile through a conjugate addition reaction. This time, we see conjugate addition onto an unsaturated imine rather than an unsaturated ketone. Re-aromatization produces a non-toxic paracetamol–glutathione adduct. Unfortunately, if someone takes a large overdose of paracetamol, there may be insufficient glutathione available to detoxify all the metabolite. This can precipitate cell damage, particularly to the liver. Paracetamol is a safe analgesic unless taken in overdose.

Box 10.21

Multiple conjugate additions: anionic polymerization and superglue

We have seen a number of reactions in which alkene derivatives can be polymerized. Radical polymerization (see Section 9.4.2) is the usual process by which industrial polymers are produced, but we also saw the implications of cationic polymerization (see Section 8.3). Here we see how an anionic process can lead to polymerization, and that this is really an example of **multiple conjugate additions**.

Alkene polymers such as poly(methyl methacrylate) and polyacrylonitrile are easily formed via **anionic polymerization** because the intermediate anions are resonance stabilized by the additional functional group, the ester or the nitrile. The process is initiated by a suitable anionic species, a nucleophile that can add to the monomer through conjugate addition in Michael fashion. The intermediate resonance-stabilized addition anion can then act as a nucleophile in further conjugate addition processes, eventually giving a polymer. The process will terminate by proton abstraction, probably from solvent.

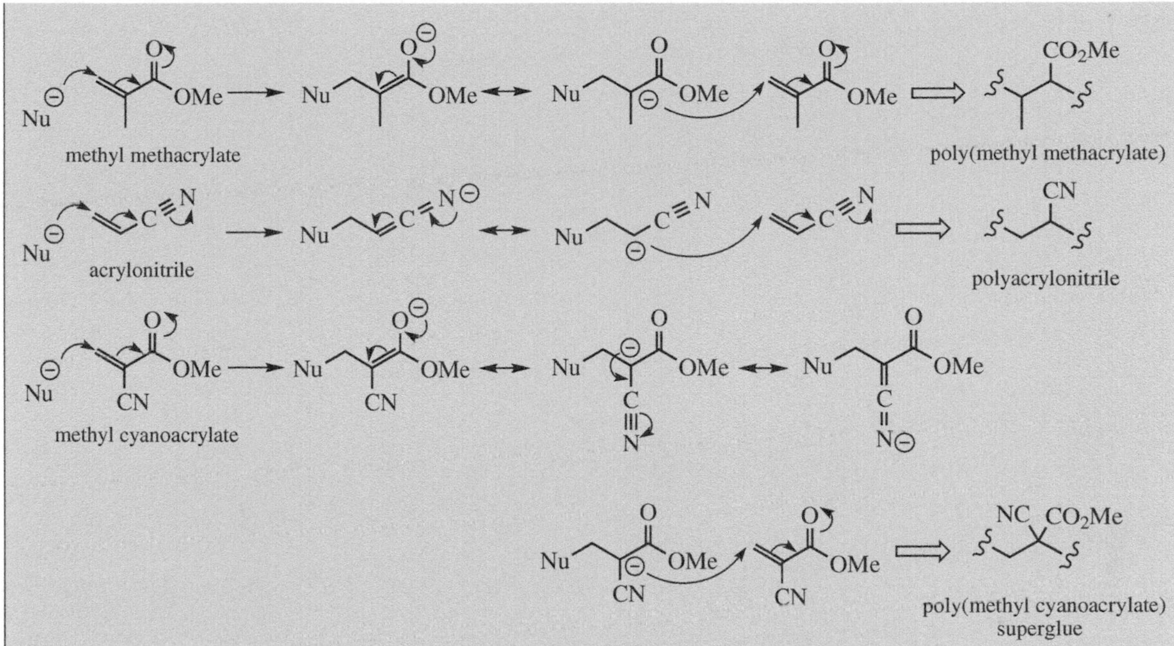

methyl methacrylate

poly(methyl methacrylate)

acrylonitrile

polyacrylonitrile

methyl cyanoacrylate

poly(methyl cyanoacrylate)
superglue

Methyl cyanoacrylate combines the anion-stabilizing features of both an ester group and a nitrile. Addition anions form very easily because of their enhanced resonance stability, and this polymerization process forms the basis of **superglue**. Traces of moisture on surfaces (including fingers) initiate anionic polymerization and the bonding together of almost any materials. Superglue now has some value in surgery, bonding tissue without the need for stitches.

11
Heterocycles

11.1 Heterocycles

Cyclic compounds in which one or more of the ring atoms is not carbon are termed **heterocycles**; the non-carbon atoms are referred to as **heteroatoms**. We shall limit our discussions to compounds in which the heteroatoms are nitrogen, oxygen, or sulfur. For the purposes of studying and understanding their properties, heterocycles are conveniently grouped into two classes, i.e. non-aromatic and aromatic.

11.2 Non-aromatic heterocycles

We have already met many examples of non-aromatic heterocycles in earlier chapters, e.g. cyclic ethers (see Section 6.3.6), including epoxides (see Section 8.1.2), and cyclic amines (see Section 7.7.1), as well as lactones (see Section 7.9.1), lactams (see Section 7.10), and cyclic acetals and ketals (see Section 7.2). From the familiar examples shown below, it should be clear that the standard approach to generating heterocyclic systems requires a difunctional compound containing a leaving group or electrophilic centre, together with a nucleophilic species that provides the heteroatom.

The **nomenclature** of simple heterocyclic ring systems containing one heteroatom is indicated overleaf. These form a useful reference, but there is little to be gained in committing them to memory.

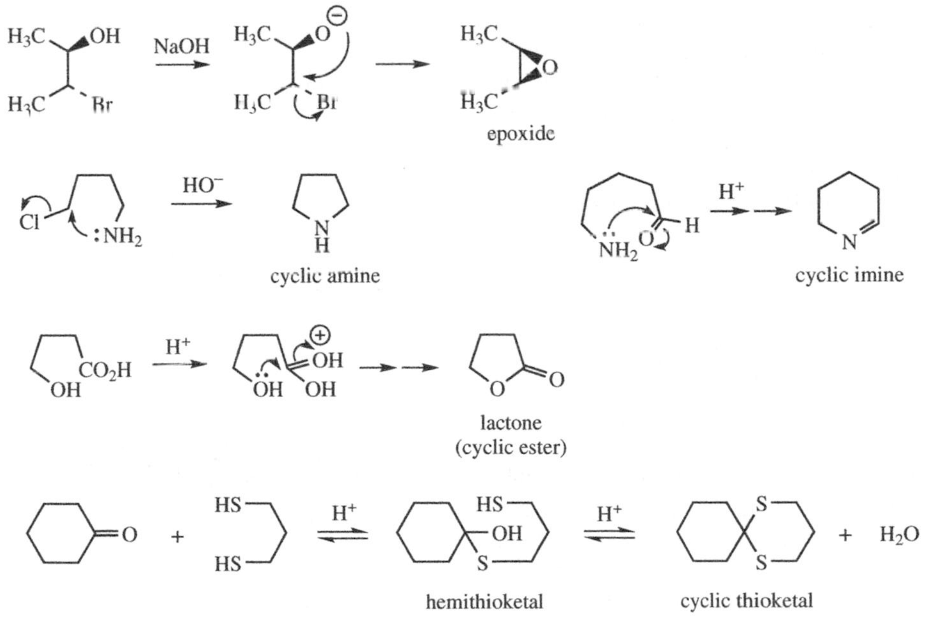

epoxide

cyclic amine

cyclic imine

lactone
(cyclic ester)

hemithioketal

cyclic thioketal

Essentials of Organic Chemistry   Paul M Dewick
© 2006 John Wiley & Sons, Ltd

Note, however, that the most important of these structures tend to have a trivial rather than systematic name, a consequence of long-standing common usage. Some of these, e.g. tetrahydrofuran and tetrahydropyran, are derived from the name of the corresponding aromatic heterocycle by the concept of reduction. A few examples of commonly encountered heterocycles with two heteroatoms are also shown.

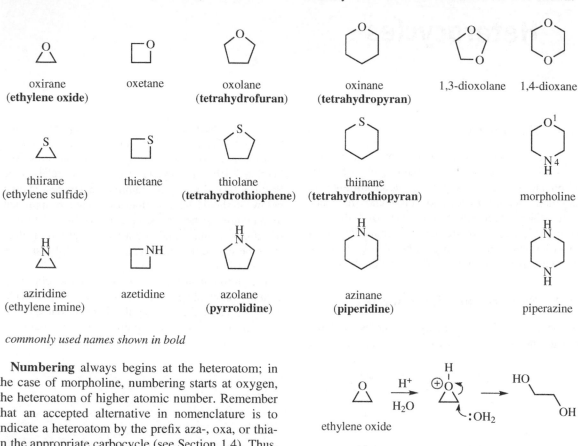

| oxirane **(ethylene oxide)** | oxetane | oxolane **(tetrahydrofuran)** | oxinane **(tetrahydropyran)** | 1,3-dioxolane | 1,4-dioxane |

| thiirane (ethylene sulfide) | thietane | thiolane **(tetrahydrothiophene)** | thiinane **(tetrahydrothiopyran)** | | morpholine |

| aziridine (ethylene imine) | azetidine | azolane **(pyrrolidine)** | azinane **(piperidine)** | | piperazine |

commonly used names shown in bold

Numbering always begins at the heteroatom; in the case of morpholine, numbering starts at oxygen, the heteroatom of higher atomic number. Remember that an accepted alternative in nomenclature is to indicate a heteroatom by the prefix aza-, oxa, or thia- in the appropriate carbocycle (see Section 1.4). Thus, we could name piperidine as azacyclohexane, and tetrahydrofuran as oxacyclopentane.

The chemistry of these non-aromatic heterocycles differs little from the chemistry of their acyclic counterparts, and we emphasize only the relative reactivity of the three-membered ring systems towards ring opening, thus achieving relief of ring strain (see Section 3.3.2). We have already noted the ring opening of epoxides (oxiranes; see Section 6.3.2), and similar reactivity is found with aziridines and thiiranes. Four-membered systems are also considerably strained and reactive towards nucleophiles, though not as readily as the three-membered compounds.

Some of these heterocycles provide us with valuable laboratory solvents, e.g. the ethers tetrahydrofuran and dioxane (1,4-dioxane). Others are useful as organic bases, e.g. piperidine, pyrrolidine, and

ethylene oxide

2,2-dimethylaziridine

oxetane

morpholine. The basicities of the nitrogen derivatives are comparable to those of similar acyclic amines, but physical properties, e.g. higher boiling point, make them more versatile than the simple amines.

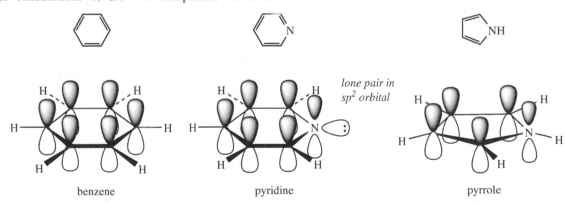

	pyrrolidine	piperidine	morpholine	piperazine	diethylamine
pK_a (conjugate acid)	11.3	11.1	8.5	9.7, 5.3	10.8

From the pK_a values shown, there is relatively little difference in basicities for diethylamine, pyrrolidine, or piperidine. Note, however, that morpholine and piperazine are weaker bases than piperidine. This is the result of an electron-withdrawing inductive effect from the second heteroatom, making the nitrogen atom both less basic and also less nucleophilic. This makes morpholine a useful base with basicity between that of piperidine and pyridine (pK_a 5.2) (see Section 4.6). The second pK_a value for the diamine piperazine is substantially lower than the first, since the inductive effect from the protonated amine will withdraw electrons away from the unprotonated amine (see Section 4.7).

11.3 Aromaticity and heteroaromaticity

Pyridine is structurally related to benzene: one CH unit has been replaced by N. If we consider the constitutions of the two compounds in more detail, we shall see even closer similarity. Thus, we have seen that the ring atoms in benzene are sp^2 hybridized (see Section 2.9.1). The remaining singly occupied p orbitals are oriented at right angles to the plane of the ring, and overlap to form a delocalized π system, extending to form a closed loop above and below the ring (see Section 2.9.1). Compared with what we might expect for the hypothetical cyclohexatriene, this results in a considerable stabilization, with significantly modified structure and reactivity in benzene. We termed this **aromaticity** (see Section 2.9).

Benzene conforms to **Hückel's rule**, which predicts that planar cyclic polyenes containing $4n + 2$ π electrons show enhanced stability associated with aromaticity (see Section 2.9.3). Pyridine is also aromatic: nitrogen contributes one electron in a p orbital to the π electron system, and its lone pair is located in an sp^2 orbital that is in the plane of the ring and perpendicular to the π electron system. It also conforms to Hückel's rule, in that we still have an aromatic sextet of π electrons.

lone pair in sp^2 orbital

benzene   pyridine   pyrrole

One of the structural features of benzene that derives from aromaticity is the equal length of the C–C bonds (1.40 Å), which lies between that for normal single (1.54 Å) and double (1.34 Å) bonds. Nevertheless, we continue to draw benzene with single and double bonds because this allows us to represent reaction mechanisms in terms of electron movements (see Section 5.1). Pyridine does not have a perfect hexagon shape; the symmetry is distorted because the C–N bonds are slightly shorter (1.34 Å) than the C–C bonds (1.39–1.40 Å).

bond lengths in benzene and pyridine

Nitrogen is more electronegative than carbon, and this influences the electron distribution in the π-electron system in pyridine through inductive effects, such that nitrogen is electron rich. In addition, the nitrogen will also become electron rich through a resonance effect: several resonance forms may be drawn that have a negative charge on nitrogen. These effects thus reinforce each other. The heteroatom thus distorts the π electron cloud of the aromatic ring system, drawing electrons towards the nitrogen and away from the carbons. The consequences of this are that we can predict that the pyridine nitrogen will react readily with electrophiles, whereas the remainder of the ring system will be resistant to electrophilic attack.

inductive effect   resonance effect

resonance effect reinforces inductive effect   dipole   *dipole is substantially greater than in piperidine*   dipole

An experimental probe for aromaticity is the chemical shift of the hydrogen signals in NMR spectroscopy (see Section 2.9.5). The substantially greater δ values for benzene protons (δ 7.27 ppm) compared with those in alkenes (δ 5–6 ppm) have been ascribed to the presence of a **ring current** that creates its own magnetic field opposing the applied magnetic field. This ring current is the result of circulating electrons in the π system of the aromatic ring. The hydrogen NMR signals for pyridine also appear at relatively large δ values, in the range 7.1–8.5 ppm, typical of aromatic systems. The signals do not all appear at the same chemical shift; the heteroatom distorts the π electron distribution and affects the 2/6, 3/4, and 5 positions to different extents.

Now let us now consider **pyrrole**, where we have a five-membered ring containing nitrogen. Pyrrole is also aromatic. This is somewhat unexpected: how can we get six π electrons from just five atoms? The answer is that each carbon contributes one electron as before, but nitrogen now contributes two electrons, its lone pair, to the π electron system.

We can draw **Frost circles** (see Section 2.9.3) to show the relative energies of the molecular orbitals for pyridine and pyrrole. The picture for pyridine is essentially the same as for benzene, six π electrons forming an energetically favourable closed shell (Figure 11.1). For pyrrole, we also get a closed shell, and there is considerable aromatic stabilization over electrons in the six atomic orbitals.

However, the contribution of the nitrogen lone pair to the aromatic sextet in pyrrole makes the nitrogen atom relatively electron deficient. The nitrogen atom should create an inductive effect, as in pyridine, drawing electrons towards the heteroatom. However, a consideration of the resonance structures leads to several resonance forms with a positive charge on nitrogen. The resonance effect is opposite to the inductive effect, and of greater magnitude. Overall, the heteroatom distorts the π electron cloud of the aromatic ring system by pushing electrons away from the nitrogen and towards the carbons.

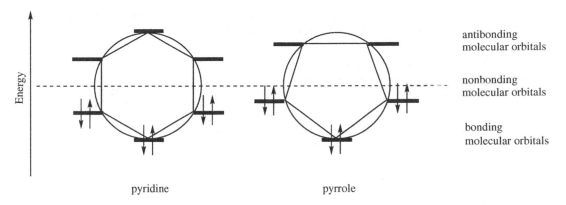

Figure 11.1 Relative energies of pyridine and pyrrole molecular orbitals from Frost circles

The difference in electron distribution in pyridine and pyrrole manifests itself via the measured **dipole moments**. More importantly, we shall see that this electron distribution influences the chemical reactivity of the two systems. In broad terms, ring systems where the carbons are electron deficient because of the electron-withdrawing effect of the heteroatom, e.g. pyridine, are more reactive towards nucleophiles than benzene. On the other hand, ring systems where the carbons are electron rich because of the electron-donating heteroatom, e.g. pyrrole, are more reactive towards electrophiles than benzene. Note the deliberate choice of terminology here: *ring systems where the carbons are electron deficient or electron rich*. You may meet the older terminology of π-deficient heterocycles and π-excessive heterocycles, but these can give a false impression. Each heterocycle contains six π electrons, so it is not the heterocycle that is electron deficient or electron rich, but the carbons that receive less or more than their equal share because of the effect of the heteroatom.

Though we shall return to this again, one critical difference between pyridine and pyrrole to note here relates to **basicity**. Pyridine is a base because its nitrogen still carries a lone pair able to accept a proton. Pyrrole is not basic: it has already used up its lone pair in contributing to the aromatic sextet.

11.4 Six-membered aromatic heterocycles

11.4.1 Pyridine

From our discussions in the last section, we might expect pyridine to display properties associated with the nitrogen function and also with the aromatic ring. Not surprisingly, it turns out that the aromatic ring affects the properties of the amine; but, more significantly, the aromatic properties are greatly influenced by the presence of the heteroatom.

Based on our earlier knowledge, and from a simple inspection of its structure, we might expect to observe

- Reaction at the heteroatom – the non-bonding electrons on the nitrogen might coordinate to H$^+$ or another suitable electrophile.

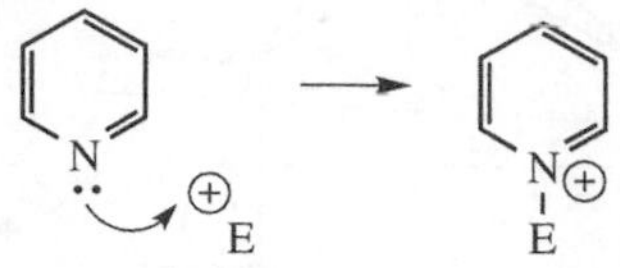

- Reaction of the aromatic π system – typical electrophilic substitution as seen for benzene might be

- Reaction of the C=N 'imine' function – though this is not an isolated imine function but is part

three types of general reactivity in pyridine. We might expect to see:

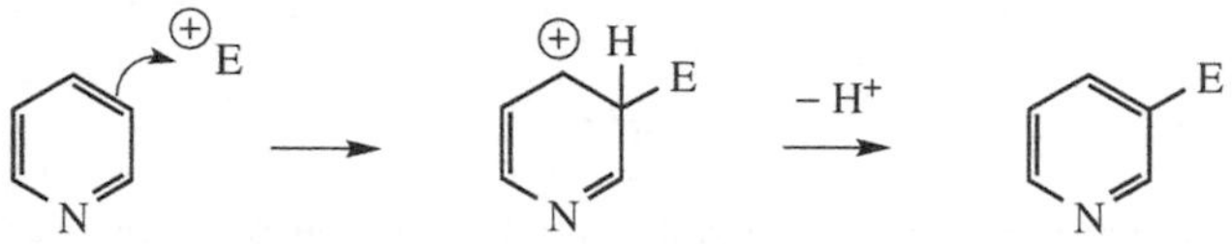

N-coordination

expected.

electrophilic substitution

of the aromatic ring, its polarization might make it susceptible to nucleophilic attack.

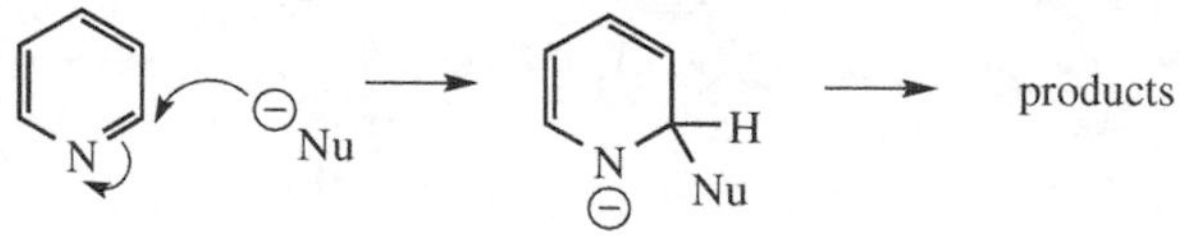

products   *nucleophilic attack on C=N*

Reassuringly, our predictions turn out to be well founded.

Pyridine is a base (pK_a pyridinium cation 5.2), but it is a considerably weaker base than a typical non-aromatic heterocyclic amine such as piperidine (pK_a piperidinium cation 11.2). This is because the lone pair electrons in pyridine are held in an sp^2 orbital. The increased s character of this orbital, compared with the sp^3 orbital in piperidine, means

that the lone pair electrons are held closer to the nitrogen, and are consequently less available for protonation. The lower basicity of pyridine compared with piperidine is thus a **hybridization effect** (see Section 4.6). Although pyridine is a weak base, it can form salts with acids and is widely used in chemical reactions as an acid scavenger and as a very good polar solvent.

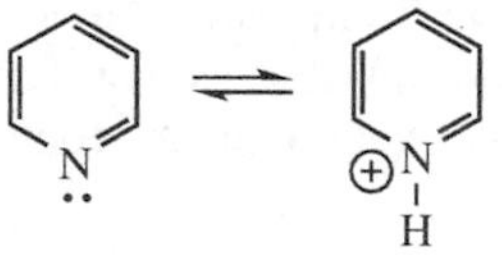

lone pair in   pK_a 5.2
sp^2 orbital

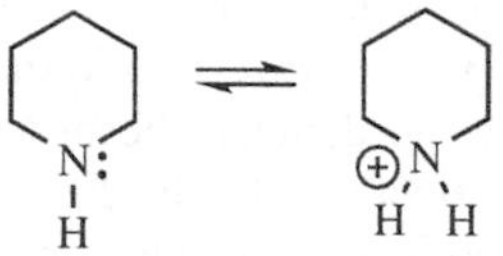

lone pair in   pK_a 11.2
sp^3 orbital

Just as pyridine is a weaker base than piperidine, it is also a poorer nucleophile. Nevertheless, it reacts with electrophiles to form stable **pyridinium salts**. In the examples shown, primary alkyl halides form *N*-alkylpyridinium salts, whereas acyl halides and anhydrides react to give *N*-acylpyridinium salts.

We have already seen the latter compounds involved in esterification reactions (see Section 7.9.1), and seen the value of pyridine in removing acidic by-products, e.g. HCl. Of course, *N*-acylpyridinium salts will easily be hydrolysed under aqueous conditions.

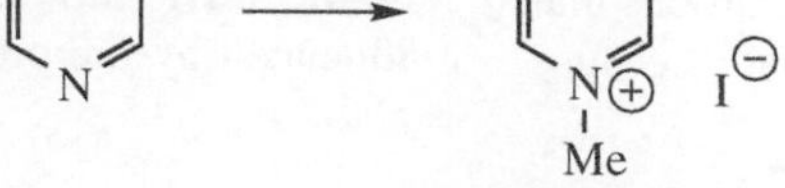

N-methylpyridinium iodide

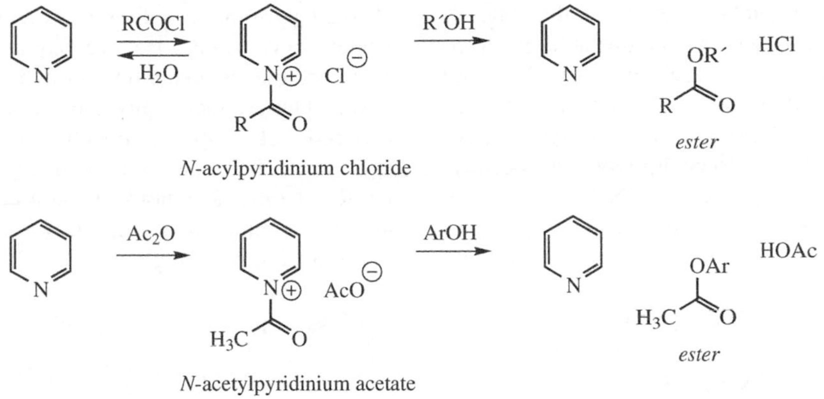

N-acylpyridinium chloride

N-acetylpyridinium acetate

Even better than pyridine in such reactions is the derivative **4-*N,N*-dimethylaminopyridine** (DMAP), where a resonance effect from the dimethylamino substituent reinforces the nucleophilicity of the pyridine nitrogen. It then also promotes the acylation step by improving the nature of the leaving group. The example shows its function in a typical esterification process, i.e. acylation of an alcohol.

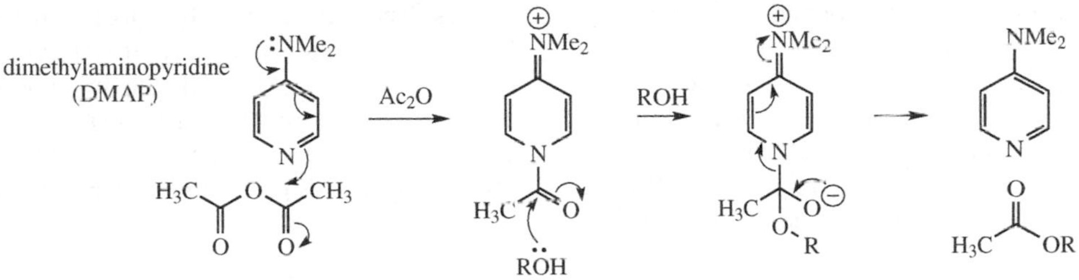

There are other gains, as well. Pyridine as a solvent is difficult to remove from the products, and it smells quite awful. In this reaction, a catalytic amount of DMAP is all that is necessary, and a more acceptable solvent can be employed.

The pre-eminent reactivity associated with aromatic compounds is the ease of **electrophilic substitution** (see Section 8.4). As we have already predicted, the pyridine ring is rather unreactive towards electrophilic reagents, and these tend to be attacked by the nitrogen instead, making the ring even less reactive.

It is readily seen from the intermediate addition cations and their resonance structures that attack at

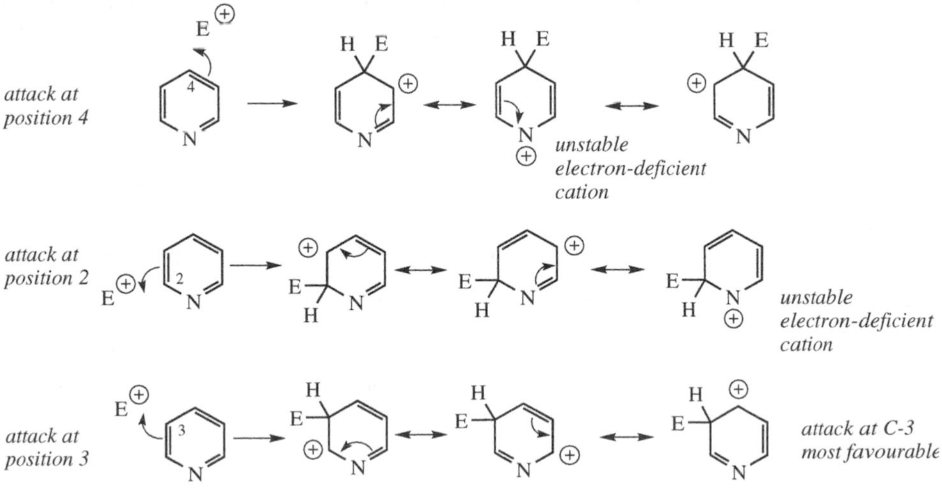

C-2 or C-4 will be unfavourable, in that one of the resonance forms features an unstable electron-deficient nitrogen cation. Attack at C-3 is the more likely, simply based on an inspection of resonance structures for the addition cation. However, electrophilic attack still tends to be unfavourable, because many electrophilic reagents, e.g. HNO_3–H_2SO_4, are strongly acidic, and the first effect is protonation on nitrogen. Attack of E^+ on to a positively charged pyridinium cation is even less favourable. Under acidic conditions, we require attack on free pyridine, the concentration of which will be very small. Thus, under equivalent conditions, pyridine undergoes electrophilic substitution very much more slowly than benzene, by a factor of about 10^6. Even Friedel–Crafts acylations are inhibited, because the nitrogen complexes with the Lewis acid, again leading to a cationic nitrogen.

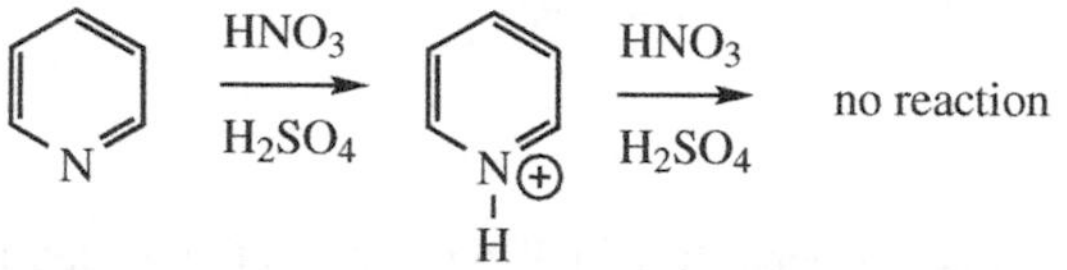

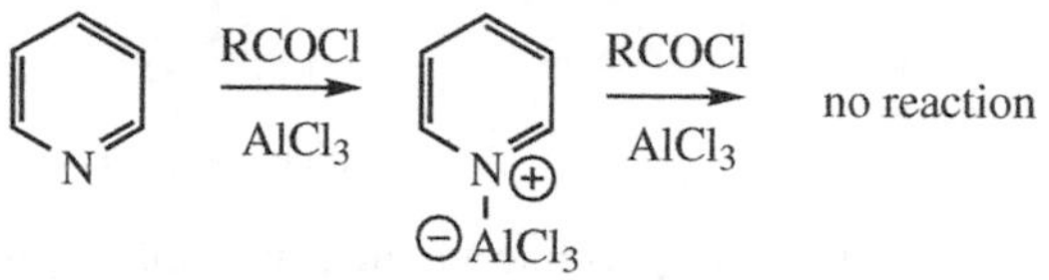

A striking demonstration of the reduced activity towards electrophiles for the pyridine ring compared with the benzene ring will be seen later when we consider the fused heterocycles quinoline and isoquinoline (see Section 11.8.1). These contain a benzene ring fused to a pyridine ring; electrophilic substitution occurs exclusively in the benzene ring.

To facilitate electrophilic substitution, it is possible to first convert pyridine into pyridine N-oxide by the action of a peracid such as peracetic acid or m-chloroperbenzoic acid (MCPBA; see Section 8.1.2). N-Oxide formation is not peculiar to pyridine, but it is a general property of tertiary amines. There is no overall charge in the molecule, but it is not possible to draw the structure without charge separation. Although the introduced oxygen atom causes electron withdrawal through an inductive effect, there is a greater and opposing resonance effect that donates electrons into the ring system.

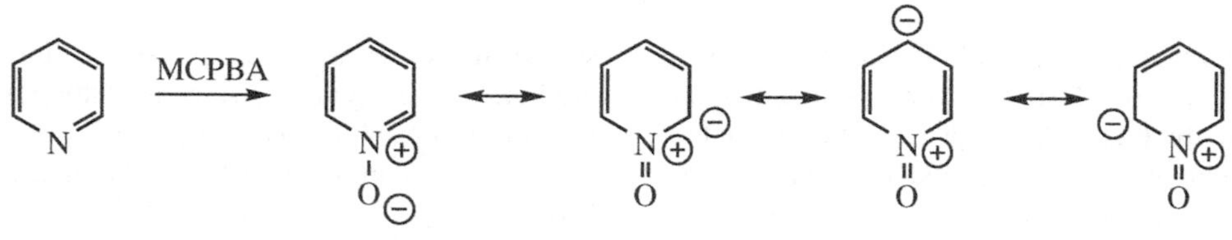

pyridine N-oxide

This improves reactivity towards electrophiles. Consideration of resonance structures shows positions 2, 4, and 6 are now electron rich. Nitration of pyridine N-oxide occurs at C-4; very little 2-nitration is observed. The pyridine compound can then be regenerated by deoxygenation with triphenylphosphine.

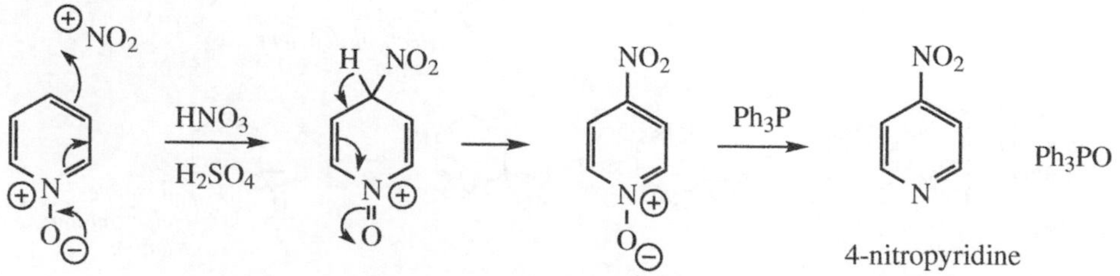

pyridine N-oxide   4-nitropyridine
                                                N-oxide

4-nitropyridine

Pyridine, on the other hand, is more reactive than benzene towards **nucleophilic aromatic substitution**. This is effectively reaction towards the C=N 'imine' function, as described above. Attack is principally at positions 2 and 4, as predictable from resonance structures of reaction intermediates. Attack at the 3 position does not allow the nitrogen to help stabilize the negative charge.

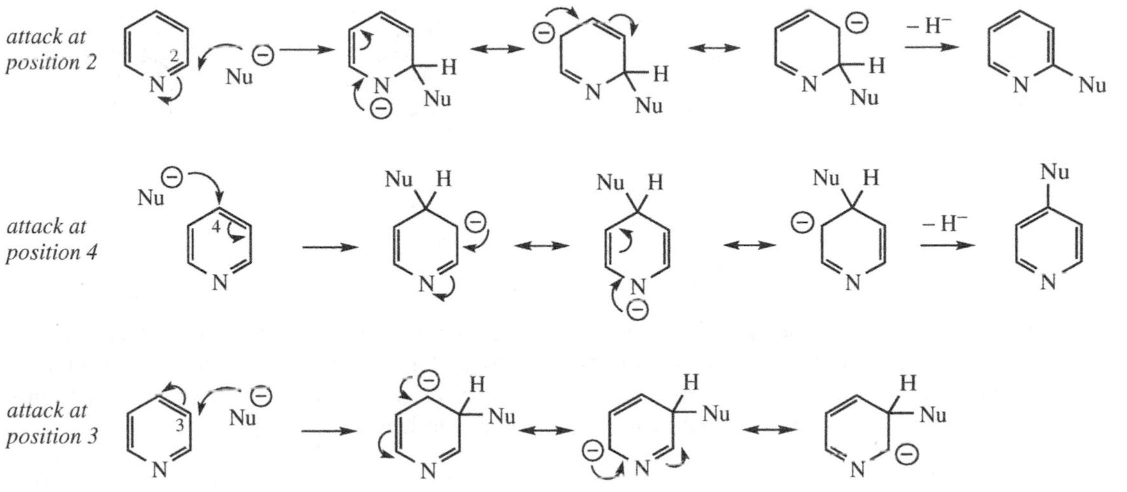

attack at position 2

attack at position 4

attack at position 3

However, for an unsubstituted pyridine, the leaving group to finish off this reaction is hydride, which is a strong base and thus a poor leaving group (see Section 6.1.4). It may be necessary to use an oxidizing agent to function as hydride acceptor to facilitate this type of hydride transfer. Nevertheless, there is a classic example of this process, known as the **Chichibabin reaction**, in which pyridine is converted into 2-aminopyridine through heating with sodium amide.

Chichibabin reaction

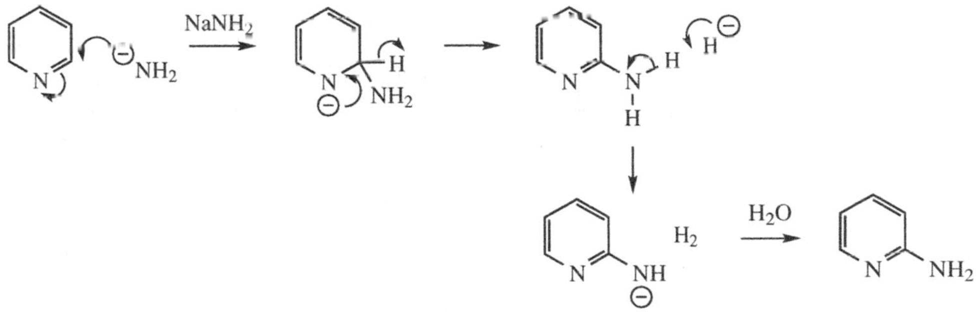

The hydride released appears to abstract a proton from the product since the other product of the reaction is gaseous hydrogen. The aminopyridine anion is finally quenched with water. The product is mainly 2-aminopyridine, probably the result of the enhanced inductive effect on carbons immediately adjacent to the electronegative nitrogen.

It is much more effective to have a better leaving group in the pyridine system. Thus 2- or 4-**chloropyridines** react with a number of nucleophiles to generate substituted products. Note that one can predict from the resonance structures that 3-chloropyridine, despite having a satisfactory leaving group, would not be susceptible to nucleophilic substitution at position 3. It is not possible in the addition anion to share the charge with nitrogen.

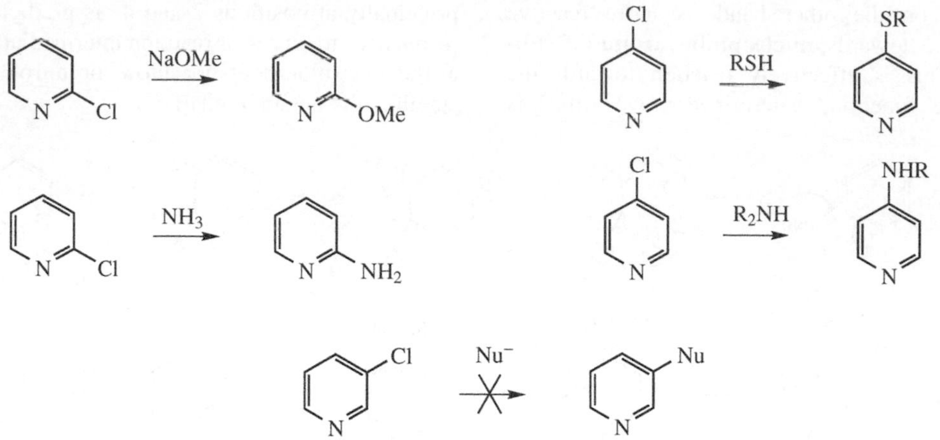

Methylpyridines are called **picolines**. 2-Picoline and 4-picoline may be deprotonated by treatment with a strong base, giving useful anions. The methyl acidity results because of resonance stabilization in the conjugate base, providing an enolate anion analogue. However, pK_a values for 2-picoline (32) and 4-picoline (34) show that they are somewhat less acidic than ketones.

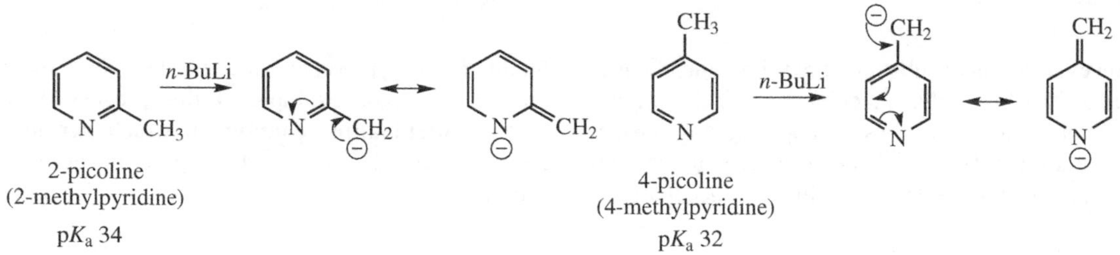

2-picoline
(2-methylpyridine)
pK_a 34

4-picoline
(4-methylpyridine)
pK_a 32

These anions can now be used as nucleophiles in a number of familiar reactions, e.g. S_N2 reactions with alkyl halides, or aldol reactions with carbonyl compounds.

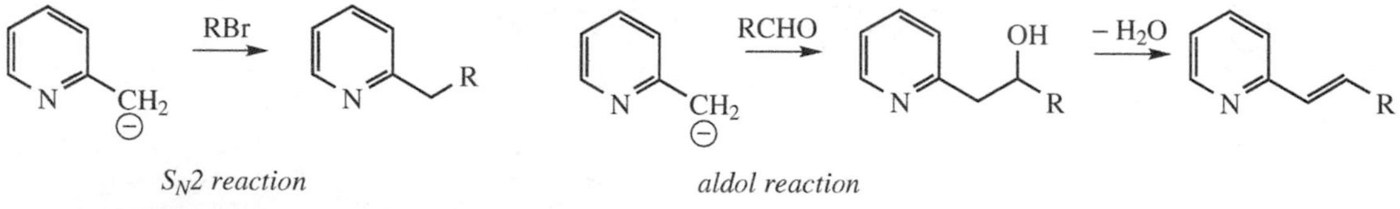

S_N2 reaction

aldol reaction

It is worthwhile here to relate the behaviour of 2-chloropyridine and 2-methylpyridine to carbonyl chemistry. If we consider the pyridine ring as an imine, and therefore a carbonyl analogue (see Section 7.7.1), then with 2-chloropyridine we are seeing reactions that parallel nucleophilic substitution of an acyl halide through an addition–elimination mechanism. With 2-methylpyridine we are seeing typical aldol reactions with activated methyl derivatives.

Box 11.1

Nicotine, nicotinic acid, and nicotinamide

Nicotine is an oily, volatile liquid and is the principal alkaloid found in **tobacco** (*Nicotiana tabacum*). It can be seen to be a combination of two types of heterocycle, i.e. the aromatic pyridine and the non-aromatic *N*-methylpyrrolidine.

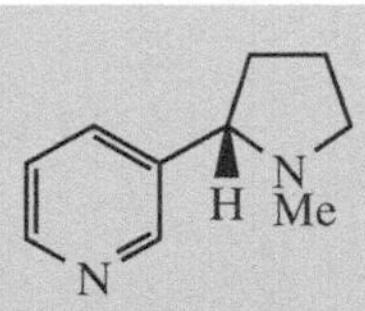

nicotine

In small doses, nicotine can act as a respiratory stimulant, though in larger doses it causes respiratory depression. Nicotine is the only pharmacologically active component in tobacco, and it is highly addictive. On the other hand, tobacco smoke contains a number of highly carcinogenic chemicals formed by incomplete combustion. Tobacco smoking also contributes to atherosclerosis, chronic bronchitis and emphysema, and is regarded as the single most preventable cause of death in modern society. Nicotine, in the form of chewing gum, nasal sprays, or trans-dermal patches, is available for use by smokers who wish to stop the habit.

Nicotine affects the nervous system, interacting with the **nicotinic acetylcholine receptors**, and the tight binding is partially accounted for by the structural similarity between **acetylcholine** and nicotine. Curare-like antagonists also block nicotinic acetylcholine receptors (see Box 6.7). There are other acetylcholine receptors, termed muscarinic, that are triggered by the alkaloid **muscarine**. The tropane alkaloid hyoscyamine (see Box 10.9) binds to **muscarinic acetylcholine receptors**.

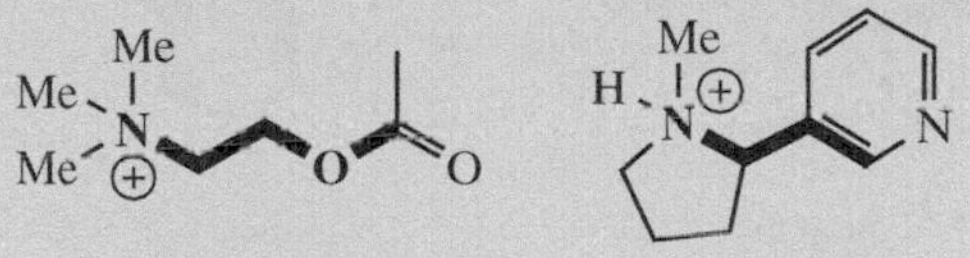

acetylcholine

nicotine
(as conjugate acid)

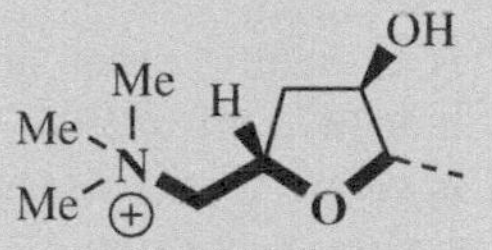

muscarine

Oxidation of nicotine with chromic acid led to the isolation of pyridine-3-carboxylic acid, which was given the trivial name **nicotinic acid**. We now find that nicotinic acid derivatives, especially **nicotinamide**, are biochemically important. Nicotinic acid (niacin) is termed vitamin B_3, though nicotinamide is also included under the umbrella term vitamin B_3 and is the preferred material for dietary supplements. It is common practice to enrich many foodstuffs, including bread, flour, corn, and rice products. Deficiency in nicotinamide leads to pellagra, which manifests itself in diarrhoea, dermatitis, and dementia.

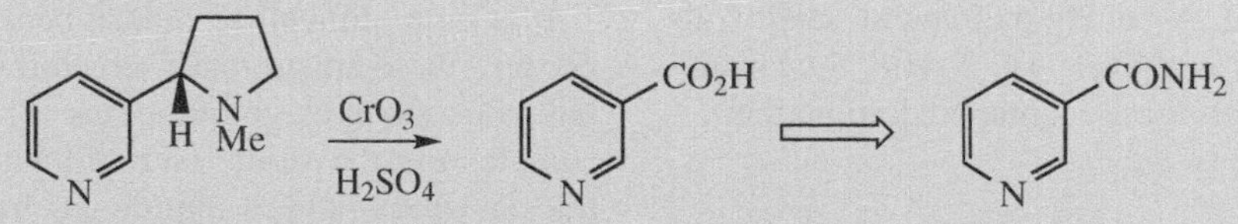

nicotine nicotinic acid nicotinamide

Nicotinic acid and nicotinamide are precursors of the coenzymes **NAD⁺** and **NADP⁺**, which play a vital role in oxidation–reduction reactions (see Box 7.6), and are the most important electron carriers in intermediary metabolism (see Section 15.1.1). We shall look further at the chemistry of NAD⁺ and NADP⁺ shortly (see Box 11.2), but note that, in these compounds, nicotinamide is bound to the rest of the molecule as an *N*-pyridinium salt.

Box 11.1 (continued)

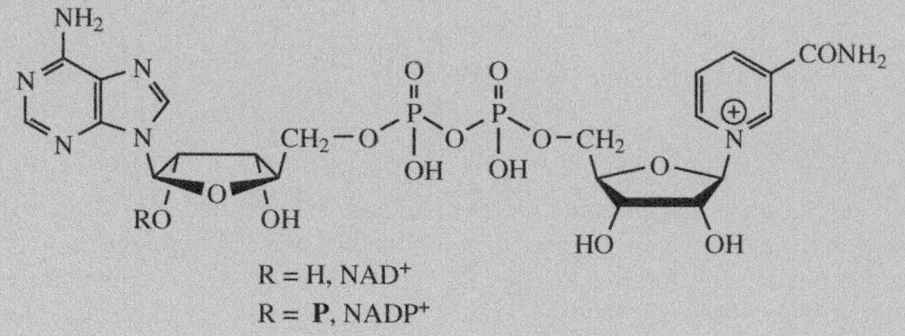

R = H, NAD⁺
R = **P**, NADP⁺

An intriguing feature of nicotinic acid formation in animals is that it is a metabolite produced from the amino acid tryptophan. This means the pyridine ring is actually formed by biochemical modification of the indole fused-ring system (see Section 11.8.2), and, as you might imagine, it involves a substantial sequence of transformations.

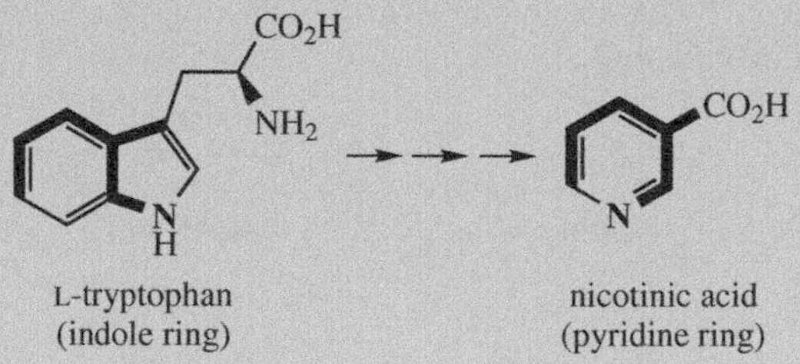

L-tryptophan
(indole ring)

nicotinic acid
(pyridine ring)

11.4.2 Nucleophilic addition to pyridinium salts

The reaction of nucleophiles with pyridinium salts leads to addition, giving dihydropyridines. Attack is normally easier at positions 2 or 6, where the inductive effect from the positively charged nitrogen is greatest; but, if these sites are blocked, then attack occurs at position 4. This is easily predicted from a consideration of resonance structures.

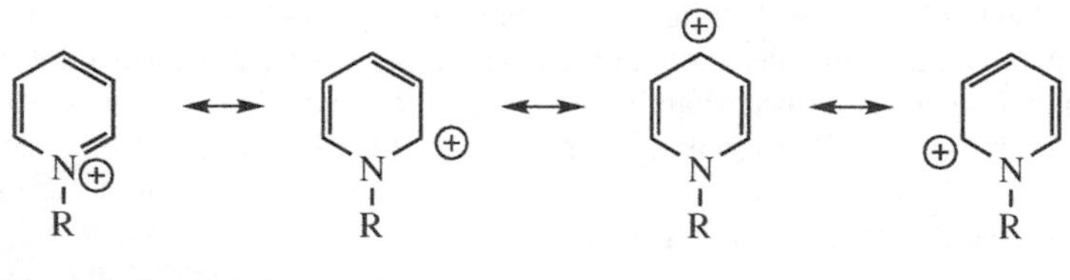

Thus, treatment of *N*-methylpyridinium salts with cyanide produces a mixture of 2- and 4-cyanodihydropyridines, with the 2-isomer predominating.

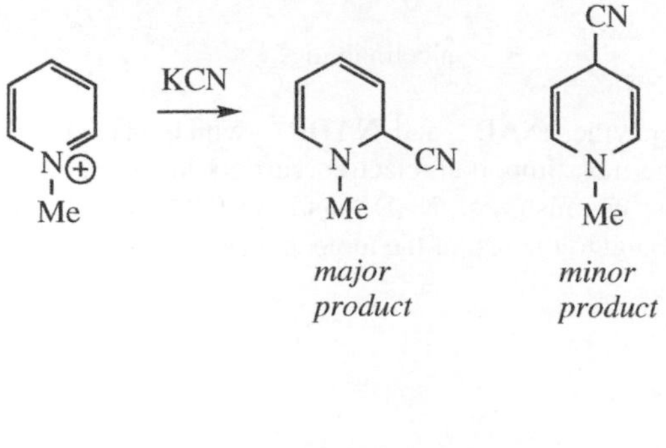

major
product

minor
product

It is quite difficult to reduce benzene or pyridine, because these are aromatic structures. However, partial reduction of the pyridine ring is possible by using complex metal hydrides on pyridinium salts. Hydride transfer from lithium aluminium hydride gives the 1,2-dihydro derivative, as predictable from the above comments. Sodium borohydride under aqueous conditions achieves a double reduction, giving the 1,2,5,6-tetrahydro derivative, because protonation through the unsaturated system is possible. The final reduction step requires catalytic hydrogenation (see Section 9.4.3). The reduction of pyridinium salts is of considerable biological importance (see Box 11.2).

Note the way we can refer to the unsaturated hetero-cycle by considering it as a reduced pyridine, e.g. a dihydropyridine (two double bonds) or tetrahydropy-ridine (one double bond).

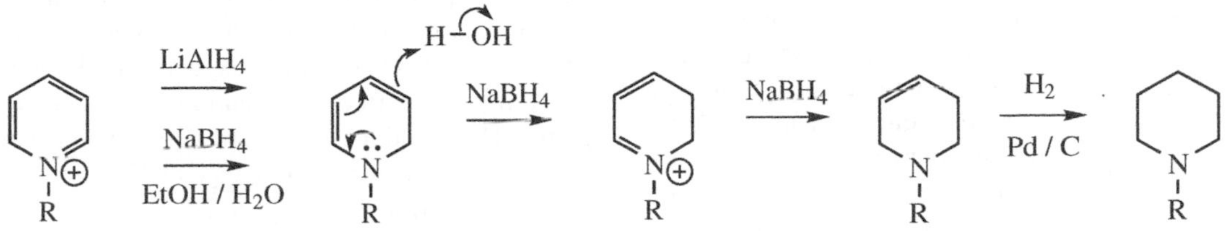

1,2-dihydro derivative

1,2,5,6-tetrahydro derivative

Box 11.2

Nicotinamide adenine dinucleotide: reduction of a pyridinium salt

Nicotinamide adenine dinucleotide (NAD$^+$) is a complex molecule in which a pyridinium salt provides the reactive functional group, hence the superscript + in its abbreviation. NAD$^+$ acts as a biological oxidizing agent, and in so doing is reduced to **NADH (reduced nicotinamide adenine dinucleotide)**. An enzyme, a dehydrogenase, catalyses the process and NAD$^+$ is the cofactor for the enzyme. The reaction can be regarded as directly analogous to the hydride reduction of a pyridinium system to a dihydropyridine, as described above. We have already seen that NADH can act as a reducing agent, delivering the equivalent of hydride to a carbonyl compound (see Box 7.6). In the oxidizing mode, the enzyme is able to extract hydride from the substrate, and use it to reduce the pyridinium salt NAD$^+$, producing the dihydropyridine NADH.

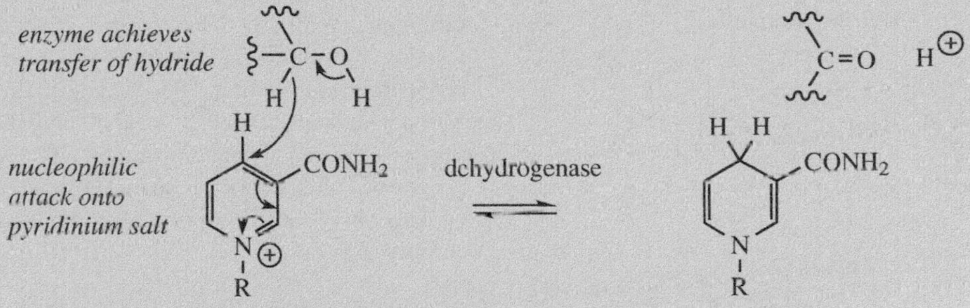

enzyme achieves transfer of hydride

nucleophilic attack onto pyridinium salt

dehydrogenase

NAD$^+$
nicotinamide adenine dinucleotide

NADH
nicotinamide adenine dinucleotide (reduced)

oxidizing agent; can remove hydride

reducing agent; in reverse reaction can supply hydride

The substrate in most reactions of this type is an alcohol, which becomes oxidized to an aldehyde or ketone, e.g. ethanol is oxidized to acetaldehyde. Some reactions employ the alternative phosphorylated cofactor **NADP$^+$**; the phosphate does not function in the oxidation step, but is merely a recognition feature helping to bind the compound to the enzyme. The full structures of NAD$^+$ and NADP$^+$ are shown in Box 11.1.

Note that attack of hydride is at position 4 of the dihydropyridine ring. This is controlled by the enzyme, but it is also probably the only site accessible, since the rest of the complex molecule hinders approach to positions 2 and 6.

11.4.3 Tautomerism: pyridones

The pyridine ring system may carry substituents, just as we have seen with benzene rings. We have encountered a number of such derivatives in the previous section. Hydroxy or amino heterocycles, however, may sometimes exist in **tautomeric forms**. We have met the concept of tautomerism primarily with carbonyl compounds, and have seen the isomerization of keto and enol tautomers (see Section 10.1). In certain cases, e.g. 1,3-dicarbonyl compounds, the enol form is a major component of the equilibrium mixture. In the example shown, liquid acetylacetone contains about 76% of the enol tautomer.

acetylacetone

keto form   *enol form*

2-Hydroxy- and 4-hydroxy-pyridines are in equilibrium with their tautomeric 'amide' structures containing a carbonyl. These tautomers are called **2-pyridone** and **4-pyridone** respectively. This type of tautomerism does not occur with the corresponding benzene derivative phenol, since it would destroy the stabilization conferred by aromaticity.

2-hydroxypyridine   2-pyridone

4-hydroxypyridine   4-pyridone

phenol

So why can tautomerism occur with a hydroxypyridine? It is because 2-pyridone and 4-pyridone still retain aromaticity, with the nitrogen atom donating its lone pair electrons to the aromatic sextet. This is more easily seen in the resonance structures, and should remind us of the resonance stabilization in

amides (see Section 4.5.4). This was used to explain why amides are very weak bases. Note that such resonance forms of pyridones are favourable, having a positive charge on the nitrogen and a negative charge on the more electronegative oxygen. In addition, the structure gains further stabilization from the carbonyl group. The pyridone forms are very much favoured over the phenol forms, and typical C=O peaks are seen in the infrared (IR) spectra.

2-pyridone

resonance stabilization in amides

Note, however, that we cannot get the same type of tautomerism with **3-hydroxypyridine**. In polar solvents, 3-hydroxypyridine may adopt a dipolar zwitterionic form. This may look analogous to the previous structure, but appreciate that there is a difference. With 3-hydroxypyridine, the zwitterion is a major contributor, and arises simply from acid–base properties (see Section 4.11.3). The hydroxyl group acts as an acid, losing a proton, and the nitrogen acts as a base, gaining a proton. The structure from 2-pyridone is a minor resonance form that helps to explain charge distribution; the compound is almost entirely 2-pyridone.

3-hydroxypyridine   *zwitterionic form*

Like amides, 2- and 4-pyridones are also very weak bases, much weaker than amines. Like amides, they actually protonate on oxygen rather than nitrogen (see Section 4.5.4). This further emphasizes that the nitrogen lone pair is already in use and not available for protonation. On the other hand, the N–H can readily be deprotonated; pyridones are appreciably acidic (pK_a about 11). The conjugate base benefits from considerable resonance stabilization, both via the carbonyl group (compare amides, Section 10.7) and also via the ring. The main contributors will be those structures in which charge is associated with the electronegative N or O atoms.

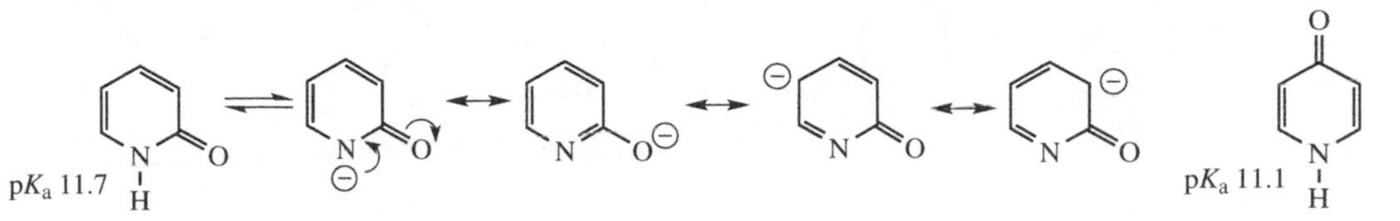

2-pyridone   pKa 11.7

4-pyridone   pKa 11.1

It is thus possible to *N*-alkylate a pyridone by exploiting its acidity. As with enolate anions (see Section 10.2), there is the possibility for *O*-alkylation and *N*-alkylation. Although it depends upon the conditions and the nature of the electrophile, carbon electrophiles tend to react on nitrogen rather than oxygen.

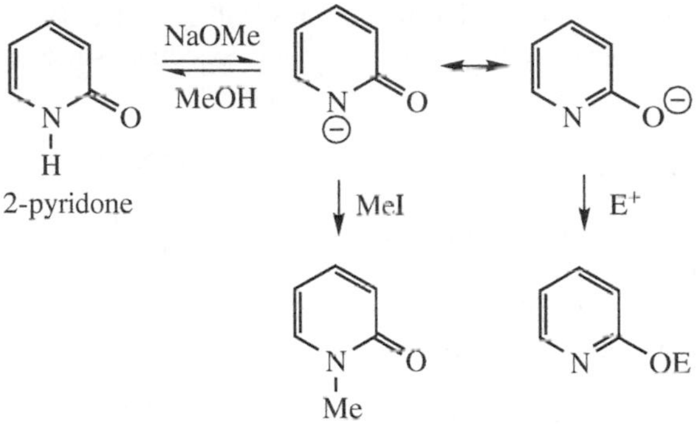

2-pyridone

A useful reaction of pyridones is conversion into chloropyridines by the use of phosphorus oxychloride $POCl_3$ in the presence of PCl_5. This appears to react initially on oxygen, forming a good leaving group, which is subsequently displaced by chloride.

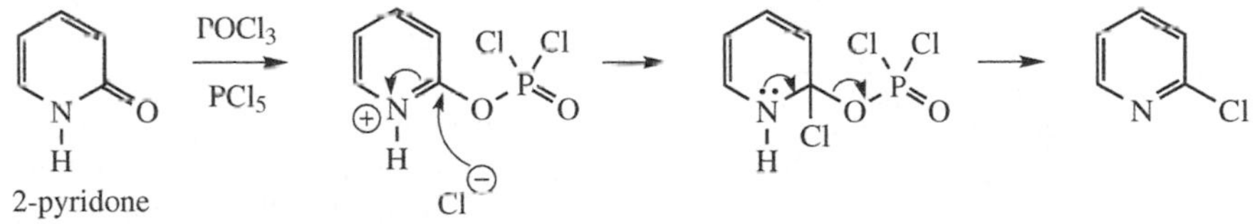

2-pyridone

Aminopyridines are also potentially tautomeric with corresponding imino forms.

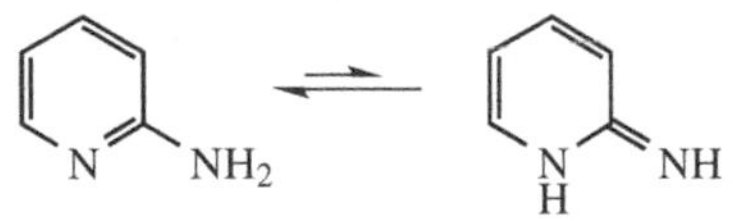

2-aminopyridine   *imino form*

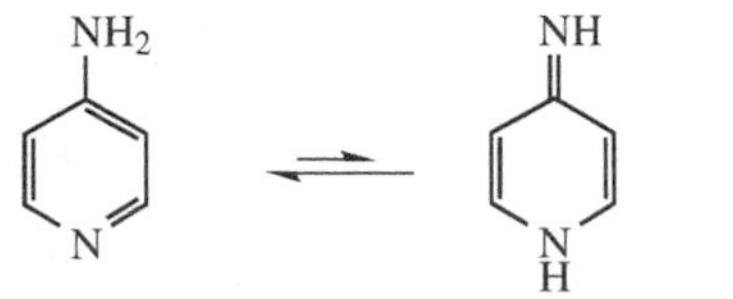

4-aminopyridine   *imino form*

However, 2-aminopyridine and 4-aminopyridine exist almost entirely as the amino tautomers – indeed, we have just seen 2-aminopyridine as a product of the Chichibabin reaction. Which tautomer is preferred for hydroxy and amino heterocycles is not always easily explained; but, as a generalization, we find that the oxygen derivatives exist as carbonyl tautomers and amino heterocycles favour the amino tautomers. At this stage, we should just register the potential for tautomerism in aminopyridines; we shall see important examples with other heterocycles (see Section 11.6.2).

Aminopyridines protonate on the ring nitrogen, and 4-aminopyridine is a stronger base than 2-aminopyridine.

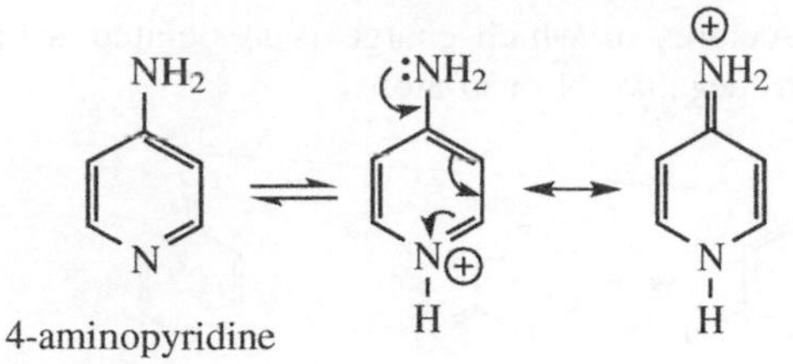

4-aminopyridine

pK_a 9.1

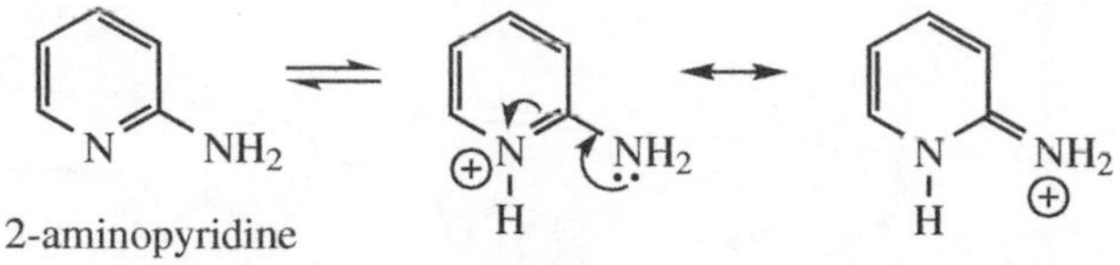

2-aminopyridine

pK_a 6.8

This may be rationalized from a consideration of resonance in the conjugate acids. The conjugate acids from ring protonation benefit from charge delocalization, which is greater in 4-aminopyridinium that in 2-aminopyridinium. This type of delocalization is not possible in 3-aminopyridinium; 3-aminopyridine (pK_a 6.0) is the weakest base of the three aminopyridines, and has basicity more comparable to that of pyridine (pK_a 5.2).

11.4.4 Pyrylium cation and pyrones

The **pyrylium cation** is isoelectronic with pyridine: it has the same number of electrons and, therefore, we also have aromaticity. Oxygen is normally divalent and carries two lone pairs. If we insert oxygen into the benzene ring structure, then it follows that, by having one electron in a p orbital contributing to the aromatic sextet, there is a lone pair in an sp^2 orbital,

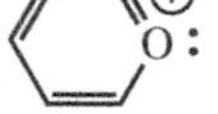

pyrylium cation

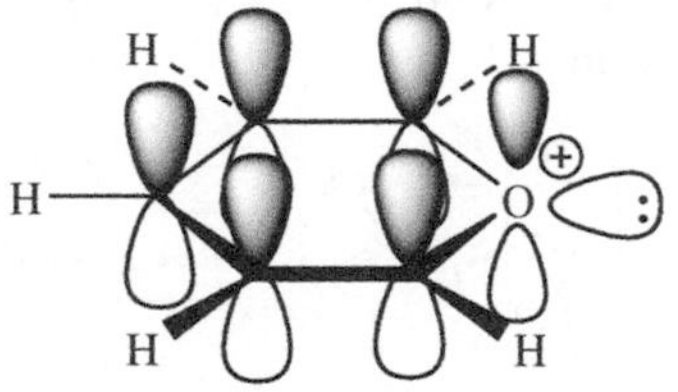

and the remaining electron needs to be removed, hence the pyrylium cation. However, oxygen tolerates a positive charge less readily than nitrogen, and aromatic stabilization is less than with pyridine.

Pyrones are oxygen analogues of pyridones, and potentially aromatic. However, there is little evidence that the dipolar resonance forms of either 2-pyrone or 4-pyrone make any significant contribution. Their chemical behaviour suggests they should be viewed more as conjugated lactones (2-pyrones) or vinylogous lactones (4-pyrones) rather than aromatic systems, since many reactions lead to ring opening.

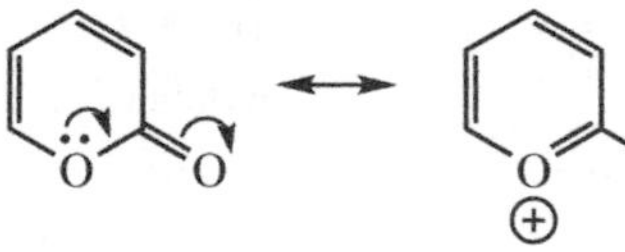

2-pyrone

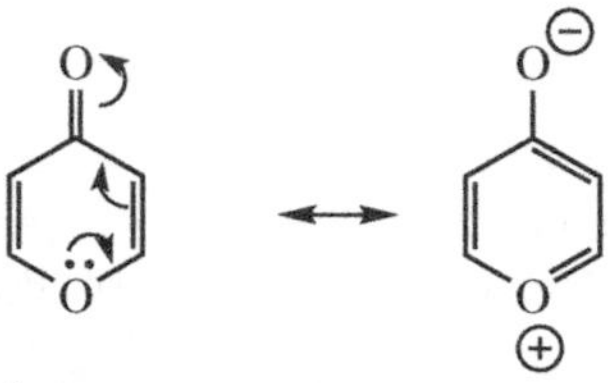

4-pyrone

Flavonoids are natural phenolic systems containing pyrylium and pyrone rings, and provide the most prominent examples. We have met some of these systems under antioxidants (see Box 9.2). **Coumarins** contain a 2-pyrone system. Note that all of these compounds are fused to a benzene ring and are strictly benzopyran or benzopyrylium systems.

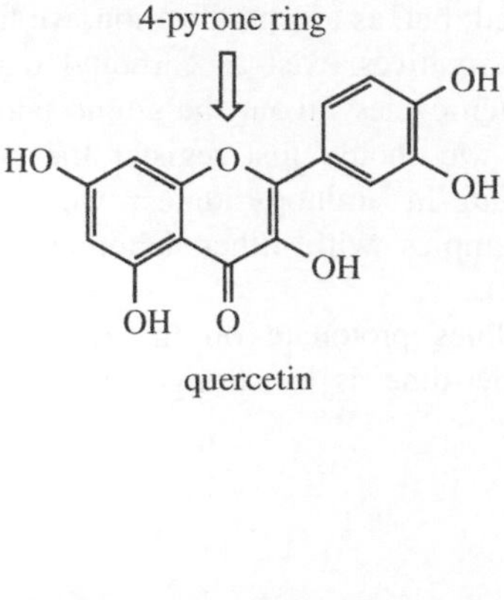

4-pyrone ring

quercetin

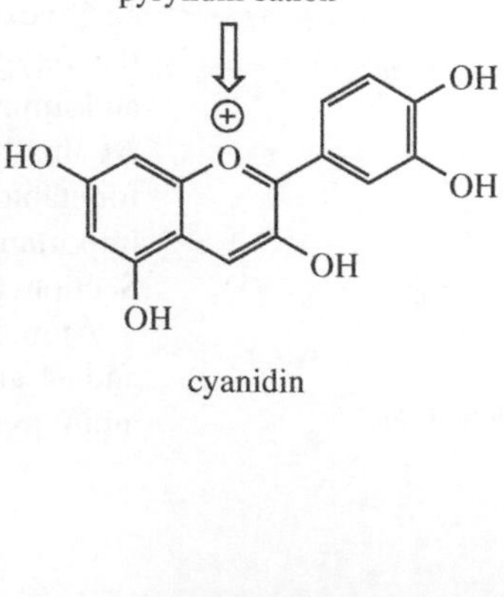

pyrylium cation

cyanidin

2-pyrone ring

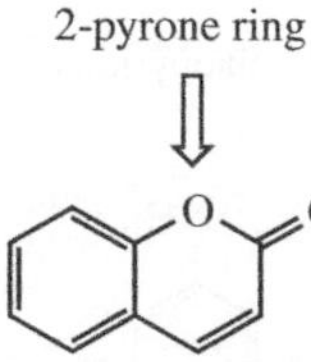

coumarin

Box 11.3

Dicoumarol and warfarin

Warfarin provides us with a slightly incongruous state of affairs: it is used as a drug and also as a rat poison. It was developed from a natural product, **dicoumarol**, and provides us with a nice example of how pyrone chemistry resembles that of conjugated lactones rather than aromatic systems.

Many plants produce coumarins; **coumarin** itself is found in sweet clover and contributes to the smell of new-mown hay. However, if sweet clover is allowed to ferment, oxidative processes initiated by the microorganisms lead to the formation of 4-hydroxycoumarin rather than coumarin. 4-Hydroxycoumarin then reacts with formaldehyde, also produced via the microbial degradative reactions, and provides dicoumarol.

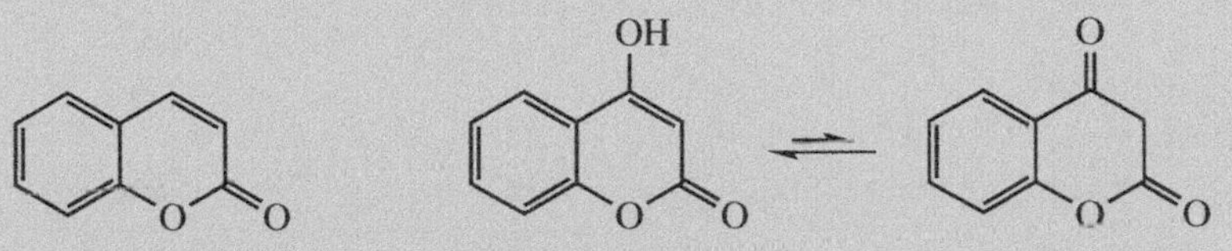

coumarin   4-hydroxycoumarin   *diketo tautomer*

4-Hydroxycoumarin can be considered as an enol tautomer of a 1,3-dicarbonyl compound; conjugation with the aromatic ring favours the enol tautomer. This now exposes its potential as a nucleophile. Whilst we may begin to consider enolate anion chemistry, no strong base is required and we may formulate a mechanism in which the enol acts as the nucleophile, in a simple aldol reaction with formaldehyde. Dehydration follows and produces an unsaturated ketone, which then becomes the electrophile in a Michael reaction (see Section 10.10). The nucleophile is a second molecule of 4-hydroxycoumarin.

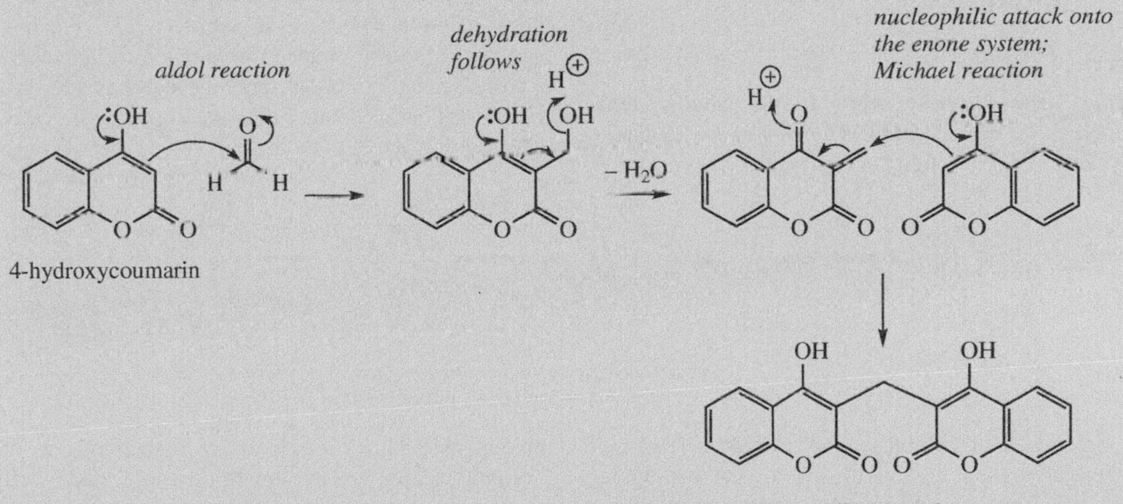

dicoumarol

Animals fed spoiled sweet clover were prone to fatal haemorrhages. The cause was traced to the presence of dicoumarol. This compound interferes with the effects of vitamin K in blood coagulation, the blood loses its ability to clot, and minor injuries can lead to severe internal bleeding. Synthetic dicoumarol has been used as an oral blood anticoagulant in the treatment of thrombosis, where the risk of blood clots becomes life threatening. It has since been superseded by **warfarin**, a synthetic development based on the natural product.

Box 11.3 (continued)

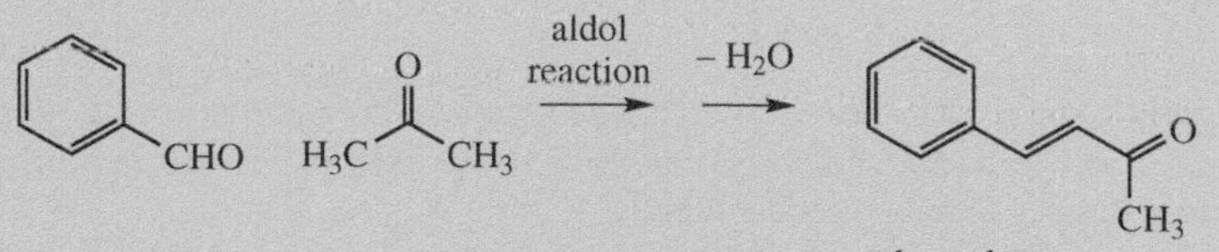

benzalacetone

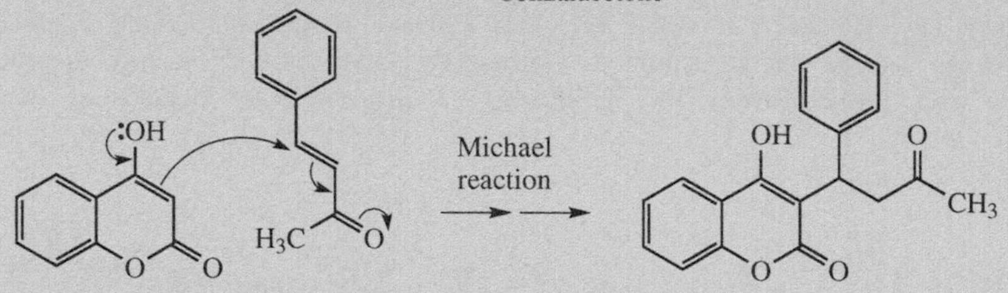

4-hydroxycoumarin   warfarin

Warfarin was initially developed as a rodenticide, and has been widely employed for many years as the first-choice agent, particularly for destruction of rats. After consumption of warfarin-treated bait, rats die from internal haemorrhage. Warfarin is synthesized from 4-hydroxycoumarin by a Michael reaction on benzalacetone, again exploiting the nucleophilicity of the hydroxypyrone. Benzalacetone is the product from an aldol reaction between benzaldehyde and acetone (see Section 10.3).

11.5 Five-membered aromatic heterocycles

11.5.1 Pyrrole

Pyrrole (azacyclopentadiene) is the ring system obtained if we replace the CH_2 group of cyclopentadiene with NH. Although cyclopentadiene is certainly not aromatic, pyrrole has aromatic character because nitrogen contributes two electrons, its lone pair, to the π-electron system (see Section 11.3). We have also noted from resonance forms that nitrogen carries a partial positive charge, and the carbons are electron rich. This is stronger than the opposing inductive effect.

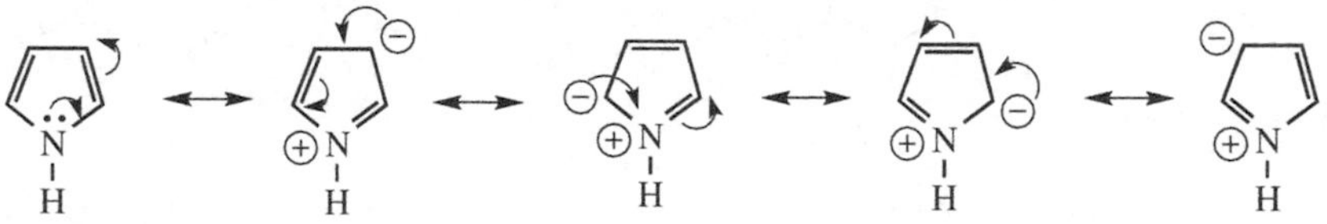

pyrrole

resonance forms predict nitrogen carries partial positive charge, and carbons are electron rich

This is reflected in the basicity of pyrrole. Pyrrole is a particularly weak base, with pK_a of the conjugate acid -3.8. First, we should realize that protonation of pyrrole will not occur on nitrogen: nitrogen has already used up its lone pair by contributing to the aromatic sextet, so protonation would necessarily destroy aromaticity.

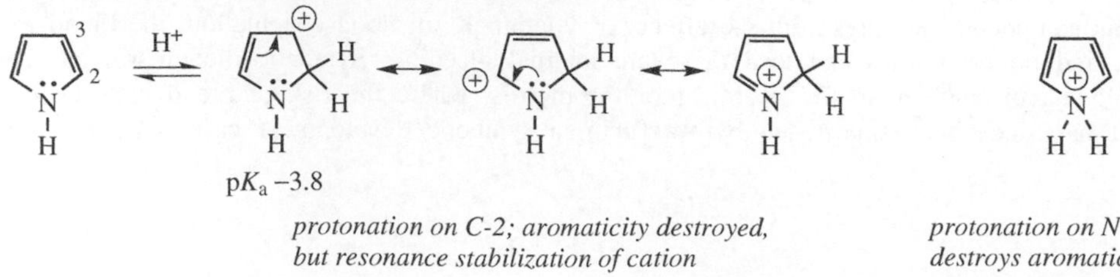

$pK_a -3.8$

protonation on C-2; aromaticity destroyed, but resonance stabilization of cation

protonation on N not favoured; destroys aromaticity

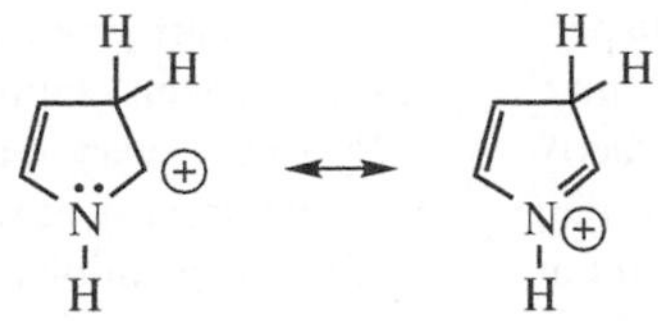

protonation on C-3 gives less resonance forms

It is possible to protonate pyrrole using a strong acid, but even then the protonation occurs on C-2 and not on the nitrogen. Although this still destroys aromaticity, there is some favourable resonance stabilization in the conjugate acid. Protonation on C-3 is not as favourable, in that there is less resonance stabilization in the conjugate acid. It turns out that, as opposed to acting as a base, pyrrole is potentially an acid (pK_a 17.5); it is not a particularly strong acid, but stronger than we might expect for a secondary amine system (pK_a about 36). This is because the anion formed by losing the proton from nitrogen has a negative charge on the relatively electronegative nitrogen, but maintains its aromaticity. Unlike in pyrrole, the anion resonance structures do not involve charge separation.

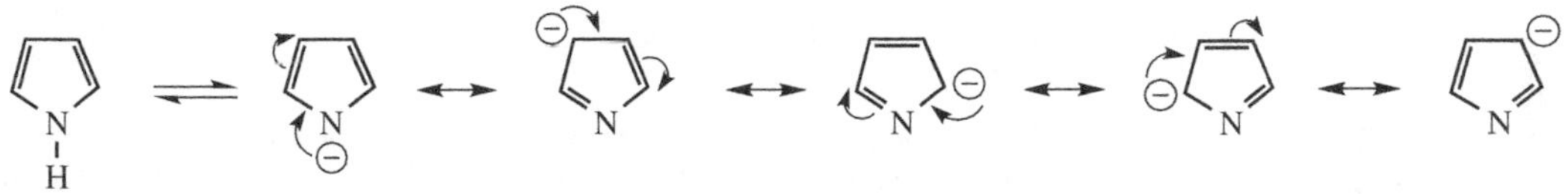

pK_a 17.5

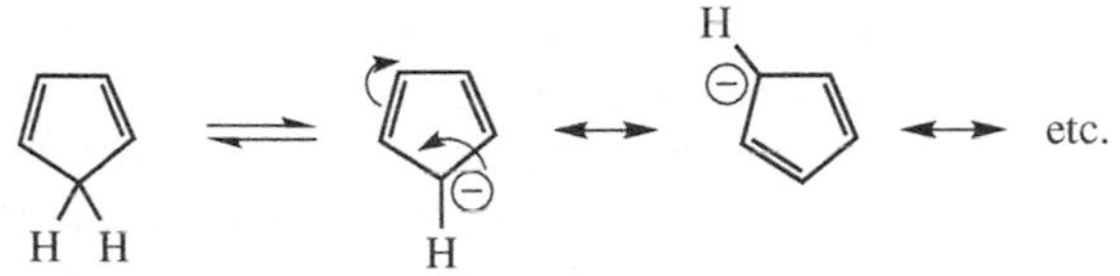

pK_a 16

cyclopentadiene

It is appropriate here to compare the acidity of **cyclopentadiene**, which has pK_a 16, considerably more acidic than most hydrocarbon systems and comparable to water and alcohols. Removal of one of the CH_2 protons from the non-aromatic cyclopentadiene generates the cyclopentadienyl anion. This anion has an aromatic sextet of electrons, two electrons being contributed by the negatively charged carbon (see Section 2.9.3).

The charge distribution in pyrrole leads us to predict that it will react readily with electrophiles; or,

put another way, pyrrole will behave as a nucleophile. This is indeed the case, and the ease of **electrophilic substitution** contrasts with the behaviour of pyridine above, where charge distribution favoured nucleophilic attack on to the heterocycle. Although our resonance description of pyrrole shows negative charge can be dispersed to any ring carbon, pyrrole reacts with electrophiles preferentially at C-2 rather than C-3, unless the 2-position is already substituted.

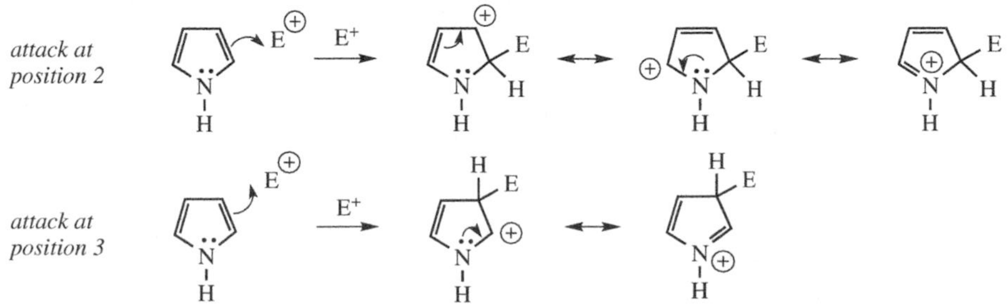

attack at position 2

attack at position 3

This may reflect that there is more charge dispersion in the addition cation from attack at C-2 than there is from attack at C-3. This is, of course, exactly the same argument as used above for *C*-protonation; protonation (pyrrole acting as a base) also occurs at C-2. As with protonation, electrophiles do not react at the nitrogen centre.

Pyrrole is very reactive towards electrophiles. For example, treatment with bromine leads to substitution of all four positions.

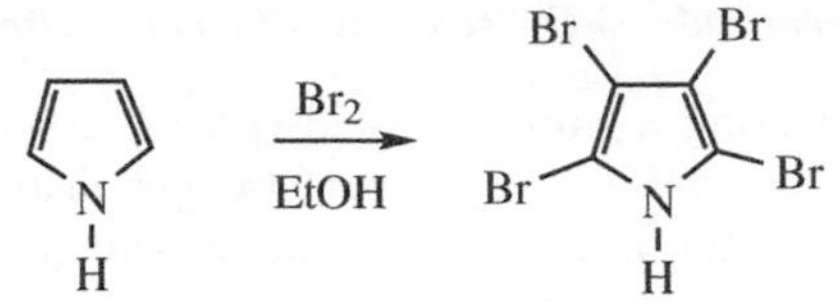

Indeed, it is often difficult to control electrophilic attack so that monosubstitution occurs. A further problem is that pyrrole polymerizes in the presence of strong acids and Lewis acids, so that typical electrophilic reagents, e.g. $HNO_3-H_2SO_4$ and $RCOCl-AlCl_3$, cannot be used. Polymerization involves the conjugate acid functioning as the electrophile.

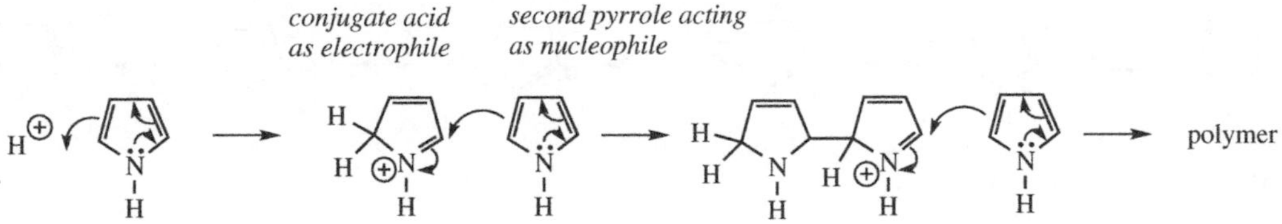

To achieve useful monosubstitution it is necessary to employ relatively mild conditions, often without a catalyst. **Nitration** may be accomplished with the reagent acetyl nitrate, giving mainly 2-nitropyrrole. Acetyl nitrate is formed by reacting acetic anhydride with fuming nitric acid. Since the other product is acetic acid, there is no strong mineral acid present to cause polymerization. It is also possible to synthesize 2-acetylpyrrole simply by using acetic anhydride, and pyrrole can act as the nucleophile in the **Mannich reaction** (see Section 10.6).

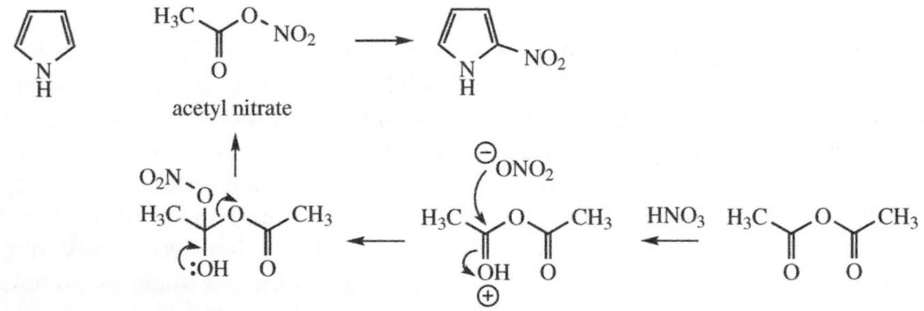

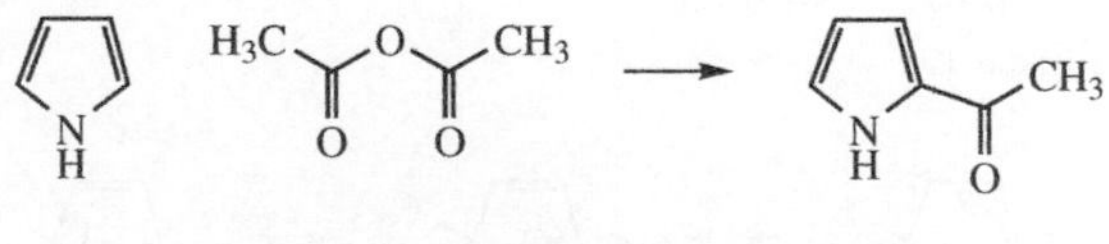

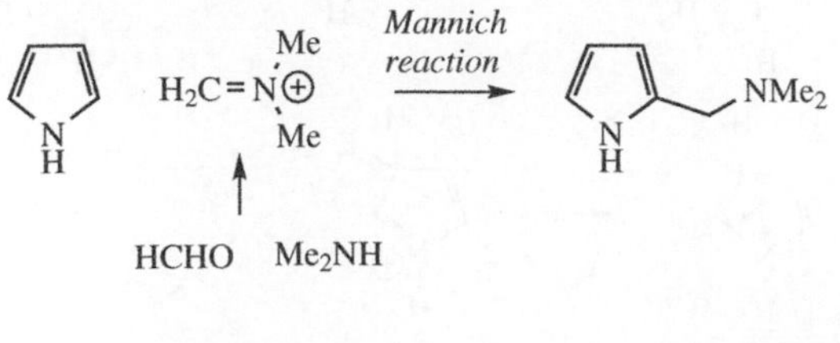

Although pyrrole is a weak acid, it can be deprotonated by using a strong base, e.g. sodium hydride, and the anion can be used in typical nucleophilic reactions. This allows simple transformations such as

N-alkylation, *N*-acylation, and *N*-sulfonation. Note particularly that whereas pyrrole reacts with electrophiles at carbon, usually C-2, the pyrrole anion reacts at the nitrogen atom.

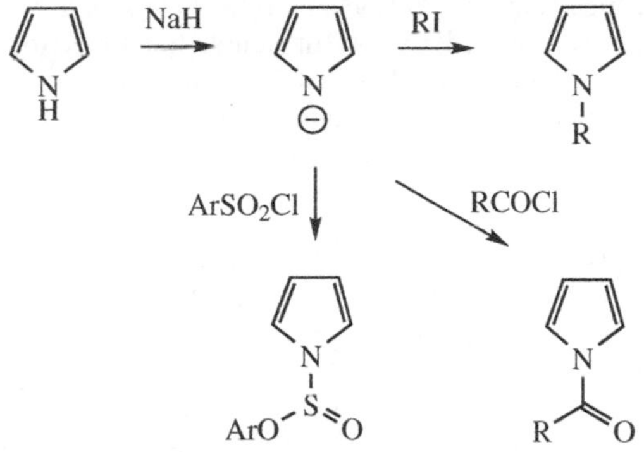

Box 11.4

Porphyrins and corrins

Pyrrole reacts with aldehydes and ketones under acidic conditions to form polymeric compounds. In many cases these are intractable resin-like materials; however, with appropriate carbonyl compounds, interesting cyclic tetramers can be formed in very good yields.

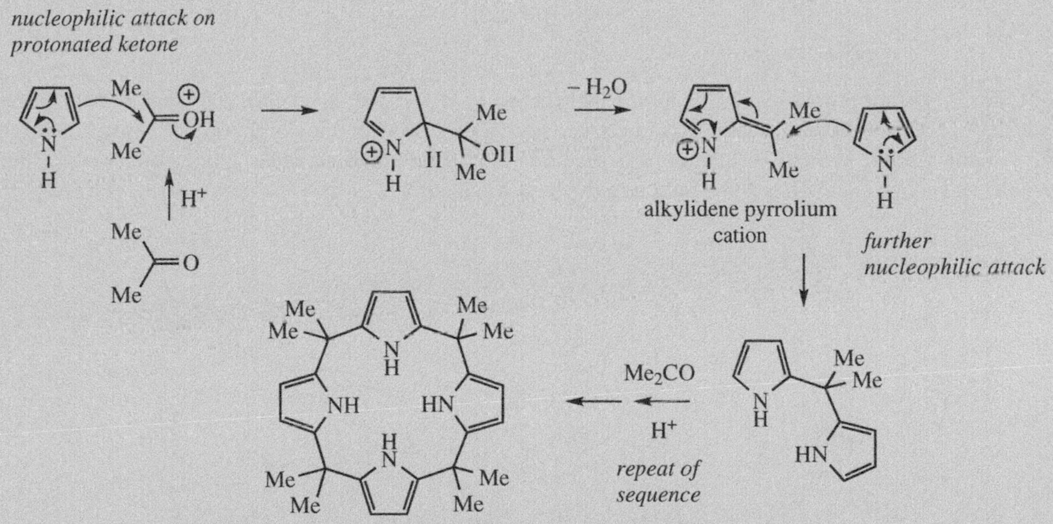

Thus, pyrrole and acetone react as shown above. This involves pyrrole acting as the nucleophile to attack the protonated ketone in an aldol-like reaction. This is followed by elimination of water, facilitated by the acidic conditions. This gives an intermediate alkylidene pyrrolium cation, a highly reactive electrophile that reacts with another molecule of nucleophilic pyrrole. We then have a repeat sequence of reactions, in which further acetone and pyrrole molecules are incorporated. The presence of the two methyl substituents from acetone forces the growing polymer to adopt a planar array, and this eventually leads to a cyclic tetramer, the terminal pyrrole attacking the alkylidene pyrrolium cation at the other end of the chain.

Box 11.4 (continued)

The cyclic tetramer shown is structurally related to the **porphyrins**. The basic ring system in porphyrins is **porphin**, which is more oxidized than the tetramer from the pyrrole–acetone reaction, and has four pyrrole rings linked together by methine (–CH=) bridges. One of the features of porphin is that it is aromatic. It contains an aromatic 18 π-electron system, which conforms to Hückel's rule, $4n + 2$ with $n = 4$. The aromatic ring weaves around the porphin structure, and is composed entirely of double-bond electrons; it does not incorporate any nitrogen lone pairs. Note that we can draw Kekulé-like resonance structures for porphin. Do not be confused by seeing alternative structures with the double bonds arranged differently.

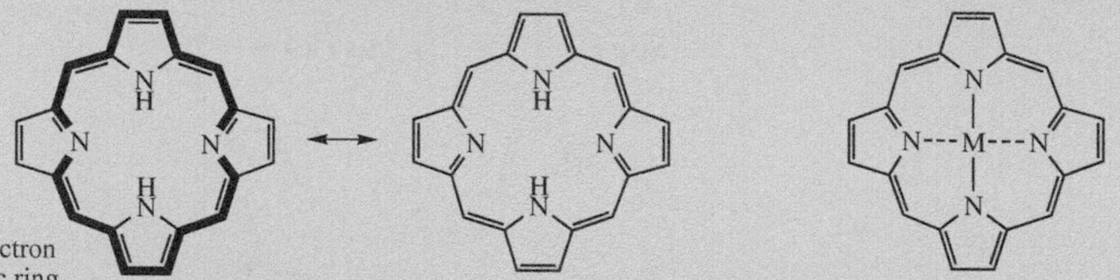

18 π electron
aromatic ring

porphin

metal–porphin complex

Porphyrin rings are formed in nature by a process that is remarkably similar to that shown above. Though the sequence contains some rather unusual features, the coupling process also involves nucleophilic attack on to an alkylidene pyrrolium cation. This may be generated from the precursor porphobilinogen by elimination of ammonia.

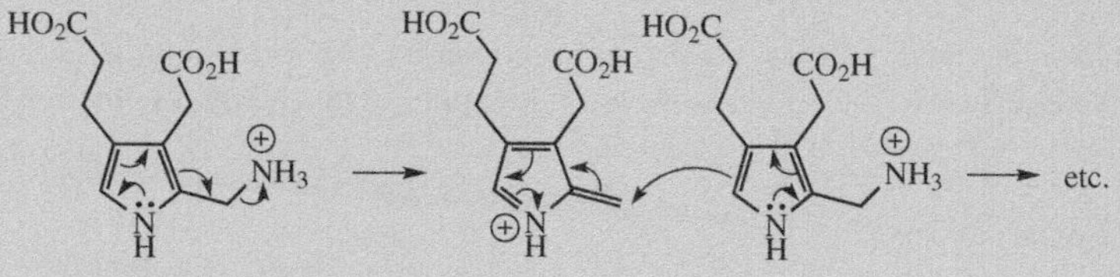

porphobilinogen

One of the important properties of porphyrins is that they complex with divalent metals, the pyrrole nitrogens being ideally spaced to allow this. Of vital importance to life processes are the porphyrin derivatives chlorophyll and haem. **Chlorophyll** (actually a mixture of structurally similar porphyrins; chlorophyll a is shown) contains magnesium, and is, of course, the light-gathering pigment in plants that permits photosynthesis.

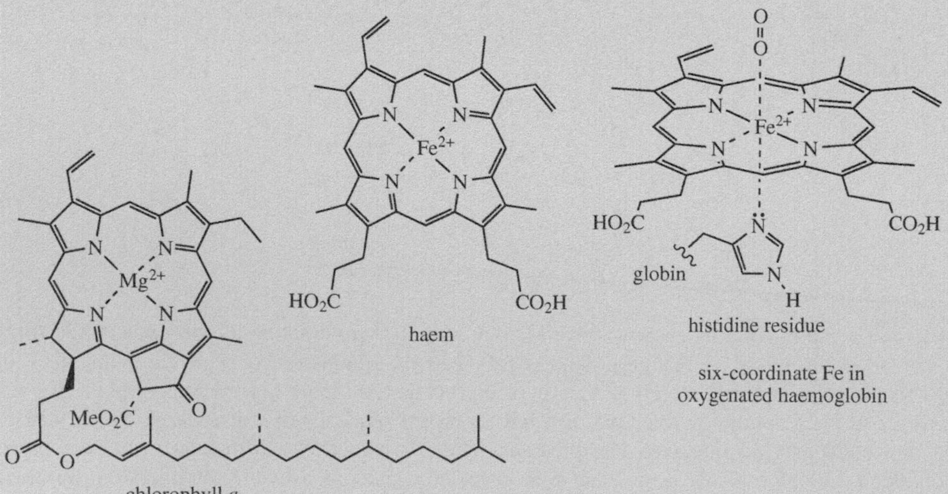

chlorophyll a

haem

globin

histidine residue

six-coordinate Fe in
oxygenated haemoglobin

Plants and a few microorganisms use photosynthesis to produce organic compounds from inorganic materials found in the environment, whereas other organisms, such as animals and most microorganisms, rely on obtaining their raw materials in their diet, e.g. by consuming plants. **Haemoglobin**, the red pigment in blood, serves to carry oxygen from the lungs to other parts of the body tissue. This material is made up of the porphyrin haem and the water-soluble protein globin. The **haem** component shares many structural features with chlorophyll, one of the main differences being the use of Fe^{2+} as the metal rather than Mg^{2+} as in chlorophyll. The oxygen-carrying ability of haemoglobin involves a six-coordinate iron, with an imidazole ring from the protein (a histidine residue) occupying the sixth position.

Porphyrin rings containing iron are also a feature of the **cytochromes**. Several cytochromes are responsible for the latter part of the electron transport chain of oxidative phosphorylation that provides the principal source of ATP for an aerobic cell (see Section 15.1.2). Their function involves alternate oxidation–reduction of the iron between Fe^{2+} (reduced form) and Fe^{3+} (oxidized form). The individual cytochromes vary structurally, and their classification (a, b, c, etc.) is related to their absorption maxima in the visible spectrum. They contain a haem system that is covalently bound to protein through thiol groups.

cytochrome c

typical oxidations achieved by cytochrome P-450-dependent mono-oxygenases:

aliphatic hydroxylation

aromatic hydroxylation

epoxidation of alkene

An especially important example is cytochrome P-450, a coenzyme of the so-called **cytochrome P-450-dependent mono-oxygenases**. These enzymes are frequently involved in biological hydroxylations, either in biosynthesis, or in the mammalian detoxification and metabolism of foreign compounds such as drugs. Cytochrome P-450 is named after its intense absorption band at 450 nm when exposed to CO, which is a powerful inhibitor of these enzymes. A redox change involving the Fe atom allows binding and the cleavage of molecular oxygen to oxygen atoms, with subsequent transfer of one atom to the substrate. In most cases, NADPH features as hydrogen donor, reducing the other oxygen atom to water. Many such systems have been identified, capable of hydroxylating aliphatic or aromatic systems, as well as producing epoxides from alkenes.

A related ring system containing four pyrroles is seen in vitamin B_{12}, but this has two pyrroles directly bonded, and is termed a **corrin** ring. **Vitamin B_{12}** is extremely complex, and features six-coordinate Co^{2+} as the metal component. Four of the six coordinations are provided by the corrin ring nitrogens, and a fifth by a dimethylbenzimidazole moiety. The sixth is variable, being cyano in cyanocobalamin (vitamin B_{12}), but other anions may feature in vitamin B_{12} analogues. Vitamin B_{12} appears to be entirely of microbial origin, with intestinal flora contributing towards human dietary needs. Insufficient vitamin B_{12} leads to pernicious anaemia, a disease that results in nervous disturbances and low production of red blood cells.

Box 11.4 (continued)

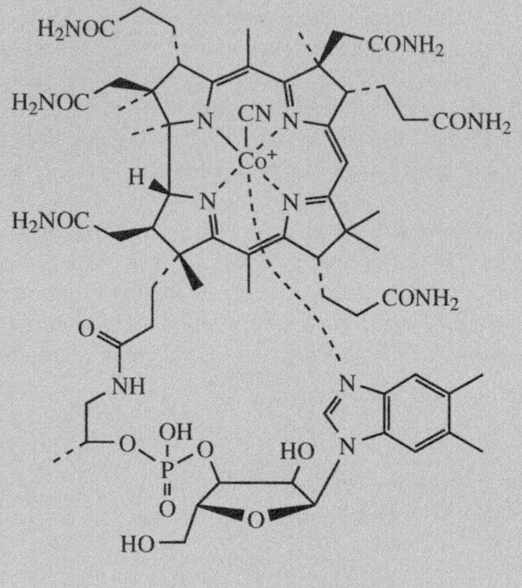

vitamin B$_{12}$ (cyanocobalamin)

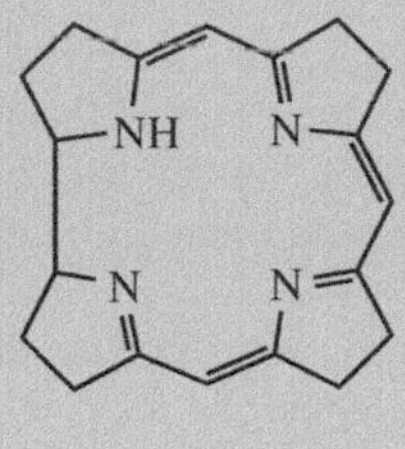

corrin ring system

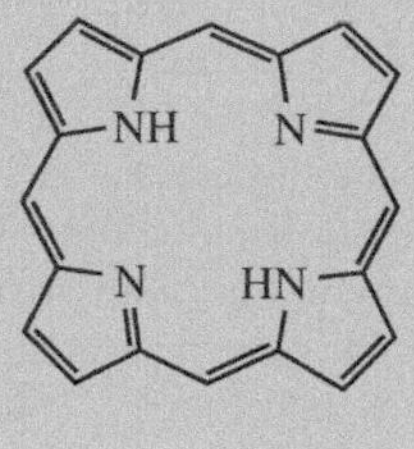

porphyrin ring system

11.5.2 Furan and thiophene

Furan and **thiophene** are the oxygen and sulfur analogues respectively of pyrrole. Oxygen and sulfur contribute two electrons to the aromatic sextet, but still retain lone pair electrons. There is one significant difference, however, in that oxygen uses electrons from a $2p$ orbital, whereas the electrons that sulfur contributes originate from a $3p$ orbital. In the case

furan

of thiophene, this reduces orbital overlap with the carbon $2p$ orbitals.

Both compounds are thus aromatic, and their chemical reactivity reflects what we have learnt about pyrrole. The most typical reaction is electrophilic substitution. However, we find that pyrrole is more reactive than furan towards electrophiles and thiophene is the least reactive; all are more reactive than benzene. This relates to the relative stability of positive charges located on nitrogen, oxygen and sulfur. We used similar electronegativity reasoning to explain the relative basic strengths of nitrogen, oxygen and sulfur derivatives (see Section 4.5.1). Furan is also the 'least aromatic' of the three, i.e. it has the least resonance stabilization, and undergoes many reactions in which the aromatic character is lost, e.g. addition reactions or ring opening.

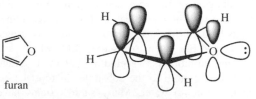

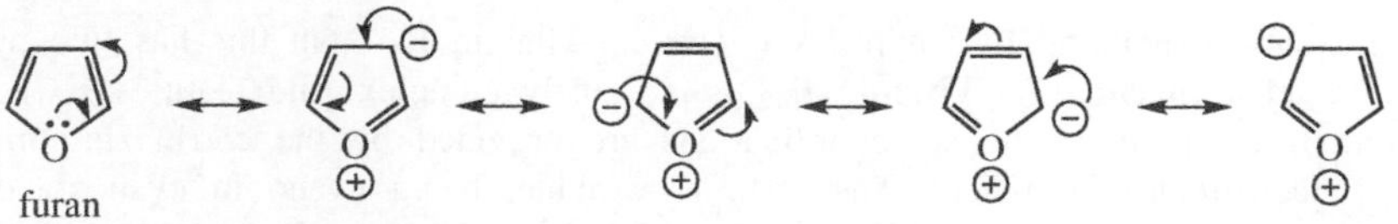

furan

Note that the dipoles of furan and thiophene are opposite in direction to that in pyrrole. In furan and thiophene, there is a greater inductive effect opposing the resonance effect, whereas in pyrrole the resonance

contribution was greater (see Section 11.3). In the non-aromatic analogues, the heteroatom is at the negative end of the dipole in all cases.

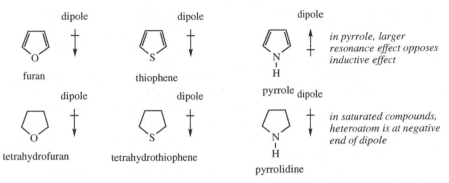

Nevertheless, we can interpret the reactions of furan and thiophene by logical consideration as we did for pyrrole. In electrophilic substitutions, there is again a preference for 2- rather than 3-substitution, and typical electrophilic reactions carried out under acidic conditions are difficult to control. However, because of lower reactivity compared with pyrrole, it is possible to exploit Friedel–Crafts acylations, though using less-reactive anhydrides rather than

acyl chlorides, and weaker Lewis acids than $AlCl_3$. Nitration can be achieved with acetyl nitrate rather than nitric acid. In the case of furan, this is slightly anomalous, in that it involves an addition intermediate by combination of the carbocation with acetate. This subsequently aromatizes by loss of acetic acid. The less-reactive thiophene can even be nitrated with concentrated nitric acid, when it yields a mixture of 2- and 3-nitrothiophene.

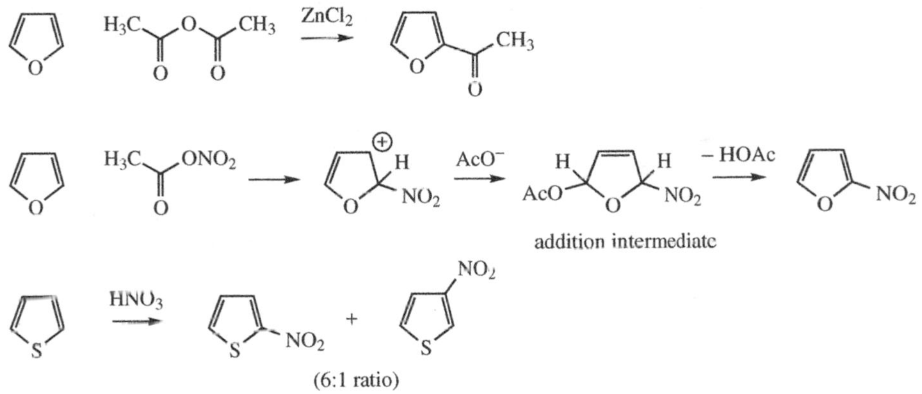

addition intermediate

(6:1 ratio)

11.6   Six-membered rings with two heteroatoms

11.6.1   Diazines

A diazabenzene, i.e. a benzene ring in which two of the CH functions have been replaced with nitrogen,

is termed a **diazine**. Three isomeric variants are possible; these are called **pyridazine, pyrimidine**, and **pyrazine**.

These structures are all aromatic, the nitrogen atoms functioning in the same way as the pyridine nitrogen, each contributing one p electron to the aromatic sextet, with a lone pair in an sp^2 orbital. The

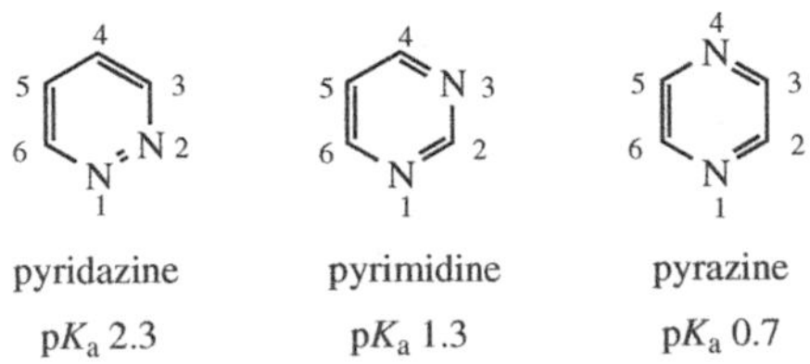

pyridazine

pK_a 2.3

pyrimidine

pK_a 1.3

pyrazine

pK_a 0.7

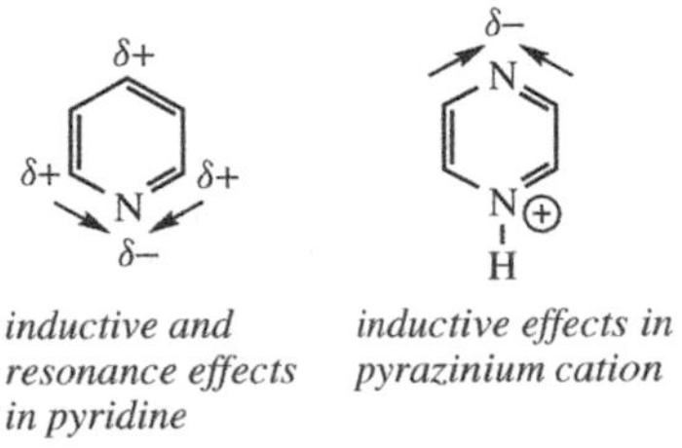

inductive and resonance effects in pyridine

inductive effects in pyrazinium cation

diazines are much weaker bases than pyridine (pK_a 5.2). If we consider the inductive and resonance effects in pyridine, we have seen that these both draw electrons towards the nitrogen (see Section 11.3). Therefore, a second nitrogen will have destabilizing effects on the conjugate acid formed by protonation of the first nitrogen. The order of basicity in pyridazine, pyrimidine, and pyrazine is influenced by secondary effects, which will not be considered here. Diprotonation is very difficult and would require extremely strong acids; in the case of pyridazine, it is essentially impossible because of the need to establish positive charges on adjacent atoms.

In general, we can consider that the extra nitrogen, through its combined effects, makes the other ring atoms more electron deficient than they would be in pyridine; as a result, the diazines are more susceptible to nucleophilic attack than pyridine. In pyrazines and pyridazines, the second nitrogen helps by withdrawing electrons from atoms that would carry a negative charge in the addition anion. In pyrimidines, the two nitrogens share the negative charge of the addition anion, and pyrimidines are the more reactive towards nucleophiles.

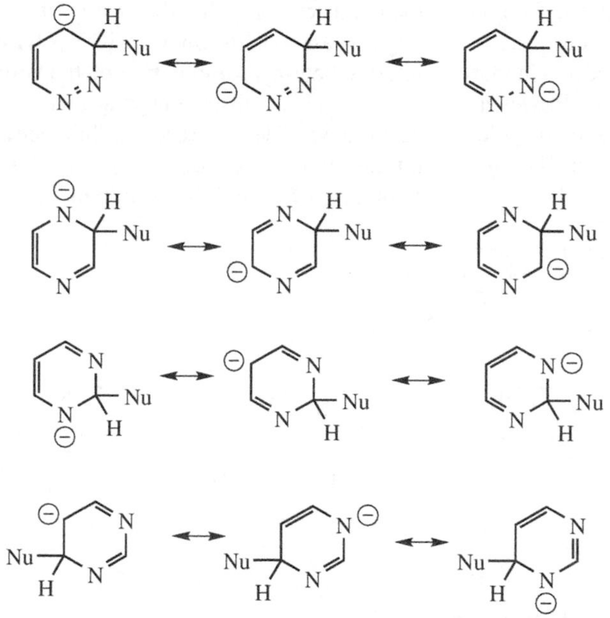

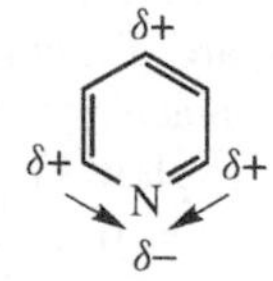

a second nitrogen will generate electron deficiencies at atoms where we are locating the negative charge

both nitrogens share the negative charge

Halodiazines react readily with nucleophiles with displacement of the halide leaving group. This follows what we have seen with halopyridines (see Section 11.4.1), but the halodiazines are more reactive because of the influence of the extra nitrogen. Thus, 2-chloropyrazine and 3-chloropyridazine easily yield the corresponding amino derivatives on heating with ammonia in alcohol solution.

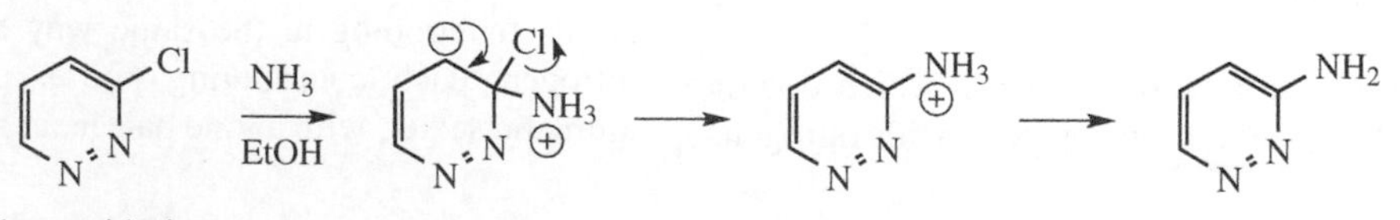

3-chloropyridazine

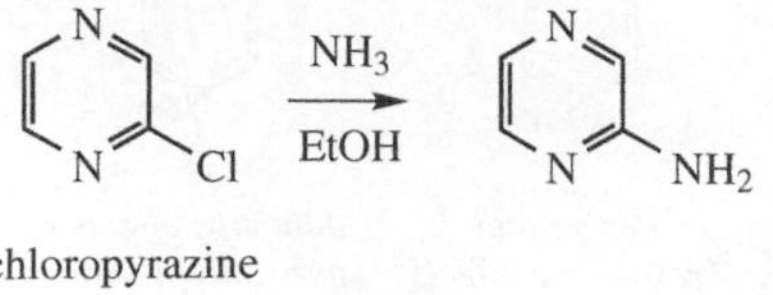

2-chloropyrazine

The 2- and 4-halopyrimidines are even more reactive, and substitute at room temperature. This is because of the improved delocalization of negative charge in the addition anion. 5-Halopyrimidines are the least susceptible to nucleophilic displacement: the halogen is neither α nor γ to a nitrogen, and cannot benefit from any favourable charge localization on nitrogen.

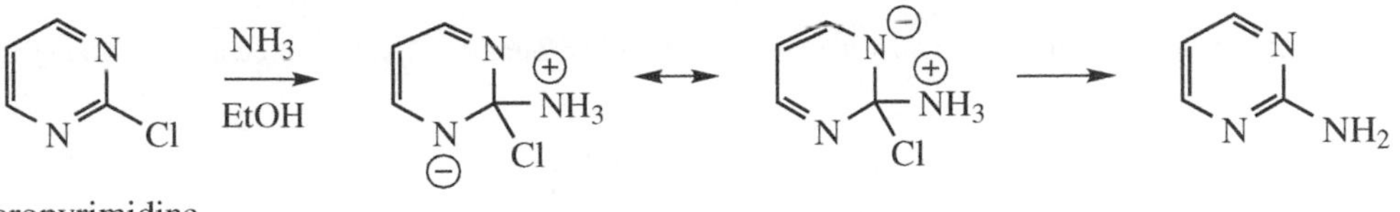

2-chloropyrimidine

Diazines are generally resistant to electrophilic attack on carbon, and, as for pyridine, addition on nitrogen is observed. Alkyl halides give mono-quaternary salts; di-quaternary salts are not formed under normal conditions. Of course, if the diazine ring carries a substituent that makes the starting material non-symmetric, then the product will almost always be a mixture of two isomeric quaternary salts. Steric and inductive effects rather than resonance effects appear to influence the reaction and formation of the major product.

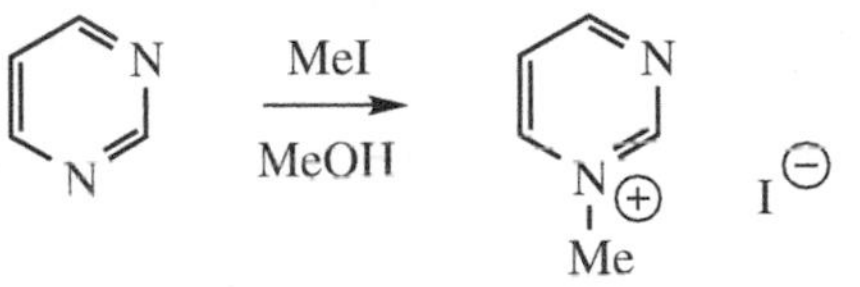

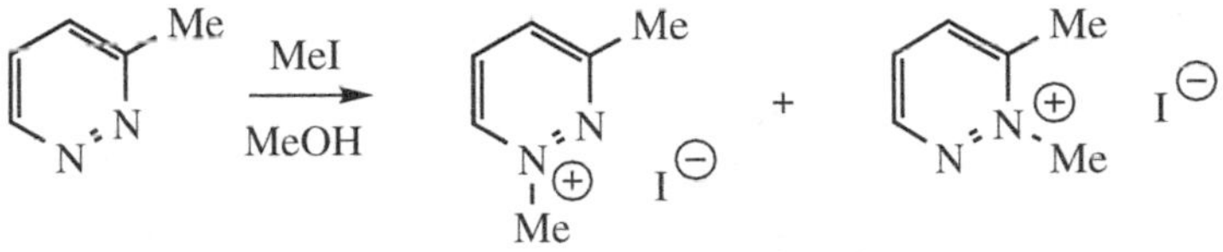

ratio 7:3

11.6.2 Tautomerism in hydroxy- and amino-diazines

We have seen that 2- and 4-hydroxypyridines exist primarily in their tautomeric 'amide-like' pyridone forms (see Section 11.4.3). This preference over the 'phenolic' tautomer was related to these compounds still retaining their aromatic character, with further stabilization from the carbonyl group. 3-Hydroxypyridine cannot benefit from this additional stabilization. In contrast, 2-aminopyridine and 4-aminopyridine exist almost entirely as the amino tautomers, although they are potentially tautomeric with imino forms (see Section 11.4.3).

We also encounter tautomerism in hydroxy- and amino-diazines, and the preference for one tautomeric form over the other follows what we have seen with the pyridine derivatives. Thus, with the exception of 5-hydroxypyrimidine, all the mono-oxygenated diazines exist predominantly in the carbonyl tautomeric form. We term these 'amide-like' tautomers **diazinones**. 5-Hydroxypyrimidine is analogous to 3-hydroxypyridine, in that the hydroxyl is wrongly positioned for tautomerism.

3-hydroxypyridazine   3(2H)-pyridazinone 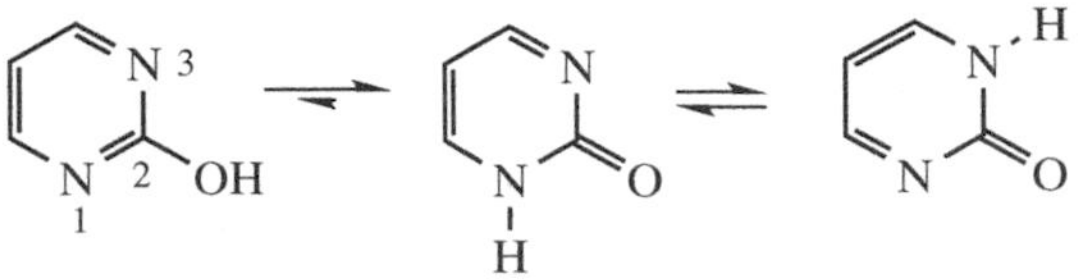   2-hydroxypyrimidine   2(1H)-pyrimidone   2(3H)-pyrimidone

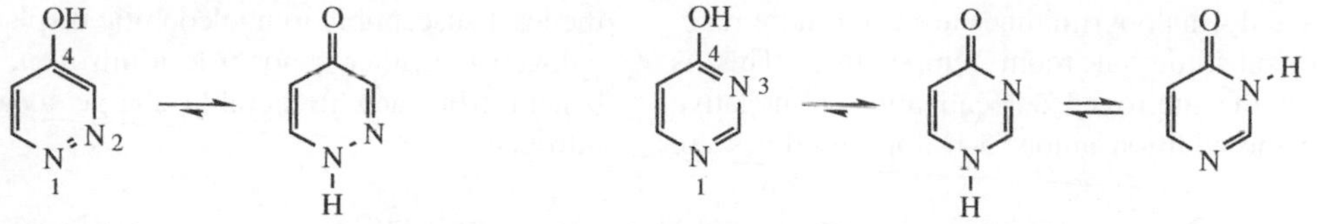

4-hydroxypyridazine   4(1H)-pyridazinone   4-hydroxypyrimidine   4(1H)-pyrimidone   4(3H)-pyrimidone

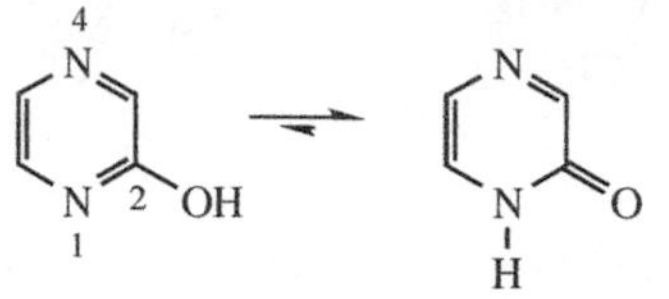

2-hydroxypyrazine   2(1H)-pyrazinone

The diazinone tautomers are identified by using terminology such as 3(2H)-pyridazinone for the carbonyl tautomer of 3-hydroxypyrazine. The 3(2H) prefix signifies the position of the oxygen (3-pyridazinone) and specifies the NH is at position 2. Note that, in addition to the diazine–diazinone tautomerism, when the nitrogens have a 1,3-relationship there is further tautomerism possible, e.g. 4(1H)-pyrimidone ⇆ 4(3H)-pyrimidone.

Diazinones may be converted into chlorodiazines by the use of phosphorus oxychloride, just as pyridones yield chloropyridines (see Section 11.4.3).

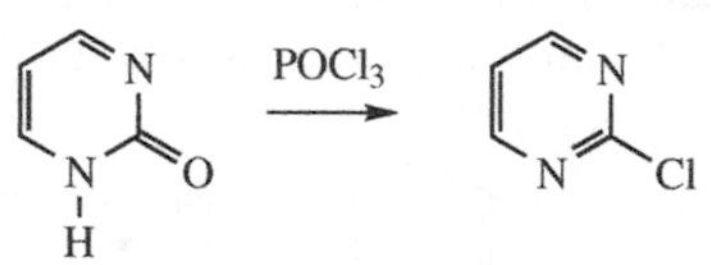

2-pyrimidone   2-chloropyrimidine

Aminodiazines exist in the amino form. These compounds contain two ring nitrogens and a primary amino group.

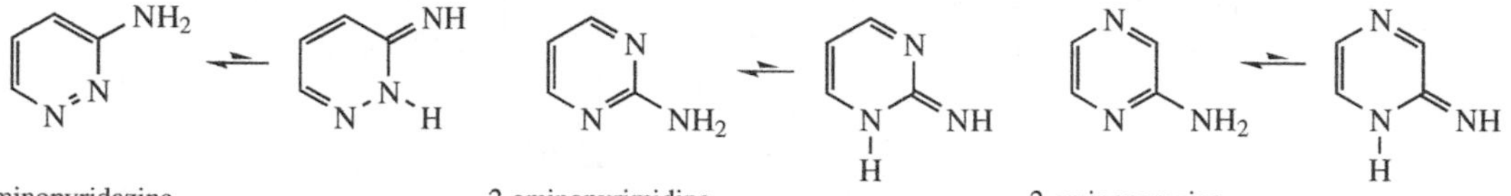

3-aminopyridazine   2-aminopyrimidine   2-aminopyrazine

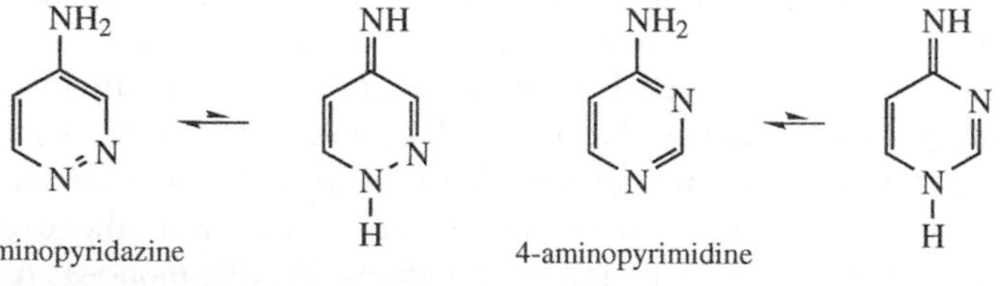

4-aminopyridazine   4-aminopyrimidine

Interestingly, they are more basic than the unsubstituted diazine, and always protonate on a ring nitrogen. This allows resonance stabilization of the conjugate acid utilizing the lone pair of the amino substituent. It has been found that one can predict which nitrogen is protonated from the ring nitrogen–amino substituent relationship, which follows the preference sequence $\gamma > \alpha > \beta$, as in the examples shown. This can be related to achieving maximum charge distribution over the molecule.

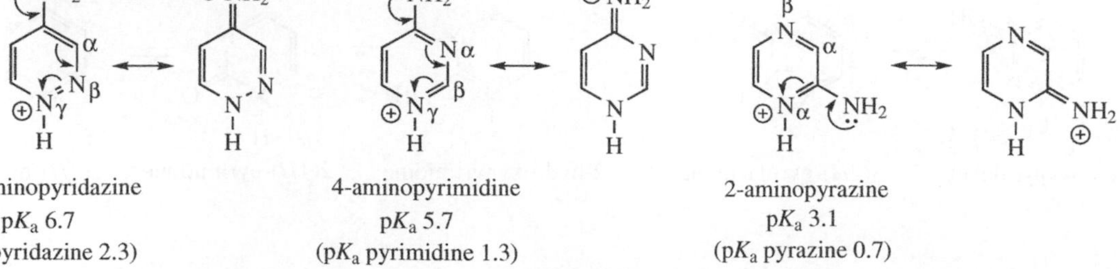

4-aminopyridazine
pK_a 6.7
(pK_a pyridazine 2.3)

4-aminopyrimidine
pK_a 5.7
(pK_a pyrimidine 1.3)

2-aminopyrazine
pK_a 3.1
(pK_a pyrazine 0.7)

Box 11.5

Pyrimidines and nucleic acids

The storage of genetic information and the transcription and translation of this information are functions of the nucleic acids **deoxyribonucleic acid** (**DNA**) and **ribonucleic acid** (**RNA**). They are polymers whose building blocks are nucleotides, which are themselves combinations of three parts, i.e. a heterocyclic base, a sugar, and phosphate (see Section 14.1).

The bases are either monocyclic pyrimidines or bicyclic purines (see Section 14.1). Three pyrimidine bases are encountered in DNA and RNA, **cytosine** (C), **thymine** (T) and **uracil** (U). Cytosine is common to both DNA and RNA, but uracil is found only in RNA and thymine is found only in DNA. In the nucleic acid, the bases are linked through an *N*-glycoside bond to a sugar, either ribose or deoxyribose; the combination base plus sugar is termed a nucleoside. The nitrogen bonded to the sugar is that shown.

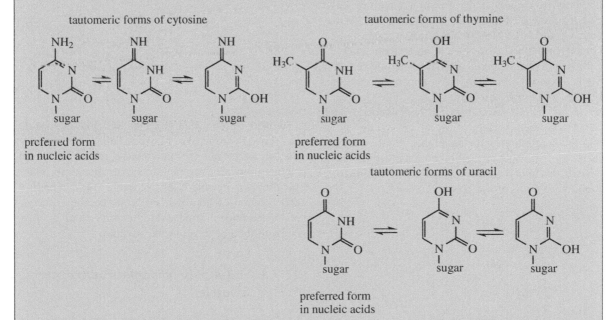

We should note particularly that uracil and thymine are dioxypyrimidines, whereas cytosine is an amino-oxypyrimidine. All three pyrimidines are thus capable of existing in several tautomeric forms (see Section 11.6.2).

The number of possible forms is reduced somewhat by the fact that one of the nitrogens is bonded to the sugar in the nucleic acid; it no longer carries a hydrogen to participate in tautomerism. The tautomeric forms indicated are found to predominate in nucleic acids. The oxygen substituents exist almost entirely as carbonyl groups, whereas the amino group is preferred over possible imino forms. Although we are accustomed to thinking of nucleic acids containing 'pyrimidine' bases, this is not strictly correct. In fact, cytosine exists as an aminopyrimidone, and thymine and uracil are pyrimidiones. Further, they are not particularly basic. Cytosine is the most basic of the three (pK_a 4.6), in that the amino group by a resonance effect can stabilize the conjugate acid (compare 4-aminopyrimidine pK_a 5.7 above). Thymine and uracil are very weak bases, in that they are 'amide-like'.

The most far-reaching feature of nucleic acids is the ability of the bases to hydrogen bond to other bases (see Box 2.2). This property is fundamental to the double helix arrangement of the DNA molecule, and the translation and transcription via RNA of the genetic information present in the DNA molecule. Hydrogen bonding occurs between complementary purine and pyrimidine bases and involves either two or three hydrogen bonds. In DNA, the base pairs are adenine–thymine and guanine–cytosine. In RNA, base pairing involves guanine–cytosine and adenine–uracil. This property will be discussed in detail in Section 14.2, but it is worth noting at this stage that hydrogen bonding is achieved between amino substituents (N–H) and the oxygen of carbonyl groups. These functions in the pyrimidine bases (and also for that matter in the purine bases; see Section 11.9.2) arise directly from the tautomeric preferences.

11.7 Five-membered rings with two heteroatoms

We have looked at the five-membered aromatic heterocycles pyrrole, furan and thiophene in Section 11.5. Introduction of a second heteroatom creates **azoles**. This name immediately suggests that nitrogen is one of the heteroatoms. As soon as we consider valencies, we discover that in order to draw a five-membered aromatic heterocycle with two heteroatoms, it *must* contain nitrogen! A neutral oxygen or sulfur atom can have only two bonds, and we cannot, therefore, have more than one of these atoms in any aromatic heterocycle. On the other hand, there is potential for having as many nitrogens as we like in an aromatic ring.

Thus, in five-membered aromatic heterocycles with two heteroatoms, we can have two nitrogens, one nitrogen plus one oxygen, or one nitrogen plus one sulfur. The heteroatoms can be positioned only 1,2 or 1,3. Numbering of the ring system starts from the heteroatom with the higher atomic number; nitrogen will always be the higher of the two numbers in **oxazole** and **thiazole** systems. In **imidazole**, numbering begins at the NH.

We can visualize these heterocycles as similar to the simpler aromatic systems pyrrole, furan and thiophene. For example, in imidazole, each carbon and nitrogen will be sp^2 hybridized, with p orbitals contributing to the aromatic π system. The carbon atoms will each donate one electron to the π system. Then, as in pyrrole, the NH nitrogen supplies two electrons, and, as in pyridine, the =N– supplies one electron and retains a lone pair. Oxygen or sulfur would also supply two electrons, as we saw in furan and thiophene.

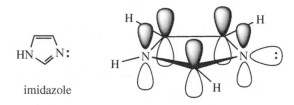

imidazole

It also follows that a compound like imidazole has one pyridine-like nitrogen, and one pyrrole-like nitrogen. We may thus expect to see imidazole having properties resembling a combination of either pyridine- or pyrrole-like reactivity. The availability and location of lone pair electrons is crucial to our understanding of imidazole chemistry, and it often helps to include these in the structure.

11.7.1 1,3-Azoles: imidazole, oxazole, and thiazole

Imidazole (pK_a 7.0) is a stronger base than either pyridine (pK_a 5.2) or pyrrole (pK_a − 3.8). When we compared the basicity of pyridine with that of the

1,2-azoles

pyrazole   isoxazole   isothiazole

1,3-azoles

imidazole   oxazole   thiazole
1*H*-imidazole

aliphatic amine piperidine (pK_a 11.1), we implicated the higher s character of the pyridine lone pair (sp^2) compared with that in piperidine (sp^3) to account for pyridine's lower basicity (see Section 11.4.1). Even so, imidazole seems abnormally basic for a compound with sp^2-hybridized nitrogen. The enhanced basicity

of imidazole appears to stem from the symmetry of the conjugate acid, and the resonance stability conferred by this. The 1,3-relationship allows the two nitrogen atoms to share the charge equally. Note that pK_a 7.0 means imidazole is 50% protonated in water (see Section 4.9).

equivalent resonance structures   imidazole   *equivalent resonance structures*

Imidazole (pK_a 14.2) is also more acidic than pyrrole (pK_a 17.5); this, again, is a feature conferred by symmetry and the enhanced resonance stabilization in the conjugate base.

effect, compared with nitrogen, but a much weaker electron-releasing resonance effect.

11.7.2 Tautomerism in imidazoles

pK_a 7.0 pK_a 0.8 pK_a 2.5

Oxazole (pK_a 0.8) and thiazole (pK_a 2.5) are weak bases. The basicity of the nitrogen is reduced by the presence of the other heteroatom. Oxygen and sulfur provide a stronger electron-withdrawing inductive

A complicating factor in imidazoles is **tautomerism. Imidazole** tautomerizes rapidly in solution and consists of two identical tautomers. This becomes a problem, though, in an unsymmetrically substituted imidazole, and tautomerism means 4-methylimidazole is in equilibrium with 5-methylimidazole. Depending upon substituents, one tautomer may predominate. Tautomerism of this kind cannot occur with N-substituted imidazoles; it is totally dependent upon the presence of an N–H group. Tautomerism is also not possible with oxazoles or thiazoles.

identical tautomers of imidazole   *tautomerism not possible*   4-methyl-$1H$-imidazole   5-methyl-$1H$-imidazole

non-identical tautomers of methylimidazole; this compound may be termed 4(5)-methylimidazole

Therefore, when we meet structures for the imidazole-containing amino acid **histidine**, we may encounter either of the tautomeric forms shown. Though there

will usually be no indication that tautomers exist, do not think there is a discrepancy in structures.

tautomeric forms of histidine

L-histidine

Box 11.6

The imidazole ring of histidine: acid–base properties

The amino acid histidine contains an imidazole ring. We have just seen that unsubstituted imidazole as a base has pK_a 7.0. From the Henderson–Hasselbalch equation

$$pH = pK_a + \log \frac{[base]}{[acid]}$$

we can deduce that in water, at pH 7, the concentrations of acid and conjugate base are equal, i.e. imidazole is 50% protonated (see Section 4.9).

The imidazole side-chain of histidine has a pK_a value of 6.0, making it a weaker base than the unsubstituted imidazole. This reflects the electron-withdrawing inductive effect of the amino group, or, more correctly the ammonium ion, since amino acids at pH values around neutrality exist as doubly charged zwitterionic forms (see Box 4.7). Using the Henderson–Hasselbalch equation, this translates to approximately 9% ionization of the heterocyclic side-chain of histidine at pH 7 (see Box 4.7). In proteins, pK_a values for histidine side-chains are estimated to be in range 6–7, so that the level of ionization will, therefore, be somewhere between 9 and 50%, depending upon the protein.

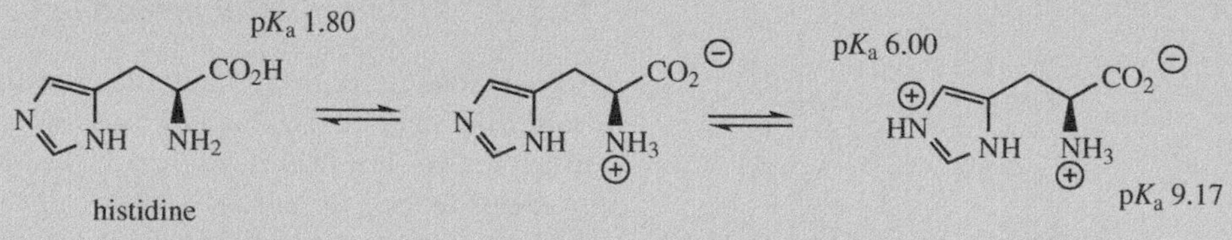

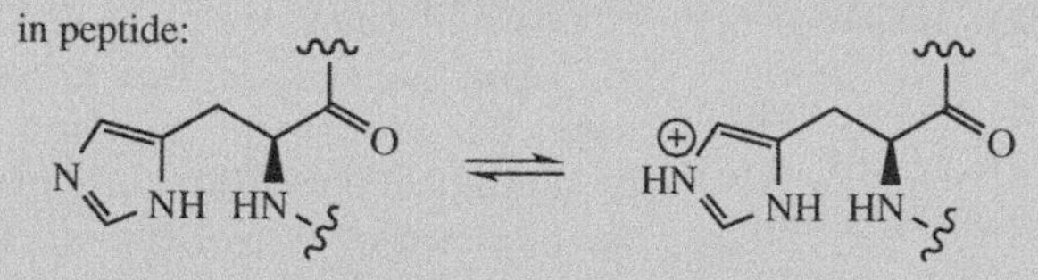

This level of ionization is particularly relevant in some enzymic reactions where histidine residues play an important role (see Section 13.4.1). This means that the imidazole ring of a histidine residue can act as a base, assisting in the removal of protons, or, alternatively, that the imidazolium cation can act as an acid, donating protons as required. The terminology used for such donors and acceptors of protons is **general acid catalyst** and **general base catalyst** respectively.

A typical role for the histidine imidazole ring is shown below, in the enzyme mechanism for a general base-catalysed hydrolysis of an ester. The imidazole nitrogen acts as a base to remove a proton from water, generating hydroxide that attacks the carbonyl. Subsequently, the alkoxide leaving group is reprotonated by the imidazolium

General base-catalysed ester hydrolysis

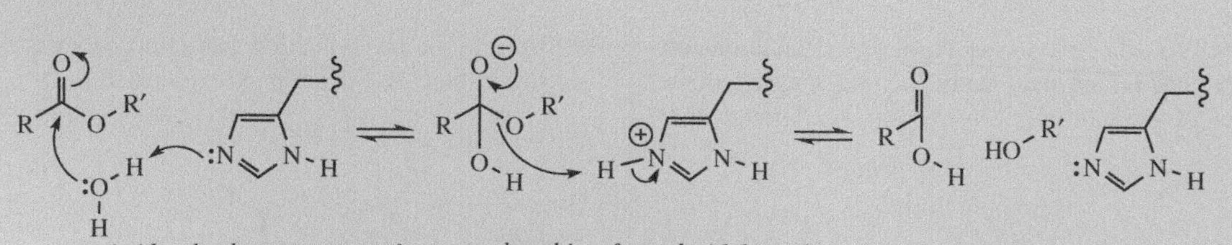

imidazole abstracts proton from water and resulting hydroxide attacks carbonyl group

carbonyl is reformed with loss of leaving group; alkoxide is protonated by imidazolium ion

ion. The beauty of this is that we effectively have the same mechanism as in the hydrolysis of an ester using aqueous sodium hydroxide (see Section 7.9.2). However, with the enzyme catalyst, this is all taking place at pH 7 or thereabouts.

Implicit in the above mechanism, though not emphasized, is the pronounced ability of imidazole rings to hydrogen bond. Imidazole resembles water, in that it is both a very good donor and a very good acceptor for **hydrogen bonding**. Imidazole (and also pyrazole) has a higher than expected boiling point, ascribed to intermolecular hydrogen bonding. This leads to polymeric-like structures for imidazole, and dimers for pyrazole.

H-bonded polymer: imidazole

H-bonded dimer: pyrazole

In enzymic mechanisms, we are not usually going to get imidazole–imidazole hydrogen bonding, but the ability of imidazole to hydrogen bond to water, to other small molecules, and to carboxylic acid side-chains facilitates the enzyme reaction by correctly positioning the reagents. We shall see examples of this in Section 13.4.

Box 11.7

Histamine and histamine receptors

Most people have heard of **antihistamines**, even if they have little concept of the nature of histamine. **Histamine** is the decarboxylation product from histidine, and is formed from the amino acid by the action of the enzyme histidine decarboxylase. The mechanism of this pyridoxal phosphate-dependent reaction will be studied in more detail later (see Section 15.7).

L-histidine

histamine

PLP = pyridoxal phosphate

Histamine is released from mast cells during inflammatory or allergic reactions. It then produces its typical response by interaction with specific histamine receptors, of which there are several types. H_1 receptors are associated with inflammatory and allergic reactions, and H_2 receptors are found in acid-secreting cells in the stomach. Drugs to target both of these types of receptor are widely used.

The term antihistamine usually relates to H_1 **receptor antagonists**. These drugs are valuable for pain relief from insect stings, or for the treatment and prevention of allergies such as hay fever. Major effects of histamine include dilation of blood vessels, inflammation and swelling of tissues, and narrowing of airways. In serious cases, life-threatening anaphylactic shock may occur, caused by a dramatic fall in blood pressure. Remarkably, current H_1 receptor antagonists, e.g. **diphenhydramine**, bear little if any structural similarity to histamine.

The main clinical use of H_2 **receptor antagonists** is to inhibit gastric secretion in the treatment of stomach ulcers. These agents all contain features that relate to the histamine structure, in particular the heterocyclic ring. Cimetidine and ranitidine are the most widely used in this class.

Box 11.7 (continued)

H₁ receptor antagonist   *H₂ receptor antagonists*

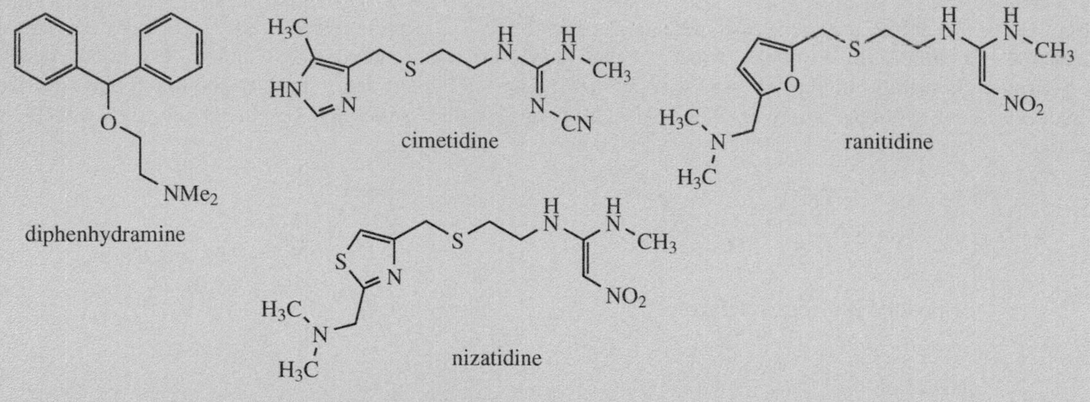

diphenhydramine

cimetidine

ranitidine

nizatidine

Cimetidine contains an imidazole ring comparable to histamine, a sulfur atom (thioether group) in the side-chain, and a terminal functional group based upon a guanidine (see Section 4.5.4). **Ranitidine** bears considerable similarity to cimetidine, but there are some important differences. The heterocycle is now furan rather than imidazole, and the guanidine has been modified to an amidine (see Section 4.5.4). A newer drug, **nizatidine**, is a variant on ranitidine with a thiazole heterocyclic ring system.

11.7.3 Reactivity of 1,3-azoles

Electrophiles can add to N-3, the azomethine =N–, of 1,3-azoles as they can to the pyridine nitrogen. *N*-**Alkylation** is complicated in the case of imidazole by the possibility of forming a dialkylimidazolium salt; the first-formed protonated *N*-alkylimidazole can be deprotonated by imidazole, then alkylated further.

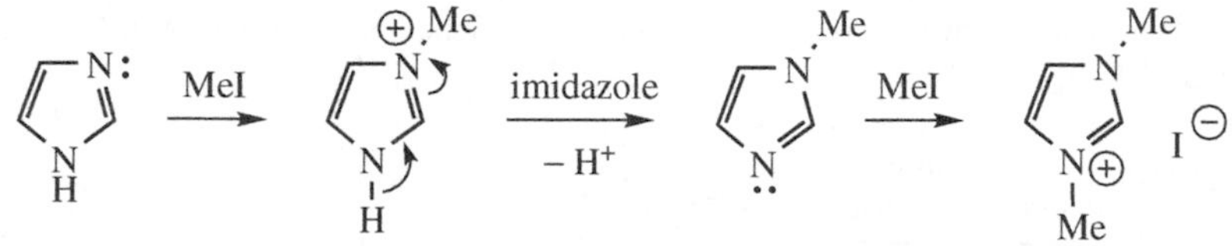

dimethylimidazolium salt

N-Acylation is mechanistically similar, and mono-acylation can be accomplished by using two molar equivalents of imidazole to one of the acylating agent, the second mole serving to deprotonate the first-formed *N*-3-acylimidazolium salt. Note that in alkylation and acylation it is the =N– that acts as the nucleophile; this carries the only lone pair. However, proton loss occurs from the other nitrogen, giving the impression that the N–H has been alkylated or acylated.

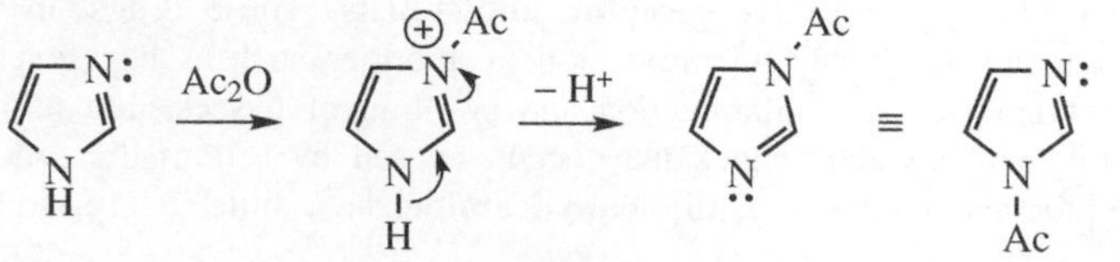

N-1-acetylimidazole

The 1,3-diazoles are much less susceptible to **electrophilic substitution** than pyrrole, furan, and thiophene, but are more reactive than pyridine. Imidazole is the most reactive, and may be nitrated readily.

Substitution occurs at C-5, but tautomerism then leads to the 4(5) mixture. The position of substitution may be predicted from a consideration of resonance structures: attack at C-5 provides maximum delocalization with no particularly unfavourable resonance forms. There is less delocalization after attack at C-2; one of the resonance forms has an unfavourable electron-deficient nitrogen.

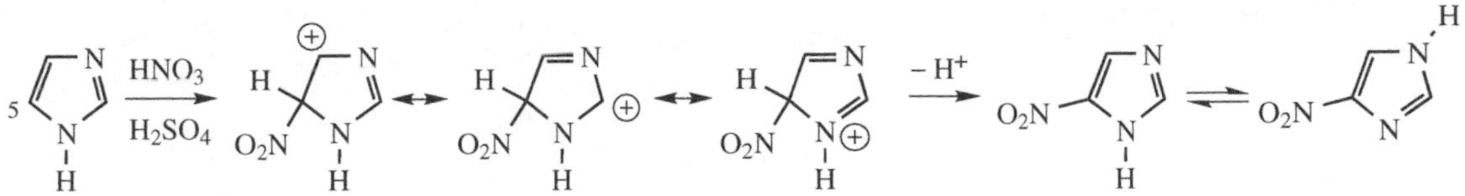

electrophilic attack at C-2 less favourable

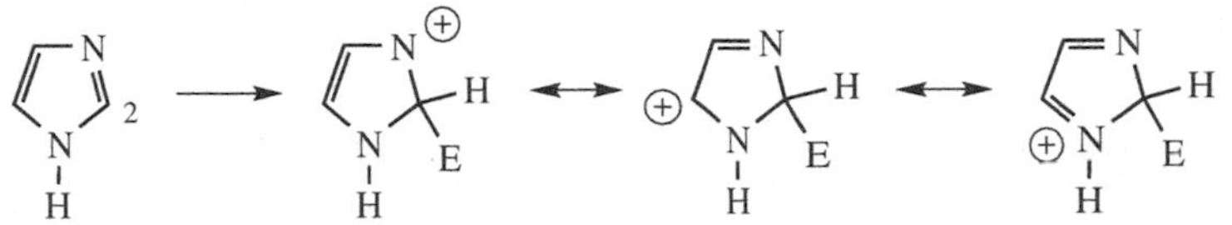

unfavourable electron-deficient cation

In general, the 1,3-diazoles do not react by nucleophilic substitution, although imidazole can participate in the Chichibabin reaction with substitution at C-2; the position of substitution is equivalent to that noted with pyridine (see Section 11.4.1). Nucleophilic species that are strong bases, like sodium amide, are more likely to remove the NH proton (pK_a 14.2) (see Section 11.7.1). However, oxazole and thiazole do not have any NH, and the most acidic proton is that at C-2. The electronegative oxygen and sulfur are able to support an adjacent negative charge.

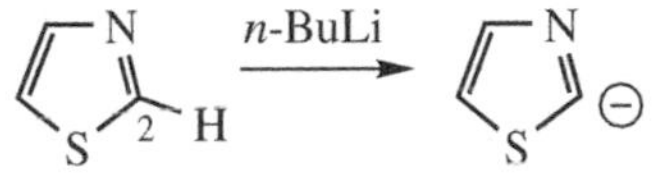

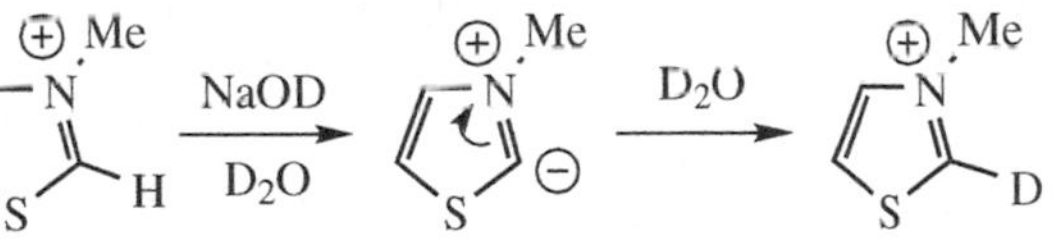

thiazolium ylid

It is found that quaternary salts of 1,3-azoles are deprotonated at C-2 in the same way. Rates of deprotonation are considerably faster because of the influence of the quaternary centre that provides a favourable inductive effect. The conjugate base bearing opposite charges on adjacent atoms is termed an ylid (or ylide; pronounced il-ide). This ylid, with negative charge on carbon, is potentially a nucleophilic species. Thus, it is found that both oxazolium and thiazolium salts undergo H–D exchange at C-2 remarkably quickly under basic conditions, illustrating very simply this nucleophilic behaviour.

Box 11.8

The thiazolium ring in thiamine

Thiamine (vitamin B$_1$), in the form of **thiamine diphosphate (TPP)**, is a coenzyme of some considerable importance in carbohydrate metabolism. Dietary deficiency leads to the condition beriberi, characterized by neurological disorders, loss of appetite, fatigue, and muscular weakness. We shall study a number of

TPP-dependent reactions in detail in Chapter 15. At this stage, we should merely examine the structure of thiamine, and correlate its properties with our knowledge of heterocycles.

Thiamine contains two heterocyclic rings, a pyrimidine and a thiazole, the latter present as a thiazolium salt. The pyrimidine portion is unimportant for our understanding of the chemistry of TPP, though it may play a role in some of the enzymic reactions.

The proton in the thiazolium ring is relatively acidic (pK_a about 18) and can be removed by even weak bases to generate the carbanion or ylid; an ylid is a species with positive and negative charges on adjacent atoms. This ylid is an ammonium ylid with extra stabilization provided by the sulfur atom.

The ylid can act as a nucleophile, and is also a reasonable leaving group. Prominent among TPP-dependent reactions is the oxidative decarboxylation of pyruvic acid to acetyl-CoA; this reaction links the glycolytic pathway to the Krebs cycle (see Section 15.3). Addition of the thiazolium ylid to the carbonyl group of pyruvic acid is the first reaction of this sequence, and this allows the necessary decarboxylation, the positive nitrogen in the ring acting as an electron sink. In due course, the thiazolium ylid is regenerated as a leaving group. We shall look at this sequence in more detail in Section 15.8.

11.7.4 1,2-Azoles: pyrazole, isoxazole, and isothiazole

As in the 1,3-azoles, the =N–nitrogen carries a lone pair of electrons and 1,2-azoles are thus potentially basic. However, the direct linking of the two heteroatoms has a base-weakening effect. Thus,

pyrazole has pK_a 2.5 and isoxazole pK_a − 3.0. The higher basicity in pyrazole is probably related to the symmetry of the contributing resonance structures. The greater electron-withdrawing effect of oxygen compared with sulfur is reflected in the basicity of isothiazole (pK_a − 0.5).

11.8 Heterocycles fused to a benzene ring

Many interesting and important heterocyclic compounds contain fused ring systems. Some of the common ones are the result of fusing a heterocycle to

a benzene ring, and these have long-established **trivial names**, e.g. indole, quinoline, and isoquinoline.

Systematic names can be derived by relating back to the parent heterocycle and using the prefix benzo to indicate its fusion to benzene. It is necessary to define which of the bonds in the heterocycle is

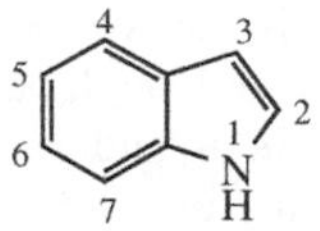

indole
benzo[b]pyrrole

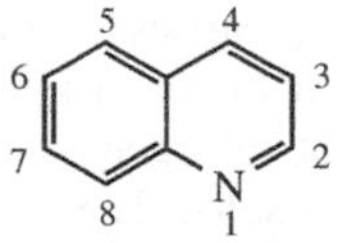

quinoline
benzo[b]pyridine

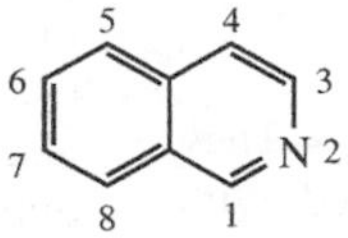

isoquinoline
benzo[c]pyridine

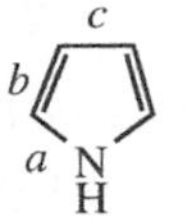

pyrrole

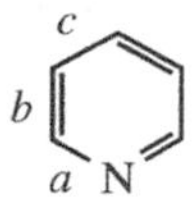

pyridine

fused to benzene, and this is accomplished through use of a bond descriptor, a lower case italic letter in square brackets. Thus, indole is benzo[b]pyrrole, quinoline is benzo[b]pyridine, and isoquinoline, an isomer of quinoline with a different type of fusion, becomes benzo[c]pyridine.

A few other examples are shown below. Note that the bonds of the heterocycle are lettered starting from the heteroatom. Where we have two similar heteroatoms, lettering is chosen to produce the lower alternative. Thus, quinazoline is benzo[d]pyrimidine, not benzo[e]pyrimidine. Where the heteroatoms are different, just as we number from the atom of higher atomic number, we also letter from the same atom. Hence, benzo[d]isoxazole is quite different from benzo[c]isoxazole.

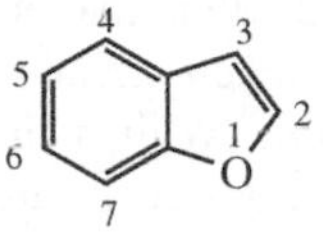

benzo[b]furan 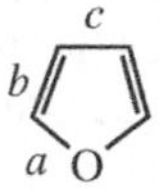   furan 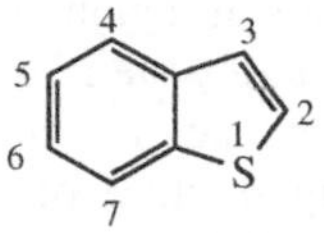   benzo[b]thiophene 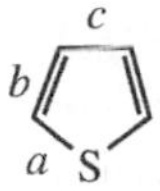   thiophene 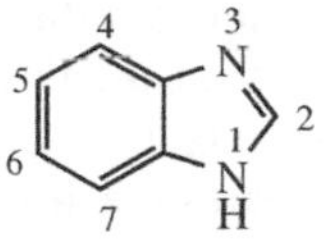   benzimidazole
benzo[d]imidazole   imidazole

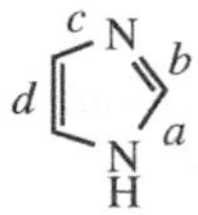

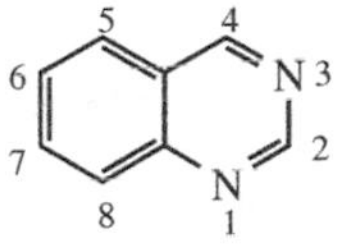

quinazoline
benzo[d]pyrimidine   pyrimidine 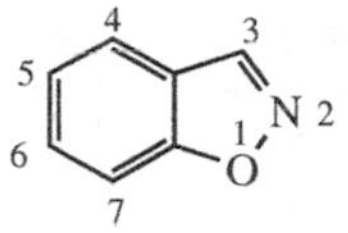   benzo[d]isoxazole 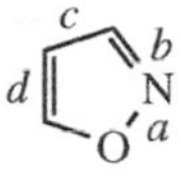   isoxazole 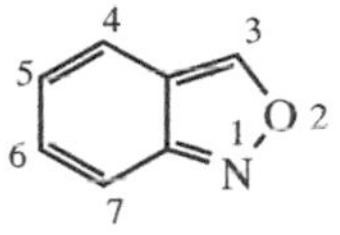   benzo[c]isoxazole

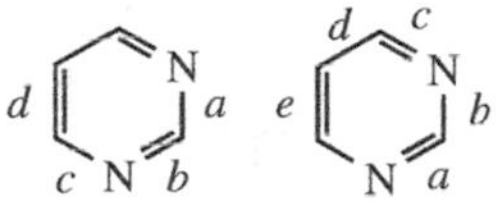

The final fused ring system is then given a completely new numbering system, different from that of the heterocycle. Typically, this starts adjacent to the bridgehead atom, then proceeds around the fused ring. The major criterion is to generate the lowest number for the first heteroatom.

Note that, in most cases, we have little regard for which Kekulé form of a benzene or pyridine ring is drawn. The three versions of quinoline shown are simply contributing resonance forms. However, some structures, such as isoindole, benzo[c]furan, or benzo[c]isoxazole above, can only be drawn in one way without invoking charge separation.

Kekulé forms of pyridine ring

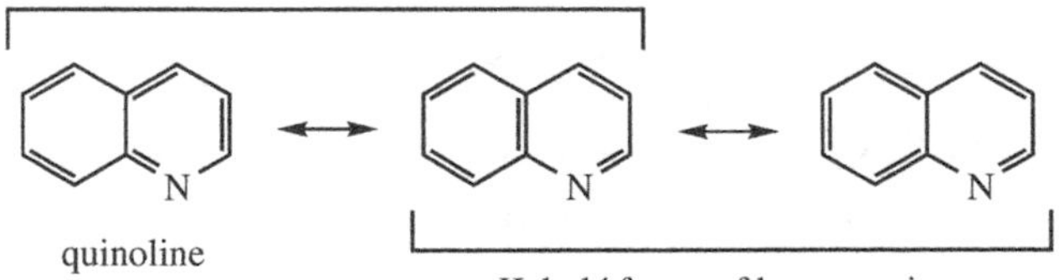

quinoline

Kekulé forms of benzene ring

indole

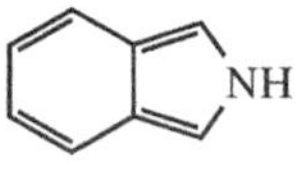

isoindole
benzo[c]pyrrole

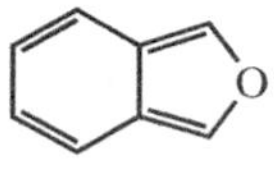

benzo[c]furan

11.8.1 Quinoline and isoquinoline

Quinoline and isoquinoline are **benzopyridines**. They behave by showing the reactivity associated with either the benzene or the pyridine rings.

Quinoline is basic with a pK_a of 4.9, similar to that of pyridine (pK_a 5.2). As with pyridine,

the nitrogen carries a lone pair in an sp^2 orbital (see Section 11.3). Alkyl halides and acyl halides also react at nitrogen to give N-alkyl- and N-acyl-quinolinium salts. The N-alkyl salts are stable, but the N-acyl salts hydrolyse rapidly in the presence of water.

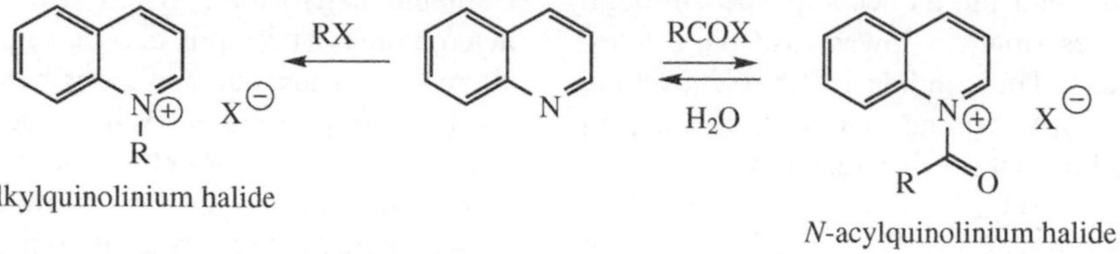

N-alkylquinolinium halide

N-acylquinolinium halide

Quinoline is much more reactive towards **electrophilic substitution** than pyridine, but this is because substitution occurs on the benzene ring, not on the pyridine. We have already seen that pyridine carbons are unreactive towards electrophilic reagents, with strongly acidic systems protonating the nitrogen first, further inhibiting reaction (see Section 11.4). This is again true in quinoline, so that the protonated system is involved in the reaction, and the benzene ring undergoes substitution. With a nitrating mixture of HNO_3–H_2SO_4, the products are 5- and 8-nitroquinoline in roughly equivalent amounts.

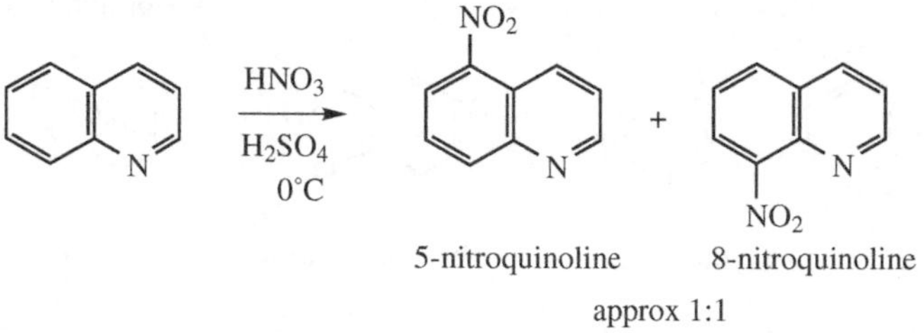

5-nitroquinoline   8-nitroquinoline

approx 1:1

This may be rationalized by considering the stability of intermediate addition cations. When the electrophile attacks at C-5 or C-8, the intermediate cation is stabilized by resonance, each having two favourable forms that do not perturb the aromaticity of the pyridinium system. In contrast, for attack at C-6 or C-7 there is only one such resonance form. We used similar reasoning to explain why naphthalene

undergoes preferential electrophilic substitution at the α-positions (see Section 8.4.4). Whilst we may be a little unhappy about protonation of a quinolinium cation to an intermediate that carries two positive charges, we find that N-methylquinolinium salts also undergo nitration at a similar rate to quinoline; so this mechanism appears correct.

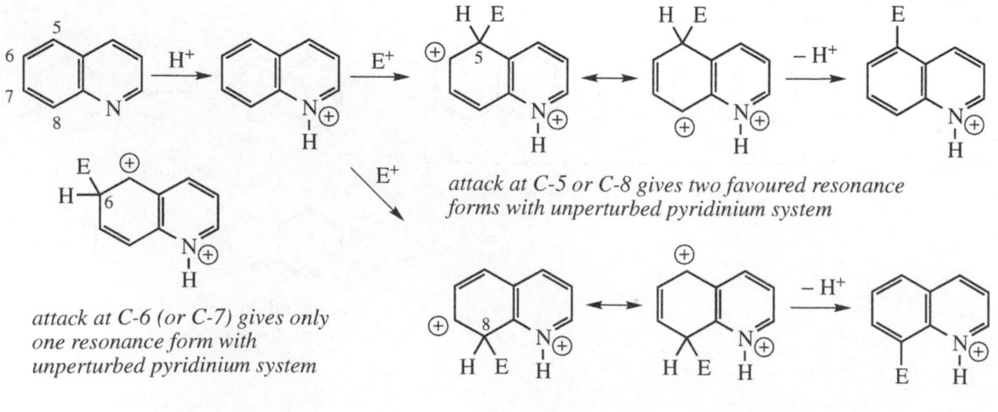

attack at C-5 or C-8 gives two favoured resonance forms with unperturbed pyridinium system

attack at C-6 (or C-7) gives only one resonance form with unperturbed pyridinium system

Nucleophilic substitution occurs at C-2, and to a lesser extent C-4, as might be predicted from similar reactions with pyridine. **Chichibabin amination** occurs rather more readily than with pyridine, giving 2-aminoquinoline. A typical hydride abstraction process occurs when quinoline is heated with sodium amide (see Section 11.4). However, better yields have been achieved by performing the reaction at low temperatures in liquid ammonia solvent, and then oxidizing the intermediate dihydroquinoline salt using potassium permanganate.

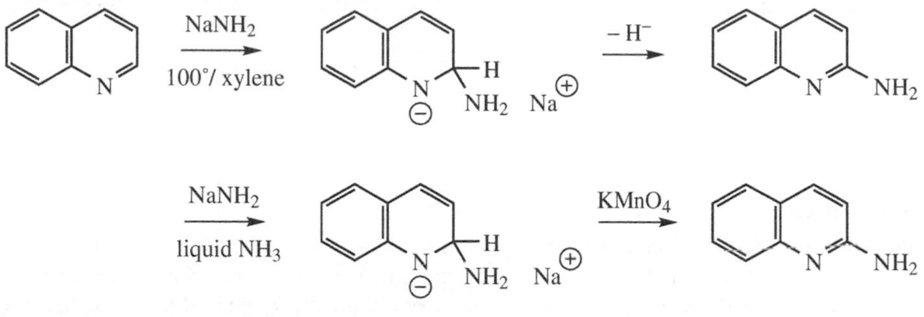

Quinolines carrying 2- or 4-halo substituents undergo nucleophilic substitution readily, in the same manner as 2- and 4-halopyridines. Hydroxyquinolines with the hydroxyl at positions 2 or 4 exist mainly in the carbonyl form, i.e. 2-quinolone and 4-quinolone.

Note, however, that hydroxyls on the benzene ring would be typical phenols. Again, aminoquinolines follow the pyridine precedent and the tautomeric imino forms are not observed.

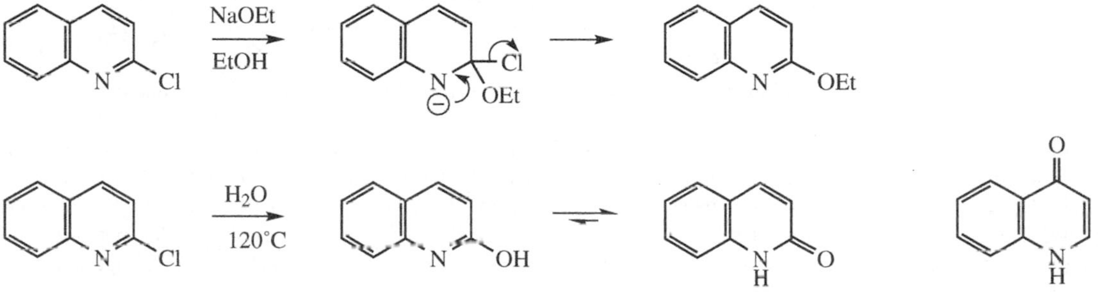

2-quinolone

4-quinolone

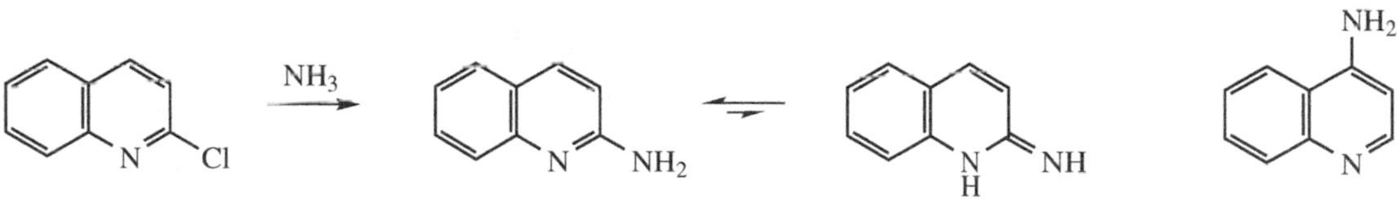

2-aminoquinoline

4-aminoquinoline

Both 2-aminoquinoline and 4-aminoquinoline protonate first on the ring nitrogen, with 4-aminoquinoline being the more basic, the conjugate acid benefiting from increased charge distribution through resonance (compare aminopyridines, Section 11.4.3). No such resonance structures can be drawn for 3-aminoquinoline, which is much less basic (pK_a 4.9).

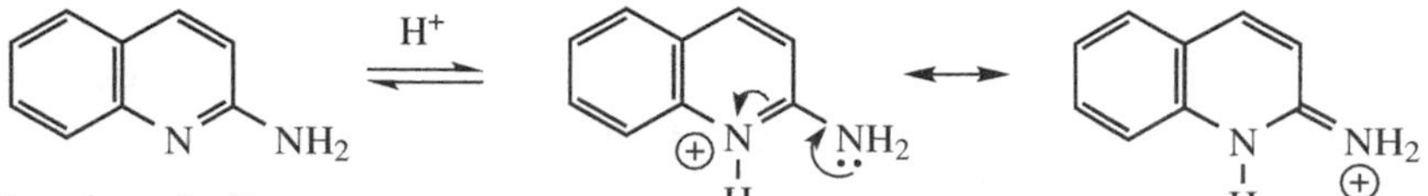

2-aminoquinoline

pK_a 7.3

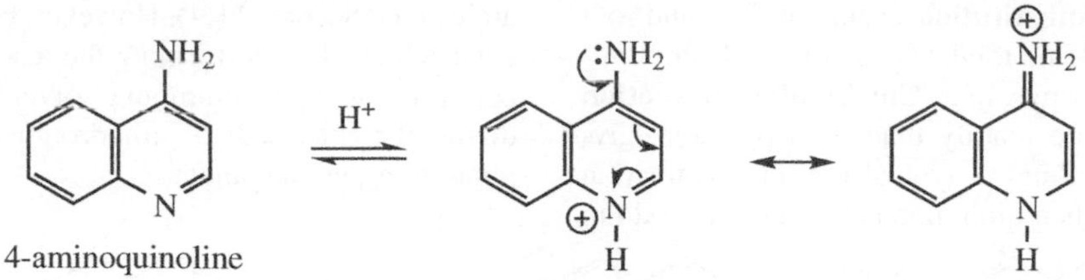

4-aminoquinoline $\xrightarrow{H^+}$ pK_a 9.2

Box 11.9

Quinolone antibiotics

The **quinolone antibiotics** feature as the one main group of antibacterial agents that is totally synthetic, and not derived from or based upon natural products, as are penicillins, cephalosporins, macrolides, tetracyclines, and aminoglycosides. The first of these compounds to be employed clinically was **nalidixic acid**; more recent drugs in current use include **ciprofloxacin, norfloxacin**, and **ofloxacin**

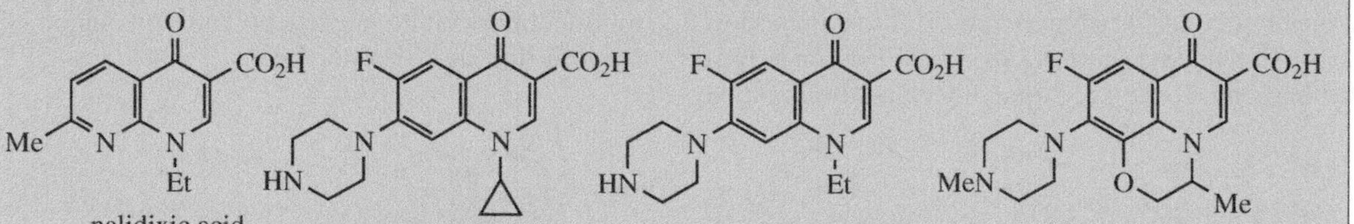

nalidixic acid   ciprofloxacin   norfloxacin   ofloxacin

'Quinolone' as a descriptor is obviously an oversimplification, since nalidixic acid contains two fused pyridine rings rather than a benzopyridine, and ofloxacin has a morpholine ring fused to the quinolone. Nevertheless, the quinolone substructure is generally used when referring to this group of antibiotics. The most important structural features for good antibacterial activity have been found to be a carboxylic acid at position 3, a small alkyl group at position 1, a 6-fluorine substituent, and a nitrogen heterocycle, often a piperazine, at position 7.

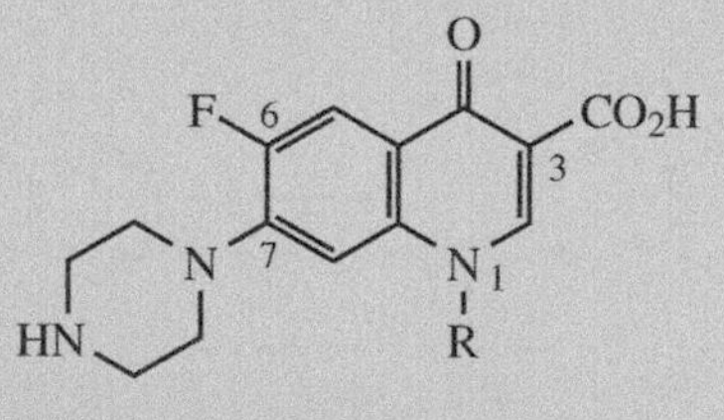

The quinolones are good general antibiotics for systemic infections, and they are particularly useful for urinary tract infections because high concentrations are excreted into the urine. The mode of action involves interference with DNA replication by inhibiting DNA gyrase, a bacterial enzyme related to mammalian topoisomerases that breaks and reseals double-stranded DNA during replication.

Isoquinoline (pK_a 5.4) has similar basicity to quinoline and pyridine, and also undergoes *N*-alkylation and *N*-acylation. **Nitration** occurs smoothly to give predominantly 5-nitroisoquinoline; the isoquinolinium cation reacts more readily than the quinolinium cation.

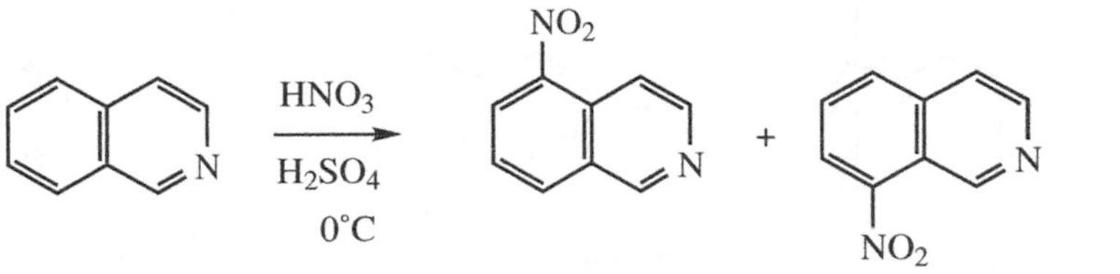

5-nitroisoquinoline   8-nitroisoquinoline

approx 9:1

Nucleophilic substitution occurs exclusively at position 1 in isoquinoline; the alternative position C-3 is quite unreactive. This is explained by the loss of benzene resonance in the intermediate anion. Thus, **Chichibabin amination** gives 1-aminoisoquinoline.

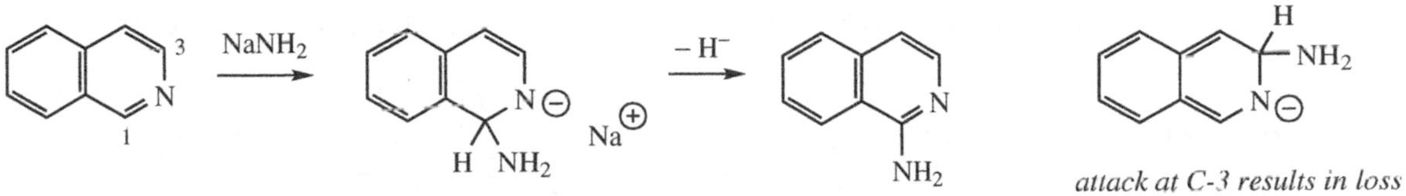

retains benzene resonance   1-aminoisoquinoline   *attack at C-3 results in loss of benzene resonance*

Substitution with displacement of halide occurs readily at C-1 and much less readily at C-3 for the same reasons, i.e. the loss of benzene resonance if C-3 is attacked. 1-Isoquinolone exists completely in the carbonyl form, whereas 1-aminoisoquinoline is the normal tautomer. The basicity of 1-aminoisoquinoline (pK_a 7.6) is similar to that of 2-aminoquinoline (pK_a 7.3).

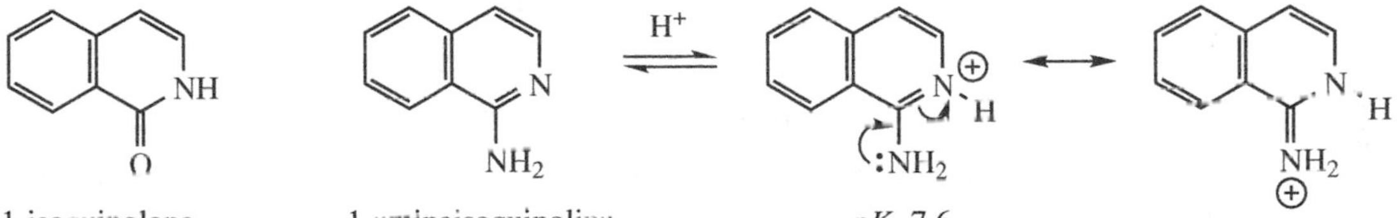

1-isoquinolone   1-aminoisoquinoline pK_a 7.6

11.8.2 Indole

Indole is the fusion of a benzene ring with a pyrrole. Like quinoline and isoquinoline, indole behaves as an aromatic compound. However, unlike quinoline and isoquinoline, where the reactivity was effectively part benzene and part pyridine, the reactivity in indole is modified by each component of the fusion. The closest similarity is between the chemistry of pyrroles and indoles.

Indoles, like pyrroles, are very weak bases. The conjugate acid of indole has pK_a − 3.5; that of pyrrole has pK_a − 3.8. As in the case of pyrrole (see Section 11.3), nitrogen has already contributed its lone pair to the aromatic sextet, so N-protonation would necessarily destroy aromaticity in the five-membered ring. Nevertheless, an equilibrium involving the N-protonated cation is undoubtedly set up, since acid-catalysed deuterium exchange of the N-hydrogen occurs rapidly, even under very mild acidic conditions. Protonation eventually occurs preferentially on carbon, as with pyrrole; but there is a difference, in that this occurs on C-3 rather than on C-2. This is the influence of the benzene ring. It can be seen that protonation on C-3 allows resonance in the five-membered ring and charge localization on nitrogen. In contrast, any resonance structure from protonation at C-2 destroys the benzene ring aromaticity.

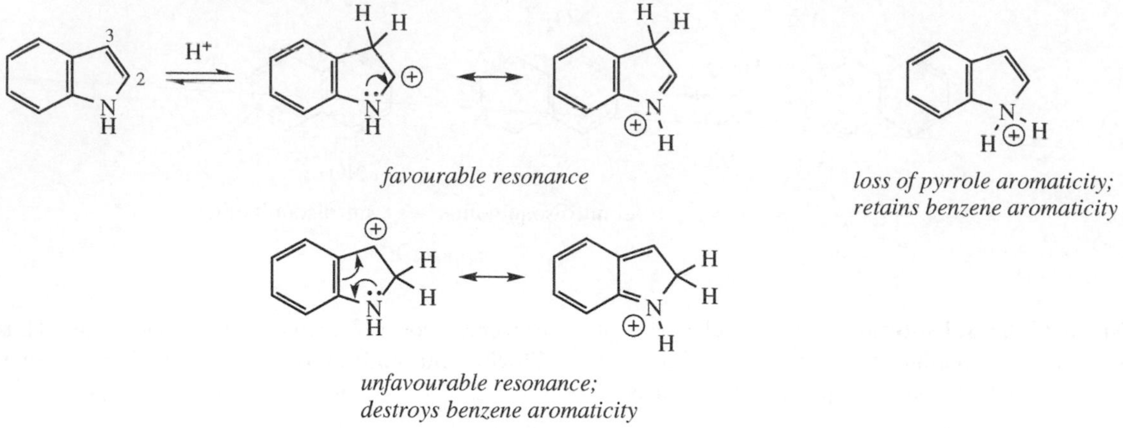

favourable resonance

*loss of pyrrole aromaticity;
retains benzene aromaticity*

*unfavourable resonance;
destroys benzene aromaticity*

Similar behaviour is encountered with other electrophiles, with substitution occurring at C-3.

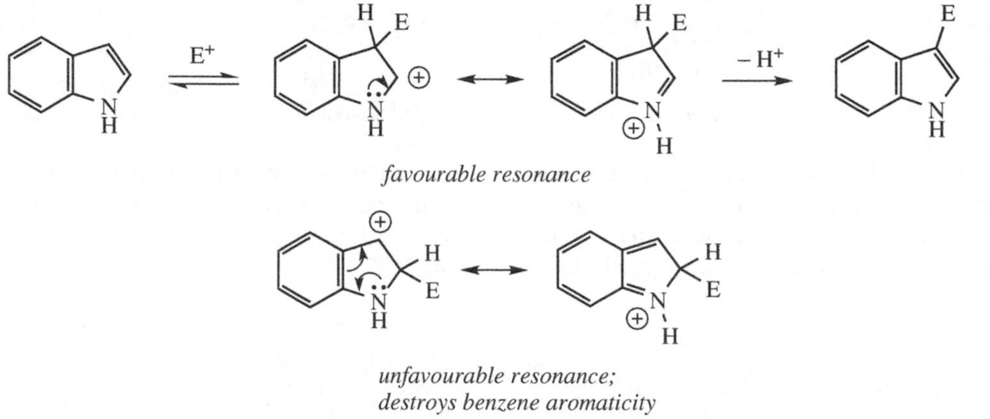

favourable resonance

*unfavourable resonance;
destroys benzene aromaticity*

Indole is very reactive towards **electrophiles**, and it is usually necessary to employ reagents of low reactivity. Nitration with $HNO_3-H_2SO_4$ is unsuccessful (compare pyrrole), but can be achieved using benzoyl nitrate.

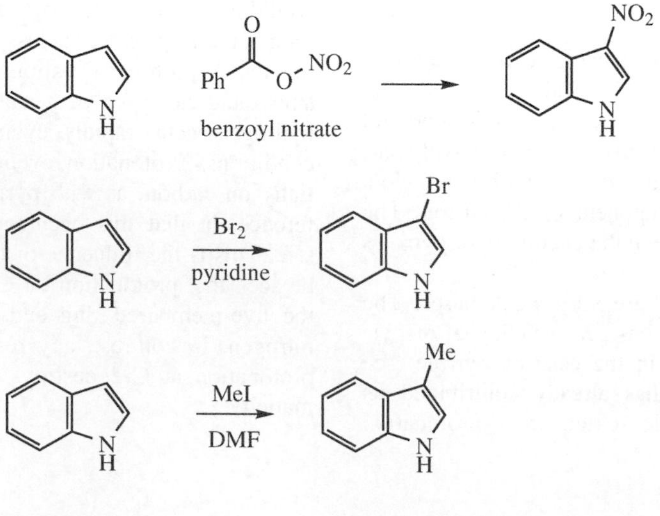

benzoyl nitrate

Br_2 / pyridine

MeI / DMF

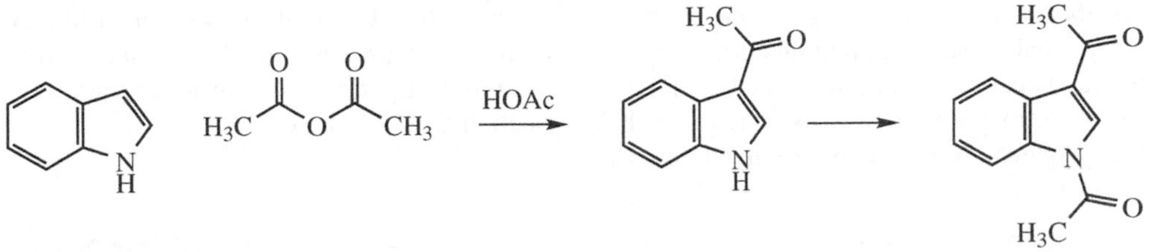

It is also possible to brominate and methylate at C-3; however, conditions must be controlled carefully, since further electrophilic reactions may then occur. Treatment with acetic anhydride leads to 1,3-diacetylindole.

Indole reacts readily as the nucleophile in **Mannich reactions**. This provides convenient access to other derivatives, as shown below.

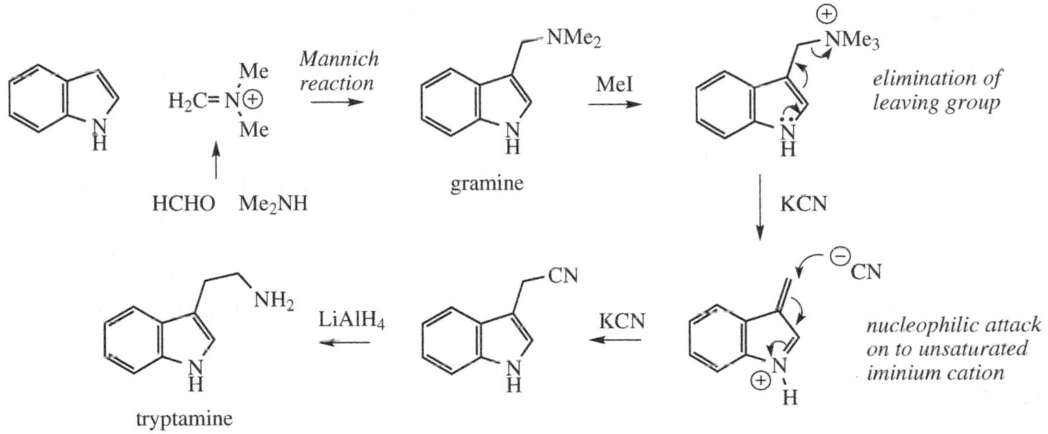

Thus, quaternization at the side-chain nitrogen allows ready elimination of trimethylamine. This is facilitated by the electron-releasing ability of the indole nitrogen, and can be brought about by mild base. By choosing KCN as the mild base, the transient 3-methyleneindoleninium salt can be trapped by cyanide nucleophile, leading to indoleacetonitrile. Reduction of the nitrile group with LAH provides a route to tryptamine.

Simple addition to carbonyl compounds occurs under mild acidic conditions. Examples given illustrate reaction with acetone, an aldol-like reaction, and conjugate addition to methyl vinyl ketone, a Michael-like reaction. The first-formed alcohol products in aldol-like reactions usually dehydrate to give a 3-alkylidene-3H-indolium cation.

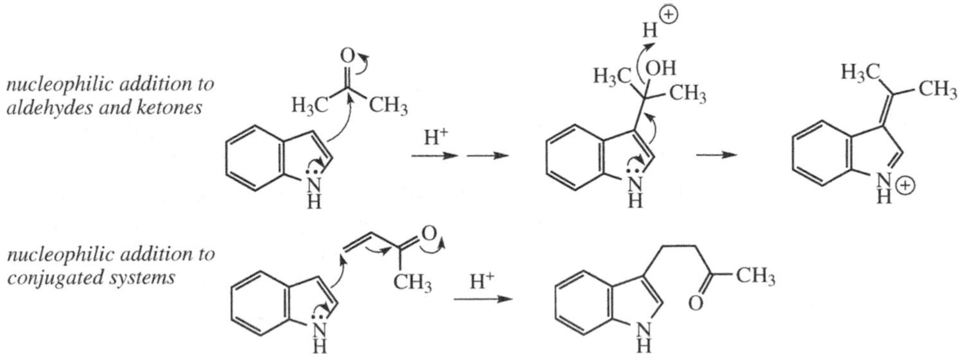

We noted above (see Section 11.5.1) that pyrrole, though a very weak base, is potentially acidic (pK_a 17.5). This was because the anion formed by losing the proton from nitrogen has a negative charge on the relatively electronegative nitrogen, but maintains its aromaticity. The indole anion is also formed by loss of the N–H proton (pK_a 16.2) using sodium amide or sodium hydride, or even a Grignard reagent (see Section 6.3.4) as base.

The indole anion is resonance stabilized, with negative charge localized mainly on nitrogen and C-3. It can now participate as a nucleophile, e.g. in alkylation reactions. However, this can lead to N-alkylation or C-alkylation at C-3. Which is the predominant product depends upon a number of variables; but, as a general rule, if the associated metal cation is sodium, then the anion is attacked at the site of highest electron density, i.e. the nitrogen. Where the cation is magnesium, i.e. the Grignard reagent, then the partial covalent bonding to nitrogen prevents attack there, and reaction occurs at C-3.

Box 11.10

Indoles in biochemistry

Some rather important indole derivatives influence our everyday lives. One of the most common ones is **tryptophan**, an indole-containing amino acid found in proteins (see Section 13.1). Only three of the protein amino acids are aromatic, the other two, phenylalanine and tyrosine being simple benzene systems (see Section 13.1). None of these aromatic amino acids is synthesized by animals and they must be obtained in the diet. Despite this, tryptophan is surprisingly central to animal metabolism. It is modified in the body by decarboxylation (see Box 15.3) and then hydroxylation to **5-hydroxytryptamine** (**5-HT, serotonin**), which acts as a neurotransmitter in the central nervous system.

L-tryptophan
(L-Trp)

5-hydroxy-L-Trp

5-hydroxytryptamine
(5-HT; serotonin)

Serotonin mediates many central and peripheral physiological functions, including contraction of smooth muscle, vasoconstriction, food intake, sleep, pain perception, and memory, a consequence of it acting on several distinct receptor types. Although 5-HT may be metabolized by monoamine oxidase, platelets and neurons possess a high-affinity mechanism for reuptake of 5-HT. This mechanism may be inhibited by the widely prescribed antidepressant drugs termed selective serotonin re-uptake inhibitors (SSRI), e.g. fluoxetine (Prozac®), thereby increasing levels of 5-HT in the central nervous system.

Migraine headaches that do not respond to analgesics may be relieved by the use of an agonist of the 5-HT$_1$ receptor, since these receptors are known to mediate vasoconstriction. Though the causes of migraine are not clear, they are characterized by dilation of cerebral blood vessels. 5-HT$_1$ agonists based on the 5-HT structure in current use include the sulfonamide derivative **sumatriptan**, and the more recent agents **naratriptan, rizatriptan** and **zolmitriptan**. These are of considerable value in treating acute attacks.

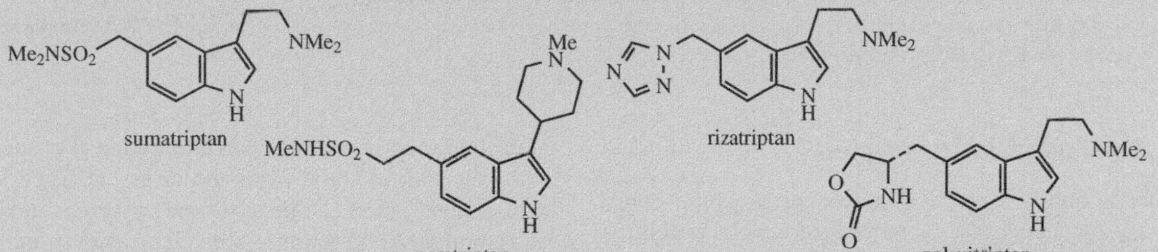

sumatriptan

naratriptan

rizatriptan

zolmitriptan

Several of the ergot alkaloids also interact with 5-HT receptors. Some are used medicinally, but the most notorious is the semi-synthetic derivative **lysergic acid diethylamide (LSD)**. This is itself an indole derivative, though the indole is part of a more complex fused-ring system. Nevertheless, from the structural similarities, it is not difficult to see why LSD might trigger 5-HT receptors. It has the additional ability to interact with noradrenaline and dopamine receptors, thus generating a complex pharmacological response. LSD is probably the most powerful pyschotomimetic known, intensifying and distorting perceptions. Experiences can vary from beautiful visions to living nightmares, and no two 'trips' are alike.

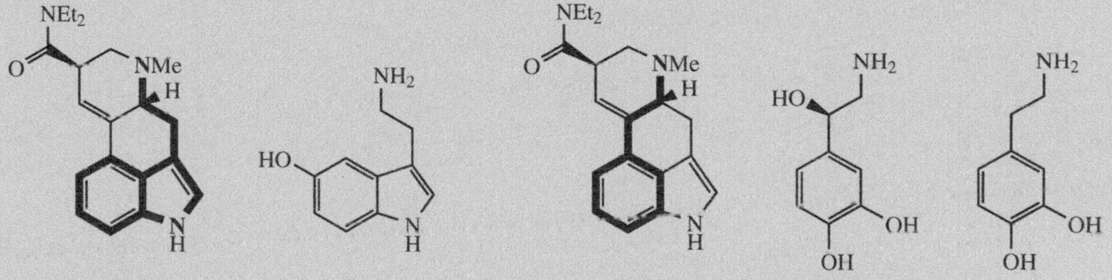

lysergic acid diethylamide
(lysergide; LSD)

5-hydroxytryptamine
(serotonin; 5-HT)

lysergic acid diethylamide
(lysergide; LSD)

noradrenaline
(norepinephrine)

dopamine

Also known to be hallucinogenic are the indole derivatives **psilocin** and **psilocybin** found in the so-called magic mushrooms, *Psilocybe* species. Ingestion of these small fungi causes visual hallucinations with rapidly changing shapes and colours. Psilocybin is the phosphate of psilocin; although based on 4-hydroxytryptamine, they also act on 5-HT receptors.

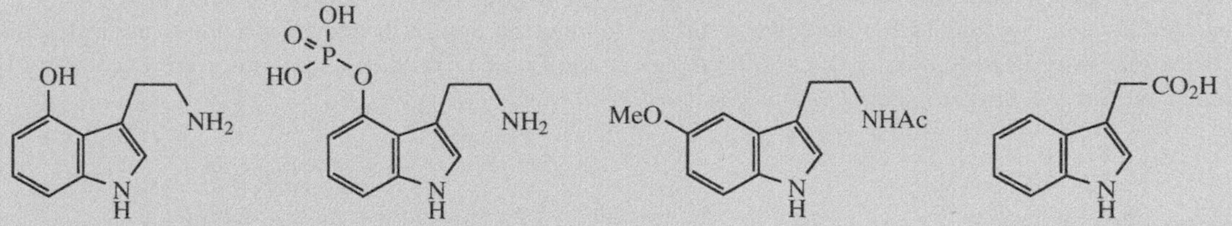

psilocin

psilocybin

melatonin

indole-3-acetic acid

Melatonin is *N*-acetyl-5-methoxytryptamine, a simple derivative of serotonin. It is a natural hormone secreted by the pineal gland in the brain during the hours of darkness. It is involved in controlling the body's day–night rhythm, the ability to sleep during the night, and to stay awake during the day. When given as a drug, melatonin induces sleep, and adjusts the internal body clock. It is now used as a means of reducing the effects of jet-lag.

Plants also require hormones to trigger their growth patterns. One of these is **indole-3-acetic acid**, which controls cell elongation and is produced in the growing shoot tips.

One of the major subdivisions of plant alkaloids is termed the indole alkaloid group. All contain the basic indole heterocycle, and many have valuable pharmacological activity that can be exploited in drug materials. The indole portion is very often fused to another heterocycle; we shall see some typical structures in Section 11.9, where we shall consider them under fused heterocycles.

11.9 Fused heterocycles

There is ample scope for increasing structural complexity by fusing two or more heterocycles together. Shown below are a few of the ring systems encountered in natural compounds, many of which have interesting, and potentially useful, biological properties. Note that in some cases the rings are fused so that the heteroatom can be at the ring junction and is thus common to both rings. This gives us even more combinations.

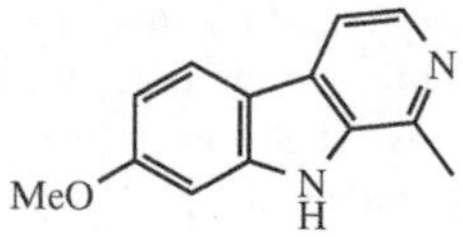

harmine
psychoactive

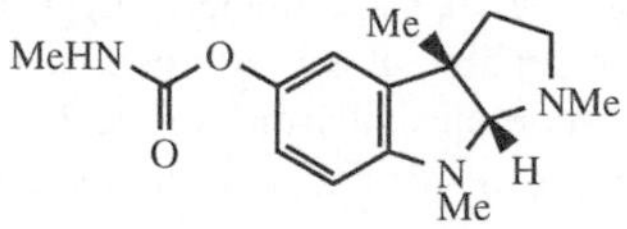

physostigmine
(eserine)
miotic; anticholinesterase

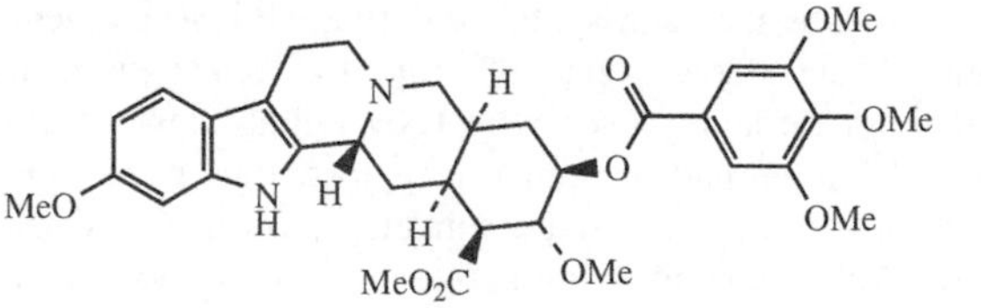

reserpine
antihypertensive

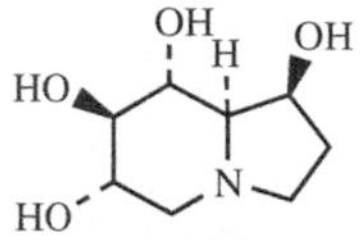

castanospermine
antiviral

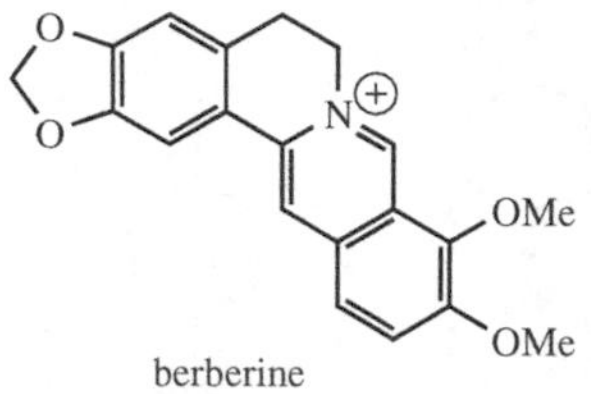

berberine
antibacterial

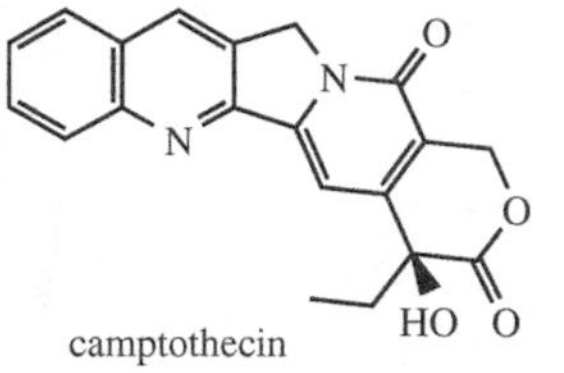

camptothecin
antitumour

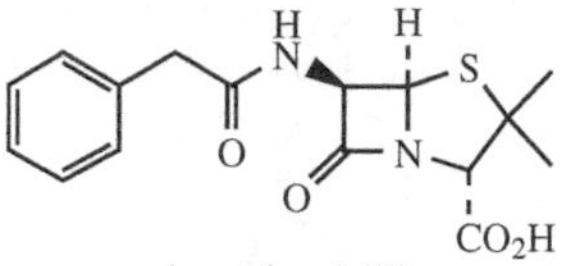

benzylpenicillin
(penicillin G)
antibacterial

We do not wish to consider these further, but instead we shall concentrate on just two groups of fused heterocycles of particular importance, the purines and pteridines.

Both purine and pteridine are parent heterocycles for **nomenclature** purposes. The systematic procedure for naming fused heterocycles is an extension of that we saw in Section 11.8 where we considered a benzene ring fused to a heterocycle. The main difference is that we have to identify bonds in two different rings to indicate the fusion. We use lettering for bonds in one heterocycle and numbers for bonds in the other. Numbering is used for the 'substituent' ring and lettering for the 'root' ring, and all are put in square brackets between substituent and root. We do not wish to include a great amount of detail, but we shall use purine and pteridine to illustrate the approach and to provide a modest level of familiarity for when such names are encountered.

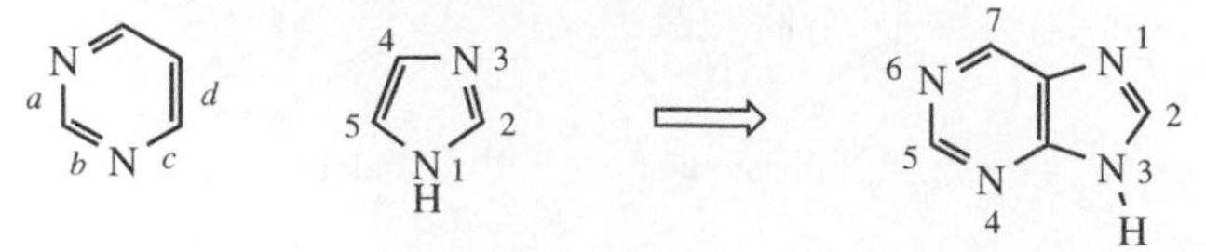

pyrimidine   imidazole

3*H*-imidazo[4,5-*d*]pyrimidine
purine

*this is the numbering appropriate
for the imidazopyrimidine;
purine has non-systematic numbering*

pyrimidine   pyrazine

pyrazino[2,3-*d*]pyrimidine
pteridine

The fused heterocycle is then given its own numbering system, starting adjacent to a bridgehead atom to generate the lowest number for the first heteroatom.

11.9.1 Purines

Purines, along with pyrimidines (see Box 11.5), feature as bases in the nucleic acids, DNA and RNA. A purine is the product of fusing a five-membered imidazole ring onto a six-membered pyrimidine ring. The accepted numbering system unfortunately is non-systematic, and treats purine as a pyrimidine derivative, the pyrimidine ring being numbered first and separately from the other ring.

purine

9*H*-purine   7*H*-purine

Purine in solution exists as a roughly equimolar mixture of two tautomeric forms, 9*H*-purine and 7*H*-purine, **tautomerism** involving the imidazole ring as we have noted earlier (see Section 11.7.2). The purine systems in nucleic acids, adenine and guanine, are **aminopurines**. The amino group is on the pyrimidine ring and, as with aminopyrimidines, these compounds exist as their amino tautomers (see Section 11.6.2). Guanine also has an oxygen substituent on the pyrimidine ring, and this adopts the carbonyl form, also following the behaviour of oxypyrimidines (see Section 11.6.2). Because adenine and guanine in nucleic acids are bonded to a sugar through N-9, the additional potential for tautomerism in the imidazole ring is no longer of concern.

adenine, A

*preferred form
in nucleic acids*

guanine, G

*preferred form
in nucleic acids*

Purines are quite weak bases. The conjugate acid of purine itself has pK_a 2.5. Protonation is found to be predominantly on N-1, though all three possible *N*-protonated forms are produced. This is perhaps unexpected, in that protonation on N-7 would provide a cation that is resonance stabilized in the imidazole ring. However, the observed pK_a more closely resembles that of pyrimidine (pK_a 1.3) rather than that of imidazole (pK_a 7.0). Amino groups increase basicity (adenine pK_a 4.3), though the oxygen substituent in guanine reduces the effect of the amino group (guanine pK_a 3.3). In the aminopurines, the position of protonation appears to be N-1 in adenine, whereas it is N-7 in guanine; this presumably reflects the opposing effects provided by amino groups (electron donating) and carbonyl groups (electron withdrawing).

protonation at N-1   *protonation at N-7*

purine pK_a 2.5

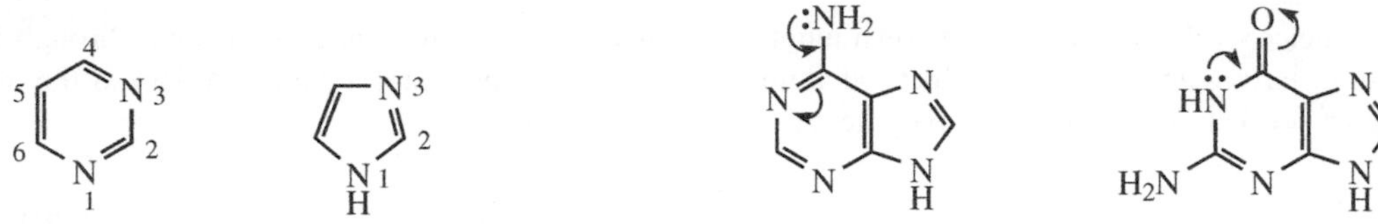

pyrimidine   imidazole   adenine   guanine
pK_a 1.3 pK_a 7.0 pK_a 4.3 pK_a 3.3

Purine has an acidic pK_a of 8.9, making it somewhat more acidic than phenol (pK_a 10), and a stronger acid than imidazole (pK_a 14.2). The

N-9 proton is lost giving an anion with substantial resonance delocalization of charge.

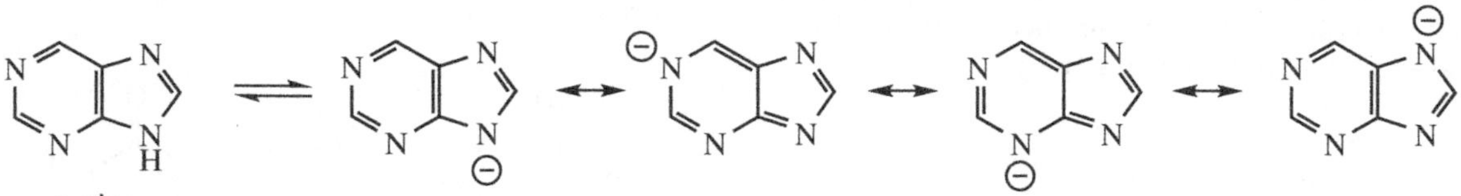

purine
pK_a 8.9

The acidities of adenine (pK_a 9.8) and guanine (pK_a 9.9) are similar, though different protons are removed. Adenine loses the N-9 proton, but guanine is ionized at N-1. N-1 is part of an amide-like system,

and charge in the conjugate base can be delocalized to the more favourable electronegative oxygen. The effect is most pronounced in uric acid, a metabolite of purines (see Box 11.11).

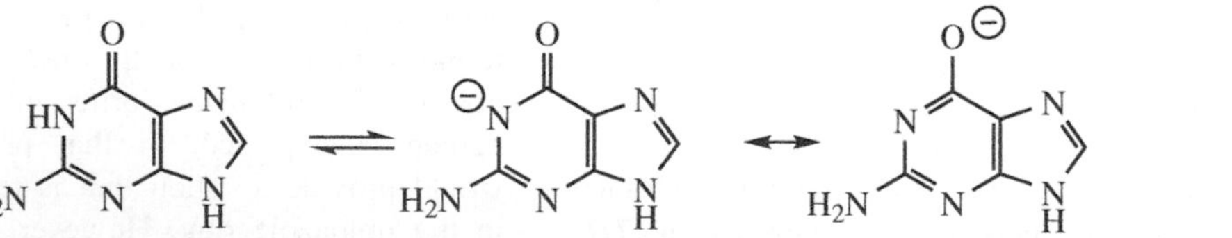

guanine
pK_a 9.9

Box 11.11

Uric acid, a purine metabolite

Nucleic acid degradation in humans and many other animals leads to production of **uric acid**, which is then excreted. The process initially involves purine nucleotides, adenosine and guanosine, which are combinations of adenine or guanine with ribose (see Section 14.1). The purine bases are subsequently modified as shown.

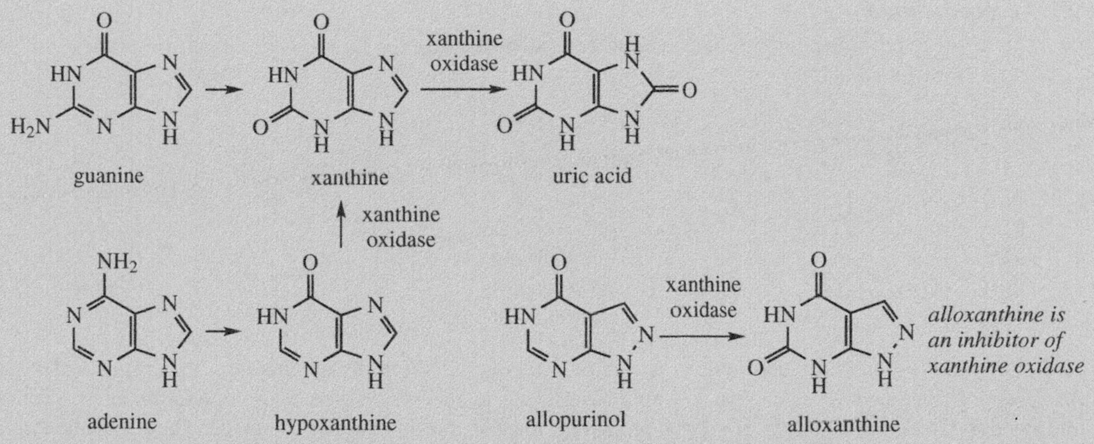

The amino groups are replaced with oxygen. Although here a biochemical reaction, the same can be achieved under acid-catalysed hydrolytic conditions, and resembles the nucleophilic substitution on pyrimidines (see Section 11.6.1). The first-formed hydroxy derivative would then tautomerize to the carbonyl structure. In the case of guanine, the product is xanthine, whereas adenine leads to hypoxanthine. The latter compound is also converted into xanthine by an oxidizing enzyme, **xanthine oxidase**. This enzyme also oxidizes xanthine at C-8, giving uric acid.

Uric acid is not a carboxylic acid, but is a relatively strong acid with pK_a 5.8. It has an 'all-amide' structure, and there are four potential sites for loss of a proton. Deprotonation occurs at N-9. Loss of this proton generates a conjugate base in which the charge can be delocalized to oxygen, giving maximum charge distribution. One resonance form is particularly favourable in having aromaticity in both rings.

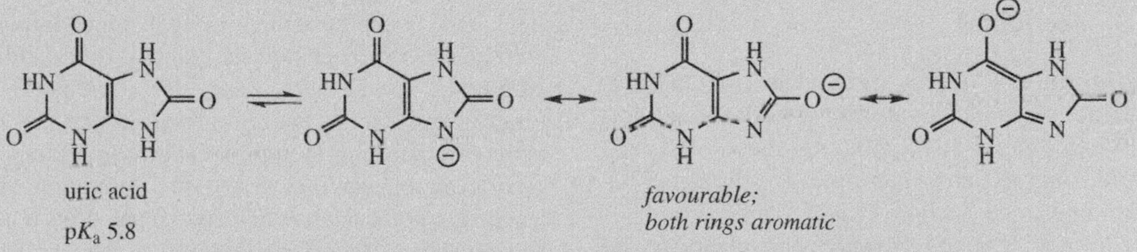

Impaired purine metabolism can lead to a build up of uric acid, and deposition of salts of uric acid as crystals in the joints. This causes the painful condition known as gout. One way of treating gout is to reduce uric acid biosynthesis by specific inhibition of the enzyme xanthine oxidase. The hypoxanthine analogue **allopurinol** is a drug that is used for this purpose. Allopurinol resembles hypoxanthine, though it contains a pyrazole ring rather than an imidazole ring. Allopurinol is oxidized by the enzyme to alloxanthine. This product then acts as an inhibitor of the enzyme, binding to the enzyme, but not being modified further and not being released.

Box 11.12

Caffeine, theobromine, and theophylline

After the nucleic acid purines adenine and guanine, the next most prominent purine in our everyday lives is probably caffeine. **Caffeine**, in the form of beverages such as tea, coffee, and cola, is one of the most widely consumed and socially accepted natural stimulants. Closely related structurally are **theobromine** and **theophylline**. Theobromine is a major constituent of cocoa, and related chocolate products. Caffeine is also used medicinally,

Box 11.12 (continued)

but theophylline is much more important as a drug compound because of its muscle relaxant properties, utilized in the relief of bronchial asthma.

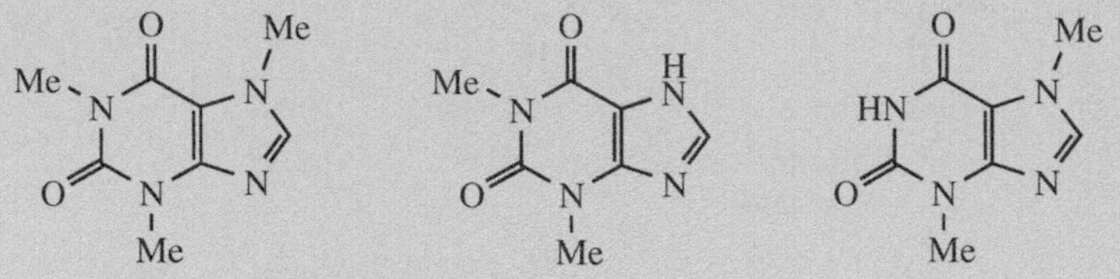

caffeine   theophylline   theobromine

These compounds competitively inhibit phosphodiesterase, resulting in an increase in **cyclic AMP** (see Box 14.3) and subsequent release of adrenaline. This leads to the major effects: a stimulation of the central nervous system (CNS), a relaxation of bronchial smooth muscle, and induction of diuresis. These effects vary in the three compounds. Caffeine is the best CNS stimulant, and has weak diuretic action. Theobromine has little stimulant action, but has more diuretic activity and also muscle relaxant properties. Theophylline also has low stimulant action and is an effective diuretic, but it relaxes smooth muscle better than caffeine or theobromine.

It has been estimated that beverage consumption may provide the following amounts of caffeine per cup or average measure: coffee, 30–150 mg (average 60–80 mg); instant coffee, 20–100 mg (average 40–60 mg); decaffeinated coffee, 2–4 mg; tea, 10–100 mg (average 40 mg); cocoa, 2–50 mg (average 5 mg); cola drink, 25–60 mg. The maximal daily intake should not exceed about 1 g to avoid unpleasant side effects, e.g. headaches, restlessness. An acute lethal dose is about 5–10 g.

Caffeine and theobromine may be obtained in large quantities from natural sources, or they may be obtained by total or partial synthesis. Theophylline is usually produced by total synthesis.

11.9.2 Pteridines

In **pteridines**, we have a pyrimidine ring fused to a pyrazine ring. There are, of course, a number of possible ways of combining these two six-membered ring systems; pteridines are pyrazino[2,3-d]pyrimidines (see Section 11.9).

We do not want to consider the chemistry of the pteridine ring system here, but instead we shall look at the structures of two rather important pteridine-based biochemicals, namely **folic acid** and **riboflavin**. In the latter case, the pteridine is also fused further to a benzene ring, giving an even more complex ring system, a benzo[g]pteridine. The accompanying diagram shows the derivation of the fused-ring nomenclature. The oxygenated form of the benzopteridine found in riboflavin is also called an isoalloxazine.

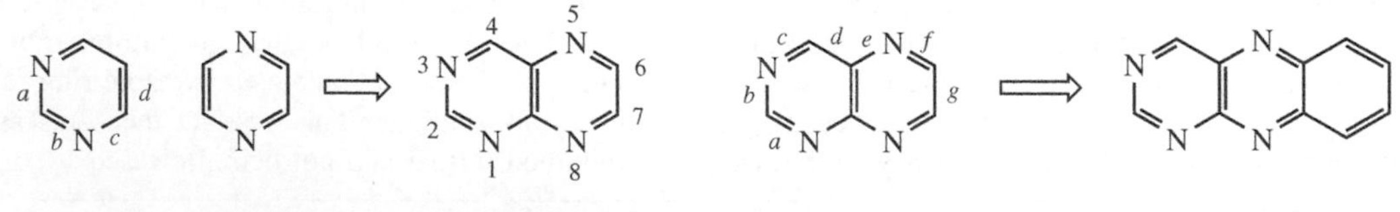

pyrimidine   pyrazine   pteridine
pyrazino[2,3-d]pyrimidine

benzo[g]pteridine

Box 11.13

Folic acid

Folic acid (vitamin B9) is a conjugate of a pteridine unit, p-aminobenzoic acid, and glutamic acid. Deficiency of folic acid leads to anaemia, and it is also standard practice to provide supplementation during pregnancy to reduce the incidence of spina bifida.

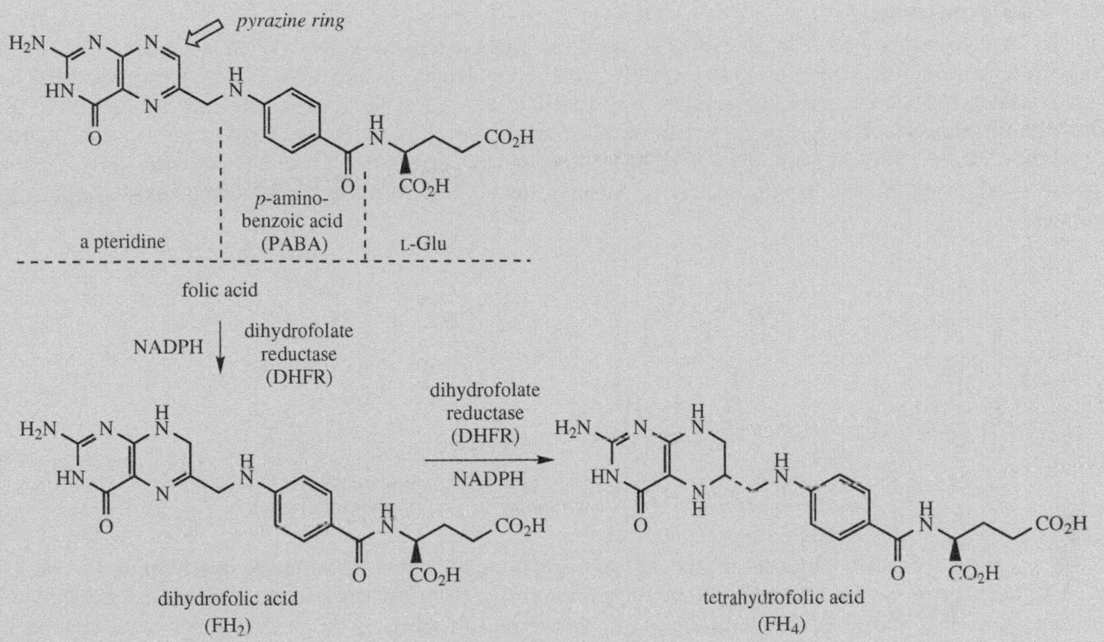

folic acid

a pteridine | *p*-amino-benzoic acid (PABA) | L-Glu

dihydrofolic acid (FH₂)

tetrahydrofolic acid (FH₄)

Folic acid becomes sequentially reduced in the body by the enzyme dihydrofolate reductase to give dihydrofolic acid (FH$_2$) and then tetrahydrofolic acid (FH$_4$). Reduction occurs in the pyrazine ring portion.

Tetrahydrofolic acid then functions as a carrier of one-carbon groups for amino acid and nucleotide metabolism. The basic ring system is able to transfer methyl, methylene, methenyl, or formyl groups, and it utilizes slightly different reagents as appropriate. These are shown here; for convenience, we have left out the benzoic acid–glutamic acid portion of the structure. These compounds are all interrelated, but we are not going to delve any deeper into the actual biochemical relationships.

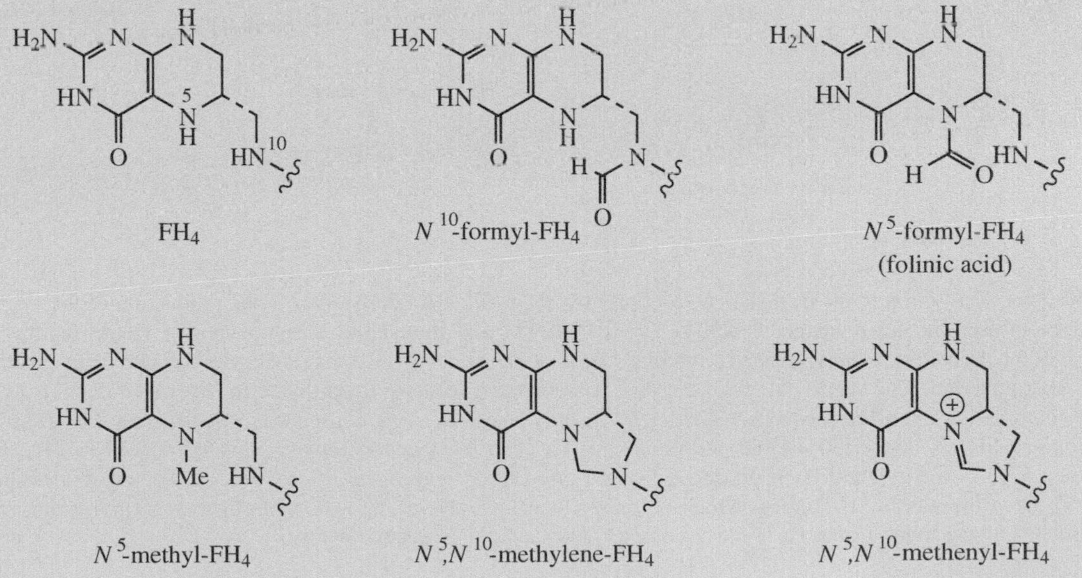

FH$_4$

N^{10}-formyl-FH$_4$

N^5-formyl-FH$_4$ (folinic acid)

N^5-methyl-FH$_4$

N^5,N^{10}-methylene-FH$_4$

N^5,N^{10}-methenyl-FH$_4$

Box 11.13 (continued)

In any case, you might be able to analyse some of the relationships on a purely chemical basis. For example, tetrahydrofolic acid reacts readily and reversibly with formaldehyde to produce N^5,N^{10}-methylene-FH$_4$. You could consider N-5 of the reduced pteridine ring reacting with formaldehyde (a one-carbon reagent) to give an iminium cation, which could then cyclize via nucleophilic attack of N-10. We might also consider reducing the iminium cation with, say, borohydride to give the *N*-methyl derivative. These are not necessarily the same as what is occurring in the enzymic reactions, but they should help to make the structures appear rather more familiar.

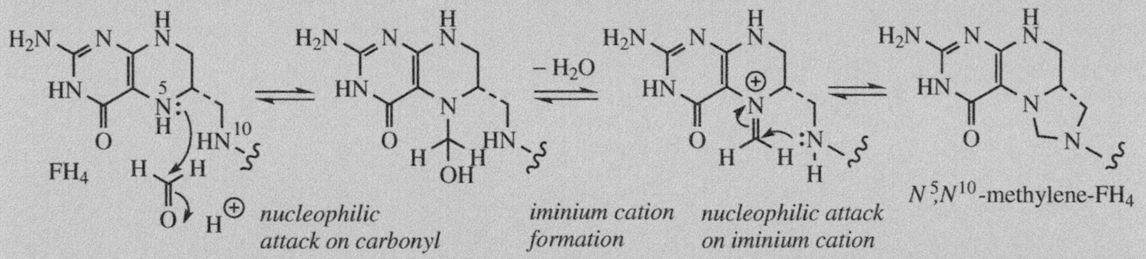

nucleophilic   iminium cation   nucleophilic attack
attack on carbonyl   formation   on iminium cation

N^5,N^{10}-methylene-FH$_4$

Now we have seen that the usual reagent for biological methylations is **S-adenosylmethionine (SAM)** (see Box 6.4). One occasion where SAM is not employed, for fairly obvious reasons, is the regeneration of methionine from homocysteine, after a SAM methylation. For this, N^5-methyl-FH$_4$ is the methyl donor, with vitamin B$_{12}$ (see Box 11.4) also playing a role as coenzyme.

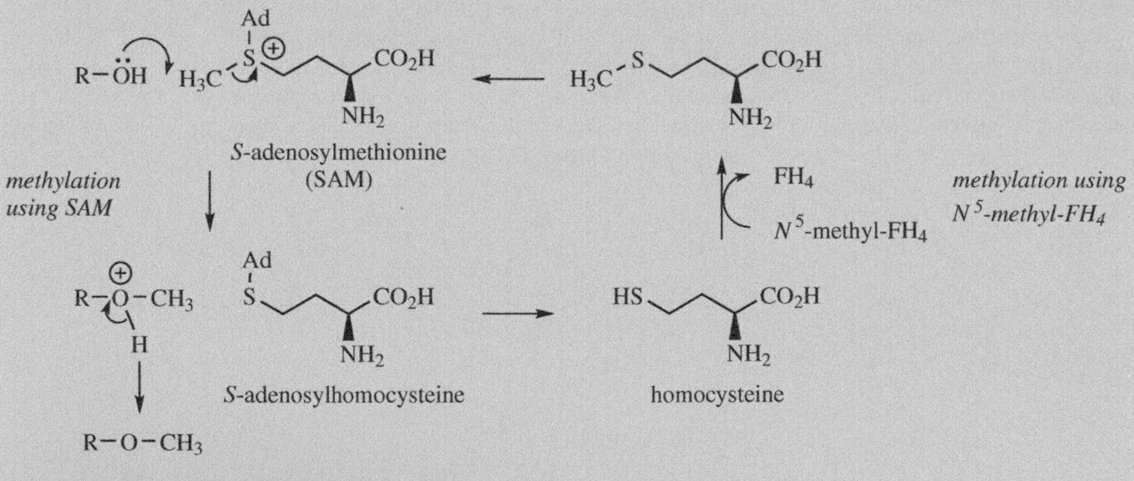

Another vitally important methylation reaction involving folic acid derivatives is the production of the nucleic acid base **thymine** from **uracil**. Uracil is found in RNA, and thymine is a component of DNA; thymine is the methyl derivative of uracil. For continuing DNA synthesis, it is necessary to methylate uracil. In practice, it is the nucleotide deoxyuridylate (dUMP) that is methylated to deoxythymidylate (dTMP) (see Section 14.1). The methylating agent employed here is N^5,N^{10}-methylene-FH$_4$. As a consequence of this reaction, N^5,N^{10}-methylene-FH$_4$ is converted into dihydrofolic acid. To keep the reaction flowing, this is reduced to FH$_4$, and further N^5,N^{10}-methylene-FH$_4$ is produced using a one-carbon reagent. In this process, the one-carbon reagent comes from the amino acid serine, which is transformed into glycine by loss of its hydroxymethyl group. The chemistry of the transformations is fairly complex and outside our requirements.

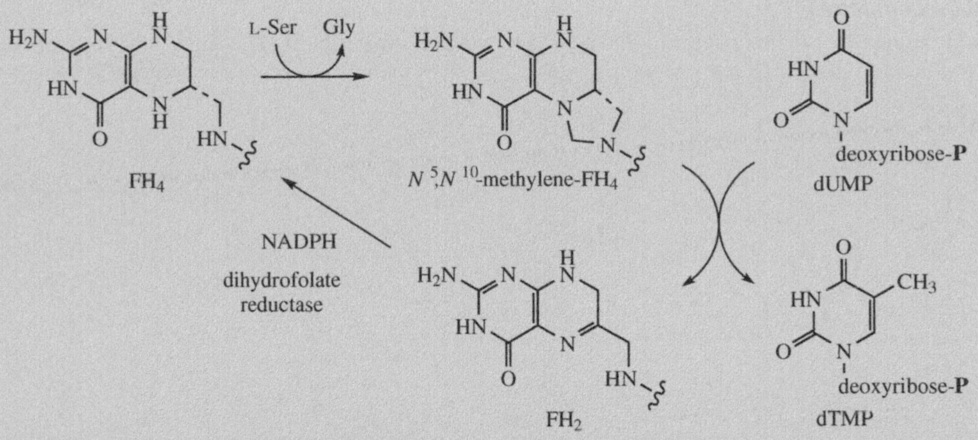

Folic acid derivatives are essential for DNA synthesis, in that they are cofactors for certain reactions in purine and pyrimidine biosynthesis, including the uracil–thymine methylation just described. They are also cofactors for several reactions relating to amino acid metabolism. The folic acid system thus offers considerable scope for drug action.

Mammals must obtain their tetrahydrofolate requirements from their diet, but microorganisms are able to synthesize this material. This offers scope for selective action and led to the use of **sulfanilamide** and other antibacterial sulfa drugs, compounds that competitively inhibit the biosynthetic enzyme (dihydropteroate synthase) that incorporates *p*-aminobenzoic acid into the structure (see Box 7.23).

Rapidly dividing cells need an abundant supply of dTMP for DNA synthesis, and this creates a need for dihydrofolate reductase activity. Specific **dihydrofolate reductase inhibitors** have become especially useful as antibacterials, e.g. **trimethoprim**, and antimalarial drugs, e.g. **pyrimethamine**.

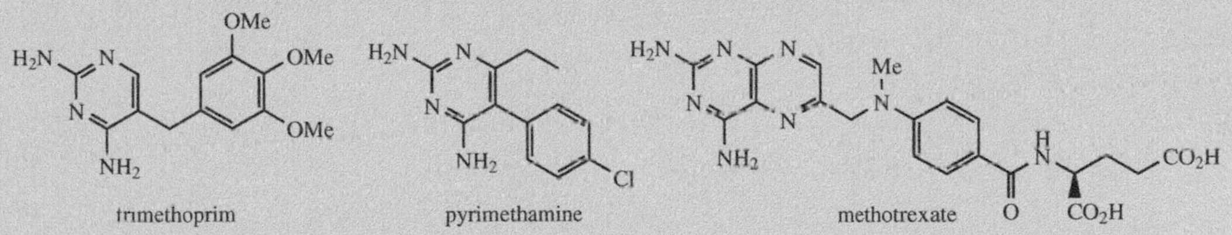

These are pyrimidine derivatives and are effective because of differences in susceptibility between the enzymes in humans and in the infective organism. Anticancer agents based on folic acid, e.g. **methotrexate**, inhibit dihydrofolate reductase, but they are less selective than the antimicrobial agents and rely on a stronger binding to the enzyme than the natural substrate has. They also block pyrimidine biosynthesis. Methotrexate treatment is potentially lethal to the patient, and is usually followed by 'rescue' with folinic acid (N^5-formyl-tetrahydrofolic acid) to counteract the folate-antagonist action. The rationale is that folinic acid 'rescues' normal cells more effectively than it does tumour cells.

Box 11.14

Riboflavin

Riboflavin (vitamin B$_2$) is a component of **flavin mononucleotide (FMN)** and **flavin adenine dinucleotide (FAD)**, coenzymes that play a major role in oxidation–reduction reactions (see Section 15.1.1). Many key enzymes involved in metabolic pathways are actually covalently bound to riboflavin, and are thus termed flavoproteins.

Box 11.14 (continued)

Riboflavin contains an isoalloxazine ring linked to the reduced sugar ribitol. The sugar unit in riboflavin is the non-cyclic ribitol, so that FAD and FMN differ somewhat from the nucleotides we encounter in nucleic acids.

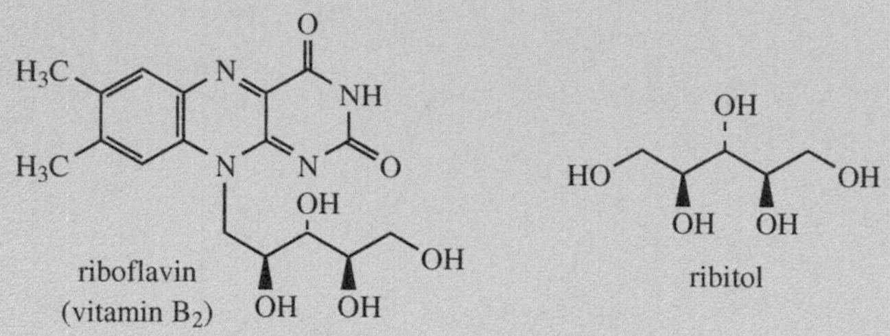

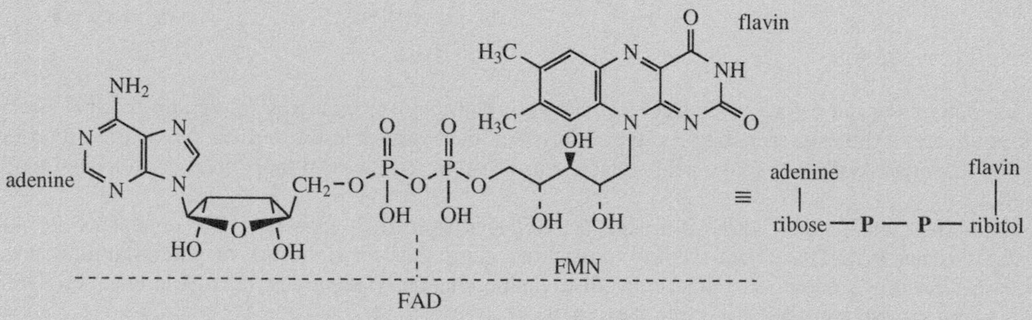

Riboflavin is widely available in foods; dietary deficiency is uncommon, but it manifests itself by skin problems and eye disturbances.

The **flavin nucleotides** are typically involved in the oxidations creating double bonds from single bonds. The flavin takes up two hydrogen atoms, represented in the figure as being derived by transfer of hydride from the substrate and a proton from the medium.

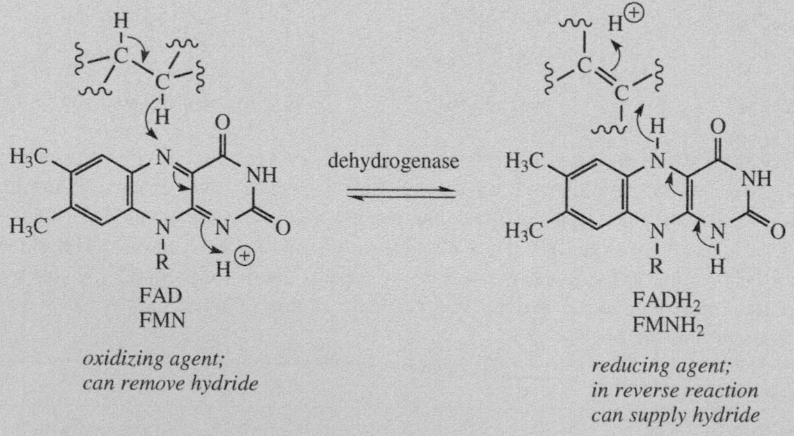

Reductive sequences involving flavoproteins may be represented as the reverse reaction, where hydride is transferred from the coenzyme, and a proton is obtained from the medium. The reaction mechanism shown here is in many ways similar to that in NAD⁺ oxidations, i.e. a combination of hydride and a proton (see Box 11.2); it is less easy to explain adequately why it occurs, and we do not consider any detailed explanation advantageous to our studies. We should register only that the reaction involves the N=C–C=N function that spans both rings of the pteridine system.

11.10 Some classic aromatic heterocycle syntheses

Our study of heterocyclic compounds is directed primarily to an understanding of their reactivity and importance in biochemistry and medicine. The synthesis of aromatic heterocycles is not, therefore, a main theme, but it is useful to consider just a few examples to underline the application of reactions we have considered in earlier chapters. From the beginning, we should appreciate that the synthesis of substituted heterocycles is probably not best achieved by carrying out substitution reactions on the simple heterocycle. It is often much easier and more convenient to design the synthesis so that the heterocycle already carries the required substituents, or has easily modified functions. We can consider two main approaches for **heterocycle synthesis**, here using pyridine and pyrrole as targets.

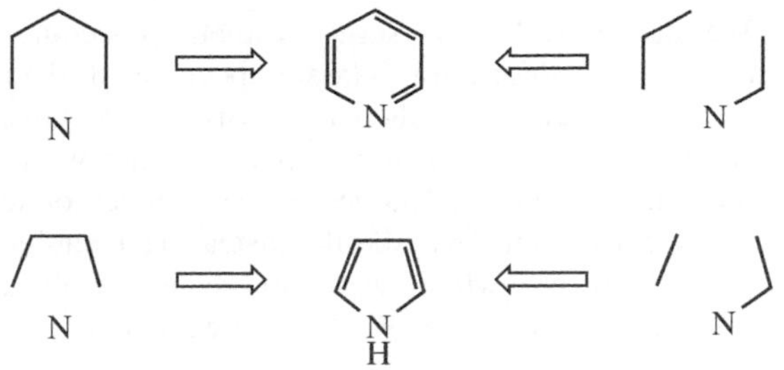

We can insert the heteroatom into the rest of the carbon skeleton, or attempt to join two units, one of which contains the heteroatom, by means of C–C and C–heteroatom linkages. To make the new bonds, two reaction types are most frequently encountered. Heteroatom–C bond formation is achieved using the **heteroatom as a nucleophile** to attack an electrophile such as a carbonyl group (see Section 7.7.1). **Aldol-type reactions** may be exploited for C–C bond formation (see Section 10.3), employing enamines and enols/enolate anions (see Section 10.5).

N–C bond formation

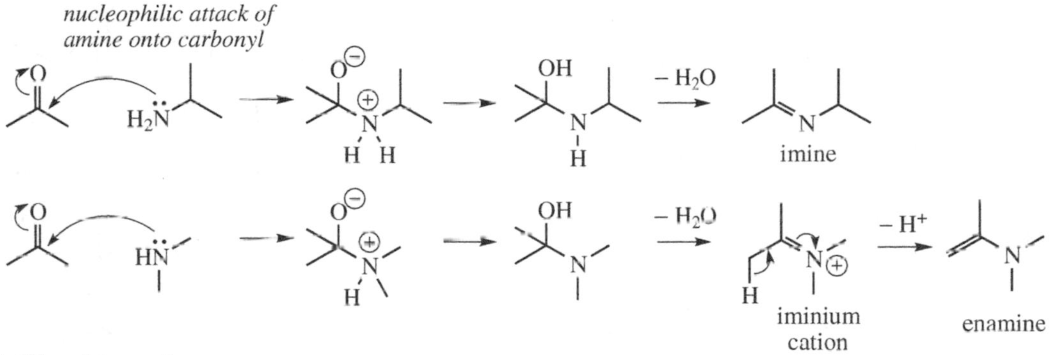

nucleophilic attack of amine onto carbonyl

C–C bond formation

attack of carbon nucleophile onto carbonyl

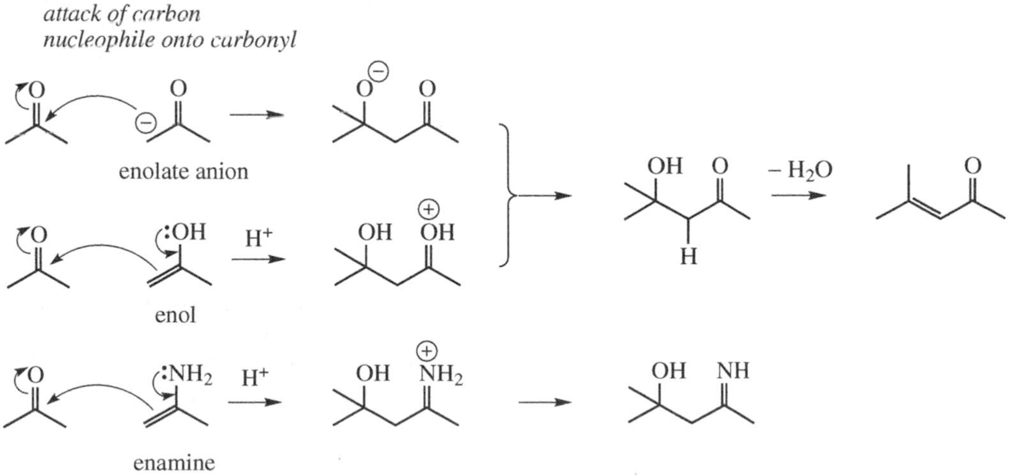

We shall now look at some synthetic procedures that merit the descriptor 'classic' because of their general application, and their longevity – some have been around for more than 100 years. Do not worry about remembering the names: these commemorate the originators, and we should instead concentrate on the chemistry, which we shall see is usually a combination of processes we have already met.

11.10.1 Hantzsch pyridine synthesis

In its simplest form, this consists of the condensation of a β-ketoester with an aldehyde and ammonia.

The product is a 1,4-dihydropyridine, which is subsequently transformed into the pyridine by oxidation. Several separate reactions occur during this synthesis, and the precise sequence of events may not be quite as shown below – they may be in a different order.

The normal Hantzsch synthesis leads to a symmetrical product. The diesters formed may be hydrolysed and decarboxylated using base to give pyridines with less substitution. Note that we are using the ester groups as activating species to facilitate enolate anion chemistry (see Section 10.9)

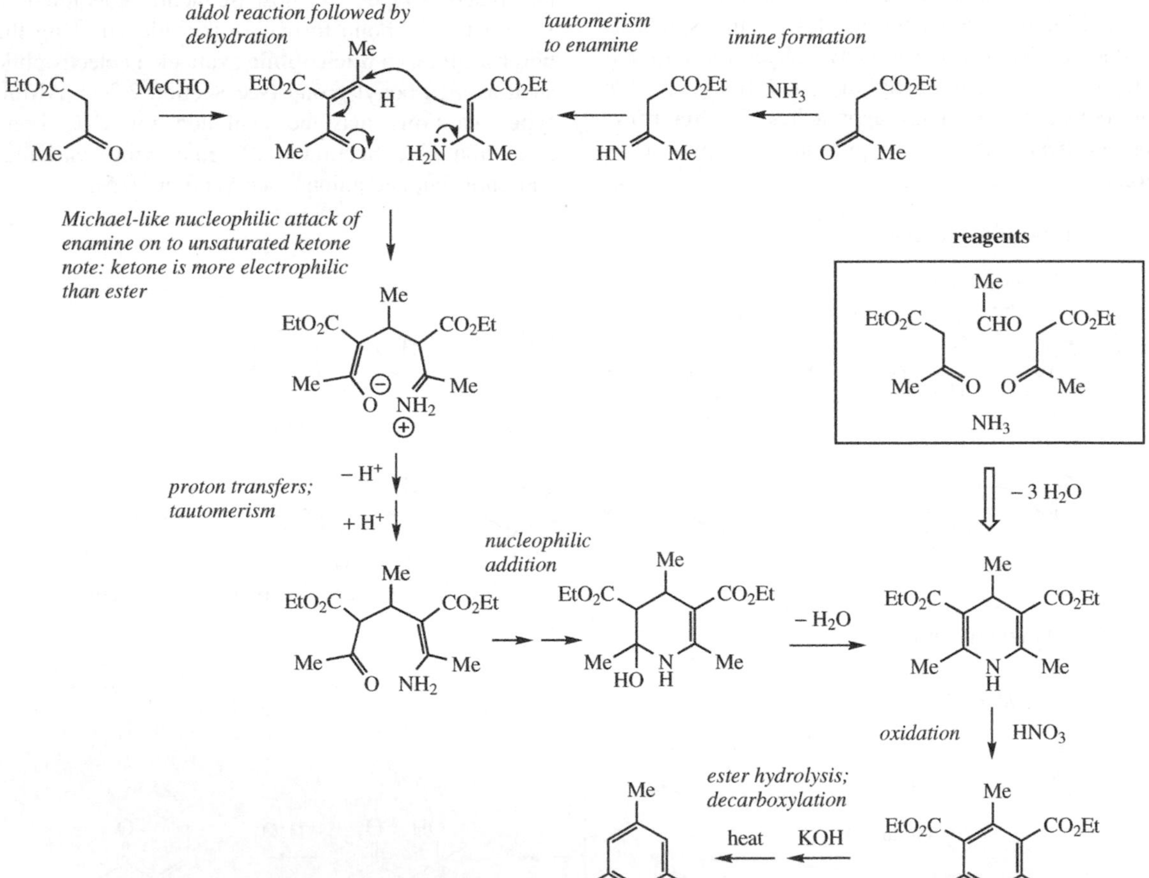

11.10.2 Skraup quinoline synthesis

The most general method for synthesizing quinolines employs aniline or a substituted aniline, glycerol, sulfuric acid, and an oxidizing agent such as a ferric

salt or nitrobenzene. The first step is acid-catalysed dehydration of glycerol to the unsaturated aldehyde acrolein. Variations of the Skraup synthesis use different acroleins instead of glycerols.

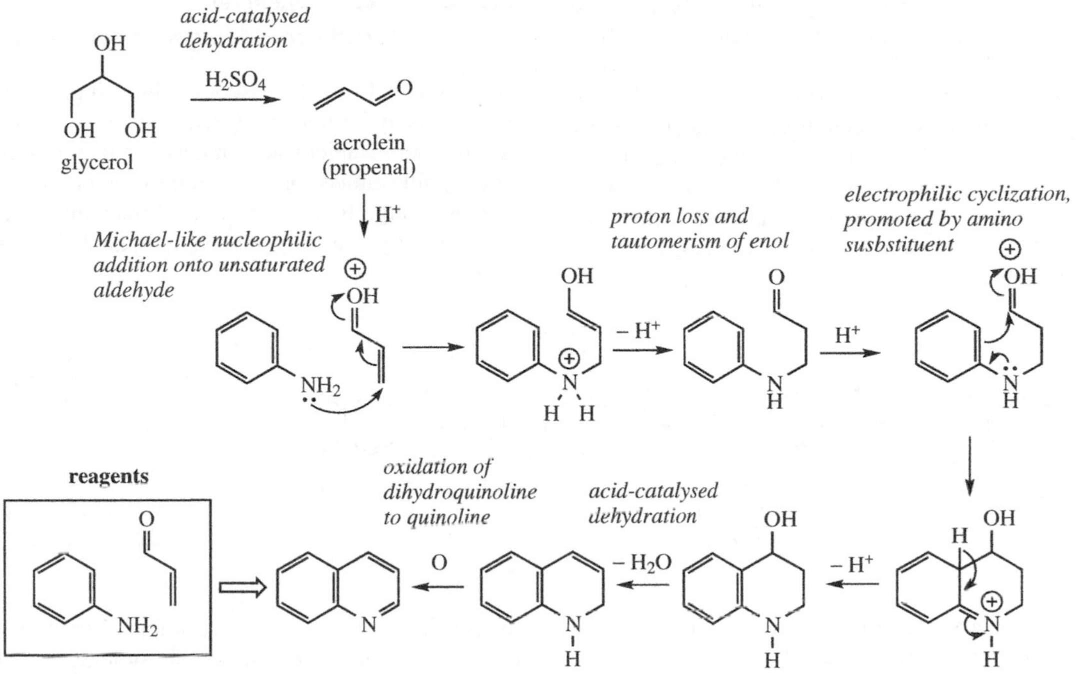

The initial product is a dihydroquinoline; it is formed via Michael-like addition, then an electrophilic aromatic substitution that is facilitated by the electron-donating amine function. A mild oxidizing agent is required to form the aromatic quinoline. The Skraup synthesis can be used with substituted anilines, provided these substituents are not strongly electron withdrawing and are not acid sensitive.

11.10.3 Bischler–Napieralski isoquinoline synthesis

Isoquinolines are easily prepared by the reaction of an acyl derivative of a β-phenylethylamine with a dehydrating agent, e.g. P_2O_5, then using a catalytic dehydrogenation to aromatize the intermediate 3,4-dihydroisoquinoline.

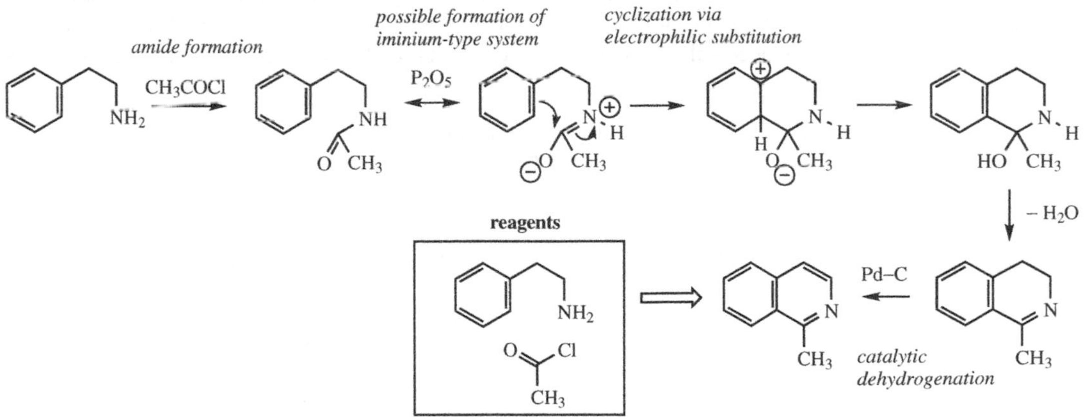

The crucial cyclization step is represented here as an electrophilic attack involving the aromatic ring and an iminium-type system, a resonance form of the amide, suitably coordinated to the phosphorus reagent. The cyclizing agent P_2O_5 also dehydrates the intermediate hydroxyamine to a dihydroisoquinoline. The isoquinoline is then obtained by heating over a catalyst, effectively reversing a catalytic hydrogenation reaction (see Section 9.4.3), facilitated by the generation of aromaticity in the product. As in the Skraup synthesis above, electron-withdrawing substituents on the aromatic ring will deactivate it towards electrophilic attack, whereas electron-donating substituents will favour the reaction.

11.10.4 Pictet–Spengler tetrahydroisoquinoline synthesis

This approach to the isoquinoline ring, albeit a reduced isoquinoline, is mechanistically similar to the Bischler–Napieralski synthesis, in that it involves electrophilic attack of an iminium cation on to an aromatic ring. In this case, the imine intermediate is formed by reacting a phenylethylamine with an aldehyde.

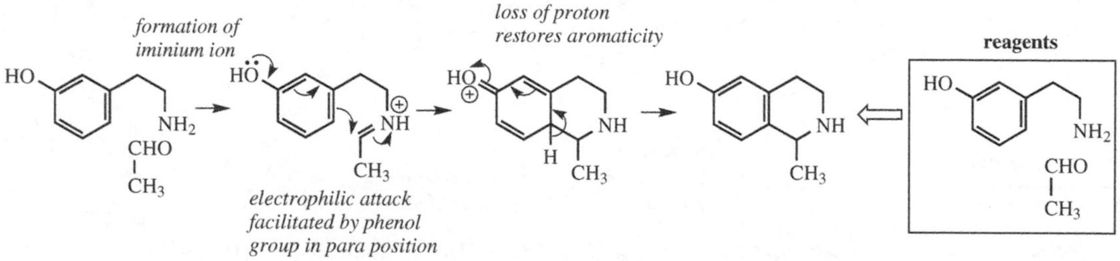

We have already met this reaction as an analogue of the Mannich reaction (see Box 10.7), which we then interpreted as nucleophilic attack of an electron-rich phenolic ring on to an iminium cation. Is it electrophilic or nucleophilic? It matters little; they are the same, though the descriptor used depends upon which species you consider the more important, the nucleophilic phenol or the electrophilic iminium cation. For effective cyclization, we need an electron-donating substituent *para* to the point of ring closure, since the Mannich-type electrophile is less reactive than the phosphorus-linked intermediates in the Bischler–Napieralski synthesis. It is also found that a similar group in the *ortho* position does not work, though we could still write an acceptable mechanism. With a good electron-donating substituent like

hydroxyl, the whole process, imine formation and cyclization, can occur under 'physiological' conditions, pH 6–7 at room temperature. In nature, this is precisely how tetrahydroisoquinoline alkaloids are biosynthesized, though the reactions are enzyme controlled.

11.10.5 Knorr pyrrole synthesis

This approach to the five-membered pyrrole ring reacts an α-aminoketone with a β-ketoester. The mechanism will probably involve imine formation then cyclization via an aldol-type reaction using the enamine nucleophile. Dehydration leads to the pyrrole. Only the key parts of this sequence are shown below.

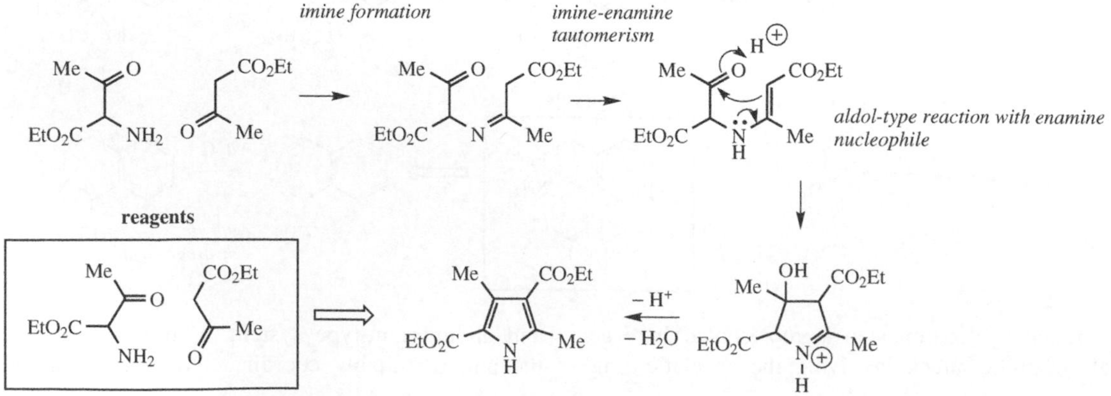

The synthesis works well only with an activated ester like ethyl acetoacetate. Otherwise, self-condensation of the α-aminoketone to a dihydropyrazine occurs more readily than the cyclization.

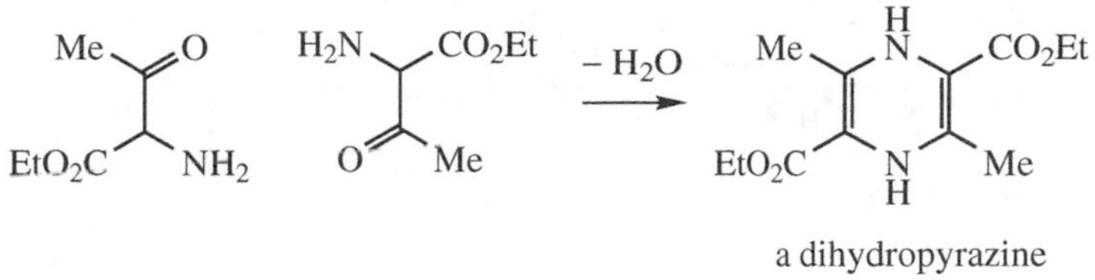

a dihydropyrazine

11.10.6 Paal–Knorr pyrrole synthesis

The other major route to pyrroles is the interaction of a 1,4-dicarbonyl compound with ammonia. The mechanism below shows successive nucleophilic additions of amino groups on to the carbonyls; but, since no intermediates have been isolated, the precise sequence of steps is speculative.

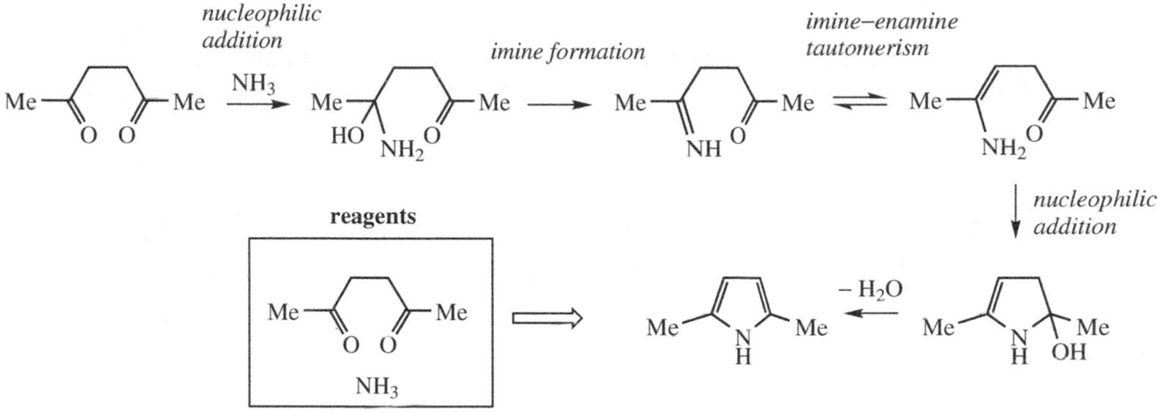

Note, however, that this synthesis gives furans if no ammonia is included. This would involve nucleophilic attack of an enol tautomer of the substrate on to the other carbonyl to give a hemiketal, followed by dehydration. The heteroatom is thus derived from a carbonyl oxygen. The procedure works well, and is usually carried out with acid catalyst under non-aqueous conditions.

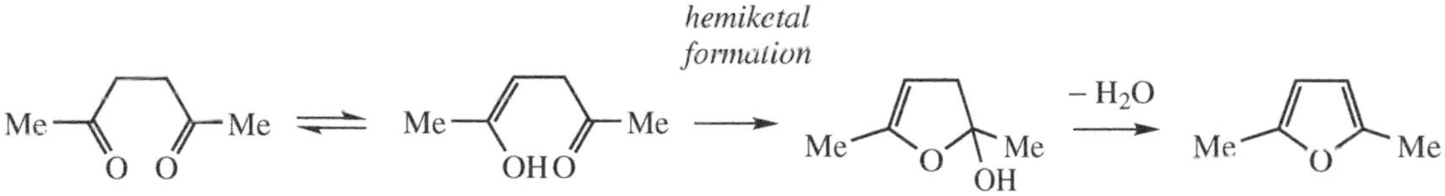

11.10.7 Fischer indole synthesis

The most useful route to indoles is the Fischer indole synthesis, in which an aromatic phenylhydrazone is heated in acid. The phenylhydrazone is the condensation product from a phenylhydrazine and an aldehyde or ketone. Ring closure involves a cyclic rearrangement process.

The hydrazine behaves as an amine towards a carbonyl compound and forms the imine-like product, a hydrazone. The cyclic rearrangement involves the enamine tautomer of this hydrazone, and proceeds because the cyclic flow of electrons forms a strong C–C bond whilst cleaving a weak N–N bond. This produces what appears to be a di-imine. One of these is involved in rearomatization and creates an aromatic amine. This then attacks the other imine function, and we get the nitrogen equivalent of a hemiketal (see Section 7.2). Finally, acid-catalysed elimination of ammonia gives the aromatic indole system.

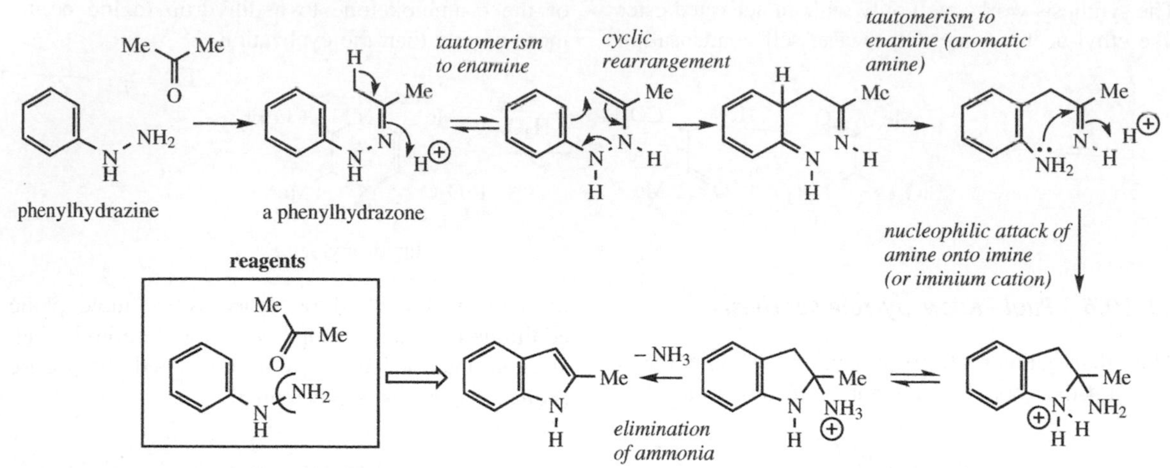

phenylhydrazine   a phenylhydrazone

tautomerism to enamine

cyclic rearrangement

tautomerism to enamine (aromatic amine)

nucleophilic attack of amine onto imine (or iminium cation)

reagents

− NH₃

elimination of ammonia

Unfortunately, the reaction fails with acetaldehyde and cannot, therefore, be used to synthesize indole itself. It is possible to use the ketoacid pyruvic acid instead and decarboxylate the product to yield indole.

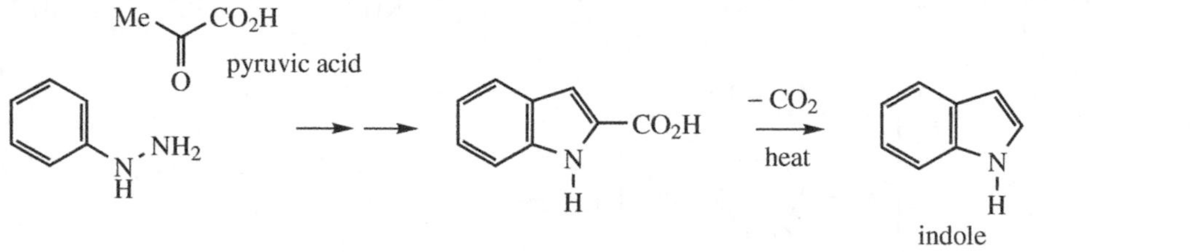

pyruvic acid

− CO₂

heat

indole

12

Carbohydrates

12.1 Carbohydrates

Many aspects of the chemistry of carbohydrates are not specific to this class of compounds, but are merely examples of the simple chemical reactions we have already met. Therefore, against usual practice, we have not attempted a full treatment of carbohydrate chemistry and biochemistry in this chapter. We want to avoid giving the impression that the reactions described here are something special to this group of compounds. Instead, we have deliberately used carbohydrates as examples of reactions in earlier chapters, and you will find suitable cross-references.

Carbohydrates are among the most abundant constituents of plants, animals, and microorganisms. Polymeric carbohydrates function as important food reserves, and as structural components in cell walls. Animals and most microorganisms are dependent upon the carbohydrates produced by plants for their very existence. Carbohydrates are the first products formed in photosynthesis, and are the products from which plants synthesize their own food reserves, as well as other chemical constituents. These materials then become the foodstuffs of other organisms. The main pathways of carbohydrate biosynthesis and degradation comprise an important component of

intermediary metabolism that is essential for all organisms (Chapter 15).

The name **carbohydrate** was introduced because many of the compounds had the general formula $C_x(H_2O)_y$, and thus appeared to be hydrates of carbon. The terminology is now commonly used in a much broader sense to denote polyhydroxy aldehydes and ketones, and their derivatives. **Sugars** or saccharides are other terms used in a rather broad sense to cover carbohydrate materials. Though these words link directly to compounds with sweetening properties, application of the terms extends considerably beyond this. A **monosaccharide** is a carbohydrate usually in the range C_3–C_9, whereas **oligosaccharide** covers small polymers comprised of 2–10 monosaccharide units. The term **polysaccharide** is used for larger polymers.

12.2 Monosaccharides

Six-carbon sugars (**hexoses**) and five-carbon sugars (**pentoses**) are the most frequently encountered monosaccharide carbohydrate units in nature. Primary examples of these two classes are the hexoses glucose and fructose, and the pentose ribose. Note the suffix -ose as a general indicator of carbohydrate nature.

Essentials of Organic Chemistry   Paul M Dewick
© 2006 John Wiley & Sons, Ltd

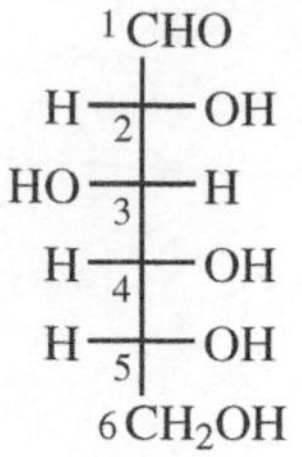

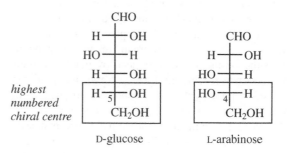

D-glucose 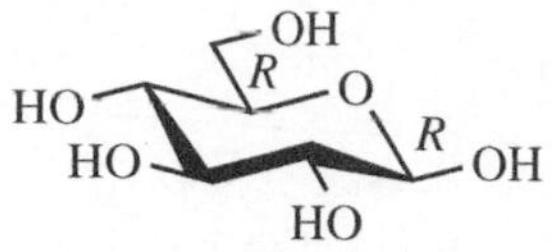   β-D-glucose (β-D-Glc)

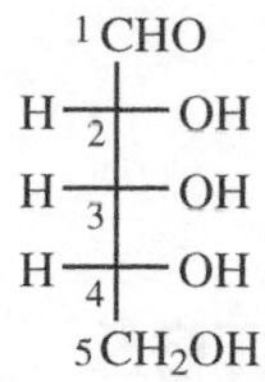

D-ribose 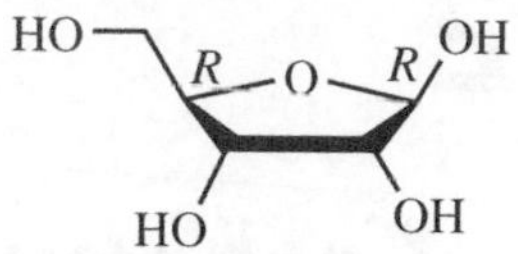   β-D-ribose (β-D-Rib)

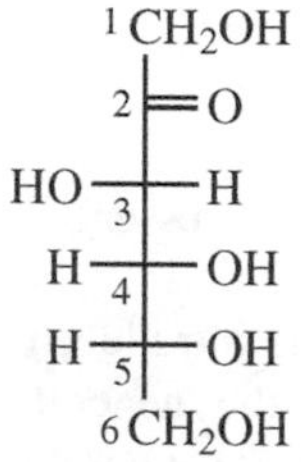

D-fructose 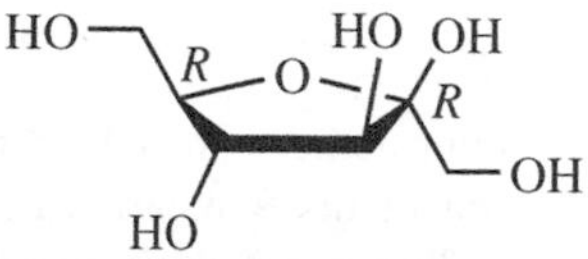   β-D-fructose (β-D-Fru)

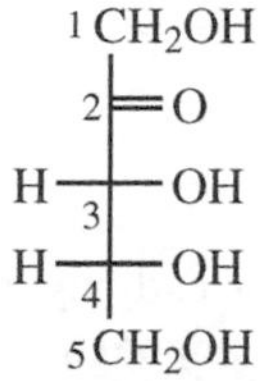

D-ribulose

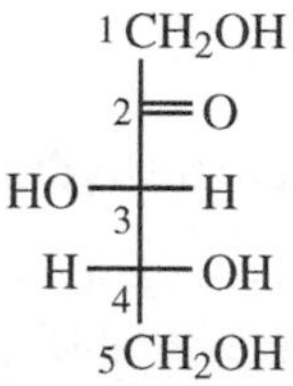

D-xylulose

The structures above show some of the fundamental features of carbohydrates. Initially, we have drawn these compounds in the form of **Fischer projections**, a depiction developed for these compounds to indicate conveniently the stereochemistry at each chiral centre (see Section 3.4.10). The Fischer projection is drawn as a vertical carbon chain with the group of highest oxidation state, i.e. the carbonyl group, closest to the top, and numbering takes place from the topmost carbon.

The carbonyl group in **glucose** and **ribose** is an aldehyde; such compounds are termed **aldoses**. **Fructose**, by contrast, has a ketone group and is therefore classified as a **ketose**. Glucose could also be termed an aldohexose and fructose a ketohexose, whereas ribose would be an aldopentose, names which indicate both the number of carbons and the nature of the carbonyl group. Another aspect of nomenclature is the use of the suffix -ulose to indicate a ketose. Fructose could thus be referred to as a hexulose, though we are more likely to see this suffix in the names of specific sugars, e.g. **ribulose** is a ketose isomer of the aldose ribose.

Each of these compounds has a prefix D- with the name. As we saw in Section 3.4.10, this indicates that the configuration at the highest numbered chiral centre is the same as that in D-(R)-(+)-glyceraldehyde; the alternative stereochemistry would be related to

L-(S)-(−)-glyceraldehyde and consequently be part of an L-sugar.

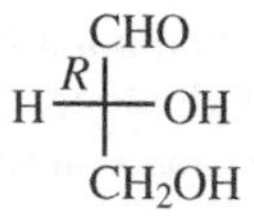

D-(+)-glyceraldehyde 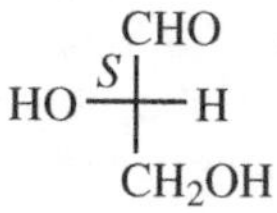   L-(−)-glyceraldehyde

highest numbered chiral centre

D-glucose   L-arabinose

Structures of the various D-aldoses in the range C_3–C_6 are shown below. These compounds are multifunctional structures, having a carbonyl group and several hydroxyls, usually with two or more chiral centres. You will notice that we are comparing the stereochemistry in the different possible diastereoisomers for compounds containing several chiral centres (see Section 3.4.4). There is a corresponding series of enantiomeric L-sugars; only a few of these are shown.

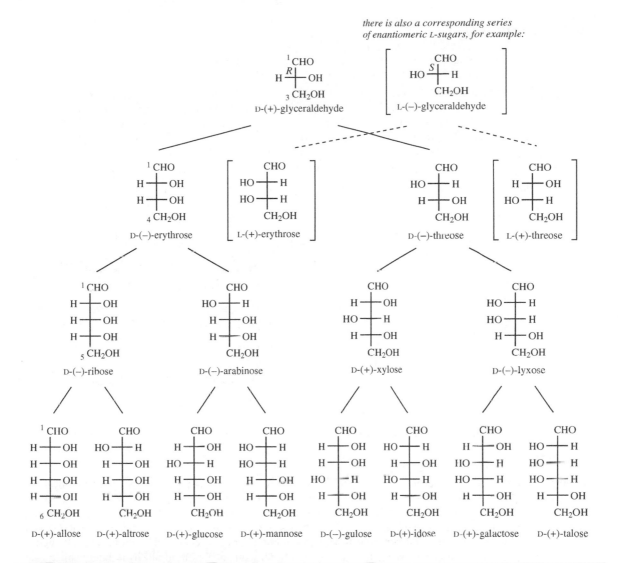

there is also a corresponding series
of enantiomeric L-sugars, for example:

D-(+)-glyceraldehyde   L-(−)-glyceraldehyde

D-(−)-erythrose   L-(+)-erythrose   D-(−)-threose   L-(+)-threose

D-(−)-ribose   D-(−)-arabinose   D-(+)-xylose   D-(−)-lyxose

D-(+)-allose   D-(+)-altrose   D-(+)-glucose   D-(+)-mannose   D-(−)-gulose   D-(+)-idose   D-(+)-galactose   D-(+)-talose

---

Box 12.1

Synthesis of ^{14}C-labelled glucose

A sequence known as the **Kiliani–Fischer synthesis** was developed primarily for extending an aldose chain by one carbon, and was one way in which configurational relationships between different sugars could be established. A major application of this sequence nowadays is to employ it for the synthesis of ^{14}C-labelled sugars, which in turn may be used to explore the role of sugars in metabolic reactions.

The synthesis of ^{14}C-labelled D-glucose starts with the pentose D-arabinose and ^{14}C-labelled potassium cyanide, which react together to form a cyanohydrin (see Section 7.6.1). Since cyanide can attack the planar carbonyl group from either side, the cyanohydrin product will be a mixture of two diastereoisomers that are epimeric at the new chiral centre. The two epimers are usually formed in unequal amounts because of a chiral influence from the rest of the arabinose structure during attack of the nucleophile.

Box 12.1 (continued)

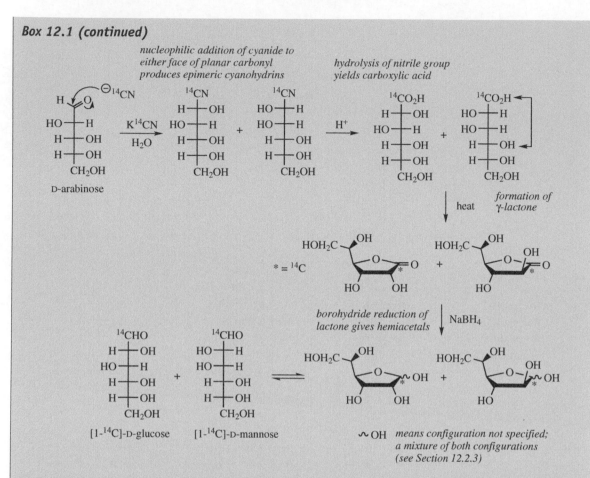

The nitrile groups in the product mixture are then hydrolysed to carboxylic acids (see Box 7.9). Upon heating, the acids readily form cyclic esters (lactones) through reaction of the hydroxyl group on C-4 with the carboxylic acid, the five-membered ring being most favoured (see Section 7.9.1). The pair of lactones is then reduced using sodium amalgam under acidic conditions to yield aldehydes, though it has been found that this reaction can also be achieved using aqueous sodium borohydride. Sodium borohydride reacts readily with lactones, though it is not usually effective in reducing esters. It is also normally difficult to stop at an aldehyde intermediate (see Section 7.11), but reduction of a lactone gives initially a hemiacetal; ring opening of the hemiacetal then leads to the aldehyde. The product will be a mixture of the two epimeric sugars D-glucose and D-mannose, which will be labelled with ^{14}C in the aldehyde function. Separation of the diastereoisomeric products may be achieved via fractional crystallization or by chromatography, and may be carried out at either the cyanohydrin stage, or at the final product stage.

Note how the process may be modified to extend its versatility. Thus, using ^{14}C-labelled potassium cyanide with D-erythrose yields a mixture of [1-^{14}C]-D-ribose and [1-^{14}C]-D-arabinose. The sequence could then be repeated on the latter product, using unlabelled KCN, to give [2-^{14}C]-D-glucose.

12.2.1 Enolization and isomerization

In common with other aldehydes or ketones that have hydrogen on the α-carbon, enolization is possible (see Section 10.1), especially when sugars are treated with base. The additional presence of a hydroxyl on the α-carbon causes further isomerization. Thus, treatment of D-glucose with dilute aqueous sodium hydroxide at room temperature leads to an equilibrium mixture also containing D-mannose and D-fructose.

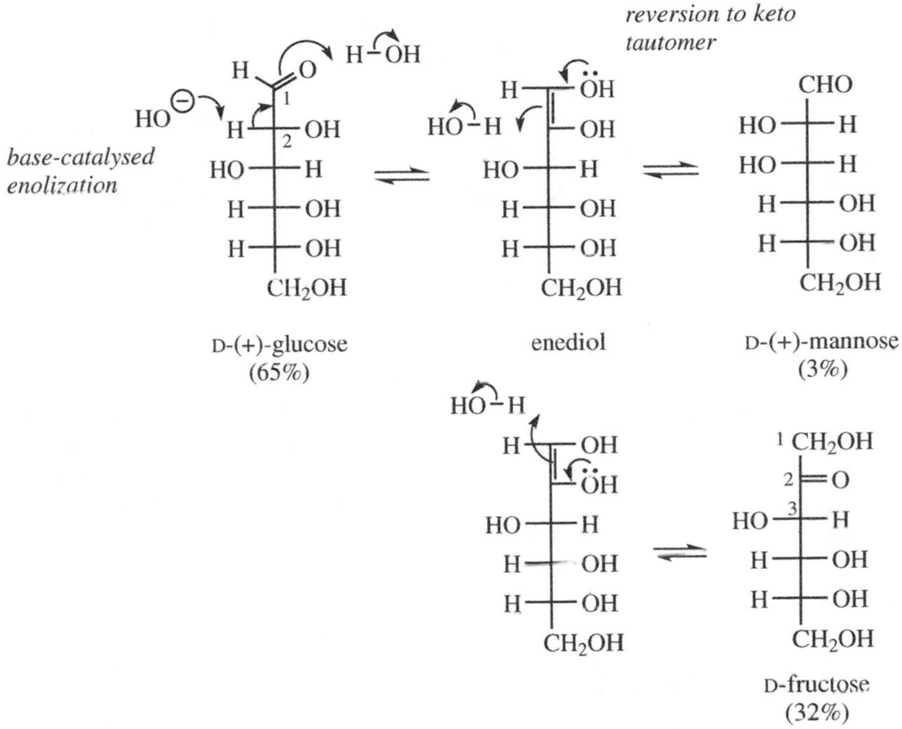

Removal of the α-hydrogen in D-glucose leads to **enolization** (we have omitted the enolate anion in the mechanism). Reversal of this process allows **epimerization** at C-2, since the enol function is planar, and a proton can be acquired from either face, giving D-mannose as well as D-glucose. Alternatively, we can get **isomerization** to D-fructose. This is because the intermediate enol is actually an enediol; restoration of the carbonyl function can, therefore, provide either a C-1 carbonyl or a C-2 carbonyl. The equilibrium mixture using dilute aqueous sodium hydroxide at room temperature consists mainly of D-glucose and D-fructose, with smaller amounts of D-mannose. The same mixture would be obtained if either D-mannose or D-fructose were treated similarly.

Note that harsher conditions may lead to further changes, e.g. epimerization at C-3 in fructose, plus isomerization, or even reverse aldol reactions (see Section 10.3). In general, basic conditions must be employed with care if isomerizations are to be avoided. To preserve stereochemistry, it is usual to ensure that free carbonyl groups are converted to acetals or ketals (glycosides, see Section 12.4) before basic reagents are used. Isomerization of sugars via enediol intermediates features prominently in the glycolytic pathway of intermediary metabolism (see Box 10.1).

12.2.2   Cyclic hemiacetals and hemiketals

Monosaccharide structures may be depicted in open-chain forms showing their carbonyl character, or in cyclic **hemiacetal** or **hemiketal** forms. Alongside the Fischer projections of glucose, ribose, and fructose shown earlier, we included an alternative representation of the compound in its cyclic form. The compounds exist predominantly in the cyclic forms, which result from nucleophilic attack of an appropriate hydroxyl onto the carbonyl (see Section 7.2).

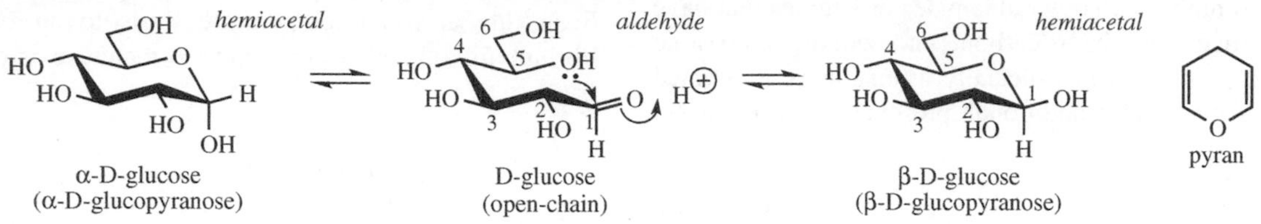

α-D-glucose
(α-D-glucopyranose)

D-glucose
(open-chain)

β-D-glucose
(β-D-glucopyranose)

pyran

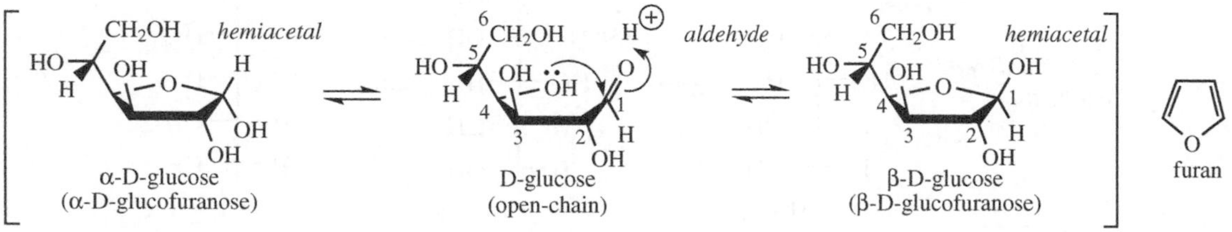

α-D-glucose
(α-D-glucofuranose)

D-glucose
(open-chain)

β-D-glucose
(β-D-glucofuranose)

furan

Both six-membered **pyranose** and five-membered **furanose** structures are encountered, a particular ring size usually being characteristic for any one sugar. Thus, although **glucose** has the potential to form both six-membered and five-membered rings, an aqueous solution consists almost completely of the six-membered hemiacetal form; five-membered rings are usually formed more rapidly, but six-membered rings are generally more stable and predominate at equilibrium. The names pyranose and furanose are derived from the oxygen heterocycles pyran and furan. Shown below is a reminder of how we can transform a Fischer projection of a sugar into a cyclic form (see Box 3.16).

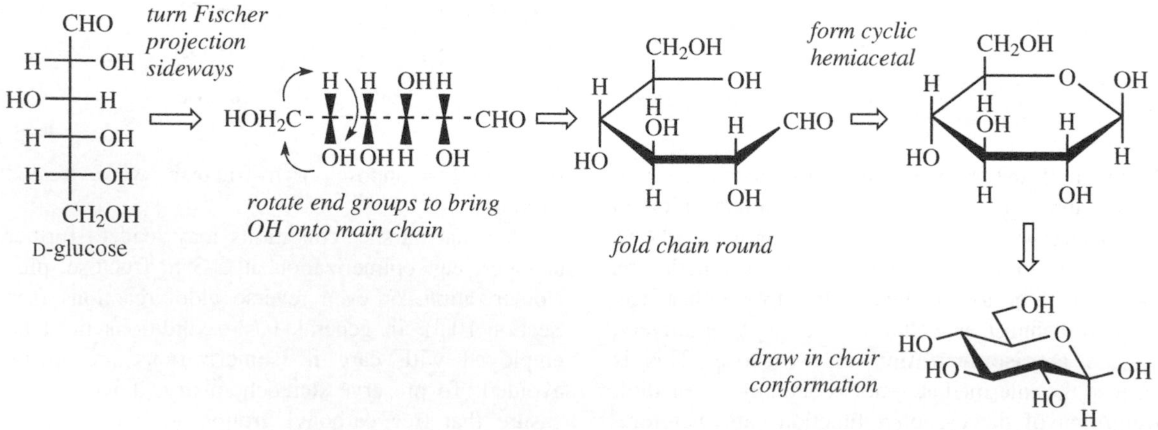

D-glucose

turn Fischer projection sideways

rotate end groups to bring OH onto main chain

form cyclic hemiacetal

fold chain round

draw in chair conformation

The pentose **ribose** is also able to form six-membered pyranose and five-membered furanose rings. In solution, ribose exists mainly (76%) in the pyranose form; interestingly, however, when we meet ribose in combination with other entities, e.g. nucleosides, it is almost always found in furanose form (see Box 7.2).

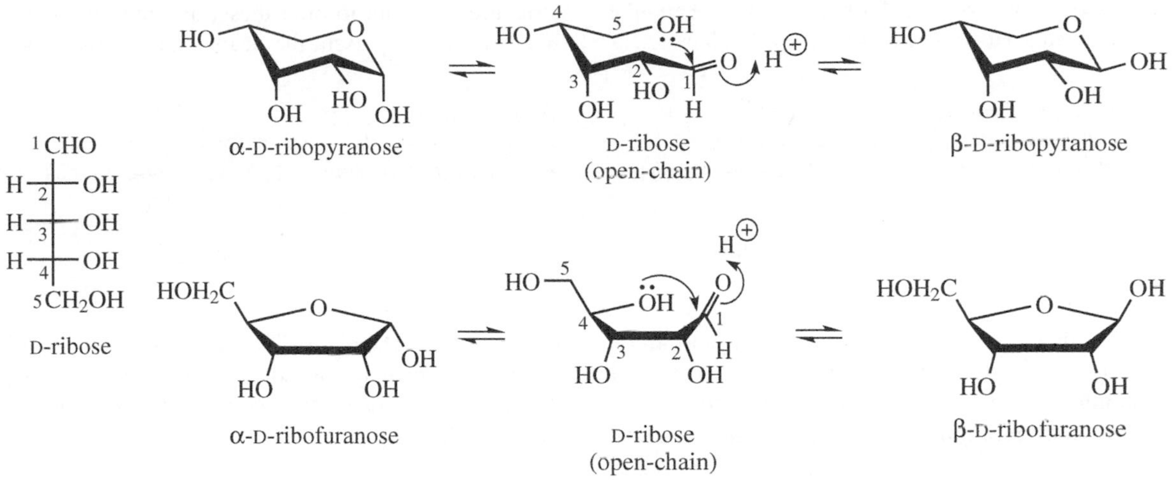

α-D-ribopyranose   D-ribose
                          (open-chain)   β-D-ribopyranose

¹CHO
H—2—OH
H—3—OH
H—4—OH
5CH₂OH

D-ribose

α-D-ribofuranose   D-ribose   β-D-ribofuranose
                          (open-chain)

Fructose is a ketose and, therefore, forms hemiketal ring structures. Like ribose, it is usually found in combination as a five-membered furanose ring, though the simple sugar in solution exists primarily in pyranose form (67%).

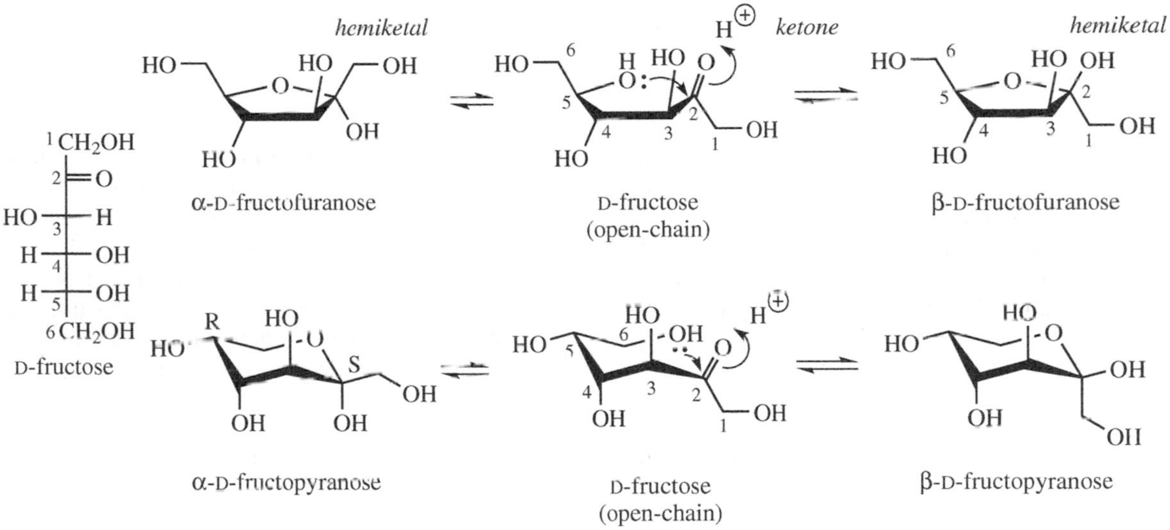

hemiketal   ketone   hemiketal

¹CH₂OH
2═O
HO—3—H
H—4—OH
H—5—OH
6CH₂OH

D-fructose

α-D-fructofuranose   D-fructose   β-D-fructofuranose
                          (open-chain)

α-D-fructopyranose   D-fructose   β-D-fructopyranose
                          (open-chain)

12.2.3 The anomeric centre

Since the carbonyl group is planar and may be attacked from either side, two epimeric structures (**anomers**) are possible in each case, and in solution the two forms are frequently in equilibrium, because hemiacetal or hemiketal formation is reversible (see Section 7.2). The two anomers are designated α or β by comparison of the chiralities at the anomeric centre and at the highest numbered chiral centre. If these are the same (*RS* convention), the anomer is termed β, or α if they are different. In practice, this translates to the anomeric hydroxyl being 'up' in the case of β-D-sugars and α-L-sugars. It is interesting to note that the descriptors α or β were originally assigned to the two forms of glucose based on the order in which they crystallized out from solution. Without changing the nomenclature for these two compounds, α or β are now assigned on a much more rigid stereochemical basis.

By convention, the ring form of sugars is drawn with the ring oxygen to the rear and the anomeric carbon furthest right. Wedges and the bold bond help to emphasize how we are looking at the chair-like pyranose ring. However, to speed up the drawing of

structures we tend to omit these, and then the lower bonds always represent the nearest part of the ring.

for ease of drawing, we usually omit bold bonds and wedges

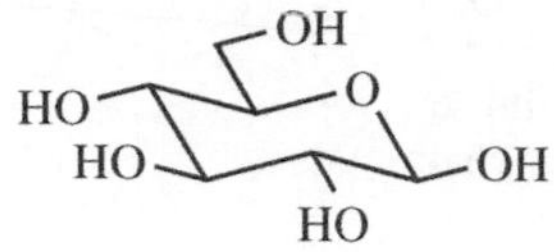

β-D-glucose

the lower bonds always represent the nearest part of the ring

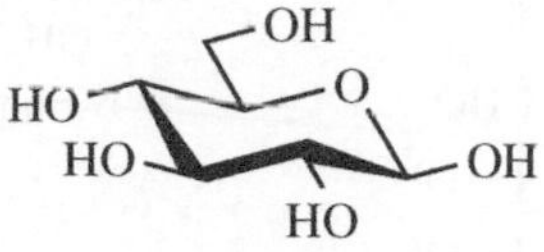

β-D-glucose

Since there are two anomeric forms, and these are often in equilibrium via the acyclic carbonyl compound, we can use a new type of bond to indicate that the configuration is not specified and could be

of either stereochemistry. This is the wavy or wiggly bond, and to display our indecision further we usually site it halfway between the two possible positions (see Section 7.2).

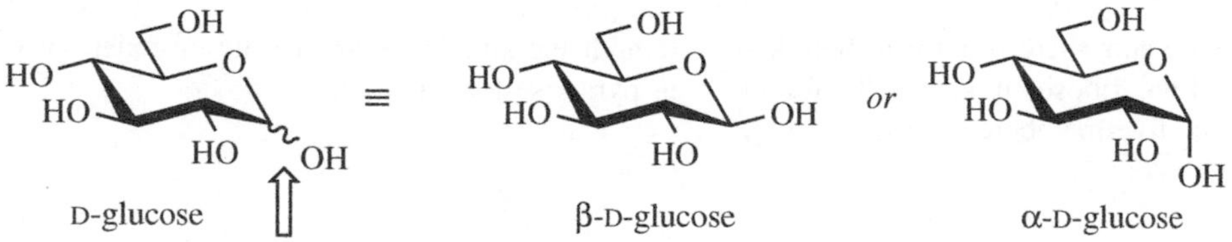

D-glucose

wavy bond; configuration not specified

≡

β-D-glucose

or

α-D-glucose

or a mixture

It follows that, when we dissolve a sugar such as glucose or ribose in water, we create a mixture of various equilibrating structures. The relative proportions of pyranose and furanose forms, and of their respective anomers for the eight **aldohexoses**, are shown in Table 12.1. In each case, the proportion of non-cyclic form is very small (<1%).

The most stable **conformation** of the cyclic sugar is mainly determined by a minimization of steric interactions, i.e. the maximum number of equatorial substituents (see Section 3.3.2). It follows that the preferred conformation for β-D-glucose will be that with all substituents equatorial; the alternative has all substituents axial. Carbohydrate chemists have introduced a neat way of referring to the two

Table 12.1 Equilibrium proportions of pyranose and furanose forms of aldohexoses in water[a]

Aldohexose	α-Pyranose form (%)	β-Pyranose form (%)	α-Furanose form (%)	β-Furanose form (%)
Allose	16	76	3	5
Altrose	27	43	17	13
Glucose	36	64	<1	<1
Mannose	66	34	<1	<1
Gulose	16	81	<1	3
Idose	39	36	11	14
Galactose	29	64	3	4
Talose	37	32	17	14

[a]The proportion of non-cyclic form is <1%.

Box 12.2

Ring size and anomeric form of common sugars

Sugars exist predominantly in cyclic hemiacetal or hemiketal forms, and whilst both six-membered pyranose and five-membered furanose structures are encountered, a particular ring size is usually characteristic for any one sugar, especially when it is found in combination with other entities in natural structures. The most commonly encountered monosaccharides and their usual anomers are shown here. By convention, the ring form is drawn with the ring oxygen to the rear and the anomeric carbon furthest right. Also shown are the accepted abbreviations for these sugars.

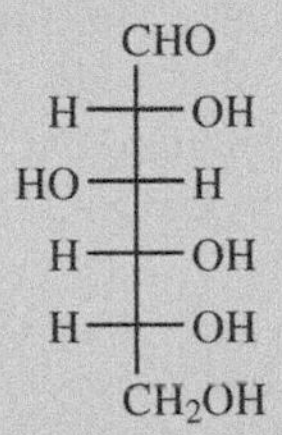

D-(+)-glucose 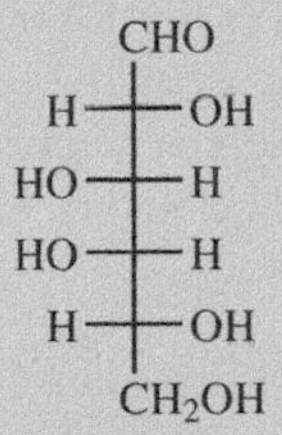 D-(+)-galactose 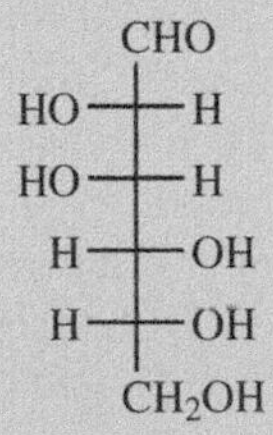 D-(+)-mannose 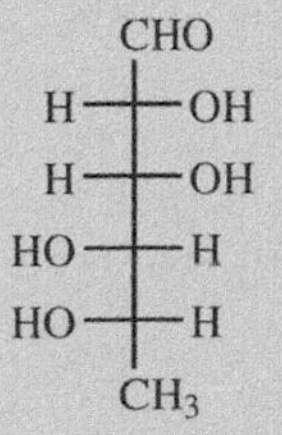 L-(−)-rhamnose

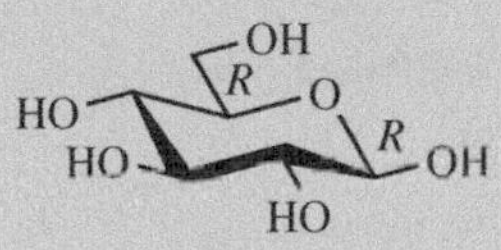

β-D-glucose (β-D-Glc) β-D-galactose (β-D-Gal) β-D-mannose (β-D-Mann) α-L rhamnose (α-L-Rha) [preferred conformation]

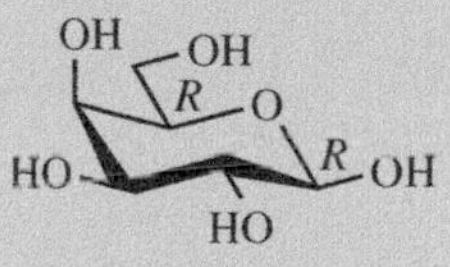

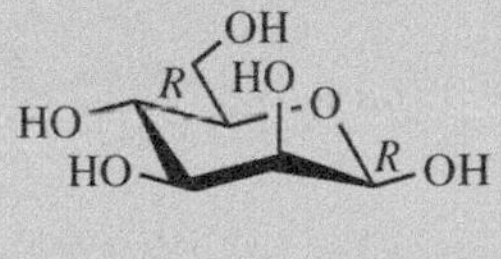

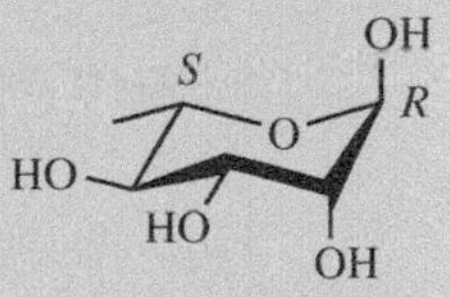

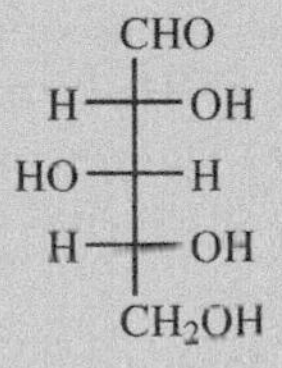

D-(+)-xylose 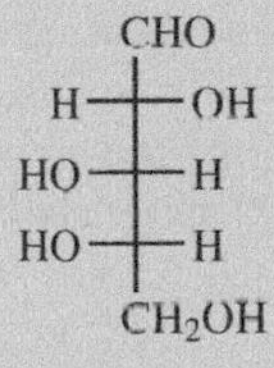 L-(+)-arabinose 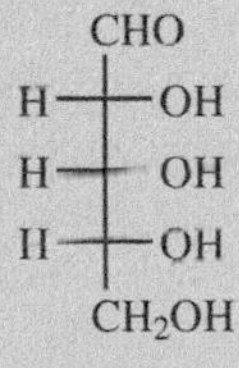 D-(−)-ribose 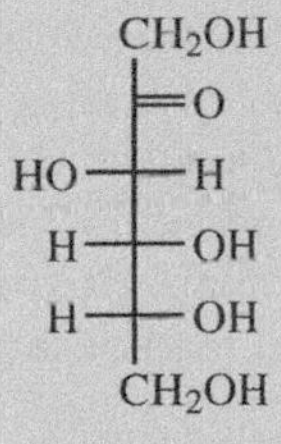 D-(−)-fructose

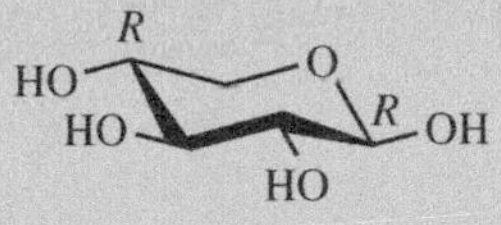

β-D-xylose (β-D-Xyl) α-L-arabinose (α-L-Ara) β-D-ribose (β-D-Rib) β-D-fructose (β-D-Fru)

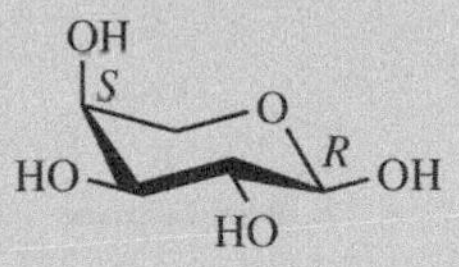

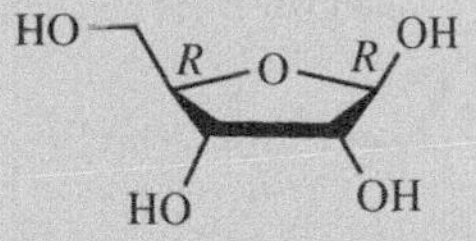

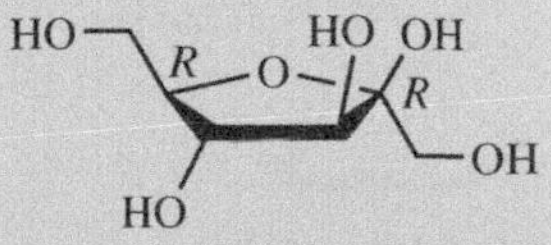

The two anomers are designated α or β by comparison of the chiralities at the anomeric centre and at the highest numbered chiral centre. If these are the same (RS convention), the anomer is termed β, or α if they are different. Note that the D and L prefixes are assigned on the basis of the chirality (as depicted in Fischer projections) at the highest numbered chiral centre and its relationship to D-(R)-(+)-glyceraldehyde or L-(S)-(−)-glyceraldehyde (see Section 3.4.10).

The stereochemistries of the various substituents may be deduced by considering the implications of the Fischer projection (see Box 3.16).

conformers, in that the left-hand conformer of glucose is termed $^4C_1$, and the right-hand one $^1C_4$. The 'C' indicates chair conformation, the superscript numeral is the carbon atom that is above the plane of the ring, and the subscript numeral is which carbon atom is below the plane of the ring. For this description, we consider the pyranose ring as originally planar, distorted to a chair by pushing carbons 1 and 4 out of the plane.

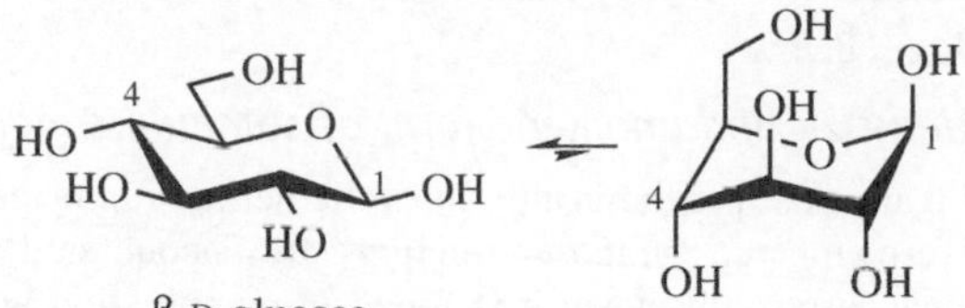

At first glance, the preferred conformation for L-hexoses, e.g. α-L-rhamnose, appears different from that of the D-hexoses. This is readily rationalized by considering the preferred conformation of α-L-glucose – the α-anomer is chosen simply because

β-D-glucose

$^4C_1$ conformer $^1C_4$ conformer

we can easily follow the anomeric substituent. Since α-L-glucose is the enantiomer of α-D-glucose, we can draw the mirror image representation, then rotate this so that the heterocyclic oxygen comes to the required position.

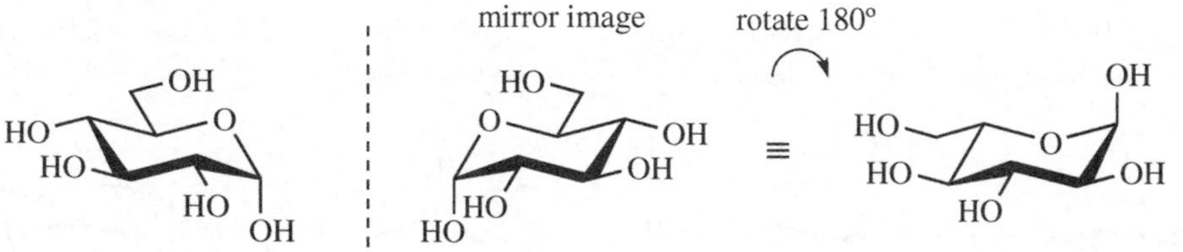

mirror image   rotate 180°

α-D-glucose   α-L-glucose   α-L-glucose
preferred conformation

Note that you may also encounter another version of the cyclic form referred to as the **Haworth representation** (see Box 3.16). This shows the ring as a planar system, and is commonly used in biochemistry books. However, we know that five-membered and six-membered rings are certainly not planar. The Haworth representation nicely reflects the up–down relationships of the various substituent

groups, but is quite uninformative about the shape of the molecule, and whether the substituents are equatorial or axial. In really bad cases, authors omit the hydrogen atoms, giving an ambiguous structure – do lines mean methyl or hydrogen? Haworth representations may be easier to draw, but you are strongly encouraged to use the more informative conformational structures.

β-D-glucose

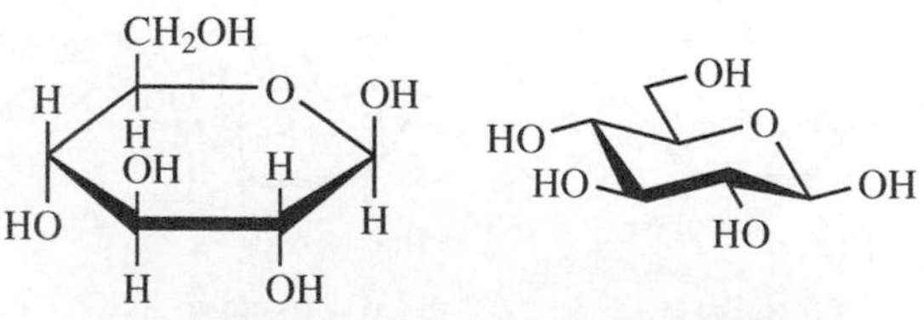

Haworth   conformational
representation   representation

β-D-ribose

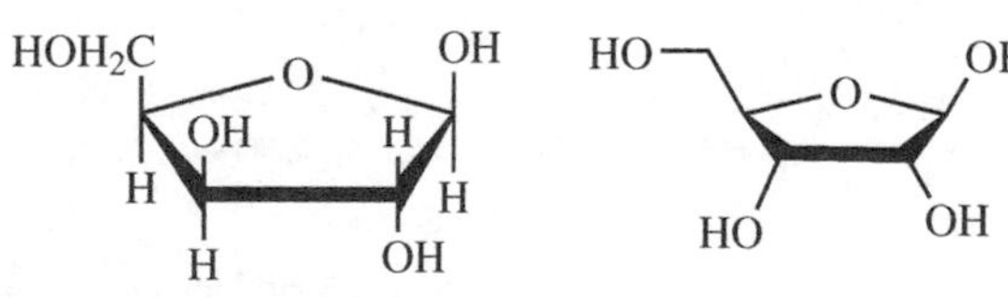

Haworth   conformational
representation   representation

often seen, but careless versions that omit hydrogens; don't think the lines mean methyls!

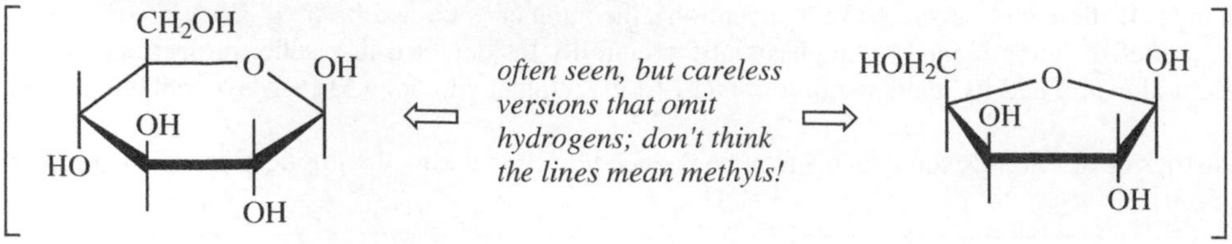

One of the consequences of forming a cyclic hemiacetal or hemiketal is that the nucleophilic hydroxyl adds to the carbonyl group and forms a new hydroxyl. This new group is susceptible to many normal chemical reactions of hydroxyls, e.g. **esterification**, and this type of reaction effectively freezes the carbohydrate into one anomeric form, since the ring-opening and equilibration can now no longer take place. Consider esterification of glucose with acetic anhydride (see Section 7.9.1). β-D-Glucose will be

acetylated to give the β-acetate, whereas α-D-glucose will specifically give the α-acetate. These two forms do not equilibrate merely by dissolving in solvent, although they can be interconverted by some other means, e.g. nucleophilic substitution reactions with acetate (compare Section 6.3.2). If we wish to consider esterification of the α–β mixture, we could use the unspecified wavy bond representation shown on the right.

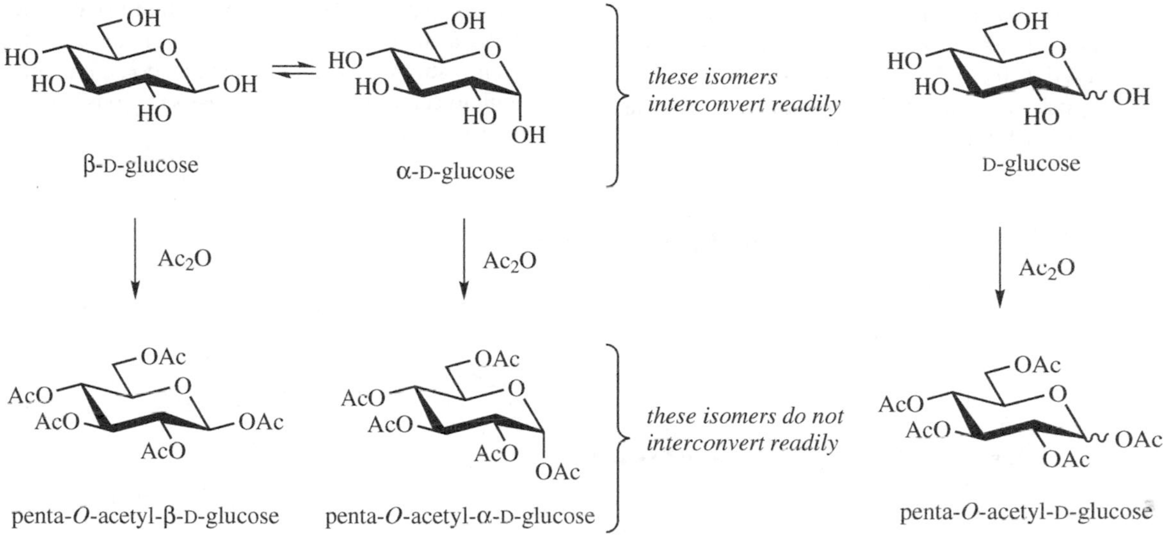

β-D-glucose   α-D-glucose   *these isomers interconvert readily*   D-glucose

penta-*O*-acetyl-β-D-glucose   penta-*O*-acetyl-α-D-glucose   *these isomers do not interconvert readily*   penta-*O*-acetyl-D-glucose

12.3 Alditols

Reduction of the aldehyde or ketone group in a sugar is readily achieved using a variety of reducing agents. Reduction occurs on the small amount of open-chain form present at equilibrium. As the open-chain form is removed, the equilibrium is disturbed until total reduction is achieved. The products are polyhydroxy compounds termed **alditols**. Reduction

of aldoses is the more satisfactory reaction, in that a single product is formed. On the other hand, reduction of ketoses generates a new chiral centre, and two epimeric alditols will result. Thus, treatment of D-glucose with sodium borohydride gives D-**glucitol**, also known as D-**sorbitol**. It should be noted that LAH is not a satisfactory reducing agent for this reaction because of the several hydroxyl groups present (see Section 7.5).

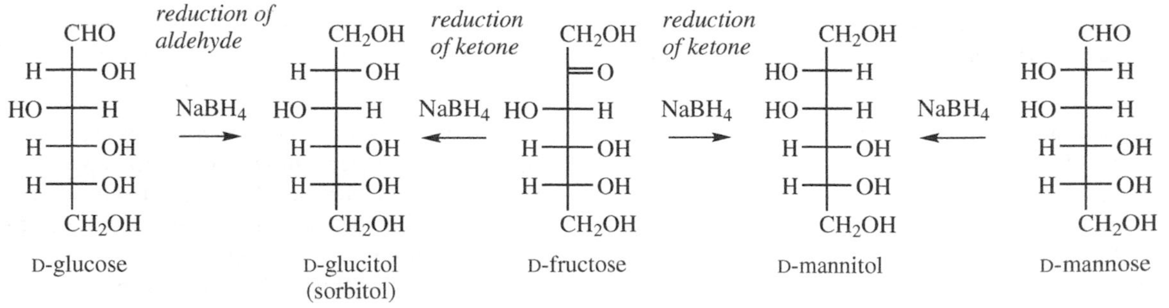

D-glucose   D-glucitol (sorbitol)   D-fructose   D-mannitol   D-mannose

On the other hand, borohydride reduction of the ketose D-fructose will give a mixture of D-glucitol and its epimer, D-**mannitol**. A better approach to D-mannitol would be reduction of the aldose D-mannose. D-Glucitol (sorbitol) is found naturally in the ripe berries of the mountain ash (*Sorbus aucuparia*), but is prepared semi-synthetically from glucose. It is half as sweet as sucrose, is not absorbed orally, and is not readily metabolized in the body. It finds particular use as a sweetener for diabetic products. D-Mannitol also occurs naturally in manna, the exudate of the manna ash *Fraxinus ornus*. This material has similar characteristics to sorbitol, but is used principally as a diuretic. It is injected intravenously, is eliminated rapidly into the urine, and removes fluid by an osmotic effect.

12.4 Glycosides

The cyclic hemiacetal and hemiketal forms of monosaccharides are capable of reacting with an alcohol to form acetals and ketals (see Section 7.2). The acetal or ketal product is termed a **glycoside**, and the non-carbohydrate portion is referred to as an **aglycone**. In the nomenclature of glycosides we replace the suffix -ose in the sugar with -oside. Simple glycosides may be synthesized by treating an alcoholic solution of the monosaccharide with an acidic catalyst, but the reaction mixture usually then contains a mixture of products. This is an accepted problem with many carbohydrate reactions; it is often difficult to carry out selective transformations because of their multifunctional nature.

β-D-glucopyranose

hemiacetals

α-D-glucopyranose

attack on either face of planar system

α- or β-D-glucofuranose

methyl α-D-glucofuranoside

methyl β-D-glucofuranoside

methyl β-D-glucopyranoside

acetals

methyl α-D-glucopyranoside

major product

Reaction of glucose with methanol and gaseous HCl yields four acetal products, the α- and β-pyranosides and α- and β-furanosides, which may be separated. The pyranosides are the predominant components, and the major product is the α-pyranoside. This is perhaps unexpected, in that the β-pyranoside has all its substituents equatorial, whereas the α-anomer has its anomeric substituent axial. This so-called **anomeric effect** apparently arises from a favourable electronic stabilization in the axial anomer that is not possible in the equatorial anomer. It involves overlap from the ring oxygen lone pair, and to achieve this the lone pair and the substituent must be antiperiplanar.

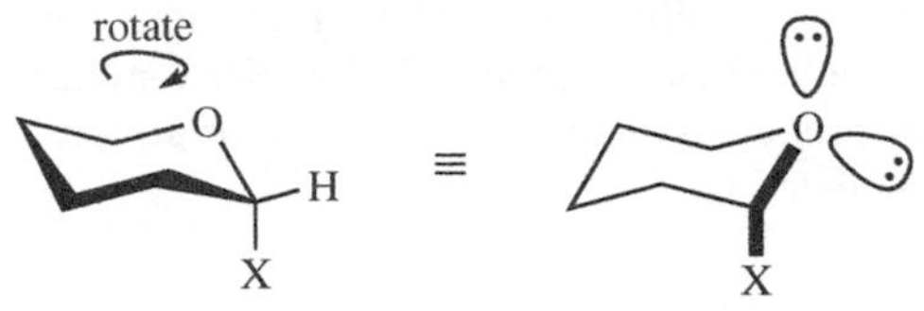

α-anomer

lone pair anti-periplanar with axial electronegative group

neither lone pair anti-periplanar with equatorial electronegative group

The anomeric effect is rather complex and will not be considered in any detail. It occurs when we have a heterocyclic ring (O, N, or S), with an electronegative substituent (halogen, OH, OR, OCOR, etc.) adjacent to the heteroatom, and favours the isomer in which the substituent is axial. Thus, with the first of the

simple acetals shown below, where we need consider only conformational isomerism, some 75% of the axial conformer is present at equilibrium. Without the ring oxygen, we would see an equatorial isomer predominating (see Section 3.3.2). In the second example, the additional stability conferred by the equatorial methyl group increases even further the proportion of the conformer with the axial methoxyl.

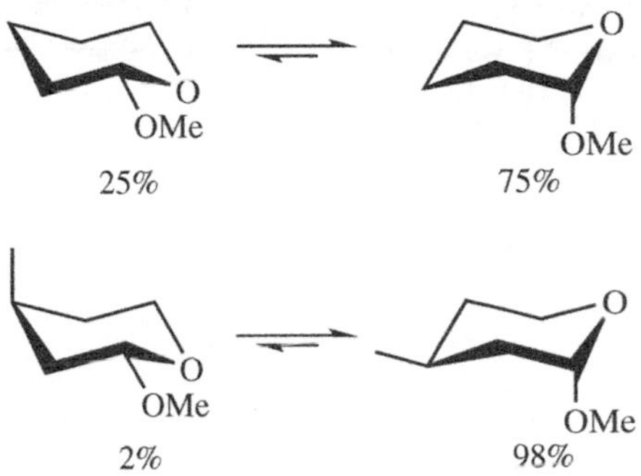

25%   75%

2%   98%

We have noted that an aqueous solution of glucose exists as an equilibrium mixture containing some 64% of the β-anomer.

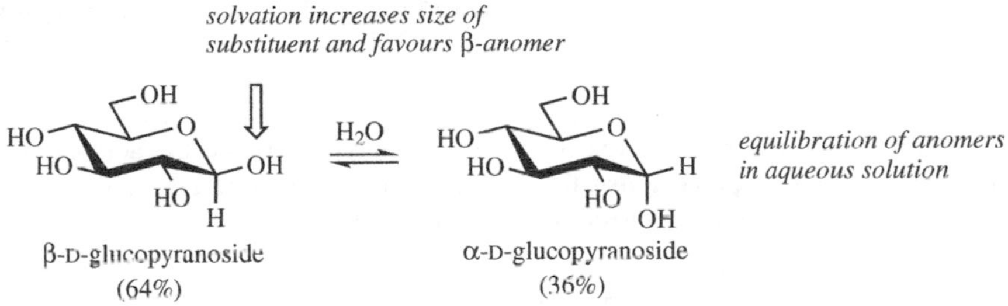

solvation increases size of substituent and favours β-anomer

β-D-glucopyranoside
(64%)

α-D-glucopyranoside
(36%)

equilibration of anomers in aqueous solution

Based simply on steric effects, this proportion appears somewhat low, whereas in view of the anomeric effect just described the proportion now seems rather high. Anomeric effects are observed to be solvent dependent, and hydroxy compounds experience considerable solvation with water through hydrogen bonding. This significantly increases the steric size of the substituent, and reinforces the steric effects.

By considering the reversibility of the acetal-forming reactions, it is apparent that treatment of either of the two methyl pyranosides with acidic methanol will produce the same equilibrium mixture. A related equilibration occurs with the anomers of glucose, as seen earlier (see Box 7.1, mutarotation of glucose).

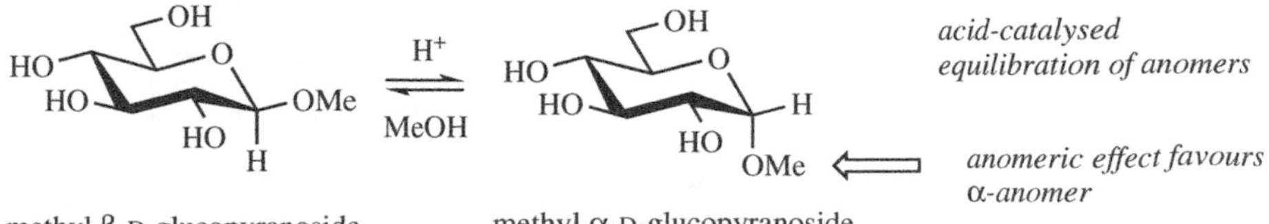

methyl β-D-glucopyranoside
(34%)

methyl α-D-glucopyranoside
(66%)

acid-catalysed equilibration of anomers

anomeric effect favours α-anomer

It should also be noted that hydrolysis of glycosides (acetals or ketals) will occur under acid-catalysed conditions if we have an excess of water present. This is a reversal of the process for glycoside formation, the equilibrium favouring the aglycone plus sugar rather than the glycoside (see Section 7.2). The sugar product will again be the equilibrium mixture of anomers.

acid-catalysed hydrolysis of glycosides

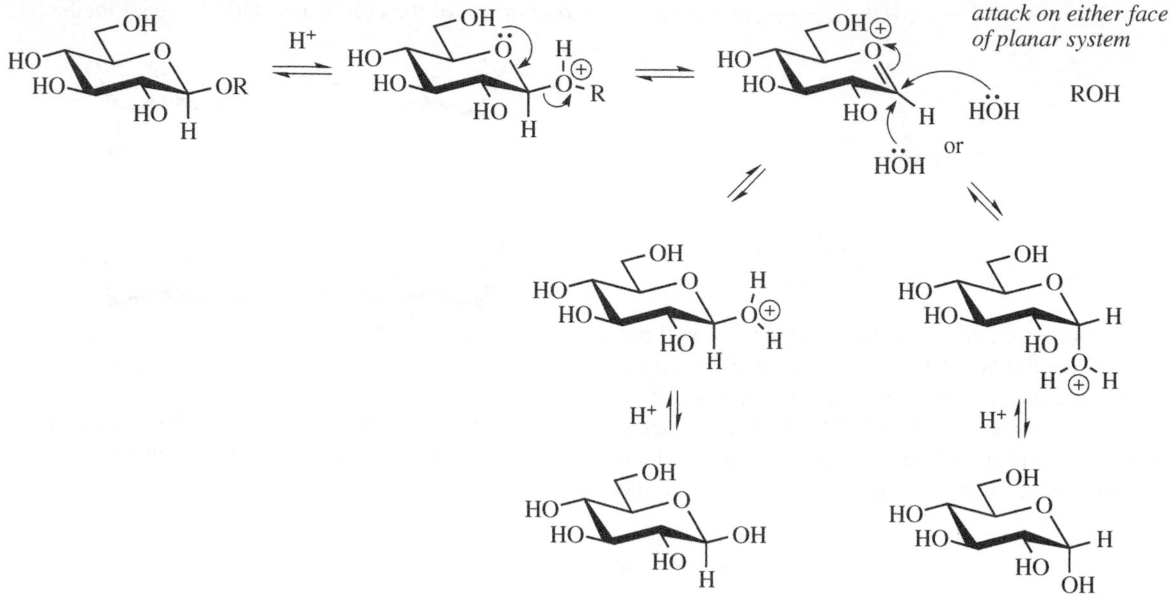

Hydrolysis of glycosides can also be achieved by the use of specific enzymes, e.g. **β-glucosidase** for β-glucosides and **β-galactosidase** for β-galactosides. These enzymes mimic the acid-catalysed processes, are commercially available, and may be used just like a chemical reagent.

Box 12.3

Some examples of natural O-, S-, C-, and N-glycosides

Many different types of glycoside structure are found in nature, especially in plants. Since the presence of a sugar unit in the structure provides polarity, it is likely that glycosylation is a means by which an organism makes an aglycone water soluble and transportable. Most of the natural glycosides are compounds in which the aglycone is an alcohol or a phenol, and such derivatives are termed **O-glycosides**. O-Glycosides are thus acetals or ketals. Less commonly, one encounters compounds in which a thiol (RSH) has been bonded to the sugar unit resulting in a thioacetal (see Section 7.4). These compounds are termed **S-glycosides**. Some examples of O- and S-glycosides are shown below.

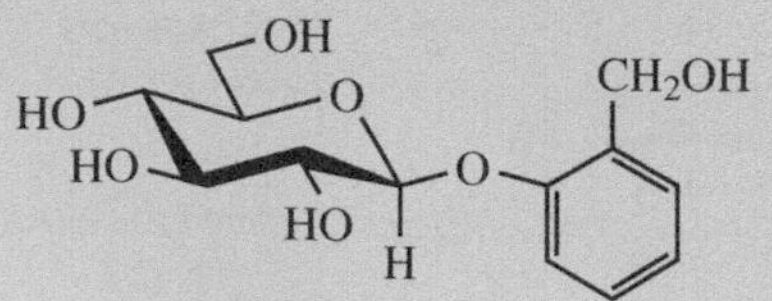

O-glycoside: salicin

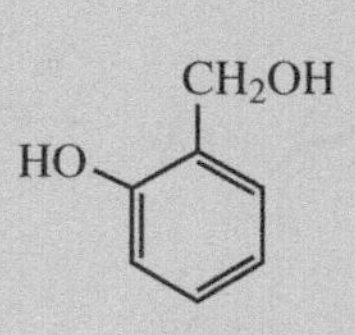

aglycone: salicyl alcohol

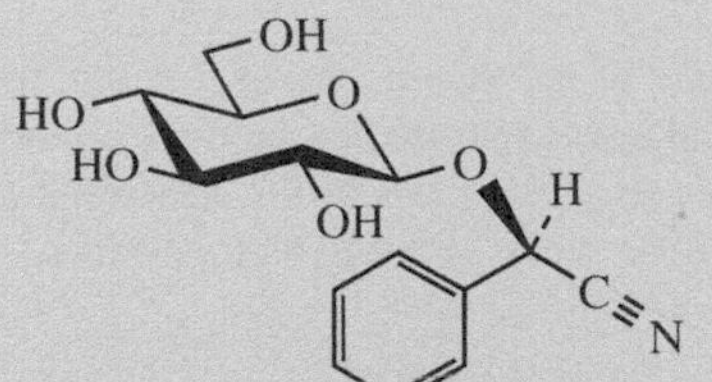

O-glycoside: prunasin

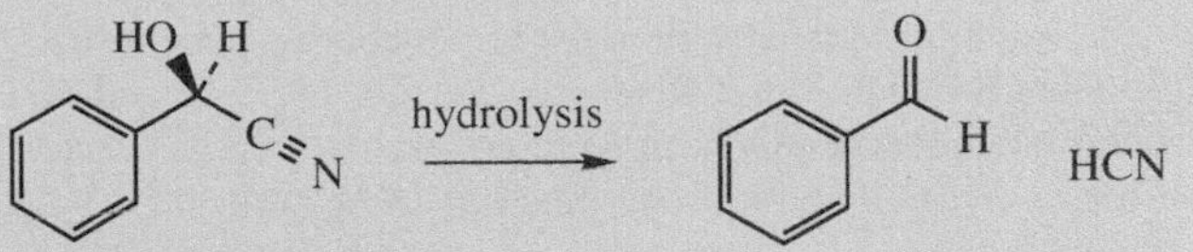

aglycone: mandelonitrile
(benzaldehyde cyanohydrin)

HCN

Salicin is an *O*-glycoside of a phenol, namely salicyl alcohol. Salicin is a natural antipyretic and analgesic found in willow bark, and is the template from which aspirin (acetylsalicylic acid, see Box 7.13) was developed. **Prunasin** from cherry laurel is an example of a **cyanogenic glycoside**, hydrolysis of which leads to release of toxic HCN (see Box 7.7). It is the *O*-glucoside of the alcohol mandelonitrile, the trivial name for the cyanohydrin of benzaldehyde. It is the further hydrolysis of mandelonitrile that liberates HCN.

S-Glycosides in nature are quite rare, but there is an important group called **glucosinolates**. These compounds are responsible for the pungent properties of mustard, horseradish and members of the cabbage family. One example is **sinigrin**, found in black mustard seeds. When seeds are crushed, enzymic hydrolysis liberates the aglycone, which subsequently rearranges to the pungent principle allylisothiocyanate.

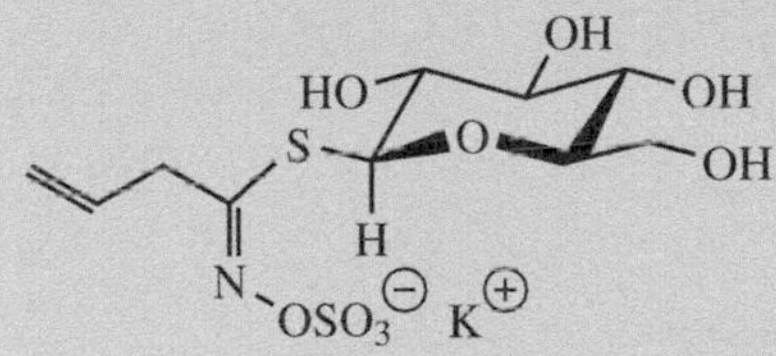

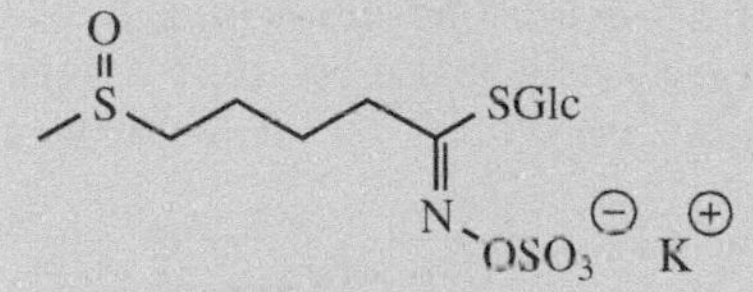

S-glycoside: sinigrin

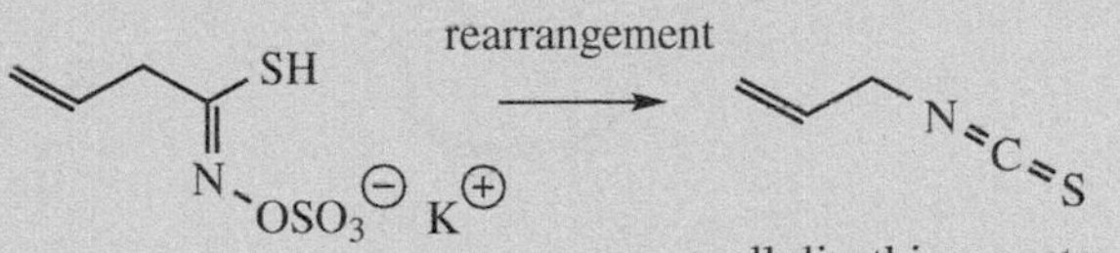

aglycone:
allyl thiohydroximate sulfonate

rearrangement

allylisothiocyanate

S-glycoside: glucoraphanin

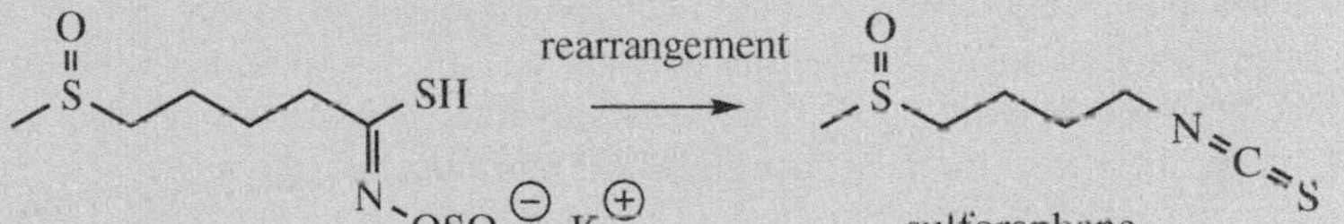

aglycone: raphanin

rearrangement

sulforaphane

A related glucosinolate **glucoraphanin** is found in broccoli, and is associated with beneficial medicinal properties of this vegetable. This is hydrolysed to the isothiocyanate sulforaphane, which is believed to induce carcinogen-detoxifying enzyme systems.

Other natural glycosides are not acetals or ketals, but analogues in which the nucleophilic species has been an amine (***N*-glycosides**), or even some carbanionic species so that the sugar becomes attached to carbon (***C*-glycosides**). It should be noted that the presence of a C–C bond between the sugar and the aglycone means that *C*-glycosides are not cleaved by simple hydrolysis, but require an oxidative process.

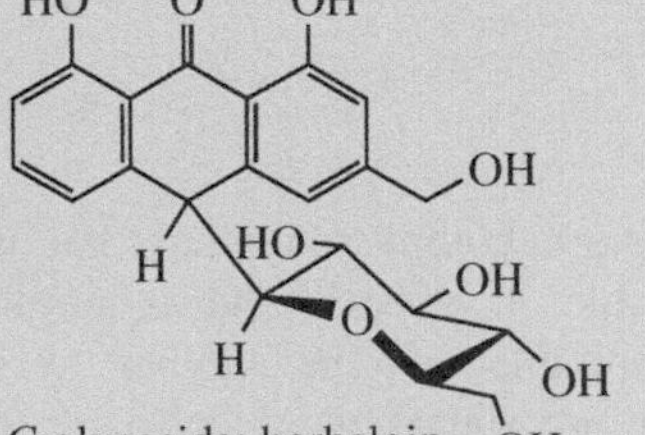

C-glycoside: barbaloin

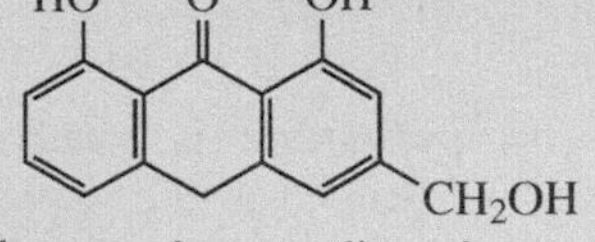

aglycone: aloe-emodin anthrone

Box 12.3 (continued)

C-Glycosides are typified by **barbaloin**, a component of the natural purgative drug cascara, but, as a group, the *N*-glycosides are perhaps the most important to biochemistry. *N*-Glycosidic linkages are found in the **nucleosides**, components of DNA and RNA (see Section 14.1). In addition, nucleosides are essential parts of the structures of crucial biochemicals such as ATP, coenzyme A, NAD$^+$, etc. The amine in these types of compound is part of a purine or pyrimidine base (see Section 14.1).

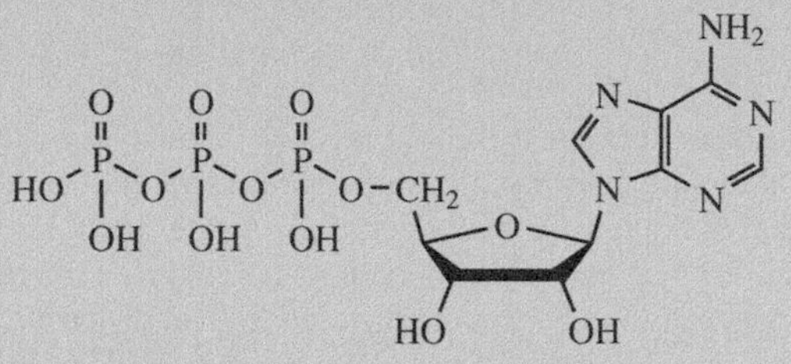

N-glycoside: ATP

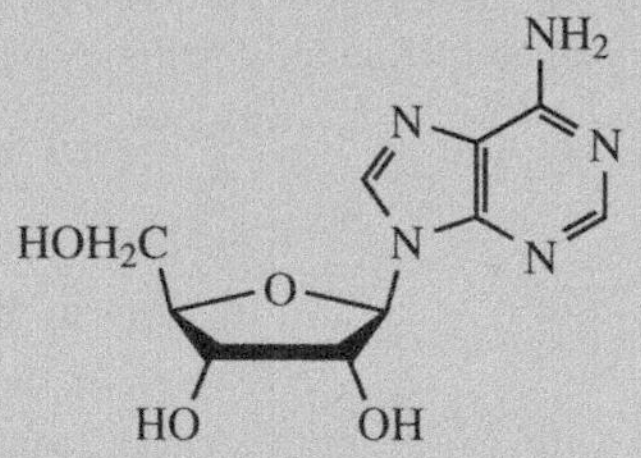

N-glycoside: adenosine

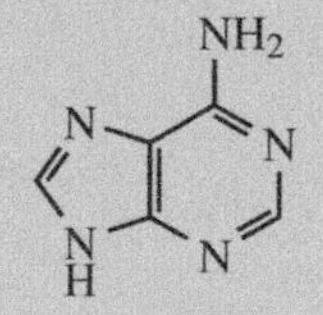

aglycone: adenine

Perhaps the most significant group of glycoside derivatives are polysaccharides. In these structures, the aglycone is itself another sugar, so that the polymer chain is composed of a series of sugar units joined by acetal or ketal linkages (see Section 12.7). Short carbohydrate polymers may also be found in some of the more complex *O*-glycosides, e.g. the heart drug **digoxin** from *Digitalis lanata*.

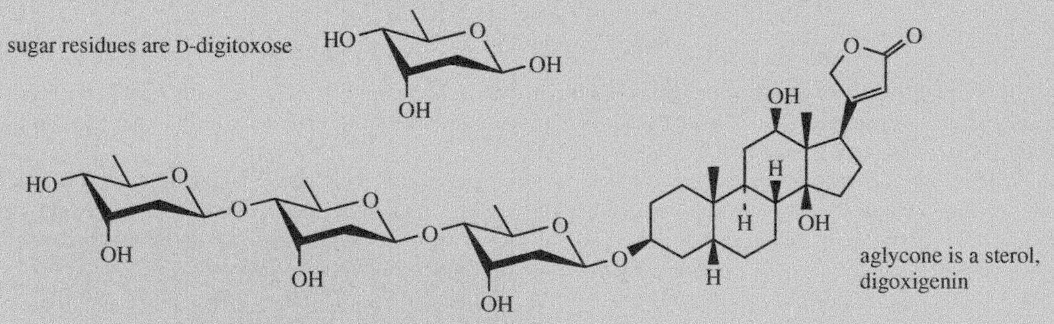

sugar residues are D-digitoxose

aglycone is a sterol, digoxigenin

digoxin

Making a methyl glucopyranoside is relatively straightforward in that we can use the alcohol methanol as solvent, and, since it is thus present in large excess, this helps to disturb the equilibrium. The process is much less attractive for a more complex alcohol that is probably not available in excess, and is unlikely to function as a suitable solvent. Trying to join together two or more sugars would also be fraught with problems, since each sugar contains several hydroxyl groups capable of acting as the nucleophile. These problems have been overcome by exploiting **nucleophilic substitution** for glycoside synthesis rather than the hemiacetal to acetal conversion we have been looking

at, combined with the use of protecting groups to avoid unwanted couplings. A valuable reagent for adding a glucose unit on to a suitable nucleophile is **acetobromoglucose**.

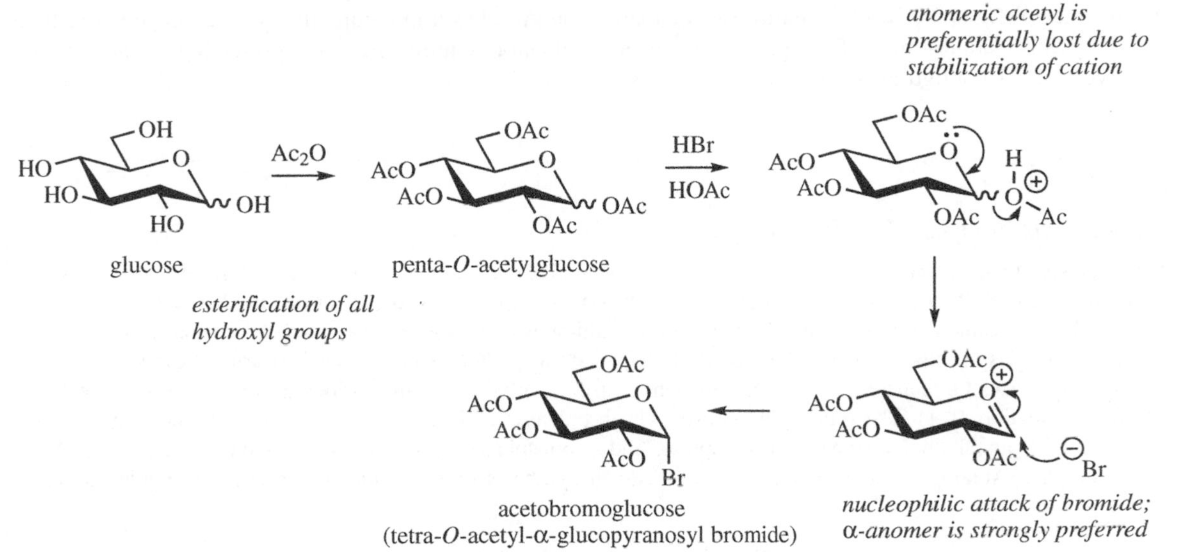

anomeric acetyl is preferentially lost due to stabilization of cation

glucose

penta-*O*-acetylglucose

esterification of all hydroxyl groups

acetobromoglucose
(tetra-*O*-acetyl-α-glucopyranosyl bromide)

nucleophilic attack of bromide; α-anomer is strongly preferred

Glucose is first esterified to penta-*O*-acetylglucose using acetic anhydride. Note that the hemiacetal hydroxyl is also esterified, and thus any equilibration with an aldehyde form is now not possible (see Section 7.2). When this penta-acetate is treated with HBr, the anomeric acetate is preferentially lost under the acidic conditions, due to the stabilization conferred by the heterocyclic oxygen. Note that this is the same type of intermediate we implicated

in the conversion of hemiacetals into acetals (see Section 7.2). Acetobromoglucose then results from nucleophilic attack of bromide onto the cationic system; in acetal formation, the nucleophile would be an alcohol. The anomeric effect is considerably larger when the substituent is halide than it is with alkoxy groups, so the product formed is almost exclusively the α-anomer.

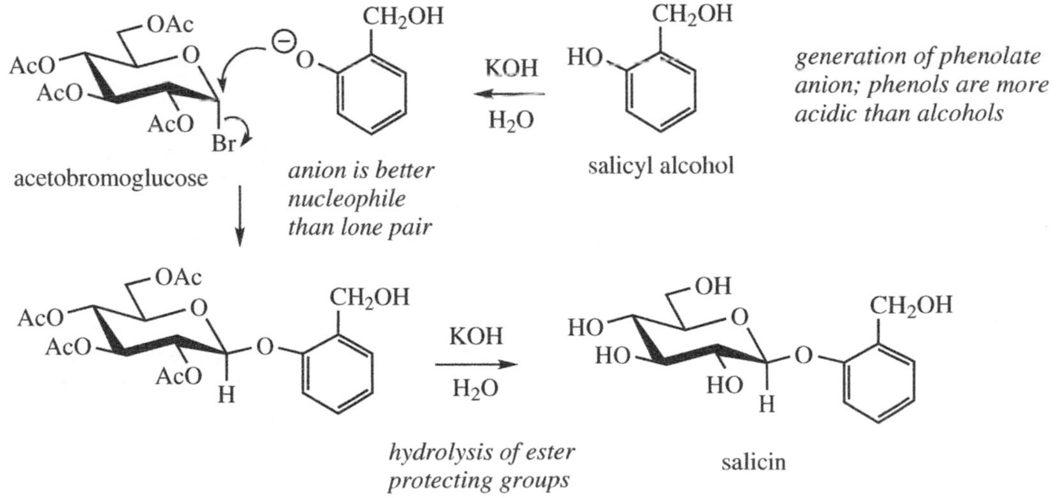

acetobromoglucose

anion is better nucleophile than lone pair

generation of phenolate anion; phenols are more acidic than alcohols

salicyl alcohol

hydrolysis of ester protecting groups

salicin

The bromide leaving group is now nicely positioned for an S_N2 reaction with an incoming nucleophile; in the example shown, this is the phenoxide anion from salicyl alcohol. In the presence of base, the phenolic group of salicyl alcohol is ionized, since phenols are very much more acidic than alcohols (see Section 4.3.5). S_N2 processes occur with inversion of configuration (see Section 6.1), so

the product is consequently the esterified β-glucoside derivative. Further base treatment then hydrolyses the ester functions, liberating the glucoside salicin. As we shall see in Box 12.4, this type of substitution process is similar to the way glucosides (and polysaccharides) are produced in nature, though the enzymic reactions do not require any ester protecting groups for the sugars.

Box 12.4

Biosynthesis of glycosides via UDPsugars

The widespread occurrence of glycosides and polysaccharides in nature demonstrates there are processes for attaching sugar units to a suitable atom of an aglycone to give a glycoside, or to another sugar to give a polysaccharide. Linkages tend to be through oxygen, although they are not restricted to oxygen, since S-, N-, and C-glycosides are also well known (see Box 12.3). The agent for glycosylation is a uridine diphosphosugar, e.g. **UDPglucose**. Of course, the uridine portion is itself a glycoside, an N-riboside of the pyrimidine base uracil (see Section 14.1). The glucosylation process can be envisaged as a simple S_N2 nucleophilic displacement reaction, with an alcohol or phenol nucleophile, and a phosphate derivative as the leaving group. This S_N2 displacement is analogous to that seen in the chemical synthesis of glycosides using acetobromoglucose (see Section 12.4).

uridine diphosphoglucose
UDPglucose

S_N2 processes occur with inversion of configuration (see Section 6.1), so since UDPglucose has its leaving group in the α-configuration, the product formed by the S_N2 process has the β-configuration. This is the configuration most commonly found in natural O-glucosides. Some natural products do possess an α-linkage, however. It appears that such compounds originate via a double S_N2 process, in which a nucleophilic group on the enzyme reacts first with the UDPglucose and then the hydroxy nucleophile displaces the enzymic group.

Other UDPsugars, e.g. UDPgalactose or UDPxylose, are utilized in the synthesis of glycosides containing different sugar units. The S-, N-, and C-glycosides are formed by a similar process with the appropriate nucleophile. This type of reaction is also that used in the biosynthesis of polysaccharides (see Section 12.7), and in the metabolism of drugs and other foreign compounds via glucuronides (see Box 12.7).

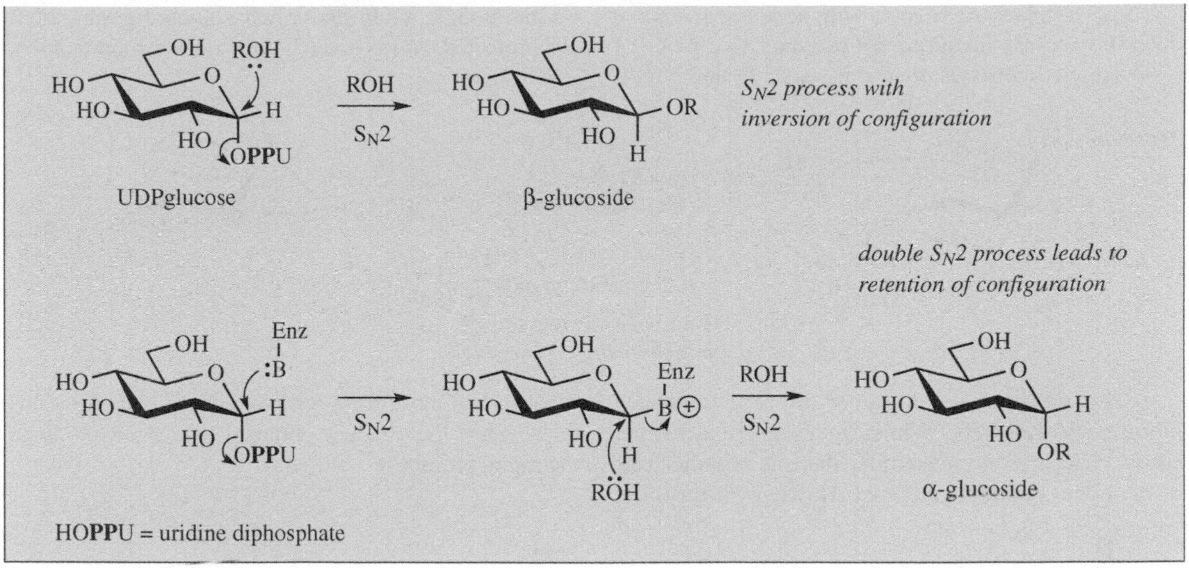

UDPglucose   β-glucoside

S_N2 process with inversion of configuration

double S_N2 process leads to retention of configuration

α-glucoside

HOPPU = uridine diphosphate

12.5 Cyclic acetals and ketals: protecting groups

We have just seen that intramolecular reactions between the carbonyl group and one or other of the hydroxyl functions readily leads to the formation of cyclic hemiacetal or hemiketal forms. Further, these products may then be converted into acetals or ketals by an intermolecular reaction with another alcohol molecule, giving us glycosides. We could also form an acetal or ketal by supplying a carbonyl compound and exploiting the hydroxyl groups of the sugar. This provides a particularly useful means of protecting some of the hydroxyl groups whilst other reactions are carried out (see Box 7.21); the protecting group is then easily removed by effectively reversing the acetal/ketal reaction using hydrolytic conditions (see Section 7.2).

In principle, a number of different types of acetal or ketal might be produced. In this section, we want to exemplify a small number of useful reactions in which two of the hydroxyl groups on the sugar are bound up by forming a cyclic acetal or ketal with a suitable aldehyde or ketone reagent. Aldehydes or ketones react with 1,2- or 1,3-diols under acidic conditions to form cyclic acetals or ketals. If the diol is itself cyclic, then the two hydroxyl groups need to be *cis*-oriented to allow the thermodynamically favourable fused-ring system to form (see Section 3.5.2). Thus, *cis*-cyclohexan-1,2-diol reacts with acetone to form a cyclic ketal, a 1,2-*O*-isopropylidene derivative usually termed, for convenience, an **acetonide**.

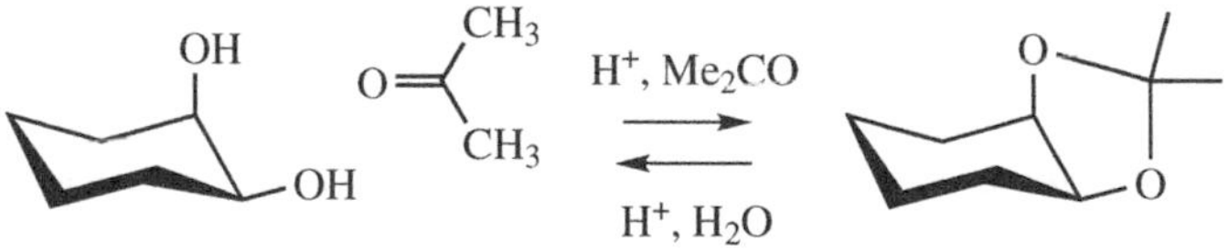

acetonide

When required, the original diol may be regenerated by acid hydrolysis.

Sugars are polyhydroxy compounds, and it is not always easy to predict which of the hydroxyls will react in this way. There are other complicating factors too. The ring size (pyranose/furanose) of the product may differ from that of the starting sugar. It may be that a more stable pyranose form does not have *cis*-oriented hydroxyl groups, whereas a less favoured furanose form does, so that the latter can form cyclic acetals/ketals. The equilibration of pyranose/furanose forms (see Section 12.2.2) allows this type of change to occur.

Thus, D-galactose reacts with acetone to give a diketal: the less-favoured α-form has two pairs of *cis*-oriented hydroxyls that can react. It thus yields a diacetonide. Only the primary alcohol group is left unprotected, and is available for further modification, if desired.

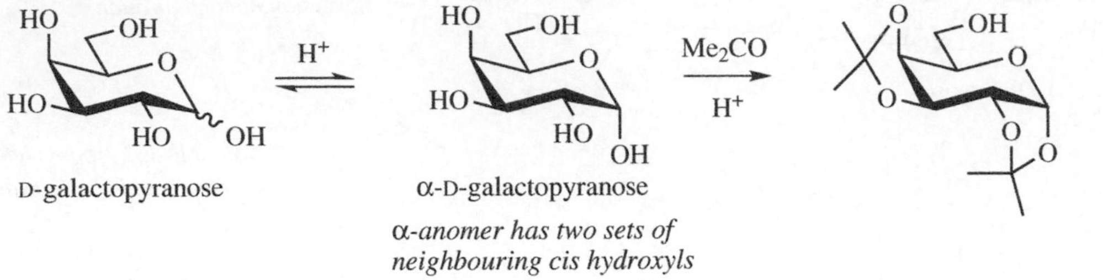

D-galactopyranose   α-D-galactopyranose

α-anomer has two sets of
neighbouring cis hydroxyls

D-Glucose provides a rather more complicated picture, unfortunately. Whilst the pyranose α-anomer could yield a mono-acetonide, there is no other pair of *cis*-hydroxyls that can react. However, it turns out that the furanose form has two sets of hydroxyls that can react; the product obtained is a diacetonide of α-D-glucofuranose.

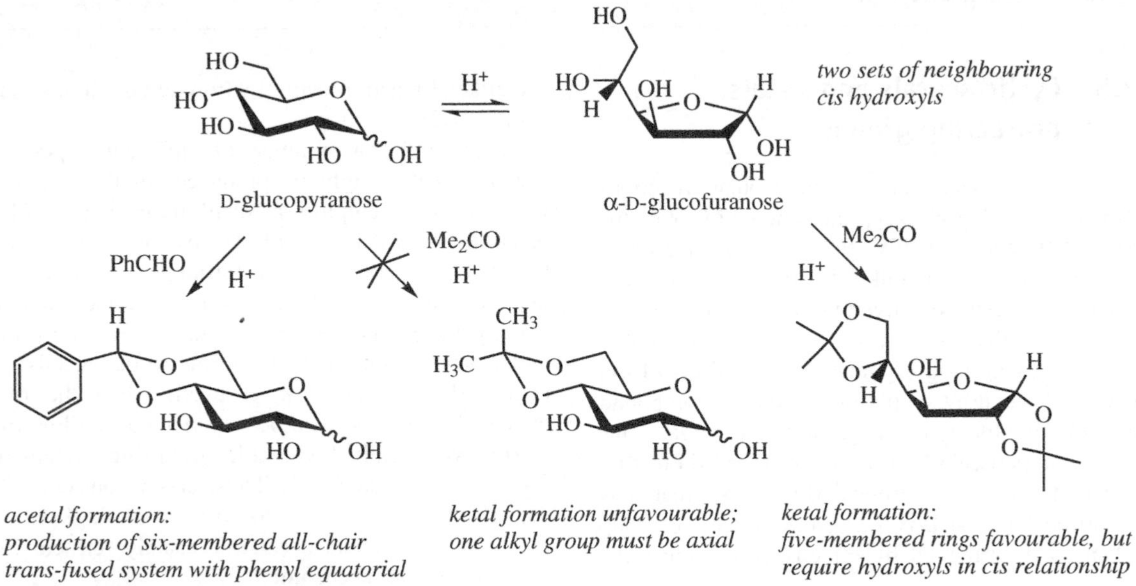

D-glucopyranose   α-D-glucofuranose   *two sets of neighbouring*
                                                                        cis hydroxyls

acetal formation:
production of six-membered all-chair
trans-fused system with phenyl equatorial

ketal formation unfavourable;
one alkyl group must be axial

ketal formation:
five-membered rings favourable, but
require hydroxyls in cis relationship

Note that a six-membered ketal ring involving the hydroxyls at 4 and 6 is not favoured; this is because such a ring would necessarily force one of the two methyls into an axial position. On the other hand, these two hydroxyls can be employed in forming a cyclic acetal with benzaldehyde. Benzaldehyde shows a tendency to form six-membered ring acetals, and because the two substituents are phenyl and hydrogen, we can have a favourable chair system with the phenyl equatorial.

It is not the intention to explain all such variations and add to potential confusion. The behaviour of most sugars with respect to cyclic acetal and ketal formation is well documented for those wishing to work with these compounds. The objective here is merely to illustrate the potential for selective protection of the hydroxyl groups.

12.6   Oligosaccharides

The term **oligosaccharide** is frequently used to classify a small polysaccharide comprised of some two to five monomer units, a name derived from the Greek *oligos*, meaning a few. A pre-eminent example is the disaccharide sucrose, which we commonly call 'sugar' and utilize widely as a sweetening agent and as the raw material for sweets and other confectionary. Other important disaccharides are **maltose**, a hydrolysis product from starch, and **lactose**, the main sugar component of cow's milk.

maltose
D-Glc(α1→4)D-Glc
4-*O*-(α-D-glucopyranosyl)-D-glucopyranose

lactose
D-Gal(β1→4)D-Glc
4-*O*-(β-D-galactopyranosyl)-D-glucopyranose

If we inspect these structures, we can see that they are acetals or ketals equivalent to the glycosides described above, though the alcohol portion is actually one of the hydroxyl groups of a second monosaccharide structure. The linkages are conveniently defined by a shorthand system of nomenclature; this indicates the carbons that are joined by the acetal/ketal bond through the use of numbers and an arrow, together with the configuration α or β at the anomeric carbon. Note that each monosaccharide is numbered separately and there is no unique numbering system for the combined structure. Thus, **maltose** becomes D-Glc(α1 → 4)D-Glc, which conveys the information that two molecules of D-glucose are bonded between carbon-1 of one molecule and carbon-4 of the second, and that the configuration at the anomeric centre (C-1 of the first glucose residue) is α.

Similarly, **lactose**, a combination of D-galactose and D-glucose, is D-Gal(β1 → 4)D-Glc, the configuration at the anomeric centre of galactose being β. Note that the configuration at the hemiacetal anomeric

centre in the second sugar (glucose) is not indicated; it could be α or β, as with a monosaccharide (see Section 12.2.3). Longhand systematic nomenclature that treats one sugar as a substituent on the other can also be used. In the systematic names, the ring size (pyranose or furanose) is also indicated. Thus maltose is 4-*O*-(α-D-glucopyranosyl)-D-glucopyranose, and lactose becomes 4-*O*-(β-D-galactopyranosyl)-D-glucopyranose.

Lactulose is a semi-synthetic disaccharide prepared from lactose, and is composed of galactose linked β1 → 4 to fructose. Galactose is an aldose and exists as a six-membered pyranose ring, whereas fructose is a ketose and forms a five-membered furanose ring. Systematically, lactulose is called 4-*O*-(β-D-galactopyranosyl)-D-fructofuranose; again, the configuration at the anomeric centre of fructose is unspecified. In abbreviated form, this becomes D-Gal(β1 → 4)D-Fru. Lactulose is widely employed as a laxative. It is not absorbed from the gastrointestinal tract, is predominantly excreted unchanged, and helps to retain fluid in the bowel by osmosis.

lactulose
D-Gal(β1 →4)D-Fru
4-*O*-(β-D-galactopyranosyl)-D-fructofuranose

sucrose
D-Glc(α1→β2)D-Fru
α-D-glucopyranosyl-(1→2)-β-D-fructofuranoside

Sucrose is composed of glucose and fructose, and again we have a six-membered pyranose ring coupled to a five-membered furanose ring. However, there is a significant difference when we compare its structure with that of lactulose: in sucrose, the two sugars are both linked through their anomeric centres. In the shorthand representation we thus have to indicate the configuration at each anomeric centre, so the linkage becomes $\alpha1 \to \beta2$. Sucrose is thus abbreviated to D-Glc($\alpha1 \to \beta2$)D-Fru. Systematic nomenclature for sucrose is α-D-glucopyranosyl-$(1 \to 2)$-β-D-fructofuranoside, which also includes the arrow to avoid confusion. Since the two sugars in sucrose are both linked through their anomeric centres, this means that both the hemiacetal/hemiketal structures are prevented from opening; and, in contrast to maltose, lactose, and lactulose, there can be no open-chain form in equilibrium with the cyclic form. Therefore, sucrose does not display any of the properties usually associated with the masked carbonyl group.

In nature, the formation of oligosaccharides, and also of polysaccharides (see Section 12.4), is dependent upon the generation of an activated sugar bound to a nucleoside diphosphate, typically a UDPsugar. As outlined above (see Box 12.4), nucleophilic displacement of the UDP leaving group by a suitable nucleophile generates the new sugar derivative. This will be a glycoside if the nucleophile is a suitable aglycone molecule, or an oligosaccharide if the nucleophile is another sugar molecule. This reaction, mechanistically of S_N2 type, should give an inversion of configuration at C-1 in the electrophile, generating a product with the β-configuration in the case of UDPglucose, as shown. Many of the linkages formed between glucose monomers actually have the α-configuration, and it is believed that a double S_N2 mechanism operates, which initially involves a nucleophilic group on the enzyme (see Box 12.4).

12.7 Polysaccharides

12.7.1 Structural aspects

Polysaccharides fulfil two main functions in living organisms, as food reserves and as structural elements. Plants accumulate **starch** as their main food reserve, a material that is composed entirely of glucopyranose units, but in two different types of polymer, namely amylose and amylopectin. **Amylose** is a linear polymer containing some 1000–2000 glucopyranose units linked $\alpha1 \to 4$. **Amylopectin** is a much larger molecule than amylose (the number of glucose residues varies widely, but may be as high as 10^6), and it is a branched-chain molecule. In addition to $\alpha1 \to 4$ linkages, amylopectin has branches at about every 20 units through $\alpha1 \to 6$ linkages. These branches then also continue with $\alpha1 \to 4$ linkages, but may have subsidiary $\alpha1 \to 6$ branching, giving a tree-like structure.

The mammalian carbohydrate storage molecule **glycogen** is analogous to amylopectin in structure, but is larger and contains more frequent branching, about every 10 residues. The branching in amylopectin and glycogen is achieved by the enzymic removal of a portion of the $\alpha1 \to 4$-linked straight chain containing several glucose residues, then transferring this short chain to a suitable 6-hydroxyl group. A less common storage polysaccharide found in certain plants is **inulin**, which is a relatively small polymer of fructofuranose, linked through $\beta2 \to 1$ bonds.

amylose
(1000–2000 residues)

amylopectin
(up to 10^6 residues; branching
about every 20 residues)

glycogen
(>10^6 residues; branching
about every 10 residues)

inulin
(30–35 residues)

cellulose
(~8000 residues)

Cellulose is reputedly the most abundant organic material on Earth, being the main constituent in plant cell walls. It is composed of glucopyranose units linked β1 → 4 in a linear chain. Alternate residues are 'rotated' in the structure, allowing hydrogen bonding between adjacent molecules, and construction of the strong fibres characteristic of cellulose, as for example in cotton.

12.7.2 Hydrolysis of polysaccharides

Hydrolysis of polysaccharides (and oligosaccharides) follows the comments under glycosides above. Thus, treatment of amylose, amylopectin, or cellulose with hot aqueous acid will result in the formation of glucose as the sole product, through hydrolysis of acetal linkages. Under milder, less-forcing conditions, it is possible to isolate short-chain oligosaccharides as a result of random hydrolysis of linkages.

More specific hydrolysis may be achieved by the use of enzymes. Thus, the enzyme **α-amylase** in saliva and in the gut is able to catalyse hydrolysis of α1 → 4 bonds throughout the starch molecule to give mainly maltose, with some glucose and maltotriose, the trisaccharide of glucose. Amylose is hydrolysed completely by this enzyme, but the α1 → 6 bonds of amylopectin are not affected. Another digestive enzyme, **α-1,6-glucosidase**, is required for this reaction. Finally, pancreatic **maltase** completes the hydrolysis by hydrolysing maltose and maltotriose.

The milk of mammals contains the disaccharide lactose as the predominant carbohydrate, to the extent of about 4–8%. Lactose, therefore, provides the basic carbohydrate nutrition for infants, who metabolize it via the hydrolytic enzyme **lactase**. Lactase enzyme activity in adult humans is usually considerably lower than in infants. Lactose intolerance is a condition in certain adults who are unable to tolerate milk products in the diet. This is a consequence of very low lactase levels, such that ingestion of lactose can lead to adverse reactions, typically gastric upsets.

Cellulose differs from amylose principally in the stereochemistry of the acetal linkages, which are α in amylose but β in cellulose. α-Amylase is specific for α1 → 4 bonds and is not able to hydrolyse β1 → 4 bonds. An alternative enzyme, termed **cellulase**, is required. Animals do not possess cellulase enzymes, and thus cannot digest wood and vegetable fibres that are predominantly composed of cellulose. Ruminants, such as cattle, are equipped to carry out cellulose hydrolysis, though this is dependent upon cellulase-producing bacteria in their digestive tracts.

12.8 Oxidation of sugars: uronic acids

Sugars may be oxidized by a variety of reagents, and the most susceptible group in aldoses is the aldehyde. Use of aqueous bromine as a mild oxidizing agent achieves oxidation of the aldehyde group in D-glucose, and the product is the corresponding carboxylic acid D-gluconic acid. The general term used for such a polyhydroxy carboxylic acid is an **aldonic acid**. These are named by substituting -onic acid for -ose of the sugar. Polyhydroxy carboxylic acids have the potential to form lactones (cyclic esters, see Section 7.9.1), and D-gluconic acid readily forms a 1,4-lactone in solution. In principle, both five- and six-membered rings might be produced, but the five-membered system is favoured (see Section 7.9.1).

More vigorous oxidation results in oxidation of one or more hydroxy groups, with the primary alcohol group being attacked most readily. Thus, oxidizing either D-glucose or D-gluconic acid with aqueous nitric acid leads to a dicarboxylic acid, D-glucaric acid. Dicarboxylic acids of this type are termed **aldaric acids**. Again, aldaric acids readily form five-membered lactones, which may be the 1,4- or 3,6-lactones, or the dilactone shown.

D-glucose

D-gluconic acid

D-gluconic acid 1,4-lactone

D-glucaric acid 1,4:3,6-dilactone

D-glucaric acid 3,6-lactone

D-glucaric acid

D-glucaric acid 1,4-lactone

lactone formation

Box 12.5

Determination of blood glucose levels

The peptide hormone **insulin** (see Box 13.1) is produced by the pancreas and plays a key role in the regulation of carbohydrate, fat, and protein metabolism. In particular, it has a hypoglycaemic effect, lowering the levels of glucose in the blood. A malfunctioning pancreas may produce a deficiency in insulin synthesis or secretion, leading to the condition known as **diabetes mellitus**. This results in increased amounts of glucose in the blood and urine, diuresis, depletion of carbohydrate stores, and subsequent breakdown of fat and protein. Incomplete breakdown of fat leads to the accumulation of ketones in the blood, severe acidosis, coma, and death.

Where the pancreas is still functioning, albeit less efficiently, the condition is known as type 2 diabetes (non-insulin-dependent diabetes, NIDDM),

and can be managed satisfactorily by a controlled diet or oral antidiabetic drugs. In type 1 diabetes (insulin-dependent diabetes, IDDM), pancreatic cells no longer function, and injections of insulin are necessary, one to four times daily, depending on the severity of the condition. These treatments need to be combined with a controlled diet and regular monitoring of glucose levels, but do not cure the disease, so treatment is lifelong.

Quick and easy methods have been developed so that patients can monitor their own blood glucose levels on a regular basis. One such method depends upon the oxidation of glucose to gluconic acid in a reaction catalysed by the enzyme glucose oxidase. This enzyme can be obtained from a number of microorganisms, e.g. *Aspergillus* and *Penicillium* species, and for convenience is usually immobilized onto a suitable support. The microbial enzyme converts glucose into gluconic acid utilizing molecular oxygen as oxidant, but detection of the process is dependent upon the simultaneous production of hydrogen peroxide.

$$\text{glucose} + O_2 \xrightarrow{\text{glucose oxidase}} \text{gluconic acid} + H_2O_2$$

Hydrogen peroxide may be detected by exploiting a secondary chemical reaction that produces a coloured product; this is compared with a standard colour chart to indicate the colour intensity and, therefore, give a measure of the glucose concentration. Alternatively, it may be scanned in a

colorimeter to give a more accurate assay. Even more accuracy can be obtained by using a voltammetric sensor, in which the hydrogen peroxide is oxidized to oxygen on an electrode surface, thus generating an electrical current that is directly proportional to the glucose concentration.

$$H_2O_2 + 2\,HO^- \rightleftharpoons O_2 + H_2O + 2\,e^-$$

The method is highly specific for glucose. Related sugars, such as mannose, xylose and galactose, are not oxidized by this enzyme, or react only in trace amounts.

Uronic acids are produced from aldoses when just the terminal $-CH_2OH$ group has been oxidized to a carboxylic acid. They are named after the parent sugar, substituting -uronic acid for -ose; thus, D-glucuronic acid is the 6-carboxylic acid analogue of D-glucose. It should be apparent from the preceding comments that it will not be possible to oxidize the primary alcohol function selectively in the presence

of the more reactive aldehyde group, so it becomes necessary to protect the aldehyde by an appropriate means. It may also be desirable to protect other hydroxyls. For example, formation of the diacetonide of galactose protects the aldehyde and all hydroxyls except that at position 6 (see Section 12.5). It now remains to oxidize the primary alcohol and remove the protecting groups.

D-galactopyranose   D-galacturonic acid

An alternative approach is to oxidize both the carbonyl and primary alcohol functions to carboxylic acids, then selectively reduce that corresponding to the required aldehyde. This may be achieved by

reducing the 1,4-lactone of D-glucaric acid, using the same reaction as in Box 12.1, where it was employed in the synthesis of labelled glucose.

D-glucaric acid
1,4-lactone   D-glucuronic acid

Uronic acids are found in nature, but they are formed enzymically by selective oxidation of the primary alcohol function of a sugar. Oxidation takes place not on the free sugar, but on UDPsugar

derivatives, as utilized in glycoside biosynthesis (see Box 12.4). **UDPglucuronic acid** is an important carrier in the metabolism of drug molecules (see Box 12.7).

UDPglucose   UDPglucuronic acid   HOPPU = uridine diphosphate

Box 12.6

Some examples of natural uronic acid derivatives

Polymers of uronic acids are encountered in nature in structures known as **pectins**, which are essentially chains of D-galacturonic acid residues linked $\alpha 1 \rightarrow 4$, though some of the carboxyl groups are present as methyl esters. These materials are present in the cell walls of fruit, and the property that aqueous solutions under acid conditions form gels is the basis of jam making.

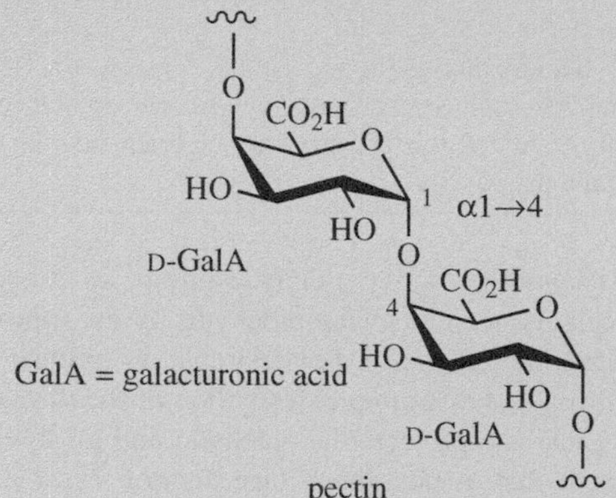

GalA = galacturonic acid

pectin
(400–1000 residues)

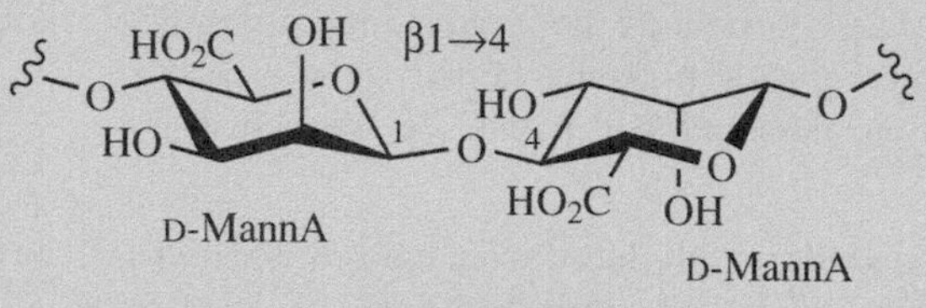

MannA = mannuronic acid

alginic acid
(200–900 residues)

Alginic acid is a polymer of D-mannuronic acid residues joined by $\beta 1 \rightarrow 4$ linkages. It is the main cell-wall constituent of brown algae (seaweeds). Salts of alginic acid are valuable thickening agents in the food industry, and the insoluble calcium salt is the basis of absorbable alginate surgical dressings.

The intensely sweet constituent in the root of liquorice (*Glycyrrhiza glabra*) is **glycyrrhizin**, a mixture of potassium and calcium salts of glycyrrhizic acid. It is said to be 50–150 times as sweet as sucrose. Glycyrrhizic acid is a glycoside of the triterpene aglycone glycyrrhetic acid. The sugar portion is a disaccharide comprised of two molecules of D-glucuronic acid, so is termed a diglucuronide. Liquorice is used in confectionary and as a flavouring agent for beers and stouts. It also finds considerable use in drug formulations to mask the taste of bitter drugs and for its emulsifying surfactant properties.

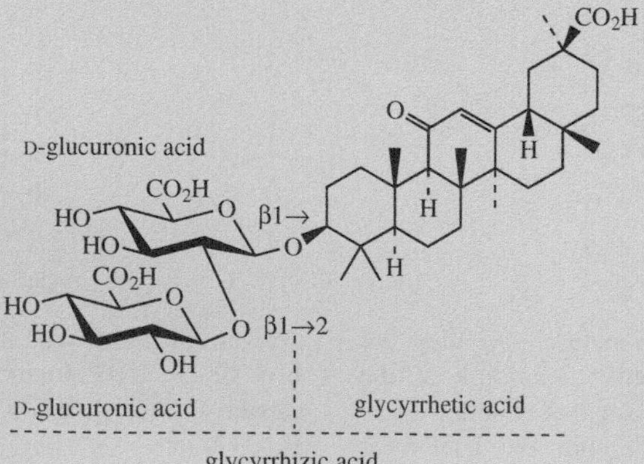

glycyrrhizic acid

Box 12.7

Glucuronides in drug metabolism

One of the principal ways by which foreign compounds are removed from the body is to conjugate them to **glucuronic acid**. This conjugation process not only binds the unwanted compound, but also converts it into a highly polar material that is water soluble and can be excreted in aqueous solution, typically via the kidneys. The polarity is provided both by the hydroxyl groups and by the ionizable carboxylic acid group. Typical chemicals that may become conjugated with glucuronic acid include alcohols, phenols, carboxylic acids, amines, and thiols.

Drugs must also be considered as foreign compounds, and an essential part of drug treatment is to understand how they are removed from the body after their work is completed. Glucuronide formation is the most important of so-called phase II metabolism reactions. Aspirin, paracetamol, morphine, and chloramphenicol are examples of drugs excreted as glucuronides.

Glucuronides are formed in mammals by reaction with uridine diphosphoglucuronic acid (UDPglucuronic acid; UDP-GA) in processes catalysed by uridine diphosphoglucuronyltransferase enzymes. This reaction is entirely analogous to the enzymic glycosylation process we looked at above (see Box 12.4). The reaction with UDP-GA can be envisaged as a simple S_N2 nucleophilic displacement reaction, with an appropriate nucleophile, e.g. an alcohol or amine, and a phosphate derivative as the leaving group. UDP glucuronyltransferase enzymes have very broad substrate specificity, and can catalyse reactions with a wide variety of foreign molecules and drugs.

UDP-GA is formed from UDPglucose (see Box 12.4) by enzymic oxidation of the primary alcohol group. We have already noted in Section 12.6 that UDPglucose is also the biochemical precursor of glucose-containing polysaccharides, e.g. starch and glycogen (see Section 12.7).

The opium alkaloid **morphine** is one of the most valuable analgesics for relief of severe pain. It is known to be metabolized in the body to O-glucuronides, by reaction at the phenolic and alcoholic hydroxyls. The glucuronides formed are water soluble and readily excreted. An interesting feature is that the two monoglucuronides have significantly different pharmacological activities. Although morphine 3-O-glucuronide is antagonistic to the analgesic effects of morphine, morphine 6-O-glucuronide is actually a more effective and longer lasting analgesic than morphine itself, and with fewer side-effects.

Box 12.7 (continued)

morphine 3-O-glucuronide   morphine   morphine 6-O-glucuronide

Box 12.8

Vitamin C

Vitamin C, also known as L-**ascorbic acid**, clearly appears to be of carbohydrate nature. Its most obvious functional group is the lactone ring system, and, although termed ascorbic acid, it is certainly not a carboxylic acid. Nevertheless, it shows acidic properties, since it is an enol, in fact an enediol. It is easy to predict which enol hydroxyl group is going to ionize more readily. It must be the one β to the carbonyl, ionization of which produces a conjugate base that is nicely resonance stabilized (see Section 4.3.5). Indeed, note that these resonance forms correspond to those of an enolate anion derived from a 1,3-dicarbonyl compound (see Section 10.1). Ionization of the α-hydroxyl provides less favourable resonance, and the remaining hydroxyls are typical non-acidic alcohols (see Section 4.3.3). Thus, the pK_a of vitamin C is 4.0, and is comparable to that of a carboxylic acid.

L-ascorbic acid

*favourable ionization; resonance stabilization
of conjugate base (enolate anion)*

unfavourable ionization

   Vitamin C is essential for the formation of collagen, the principal structural protein in skin, bone, tendons, and ligaments, being a cofactor in the hydroxylation of the amino acids proline to 4-hydroxyproline, and of lysine to 5-hydroxylysine. These hydroxyamino acids account for up to 25% of the collagen structure. Vitamin C is also associated with some other hydroxylation reactions, e.g. the hydroxylation of tyrosine to dopa (dihydroxyphenylalanine) in the pathway to catecholamines (see Box 15.3). Deficiency leads to scurvy, a condition characterized by muscular pain, skin lesions, fragile blood vessels, bleeding gums, and tooth loss. Vitamin C also has valuable antioxidant properties (see Box 9.2), and these are exploited commercially in the food industries.

   Most animals can synthesize vitamin C, though humans and primates cannot and must obtain it via the diet. Citrus fruits, peppers, guavas, rose hips, and blackcurrants are especially rich sources, but it is present in most fresh fruit and vegetables.

   In animals, ascorbic acid is synthesized in the liver from D-glucose, by a pathway that initially involves specific enzymic oxidation of the primary alcohol function, giving D-glucuronic acid (see Section 12.8). This is followed by reduction to L-gulonic acid, which is effectively reduction of the carbonyl function in the ring-opened hemiacetal.

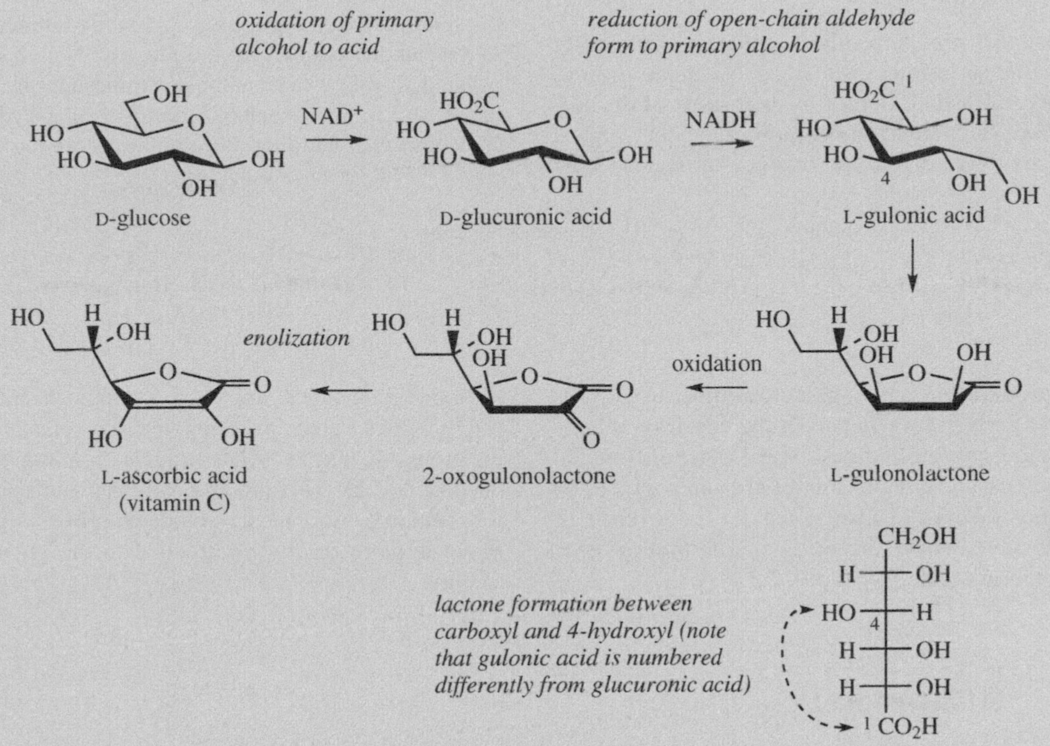

oxidation of primary alcohol to acid

reduction of open-chain aldehyde form to primary alcohol

D-glucose → (NAD⁺) → D-glucuronic acid → (NADH) → L-gulonic acid

L-gulonic acid → L-gulonolactone

L-gulonolactone → (oxidation) → 2-oxogulonolactone

2-oxogulonolactone → (enolization) → L-ascorbic acid (vitamin C)

lactone formation between carboxyl and 4-hydroxyl (note that gulonic acid is numbered differently from glucuronic acid)

Lactone formation in gulonic acid leads to the favourable five-membered system (see Section 7.9.1), and then oxidation of the secondary alcohol to a carbonyl effectively gives ascorbic acid. However, the more favourable structure of ascorbic acid is the enol tautomer with the conjugated α,β-unsaturated lactone. Ascorbic acid formation in plants follows an analogous pathway, starting from either D-glucose or D-galactose. Man and other primates appear to be deficient in the enzyme that oxidizes gulonolactone to the ketolactone, and we are thus dependent on a dietary source of vitamin C.

An unfortunate twist here is the apparent configurational change from D to L in going from glucuronic acid to gulonic acid. This is actually a consequence of renumbering. In gulonic acid the carboxylic acid group has the higher oxidation state, becomes the topmost substituent in the Fischer projection, and is numbered carbon-1 (see Section 3.4.9). As a result, the D descriptors for glucuronic acid and the L descriptor for gulonic acid are now referring to two different chiral centres. You can see why we were rather unenthusiastic about the value of D and L in Section 3.4.10. We are also uncomfortable that the –CH$_2$OH → –CO$_2$H change inferred in the glucose → glucuronic acid by convention maintains the same configuration in the two compounds. From the gulonic acid example, we might reasonably expect to apply the same type of renumbering in glucuronic acid.

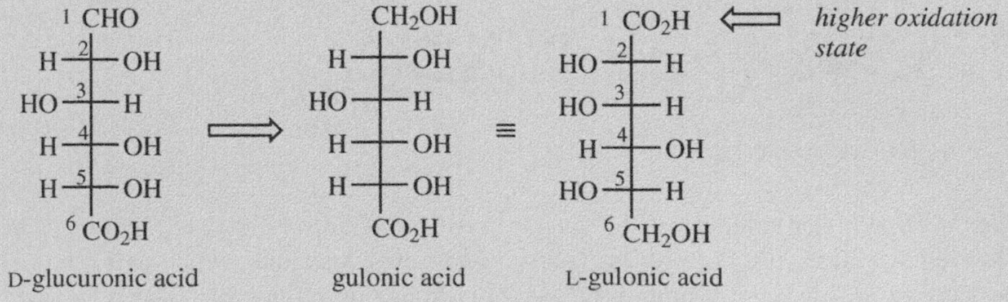

D-glucuronic acid → gulonic acid ≡ L-gulonic acid

higher oxidation state

12.9   Aminosugars

Aminosugars are the result of replacement of one or more hydroxyl groups in a sugar by amino groups. They are formed in nature by **transamination** processes (see Section 15.6) on appropriate keto sugars, which are themselves the product of regiospecific enzymic oxidation processes. Thus, D-glucosamine (2-amino-2-deoxy-D-glucose) is readily appreciated as a metabolic product from D-glucose. Note here the convenient way we can name an aminosugar by relating it to a normal sugar via removal of a hydroxyl (2-deoxy) then addition of an amino (2-amino).

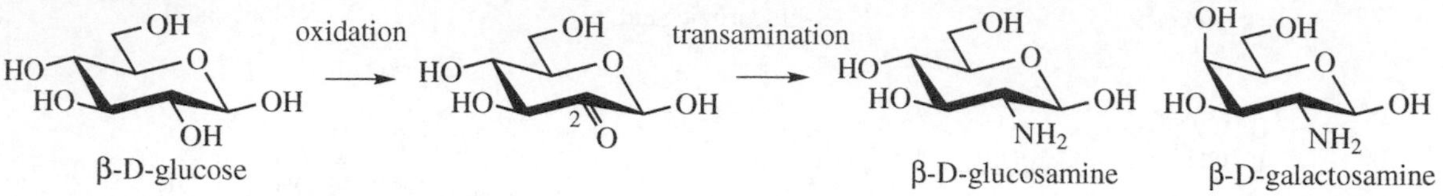

β-D-glucose   →(oxidation)   →(transamination)   β-D-glucosamine   β-D-galactosamine

D-**Glucosamine** and D-**galactosamine**, usually as *N*-acetyl derivatives, are part of the structures of several natural polysaccharides, whilst other uncommon aminosugars are components of the aminoglycoside antibiotics. We have also noted the occurrence of *N*-glycosides, where the nitrogen substitution is at the anomeric centre (see Box 12.3).

A simple chemical approach to aminosugars is to use S_N2 displacement by ammonia of a suitable leaving group, such as a tosylate (toluene *p*-sulfonate, see Section 7.13.1). This process can be made selective for position 6, since the less-hindered primary alcohol group is more readily esterified than the secondary alcohols.

D-glucose

H^+ | MeOH

TsCl ≡   H₃C—⟨⟩—SO₂Cl

methyl α-D-glucopyranoside   →   methyl 6-*O*-tosyl-α-D-glucopyranoside

since the primary alcohol reacts more readily than the secondary alcohols, selective esterification is possible

↓ NH₃

S_N2 *reaction*

$H_3N:$ ⟶ OTs

6-amino-6-deoxy-D-glucose   ←(H_2O / H^+)—

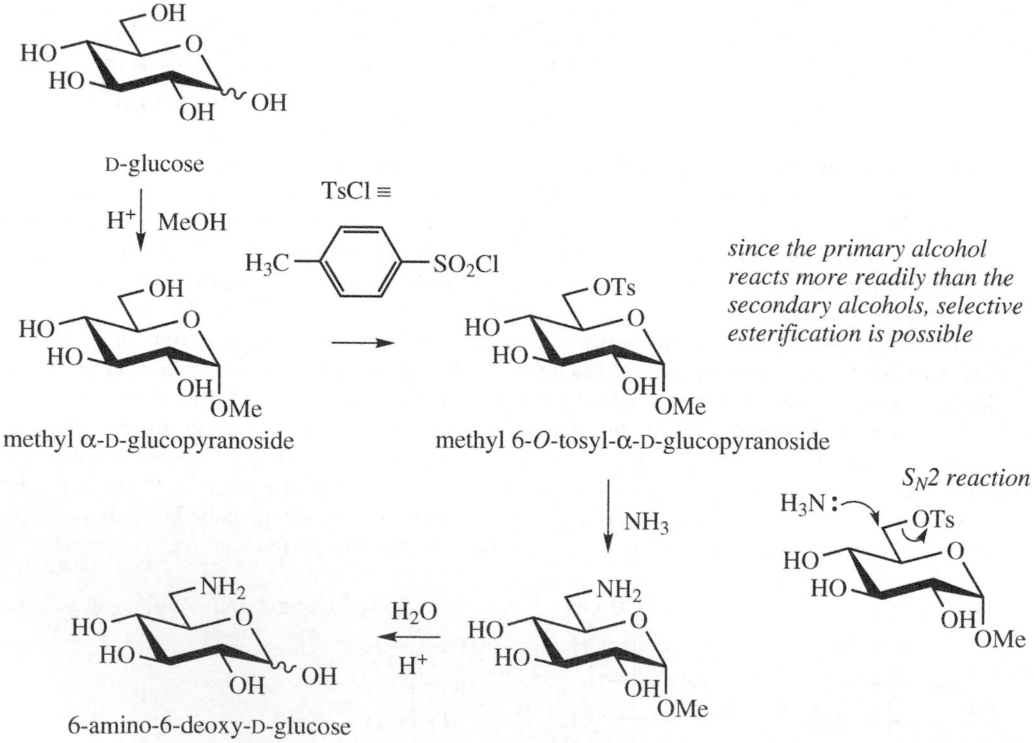

2-Aminosugars such as glucosamine may be synthesized by a modified Kiliani–Fischer process (see Box 12.1). The starting aldose, here D-arabinose, is treated with ammonia, producing an imine, and then with HCN to yield epimeric 2-aminonitriles. The remaining steps lead to a mixture of D-glucosamine and D-mannosamine, which will need to be separated.

Protecting groups such as cyclic acetals and ketals may also be employed to achieve selective reactions (see Section 12.5).

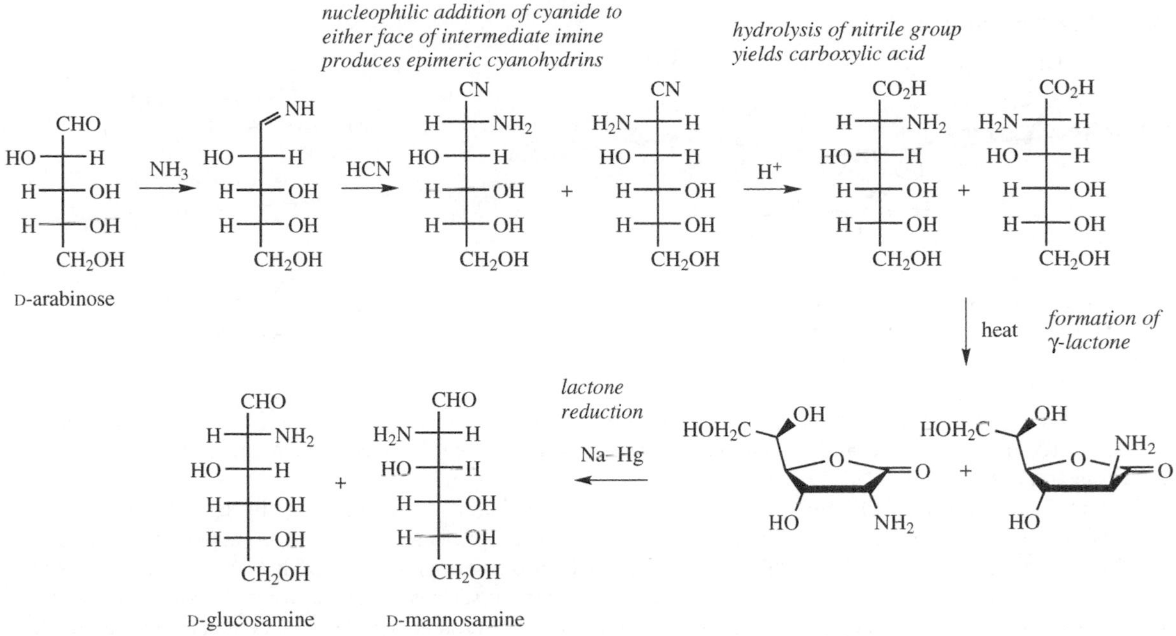

nucleophilic addition of cyanide to either face of intermediate imine produces epimeric cyanohydrins

hydrolysis of nitrile group yields carboxylic acid

formation of γ-lactone

lactone reduction

D-arabinose

D-glucosamine   D-mannosamine

Box 12.9

Aminosugars and aminoglycoside antibiotics

The **aminoglycosides** form an important group of antibiotic agents and are immediately recognizable as modified carbohydrate molecules. Typically, they have two or three uncommon sugars attached through glycoside linkages to an aminocyclitol, i.e. an amino-substituted hydroxycyclohexane system. The first of these agents to be discovered was **streptomycin** from *Streptomyces griseus*. Its structure contains the aminocyclitol **streptamine**, though both amino groups are bound as guanidino substituents in the derivative streptidine.

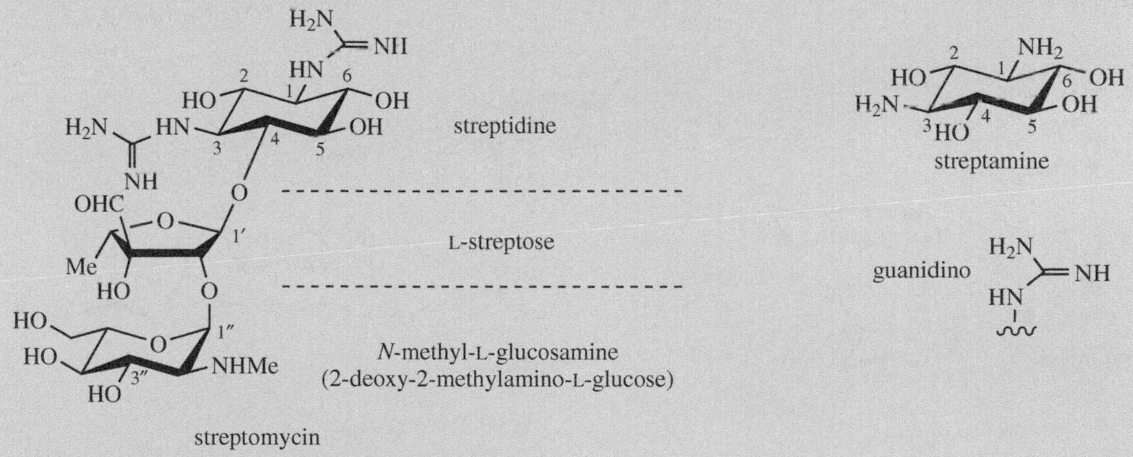

streptidine

L-streptose

N-methyl-L-glucosamine
(2-deoxy-2-methylamino-L-glucose)

streptomycin

streptamine

guanidino

Other medicinally useful aminoglycoside antibiotics are based on the aminocyclitol **2-deoxystreptamine**, e.g. **gentamicin C₁** from *Micromonospora purpurea*. Although streptamine and 2-deoxystreptamine are actually cyclohexane derivatives, they are both of carbohydrate origin and derived naturally from glucose.

Box 12.9 (continued)

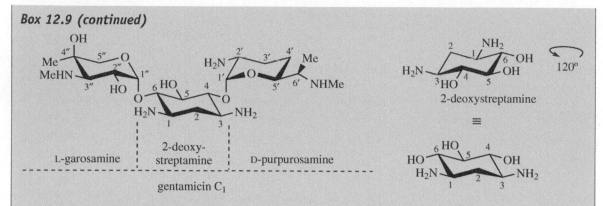

gentamicin C_1

The other component parts of streptomycin, namely L-streptose and the aminosugar 2-deoxy-2-methylamino-L-glucose (*N*-methyl-L-glucosamine), are also ultimately derived from D-glucose. Gentamicin C_1 contains two aminosugars, L-garosamine and D-purpurosamine.

The **aminoglycoside antibiotics** have a wide spectrum of activity, including activity against some Gram-positive and many Gram-negative bacteria. However, their widespread use is limited by nephrotoxicity, which results in impaired kidney function, and by ototoxicity, which is a serious side-effect that can lead to irreversible loss of hearing. These antibiotics are thus reserved for treatment of serious infections where less-toxic drugs have proved ineffective. The aminoglycoside antibiotics interfere with protein biosynthesis by acting on the smaller 30S subunit of the bacterial ribosome (see Box 14.1).

Aminosugars are also components of many **macrolide antibiotics**. These are macrocyclic lactones with a ring size typically of 12–16 atoms (see Box 7.14). Two or more sugar units are attached through glycoside linkages, these sugars tending to be unusual 6-deoxy structures often not found outside of this class of compounds, e.g. L-cladinose. At least one sugar is an amino sugar, e.g. D-desosamine. These antibiotics have a narrow spectrum of antibacterial activity, principally against Gram-positive microorganisms. Their antibacterial spectrum resembles, but is not identical to, that of the penicillins, so they provide a valuable alternative for patients allergic to the penicillins (see Box 13.12). **Erythromycin** produced by cultures of *Saccharopolyspora erythraea* is the principal macrolide antibacterial currently used in medicine. It exerts its antibacterial action by inhibiting protein biosynthesis, binding to the larger 50S subunit of bacterial ribosomes and blocking the translocation step (see Section 14.2.6).

erythronolide — erythromycin A — L-cladinose — D-desosamine

leuconolide A_1 — spiramycin I — D-forosamine — D-mycaminose — L-mycarose

Spiramycin is another macrolide, recently introduced into medicine for the treatment of toxoplasmosis, infections caused by the protozoan *Toxoplasma gondii*. This contains a 16-membered lactone ring (erythromycin has a 14-membered ring), and two aminosugars, D-mycaminose and D-forosamine. D-Forosamine is remarkable in having only one hydroxyl group, and that is bound up in the hemiacetal ring system.

12.10 Polymers containing aminosugars

The structure of **chitin** is rather similar to that of cellulose, though it is composed of $\beta1 \to 4$-linked *N*-acetylglucosamine residues. Chitin is a major constituent in insect skeletons and the shells of crustaceans, e.g. crabs and lobsters; as with cellulose, its strength again depends on hydrogen bonding between adjacent molecules, producing rigid sheets. Chemical deacetylation of chitin provides **chitosan**, a valuable industrial material used for water purification because of its chelating properties, and in wound-healing preparations.

GlcN = glucosamine
GlcNAc = *N*-acetylglucosamine

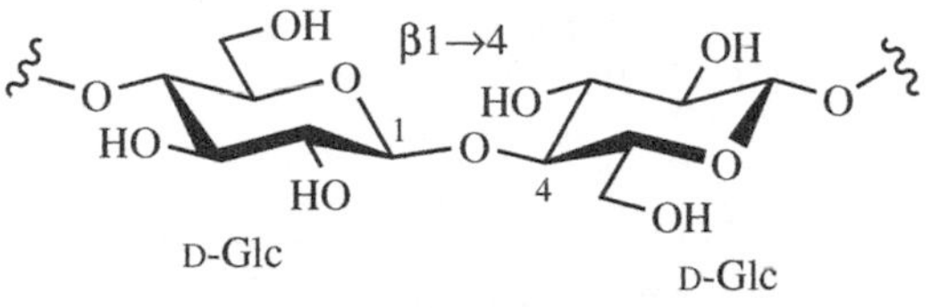

cellulose
(~8000 residues)

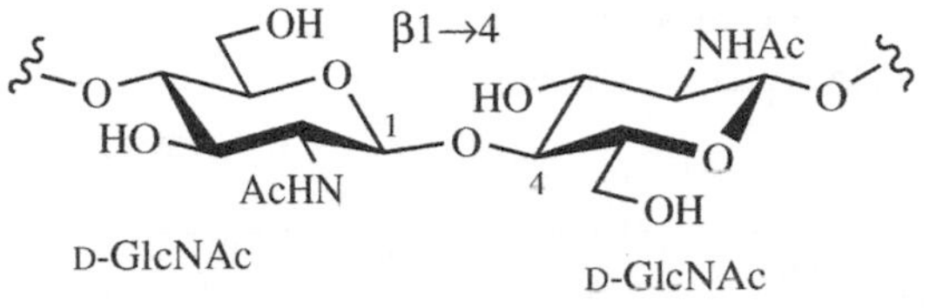

chitin
(500–5000 residues)

Bacterial cell walls contain **peptidoglycan** structures in which the carbohydrate chains are composed of alternating $\beta1 \to 4$-linked *N*-acetylglucosamine and *O*-lactyl-*N*-acetylglucosamine (also called *N*-acetylmuramic acid) residues. These chains are cross-linked via peptide structures. Part of the peptidoglycan of *Staphylococcus aureus* is shown here. Amino acid abbreviations are given in Chapter 13.

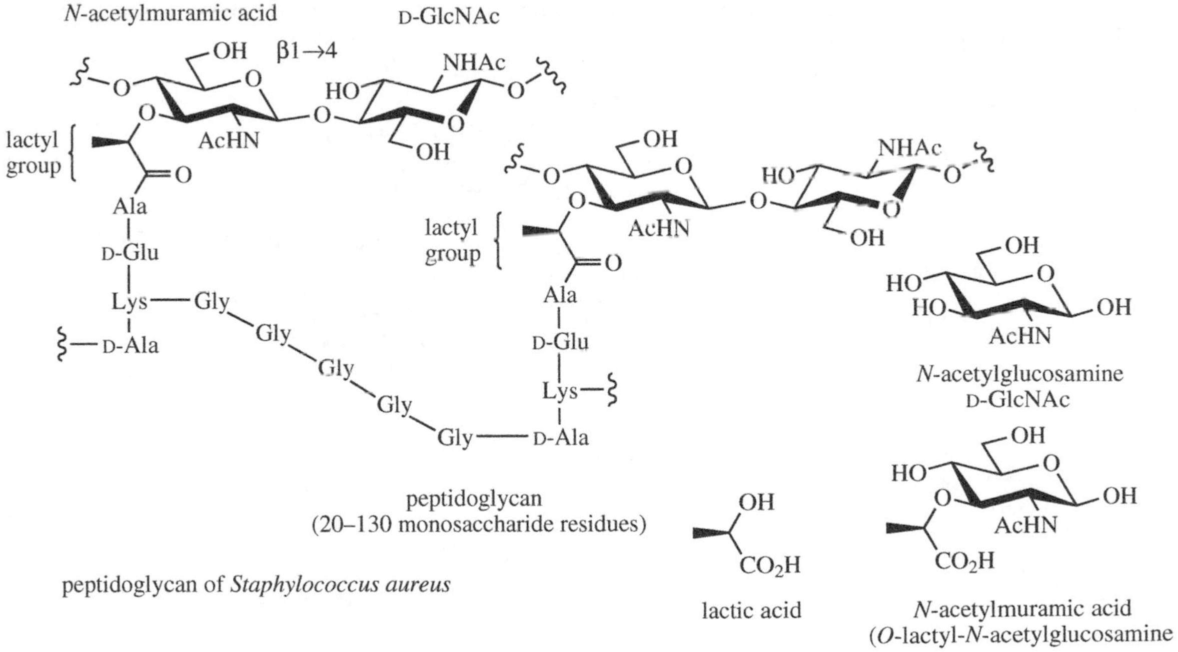

peptidoglycan
(20–130 monosaccharide residues)

peptidoglycan of *Staphylococcus aureus*

lactic acid

N-acetylglucosamine
D-GlcNAc

N-acetylmuramic acid
(*O*-lactyl-*N*-acetylglucosamine

This shows the involvement of the lactyl group of the *N*-acetylmuramic acid in linking the peptide with the carbohydrate via an amide/peptide bond. The biological activities of the β-lactam antibiotics, e.g. penicillins and cephalosporins (see Box 13.11), stem from an inhibition of the cross-linking mechanism during the biosynthesis of the bacterial cell wall (see Box 13.12).

The mammalian blood anticoagulant **heparin** is also a carbohydrate polymer in which amino sugars (glucosamine) alternate with uronic acid residues (see

Section 12.8). Polymers of this kind are known as anionic **mucopolysaccharides**, or **glycosaminoglycans**. Heparin consists of two repeating disaccharide units, in which the amino functions and some of the hydroxyls are sulfated, producing a heterogeneous polymer. The carboxyls and sulfates together make heparin a strongly acidic water-soluble material.

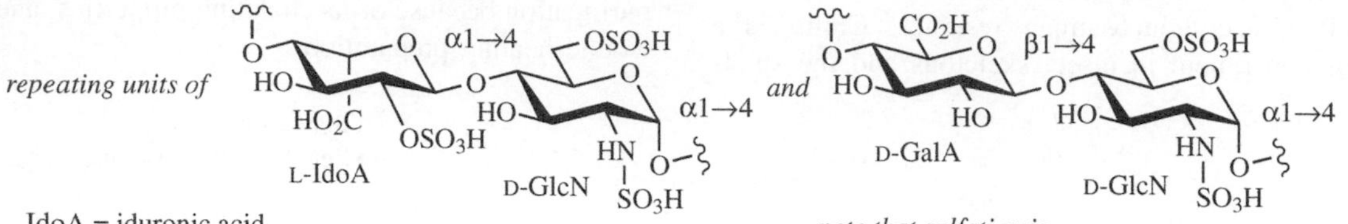

repeating units of

L-IdoA

D-GlcN

IdoA = iduronic acid

GalA = galacturonic acid

and

D-GalA

D-GlcN

note that sulfation is partial and variable

heparin
(10–30 disaccharide residues)

Box 12.10

Carbohydrate determinants of blood groups

Most people are aware that blood is classified into several types, the **blood groups**. These are termed A, B, O, etc. It is absolutely essential in blood transfusions that the donor blood matches that of the recipient, otherwise antibodies are produced to the new blood. This leads to aggregation of red blood cells, with potentially fatal results through blockage of blood vessels. The blood group antigens are actually **glycoproteins**, carbohydrates having an attached protein chain, and the various blood groups can be correlated with a single monosaccharide residue in the carbohydrate portion.

At the end of the carbohydrate section in type O blood antigens, there is a D-galactopyranose ring to which is attached an L-fucopyranose sugar through an $\alpha 1 \rightarrow 2$ linkage.

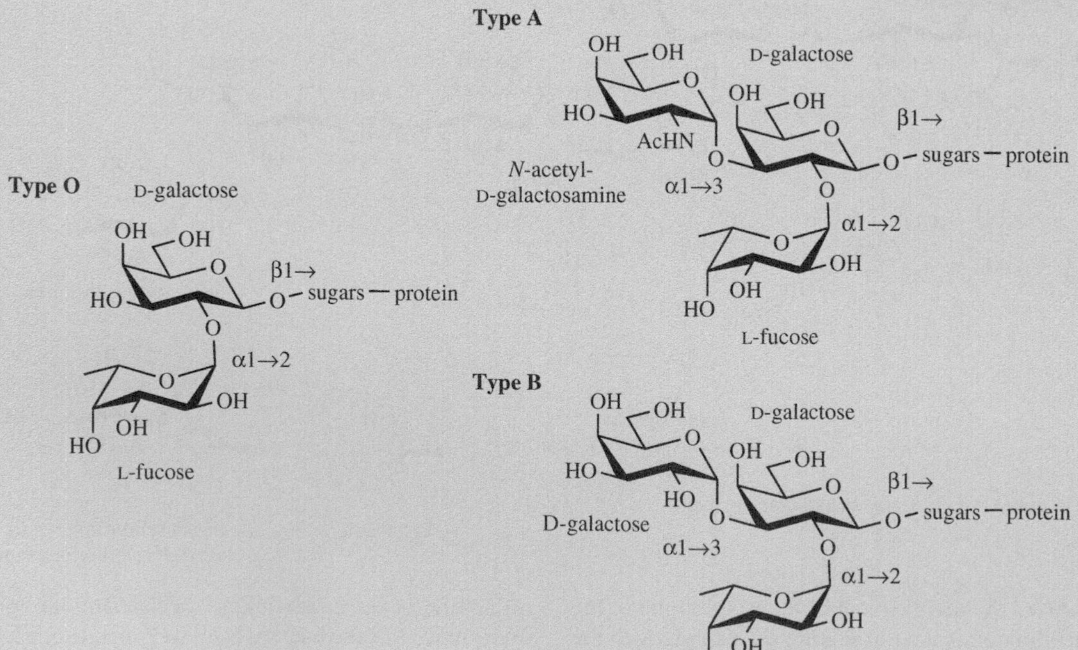

Type A

Type O

Type B

In type B blood antigens, the galactose residue has a second D-galactose residue attached, through an α1 → 3 linkage. In type A blood antigens, the second sugar residue is now N-acetyl-D-galactosamine, again attached through an α1 → 3 linkage. It has been found that enzymic removal of the terminal galactose residue from type B or of the N-acetyl-galactosamine residue from type A converts the B or A antigens into O antigens. It is also known that individuals with type B or type A antigens possess additional enzyme systems that specifically add the extra terminal carbohydrate unit to the type O antigen.

13

Amino acids, peptides and proteins

13.1 Amino acids

Numerous amino acids are found in nature, but in this chapter we are concerned primarily with those that make up the structures known as peptides and proteins. Peptides and proteins are both **polyamides** (see Section 7.10) composed predominantly of **α-amino acids** linked through their carboxyl and α-amino functions.

α-amino acids

peptide

peptide bond
=
amide bond

In biochemistry, the amide linkage is traditionally referred to as a **peptide bond**. Whether the resultant polymer is classified as a **peptide** or a **protein** is not clearly defined; generally, a chain length of more than 40 residues confers protein status, whereas the term **polypeptide** can be used to cover all chain lengths.

Proteins in all organisms are made up from the same set of 20 α-amino acids, though the organism is not necessarily capable of synthesizing all of these. Some amino acids are obtained from the diet. The amino acids are combined in a sequence that is defined by the genetic code, the sequence of bases in DNA (see Section 14.2.4). Table 13.1 gives the structures of these 20 amino acids together with the standard three-letter and one-letter abbreviations used to represent them. Proline is strictly an imino acid rather than an amino acid, but it is normally included as one of the 20 amino acids. The amino acids are also subclassified according to the chemical and physical characteristics of their R substituent. Since the polypeptide structure combines both the amino and carboxylic acid functions of an amino acid into amide linkages, the overall properties of the polypeptide are going to be defined predominantly by the characteristics of these R substituents.

The amino acid components of proteins have the L configuration (see Section 3.4.10), but many peptides are known that contain one or more D-amino acids in their structures. D-Amino acids are not encoded by DNA, and peptides containing them are produced by what is termed 'non-ribosomal peptide biosynthesis' (see Section 13.5.2). D-Amino acids generally arise by epimerization of L-amino acids (see Box 10.10). All the protein L-amino acids have the S configuration, except for glycine, which is not chiral, and L-methionine which is R, a consequence of the priority rules for systematic descriptors of configuration (see Section 3.4.10).

L-amino acid   D-amino acid

Essentials of Organic Chemistry   Paul M Dewick
© 2006 John Wiley & Sons, Ltd

Table 13.1   Amino acids: structures and standard abbreviations

Amino acid	Structure	Abbreviations		R group classification	pK_a		
					CO_2H	NH_2	Side-chain
Alanine	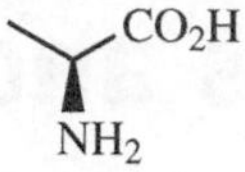	Ala	A	Alkyl non-polar	2.34	9.69	
Arginine	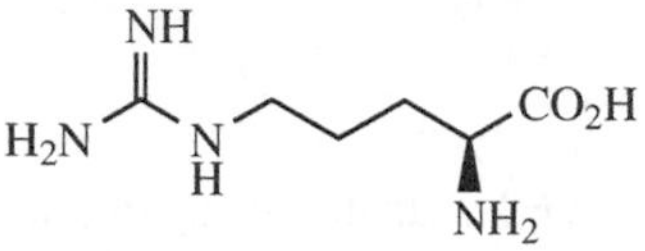	Arg	R	Guanidine basic polar	2.17	9.04	12.48
Asparagine	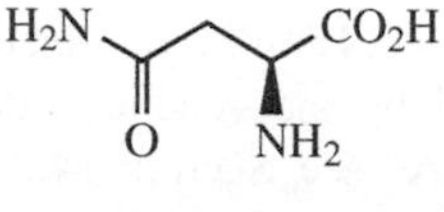	Asn	N	Amide polar	2.01	8.8	
Aspartic acid	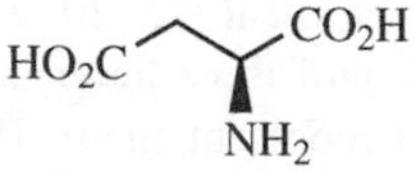	Asp	D	Carboxylic acid acidic polar	1.89	9.60	3.65
Cysteine	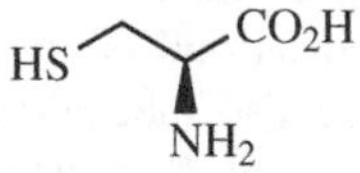	Cys	C	Thiol polar	1.96	8.18	10.29
Glutamic acid	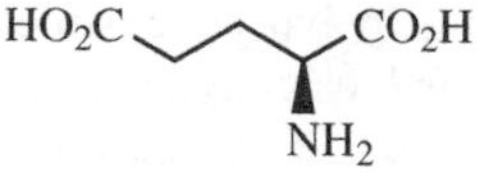	Glu	E	Carboxylic acid acidic polar	2.19	9.67	4.25
Glutamine	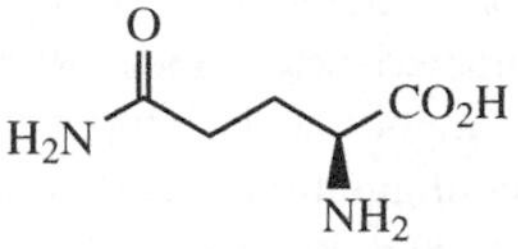	Gln	Q	Amide polar	2.17	9.13	
Glycine	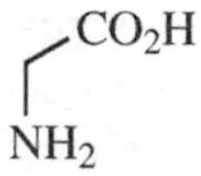	Gly	G	Hydrogen non-polar	2.34	9.6	
Histidine	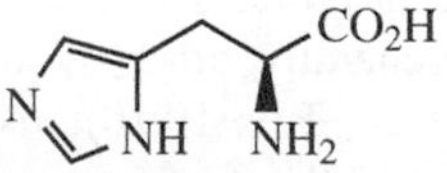	His	H	Imidazole basic polar	1.8	9.17	6.00
Isoleucine	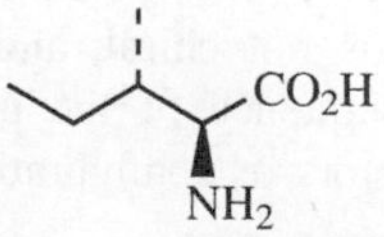	Ile	I	Alkyl non-polar	2.35	9.68	
Leucine	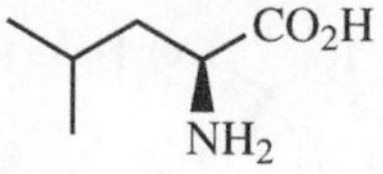	Leu	L	Alkyl non-polar	2.36	9.60	
Lysine	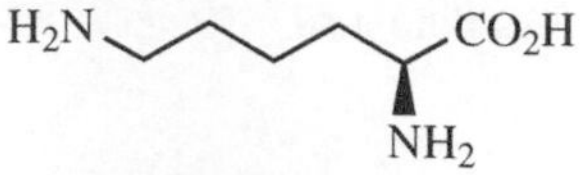	Lys	K	Amine basic polar	2.18	8.95	10.52

Table 13.1 (*continued*)

Amino acid	Structure	Abbreviations		R group classification	pKa		
					CO$_2$H	NH$_2$	Side-chain
Methionine	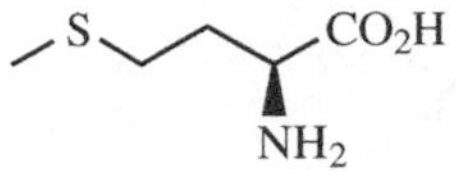	Met	M	Thioether non-polar	2.28	9.2	
Phenylalanine	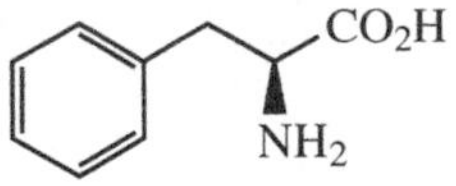	Phe	F	Aromatic non-polar	1.83	9.12	
Proline[a]	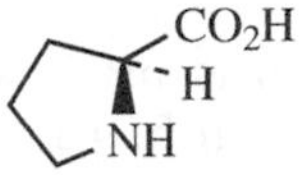	Pro	P	Non-polar	1.99	10.96	
Serine	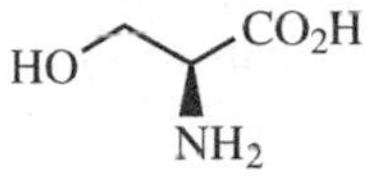	Ser	S	Alcohol polar	2.21	9.95	13.6
Threonine	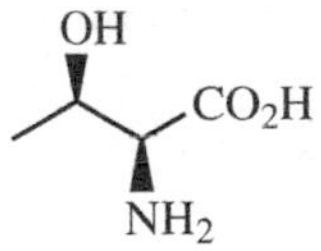	Thr	T	Alcohol polar	2.11	9.62	13.6
Tryptophan	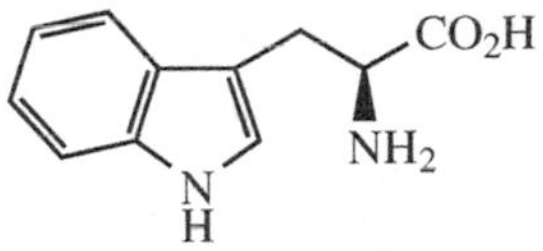	Trp	W	Aromatic non-polar	2.38	9.39	
Tyrosine	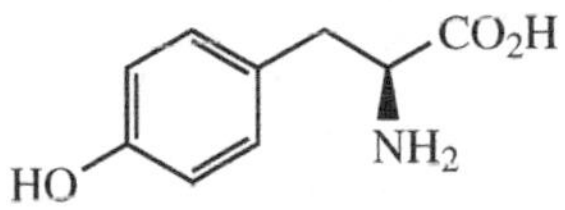	Try	Y	Phenol polar	2.2	9.11	10.06
Valine	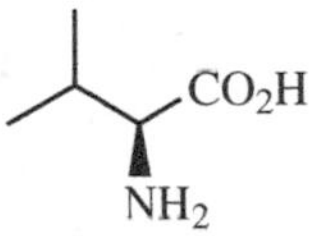	Val	V	Alkyl non-polar	2.32	9.61	

[a]Proline is actually an *imino* acid.

Two of the protein amino acids, threonine and isoleucine, have two chiral centres; therefore, diastereoisomeric forms are possible. In proteins, each of these amino acids exists in a single diastereoisomeric form.

The pKa of the carboxylic acid group of amino acids is around 2, and that of the amino group (as conjugate acid) is around 9. As we saw in Section 4.11.3, this means that the carboxylic acid group (a stronger acid than the ammonium cation) will protonate the amino group (a stronger base than

the carboxylate anion). At pH 7, therefore, amino acids with neutral R groups will exist mainly as the overall neutral, but doubly charged zwitterionic form (the weaker acid and weaker base).

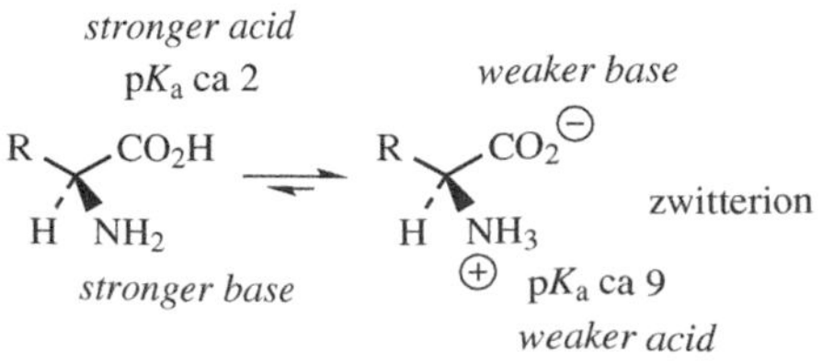

The carboxylate group becomes protonated as the pH decreases, whereas at higher pH the ammonium ion becomes deprotonated, in both cases yielding a singly charged species. The uncharged amino acid (as we almost always draw it!) is actually a negligible contributor at any pH (see Section 4.11.3).

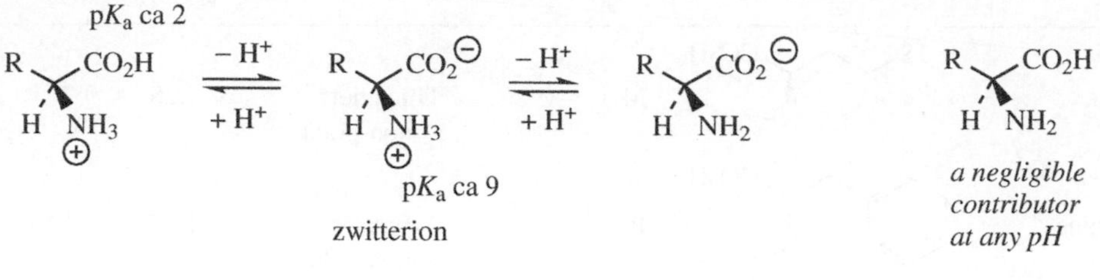

When the R group contains another ionizable group, the amino acid will have more than two dissociation constants. The carboxylic acid groups of aspartic acid and glutamic acid, the amine of lysine, and the guanidino group of arginine will all be ionized at pH 7, and the imidazole nitrogen of histidine will be partially protonated (see Box 4.7). However, neither the phenolic group of tyrosine nor the thiol group of cysteine will be ionized at this pH.

Amino acids with ionizable side-chains   **ionic form at pH 7**

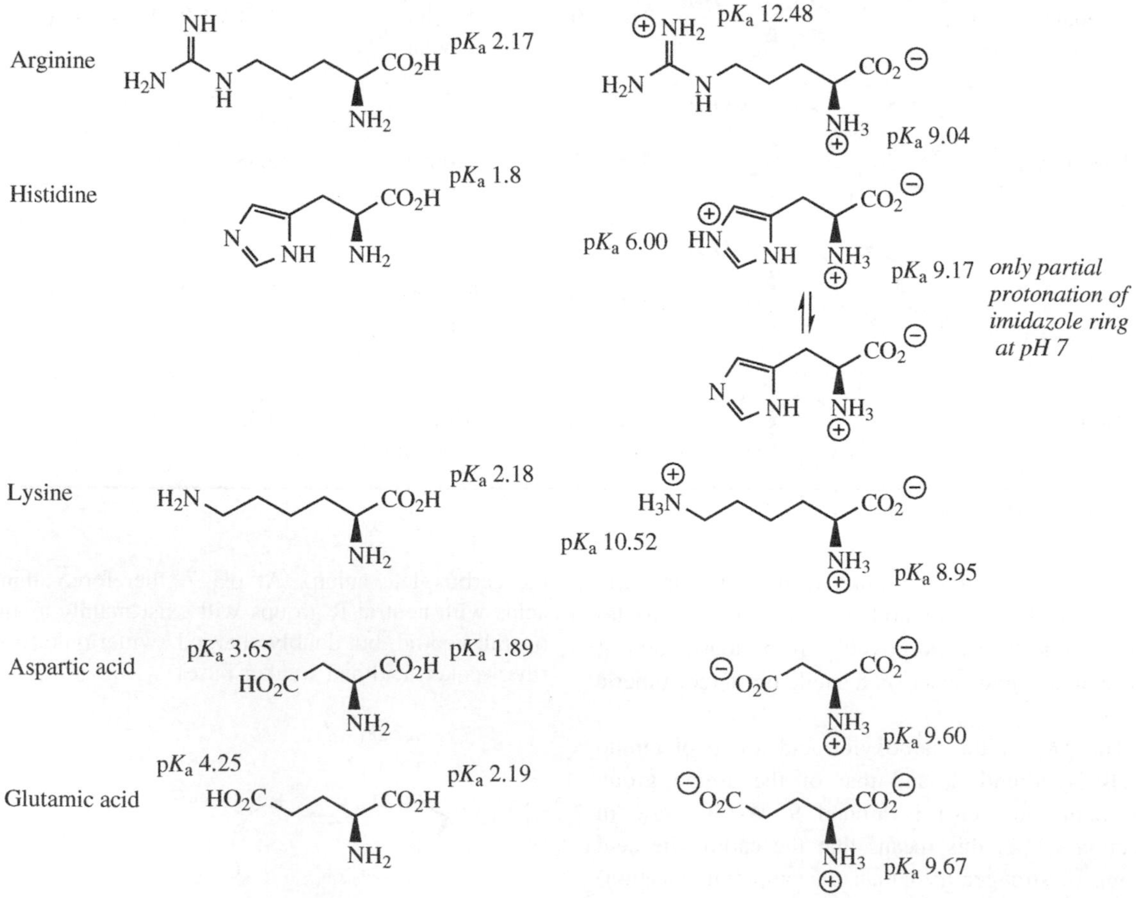

Amino acids with ionizable side-chains

ionic form at pH 7

Cysteine

Tyrosine

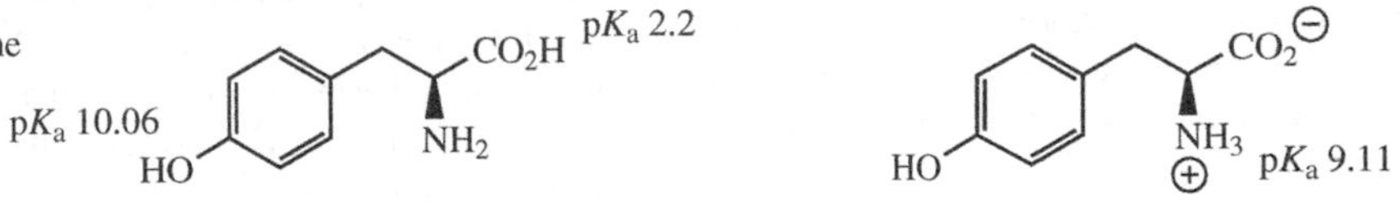

The ionization properties of side-chain substituents will usually carry through into the peptide or protein and influence the behaviour of the polymer. However, the actual pK_a values of the amino acid side-chains in the protein are modified somewhat by the position of the amino acid in the chain, and the environment created by other substituents. Typical pK_a values are shown in Table 13.2.

Note that the side-chains of glutamine and asparagine are not basic; these side-chains contain amide functions, which do not have basic properties (see Section 4.5.4). The heterocyclic ring in trypto-phan can also be considered as non-basic, since the nitrogen lone pair electrons form part of the aromatic π electrons and are unavailable for bonding to a pro-ton (see Section 11.8.2).

In addition to the 20 amino acids described, there are also a few amino acids quite frequently encountered that are not encoded by DNA. These are mainly found in peptides, and are typically slightly modified versions of the common amino acids, such as N-methyl amino acids. These components are represented by an appropriate variation of the normal abbreviation, e.g. N-methyl amino acids such as Tyr(Me) or Leu(Me), though N-methylglycine is often referred to as sarcosine (Sar).

Table 13.2 pK_a values for free and protein-bound amino acids

Amino acid	pK_a	
	Free amino acid	In proteins
Aspartic acid	3.65	3.5–4.5
Glutamic acid	4.25	3.5–4.5
Lysine	10.52	9.5–10.5
Arginine	12.48	12–13
Histidine	6.00	6–7

A frequently encountered modification is the conversion of the C-terminal carboxyl into an amide. This is represented as Phe–NH_2, for example, which must be considered carefully, and not be interpreted as an indication of the N-terminus (see Section 13.2).

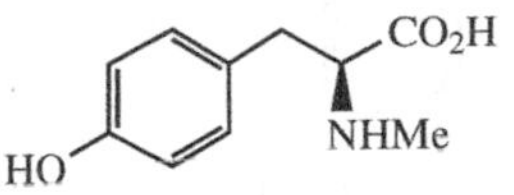

Tyr(Me)

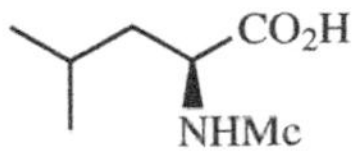

Leu(Me)

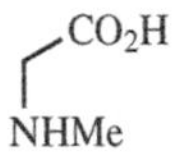

Sar
(Sarcosine;
N-methylglycine)

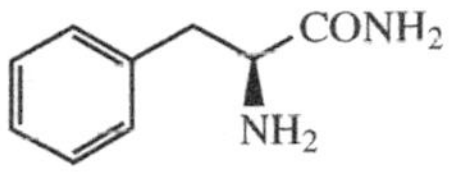

Phe–NH$_2$

Some other variants are shown below, with their abbreviations. Pyroglutamic acid may be found where a terminal glutamic acid residue, linked to the chain through its carboxyl, forms a cyclic amide (lactam) (see Section 7.10).

Some common amino acids not encoded by DNA

Pyroglutamic acid (5-oxoproline)	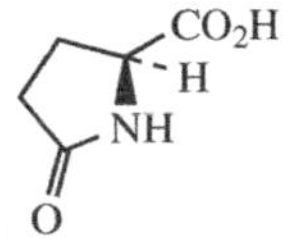	Glp oxoPro <Glu
Hydroxyproline		HPro
Hydroxyleucine	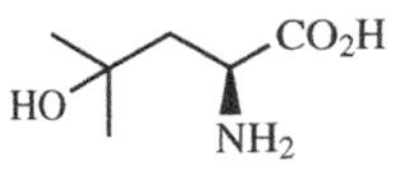	HLeu
Ornithine	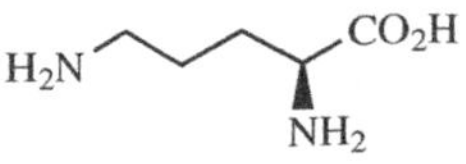	Orn

13.2　Peptides and proteins

Although superficially similar, peptides and proteins display a wide variety of biological functions, and many have marked physiological properties. For example, they may function as structural molecules in tissues, as enzymes, as antibodies, or as neurotransmitters. Acting as hormones, they can control many physiological processes, ranging from gastric acid secretion and carbohydrate metabolism to growth itself. The toxic components of snake and spider venoms are usually peptide in nature, as are some plant toxins. These different activities arise as a consequence of the sequence of amino acids in the peptide or protein (the primary structure), the

three-dimensional structure that the molecule then adopts as a result of this sequence (the secondary and tertiary structures), and the specific nature of individual side-chains in the molecule. Many structures have additional modifications to the basic polyamide system shown, and these features may also contribute significantly to their biological activity.

The tripeptide formed from L-alanine, L-phenylalanine and L-serine by two condensation reactions is alanyl–phenylalanyl–serine, considering each additional amino acid residue as a substituent on the previous. This would be more commonly represented as Ala–Phe–Ser, using the standard three-letter abbreviations for amino acids shown in Table 13.1.

L-Ala　　　L-Phe　　　L-Ser

amino- or N-terminus

carboxyl- or C-terminus

Ala–Phe–Ser

or　H–Ala–Phe–Ser–OH

or　Ala→Phe→Ser

or　AFS

N-terminus ⟶ C-terminus

Ser–Phe–Ala

By convention, the left-hand amino acid in this sequence is the one with a free amino group, the **N-terminus**, and the right-hand amino acid has the free carboxyl, the **C-terminus**. Thus, Ser–Phe–Ala is different from Ala–Phe–Ser, and represents a quite different molecule. Sometimes, the termini identities are emphasized by showing H– and –OH; H– represents the amino group and –OH the carboxyl group. Some peptides are cyclic, and this convention can have no significance, so arrows are incorporated into the sequence to indicate peptide bonds in the direction CO→NH. As sequences become longer, one-letter abbreviations for amino

acids are commonly used instead of the three letter abbreviations, thus Ala–Phe–Ser becomes AFS. Abbreviations assume the L-configuration applies throughout, and any D-amino acids would be specifically noted, e.g. Ala–D-Phe–Ser.

Glutathione is an important tripeptide (see Boxes 6.6 and 10.20); but it is a slightly unusual one, in that it has an amide linkage that involves the γ-carboxyl of glutamic acid rather than a normal amide bond utilizing the C-1 carboxyl group. To specify this bonding, the glutathione structure is written as γ-Glu–Cys–Gly.

glutathione: γ-Glu–Cys–Gly   L-Glu   L-Cys   Gly

Peptides and proteins may be hydrolysed to their constituent amino acids by either acid or base hydrolysis (see Box 7.17). However, because of its nature, the amide bond is quite resistant to hydrolytic conditions (see Section 7.9.2), a very important feature for natural proteins. **Hydrolysis of peptides** and proteins, therefore, requires heating with quite concentrated strong acid or strong base. Neither acid nor base hydrolysis provides the ideal hydrolytic conditions, however, since some of the constituent amino acids are found to be sensitive to the reagents. Acid hydrolysis is preferred, but the indole system of tryptophan is largely degraded in strong acid, and the sulfur-containing amino acid cysteine is also unstable. Serine, threonine, and tyrosine may also suffer partial degradation. Those amino acids containing amide side-chains, e.g. asparagine and glutamine, will be hydrolysed further, giving the corresponding structures with acidic side-chains, namely aspartic acid and glutamic acid.

13.3 Molecular shape of proteins: primary, secondary and tertiary structures

Peptides and proteins are composed of amino acids linked together via amide (peptide) bonds, the amino group of one condensing with the carboxylic acid of another. The sequence of amino acids in a peptide or protein is closely controlled by genetic factors. Some peptides are synthesized via a multi-functional enzyme complex (non-ribosomal peptide synthesis), whereas others, including the larger proteins, are produced on the ribosome, and the sequence can be related directly to the nucleotide sequence of DNA. This amino acid sequence provides what we term the **primary structure** of the protein, although this term also includes the position of **disulfide bridges**, the result of covalent bonding between pairs of cysteine residues. Disulfide bridges produce cross-linking in the polypeptide chain.

Cys residues   disulfide bridge   cystine (Cys–Cys)

This covalent bonding arises as a result of biochemical oxidation of the thiol groups in two cysteine residues, and it may also be achieved chemically with the use of mild oxidizing agents. This modification of thiol groups may thus loop a polypeptide chain or cross-link two separate chains. It also significantly modifies the properties of a protein by removing two polar and potentially acidic (pK_a 10.3) groups, replacing them with a non-polar disulfide function. Under suitable hydrolytic conditions, a protein containing one or more disulfide bridges will yield cysteine residues still joined by this type of bonding. This amino acid 'dimer' is called cystine (Cys–Cys). Because of the similarity in names, it is usual practice to differentiate them in speech by pronouncing cysteine as sis-tay-een, whereas cystine is pronounced sis-teen. The disulfide bridge is easily formed, and is just as easily broken. It may be cleaved to thiol groups by reduction with reagents such as sodium borohydride or by the use of other thiol reagents. For example, mercaptoethanol ($HSCH_2CH_2OH$) is routinely used in protein analysis to help locate disulfide bridges through an equilibration reaction.

$$RS-SR \xrightarrow{\text{NaBH}_4} RSH + HSR$$

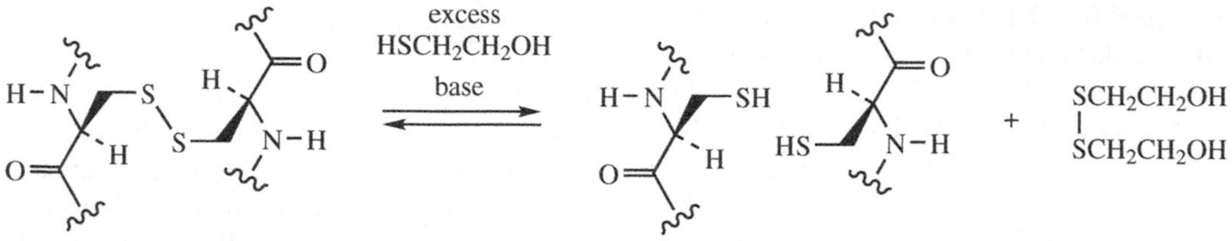

The mechanism for this reaction is shown below.

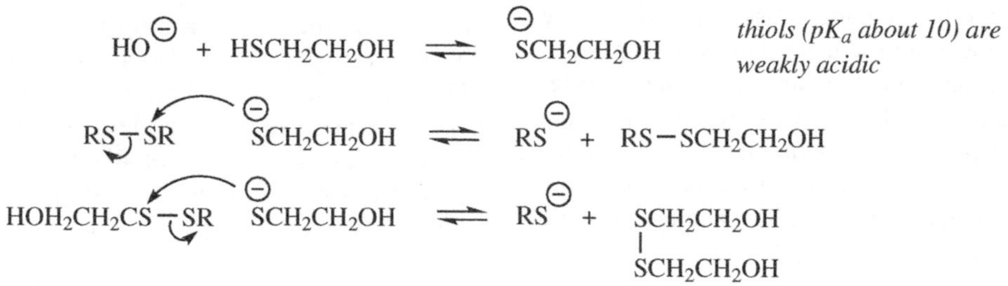

$HO^{\ominus} + HSCH_2CH_2OH \rightleftharpoons {}^{\ominus}SCH_2CH_2OH$ *thiols (pK$_a$ about 10) are weakly acidic*

Oxidation with stronger oxidizing agents, e.g. potassium permanganate or performic acid, converts the disulfide to two molecules of a sulfonic acid (see Section 7.13.1), namely cysteic acid. This reaction may be of value in sequence analysis, to determine the position of disulfide bridges (as opposed to unmodified cysteine residues) in the primary structure.

$$RS-SR \xrightarrow[\text{or} \atop \text{HCO}_3\text{H}]{\text{KMnO}_4} RSO_3H + HO_3SR$$

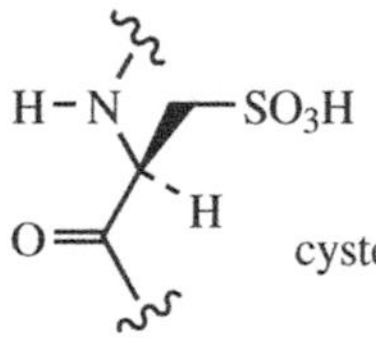

cysteic acid residue

Box 13.1
Disulfide bridges in insulin

The peptide hormone **insulin** is produced by the pancreas and plays a key role in the regulation of carbohydrate, fat, and protein metabolism. In particular, it has a hypoglycaemic effect, lowering the levels of glucose in the blood. A malfunctioning pancreas leads to a deficiency in insulin synthesis and the condition known as **diabetes**. This results in increased amounts of glucose in the blood and urine, diuresis, depletion of carbohydrate stores, and subsequent breakdown of fat and protein. Incomplete breakdown of fat leads to the accumulation of ketones in the blood, severe acidosis, coma, and death. Treatment for diabetes requires daily injections of insulin; since insulin is a peptide, it would be degraded by stomach acid if taken orally. Insulin does not cure the disease, so treatment is lifelong.

Human insulin is composed of two straight-chain polypeptides joined by disulfide bridges.

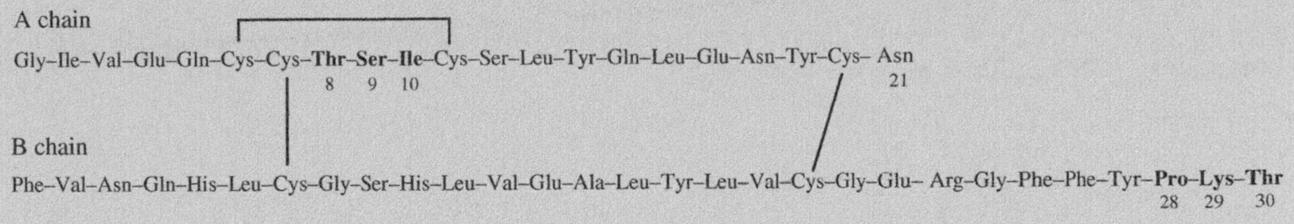

A chain
Gly–Ile–Val–Glu–Gln–Cys–Cys–Thr–Ser–Ile–Cys–Ser–Leu–Tyr–Gln–Leu–Glu–Asn–Tyr–Cys– Asn
8 9 10
21
B chain
Phe–Val–Asn–Gln–His–Leu–Cys–Gly–Ser–His–Leu–Val–Glu–Ala–Leu–Tyr–Leu–Val–Cys–Gly–Glu– Arg–Gly–Phe–Phe–Tyr–Pro–Lys–Thr
28 29 30

human insulin

This structure is known to arise from a single straight-chain polypeptide, preproinsulin, containing 100 amino acid residues. This loses a 16-residue portion of its chain and forms proinsulin, in which disulfide bridges connect the terminal portions of the chain in a loop. A central portion of the loop (the C chain) is then cleaved out, leaving the A chain (21 residues) bonded to the B chain (30 residues) by two disulfide bridges. There is also a third disulfide bridge interconnecting two cysteine residues in the A chain. This is the resultant insulin.

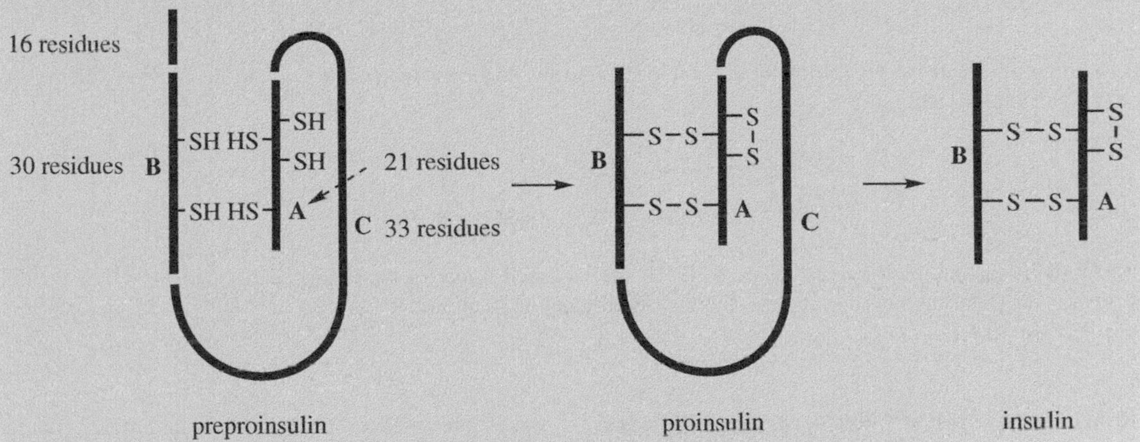

16 residues
30 residues B
SH HS
SH HS
SH
SH
A
21 residues
C 33 residues
B
S–S
S–S
S
S
A
C
B
S–S
S–S
S
S
A
preproinsulin
proinsulin
insulin

Mammalian insulins from different sources are very similar and may be used to treat diabetes. The compounds show variations in the sequence of amino acid residues 8–10 in chain A, and at amino acid 30 in chain B.

	amino acid residues					
	A chain			B chain		
	8	9	10	28	29	30
insulin (human)	Thr	Ser	Ile	Pro	Lys	Thr
insulin (porcine)	Thr	Ser	Ile	Pro	Lys	**Ala**
insulin (bovine)	Ala	Ser	**Val**	Pro	Lys	**Ala**

Porcine insulin and **bovine insulin** for drug use are extracted from the pancreas of pigs and cattle respectively. More frequently, **human insulin** is now employed. This is produced by the use of recombinant DNA technology to obtain the two polypeptide chains, and then linking these chemically to form the disulfide bridges

Box 13.2
Disulfide bridges: glutathione protects cells against damage by peroxides

Peroxides, including hydrogen peroxide (H_2O_2), can damage cells by causing unwanted oxidation reactions. The tripeptide **glutathione** (GSH) is able to participate in a cellular protection mechanism via its ability to form disulfide bridges.

In an enzymic reaction catalysed by glutathione peroxidase, GSH reacts with peroxides and becomes oxidized to form a dimer (GSSG) linked by a disulfide bridge.

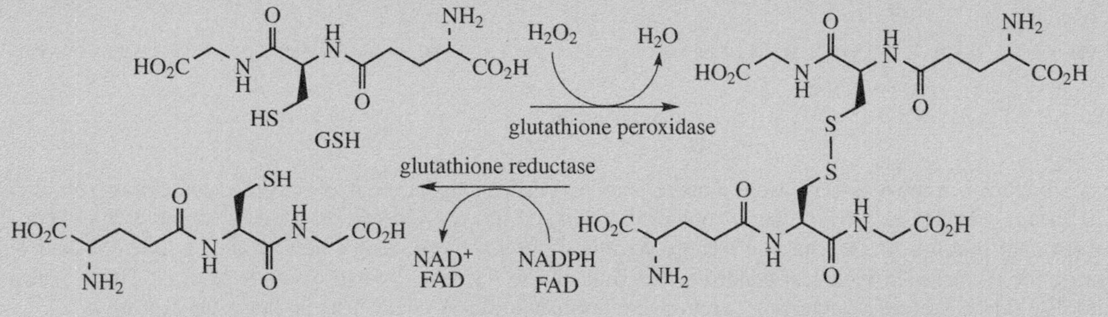

glutathione: γ-Glu–Cys–Gly
(GSH)

In so doing, it reduces the peroxide. In the case of H_2O_2, this generates water, whereas an organic peroxide would yield water and an alcohol.

$$H_2O_2 + 2GSH \longrightarrow GSSG + 2H_2O$$

$$ROOH + 2GSH \longrightarrow GSSG + ROH + H_2O$$

In order to maintain adequate levels of GSH, the oxidized dimer is then reduced back to the original thiol components. This is achieved using the enzyme GSH reductase in a reaction involving NADPH and FAD cofactors (see Section 15.1.1).

Protein chains are not the sprawling, ill-defined structures that might be expected from a single polypeptide chain. Most proteins are compact molecules, and the relative positions of atoms in the molecule contribute significantly to its biological role. A particularly important contributor to the shape of proteins is provided by the **peptide bond** itself. Drawn in its simplest form, one might expect free rotation about single bonds, with a variety of conformations possible (see Section 3.3.1). However, there is resonance stabilization in an amide, via electron movement from the lone pair on the nitrogen to the carbonyl oxygen. We have already noted this type of resonance stabilization in amides (see Section 7.9.2), and also in esters (see Section 10.7). It was invoked in explaining reactivity and pK_a values compared with other types of carbonyl compound, and the non-basic behaviour of the nitrogen atom in amides (see Section 4.5.4).

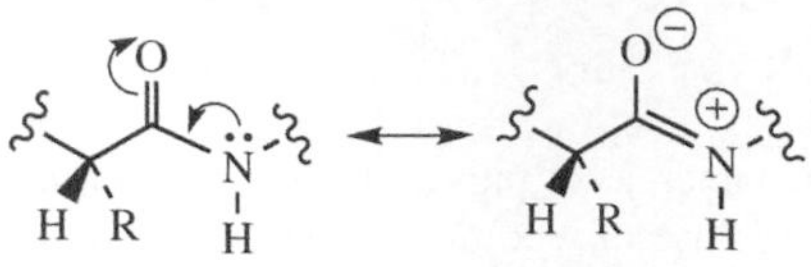

resonance stabilization of peptide bond

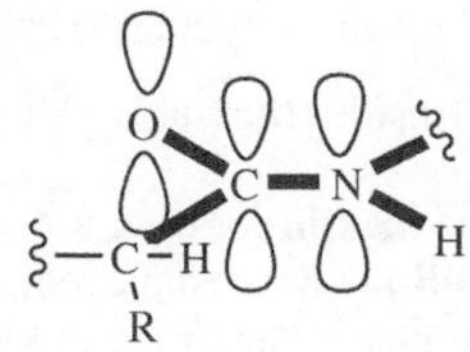

resonance stabilization demands coplanarity of p orbitals; bold bonds must be coplanar

To achieve this stabilization, the *p* orbital on nitrogen needs to be lined up with the carbonyl π bond. The immediate consequences of this are that five bonds in the peptide linkage must be coplanar. There is no free rotation about the N–C bond, because it is involved with a partial double-bond system. Of course, there are potentially two configurations with respect to this N–C bond, corresponding to *cis* and *trans* versions if it were a true double bond. It is not surprising that the *trans* form is energetically favoured, where we have the large groups, i.e. the rest of the chain, arranged to give minimum interaction.

'trans'   'cis'

zig-zag conformation with main chain *trans* oriented

rotational freedom about single bonds

rotation about the C–N bond is restricted

We now see good reasons for drawing a polypeptide chain in the accepted zigzag form. Note, however, that the remaining single bonds in the chain do allow rotation, and this is why we see a wide variety of different shapes in proteins. We can also appreciate that, in general, the carbonyl groups and N–H groups are all going to be coplanar. This leads to the **secondary structure** of proteins, a consequence of hydrogen bonding possible because of the regular array of carbonyl and N–H groups.

The most easily appreciated example of this is the **β-pleated sheet**, one of the ways in which a polypeptide chain can be arranged in an ordered fashion (Figure 13.1). Polypeptide chains align themselves side-by-side, stabilized by multiple hydrogen bonding (see Section 2.11), allowed by the regular array of carbonyl and N–H bonds. The alignment may be **parallel**, such that all the carbonyl to amino peptide linkages are in same direction, or **antiparallel** where carbonyl to amino peptide linkages run in opposite directions. Although there are going to be groups of atoms that are planar, the whole chain is not planar. Instead, these arrangements take up a pleated array, which helps to minimize interaction between the large R groups.

Parallel sheets may involve different polypeptide chains via intermolecular hydrogen bonds, or the same chain via intramolecular hydrogen bonds. For intramolecular interactions, the chain length needs to be substantial, i.e. proteins rather than peptides, and it will be necessary for the polypeptide chain to bend back upon itself. The commonest type of arrangement for bending back a chain is called the **β-turn**, resulting in hydrogen bonding between residues *n* and *n* + 3.

residue *n* + 3

β-turn

residue *n*

Note also that the imino acid **proline** must distort the regular zigzag array and introduce a bend into the chain; two configurations may be considered, and

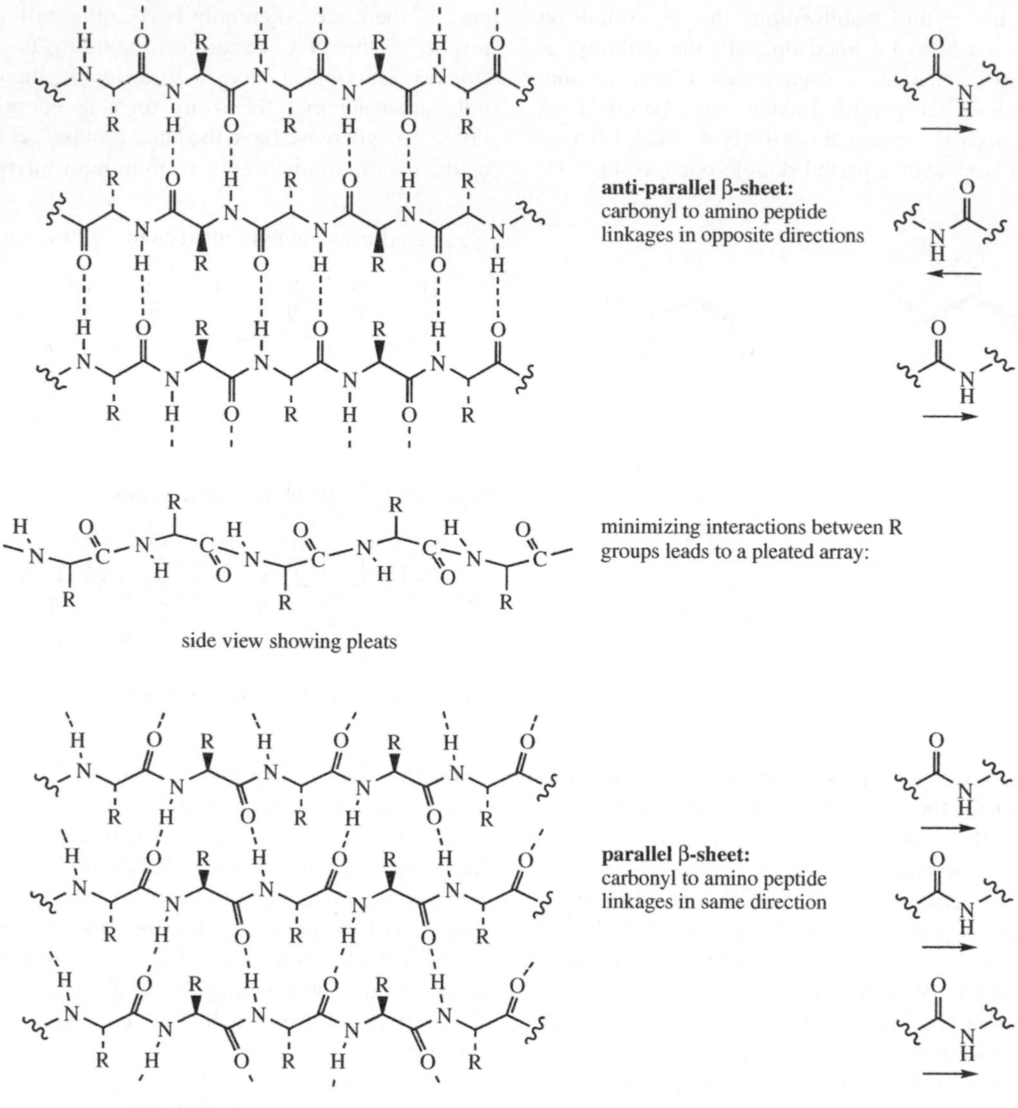

anti-parallel β-sheet:
carbonyl to amino peptide
linkages in opposite directions

minimizing interactions between R
groups leads to a pleated array:

side view showing pleats

parallel β-sheet:
carbonyl to amino peptide
linkages in same direction

Figure 13.1 Secondary structure of proteins: hydrogen bonding in β-sheets

both are possible since there is little difference in energy between them. Further, there is no N–H in proline, so hydrogen bonding involving this residue is not possible.

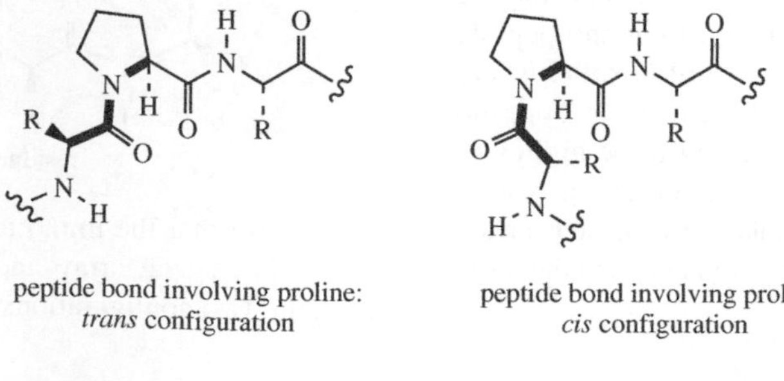

peptide bond involving proline:
trans configuration

peptide bond involving proline:
cis configuration

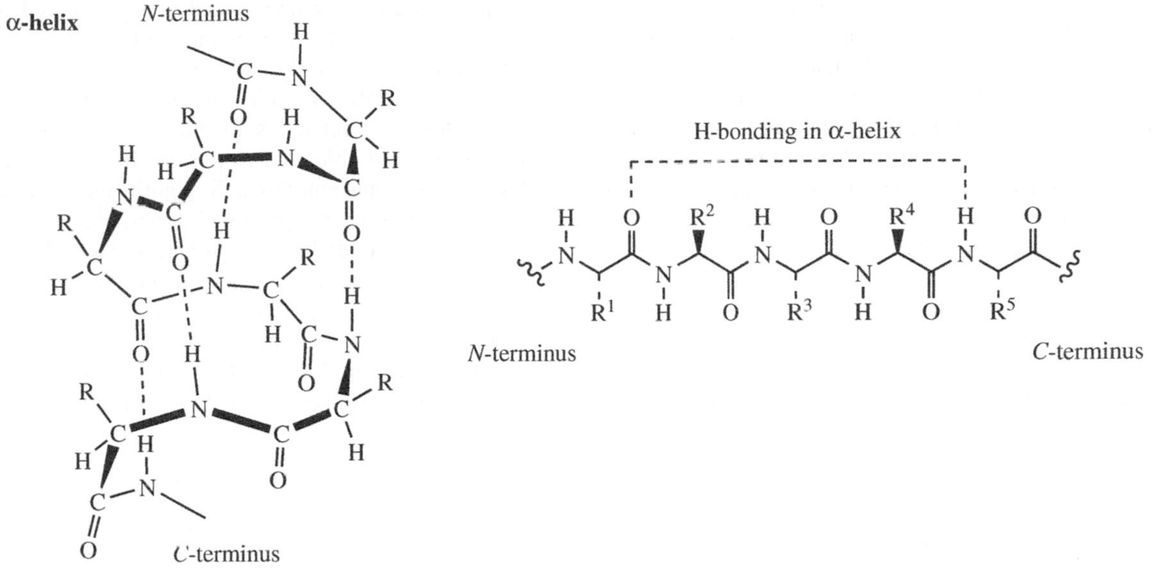

Figure 13.2 Secondary structure of proteins: hydrogen bonding in α-helix

Whereas the β-pleated sheet provides a particularly nice and easily appreciated example of regular hydrogen bonding in polypeptide chains, the most common arrangement found in proteins is actually the α-helix (Figure 13.2). Do not worry about the α or β used in the nomenclature; this merely signifies that the helical structure (α) was deduced earlier than that of the pleated sheet (β).

The **α-helix** is a right-handed helix, an ordered coil array stabilized by hydrogen bonding between carbonyl and N–H groups in the same chain. In a right-handed helix, movement along the chain involves a clockwise or right-handed twist, just like an ordinary screw – you turn the screwdriver clockwise. Hydrogen bonds link carbonyl and N–H bonds in amino acids that are separated by three other residues, and each turn of the helix is found to take up 3.6 amino acid residues. Note that all of the R groups, which in the majority of amino acids are quite bulky, are accommodated on the outside of the helix. Only the imino acid proline cannot fit into the regular array of the α-helix. We have just seen that **proline** must distort the regular array and introduce a bend into the chain, and that there is no N–H for hydrogen bonding.

The secondary structure is responsible for some of the physical properties of proteins. For example, structural proteins such as α-keratins in skin and hair are fibrous in nature, and have good elastic

properties. This elasticity can be traced back to the α-helix structure, in which weak hydrogen bonds are parallel to the direction of stretching, i.e. a spring-like structure. On the other hand, proteins such as α-fibroin in silk are relatively inelastic since they contain the β-pleated sheet structure, where extension is resisted by the full strength of covalent bonds. In the β-pleated sheet, the weaker hydrogen bonds would be perpendicular to the direction of stretching. However, most proteins actually have a roughly spherical shape and are thus termed **globular proteins**. Globular proteins are likely to contain portions of the polypeptide chain that adopt both helical and sheet structures. In contrast to the structural proteins like α-keratin or α-fibroin, the helical or sheet fragments in globular proteins are rather short and do not extend far without a change in direction. The overall folding of the polypeptide chain and the three-dimensional arrangement produced provide the **tertiary structure** of the protein. The globular shape is facilitated, however, by a number of other non-covalent interactions.

13.3.1 Tertiary structure: intramolecular interactions

The conformation of a protein is determined and maintained by a range of intramolecular interactions that arise from some of the amino acid side-chain substituents. These are non-covalent inter-actions, although we must also remember that a disulfide bond formed between pairs of cysteine residues also contributes to the three-dimensional shape of the protein by providing cross-chain links. **Non-covalent interactions** are relatively weak when compared with covalent bonds, but there are usually many such interactions in a protein and, overall, a substantial degree of stabilization is attained.

- *Hydrophobic interactions.* Many of the amino acids contain side-chains that are hydrocarbon in nature, either aliphatic or aromatic.

In an aqueous environment, such groups are hydro-phobic, and any folding in the protein that con-centrates these hydrophobic areas together, and away from water, is going to be favoured. This hydrophobic effect tends to encourage the bury-ing of hydrophobic side-chains in the interior of the protein, and provides a significant part of the driving force for protein folding. Because so many amino acids have hydrocarbon side-chains, not all can be accommodated in the interior, and hydrophobic groups tend be about equally dis-tributed between the interior and the surface of the molecule. On the other hand, hydrophilic side-chains are more likely to be found on the surface of a protein.

non-polar, hydrophobic side-chains

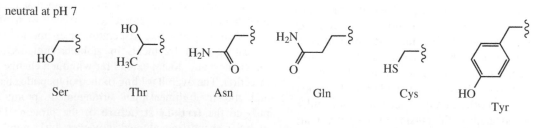

polar, hydrophilic side-chains

neutral at pH 7

- *Hydrogen bonds.* Hydrogen bonds (see Section 13.3) are responsible for the fundamental characteristics of the α-helix and β-pleated sheet; in addition, they contribute to the final shape of a globular protein. Hydrogen bonds can form in a variety of ways, involving the peptide backbone, polar amino acid side-chains, and also water molecules. Some of the hydrogen-bonding situations are shown below; others can be deduced.

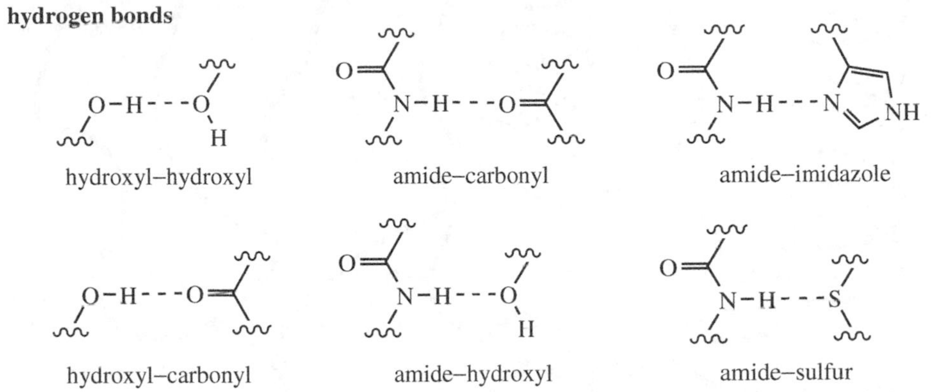

hydrogen bonds

hydroxyl–hydroxyl

amide–carbonyl

amide–imidazole

hydroxyl–carbonyl

amide–hydroxyl

amide–sulfur

It should be appreciated that amino acids such as serine, threonine, tyrosine, and cysteine all contain side-chain alcohol or thiol groups that may participate in hydrogen bonding and stabilize a particular protein conformation.

- *Ionic bonds.* Carboxylic acid groups in amino acid side-chains (aspartic acid, glutamic acid) will be ionized at pH 7, and nitrogen-containing groups (lysine, arginine) will similarly be protonated (see Box 4.7). Isolated hydrophilic groups such as these will never be found in the hydrophobic interior of a globular protein, but will be positioned on the outer surface in proximity to water molecules. However, pairs of oppositely charged ions may be found in the interior since electrostatic interactions can provide the necessary attractive forces.

Thus, non-covalent hydrophobic interactions, hydrogen bonds, and electrostatic bonds all contribute to the overall shape of a protein (Figure 13.3). As we shall see (Section 13.3.2), many pertinent properties of a protein are then provided by the appropriate combination of the remaining amino acid side-chains that reside on the surface, allowing specific binding to various molecules. This is the essence of enzymic activity and drug–receptor interactions.

With some proteins, there is a further level of structure, i.e. **quaternary structure**, which may need to be considered. This arises because two or more protein chains aggregate to form the normal functional protein. Typically, the separate subunits, often the same, are held together by non-covalent interactions, as seen in the consideration of tertiary structure. Not all proteins have quaternary structure.

13.3.2 Protein binding sites

From our considerations above, we can see just how important the interactions of various amino acid side-chains are to the structure and shape of proteins. These interactions tend to be located inside the protein molecule, stabilizing a particular conformation and generating the overall shape as in a globular protein. However, it is obvious that there are also going to be many amino acid side-chains located on the surface of a protein, and these in turn will be capable of interacting with other molecules. These interactions will be intermolecular, rather than the intramolecular interactions that contribute to protein structure.

Because there will also be several amino acid side-chains in close proximity, a combination of interactions may generate a site that has a specific shape, and a specific array of forces. The site will then be able to bind a particular molecule or part of a molecule. These amino acid side-chains, therefore, allow strong binding to specific molecules, and the

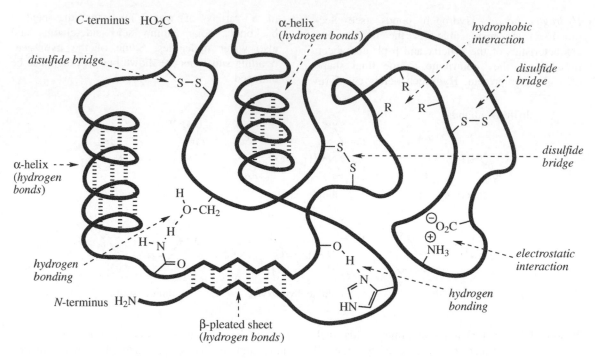

Figure 13.3 Secondary and tertiary interactions contributing to a protein structure

particular molecule can be regarded as being a perfect fit both in terms of the forces involved and in a geometric sense. As a result, even a small change in the structure of the molecule could well spoil the fit and upset the interplay of forces – the binding is powerful and reasonably specific.

The sole function of many proteins is simply to bind other molecules. Thus, **immunoproteins** are able to bind to alien molecules (**antigens**) and destroy their activity by complexation. Some proteins act as **hormones**, influencing metabolic rates by binding to another molecule or structure. Binding to a protein may be the way an organism transports a

molecule or even an ion to a different part of the organism, e.g. haemoglobin (see Box 11.4) transports molecular oxygen around the body, and cytochromes (see Section 15.1.2) transport electrons within a cell. Drug–receptor interactions and enzymic activity are also a consequence of binding of molecules to a protein. A substance that elicits a particular biological response by interaction at a receptor site is termed an **agonist**. An **antagonist** is a substance that inhibits the action of an agonist, often by competing for the same receptor site. The binding site on an enzyme is usually termed the **active site**.

Box 13.3

Pain relief: morphine mimics natural peptides called endorphins

Although the pain-killing properties of the opium alkaloid **morphine** and related compounds have been known for a considerable time, the existence of endogenous peptide ligands for the receptors to which these compounds bind is a more recent discovery. It is now appreciated that the body produces a family of endogenous opioid peptides that bind to a series of receptors in different locations. These peptides include **enkephalins, endorphins**, and **dynorphins**, and are produced primarily, but not exclusively, in the pituitary gland. The pentapeptides Met-enkephalin and Leu-enkephalin were the first to be characterized. The largest peptide is β-endorphin ('*end*ogenous m*orphine*'), which is several times more potent than morphine in relieving pain.

Box 13.3 (continued)

Tyr–Gly–Gly–Phe–Met–Thr–Ser–Glu–Lys–Ser–
Gln–Thr–Pro–Leu–Val–Thr–Leu–Phe–Lys–Asn–
Ala–Ile–Val–Lys–Asn–Ala–His–Lys–Lys–Gly–
Gln

β-endorphin

amino acid sequences listed in blocks of 10 residues

Tyr–Gly–Gly–Phe–Met   Tyr–Gly–Gly–Phe–Leu

Met-enkephalin   Leu-enkephalin

Tyr–Gly–Gly–Phe–Leu–Arg–Arg–Ile–Arg–Pro–
Lys–Leu–Lys–Trp–Asp–Asn–Gln

dynorphin A

Although β-endorphin at its *N*-terminus contains the sequence for Met-enkephalin, the latter peptide and Leu-enkephalin are derived from a larger peptide, namely proenkephalin A, and β-endorphin itself is formed by cleavage of the peptide pro-opiomelanocortin. The proenkephalin A structure contains four Met-enkephalin sequences and one of Leu-enkephalin. The dynorphins, e.g. dynorphin A, are also produced by cleavage of a larger precursor, namely proenkephalin B (prodynorphin), and all contain the Leu-enkephalin sequence. Some 20 opioid ligands have now been characterized. When released, these endogenous opioids act upon specific receptors, inducing analgesia and depressing respiratory function and several other processes. The individual peptides have relatively high specificity towards different receptors. It is known that morphine, β-endorphin, and Met-enkephalin are agonists for the same site. The opioid peptides are implicated in analgesia brought about by acupuncture, since opiate antagonists can reverse the effects. The hope of exploiting similar peptides as ideal, non-addictive analgesics has yet to be attained; repeated doses of endorphin or enkephalin produce addiction and withdrawal symptoms.

A common structural feature required for centrally acting analgesic activity in the opioids is the combination of an aromatic ring and a piperidine ring that maintains the stereochemistry at the chiral centre, as shown below.

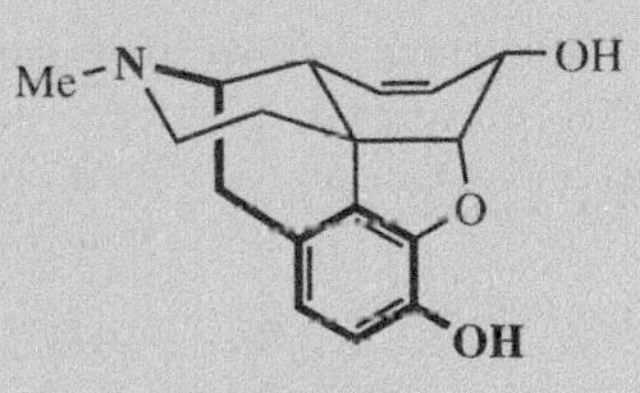

morphine

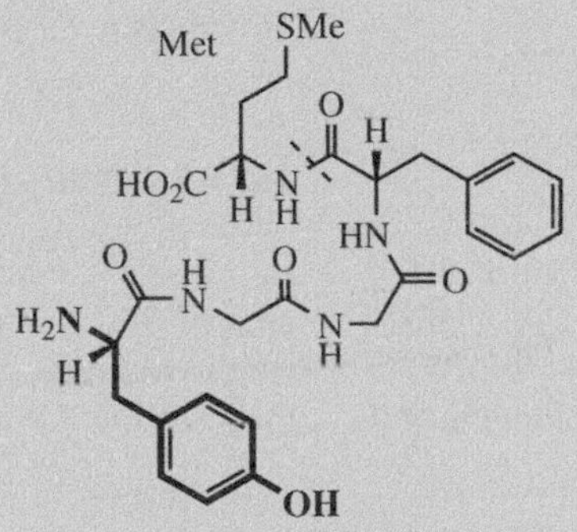

Met-enkephalin
(Tyr–Gly–Gly–Phe–Met)

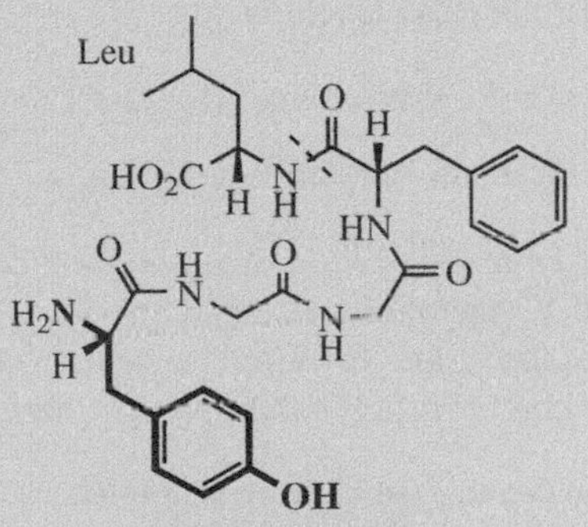

Leu-enkephalin
(Tyr–Gly–Gly–Phe–Leu)

The three-dimensional disposition of the nitrogen function to the aromatic ring allows morphine and other analgesics to bind to a pain-reducing receptor in the brain. The terminal tyrosine residue in the natural agonists Met-enkephalin and Leu-enkephalin is mimicked by portions of the morphine structure.

13.4 The chemistry of enzyme action

Enzymes are proteins that act as biological catalysts. They not only bind molecules, but also provide a special environment in which the molecules are chemically modified. Enzymes cannot promote a reaction that is not energetically favourable, but by binding the reagents in the necessary orientation and in close proximity, they significantly reduce the activation energy for the transformation.

By virtue of the amino acid side-chains, enzymes are able to provide a highly specific binding site for their substrates, anchoring these reagents in an appropriate manner and in suitable proximity so that reaction can occur, as well as providing any necessary acid or base catalyst for the reaction. In some cases, a further reagent, a **coenzyme**, must also be bound for the reaction to occur. After the reaction is completed, the products are then released from the enzyme, so that the reaction can be repeated on further molecules of the substrates. These amino acid side-chains confer immense catalytic power to the enzyme, giving it an ability to carry out reactions that organic chemists can only dream of.

Several types of bonding might be utilized to bind substrates to enzymes. These are analogous to the bondings that contribute to the secondary and tertiary structures of a protein and include the following non-covalent interactions:

- electrostatic bonding, via acids, bases, phosphates;

- hydrophobic interactions, via alkyl groups, aromatic rings;

- hydrogen bonding, via NH, OH, SH, C=O.

In addition, we can meet examples of covalent bonding that are responsible for binding a substrate, where a functional group in the substrate reacts chemically with a protein side-chain functional group. Two important reactions are:

- imine formation, via NH_2 and C=O;

- thioester formation, via SH and C=O.

A large proportion of the substrates used in intermediary metabolism are in the form of phosphates. Phosphates are favoured in nature since they usually confer water solubility on the compound, and they provide a functional group that is able to bind to enzymes through simple electrostatic bonding. In many cases, the phosphate group may also feature as a chemically reactive functional group – phosphates are good leaving groups (see Section 7.13.2).

We have considered the structures of the various amino acids in terms of polarity, basicity, acidity, etc. Here is a useful reminder of the important functionalities that are pertinent to enzyme action:

- amino acids with hydroxyl groups: Ser, Thr, Tyr;

- amino acids containing thiol (sulfhydryl) groups: Cys;

- amino acids with acidic groups: Asp, Glu;

- amino acids with basic groups: Lys, Arg, His.

It is not sensible to try to cover a wide range of enzymic reactions to demonstrate how amino acid side-chains are responsible for the chemical changes the enzyme brings about. Instead, one or two suitable examples will suffice to illustrate the general principles.

13.4.1 Acid–base catalysis

Let us first remind ourselves of the two mechanisms for hydrolysis of an ester, namely acid-catalysed hydrolysis and base-catalysed hydrolysis (see Section 7.9.2).

acid-catalyzed hydrolysis of esters

protonation of carbonyl oxygen to form conjugate acid makes better electrophile

nucleophilic attack of water nucleophile onto protonated electrophile

loss of leaving group, and reformation of protonated carbonyl

acid is **catalyst** and is regenerated; this is an **equilibrium reaction**

base hydrolysis of esters

nucleophilic attack of good nucleophile hydroxide onto carbonyl

loss of leaving group with reformation of carbonyl

leaving group (strong base) abstracts proton from acid

base is **reactant** and consumed; reaction becomes essentially **irreversible** because of formation of the carboxylate anion

carboxylate anion

ionization of carboxylic acid makes reaction essentially irreversible

Since enzymic reactions proceed in aqueous solution at pH 7 or thereabouts, where the concentrations of hydronium and hydroxide ions are both approximately 10^{-7} M, we need alternatives to strong acids and strong bases to formulate comparable enzymic mechanisms. Such reagents are provided by those amino acid side-chains that are ionized at pH 7. Thus, the capacity for acid or base catalysis is actually built into the active site of many enzymes. Furthermore, the effective concentration of these groups at the active site is high, making them very effective acid and base catalysts. These donors and acceptors of protons are called **general acid catalysts** and **general base catalysts** respectively.

The amino acids in question are the basic amino acids lysine, arginine, and histidine, and the acidic amino acids aspartic acid and glutamic acid. The side-chain functions of these amino acids, ionized at pH 7 (see Box 4.7), act as acids or bases. In a reverse sequence, protons may be acquired or donated to regenerate the conjugate acids and conjugate bases.

The most effective acid–base catalyst is one whose pK_a is 7.0, since at pH 7.0 the concentrations of acid and conjugate base are equal (see Section 4.9). With just a slight decrease in pH it would become

protonated and function as a general acid catalyst, whereas with a slight increase in pH it would become unprotonated and, therefore, a general base catalyst.

Enzymes are active over a limited pH range; the pH value of maximum activity is known as the **pH optimum**, and this is characteristic of the enzyme. It typically reflects the pH necessary to achieve the appropriate ionization of amino acid side-chains at the active site.

The side-chain of **histidine** has a pK_a value of 6.0; above pH 6.0, the **imidazole ring** acts as a proton acceptor or general base catalyst, whereas below pH 6.0 it is protonated and can act as a proton donor or general acid catalyst (see Box 11.6). At pH 7.0, the imidazole ring can be considered as partially protonated, since both ionized and non-ionized forms are present in the ratio of about 1 : 10 (see Box 4.7). Consequently, we find that the imidazole ring of histidine participates in acid–base catalysis in many enzymes. As we mentioned earlier (Section 13.1), the pK_a values of histidine side-chains in a protein are modified by the neighbouring amino acid residues, and are typically in the range 6–7. These values provide a level of ionization somewhere between 9 and 50%, depending upon the protein.

Remember that tautomerism can occur in imidazole rings (see Section 11.7.2). When we meet structures for the amino acid histidine, we may encounter either of the tautomeric forms shown. Do not think there is a discrepancy in structures. On the other hand, we can write resonance structures for the protonated ring, the imidazolium cation.

> pH 6.0
< pH 6.0
His
tautomerism
+ H+
– H+
pKa 6.0
resonance

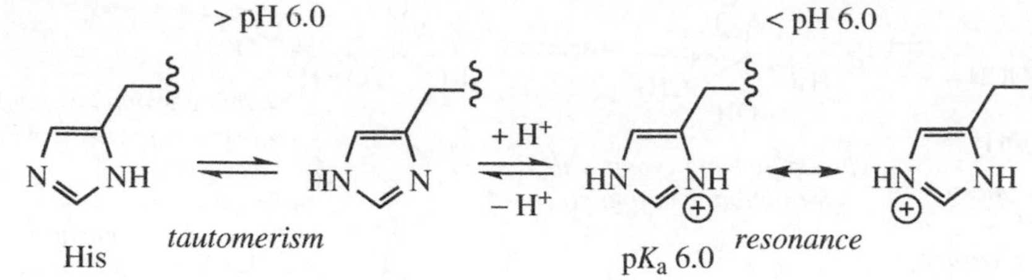

at pH 7.0, ratio non-ionized:ionized = approx 10:1

Now let us see how the imidazole grouping of histidine can be involved in the general acid-catalysed and general base-catalysed hydrolysis of esters by enzymes. In essence, the chemical and enzymic reactions are very similar. This is to be expected, since enzymes can only catalyse a reaction that is energetically favourable. The role of acid or base is largely achieved in the enzyme reactions by implicating the imidazolium/imidazole system.

general acid-catalysed ester hydrolysis

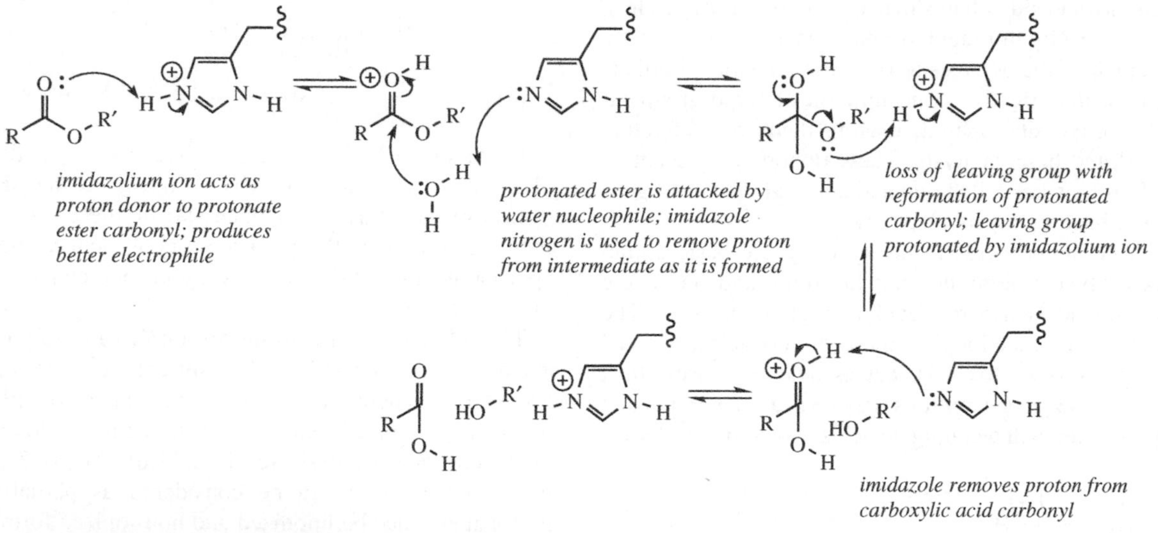

imidazolium ion acts as proton donor to protonate ester carbonyl; produces better electrophile
protonated ester is attacked by water nucleophile; imidazole nitrogen is used to remove proton from intermediate as it is formed
loss of leaving group with reformation of protonated carbonyl; leaving group protonated by imidazolium ion
imidazole removes proton from carboxylic acid carbonyl

general base-catalysed ester hydrolysis

imidazole abstracts proton from water and resulting hydroxide attacks carbonyl group
carbonyl is reformed with loss of leaving group; alkoxide is protonated by imidazolium ion

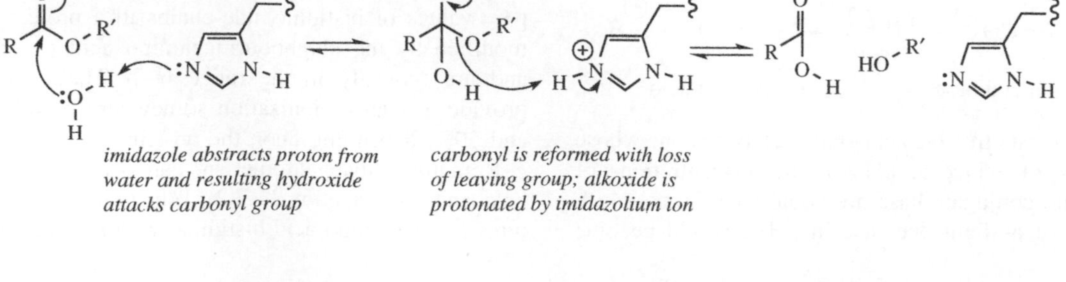

In the general acid-catalysed mechanism, the imidazolium ion acts as a proton donor to protonate the carbonyl oxygen, thus producing a better electrophile. The protonated ester is then attacked by a water nucleophile, after which the imidazole nitrogen removes the now unwanted proton from the nucleophile; the imidazole nitrogen consequently becomes reprotonated. The carboxylic acid is then formed via loss of the leaving group, and this is facilitated by protonation, so that the leaving group is an alcohol rather than alkoxide. The imidazolium proton is again a participant in this process. Finally, abstracting the proton from the protonated carbonyl regenerates the imidazolium ion. As in the chemical reactions, the general base-catalysed process is mechanistically rather simpler. The imidazole nitrogen acts as a base to remove a proton from water, generating hydroxide that attacks the carbonyl. Subsequently, the alkoxide leaving group is reprotonated by the imidazolium ion.

However, not included in the above mechanisms are other amino acid side-chains at the active site, whose special role will be to help bind the reagents in the required conformation for the reaction to occur. Examples of such interactions are found with **acetylcholinesterase** and **chymotrypsin**, representatives of a group of hydrolytic enzymes termed **serine hydrolases**, in that a specific serine amino acid residue is crucial for the mechanism of action.

The proposed enzyme mechanisms just described, and those that follow, are depicted to show how certain amino acid residues become involved. Remember that the enzyme active site is three-dimensional, whereas our representation is only two-dimensional. This means that bond angles and bond lengths sometimes look a little odd and distorted. However, such imperfect representations are actually easier to follow than if we had provided pictures that tried to emulate three-dimensional views.

Box 13.4

Acetylcholinesterase, a serine esterase

Acetylcholine is a relatively small molecule that is responsible for nerve-impulse transmission in animals. As soon as it has interacted with its receptor and triggered the nerve response, it must be degraded and released before any further interaction at the receptor is possible. Degradation is achieved by hydrolysis to acetate and choline by the action of the enzyme **acetylcholinesterase**, which is located in the synaptic cleft. Acetylcholinesterase is a **serine esterase** that has a mechanism similar to that of chymotrypsin (see Box 13.5).

Hydrolysis involves nucleophilic attack by the serine hydroxyl onto the ester carbonyl (see Box 7.26). This leads to transfer of the acetyl group from acetylcholine to the enzyme's serine hydroxyl, i.e. formation of a transient acetylated enzyme, and release of choline. We have met this type of reaction before under transesterification (see Section 7.9.1). Hydrolysis of the acetylated enzyme then occurs rapidly, releasing acetate and regenerating the free enzyme.

The active site of the enzyme contains two distinct regions: an anionic region that contains a glutamic acid residue, and a region in which a histidine imidazole ring and a serine hydroxyl group are particularly important.

Box 13.4 (continued)

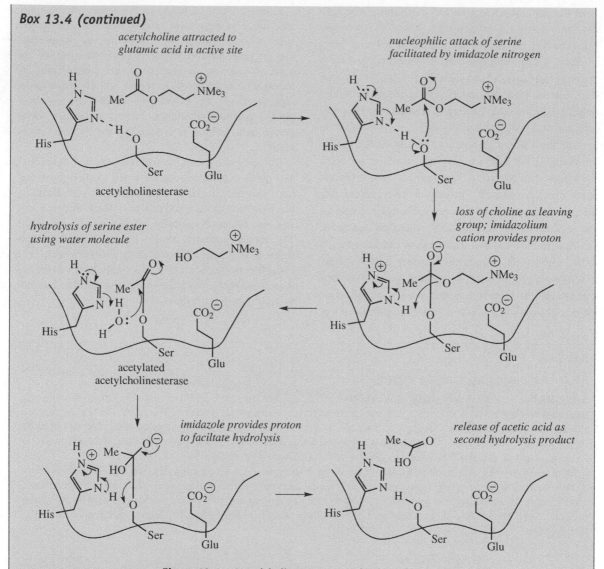

Figure 13.4 Acetylcholinesterase: mechanism of action

It is easy to see that the glutamic acid side-chain, ionized at pH 7, can attract the positively charged acetylcholine by means of ionic interactions. This allows binding, and locates the ester function close to the serine side-chain and the imidazole ring (Figure 13.4).

Serine itself would be insufficiently nucleophilic to attack the ester carbonyl, so the reaction is facilitated by participation of the imidazole ring of histidine. The basic nitrogen in this residue is oriented so that it can remove a proton from the serine hydroxyl, increasing nucleophilicity and allowing attack on the ester carbonyl. This leads to formation of the transient acetylated enzyme, and release of choline. Hydrolysis of the acetylated enzyme utilizes water as nucleophile, but again involves the imidazole ring, and regenerates the free enzyme.

Not included in this simplified description of the active site is the important role played by another amino acid residue. The basicity of the histidine nitrogen is increased because of the proximity of a neighbouring aspartate residue. This facilitates removal of a proton from the active site serine. The relationship of these three residues provides a **charge-relay network**, in which a charge, here a proton, is effectively passed from one molecule to another. In due course, the groups can be restored to their original nature by a reverse of the sequence. We shall see this feature again with chymotrypsin below.

aspartate–histidine–serine charge relay network

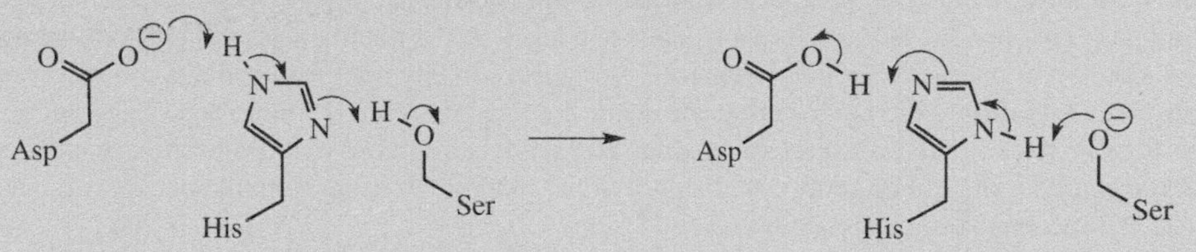

aspartate facilitates removal of proton from serine via charge-relay network

later in the sequence, the original residues can be regenerated

Acetylcholinesterase is a remarkably efficient enzyme; turnover has been estimated as over 10 000 molecules per second at a single active site. This also makes it a key target for drug action, and acetylcholinesterase inhibitors are of considerable importance. Some natural and synthetic toxins also function by inhibiting this enzyme (see Box 7.26).

Box 13.5

Chymotrypsin and other serine proteases

The **serine proteases** cleave amide (peptide) bonds in peptides and have a wide variety of functions, including food digestion, blood clotting, and hormone production. They feature as one of the best-understood groups of enzymes as far as mechanism of action is concerned. We are able to ascribe a function to many of the amino acid residues in the active site, and we also understand how they determine the specificity of the various enzymes in the group.

For example, **chymotrypsin** cleaves peptides on the *C*-terminal side of aromatic amino acid residues phenylalanine, tyrosine, and tryptophan, and to a lesser extent some other residues with bulky side-chains, e.g. Leu, Met, Asn, Gln. On the other hand, **trypsin** cleaves peptides on the *C*-terminal side of the basic residues arginine and lysine. **Elastase** usually catalyses hydrolysis of peptide bonds on the *C*-terminal side of neutral aliphatic amino acids, especially glycine or alanine. These three pancreatic enzymes are about 40% identical in their amino acid sequences, and their catalytic mechanisms are nearly identical.

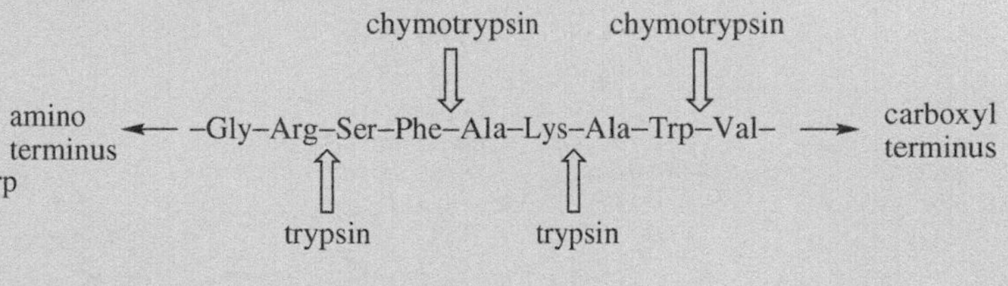

cleavage sites for serine proteases

Enzyme	**Cleavage site**
trypsin	*C*-terminal Lys, Arg
chymotrypsin	*C*-terminal Tyr, Phe, Trp

hydrolysis with trypsin ⟹ –Gly–Arg Ser–Phe–Ala–Lys Ala–Trp–Val–

hydrolysis with chymotrypsin ⟹ –Gly–Arg–Ser–Phe Ala–Lys–Ala–Trp Val–

Because of their known specificity, these enzymes, especially trypsin and chymotrypsin, have been widely utilized in helping to determine the amino acid sequences of peptides (see Section 13.7). Hydrolysis using these

Box 13.5 (continued)

enzymes generates smaller peptide fragments via hydrolysis at specific amino acid residues. The shortened chains can then be sequenced and, with a little logic and reasoning, the order in which they are attached in the larger peptide can be deduced.

The differences in specificity are known to be a consequence of the amino acid sequences at the binding sites of the enzymes; these sequences are almost identical. Thus, trypsin and chymotrypsin differ in only one residue at the binding site. This residue is located in a so-called 'pocket' in the binding site, and allows binding of substrates containing specific amino acids in their structure. The pocket in chymotrypsin contains a serine residue, and the pocket provides a hydrophobic environment allowing binding of aromatic amino acid side-chains. On the other hand, the trypsin pocket has an aspartate residue, and binds substrates with the positively charged amino acid residues lysine and arginine. For simplicity, the additional binding resulting from the pocket residues has not been included in the mechanistic interpretation below.

The mechanism of action of **chymotrypsin** can be rationalized as follows (Figure 13.5). The enzyme–substrate complex forms, with the substrate being positioned correctly through hydrogen bonding and interaction with the 'pocket' as described above. The nucleophilicity of a serine residue is only modest, but here it is improved by

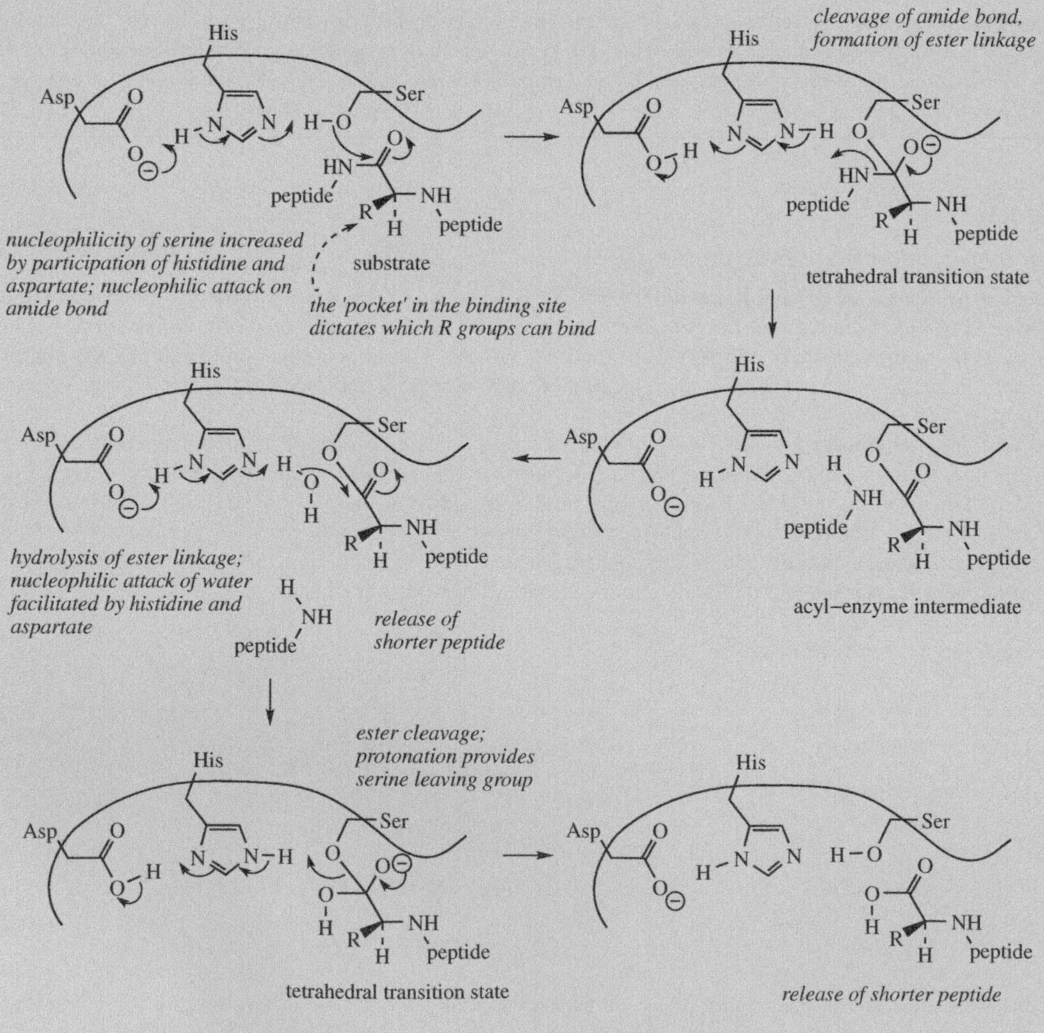

Figure 13.5   Chymotrypsin: mechanism of action

the participation of the histidine group, with the basicity of the histidine nitrogen also being increased because of the proximity of a neighbouring aspartate residue – a **charge-relay network**, as seen with acetylcholinesterase (see Box 13.4). This allows nucleophilic attack on to the peptide carbonyl, giving an initial tetrahedral transition state. We also know that specific amino acid residues are positioned so that they help to stabilize this anionic transition state. Reformation of the carbonyl group is followed by cleavage of the peptide bond. The proton required to form the amino group is acquired from the imidazole. The product is now an acyl–enzyme intermediate, actually an ester involving the serine hydroxyl. This ester is hydrolysed by a water nucleophile, and deprotonation is achieved via the aspartate–histidine system once again. This generates another tetrahedral transition state, which collapses and allows release of the carboxylic acid and regeneration of the serine hydroxyl by protonation from the imidazole system.

Note that **penicillins** and structurally related antibiotics are frequently deactivated by the action of bacterial **β-lactamase** enzymes. These enzymes also contain a serine residue in the active site, and this is the nucleophile that attacks and cleaves the β-lactam ring (see Box 7.20). The β-lactam (amide) linkage is hydrolysed, and then the inactivated penicillin derivative is released from the enzyme by further hydrolysis of the ester linkage, restoring the functional enzyme. The mode of action of these enzymes thus closely resembles that of the serine proteases; there is further discussion in Box 7.20.

Whilst chymotrypsin and trypsin are especially useful in peptide sequence analysis, they also have medicinal applications. Their ability to hydrolyse proteins makes them valuable for wound and ulcer cleansing (trypsin) or during cataract removal (chymotrypsin).

13.4.2 Enolization and enolate anion biochemistry

Let us now look at an example of how nature exploits the equivalent of enol and enolate anion chemistry. **Enolization** provides another application of acid–base catalysis. We saw that the chemical process for enolization could be either acid- or base-catalysed (see Section 10.1), and the following scheme should remind us of the mechanism for base-catalysed enolization of acetone.

base-catalysed enolization

abstraction of proton

resonance

conjugate base enolate anion

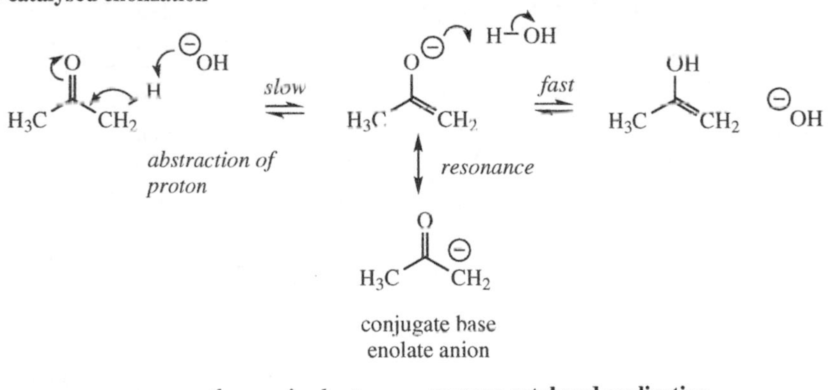

The enzymic processes appear exactly equivalent, except that protons are removed and supplied through the involvement of peptide side-chains. It is unlikely that a distinct enolate anion is formed; instead, we should consider the process as concerted with a smooth flow of electrons. Thus, as a basic group removes a proton from one part of the molecule, an acidic group supplies a proton at another.

The example of triose phosphate isomerase in Box 13.6 provides us with an easily understood analogy.

enzyme-catalysed enolization

forward reaction reverse reaction

A and B are part of enzyme

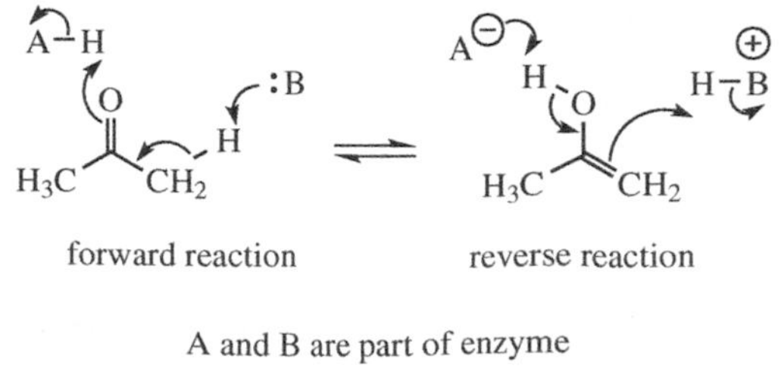

Box 13.6

Triose phosphate isomerase: enolization via acid–base catalysis

Triose phosphate isomerase is one of the enzymes of **glycolysis** (see Section 15.2) and is responsible for converting dihydroxyacetone phosphate into glyceraldehyde 3-phosphate by a two-stage enolization process. An intermediate enediol is involved – this common enol can revert to a keto form in two ways, thus providing the means of isomerization.

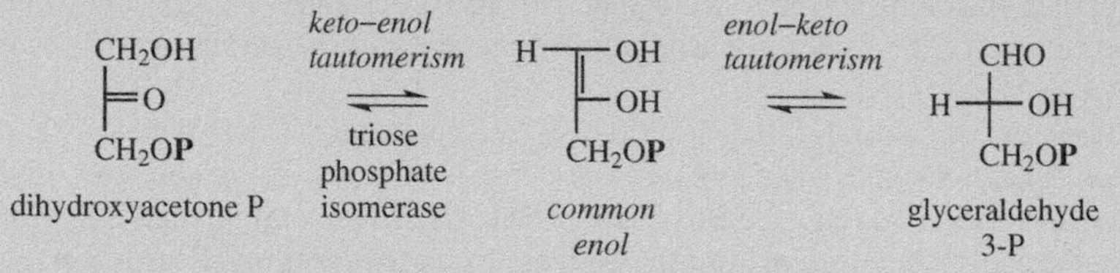

$$
\begin{array}{ccc}
\text{CH}_2\text{OH} & & \text{H}\!-\!\text{OH} \\
|\!\!=\!\!\text{O} & \underset{\substack{\text{triose} \\ \text{phosphate} \\ \text{isomerase}}}{\overset{\substack{\text{keto–enol} \\ \text{tautomerism}}}{\rightleftharpoons}} & |\!\!-\!\!\text{OH} \\
\text{CH}_2\text{OP} & & \text{CH}_2\text{OP}
\end{array}
\qquad
\begin{array}{c}
\text{enol–keto} \\ \text{tautomerism} \\ \rightleftharpoons
\end{array}
\qquad
\begin{array}{c}
\text{CHO} \\
\text{H}\!-\!\text{OH} \\
\text{CH}_2\text{OP}
\end{array}
$$

dihydroxyacetone P   *common enol*   glyceraldehyde 3-P

The active site of the enzyme contains a glutamic acid residue that is ionized at pH 7 and supplies the base. A histidine residue, partially protonated at pH 7, in turn supplies the proton necessary to form the common enol (Figure 13.6).

The process continues, in that the now uncharged histidine is suitably placed to remove a proton from the second of the two hydroxyls, and tautomerization is achieved by abstraction of a proton from the now non-ionized

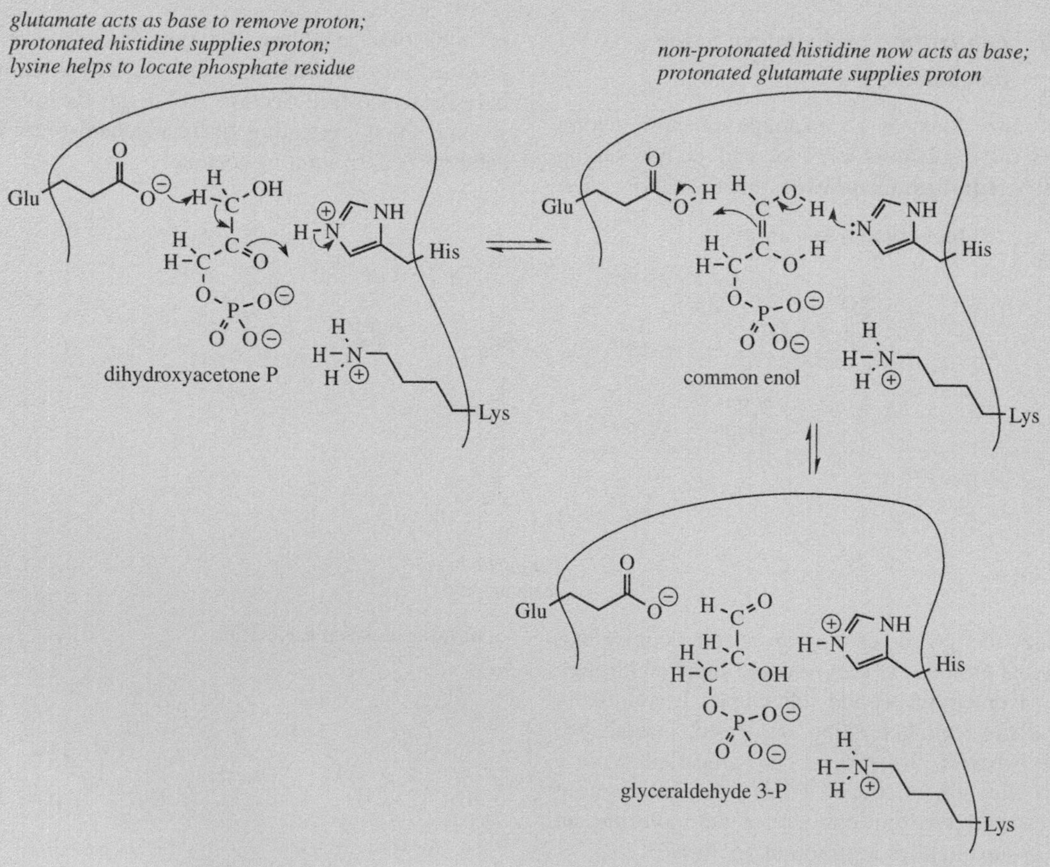

glutamate acts as base to remove proton;
protonated histidine supplies proton;
lysine helps to locate phosphate residue

non-protonated histidine now acts as base;
protonated glutamate supplies proton

Figure 13.6   Triose phosphate isomerase: mechanism of action

glutamic acid. These two enolization processes thus account for the observed isomerization. Note that the glutamic acid and histidine side-chains are suitably positioned so that they can remove/supply protons at either of two different atoms. The only additional feature to mention is the known involvement of a lysine residue, which acts to provide ionic bonding at the active site through an ammonium–phosphate interaction.

We have seen many examples of chemical reactions involving enolate anions, and should now realize just how versatile they are in chemical synthesis (see Chapter 10). We have also seen several examples of how equivalent reactions are utilized in nature. For the triose phosphate isomerase mechanism above, we did not actually invoke a distinct enolate anion intermediate in the enolization process, but proposed that there was a smooth flow of electrons. For other reactions, we shall also need to consider whether enolate anions are actually involved, or whether a more favourable alternative exists. The aldol-type reaction catalysed by the enzyme aldolase is an excellent illustration of nature's approach to enolate anion chemistry.

Aldolase catalyses both aldol and reverse aldol reactions according to an organism's needs. In glycolysis, the substrate fructose 1,6-diphosphate is cleaved by a reverse aldol reaction to provide one molecule of glyceraldehyde 3-phosphate and one molecule of dihydroxyacetone phosphate. In carbohydrate synthesis, these two compounds can be coupled in an aldol reaction to produce fructose 1,6-diphosphate.

It is conceptually easier to consider initially the aldol reaction rather than the reverse aldol reaction. This involves generating an enolate anion from the dihydroxyacetone phosphate by removing a proton from the position α to the ketone group. This enolate anion then behaves as a nucleophile towards the aldehyde group of glyceraldehyde 3-phosphate, and an addition reaction occurs, which is completed by abstraction of a proton, typically from solvent. In the reverse reaction, the leaving group would be the enolate anion of dihydroxyacetone phosphate.

Now let us consider the difficulties associated with this reaction, should we attempt it using chemical reagents (see Section 10.3). In contrast to the chemical aldol reaction, the enzymic reaction has several remarkable advantages:

- the reaction is conducted at room temperature;

- it is conducted at pH 7 without the need for a strong base to generate the enolate anion;

- although it is a mixed aldol reaction, it is quite specific, giving a single product;

- both substrates have the potential to form an enolate anion;

- both substrates have the potential to act as an electrophile;

- dihydroxyacetone phosphate has the potential to form two enolate anions;

- other functional groups in the substrates remain unchanged;

- the reaction is reversible, and can be employed in either direction under similar conditions.

How this is achieved with the enzyme and the role played by the some of the amino acid side-chains can now be considered.

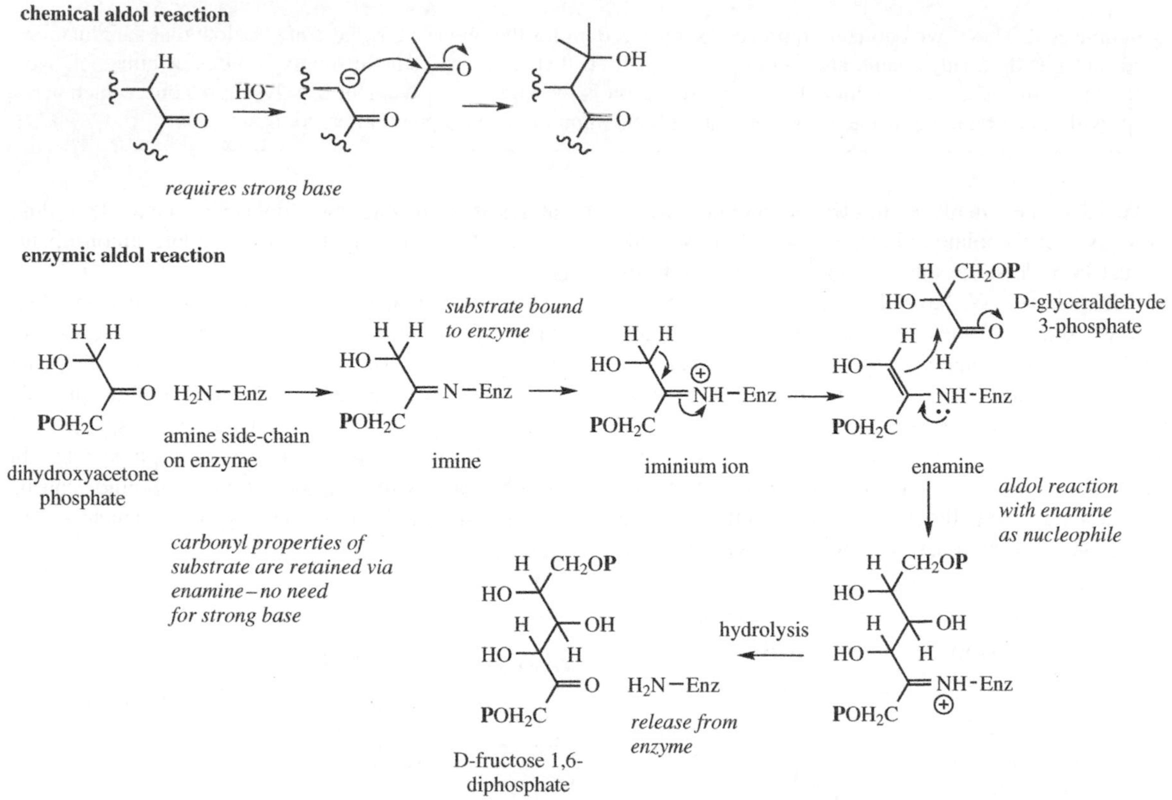

chemical aldol reaction

requires strong base

enzymic aldol reaction

dihydroxyacetone phosphate

H_2N—Enz
amine side-chain on enzyme

substrate bound to enzyme

imine

iminium ion

enamine

D-glyceraldehyde 3-phosphate

aldol reaction with enamine as nucleophile

carbonyl properties of substrate are retained via enamine—no need for strong base

hydrolysis

H_2N—Enz

release from enzyme

D-fructose 1,6-diphosphate

A particularly important interaction with the enzyme is that dihydroxyacetone phosphate is bound to the protein by means of an **imine** linkage between the ketone group and an amino group on the enzyme. This produces a twofold advantage. First, it anchors the substrate to the enzyme through a covalent linkage; second, it allows formation of an **enamine** by removal of the proton originally α to the ketone. An enamine is the equivalent of an enolate anion, but enamine formation is much easier than enolate anion formation and can occur without the need for a strong base (see Section 10.5). Proton removal is achieved by participation of one of the basic groups on the enzyme. With the second substrate glyceraldehyde 3-phosphate appropriately positioned, the aldol addition can then take place, completion of which requires supply of a proton from the enzyme. The product can then be released from the enzyme by hydrolysis of the imine bond, restoring the original ketone of the substrate and the amino group on the enzyme. The reverse aldol reaction can be rationalized in a similar way.

Box 13.7

The active site of aldolase

The active site of the aldolase enzyme is believed to be as shown (Figure 13.7). Although several amino acid residues are involved with bonding the substrates at the active site, the critical amino acid residues are a lysine and an aspartic acid residue. The lysine forms a substrate–enzyme bond via an imine linkage, and the aspartic acid residue functions as a general acid–base.

- Basic amino acid residues are involved in binding the phosphate substrates; to simplify the overall picture, these are not specified here.

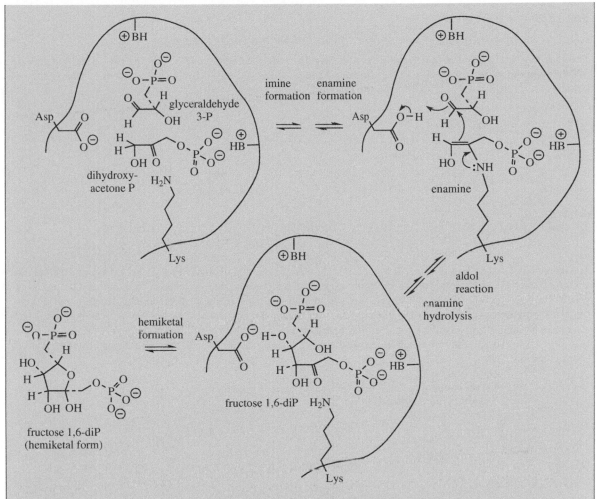

Figure 13.7 Active site of aldolase

- A lysine residue reacts with the carbonyl of dihydroxyacetone phosphate, forming first an addition product that dehydrates to give an imine linkage.

- An aspartate residue is suitably positioned to function as the active site base that removes a proton from the imine and generates the enamine.

- The resultant aspartic acid residue is then involved again in providing a proton to complete the aldol addition.

- The active site also facilitates the ketone–hemiketal interconversion, so that the product liberated is the hemiketal form of fructose 1,6-diphosphate.

- In the reverse reaction, aspartate removes a proton from the alcohol, which allows formation of a transient carbanion or enamine.

- The carbanion/enamine is subsequently protonated via aspartic acid.

Box 13.8

Citrate synthase catalyses an aldol reaction rather than a Claisen reaction

The reaction catalysed by **citrate synthase** in the **Krebs cycle** (see Section 15.3) is primarily an aldol reaction, but the subsequent step, hydrolysis of a thioester linkage, is also catalysed by the same enzyme. This is shown below.

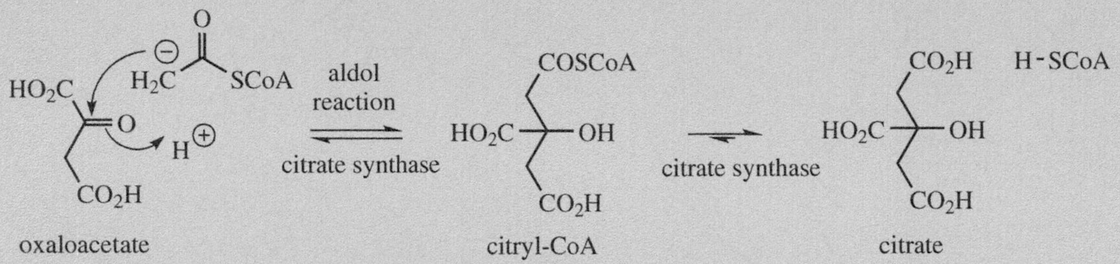

Mechanistically, we can consider it as attack of an enolate anion equivalent from acetyl-CoA on to the ketone group of oxaloacetate. However, if we think carefully, we come to the conclusion that this is not what we would really predict. Of the two substrates, oxaloacetate is the more acidic reagent, in that two carbonyl groups flank a methylene. According to the enolate anion chemistry we studied in Chapter 10, we would predict that oxaloacetate should provide the enolate anion, and that this might then attack acetyl-CoA in a Claisen reaction (see Box 10.4). The product expected in a typical base-catalysed reaction would, therefore, be an acetyl derivative of oxaloacetate.

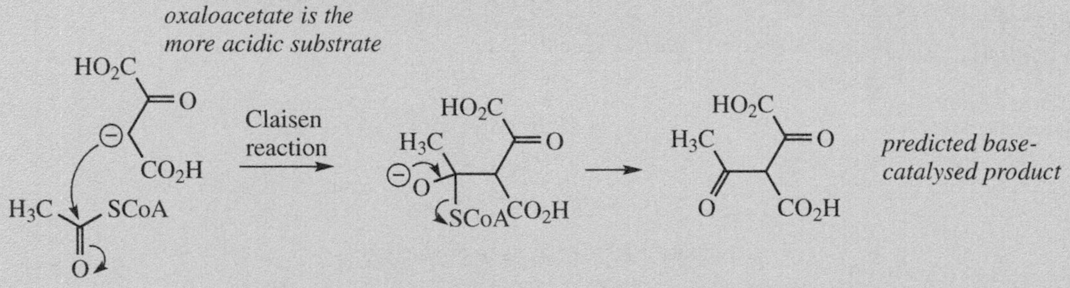

That this is not the case for the enzyme citrate synthase suggests we must look at the enzyme binding site to rationalize the different reaction sequence. It becomes clear that the enzyme binding site positions the substrates so that there are acidic and basic amino acid residues available to produce the enolate anion equivalent of acetyl-CoA (shown here as the enol), but not for the oxaloacetate (Figure 13.8).

Imidazole rings of histidine residues are suitably oriented to participate in the aldol reaction. A histidine residue is also involved in the next step, the hydrolysis of citryl-CoA, and release of citric acid as the final product. It is the hydrolysis of the thioester that disturbs the equilibrium and drives the reaction to completion.

As with other examples of enzyme mechanisms, we can see that the exact array of amino acid residues in the binding site dictates binding of substrates and their chemical interaction to yield products.

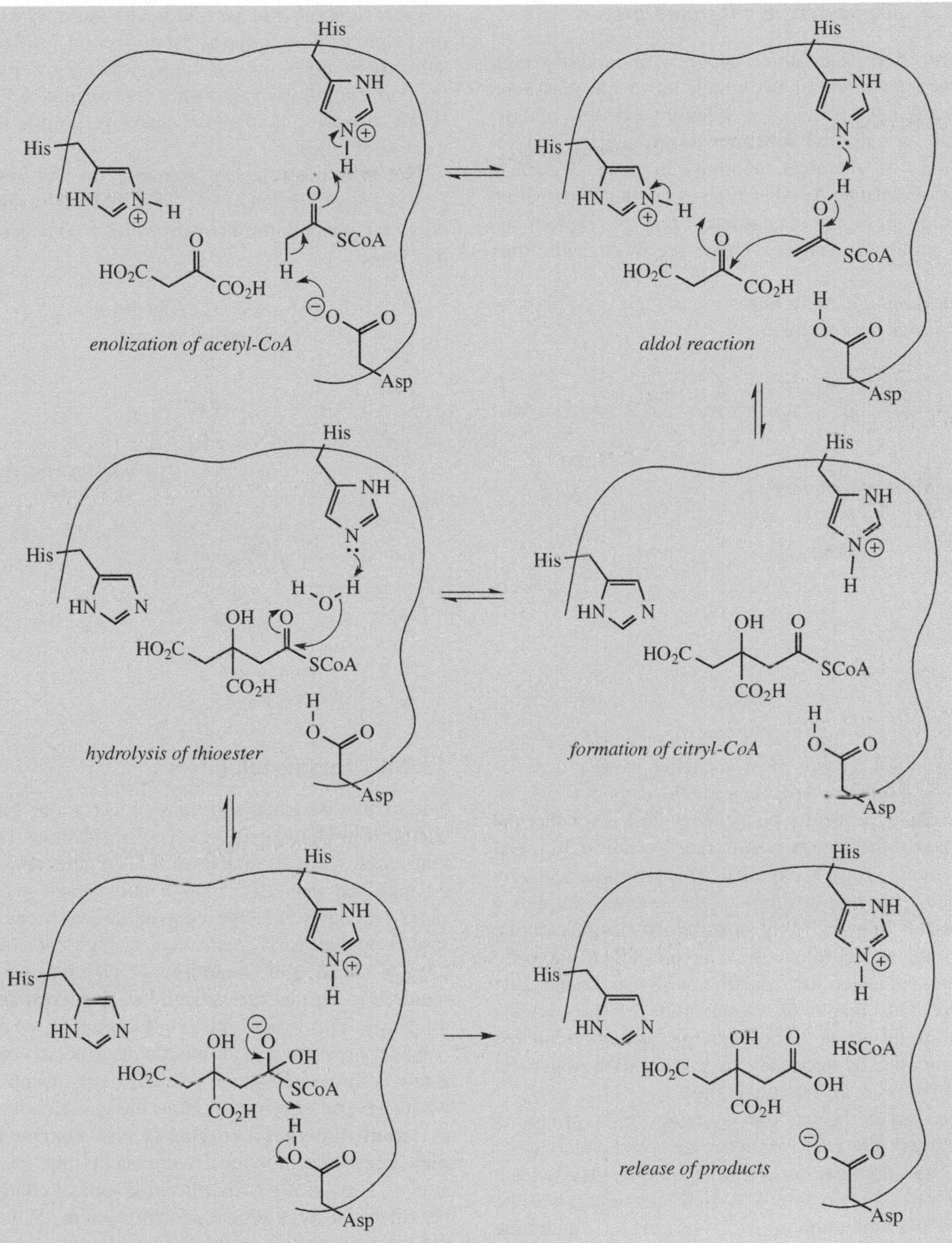

Figure 13.8   Citrate synthase: mechanism of action

13.4.3 Thioesters as intermediates

The reaction of an amino group with an aldehyde or ketone leads to an imine, which, as we have just seen with aldolase, provides a splendid example of how to bond a carbonyl substrate to an enzyme, and yet maintain its chemical reactivity in terms of enolate anion chemistry. Another type of covalent interaction is quite commonly encountered, and this exploits the thiol group of cysteine. Thiols are more acidic than oxygen alcohols (see Section 4.3.2), sulfur is a better nucleophile than oxygen (see Section 6.1.2), and sulfur derivatives provide better leaving groups than the corresponding oxygen ones (see Section 6.1.4). It is not surprising that nature makes very good use of these properties.

We shall meet several examples of this type of process (see Sections 15.2, 15.4 and 15.5), and so only the general mechanism will be considered at this stage.

The thiol group of a cysteine residue acts as a nucleophile towards a suitable carbonyl system, which may frequently be a coenzyme A ester. Normal addition–elimination occurs (see Section 6.3.2), and the leaving group is expelled. This process effectively anchors the acyl residue to the enzyme through a **thioester** linkage. This now allows a nucleophilic substrate to approach the enzyme active site, and become acylated by reacting with the bound acyl group. This results in regeneration of the cysteine thiol group. It is likely that protons are removed and supplied as necessary by participation of general acids or bases at the active site.

We have shown the cysteine thiol group as uncharged. The pK_a for this group in cysteine is about 10.3, and application of the Henderson–Hasselbach equation (see Section 4.9) indicates there will be negligible ionization at pH 7. Nevertheless, under the influence of a suitable basic group, e.g. arginine pK_a 12.5, ionization to thiolate may be possible. In such an environment, thiolate may act as the nucleophile in the mechanism.

13.4.4 Enzyme inhibitors

Nature has designed enzymes to carry out modest chemical modifications on a specific substrate. In certain cases, a small number of related substrates may be modified similarly, though not always with the same efficiency, i.e. the enzyme shows broad substrate specificity. The chemical change catalysed is usually small, and a number of enzymes will be required to change the structure of the substrate significantly. This is made clear in Chapter 15, when we consider the pathways of intermediary metabolism. In a few of these pathways we shall meet examples of where several enzyme activities are combined, either as a **multi-functional enzyme** or as an **enzyme complex** where the individual components may be separated. This allows a significant chemical change to be catalysed by a single protein system. Whatever the arrangement of enzymes, it is clear to see that a single enzyme activity functions as a link in a chain and, therefore, can be used to control whether or not a sequence of reactions proceeds. We can thus exploit a chain's weakest link.

Enzyme inhibitors are chemicals that may serve as a natural means of controlling metabolic activity by reducing the number of enzyme molecules available for catalysis. In many cases, natural or synthetic inhibitors have allowed us to unravel the pathways and mechanisms of intermediary metabolism. Enzyme inhibitors may also be used as pesticides or drugs. Such materials are designed so that they inhibit a specific enzyme that is peculiar to an organism or a disease state. For example, a good antibiotic may inhibit a bacterial enzyme, but it should have no effect on the host person or animal.

We may consider enzyme inhibitors as either irreversible or reversible inhibitors. Some inhibitors become covalently linked to the enzyme and are bound so strongly that they cannot be removed. As a result, the enzyme activity decreases and eventually becomes zero.

irreversible inhibitor: E + I $\longrightarrow$ EI

reversible inhibitor: E + I $\rightleftharpoons$ EI

Irreversible inhibition in an organism usually results in a toxic effect. Examples of this type of inhibitor are the organophosphorus compounds that interfere with acetylcholinesterase (see Box 7.26). The organophosphorus derivative reacts with the enzyme in the normal way, but the phosphorylated intermediate produced is resistant to normal hydrolysis and is not released from the enzyme.

The enzyme becomes inactivated, and a toxic level of acetylcholine builds up. Organophosphorus compounds provide a range of insecticides and nerve gases.

Reversible inhibitors are potentially less damaging. In the presence of a reversible inhibitor, the enzyme activity decreases, but to a constant level as equilibrium is reached. The enzyme activity reflects the lower level of enzyme available for catalysis. We can subdivide the reversible inhibition into three types, i.e. competitive, non-competitive, and allosteric inhibition.

Competitive inhibitors bind to specific groups in the enzyme active site to form an enzyme–inhibitor complex. The inhibitor and substrate compete for the same site, so that the substrate is prevented from binding. This is usually because the substrate and inhibitor share considerable structural similarity. Catalysis is diminished because a lower proportion of molecules have a bound substrate. Inhibition can be relieved by increasing the concentration of substrate. Some simple examples are shown below. Thus, sulfanilamide is an inhibitor of the enzyme that incorporates *p*-aminobenzoic acid into folic acid, and has antibacterial properties by restricting folic acid biosynthesis in the bacterium (see Box 11.13). Some phenylethylamine derivatives, e.g. phenelzine, provide useful antidepressant drugs by inhibiting the enzyme monoamine oxidase. The *cis*-isomer maleic acid is a powerful inhibitor of the enzyme that utilizes the *trans*-isomer fumaric acid in the Krebs cycle.

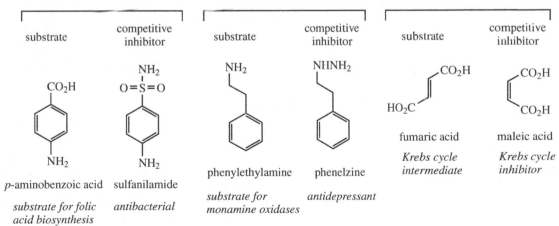

substrate — competitive inhibitor | substrate — competitive inhibitor | substrate — competitive inhibitor

p-aminobenzoic acid   sulfanilamide

substrate for folic acid biosynthesis   *antibacterial*

phenylethylamine   phenelzine

substrate for monamine oxidases   *antidepressant*

fumaric acid   maleic acid

Krebs cycle intermediate   *Krebs cycle inhibitor*

Non-competitive inhibitors do not bind to the active site, but bind at another site on the enzyme and distort the shape of the protein, resulting in a lowering of activity. Both inhibitor and substrate can bind simultaneously to the enzyme. A non-competitive inhibitor decreases the activity of the enzyme rather than lowering the proportion of molecules with a bound substrate. In contrast to competitive inhibition, increasing the concentration of substrate has no effect on the level of inhibition. The chemical structures of non-competitive inhibitors frequently bear no similarity to the natural substrate structures. For example, heavy metal ions, such as Pb^{2+} and Hg^{2+}, inhibit the activity of some enzymes by binding to thiol groups, and cyanide reacts with and inhibits iron–porphyrin enzymes (see Box 11.4).

The third type of inhibition is called **allosteric inhibition**, and is particularly important in the control of intermediary metabolism. This refers to the ability of enzymes to change their shape (tertiary and quaternary structure, sec Section 13.3) when exposed to certain molecules. This sometimes leads to inhibition, whereas in other cases it may actually activate the enzyme. The process allows subtle control of enzyme activity according to an organism's demands. Further consideration of this complex phenomenon is outside our immediate needs.

Box 13.9
Angiotensin-converting enzyme (ACE) inhibitors: captopril

Captopril was the first of a range of orally active drugs to counter high blood pressure, a group known collectively as **ACE inhibitors**. ACE is the abbreviation for **angiotensin-converting enzyme**, a protein that converts the decapeptide angiotensin I into the octapeptide angiotensin II by hydrolytic removal of a pair of amino acids. Angiotensin II has a powerful vasoconstrictor effect, so increases blood pressure. By inhibiting the action of ACE, angiotensin II levels are limited, blood vessels dilate, and blood pressure is reduced. This is of particular value in reducing the risk of heart attacks in patients prone to high blood pressure.

ACE is a carboxypeptidase enzyme that splits off a pair of amino acids from the *C*-terminal end; its active site is known to contain a zinc atom.

The development of captopril was one of the first examples of successful drug design based upon knowledge of the active site of the target enzyme. It was designed to fit the known active site of carboxypeptidase A, an enzyme very similar to ACE. Captopril resembles the terminal dipeptide cleaved from angiotensin I, in that the proline carboxylate can bind to a positive centre, the amide carbonyl can hydrogen bond, and the thiol group is a good ligand for the Zn^{2+} component. Captopril is thus a competitive inhibitor of the enzyme; it can bind to the enzyme, but, in so doing, inhibits its hydrolytic action.

Several captopril-like drugs are now available, their main advantages over captopril being their increased duration of action, e.g. enalapril.

13.5 Peptide biosynthesis

Synthesis and biosynthesis of peptides and proteins requires the combination of amino acids via amide bonds. We have seen earlier that the chemical reaction of amines and acids to produce a simple amide is severely hindered by initial salt formation, and that a more efficient way of making amides is to employ a carboxylic derivative that is non-acidic and has a better leaving group (see Section 7.10). Thus, acyl halides, anhydrides, or even esters provide better substrates. In nature, we find that esters or thioesters are the reactive species employed.

$$RCO_2H \;+\; H_2NR' \longrightarrow RCO_2^{\ominus} \; \overset{\oplus}{N}H_2R' \quad \textit{ammonium salt}$$

high temp
$- H_2O$

$$RCONHR'$$
amide

$$RCOCl \;+\; H_2NR' \longrightarrow RCONHR' \;+\; HCl$$

$$(RCO)_2O \;+\; H_2NR' \longrightarrow RCONHR' \;+\; RCO_2H$$

$$RCO_2R \;+\; H_2NR' \longrightarrow RCONHR' \;+\; ROH$$

$$RCOSR \;+\; H_2NR' \longrightarrow RCONHR' \;+\; RSH$$

esters and thioesters
are used in nature

A further requirement for chemical synthesis of peptides would be to take steps to avoid any side-chain functional groups reacting under the conditions used for amide bond formation. This can be accomplished by the use of appropriate protecting groups, though these will then have to be removed at a later stage in the synthesis (see Section 13.6.1). Nature employs enzymic reactions that position the functional groups in an appropriate orientation to react. Consequently, any side-chain functionalities are kept well away and do not interfere with the processes of amide bond formation. The final consideration is to assemble the amino acids in the correct order. In all cases, we need to choose the correct amino acid at each step, but we shall see that nature uses quite sophisticated techniques, and specificity is conferred by nucleic acids and enzymes (see Section 14.2.6). In the laboratory, we must pick up reagent bottles in the correct sequence.

Peptides are produced in nature by one of two methods, termed **ribosomal peptide biosynthesis** and **non-ribosomal peptide biosynthesis**. In the former process, peptide biosynthesis takes place on the ribosomes, and the amino acid precursors are combined in a sequence that is defined by the genetic code, the sequence of bases in DNA. In non-ribosomal peptide biosynthesis, peptides are synthesized by a more individualistic sequence of enzyme-controlled reactions. Despite the differences in programming the sequence, the chemical linkage of amino acid residues is achieved in a rather similar fashion.

13.5.1 Ribosomal peptide biosynthesis

A simplified representation of peptide biosynthesis, as characterized in the bacterium *Escherichia coli*, is discussed in Section 14.2.6. The major aspect to be considered here relates to the bond forming processes involved in linking the amino acids.

activation of amino acid by
formation of mixed anhydride

aminoacyl group transferred to ribose
hydroxyl in 3′-terminal adenosine of
tRNA; formation of ester linkage

aminoacyl-AMP

aminoacyl-tRNA

Initially, the amino acid is activated by an ATP-dependent process, producing an **aminoacyl-AMP**. This may be considered to be nucleophilic attack of the amino acid carboxylate group on to the P=O system of ATP with expulsion of diphosphate as the leaving group.

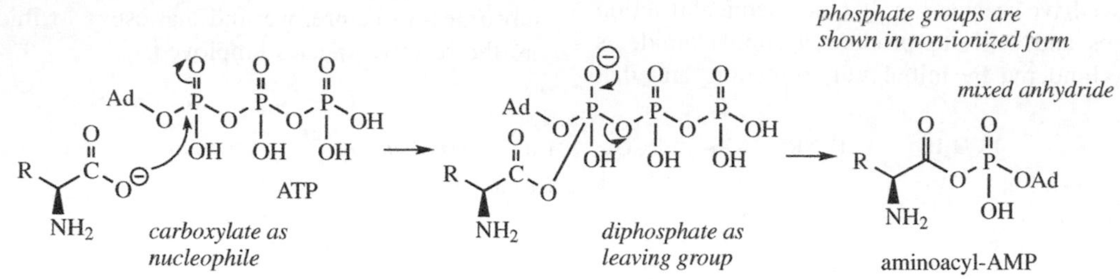

phosphate groups are shown in non-ionized form

Carboxylate is not an especially good nucleophile, but we have seen it used in S_N2 reactions to synthesize esters (see Section 6.3.2). Here, attack is on a reactive anhydride; a similar type of reaction is seen in fatty acid degradation (see Section 15.4.1).

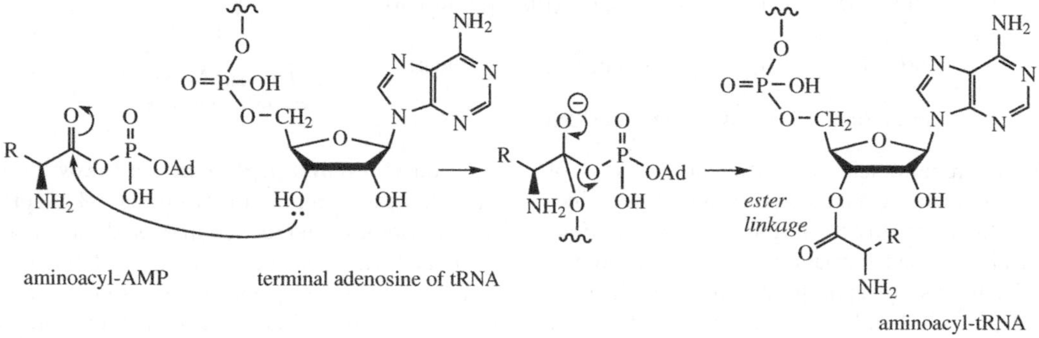

The intermediate aminoacyl-AMP can also be seen to be an anhydride, but in this case a mixed anhydride of carboxylic and phosphoric acids (see Box 7.27). This can react with a hydroxyl group in ribose, part of a terminal adenosine group of transfer-RNA (tRNA). This then binds the amino acid via an ester linkage, giving an **aminoacyl-tRNA**. The tRNA involved will be specific for the particular amino acid.

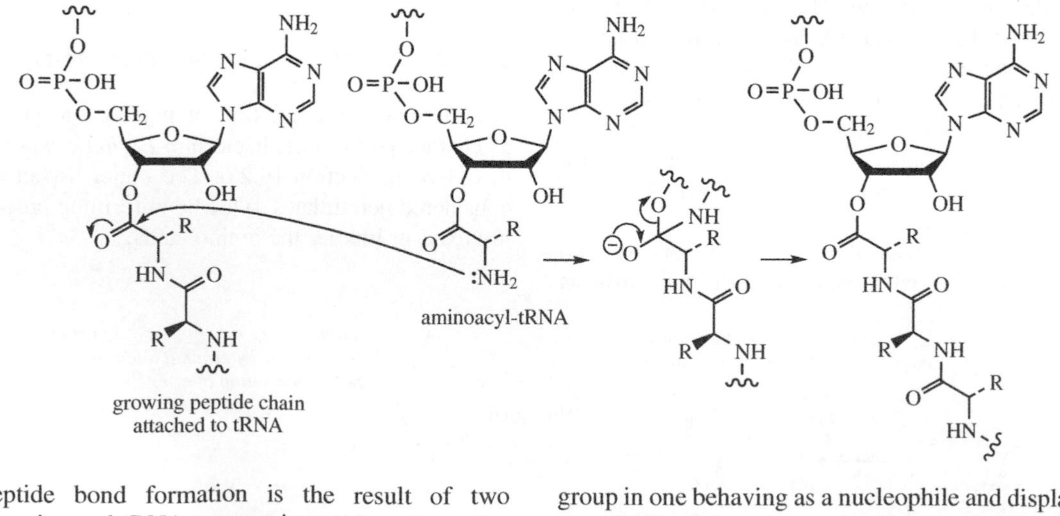

Peptide bond formation is the result of two such aminoacyl-tRNA systems interacting, the amino group in one behaving as a nucleophile and displacing the tRNA from the second, i.e. simply amide formation utilizing an ester substrate. The process is repeated as required. The sequence of amino acids is controlled by messenger RNA (mRNA), the message being stored as a series of three-base sequences (codons) in its nucleotides (see Section 14.2.4). Elongation of the peptide continues until a termination codon is reached, and the peptide or protein is then hydrolysed and released from the tRNA carrier.

13.5.2 Non-ribosomal peptide biosynthesis

In marked contrast to the ribosomal biosynthesis of peptides and proteins where a biological production line interprets the genetic code of mRNA, many natural peptides are known to be synthesized by a more individualistic sequence of enzyme-controlled processes, in which each amino acid is added as a result of the specificity of each enzyme involved. The many stages of the whole process appear to be carried out by a multi-functional enzyme **non-ribosomal peptide synthase** (NRPS) comprised of a linear sequence of modules. Each **module** is responsible for inserting a particular amino acid to generate the sequence in the peptide product. The amino acids are first activated to **aminoacyl-AMP** derivatives as for ribosomal peptide biosynthesis. These are then converted into **thioesters**, by reaction with thiol functions in the enzyme. The process is exactly analogous to forming aminoacyl-tRNA units, but utilizes SH rather than OH as nucleophile.

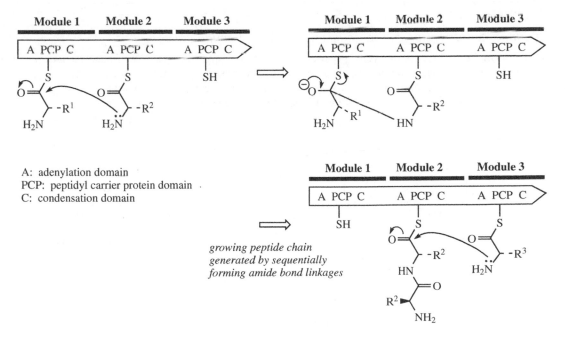

The residues are held so as to allow a sequential series of peptide bond formations (Figure 13.9 gives a simplified representation), until the peptide is finally released from the enzyme.

A typical **module** consists of an adenylation (A) domain, a peptidyl carrier protein (PCP) domain, and a condensation (C) or elongation domain. The A domain activates a specific amino acid as an

A: adenylation domain
PCP: peptidyl carrier protein domain
C: condensation domain

Figure 13.9 Modular non-ribosomal peptide synthase

536AMINO ACIDS, PEPTIDES AND PROTEINS

aminoacyl-AMP mixed anhydride, which is then transferred to the PCP domain to form an aminoacyl thioester. The thioester linkage is not to a cysteine residue in the protein, as we described in Section 13.4.3. Instead, it involves pantothenic acid (vitamin B$_5$) bound to the enzyme as pantetheine, and

this is used to carry the growing peptide chain via its thiol group. The important significance of this is that the long '**pantetheinyl arm**' allows different active sites on the multi-functional enzyme to be reached in the chain assembly process (compare biosynthesis of fatty acids, see Box 15.2).

'pantetheinyl arm'

attached to growing peptide chain

bound to enzyme through phosphate

cysteamine | pantothenic acid

pantetheine

Nucleophilic attack by the amino group of the neighbouring aminoacyl thioester is catalysed by the C domain, and this results in amide (peptide) bond formation. Enzyme-controlled biosynthesis in this manner is a feature of many microbial peptides, especially those containing unusual amino acids not encoded by DNA and where post-translational modification (see Section 13.1) is unlikely.

As well as activating the amino acids and catalysing formation of the peptide linkages, the enzyme may possess other domains that are responsible for epimerizing L-amino acids to D-amino acids,

probably through enol-like tautomers in the peptide (see Box 10.10). A terminal **thioesterase** domain is also required. This is responsible for terminating the chain extension process by hydrolysing the thioester and releasing the peptide from the enzyme.

Many medicinally useful peptides have cyclic structures. Cyclization may result if the amino acids at the two termini of a linear peptide link up to form another peptide bond. Alternatively, ring formation can very often be the result of ester or amide linkages that utilize side-chain functionalities (CO$_2$H, NH$_2$, OH) in the constituent amino acids.

Box 13.10

Ciclosporin, a cyclic peptide composed mainly of unusual amino acids

The **cyclosporins** are a group of cyclic peptides produced by fungi such as *Cylindrocarpon lucidum* and *Tolypocladium inflatum*. These agents show a rather narrow range of antifungal activity, but high levels of immunosuppressive and anti-inflammatory activities. The main component from the culture extracts is cyclosporin A, but some 25 naturally occurring cyclosporins have been characterized.

(Me)Bmt = 4-(2-butenyl)-4,N-dimethyl-L-threonine

Abu = L-α-aminobutyric acid

Sar = sarcosine (N-methylglycine)

ıııııı amide linkages (peptide bonds)

ciclosporin (cyclosporin A)

Cyclosporin A contains 11 amino acids, joined in a cyclic structure by peptide bonds. The structure is also stabilized by intramolecular hydrogen bonds. Only two of the amino acids, i.e. alanine and valine, are typical of proteins. The compound contains several N-methylated amino acid residues, together with the even less common L-α-aminobutyric acid and an N-methylated butenylmethylthreonine. There is one D-amino acid, i.e. D-alanine, and the assembly of the polypeptide chain is known to start from this residue. Many of the other natural cyclosporin structures differ only with respect to a single amino acid (the α-aminobutyric acid residue) or the number of amino acids that have the extra N-methyl group.

Of all the natural analogues, and many synthetic ones produced, cyclosporin A is the most valuable for drug use, under the drug name **ciclosporin**. It is now widely exploited in organ and tissue transplant surgery, to prevent rejection following bone marrow, kidney, liver, and heart transplants. It has revolutionized organ transplant surgery, substantially increasing survival rates in transplant patients. It is believed to inhibit T-cell activation in the immunosuppressive mechanism by first binding to a receptor protein, giving a complex that then inhibits a phosphatase enzyme called calcineurin. The resultant aberrant phosphorylation reactions prevent appropriate gene transcription and subsequent T-cell activation.

Box 13.11

Penicillins and cephalosporins are modified tripeptides

Penicillin and **cephalosporin** antibiotics are usually classed as **β-lactam antibiotics**, since their common feature is a lactam function in a four-membered ring, typically fused to another ring system. This second ring takes in the β-lactam nitrogen atom and also contains sulfur. In the case of penicillins, e.g. benzylpenicillin, the second ring is a thiazolidine, and in the cephalosporins, e.g. cephalosporin C, this ring is a dihydrothiazine. What is not readily apparent from these structures is that they are both modified tripeptides and their biosyntheses share a common tripeptide precursor.

The tripeptide precursor is called ACV, an abbreviation for δ-(L-α-aminoadipyl)-L-cysteinyl-D-valine. ACV is an acronym, and does not refer to the systematic abbreviations for amino acids described in Table 13.1. ACV is the linear tripeptide that leads to isopenicillin N, the first intermediate with the fused-ring system found in the penicillins.

ACV is produced by the modular system for non-ribosomal peptide biosynthesis. The amino acid precursors are L-α-aminoadipic acid (an unusual amino acid derived by modification of L-lysine), L-cysteine, and L-valine; during tripeptide formation, the L-valine is epimerized to D-valine (see Box 10.10).

Box 13.11 (continued)

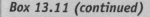

A: adenylation domain
PCP: peptidyl carrier protein domain
C: condensation domain
E: epimerization domain (Val module only)
TE: thioesterase domain

peptide bond formation; during assembly, the L-valine residue is epimerized to the D-configuration

ACV

L-α-aminoadipic acid

Medicinally useful penicillins are formed by replacing the acyl group of the side-chain amide in isopenicillin N with an alternative acyl group. This is sometimes achieved biochemically in the fungal culture, but more frequently it is accomplished through semi-synthetic procedures (see Box 7.20). Isopenicillin N is also the precursor of the cephalosporins, formation of which requires a ring expansion. The five-membered thiazolidine ring of the penicillin is expanded, taking in one of the methyl groups, to produce a six-membered heterocycle.

Box 13.12

Bacterial peptidoglycans: D-amino acids and the antibacterial action of penicillins

Bacterial cell walls contain **peptidoglycan** structures in which carbohydrate chains (composed of alternating β1 → 4-linked *N*-acetylglucosamine and *O*-lactyl-*N*-acetylglucosamine residues) are cross-linked via peptide structures (see Section 12.10). Part of the peptidoglycan of *Staphylococcus aureus* is shown here, illustrating the involvement of the lactyl group of the *O*-lactyl-*N*-acetylglucosamine (also called *N*-acetylmuramic acid) in linking the peptide with the carbohydrate via an amide/peptide bond. The peptide cross-links include some D-amino acids, namely D-alanine and D-glutamic acid.

At the start of the cross-linking process, the peptide chains from the *N*-acetylmuramic acid residues have a terminal –Lys–D-Ala–D-Ala sequence. The lysine from one chain then becomes bonded to the penultimate D-alanine of another chain through five glycine residues, at the same time displacing the terminal D-alanine. The mechanism involves a serine residue at the active site of the enzyme. This residue is used to convert an amide linkage into an ester, and a reversal of this sequence provides the new peptide bond.

cross-linking in peptidoglycan biosynthesis:

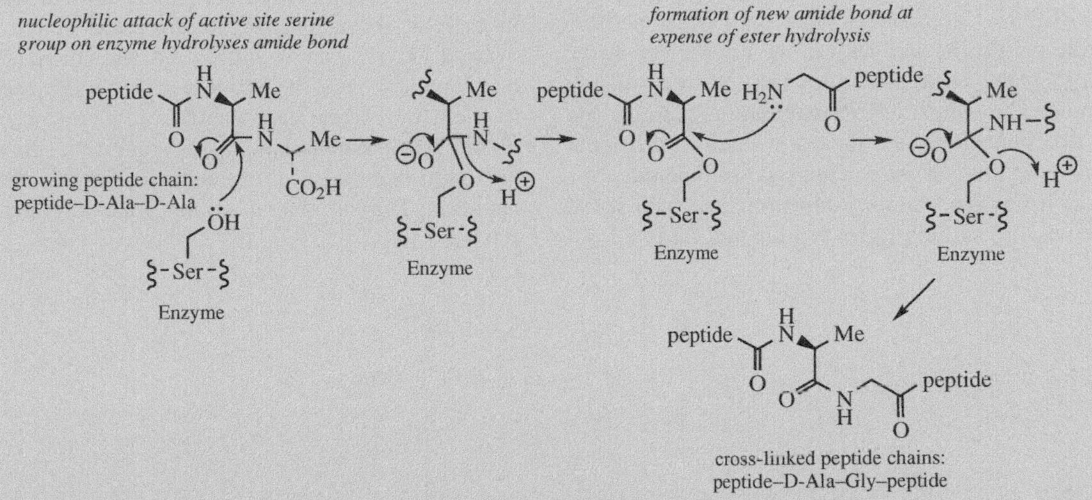

nucleophilic attack of active site serine group on enzyme hydrolyses amide bond

formation of new amide bond at expense of ester hydrolysis

cross-linked peptide chains:
peptide–D-Ala–Gly–peptide

The biological activities of the **β-lactam antibiotics**, e.g. **penicillins** and **cephalosporins** (see Box 13.11), stem from an inhibition of the cross-linking mechanism during the biosynthesis of the bacterial cell wall. The β-lactam drugs bind to enzymes (penicillin-binding proteins) that are involved in the late stages of the biosynthesis of the bacterial cell wall. During the cross-linking process, the peptide–D-Ala–D-Ala intermediate in its transition state conformation closely resembles the penicillin molecule.

enzyme inhibition by β-lactams:

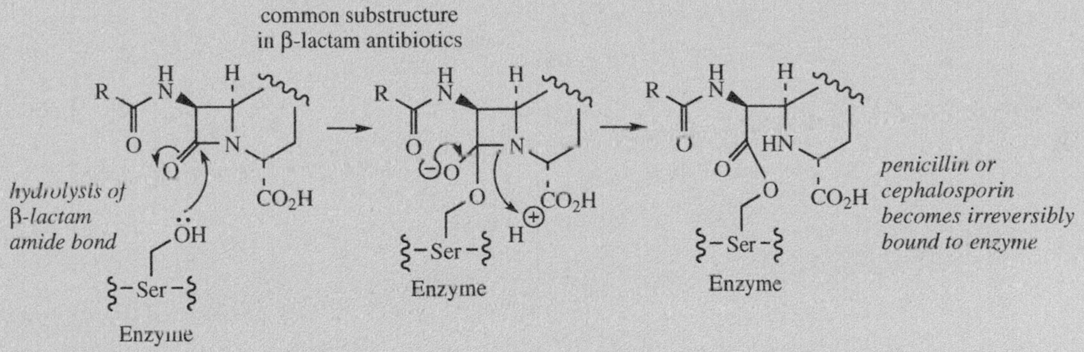

common substructure
in β-lactam antibiotics

penicillin or cephalosporin becomes irreversibly bound to enzyme

As a result, the penicillin occupies the active site of the enzyme, and becomes bound via the active-site serine residue. This binding causes irreversible enzyme inhibition, and stops cell-wall biosynthesis. Growing cells are killed due to rupture of the cell membrane and loss of cellular contents. The binding reaction between penicillin-binding proteins and penicillins is chemically analogous to the action of **β-lactamases** (see Boxes 7.20 and 13.5); however, in the latter case, penicilloic acid is subsequently released from the β-lactamase, and the enzyme can continue to function. Inhibitors of acetylcholinesterase (see Box 7.26) also bind irreversibly to the enzyme through a serine hydroxyl.

The penicillins are very safe antibiotics for most individuals. The bacterial cell wall has no counterpart in mammalian cells, and the action is thus very specific. However, a significant proportion of patients can experience allergic responses, ranging from a mild rash to fatal anaphylactic shock. Cleavage of the β-lactam ring through nucleophilic attack of an amino group in a protein is believed to lead to the formation of antigenic substances that then cause the allergic response.

13.6 Peptide synthesis

Many different approaches have been developed for **peptide synthesis**, and it is not the intention to cover more than the basic principles here, with a suitable example. The philosophy to convert two amino acids into a **dipeptide** is to transform each difunctional amino acid into a monofunctional compound, one of which has the amino group protected, whilst the other has the carboxyl group protected. This allows

the remaining amino and carboxyl groups to react, provided the carboxyl group is suitably activated to make it more reactive, as discussed above (see Section 13.5). After coupling and formation of the new amide bond, the product can be deprotected to yield the dipeptide. Alternatively, one or other of the protecting groups can be removed, allowing the sequence to be repeated, leading to larger peptides. This is shown in the following general scheme.

13.6.1 Protecting groups

What is not included here is the need also to protect any vulnerable functional groups in the amino acid side-chains. A range of methods is available to protect amino, carboxyl, thiol, and hydroxyl groups and prevent them reacting during the amide bond synthesis. Such groups also have to be removed after their job is done, using conditions that do not

destroy the new amide bonds. Where amino acid side-chains have carboxylic acid or amino groups, you will readily appreciate that manipulating protecting groups on these groups separately from those related to making the peptide linkage can turn out to be a highly delicate operation.

Let us consider one method to synthesize the dipeptide Ala–Leu. It is necessary to protect the amino group of Ala and the carboxyl group of Leu.

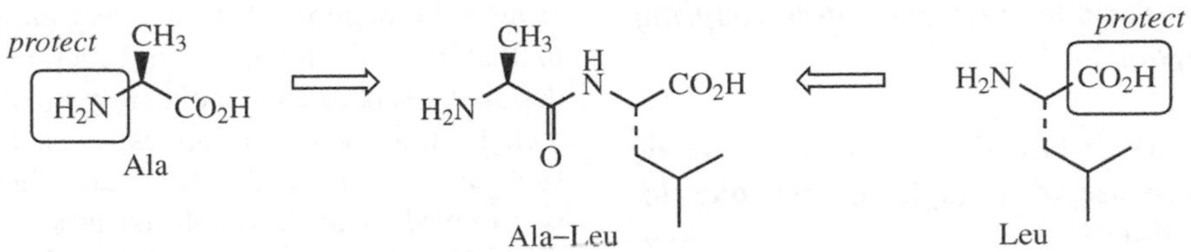

Ala

Ala–Leu

Leu

Amino group protection may be achieved by converting the amine into its *N-tert*-butyloxycarbonyl (tBOC or just BOC) derivative, by reaction with di-*tert*-butyl dicarbonate. This reagent should be considered as a variant of a carboxylic acid anhydride; it reacts in just the same way (see Section 7.10). The product is termed BOC-Ala, and is strictly a carbamate, a half ester–half amide of carbonic acid.

protection of amino group: tBOC

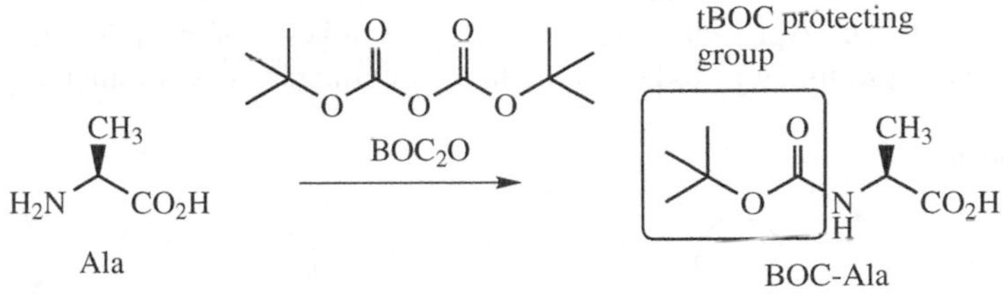

BOC$_2$O

tBOC protecting group

Ala

BOC-Ala

Carbamates behave like amides; the amino group is no longer basic or nucleophilic (see Section 4.5.4). The BOC protecting group can thus be removed readily by treating with dilute aqueous acid. The process involves protonation, loss of the *tert*-butyl cation, and then decarboxylation. On the other hand, the carbonyl group is too hindered to be attacked by base.

removal of tBOC protecting group

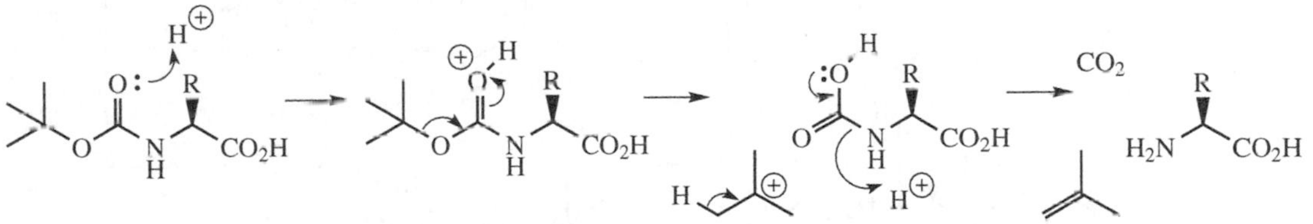

CO$_2$

Carboxyl protection of the second amino acid is usually achieved by conversion to an ester using an appropriate alcohol and acidic catalyst (see Section 7.9.1). Although methyl and ethyl esters work perfectly well, their removal typically requires alkaline hydrolysis, which may be undesirable. More acceptable are esters that can be removed via catalytic hydrogenolysis, e.g. benzyl esters.

protection of carboxyl group: benzyl ester

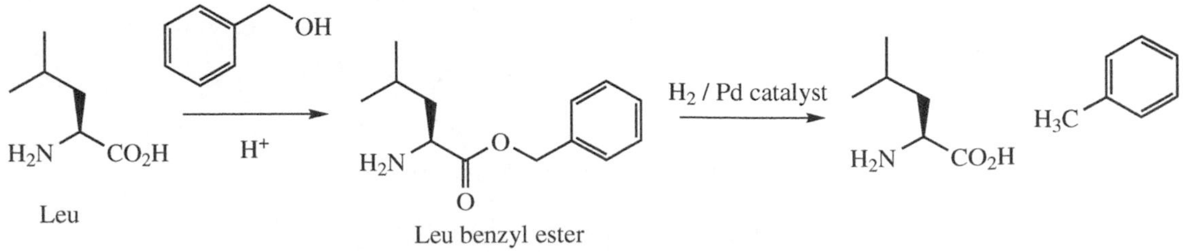

Leu

H$^+$

Leu benzyl ester

H$_2$ / Pd catalyst

13.6.2 The dicyclohexylcarbodiimide coupling reaction

Activation of the carboxyl and coupling may be achieved through use of a single reagent, **dicyclohexylcarbodiimide** (DCC).

This compound removes a proton from the carboxylic acid, producing a cation that is readily attacked by the carboxylate nucleophile across one of the C–N double bonds – the protonated imine behaves as a good electrophile (see Section 7.7.1). The product is now an activated ester (an *O*-acylisourea) that can be attacked by any available nucleophile. The amino group of the second amino acid derivative provides the nucleophile, resulting in expulsion of a very stable urea as the leaving group, and production of the

protected dipeptide. DCC is a very attractive reagent, in that there is no need to generate the activated derivative separately. One merely mixes the two protected amino acid derivatives in an aprotic solvent such as CH_2Cl_2, adds DCC, and dicyclohexylurea is removed as an insoluble by-product. The desired dipeptide can then be obtained by removal of the protecting groups, as already outlined. Note that, in the example shown, we are extending the chain by adding new amino acid residues to the carboxyl terminus.

13.6.3 Peptide synthesis on polymeric supports

Synthesis of peptides in solution using the method outlined above, or alternative procedures, is laborious

DCC coupling

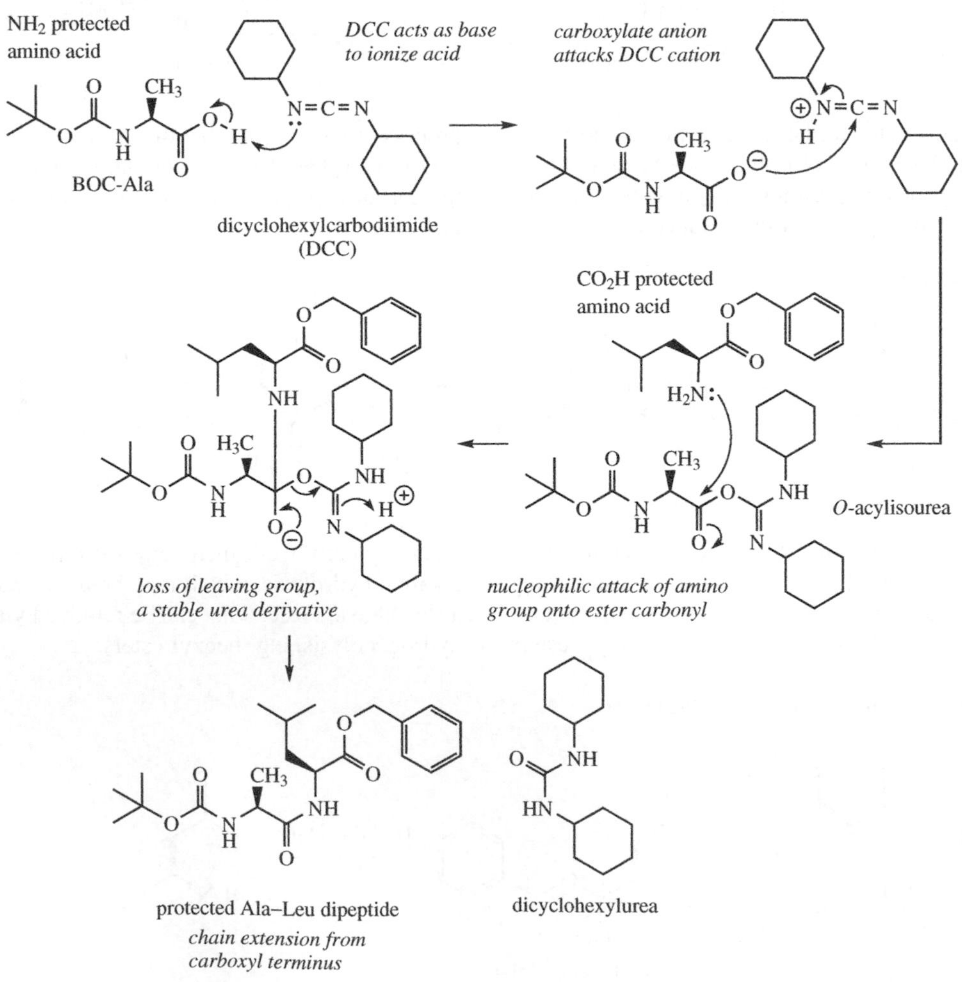

protected Ala–Leu dipeptide

chain extension from carboxyl terminus

dicyclohexylurea

and often low yielding, since each intermediate needs isolating and purifying at each stage of the synthesis. An alternative approach developed by Merrifield is to attach the growing peptide chain to a polymer, which renders it insoluble. This allows the use of excess reagents, and the removal of impurities merely by washing the polymer, which is usually in the form of beads. This approach is the basis of automated peptide synthesizers, since the process can be fast, simple and readily repeated.

In the initial step, the first BOC-protected amino acid is bound to the polymer, e.g. polystyrene in which a proportion of the phenyl rings have chloromethyl substitution. Attachment to these residues is through the carboxyl via an ester linkage. This involves a simple nucleophilic substitution reaction, with the carboxylate as nucleophile and chloride as leaving group (see Section 6.3.2). After each stage, the insoluble polymer–product combination is washed free of impurities.

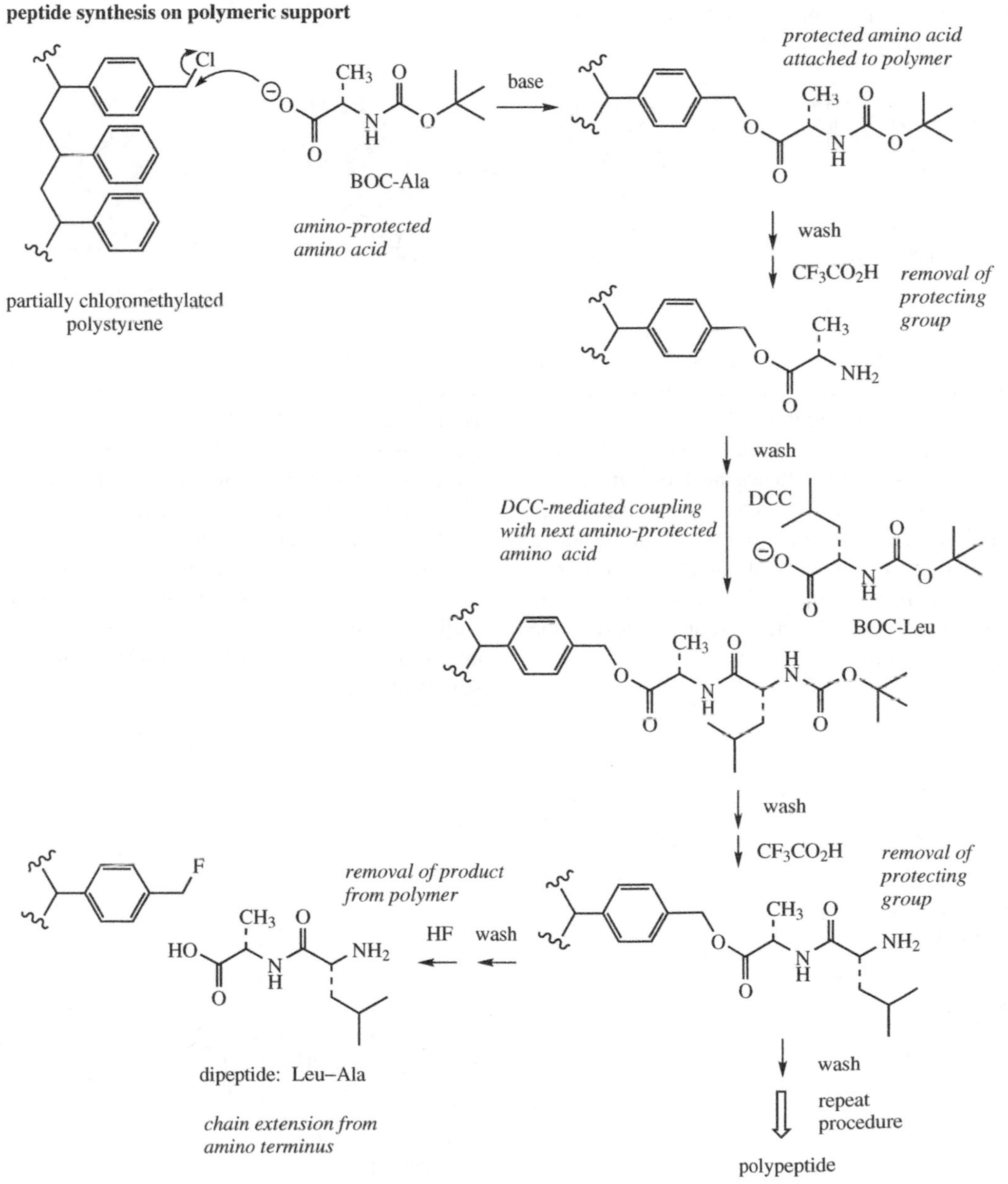

peptide synthesis on polymeric support

partially chloromethylated polystyrene

amino-protected amino acid

BOC-Ala

protected amino acid attached to polymer

removal of protecting group

DCC-mediated coupling with next amino-protected amino acid

BOC-Leu

removal of protecting group

removal of product from polymer

dipeptide: Leu–Ala

chain extension from amino terminus

repeat procedure

polypeptide

The BOC protecting group is then removed from the amino acid, allowing the next protected amino acid to be bonded to the polymer-bound substrate via the DCC coupling reaction. The processes of BOC removal and DCC coupling are then repeated with as many amino acid residues as required. This procedure extends the chain by adding new amino acid residues to the amino terminus. Finally, the polypeptide is released from the polymer by treatment with HF. All the steps are carried out without isolating any intermediate. An early peptide synthesizer produced the 125 amino acid protein ribonuclease in an overall yield of 17%, a quite staggering achievement.

13.7 Determination of peptide sequence

Chemical methods for determining the **amino acid sequence** of a peptide or protein have been developed, and the normal approach is to exploit the properties of the amino group at the *N*-terminus. A long-established procedure for identifying the ***N*-terminal amino acid** is use of the **Sanger reagent** 2,4-dinitrofluorobenzene. This reacts with an amine by nucleophilic displacement of the fluorine.

2,4-dinitrofluorobenzene

S_NAr mechanism

Normally, substitution on a benzene ring is achieved by electrophilic attack, with subsequent loss of a proton (see Section 8.4). With the Sanger reagent, the presence of three strongly electron-withdrawing substituents allows nucleophilic attack and then displacement of fluoride as a leaving group. The initial addition of a nucleophile to the aromatic system generates a transient carbanion, which is stabilized by the nitro groups. Charge is then lost by expelling fluoride as a leaving group, restoring the aromatic ring system.

We have already noted that fluoride is not normally a very effective leaving group (see Section 6.1.4). Here, the nucleophilic addition is the rate-determining step, though it is favoured by the very large inductive effect from the fluorine and the stabilization from the nitro groups. This allows formation of the addition carbanion, and, even though fluoride is a poor leaving group, it can be lost from the anion to restore aromaticity. This type of reaction is strictly an addition–elimination mechanism, but is referred to as an S_NAr mechanism, or **nucleophilic aromatic substitution**.

use of Sanger reagent

2,4-dinitrofluorobenzene
(Sanger reagent)

peptide

mild base to remove HF | NaHCO₃

acid hydrolysis | H⁺

N-terminal amino acid carries 2,4-dinitrophenyl group

mixture of amino acids

After treatment of the peptide with the Sanger reagent, all peptide bonds are then cleaved by hydrolysis, giving a mixture of amino acids, with the N-terminal one carrying a 2,4-dinitrophenyl group. Being yellow, this compound is readily detected and can be characterized easily by chromatographic comparison with standards. Although 2,4-dinitrofluorobenzene will also react with any free amino group in an amino acid side-chain, e.g. that in

lysine, only the N-terminal amino acid will carry the 2,4-dinitrophenyl residue in its α-amino group.

A more useful procedure, in that it allows sequential determination of the N-terminal amino acids in a peptide, is the **Edman degradation**. This process removes the N-terminal amino acid, but leaves the rest of the chain intact, so allowing further reactions to be applied. The reagent used here is phenyl isothiocyanate.

Edman degradation

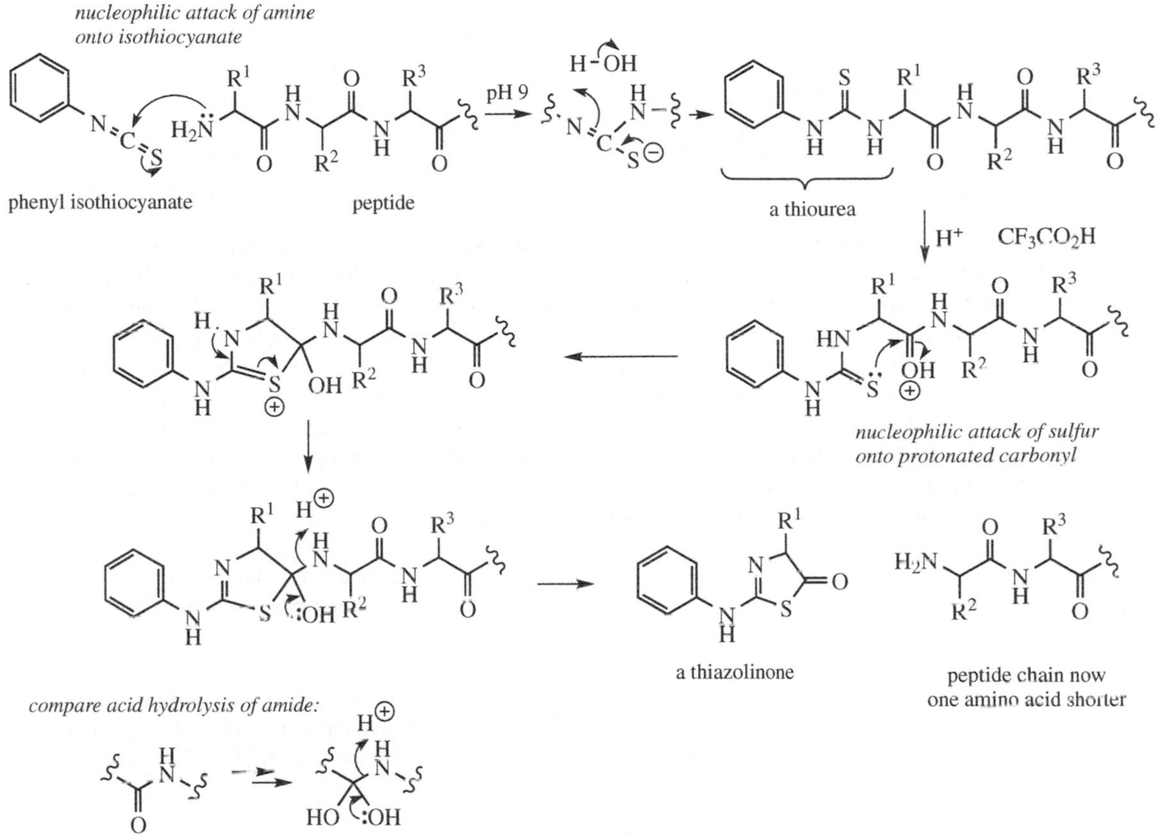

The carbon in the isothiocyanate grouping is highly susceptible to nucleophilic attack by the peptide's free amino group. Overall addition to the C=N creates a thiourea derivative. Making the conditions strongly acidic then promotes nucleophilic attack by the sulfur of the thiourea on to the carbonyl of the first peptide bond, producing a five-membered thiazoline heterocycle. Proton loss occurs from the nitrogen, and this creates an intermediate that is equivalent to the addition product in simple acid-catalysed amide

hydrolysis, though here we have employed a sulfur rather than an oxygen nucleophile. Bond cleavage follows, leaving the first amino acid as part of a thiazolinone system. The rest of the peptide chain is unaffected.

Thus, the N-terminal amino acid can be identified by analysis of the thiazolinone, and the process can be repeated on the one-unit-shortened polypeptide chain. Under the acidic conditions, the thiazolinone is actually unstable, and rearranges to a phenylthiohydantoin. The reasons for the rearrangement need not concern us; a mechanism is shown merely to demonstrate that it can be rationalized. The phenylthiohydantoin derivative produced can be identified simply by chromatographic comparison with authentic standards.

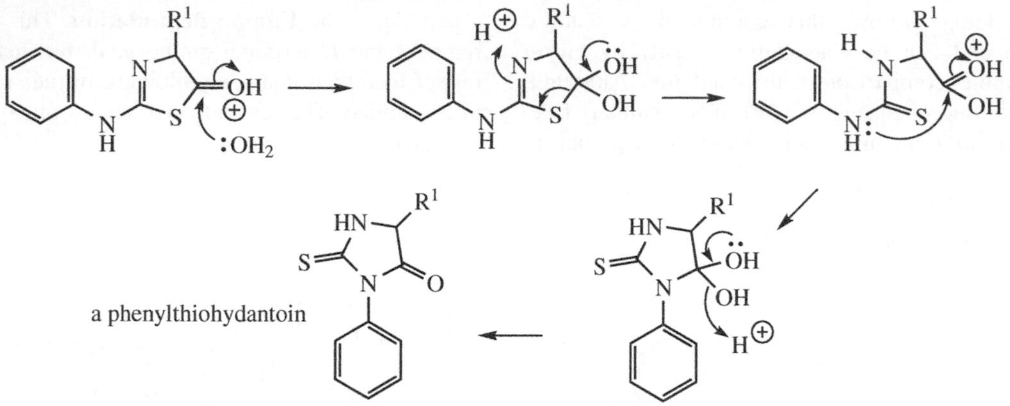

a phenylthiohydantoin

The repetitive cycle to identify a sequence of N-terminal amino acids has been automated. In practice, it is limited to about 20–30 amino acids, since impurities build up and the reaction mixture becomes too complex to yield unequivocal results. The usual approach is to break the polypeptide chain into smaller fragments by partial hydrolysis, preferably at positions relating to specific amino acid residues in the peptide chain. There are ways of doing this chemically, and the enzymes chymotrypsin and trypsin are also routinely used for this purpose (see Box 13.5). The shortened chains can then be sequenced and, with a little logic and reasoning, the order in which they are attached can be deduced, leading us to the entire amino acid sequence. The process can be exemplified using a simple hypothetical example containing 12 amino acid residues, although, in practice, this is small enough to be achieved by an automatic amino acid sequencer.

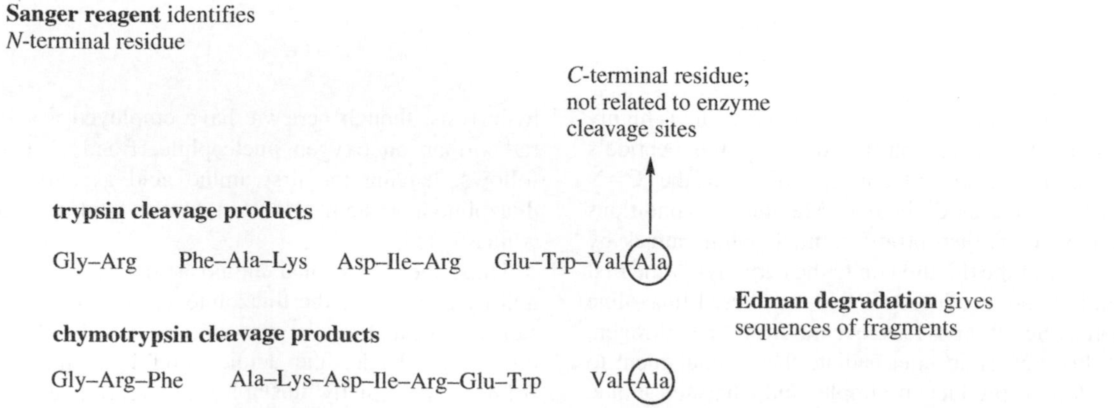

trypsin cleavage sites

Gly–Arg–Phe–Ala–Lys–Asp–Ile–Arg–Glu–Trp–Val–Ala

chymotrypsin cleavage sites

Sanger reagent identifies N-terminal residue

trypsin cleaves on C-terminal side of basic amino acids Arg, Lys

chymotrypsin cleaves on C-terminal side of aromatic amino acids Phe, Tyr, Trp, and to lesser extent Leu, Met, Asn, Gln

C-terminal residue; not related to enzyme cleavage sites

trypsin cleavage products

Gly–Arg   Phe–Ala–Lys   Asp–Ile–Arg   Glu–Trp–Val–Ala

Edman degradation gives sequences of fragments

chymotrypsin cleavage products

Gly–Arg–Phe   Ala–Lys–Asp–Ile–Arg–Glu–Trp   Val–Ala

The *N*-terminal amino acid can be ascertained by the Sanger method. Enzymic cleavage using either chymotrypsin or trypsin will break the peptide into smaller fragments. The fragments obtained will be different, depending on the enzyme and its specificity. The smaller fragments are then each sequenced by the Edman technique. *C*-Terminal residues in the smaller peptides can be related to knowledge of the enzyme cleavage sites; this may point to the *C*-terminal residue of the full peptide if it does not correspond to an enzymic cleavage site. The full sequence can be deduced from these fragments by lining up matching sequences of overlapping portions.

14

Nucleosides, nucleotides and nucleic acids

14.1 Nucleosides and nucleotides

The nucleic acids **DNA (deoxyribonucleic acid)** and **RNA (ribonucleic acid)** are the molecules that play a fundamental role in the storage of genetic information, and the subsequent manipulation of this information. They are polymers whose building blocks are **nucleotides**, which are themselves combinations of three parts: a heterocyclic base, a sugar, and phosphate. The most significant difference in the nucleotides comprising DNA and RNA is the sugar unit, which is **deoxyribose** in DNA and **ribose** in RNA. The term **nucleoside** is used to represent a nucleotide lacking the phosphate group, i.e. the base–sugar combination. The general structure of nucleotides and nucleosides is shown below.

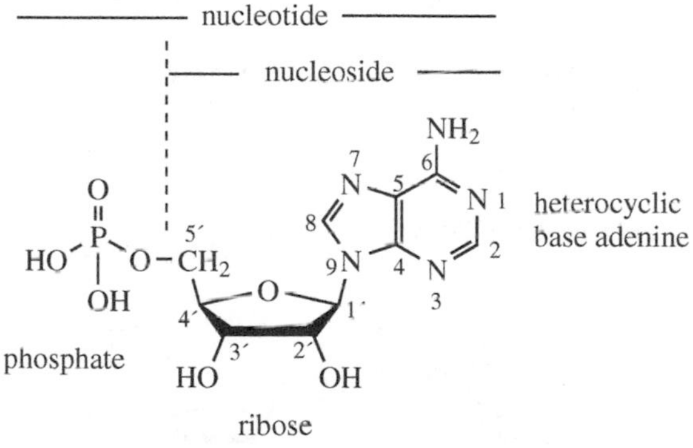

adenylic acid
(adenosine 5′-monophosphate, AMP)

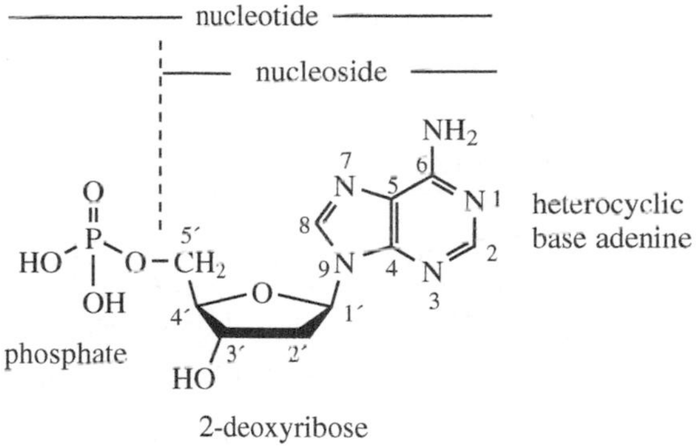

deoxyadenylic acid
(2′-deoxyadenosine 5′-monophosphate, dAMP)

Before we analyse nucleotide structure in detail, it is perhaps best that we consider the nature of the various component parts. In nucleic acid structures, there are five different bases and two different sugars.

Essentials of Organic Chemistry Paul M Dewick

bases

in both DNA and RNA		*in DNA only*	*in RNA only*

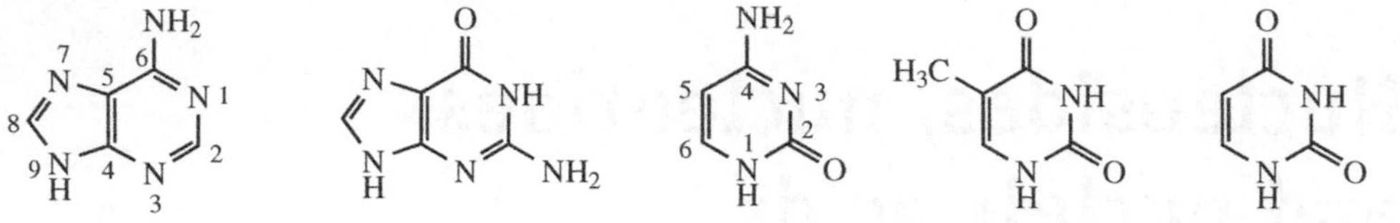

adenine, A   guanine, G   cytosine, C   thymine, T   uracil, U

purines   pyrimidines

The **bases** are monocyclic **pyrimidines** (see Box 11.5) or bicyclic **purines** (see Section 11.9.1), and all are aromatic. The two purine bases are **adenine** (A) and **guanine** (G), and the three pyrimidines are **cytosine** (C), **thymine** (T) and **uracil** (U). Uracil is found only in RNA, and thymine is found only in DNA. The other three bases are common to both DNA and RNA. The heterocyclic bases are capable of existing in more than one tautomeric form (see Sections 11.6.2 and 11.9.1). The forms shown here are found to predominate in nucleic acids. Thus, the oxygen substituents are in keto form, and the nitrogen substituents exist as amino groups.

sugars

in RNA only	*in DNA only*

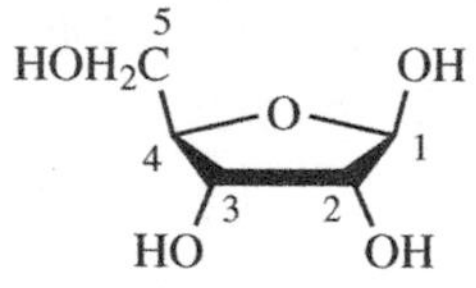

β-D-ribose   β-D-2-deoxyribose

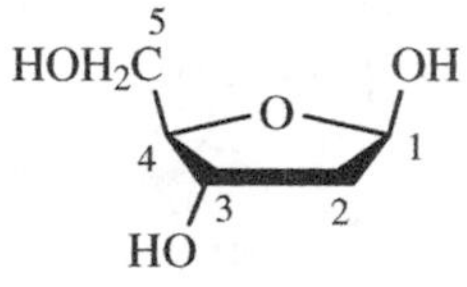

The two **sugars** are pentoses, D-**ribose** in RNA and **2-deoxy-D-ribose** in DNA. In all cases, the sugar is present in five-membered acetal ring form, i.e. a furanoside (see Section 12.4). The base is combined with the sugar through an *N*-glycoside linkage at C-1, and this linkage is always β. Purine bases are linked through N-9, and pyrimidines through N-1. When numbering nucleosides and nucleotides, we use primed numbers for the sugar, since non-primed numbers are already employed in the base part. There are thus four different nucleosides for each type of nucleic acid, as shown.

nucleosides in RNA

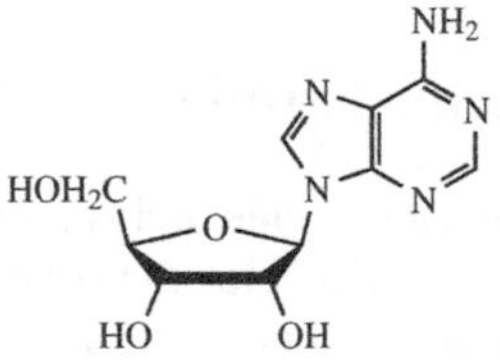

adenosine 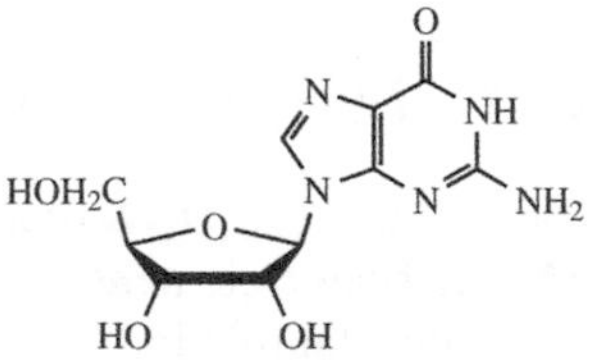   guanosine

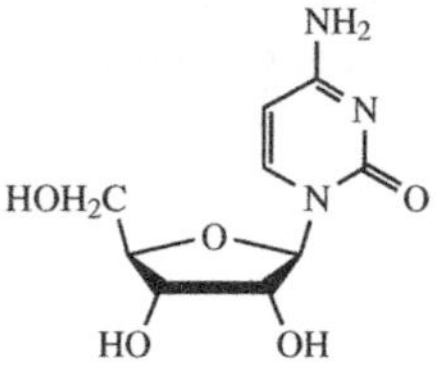

cytidine 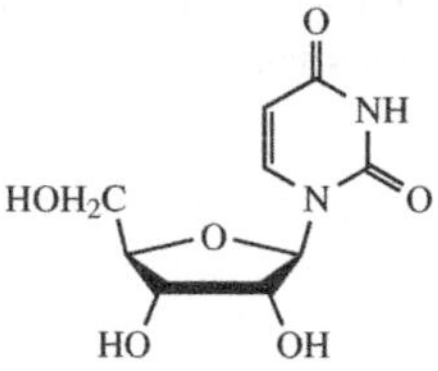   uridine

nucleosides in DNA

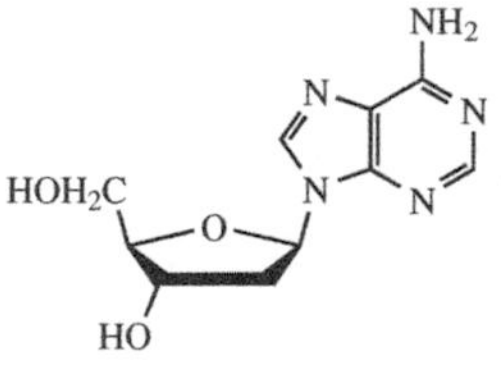

2′-deoxyadenosine 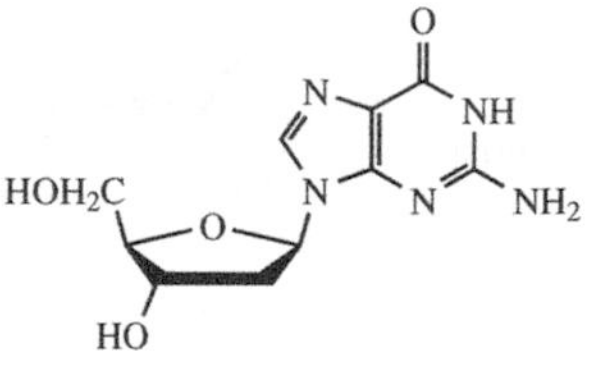   2′-deoxyguanosine

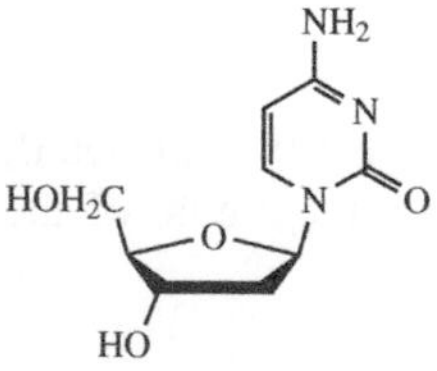

2′-deoxycytidine 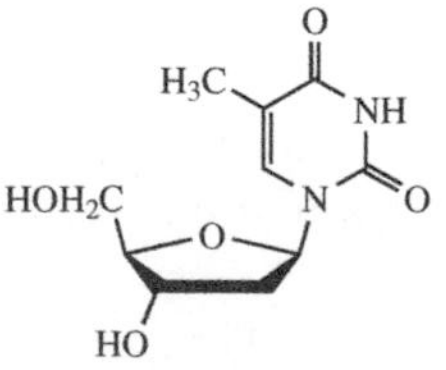   2′-deoxythymidine

The phosphate group of nucleotides is attached via a phosphate ester linkage, and may be attached to

Table 14.1 Nomenclature of bases, nucleosides, and nucleotides

Base	Ribonucleoside/ deoxyribonucleoside	Ribonucleotide/ deoxyribonucleotide (5′-monophosphate)
RNA		
Adenine (A)	Adenosine	Adenylate (AMP)
Guanine (G)	Guanosine	Guanylate (GMP)
Uracil (U)	Uridine	Uridylate (UMP)
Cytosine (C)	Cytidine	Cytidylate (CMP)
DNA		
Adenine (A)	Deoxyadenosine	Deoxyadenylate (dAMP)
Guanine (G)	Deoxyguanosine	Deoxyguanylate (dGMP)
Thymine (T)	Deoxythymidine	Deoxythymidylate (dTMP)
Cytosine (C)	Deoxycytidine	Deoxycytidylate (dCMP)

either C-5′ or C-3′. As we shall see, nucleosides in nucleic acids are joined together through a phosphate linkage between the 3′-hydroxyl of one sugar and the 5′-hydroxyl of another. As a result, hydrolysis of nucleic acid could give us nucleotides containing either 5′- or 3′-phosphate groups. It is usual, however, to consider nucleic acids as being composed of nucleotides that contain a 5′-phosphate group.

The accepted nomenclature for the various components in RNA and DNA is shown in Table 14.1.

14.2 Nucleic acids

14.2.1 DNA

The **nucleic acids** are composed of a long unbranched chain of **nucleotide** monomeric units. The nucleotides are linked together via the phosphate group, which joins the sugar units through ester linkages, usually referred to as a **phosphodiester bond**. The phosphodiester bond links the 5′ position of one sugar with the 3′ position of the next. A short portion of a DNA molecule is shown in Figure 14.1.

The nucleic acid chain is thus composed of alternating units of sugar and phosphate, with the bases appearing as side-chains from the sugar components. The nucleotide chain has ends, referred to as the 5′- and 3′-ends, according to the sugar hydroxyl that is available for further bonding. Though we shall not be considering this aspect further, in some organisms,

especially bacteria and some viruses, the two ends of the DNA chain are joined together so that we encounter a circular form of DNA. Nucleic acid structures generally need to be written in a much abbreviated form. The sugar–phosphate backbone is taken for granted; it can be indicated by a line, with the attached bases defined. Even this is tedious. It is thus reduced further to the sequence of attached bases. The **base sequence** of the nucleic acid is the standard way of defining its structure; strictly, the structure is a sequence of nucleotides. By convention, the base sequence is written from the 5′-end to the 3′-end, so that the short strand of DNA shown in Figure 14.1 would be given as –ACGT–.

Perhaps the most far-reaching feature of nucleic acids is the ability of the bases to **hydrogen bond** (see Box 2.2) to other bases. This property is fundamental to the **double helix** arrangement of the DNA molecule, and the translation and transcription via RNA of the genetic information present in the DNA molecule. The polymeric strand of DNA coils into a helix, and it is bonded to a second helical strand by hydrogen bonds between appropriate **base pairs**. In DNA, the base pairs are adenine–thymine and guanine–cytosine. It should be appreciated that each of these bases is planar, and that the hydrogen-bonded base pair is also planar. The hydrogen-bonded N–H–N and N–H–O interatomic distances are in the range 2.8–3.0 Å. By comparison, N–H and O–H bonds are typically about 1.0 Å.

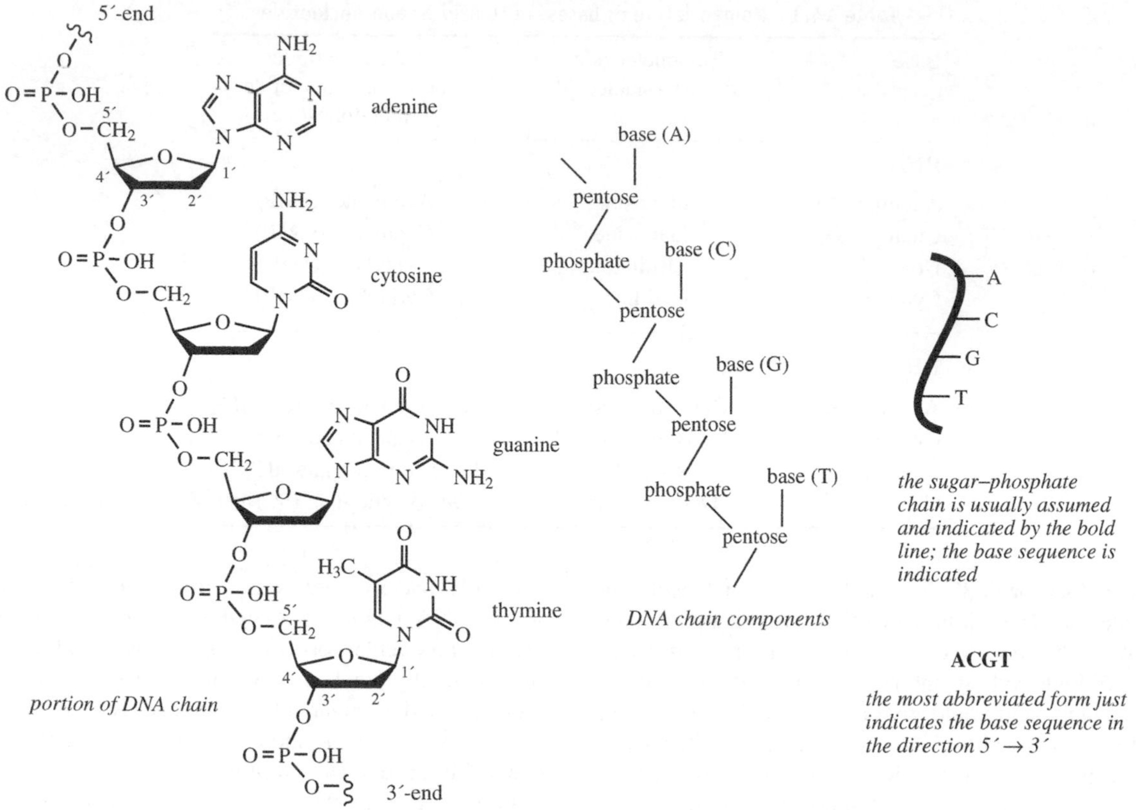

Figure 14.1 Portion of DNA molecule

Thus, each purine is specifically linked to a pyrimidine by either two or three hydrogen bonds. The result of these interactions is that each nucleotide recognizes and bonds with its complementary partner. This specific base-pairing means that the two strands in the DNA double helix are complementary. Wherever adenine appears in one strand, thymine appears opposite it in the other; wherever cytosine appears in one strand, guanine appears opposite it in the other. We shall see later the significance of base pairing

between adenine and uracil (see Section 14.2.3). The latter base is found in RNA instead of thymine.

The **DNA double helix** has both chains twisting on a common axis. The bases are directed inwards to allow hydrogen bonding, and the sugar and phosphodiester parts of the main chain form the outside portion. The planes of the base pairs are perpendicular to the helix axis, so that the molecule looks like a spiral staircase with the base-pair combinations forming the treads. The helix makes

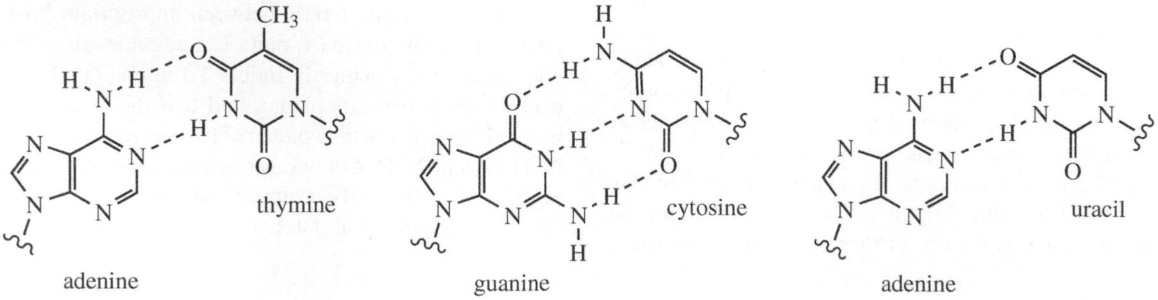

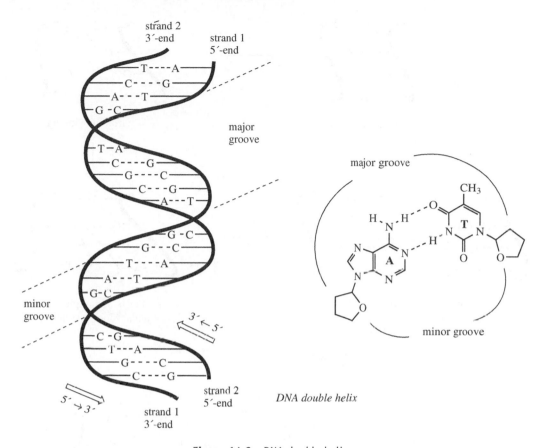

Figure 14.2 DNA double helix

a complete turn every 10 base pairs along the chain. The two strands are complementary: if you know the sequence along one chain, you can write down the sequence along the other via the base-pairing relationship. Note, however, that the chains are **antiparallel**, i.e. they run in opposite directions. This is indicated in the schematic diagram in Figure 14.2.

One further point arises because the glycoside bonds between the sugars and bases of a particular base pair are not directly opposite each other. This is easily appreciated from the illustrations of hydrogen-bonded base pairings. The consequence of this is that the grooves along the outside of the double helix array are of unequal width, and are termed the **major groove** and the **minor groove**. These grooves contain many water molecules through interaction with amino and carbonyl groups of the bases, and are distinguishable to agents that bind to DNA, e.g. some anticancer drugs.

14.2.2 Replication of DNA

During cell division, the DNA molecule is replicated so that each daughter cell will carry its own DNA molecule. During the process, the two strands of DNA unwind, and each strand then acts as the template for synthesis of a new strand; in each case, the new strand is complementary to the original because of the base-pairing restrictions (Figure 14.3).

Each new double helix is comprised of one strand that was part of the original molecule and one strand that is newly synthesized. Not surprisingly, this is a very simplistic description of a quite complex process, catalysed by enzymes known as **DNA polymerases**. The precursors for synthesis of the new chain are the **nucleoside triphosphates**, dATP, dGTP, dTTP, and dCTP. We have already met ATP when we considered anhydrides of phosphoric acid (see Box 7.25); these compounds are analogues of ATP, though the sugar is deoxyribose rather than ribose.

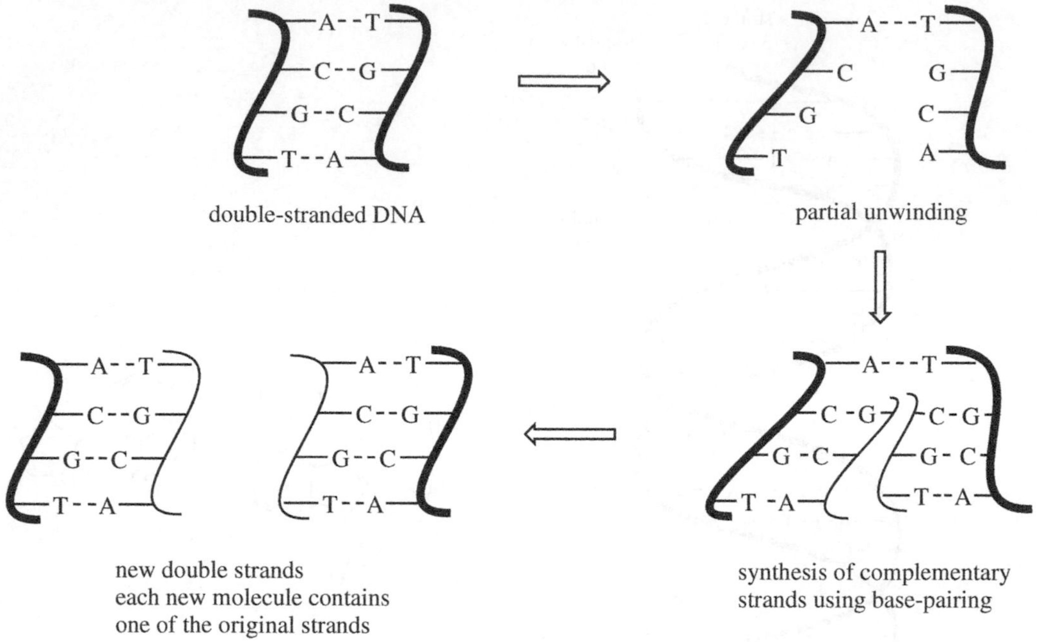

double-stranded DNA

partial unwinding

new double strands
each new molecule contains
one of the original strands

synthesis of complementary
strands using base-pairing

Figure 14.3 Replication of DNA

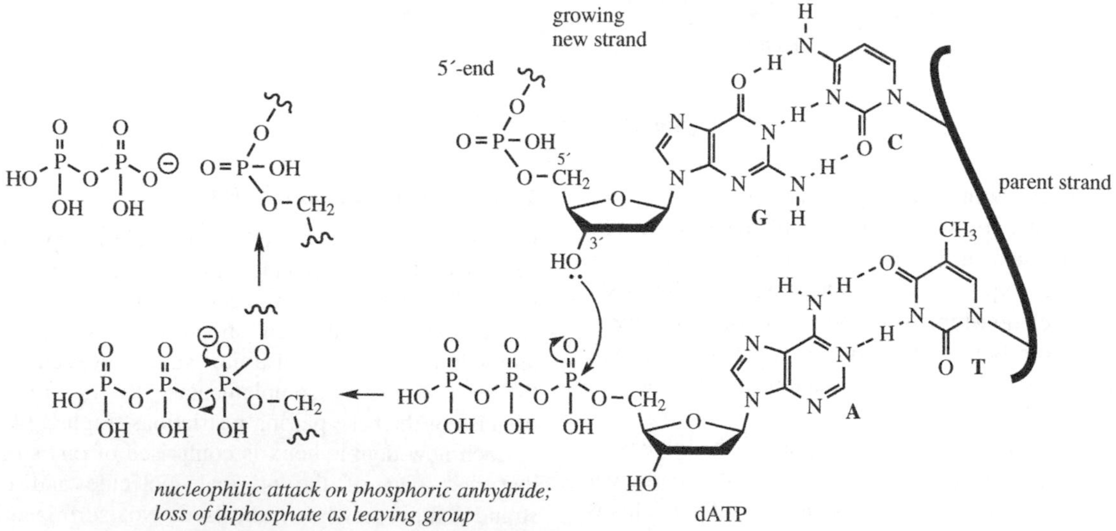

growing
new strand

5′-end

parent strand

G

C

A

T

dATP

nucleophilic attack on phosphoric anhydride;
loss of diphosphate as leaving group

These **triphosphate anhydrides** are susceptible to nucleophilic attack by hydroxyl groups. Chain extension is simply an **esterification** reaction utilizing the 3′-hydroxyl of the sugar in the growing chain, with diphosphate as a good leaving group. The correct nucleoside triphosphate is selected because of the hydrogen bonding properties of base pairs. This also provides the correct alignment so that the reaction can occur. In the illustration, the next base in the original DNA strand is thymine, which dictates that only an adenine nucleotide can hydrogen bond and form the complementary base pair. The esterification occurs, with loss of diphosphate as leaving group, and the new daughter strand is extended by one nucleotide. The process repeats as the enzyme moves on to the next position on the original DNA strand. Hydrolysis of diphosphate to two molecules of phosphate provides some of the driving force to facilitate the reaction (compare Box 14.3 and Section 15.4.1).

14.2.3   RNA

RNA differs structurally from DNA in three important ways. First, as indicated above, the sugar in RNA is **ribose**, not 2-deoxyribose. Second, **thymine** is replaced by uracil, so that the four bases are adenine, uracil, guanine, and cytosine. The third difference is that RNA is usually single stranded. Although an RNA molecule may be single stranded, it does not exclude the possibility of partial double-stranded sequences being present. In such cases, the molecule doubles back on itself, and coils up with a complementary base sequence elsewhere. Remember that complementary sequences now involve A–U rather than A–T hydrogen-bonding interactions.

DNA stores the genetic information for a cell, but it is RNA that participates in the processes by which this information is used. RNA molecules are classified according to their function or cellular location. Three major forms are found in prokaryotic cells:

- **messenger RNA** (mRNA) carries genetic information from DNA to ribosomes, the organelles responsible for protein synthesis;

- **ribosomal RNA** (rRNA) is an integral part of the ribosomes;

- **transfer RNA** (tRNA) carries the amino acid residues that are added to the growing peptide chain during protein synthesis.

14.2.4   The genetic code

It is the sequence of bases along one of the strands of the DNA molecule, the **coding strand**, that provides the information for the synthesis of proteins, especially enzymes, in an organism. A complementary sequence exists along the second strand, and this is termed the **template strand**. A **gene** is that segment of DNA that contains the information necessary for the synthesis of one protein.

Each amino acid in a protein is specified by a sequence of three nucleotides, termed a codon. A **codon** is usually designated in terms of the base sequence, however, just as we saw with nucleic acid sequences above. With four different bases, there are $4^3 = 64$ different combinations of three bases (codons) available, more than enough for the 20 different amino acids found in proteins (see Section 13.1). Most amino acids can be specified by two or more different codons, and three particular codons are known to carry the signal for stop, i.e. chain termination. The signal for start is the same as for methionine (unusual in having only one codon rather than several), and means that all proteins should begin with a methionine residue. Since this is obviously not the case, the inference is that many proteins are subsequently modified by cleaving off a fragment that contains this starter amino acid residue.

The codon combinations are shown in Table 14.2. A codon can be the DNA sequence in the coding strand or, alternatively, the related sequence found in mRNA. The table shows the mRNA sequences, since we shall be using these during consideration of protein synthesis. The DNA sequences merely have thymine (T) in place of uracil (U), as appropriate. The sequence is always listed from the 5'-end to the 3'-end.

Table 14.2   The genetic code: mRNA sequences

First position, 5'-end	Second position				Third position, 3'-end
	U	C	A	G	
U	Phe	Ser	Tyr	Cys	U
	Phe	Ser	Tyr	Cys	C
	Leu	Ser	STOP	STOP	A
	Leu	Ser	STOP	Trp	G
C	Leu	Pro	His	Arg	U
	Leu	Pro	His	Arg	C
	Leu	Pro	Gln	Arg	A
	Leu	Pro	Gln	Arg	G
A	Ile	Thr	Asn	Ser	U
	Ile	Thr	Asn	Ser	C
	Ile	Thr	Lys	Arg	A
	Met	Thr	Lys	Arg	G
G	Val	Ala	Asp	Gly	U
	Val	Ala	Asp	Gly	C
	Val	Ala	Glu	Gly	A
	Val	Ala	Glu	Gly	G

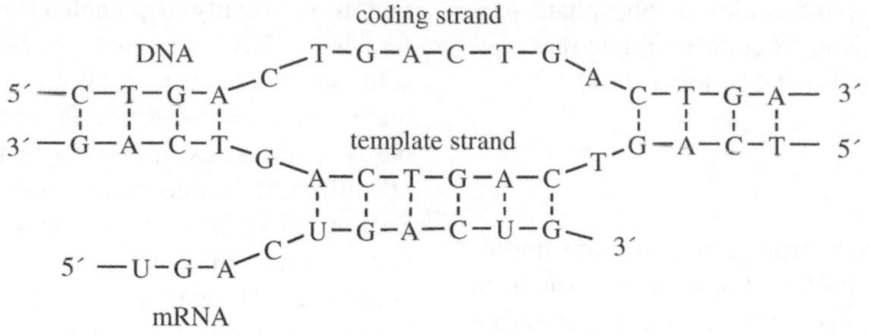

Figure 14.4   Transcription of DNA to mRNA

14.2.5   Messenger RNA synthesis: transcription

Although the amino acid sequence of a protein is defined by the sequence of codons in DNA, it is RNA that participates in the interpretation of this sequence and the subsequent joining together of amino acids. The process starts with the synthesis of **mRNA**, in a process called **transcription** (Figure 14.4).

Part of the DNA double helix, corresponding to the gene in question, is unwound. Rather like in the replication of DNA, the base sequence is used to synthesize a new nucleic acid strand. However, this time only one strand is interpreted, the **template strand**, and ribonucleotides (ATP, GTP, CTP, and UTP) are used in the new chain assembly instead of deoxyribonucleotides. The sequence of ribonucleotides incorporated is dictated by the sequence of nucleotides in DNA, and depends on hydrogen bonding between pairs of bases. In RNA synthesis, uracil nucleotides are employed rather than thymine nucleotides. The result is synthesis of a single strand of RNA with a sequence analogous to the coding strand of DNA, except that U replaces T. Coupling of the ribonucleotide units is catalysed by the enzyme **RNA polymerase**, and is mechanistically the same as with DNA replication above, i.e. esterification of a hydroxyl via a phosphoric anhydride.

14.2.6   Transfer RNA and translation

Although messenger RNA is synthesized in the cell nucleus, it then moves to the cytoplasm and to the **ribosomes**, where protein biosynthesis occurs. These particles are composed of two subunits, termed 50S and 30S, and are combinations of **rRNA** and protein. The ribosomes are responsible for binding the two other types of RNA, mRNA (which contains the genetic code) and **tRNA** (which carries the individual amino acids). tRNA molecules are very small compared with the other forms of RNA, being less than 100 nucleotides. The size of mRNA reflects the number of amino acid residues in the protein being synthesized, but could be a thousand or more nucleotides. rRNA is the most abundant of the three types of RNA, and in size covers a range from about 75 to 3700 nucleotides.

A tRNA molecule is specific for a particular amino acid, though there may be several different forms for each amino acid. Although relatively small, the polynucleotide chain may show several loops or arms because of base pairing along the chain. One arm always ends in the sequence cytosine–cytosine–adenosine. The 3'-hydroxyl of this terminal adenosine unit is used to attach the amino acid via an ester linkage. However, it is now a section of the nucleotide sequence that identifies the tRNA–amino acid combination, and not the amino acid itself. A loop in the RNA molecule contains a specific sequence of bases, termed an anticodon, and this sequence allows the tRNA to bind to a complementary sequence of bases, a codon, on mRNA. The synthesis of a protein from the message carried in mRNA is called **translation**, and a simplified representation of the process as characterized in the bacterium *Escherichia coli* is shown below.

Initially, the amino acid is activated by an ATP-dependent process, producing an aminoacyl-AMP. A hydroxyl group in ribose, part of a terminal adenosine group of tRNA, then reacts with this mixed anhydride. In this way, the amino acid is bound to tRNA via an ester linkage as an aminoacyl-tRNA. The tRNA involved will be specific for the particular amino acid. A detailed mechanism for this process has been considered in Section 13.5.1.

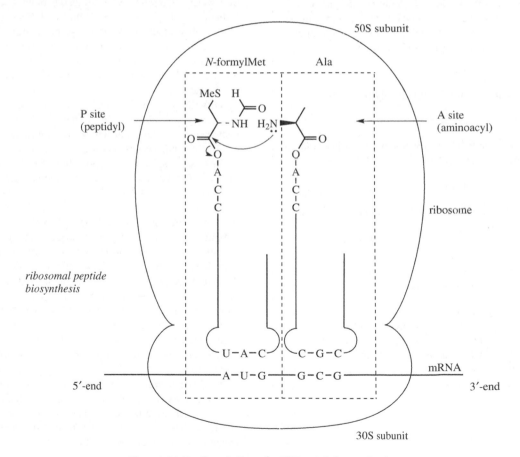

activation of amino acid by formation of mixed anhydride

aminoacyl group transferred to ribose hydroxyl in 3′-terminal adenosine of tRNA; formation of ester linkage

aminoacyl-AMP

aminoacyl-tRNA

The mRNA is bound to the smaller 30S subunit of the bacterial ribosome. The mRNA is a transcription of one of the genes of DNA, and carries the information as a series of three-base codons. The message is read (translated) in the 5′ to 3′ direction along the mRNA molecule. The aminoacyl-tRNA anticodon (UAC) allows binding via hydrogen bonding to the appropriate codon (AUG) on mRNA. In prokaryotes, the first amino acid encoded in the sequence is *N*-formylmethionine (fMet). Although the codon for initiation (*N*-formylmethionine) is the same as

that for methionine (see Section 14.2.4), the initiator tRNA used is different to that employed for the incorporation of methionine elsewhere in the peptide chain. The initiator aminoacyl-tRNA is thus bound and positioned at the P (for peptidyl) site on the ribosome (Figure 14.5).

The next aminoacyl-tRNA (Figure 14.5 shows a tRNA specific for alanine) is also bound via a codon (GCG)–anticodon (CGC) interaction and is positioned at an adjacent A (for aminoacyl) site on the ribosome. This allows peptide bond formation to

Figure 14.5   Translation of mRNA: protein synthesis

occur, the amino group of the amino acid in the A site attacking the activated ester in the P site. The peptide chain is thus initiated and has become attached to the tRNA located in the A site. The tRNA at the P site is no longer required and is released from the ribosome.

Then the peptidyl-tRNA at the A site is translocated to the P site by the ribosome moving along the mRNA a codon at a time, exposing the A site for a new aminoacyl-tRNA appropriate for the particular codon, and a repeat of the elongation process occurs. The cycles of elongation and **translocation** continue until a termination codon is reached, and the peptide or protein is then hydrolysed and released from the ribosome. Note that the protein is synthesized from the N-terminus towards the C-terminus (see Section 13.2).

Some special features of proteins are elaborated by **secondary transformations** that are not part of the translation process. The N-formylmethionine initiator may be hydrolysed to methionine, or, as we have already indicated, the methionine unit may be removed altogether. Other **post-translational changes** to individual amino acids may be seen, e.g. the hydroxylation of proline to hydroxyproline (see Section 13.1) or the generation of disulfide bridges between cysteine residues (see Section 13.3).

Box 14.1

Antibiotics that interfere with ribosomal peptide biosynthesis

Many of the antibiotics used clinically are active by their ability to **inhibit protein biosynthesis** in bacteria. The individual steps of protein biosynthesis all seem susceptible to disruption by specific agents. Some specific examples are listed below:

Inhibitors of transcription

rifampicin (inhibits RNA polymerase)

Inhibitors of aminoacyl-tRNA binding to ribosome

tetracyclines (bind to 30S subunit of ribosome and prevent attachment of aminoacyl-tRNA)

Inhibitors of translation

streptomycin (binds to 30S subunit of ribosome, causes mRNA to be misread)
erythromycin (binds to 50S subunit of ribosome, inhibits translocation)
chloramphenicol (binds to 50S subunit, inhibits peptidyltransferase activity)

Naturally, if such materials are going to be useful as antibiotic drugs, we require a selective action. We need to be able to inhibit protein biosynthesis in bacteria, whilst producing no untoward effects in man or animals. Although the mechanisms for protein biosynthesis are essentially the same in prokaryotes and eukaryotes, there are some subtle differences, e.g. in the nature of the ribosome and how the process is initiated. Without such differences, the agent would be toxic to man as well as to bacteria.

Box 14.2

Nucleosides as antiviral agents

Viruses are responsible for many human and animal diseases, with a variety of symptoms and levels of severity. Common viral illnesses include colds, influenza, cold-sores (herpes), and childhood infections such as chickenpox, measles, and mumps. More serious conditions include meningitis, poliomyelitis, and human immunodeficiency virus (HIV), the latter potentially leading to acquired immune deficiency syndrome (AIDS).

Viruses are simpler than bacteria and consist essentially of nucleic acid (either DNA or RNA) enclosed in a protein coat. Those causing chickenpox, smallpox, and herpes belong to the DNA virus group, whereas those responsible for influenza, measles, mumps, meningitis, poliomyelitis, and HIV are classified as RNA viruses. Viruses have no metabolic machinery of their own, and for their very existence are intracellular parasites of other organisms. To survive and reproduce, they have to tap into the metabolic processes of the host organism. For this reason, it is difficult to find drugs that are selective towards viruses without damaging the host. Most antiviral agents are only effective whilst the virus is replicating, and viral replication is very far advanced by the time the infection is detectable. There are relatively few effective antiviral drugs, and most of these are nucleoside derivatives.

Aciclovir (acyclovir) was one of the first effective selective antiviral agents. It is a guanine derivative of value in treating herpes viruses, though it does not eradicate them, and is only useful if drug treatment is started at the onset of infection.

2'-deoxyguanosine

aciclovir

Aciclovir is a member of a group of nucleoside derivatives termed **acyclonucleosides**, in that there is an incomplete sugar ring. The structural relationship to 2'-deoxyguanosine should be very clear. Aciclovir is converted into its monophosphate by the viral enzyme thymidine kinase – some viruses also possess enzymes that facilitate their replication in the host cell. The viral enzyme turns out to be much more effective than that of the host cell, and conversion is, therefore, mainly in infected cells. The monophosphate is subsequently converted into the triphosphate by the host cell enzymes. Aciclovir triphosphate inhibits viral DNA polymerase, much more so than it does the host enzyme, and so terminates DNA replication.

Zidovudine is 3'-azido-3'-deoxythymidine, and is a derivative of deoxythymidine in which an azide group replaces the 3'-hydroxyl. It is better

2'-deoxythymidine

zidovudine; AZT

known as the anti-AIDS drug **AZT**. The AIDS virus is an RNA retrovirus. In **retroviruses**, an enzyme **reverse transcriptase** makes a DNA copy of viral RNA (contrast transcription: making an RNA copy of DNA). This DNA copy is then integrated into the host genome, and gets transcribed into both new viral RNA and mRNA for translation into viral proteins. AZT is an inhibitor of reverse transcriptase.

AZT is phosphorylated by cellular enzymes to the triphosphate, which competes with normal substrates for formation of DNA by reverse transcriptase, and blocks viral DNA synthesis. Mammalian DNA polymerase is relatively unaffected, but there can be some toxic effects. AZT is used in AIDS treatment along with other antiretroviral drugs.

14.3 Some other important nucleosides and nucleotides: ATP, SAM, Coenzyme A, NAD, FAD

The terminology nucleotide or nucleoside immediately directs our thoughts towards nucleic acids. Remarkably, nucleosides and nucleotides play other roles in biochemical reactions that are no less important than their function as part of nucleic acids. We also encounter more structural diversity. It is rare that the chemical and biochemical reactivities of these derivatives relate specifically to the base plus sugar part of the structure, and usually reside elsewhere in the molecule. Almost certainly, it is this base plus sugar part of the structure that provides a recognition

feature for the necessary enzymes that utilize these compounds.

adenosine triphosphate; ATP

Ad = adenosine

S-adenosylmethionine
(SAM)

ATP, adenosine triphosphate, provides the currency unit for energy in biochemical reactions (see Section 15.1.1) and is simply a triphosphate variant of a standard RNA nucleotide. It is, of course, the biosynthetic precursor for adenine-based units in RNA (see Section 14.2.5). As we have already seen (see Box 7.25), the functions of ATP can be related to hydrolytic reactions in the triphosphate (anhydride) part of the molecule.

SAM, S-adenosylmethionine, has been encountered as a biological **methylating agent**, carrying out its function via a simple S_N2 reaction (see Box 6.5). This material is a nucleoside derivative formed by nucleophilic attack of the thiol group of methionine on to ATP (see Box 6.5). It provides in its structure an excellent leaving group, the neutral S-adenosylhomocysteine.

Coenzyme A
HSCoA

Coenzyme A is another adenine nucleotide derivative, with its primary functional group, a thiol, some distance away from the nucleotide end of the molecule. This thiol plays an important role in biochemistry via its ability to form thioesters with suitable acyl compounds (see Box 7.18). We have seen how **thioesters** are considerably more reactive than oxygen esters, with particular attention being paid to their improved ability to form enolate anions, coupled with thiolates being excellent leaving groups (see Box 10.8).

Nature's oxidizing agents **NAD⁺** and **NADP⁺**, and the corresponding reducing agents **NADH** and **NADPH**, are all dinucleotide derivatives (see Boxes 11.2 and 7.6). Indeed, the full names betray this: NAD is **nicotinamide adenine dinucleotide**. From the structures of nucleic acids, one interprets a dinucleotide as a repeated nucleotide. This would have two bases attached to a chain that reads phosphate–sugar–phosphate–sugar: a phosphodiester linkage. Note that these NAD derivatives have a sugar–phosphate–phosphate–sugar sequence, a broader interpretation of dinucleotide terminology. The reactive centre in these compounds relates to the pyridine ring in nicotinamide, which is capable of accepting or donating hydride equivalents according to its oxidation state. We have seen that, in biochemical reactions, NADH and NADPH may be considered analogues of complex metal hydride reagents (see Box 7.6). Here is our first example, then, of a nucleotide where the base, **nicotinamide**, is different from those in nucleic acids.

FAD shares a lot of features with NAD⁺ and NADP⁺, but contains two new variants: a sugar that is neither ribose nor deoxyribose, and a fairly complex heterocyclic base flavin. The new sugar is **ribitol**, non-cyclic because it contains no carbonyl group (see Section 12.3). The chemistry of FAD is concentrated in the flavin part, and features oxidation/reduction processes (see Box 11.14).

FMN, flavin mononucleotide, is simply the flavin-containing structure from the dinucleotide FAD.

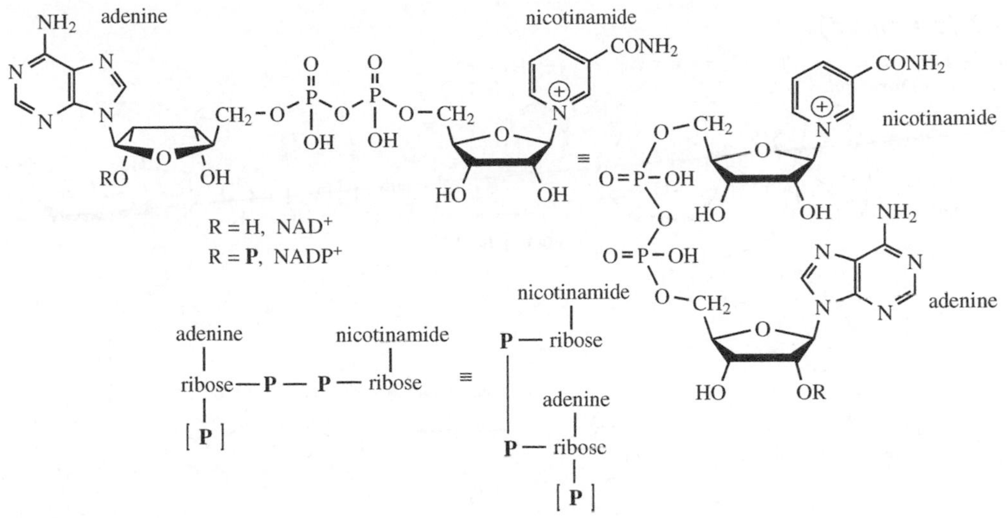

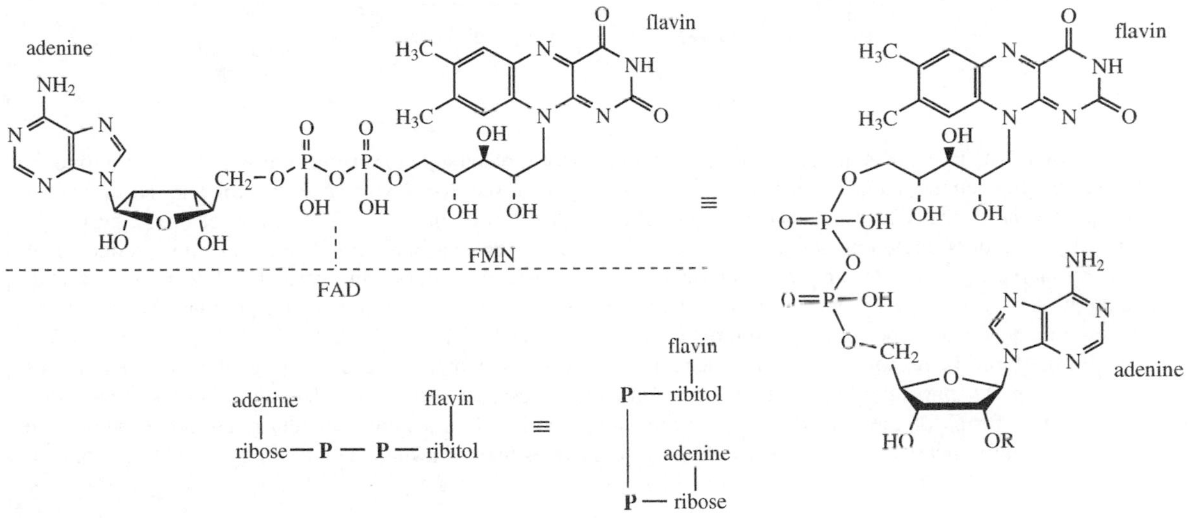

R = H, NAD$^+$
R = P, NADP$^+$

Box 14.3

Cyclic AMP

The nucleotide **cyclic AMP** (3',5'-cyclic adenosine monophosphate, **cAMP**) is a cyclic phosphate ester of particular biochemical significance. It is formed from the triester ATP by the action of the enzyme **adenylate cyclase**, via nucleophilic attack of the ribose 3'-hydroxyl onto the nearest P=O group, displacing diphosphate as leaving group. It is subsequently inactivated by hydrolysis to 5'-AMP through the action of a phosphodiesterase enzyme.

Box 14.3 (continued)

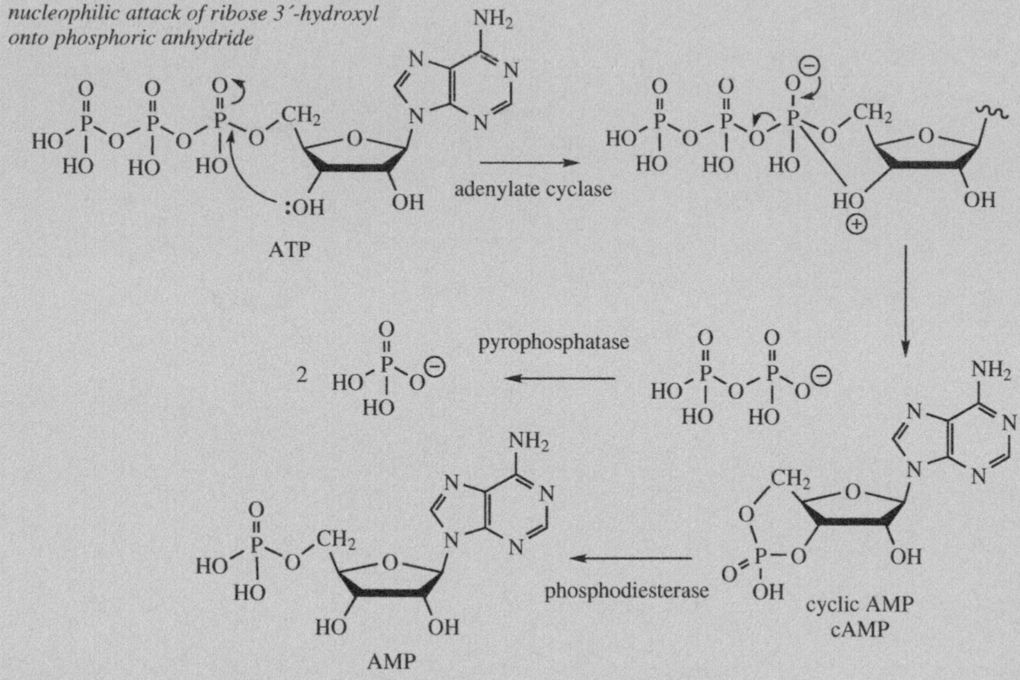

nucleophilic attack of ribose 3′-hydroxyl onto phosphoric anhydride

cAMP functions in cells as a second messenger, a mediator molecule that transmits the signal from a hormone. Other second messengers identified include Ca^{2+}, prostaglandins (see Box 9.3), diacyl glycerol, and the equivalent cyclic phosphate derivative of guanosine, cyclic GMP. cAMP is the mediator for a variety of drugs, hormones and neurotransmitters, including adrenaline, glucagon, calcitonin, and vasopressin. Such compounds produce their effects by increasing or decreasing the catalytic activity of adenylate cyclase, thus raising or lowering the cAMP concentration in a cell. A pyrophosphatase activity rapidly removes the other reaction product, disturbing the equilibrium, and making the reaction unidirectional (see Section 14.2.2).

cAMP, in turn, is responsible for the activation of various protein kinases that regulate the activity of cellular proteins by phosphorylation of serine and threonine residues using ATP. The phosphorylated and non-phosphorylated forms of the enzymes catalyse the same reaction, but at quite different rates; often, one of the forms is essentially inactive. This means the activity of enzymes may be switched on or off by addition or removal of phosphate groups, and can thus be controlled by hormones.

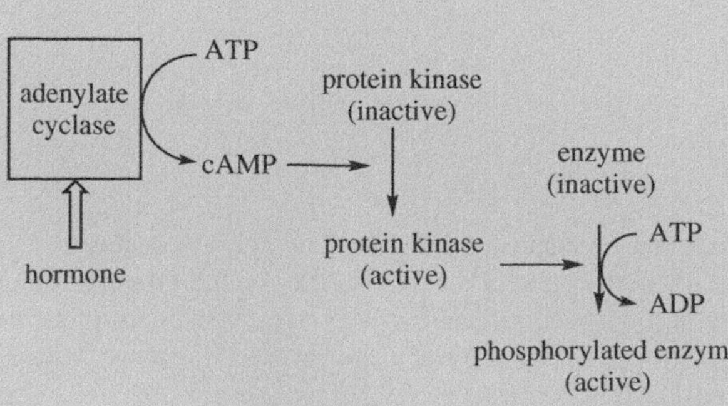

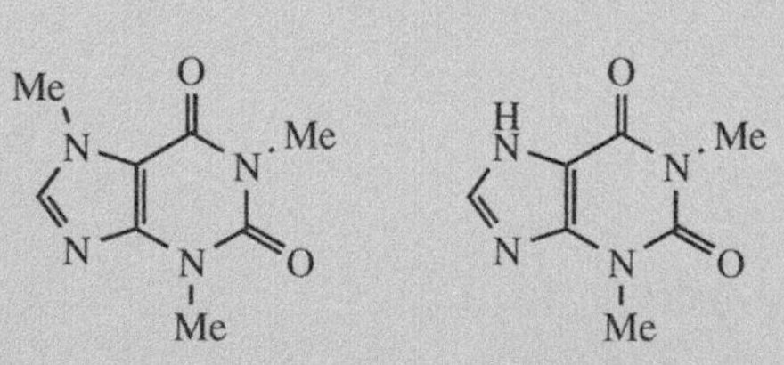

Caffeine in tea and coffee inhibits the phosphodiesterase that degrades cAMP. The resultant increase in cAMP levels, therefore, mimics the action of mediators such as the catecholamines that modulate adenylate cyclase. Caffeine and the related **theophylline** (both purine alkaloids, see Box 11.12) are thus effective stimulants of the CNS.

14.4 Nucleotide biosynthesis

Nucleic acids are synthesized in nature from nucleoside triphosphates, which are coupled by a chain extension process. We have seen that coupling is simply an esterification reaction utilizing the 3′-hydroxyl of the sugar of the growing chain, with diphosphate as a good leaving group (see Section 14.2.2). Nucleoside triphosphates, especially ATP, have other major biochemical roles (see Section 15.1.1). A full discussion of the origins of these compounds is outside our requirements, but there are some features of particular interest pertinent to our understanding of these compounds. One of these is that several of the biosynthetic reactions require the involvement of ATP, demonstrating that nucleotide production requires input from other nucleotides. Another interesting aspect is the quite different approach nature adopts for synthesis of pyrimidine or purine nucleotides.

Pyrimidine nucleotides are made by adding a preformed pyrimidine ring to the sugar phosphate. On the other hand, the purine ring of purine nucleotides is built up gradually, and assembly occurs with the growing ring attached to the sugar phosphate.

A common intermediate for all the nucleotides is **5-phosphoribosyl-1-diphosphate** (PRPP), produced by successive ATP-dependent phosphorylations of ribose. This has an α-diphosphate leaving group that can be displaced in S_N2 reactions. Similar S_N2 reactions have been seen in glycoside synthesis (see Section 12.4) and biosynthesis (see Box 12.4), and for the synthesis of aminosugars (see Section 12.9). For **pyrimidine nucleotide biosynthesis**, the nucleophile is the 1-nitrogen of uracil-6-carboxylic acid, usually called **orotic acid**. The product is the nucleotide orotidylic acid, which is subsequently decarboxylated to the now recognizable **uridylic acid** (UMP).

Formation of UTP requires successive phosphorylations using ATP. CTP is, in turn, formed from UTP by an amination reaction in the pyrimidine ring, with the amino acid glutamine supplying the nitrogen; this is also an ATP-dependent reaction.

Glutamine also supplies an amino function to start off **purine nucleotide biosynthesis**. This complex little reaction is again an S_N2 reaction on PRPP, but only an amino group from the amide of glutamine is transferred. The product of the enzymic reaction is thus **5-phosphoribosylamine**.

The amino group now provides the nucleus for purine ring formation, an extended series of reactions we shall not describe. The first-formed purine product is **inosine 5'-phosphate** (IMP), which leads to either AMP or GMP; these require amination at alternative sites, and utilize either GTP- or ATP-dependent reactions for amination. GTP or ATP (as appropriate) will also be required for further phosphorylations to produce the nucleotide triphosphates.

2'-Deoxyribonucleotides are generally formed by reduction of ribonucleoside diphosphates. This involves a series of redox reactions in which $NADP^+$ and FAD play a role (see Section 15.1.1), with a subsequent electron transport chain. DNA contains thymine rather than uracil, so **thymidine triphosphate** (dTTP) is a requirement. Methylation of dUMP to dTMP is a major route to thymine nucleotides, and is dependent upon N^5, N^{10}-methylenetetrahydrofolate as the source of the methyl group (see Box 11.13).

14.5 Determination of nucleotide sequence

14.5.1 Restriction endonucleases

Natural DNA molecules are extremely large, and for sequence determination it is necessary to cleave them into manageable fragments. This may be accomplished by using enzymes, called **restriction endonucleases**, that are obtained mainly from bacterial sources. These enzymes appear to have developed so that a cell can destroy foreign, particularly viral, DNA. The enzymes, of which several hundred are available, cleave the DNA at specific points in the chain, dictated by a series of nucleotides, typically three to six nucleotides. For example, the enzyme *Eco*RI (from *Escherichia coli*) cleaves a GAATTC sequence between G and A. Also important is the property that most restriction endonucleases cleave both strands of DNA, because the recognition sequence reads the same both ways: the complementary strand to GAATTC (in 5' → 3' direction) is CTTAAG (in 3' → 5' direction) (see Section 14.2.1).

The restriction endonuclease recognizes a specific sequence, but the probability of these sequences occurring in a given DNA molecule is usually quite low; therefore, cleavage produces only a few fragments. Use of a different enzyme on the same DNA will produce different fragments, but there then will be overlap of sequences. Hence, sequencing of both sets of fragments should allow the full sequence to be deduced. This deductive approach is thus similar to that used in amino acid sequencing (see Section 13.7).

14.5.2 Chemical sequencing

Before separation, double-stranded restriction fragments are labelled chemically, by attaching a radioactive or fluorescent marker to the 5'-end of the chain. For example, radioactive ^{32}P-labelled phosphate may be added using labelled ATP in an enzymic reaction. The labelled fragments are then separated chromatographically using conditions that are known to cause strand separation into single-stranded DNA molecules. The separated fragments are then split into four portions, and each portion is treated chemically with a suitable reagent. The reagent needs to be one that induces cleavage reactions, but which shows selectivity for the different nucleotides. Now this could potentially lead to almost total cleavage, but the trick is to use reagents at concentrations so low that, statistically, only one cleavage occurs per chain.

The reagents are **dimethyl sulfate** and **hydrazine** (only two reagents, but read on), and though we shall not consider the full mechanisms of the reactions here, they may be summarized as follows:

- Me_2SO_4, then aqueous piperidine; cleavage at G;
- Me_2SO_4 and aqueous formic acid, then aqueous piperidine; cleavage at A and G;
- aqueous hydrazine (H_2NNH_2), then aqueous piperidine; cleavage at C and T;
- aqueous hydrazine (H_2NNH_2) and NaCl, then aqueous piperidine; cleavage at C.

Dimethyl sulfate is an effective methylating agent (see Section 7.13.1). Methylation of the purine rings in guanine and adenine makes them susceptible to hydrolysis and subsequent rupture. This, in turn, makes the glycosidic bond vulnerable to attack, and the heterocycle is displaced from the phosphodiester. The phosphodiester bond can then cleaved by basic hydrolysis (aqueous piperidine).

Guanine is methylated on the imidazole ring at N-7, whereas adenine undergoes methylation at N-3. Under the conditions used, guanine is methylated more readily than adenine; therefore, cleavage of the DNA occurs predominantly where there a guanine residues. However, by treating the methylated DNA with acid, cleavage at the methylated adenine sites becomes enhanced, and the chain is broken at sites that originally contained either adenine or guanine.

The pyrimidines cytosine and thymine both react with **hydrazine**, which initially attacks the unsaturated carbonyl system and then leads to ring opening. Again, base treatment is used to hydrolyse the phosphodiester bond. This reaction becomes selective for cytosine in the presence of NaCl, which suppresses reaction with thymine.

The reaction products from the four reactions are then separated by gel electrophoresis in parallel lanes. This procedure will separate the components according to their charge (mainly from phosphate groups) and their size. The smallest species will migrate furthest. After chromatography, the gel is visualized by autoradiography, detecting bands via the radioactive tracer used. The **base sequence** can be read directly from the gel by the pattern of bands produced using the following reasoning.

Consider a short sequence as shown (by convention written from 5'-end to 3'-end):

$$AGTCGGAACGTA$$

This is labelled at the 5'-end with ^{32}P to give

$$^{32}P-AGTCGGAACGTA$$

cleavage at the 5'-side of G residues using the first reagent (Me_2SO_4, then aqueous piperidine) leads to fragments:

$$^{32}P-A$$

$$^{32}P-AGTC$$

$$^{32}P-AGTCG$$

$$^{32}P-AGTCGGAAC$$

Of course, there will be other fragments that do not contain the 5'-end with its ^{32}P label, but we shall not detect any of these since they contain no radioactive label.

Corresponding fragments will be produced when we use the other three types of cleavage reaction. The resultant chromatogram with the four reaction mixtures will then look something like Figure 14.6, though the bands will be much closer together in practice.

Bands that occur in the left-hand lane represent guanine, and bands that occur in the second lane but not the first lane represent adenine. Similarly, bands in the third lane but not the fourth lane represent thymine, and bands that occur in the fourth lane represent cytosine. By reading up the

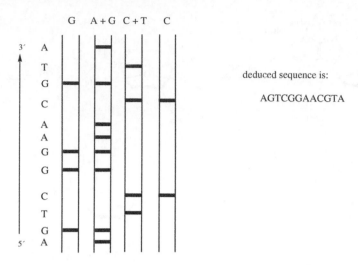

Figure 14.6   Representation of a DNA sequencing gel

chromatogram, the sequence AGTCGGAAC may be deduced. It is possible to distinguish about 200 bands on a single gel. The process is so reliable that automated equipment is available to perform routine analyses. As an alternative to using radioactive labelling, a modification uses differently coloured fluorescent dyes, one for each base-selective reaction. All samples are then applied in one lane, and the base sequence can then be read automatically from the colour of the bands along the gel.

Similar sequencing methodology can be applied to RNA samples.

14.6   Oligonucleotide synthesis: the phosphoramidite method

The ability to synthesize chemically short sequences of single-stranded DNA (**oligonucleotides**) is an essential part of many aspects of genetic engineering. The method most frequently employed is that of **solid-phase synthesis**, where the basic philosophy is the same as that in solid-phase peptide synthesis (see Section 13.6.3). In other words, the growing nucleic acid is attached to a suitable solid support, protected nucleotides are supplied in the appropriate sequence, and each addition is followed by repeated coupling and deprotection cycles.

As with peptide synthesis, similar considerations must to be incorporated into the methodology. Vulnerable functional groups in the base, the sugar,

and the phosphates will need to be protected. The groups to be coupled may need suitable activation, and after the coupling reaction the protecting groups must be removed under mild conditions. In addition, we need to attach the starting material to the support, and eventually the product will need to be released from the support. Nevertheless, the procedure is efficient and has allowed development of automatic **DNA synthesizers** capable of preparing oligonucleotides of up to about 150 residues.

In solid-phase syntheses, oligonucleotides are usually synthesized in the 5′-direction from an immobilized 3′-terminus. The solid phase is generally silica or controlled pore glass (CPG), which has been derivatized to provide a spacer molecule carrying a primary amino group. This **spacer group** is used to bring the nucleotide away from the support and allow the reagents free access. The first residue, as a nucleoside (i.e. without phosphate), is affixed to the support via its 3'-hydroxyl, using a succinic acid residue to achieve bonding, and also extend the spacer further. The succinic acid residue thus has an amide link at one end and an ester link to the sugar of the nucleoside. In practice, the ester linkage is performed first.

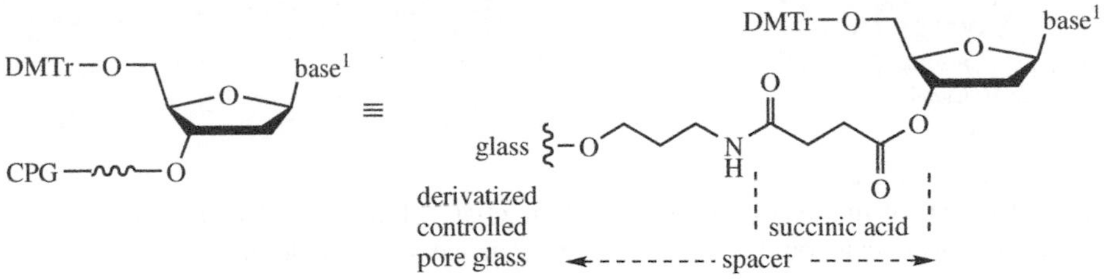

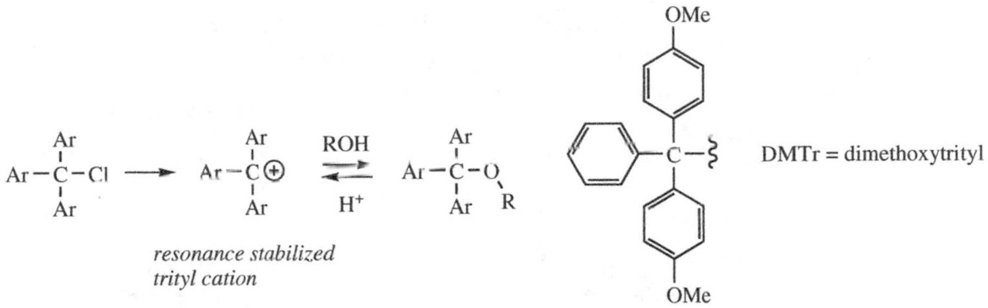

Protection of the 5'-hydroxyl of the sugar unit is usually as a dimethoxytrityl ether (trityl: triphenylmethyl), by reaction with dimethoxytrityl chloride. The dimethoxytrityl group is bulky, and reaction only occurs at the primary 5'-hydroxyl of the sugar group, the secondary 3'-hydroxyl being too hindered to react. This protecting group is easily removed by treatment with acid, even more easily than trityl groups, since the electron-donating methoxy groups stabilize the triarylmethyl carbocation that is an intermediate in the deprotection reaction (see Section 6.2.1).

The bases adenine, guanine, and cytosine all contain exocyclic amino substituents that require protection, since these are potential nucleophiles. They are converted into amides that are stable to the other reagents used in the process, yet can be removed readily by basic hydrolysis. The most effective protecting groups have been found to be isobutyryl for the amino group of guanine, and benzoyl for adenine and cytosine. Thymine has no exocyclic nitrogen and does not need protection.

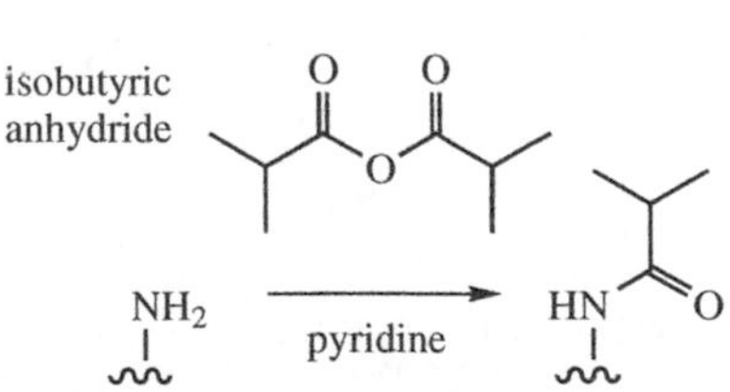

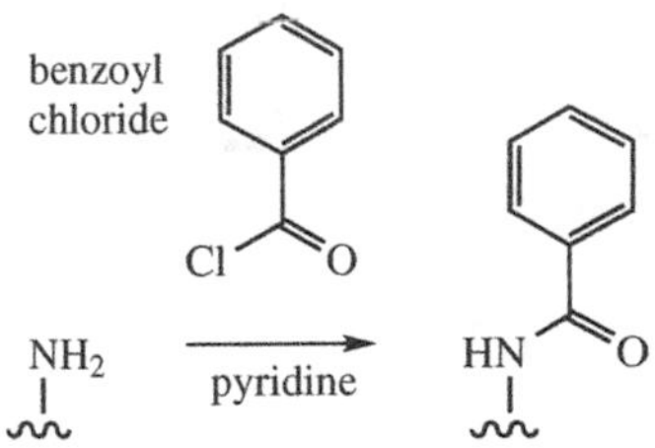

Protection and **activation** of the phosphate moiety is achieved by employing a **phosphoramidite** derivative, $-P(OR)NR_2$. This reagent has phosphorus in its P^{III} oxidation state; the phosphate that we finally require contains P^V. Favoured R groups in the phosphoramidite are 2-cyanoethyl for OR and 2-propyl (isopropyl) for NR_2. The reagent used to attach this to the 3'-hydroxyl is the phosphorodiamidite shown, the hydroxyl displacing an NR_2 group in the presence of tetrazole as a mild acidic catalyst (pK_a 4.9).

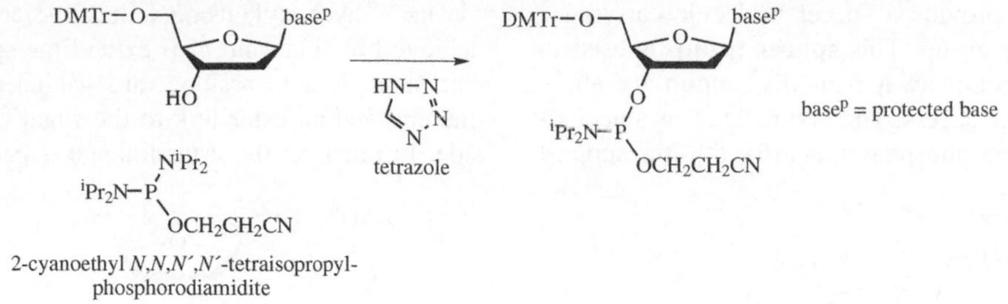
DMTr–O
base^p
HO
N^iPr_2
^iPr_2N–P
OCH_2CH_2CN
2-cyanoethyl N,N,N′,N′-tetraisopropyl-phosphorodiamidite
HN–N
tetrazole
^iPr_2N–P
base^p = protected base

In what is essentially a repeat of this reaction, the 5′-hydroxyl of a second nucleoside can couple to this intermediate; this is the crucial **coupling** reaction in the sequence shown below.

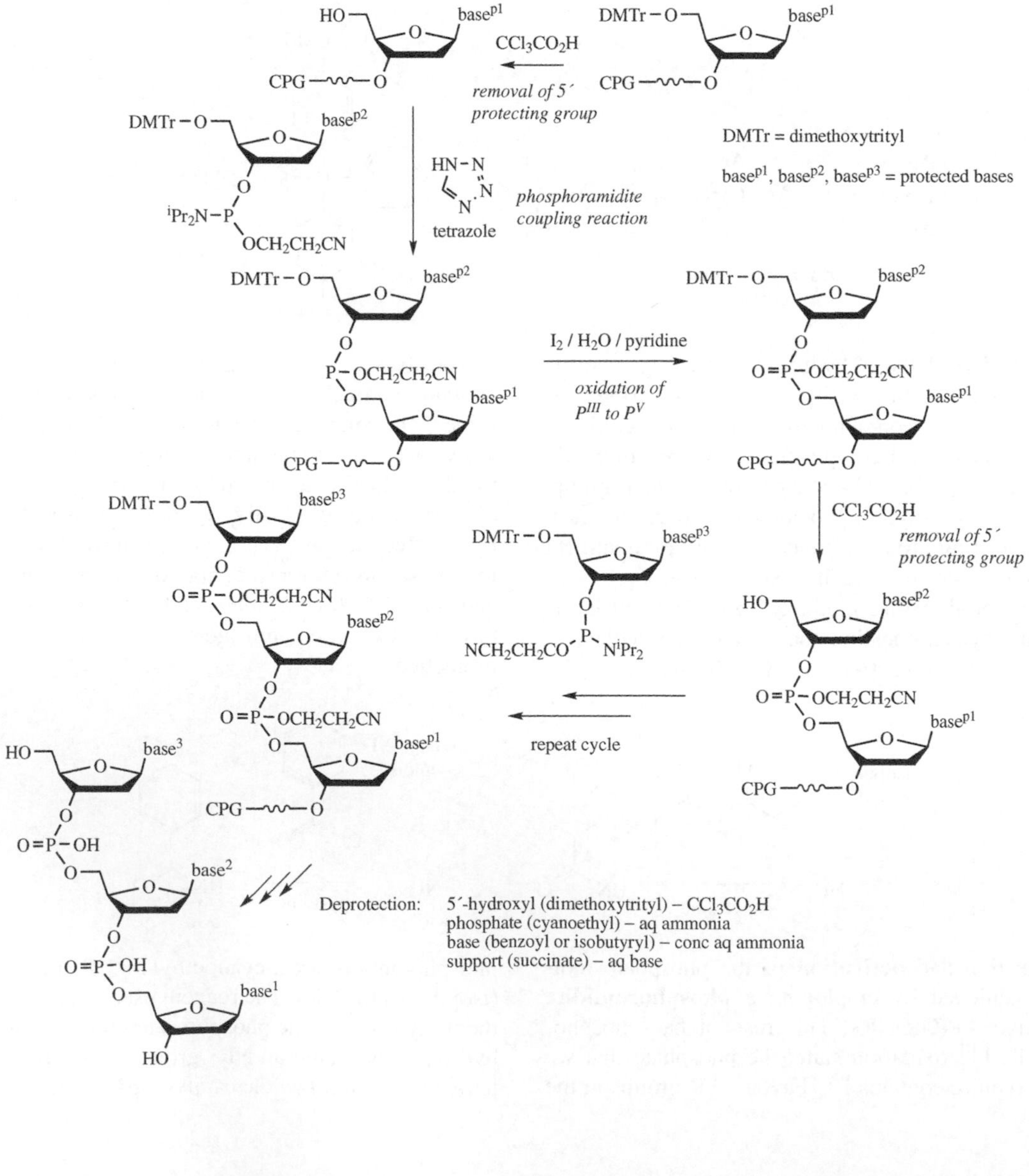
CCl_3CO_2H
removal of 5′ protecting group
DMTr = dimethoxytrityl
base^p1, base^p2, base^p3 = protected bases
tetrazole
phosphoramidite coupling reaction
I_2 / H_2O / pyridine
oxidation of P^III to P^V
CCl_3CO_2H
removal of 5′ protecting group
repeat cycle
Deprotection: 5′-hydroxyl (dimethoxytrityl) – CCl_3CO_2H
phosphate (cyanoethyl) – aq ammonia
base (benzoyl or isobutyryl) – conc aq ammonia
support (succinate) – aq base

Of course, the product does not have a phosphate linker between the two nucleosides, and phosphorus is still in the wrong oxidation state. This is remedied by oxidation of the dinucleotide phosphite to a phosphotriester using iodine. We now have the required phosphate linker, though it is still protected with the cyanoethyl group. This is retained at this stage.

The dimethoxytrityl ester protecting group is now removed by treatment with mild acid (CCl_3CO_2H), which is insufficiently reactive to hydrolyse the amide protection of bases, or the cyanoethyl protection of the phosphate. The coupling cycle can now be repeated using a phosphoramidite derivative of the next appropriate nucleoside. The sequences will be continued as necessary until the desired oligonucleotide is obtained.

It then remains to remove protecting groups and release the product from the support. All of these tasks, except for the removal of the dimethoxytrityl group, are achieved by use of a single **deprotection** reagent, aqueous base (ammonia). The cyanoethyl groups are lost from the phosphates by base-catalysed elimination, and amide protection of the bases is removed by base-catalysed hydrolysis. The latter process also achieves hydrolysis of the succinate ester link to the support.

14.7 Copying DNA: the polymerase chain reaction

The **polymerase chain reaction** (PCR), developed by Mullis, is a simple and most effective way of amplifying, i.e. producing multiple copies of, a DNA sequence. It finds applications in all sorts of areas not immediately associated with nucleic acid biochemistry, e.g. genetic screening, medical diagnostics, forensic science, and evolutionary biology. The general public is now well aware of the importance of some of these topics, e.g. the ability to identify a person by DNA analysis, but perhaps does not realize that tiny samples of DNA must be copied millions of times to provide a sample large enough for chromatographic analysis.

PCR makes use of the heat-stable enzyme **DNA polymerase** from the bacterium *Thermus aquaticus* and its ability to synthesize complementary strands of DNA when supplied with the necessary deoxyribonucleoside triphosphates. We have already looked at the chemistry of DNA replication (see Section 14.2.2), and this process is exactly the same, though it is carried out in the laboratory and has been automated.

Although knowledge of the whole nucleotide sequence of the target area of DNA is not required, one must know the sequence of some small stretch on either side of the target area. These data may be known from other sequencing studies; or, surprisingly, it can even be predictable from knowledge of related genes. Two single-stranded oligonucleotides, one for each sequence, are then synthesized to act as primers. Typically, the primers should contain about 20 nucleotides, and they must be complementary to the DNA sequences of opposite strands. In the schematic illustration of the process (Figure 14.7), the central target area is indicated, and the primers are depicted as short complementary sequences.

Initially, the double-stranded DNA is heated to separate the strands. The primers are then added and the temperature lowered so that the primers anneal to the complementary sequences of each strand. In the presence of nucleoside triphosphates, the DNA polymerase enzyme will replicate a length of DNA starting from the 3'-end of a nucleotide, extending the chain towards the 5'-end (see Section 14.2.2). It will thus start chain extension from the 3'-ends of the primers and continue to the end of the DNA strands. This will lead to two double-stranded DNA

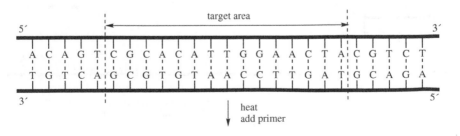

Figure 14.7 Representation of DNA amplification via the PCR

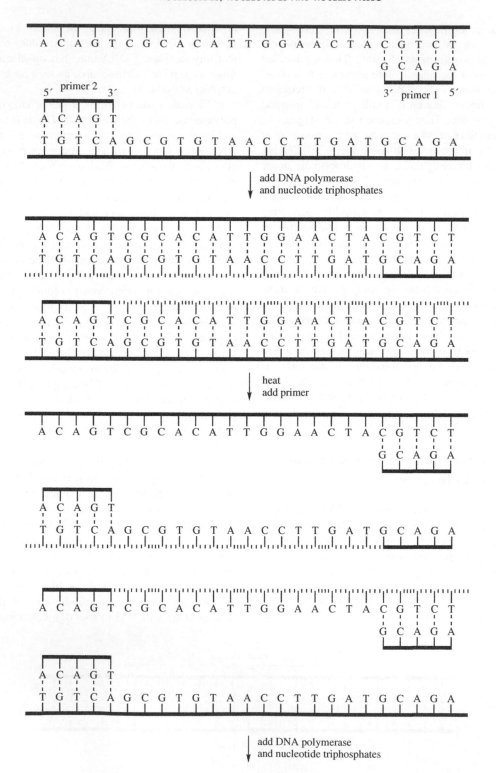

Figure 14.7 (*continued*)

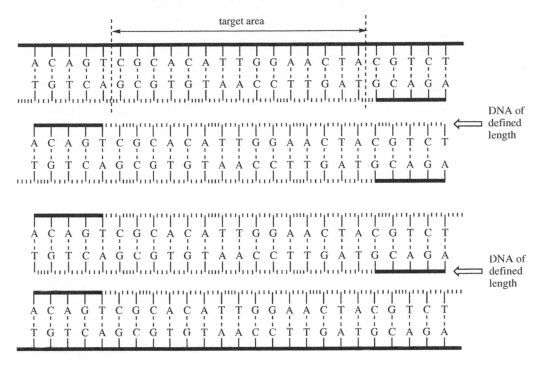

Figure 14.7   *(continued)*

molecules, composed of initial strands, and primer plus newly synthesized DNA, as shown.

The process is repeated. Heating causes separation of strands, and cooling allows primer to attach to the appropriate nucleotide sequence. Enzymic chain extension then produces four double stranded DNA molecules.

The number of DNA molecules doubles in each cycle of the process, so that after 30 cycles, say, we have 2^{30} molecules (approximately 10^9 copies). However, there is another, less obvious feature that makes the PCR even more useful. In the second cycle, two of the newly synthesized single-stranded chains will be of defined length. They will consist of the target area plus two primers; the 5'-ends of the primers define the length of DNA. Other molecules will be much longer, because replication goes on to the end of the template. Should you wish to follow this through, you will find that, after the third cycle, there will be eight single-stranded DNA molecules of defined length and eight that are longer. After each cycle, the number of defined-length DNA molecules increases geometrically, whereas the number of DNA strands containing sequences outside of the primers only increases arithmetically. This means that, after about 20 cycles, the DNA synthesized is almost entirely composed of molecules whose length is defined by the primers, i.e. the target area plus a short extra length defined by the primers.

Related work

15

The organic chemistry of intermediary metabolism

15.1   Intermediary metabolism

Intermediary metabolism is the all-encompassing name given to the highly integrated network of chemical reactions by which organisms obtain energy from their environment and synthesize those molecules necessary for their continued well being and existence. These are the reactions that we consider to comprise 'biochemistry', and they are usually studied as part of a biochemistry course. However, if we choose to study these reactions as part of organic chemistry, we shall see some interesting, elaborate, and often quite complex transformations taking place. Two very important characteristics differentiate these reactions from those we have already encountered. First, they are almost always **enzyme-mediated reactions**, and thus take place readily at near-neutral pH and at ambient temperatures. Second, they are also highly regulated, and their participation can be switched on or off, or otherwise finely controlled by the organism according to its needs.

As we look at some of the reactions of intermediary metabolism, we shall rationalize them in terms of the chemistry that is taking place. In general, we shall not consider here the involvement of the enzyme itself, the binding of substrates to the enzyme, or the role played by the enzyme's amino acid side-chains. In Chapter 13 we looked at specific examples where we know just how an enzyme is able to catalyse a reaction. Examples such as aldolase and triose phosphate isomerase, enzymes of the glycolytic pathway, and citrate synthase from the Krebs cycle were considered in some detail. It may be advantageous to look back at these examples in order to underline the participation of an enzyme.

A large proportion of the substrates used in intermediary metabolism are in the form of phosphates. **Phosphates** are favoured in nature since they usually confer water solubility on the compound, and provide a functional group that is able to bind to enzymes through simple electrostatic bonding. In many cases, the phosphate group may also feature as a chemically reactive functional group – phosphates are good leaving groups (see Section 7.13.2). In many structures, the abbreviation **P** is used to represent the phosphate group and **PP** the diphosphate (or pyrophosphate) group:

ionized forms

ROP (phosphate)   ROPP (diphosphate)

non-ionized forms

ROP (phosphate)   ROPP (diphosphate)

At physiological pH values, these groups will be ionized as shown, but in schemes where structures are given in full, the non-ionized acids are

Essentials of Organic Chemistry   Paul M Dewick
© 2006 John Wiley & Sons, Ltd

usually depicted. This is done primarily to simplify structures, to eliminate the need for counterions, and to avoid mechanistic confusion. Likewise, amino acids are usually shown in non-ionized form, although they will typically exist as zwitterions (see Section 4.11.3):

$$R \overset{CO_2H}{\underset{H\ \ NH_2}{\diagup}} \qquad R \overset{CO_2^{\ominus}}{\underset{\underset{\oplus}{H\ \ NH_3}}{\diagup}}$$

zwitterion

Ionized and non-ionized forms of many compounds are regarded as synonymous in the text, thus citrate/citric acid, acetate/acetic acid or others may be used according to the author's whim and context, and should not be considered as having any especial relevance.

15.1.1 Oxidation reactions and ATP

The currency unit for energy in biochemical reactions is the nucleotide derivative **ATP, adenosine triphosphate**. We have already discussed this molecule in Chapter 7, where we rationalized many of the reactions of phosphates in terms of them being analogues of carbonyl compounds. Only the triphosphate portion of ATP is involved chemically in the energy processes; the remaining complex part of the molecule is a recognizable feature that allows binding to the enzyme.

The triphosphate portion can be visualized as containing two anhydride functions and one ester function. We have seen that hydrolysis of anhydrides is achieved much more easily than is hydrolysis of esters, an observation that can be related to the nature of the leaving group (see Section 7.8). Accordingly, hydrolysis of the anhydride bond liberates considerably more energy than does hydrolysis of the ester bond, and it is anhydride hydrolysis that is crucial to ATP's role in biochemistry. Hydrolysis of ATP to ADP liberates energy, which can be coupled to energy-requiring processes. Alternatively, energy-releasing processes can be coupled to the synthesis of ATP from ADP.

currency unit for energy is **ATP**, adenosine triphosphate

adenosine triphosphate; ATP

ATP $\longrightarrow$ ADP + **P** ⎫ both represent hydrolysis of anhydride bond
ADP $\longrightarrow$ AMP + **P** ⎬ – release significant amount of energy ΔG –34 kJ / mole

hydrolysis of ATP to ADP
– coupled to **energy-requiring processes**

synthesis of ATP from ADP
– coupled to **energy-releasing processes**

AMP $\longrightarrow$ A + **P** hydrolysis of ester bond
– releases smaller amount of energy ΔG –9 kJ / mole

Hydrolysis of ATP to ADP is rationalized simply as nucleophilic attack of water on to the terminal P=O double bond, followed by cleavage of the anhydride bond and expulsion of ADP as the leaving group (see Box 7.25).

There are two anhydride linkages in ATP, but nucleophilic attack in the enzyme-controlled reaction usually occurs on the terminal P=O (hydrolysis of ATP to ADP), and only occasionally do we encounter attack on the central P=O (hydrolysis of ATP to adenosine monophosphate, AMP). Both reactions yield the same amount of energy, $\Delta G - 34$ kJ mol^{-1}. This is not surprising, since in each case the same type of bond is being hydrolysed. The further hydrolysis of AMP to adenosine breaks an ester linkage and would liberate only a fraction of the energy, $\Delta G - 9$ kJ mol^{-1}, and this reaction is not biochemically important.

Oxidation reactions are the main providers of energy for ATP synthesis. Whilst oxidation usually involves incorporation of one or more oxygen atoms, in its simplest form it can be thought of as a loss of electrons. Thus, the transformation of ferrous ion to ferric ion is an oxidation reaction and involves loss of one electron. Such electrons can be considered as carrying the energy released from the oxidation reactions.

$$Fe^{2+} \xrightarrow{\text{oxidation}} Fe^{3+} + e^-$$

packets of energy

In biochemical reactions, these electrons are eventually passed to oxygen, which becomes reduced to water. Overall, the oxidation of a substrate AH$_2$ could be represented by the equation

$$AH_2 + \tfrac{1}{2}O_2 \rightleftharpoons A + H_2O \quad \text{large negative } \Delta G$$

and this reaction has the potential to liberate energy, i.e. it has a large negative ΔG.

Now this reaction is not possible directly. We are not accustomed to seeing our food spontaneously reacting with atmospheric oxygen and igniting because of the energy released! However, food such as carbohydrate, fat and protein is oxidized after we have eaten it, and energy is released and utilized by our bodies. The secret is to react AH$_2$ through the involvement of a suitable **coenzyme**, not directly with oxygen. This reaction can be considered as

$$AH_2 + \underset{coenzyme}{X \text{ (oxidized)}} \rightleftharpoons A + \underset{coenzyme}{X \text{ (reduced)}}$$

where X is the coenzyme. The reaction is catalysed by an enzyme termed a dehydrogenase, which removes two hydrogen atoms from the substrate. The coenzyme system involved can generally be related to the functional group being oxidized in the substrate. If the oxidation process is then a pyridine nucleotide, **nicotinamide adenine dinucleotide (NAD$^+$)** or **nicotinamide adenine dinucleotide phosphate (NADP$^+$)**, tends to be utilized as hydrogen acceptor. One hydrogen from the substrate (that bonded to carbon) is transferred as hydride to the coenzyme, and the other, as a proton, is passed to the medium (see Box 11.2).

NAD$^+$ and NADP$^+$

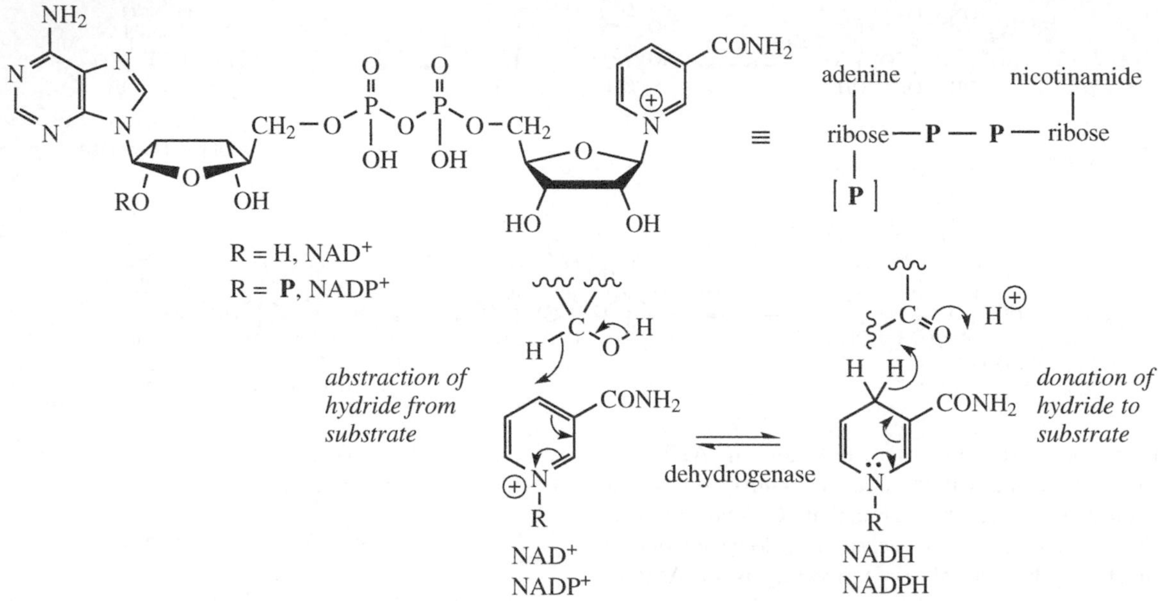

$$R = H, NAD^+$$
$$R = P, NADP^+$$

abstraction of hydride from substrate

donation of hydride to substrate

dehydrogenase

NAD$^+$ NADH
NADP$^+$ NADPH

NAD$^+$ and NADP$^+$ may also be used in the oxidations

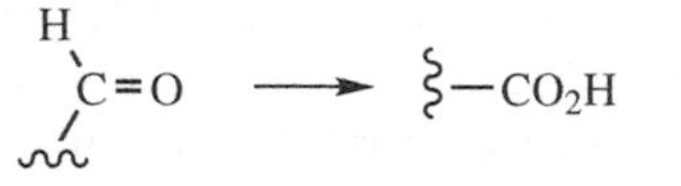

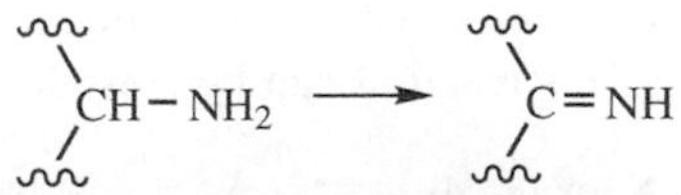

The reverse reaction, i.e. reduction, is also indicated in the scheme, and may be compared with the chemical reduction process using complex metal hydrides, e.g. LiAlH$_4$ or NaBH$_4$, namely nucleophilic addition of hydride and subsequent protonation (see Section 7.5). The reduced forms NADH and NADPH are conveniently regarded as hydride-donating reducing agents (see Box 7.6). We also noted that there were stereochemical features associated with these coenzymes (see Box 3.14). During a reduction sequence, there is stereospecific transfer of hydride from a prochiral centre on the dihydropyridine ring, and it is delivered to the carbonyl compound also in a stereospecific manner. In practice, NADPH is generally employed in reductive processes, whereas NAD$^+$ is used in oxidations.

Should the oxidative process be the conversion

$$\text{-CH}_2\text{-CH}_2\text{-} \longrightarrow \text{-CH=CH-}$$

then the coenzyme used as acceptor is usually a flavin nucleotide: **flavin adenine dinucleotide (FAD)** or **flavin mononucleotide (FMN)**. These entities are bound to the enzyme in the form of a **flavoprotein**, and take up two hydrogen atoms, represented in the figure as being derived by addition of hydride from the substrate and a proton from the medium. Reductive sequences involving flavoproteins may be represented as the reverse reaction, where hydride is transferred from the coenzyme, and a proton is obtained from the medium.

FAD and FMN

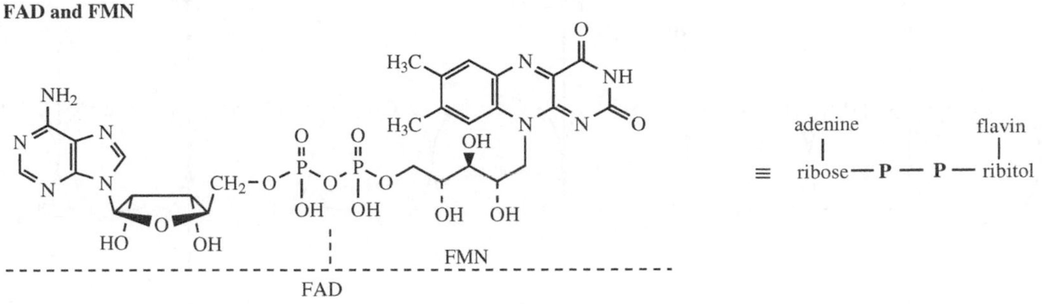

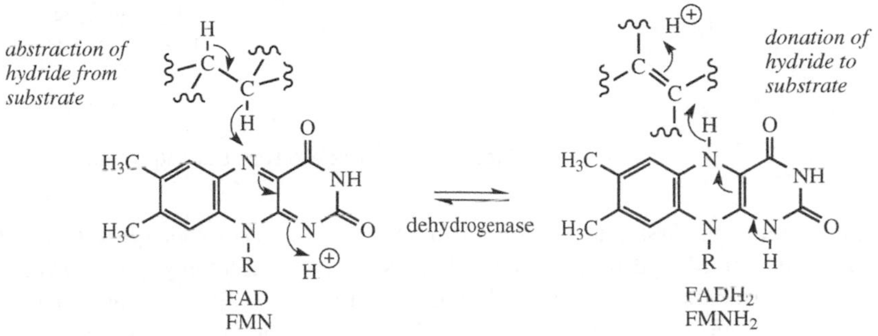

$$\equiv \quad \underset{ribose}{\overset{adenine}{|}} — P — P — \underset{ribitol}{\overset{flavin}{|}}$$

After the substrates containing either CH–OH or CH$_2$–CH$_2$ functional groups have been oxidized by the dehydrogenase enzyme–coenzyme system, energy abstraction from the oxidative transformation now depends upon reoxidation of the reduced coenzyme.

$$
\begin{array}{lllll}
\text{NAD(P)H} & + \; \tfrac{1}{2}\,O_2 & \rightleftharpoons & \text{NAD(P)}^+ & + \; H_2O \\
\text{FADH}_2 & + \; \tfrac{1}{2}\,O_2 & \rightleftharpoons & \text{FAD} & + \; H_2O \\
\text{FMNH}_2 & + \; \tfrac{1}{2}\,O_2 & \rightleftharpoons & \text{FMN} & + \; H_2O
\end{array}
\right\} \; \text{large negative } \Delta G
$$

This will be one of the processes shown, all of which have a large negative ΔG and are capable of harnessing this energy via synthesis of ATP molecules. However, even these are not achievable directly, and the electron transport chain of oxidative phosphorylation is utilized.

15.1.2 Oxidative phosphorylation and the electron transport chain

The total oxidation of an organic compound using molecular oxygen as the electron acceptor has the potential to yield a very large amount of energy, sufficient for the synthesis of several molecules of ATP from ADP, if there could be one efficiently coupled oxidation process. It is quite unrealistic to achieve this in one step; instead, a multi-stage process termed **oxidative phosphorylation** is employed. This process removes packets of energy, more nearly corresponding to the amounts required for the synthesis of single ATP molecules from ADP. Oxidation of a compound, per atom of oxygen used, can yield up to three molecules of ATP, representing an energy efficiency of about 50%. It takes place in the mitochondria via a sequence of redox (reduction–oxidation) reactions known as the **electron transport chain** or the respiratory chain, and provides the principal source of ATP for an aerobic cell.

The electron transport chain (Figure 15.1) involves a series of compounds acting together, achieving the removal of hydrogen equivalents from organic molecules and eventually reacting them with oxygen

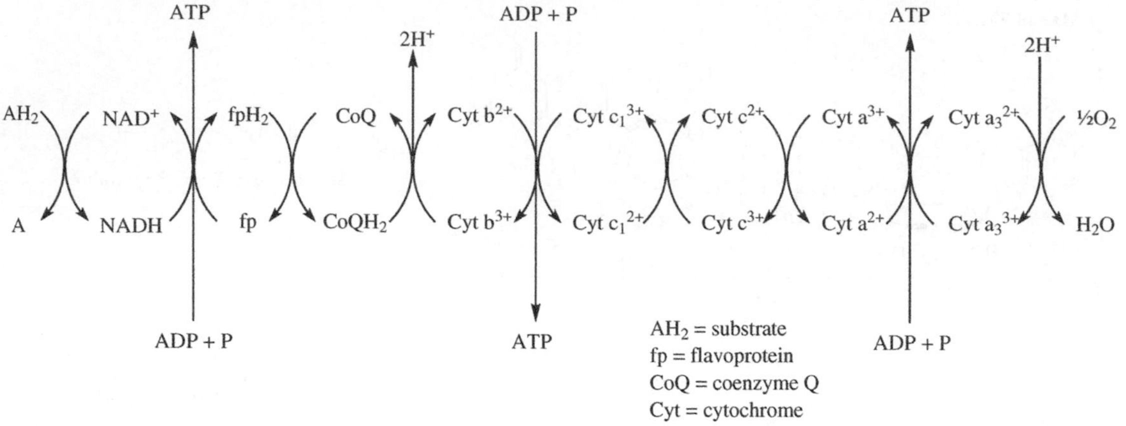

AH$_2$ = substrate
fp = flavoprotein
CoQ = coenzyme Q
Cyt = cytochrome

Figure 15.1   The electron transport chain

to form water. The first part of the chain involves transfer of a pair of hydrogen atoms, whereas only electrons are transferred in the final stages of the pathway. The individual molecules involved are NAD$^+$, flavoprotein, coenzyme Q (ubiquinone), and a number of cytochromes.

We have already seen the redox reactions of NAD$^+$ and flavoproteins containing FAD or FMN. The term **coenzyme Q** or **ubiquinone** covers a range of structures (as shown), depending on the length of the hydrocarbon side-chain, which varies according to species. In humans, the redox carrier is coenzyme Q$_{10}$ ($n = 10$).

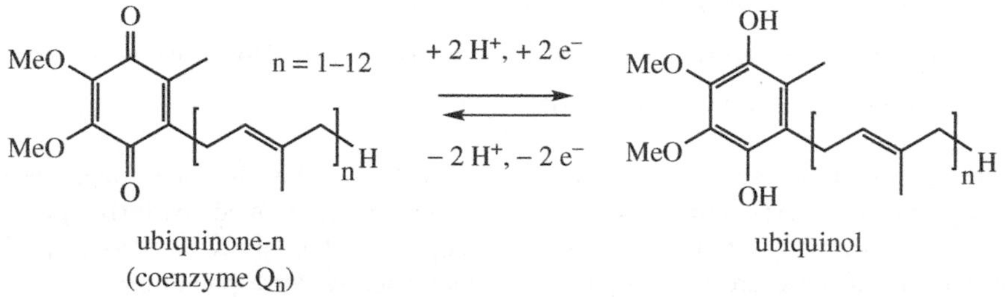

ubiquinone-n
(coenzyme Q$_n$)

ubiquinol

Ubiquinone is readily reduced to ubiquinol, a process requiring two protons and two electrons; similarly, ubiquinol is readily oxidized back to ubiquinone. This redox process is important in oxidative phosphorylation, in that it links hydrogen transfer to electron transfer. The **cytochromes** are haem-containing proteins (see Box 11.4). As we have seen, haem is an iron–porphyrin complex. Alternate oxidation–reduction of the iron between Fe^{2+} (reduced form) and Fe^{3+} (oxidized form) in the various cytochromes is responsible for the latter part of the electron transport chain. The individual cytochromes vary structurally, and their classification

(a, b, c, etc.) is related to their absorption maxima in the visible spectrum.

Most compounds oxidized by the electron transport chain donate hydrogen to NAD$^+$, and then NADH is reoxidized in a reaction coupled to reduction of a flavoprotein. During this transformation, sufficient energy is released to enable synthesis of ATP from ADP. The reduced flavoprotein is reoxidized via reduction of coenzyme Q; subsequent redox reactions then involve cytochromes and electron transfer processes rather than hydrogen transfer. In two of these cytochrome redox reactions, there is sufficient energy release to allow ATP synthesis. In due course, electrons are passed to oxygen, which is converted into water in the presence of protons. The total process whereby hydrogen atoms are passed to NAD^+ generates three molecules of ATP per pair of hydrogen atoms. However, substrates with the CH_2–CH_2 grouping that are oxidized by flavoproteins effectively bypass the first ATP generation step, so only produce two molecules of ATP per pair of hydrogen atoms.

The electron transport chain is vital to aerobic organisms. Interference with its action may be life threatening. Thus, cyanide and carbon monoxide bind to haem groups and inhibit the action of the enzyme cytochrome c oxidase, a protein complex that is effectively responsible for the terminal part of the electron transport sequence and the reduction of oxygen to water.

What has been achieved by the participation of coenzyme systems and the electron transport chain is twofold. First, there is no need for the substrate AH_2 to react with oxygen. Second, it provides common routes for the oxidation of many different organic compounds, rather than a specific route for every compound, a vast variety of which will be present in the normal diet. Although it is somewhat simplistic, many of the non-oxidative reactions of intermediary metabolism can be viewed as additional chemical transformations designed to provide substrates containing either CH–OH or CH_2–CH_2 that may then be subjected to dehydrogenation.

15.2 The glycolytic pathway

The **glycolytic pathway**, or **glycolysis**, is a metabolic sequence in which **glucose** is broken down to **pyruvic acid**. The subsequent fate of pyruvate then depends upon whether or not the organism is aerobic or anaerobic: Under aerobic conditions, pyruvate is oxidized via oxidative phosphorylation; under anaerobic conditions, pyruvate is converted further into compounds such as lactate or ethanol, depending upon the organism.

The first step in glycolysis is the phosphorylation of glucose to give the ester **glucose 6-phosphate**. The glucose starting material may well have come from hydrolysis of starch obtained in the diet, or by utilization of glycogen reserves.

glucose is usually shown in α-pyranose form as found in starch and glycogen; the configuration α or β has no bearing on the subsequent chemistry

OP = (*ionized at physiological pH*)

activation of glucose by formation of phosphate ester; driven by energy released from hydrolysis of ATP to ADP (anhydride bond)

phosphate ester

This phosphorylation step is achieved by reaction of the 6-hydroxyl with the anhydride ATP, during which process ATP is converted into ADP. This process is driven by the energy contained in the anhydride function of ATP, and represents an expenditure of energy to get the metabolic process started, though the overall objective of glycolysis is to acquire energy via synthesis of ATP molecules.

Glucose 6-phosphate is then isomerized to **fructose 6-phosphate**. This conversion of an aldose sugar to a ketose sugar is easy to rationalize in terms of keto–enol tautomerism (see Box 10.1).

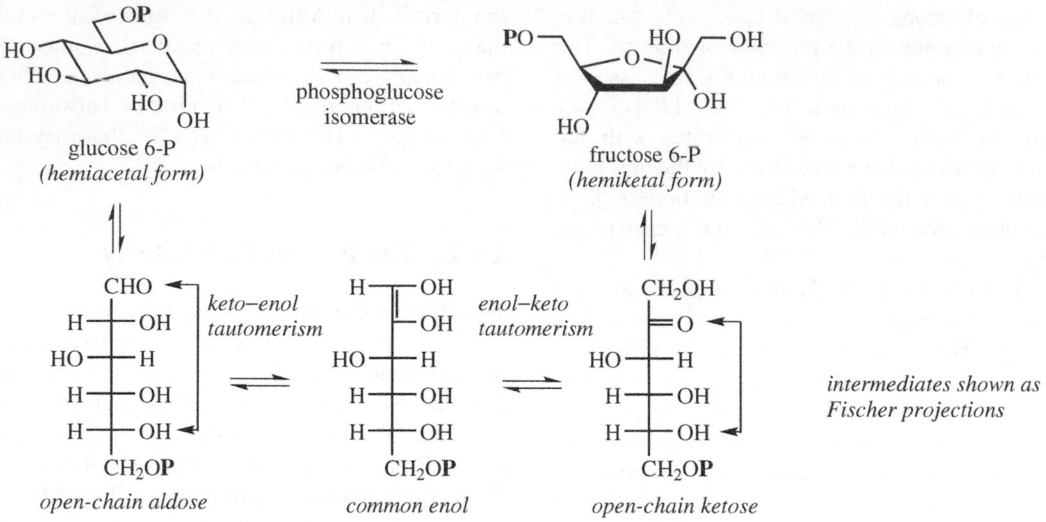

glucose 6-P
(*hemiacetal form*)

phosphoglucose
isomerase

fructose 6-P
(*hemiketal form*)

CHO — H—OH — HO—H — H—OH — H—OH — CH₂OP
open-chain aldose

keto–enol tautomerism

H—OH — OH — HO—H — H—OH — H—OH — CH₂OP
common enol

enol–keto tautomerism

CH₂OH — O — HO—H — H—OH — H—OH — CH₂OP
open-chain ketose

intermediates shown as Fischer projections

isomerism of aldose to ketose via keto–enol tautomerism

We should first consider the open-chain form of glucose 6-phosphate, rather than its pyranose hemiacetal form (see Section 12.2.1). The open-chain aldose has the requirements for enolization, namely a hydrogen α to the aldehyde carbonyl group. Enolization produces in this case an enediol, which can revert to a keto form in two ways, i.e. reforming the open-chain aldose or, alternatively, producing the ketose fructose 6-phosphate. The enediol may be considered a common enol for the two enolization processes. The open-chain form of fructose 6-phosphate may then form a hemiketal, as shown, generating a furanose ring.

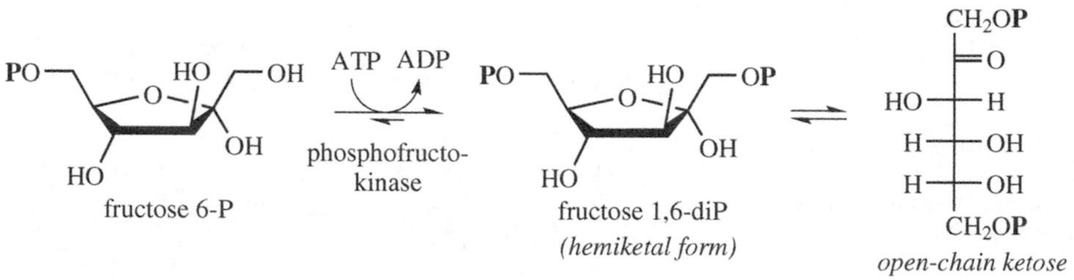

fructose 6-P

ATP ADP

phosphofructo-kinase

fructose 1,6-diP
(*hemiketal form*)

CH₂OP — O — HO—H — H—OH — H—OH — CH₂OP
open-chain ketose

further phosphorylation driven by hydrolysis of ATP

Further phosphorylation, again using ATP as in the first reaction, converts fructose 6-phosphate into **fructose 1,6-diphosphate**. Again, there is the expenditure of energy by the use of ATP; we have now used two molecules of ATP, and there has been no net generation of energy. This represents a significant investment before any rewards are forthcoming.

For the subsequent reactions we need to consider fructose 1,6-diphosphate in its open-chain form rather than the hemiketal originally drawn. Now follows the reverse aldol reaction catalysed by aldolase, as we have already discussed in some detail elsewhere (see Boxes 10.4 and 10.5). For a simple chemical interpretation, we can write this as involving enolate anions, either as leaving group in the forward reaction or as nucleophile in the reverse reaction, but the enzymic reaction is known to utilize enamine derivatives. It is worth emphasizing again that organisms make use of this reaction both in its forward direction for carbohydrate metabolism and in its reverse direction for carbohydrate synthesis, according to requirements.

cleavage of hexose via reverse aldol reaction

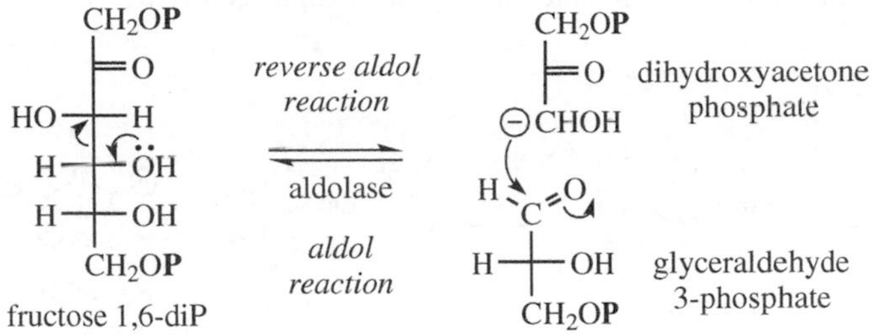

important in carbohydrate metabolism – reverse aldol reaction
important in carbohydrate synthesis – aldol reaction

The reverse aldol reaction results in the formation of **dihydroxyacetone phosphate** and **glyceraldehyde 3-phosphate**. Dihydroxyacetone phosphate is not on the direct pathway, and is converted into a second molecule of glyceraldehyde 3-phosphate by the enzyme triose phosphate isomerase.

dihydroxyacetone phosphate is not on the direct pathway – it is converted into a second molecule of glyceraldehyde 3 phosphate via a common enol

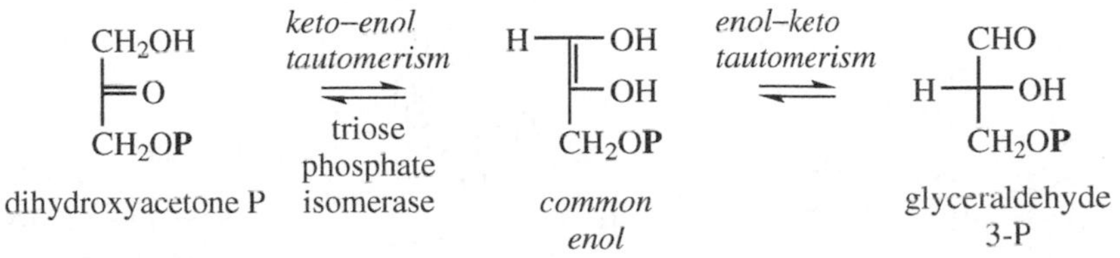

This is achieved by two keto–enol tautomerism reactions and a common enol (see Box 10.1). Mechanistically, it is identical to the isomerization of glucose 6-phosphate to fructose 6-phosphate seen earlier in the sequence, so we can move on to the next step of the pathway.

In this step, the aldehyde group of glyceraldehyde 3-phosphate appears to be oxidized to an acid, which becomes phosphorylated, and hydrogen is passed to NAD^+, which becomes reduced to NADH. We shall see shortly that the fate of this NADH is quite significant.

aldehyde group is oxidized and phosphorylated

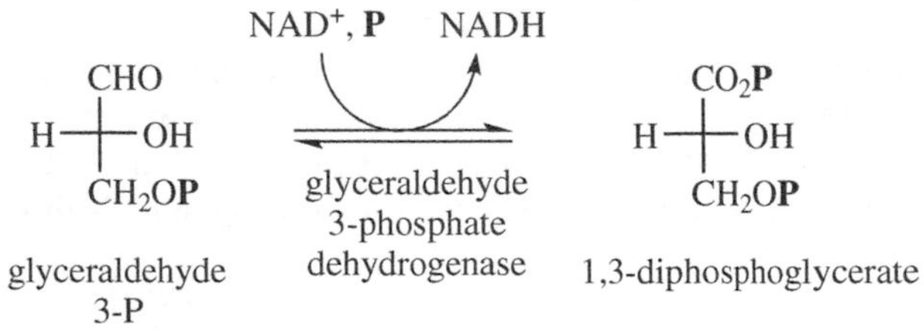

At first glance, this oxidation–phosphorylation reaction seems rather obscure. It becomes much more logical when we see that the enzyme achieves this via a multi-stage process. Critical to the reaction is the involvement of a thiol group on the enzyme. This reacts with the aldehyde group of the substrate glyceraldehyde 3-phosphate to form a hemithioacetal (see Section 7.4). It is this intermediate that reacts with NAD^+, since it contains an oxidizable CH–OH function. The product is then a thioester. The thioester is attacked by a phosphate nucleophile, and

since it contains a good leaving group, $EnzS^-$, the reaction product is released from the enzyme as **1,3-diphosphoglycerate**.

stepwise sequence involving thiol-containing enzyme

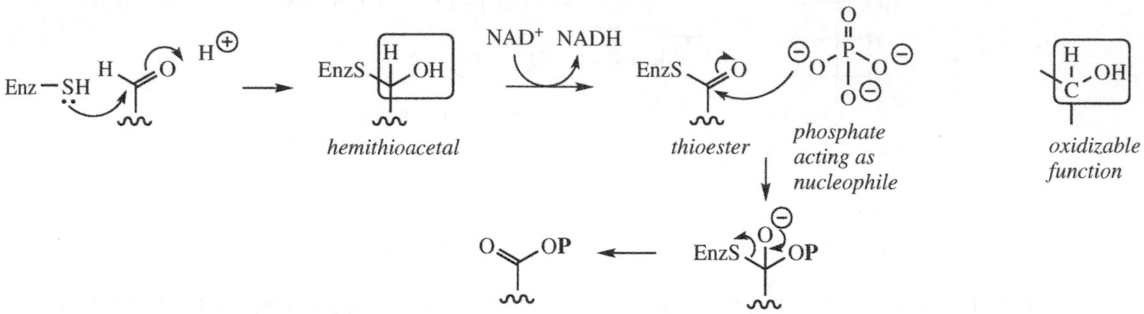

hemithioacetal   *thioester*   *phosphate acting as nucleophile*   *oxidizable function*

If we look at the structure of 1,3-diphosphoglycerate, we can see that it is actually an anhydride, albeit a mixed anhydride of a carboxylic acid and phosphoric acid (see Box 7.27). Accordingly, we expect it to

be fairly reactive towards nucleophiles, and indeed it is. It is sufficiently reactive that hydrolysis liberates enough energy to synthesize ATP from ADP.

1,3-diphosphoglycerate is a mixed anhydride of carboxylic and phosphoric acid

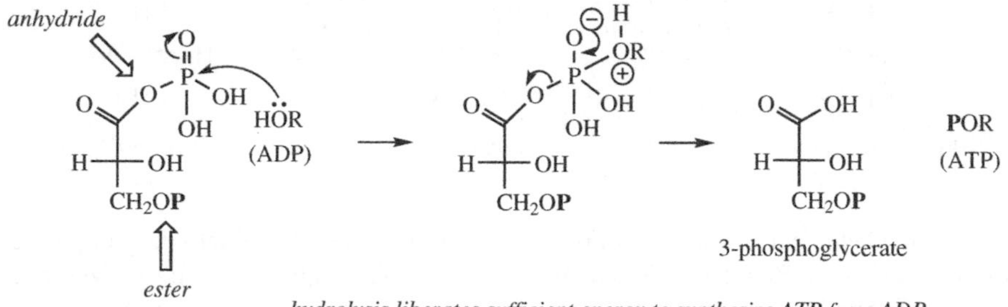

anhydride   *ester*   3-phosphoglycerate

hydrolysis liberates sufficient energy to synthesize ATP from ADP

substrate-level phosphorylation

ATP synthesis is achieved by ADP acting as the nucleophile towards this mixed anhydride, attacking the P=O bond, with the carboxylate being the leaving group. Note that this reaction is favoured, whereas the alternative possibility involving hydrolysis of the phosphate ester does not occur. This is precisely what we would predict knowing the different reactivities of anhydrides and esters (see Section 7.8). This direct synthesis of ATP by a process in which

ADP acquires an additional phosphate from a suitable donor molecule is often termed **substrate-level phosphorylation**, differentiating it from ATP synthesis that is achieved through oxidative phosphorylation.

After donating its phosphate group to ADP, 1,3-diphosphoglycerate is converted into **3-phosphoglycerate**. This reaction is followed by enzymic modification to **2-phosphoglycerate**.

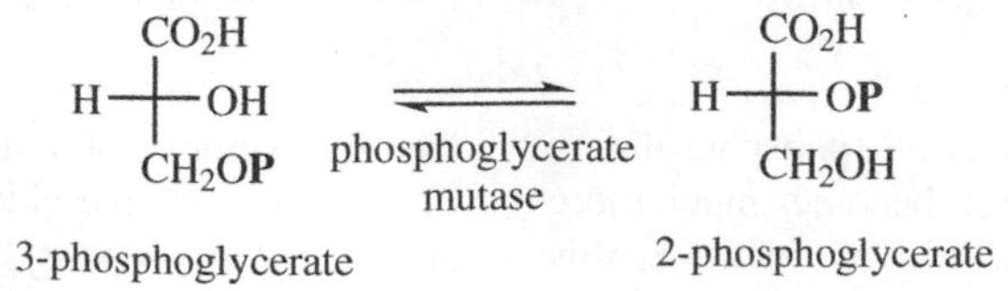

3-phosphoglycerate   phosphoglycerate mutase   2-phosphoglycerate

mutase adds a second phosphate group to position 2, then
removes that from position 3, i.e. proceeds via the diphosphate

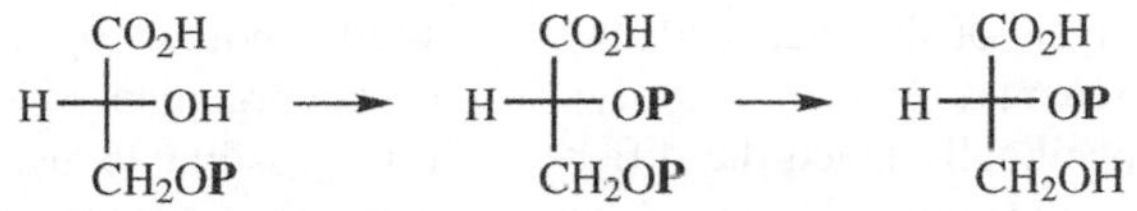

$$
\begin{array}{ccc}
CO_2H & CO_2H & CO_2H \\
H\!\!-\!\!OH & H\!\!-\!\!OP & H\!\!-\!\!OP \\
CH_2OP & CH_2OP & CH_2OH
\end{array}
$$

the phosphate group is derived from the enzyme, not ATP

Although this reaction is catalysed by a mutase, which perhaps suggests this is a rearrangement reaction, there is no transfer of the phosphate group to the adjacent hydroxyl. Instead, this reactions proceeds via an intermediate diphosphate, so we are actually seeing a phosphorylation–dephosphorylation or esterification–hydrolysis sequence. There is an unusual aspect of this extra phosphorylation, and that is that no ATP is involved. Instead, the new phosphate group is derived from the enzyme itself. Although this is unexpected, it does avoid another energy-requiring reaction and the use of precious ATP.

Then follows an elimination reaction, in which water is removed from 2-phosphoglycerate to yield **phosphoenolpyruvate**.

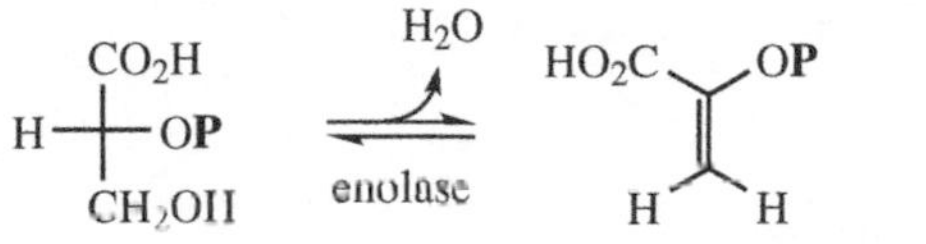

2-phosphoglycerate   phosphoenolpyruvate

an elimination reaction (dehydration); although phosphate is the better leaving group, the elimination is enzyme-controlled

postulated elimination reaction

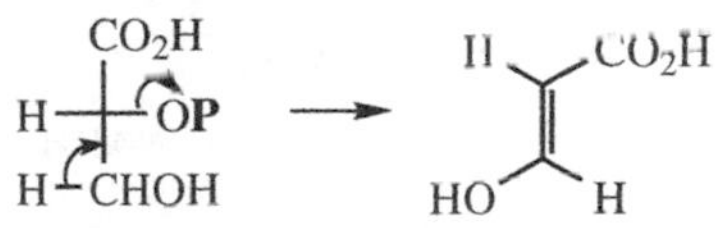

phosphate is the better leaving group

This reaction is catalysed by an enzyme called enolase; though this may appear quite straightforward, it is chemically unusual. Eliminations depend upon the presence of a suitable leaving group (see Section 6.4.1), and by far the better leaving group in 2-phosphoglycerate is the phosphate. We might predict that the product from an elimination reaction on 2-phosphoglycerate would logically be the alternative enol system. That this does not occur indicates and emphasizes the enzyme's special contribution to the reaction.

The product phosphoenolpyruvate is able to donate its phosphate group directly to ADP, resulting in ATP synthesis.

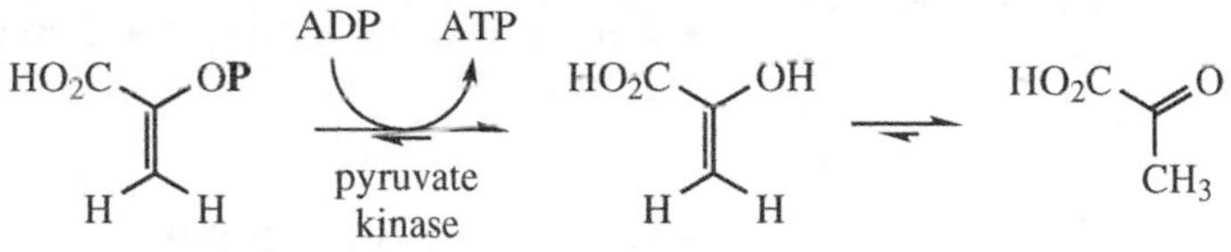

phosphoenolpyruvate   enolpyruvate   pyruvate

substrate-level phosphorylation
*although phosphoenolpyruvate is only an enol **ester**, hydrolysis gives an unfavoured enol;*
tautomerism to the keto form is the driving force for the reaction and results in a large negative ΔG

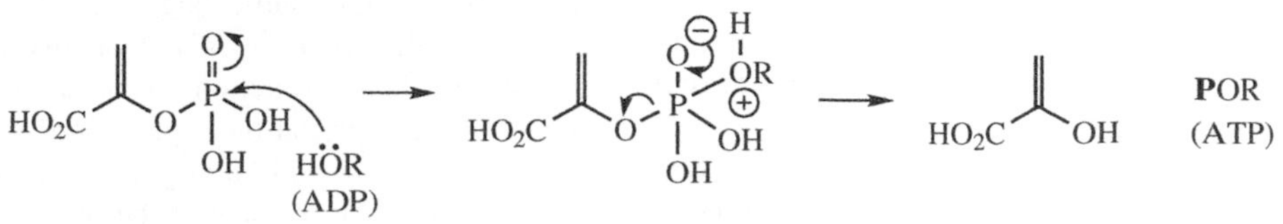

This is another example of **substrate-level phosphorylation**, but differs from the earlier example that involved hydrolysis of a mixed anhydride. Here, we have merely the hydrolysis of an ester, and thus a much lower release of energy. In fact, with 1,3-diphosphoglycerate, we specifically noted the difference in reactivity between the anhydride and ester groups. So how can this reaction lead to ATP synthesis? The answer lies in the stability of the hydrolysis product, **enolpyruvic acid**. Once formed, this enol is rapidly isomerized to its keto tautomer, **pyruvic acid**, with the equilibrium heavily favouring the keto tautomer (see Section 10.1). The driving force for the substrate-level phosphorylation reaction is actually the position of equilibrium in the subsequent tautomerization.

This completes the glycolytic pathway; well, almost. To maintain operation of the pathway, the NAD$^+$ used in the conversion of glyceraldehyde 3-phosphate into 1,3-diphosphoglycerate must be regenerated from its reduced form NADH, since only small amounts of the coenzyme will be available to the organism. If the organism is aerobic, then it is possible to use the oxidative phosphorylation processes to regenerate NAD$^+$ from NADH, and in so doing also achieve the synthesis of ATP. However, for anaerobes, or for aerobes under temporary anaerobic conditions, pyruvate synthesized in the last reaction is modified further to achieve this end. For example, certain organisms use NADH to reduce pyruvate to **lactate**, and regenerate NAD$^+$. This process might occur in actively exercised muscle, when there is a temporary shortage of oxygen, leading to a build up of lactic acid and ensuing cramp pains.

regeneration of NAD$^+$

Other organisms are equipped to produce **ethanol**, by employing a thiamine diphosphate-dependent decarboxylation of pyruvate to acetaldehyde (see Section 15.8) and NAD$^+$ is regenerated by reducing the acetaldehyde to ethanol. This is a characteristic of baker's yeast, and forms the essential process for both bread making (production of CO_2) and the brewing industry (formation of ethanol).

The glycolytic pathway is crucial to anaerobes for ATP production; this is reflected in the fact that ATP synthesis is achieved via substrate-level phosphorylation, and does not depend on the availability of oxidative phosphorylation.

The **energy yield from glycolysis** for the anaerobic decomposition of glucose to 2 mol of lactic acid may be calculated as follows:

- 2 mol of ATP are used up in phosphorylations;

- 2 mol of ATP are gained per half molecule of glucose, i.e. a total of 4 mol ATP;

- Net yield from glucose $\rightarrow$ 2 mol lactic acid = 2 mol ATP.

15.3 The Krebs cycle

The **Krebs cycle** is sometimes still referred to as the **citric acid cycle**, citric acid being one of the intermediates involved, and even the **tricarboxylic acid cycle**, in that several of the intermediates are tri-acids. As the name suggests, the process is a cycle, so that there is a reasonably constant pool of intermediates functioning in an organism, and material for degradation is processed via this pool of intermediates. Overall, though, the material processed does not increase the size of the pool. The compound that enters the cycle is the thioester **acetyl-coenzyme A (acetyl-CoA)**.

We have just seen that anaerobic organisms metabolize pyruvate from the glycolytic pathway by various means, but that the prime objective is to reoxidize NADH to NAD$^+$. In aerobic organisms, reoxidation of NADH is achieved via oxidative phosphorylation, generating ATP in the process, and there is no longer any need to sacrifice pyruvate for this purpose. Accordingly, pyruvate from glycolysis is converted into acetyl-CoA by a process known as **oxidative decarboxylation**. This transformation is dealt with in more detail in Section 15.8; but, for the moment, it can be represented by the equation

$$\text{pyruvate} + \text{HS-CoA} + \text{NAD}^+ \xrightleftharpoons[\text{3 enzyme activities}]{\text{pyruvate dehydrogenase complex}} \text{acetyl-CoA} + CO_2 + \text{NADH}$$

The whole process is multi-step, and catalysed by the **pyruvate dehydrogenase** enzyme complex, which has three separate enzyme activities. During the transformation, an acetyl group is effectively removed from pyruvate, and passed via carriers **thiamine diphosphate** (TPP) and **lipoic acid** eventually to coenzyme A, a complex material whose principal functional group involved in metabolic reactions is a thiol (see Box 10.8).

Coenzyme A
HSCoA

The requirement for NAD$^+$ is to reoxidize the lipoic acid carrier. It is worth mentioning that the pyruvate → acetaldehyde conversion we considered at the end of the glycolytic pathway involves the same initial sequence, and pyruvate decarboxylase is another thiamine diphosphate-dependent enzyme.

Acetyl-CoA (see Box 10.8) is a thioester of acetic acid with coenzyme A. It is a remarkably common intermediate in many metabolic degradative and synthetic pathways, for which the reactivity of the thioester function plays a critical role. There are two major sources of the acetyl-CoA entering the Krebs cycle: glycolysis via the oxidative decarboxylation of pyruvate and fatty acid degradation (see Section 15.4). There are other minor sources of acetyl-CoA, including metabolism of amino acids from protein.

The Krebs cycle intermediate that reacts with acetyl-CoA is **oxaloacetate**, and this reacts via an aldol reaction, giving **citryl-CoA**. However, the enzyme citrate synthase also carries out hydrolysis of the thioester linkage, so that the product is **citrate**; hence the terminology 'citric acid cycle'. The hydrolysis of the thioester is actually responsible for disturbing the equilibrium and driving the reaction to completion.

aldol reaction

HO$_2$C   H$_2$C   SCoA
   O
   H$^{\oplus}$
   CO$_2$H

oxaloacetate

$\rightleftharpoons$ citrate synthase

COSCoA
HO$_2$C—OH
   CO$_2$H

citryl-CoA

$\rightleftharpoons$ citrate synthase

CO$_2$H   H–SCoA
HO$_2$C—OH
   CO$_2$H

citrate

also involves hydrolysis of thioester which drives reaction from oxaloacetate to citrate

The aldol reaction is easily rationalized, with acetyl-CoA providing an enolate anion nucleophile that adds to the carbonyl of oxaloacetate – easily rationalized, but surprising. Oxaloacetate is more acidic than acetyl-CoA, in that there are two carbonyl groups flanking the methylene. If one were to consider a potential base-catalysed reaction between these two susbstrates, then logic suggests that oxaloacetate would be preferentially converted into

an enolate anion nucleophile. This could then attack the carbonyl of acetyl-CoA, but via a Claisen reaction, since there is a thiolate leaving group. That citrate synthase achieves an aldol reaction (as shown) reflects that the enzyme active site must have a basic residue appropriately positioned to abstract a proton from acetyl-CoA allowing it to act as the nucleophile (see Box 13.8).

oxaloacetate is the more acidic substrate

HO$_2$C
   O
   CO$_2$H
H$_3$C   SCoA
   O

Claisen reaction $\longrightarrow$

HO$_2$C
H$_3$C   O
$^{\ominus}$O
   SCoA   CO$_2$H

$\longrightarrow$

HO$_2$C
H$_3$C   O
   O   CO$_2$H

predicted base-catalysed product

Citrate is subsequently isomerized to **isocitrate**; this involves dehydration and rehydration via the intermediate *cis*-**aconitate**. Both reactions are

catalysed by the single enzyme aconitase. They may be considered simply as acid-catalysed elimination followed by acid-catalysed addition reactions.

HO$_2$C
HO—CO$_2$H
HO$_2$C

citrate

$\rightleftharpoons$ H$_2$O
aconitase

CO$_2$H
   CO$_2$H
   CO$_2$H

cis-**aconitate**

$\rightleftharpoons$ H$_2$O
aconitase

CO$_2$H
   CO$_2$H
H
HO   CO$_2$H

isocitrate

H—C
HO—C

oxidizable function

Worthy of note in this reaction is that citrate displays prochirality (see Section 3.4.7). The methylene carbons may be considered prochiral, in that enzymic elimination of a proton is likely to be entirely stereospecific. In addition, the apparently equivalent side-chains on the central carbon are also prochiral and going to be positioned quite differently on the enzyme. This means that only one of these side-chains is involved in the dehydration–rehydration

sequence, and it can be shown from labelling studies that the side-chain modified is not the one that was recently derived from acetyl-CoA as nucleophile.

In isocitrate, there is a CHOH group that is available for oxidation via the coenzyme NAD$^+$ and the enzyme isocitrate dehydrogenase. NADH will then be reoxidized via oxidative phosphorylation, and lead to ATP synthesis. The oxidation product from isocitrate is **oxalosuccinate**, a β-ketoacid that easily decarboxylates through an intramolecular hydrogen-bonded system (see Section 10.9). Although thermal (non-enzymic) decarboxylation would probably occur readily, it turns out that the enzyme isocitrate dehydrogenase also catalyses this reaction.

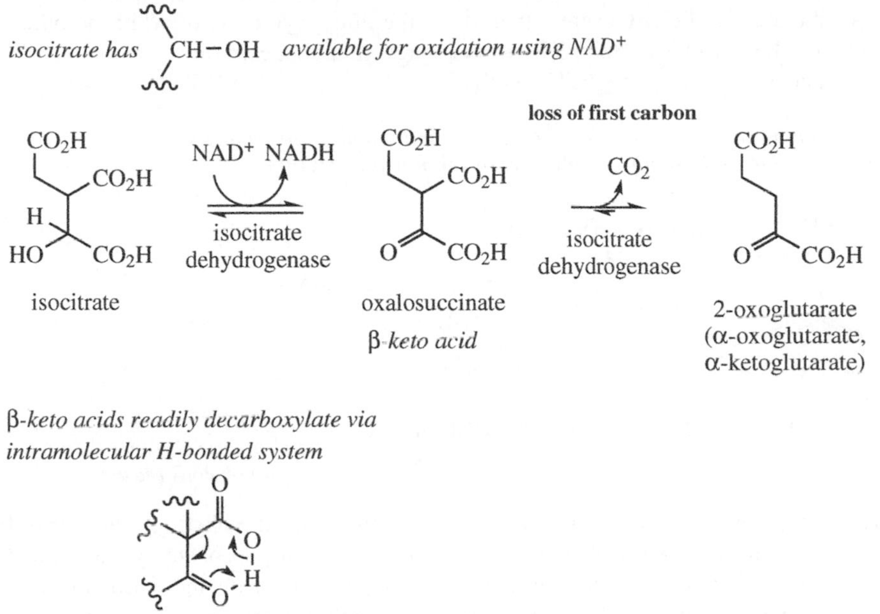

isocitrate has CH–OH *available for oxidation using NAD^+*

loss of first carbon

isocitrate

oxalosuccinate

β-*keto acid*

2-oxoglutarate
(α-oxoglutarate,
α-ketoglutarate)

β-*keto acids readily decarboxylate via intramolecular H-bonded system*

The product is **2-oxoglutarate**, sometimes referred to as α-oxoglutarate or α-ketoglutarate. Note specifically that we have just lost one of the carbon atoms. Oxaloacetate (a C_4 compound) reacted with acetyl-CoA (C_2) to give citrate (C_6), and this reaction now gives us a C_5 compound; to complete the cycle and get back to C_4, we shall need to lose another carbon atom. This is achieved in the next reaction catalysed by **2-oxoglutarate dehydrogenase**.

a repeat of the pyruvate → acetyl-CoA oxidative decarboxylation then occurs - similarly requires thiamine diphosphate, lipoic acid, coenzyme A and NAD^+

2-oxoglutarate

+ HS–CoA + NAD^+

2-oxoglutarate dehydrogenase complex 3 enzyme activities

loss of second carbon

+ CO_2 + NADH

succinyl-CoA

note that neither carbon lost originates from the acetyl-CoA added in the first reaction

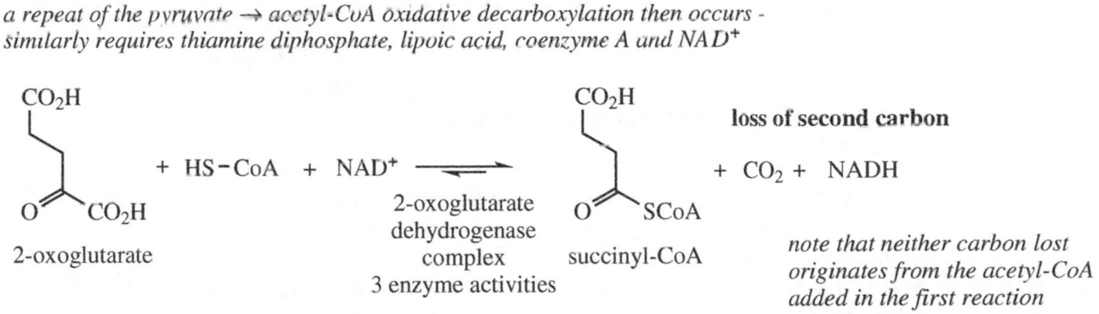

Now this reaction is effectively a repeat of the pyruvate → acetyl-CoA oxidative decarboxylation we saw at the beginning of the Krebs cycle. It similarly requires thiamine diphosphate, lipoic acid, coenzyme A and NAD^+. A further feature in common with that reaction is that 2-oxoglutarate dehydrogenase is also an enzyme complex comprised of three separate enzyme activities. 2-Oxoglutarate is thus transformed into **succinyl-CoA**, with the loss of a further carbon as CO_2, and producing NADH that can be exploited in ATP synthesis via oxidative phosphorylation. Note that, because of the prochirality in citric acid and subsequent enzymic selectivity, neither of the carbon atoms lost in the two decarboxylations originates from the acetyl-CoA molecule added in the first reaction, the aldol addition. These carbons are not lost until further cycles of the pathway have been completed.

The product succinyl-CoA is able to participate in ATP synthesis as an example of **substrate-level phosphorylation** – we met some other examples in the glycolytic pathway. Essentially, hydrolysis of succinyl-CoA liberates sufficient energy that it can be coupled to the synthesis of ATP from ADP. However, guanosine triphosphate (GTP) is the nucleoside phosphate produced, rather than ATP. ATP is then produced indirectly from GTP. There appears to be no obvious reason why this reaction should be coupled to the synthesis of GTP, rather than to the direct synthesis of ATP; the other product of the reaction is **succinate**.

hydrolysis of thioester has large negative ΔG and can be coupled to synthesis of GTP (guanosine triphosphate) – **substrate-level phosphorylation**

succinyl-CoA → succinate + HSCoA (succinyl-CoA synthetase, GDP, **P**, GTP)

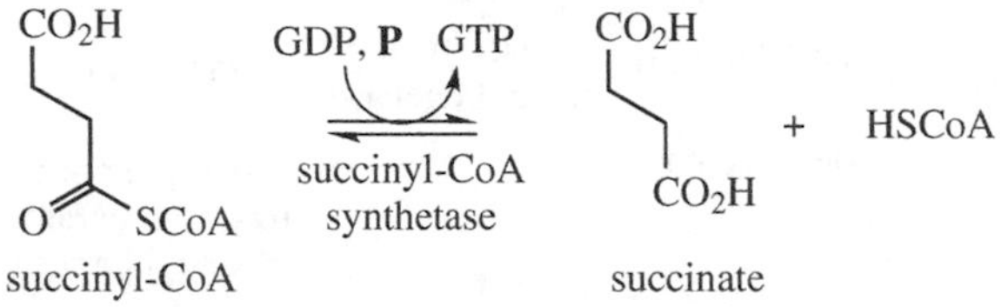

then   GTP + ADP ⇌ GDP + ATP

query?
why use G rather than A directly?
no obvious reason

When we investigate this substrate-level phosphorylation reaction in detail, we find it also involves a molecule of phosphate. Phosphate reacts initially with succinyl-CoA, converting the thioester into an acyl phosphate, which is, of course, a mixed anhydride (see Box 7.27). It is actually hydrolysis of this mixed anhydride that can be coupled to nucleoside triphosphate synthesis, and it is fitting to compare this with the formation and hydrolysis of 1,3-diphosphoglycerate in the glycolytic pathway (see Section 15.2).

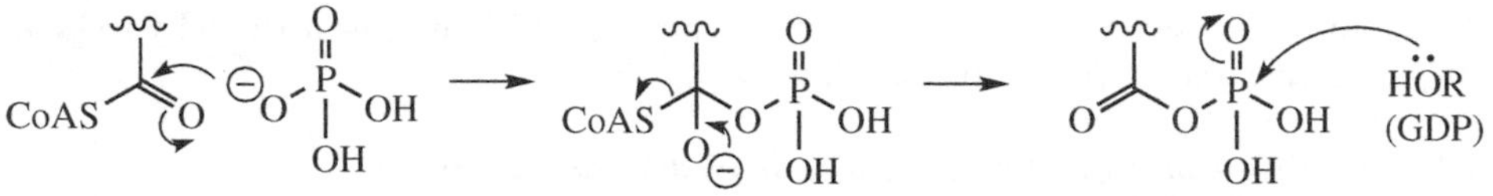

acyl phosphate (mixed anhydride)

this sequence is analogous to the formation / hydrolysis of 1,3-diphosphoglycerate in glycolysis

As we move on in the Krebs cycle, the next reaction is oxidation of the CH_2–CH_2 grouping in succinate to give the unsaturated diacid **fumarate**. We have already looked at this type of oxidation and seen that it involves a dehydrogenase enzyme coupled to a flavin nucleotide coenzyme (see Section 15.1.2). For this reaction, the coenzyme is FAD. The reduced form of FAD can then be reoxidized to FAD via oxidative phosphorylation, generating energy in the form of ATP in the process.

succinate contains –CH_2CH_2– *; oxidation using FAD*

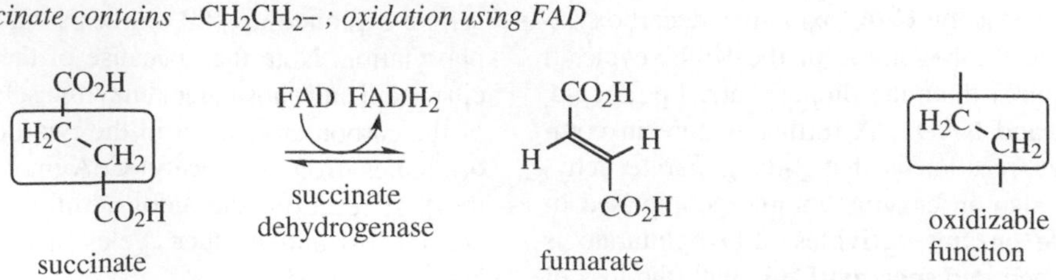

succinate → fumarate (succinate dehydrogenase, FAD, FADH₂)   oxidizable function

The sequence continues with hydration, addition of water, to produce **malate**, which contains an oxidizable CHOH group. Oxidation involves NAD^+,

and results in the formation of **oxaloacetate**, which completes the cycle and regenerates the substrate to react with further acetyl-CoA.

addition of water produces malate, which contains $CH-OH$

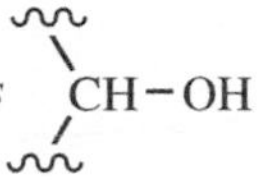

oxidation using using NAD^+ gives oxaloacetate

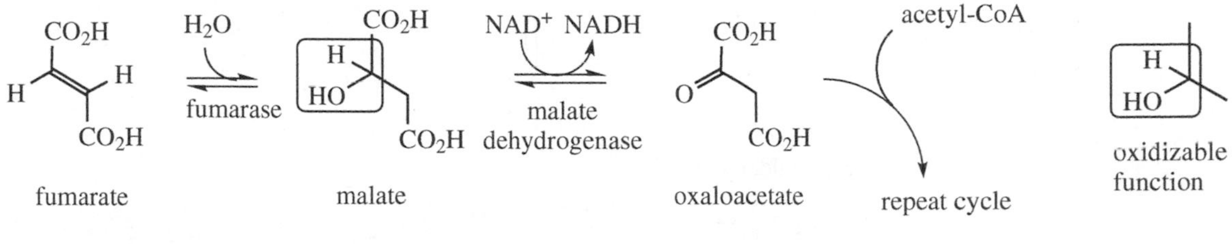

fumarate   malate   oxaloacetate   repeat cycle   oxidizable function

this sequence of reactions is also seen in β-oxidation / fatty acid metabolism

This sequence of reactions, namely oxidation of $CH_2–CH_2$ to $CH=CH$, then hydration to $CH_2–CHOH$, followed by oxidation to $CH_2–CO$, is a sequence we shall meet again in the β-oxidation of fatty acids (see Section 15.4.1). The first oxidation utilizes FAD as coenzyme, the second NAD^+. In both cases, participation of the oxidative phosphorylation system allows regeneration of the oxidized coenzyme and the subsequent generation of energy in the form of ATP.

The **energy yield from the Krebs cycle** by the aerobic breakdown of pyruvate may be calculated as follows. overall:

$$CH_3COCO_2H + 3 H_2O \longrightarrow 3 CO_2 + 5 \times 2H$$

- five pairs of hydrogen atoms are available for oxidation;
- four pairs are passed to NAD^+ and via the respiratory chain yield $4 \times 3 = 12$ mol ATP;
- one pair is passed to FAD and via the respiratory chain yields 2 mol ATP;
- there is also the gain of one ATP via GTP;
- therefore, there will be a total yield of 15 mol ATP.

By combining the glycolytic pathway, the Krebs cycle, and oxidative phosphorylation, the energy yield from the aerobic degradation of glucose will be

glycolysis

glucose $\longrightarrow$ $2 CH_3COCO_2H$ + 2 ATP + 2 NADH

Krebs cycle

respiratory chain

2×15 mols ATP

2×3 mols ATP

Total = 38 mols ATP

The total yield of 38 mol ATP by aerobic degradation of glucose may not be achieved under all circumstances, but it is, nevertheless, considerably more efficient than that from the anaerobic breakdown, namely 2 mol ATP (see Section 15.2).

15.4 Oxidation of fatty acids

Fat degradation provides a major source of energy for most organisms. **Fats** are esters of **glycerol** with long-chain **fatty acids** (see Box 7.16) and are hydrolysed by the action of enzymes called **lipases**. This gives the alcohol portion, glycerol, together with a range of fatty acids, such as stearic acid. Most fats taken in the diet provide a range of fatty acids of varying chain length and different levels of unsaturation, according to source. Because of their biosynthetic origin (see Section 15.5), the vast majority of fatty acids have an even number of carbon atoms.

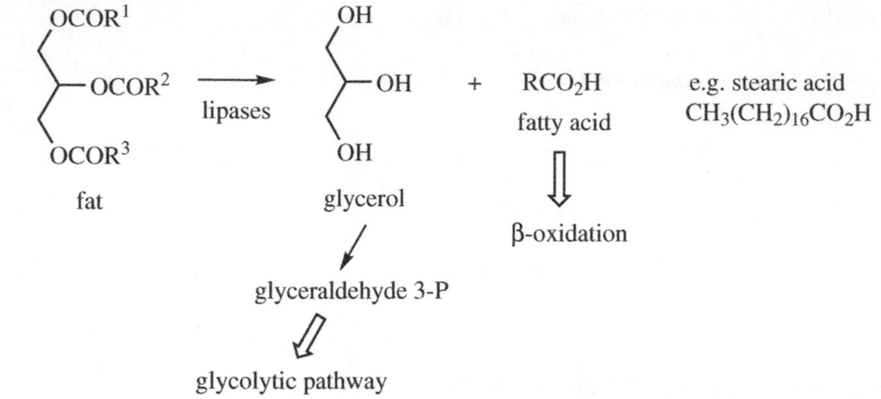

β-**oxidation** *involves the sequential removal of two-carbon units via oxidation at th e β position*

Glycerol provides a minor source of energy, in that it can be modified readily to **glyceraldehyde 3-phosphate**, one of the intermediates in the glycolytic pathway. The fatty acids are metabolized by a process termed **β-oxidation**, which involves the sequential removal of two-carbon units via oxidation at the β-position. The process for saturated fatty acids will now be described.

15.4.1 Metabolism of saturated fatty acids

The free fatty acid needs activating before it can be metabolized. This is achieved by conversion into its **thioester** by esterification with **coenzyme A**. We have already seen that thioesters are reactive entities, and it is reasonable, therefore, to suppose that such activation will cost energy. It is achieved in a two-stage reaction catalysed by a single enzyme, an acyl-CoA synthetase. Energy is supplied in the form of ATP.

The fatty acid is initially converted into an acyl-AMP derivative by attack of the carboxylate as a nucleophile onto the P=O system of ATP, with loss of diphosphate as a leaving group. This reaction is far from favourable, and the equilibrium is disturbed by subsequent pyrophosphatase-catalysed hydrolysis of diphosphate into two molecules of phosphate.

fatty acid is activated by conversion into thioester with coenzyme A; a two-stage reaction catalysed by single enzyme

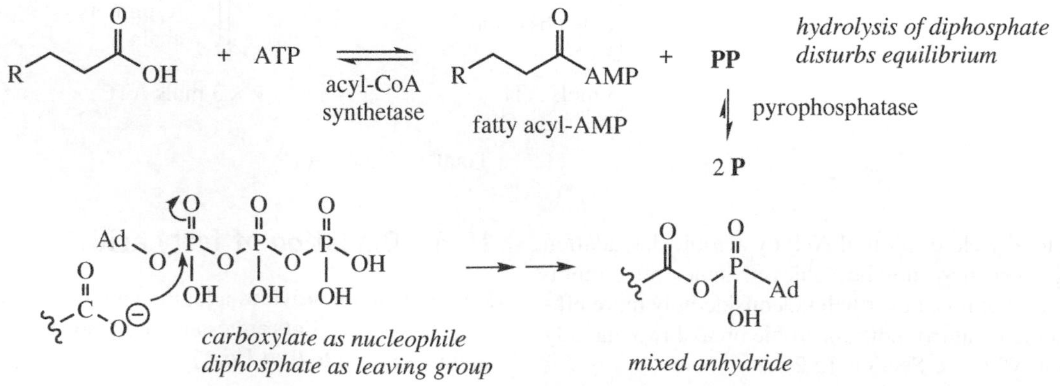

This means that the energy demands (ATP → AMP) are equivalent to two ATP → ADP transformations. However, the product fatty acyl-AMP is actually a reactive mixed anhydride and may be attacked by the thiol group of coenzyme A, giving the required thioester. We have met an analogous series of reactions in non-ribosomal peptide biosynthesis (see Section 13.5.2).

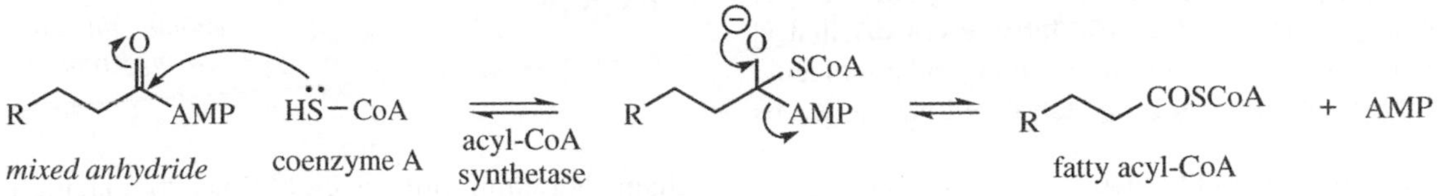

mixed anhydride   coenzyme A   acyl-CoA synthetase   fatty acyl-CoA   + AMP

Oxidation at the β-position is then achieved by the same sequence of dehydrogenation, hydration, and dehydrogenation reactions that we have seen earlier in the succinate → fumarate → malate → oxaloacetate transformations in the Krebs cycle (see Section 15.3).

β-oxidation is achieved by a sequence of dehydrogenation, hydration, and dehydrogenation

FAD FADH$_2$ acyl-CoA dehydrogenase

oxidizable function

H$_2$O enoyl-CoA hydratase

NADH NAD$^+$ β-hydroxyacyl-CoA dehydrogenase

oxidizable function

two dehydrogenation reactions; both FADH$_2$ and NADH can then yield ATP via oxidative phosphorylation

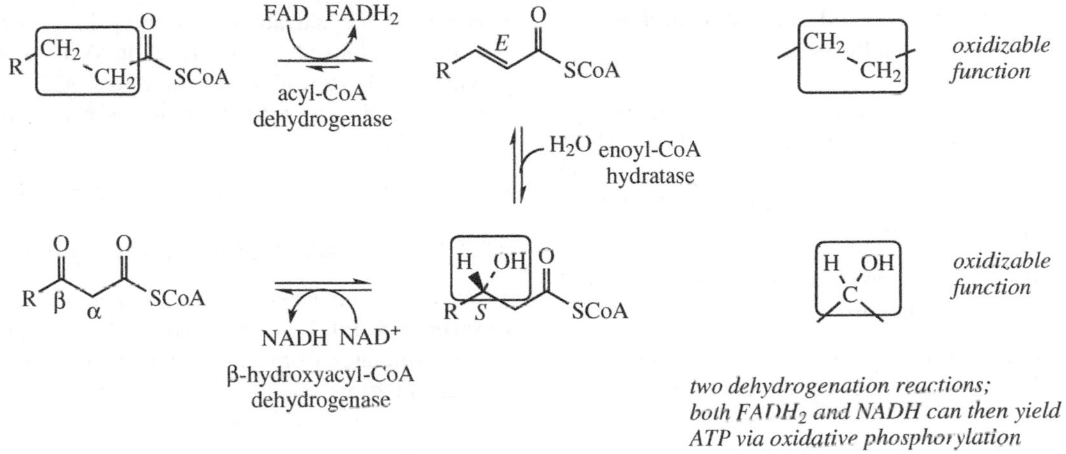

compare the equivalent sequence in the Krebs cycle:

succinate → fumarate → malate → oxaloacetate

Because of the enzyme specificity in the hydration step, the new carbonyl group is introduced β to the original thioester carbonyl. The sequence includes two dehydrogenation reactions, and involves both FAD and NAD$^+$ as coenzymes. The reduced forms of these coenzymes can be reoxidized by means of oxidative phosphorylation, and can, therefore, yield ATP. Although the reactions just described form the basis of β-oxidation, the terminology β-oxidation when applied to fatty acid metabolism is usually understood to include the next step, the sequential chain shortening.

then follows cleavage of acetyl-CoA from the end of the chain via a reverse Claisen reaction

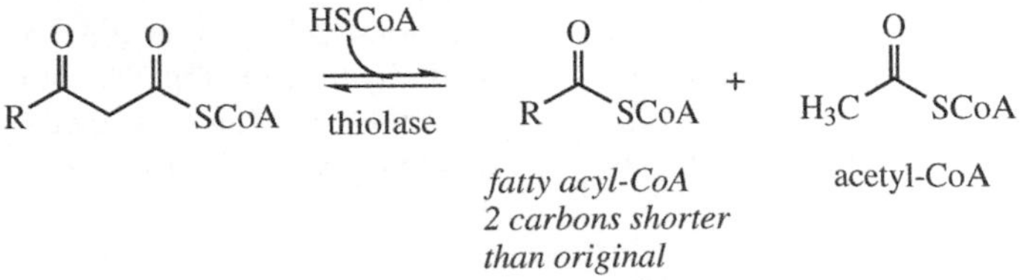

HSCoA   thiolase

fatty acyl-CoA 2 carbons shorter than original   acetyl-CoA

There follows cleavage of acetyl-CoA from the end of the chain via a **reverse Claisen reaction** (see Box 10.15). This requires use of a molecule of coenzyme A as nucleophile, with the loss of the enolate anion of acetyl-CoA as leaving group. The net result is production of a new fatty acyl-CoA that is two carbons shorter than the original, and a molecule of acetyl-CoA that can be metabolized via the Krebs cycle.

Thus, a fatty acid such as stearic acid (C_{18}), after activation, can undergo the β-oxidation and

enolate anion as leaving group

chain shortening process eight times, producing nine molecules of acetyl-CoA for further metabolism.

stearic acid can undergo β-oxidation / chain shortening 8×

$$CH_3(CH_2)_{16}CO_2H \longrightarrow 9 \ CH_3CO{-}SCoA \Longrightarrow Krebs \ cycle$$

The overall **energy yield from β-oxidation** may thus be calculated as follows:

- Each sequence of β-oxidation involves the passage of one pair of hydrogen atoms to FAD (which yields 2 mol ATP via the respiratory chain) and one pair to NAD^+ (which yields 3 mol ATP via the respiratory chain).

- Stearoyl-CoA thus produces $8 \times 5 = 40$ mol ATP from eight β-oxidations.

- The nine acetyl-CoA moles generated will yield $9 \times 12 = 108$ mol ATP via the Krebs cycle.

- However, the activation of stearic acid to stearoyl-CoA is achieved by the reaction ATP → AMP, which is the effective loss of 2 mol ATP; nevertheless, only one activation step is necessary per fatty acid.

- Total yield is thus $40 + 108 - 2 = 146$ mol ATP.

15.4.2 Metabolism of unsaturated fatty acids

Much of the fat taken in via the diet will contain **unsaturated fatty acids**, particularly that portion which originates from plant material. For example, the fats in olive oil contain up to 85% oleic acid

---

Box 15.1

Comparison of fat and carbohydrate as energy stores

It is instructive to compare the energy yield from three molecules of **glucose** (C_{18}) with that from one molecule of **stearic acid** (also C_{18}).

We saw that aerobic degradation of each molecule of glucose via glycolysis and the Krebs cycle gave 38 mol ATP; three molecules would thus give $3 \times 38 = 114$ mol ATP. Stearic acid gives 146 mol ATP.

The higher energy yield per carbon atom from fatty acid compared with carbohydrate reflects its higher level of reduction, which consequently allows more oxidation. Thus, fat is logically the preferred storage molecule to carbohydrate. This is

borne out in practice. A 70 kg man would typically have fat reserves of about 7 kg, equivalent to his energy needs for 1 month, and carbohydrate reserves of about 0.35 kg, equivalent to his energy needs for only about 1 day.

This is undoubtedly why low-carbohydrate diets have proved so effective for rapid weight loss. As soon as the reserves of carbohydrate are used up, the body resorts to metabolizing fat for its energy needs. This continues whenever carbohydrate intake is limited. It should also be appreciated that although carbohydrate can readily be converted into fat (via acetyl-CoA; see Section 15.5), fat is not readily converted into carbohydrate in animals. Fat metabolism produces acetyl-CoA, which is then usually metabolized completely via the Krebs cycle.

(C$_{18}$ unsaturated), and only relatively small amounts of saturated fatty acids. Animal fats have a much higher proportion of saturated fatty acid derivatives, but they still contain a substantial level of unsaturated fatty acids. The fatty acid analysis of butterfat, for example, shows it contains about 28% oleic acid,

and most of the remainder is composed of saturated fatty acids: 13% stearic acid (C$_{18}$), 29% palmitic acid (C$_{16}$), 12% myristic acid (C$_{14}$), and other shorter-chain saturated fatty acids. The vast majority of natural unsaturated fatty acids have one or more double bonds with the Z or *cis* configuration.

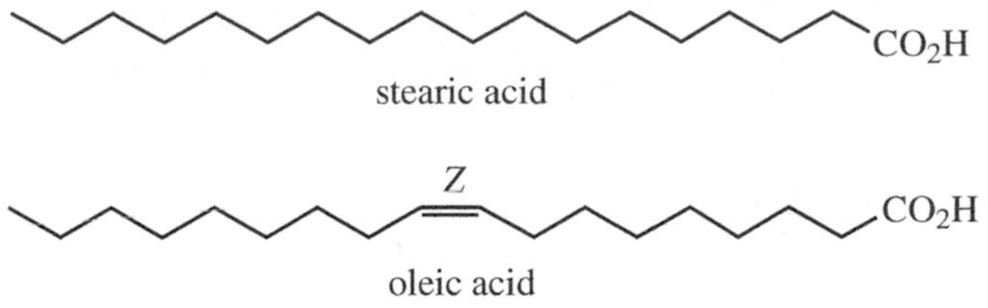

stearic acid

oleic acid

Metabolism of unsaturated fatty acids is similar to that of the saturated compounds just described, but additional enzymic reactions are necessary.

Thus, oleoyl-CoA, the CoA ester of oleic acid, will undergo β-oxidation three times, until the C$_{12}$ derivative is reached. The 3,4-Z-double bond in this

compound now prevents the normal dehydrogenation step that should introduce a 2,3-double bond. As a result, the normal degradative process stops until this compound is isomerized to the normal intermediate with a 2,3-*E*-double bond by the action of an **isomerase** enzyme.

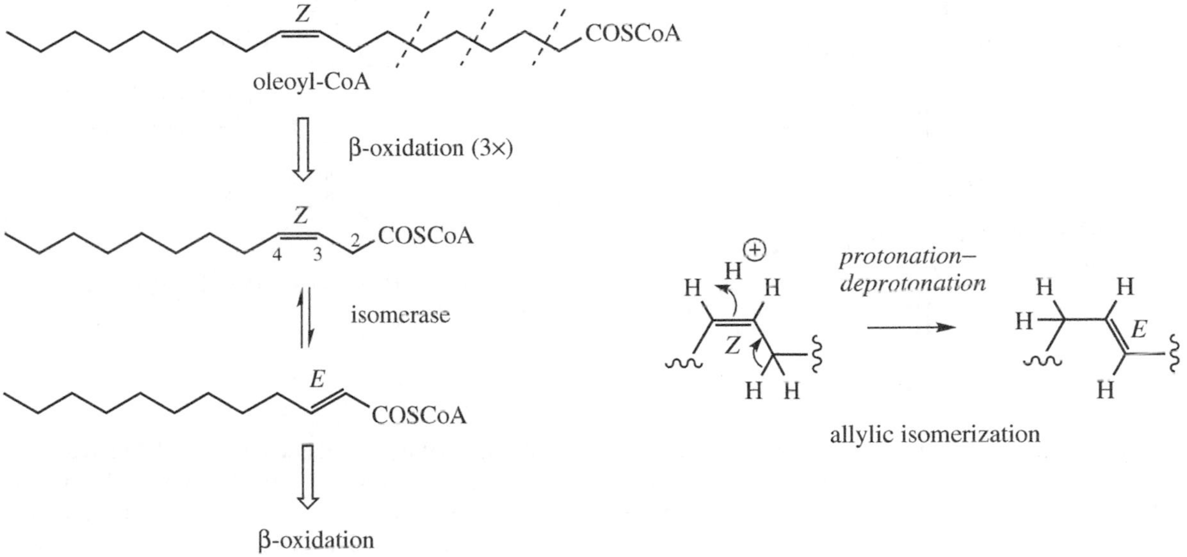

oleoyl-CoA

β-oxidation (3×)

isomerase

protonation–deprotonation

allylic isomerization

β-oxidation

By means of this additional step, the β-oxidation process can then continue as normal. The energy yield will be only slightly less than that for stearoyl-CoA, since there is omission of transfer of one pair of hydrogen atoms to FAD, and consequently loss of 2 mol ATP.

Of course, the double bond in the starting ester may end up in the correct position for the β-oxidation

processes, but it turns out that the usual Z or *cis* configuration of this double bond is wrong for the normal enzymes. Although hydration of the Z double bond occurs, the configuration of the hydroxy derivative is wrong for the subsequent dehydrogenase, so an inversion to the required configuration is achieved by the action of an **epimerase** enzyme. β-Oxidation processes can then continue normally.

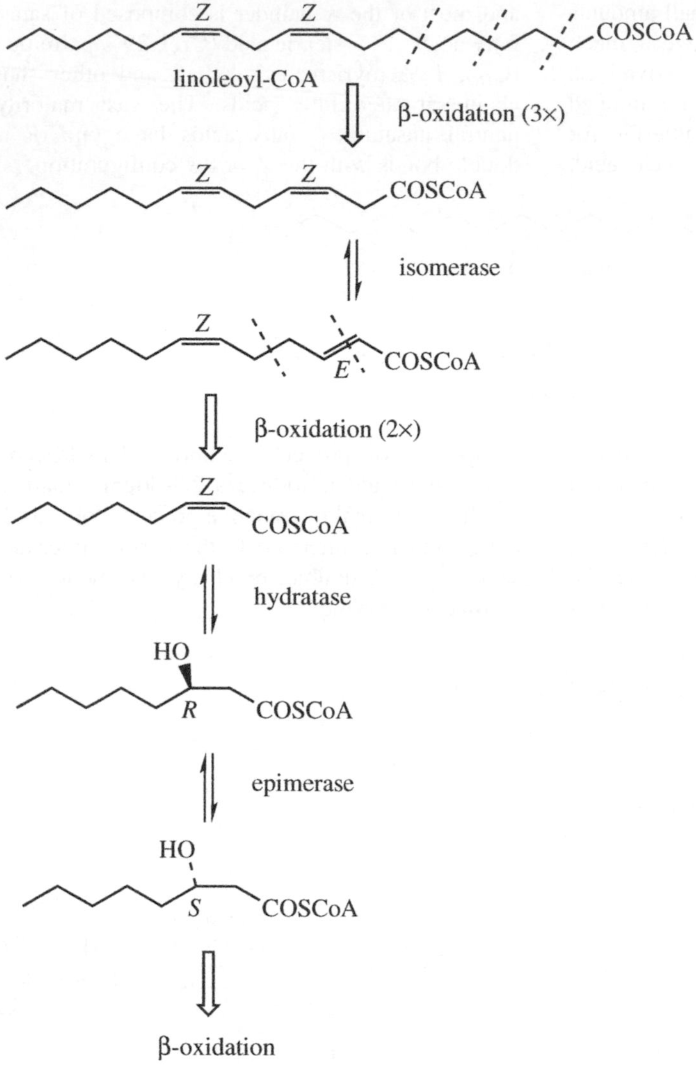

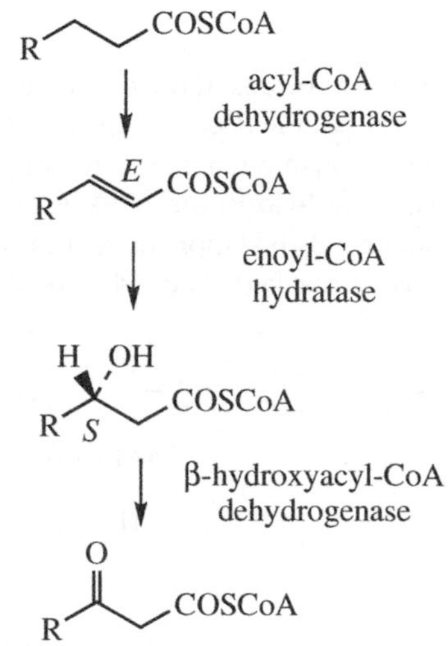

stereochemistry of β-oxidation

These processes are shown for the CoA ester of linoleic acid, the most common of the polyunsaturated acids.

15.5 Synthesis of fatty acids

Fatty acid synthesis provides an organism with a means of storing energy in the form of an organic

molecule that can be degraded by oxidative reactions whenever necessary. In principle, fatty acid synthesis is the reverse of fatty acid metabolism, though there are some fundamental differences, which are quite logical when we consider the chemical reactivity of the intermediate reagents.

Fatty acid degradation involves a reverse Claisen reaction

reverse Claisen reaction

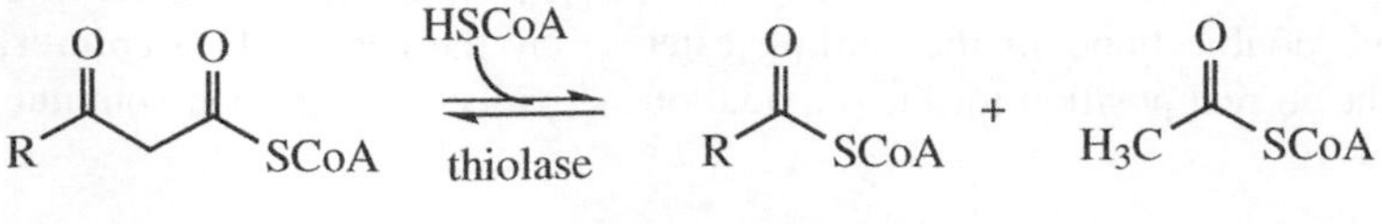

Therefore, we could consider using the **Claisen   reaction** in fatty acid synthesis.

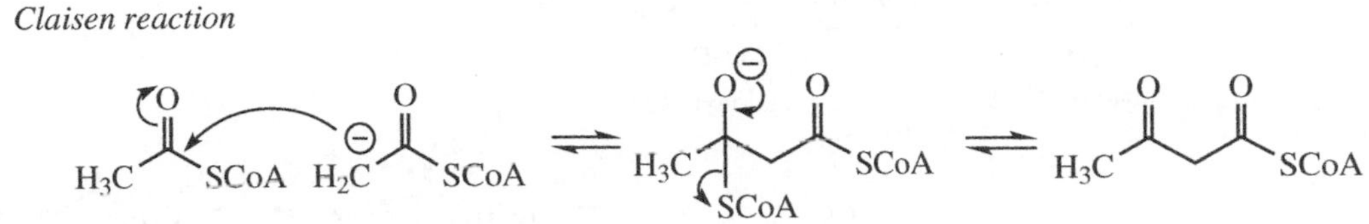

Claisen reaction

However, a more favourable pathway is used, employing a more reactive nucleophile. Rather than using the enolate anion derived from **acetyl-CoA**, nature uses the enolate anion derived from malonyl-CoA. **Malonyl-CoA** is obtained from acetyl-CoA by means of an enzymic carboxylation reaction, incorporating CO_2 (usually from the soluble form bicarbonate). Now CO_2 is a particularly unreactive material, so this reaction requires the input of energy (from ATP) and the presence of a suitable coenzyme, **biotin**, as the carrier of CO_2 (see Section 15.9). The

conversion of acetyl-CoA into malonyl-CoA increases the acidity of the α-hydrogens, since the acidic protons are flanked by two carbonyl groups, and thus it is easier to generate a nucleophile for the Claisen condensation. We should relate this to the use of diethyl malonate rather than ethyl acetate as a nucleophile in Section 10.9.

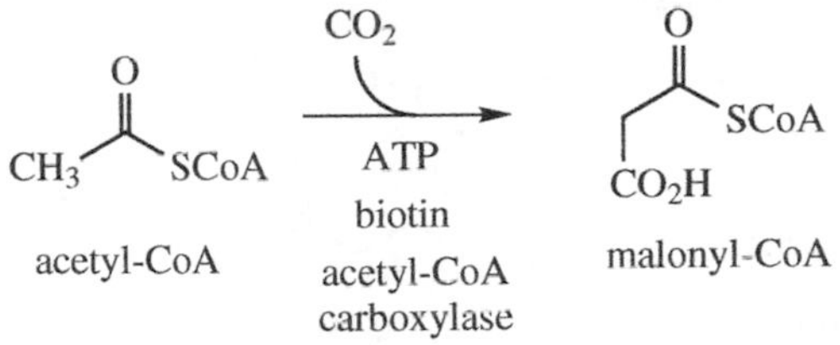

The Claisen reaction can now proceed smoothly, but nature introduces another little twist. The carboxyl group introduced into malonyl-CoA is simultaneously lost by a decarboxylation reaction during the Claisen condensation. Accordingly, we now see that the carboxylation step helps to activate the α-carbon and facilitate Claisen condensation, and the carboxyl is immediately removed on completion of this task. An alternative rationalization is that decarboxylation of the malonyl ester is used to generate the acetyl enolate anion without any requirement for a strong base (see Box 10.17).

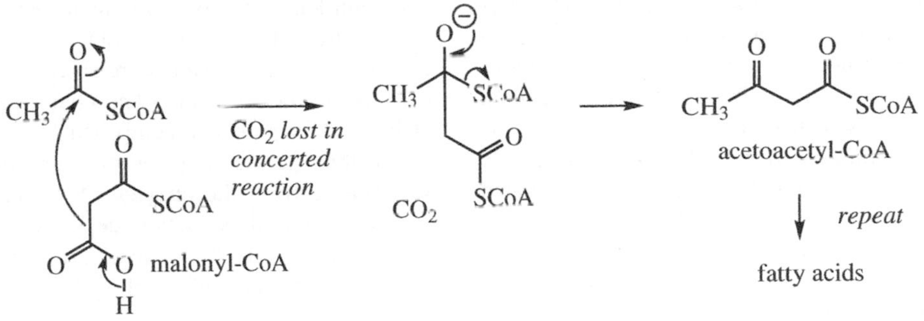

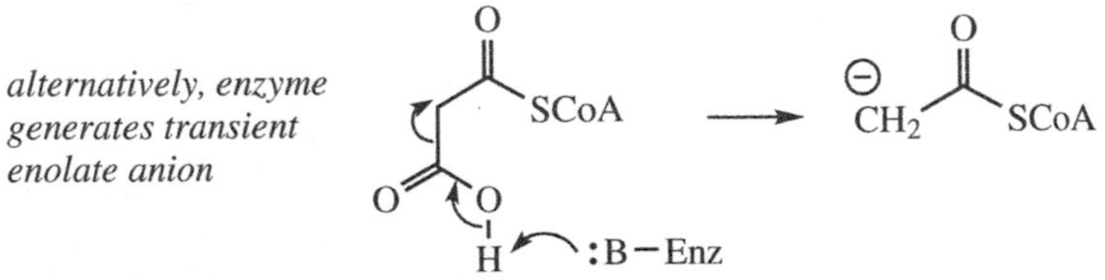

The processes of fatty acid biosynthesis are catalysed by the enzyme **fatty acid synthase**. In animals, this is a multifunctional protein containing all of the catalytic activities required, whereas in

bacteria and plants it is an assembly of enzymes that can be separated. Acetyl-CoA and malonyl-CoA themselves are not involved in the condensation step: they are converted into enzyme-bound **thioesters**.

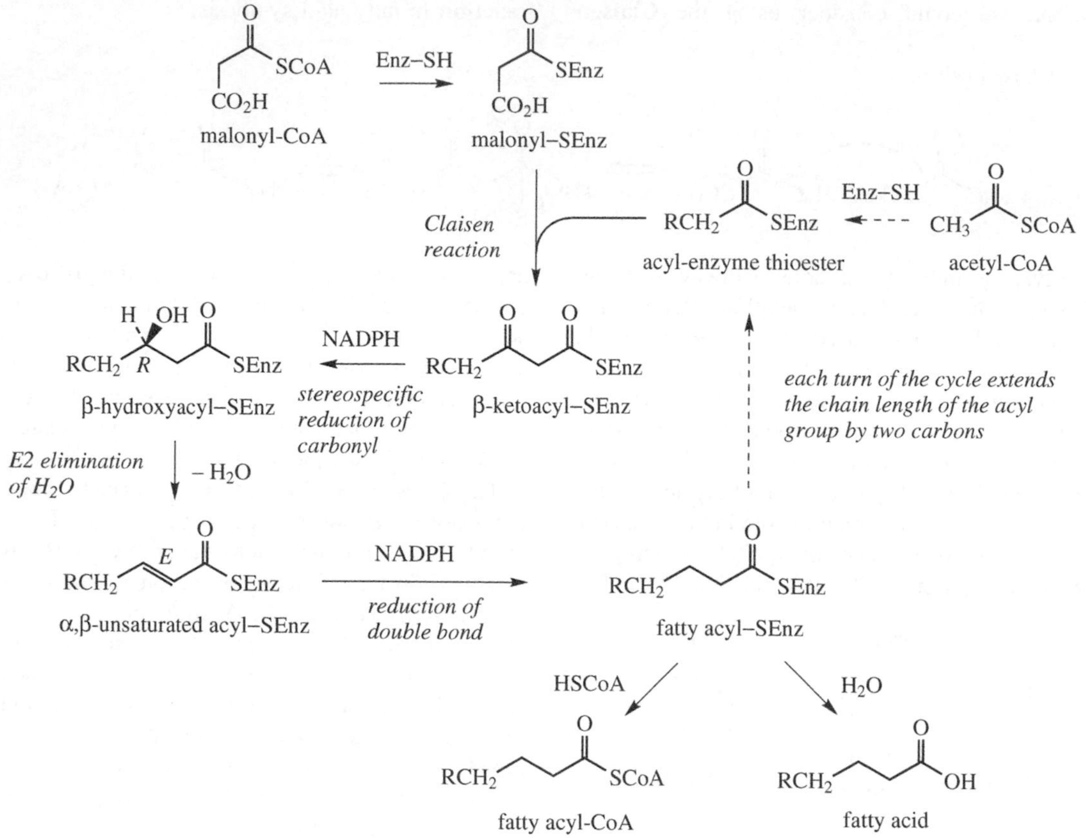

The Claisen reaction follows, giving the acetoacetyl thioester (β-ketoacyl–SEnz; R=H), which is reduced stereospecifically to the corresponding β-hydroxy ester, consuming NADPH in the reaction. Then follows elimination of water, giving the E (*trans*) α,β-unsaturated ester. Reduction of the double bond again utilizes NADPH and generates a saturated acyl–SEnz (fatty acyl–SEnz; R=H) that is two carbons longer than the starting material. This can feed back into the system, condensing again with malonyl thioester, and going through successive reduction, dehydration and reduction steps, gradually increasing the chain length by two carbons for each cycle, until the required chain length is obtained. At that point, the fatty acyl chain can be released as a fatty acyl-CoA or as the free acid. The chain length actually elaborated is probably controlled by the specificity of the thioesterase enzymes that subsequently catalyse release from the enzyme. Note that the reduction, dehydration, reduction steps are essentially the reverse of the oxidation, hydration, oxidation steps in fatty acid metabolism, though the enzymes and coenzymes involved are different.

Box 15.2

Fatty acid synthase

Fatty acid synthesis is catalysed in animals by the enzyme **fatty acid synthase**, which is a multifunctional protein containing all of the catalytic activities required. Bearing in mind the necessity to provide a specific binding site for the various substrates involved, and then the fairly complex sequence of reactions carried out, it raises the question of just how it is possible for this process to be achieved at the enzymic level. Nature has devised an elaborate but satisfyingly simple answer to this problem.

The fatty acid synthase protein is known to contain an **acyl carrier protein** (ACP) binding site, and also an active-site cysteine residue in the β-ketoacyl synthase domain. Acetyl and malonyl groups are successively transferred from coenzyme A esters and attached to the thiol groups of Cys and ACP.

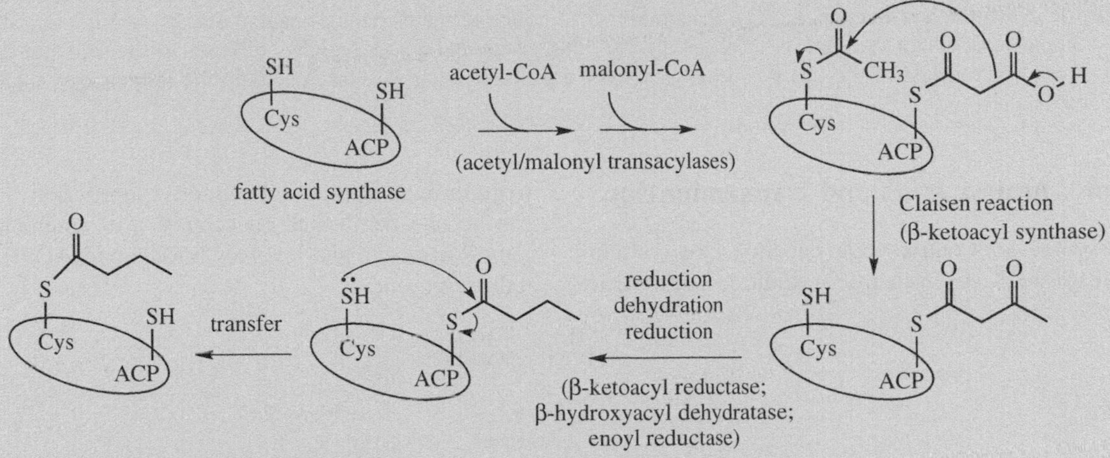

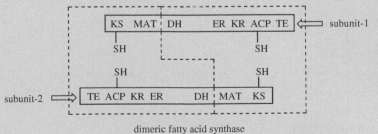

The Claisen condensation occurs, and the processes of reduction, dehydration, and reduction then occur whilst the growing chain is attached to ACP. The ACP carries a **phosphopantetheine** group exactly analogous to that in coenzyme A (pantothenic acid: vitamin B_5). This phosphopantetheine group provides a long flexible arm, enabling the growing fatty acid chain to reach the active site of each enzyme in the complex, and allowing the different chemical reactions to be performed without releasing intermediates from the enzyme. The chain is then transferred to the thiol of Cys, and the process can continue.

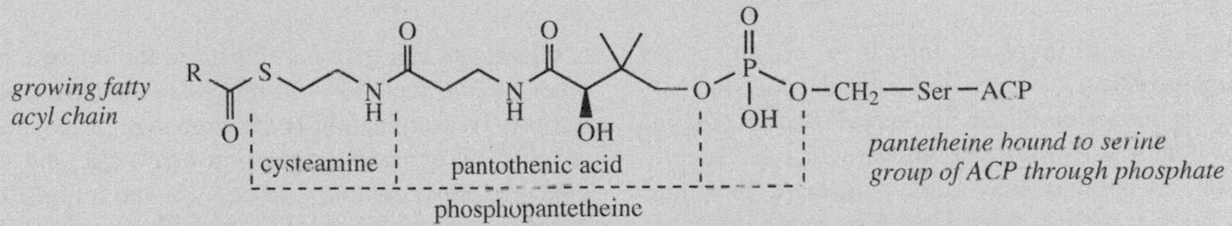

Making the process even more efficient, animal fatty acid synthase is a dimeric protein containing two catalytic centres, and it is able to generate two growing chains at the same time. The monomeric subunits are also arranged head to tail, so that the acyl group of one unit actually picks up a malonyl extender from the other unit. It is interesting that the sequence of enzyme activities along the protein chain of the enzyme complex does not correspond with the order in which they are employed.

dotted lines indicate two sites for fatty acid synthesis, utilizing enzyme activities from both subunits

ACP: acyl carrier protein
DH: dehydratase
ER: enoylreductase
KR: β-ketoacylreductase
KS: β-ketoacylsynthase
MAT: malonyl/acetyltransferase
TE: thioesterase

dimeric fatty acid synthase

A similar approach is employed in the formation of peptides such as peptide antibiotics (see Box 13.11). In marked contrast to the ribosomal biosynthesis of proteins, where a biological production line interprets the genetic

---

Box 15.2 (continued)

code, many natural peptides are known to be synthesized by a more individualistic sequence of enzyme-controlled processes, in which each amino acid is added as a result of the specificity of the enzyme involved. The many stages of the whole process appear to be carried out by a multi-functional enzyme termed a **non-ribosomal peptide synthase** (see Section 13.5.2). Pantothenic acid bound to the enzyme as pantotheine is used to carry the growing peptide chain through its thiol group. The long 'pantotheinyl arm' allows different active sites on the enzyme to be reached in the chain assembly process, a process remarkably analogous to the fatty acid synthase mechanism.

15.6   Amino acids and transamination

The synthesis of amino acids depends upon the amination of the Krebs cycle intermediate **2-oxoglutarate** to **glutamate**, a process of **reductive amination**. This can occur when a high concentration of ammonium ions is available and involves NADH or NADPH as reducing agent.

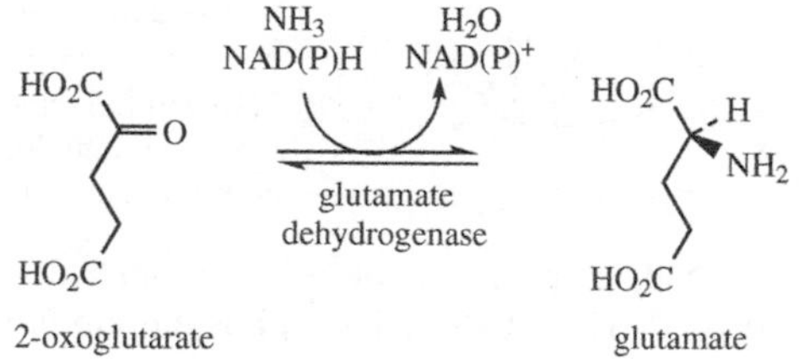

2-oxoglutarate   glutamate

this involves formation and reduction of an intermediate imine

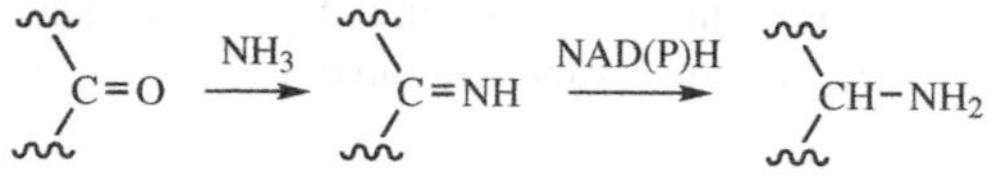

The reaction involves formation of an imine through reaction of ammonia with the ketone, followed by reduction of this imine (see Section 7.7.1). As we noted earlier (see Section 15.1.1), nicotinamide coenzymes may also participate in imine reductions as well as aldehyde/ketone reductions, further emphasizing the imine–carbonyl analogy (see Section 7.7.1). The reverse reaction, removal of ammonia from glutamate, is also of importance in amino acid catabolism.

Glutamate can then participate in the formation of other amino acids via the process called **transamination**. Transamination is the exchange of the amino group from an amino acid to a keto acid, and provides the most common process for the introduction of nitrogen into amino acids, and for the removal of nitrogen from them. The reaction is catalysed by a **transaminase** enzyme, and the coenzyme **pyridoxal phosphate** (**PLP**) is required.

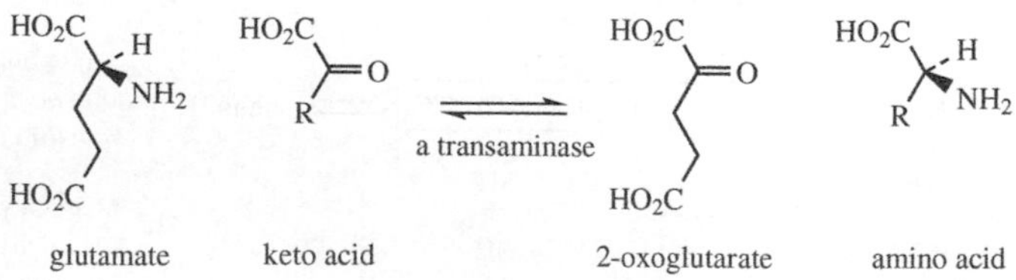

glutamate   keto acid   2-oxoglutarate   amino acid

coenzyme **pyridoxal phosphate** involved

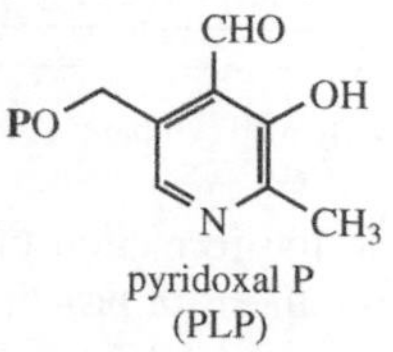

pyridoxal P
(PLP)

The process initially features formation of an imine intermediate (aldimine) using the amine group of the amino acid with the aldehyde group of PLP. The imine function formed is conjugated with the aromatic pyridine ring. Accordingly, protonation of the pyridine nitrogen (as would occur at physiological pH) makes the α-hydrogen of the original amino acid considerably more acidic. This can be removed, in a process rather like that seen with conjugated carbonyl compounds (see Section 10.1.3). This generates a dihydropyridine ring system; the process is effectively imine–enamine tautomerization (see Section 10.5), though in an extended conjugated system.

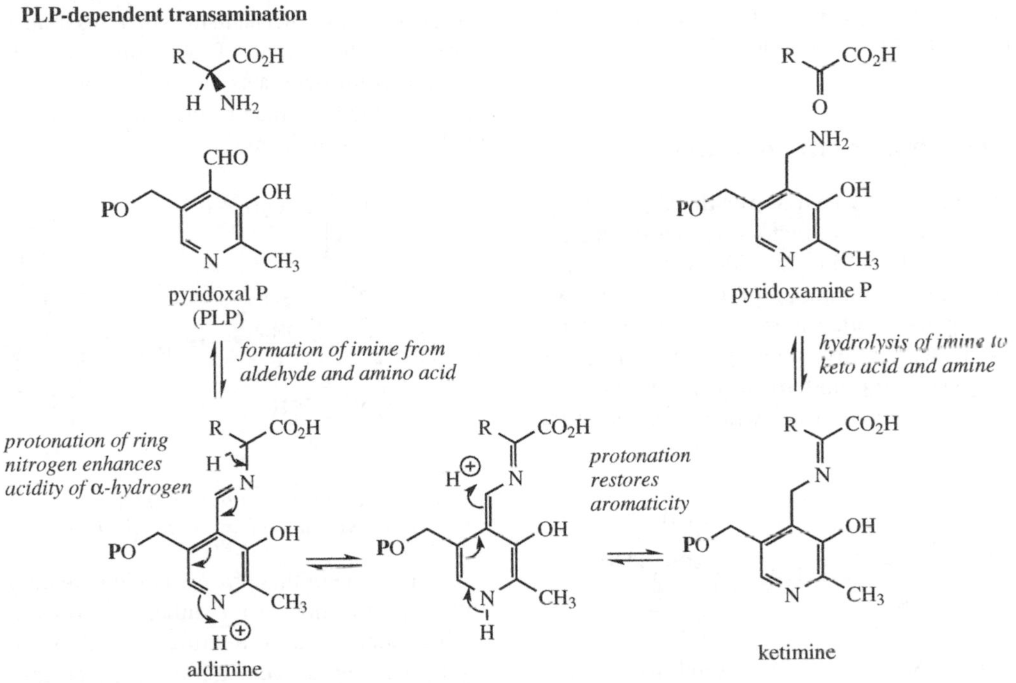

PLP-dependent transamination

Reprotonation then produces a new imine (ketimine), and also restores aromaticity in the pyridine ring. However, because of the conjugation, it allows protonation at a position that is different from where the proton was originally lost. The net result is that the imine double bond has effectively moved to a position adjacent to its original position. Hydrolysis of this new imine group generates a keto acid and **pyridoxamine phosphate**. The remainder of the sequence is now a reversal of this process. This now transfers the amine function from pyridoxamine phosphate to another keto acid.

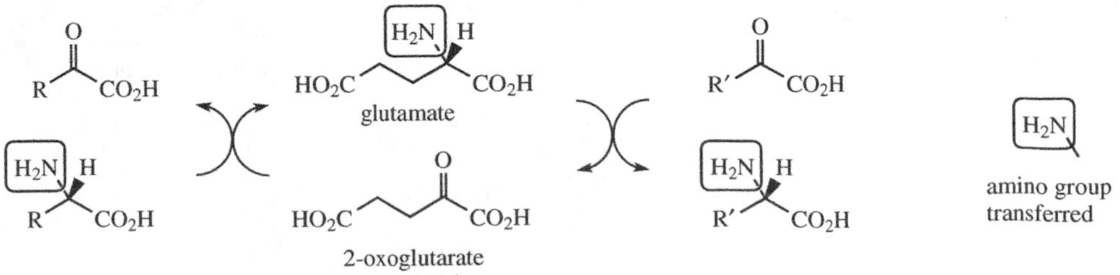

The glutamic acid–2 oxoglutaric acid couple features as the usual donor–acceptor molecules for the amino group, and **glutamate transaminase** is thus the most important of the transaminases. Transamination allows the amino group to be transferred from glutamic acid to a suitable keto acid. In the reverse mode, the amino group can be transferred from an amino acid to 2-oxoglutaric acid. Equilibrium constants for the reactions catalysed by transaminases are close to unity, so the reactions proceed readily in either direction. It now becomes possible, as shown, to transfer the amino group of one amino acid, which may be readily available, to provide another amino acid, which could be in short supply.

This has obvious advantages over the process seen for glutamate synthesis via the reductive amination of 2-oxoglutarate, in that it no longer requires the intervention of free ammonia. We thus have the situation that some organisms are able to carry out the fixation of ammonia via reductive amination, whereas others manipulate via transamination the amino acid structures obtained from protein in the diet.

15.7 PLP-dependent reactions

We have just noted the role that **pyridoxal phosphate** plays as a coenzyme (cofactor) in transamination reactions (see section 15.6). Pyridoxal 5′-phosphate (PLP) is crucial to a number of biochemical reactions. PLP, together with a number of closely related materials that are readily converted into PLP, e.g. **pyridoxal, pyridoxine** and **pyridoxamine**, are collectively known as vitamin B_6, which is essential for good health.

At neutral pH, PLP is considerably ionized, so that the phenol group loses a proton and the pyridine nitrogen is protonated. Of course, the phosphate will also be ionized. These ionic centres facilitate binding to the enzyme, but, for clarity, they are omitted from the mechanisms shown. However, the positively charged nitrogen is essential for the cofactor's chemical reactivity, and we need to invoke it in the mechanisms. In transamination, we have seen formation of an imine, in which the protonated nitrogen acts as an electron sink, making the α-hydrogen of the original amino acid acidic and facilitating its removal.

acidity of α-proton: resonance stabilization gives extended enamine

The reversal of this process could potentially occur with reprotonation from either face of the C=N double bond, and a mixture of aldimines would result, leading to generation of a racemic amino acid. This accounts for the mode of action of PLP-dependent amino acid **racemase** enzymes. Of course, the enzyme controls removal and supply of protons; this is not a random event. One important example of this reaction is alanine racemase, employed by bacteria to convert L-alanine into D-alanine for cell-wall synthesis (see Box 13.12).

pyridoxine
(pyridoxol)

pyridoxal

pyridoxamine

PLP-dependent racemization

L-amino acid

reprotonation from opposite face of imine

D-amino acid

imine formation

hydrolysis of imine

In transamination and racemization reactions, we have seen loss of a proton from the aldimine, i.e. breaking of bond *a*. Let us now consider the two alternatives, namely the breaking of bonds *b* or *c*, to explore further the scope for PLP-dependent reactions.

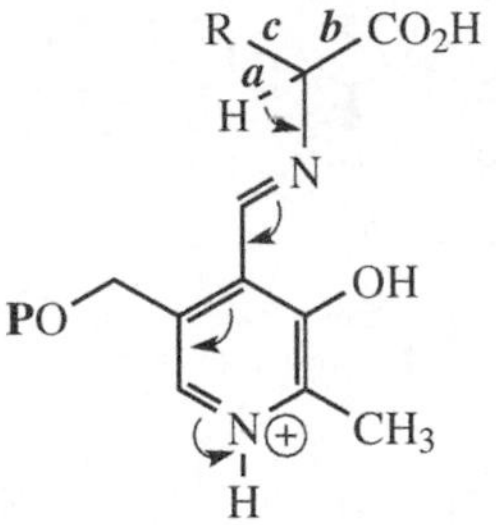

Breaking of bond *b* accounts for **PLP-dependent decarboxylations**. Decarboxylation of the intermediate aldimine is facilitated in the same way as loss of a proton in the transamination sequence. The protonated nitrogen acts as an electron sink, and the conjugated system allows loss of the carboxyl proton, with subsequent bond breaking and loss of CO_2. The resultant imine may subsequently be hydrolysed, releasing an amine (the decarboxylated amino acid) and regenerating PLP. There are many examples for decarboxylation of amino acids (see Box 15.3).

PLP-dependent decarboxylation

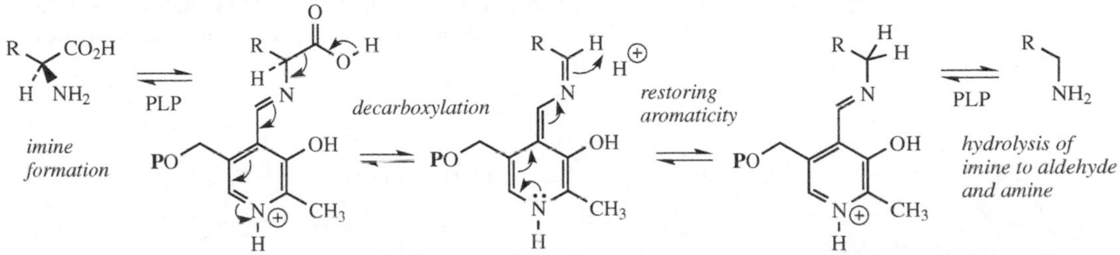

Box 15.3

PLP-dependent amino acid decarboxylations

An important example of **PLP-dependent amino acid decarboxylation** is the conversion of histidine into histamine. **Histamine** is often involved in human allergic responses, e.g. to insect bites or pollens. Stress stimulates the action of the enzyme histidine decarboxylase and histamine is released from mast cells. Topical antihistamine creams are valuable for pain relief, and oral antihistamines are widely prescribed for nasal allergies such as hay fever. Major effects of histamine include dilation of blood vessels, inflammation and swelling of tissues, and narrowing of airways. In serious cases, life-threatening anaphylactic shock may occur, caused by a dramatic fall in blood pressure.

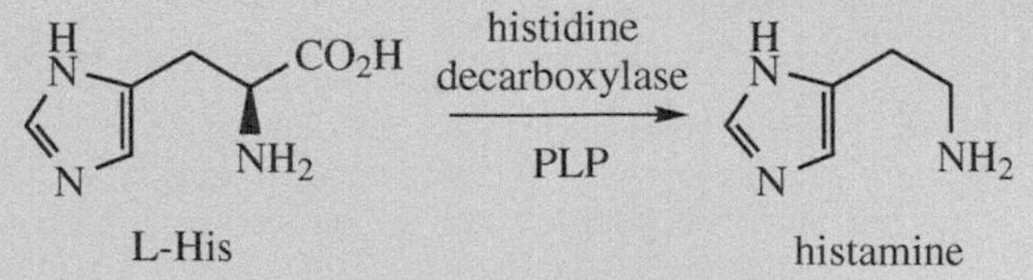

The **catecholamines noradrenaline (norepinephrine)** and **adrenaline (epinephrine)** are amines derived via decarboxylation of amino acids. Noradrenaline is a mammalian neurotransmitter, and adrenaline, the

Box 15.3 (continued)

'fight or flight' hormone, is released in animals from the adrenal gland as a result of stress. These compounds are synthesized by successive hydroxylation and N-methylation reactions on dopamine.

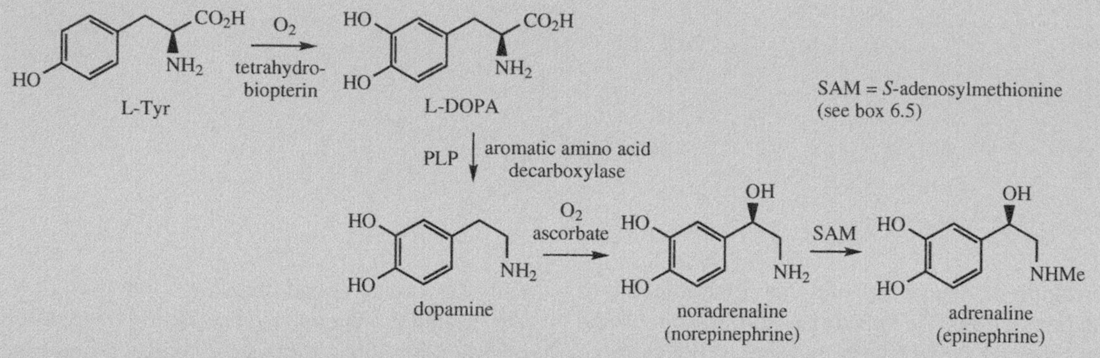

SAM = S-adenosylmethionine (see box 6.5)

Dopamine is the decarboxylation product of DOPA, dihydroxyphenylalanine, and is formed in a reaction catalysed by DOPA decarboxylase. This enzyme is sometimes referred to as aromatic amino acid decarboxylase, since it is relatively non-specific in its action and can catalyse decarboxylation of other aromatic amino acids, e.g. tryptophan and histidine. DOPA is itself derived by aromatic hydroxylation of tyrosine, using tetrahydrobiopterin (a pteridine derivative; see Section 11.9.2) as cofactor.

The neurotransmitter **5-hydroxytryptamine (5-HT, serotonin)** is formed from tryptophan by hydroxylation then decarboxylation, paralleling the tyrosine → dopamine pathway. The non-specific enzyme aromatic amino acid decarboxylase again catalyses the decarboxylation.

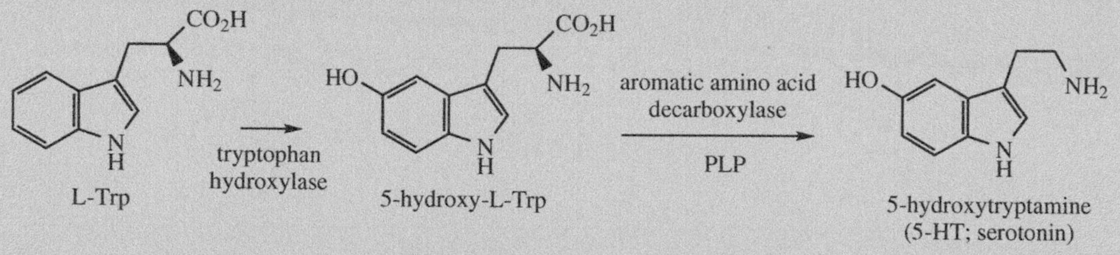

5-HT is a neurotransmitter found in cardiovascular tissue, the peripheral nervous system, blood cells, and the CNS. It mediates many central and peripheral physiological functions, including contraction of smooth muscle, vasoconstriction, food intake, sleep, pain perception, and memory, a consequence of it acting on several distinct receptor types (see Box 11.10). Although 5-HT may be metabolized by monoamine oxidase, platelets and neurons possess a high-affinity 5-HT re-uptake mechanism. This mechanism may be inhibited by the widely prescribed antidepressant drugs termed selective serotonin re-uptake inhibitors (SSRI), e.g. fluoxetine (Prozac®), thereby increasing levels of 5-HT in the CNS.

Yet another neurotransmitter, **γ-aminobutyric acid** or **GABA**, is formed by PLP-dependent decarboxylation of an amino acid, in this case glutamic acid.

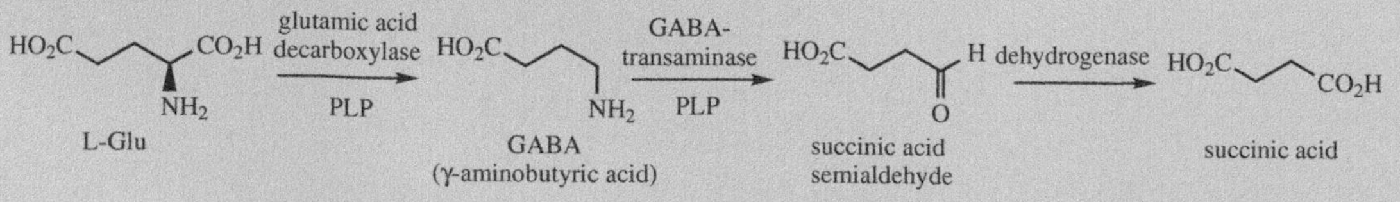

GABA acts as an inhibitory transmitter in many different CNS pathways. It is subsequently destroyed by a transamination reaction (see Section 15.6) in which the amino group is transferred to 2-oxoglutaric acid, giving glutaric acid and succinic semialdehyde. This also requires PLP as a cofactor. Oxidation of the aldehyde group produces succinic acid, a Krebs cycle intermediate.

The breaking of bond *c* is going to be less common than deprotonation or decarboxylation. In most amino acids, R is an alkyl group, so there is little chance of losing R as a cation. Indeed, the only occasions on which we can break bond *c* are when R is hydroxymethyl (as in serine) or a similar grouping (as in threonine). In both cases, bond breaking is facilitated by the hydroxyl, in that the lone pair can feed into the conjugated system. The result in the case of serine is the loss of formaldehyde, whereas in the case of threonine it is loss of acetaldehyde.

PLP-dependent reverse aldol reactions

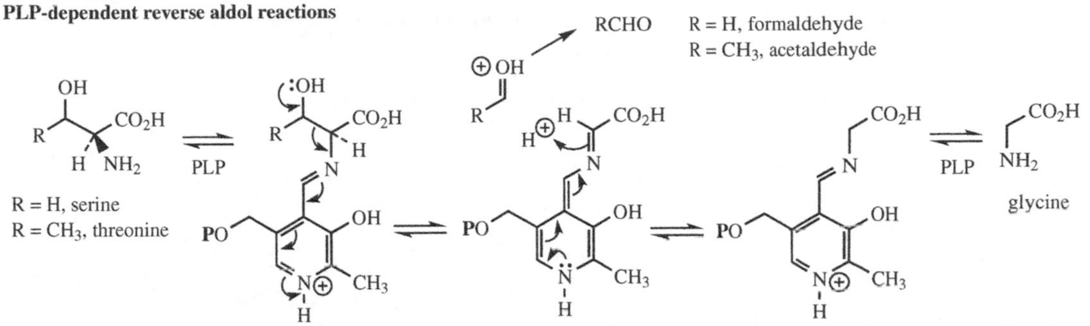

The fragment attached to pyridoxal will be the same in both cases, and after hydrolysis it is released as the amino acid glycine. In case this seems a bit complicated, consider the reverse reaction, which would be attack of an electron-rich system on to the carbonyl group of an aldehyde, i.e. an aldol reaction. Therefore, what we are seeing here is merely a **reverse aldol-type reaction** (see Section 10.3).

It is interesting to see how different products are formed according to which of the three different bonds is cleaved in the aldimine derived from an amino acid and PLP. There is one other point to ponder though. What determines the type of cleavage that occurs? The answer must lie in the enzyme and how it binds the substrate, and it merely becomes a consideration of stereochemistry. By considering the shape of the aldimine, we see that the pyridine ring and the adjacent C=N double bond must be planar to achieve maximum orbital overlap and conjugation. Electrons from the bond that is broken should feed smoothly into this planar conjugated system. This requires the bond to be positioned at right angles to the plane and, therefore, parallel to the *p* orbitals. As shown in the accompanying diagrams, rotation about the N–C bond positions the vulnerable group towards the exterior face of the enzyme so that it can be attacked.

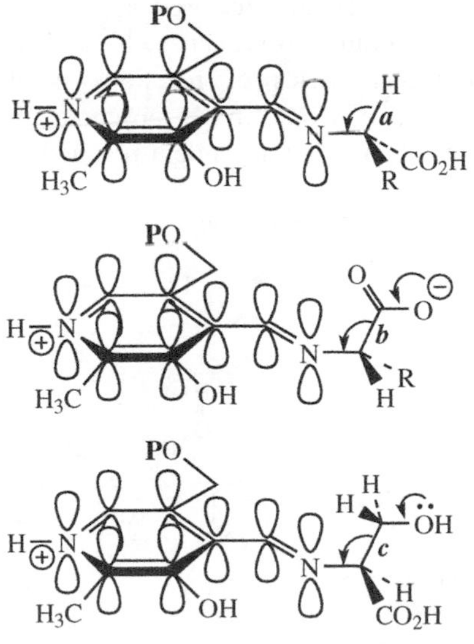

bond being cleaved is aligned perpendicular to plane of conjugated system

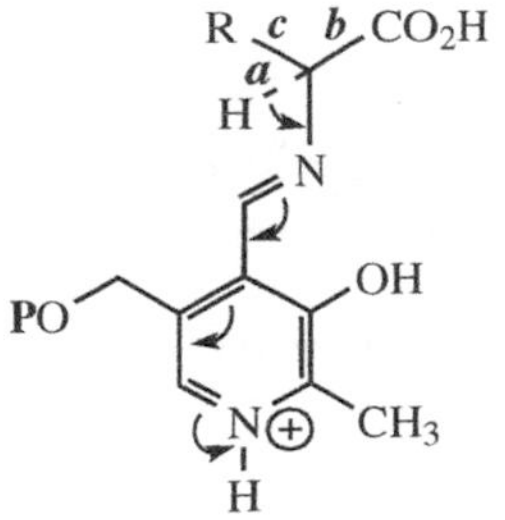

Do appreciate that one enzyme will not catalyse all three types of reaction. We need different enzymes to accomplish a particular reaction on a particular substrate. Rotation about the N–C bond positions the vulnerable group for reaction; it also positions the R and/or CO_2^- groups so that they can interact and bind to the enzyme, thus providing the specificity.

One further point for the sake of accuracy; we have omitted it to simplify the mechanistic features.

PLP does not exist as the free aldehyde when it is bound to the enzyme, but actually uses the aldehyde group in its binding. An imine linkage is formed between the aldehyde and the primary amine group of a lysine residue in the enzyme active site. When the substrate RNH_2 binds to the enzyme to produce the intermediates shown above, it achieves this by a **transimination** reaction.

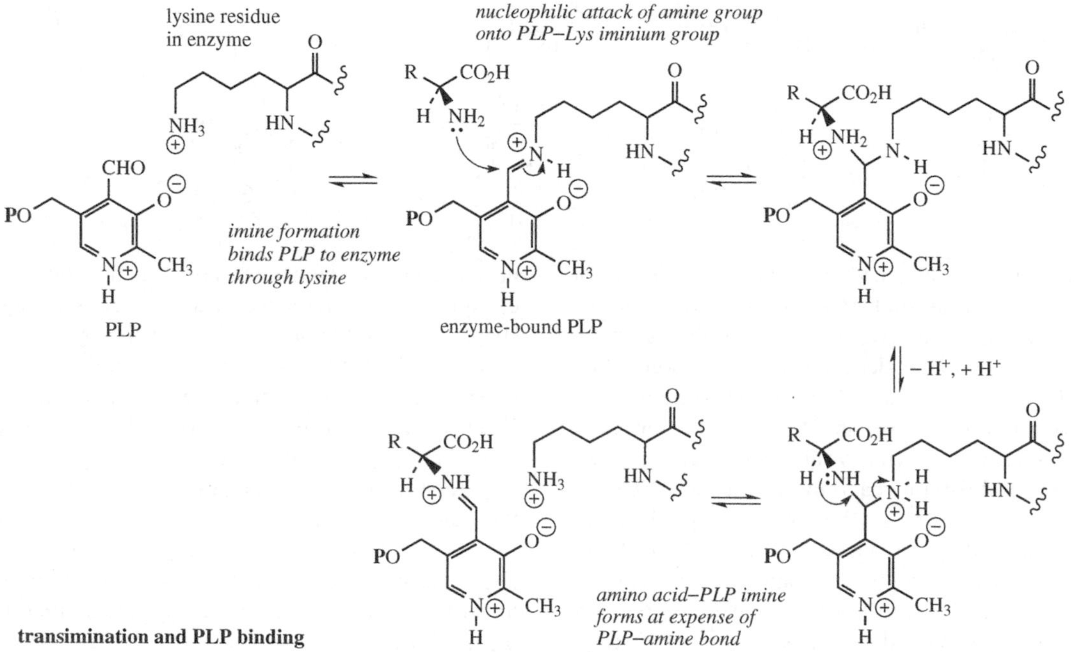

transimination and PLP binding

A similar transimination, in the reverse sense, takes place at the end of the reaction sequence to displace the product, but still retains PLP bound to the enzyme via the lysine group. Should the chemistry seem a little complicated, remind yourself

of the mechanism for hemiacetal and acetal formation (see Section 7.2); that featured an oxygen analogy for the transimination sequence. If we had included this additional series of reactions in the above descriptions, it would have obscured the simple principles of the PLP-dependent reactions.

transimination mechanism

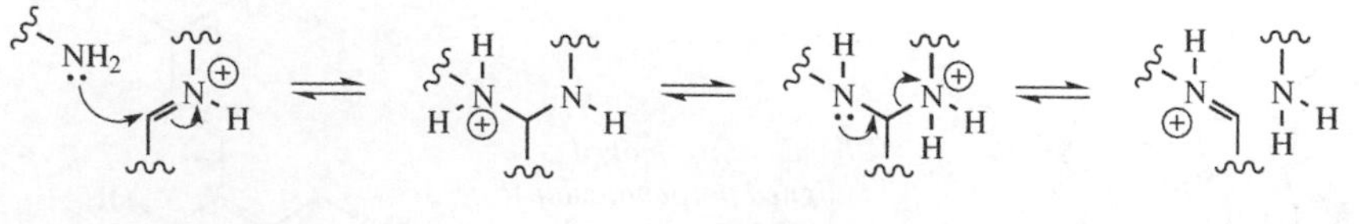

synthesis of hemiacetals and acetals

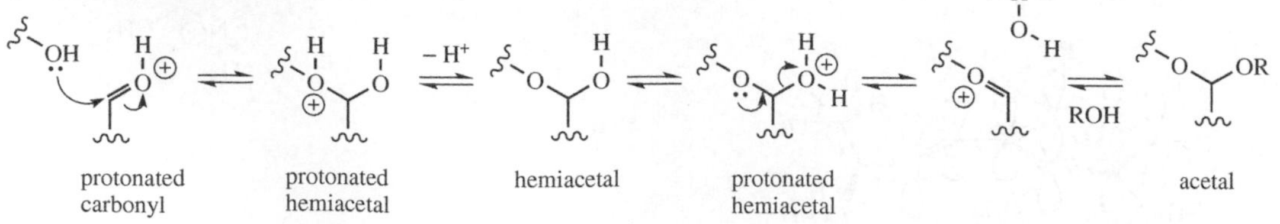

protonated carbonyl   protonated hemiacetal   hemiacetal   protonated hemiacetal   acetal

15.8 TPP-dependent reactions

Thiamine diphosphate (TPP) is the coenzyme for the **pyruvate dehydrogenase** complex that catalyses the oxidative decarboxylation of pyruvate to acetyl-CoA, and thus links the glycolytic pathway to the Krebs cycle (see Section 15.3). Later in the Krebs cycle, TPP is the cofactor for the **2-oxoglutarate dehydrogenase** complex, which catalyses a similar reaction on 2-oxoglutarate. In the glycolytic pathway, the pyruvic acid → acetaldehyde conversion also features TPP (see Section 15.2). A further enzyme, transketolase, has the unexpected property of transferring a two-carbon fragment between carbohydrates in the pentose phosphate pathway, and TPP is again involved. The biosynthetic pathways to two amino acids, valine and isoleucine, also involve TPP-dependent enzymes. All of these reactions employ TPP as a carrier of an acyl anion equivalent.

TPP thus plays a very important role in carbohydrate metabolism. The parent alcohol thiamine is one of the B group vitamins, namely vitamin B_1; dietary deficiency leads to the condition beriberi, characterized by neurological disorders, loss of appetite, fatigue, and muscular weakness.

thiamine diphosphate
(TPP)

thiamine
(vitamin B_1)

The conversion of pyruvic acid into acetyl-CoA is conveniently written according to the equation

pyruvate + HS–CoA + NAD⁺ ⇌ (pyruvate dehydrogenase complex, 3 enzyme activities) acetyl-CoA + CO_2 + NADH

This simple equation conceals a quite complex reaction sequence involving not just TPP, but other coenzymes, including lipoic acid. Let us first inspect the nature of TPP. With its pyrimidine ring and diphosphate grouping, TPP looks rather like a nucleotide, but the central ring system is a **thiazole** rather than a sugar. This heterocyclic ring is alkylated on nitrogen, and is thus a **thiazolium salt** (see Box 11.8). This plays the key role in the reactivity of TPP.

The proton in the thiazolium ring is relatively acidic (pK_a about 18) and can be removed by even weak bases to generate the carbanion or **ylid** (see Box 11.8). An ylid (also ylide) is a species with positive and negative charges on adjacent atoms; this ylid is an ammonium ylid with extra stabilization from the sulfur atom.

stabilization of ylid from sulfur

thiazolium salt

ylid

This ylid can act as a nucleophile, and is also a reasonable leaving group. Addition to the carbonyl group of pyruvic acid is followed by decarboxylation, the positive nitrogen in the ring acting as an electron sink (compare PLP-dependent decarboxylation; Section 15.7). The resulting molecule is an enamine, but because of the neighbouring heteroatoms it is extremely electron rich. It accepts a proton, and this achieves tautomerism of the enamine to the iminium ion. This is followed by a reverse aldol reaction, which also regenerates the ylid as the leaving group. This sequence would thus accommodate the pyruvic acid → acetaldehyde conversion catalysed by pyruvate decarboxylase in the glycolytic pathway (see Section 15.2).

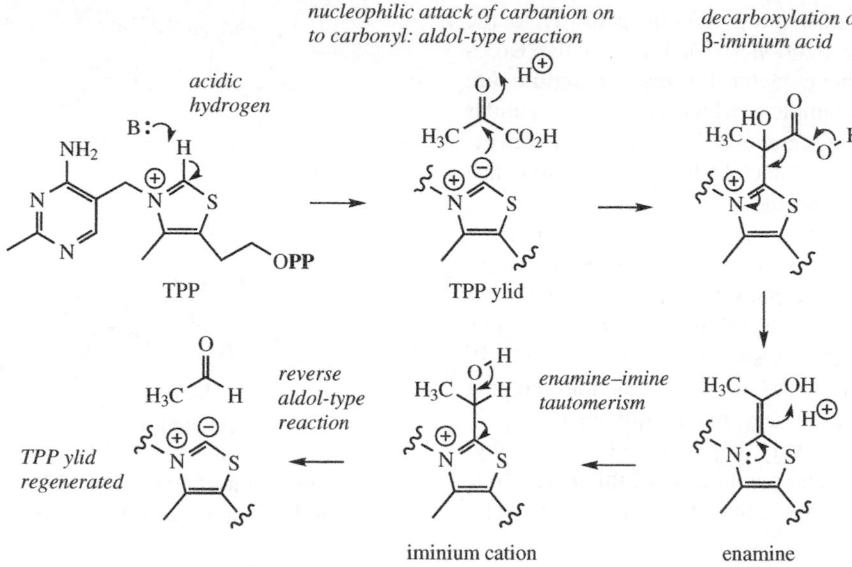

In oxidative decarboxylation of pyruvate to acetyl-CoA, the enzyme-bound disulfide-containing coenzyme **lipoic acid** is also involved. The electron-rich enamine intermediate, instead of accepting a proton, is used to attack a sulfur in the lipoic acid moiety. This leads to fission of the S–S bond, and thereby effectively reduces the lipoic acid fragment. Regeneration of the TPP ylid via the reverse aldol-type reaction leaves the acetyl group bound to the dihydrolipoic acid as a thioester. This acetyl group is then released as acetyl-CoA by displacement with the thiol coenzyme A. The bound dihydrolipoic acid fragment must then be reoxidized to restore its function. An exactly equivalent reaction is encountered in the Krebs cycle in the conversion of 2-oxoglutaric acid into succinyl-CoA.

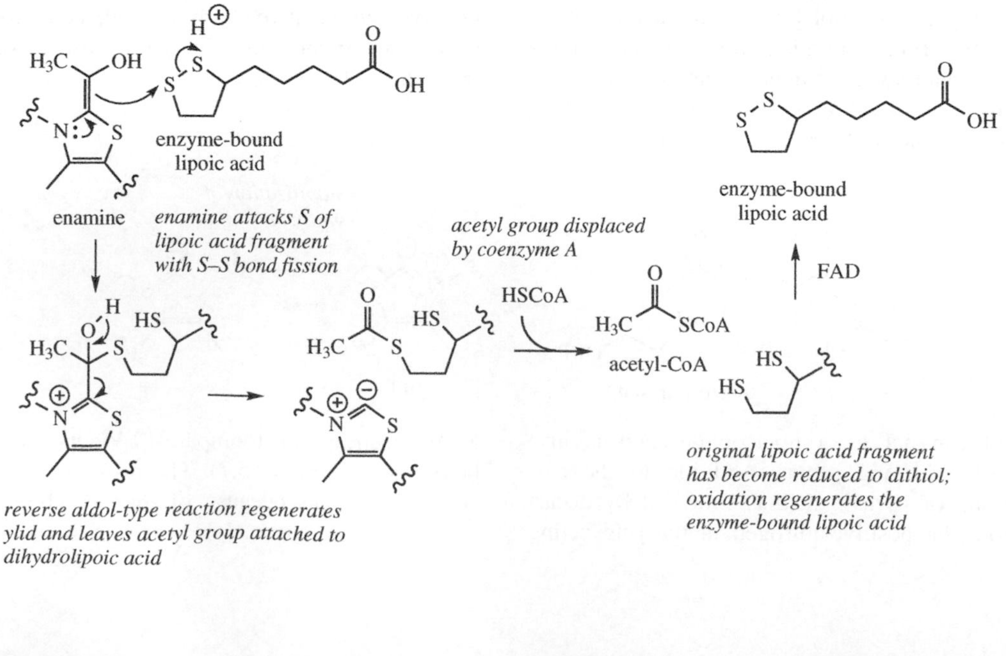

Each of the complexes **pyruvate dehydrogenase** and **2-oxoglutarate dehydrogenase** actually contains three enzyme activities. We can now readily appreciate what the separate activities might be. In the case of pyruvate dehydrogenase, the individual activities are pyruvate dehydrogenase (requires cofactor TPP), dihydrolipoamide acyltransferase (requires cofactors lipoic acid and coenzyme A), and dihydrolipoamide dehydrogenase (requires cofactors FAD and NAD$^+$). The requirement for NAD$^+$ is to reoxidize FADH$_2$ after regeneration of lipoic acid. The 'lipoamide' terminology indicates that the lipoic acid is enzyme bound through its carboxylic acid group to an amino group of lysine via an amide linkage.

At the end of the first paragraph in this section we stated 'All these reactions employ TPP as a carrier of an acyl anion equivalent'. An acyl anion is an unlikely species, since we would consider locating a negative charge on the carbon of a carbonyl group as definitely unfavourable. However, the following scheme should emphasize how we can consider that TPP is effectively removing and transferring an acetyl anion equivalent in the above reactions.

TPP and acyl anion equivalents

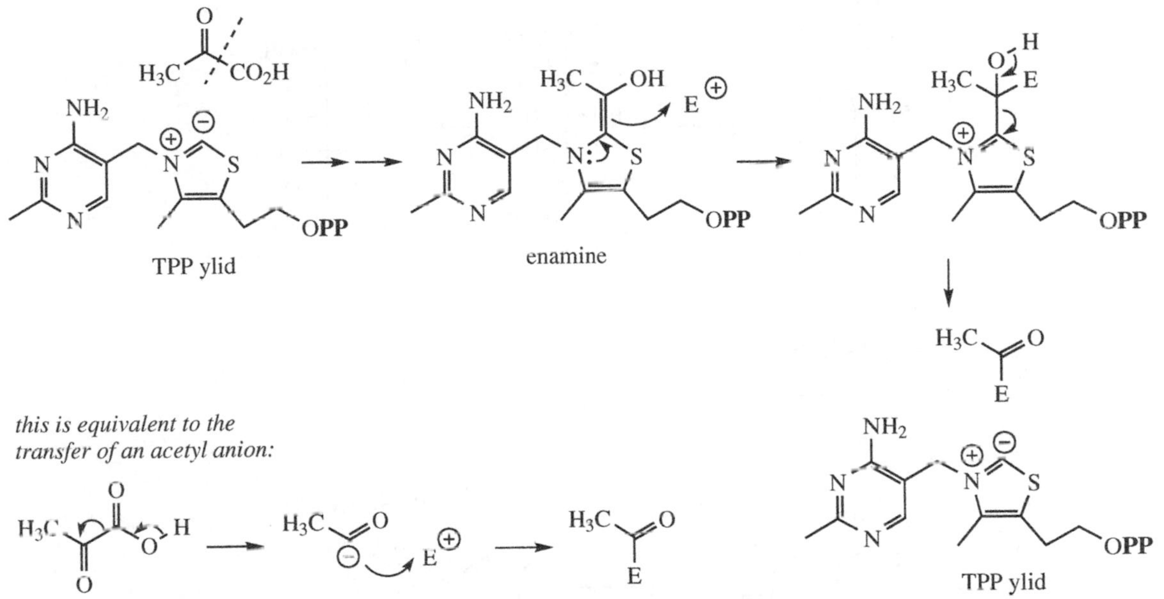

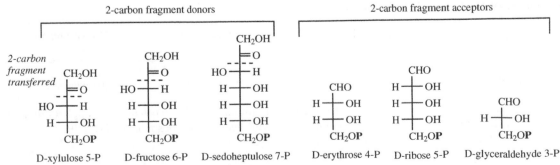

A slightly different acyl anion equivalent is transferred in transketolase reactions, and this anion is then used in a subsequent aldol reaction. **Transketolase** removes a two-carbon fragment from keto sugars such as xylulose 5-phosphate (alternatively fructose 6-phosphate or sedoheptulose 7-phosphate) through the participation of the thiamine diphosphate ylid.

transketolase

| | 2-carbon fragment donors | | | 2-carbon fragment acceptors | | |

2-carbon fragment transferred

| D-xylulose 5-P | D-fructose 6-P | D-sedoheptulose 7-P | D-erythrose 4-P | D-ribose 5-P | D-glyceraldehyde 3-P |

Nucleophilic attack of this ylid on to the ketone carbonyl results in an addition product that then fragments by a reverse aldol reaction. This generates a chain-shortened aldose, e.g. glyceraldehyde 3-phosphate from xylulose 5-phosphate, and the two-carbon acyl anion equivalent attached to TPP.

transketolase

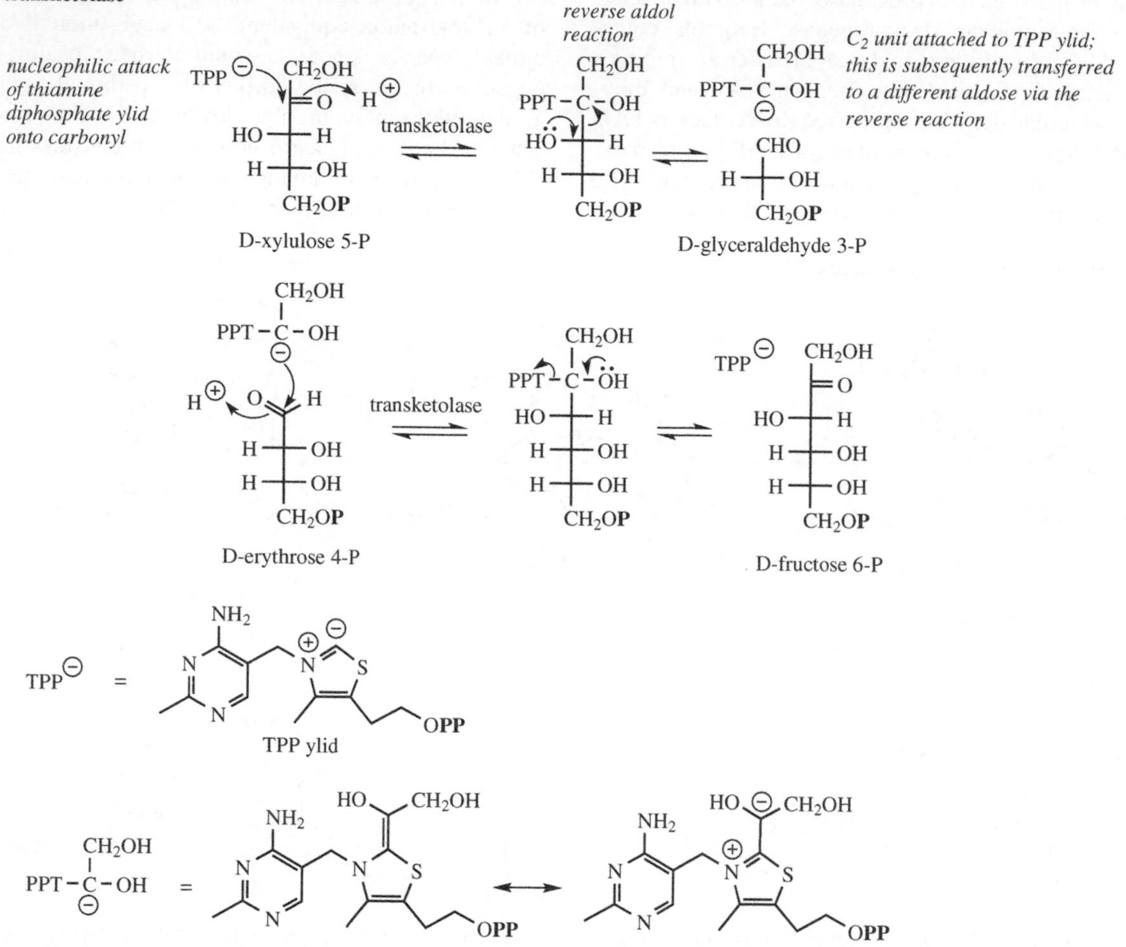

nucleophilic attack of thiamine diphosphate ylid onto carbonyl

reverse aldol reaction

C_2 *unit attached to TPP ylid; this is subsequently transferred to a different aldose via the reverse reaction*

D-xylulose 5-P

D-glyceraldehyde 3-P

D-erythrose 4-P

D-fructose 6-P

TPP ylid

$PPT-\overset{CH_2OH}{\underset{O^{\ominus}}{C}}-OH$ =

Then, in what is formally a reverse of this reaction, this carbanion equivalent can attack another aldose, such as erythrose 4-phosphate, extending its chain length by two carbons. Transketolase is crucial to metabolism in creating a link between the pentose phosphate pathway and glycolysis.

reactions catalysed by transketolase enzymes:

D-xylulose 5-P + D-erythrose 4-P ⇌ D-glyceraldehyde 3-P + D-fructose 6-P

D-xylulose 5-P + D-ribose 5-P ⇌ D-glyceraldehyde 3-P + D-sedoheptulose 7-P

An additional enzyme that transfers C_3 rather than C_2 units is called **transaldolase**, but, in common with aldolase (see Box 10.5), this enzyme utilizes an imine–enamine mechanism through an imine link with lysine, and does not involve TPP.

15.9 Biotin-dependent carboxylations

We have briefly noted the role of **biotin** when we considered the biosynthesis of fatty acids (see Section 15.5). Biotin is a carrier of carbon dioxide and involved in **carboxylation reactions**. In fatty acid biosynthesis, we noted how acetyl-CoA was transformed by carboxylation into the more effective nucleophilic agent malonyl-CoA, thus facilitating the Claisen reaction. Structurally, biotin (vitamin H) is composed of two fused five-membered heterocycles, a cyclic urea and a cyclic sulfide (tetrahydrothiophene).

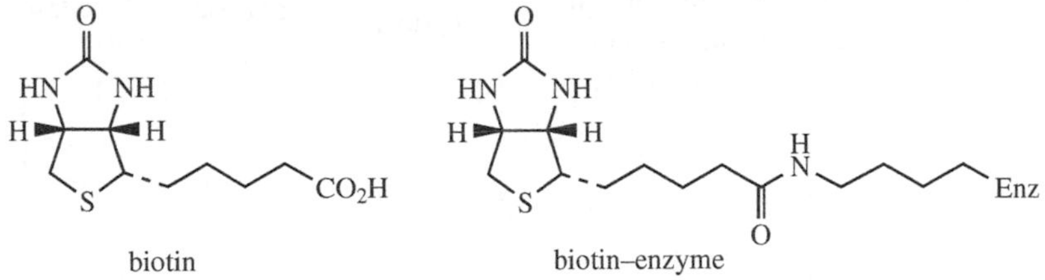

biotin

biotin–enzyme

Carbon dioxide is a normally unreactive material, and combination with biotin requires the input of energy (from ATP). Carbon dioxide is usually present as the soluble form bicarbonate, and this reacts with ATP to form a mixed anhydride, as part of the reaction catalysed by the carboxylase.

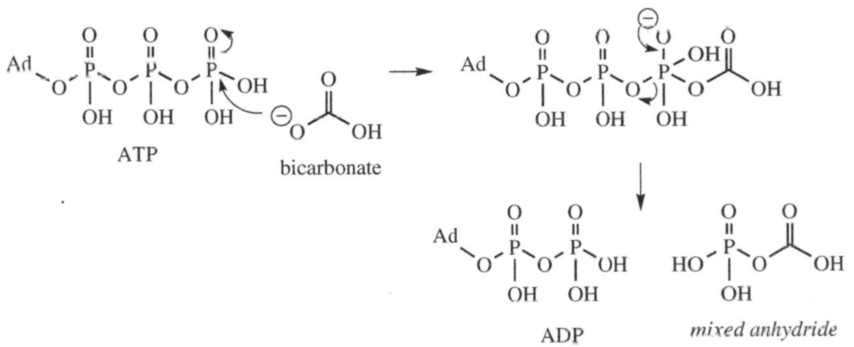

ATP

bicarbonate

ADP

mixed anhydride

This mixed anhydride carboxylates the coenzyme in a biotin–enzyme complex. Biotin is bound to a lysine residue in the enzyme as an amide. The carboxylation reaction is effectively a nucleophilic attack of the cyclic urea on the mixed anhydride.

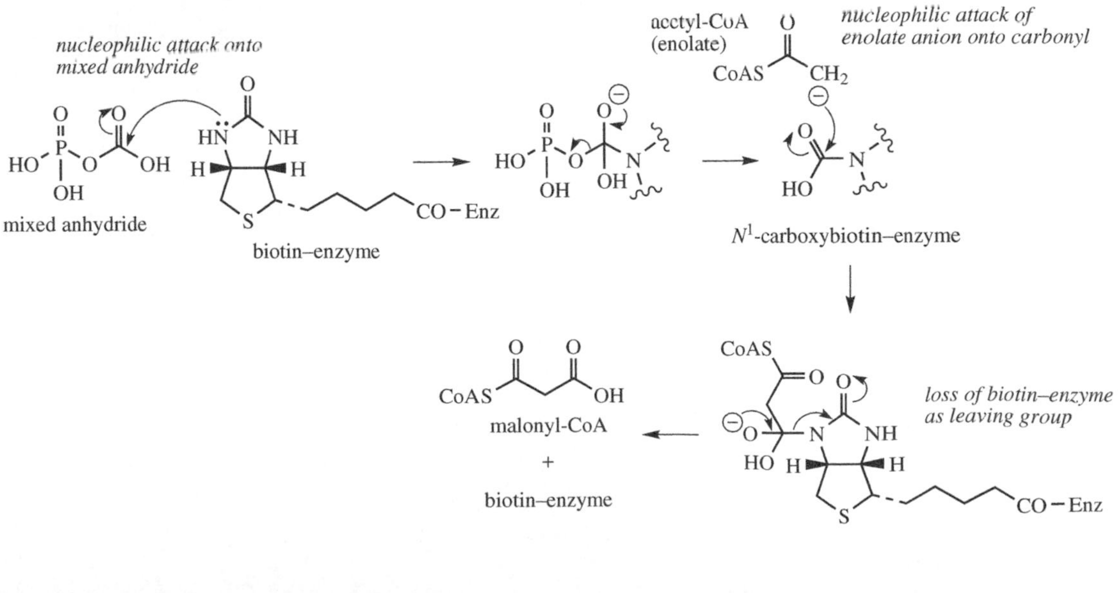

nucleophilic attack onto mixed anhydride

mixed anhydride

biotin–enzyme

acetyl-CoA (enolate)

nucleophilic attack of enolate anion onto carbonyl

N^1-carboxybiotin–enzyme

malonyl-CoA

+

biotin–enzyme

loss of biotin–enzyme as leaving group

In what can be considered a reversal of this sequence, the acetyl-CoA acts as the nucleophile and is carboxylated to malonyl-CoA with displacement of the biotin–enzyme system.

Fixation of carbon dioxide by biotin–enzyme complexes is not unique to acetyl-CoA, and another important example occurs in the generation of oxaloacetate from pyruvate in the synthesis of glucose from non-carbohydrate sources (gluconeogenesis). This reaction also allows replenishment of Krebs cycle intermediates when compounds are drawn off for biosynthetic purposes, e.g. amino acid synthesis (see Section 15.6).

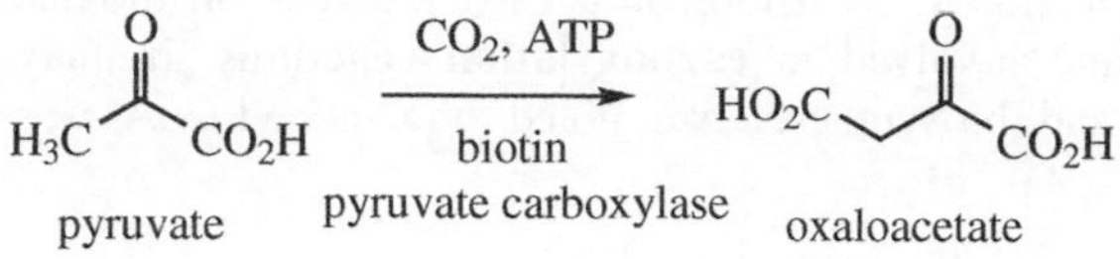

16

How to approach examination questions: selected problems and answers

No matter how intrinsically interesting students find the study of organic chemistry, there will usually be one or more hurdles that need to be overcome, the examinations. Examination technique is undoubtedly a skill, and this chapter is aimed at developing such skills for organic chemistry. Throughout the book, we have tried to convince the reader that, by applying principles and deductive reasoning, we can reduce to a minimal level the amount of material that needs be committed to memory.

This section contains a selection of typical examination questions, many of them based on real ones used at Nottingham, together with the answers. However, it is not so much the answers themselves that are important, but developing the skills to answer them. We wish to show that, in many questions, the information given can actually direct us toward the answer. Thus, emphasis has been placed on what to look for in the question, how to approach the problem, and how to develop the answer in a logical manner. We are not looking at questions that ask for essay-style answers, and thus require a lot of memorized material, but are concerned predominantly with those questions that start out 'Propose a mechanism for . . .' or 'Explain the following observations . . .'

Rather than having questions at the end of each chapter, we have purposely decided to put all these problems together in a separate chapter to help emphasize links and the integration of different ideas. One of the disadvantages of the almost-universal modular system as a study method is that particular themes become imprisoned in self-contained packages. For real skill and understanding, it is necessary to be able to interrelate ideas from different modules. The problems used here are usually going to require integration of knowledge from one or more chapters to provide the answer. The source material required for each problem is broadly indicated, so that they can be used as self-test examples after the appropriate chapters have been covered. The questions are targeted predominantly towards the general material covered in Chapters 1–11. It may well be that these questions become more relevant towards the end of a particular course, when examinations are no longer a distant consideration but are becoming an imminent prospect.

16.1 Examination questions: useful advice

Read the instructions very carefully. Make sure you are clear how many questions you need to answer. Do not answer more questions than are required. You may be penalized, in that only those questions answered first will be marked; these may not be your highest scoring answers.

Read the entire paper. Choose the questions that you are going to answer and decide the order in which you will tackle them. Start with a question which you think that you can answer well, to build up your confidence for the less-appealing questions, but do not spend more than the allocated time on this question.

Essentials of Organic Chemistry   Paul M Dewick
© 2006 John Wiley & Sons, Ltd

Read each question carefully before beginning your answer. If the question is divided into parts, be clear how many parts you must answer. Understand clearly what is being asked. Jot down headings to help you structure your answer. To be sure to get maximum credit, organize your answer, setting the solution neatly and legibly. Cross out any rough work or mistakes. Take special care with multiple-choice questions, and do not make guesses if you are going to be penalized for wrong answers.

Answer the question asked. Easy marks are lost by not answering the question asked. Do not spend time giving information that is not required whilst missing out other aspects that are clearly asked for. Do not try to copy out your lecture notes. The examiner expects you to be selective and to formulate a critical answer to the question.

Be quite clear you fully understand the meanings of the terms 'discuss', 'evaluate', 'illustrate', and 'outline', which are frequently used in examination questions. You must appreciate the subtle differences, and provide the appropriate level of detail to answer the question effectively.

A title that includes 'discuss' or 'evaluate' requires you to sum up both positive and negative evidence or aspects. A title that says 'illustrate' or 'outline' is only asking for evidence or aspects that are consistent with the proposition.

Most organic chemistry questions do not require descriptive answers, but can be adequately answered by generous use of structures and mechanisms. These can be annotated with pertinent points as appropriate.

Budget your time. Keep an eye on the clock, and do not spend more than the appropriate amount of time on any question. It may not be worthwhile trying to get a few extra marks on one question at the expense of gaining easy marks by starting another. You must attempt the full number of questions specified on the paper. Note any breakdown of marks allocated for each part of a question, and use this to apportion the time and effort spent on the component parts.

Check your answers. Try to finish a little early so that you can read through your answers. Make sure your answers are complete. It is easy to make simple errors. Check for any missing answers to parts of questions, especially multiple-choice questions. Make sure your answers make scientific sense, and in numerical answers check that you have included units and considered significant figures.

Mechanisms or calculations that appear unnecessarily complicated are probably wrong and should be checked for obvious errors. Incorrect transcription of structures or figures is a common cause of problems.

16.2   How to approach the problem: 'Propose a mechanism for . . .'

You will usually be provided with the reactants, reagents, conditions, and the products formed, though in some questions you may be required to predict the nature of the products.

- Carry out a quick analysis of the carbon skeletons of reactants and products to ascertain what the reaction represents.

    Is it merely modification of substituents, e.g. as in substitution or elimination?
    Is it formation of a bigger skeleton, e.g. as in an addition reaction?
    Is it formation of a smaller skeleton, e.g. as in a cleavage reaction?
    Is it a modification that has involved a rearrangement of the basic skeleton?
    Register this change, but do not try to accommodate the reaction at this stage. Instead, move on to the reagents and conditions, as follows.

- Consider the essential principles of mechanism. Look for nucleophiles, electrophiles, and leaving groups.

- Consider the conditions. If the reaction requires a catalyst, e.g. acid or base, it is almost certain that this needs to be used in the first step. For example, acid may protonate an electronegative atom, making a more reactive species. Thus, a protonated carbonyl becomes a better electrophile, and a protonated alcohol now has a better leaving group. Base may remove an acidic proton to generate a better nucleophile, although in some reactions it may itself act as the nucleophile.

- At this stage, the mechanism should develop logically by interaction of a nucleophile with an electrophile, or by displacement of a leaving group. If your mechanism is to make sense though, certain aspects, which are really rather obvious, need to be considered. You cannot sensibly use X^- as a nucleophilic reagent under acid conditions, and there is not going to be any H^+ under basic conditions. In the former case, you must use HX as the nucleophilic reagent. In the latter case, a requirement for a proton, perhaps to finish off a mechanism, will probably be met by abstracting it from a solvent molecule, such as water or alcohol.

- There are stability considerations with carbocations, with tertiary carbocations being more stable than secondary ones; if your mechanism includes CH_3^+ or RCH_2^+, it is almost certainly wrong! Carbocation mechanisms are also going to be much more likely under acidic conditions (H^+) rather than under basic conditions (HO^-).

- Do not even consider radical reactions unless there is some obvious pointer from the reagents, e.g. a radical initiator such as a peroxide or electromagnetic radiation.

- After each step of the mechanism, write out the intermediate structure that will be produced. Do not try to include more than one such change per step. Always try to develop the mechanism logically, working forwards so that the product you end up with turns out to be identical to that in the question. It is rather less satisfactory, and definitely more risky, to apply the reverse strategy, i.e. aiming for the product and trying to manipulate the reagents. If a proposed mechanism starts to get complex or overcomplicated, then stop – it is probably wrong. Check it through, first ensuring

that the structure is transposed correctly at each stage. If this does not show up an error, go back to the essentials – electronegativity, nucleophiles, electrophiles, leaving groups.

16.3 Worked problems

These problems are typical examination questions, though perhaps longer than most, and have been graded as level 1 or level 2. Based on a 2-year period of study to cover most of the content of this book, level-1 questions might be answered after year-1 studies, whereas the level-2 questions are more appropriate for year-2 studies.

Purists might criticize the avoidance of equilibrium arrows in the mechanisms shown. Some reactions, e.g. hemiacetal formation or acid-catalysed ester hydrolysis, are undoubtedly reversible, yet we have shown them as proceeding only in the forward direction. We believe it is more important to develop the skills for predicting a rational mechanism rather than remembering whether the reaction is reversible or not. Unless there is any specific comment regarding reversible reactions, we should concentrate on the reaction in the sense given in the question.

Note also that we illustrated some common mistakes in drawing mechanisms, all taken from students' examination answers, in Box 5.1.

Problem 16.01 (level 1; chapters 1, 3 and 7)

Taxol is an anticancer drug obtained from species of yew, and has the partial structure shown. R represents a splendidly complex terpenoid group.

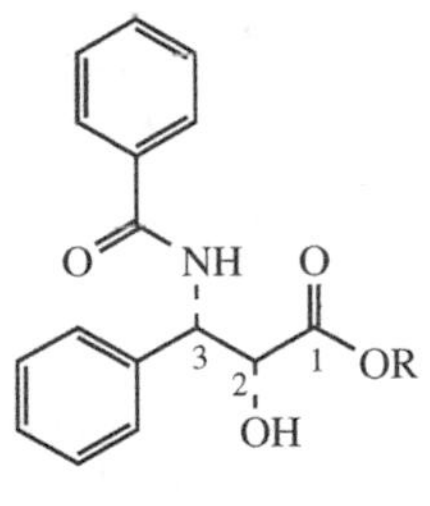

taxol

(a) From the following terms, indicate those which may be used to describe the chemical structure of taxol: (i) ketone; (ii) lactone; (iii) ester; (iv) amine; (v) amide; (vi) lactam; (vii) acetal; (viii) ether.

(b) Identify the chiral centres in taxol, and designate their configurations as *R* or *S*. Show the priorities assigned to the groups in establishing the configurations.

(c) Draw taxol in the form of a Newman projection looking down the 2–3 bond, and showing the stereochemically preferred conformation.

(d) Draw the enantiomer of taxol, using wedge–dot convention.

(e) Give a mechanism for the reaction of taxol with acetic anhydride [$(CH_3CO)_2O$].

(f) What products would be produced if taxol were treated with hot, aqueous HCl?

Answer 16.01

This is a relatively complex molecule, but these are simple questions requiring only standard procedures and reasoning.

(a) Ester, amide.

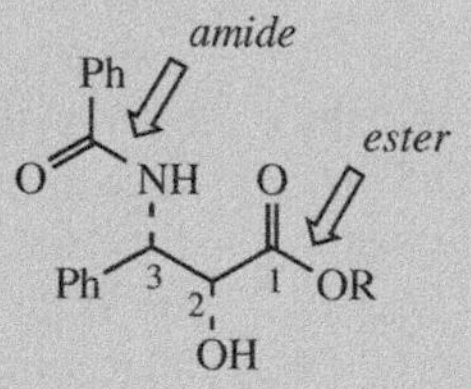

(b) 2R, 3S.

chiral centres have four different groups attached

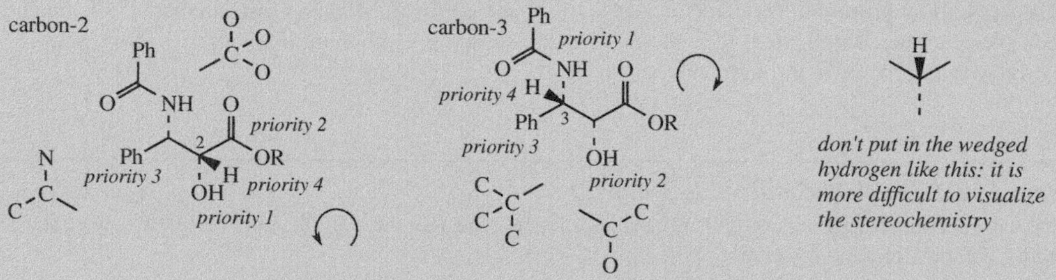

the sequence $1 \rightarrow 2 \rightarrow 3$ looks anticlockwise from front, but must be viewed from the rear towards H; therefore it is clockwise: (R)

the sequence $1 \rightarrow 2 \rightarrow 3$ looks clockwise from front, but must be anticlockwise looking from the rear towards H: (S)

don't put in the wedged hydrogen like this: it is more difficult to visualize the stereochemistry

(c)

view from this side

rotate front group 120°

3 interactions

2 interactions

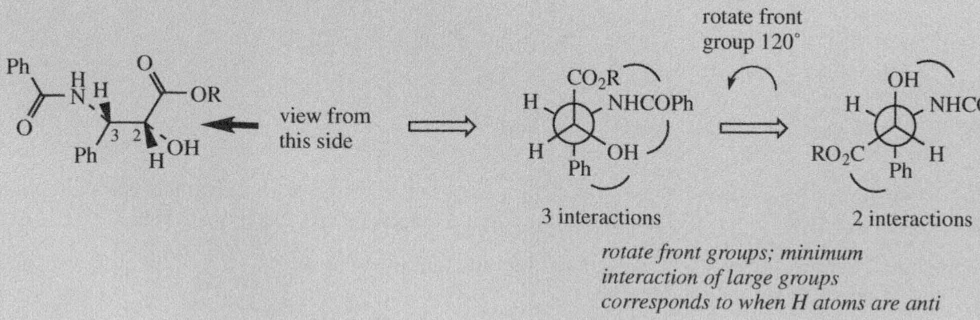

rotate front groups; minimum interaction of large groups corresponds to when H atoms are anti

(d) Enantiomer.

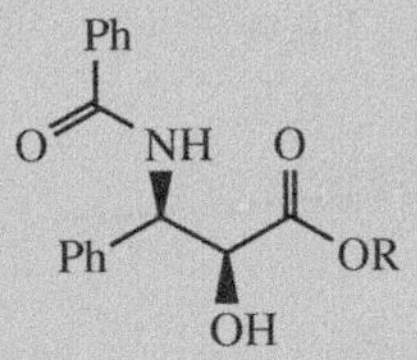
Ph
O
NH
O
Ph
OR
OH

change the configuration at both centres; change dot-wedge representation rather than trying to draw mirror image

(e) Alcohol plus acetic anhydride gives acetyl ester.

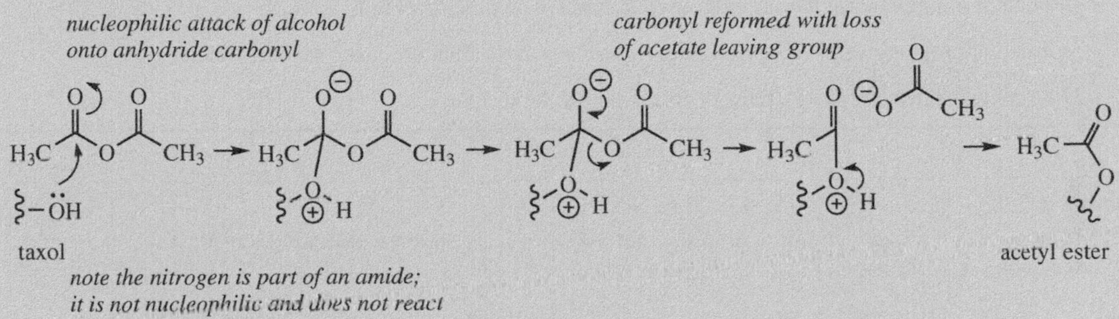
nucleophilic attack of alcohol onto anhydride carbonyl
carbonyl reformed with loss of acetate leaving group
H3C
CH3
OH
H
taxol
acetyl ester
note the nitrogen is part of an amide; it is not nucleophilic and does not react

(f) Note that ester and amide functions were identified in section (a); acidic hydrolysis (hot aqueous HCl) would lead to acid plus alcohol from ester, acid plus amine from amide. There will be three products. You are not asked for any mechanisms.

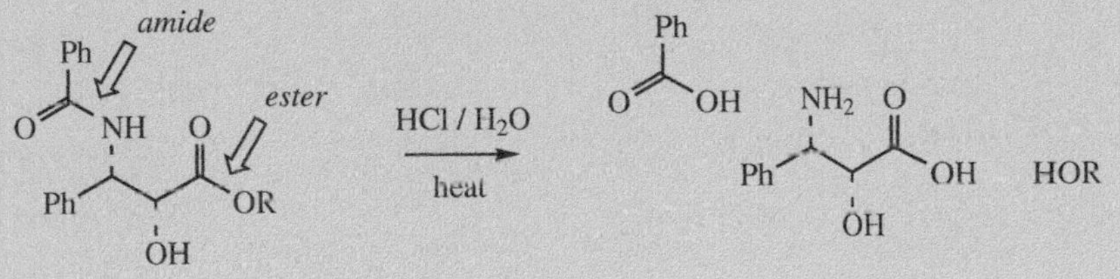
amide
ester
Ph
NH
OR
OH
HCl / H2O
heat
OH
NH2
HOR

Problem 16.02 (level 1; chapters 3 and 7)

Fluvastatin is an inhibitor of the cholesterol biosynthetic pathway and may be used to reduce the risk of coronary conditions, e.g. strokes and heart attacks, in susceptible patients. It has the structure shown, though drug material is supplied as the racemic form. A partial numbering system is given, and the rest of the molecule may be abbreviated to 'aryl' in answers.

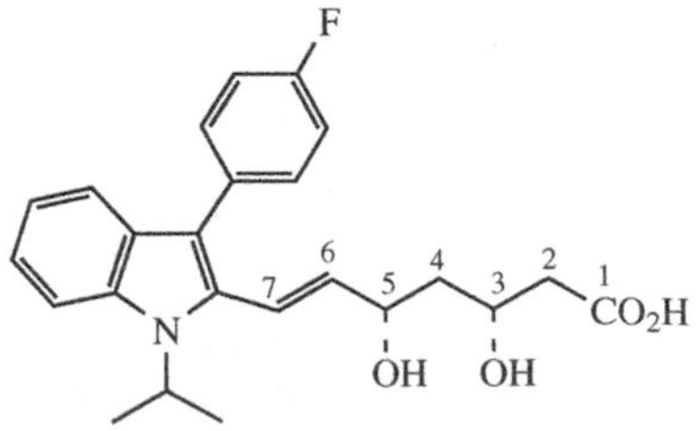
F
N
7
6
5
4
3
2
1
CO2H
OH
OH

fluvastatin

(a) Identify the chiral centres in fluvastatin, and designate their configuration as *R* or *S*. Show the priorities assigned to groups in establishing the configurations.

(b) Define the configuration of the double bond as *E* or *Z*. Show the priorities assigned to groups in establishing the configuration.

(c) Draw the structure of the other component of racemic fluvastatin.

(d) Draw the structure of a diastereoisomer of fluvastatin.

(e) Fluvastatin can easily form a cyclic ester (lactone):

- Show the structures, including stereochemistry, of two possible lactone products.
- Indicate, giving reasons, which of the two lactones would be favoured.
- With an appropriate drawing, predict the most likely conformation of the favoured lactone.
- Give a mechanism for acid-catalysed formation of the lactone.

Answer 16.02

This problem shows a fairly complex structure, but asks relatively simple questions about it. It is important not to be discouraged by the structure; simply apply standard procedures and reasoning.

(a) 3*R* and 5*S*.

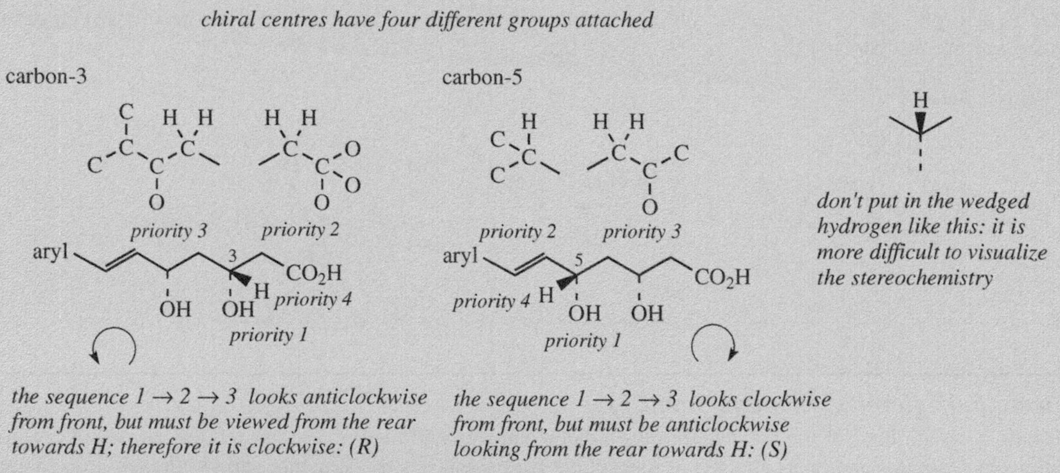
chiral centres have four different groups attached
carbon-3
carbon-5
priority 3
priority 2
priority 4
priority 1
priority 2
priority 3
priority 4
priority 1
aryl
CO_2H
OH
H
the sequence 1 → 2 → 3 looks anticlockwise from front, but must be viewed from the rear towards H; therefore it is clockwise: (R)
the sequence 1 → 2 → 3 looks clockwise from front, but must be anticlockwise looking from the rear towards H: (S)
don't put in the wedged hydrogen like this: it is more difficult to visualize the stereochemistry

(b) *E*.

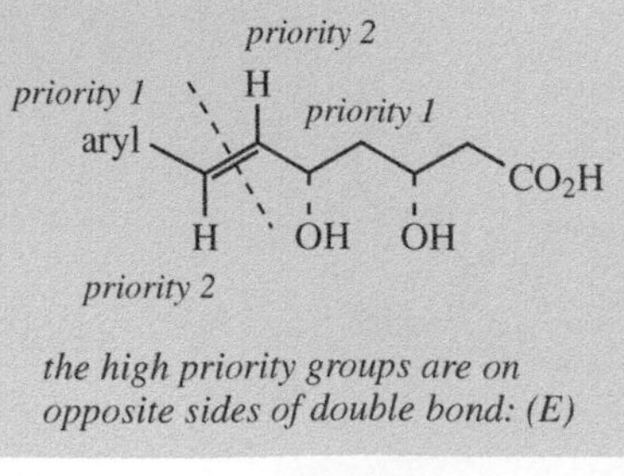
priority 2
priority 1
priority 1
priority 2
aryl
H
CO_2H
OH
OH
the high priority groups are on opposite sides of double bond: (E)

(c) The other component of racemic fluvastatin is its enantiomer.

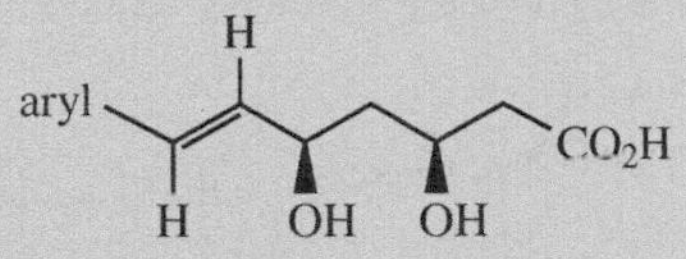
H
aryl
CO_2H
H
OH
OH

change the configuration at both centres; change dot-wedge representation rather than trying to draw mirror image

(d) Diastereoisomers.

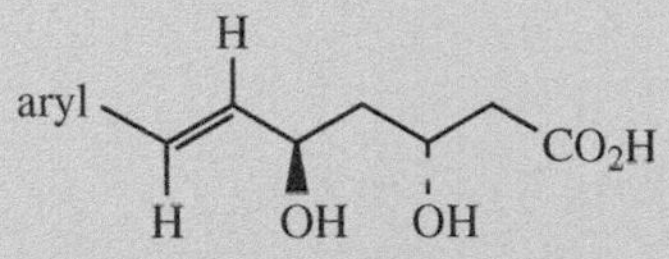
H
aryl
CO_2H
H
OH
OH

or

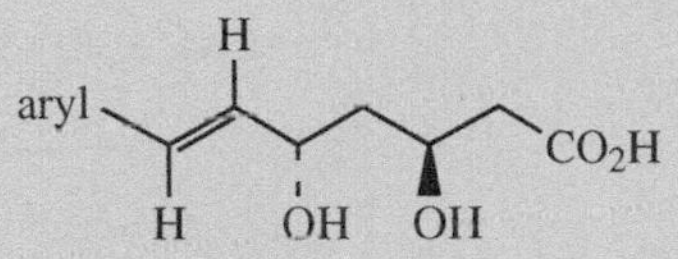
H
aryl
CO_2H
H
OH
OH

change the configuration at just one of the chiral centres

(e) Make ester from acid plus one of the two alcohol groups.

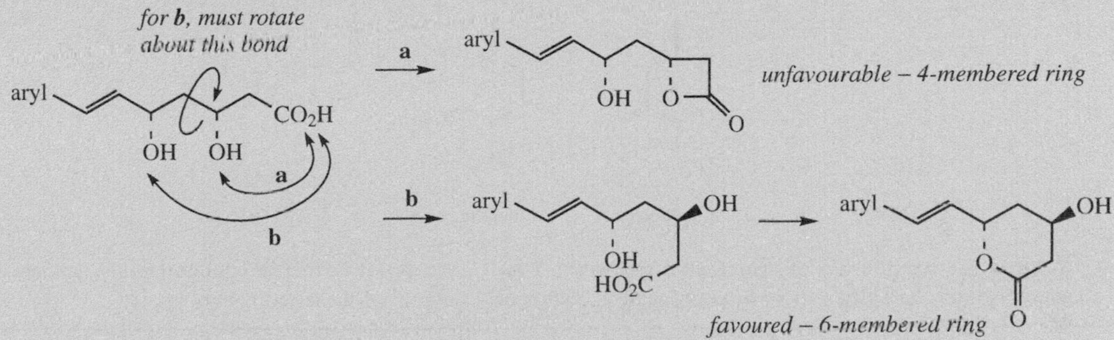
for b, must rotate about this bond
aryl
CO_2H
OH
OH
a
b
a
aryl
OH
O
O
unfavourable – 4-membered ring
b
aryl
OH
OH
HO_2C
aryl
OH
O
O
favoured – 6-membered ring

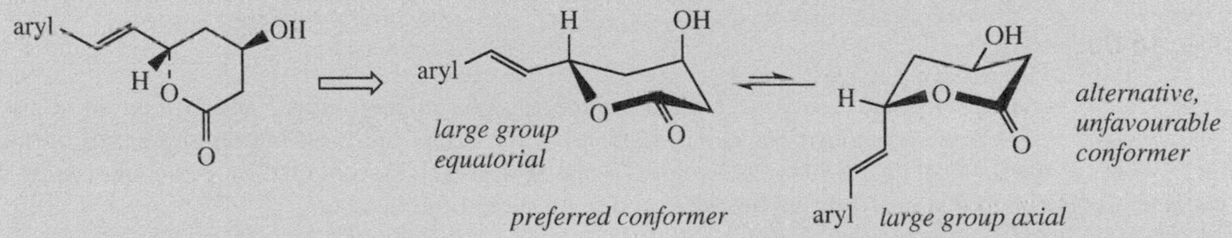
aryl
OH
H
O
O
H
OH
aryl
large group equatorial
O
O
preferred conformer
OH
H
O
O
aryl
large group axial
alternative, unfavourable conformer

Acid catalysed lactone (ester) formation.

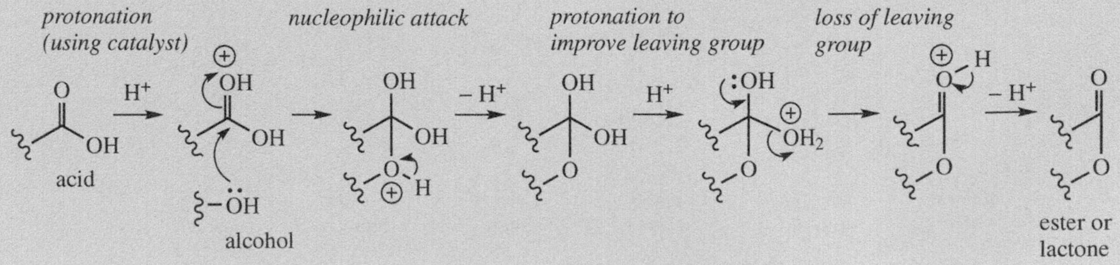
protonation (using catalyst)
nucleophilic attack
protonation to improve leaving group
loss of leaving group
O
H^+
OH
acid
OH
OH
OH
alcohol
OH
– H^+
OH
O
H
OH
H^+
OH
O
:OH
OH_2
O
O
H
O
– H^+
O
O
ester or lactone

Problem 16.03 (level 1; chapters 3 and 7)

(a) Indicate, with appropriate comments, which of the following compounds could exist in optically active form:

(b) Draw the compound cinerolone to show the shape of its ring and other stereochemical features.

$$HO\diagdown \overset{CH_3}{\underset{O}{\diagup}} \quad \overset{Z}{-CH_2CH=CHCH_3}$$

cinerolone

(c) Reduction of menthone with sodium borohydride gives menthol and a second isomeric material, neomenthol. Give a mechanism for the reaction and suggest a structure for neomenthol.

menthone   menthol

What is the relationship of neomenthol to menthol? Is it a structural isomer, a configurational isomer, a diastereoisomer, an epimer, a *meso* compound, a conformer? Several of these terms may apply.

Answer 16.03

(a) This is a 'spot the chiral centre' problem. A chiral centre has four different groups attached; we must also bear in mind that a *meso* compound has chiral centres, but the compound itself is optically inactive. This arises because the molecule has a plane of symmetry, and optical activity conferred by one chiral centre is equal and opposite to that conferred by the other, and is therefore cancelled out.

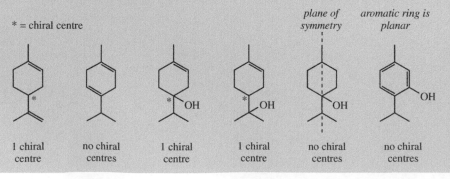

| 1 chiral centre | no chiral centres | 1 chiral centre | 1 chiral centre | no chiral centres | no chiral centres |

Only three of the compounds shown have chiral centres; these are indicated. These three compounds could, therefore, exist in optically active form. None of the compounds shown is a *meso* compound. Note that we have to consider part of the ring system as a 'group' attached to the centre in question. Follow this around until a decision can be made. The last two are not exactly trick questions, but require care. One has a plane of symmetry, and the carbon carrying the hydroxyl has two of its attached groups the same; the benzene ring is planar, so none of its carbons has the potential to be chiral.

(b) There are two parts of this molecule that require attention, the ring and the side-chain. The latter is the easier; we just have to interpret Z. In simple systems, as we have here, Z is the equivalent of *cis*. Systematically, we need to consider priorities of groups attached to the double bond. The Z configuration has groups of higher priority on the same side of the double bond, as shown.

cinerolone

planar bonds   planar bonds   cinerolone

Now for the ring, and you are advised to make some deductions before you start thinking about the conformations of five-membered ring systems. We have a double bond and a ketone in this structure. These both confer planarity on adjacent atoms. As a result, we must conclude that the carbon atoms of the ring must all be in one plane. Only the hydrogen and hydroxyl substituents will be out of plane. The stereodrawing, therefore, requires a planar five-membered ring seen in perspective, as shown.

(c) Sodium borohydride is a reducing agent that reacts with a ketone to give an alcohol through nucleophilic attack of hydride (it is not actually hydride that attacks, but we can formulate it as such). One of the products is menthol, and the second is an isomer, neomenthol. The carbonyl group is planar, so hydride can be delivered from either face. The isomer must be the alcohol in which the hydroxyl substituent has the alternative stereochemistry.

menthone   menthol   isomeric product:
                                                                       neomenthol

Neomenthol differs from menthol in having different stereochemistry at a single position. This makes it a configurational isomer, a diastereoisomer (there are two other chiral centres that are unchanged), and an epimer. The other terms are not applicable.

Problem 16.04 (level 1; chapters 1, 3 and 7)

The enzyme fumarase catalyses the stereospecific *trans*-addition of water to fumaric acid giving (S)-malic acid, and the reverse reaction, the *trans*-elimination of water from (S)-malic acid:

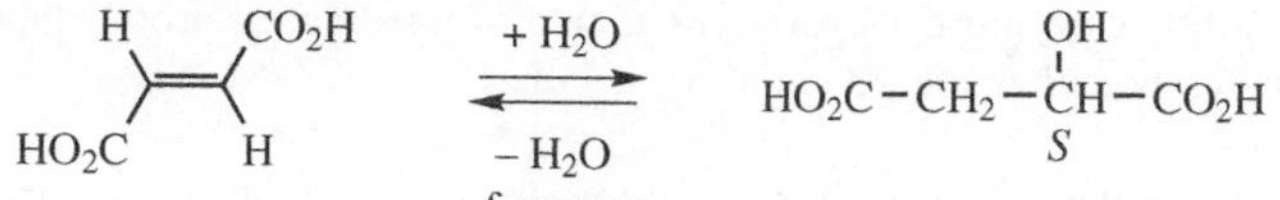

fumaric acid   fumarase   (S)-malic acid

(a) Give systematic names with stereochemical descriptors for fumaric acid and malic acid.

(b) Show the stereochemistry of (S)-malic acid using wedge–dot conventions.

(c) Deduce the structure of the product from the action of fumarase on ($2S,3S$)-[3-^{2}H]malic acid.

The enzyme malate synthase also synthesises (S)-malic acid from glyoxylic acid ($OHC–CO_2H$) and the thioester acetyl-coenzyme A ($CH_3CO.SCoA$; this may be regarded as a source of the nucleophile $^-CH_2CO.SCoA$).

(d) Draw a mechanism for the addition reaction.

(e) Draw a mechanism for the hydrolysis of the intermediate thioester, assuming base catalysis.

(f) Indicate the stereochemistry implicit in the addition reaction. What face of the carbonyl group is attacked?

Answer 16.04

Whilst this may look like an intermediary metabolism question, it requires no knowledge of biochemistry, and is merely a simple chemical analysis of two reactions from biochemistry.

(a) The nomenclature for a dicarboxylic acid follows from that of a monocarboxylic acid, e.g. butanoic acid becomes butandioic acid. The 1,4-numbering is strictly redundant; the carboxylic acids must be at the ends of the chain. Stereochemical descriptors are (E) for the double bond in fumaric acid, and the (S) configuration stated for malic acid.

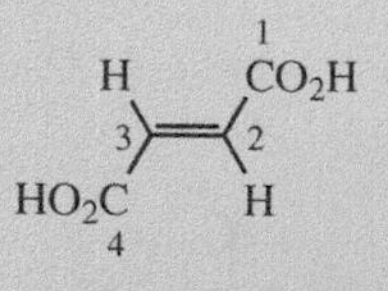

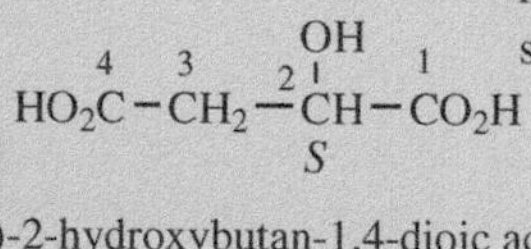

(b) Draw the carbon chain in zigzag form, put in the large carboxylic acid groups *anti*, then the substituents. The configuration at C-2 is determined by trial and error.

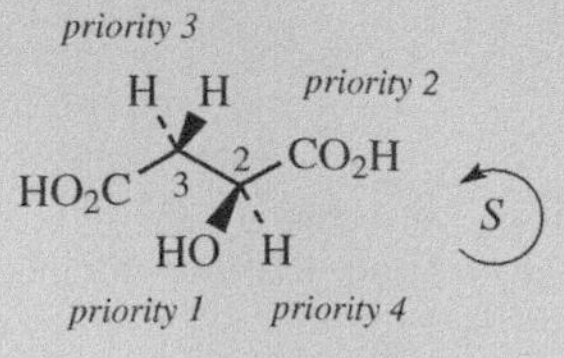

(c) The *trans* elimination of water from (S)-malic acid must remove the groups shown in bold. We then substitute one of the hydrogens at position 3 with ^2H (deuterium; D) to get the $3S$ configuration by the trial-and-error procedure in (b). ($2S,3S$)-[3-^2H]malic acid is thus labelled as shown, and *trans* elimination of water must result in the deuterium label ending up as shown in fumaric acid.

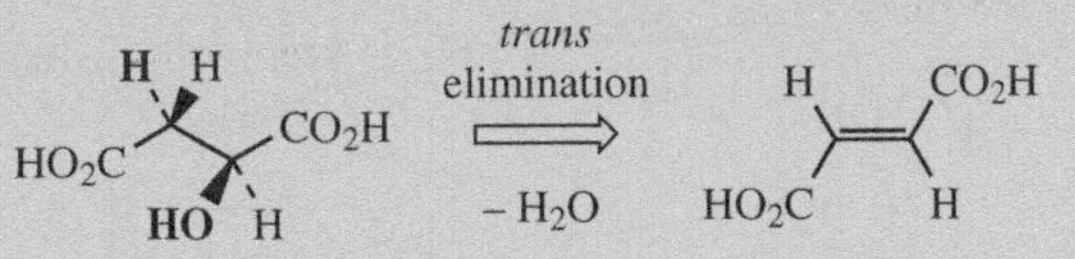

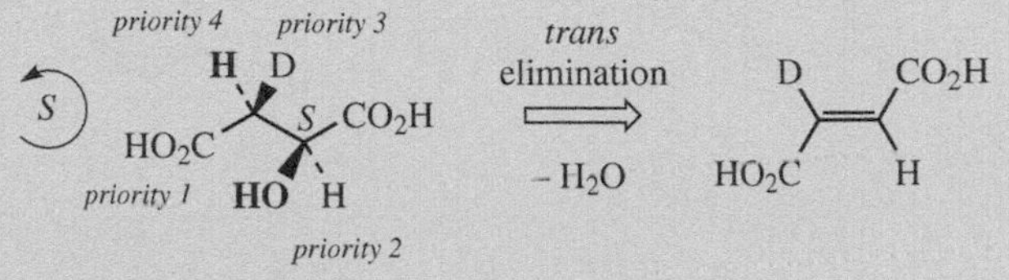

(2S,3S)-[3-^2H]malic acid

(d) As soon as you have written down the reagents, you will find the reaction follows quite readily. We have simple nucleophilic addition to an aldehyde.

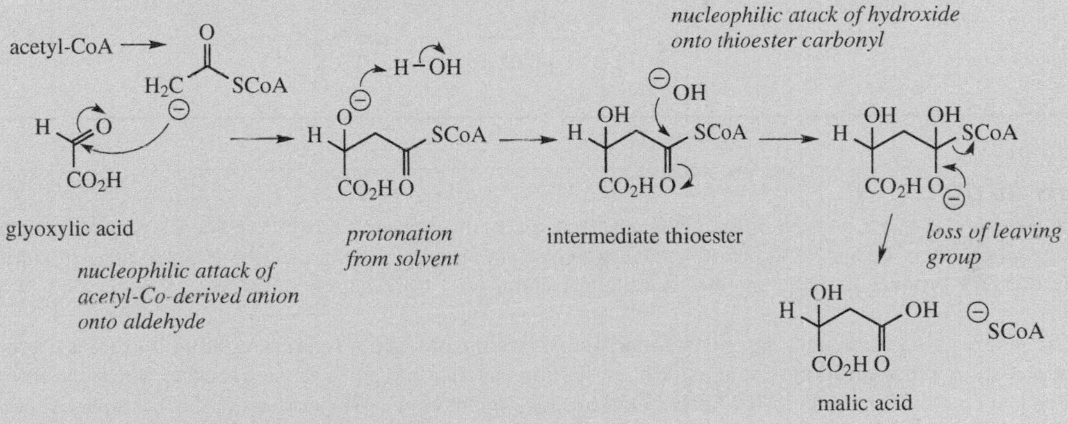

malic acid

(e) The intermediate thioester is then hydrolysed by base as included above.

(f) Now for the stereochemistry. Draw the malic acid product with the 2S configuration. It is also possible to turn around the structure obtained in part (b). You can only get this configuration if the planar aldehyde is attacked from the face shown. This is the Si face, using the priority rules for enantiotopic faces, which are an extension of the Cahn–Ingold–Prelog R and S system.

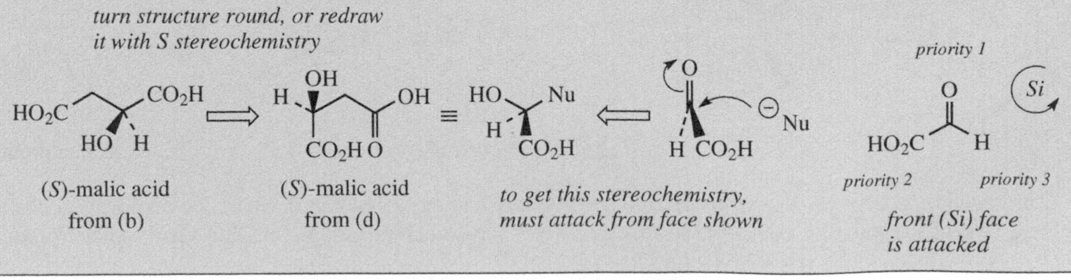

Problem 16.05 (level 1; chapters 3, 6, 7 and 10)

In each of the following reactions, an optically active starting material is converted into a product that is optically inactive. Suggest mechanisms for the reactions and explain the loss of optical activity.

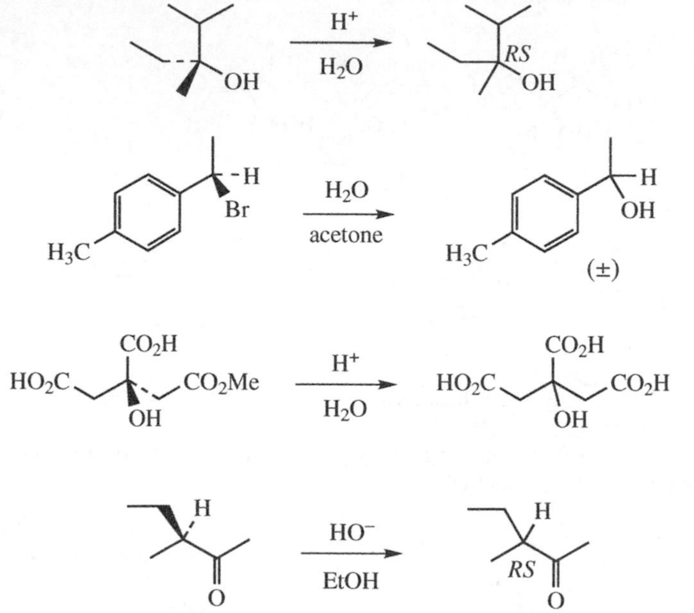

Answer 16.05

Whilst this may appear a problem in stereochemistry, it is actually mechanism based and depends upon the nature of the reaction intermediates. The loss of optical activity is because we form a racemic product, i.e. a mixture of enantiomers, or produce a compound that is no longer chiral.

(a) The first reaction converts a single enantiomer into a racemic product; it is a racemization. The logical way of achieving any racemization is to transform the starting material into a planar intermediate, which can then be attacked equally from either side. Such an intermediate might be a carbocation, as in this example. Although this might appear a non-reaction, it is actually S_N1 chemistry where the leaving group and nucleophile are the same, namely water.

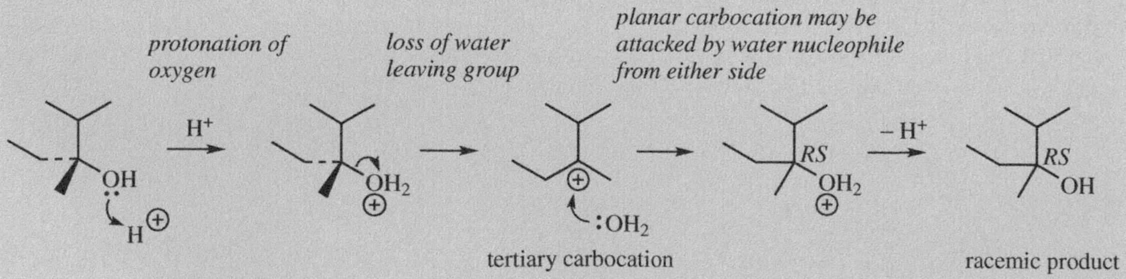

protonation of oxygen loss of water leaving group planar carbocation may be attacked by water nucleophile from either side

tertiary carbocation racemic product

The reaction occurs because a favourable tertiary carbocation is generated. Since the carbocation also has three different substituents, nucleophilic attack of water forms a chiral centre, and thus enantiomeric products.

(b) This reaction differs little from (a), though the leaving group and nucleophile are no longer the same; it is a standard S_N1 reaction. We show the alternative way of depicting a racemic product: RS or $(\pm)$- may be used for racemates.

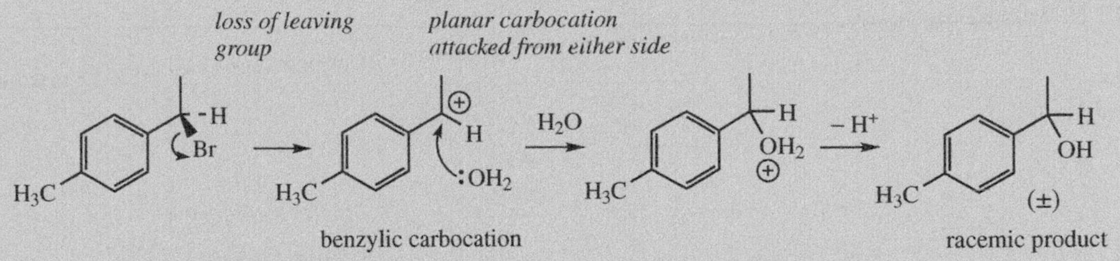

loss of leaving group

planar carbocation attacked from either side

benzylic carbocation

racemic product

The intermediate carbocation initially appears secondary, but it is also benzylic and, therefore, resonance stabilized by the aromatic ring.

(c) There is no need even to consider the chiral centre in this problem; hydrolysis of the ester grouping to an acid generates a molecule that is achiral because it has two equivalent substituents. The answer is simply acid-catalysed ester hydrolysis.

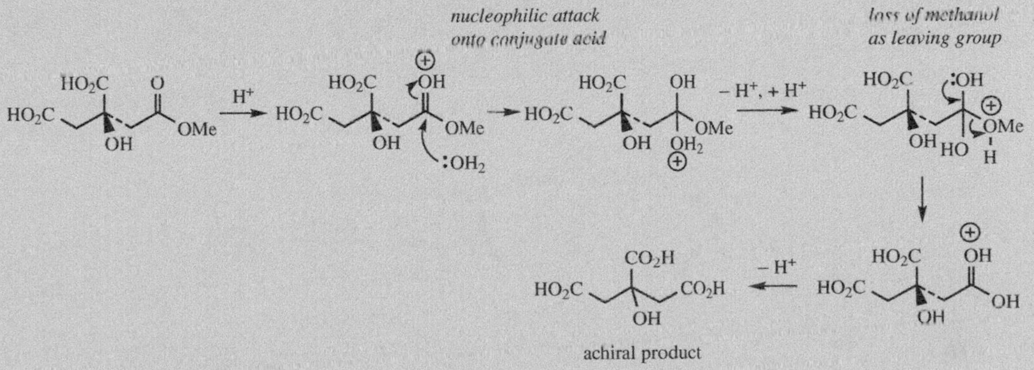

nucleophilic attack onto conjugate acid

loss of methanol as leaving group

achiral product

(d) This is a base-catalysed racemization that is dependent upon the formation of an enolate anion. The resonance form shown first is planar, so reprotonation can occur from either face, producing the racemic ketone.

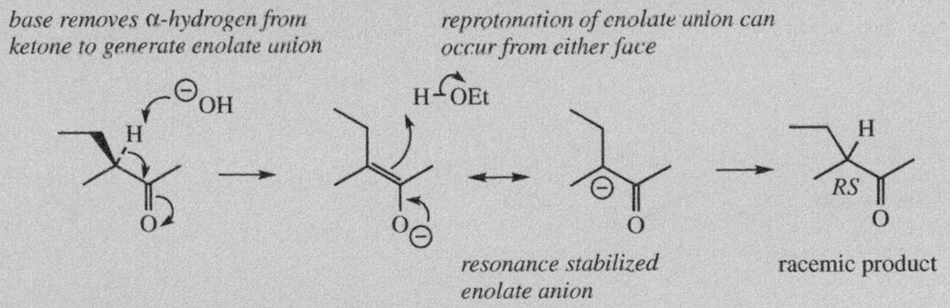

base removes α-hydrogen from ketone to generate enolate anion

reprotonation of enolate anion can occur from either face

resonance stabilized enolate anion

racemic product

This is an equilibrium reaction, and it raises a couple of points. First, there are two α-positions in the ketone, so what about the $COCH_3$-derived enolate anion? The answer is that it is formed, but since the CH_3 group is not chiral, proton removal and reprotonation have no consequence. Racemization only occurs where we have a chiral α-carbon carrying a hydrogen substituent. Second, the enolate anion resonance structure with charge on carbon is not planar, but roughly tetrahedral. If we reprotonate this, it must occur from just one side. Yes, but both enantiomeric forms of the carbanion will be produced, so we shall still get the racemic mixture.

Problem 16.06 (level 1; chapters 3 and 6)

Treatment of (2*R*,3*R*)-2-bromo-3-methylpentane with sodium ethoxide in ethanol gives (*E*)-3-methylpent-2-ene as the major product. What would be the major product from similar base treatment of (2*S*,3*S*)- and (2*R*,3*S*)-2-bromo-3-methylpentane? Hint: Newman projections will be useful here.

Answer 16.06

(2*R*,3*R*)-2-bromo-3-methylpentane →[NaOEt] (*E*)-3-methylpent-2-ene (major product)

Draw out the structures with stereochemistry. Since it is not always easy to visualize a stereoisomer with the required configuration, draw one tentative arrangement, work out the chirality, then adjust if necessary. It is always easiest to put the group of lowest priority (usually H) away from you, i.e. dotted bond.

Draw the starting material as a Newman projection

The major products from similar base treatment of (2*S*,3*S*)- and (2*R*,3*S*)-2-bromo-3-methylpentane can be deduced as below

(2*S*,3*S*) is the enantiomer; opposite configuration at each centre

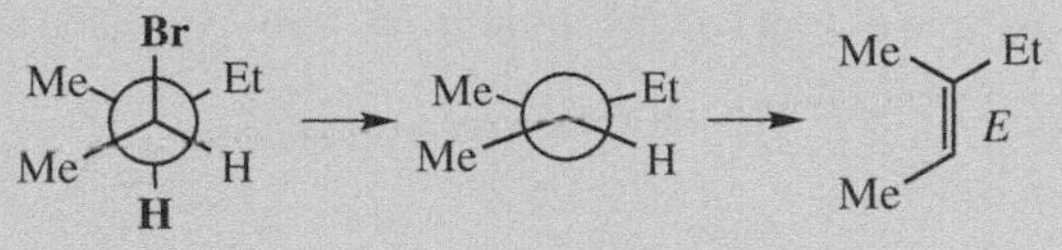

2*S*,3*S*

it gives the same product

(2*R*,3*S*) is a diastereoisomer; opposite configuration at C-3 (C-3 is rear chiral centre)

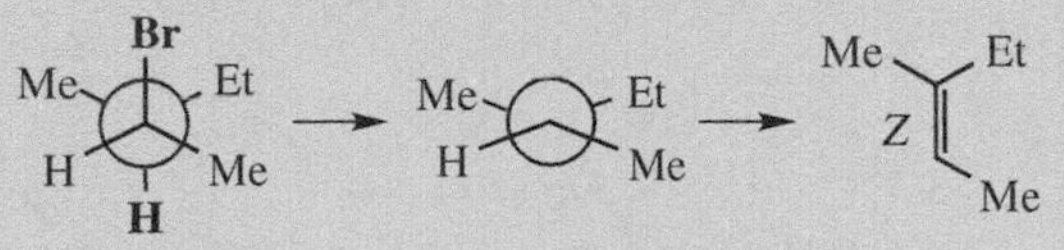

2*R*,3*S*

it gives the *Z* product

Problem 16.07 (level 1; chapters 6, 7 and 8)

In the following four transformations, give detailed mechanisms for the steps indicated.

(a)

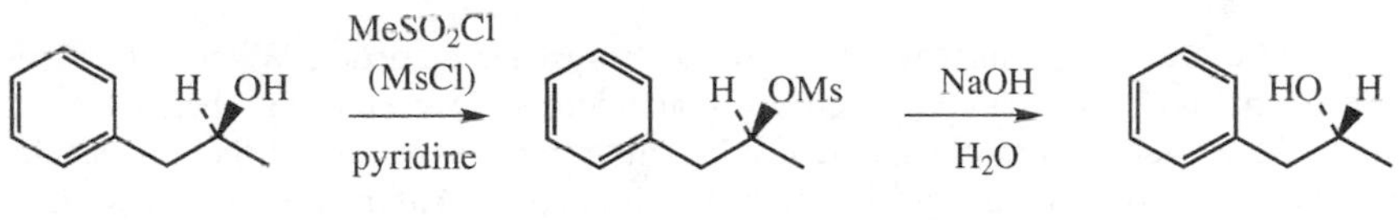

(b)

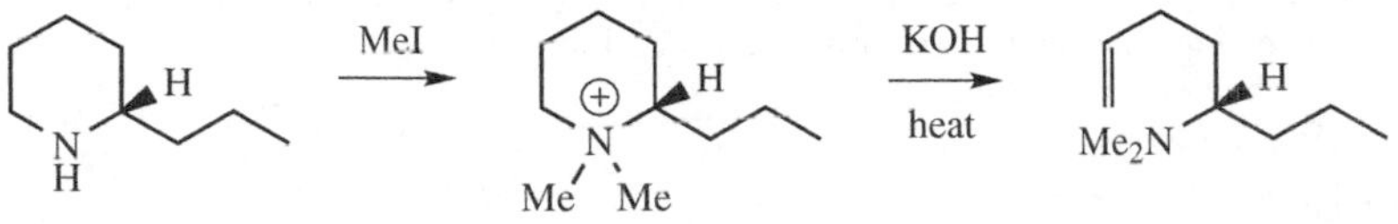

(c)

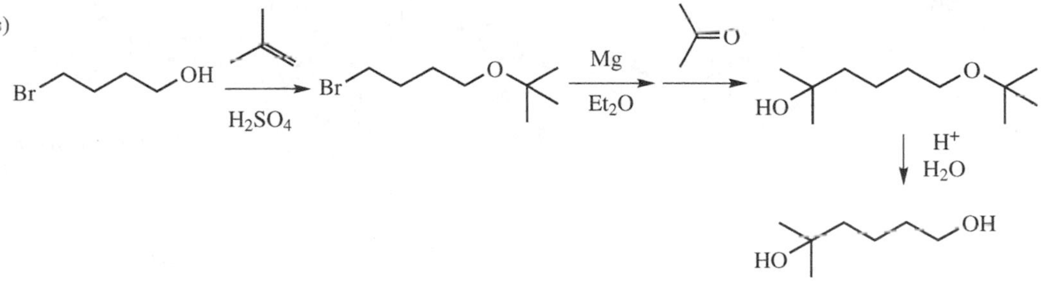

(d)

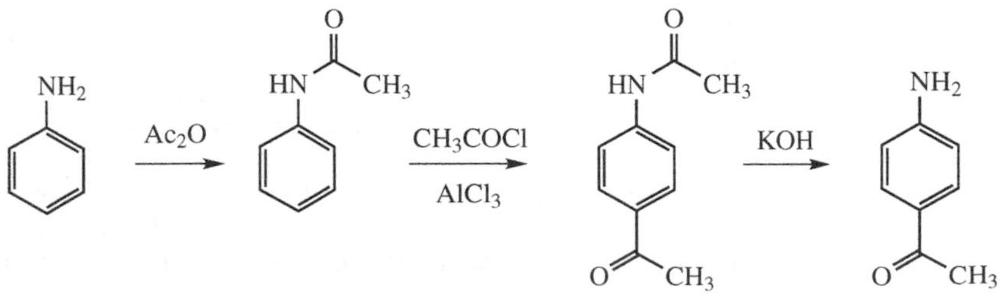

For each sequence, explain why the intermediary steps are advantageous in obtaining the product shown.

Answer 16.07

(a) This achieves a change in stereochemistry and, therefore, requires an S_N2 reaction. An alternative S_N1 mechanism is likely to lead to a racemic product, so may be discounted.

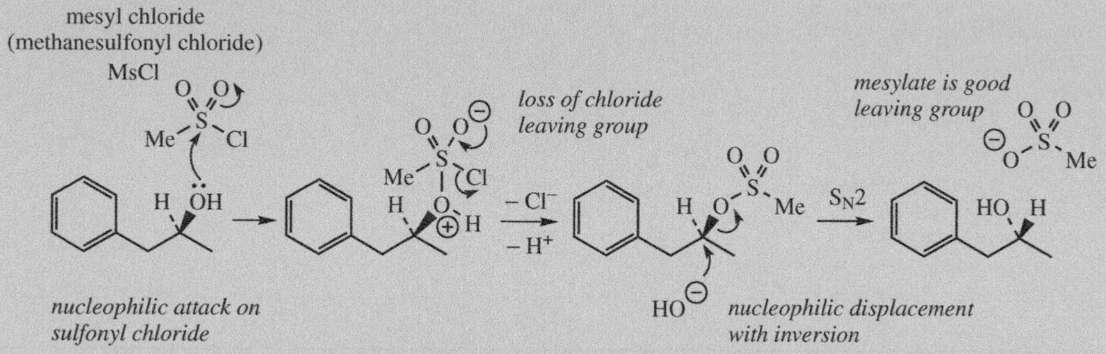

mesyl chloride
(methanesulfonyl chloride)
MsCl

loss of chloride leaving group

mesylate is good leaving group

$-Cl^-$
$-H^+$

S_N2

nucleophilic attack on sulfonyl chloride

nucleophilic displacement with inversion

The purpose of the first reaction with methanesulfonyl chloride (mesyl chloride) is to form a mesylate ester, a derivative with a better leaving group than hydroxide. Then, the S_N2 reaction can proceed more favourably. Esterification with mesyl chloride is mechanistically analogous to esterification with acetyl chloride.

(b) The extra steps here are also to improve the leaving group characteristics. We require a base-initiated E2 elimination reaction, and by forming a quaternary ammonium salt we produce a better leaving group. Since we are starting with a cyclic amine, the amine leaving group remains as part of the product. In this particular example, we also need to consider the direction of elimination. Alternatives are possible, as shown. We actually obtain the Hofmann-type less-substituted alkene as major product, since the elimination involves a relatively large leaving group, the quaternary ammonium cation.

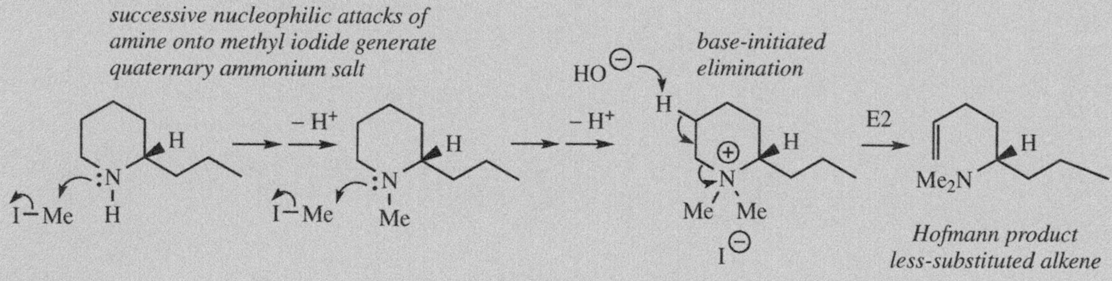

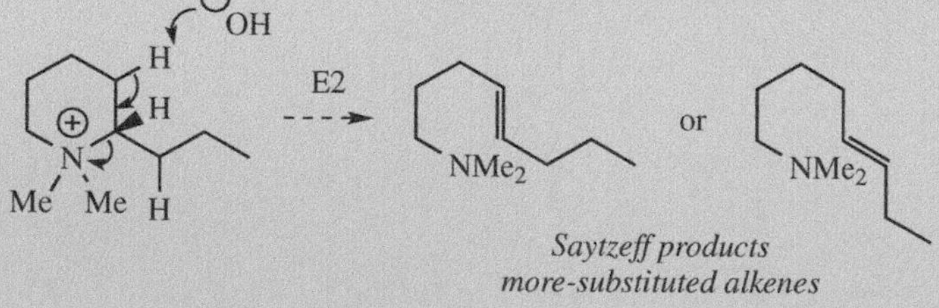

successive nucleophilic attacks of amine onto methyl iodide generate quaternary ammonium salt

base-initiated elimination

Hofmann product less-substituted alkene

Saytzeff products more-substituted alkenes

(c) The overall reaction is generation of a Grignard reagent from the alkyl bromide, followed by nucleophilic addition of this carbanion equivalent on to the ketone acetone. However, there is also a hydroxyl group in the starting material, and we must use a protecting group to ensure this does not react with and destroy the OH-sensitive Grignard agent. It is protected here through formation of an ether. This is an electrophilic addition reaction, using a *tert*-butyl carbocation formed by protonation of an alkene.

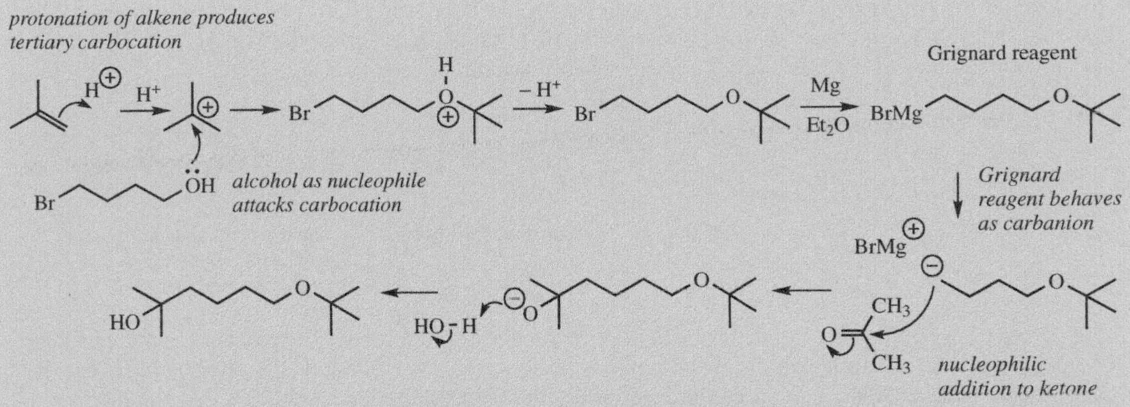

protonation of alkene produces tertiary carbocation

alcohol as nucleophile attacks carbocation

Grignard reagent

Grignard reagent behaves as carbanion

nucleophilic addition to ketone

With the alcohol protected, preparation of the Grignard reagent can proceed, and this can then react with the ketone carbonyl in a nucleophilic addition. The protecting group can then be removed by treatment with acid, to restore the hydroxyl function. This also involves a tertiary carbocation that is subsequently quenched with water.

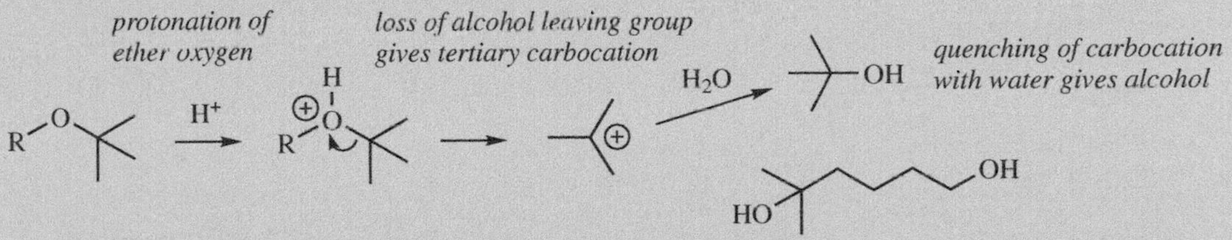

protonation of ether oxygen

loss of alcohol leaving group gives tertiary carbocation

quenching of carbocation with water gives alcohol

(d) In this sequence of reactions, we achieve Friedel–Crafts acylation of the aromatic amine. This could not be performed directly. Although an amino group is *ortho* and *para* directing, and should give mainly the *para* product because steric considerations favour this over the *ortho* product, this is because the amine group is an electron donor. As such, this means it will preferentially complex with the Friedel–Crafts reactive intermediate, the acylium cation, preventing further reaction. Converting the amine to an amide stops this behaviour. Amides are non-basic and poor electron donors; this is because the nitrogen lone pair is utilized with the carbonyl group to provide resonance stabilization. The *N*-acetyl is thus a protecting group for the amine.

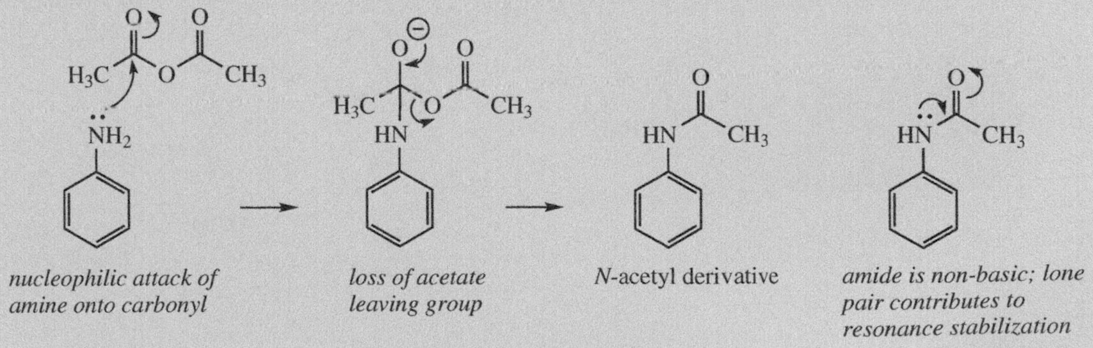

nucleophilic attack of amine onto carbonyl

loss of acetate leaving group

N-acetyl derivative

amide is non-basic; lone pair contributes to resonance stabilization

The sequence requires first simple *N*-acetylation with acetic anhydride. This product is then the substrate for the Friedel–Crafts reaction. The acylium cation is generated from acetyl chloride and aluminium chloride.

You may perhaps remember that the acylium cation is the acylating species or, alternatively, you can deduce that the acyl chloride interacts with the Lewis acid $AlCl_3$, and is most likely to give $AlCl_4^-$. The acylium cation then emerges as the other part. This is resonance stabilized, as shown.

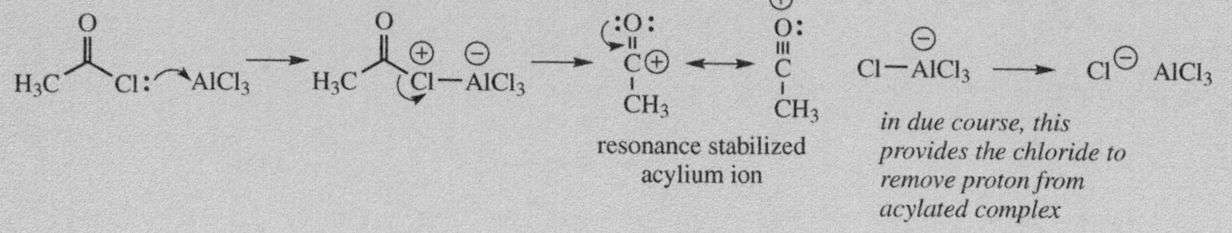

resonance stabilized
acylium ion

*in due course, this
provides the chloride to
remove proton from
acylated complex*

Now the electrophilic attack of the aromatic system onto the acylium cation. The product is *para* substituted, so go for that site of attack. We then need removal of a proton, and chloride is the obvious base, but since this is complexed as $AlCl_4^-$, we need the latter to decompose to chloride and $AlCl_3$.

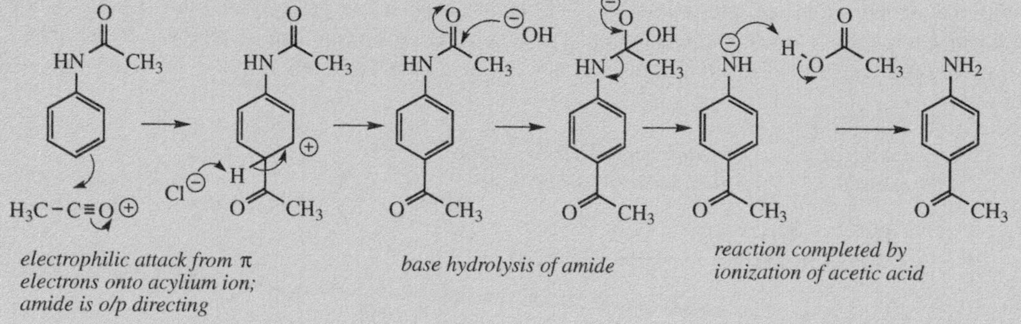

*electrophilic attack from π
electrons onto acylium ion;
amide is o/p directing*

base hydrolysis of amide

*reaction completed by
ionization of acetic acid*

The sequence is completed by base hydrolysis of the amide and removal of the protecting group. This is much the same as an ester hydrolysis, and needs to include as last stage the ionisation of acetic acid; since RNH^- is a poor leaving group, it is this ionisation that allows the reaction to proceed.

Problem 16.08 (level 1; chapters 6 and 7)

(a) The natural alkaloid diester cinnamoylcocaine can be converted into the local anaesthetic and much abused drug cocaine by the sequence shown.

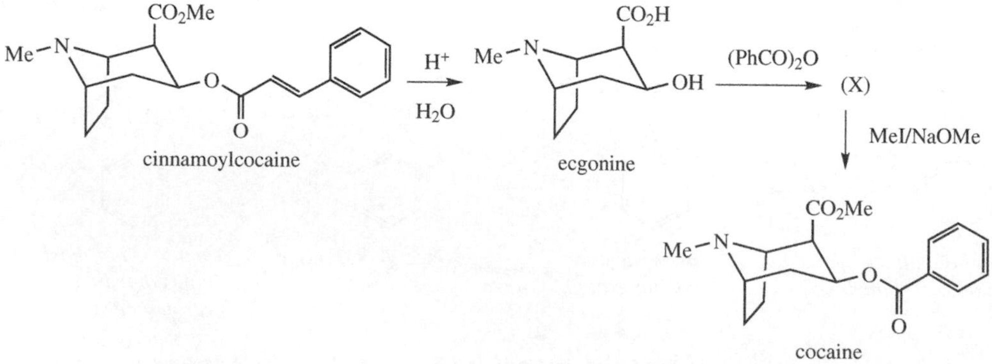

cinnamoylcocaine

ecgonine

cocaine

Give a mechanistic interpretation for each of the reactions, and give a structure for the intermediate (X).

(b) Give mechanisms for conversion of estrone into the oestrogen drugs estradiol and ethinylestradiol.

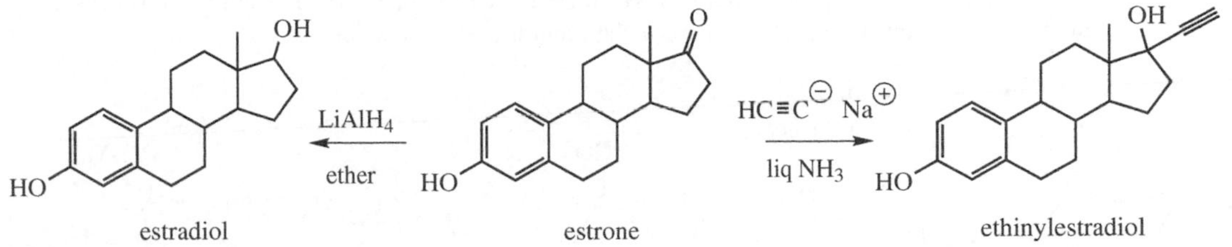

estradiol   estrone   ethinylestradiol

Answer 16.08

Do not be put off by the structures; instead, you should be looking at the functional groups and the changes involved. The reactions are relatively simple.

(a) You are told cinnamoylcocaine is a diester. The first reaction is merely acid-catalysed hydrolysis of these two groups. Abbreviate groups as necessary; and provided you make things clear, you will not be required to give the same mechanism for the second group.

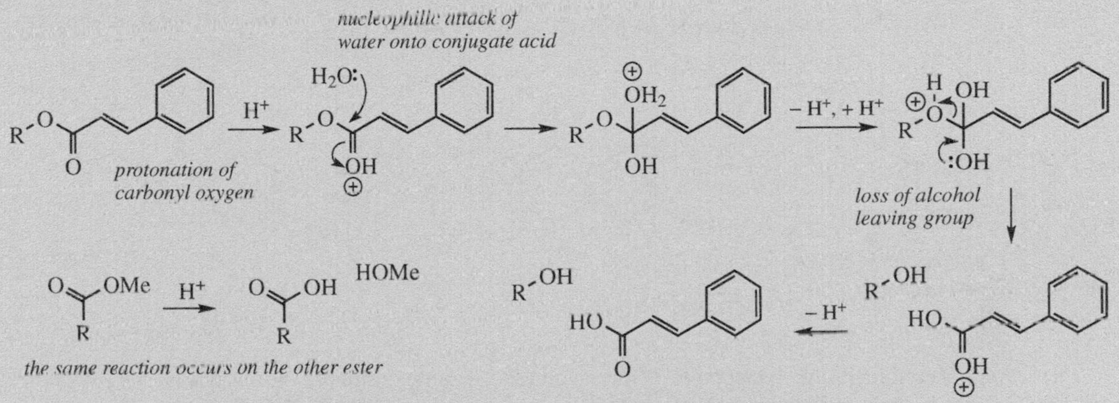

Ecgonine now reacts with (PhCO)$_2$O to give (X). This reagent is benzoyl anhydride, an analogue of acetic anhydride. The alcohol group reacts, displacing benzoate as leaving group and forming (X), the benzoyl ester of ecgonine.

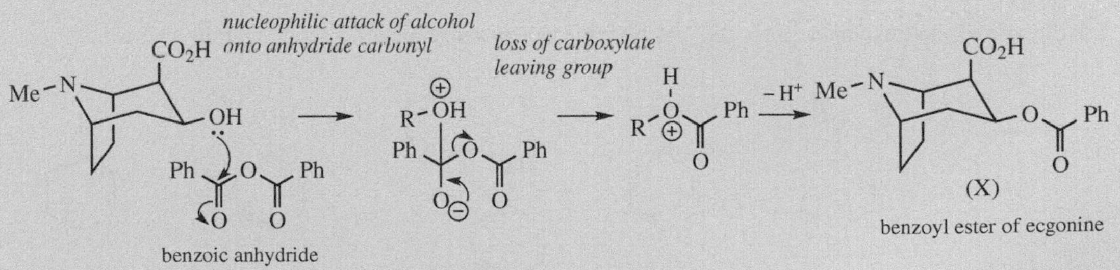

The next step is an unusual approach to making an ester via an S$_N$2 reaction. We usually make esters by the reaction of alcohols with activated carboxylic acid derivatives like acyl chlorides, or anhydrides as we have just seen. We do not particularly want to have to convert the carboxylic acid (X) into a more reactive derivative, and we can not just react with methanol and acid because that would be likely to transesterify the existing ester. The clue is in the use of base; this will form the carboxylate anion. The reaction then becomes a base-initiated S_N2 reaction with MeI; the carboxylate anion is the nucleophile.

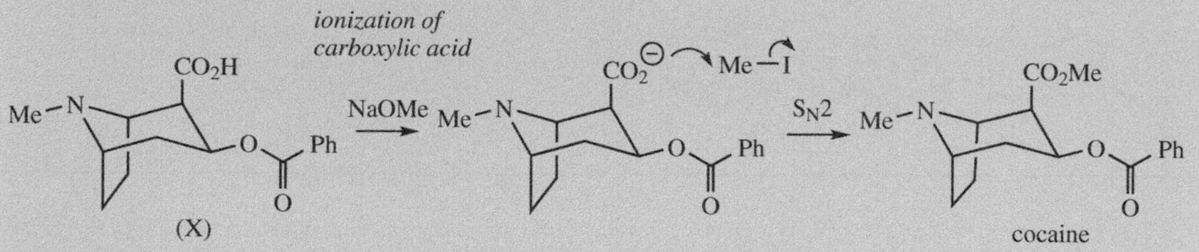

(b) This problem provides two examples of nucleophilic addition to a ketone; the ketone is the only reacting functional group in this complex steroidal structure.

Estradiol is formed by lithium aluminium hydride reduction of the ketone. We can formulate this simply as hydride acting as the nucleophile, though hydride delivery by LAH is strictly more complex than this. Unless you are specifically asked for details, treat LAH as a source of hydride ion.

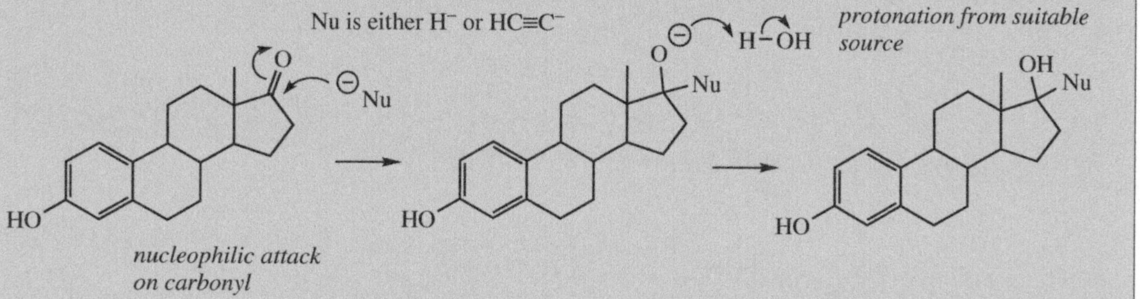

Similarly, acetylide is the nucleophile in the synthesis of ethinylestradiol. In each case the reaction is finished off by protonation from a suitable source, e.g. water.

Problem 16.09 (level 1; chapter 4)

(a) Provide a reasoned explanation for the relative pK_a values determined for the five amines shown.

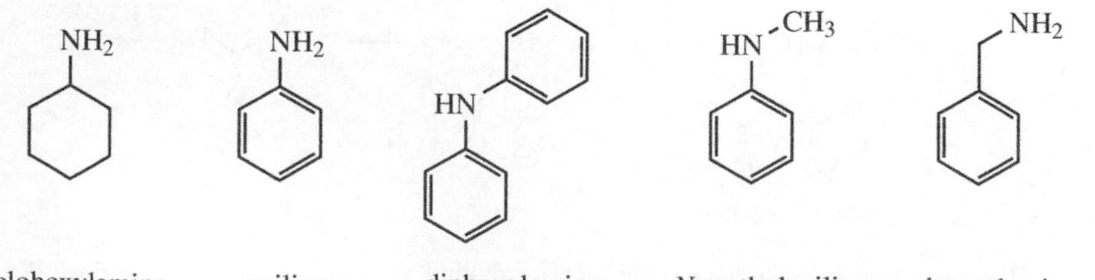

	cyclohexylamine	aniline	diphenylamine	N-methylaniline	benzylamine
pK_a (conjugate acid)	10.6	4.6	0.8	4.9	9.3

(b) Comment on the relative pK_a values determined for the following five carboxylic acids.

	cyclohexane-carboxylic acid	benzoic acid	4-methyl-benzoic acid	4-hydroxy-benzoic acid	4-acetyl-benzoic acid
pK_a	4.9	4.2	4.4	4.6	3.7

Answer 16.09

The relative acidities or basicities of a range of similar organic compounds can usually be rationalized quite readily by considering the implications of electron-withdrawing and electron-donating substituents in the molecule. Consideration of inductive and resonance effects is usually adequate, and although other effects, e.g. electronegativity or hybridization, may come into play, they should not form the first line of reasoning. Remember also that, where inductive and resonance effects are in opposition, the resonance effect is usually of greater magnitude and will prevail.

One of the compounds shown should be used as the point of reference from which comparisons can be made. There is no way we would encourage memorising of pK_a values, but two easily remembered figures can be valuable for comparisons. These are pK_a around 5 for a typical aliphatic carboxylic acid, and pK_a around 10 for a typical aliphatic amine. These then allow us to consider whether the compound in question is more acidic, more basic, etc.

For acids, the predominant consideration should be stabilization or destabilization of the conjugate base. The stability of negatively charged conjugate bases will tend to be enhanced by electron-withdrawing substituents that can help delocalize the charge. Electron-donating substituents will do the opposite, and thus reduce the likelihood of conjugate base formation.

Conversely, we can reason that the formation of conjugate acids from bases will be favoured by electron-donating substituents and inhibited by electron-withdrawing groups. However, the feature of bases is that they have a lone pair of electrons that are able to coordinate with a proton. Sometimes, this lone pair may feed into the molecule via a resonance effect, and this can stabilize the free base and inhibit conjugate acid formation. With bases, therefore, we normally consider two approaches, either stabilization of the conjugate acid, which increases basicity, or stabilization of the free base, which decreases basicity.

Most substituents are easily recognized as potentially electron withdrawing or electron donating, e.g. halogens, carbonyls, amines, phenols. We also need to remember that phenyl groups provide an electron-withdrawing inductive effect (a consequence of sp^2 hybridization), whilst alkyl groups are weakly electron donating.

(a) We have here a mix of aliphatic and aromatic amines, and two benchmark compounds, cyclohexylamine and aniline. Remember, of course, that the pK_a values refer to the acidity of the conjugate acids, and not to the acidity of the amines themselves, which could be relevant in different circumstances. Nevertheless, it is usually acceptable to talk of the pK_a of aniline, when we should strictly say the pK_a of the conjugate acid of aniline.

The pK_a of cyclohexylamine (10.6) is typical of an aliphatic amine (methylamine also has pK_a 10.6). Aniline is a much weaker base (the smaller the pK_a, the weaker the base) than cyclohexylamine. This is a combination of an inductive effect and a resonance effect, which reinforce each other. The phenyl ring provides an electron-withdrawing inductive effect that destabilizes the conjugate acid. More important, though, is the resonance effect that stabilizes the uncharged amine.

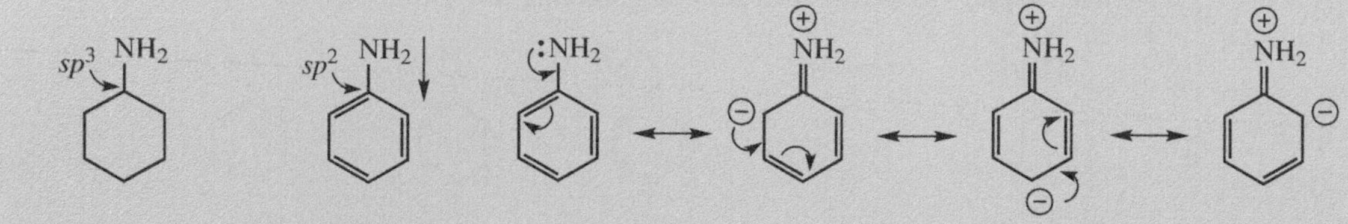

typical aliphatic amine   *electron-withdrawing inductive effect*   *resonance stabilization possible in the base, but not the conjugate acid*

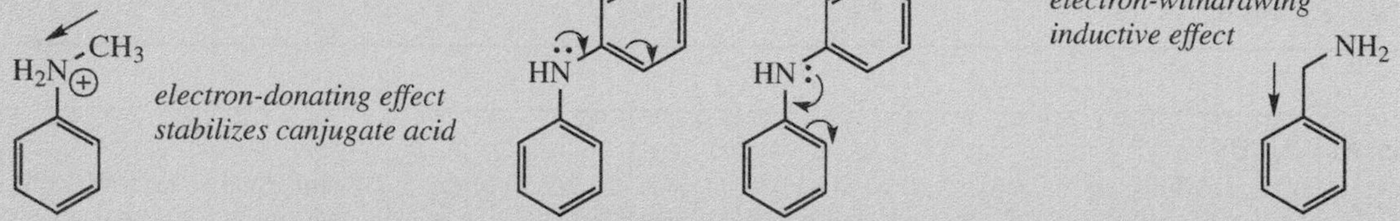

electron-donating effect stabilizes canjugate acid   *resonance effect from both aromatic rings*   *electron-withdrawing inductive effect*

As we move to *N*-methylaniline, we see only a modest change in pK_a. This is undoubtedly due to the electron-donating effect of the methyl group, and this would be expected to stabilized the conjugate acid, increasing observed basicity. There is a modest increase in basicity, but it is apparent that the resonance effect, as in aniline, is also paramount here, and this compound is also a weak base. However, diphenylamine (*N*-phenylaniline) is an extremely weak base; this can be ascribed to the resonance effect allowing electron delocalization into two rings.

On the other hand, benzylamine (pK_a 9.3) is a much stronger base than aniline, somewhat weaker than cyclohexylamine. Benzylamine is merely an aliphatic amine, with an electron-withdrawing aromatic ring separated by one carbon from the nitrogen. There is no chance of a resonance effect as in aniline, and the inductive effect is smaller than if the phenyl were directly bonded to nitrogen. The net result is a small lowering of charge on nitrogen with destabilization of the conjugate acid, and hence a lower basicity than cyclohexylamine.

(b) The reference point here is the acidity of cyclohexanecarboxylic acid (pK_a 4.9), a typical aliphatic acid. Benzoic acid is a stronger acid (the smaller the pK_a, the stronger the acid), and this can be ascribed to the hybridization-derived electron-withdrawing inductive effect of the phenyl ring. This helps to delocalize the charge on the conjugate acid and is a stabilizing factor. Inductive effects in the opposite direction destabilize the conjugate base, and we see a weakening of acidic strength. This is why 4-methylbenzoic acid is not as strong an acid as benzoic acid, though the inductive effect of the methyl is small.

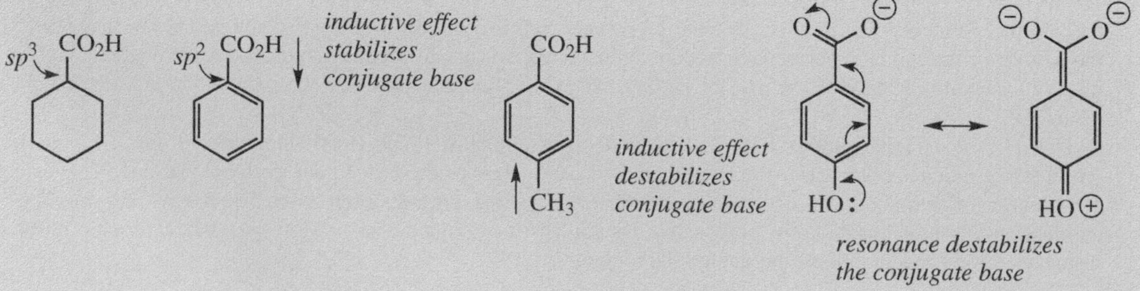

inductive effect stabilizes conjugate base   *inductive effect destabilizes conjugate base*   *resonance destabilizes the conjugate base*

*resonance effect
stabilizes conjugate base*

The inductive effect in 4-hydroxybenzoic acid is electron withdrawing, but this is a modest effect compared with the electron-donating resonance effect. This significantly destabilizes the conjugate base, and 4-hydroxybenzoic acid (pK_a 4.6) is less acidic than benzoic acid. The last example, 4-acetylbenzoic acid, is the most acidic of the group, and this is primarily the result of an electron-withdrawing resonance effect, though there is also a favourable inductive effect.

Problem 16.10 (level 1; chapter 4)

(a) Ignoring possible dilution effects, calculate the weight of sodium acetate (MW 82.03) that should be added to 1 litre of an aqueous solution containing 3.85 g of acetic acid (MW 60.05) to produce a solution of pH 4.80. The pK_a of acetic acid is 4.75.

(b) Calculate the pH of the solution in (a) following the addition of 1 ml of 1.0 M aqueous solution of sodium hydroxide.

(c) Calculate the pH of a solution containing 6.50 g of dimethylamine (MW 45; pK_a 10.8) in 600 ml distilled water.

(d) Calculate the percentage ionized of an acidic drug (pK_a 5.70) dissolved in an aqueous buffer solution of pH 4.90.

Answer 16.10

Calculations of this type require the application of standard equations that should be committed to memory, unless you feel up to deriving them from first principles.

The first is the Henderson–Hasselbalch equation:

$$pH = pK_a + \log \frac{[base]}{[acid]}$$

The second pair of equations relate pH to pK_a for weak acids and weak bases, though they are actually variants of the Henderson–Hasselbalch equation:

for acids

$$pH = \tfrac{1}{2}pK_a - \tfrac{1}{2}\log[HA]$$

for bases

$$pH = \tfrac{1}{2}pK_w + \tfrac{1}{2}pK_a + \tfrac{1}{2}\log[B]$$

It follows that we use the Henderson–Hasselbalch equation when both acid and base concentrations are applicable.

With these types of calculation, two general points should be made. Make sure your result makes sense. For example, do not forget the painfully obvious fact that bases should yield pH values >7, and acids <7. The second point relates to the number of significant figures you present in your result. Do not write down every figure your calculator provides, but round up to an appropriate level, for which the question will give guidelines. If a pH or pK is given to two decimal places, do likewise in your answers.

(a) This is preparation of a buffer solution, using acetic acid (acid) and sodium acetate (base). Putting values into the Henderson–Hasselbalch equation

$$pH = pK_a + \log \frac{[base]}{[acid]}$$

we get

$$4.80 = 4.75 + \log \frac{[base]}{[acid]}$$

so that

$$\log \frac{[base]}{[acid]} = 4.80 - 4.75 = 0.05$$

and

$$\frac{[\text{base}]}{[\text{acid}]} = 10^{0.05} = 1.12$$

The [acid] is known; it is

$$[\text{acid}] = \frac{3.85}{60.05} = 0.0641\text{M}.$$

Therefore, $[\text{base}] = 1.12 \times 0.0641 = 0.0718$ M.
Amount of sodium acetate required
$= 0.0718 \times 82.03 = \textbf{5.89 g}$.

(b) We again use the Henderson–Hasselbalch equation; we are adding extra base, which in turn removes some acid, so the [base]/[acid] ratio will alter accordingly.

The amount of base added is 1 ml of 0.1 M NaOH $= 0.001$ M; this will also remove 0.001 M of acid

$$pH = pK_a + \log \frac{[\text{base}]}{[\text{acid}]}$$

so

$$pH = 4.75 + \log \frac{0.0718 + 0.001}{0.0641 - 0.001}$$

$$= 4.75 + \log \frac{0.0728}{0.0631}$$

$$pH = 4.75 + \log 1.15 = 4.75 + 0.06 = 4.81$$

New pH $= \textbf{4.81}$, a change of just 0.01 pH units

(c) The pH of a solution of a weak base may be determined from the equation:

$$pH = \tfrac{1}{2}pK_w + \tfrac{1}{2}pK_a + \tfrac{1}{2}\log[B]$$

$$[\text{base}] = \frac{6.50}{45} \times \frac{1000}{600} = 0.241\text{M}$$

$$pH = \tfrac{1}{2}(14) + \tfrac{1}{2}(10.8) + \tfrac{1}{2}\log(0.241)$$

$$= 7 + 5.4 + \tfrac{1}{2}(-0.618)$$

$$pH = 7 + 5.4 - 0.31 = 12.09$$

pH of solution $= \textbf{12.09}$

(d) This again requires use of the Henderson–Hasselbalch equation; we have a partially ionized acid, so the ionized fraction represents [base] and the non-ionized fraction represents [acid].

Let us call the fraction ionized I; then the fraction non-ionized is $1 - I$.

$$pH = pK_a + \log \frac{[\text{base}]}{[\text{acid}]}$$

$$4.90 = 5.70 + \log \frac{I}{1 - I}$$

so that

$$\log \frac{I}{1 - I} = 4.90 - 5.70 = -0.80$$

and

$$\frac{I}{1 - I} = 10^{-0.80} = 0.158$$

from this

$$I = 0.158 - 0.158I$$

and

$$I = \frac{0.158}{1.158} = 0.136$$

This means the percentage ionized is $\textbf{13.6\%}$.

Problem 16.11 (level 1; chapters 7 and 3)

(a) Give a mechanism for the acid-catalysed formation of a hemiacetal:

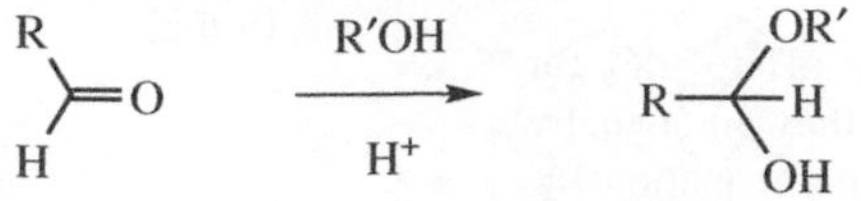

The heart drug digoxin contains in its structure three molecules of digitoxose as the trisaccharide (digitoxose)$_3$:

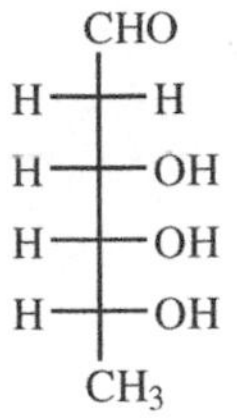

digitoxose

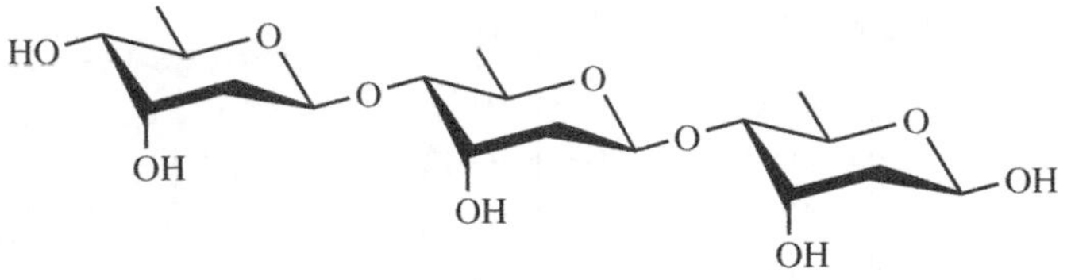

(digitoxose)$_3$

(b) Explain how a digitoxose molecule is capable of forming a six-membered cyclic form as seen in the trisaccharide with the stereochemistry shown.

(c) Indicate what information is conveyed by the four underlined parts of the nomenclature (+)-β-D-digitoxopyranose for the digitoxose units in the trisaccharide.

(d) Give a mechanism to account for the acid-catalysed formation of the dimer (digitoxose)$_2$ from two molecules of digitoxose (cyclic form). It may be assumed that no other functional groups in the molecules interfere with this reaction.

(e) Explain mechanistically the following two observations:

• digitoxose reacts with HCN to give a mixture of two diastereoisomeric products;

• sodium borohydride reduction of digitoxose gives a single product.

Answer 16.11

(a) This multi-part question starts with a standard general mechanism. It is the lead-in to other questions based on this mechanism. This is a textbook reaction; but, rather than remembering it in detail, work it out. The acid catalyst is used in the very first step to produce the conjugate acid of the carbonyl compound, increasing its electrophilicity. The standard nucleophilic addition is followed by proton loss, giving the product and regenerating the catalyst.

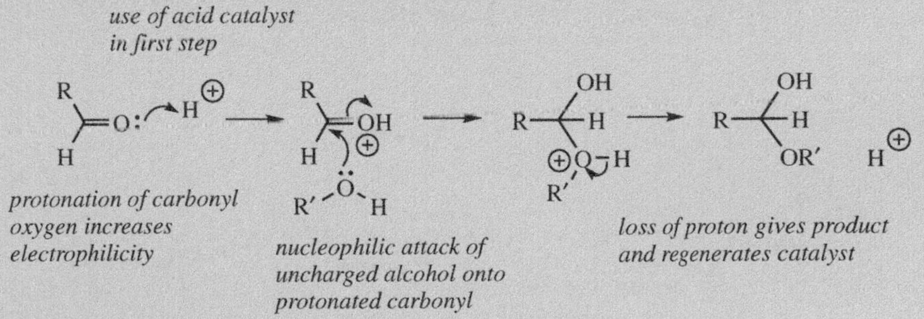

use of acid catalyst in first step

protonation of carbonyl oxygen increases electrophilicity

nucleophilic attack of uncharged alcohol onto protonated carbonyl

loss of proton gives product and regenerates catalyst

(b) The sugar digitoxose is shown in the form of a Fischer projection. The cyclic form is merely a hemiacetal; we need to react one of the hydroxyls with the carbonyl to form a six-membered ring. We must also manipulate the Fischer projection to deduce the stereochemistry in the product, which should end up the same as shown in the question. Note the approach here: deduce the stereochemistry rather than trying to obtain the stereochemistry shown. We are only asked for a single digitoxose molecule, despite the structure shown being the trimer.

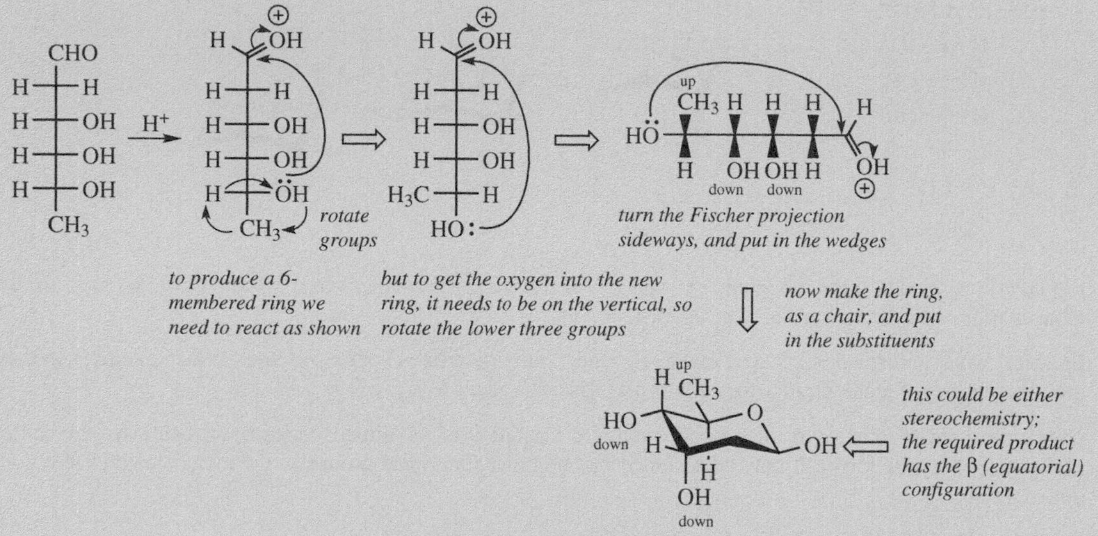

to produce a 6-membered ring we need to react as shown

but to get the oxygen into the new ring, it needs to be on the vertical, so rotate the lower three groups

turn the Fischer projection sideways, and put in the wedges

now make the ring, as a chair, and put in the substituents

this could be either stereochemistry; the required product has the β (equatorial) configuration

(c) The underlined parts in the name (+)-β-D-digitoxopyranose convey the following information:

(+) means the compound is in one enantiomeric form and is dextrorotatory; β defines the configuration at the new chiral centre formed during hemiacetal formation (the anomeric centre) and it has the OH group equatorial; D conveys the configuration at the highest numbered chiral centre (C-5); pyranose means the hemiacetal ring is a six-membered oxygen heterocycle.

(d) The mechanism for formation of the dimer (digitoxose)₂ from two molecules of cyclic digitoxose is the transformation of a hemiacetal into an acetal. We are told that no other functional groups interfere in the reaction, so let us miss them out of the structures. It is a textbook reaction, the continuation of part (a), but on specified molecules. We need to join position 1 of one molecule to position 4 of a second, and the reaction is acid catalysed.

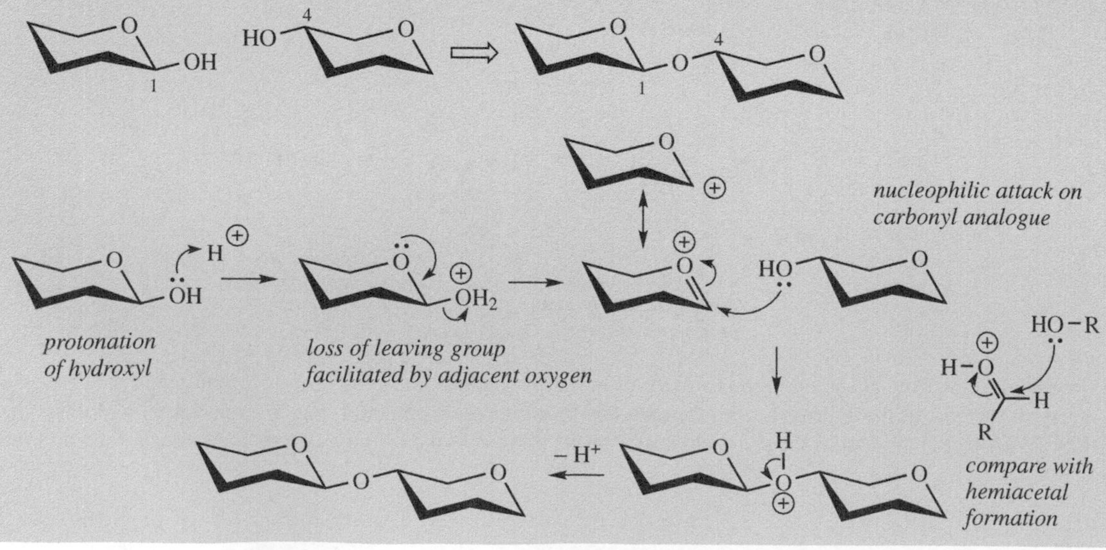

protonation of hydroxyl

loss of leaving group facilitated by adjacent oxygen

nucleophilic attack on carbonyl analogue

compare with hemiacetal formation

The reaction starts of with a protonation – use the catalyst. Resist the urge to protonate the 4-hydroxyl, but go for the one at position 1 that has the added functionality of the hemiacetal linkage. It is going to be the more reactive one. Protonation is followed by loss of water as leaving group. The intermediate oxonium cation shown is actually a resonance form of the simpler carbocation; now you can see the role of the adjacent oxygen. The reaction is completed by attack of the nucleophile, the 4-hydroxyl of another molecule. This is not special, but is merely another version of the hemiacetal synthesis done in part (a).

(e) This looks more complicated than it is. HCN reacts with an aldehyde to give a cyanohydrin through nucleophilic attack of cyanide; NaBH$_4$ reacts to give an alcohol through nucleophilic attack of hydride (it is not actually hydride that attacks, but we can formulate it as such). Furthermore, attack on the planar carbonyl group may be from either face. Now just follow that through.

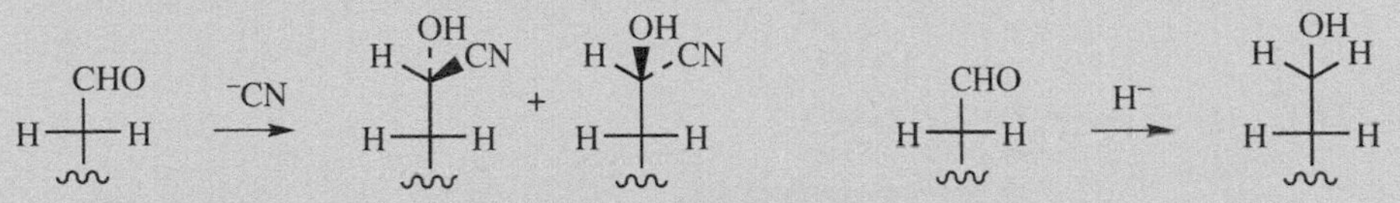

It can be seen that addition of cyanide creates a new chiral centre, whereas addition of hydride does not. Therefore, cyanide reacts to give two epimeric products. We already have chirality in the rest of the digitoxose chain, so the cyanide reaction must yield two diastereoisomers. On the other hand, borohydride reduction gives just a single product.

Problem 16.12 (level 1; chapters 8 and 9)

(a) Explain why addition of HCl to the conjugated diene at low temperature gives two main products, in the proportions shown.

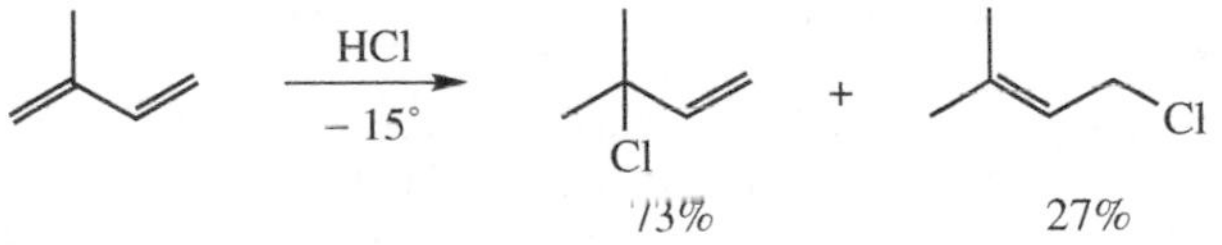

(b) Suggest mechanisms for the following two reactions:

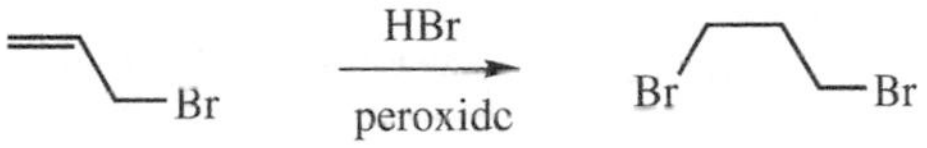

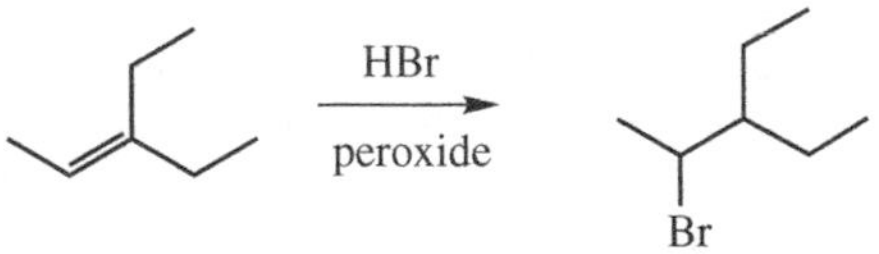

(c) Give a mechanism to explain the formation of two products in the following reaction:

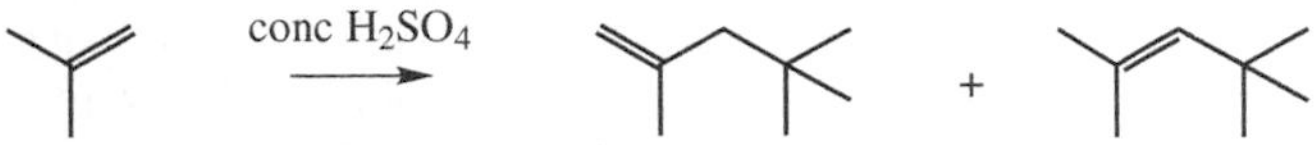

(d) Explain why HBr adds to the trichlorinated alkene with anti-Markovnikov regiochemistry.

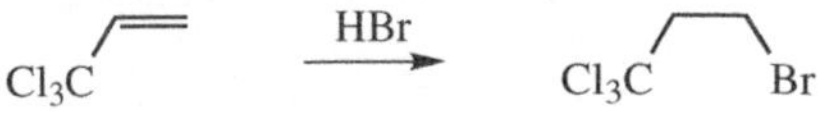

Answer 16.12

We have here a mixture of electrophilic and radical addition reactions to alkenes. Remember the guidelines that radical reactions are characterized by the inclusion of radical initiators, such as light or peroxides. In the absence of such initiators, consider only the alternative electrophilic mechanisms.

(a) Electrophilic addition to conjugated dienes is complicated by the concepts of kinetic and thermodynamic control. Briefly, you should think of kinetic control being related to the stability of the intermediate carbocations, whereas thermodynamic control is dependent on the stability of the products. Here, we have low temperature conditions, so we need to consider protonation of the alkene, and the relative stability of the carbocations formed. Draw the mechanisms for the two possible scenarios where we protonate either end of one double bond.

primary carbocation;
no resonance stabilization

resonance stabilized
allylic cation

tertiary   primary

major product   minor product

Protonation on C-2 gives an unfavourable primary carbocation. On the other hand, protonation on C-1 gives a favourable resonance-stabilized allylic carbocation. The two products are then formed by capture of chloride in a ratio that reflects the relative contribution of the limiting structures for the allylic carbocation; one is tertiary and the other is primary.

resonance stabilized
allylic cation

secondary   primary

primary
carbocation

You may think that is the end of the problem; but, since we have an unsymmetrical diene, it is also necessary to consider protonation of the other double bond. Protonation on C-4 also gives a favourable resonance-stabilized allylic carbocation, this time with primary and secondary limiting structures. Protonation on C-3 gives an unfavourable primary carbocation with no resonance stabilization. Since the products formed are related to initial protonation at C-1, it is apparent that, despite the stability associated with an allylic cation, a tertiary limiting structure is formed in preference to that with a secondary limiting structure.

(b) The inclusion of 'peroxides' indicates a radical reaction. Jot down the general sequence for radical reactions involving peroxides and HBr. This requires homolysis of a general peroxide, hydrogen abstraction from HBr, and generation of a bromine radical. Remember to use the fish-hook curly arrows here, representing the movement of a single electron. You can use pairs of arrows if you prefer, or single ones as in a two-electron mechanism. Now we can attack the double bond, and the carbon attacked is the one that leads to the more stable radical, in this case a secondary radical.

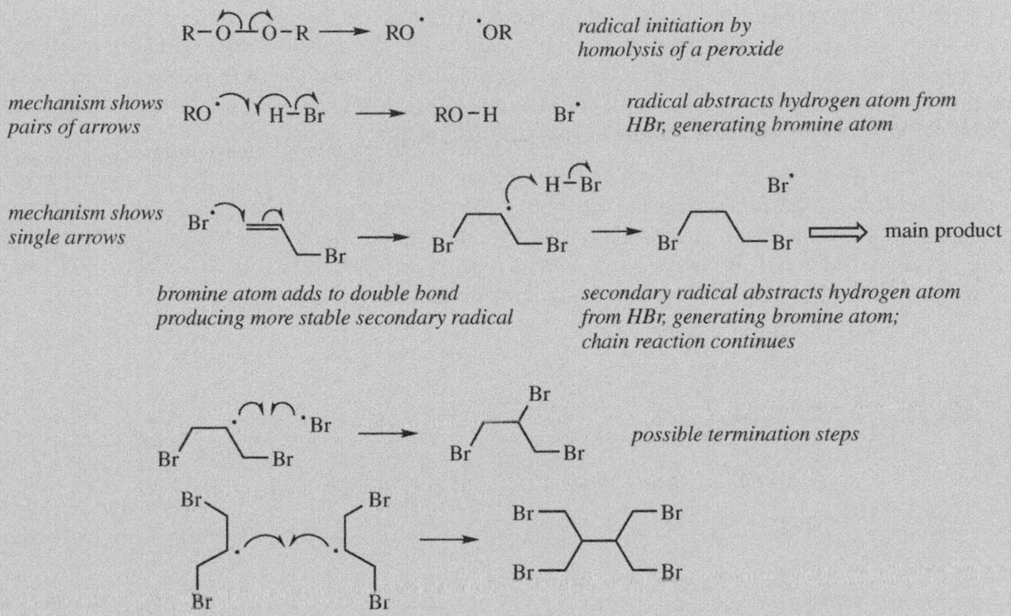

radical initiation by homolysis of a peroxide

mechanism shows pairs of arrows

radical abstracts hydrogen atom from HBr, generating bromine atom

mechanism shows single arrows

main product

bromine atom adds to double bond producing more stable secondary radical

secondary radical abstracts hydrogen atom from HBr, generating bromine atom; chain reaction continues

possible termination steps

This new radical then abstracts hydrogen from another molecule of HBr, and the chain reaction can continue. You may wish to show typical chain termination steps.

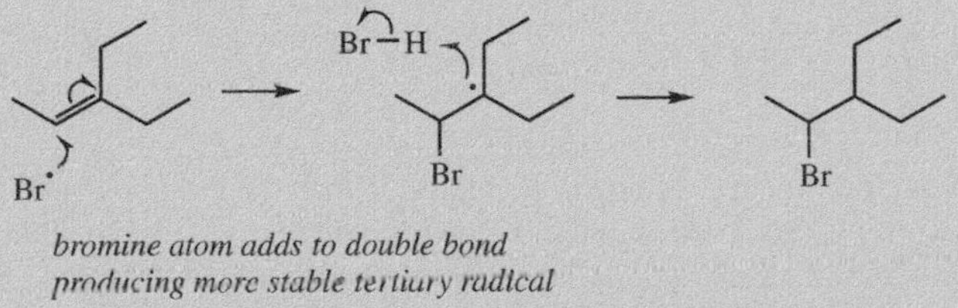

bromine atom adds to double bond producing more stable tertiary radical

The main product in the second reaction is easily rationalized, in that addition of a bromine atom to the alkene gives the favourable tertiary radical rather than the less attractive secondary radical.

(c) The obvious change here is that two molecules of the alkene are combining under acidic conditions. We should consider an electrophilic reaction initiated by protonation of the alkene to a carbocation. Protonation is governed by the stability of the carbocation; this is straightforward here, in that we could get a favourable tertiary carbocation or a very unfavourable primary one.

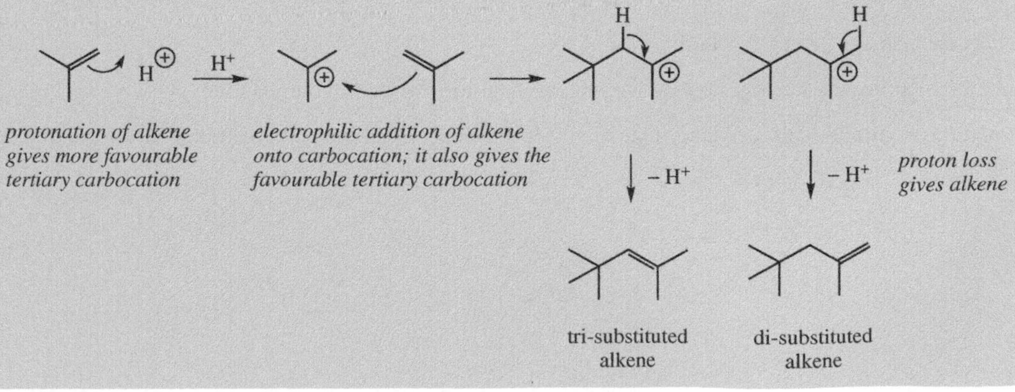

protonation of alkene gives more favourable tertiary carbocation

electrophilic addition of alkene onto carbocation; it also gives the favourable tertiary carbocation

proton loss gives alkene

tri-substituted alkene

di-substituted alkene

The next step is merely a repeat; instead of an electrophilic reaction with a proton, we have an electrophilic reaction with a carbocation. The reasoning is the same; we get the tertiary carbocation intermediate. What then follows is loss of a proton to give an alkene, and there are two possible products, depending upon which proton is released. Both products are formed, though we would expect the more-substituted alkene to predominate. We are not asked for, or given, any information about product ratios.

(d) According to the reasoning we have so far used, protonation of the double bond in this alkene should give the secondary carbocation rather than the alternative primary carbocation. Based on the information given in the question, it does not. We must get the primary carbocation, which is then quenched by bromide, i.e. anti-Markovnikov addition. It is not necessary to remember what Markovnikov or anti-Markovnikov additions mean, we just need to consider the carbocation intermediates and their relative stability.

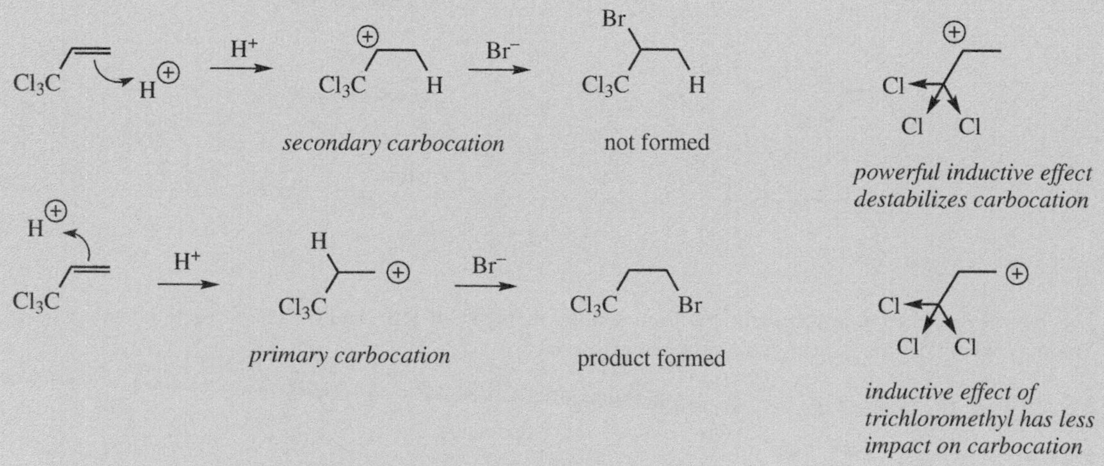

secondary carbocation   not formed   *powerful inductive effect destabilizes carbocation*

primary carbocation   product formed   *inductive effect of trichloromethyl has less impact on carbocation*

We can assume the trichloromethyl group is to blame. What do we know abut this? It has three very electronegative chlorine atoms, so there will be a powerful inductive effect, withdrawing electrons. Now alkyl groups, which are weakly electron donating, help to stabilized a carbocation; and the more there are, the more stabilization we get. At the other extreme, electron-withdrawing groups destabilize a carbocation. It appears from this reaction that the primary carbocation is more favourable than one adjacent to the powerful electron-withdrawing trichloromethyl group. Hence the observed regiochemistry.

Problem 16.13 (level 1; chapters 7, 8 and 9)

For the transformations shown, provide the following information:

- a detailed mechanism for the reaction;

- resonance structures for the organic reactive intermediate formed during the reaction;

- the structure of an alternative product that might be formed in the reaction, with a brief explanation for its formation.

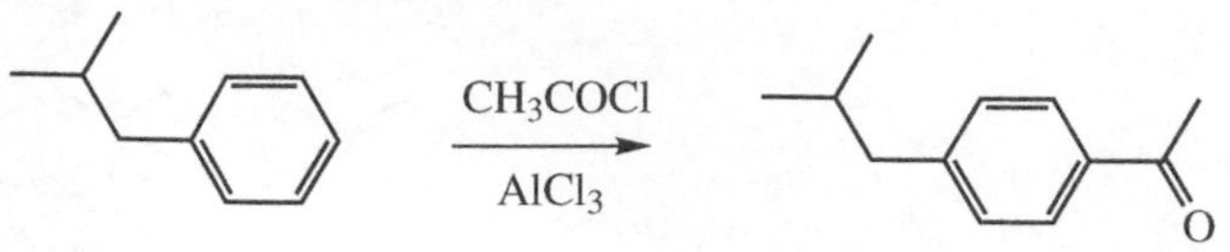

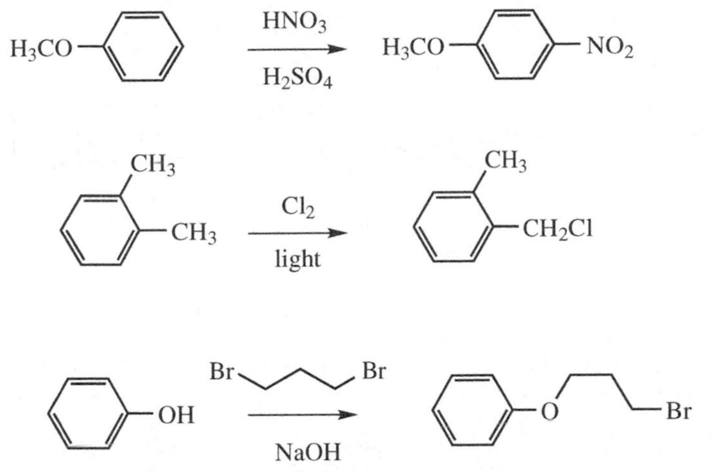

Answer 16.13

(a) This should be recognized as acylation of an aromatic ring, and from the use of aluminium chloride it is a standard Friedel–Crafts acylation reaction. You may remember that the acylium cation is the acylating species; alternatively, you can deduce that the acyl chloride interacts with the Lewis acid $AlCl_3$, and is most likely to give $AlCl_4^-$. The acylium cation then emerges as the other part. This is resonance stabilized as shown.

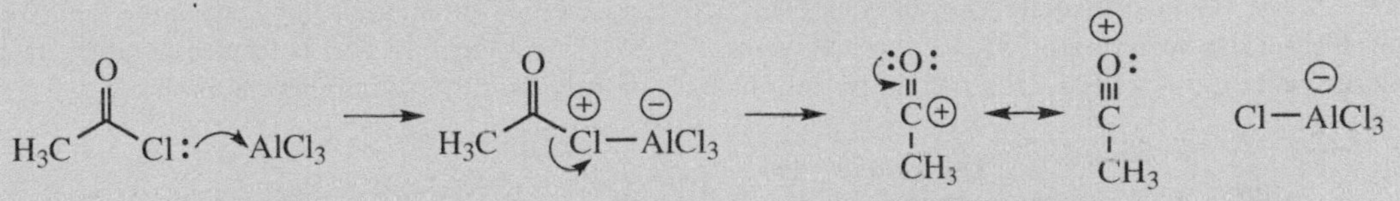

resonance stabilized
acylium ion

Now the electrophilic attack of the aromatic system onto the acylium cation. The product is *para* substituted, so go for that site of attack. We then need removal of a proton, and chloride is the obvious base, but since this is complexed as $AlCl_4^-$, we need the latter to decompose to chloride and $AlCl_3$. We should also show resonance stabilization of the addition cation, delocalizing the charge around the ring; this helps with the alternative product question.

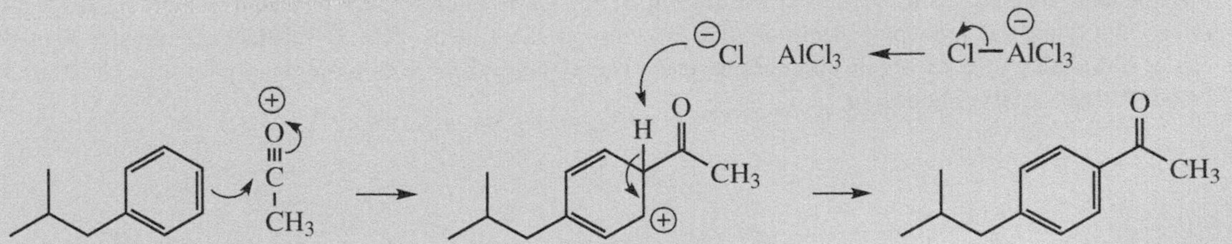

electrophilic attack from π
electrons onto acylium ion

loss of proton to
regenerate aromaticity

resonance stabilized arenium cation

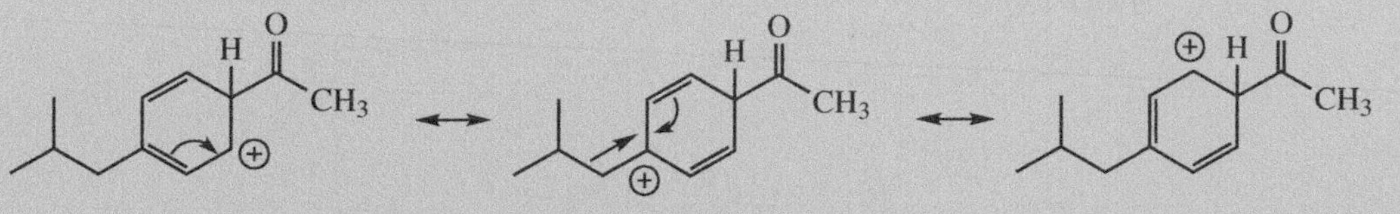

*electron donating effect of alkyl
group stabilizes carbocation*

An alternative product might be the *ortho*-substituted analogue. Alkyl groups are *ortho* and *para* directors for further electrophilic substitution. This follows from stabilization of one of the resonance forms by the electron-donating effect of an alkyl group. This is seen in the *para* substitution case, and extrapolation to *ortho* substitution shows a similar stabilization.

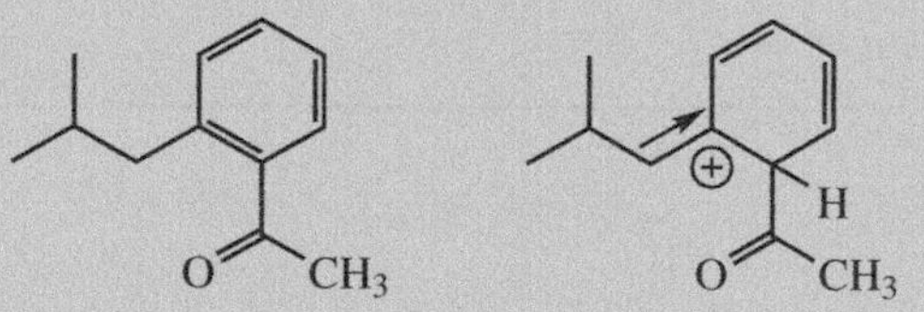

*electron donating effect of alkyl
group stabilizes carbocation*

alternative product

Though not asked for, it might be useful to indicate that, because we have relatively bulky substituents in this example, the *para* product will be formed in preference to the *ortho* product.

(b) This is another electrophilic aromatic substitution, though a different reagent, and a different type of stabilization for intermediate carbocations. The active species in nitration is the nitronium ion. We are introducing the nitro group NO_2, so the involvement of NO_2^+ is logical. How it is formed is rather unusual, but aromatic nitration is such a routine transformation that it is worth remembering the basics.

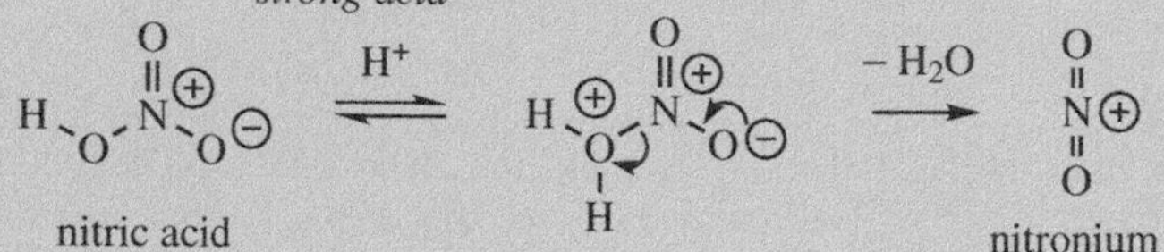

*protonation from
strong acid*

H^+

$- H_2O$

nitric acid

nitronium ion

Electrophilic attack is much the same as in the Friedel–Crafts reaction above; note that we have to write the nitro group with charge separation in line drawings. Again, we can draw resonance structures for the intermediate arenium cation. The extra stabilization this time comes from an electron-donating resonance effect involving the lone pair electrons of the methoxyl substituent. This is substantially greater than the electron-donating effect from an alkyl group; phenols and their ethers react more readily towards electrophilic reagents than do alkyl benzenes.

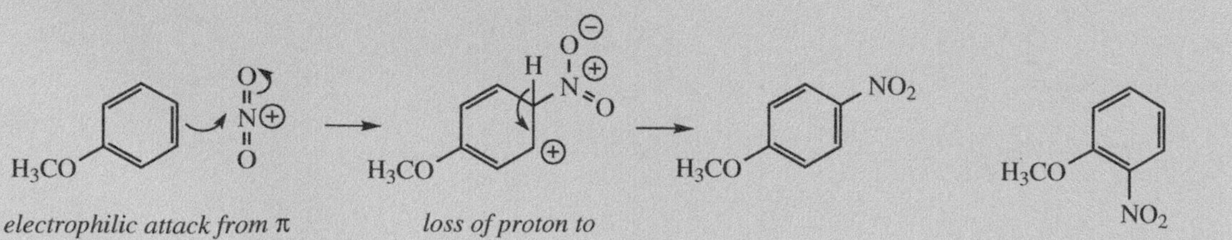

electrophilic attack from π
electrons onto nitronium ion

*loss of proton to
regenerate aromaticity*

alternative product

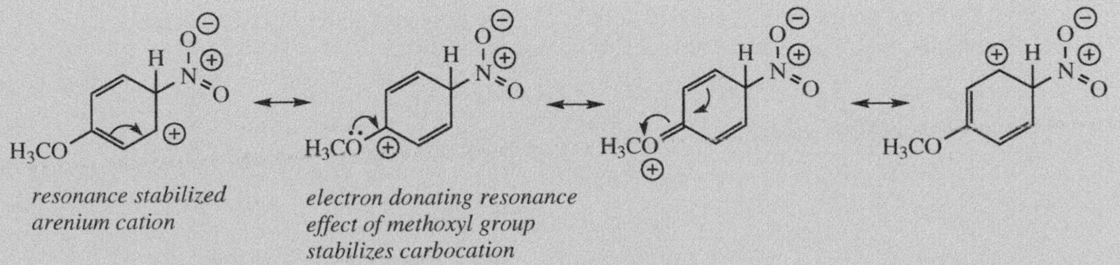

resonance stabilized
arenium cation

electron donating resonance
effect of methoxyl group
stabilizes carbocation

By the same reasoning as in part (a), an alternative product is likely to be the corresponding *ortho* derivative. The intermediate carbocation can be stabilized in the same way.

(c) This is radical reaction. The pointer is that the reaction also requires light energy. Accordingly, we need to formulate a sequence starting from homolytic fission of a chlorine molecule to chlorine atoms, then a chain reaction stemming from these reactive species.

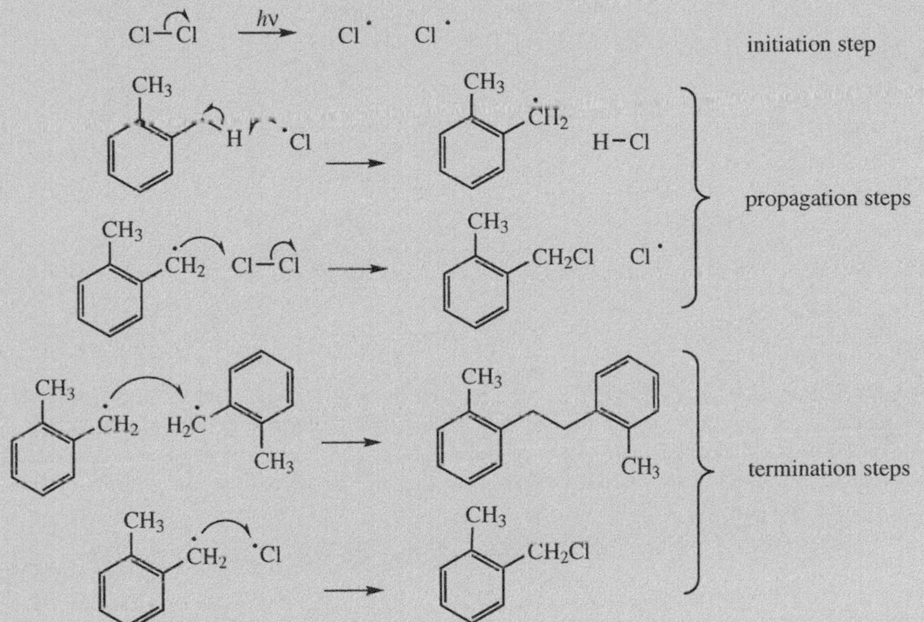

The propagation steps involve removal of a hydrogen atom from one of the methyl substituents on the benzene ring. Abstraction from the methyl group is favourable because it generates a resonance-stabilized benzylic radical, in which the unpaired electron can be delocalized into the aromatic ring system.

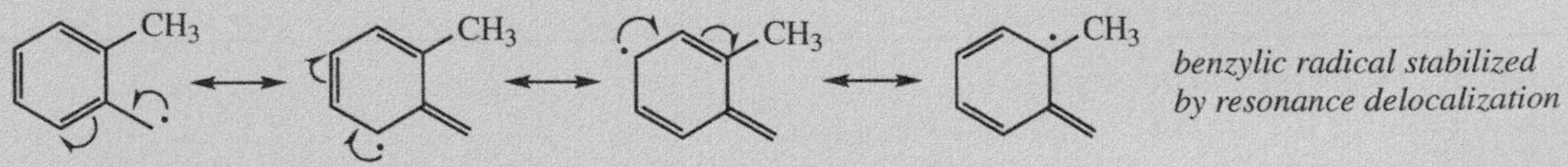

benzylic radical stabilized
by resonance delocalization

Two possible termination steps are shown, one of which produces the same product as the chain reaction. Recombination of two chlorine radicals is feasible, but less likely. Either of the termination products could be suggested as an alternative product; but, in practice, what we are going to get is derivatives with more than one chlorine substituent. It is difficult to control a radical process once the chain reaction is under way, so we usually get a mixture of products. Note particularly that further substitution is on the original or alternative methyl group, and not on the ring. Again, the formation of a resonance-stabilized benzylic radical directs the reaction. Eventually, we might get the hexachloro compound shown.

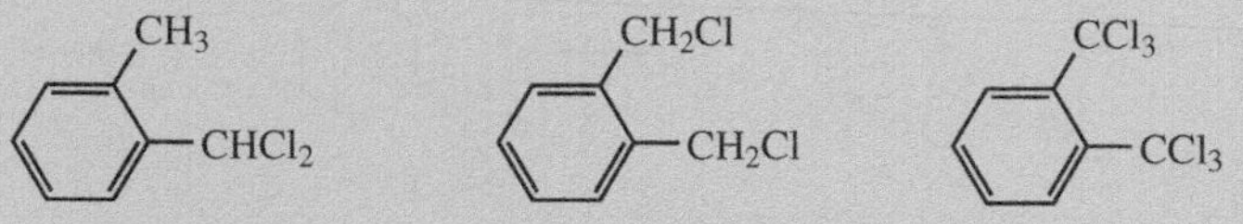

alternative products

(d) This is a simple S_N2 displacement of a bromide leaving group by the phenolate anion. Phenols are relatively acidic, and treatment with base generates the resonance-stabilized conjugate base. The phenoxide anion with charge on the electronegative oxygen is the preferred nucleophile. The reaction is then completed by displacement of bromide to produce the ether.

Note the use of the reagent NaOH in the first step. Do not consider hydroxide as the nucleophile; the immediate consequence of mixing the reagents is ionization of the phenol.

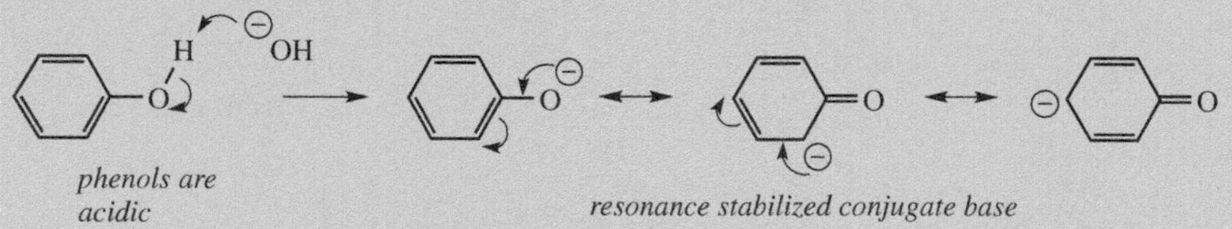

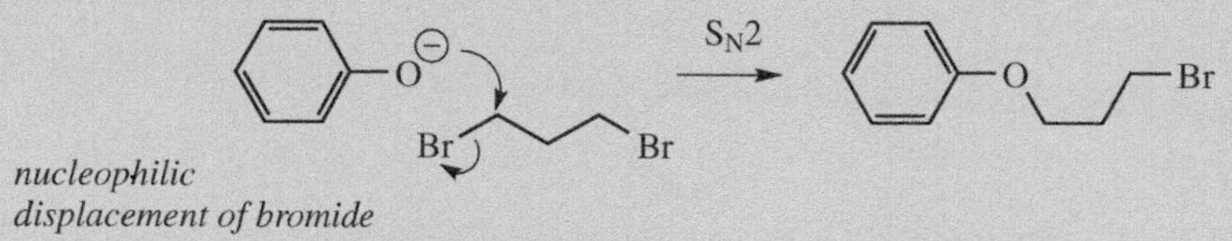

phenols are acidic   *resonance stabilized conjugate base*

nucleophilic displacement of bromide

And what about an alternative product? There are two lines of thought, and the most obvious is that the reaction is repeated, since we are using a dibromide as substrate. Alternatively, we could consider one of the other resonance forms of the phenolate anion as nucleophile. This would generate a *C*-alkylated phenol. In the majority of cases, *C*-alkylation is not observed, in that the preferred resonance structure has charge on the electronegative oxygen.

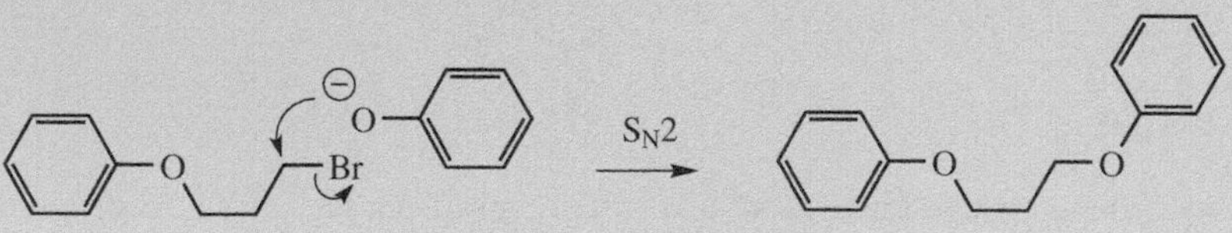

alternative product

Problem 16.14 (level 2; chapter 6)

Provide mechanistic explanations for the following observations:

(a) Treatment of 3,3-dimethylbutan-2-ol with hot aqueous HCl gives principally 2-chloro-2,3-dimethylbutane; similar treatment of 3-methylbutan-2-ol gives principally 2-chloro-2-methylbutane.

(b) During nucleophilic substitution, the alcohol (A) undergoes rearrangement. Under the same conditions, the alcohol (B) does not suffer rearrangement.

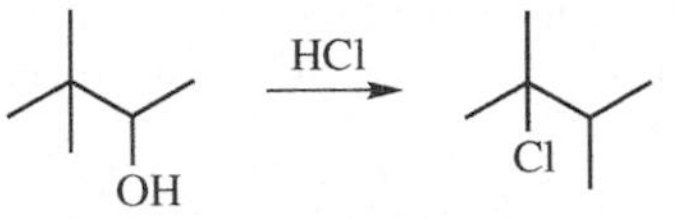

(A) 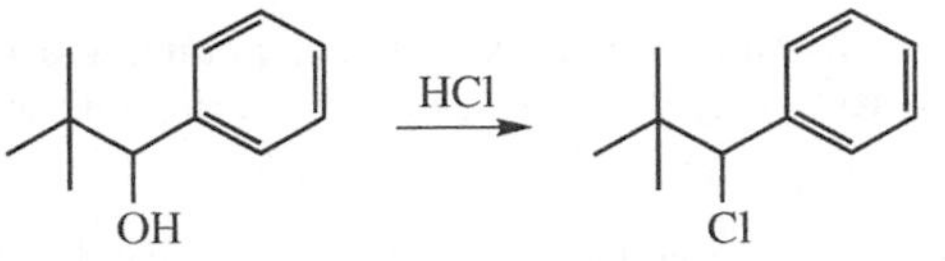 (B)

(c) Treatment of 3,3-dimethylbutan-2-ol with hot sulfuric acid gives three products: 3,3-dimethylbut-1-ene (3%), 2,3-dimethylbut-2-ene (61%), and 2,3-dimethylbut-1-ene (31%).

(d) Under acidic conditions, the unsaturated ketone (C) rearranges to a phenol (D).

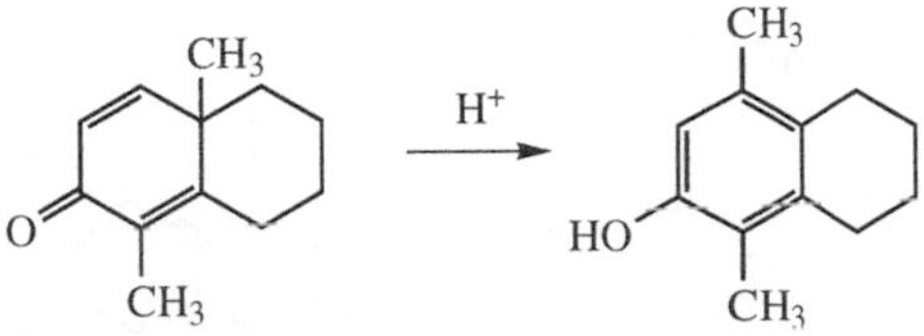

(C) (D)

Answer 16.14

These are all examples of carbocation rearrangements. The approach here is to spot the potential for carbocation formation, then see if an alternative carbocation might be formed by migration of an alkyl group, or perhaps hydride, to produce a more favourable carbocation. Favourable carbocations are typically tertiary, allylic, or benzylic.

(a) Draw out the structures, which are both secondary alcohols. Draw out the products and observe that, at least in the first example, we must have a rearrangement reaction because the carbon skeleton is different. There is also something unusual about the second reaction; the nucleophile is not attached at the same carbon that had the potential leaving group.

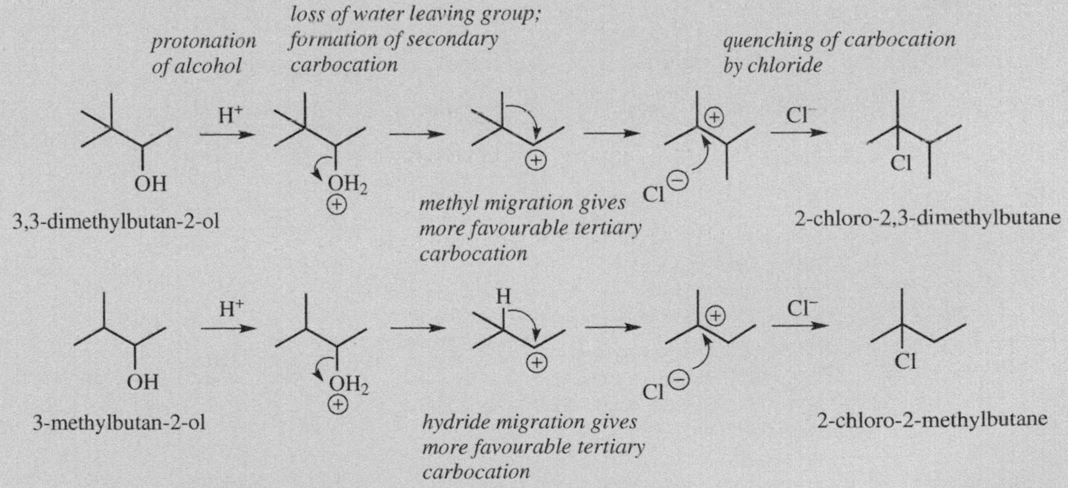

Since we can spot a rearrangement in the first example, formulate a mechanism for that first. The sequence is quite simple, protonation of the hydroxyl to make a better leaving group, loss of water and generation of a secondary carbocation. Now we should be able to see that a methyl migration would produce a tertiary carbocation, a rearranged skeleton, and following through we get the required product.

Now do the same for the second compound, and keep thinking about rearrangements. We can formulate this in the same way once we see that a hydride migration would achieve the secondary to tertiary carbocation change.

(b) The first example is actually the same as we have discussed in part (a), though here we are given the structure. What is different about the behaviour of the second compound? Draw out the same type of mechanism as far as the secondary carbocation.

benzylic carbocation; stabilized by resonance

etc

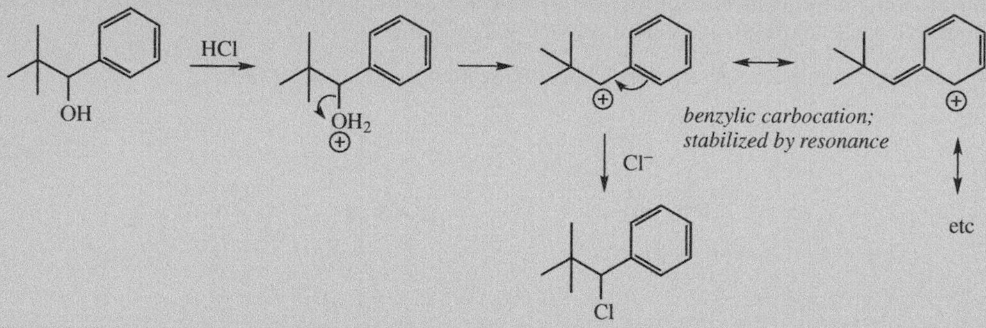

Why doesn't this secondary carbocation rearrange? Why doesn't it want to become tertiary? There must be something extra that is stabilizing it so that there is no need for rearrangement. There is, and it is the aromatic ring. This carbocation is benzylic and stabilized by resonance. There is no need for rearrangement to acquire improved stability; accordingly, chloride attacks at the same carbon the leaving group was attached to, an S_N1 reaction.

(c) The starting material is one of those from part (a), but because of the conditions, probably the high temperature, we are getting eliminations rather than substitutions. On drawing out the products, we can see two of them are produced as a result of rearrangements. The first product is not rearranged and is a simple elimination. Formulate it as E1, because the rearrangements are carbocation related.

3,3-dimethylbutan-2-ol   3,3-dimethylbut-1-ene   2,3-dimethylbut-2-ene   2,3-dimethylbut-1-ene

no rearrangement   *rearrangement*   *rearrangement*

E1 elimination

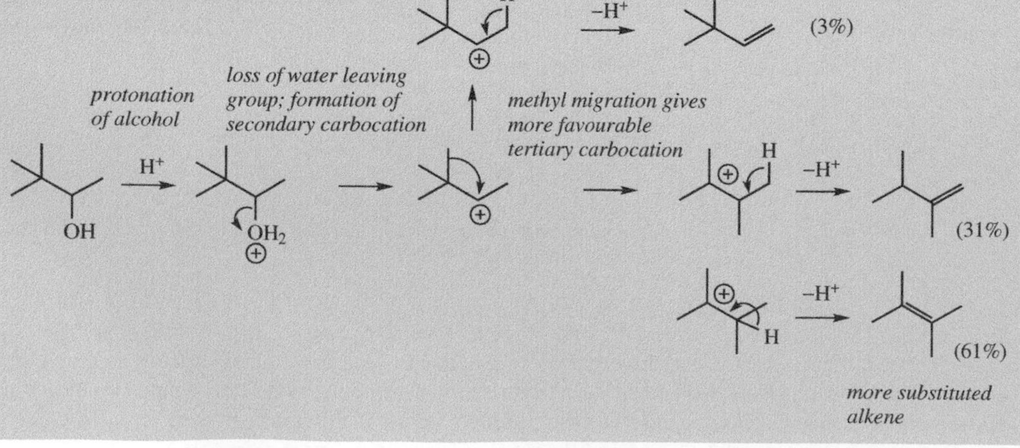

Methyl migration as in part (a) converts the first-formed secondary carbocation into a more favourable tertiary carbocation, and we see proton loss from this to give the other two products. These are unexceptional; there are two possible sites for proton loss.

Now for the relative proportions of products. Only 3% of the unrearranged product shows just how unfavourable the secondary carbocation is compared with the rearranged tertiary carbocation. The relative proportions of the other two alkenes are explained by the increased thermodynamic stability of the more-substituted alkene, though this is not sufficient to produce just the single product.

(d) This is clearly a rearrangement reaction; the methyl has migrated to the adjacent carbon in the transformation. There is no leaving group to generate a carbocation. We might consider protonation of the alkene; but be realistic, acid is going to protonate the carbonyl oxygen rather than the alkene. Let us see where that might lead.

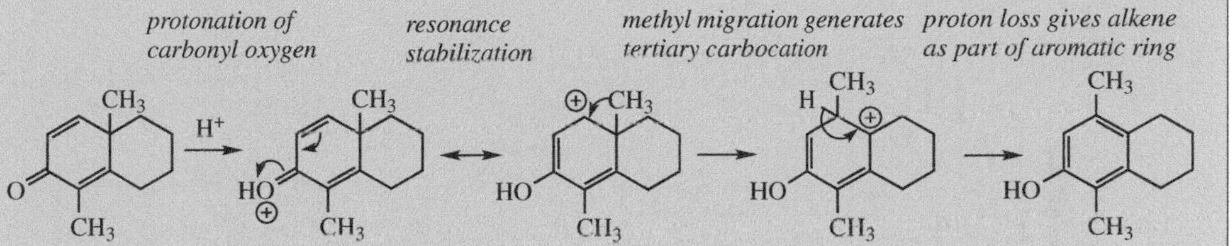

protonation of carbonyl oxygen   *resonance stabilization*   *methyl migration generates tertiary carbocation*   *proton loss gives alkene as part of aromatic ring*

Protonation of the carbonyl oxygen gives the conjugate acid, and because there is a double bond conjugated with the carbonyl, we are able to draw resonance structures. The pertinent resonance structure is the one that puts the positive charge adjacent to the migratory methyl group. Methyl migration then leads to a tertiary carbocation, and proton loss introduces further stability through generation of an aromatic ring.

Problem 16.15 (level 2; chapter 3)
Give stereodrawings for the natural products illustrated, in order to indicate their molecular shape.

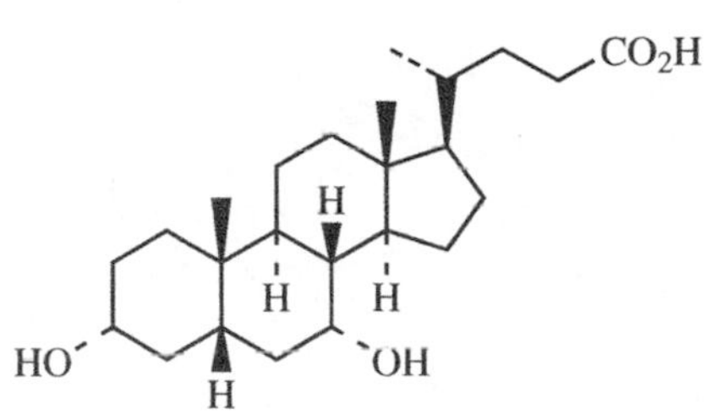

chenodeoxycholic acid 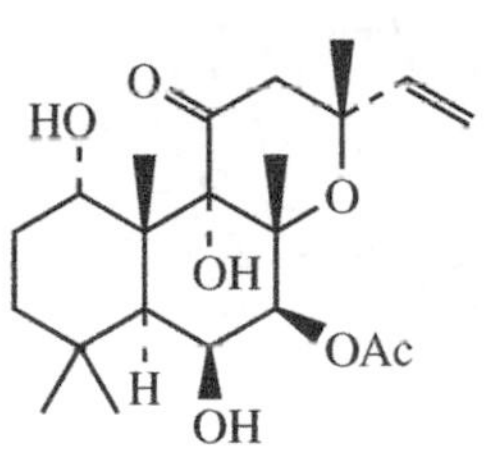   forskolin 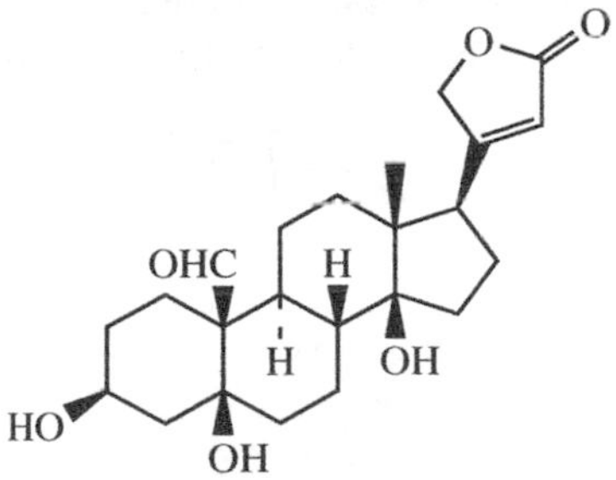   strophanthidin

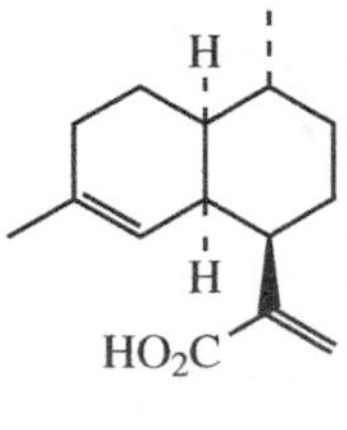

artemisinic acid 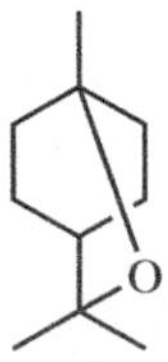   cineole 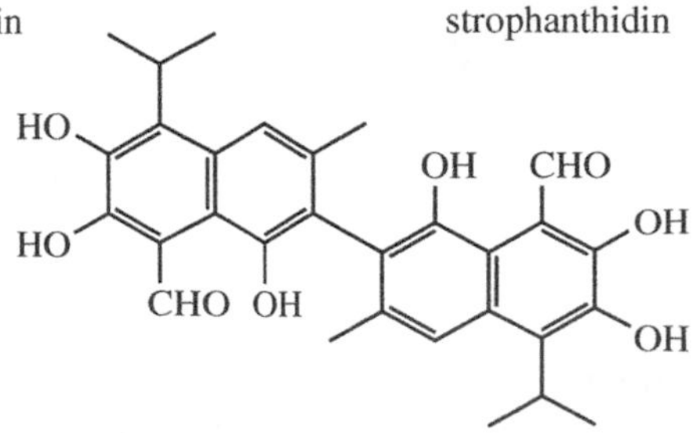   gossypol

Answer 16.15

The shape of fused ring structures can be deduced readily by extrapolation from a standard ring system. The most useful starting point is the all-*trans*-fused steroid system. The approach is to modify this according to requirements.

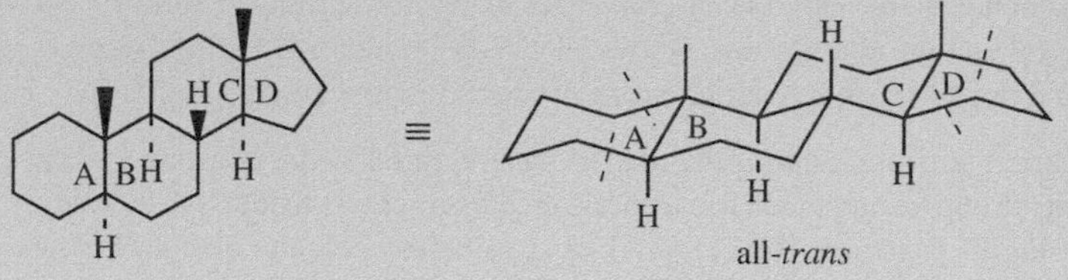

Chenodeoxycholic acid typifies a steroid with A/B *cis*-fusion; the remaining rings are the same as the all-*trans* model.

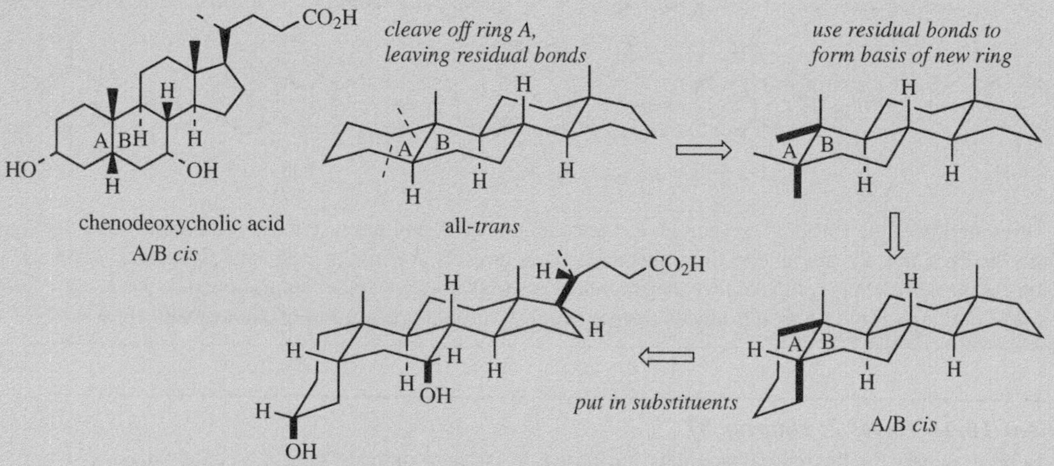

Forskolin is not a steroid; however, if we disregard substituents and that one ring is a heterocycle, then all we have is the first three rings A, B, and C of an all-*trans* steroid.

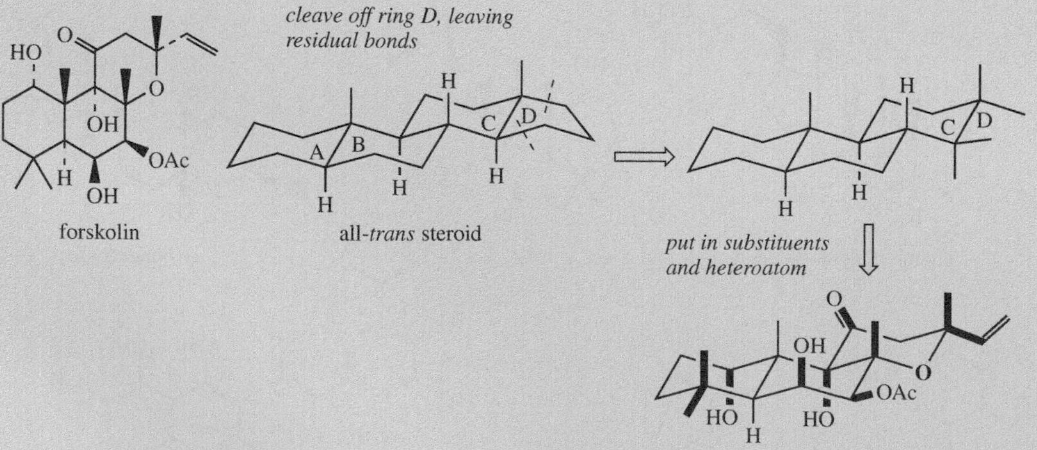

Strophanthidin is a steroid with rings A/B *cis*-fused, and rings C/D also *cis*-fused. The latter feature is approached in the same way as the A/B *cis*-fusion shown with chenodeoxycholic acid above.

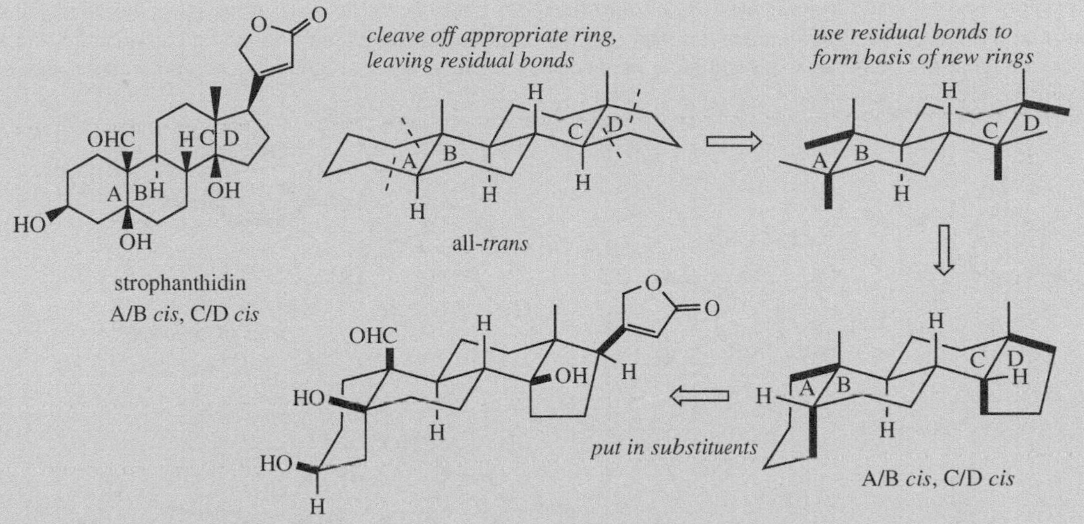

cleave off appropriate ring, leaving residual bonds

use residual bonds to form basis of new rings

strophanthidin
A/B *cis*, C/D *cis*

all-*trans*

put in substituents

A/B *cis*, C/D *cis*

Artemisinic acid is a *cis* fused decalin; we can still deduce its shape from the steroid approach. Note that both hydrogens at the ring fusion are down, whereas we have been looking at systems where the ring fusion substituents are up. We shall need to change ring B rather than ring A.

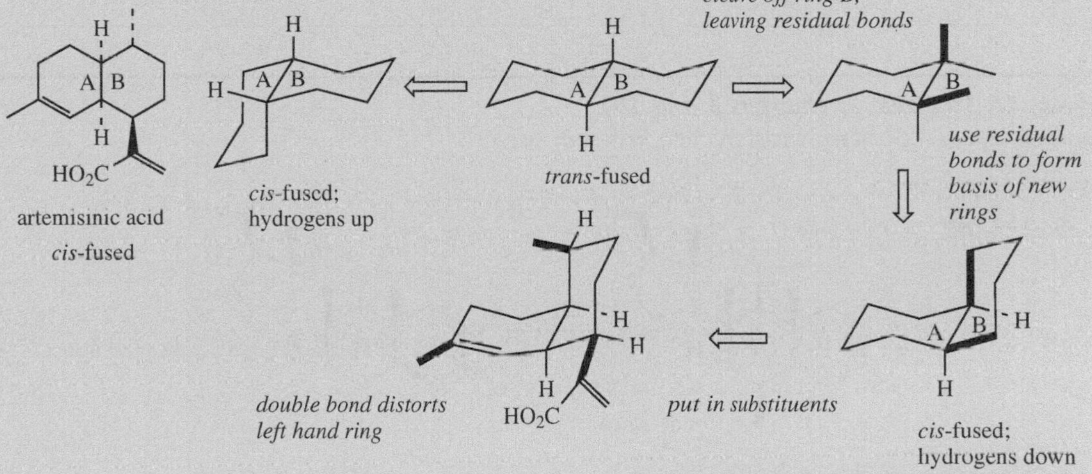

cleave off ring B, leaving residual bonds

artemisinic acid
cis-fused

cis-fused;
hydrogens up

trans-fused

use residual bonds to form basis of new rings

double bond distorts left hand ring

put in substituents

cis-fused;
hydrogens down

Cineole provides an example of a bridged ring system. The essence of this problem is to spot that we cannot bridge the opposite carbons of a cyclohexane chair conformation; it is necessary to have the cyclohexane in a boat form. Then the answer is readily obtained.

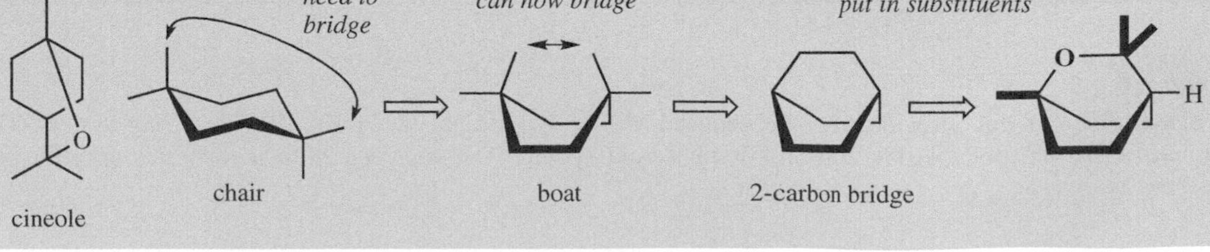

need to bridge

can now bridge

put in substituents

cineole

chair

boat

2-carbon bridge

Gossypol is a biphenyl system with *ortho* substituents, so there is restricted rotation about the bond joining the aromatic rings, creating torsional asymmetry and the existence of enantiomers without a chiral centre. We are only asked about the shape, and this will have the planar aromatic systems at right angles to minimize interaction between the *ortho* substituents.

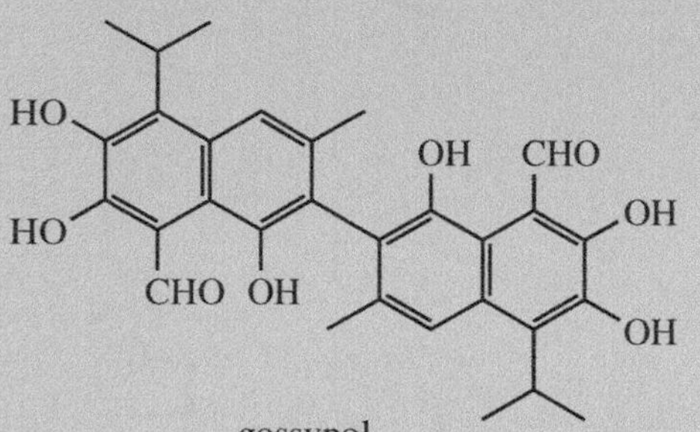

gossypol

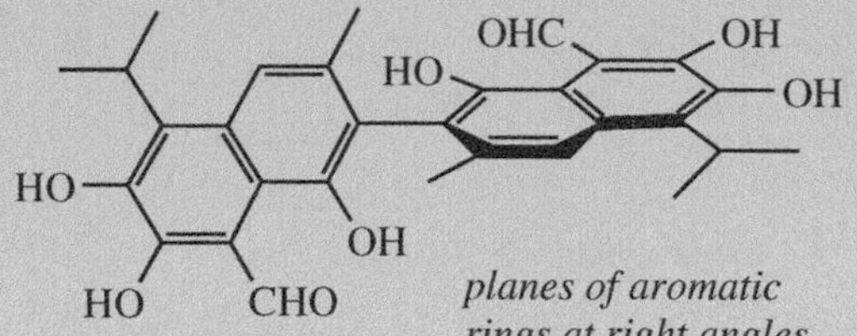

planes of aromatic rings at right angles

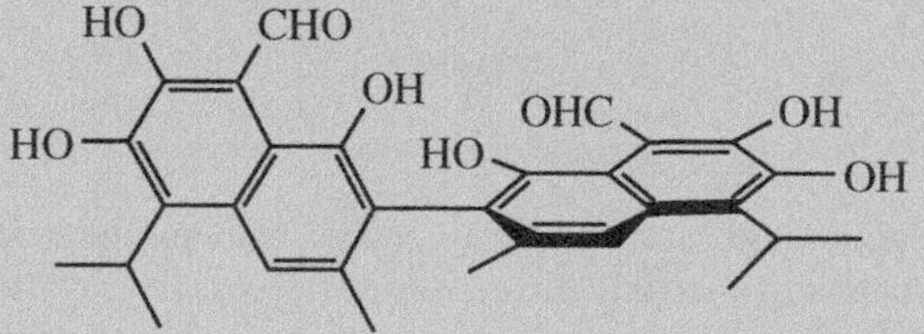

this would be the alternative enantiomeric form

Problem 16.16 (level 2; chapters 3 and 10)

Treatment with acid converts steroid (A) into its isomer (B).

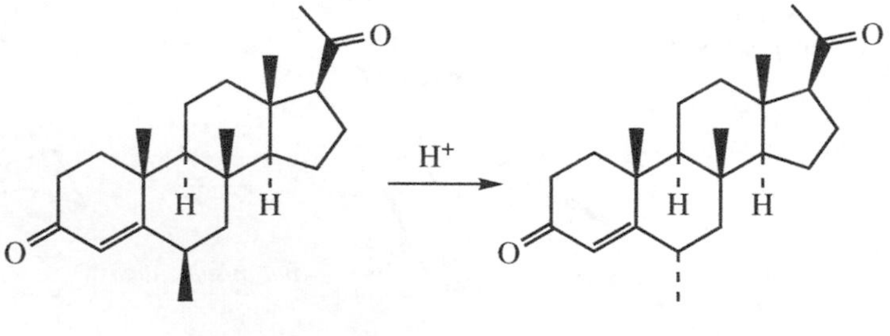

(A)   (B)

Use a stereodrawing to indicate the molecular shape of (A).

Give a mechanistic explanation for the acid-catalysed isomerization of (A) into (B).

Answer 16.16

The shape of fused ring structures can be deduced readily by extrapolation from a standard ring system. The most useful starting point is the all-*trans*-fused steroid system. The approach is to modify this according to requirements.

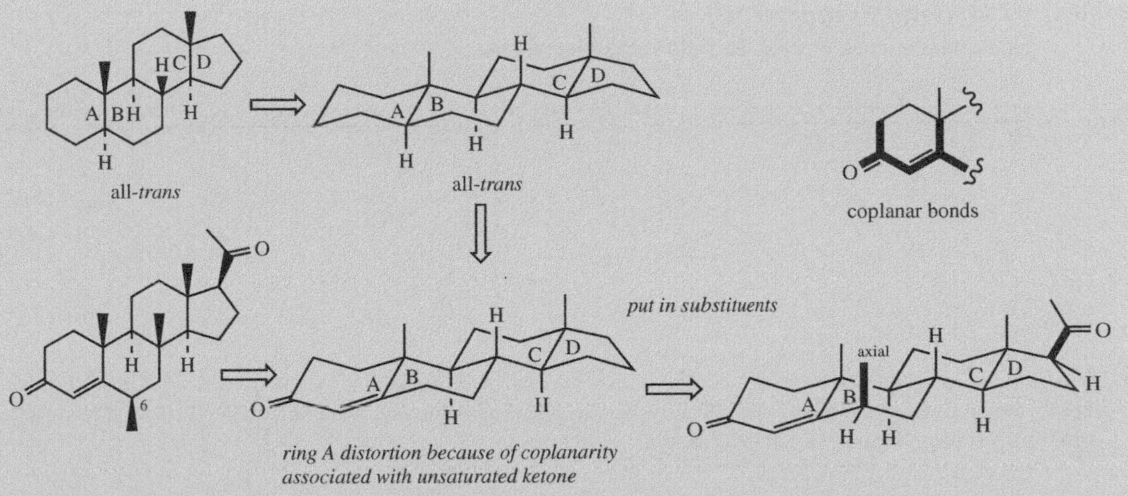

all-*trans*

all-*trans*

coplanar bonds

put in substituents

axial

ring A distortion because of coplanarity
associated with unsaturated ketone

The steroid skeleton in question differs from the all-*trans*-fused system only in respect of ring A, which has
an unsaturated ketone. Because of the coplanarity of bonds in this system, there will be some distortion in ring
A, which can be represented (approximately) as shown.

We then need to put in the substituents. The critical one to note is the methyl in ring B, because this must be
axial. In the following part of the question, the stereochemistry at this position changes, so that the methyl becomes
equatorial. This is not a conformational change, but a configurational change, so requires appropriate chemistry.
We should be formulating a proposal where this centre first becomes planar, and then reverts to tetrahedral, but
with the substituent then taking up the more favourable equatorial position.

protonation of carbonyl

*enolization through
conjugated system*

enol tautomer

*reversion to keto tautomer allows proton to be
acquired from upper face, so that substituent
takes up favourable equatorial position*

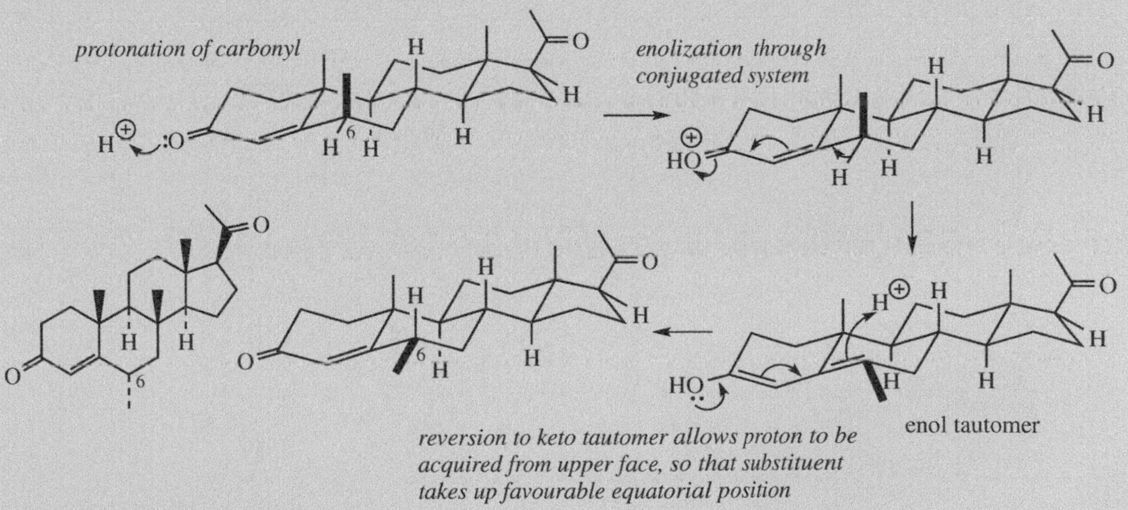

The functional group in ring A is a ketone; a planar system that takes in carbons adjacent to a ketone is the
enol tautomer. We are using acid catalysis here, and that is appropriate for enolization. If we first protonate the
carbonyl, then try moving electrons, we shall soon find that the proton at position 6 (standard steroid numbering)
can be lost in generating a conjugated enol tautomer. In a reversal of this process back to the ketone, we can
pick up a proton at position 6 from either face. In this case, protonation on the upper face allows the methyl
substituent to take up the more favourable equatorial position.

Problem 16.17 (level 2; chapter 10)

Provide mechanistic explanations for the following observations:

(a) In a base-catalysed aldol reaction, compound (A) yields an α,β-unsaturated aldehyde, whereas compound (B) gives a β-hydroxyaldehyde.

(A)   (B)

(b) Two possible products (D) and (E) can be considered when bromoketone (C) is treated with base. Product (D) is favoured when the base is potassium *tert*-butoxide in *tert*-butanol, and compound (E) is produced using lithium di-isopropylamide in tetrahydrofuran.

(C)   (D)   (E)

(c) Treatment of acetaldehyde with an excess of formaldehyde under basic conditions produces the trihydroxyaldehyde (F).

$$CH_3CHO \quad + \quad \underset{\text{excess}}{HCHO} \quad \xrightarrow[\text{H}_2\text{O}]{\text{Na}_2\text{CO}_3} \quad \underset{\text{(F)}}{HOH_2C} \overset{\displaystyle CH_2OH}{\underset{\displaystyle CHO}{\overset{\displaystyle |}{-}} CH_2OH}$$

(d) The cyclic ketone (I) is formed when ketones (G) and (H) are treated with base.

(G)   (H)   (I)

Answer 16.17

(a) A base-catalysed aldol reaction involves an enolate anion acting as nucleophile and adding to another carbonyl compound. In each of these cases, the enolate anion derived from the aldehyde will react with a second molecule of the same aldehyde.

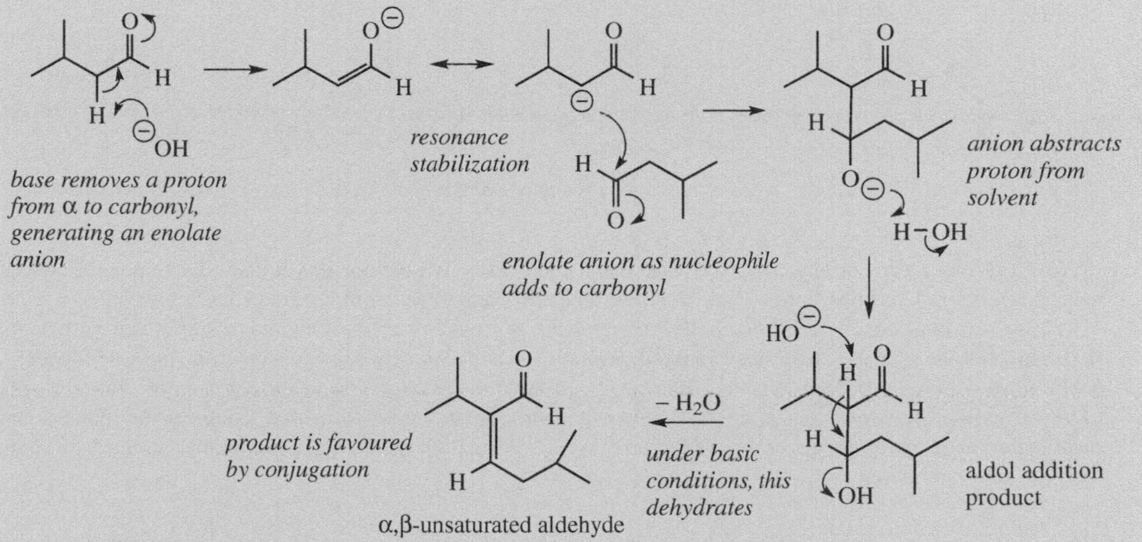

base removes a proton
from α to carbonyl,
generating an enolate
anion

*resonance
stabilization*

*enolate anion as nucleophile
adds to carbonyl*

*anion abstracts
proton from
solvent*

*product is favoured
by conjugation*

− H₂O

*under basic
conditions, this
dehydrates*

*aldol addition
product*

α,β-unsaturated aldehyde

Aldehyde (B) also has hydrogen α to the carbonyl and can generate an enolate anion; the aldol reaction follows as with the first aldehyde to give the product shown. However, dehydration of the β-hydroxyaldehyde is not favoured.

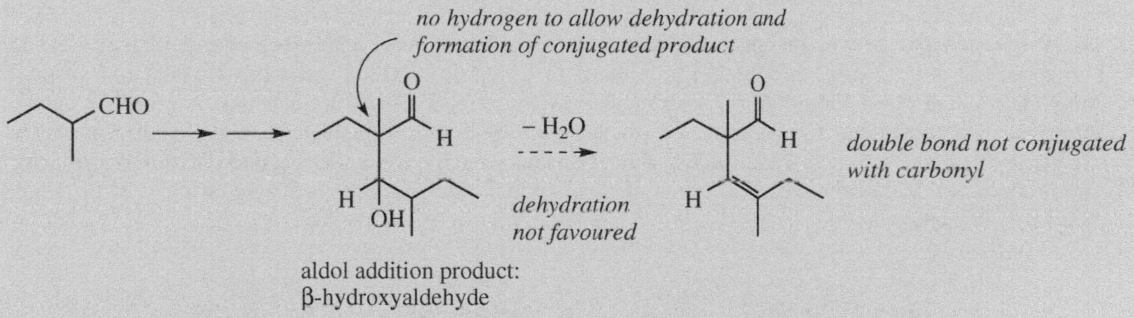

*no hydrogen to allow dehydration and
formation of conjugated product*

− H₂O

*dehydration
not favoured*

*double bond not conjugated
with carbonyl*

aldol addition product:
β-hydroxyaldehyde

(b) Structural analysis suggests we have intramolecular reactions with displacement of bromide and formation of a new ring. Since it is a base-initiated reaction, we should consider enolate anions as the nucleophiles. There are two possible sites to remove protons in the starting material, and the likely substitution reactions follow.

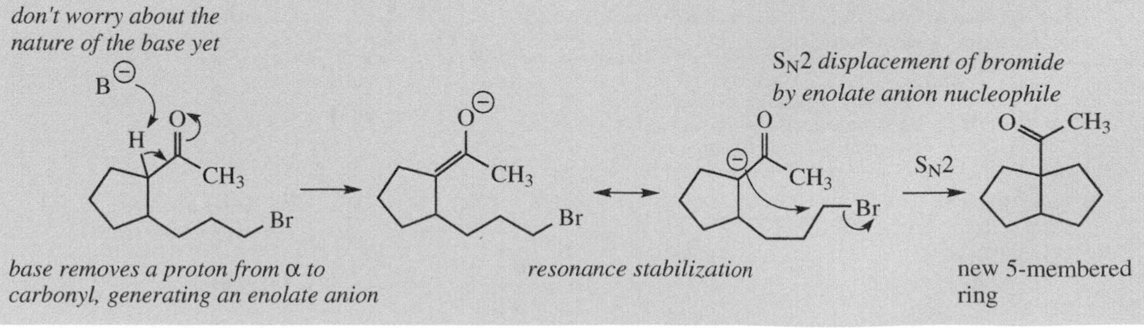

*don't worry about the
nature of the base yet*

S_N2 *displacement of bromide
by enolate anion nucleophile*

base removes a proton from α to
carbonyl, generating an enolate anion

resonance stabilization

S_N2

new 5-membered
ring

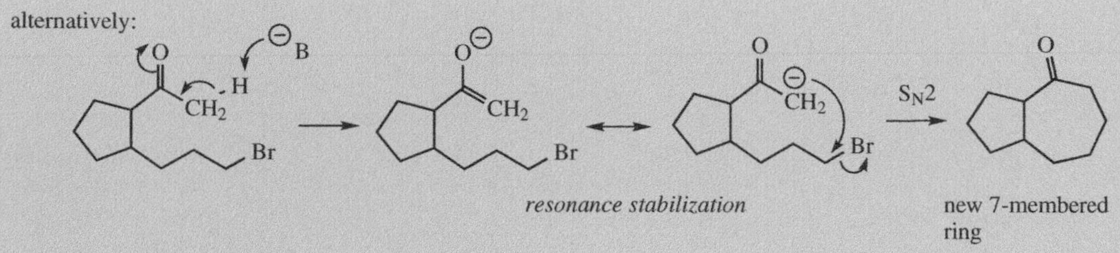

alternatively:

resonance stabilization   new 7-membered
                                                               ring

Now LDA is a very strong base, and it is also a large base. We deduce that it can only remove a proton from the less-hindered position, i.e. the $COCH_3$, and this leads to seven-membered ring formation.

To say *tert*-butoxide is a smaller base and removes a proton from the alternative site is not sufficient. If it can remove a proton from the hindered position, then it can also remove one from the less-hindered position, so we ought to get a mixture of the two possible products. Here, we need to remember that although LDA forms the enolate anion in an irreversible manner, enolate anion formation using *tert*-butoxide is an equilibrium and, therefore, reversible. Therefore, we see formation of the more favourable product, i.e. that with the five-membered ring.

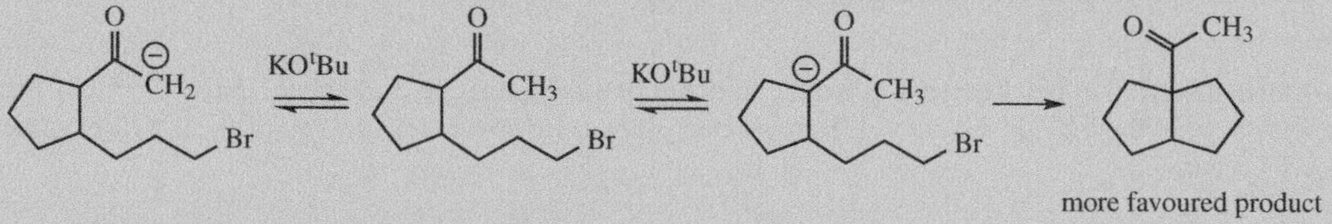

more favoured product

(c) The obvious change here is that three molecules of formaldehyde react with one molecule of acetaldehyde. Formaldehyde is in excess; the reaction is initiated by mild base, sodium carbonate. Combination of these compounds under basic conditions suggests enolate anion chemistry. Only acetaldehyde can form an enolate anion; the less-substituted formaldehyde is a better electrophile than acetaldehyde, and it is also present in excess. In the product, we can visualize the original acetaldehyde fragment carrying three CH_2OH substituents. The product is readily rationalized in terms of three similar aldol reactions, utilizing all three of the acidic hydrogens in acetaldehyde.

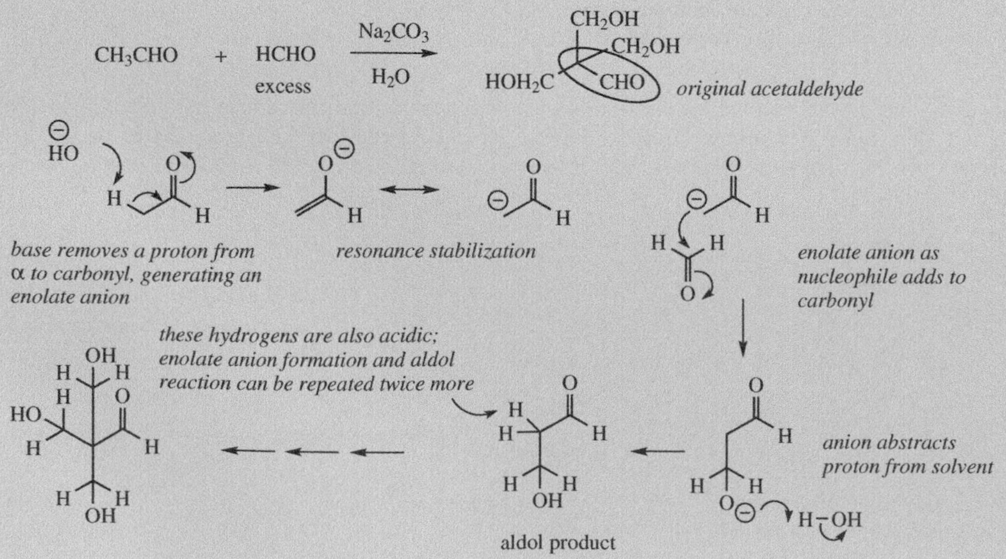

$$CH_3CHO \quad + \quad HCHO \xrightarrow[H_2O]{Na_2CO_3}$$
                           excess

original acetaldehyde

base removes a proton from   resonance stabilization   enolate anion as
α to carbonyl, generating an   nucleophile adds to
enolate anion   carbonyl

these hydrogens are also acidic;
enolate anion formation and aldol
reaction can be repeated twice more   anion abstracts
                                                                               proton from solvent

aldol product

(d) Redrawing this equation allows us to see how the two starting materials are required to combine. We can also think in terms of enolate anion chemistry to achieve the necessary bonding; we have a carbonyl compound and basic conditions.

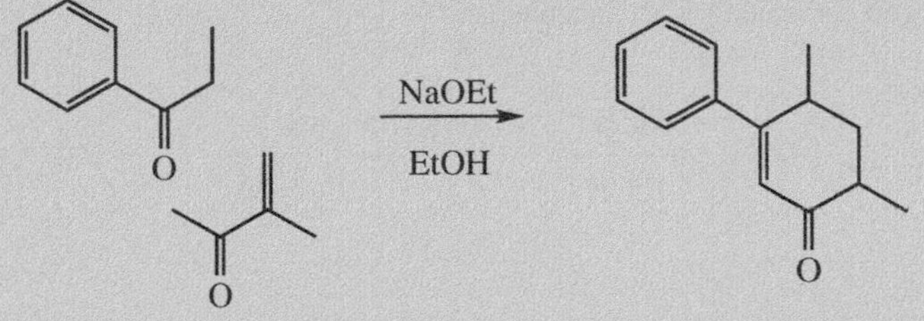

The reaction is considered as a combination of a Michael reaction, the conjugate addition of an enolate anion on to an unsaturated carbonyl compound, plus an aldol reaction followed by elimination of water.

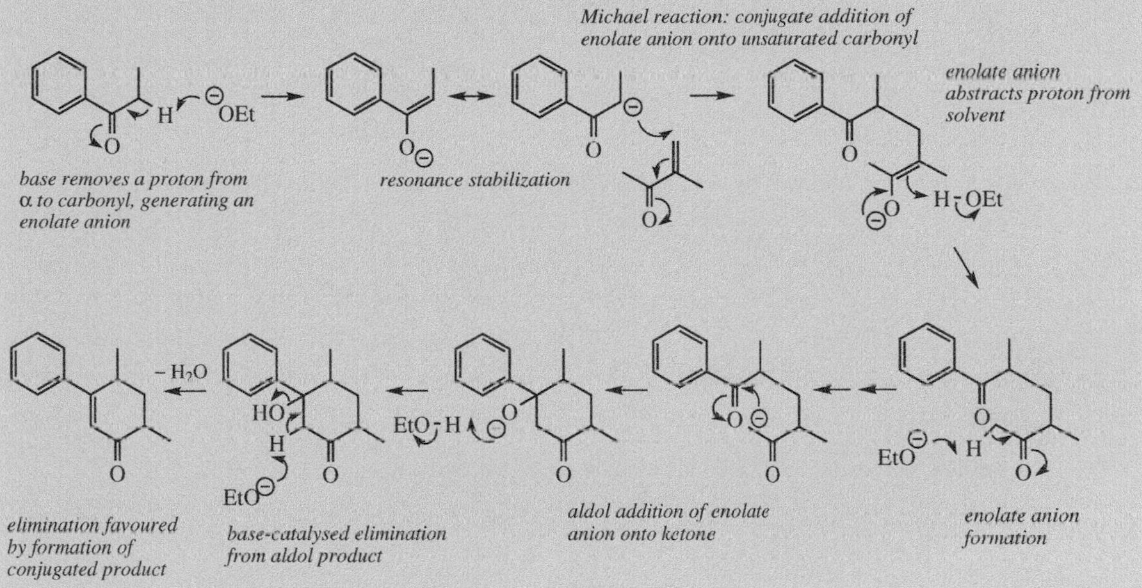

Michael reaction: conjugate addition of enolate anion onto unsaturated carbonyl

enolate anion abstracts proton from solvent

base removes a proton from α to carbonyl, generating an enolate anion

resonance stabilization

elimination favoured by formation of conjugated product

base-catalysed elimination from aldol product

aldol addition of enolate anion onto ketone

enolate anion formation

The irreversible elimination drives the reversible aldol reaction and gives a favourable conjugated ketone in a favourable six-membered ring. On paper, one could also draw an acceptable mechanism in which the order of events was reversed. This is not so neat, and would require generating an enolate anion γ to the α,β-unsaturated ketone formed by the first aldol–dehydration sequence.

the alternative approach:

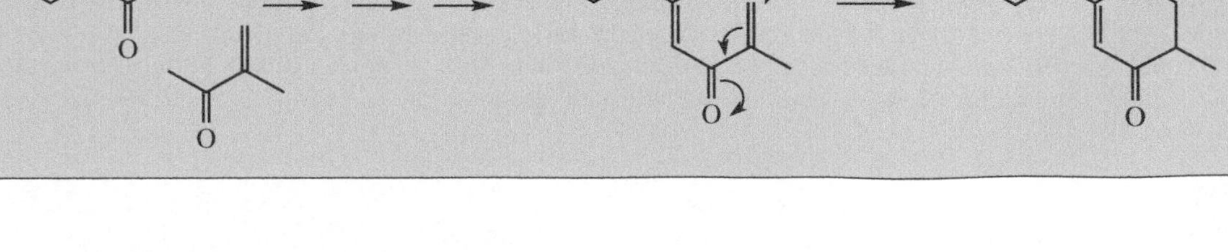

Problem 16.18 (level 2; chapters 7 and 10)

Provide mechanistic explanations for the following observations:

(a) The ketone (A) may be synthesized from ethyl acetoacetate by base-catalysed reaction with 1,4-dibromobutane, followed by heating with aqueous acid.

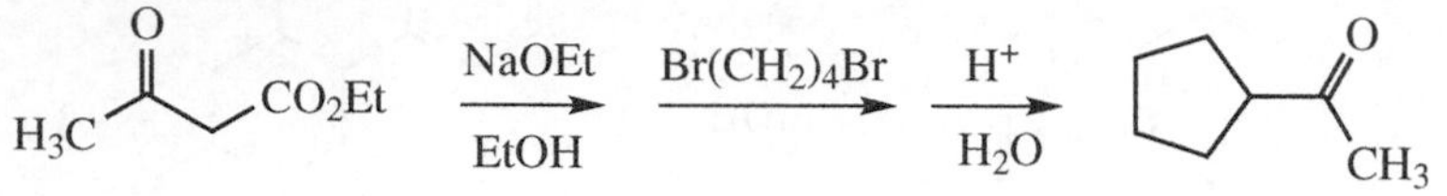

(A)

(b) The diketone (B) is transformed into the bicyclic ketone (C) on treatment with mild base.

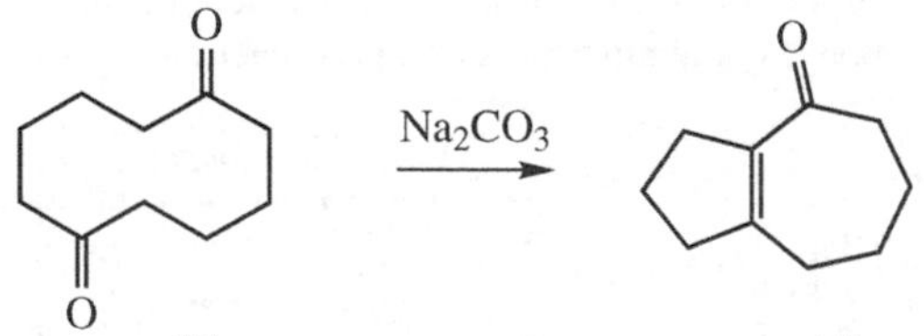

(B)　　　　　(C)

(c) Base-catalysed reaction, followed by acid hydrolysis converts the diester (D) into the cyclic ketone (E).

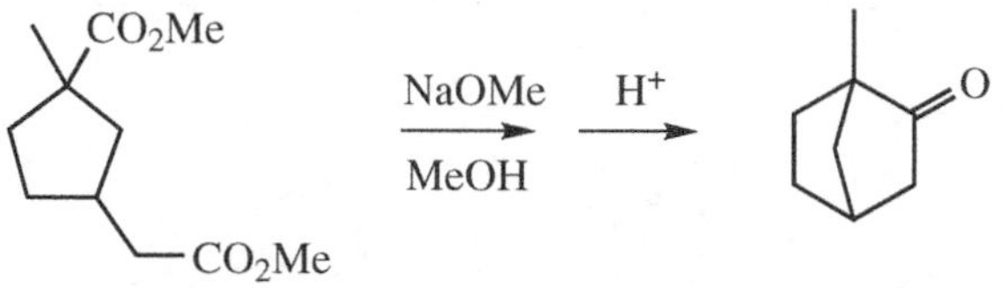

(D)　　　　　(E)

(d) The ketoester (F), on heating with aqueous base, is converted into a mixture of two acids.

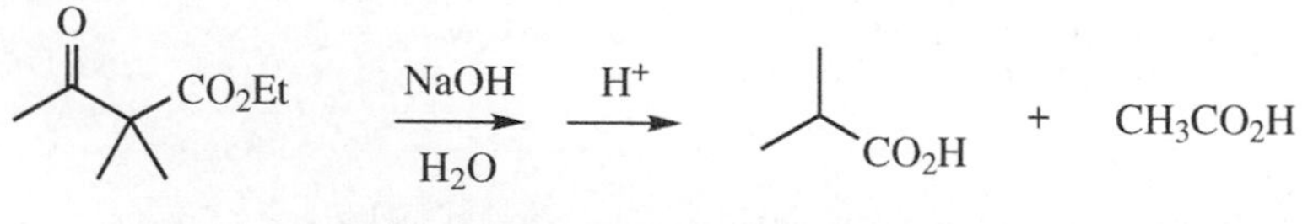

(F)

Answer 16.18

These provide examples of enolate anion chemistry. This can be surmised from the variety of bond-forming reactions on ketone or ester substrates all achieved under basic conditions.

(a) This shows a three-step sequence; do not be put off, but consider the implications of the first two. We start with ethyl acetoacetate, a β-keto ester. Under the basic conditions, this can give an enolate anion that presumably then reacts with the halide. No halogen is in the product, so think about S_N2 displacement; since we actually use a dihalide, we might consider two such reactions. Put this down, and you will see we are well on the way.

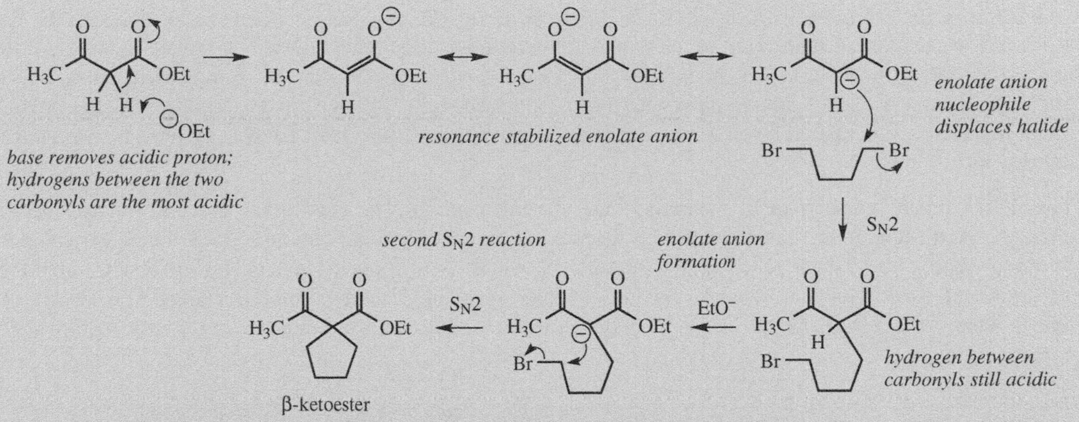

base removes acidic proton; hydrogens between the two carbonyls are the most acidic

resonance stabilized enolate anion

enolate anion nucleophile displaces halide

second S$_N$2 reaction   *enolate anion formation*

hydrogen between carbonyls still acidic

β-ketoester

The methylene hydrogens between the two carbonyls are the most acidic, so this is where enolate anion formation occurs. Now follows an S$_N$2 reaction with the dibromide reagent. It is soon apparent that this sequence of enolate anion formation and S$_N$2 displacement can be repeated, since the substrate still contains an acidic hydrogen. We soon end up with an alkylated ketoester.

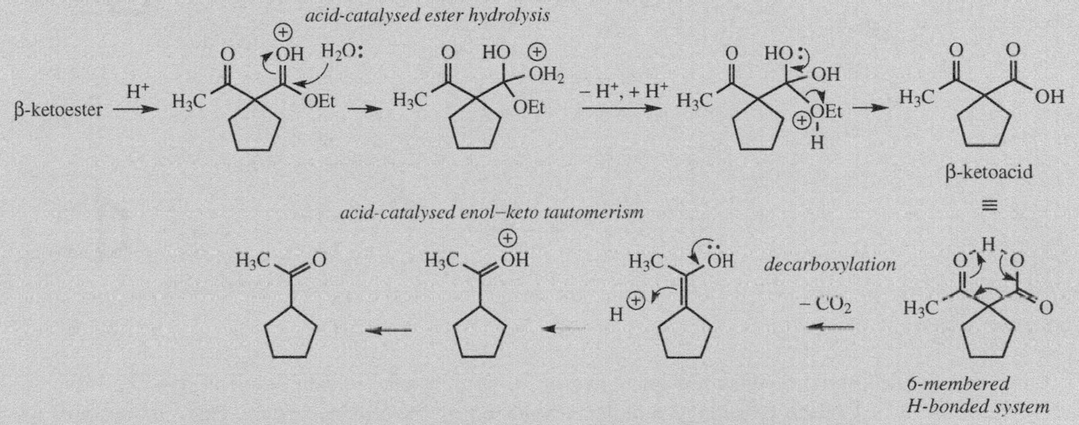

acid-catalysed ester hydrolysis

β-ketoester

β-ketoacid

acid-catalysed enol–keto tautomerism

decarboxylation

6-membered H-bonded system

To get the final product we need to lose the ester function. This is a standard combination of acid-catalysed ester hydrolysis followed by heating. The β-ketoacid forms a hydrogen-bonded six-membered ring that facilitates decarboxylation.

(b) The base used in this reaction is quite mild, namely sodium carbonate, but this does not change the approach through enolate anions. Overall, we have modified a system with two ketone functions, lost one, and produced an alkene. This is an aldol reaction followed by dehydration, but it is an intramolecular reaction. Do not worry about the five-membered/seven-membered ring combination; that will fall into place as we consider the mechanism.

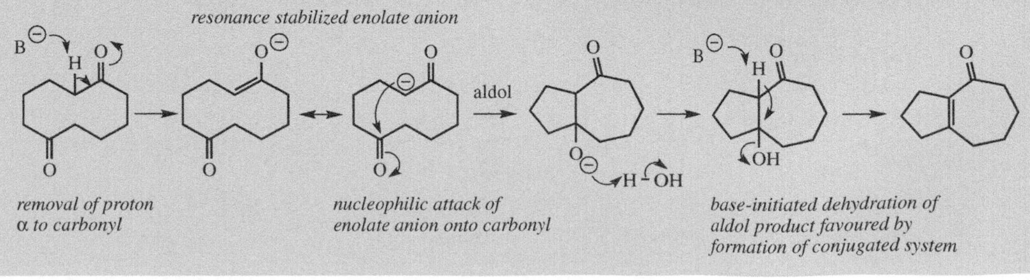

resonance stabilized enolate anion

aldol

removal of proton α to carbonyl

nucleophilic attack of enolate anion onto carbonyl

base-initiated dehydration of aldol product favoured by formation of conjugated system

Base removes a proton from adjacent to a carbonyl group. We can use a general B⁻ to represent the base; however, we could use hydroxide, since that is released upon dissolving sodium carbonate in water. It does not matter which α-position we choose; they are actually all equivalent in this symmetrical substrate. The nucleophilic attack of the enolate anion onto the second carbonyl is followed by base-initiated dehydration, so that a favourable conjugated ketone is the product. As you can see, the size of the ring systems is automatically defined by the reaction.

(c) This looks much worse than it is. Analyse the starting material, the conditions, and the functional group changes. A diester under basic conditions leads us to consider enolate anions. From there, we should be thinking about the Claisen reaction, the product of which is a β-ketoester. The loss of ester functions is accommodated by typical follow-up reactions of hydrolysis and decarboxylation, leading to a ketone. And that is what we get here. As in (b), the ring systems will be automatically defined and do not require special consideration.

nucleophilic attack of enolate anion onto carbonyl   *loss of leaving group*

no acidic hydrogen   CO₂Me   CO₂Me   OMe   OMe

Claisen

OMe   OMe   OMe

H   H   CO₂Me   CO₂Me

removal of proton α to carbonyl   OMe   *resonance stabilized enolate anion*   β-ketoester

6-membered H-bonded system   *redraw it*

– CO₂   O   H   O   *acid-catalysed ester hydrolysis*   CO₂Me

decarboxylation   O

β-ketoacid   β-ketoester

Only one of the ester functions has an α hydrogen, so only one enolate anion is possible. This attacks the other ester in a Claisen reaction, leading to expulsion of the leaving group. This product will need a bit of redrawing to get it into the format of the final product; you see it is exactly right and follows from the mechanistic approach. All that remains is acid-catalysed hydrolysis of the ester, and decarboxylation of the β-ketoacid; this is exactly as in part (a), so is not shown again. It involves a hydrogen-bonded cyclic intermediate.

(d) This little reaction is a reverse Claisen reaction on a β-ketoester. We normally think of a Claisen reaction occurring between two esters; the products here are acids, but this is because the aqueous basic conditions also lead to ester hydrolysis.

When you are confronted with a reverse reaction, it is a good idea to write down the forward reaction, which you are probably more familiar with, then just reverse the mechanistic sequence.

Claisen reaction

O   CO₂Et   O   CO₂Et   O   CO₂Et

OEt   EtO   EtO

nucleophilic attack of enolate anion onto carbonyl   *loss of leaving group*

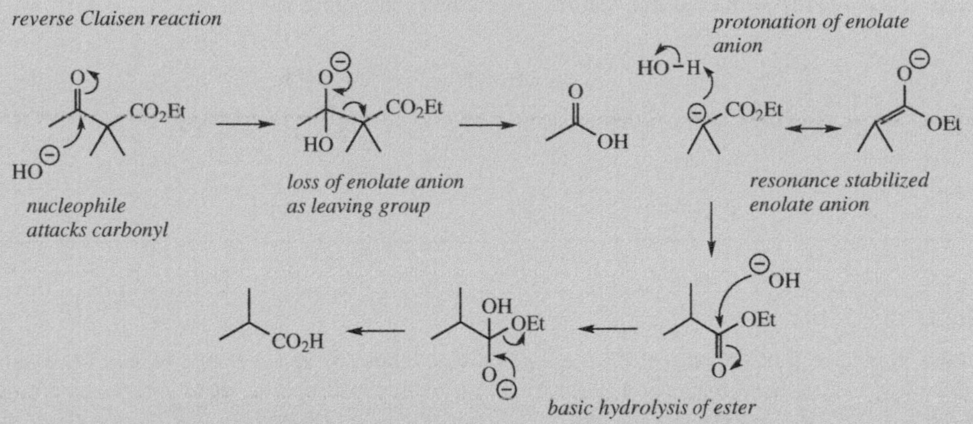

reverse Claisen reaction

protonation of enolate anion

nucleophile attacks carbonyl

loss of enolate anion as leaving group

resonance stabilized enolate anion

basic hydrolysis of ester

In this case, we formulate the Claisen reaction between two ester molecules as enolate anion formation, nucleophilic attack, then loss of the leaving group. Now reverse it. Use hydroxide as the nucleophile to attack the ketone carbonyl, then expel the enolate anion as the leaving group. All that remains is protonation of the enolate anion, and base hydrolysis of its ester function.

Problem 16.19 (level 2; chapters 7 and 10)

Provide mechanistic explanations for the following observations:

(a) The antidepressant diazepam may be synthesized by reaction of the aminoketone (A) with glycine.

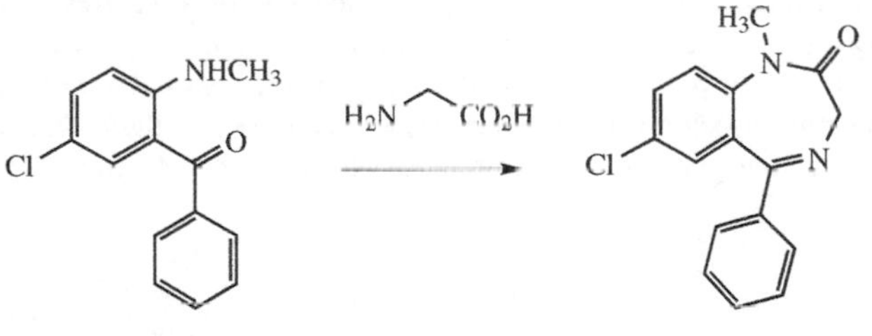

(A)   diazepam

(b) The amino acid valine may be synthesized by the reaction of aldehyde (B) with ammonia and HCN, followed by acid hydrolysis.

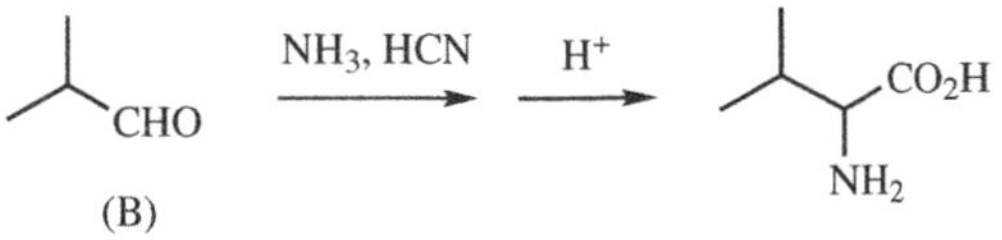

(B)   valine

(c) The aminoketone (D) is produced when ketone (C) is reacted with piperidine and formaldehyde under mild acidic conditions.

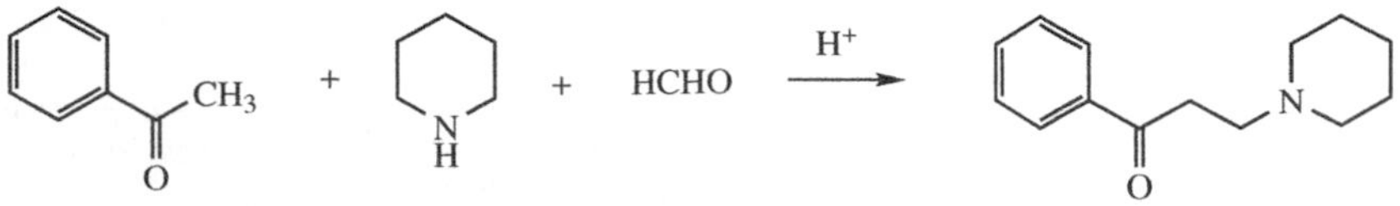

(C)   (D)

(d) Dopamine reacts readily with acetaldehyde to produce the cyclic amine salsolinol.

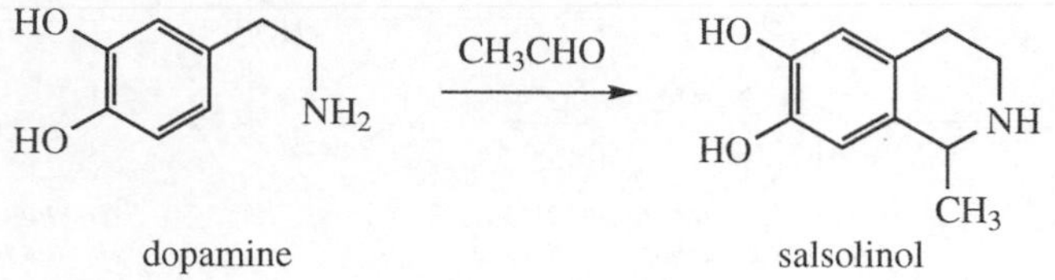

dopamine   salsolinol

Answer 16.19

(a) Do a quick comparison of the reagents and the product; it should be apparent that we need to synthesize a new amide bond and a new imine bond. Both reactions involve nucleophilic attack of an amino group onto a carbonyl.

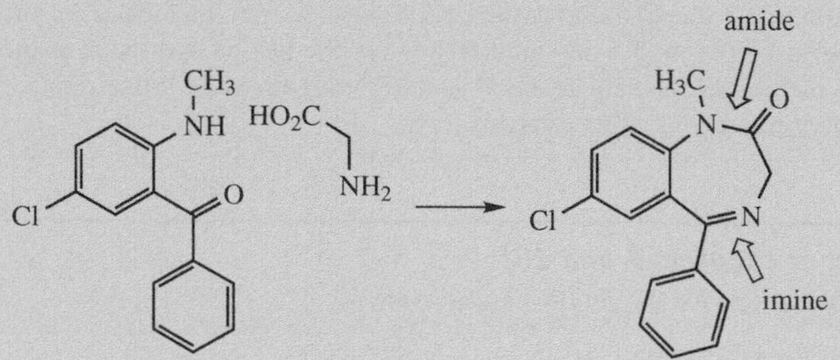

These are standard reactions, and it does not matter which you do first. Resist the urge to do both at the same time; it's unlikely to occur, and will certainly complicate your structures and mechanisms. Note that we have to include a number of protonations to make the mechanisms work. This is quite legitimate, and although the question has not specified any conditions, you can assume there will be a source of protons. In any case, you will observe that protons are first released to solvent, then transferred back to an alternative site in the molecule.

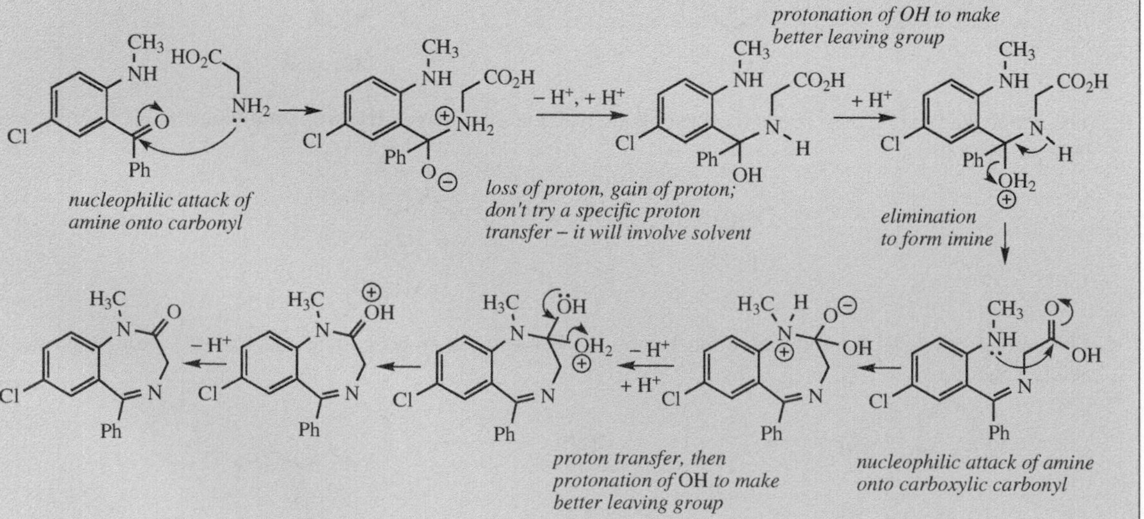

nucleophilic attack of
amine onto carbonyl

loss of proton, gain of proton;
don't try a specific proton
transfer – it will involve solvent

protonation of OH to make
better leaving group

elimination
to form imine

proton transfer, then
protonation of OH to make
better leaving group

nucleophilic attack of amine
onto carboxylic carbonyl

(b) This is a standard Strecker synthesis of an amino acid; it is not important that you remember the name in 'named' reactions. An analysis of the transformation should lend some clues. You are informed it is a two-stage process; acid hydrolysis is a necessary part, and you can relate this to hydrolysis of a nitrile, supplied by the cyanide in the first part. The initial reaction is between the aldehyde and the ammonia to form an imine. This then acts as carbonyl analogue for attack of the cyanide. Alternative thoughts starting with aldehyde plus cyanide giving a cyanohydrin are unproductive, in that you then wonder how to use the ammonia.

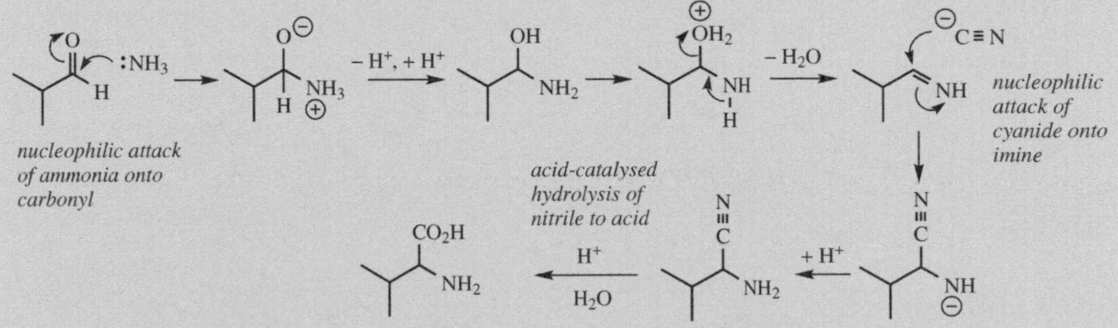

nucleophilic attack of ammonia onto carbonyl

acid-catalysed hydrolysis of nitrile to acid

nucleophilic attack of cyanide onto imine

The initial part of the reaction is the same as in part (a), and leads to the imine intermediate. Nucleophilic attack by the cyanide onto this carbonyl analogue provides the next intermediate, an aminonitrile. The last part of the synthesis is hydrolysis of the nitrile function. It involves another intermediate, an amide, and this also has to be subjected to acid hydrolysis. As a result, it all adds up to a lengthy mechanism that can become rather tedious, because it contains so many repetitive steps. It is shown in general terms below.

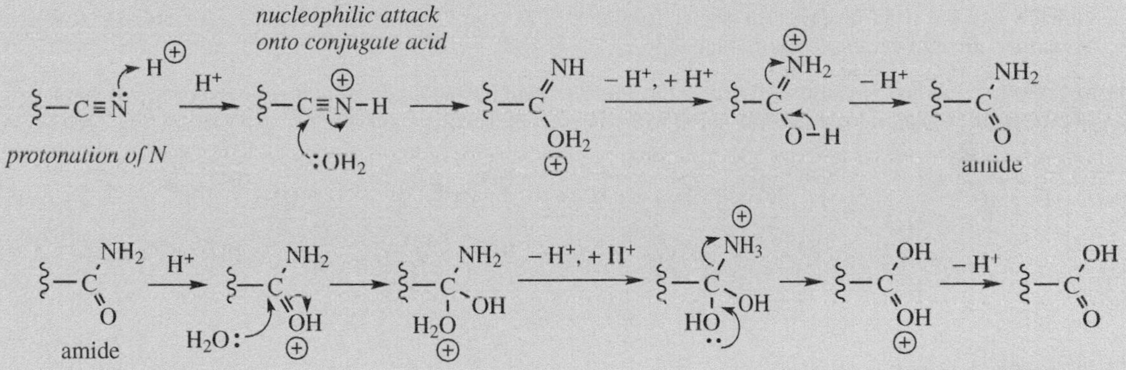

nucleophilic attack onto conjugate acid

protonation of N

amide

amide

(c) This is another 'named' reaction, the Mannich reaction. Three reagents join together in a two-stage process.

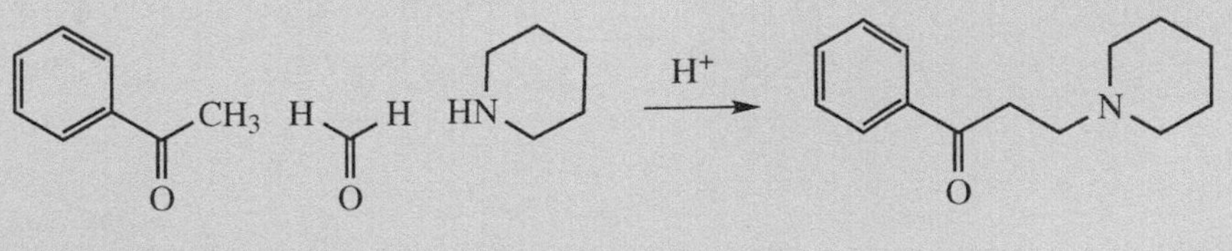

We have an aldehyde, an amine, and a ketone. As in part (b), the amine reacts first to give an imine, and this behaves as a carbonyl analogue, which in the Mannich reaction is then the electrophile for an enolate anion equivalent. How can we remember the sequence of events? The most common mistake is to react the aldehyde and ketone via an aldol reaction, but this then leads to an alcohol and one is faced with a substitution reaction to incorporate the amine. It is the mild acidic conditions that help us to avoid wrong proposals. Such conditions are not especially conducive to aldol reactions and substitution reactions. Instead, they are favourable for imine formation. And which of the carbonyl reagents reacts? That is easy, it is the most reactive electrophile, the aldehyde.

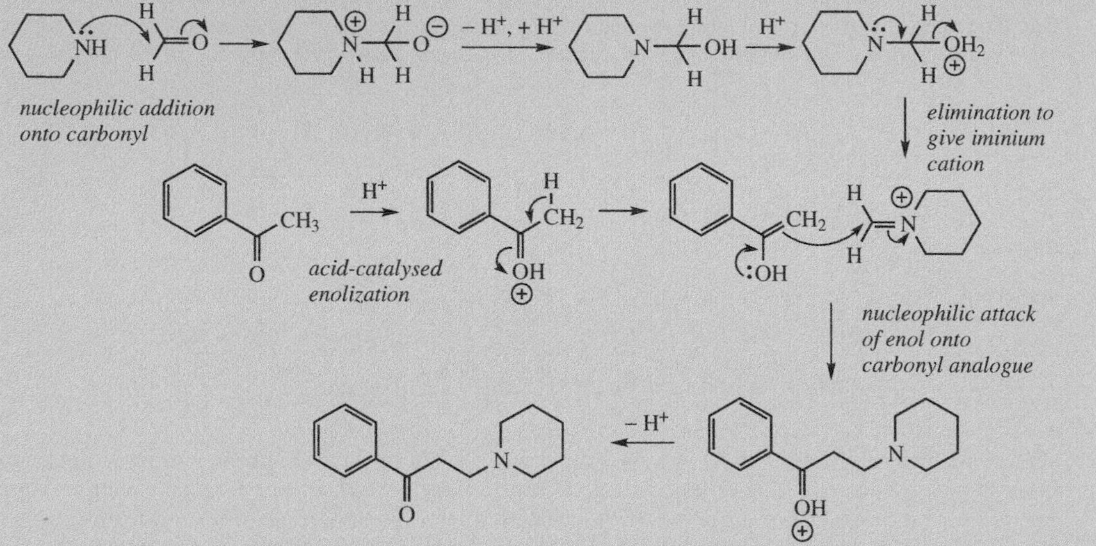

nucleophilic addition
onto carbonyl

elimination to
give iminium
cation

acid-catalysed
enolization

nucleophilic attack
of enol onto
carbonyl analogue

We thus have standard imine formation; in this case, the secondary amine leads to an iminium cation. This is attacked by the ketone nucleophile. This cannot be an enolate anion because of the mild acidic conditions under which the reaction proceeds, so we formulate it as involving the enol tautomer. Accordingly, we need to include an acid-catalysed enolization step.

(d) Yet another 'named' reaction, this time the Pictet–Spengler tetrahydroisoquinoline synthesis. It is really a Mannich-like reaction as in part (c), with the enol nucleophile being a phenol, and achieving a ring cyclization.

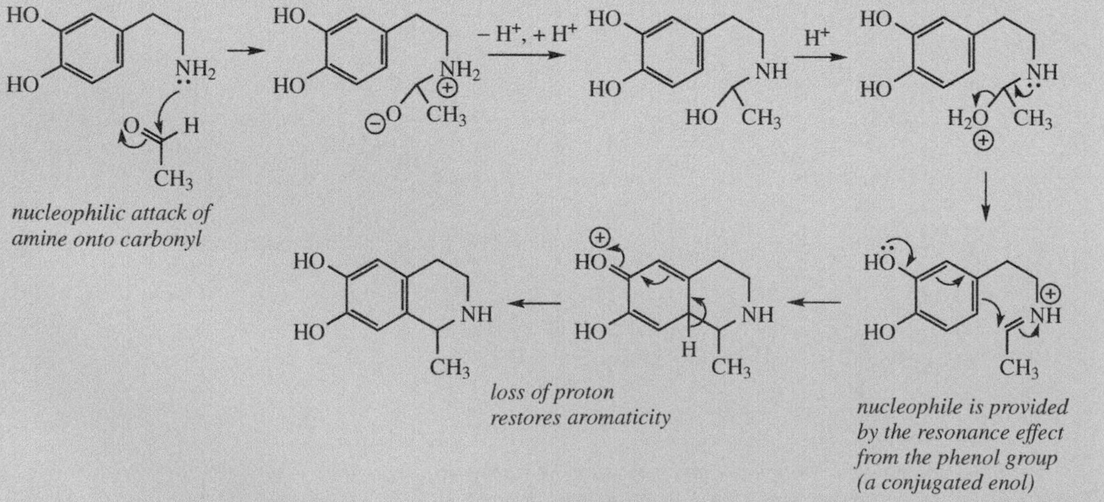

nucleophilic attack of
amine onto carbonyl

loss of proton
restores aromaticity

nucleophile is provided
by the resonance effect
from the phenol group
(a conjugated enol)

The mechanism of imine formation is standard, as seen in the other examples. The cyclization reaction is then like the Mannich reaction, attack of an enol on to the iminium cation. This time though, the nucleophile is provided by the resonance effect from the phenol system.

Problem 16.20 (level 2; chapters 6, 7, 8 and 10)

A synthetic pathway to vitamin A involves the following partial sequence:

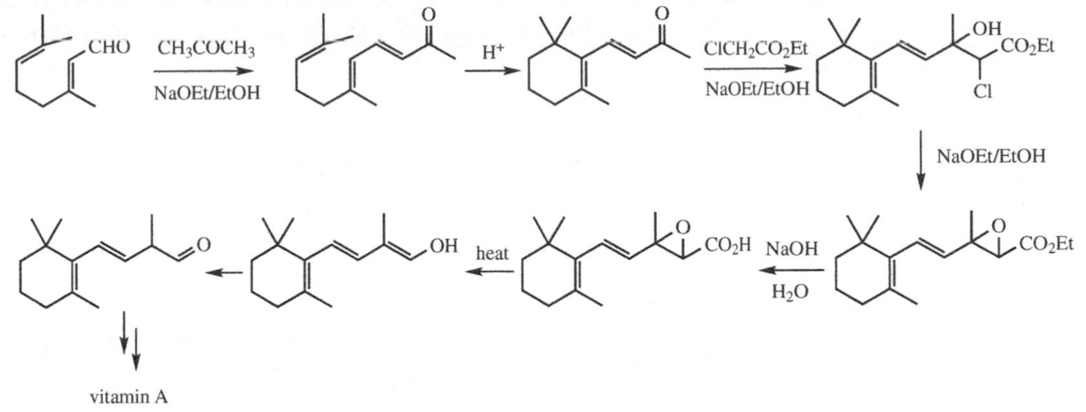

Give detailed mechanisms for the reactions in this sequence.

Answer 16.20

This problem covers a reaction sequence and a variety of different reactions, some easier than others. This one includes enolate anions, electrophilic cyclization, nucleophilic substitution, and simple carboxylic acid chemistry.

The first reaction involves a ketone reaction with an aldehyde under basic conditions, so enolate anion chemistry is likely. This is a mixed aldol reaction; the acetone has acidic α-hydrogens to form an enolate anion, and the aldehyde is the more reactive electrophile. The reaction is then driven by the ability of the intermediate alcohol to dehydrate to a conjugated ketone.

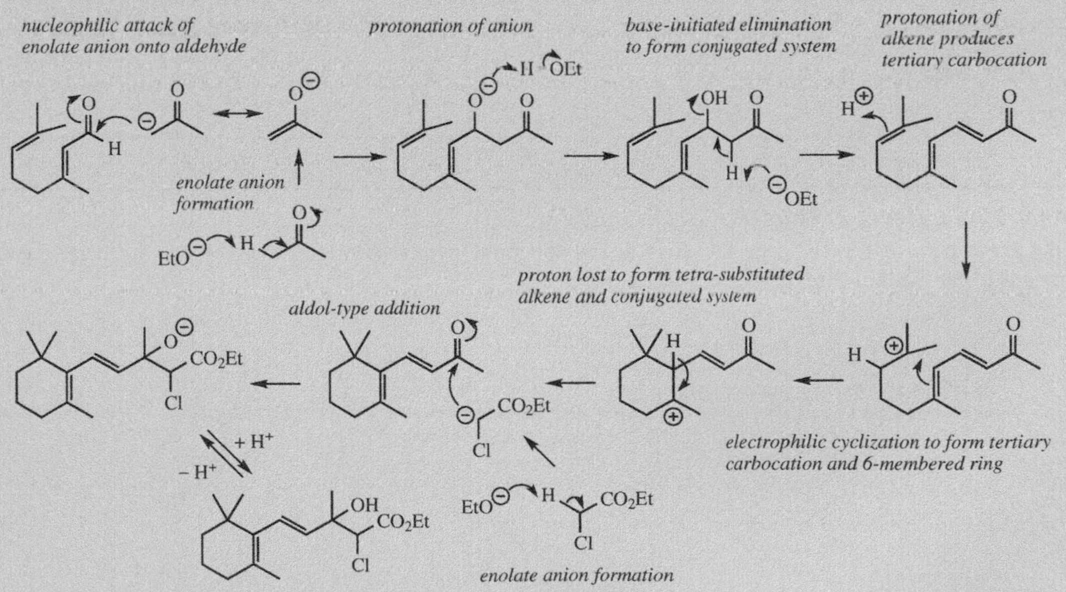

The next transformation is an electrophilic cyclization. Protonation of the terminal alkene produces the more favourable tertiary carbocation, then ring formation occurs by electrophilic addition to the neighbouring alkene.

Again, this produces a favourable tertiary carbocation. Loss of a proton gives the required alkene. Note that potentially three different carbons could lose a proton. The reaction shown generates the most stable product; this has the maximum number of alkyl substituents and also benefits from extended conjugation. We then get another aldol-type reaction. The enolate anion is produced from the ethyl chloroacetate, and simple addition yields an anion that is subsequently protonated.

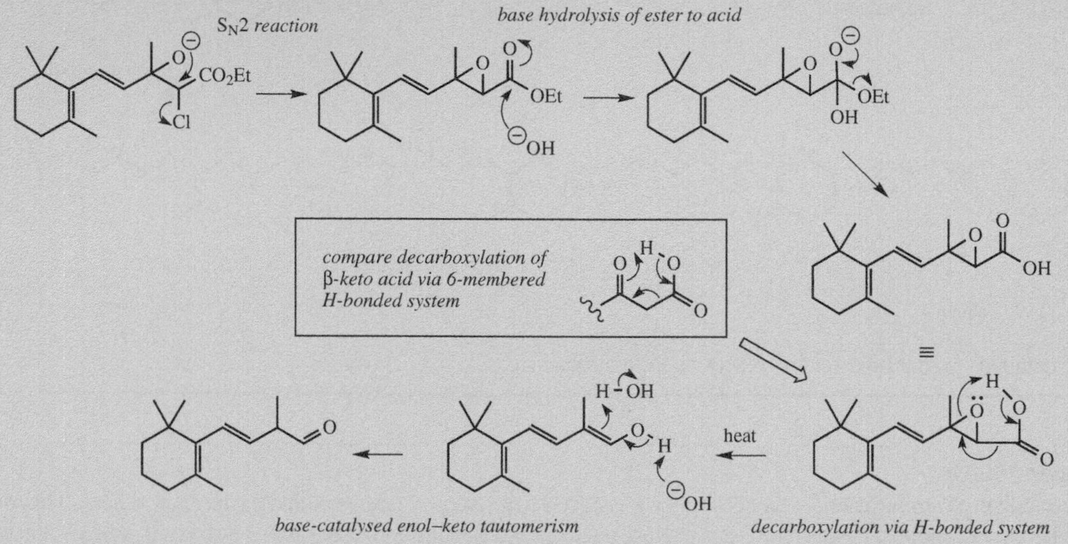

base-catalysed enol–keto tautomerism   *decarboxylation via H-bonded system*

However, the same anion has to be regenerated for the next reaction, in which an S_N2 displacement leads to epoxide formation. Base hydrolysis of the ester produces the carboxylic acid, which decarboxylates on heating. The mechanism for this reaction should be suggested by the structural change (loss of a carbon), and that this is achieved merely by heating. What other systems do this? It is actually analogous to thermal decarboxylation of a β-keto acid, a process where we invoke a cyclic mechanism within a hydrogen-bonded system. Think laterally to see whether we can do the same here. We can, though any hydrogen bonding must involve a five-membered ring rather than a six-membered ring. The sequence is completed by base-catalysed tautomerism of the enol to a ketone.

Problem 16.21 (level 2; chapters 10, 6, and 7)

Propose a mechanism for all steps involved in the following synthesis of the ester of chrysanthemic acid, a component of the natural insecticides, the pyrethrins:

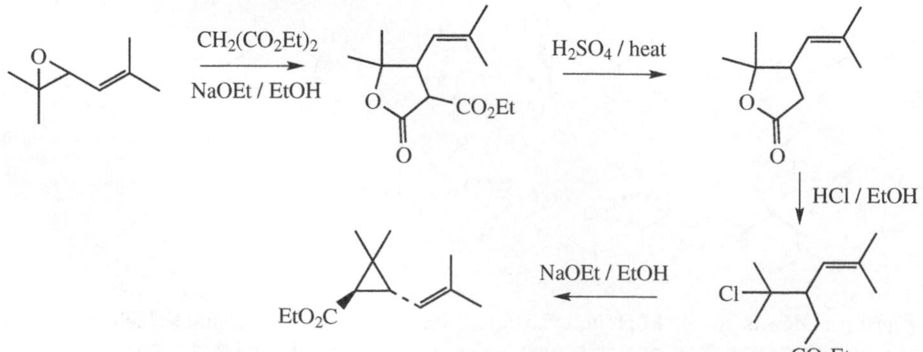

ethyl chrysanthemate

Answer 16.21

This problem brings in a wide range of reactions. It includes enolate anions as nucleophiles, with S_N2 reactions opening an epoxide ring, and displacing halide. There is a simple S_N1 reaction and the decarboxylation of a β-keto acid through a cyclic hydrogen-bonded system. Most of the remaining steps involve ester formation and hydrolysis. Although a lactone intermediate can be isolated in the sequence, it need not be specifically implicated in the mechanism; it is formed under the acidic conditions from a hydroxy acid, but we could utilize the hydroxy acid in the sequence. Some of the steps might then occur in a different order from that shown.

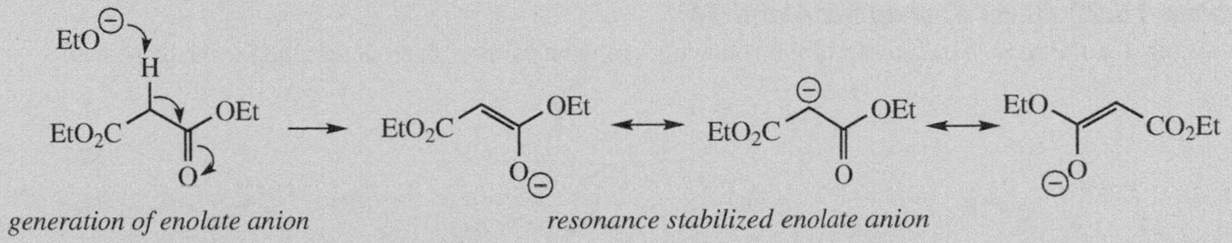

generation of enolate anion

resonance stabilized enolate anion

The mechanistic steps can be deduced by inspection of structures and conditions. Enolate anion formation from diethyl malonate under basic conditions is indicated, and that this must attack the epoxide in an S_N2 reaction is implicated by the addition of the malonate moiety and disappearance of the epoxide. The subsequent ring formation follows logically from the addition anion, and is analogous to base hydrolysis of an ester. Ester hydrolysis followed by decarboxylation of the β-keto acid is then implicated by the acidic conditions and structural relationships.

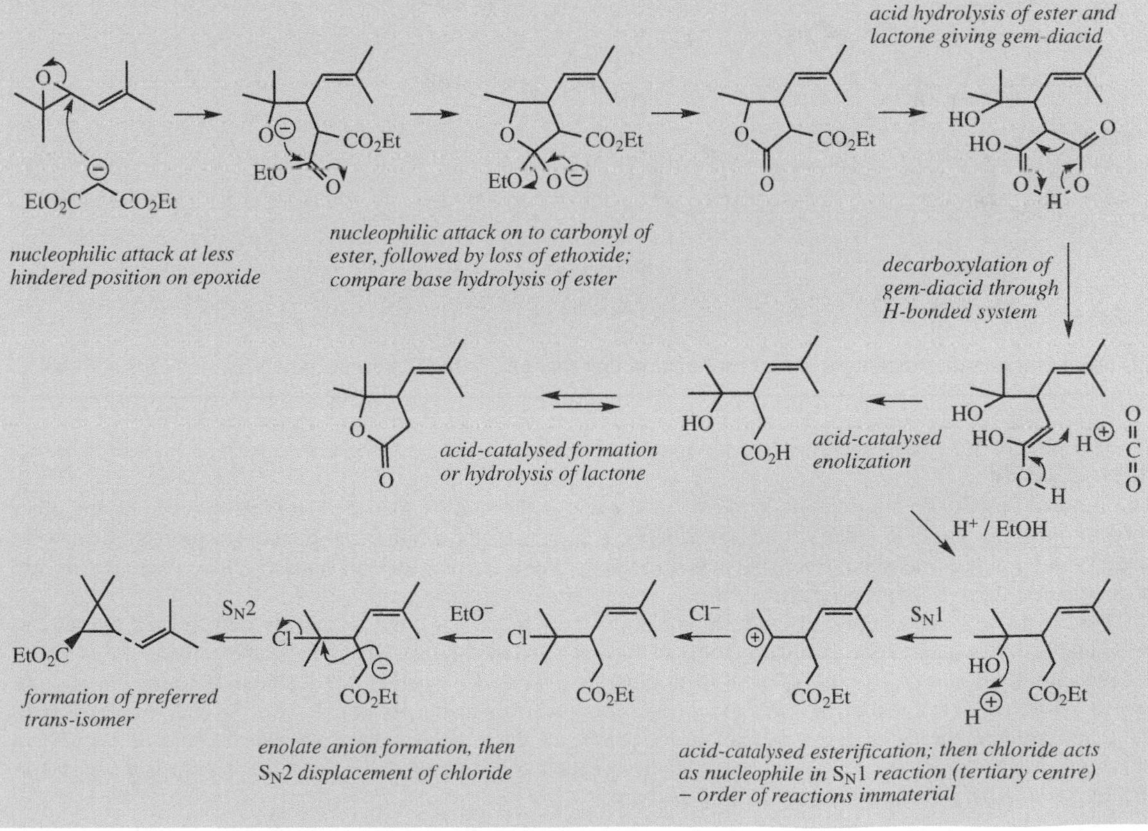

acid hydrolysis of ester and lactone giving gem-diacid

nucleophilic attack at less hindered position on epoxide

nucleophilic attack on to carbonyl of ester, followed by loss of ethoxide; compare base hydrolysis of ester

decarboxylation of gem-diacid through H-bonded system

acid-catalysed formation or hydrolysis of lactone

acid-catalysed enolization

$H^+ / EtOH$

formation of preferred trans-isomer

S_N2 EtO^- Cl^- S_N1

enolate anion formation, then S_N2 displacement of chloride

acid-catalysed esterification; then chloride acts as nucleophile in S_N1 reaction (tertiary centre) – order of reactions immaterial

The next step is not immediately obvious. The generation of an ethyl ester from a lactone can be accommodated by transesterification (we might alternatively consider esterification of the free hydroxyacid). The incorporation of chlorine where we effectively had the alcohol part of the lactone leads us to nucleophilic substitution. That it can be S_N1 is a consequence of the tertiary site. Cyclopropane ring formation from an S_N2 reaction in which an enolate anion displaces a halide should be deducible from the structural relationships and basic conditions.

Problem 16.22 (level 2; chapters 4 and 11)

pK_a values for the most basic group in the following compounds have been determined to be as shown:

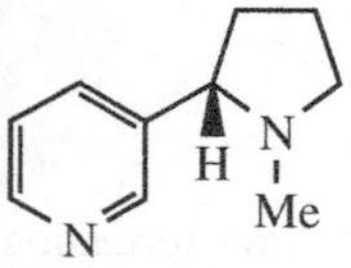

nicotine
pK_a 8.1

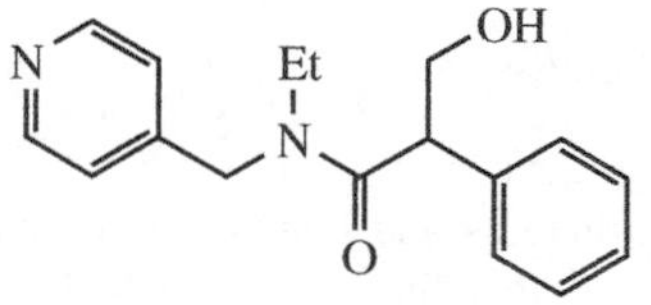

tropicamide
pK_a 5.2

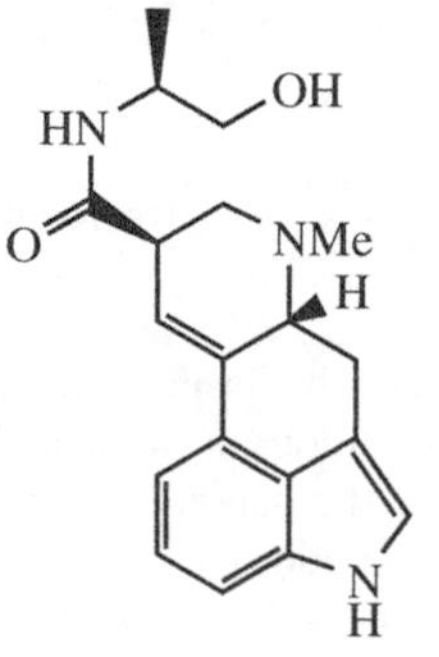

ergometrine
pK_a 6.8

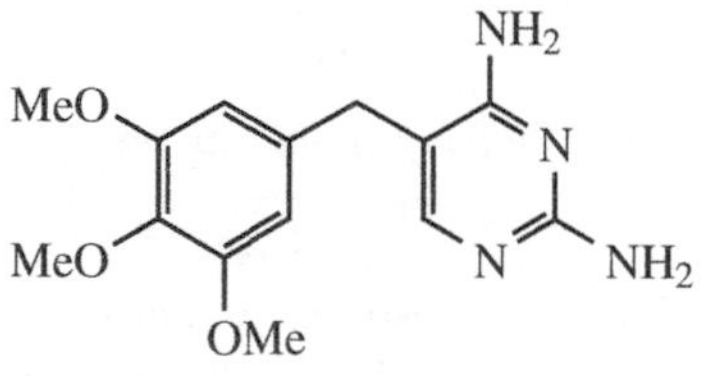

trimethoprim
pK_a 7.2

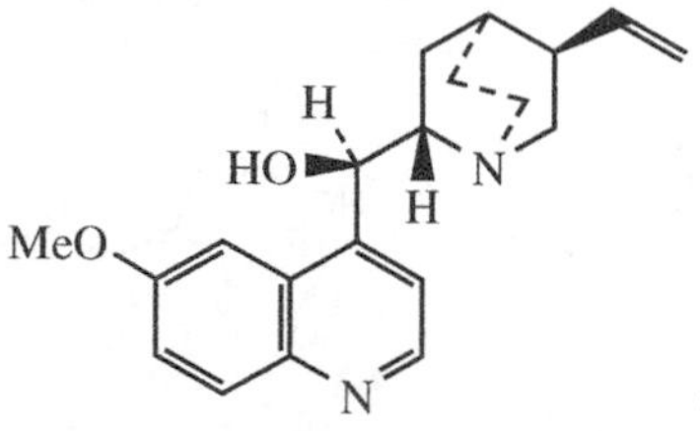

quinine
pK_a 8.5

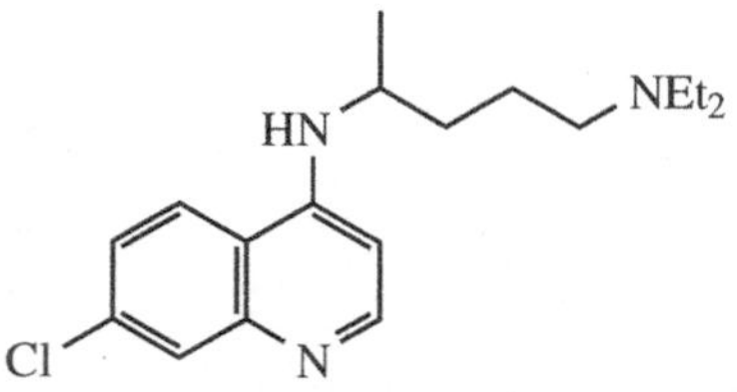

chloroquine
pK_a 10.8

For each compound, predict the site of protonation that this pK_a value represents. Give reasons for your choice.

Answer 16.22

The relative basicity of an aromatic nitrogen heterocycle is dictated by the ring size, the presence of any other heteroatoms, and possible effects from substituents. It is potentially a rather more complex problem than with, say, simple amines, and should be approached logically and systematically. In practice, these examples do not present particularly difficult problems.

There is no way we would encourage the memorising of pK_a values, but two easily remembered figures can be valuable for comparisons. Conjugate acids of typical aliphatic amines have pK_a values around 10, whereas for pyridine the figure is around 5. These then allow us to consider whether the compound in question is more or less basic, etc. The smaller the pK_a, the weaker the base. Remember, of course, that the pK_a values refer to the acidity of the conjugate acids, and not to the acidity of the amines themselves, which could be relevant in different circumstances. Nevertheless, it is usually acceptable to talk of the pK_a of the base, when we should strictly say the pK_a of the conjugate acid of the base.

Nicotine has two nitrogen atoms, one as a cyclic tertiary amine and one in a pyridine ring. The basicities are easily distinguished, in that a pyridine system is much less basic than a simple amine. This is essentially a hybridization effect, the nitrogen lone pair in pyridine being held in an sp^2 orbital. This means the lone pair electrons are held closer to the nitrogen, and are consequently less available for protonation than in an sp^3-hybridized aliphatic amine. Hence, as mentioned above, pyridine has pK_a approximately 5. It follows that pK_a 8.1 is more appropriate for the pyrrolidine nitrogen.

Tropicamide has pK_a appropriate for a pyridine ring. The second nitrogen is part of an amide group, and is thus a very weak base. Amides are stabilized through resonance, a stabilization that is not possible in the conjugate acid.

electron overlap from nitrogen lone pair allows resonance stabilization of amide

protonation on nitrogen does not occur; destroys resonance stabilization

Ergometrine has three nitrogen atoms, a cyclic tertiary amine, an amide (very weak base as just described for tropicamide), and one in an indole ring. Indoles, like pyrroles, are very weak bases, with negative pK_a values for the conjugate acid. This is because nitrogen has already used its lone pair electrons by contributing to the aromatic sextet. Consequently, protonation on nitrogen destroys aromaticity in the five-membered ring. In fact, like pyrrole, indole preferentially protonates on carbon rather than nitrogen. It follows that pK_a 6.8 for ergometrine is not appropriate for the indole nitrogen or the amide nitrogen, but must be the figure for the cyclic tertiary amine. It is somewhat less basic than we might predict for a cyclic tertiary amine, and this must relate to its position in a fused-ring system.

Trimethoprim is a diaminopyrimidine derivative. It is reasonably basic (pK_a 7.2) and we should remember here that amino substituents are able to utilize their lone pairs and provide resonance stabilization to a conjugate acid. Consequently, aminopyrimidines protonate on a ring nitrogen. If we consider protonation of the two ring nitrogens separately, and then think about potential resonance stabilization, we can predict the site of protonation.

this type of stabilization involving adjacent nitrogens is the same in both 1- and 3-protonated structures

this provides the most favourable resonance stabilization: maximum charge distribution

If we protonate on N-3, we can predict resonance stabilization involving either of the adjacent amino functions. In each case we see charge distribution over three atoms using an N–C=N system. However, protonation on N-1 allows similar resonance stabilization, but increased charge dispersal emanating from the

4-amino substituent. We do not need to consider any further resonance structures. Effects involving the 2-amino substituent will be the same whether N-1 or N-3 is protonated. We can predict that protonation occurs on N-1.

Quinine has two potentially basic centres, a cyclic tertiary amine at a ring junction, and one in a quinoline ring system. pK_a 8.5 is reasonably basic, and this is most likely from the aliphatic tertiary amine. We need to convince ourselves that the quinoline nitrogen is less basic. This is true. As far as reactivity is concerned, a quinoline ring system behaves as two separate parts, either pyridine or benzene, depending upon the reagent. Thus, quinoline has pK_a very similar to that of pyridine, i.e. around 5.

Chloroquinine is more basic than quinine, and the pK_a suggests we must be looking at an aliphatic amine. There are two; so which? Note though that one of the amine functions is *para* to the quinoline nitrogen, so we may expect resonance stabilization when the quinoline nitrogen is protonated.

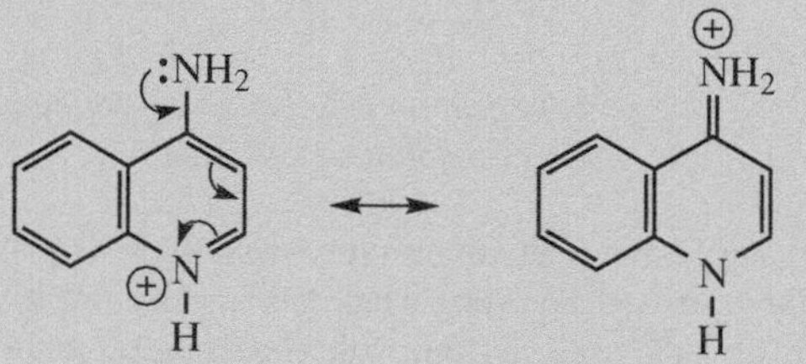

resonance effect stabilizes conjugate acid and provides maximum charge distribution

This will increase the basicity of the quinoline system from about pK_a 5, but almost certainly not as far as represented by pK_a 10.8. However, it solves our problem, since it means the 4-amino substituent is donating its lone pair into the aromatic ring system and is not, therefore, available for bonding to a proton. This site is going to be less basic than a tertiary amine. pK_a 10.8 must represent the terminal $-NEt_2$.

Problem 16.23 (level 2; chapters 11 and 7)

(a) During porphyrin biosynthesis, porphobilinogen is produced from two molecules of 5-aminolaevulinic acid and is subsequently converted into uroporphyrinogen I. Formulate a chemical mechanism to account for the transformations.

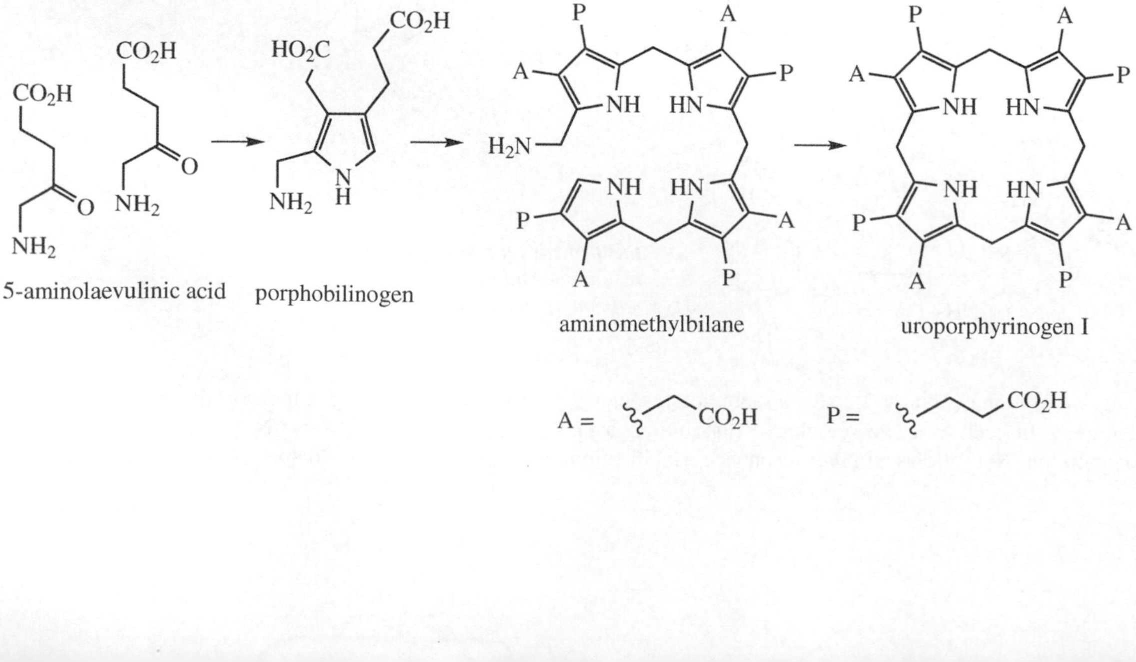

5-aminolaevulinic acid   porphobilinogen   aminomethylbilane   uroporphyrinogen I

$A = $ CO_2H $P = $ CO_2H

(b) The Reissert indole synthesis is shown below. Propose rational mechanisms to account for the first and third steps. Explain why the nitro group is essential for the success of the first step.

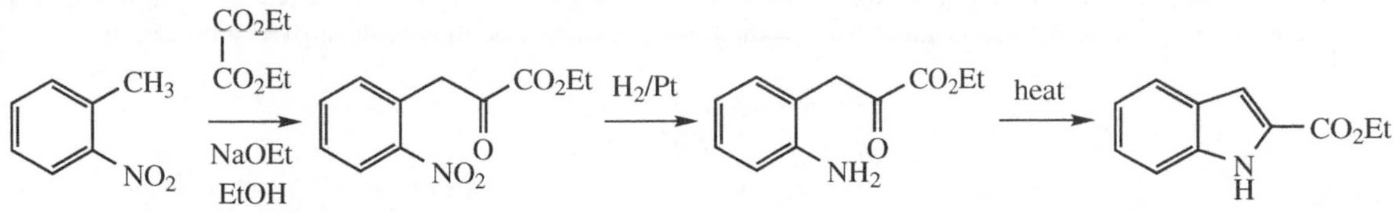

CO_2Et | CO_2Et

NaOEt / EtOH

H_2/Pt

heat

Answer 16.23

(a) Ignore the fact that this is a biosynthetic sequence. We are required to formulate a mechanism that treats it as a chemical transformation.

The first step is formation of a pyrrole ring system from two identical aminoketones. It is actually a Knorr pyrrole synthesis, but we do not need to identify it as such, just approach it logically. In fact, if we look back at the Knorr pyrrole synthesis, we shall see that, under chemical conditions, the reagents used here are not sufficiently reactive for the pyrrole synthesis; we need a more activated compound, like ethyl acetoacetate. Furthermore, we could not possibly proceed without masking the carboxyls as esters. This underlines how a biosynthetic sequence might differ somewhat from a purely chemical synthesis.

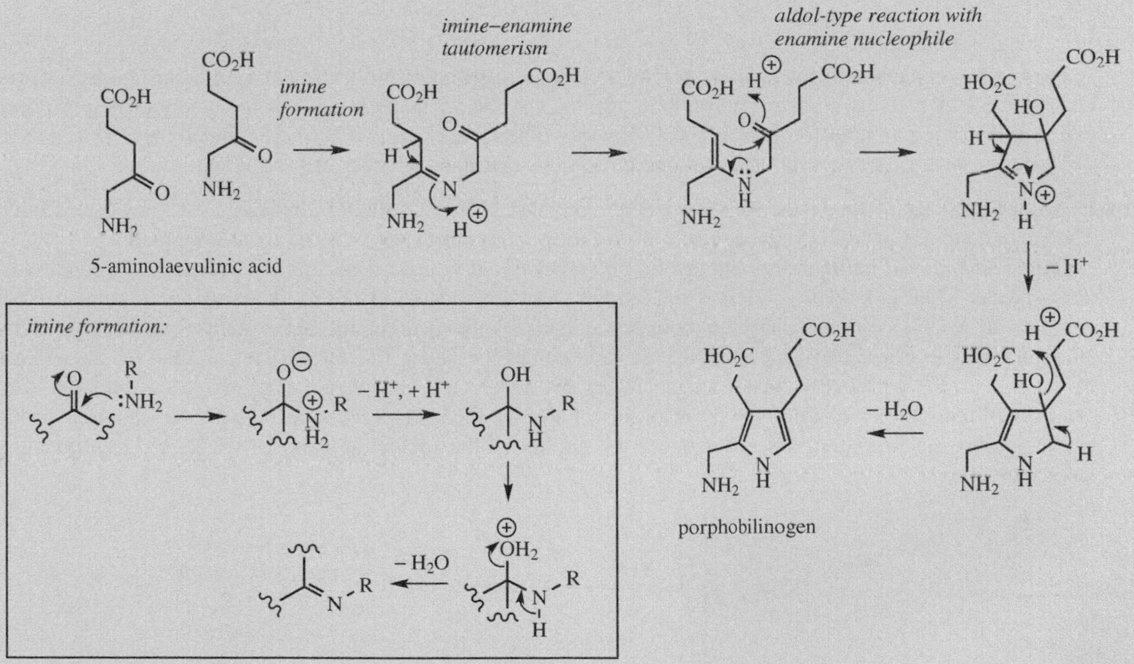

imine formation

imine–enamine tautomerism

aldol-type reaction with enamine nucleophile

5-aminolaevulinic acid

imine formation:

$-H^+, +H^+$

$-H_2O$

porphobilinogen

$-H_2O$

$-H^+$

We shall consider the sequence as firstly imine formation (an abbreviated form of this mechanism is shown), followed by imine–enamine tautomerism. This provides a nucleophilic centre and allows a subsequent aldol-type reaction with enamine plus ketone. The pyrrole ring is produced by proton loss and a dehydration.

The linear tetrapyrrole has methylene bridges between the pyrrole rings; we start from porphobilinogen that has either –H or –CH$_2$NH$_2$ as the ring substituents at these positions. Since the nitrogens are lost, we should consider an elimination, and this is assisted by the pyrrole nitrogen. We can consider protonation of the amine to facilitate the elimination. The product is an electrophilic methylidene pyrrolium cation.

conjugate addition
electrophile   *nucleophile*

elimination of ammonia
facilitated by pyrrole nitrogen   *methylidene pyrrolium cation*

etc

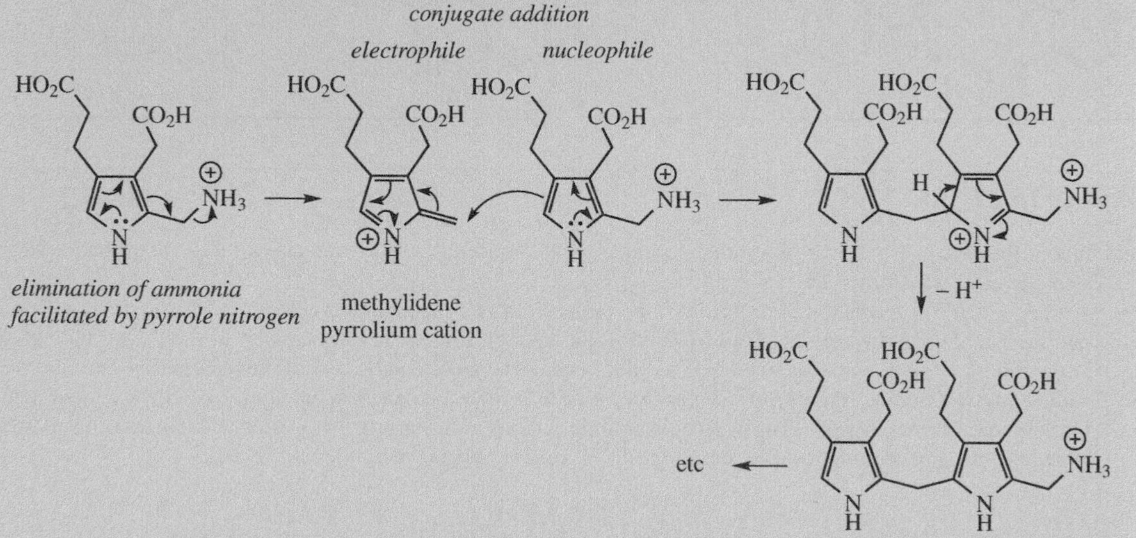

Pyrrole is very reactive towards electrophiles; charge distribution from the nitrogen makes either C-2 (or C-3) electron rich. Thus, a second porphobilinogen acts as the nucleophile towards the methylidene pyrrolium cation in a conjugate addition reaction. It is now possible to see that two further identical steps will give us the required linear tetrapyrrole, and that one more time will then achieve ring formation.

(b) It is not necessary to know the Reissert indole synthesis; this is a rational application of known reactivity. The second step is catalytic reduction of a nitro group to an amine, and no details are required.

Step 1 has all the hallmarks of enolate anion chemistry. It is carried out under strongly basic conditions, and a diester (diethyl oxalate) suffers displacement of an ethoxide in a Claisen-like reaction. We thus need to generate an ArCH$_2^-$ anion. Toluene is more acidic than, say, ethane, but not that acidic for proton removal by ethoxide. This is where the last part of the question comes in: why is the nitro group essential for the success of the first step? It probably relates to generating the anion; and indeed, we can make a case for the nitro group stabilizing the conjugate base by resonance. Later on, the nitro is reduced to an amine; an amino group would inhibit the first step, since it reduces the acidity of the methyl substituent via its electron-donating resonance effect.

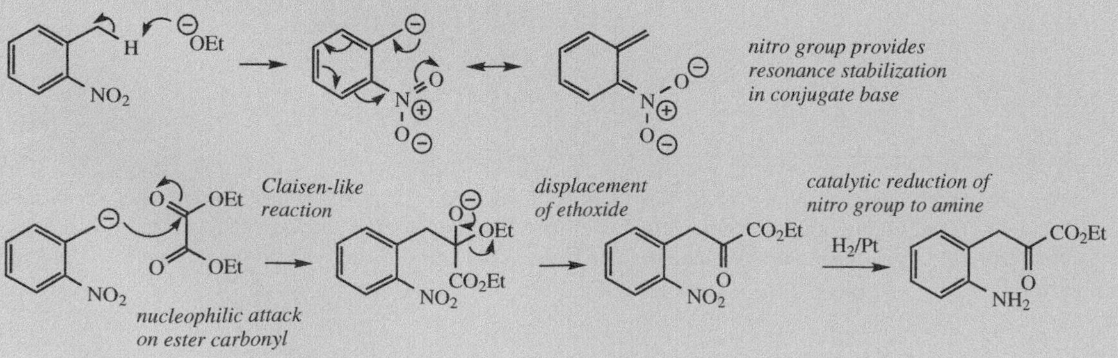

nitro group provides resonance stabilization in conjugate base

Claisen-like reaction   *displacement of ethoxide*   *catalytic reduction of nitro group to amine*

nucleophilic attack on ester carbonyl

Finally, indole ring formation is via condensation of the amino and keto functions. This is analogous to imine formation, as seen in part (a), but dehydration produces the aromatic pyrrole ring rather than an imine. Alternatively, one could write imine formation followed by tautomerism to the aromatic enamine.

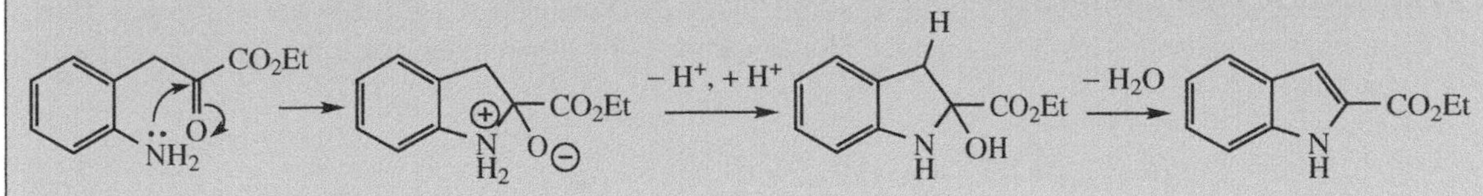

Problem 16.24 (level 2; chapters 11 and 7)

(a) Treatment of 2-methylpyridine with benzaldehyde under basic conditions gives the compound stilbazole. Give a mechanism for the reaction and predict the product formed when 2,3-dimethylpyridine is reacted with benzaldehyde under the same conditions.

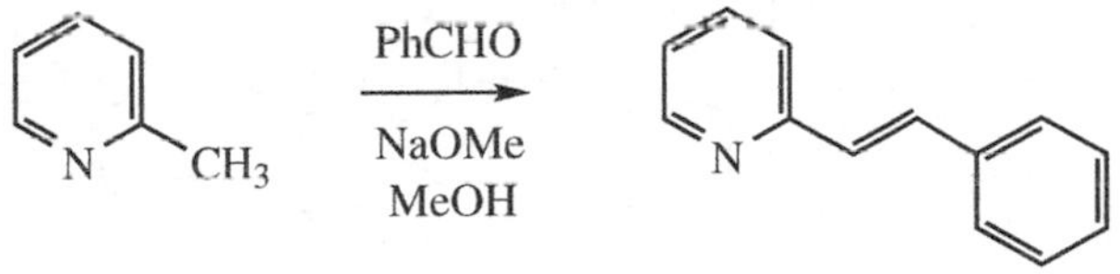

stilbazole

(b) What product would you expect to obtain if 2-chloropyridine was treated with sodium methoxide in methanol? What product would be obtained upon similar treatment of 3,4-dichoropyridine?

(c) Give mechanisms to account for the following two reactions:

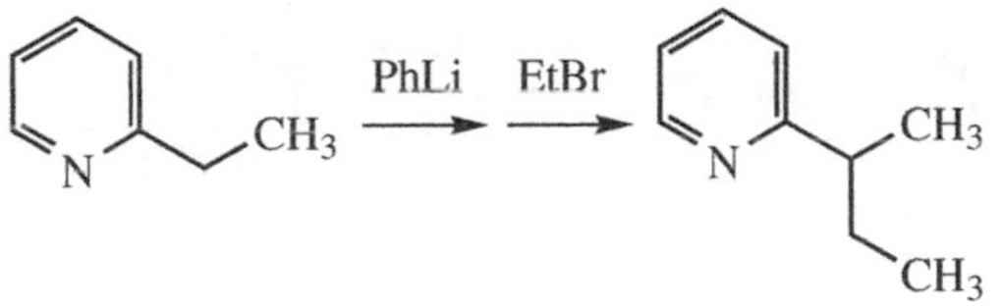

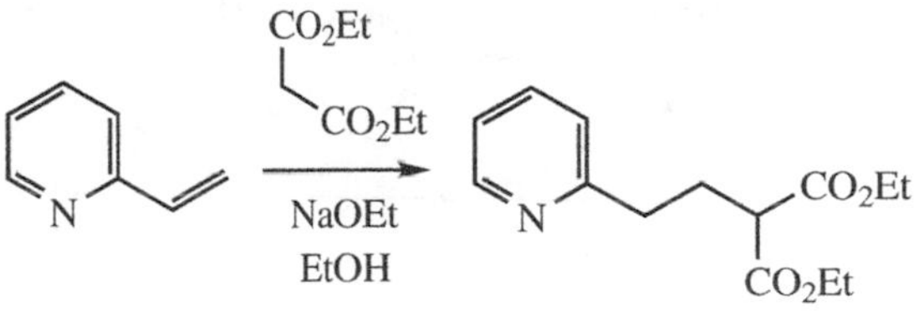

(d) Treating indole with the unsaturated ketone shown (mesityl oxide) under acid conditions gives a carbazole derivative. Formulate a rational mechanism for the transformation, which proceeds through an indolenium cation intermediate (not isolated).

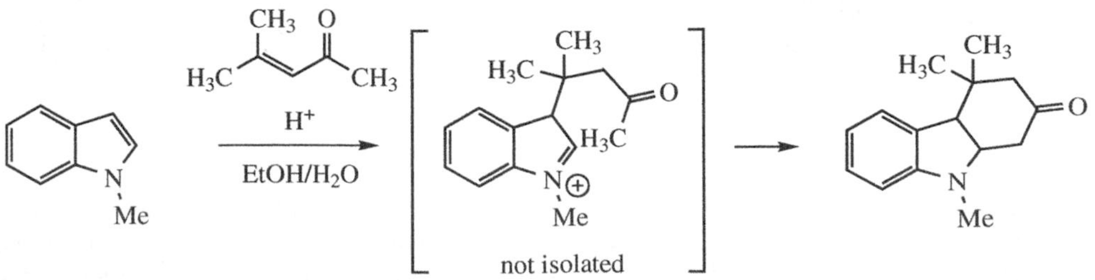

not isolated

Answer 16.24

(a) The basic conditions point to initial generation of an anion that is subsequently used in reaction with benzaldehyde. Since the product is an alkene conjugated to the aromatic rings, an addition–elimination sequence seems a likely proposal. To generate the anion, the 2-methylpyridine needs to lose a proton from the methyl substituent. This is feasible; the resultant anion is resonance stabilized and behaves much as an enolate anion.

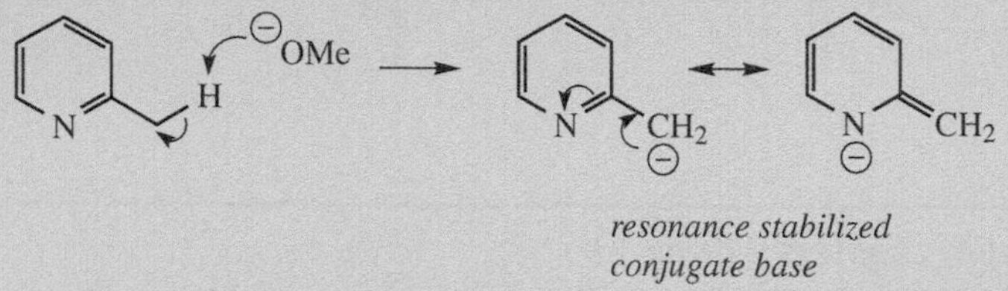

*resonance stabilized
conjugate base*

Then follows an aldol-like reaction with subsequent dehydration, favoured by the conjugation in the product.

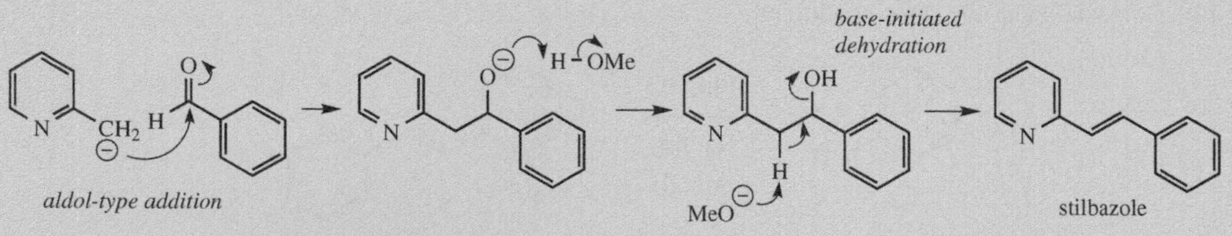

aldol-type addition   *stilbazole*

Removal of a proton from 3-methylpyridine does not provide the same resonance stabilization in the anion. Charge cannot be dispersed to the nitrogen, only to carbons in the aromatic ring. Hence, 3-methylpyridine is a weaker acid than either 2- or 4-methylpyridine. We predict, then, that similar treatment of 2,3-dimethylpyridine will produce the methylstilbazole derivative shown.

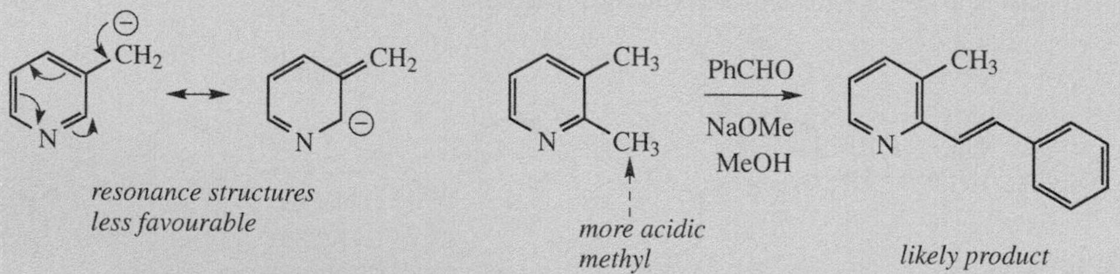

*resonance structures
less favourable*

*more acidic
methyl*

likely product

(b) The reaction of 2-chloropyridine with sodium methoxide is a nucleophilic substitution. This is represented as addition to the imine function, followed by loss of the chloride leaving group. The reaction is facilitated by the resonance stabilization in the intermediate anion.

*nucleophilic attack
on imine function*

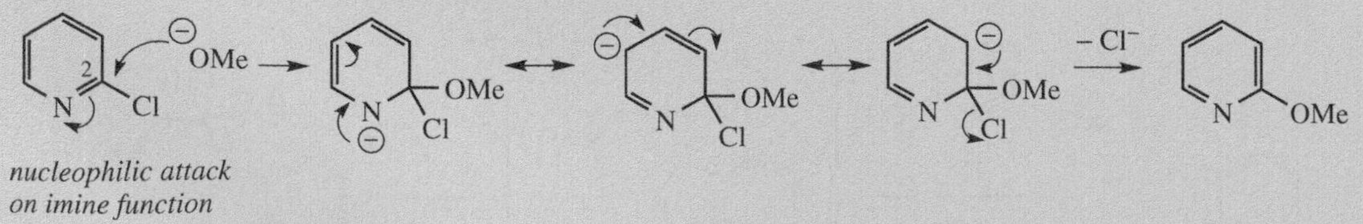

Similar resonance structures may be drawn in the case of 4-chloropyridine, and this will also undergo nucleophilic substitution. However, 3-chloropyridine, despite having the same favourable leaving group, does not undergo substitution. This may be deduced from consideration of the intermediate anion; resonance structures involve only aromatic carbons, with no stabilization contributed by the nitrogen. It follows, therefore, that the product from treatment of 3,4-dichoropyridine with sodium methoxide will be 3-chloro-4-methoxypyridine. Substitution will occur only at position 4.

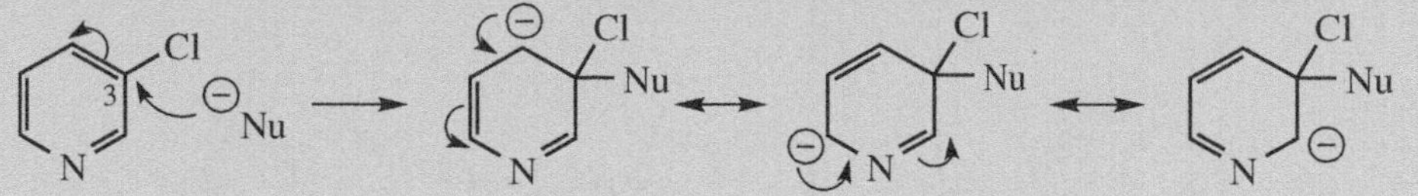

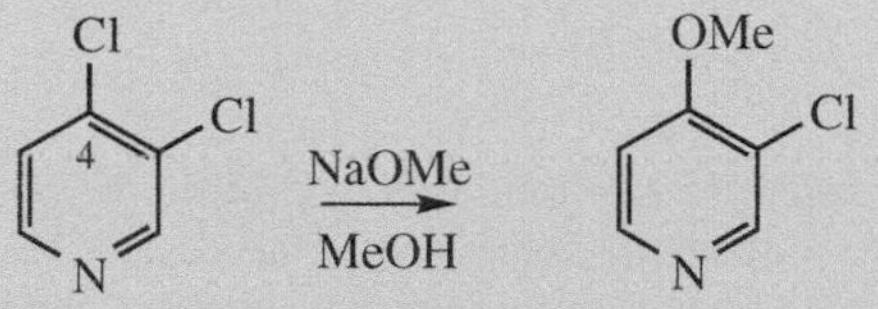

(c) The first reaction follows from part (a), in that we can generate the conjugate base of the alkylated pyridine derivative, and use this anion as a nucleophile in an S_N2 reaction. Note that we remove a proton from the position adjacent to the pyridine ring to obtain the resonance-stabilized anion.

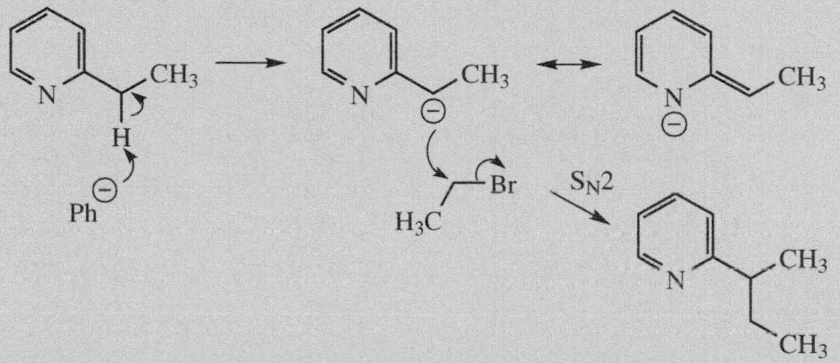

In the second reaction, we also have a base-generated anion, this time an enolate anion from diethyl malonate. This reagent is more acidic than an alkylpyridine; and in any case, the unsaturated side-chain does not allow proton removal in the way we have just seen. It can be seen that the enolate anion joins up with the heterocyclic reagent, but now the heterocycle must be responsible for providing the electrophile. The double bond in the side-chain gives a clue, and we note this is conjugated with the pyridine ring; the reaction is nucleophilic attack on a conjugated imine, a Michael-like reaction.

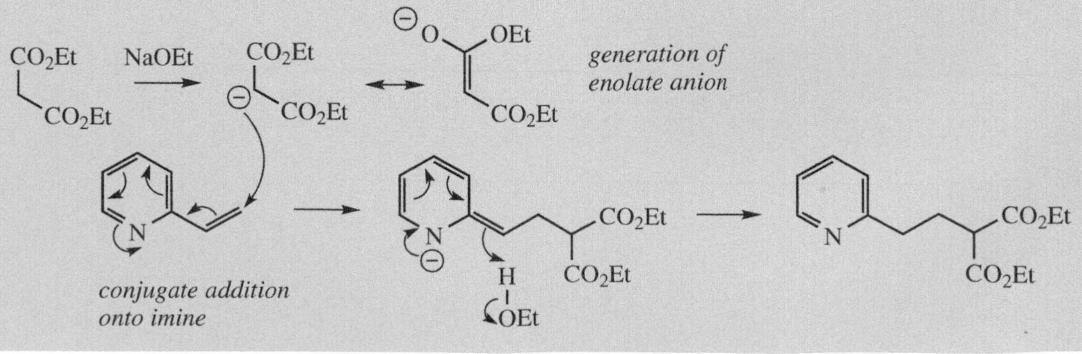

(d) This is a two-step reaction, where the intermediate is not isolated; it is presented to give you a clue. The clue is that we have Michael-like conjugate addition on to the unsaturated ketone, and the heterocycle must provide the nucleophile. Since the heterocycle is indole, this is quite reasonable; indoles are very reactive towards electrophiles with attack occurring at C-3.

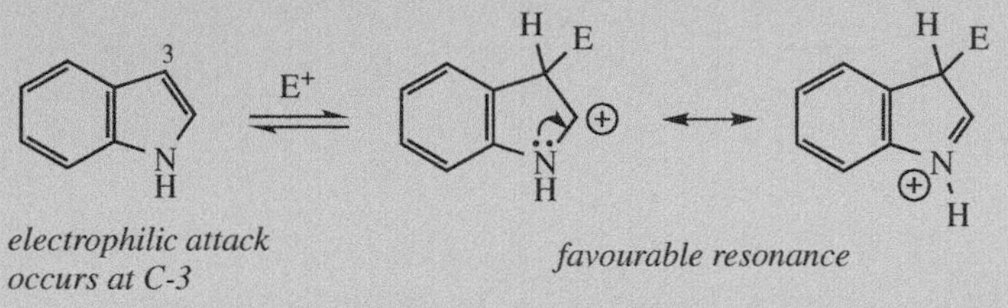

electrophilic attack occurs at C-3

favourable resonance

We can now proceed. Indole attacks the unsaturated ketone in Michael fashion; because the conditions are acidic, we protonate the ketone. The product is an enol, so that is tautomerized to the ketone. This gives the 'not isolated' intermediate. The next step is ring closure, and we have the iminium system of the heterocycle as electrophile. The nucleophile should be distinguishable as the α-position of the ketone, an enolate anion equivalent. Under acidic conditions, we cannot have an enolate anion, so we must employ the enol. This means a second tautomerization step. Ring closure follows; you may recognize it as equivalent to a Mannich reaction: attack of an enol onto an iminium cation. Loss of a proton and the job is done.

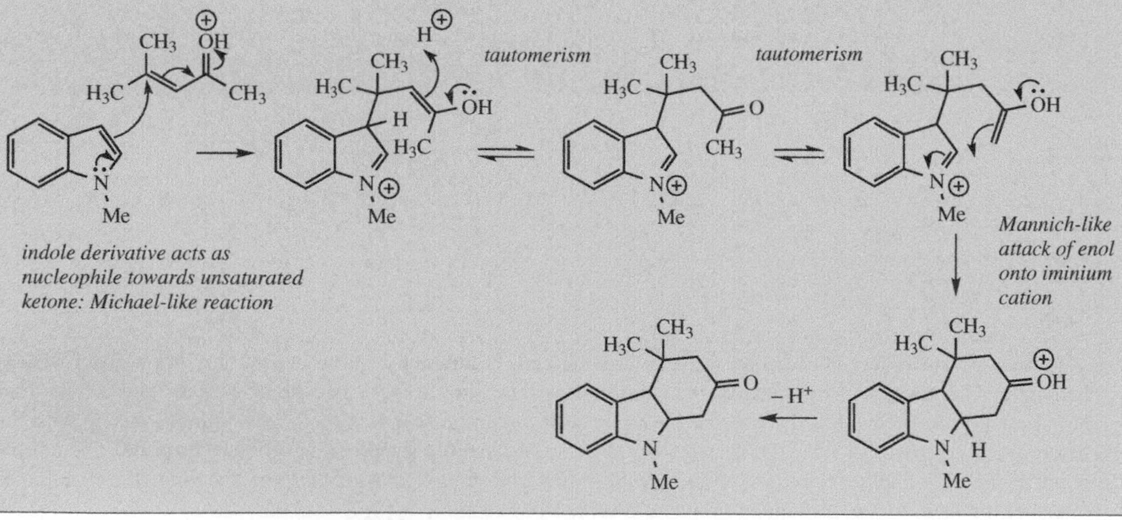

indole derivative acts as nucleophile towards unsaturated ketone: Michael-like reaction

Mannich-like attack of enol onto iminium cation

Index

Essentials of Organic Chemistry   Paul M Dewick
© 2006 John Wiley & Sons, Ltd

Printed and bound by CPI Group (UK) Ltd, Croydon, CR0 4YY